Physical Constants[a]

Velocity of light	$c = 2.99792458 \times 10^8$ m/s (exactly)
Planck's constant	$h = 6.6260755 \times 10^{-34}$ J s
	$\hbar = 1.054573 \times 10^{-34}$ J s
	$= h/2\pi$
Boltzmann constant (gas constant per molecule)	$k = 1.380658 \times 10^{-23}$ J/K
Gas constant	$R = 8.314510$ J/mol K
	$= 1.98721$ cal/mol K
	$= 0.082051$ L atm/mol K
Avogadro's number	$N_A = 6.0221367 \times 10^{23}$/mol
Electron charge	$e = 1.602177 \times 10^{-19}$ C
	$= 4.80320 \times 10^{-10}$ esu
Faraday	$\mathscr{F} = 9.6485309 \times 10^4$ C/mol
Atomic mass unit	amu $= 1.6605402 \times 10^{-27}$ kg
Electron rest mass	$m_e = 9.109390 \times 10^{-31}$ kg
	$= 5.485716 \times 10^{-4}$ amu
Proton rest mass	$m_p = 1.672623 \times 10^{-27}$ kg
	$= 1.007261$ amu
Neutron rest mass	$m_n = 1.6749286 \times 10^{-27}$ kg
	$= 1.0086495$ amu
Bohr magneton	$\mu_B = e\hbar/2m_e = 9.274015 \times 10^{-24}$ J/T
Nuclear magneton	$\mu_N = (m_e/m_p)\mu_B = 5.0507866 \times 10^{-27}$ J/T
Debye	$D = 1.0 \times 10^{-18}$ esu cm
	$= 3.33 \times 10^{-32}$ C cm
Gravitational constant	$G = 6.67259 \times 10^{-11}$ m^3/s^2 kg
Acceleration of gravity	$g = 9.80665$ m/s^2 (exactly)
Rydberg constant	$R = 1.097373153 \times 10^7$/m
Pi	$\pi = 3.1415926536$

[a] Values from *Quantities, Units, and Symbols in Physical Chemistry*, I. Mills, Ed., Blackwell Scientific, Oxford, 1988, p.81.

Concepts and Models of Inorganic Chemistry

Concepts and Models
of Inorganic Chemistry

Third Edition

Bodie E. Douglas
University of Pittsburgh

Darl H. McDaniel
University of Cincinnati

John J. Alexander
University of Cincinnati

John Wiley & Sons, Inc.
New York   Chichester   Brisbane   Toronto   Singapore

ACQUISITIONS EDITOR   Nedah Rose
MARKETING MANAGER   Cathy Faduska
PRODUCTION EDITOR   Bonnie Cabot
DESIGNER   Kevin Murphy
MANUFACTURING MANAGER   Andrea Price
ILLUSTRATION   Jaime Perea/Sigmund Malinowski

This book was set in 10/12 Times Roman by Interactive Composition Corp. and
printed and bound by Malloy Lithographing. The cover was printed by Lehigh Press.

The cover drawing was taken from *Physics Today*, November 1992. It is an experimental electron density plot of the 100 plane of solid C_{60}. The circular depressions at the center and corners are cross-sections of the rotating bucky balls on a face of the face-centered cubic (ccp) cell.

Library of Congress Cataloging in Publication Data:
Douglas, Bodie Eugene, 1924–
        Concepts and models of inorganic chemistry / Bodie E. Douglas,
Darl McDaniel, John Alexander.—3rd ed.
        p.   cm.
        Includes bibliographical references and index.
        ISBN 0-471-62978-2 (acid-free)
        1. Chemistry, Inorganic.   I. McDaniel, Darl Hamilton, 1928-
II. Alexander, John J.   III. Title.
QD475.D65   1993
546—dc20   93-38708
                                                                CIP

Printed in the United States of America

10   9   8   7   6

Printed and bound by Malloy Lithographing, Inc.

Preface

. . . .

Textbooks are written or revised to meet a need perceived by the authors. The goal of the first edition of *Concepts and Models* was to create an inorganic chemistry text organized around the models undergirding the field rather than the traditional group-by-group survey of the periodic table. We drew heavily on such sources as Pauling's *Nature of the Chemical Bond*, Herzberg's *Atomic Spectra*, Latimer's *Oxidation Potentials*, Basolo & Pearson's *Mechanisms of Inorganic Reactions* and other "classic" books as well as the original literature of the decade past as filtered through our eyes. We did not hesitate to speculate on structures then unknown (p. 181), numerical data not yet determined (p. 186), or unexplored mechanisms (with warnings to the reader such as "probably having units such as," "based on the assumption," "somewhat speculative explanation") since this is exactly what we hoped the student would attempt. We encouraged use of the original literature with extensive citations to journal articles as well as the general references to books and review articles at the end of chapters. We included solved example problems within chapters as well as problem sets at the end of chapters. We had an extensive set of appendices including an exhaustive update of Latimer's redox potentials. As befitted the title, we took special care with the numerous illustrations, and virtually all of the printed chemical formulas conveyed stereochemical information.

In the early sixties "advanced inorganic chemistry" courses varied widely depending on the interests of the instructor. *Concepts and Models of Inorganic Chemistry* helped to define modern upper level inorganic chemistry courses, standardized examinations at that level, and inorganic chemistry texts that followed.

The second edition of this text (1983) updated the book. The introduction of symmetry and group theory was a leadership step. Bioinorganic chemistry was added. A new coauthor, John J. Alexander, was added to broaden the expertise of the authors. The coverage of organometallic chemistry was expanded and kinetics was treated in greater depth.

Concepts and Models of Inorganic Chemistry is neither an encyclopedia of descriptive inorganic chemistry nor a textbook of structural inorganic chemistry. Instead, information regarding chemical reactions is presented within a framework of concepts and models that help students to organize and retrieve chemical knowledge. Descriptive chemistry is woven into almost all chapters and is the subject of special topics chapters (Chapters 15–18) and major parts of Chapters 9 and 13.

Revisions of this third edition have been made with the student in mind, trying to make reading clear and to anticipate questions that might arise. Boxed inserts serve several purposes: (a) to provide a more detailed treatment of a topic, sometimes at a higher level than is essential for the text; (b) to introduce interesting asides; and (c) to provide an explanation that might be unnecessary for some readers. Examples and problems illustrate the points being discussed. They should help students develop their own critical reading, raising questions of why and how. References to literature are cited throughout for pursuing topics of interest to the reader.

The book is intended for an upper level undergraduate course or a first year graduate course. Comprehensive inorganic texts have evolved into reference books. This text should be particularly suitable for a two-term inorganic chemistry course in the United States or Canada. It is suitable for general courses extending over more than a year in the United Kingdom. For the one-term inorganic chemistry course, selections of topics must be made. We have included more material than most instructors will want to cover. We feel that it is easier for instructors to omit some sections than to scurry for ways to round out too-brief presentations. Also we have striven to give explanations in enough detail so a student does not have difficulty understanding the material because of gaps or terse writing.

The text has been reorganized extensively into six parts. Core material is covered in the first eight chapters and is grouped into three parts: Basic Concepts, Bonding and Structure, and Chemical Reactions. However, parts of each of these chapters might serve as a concise review for students with good backgrounds in chemistry. Symmetry might have been covered in an earlier course or group theory might be deferred. For those who have encountered the material in other courses, terms used are defined at the end of the chapter to provide a review or fill in gaps. Terminology and notation of symmetry and group theory are essential in modern inorganic chemistry. If only symmetry is covered, Section 3.5 can be omitted along with the applications of group theory in Sections 4.5–4.7, 10.8, and 17.4. Section 4.2 can also be omitted.

Chapter 5 covers the basic concepts and models of ionic solids, including important structures. In Chapter 6, we look into some of the recent developments in graphite chemistry, superconductors, and the fullerenes. Although any of these can be covered in part without the Bloch equation, the main feature of this chapter is the development of the interelationship of real space and reciprocal space as embodied in the Bloch equation. Selections used from Chapters 5 and 6 will depend on time available and emphasis.

Part III, Chemical Reactions, includes acids and bases, and oxidation–reduction reactions. A great deal of descriptive chemistry is included to illustrate the reactions and periodic trends. Chapter 7 surveys several models from acid–base chemistry, including the hard–soft acid–base model and solid acids. Chapter 8 illustrates the use of emf and Pourbaix diagrams. Redox chemistry is surveyed for Mn, Tc, V, and some nonmetals.

Selection of chapters or sections from Parts IV, V, and VI can be made to fit the desired schedule. Each chapter can stand alone following the basic essentials from Parts I–III.

Chapter 9 covers basic coordination chemistry including valence bond and ligand field theory models, stability, isomerism, and coordination numbers 2–14. Chapter 10 is a more detailed treatment of electronic spectra and the molecular orbital model for coordination compounds. Chapter 11 covers important mechanisms and applications to reactions in coordination chemistry.

The coverage of organometallic chemistry has been expanded; however, the new organization offers added flexibility. The essentials of electron counting, bonding, and the chemistry of some typical compounds are presented in Chapter 12. For a presentation that also emphasizes catalysis, insertion reactions (Section 14.1.2) and the scope of oxidative addition and reductive elimination (Section 14.1.3) might be covered (possibly omitting mechanisms of oxidative addition) along with Section 14.2 on homogeneous catalysis. Chapter 14 does not require Chapter 13 for understanding. A greater emphasis can be achieved by also including the systematic survey of organometallic compounds in Chapter 13 and Sections 14.1.1 and part of 14.1.2, which present mechanisms as well as the topics of C—H activation (Section 14.1.4) and attack on coordinated ligands (Section 14.1.5).

Chapters 15 (metals) and 16 (nonmetals) survey the chemistry and periodic trends of most of the elements. Much more descriptive chemistry is incorporated throughout the book.

A unified treatment of boron hydrides and metal clusters is presented in Chapter 17. Sections 17.6–17.9 deal with the chemistry of the boranes; the remainder of the chapter covers structure and bonding of boranes and clusters.

Chapter 18 is a survey of bioinorganic chemistry with emphasis on important roles of metal ions. The sections of the chapters are independent so omissions will cause no problem.

The third edition of the companion book, *Problems for Inorganic Chemistry,* provides solutions for all problems in the main text plus additional problems with solutions. This is a self-study aid that stands alone and can be used with any text.

Development of a text through several editions is a cumulative effort. We would like to acknowledge friends, colleagues, and reviewers who have offered helpful suggestions including Drs. Elmer Amma, Gordon Atkinson, the late Carl H. Brubaker, Thomas B. Cameton, Henry S. Frank, and Hans H. Jaffé, James C. Carter, Joyce Y. Corey, William L. Jolly, Laxman N. Mulay, Alan Searcy, and Darel K. Straub (first edition); Edwin H. Abbott, Paul Gaus, William E. Hatfield, Rex Shepherd, Duward F. Shriver, Darel K. Straub, John S. Thayer, and Andrew A. Wojcicki (second edition); Punit Boolchand, Richard Burrows, William B. Jensen, Nemad M. Kostić, Allan R. Pinhas, Barbara E. Stout, and Kathleen S. Richardson (third edition). This edition was improved by detailed comments on the entire manuscript by Kenneth Balkus, University of Texas at Dallas, Ruth Bowen, California State Polytechnic University, Van Crawford, Mercer College, and James Golen, University of Massachusetts at Dartmouth.

We are grateful to Nedah Rose (Acquisitions Editor), Bonnie Cabot (Production Editor), and Jaime Perea and Sigmund Malinowski (Art department) for their kind assistance during the revisions and production of the text. The final book has been improved by their contributions.

We appreciate comments and advice of colleagues and users of the former editions of *Concepts and Models of Inorganic Chemistry* and *Problems for Inorganic Chemistry.* Your comments and advice are welcome.

We want to express our appreciation to Kevin Murphy for the design of the cover of this book and that of DMA, 3rd ed. and to AT&T and Arthur F. Hebard of AT&T Bell Laboratories for the drawing. The drawing, based on work at AT&T, was the cover of *Physics Today*, November, 1992. It is an experimental density plot of the 100 plane of solid C_{60}. The circular depressions at the center and corners are cross sections of the rotating "bucky balls" on a face of the face-centered cubic (*ccp*) unit cell. Another figure, for molecular orbitals of the superconducting K_3C_{60}, from A. F. Hebard, *Physics Today* **1992,** Nov. 29 was used in Chapter 6 of DMA 3rd ed.

Bodie E. Douglas
Darl H. McDaniel
John J. Alexander

Excerpts from
The Preface to
The Second Edition

•••

The title of the book still serves as a central theme. The text should help students integrate their knowledge of chemistry, enabling them to draw on the wealth of knowledge learned in the highly compartmentalized chemistry courses. The upper level inorganic course is a fun course for many of us (instructors and students) because here many students gain an overall view of chemistry and acquire the intuitive feel of a chemist. The concepts-and-models approach should foster this goal.

The first edition of this text appeared in 1965. The developments in inorganic chemistry since then have been impressive. We have made the following additions to the text to keep abreast of these developments. We use figures more extensively than in the first edition; they have been selected with care and rendered more effectively. The treatment of bonding is more sophisticated, using molecular orbital theory (Chapter 4). Symmetry and group theory (Chapter 3) are presented for applications to bonding; with this background we are able to have a more detailed discussion of ligand field spectra (Chapter 7). Because the solid state is not given proper emphasis in inorganic courses, discussion of this topic has been expanded (Chapter 6). Reaction mechanisms have increased in importance and receive greater attention (Chapters 9 and 10). The field of organometallic chemistry has grown tremendously and requires a full chapter (Chapter 10), including reaction mechanisms. The treatment of hard and soft acids and bases was added (Chapter 12), and applications to coordination chemistry are included in the same chapter. More descriptive chemistry has been added in Chapters 13, 14, 15, and elsewhere. The importance of the emerging field of metal cluster compounds was acknowledged in the first edition by using a picture of the $Re_3Cl_{12}^{3-}$ structure on the dust jacket, and metal clusters are now recognized to be much more common than had been believed. More systematic treatments of cage and cluster compounds merit regrouping of the topics (Chapter 15). Bioinorganic chemistry (Chapter 16) has become a major component of inorganic chemistry and requires introduction.

The goal of the text is to offer a reasonable balance of material for an upper level undergraduate or first year graduate course. Even reasonable, but not comprehensive, coverage of the currently important topics in inorganic chemistry leads to a book that is too large for a one semester course. Each instructor must select topics and decide on emphasis. Material is provided for courses differing significantly in the balance between theory and descriptive chemistry. Chapters 8, 10, 13, 14, and 15 are primarily descriptive, although in Chapters 6, 9, 11, 12, and 16 we have incorporated a great deal of descriptive chemistry.

Bodie E. Douglas
Darl H. McDaniel
John J. Alexander

Contents

•••••••••••••••

PART I
Some Basic Concepts

•••••••••••••••••••••••••••

CHAPTER 1
Atomic Structure and the Periodic Table

CHAPTER 2
Molecular Models

CHAPTER 3
Symmetry   105

PART II
Bonding and Structure   149

CHAPTER 4
Discrete Molecules: Molecular Orbitals   151

CHAPTER 10
Spectra and Bonding 441

CHAPTER 11
Reaction Mechanisms 482

PART V
Organometallic Chemistry   559
. . . .

PART VI
Selected Topics
• • • • • • • • • • • • • • • • • •

CHAPTER 15
Chemistry and Periodic Trends among Metals

CHAPTER 16
Chemistry of Some Nonmetals

CHAPTER 17
Cluster and Cage Compounds

CHAPTER 18
Bioinorganic Chemistry

APPENDIX A
Units

APPENDIX B
Nomenclature of Inorganic Chemistry

APPENDIX C
Character Tables

APPENDIX D
Tanabe–Sugano Diagrams

APPENDIX E
Standard Half-Cell emf Data

INDEX

PART I

Some Basic Concepts

···

▶ 1 ◀

Atomic Structure and the Periodic Table

· ·

> The development of the periodic table is one of the great scientific achievements of Western civilization. In this chapter we discuss its development and the basis for its organization, which lies in the electronic structure of atoms. Atomic structure is treated using the results of the Schrödinger equation. We begin with a brief look at the role of models in chemistry which sets out the viewpoint from which this book is written.

1.1 MODELS IN CHEMISTRY

The title *Concepts and Models of Inorganic Chemistry* reveals much about this book's content and about how you should read and understand the material in it. Chemistry consists of a seemingly endless catalogue of information about the composition, structure, reactivity, and physical properties of substances. Because inorganic chemistry deals with all the elements of the periodic table, the amount of information encompassed is particularly vast. It would be a monumental task to study inorganic chemistry in encyclopedic fashion by learning, say, the stoichiometries of all the oxides of vanadium, the structures of all the halides of selenium, and so on.

An important goal of chemistry is to be able to predict what will happen based on our store of chemical information. If we are to go beyond information connected with a particular occasion and to predict the chemical future on the basis of experience, we must store information in a structured way. A very effective and intellectually satisfying way of doing this is to create a **model.** A model can be thought of as a simpler form of whatever part of nature we wish to study. Studying a valid model enables us to understand and rationalize the behavior of physical "reality" and, ideally, to predict its behavior.

A familiar model is the atomic billiard balls of the kinetic/molecular theory of gases. Although gaseous atoms are not actually little billiard balls, we may think of a gas sample as consisting of spherical, hard billiard-ball-like molecules in constant motion undergoing perfectly elastic collisions with one another and with container walls. Applying Newtonian mechanics to the model enables us to understand gas pressure in terms of elastic collisions with container walls. This, in turn, leads to the conclusion that increasing the number of molecules confined to the same volume would increase collision frequency with the walls and hence the pressure, as observed. The separate abilities of heavy and light billiard ball molecules to collide with the container sides and so to exert pressure should be unaltered by the presence of each other—that is, the model leads to the rationalization of Dalton's Law of Partial Pressures. Thus, much of the behavior of gases can be stored and retrieved by attaching it to the model. Modifications and refinements lead to such current areas as molecular dynamics.

A major purpose of this book is to present the fundamental concepts of inorganic chemistry as well as selected topics of current interest and importance. A **concept** is "an abstract idea generalized from particular instances." Models, concepts, and theories are employed to explain and systematize experimental facts. Thus, facts are the foundation of science. Because there is little use in knowing explanations without having an idea what they explain, you will need to learn some facts of inorganic chemistry as well.

Furthermore, real advances in science occur at just the points where models and theories do not fit experimental observations. Such situations can reveal those parts of the universe which we still do not understand and can lead to modification of models and theories to include previously unexplained phenomena. Of course, it is necessary to be conversant with what the current models are, in order to recognize a discrepancy. This is another reason for our emphasis on the use of models in studying chemistry. We should not use models as a way to explain everything and discover nothing.

The term **theory** has been used above relying on your prior notions of what a theory is. Theories are sets of statements (some or all of which may be mathematical equations) which relate the part of the universe in which we are interested (the **system**) to the wider logical structure of the physical universe. Theories are often understood in terms of models which give physical interpretation to terms of the theory. In the example above, sets of equations and statements of the laws of Newtonian mechanics, along with the mathematical postulate of the relation between kinetic energy and temperature ($\frac{1}{2}mv^2 = \frac{3}{2}kT$), constitute the kinetic/molecular *theory*. The *model* is our interpretation of the theoretical terms such as mass and velocity as belonging to hard spherical billiard ball molecules. Models give a concrete physical interpretation to the more abstract terms of scientific theories.

Theories ⇔ Models

⇕

Laws

⇕

Hypotheses

⇕

Experimental facts

The use of models is deeply ingrained in the way in which science is practiced. When we are presented with some new information, it is natural to try to construct a model to systematize it or to inquire how the new information does or does not fit with accepted models. Which model is selected depends on what we want to explain. For example, the result of an electron diffraction experiment is explained by considering the electron to be a standing wave. The Millikan oil drop experiment is explained by considering the electron to be a charged particle. The wave and particle descriptions seem literally mutually contradictory. A particle has an exact location and mass whereas a wave does not, and so on. Evidently, the models we employ to understand a system need not always be logically compatible. (But the incompatibility is an unresolved chord and often leads to efforts to resolve the contradiction at a more fundamental level.) Practicing scientists must often be content with a state of knowledge which does not reflect a unified version of reality. Pending development of models and theories which present all of nature as a unified whole, we often explain various aspects of the same system by using logically incompatible models.

Thus, in this book we often describe more than one useful model of phenomena, and we often mention the contributions of particular scientists to developing and expanding currently accepted models.

Early on in any investigation of nature, *pattern recognition* takes place. Some set of natural phenomena is seen to exhibit a recognizable similarity. Scientific investigation consists in the search for an explanation of the regularities in more fundamental terms (i.e., the terms of some theory). Classification (recognizing similarities in properties which allow assignment to a related **group** such as the alkali metals, halogens, etc.) is a one-dimensional pattern recognition. The most important instance of pattern recognition in chemistry is the periodic table, which systematizes the myriad complexity of chemical behavior in a two-dimensional pattern: chemical reactivity and atomic number. It presents in graphic form the content of the periodic law: *The properties of the elements are periodic functions of their atomic numbers*. We now consider the development and structure of the periodic table and the underlying explanation for the regularities in the properties of the elements which it displays. Many experiments and results which we will describe are familiar to you from courses in physics and general chemistry. Our purpose here is to recall pertinent material to your memory, add to it, and present it in a way which emphasizes its importance for and its logical connection to the development and understanding of the periodic table.

1.2   HISTORICAL BACKGROUND

A systematic approach to inorganic chemistry is today almost synonymous with studying the periodic relationships of the elements and their compounds. This approach has an empirical foundation built during the last two centuries and has a theoretical justification of half a century.

Stanislao Cannizzaro presented a consistent set of atomic weights at the Karlsruhe Conference in 1860. Within a decade, various forms of the periodic table appeared in

France, England, Germany, and Russia. The role played by Cannizzaro's list of atomic weights in the development of the periodic table can better be appreciated by noting that an attempt in 1852 by Gladstone to find a relationship between the atomic weights and other properties of the elements failed because of the lack of a consistent set of atomic weights. The greatest share of credit for the periodic table is usually given to Dimitri Mendeleev, and properly so, because it was the realization of his bold prophecy of new elements and their properties that led to the almost immediate acceptance of the periodic law. In 1871, Mendeleev made many detailed predictions about an element that was discovered by Boisbaudran in 1875. Here are some:

> The properties of ekaaluminum, according to the periodic law, should be the following: Its atomic weight will be 68. Its oxide will have the formula El_2O_3; its salts will present the formula ElX_3. Thus, for example, the chloride of ekaaluminum will be $ElCl_3$; it will give for analysis 39% metal and 61% chlorine and will be more volatile than $ZnCl_2$. The sulfide El_2S_3, or the oxysulfide $El_2(O,S)_3$, will be precipitated by H_2S and will be insoluble in ammonium sulfide. The metal will be easily obtained by reduction: its density will be 5.9; accordingly its atomic volume will be 11.5; it will be soft and fusible at a very low temperature. It will not be oxidized on contact with air; it will decompose water when heated to redness . . . .

Examining the above predicted properties for gallium reveals that physical properties of the element and its compounds (specific gravity, hardness, melting point, boiling point, etc.), spectrographic properties, and chemical properties (formulas of possible compounds, acidic and basic properties of compounds, etc.) all reflect its position in the periodic table. In fact, properties of the elements or their compounds that cannot be correlated by means of the periodic table are somewhat exceptional.

Much of this book is devoted to attempting to understand the underlying principles that bring about these periodic relationships—that is, properties that show greater-than-chance similarity for elements which lie in the periodic table (a) in a vertical column (called a **group** or **family**), (b) in a horizontal row (called a **period**), (c) within a given area (bounded by elements of two or more groups and two or more periods), and (d) on diagonals. The following illustrate these types of relationships. Elements in a group have similar spectra, similar **valences** (capacities to combine with other elements in certain proportions), similar crystal structures both for the element and for particular series of compounds, and so on. Elements in a given period are bonded to similar maximum numbers of other atoms in their compounds. The compounds of the second-period elements Li, Be, and B show many similarities to the compounds of the third-period elements Mg, Al, and Si, to which they are diagonally related. Finally, there are numerous properties—such as classification of the elements as metals, metalloids, and nonmetals—that have an area relationship to the periodic table. Metals appear to the left and bottom, whereas nonmetals appear to the right and top. Moving to higher rows, progressively more columns contain nonmetals. A diagonal band of metalloids (B, Si, Ge, As, Sb, Te, Po) separates metals from nonmetals. Area relationships are often the most difficult to explain, because of the wider possible variation of the factors involved.

Before proceeding to the theoretical basis of the period table, let us note several steps in its evolution.

In 1829, Döbereiner had recognized one-dimensional patterns, namely, a number of cases in which three elements (triads) have similar chemical properties; he also noted that the middle triad member has properties very close to the average value of the other two—this is particularly true of the atomic masses. Two of Döbereiner's triads were (a) Li, Na, K and (b) Cl, Br, I.

Between 1860 and 1870, Newlands, Meyer, and Mendeleev prepared periodic tables by listing the elements in the order of increasing atomic mass and then grouping them according to chemical properties.[1] In Mendeleev's table the triads of Döbereiner always fell within the same group. Mendeleev reassigned the positions of a number of elements in order to obtain a fit with the chemical properties of the other elements in the group. For example, Te has a larger atomic mass than does I; placement strictly on this basis would put Te in the group with Cl and Br, with which its chemistry has little in common. For several pairs of elements [the early known cases are (a) Te and I and (b) Co and Ni] a reversal of position in the periodic table as compared to atomic masses is now established.

As the rare earth elements were discovered, difficulty was encountered in fitting them into the table. This led Basset and later Thomsen to propose the extended form of the table generally accepted today. Thomsen's table is shown in Figure 1.1. Furthermore, from considering the change of group valence from −1 for the halogens to +1 for the alkali metals, Thomsen reasoned that one should expect a group of elements lying between groups VII and I and having either infinite or zero valence (combining capacity). Because a valence of infinity is unacceptable from a chemical viewpoint, he proposed that an element group of zero valence separated the halogens from the alkali metals. Thomsen proceeded to predict the atomic masses of these elements as 4, 20, 36, 84, 132, and 212; he felt that these elements should terminate each period. Unfortunately, Thomsen did not publish these remarkable predictions until a year after argon had been discovered in 1894.

The last stage in the empirical development of the periodic table came in 1913, when Moseley found the x-ray emission from different elements had characteristic frequencies (ν) which varied in a regular fashion with the ordinal number of the elements in the table. (The ordinal number reflects the element's position in the periodic table: H = 1, He = 2, etc.) The empirical relationship is

$$\nu = k(Z - \sigma)^2 \tag{1.1}$$

where Z is the ordinal or **atomic number,** ν is the characteristic x-ray frequency, and k and σ are constants for a given series. No reversals in atomic number occur in the periodic table; hence, Moseley's results showed that atomic number is a more fundamental property than atomic mass.

The empirical evolution of the periodic table had reached its peak. It was then possible to make a strictly ordered list of the elements with definite indication of missing elements. Each period terminated with a noble gas, and it was possible to tell how many elements belonged to each period. The modern periodic table is shown inside the back cover of this book.

---

[1] It is surprising that the sequence of atomic masses worked so well because atomic mass depends on the relative abundances of light and heavy isotopes.

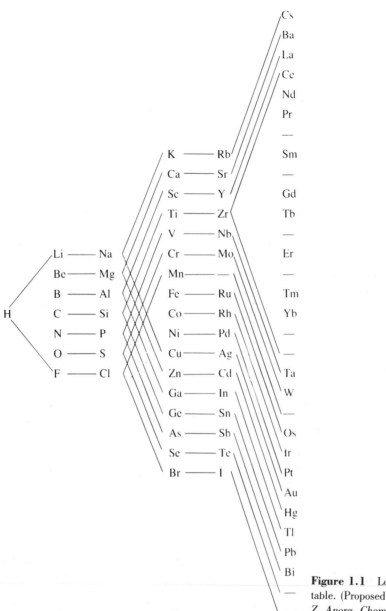

Figure 1.1 Long form of the periodic table. (Proposed by J. J. Thomsen, *Z. Anorg. Chem.* **1895,** *9,* 190.)

1.3 ATOMIC STRUCTURE AND THE BASIS OF THE PERIODIC TABLE

The theoretical basis of the periodic table had to await the development of a clearer picture of the atom. The concept of atoms as fundamental (indivisible) particles had to be abandoned at the beginning of this century. This resulted from, among other things, studies of cathode rays (electrons) and canal rays (positive ions) which appear when large voltages induce electric discharges in tubes containing very low pressures of gases. These studies led to the recognition of the existence of negative and positive charges within the atom. Further complexity of the atom could be inferred from studying the light emitted by energetically excited atoms (atomic emission spectra). It was also found that the lines (the characteristic single wavelengths emitted) in the spectra of gaseous atoms were split if the atoms were placed in a magnetic field (the Zeeman effect). The discovery of radioactivity indicated that the various kinds of particles emitted in radioactive decay must exist inside atoms. In fact, radioactivity not only indicated that the atom was not a fundamental unit, but the alpha particles emitted in radioactive decay also provided a probe with which to examine the atom. When alpha particles were propelled at thin metal foils, the pattern of scattering observed led Rutherford to a nuclear model of the atom, with a tiny nucleus carrying a number of unit positive charges equal to approximately one-half the atomic mass of the element. Moseley associated the atomic number with the nuclear charge.

▶ 1.3.1   Bohr Model of the Atom

A major advance in the understanding of the atom was Niels Bohr's development of a model that could account for the spectra of hydrogenlike atoms (i.e., one electron—H, He^+, Li^{2+}, etc.). In developing his model, Bohr accepted some past notions, rejected others, and assumed some new ones.

1. The Rutherford nuclear model of the atom was accepted.
2. The theories of Planck and of Einstein that radiant energy is quantized in units of $h\nu$, where h is Planck's constant and ν is the frequency of the radiant energy, were accepted.
3. The electron was assumed to travel in circular orbits.
4. Of all possible orbits, only certain ones were acceptable—namely those which had a specified angular momentum. (Another way of saying this is that the angular momentum is *quantized*.) The values were taken to be integral multiples of $h/2\pi$.
5. The classical physical theory from electrodynamics that a charged particle undergoing acceleration must emit electromagnetic radiation was rejected for electrons within atoms.
6. Radiation was postulated to be emitted or absorbed only when the electron jumped from one orbit to another; the energy emitted or absorbed corresponded to the difference in the energies for the initial and final states of the system.
7. Except as noted above, classical physics was assumed to be applicable to the atoms.

Before going further, we should note that Bohr's assumption of circular orbits has been shown to be too restrictive. Assumptions 1, 2, 5, and 6 are retained in the presently held description of atoms by wave mechanics, whereas assumption 4 comes as a result of the one abitrary assumption of wave mechanics. Accordingly, we will not pay too much

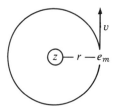

Figure 1.2   Bohr model of hydrogenlike atoms.

attention to the geometry of the Bohr model, but rather shall be more concerned with the energy states of the atoms based on Bohr's model.

From assumptions 1 and 3 above, the Bohr model for hydrogenlike atoms may be pictured as having a heavy nucleus bearing a charge of Ze (where Z is the atomic number and e is the magnitude of the charge on the electron), with an electron of charge e and mass m traveling with a velocity v in an orbit of radius r from the nucleus (see Figure 1.2).

The following relationships for the electron in H result from the assumptions listed above. They permit a calculation of the total energy of the H atom.

From classical physics the centrifugal force may be equated with the coulombic attraction:

$$\frac{mv^2}{r} = \frac{Ze^2}{r^2} \tag{1.2}$$

$$mv^2 r = Ze^2 \tag{1.3}$$

The total energy, E, is the sum of the kinetic and potential energy:

$$E = \tfrac{1}{2}mv^2 - \frac{Ze^2}{r} \tag{1.4}$$

Substituting from (1.3) for $\tfrac{1}{2}mv^2$, we obtain

$$E = -\frac{1}{2}\frac{Ze^2}{r} \tag{1.5}$$

Quantizing the angular momentum gives

$$mvr = n\left(\frac{h}{2\pi}\right) \tag{1.6}$$

where n (called the **quantum number**) must be an integer and h is Planck's constant.[2]

---

[2] Bohr did not start out by assuming quantization of angular momentum. Instead, he realized that he needed to find analytical expressions for energy states such that the differences between them would account for the frequencies of spectral lines in the same way that differences between the numerical terms (which were theoretically uninterpretable) did. Solving the equations of motion for a charged particle in the field of a nucleus gave closed orbits only if the total energy of the system were somehow given. Bohr assumed that, in attachment of the electron to the H nucleus, an amount of energy given by $\tfrac{1}{2}\pi nh\omega$ was emitted, giving a total energy for the

When (1.3) is divided by (1.6) we obtain

$$v = Ze^2 \frac{2\pi}{nh} \qquad (1.7)$$

From (1.3) and (1.7) we obtain

$$r = \frac{Ze^2}{mv^2} = \frac{n^2 h^2}{4\pi^2 m Z e^2} \qquad (1.8)$$

From (1.5) and (1.8) we obtain

$$E = \frac{-2\pi^2 m Z^2 e^4}{n^2 h^2} = \frac{E_{n=1}}{n^2} \qquad (1.9)$$

Equation (1.9) gives the energy of hydrogenlike atoms in various quantum states. For the hydrogen atom itself, the lowest possible energy state—that is, the energy state with $n = 1$—has the value of -13.6 eV or -1312 kJ/mol. The lowest energy state for an atom (or ion or molecule) is called the **ground state.** The first higher-energy state ($n = 2$) above the ground state is called the **first exited state,** the next higher state ($n = 3$) is called the **second excited state,** and so on. The amount of energy needed to promote an atom from the ground state to a given excited state is called the **excitation energy.** The amount of energy needed to remove an electron from an atom in its ground state is called the **ionization energy.** The **separation energy** is the amount of energy necessary to remove an electron from an atom in a particular excited state. These relationships are illustrated in Figure 1.3, an energy-level diagram for the hydrogen atom. In such a diagram, only the ordinate has meaning.

On the basis of the Bohr model, we may conclude about hydrogenlike atoms that (1) the ionization energy for the removal of the single electron is proportional to Z^2, (2) the radius of the hydrogen atom in the ground state is 52.9 pm, and for hydrogenlike atoms it is inversely proportional to Z, (3) the radius of the atom in excited states is proportional to n^2, and (4) in the ground state of H the electron is traveling with a velocity of 2.187×10^8 cm/s. These values have become the standard units of atomic physics. The unit of distance $a_0 = 52.9$ pm is called the **Bohr radius;** the unit of energy is either the Rydberg (13.605 eV) or the Hartree, which is e^2/a_0 and turns out to be 2×13.605 eV.

A major achievement of the Bohr model was its ability to account for the spectra of hydrogenlike atoms. By the time Bohr proposed his model of the atom, spectroscopists had formulated many empirical rules dealing with the line spectra of atoms. (That is, many regularities in the positions of spectral lines had been observed which could be formulated mathematically, but no theoretical reasons for such regularities were apparent.)

---

system of $-\frac{1}{2}\pi n h \omega$, where ω is the frequency of rotation of the electron in its orbit. This rather arbitrary assumption led to a correct fit of frequencies of spectral lines. The quantization of angular momentum in units of $h/2\pi$ results for the special case of circular orbits. His theory was in contrast to other accounts of the origin of spectral lines which explained the frequencies of emission as due to oscillations of electrons with certain natural frequencies within the atom which matched the frequencies of emitted radiation in accord with Maxwell's equations.

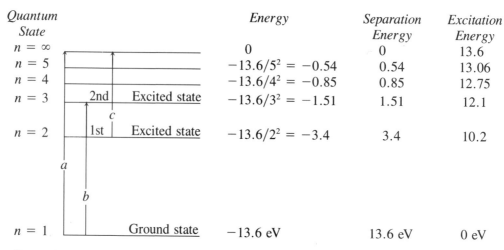

Quantum State			Energy	Separation Energy	Excitation Energy
$n = \infty$			0	0	13.6
$n = 5$			$-13.6/5^2 = -0.54$	0.54	13.06
$n = 4$			$-13.6/4^2 = -0.85$	0.85	12.75
$n = 3$	2nd	Excited state	$-13.6/3^2 = -1.51$	1.51	12.1
$n = 2$	1st	Excited state	$-13.6/2^2 = -3.4$	3.4	10.2
$n = 1$		Ground state	-13.6 eV	13.6 eV	0 eV

Figure 1.3 Energy-level diagram for the hydrogen atom. Line *a* corresponds to the ionization energy: 13.6 eV. Line *b* corresponds to the excitation energy necessary to produce the second excited state: 12.1 eV. Line *c* corresponds to the separation energy of the first excited state: 3.4 eV.

Among these rules, the one having the greatest influence on Bohr was that frequency (or wavenumber, $\tilde{\nu} \equiv 1/\lambda$) of the numerous individual lines observed in a given spectrum can be reduced to differences among a smaller number of numerical terms T. The terms were a manifold (collection) of numerical constants characteristic of each different kind of atom. For atomic H the terms take the form R/n^2, where R is a constant called the **Rydberg constant** (having a value of 109,677.581 cm^{-1} = 13.605 eV) and n is an integer. Thus, empirically, all lines of the spectrum of atomic hydrogen have wave numbers given by the equation

$$\tilde{\nu} = \frac{R}{n_1^2} - \frac{R}{n_2^2} = T_1 - T_2 = \frac{1}{\lambda} \tag{1.10}$$

According to Bohr's theory

$$h\nu = E_2 - E_1 = \frac{hc}{\lambda} \tag{1.11}$$

The term values of the spectroscopist thus are virtually identical to energy levels within the atom, differing only in sign (as a result of defining the potential energy of the ionized atom as zero) and a constant factor of hc (which essentially takes care of the difference in units). *Bohr's theory gives an interpretation to the empirically discovered integer constants in Equation (1.10): They are the quantum numbers which determine the energy of the allowed states of the H atom.* From Bohr's theory the Rydberg constant[3] R is given by

---

[3] More rigorously, the electron mass, m, in the Bohr equations should be replaced by the reduced mass, μ. The reduced mass is given by $1/\mu = 1/m + 1/M$, where M is the nuclear mass. If the nucleus is assumed to be stationary (i.e., $M \gg m$), the value of R is 109,737.31 cm^{-1}.

$2\pi^2me^4/h^3c$. The excellent agreement between the value calculated from these fundamental constants and the spectroscopically derived value gave strong support to Bohr's theory.

The Bohr theory, with some modifications, was found to be capable of explaining the Moseley relationship between characteristic x-ray spectra and atomic number [Equation (1.1)]. The excited state for emission of x-rays was postulated as one in which a low-energy electron had been knocked out of a polyelectron (many-electron) atom. An electron from a higher-energy state could then drop down to the lower empty orbit with emission of an x-ray. In a polyelectron atom, a given electron is shielded partially from the positive charge of the nucleus by the electrons between it and the nucleus. The effective nuclear charge may be taken as $Z - \sigma$, where σ represents the shielding effect of underlying electrons. For an electron in a polyelectron atom dropping from the n_2 to the n_1 state, the Bohr theory thus would predict

$$\nu = \frac{2\pi^2 m(Z - \sigma)^2 e^4}{h^3}\left(\frac{1}{n_1^2} - \frac{1}{n_2^2}\right) \tag{1.12}$$

or, for $n_2 = 2$ and $n_1 = 1$,

$$\nu = \tfrac{3}{4}Rc(Z - \sigma)^2 \tag{1.13}$$

Using the Bohr model adjusted for shielding allows us to focus on a single electron at a time, lumping effects due to all other electrons into the shielding parameter σ. This is the so-called "one-electron" model for electrons in atoms.

Despite partially successful attempts to extend the Bohr theory to atoms and ions with more than one electron, it proved impossible to treat rigorously even the two-electron case of He. The main difficulty lay in Bohr's treatment of the electron as a particle.

In classical physics a dichotomy exists between particle and wave models. A particle has measurable mass, momentum ($p = mv$), and trajectory (orbit). Certain problems such as the motions of the planets around the sun or the path of a cannonball, rocket, or satellite can be solved by considering the motion of particles subject to forces governed by Newton's laws and electrodynamics. Bohr's treatment of the H atom followed along these lines [see Equations (1.2)–(1.5)], although there were certain *ad hoc* assumptions [Equation (1.6)].

Other phenomena in classical physics were analyzed in terms of a wave model. In contrast to a particle, a wave has no definite location in space, but instead is a periodic disturbance with amplitude A which oscillates along the direction of propagation the pattern repeating over a characteristic distance λ (called the **wavelength**) and repeating with a characteristic frequency ν over a time period $\omega = 1/\nu$. The wave is propagated with a characteristic velocity. Vibrating strings and electromagnetic radiation (including visible light) can be treated using the wave model.

By the 1920s, it became apparent that matter also has wave properties and that wave behavior is especially important for matter on the atomic scale. Only by incorporating the wave description into equations governing the behavior of electrons in atoms was a satisfactory account of this behavior achieved. We now consider some of the experiments and ideas which led to the "dual" description of electrons as both particles and waves and which finally provided a more adequate treatment of polyelectron atoms and ions.

▶ *1.3.2 Wave Mechanics*

Wave Properties of Matter

Compton applied the laws of conservation of energy and conservation of momentum to the collision of a photon with a free electron and predicted the change in momentum or frequency of the photon for various angles of the scattered radiation relative to the incident radiation. Measurements of the scattering of x-rays by the "free" electron of graphite verified the predictions, thereby proving that photons have momenta and thus can behave like particles.

The Planck equation for the energy of a photon, $E = h\nu$, may be combined with the Einstein equation relating mass and energy, $E = mc^2$, to give an expression for the momentum of a photon, $mc = h\nu/c$. Using the relationship $\lambda\nu = c$ leads to a connection between wavelength λ and photon momentum $mc = h/\lambda$. DeBroglie suggested in 1924 that the particle–wave dualism is not restricted to light, but that all moving particles must have an associated wave. For particles in general, the wavelength of the associated waves is given by the deBroglie equation

$$\lambda = \frac{h}{m\upsilon} \tag{1.14}$$

where m is the mass of a particle traveling with a velocity υ. The wave properties predicted for matter have been verified for free particles by experiments in which diffraction patterns have been observed for beams of electrons, neutrons, and atoms.

In a classical wave picture, the intensity I of a light wave is proportional to the square of the maximum wave amplitude A: $I \propto A^2$. In a photon picture, intensity is proportional to the number of photons striking a given area. If both representations are valid, then the number of photons striking a given area in a given time period must be proportional to the square of the amplitude of the light wave striking that area. Because a given area may be subdivided into an infinite number of smaller areas, whereas a light beam would have a finite number of photons, the above statement may be modified to read that the *probability of finding a photon* in a given area is proportional to the square of the amplitude of the wave associated with the particle in the given volume.

In terms of deBroglie's matter waves, the Bohr quantization condition amounts to allowing just those orbits having a circumference which is a whole multiple of the wavelength associated with the electron in the orbit. That is, $n\lambda = 2\pi r$ would give rise to a standing electron wave. Replacing λ by $h/m\upsilon$ gives Equation (1.6). A circumference whose length was a nonintegral number of wavelengths would result in wave interference causing the electron wave to partially or completely cancel itself.

The Uncertainty Principle

We can see a reciprocal relationship between wave properties and particle properties by rearranging the equations $E = h\nu$ and $\lambda = h/p$ [where p stands for momentum and replaces the $m\upsilon$ of Equation (1.14)] to give

$$h = ET = p\lambda \tag{1.15}$$

(where T is the period of the wave, $T = 1/\nu$).

Energy and momentum usually are associated with particles, whereas period and wavelength are associated with waves. When one of these (E or p) is large, the other (T or λ) will be small. Thus, for long wavelengths, such as radio waves, it is difficult to show particle behavior. For heavy particles, such as a ball, the wave treatment is not very useful.

An equation very similar in form to Equation (1.15) was proposed by Heisenberg and has become known as **Heisenberg's uncertainty principle.** This equation, $(\Delta p_x)(\Delta x) \geq h$, states that the product of the uncertainty in momentum with respect to a given coordinate and the uncertainty in position with respect to the same coordinate must be equal to or greater than Planck's constant. This uncertainty is caused not by experimental errors, but rather by the inherent indeterminacy in describing simultaneously both position and momentum. A corollary, from Bohr, states that a single experiment cannot simultaneously show particle and wave properties of radiation.

The Bohr model of the atom specified both the momentum and position of the electron. According to the uncertainty principle, it is not possible to have such precise knowledge of both momentum and position. As the uncertainty in position becomes smaller (that is, as we locate an electron more precisely), Δp becomes larger so that our knowledge of the momentum loses precision and vice versa.

▶ *1.3.3   The Schrödinger Equation*

In 1926, Schrödinger proposed an equation that specified no discrete orbits, but instead described the wave associated with the electron "particle". This equation, given below, provided the basis for wave mechanics[4]:

$$-\frac{h^2}{8\pi^2 m}\left(\frac{\partial^2 \Psi}{\partial x^2} + \frac{\partial^2 \Psi}{\partial y^2} + \frac{\partial^2 \Psi}{\partial z^2}\right) + V\Psi = E\Psi \tag{1.16}$$

In this equation, Ψ is the amplitude of the wavefunction associated with the electron, E is the total energy of the system, V is the potential energy of the system (equal to $-e^2/r$ for hydrogen), m is the mass of the electron, h is Planck's constant, and x, y, and z are the Cartesian coordinates. The frequency of the wave describing the electron is related to its energy by $E = h\nu$. Compared to Bohr's postulates, given on p. 9, the last statement replaces postulate 5 and the wave form of the equation replaces postulate 4, the other postulates being retained.

The Schrödinger equation is generally written in operator form

$$\tilde{H}\Psi = E\Psi \tag{1.17}$$

(Hamiltonian) operator   **Eigenvalue** (a constant; the energy or other observable)

Eigenfunction (or operand, or trial wavefunction)

---

[4] The equation is known as the **time-independent Schrödinger wave equation,** and the solutions obtained are to be multiplied by a phase factor $e^{-i\omega t}$.

$\tilde{H}$ is an **operator**—that is, it gives directions for an operation (such as taking a derivative) to be performed on the function that follows it. If the result of performing this operation is a constant times the original trial function, the trial function is termed an **eigenfunction** of the operator and the constant is called an **eigenvalue.** The Hamiltonian operator is the one associated with energy. Other operators are associated with other observables such as angular momentum. In general, an operator may itself be several other operators applied sequentially provided that the order of application does not alter the result—that is, the operators commute. Such commutivity is necessary if the eigenvalues corresponding to two or more operators are to be simultaneously observable. In Equation (1.16), the Hamiltonian operator consists of taking the second partial derivatives of the trial function with respect to each of the Cartesian coordinates and of multiplying each by a constant and adding the results to the potential energy V.

Because the probability of finding the electron in a given volume element is proportional to $\Psi^2 d\tau$ (where $d\tau$ is the size of the volume element), Ψ itself must be a single-valued function with respect to the spatial coordinates, must be continuous, and must become zero at infinity. These conditions are imposed as boundary conditions on the solutions to the wave equation.

The transformation from Cartesian coordinates x, y, z into polar coordinates facilitates solution of the wave equation. The position variables in polar coordinates are r, θ, and ϕ, where r is the radial distance of a point from the origin, θ is the inclination of the radial line to the z axis, and ϕ is the angle made with the x axis by the projection of the radial line in the xy plane (see Figure 1.4). After this transformation the Schrödinger equation can be factored into two equations, one depending only on r and the other only on θ and ϕ.

The relationship between the Cartesian and spherical polar coordinates results using the right triangles visible in Figure 1.4:

$$\cos \theta = z/r, \qquad z = r \cos \theta \qquad (1.18)$$

$$\cos(90 - \theta) = \sin \theta = s/r, \qquad s = r \sin \theta \qquad (1.19)$$

$$\cos \phi = x/s, \qquad x = s \cos \phi = r \sin \theta \cos \phi \qquad (1.20)$$

$$\sin \phi = y/s, \qquad y = s \sin \phi = r \sin \theta \sin \phi \qquad (1.21)$$

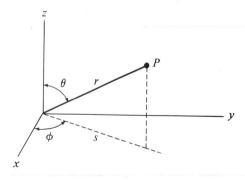

Figure 1.4 Variables of polar coordinates.

The solutions for Ψ, called **wavefunctions,** may be expressed as the product of three functions, each of which depends on only one spherical polar coordinate:

$$\Psi(r, \theta, \phi) = R(r)\Theta(\theta)\Phi(\phi) \qquad (1.22)$$

The boundary conditions require that certain constants entering into the solution of the wave equation take on only certain integral values. These constants, called **quantum numbers,** are designated by n, l, and m_l. The **principal quantum number,** n, may take on the values 1, 2, 3, . . . ; l may have the values 0, 1, 2, . . . , up to $(n - 1)$; m_l can have values ranging from $-l$ through 0 to $+l$. The **wavefunctions,** Ψ_{n,l,m_l}, which are solutions of the Schrödinger equation, are commonly called **orbitals.** Orbitals for which $l = 0, 1, 2, 3$, and 4, are called, respectively, s, p, d, f, and g orbitals.

Solving the Schrödinger equation is beyond the scope of this book. Instead, we examine the wavefunctions (orbitals) which are solutions. These Ψ's describe the energies and spatial distributions of electrons in the H atom.

Radial Functions

The radial part of the wavefunction is given below:

$$R_{n,l} = \text{constant} \times \left(\frac{2Zr}{na_0}\right)^l \cdot e^{-Zr/na_0} \cdot L_{n+l}^{2l+1}(x) \qquad (1.23)$$

R depends only on the n and l values and has an exponential term e^{-Zr/a_0} and a term involving a polynomial L of the $(n - 1)$ degree known as an **associated Laguerre polynomial** given in Table 1.1. The preexponential term consists of a normalization constant and a term $(2Zr/na_0)^l$.

Table 1.1   Associated Laguerre polynomials[a,b]

Electron	n	l	$L_{n+l}^{2l+1}(x)$
$1s$	1	0	$-1!$
$2p$	2	1	$-3!$
$3d$	3	2	$-5!$
$4f$	4	3	$-7!$
$2s$	2	0	$2x - 4$
$3p$	3	1	$24x - 96$
$4d$	4	2	$720x - 4320$
$3s$	3	0	$-3x^2 + 18x - 18$
$4p$	4	1	$-60x^2 + 600x - 1200$
$4s$	4	0	$4x^3 - 48x^2 + 144x - 96$

[a] From H. E. White, *Introduction to Atomic Spectroscopy*, McGraw–Hill, New York, 1934, p. 67.
[b] $x = 2Zr/na_0$.

The exponential term has the effect of causing R to drop off more slowly as values of the principal quantum number n increase; in other words, orbital size increases with increasing n. The $(2Zr/na_0)^l$ term causes the R value of all orbitals with $l > 0$ to go to zero as r goes to zero; that is, only s orbital electrons have any probability of being found in the vicinity of the atomic nucleus. (This has consequences in such areas as Mössbauer spectroscopy, in which only changes in electron density at the nucleus are detected.) The associated Laguerre polynomials produce sign changes in R as r varies for orbitals with $(n - l) \geq 2$. This produces orthogonality in orbitals with equal l but different n; that is, $\int \Psi_1 \Psi_2 d\tau = 0$ or $\int R_1 R_2 dr = 0$. Orthogonality guarantees the noninterference of different orbitals. Finally, the normalization constant is introduced to ensure a unity probability of finding the electron in all space—that is, $\int \Psi\Psi^* d\tau = 1$.

The probability of finding the electron in a volume element of fixed size centered about some given point at a distance r from the nucleus is given by $R^2 d\tau$ where $d\tau$ is the elemental volume unit. The probability that the electron will be found at a distance r from the nucleus is given by $(4\pi r^2)R^2 dr$; this is the probability that the electron will be found in a spherical shell of thickness dr at a distance from the nucleus ranging from r to $r + dr$. Figure 1.5 gives curves showing R^2 and $(4\pi r^2)R^2$ versus r for various hydrogen atomic orbitals (note that the scale changes with each plot). The maximum in the shaded curve occurs at the distance where the probability of finding the electron is a maximum. At some r values the probability is zero; these are called **radial nodes.**

Angular Functions

Figure 1.5 shows electron distribution curves for an atom in free space—that is, where no basis exists for assigning one or more Cartesian reference axes and where, consequently, Θ^2 and Φ^2 are constant. In other words, there is no dependence on θ or ϕ, only on r. In general, we are interested in atoms in molecules in which the symmetry is less than spherical and where the electron distribution *is* dependent on the angular part of the wavefunction as well as on the radial part. Thus we now discuss the angular functions which specify the shapes of orbitals.

The angular functions may be generated from the equation

$$A = \frac{x^a y^b z^c}{r^l} \tag{1.24}$$

where l is the l quantum number and a, b, and c may take positive (or zero) integer values such that $a + b + c = l$. The variables x, y, and z are the Cartesian coordinates, and r is the distance from the origin. In addition to the angular function of a given l value, all orbitals of lower l value of the same **parity** (even or odd l) will be generated. Linear combinations of the solutions may be made to achieve an orthogonal (linearly independent) set of solutions.

For $l = 0$, $A = x^0 y^0 z^0/r^0 = 1$. An s orbital has no angular dependence; it is spherically symmetric, and the radial functions given in Figure 1.5 completely describe the s orbital. For $l = 1$, the possibilities for A are x/r, y/r, and z/r. Making a plot of these functions to reveal their shape is more complicated than it might seem at first, because, as written, they contain both Cartesian coordinates and the polar coordinate $r = \sqrt{x^2 + y^2 + z^2}$.

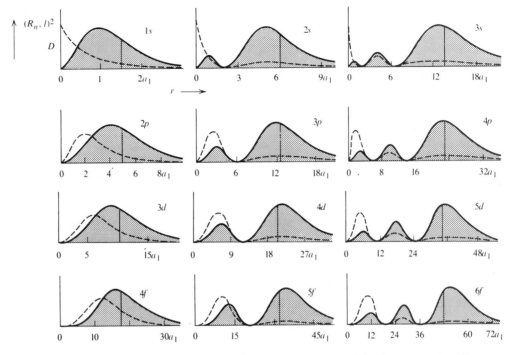

Figure 1.5 The probability-density factor $(R_{n,l})^2$ plotted as a function of the electron–nuclear distance r (r is given in units $a_0 = 53$ pm, the radius of the first Bohr circular orbit). The density distribution curves, $D = 4\pi r^2(R_{n,l})^2$, are the shaded areas. Vertical lines represent average distance over time for $(R_{n,l})^2$. (From H. E. White, *Introduction to Atomic Spectra*, McGraw–Hill, New York, 1934, p. 68.)

EXAMPLE: Make a plot of the function $A = x/r$.

Solution: It is simpler to express the function only in polar coordinates $x/r = \sin \theta \cos \phi$. Because the function depends on two variables we would need a three-dimensional plot to represent it. Selecting a particular value of θ gives the projection of the function in a plane. $\theta = \pi/2$ gives the xy plane. The data of Table 1.2 are plotted in Figure 1.6. A value of ϕ is selected and a dot is placed at that angle at a distance from the origin which is proportional to the value of the function $\cos \phi$. As might be guessed from the Cartesian form of the function x/r, the maximum in Figure 1.6 occurs along the x axis where $\phi = 0$. Going up or down from the xy plane, the value of θ increases or decreases from its value of $\pi/2$ in the plane and $\sin \theta$ becomes <1. The value of the function is reduced by the factor $\sin \theta$ for each ϕ value until it is 0 for $\theta = 0$ and π. Thus, the function is the surface of a pair of tangent spheres, with the value of the function in any direction being the distance from the origin to the surface. The sign of x/r is the same as the sign of x. The sign of the wavefunction is of particular importance in connection with the formation of molecular orbitals from atomic orbitals. The function x/r vanishes where $x = 0$—that is, in the yz plane which is an angular nodal plane for this orbital.

The plots of the functions y/r and z/r also consist of tangent spheres. These have the same appearance as Figure 1.6 except that the maximum lies along the y and z axes, respectively. The three p orbitals having the angular functions x/r, y/r, and z/r are commonly denoted as p_x, p_y, and p_z.

Table 1.2   Data for the function
$A = x/r = \sin\theta\cos\phi$ **in the xy plane**

θ	ϕ	$A = \sin\theta\cos\phi = 1\cdot\cos\phi$
$\pi/2$	0	1
$\pi/2$	$\pi/4$	0.707
$\pi/2$	$\pi/2$	0
$\pi/2$	$3\pi/4$	-0.707
$\pi/2$	π	-1
$\pi/2$	$5\pi/4$	-0.707
$\pi/2$	$3\pi/2$	0

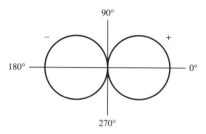

Figure 1.6   A plot of the function $A = x/r = \sin\theta\cos\phi$ in the xy plane (where $\theta = \pi/2$).

Turning to the angular functions for d orbitals ($l = 2$), the possible solutions for A are x^2/r^2, y^2/r^2, z^2/r^2, xy/r^2, xz/r^2, and yz/r^2. Knowing that the s-orbital angular function ($l = 0$) is included in the above solutions, we readily find it in the sum of the first three: $A_s = x^2/r^2 + y^2/r^2 + z^2/r^2 = 1$. The two other linear combinations we can construct from these three terms, which are orthogonal to the s orbital and to each other, are

$$d_{x^2-y^2} = x^2/r^2 - y^2/r^2$$
$$d_{z^2} = -x^2/r^2 - y^2/r^2 + 2z^2/r^2 \tag{1.25}$$

These and other angular functions are shown in Figure 1.7. Table 1.3 gives angular functions in polar coordinates.

For f orbitals ($l = 3$) A functions contain terms of third degree. Comparing x/r, xy/r^2 and the f orbital xyz/r^3 we note that each time a new coordinate is introduced, the number of lobes doubles and that the sign of each lobe remains the same on the positive side of the "new" coordinate and inverts on the negative side of the new coordinate. Similarly, we may visualize other f orbitals from knowledge of d orbital shapes; that is, $f_{z(x^2-y^2)}$ and the similar $f_{y(x^2-z^2)}$ and $f_{x(z^2-y^2)}$ are all related in shape to the $d_{x^2-y^2}$ orbital, with lobes both above and below the nodal xy plane alternating in sign by octant. The latter two f orbitals differ only in orientation, being symmetric about the xz and yz planes, respectively. The "d_{z^2}" orbital is actually the $d_{(2z^2-x^2-y^2)}$; the $-x^2 - y^2$ terms give the torus of sign -1, which is maximized in the xy plane. The "f_{z^3}" is actually $(2z^3 - zx^2 - zy^2)/r^3$ and might be visualized as $z(2z^2 - x^2 - y^2)$; that is, a new nodal plane appears at $z = 0$ (the part of the torus on the $+z$ side has a negative sign, whereas the part on the $-z$ side has a positive sign).[5]

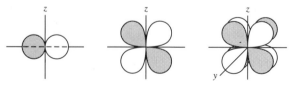

---

[5] Representations of f-orbital shapes can be found in O. Kikuchi and K. Suzuki, *J. Chem. Educ.* **1985,** *62,* 206.

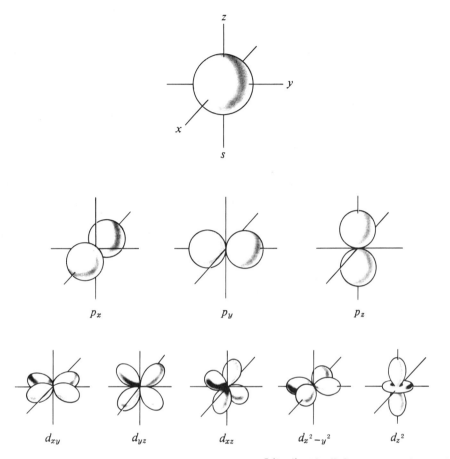

Figure 1.7 Total angular dependence of the wavefunction $\Psi(\theta, \phi)$ with all Cartesian coordinates fixed.

Table 1.3 Φ_{m_l} Functions for the hydrogen atom[a]

$$\Phi_0(\phi) = \frac{1}{\sqrt{2\pi}} \qquad \text{or} \qquad \Phi_0(\phi) = \frac{1}{\sqrt{2\pi}}$$

$$\Phi_1(\phi) = \frac{1}{\sqrt{2\pi}} e^{i\phi} \qquad \text{or} \qquad \Phi_{1\cos}(\phi) = \frac{1}{\sqrt{\pi}} \cos \phi$$

$$\Phi_{-1}(\phi) = \frac{1}{\sqrt{2\pi}} e^{-i\phi} \qquad \text{or} \qquad \Phi_{1\sin}(\phi) = \frac{1}{\sqrt{\pi}} \sin \phi$$

$$\Phi_2(\phi) = \frac{1}{\sqrt{2\pi}} e^{i2\phi} \qquad \text{or} \qquad \Phi_{2\cos}(\phi) = \frac{1}{\sqrt{\pi}} \cos 2\phi$$

$$\Phi_{-2}(\phi) = \frac{1}{\sqrt{2\pi}} e^{-i2\phi} \qquad \text{or} \qquad \Phi_{2\sin}(\phi) = \frac{1}{\sqrt{\pi}} \sin 2\phi$$

[a] From L. Pauling, *The Nature of the Chemical Bond*, 3rd ed. Copyright © 1960 by Cornell University. Used by permission of Cornell University Press.

The real angular functions presented here do not all have well-defined m_l values. When the Schrödinger equation is solved using the methods of differential equations, separation of variables leads to three separate differential equations—one each in r, θ, and ϕ. As pointed out previously, $\Psi(r, \theta, \phi)$ is the product of three separate functions $R(r)$, $\Theta(\theta)$, and $\Phi(\phi)$, which are the solutions of the three differential equations. The differential equation in ϕ is

$$\frac{\partial^2 \Phi}{\partial \phi^2} = -m_l^2 \Phi \tag{1.26}$$

Normalized solutions to this equation are $\Phi(\phi) = (2\pi)^{-1/2} e^{im_l}$ which give Φ's having well-defined m_l values. Figure 1.8 shows plots of the imaginary angular functions. Any linear combination of Φ's having the same $|m_l|$ is also a solution of Equation (1.26). Hence, it is perfectly permissible to use orbitals which are linear combinations of the Φ's chosen to give real angular functions.

EXAMPLE: Show that $\Phi_1 = e^{-im_l} + e^{im_l}$ is a solution of the differential equation (1.26).

Solution:

$$\frac{\partial^2 \Phi_1}{\partial \phi^2} = \frac{\partial}{\partial \phi}[im_l e^{im_l\phi} - im_l e^{-im_l\phi}]$$

$$= -m_l^2 e^{im_l\phi} - m_l^2 e^{-im_l\phi}$$

$$= -m_l^2 [e^{im_l\phi} + e^{-im_l\phi}] = -m_l^2 \Phi_1$$

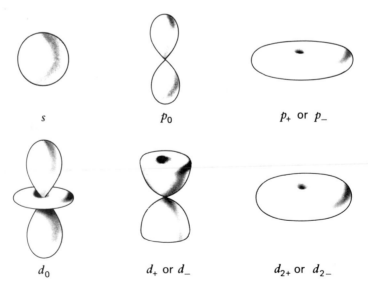

Figure 1.8   Angular dependence of $\Psi\Psi^*$ with a single designated Cartesian axis (z).

EXAMPLE: The function $\Theta_l^{\pm 1}(\theta)$ for $l = 1$ is $\sin\theta$. Show that the angular function $\frac{1}{2}\Theta_l^{\pm 1}[e^{i\phi} + e^{-i\phi}]$ gives the real p_x orbital x/r.

Solution:

$$\frac{1}{2}\sin\theta[e^{i\phi} + e^{-i\phi}] = \frac{1}{2}\sin\theta[\cos\phi + i\sin\phi + \cos\phi - i\sin\phi]$$

$$= \frac{1}{2}\sin\theta[2\cos\phi] = \sin\theta\cos\phi = x/r$$

Because any linear combination of Φ's is also a solution of the Schrödinger equation, we can obtain valid angular functions from any linear combination of these or from any linear combination of real angular functions A (because these are also no more than linear combinations of Φ's).

The following set of d orbitals of well-defined m_l values may be constructed from the "real" set:

$$d_{-2} = \sqrt{\frac{3}{8}}\frac{(x - iy)^2}{r^2}$$

$$d_{-1} = \sqrt{\frac{3}{2}}\frac{z(x - iy)}{r^2}$$

$$d_0 = \sqrt{\frac{1}{4}}\frac{3z^2 - r^2}{r^2} \qquad (1.27)$$

$$d_1 = -\sqrt{\frac{3}{2}}\frac{z(x + iy)}{r^2}$$

$$d_2 = \sqrt{\frac{3}{8}}\frac{(x + iy)^2}{r^2}$$

The information needed to determine the angular functions of the orbitals is carried in the subscript of the orbital name ($d_{x^2-y^2}$, etc.). In the few cases in which the subscript is an abbreviated expression (d_{z^2}, f_{z^3}, etc.) the orbital behaves mathematically as the polynomial given. Several different sets of orbitals may be formulated for a particular l value by linear combination; the appropriate set for any particular situation is determined by the symmetry of the environment of the atom. The A surfaces are often referred to as orbitals and, indeed, serve very satisfactorily for most arguments concerned with chemical bonding. Note, however, that the variation of Ψ in space is the product of $R(r)$ and $Y(\theta, \phi) = \Theta(\theta)\,\Phi(\phi)$.

We are often interested in knowing the electron density at each point in space, which is proportional to Ψ^2. A common way of representing this information is by a contour plot such as the ones depicted in Figure 1.9. A two-dimensional plot of the three-dimensional function $\Psi^2(x, y, z)$ is made by slicing through the three-dimensional surface and thus projecting the function only in one plane. For example, Figure 1.9a shows the plot for $2p_z$

in the xy ($z = 0$ or $\theta = \pi/2$) plane, whereas Figure 1.9b shows $3d_{z^2}$ in the xz ($y = 0$ or $\theta = 0$, $\phi = 0$) or yz ($x = 0$ or $\theta = 0$, $\phi = \pi/2$) plane (or in any plane containing the z axis). Contours having a constant value of Ψ^2 are represented.

▶ *1.3.4 Polyelectronic Atoms*

The one-electron orbitals are exact solutions of the Schrödinger equation and are determined uniquely by three quantum numbers n, l, and m_l. These quantum numbers specify the energy (n) and angular momentum (l, m_l) of an electron in the orbital. In particular, l specifies the magnitude of the orbital angular momentum vector (which has a length of $\sqrt{l(l + 1)}\,(h/2\pi)$). m_l gives the component of the orbital angular momentum in units of $h/2\pi$ along some direction arbitrarily defined (for example, the direction of an imposed magnetic field) and conventionally labeled as z.

Solving the Schrödinger equation exactly for polyelectron systems is a challenge still to be met. Wavefunctions for polyelectron atoms are currently approximated as products of a number of hydrogenlike (one-electron) wavefunctions. But how many of each kind ($1s$, $2s$, etc.) should we use? The answer lies in one of the remarkable insights into atomic structure reached just prior to Schrödinger's equation—electron spin and the Pauli exclusion principle.

The number of lines in atomic spectra increases when the atoms are placed in a magnetic field (the Zeeman effect). Splitting occurs because the different possible orientations of the orbital angular momentum l with respect to the magnetic field differ in energy. The

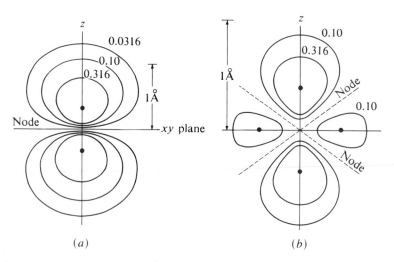

Figure 1.9   (*a*) Contours of constant Ψ^2 at 0.0316, 0.10, and 0.316 of maximum for a C $2p_z$ orbital. The xy plane is a nodal surface. (*b*) Contours of constant Ψ^2 at 0.10 and 0.316 of maximum for a Ti^{III} $3d_{z^2}$ orbital. (From E. A. Ogryzlo and G. Porter, *J. Chem. Educ.* **1963,** 40, 258.)

l component in the z direction is permitted only the values $m_l = -l, \ldots, 0, \ldots, +l$; there are $(2l + 1)$ possibilities. An electron in an l state has $(2l + 1)$ accessible energy levels in a magnetic field. Obviously, only one level is expected when $l = 0$. In 1922, Stern and Gerlach passed a beam of Ag atoms through an inhomogeneous magnetic field and found that it was split into two beams. The highest-energy electron in Ag has $l = 0$ and was not predicted to show a Zeeman splitting. The only value of l which would predict a splitting into two levels was $l = \frac{1}{2}$, an impermissible value! Furthermore, interpretation of the atomic spectrum of Ag indicated that its ground state has no orbital angular momentum and should thus be unaffected by a magnetic field. To account for these observations which did not fit then-current models, Uhlenbeck and Goudsmit proposed that the electron is itself spinning and that its spin angular momentum is $\frac{1}{2}(h/2\pi) = s(h/2\pi)$. This is just half the value of the quantized units of orbital angular momentum. The Stern–Gerloch experiment could be explained if the electron could align itself only parallel or antiparallel to an external magnetic field—that is, with $m_s = \pm\frac{1}{2}$. We now have four quantum numbers which characterize an electron in an orbital: n, l, m_l, and m_s (which can adopt only values of $\pm\frac{1}{2}$).

Wolfgang Pauli proposed in 1925 (without knowledge of the proposal of electron spin) that *four quantum numbers are required to characterize an electron in an atom and that no two electrons in an atom can have the same set of quantum numbers.*[6] This is known as the **Pauli exclusion principle.**

Because $l = 0, 1, \ldots, (n - 1)$ and m_l can have the values of $-l, \ldots, 0, \ldots, +l$, there will be n^2 orbitals for a given n, each of which can contain two electrons—one with $m_s = +\frac{1}{2}$ and one with $m_s = -\frac{1}{2}$. Table 1.4 shows the quantum numbers, orbital designations, and shell descriptions corresponding to the orbitals through $n = 3$. The shell designation stands for the orbitals of a given principal quantum number n: $K, L, M, N, \ldots$ correspond to values of n of 1, 2, 3, 4, $\ldots$, respectively. The letters s, p, d, f, g of the orbital designation indicate l values of 0, 1, 2, 3, 4, $\ldots$, respectively, and the number preceding the letter gives the principal quantum number. Thus for a $3p$ electron, $n = 3$ and $l = 1$.

Polyelectron atoms are formed by adding electrons successively to hydrogenlike orbitals. The number of electrons that may occupy each orbital is limited to two by the Pauli

Table 1.4   Orbital and shell designations for different quantum states

Value of n:	1	2	2	2	2	3	3	3	3	3	3	3	3	3
Value of l:	0	0	1	1	1	0	1	1	1	2	2	2	2	2
	s	s		p		s		p				d		
Value of m_l:	0	0	1	0	-1	0	1	0	-1	2	1	0	-1	-2
Value of m_s:	$\pm\frac{1}{2}$	$\pm\frac{1}{2}$	$\pm\frac{1}{2}$	$\pm\frac{1}{2}$	$\pm\frac{1}{2}$	$\pm\frac{1}{2}$	$\pm\frac{1}{2}$	$\pm\frac{1}{2}$	$\pm\frac{1}{2}$	$\pm\frac{1}{2}$	$\pm\frac{1}{2}$	$\pm\frac{1}{2}$	$\pm\frac{1}{2}$	$\pm\frac{1}{2}$
Shell designation:	K		L							M				

[6] We shall later look at a more general statement of this rule which is of fundamental importance in quantum mechanics.

exclusion principle.[7] Electrons enter the lowest available empty orbital, and the order of increasing energy is that of increasing $(n + l)$ or, where $(n + l)$ values are the same, of increasing n:

$$1s < 2s < 2p < 3s < 3p < 4s \geq 3d < 4p < 5s \geq 4d < 5p < 6s \approx 5d \approx 4f < 6p$$

Where approximate inequality is indicated, the order given is that for the lighter elements. At high atomic numbers the order of increasing energy is that of increasing n, and, within a given shell, of increasing l. Where more than one orbital has the same n and l value (such as the p orbitals), each individual orbital is filled with a single electron before any of the equivalent orbitals are occupied by pairs of electrons. This is a statement of one of **Hund's rules.** Table 1.5 gives the occupancy of the orbitals, usually called the **electron configuration,** for the known elements. In writing configurations, superscripts indicate the number of electrons in a given set of orbitals: For nitrogen, for example, the $1s^2 2s^2 2p^3$ indicates that there are two electrons in the $1s$ orbital, two in the $2s$ orbital, and one electron in each of the three $2p$ orbitals. A shorthand convention is to represent the part of the configuration for the preceding noble gas by its symbol in square brackets. Thus, for N, we also have $[\text{He}]2s^2 2p^3$. More than one energy state is usually associated with an electron configuration. In Section 1.6 we shall show, for example, that the $1s^2 2p^2$ configuration gives rise to three energy states which differ in the Pauli-allowed orientations of l and s. Hence, more detailed descriptions called **microstates** are sometimes shown in which m_s values are indicated by the direction of an arrow. Whereas more than one microstate may be associated with a given energy level, the ground-state energy level is always associated with the microstate having *the maximum number of electrons of like spin.* Thus for nitrogen, we have

$$\underset{2s}{\underline{\uparrow\downarrow}} \qquad \underset{2p}{\underline{\uparrow}\;\underline{\uparrow}\;\underline{\uparrow}}$$

where $\uparrow$ represents $m_s = +\frac{1}{2}$ and $\downarrow$ represents $m_s = -\frac{1}{2}$.

The periodic chemical and physical behavior of the elements results from the periodic recurrence of the same outermost electron configuration. Thus, for example, Li, Na, K, Rb, and Cs all have the configuration . . . ns^1 and are soft low-melting metals; F, Cl, Br, and I all have the configuration . . . $ns^2 np^5$ and form diatomic molecules X_2 and sodium salts with formula NaX. Outermost electrons (highest n's) are the most loosely held and can interact with those in other atoms to form chemical bonds. These electrons are called **valence electrons.** In contrast, electrons of lower n values are tightly held and are referred to as **core electrons.** Each period of the table begins with an element having an ns^1 configuration (hydrogen or the alkali metals) and ends with an element having a noble gas configuration ($1s^2$ or $ns^2 np^6$) Table 1.5 illustrates the makeup of the periodic table in terms of the orbitals being filled. Elements with an incompletely filled set of s and p orbitals are called **main-group (or representative) elements.** Within a given group of the

---

[7] The nonelectrostatic repulsive force between two electrons of the same spin is known as the **Pauli force** and comes into play very sharply at short range. The nonpenetrability of matter and the shapes of molecules are determined largely by Pauli forces.

periodic table the main-group elements have the same valence electron configuration, with the principal quantum numbers corresponding to the period number. The ionic states and the covalent states formed by the main-group elements are related directly to the electron configuration of the elements and are reasonably predictable.

Elements having an incompletely filled set of d orbitals in one or more oxidation states are called **transition elements.** If only the configuration of neutral atoms were considered, Pd (which has a d^{10} configuration) would be excluded. Despite this unique configuration, the chemistry of Pd belongs with that of Ni and Pt and Pd is properly termed a **transition metal.** Cu, Ag, and Au are often considered to be transition metals because of the electron configurations of the 2+ or 3+ ions. Although the ground-state configurations within a transition element group are not always the same (see V, Nb, Ta; Cr, Mo, W; Fe, Ru, Os; Co, Rh, Ir; Ni, Pd, Pt), the differences in energy between the ground-state configuration and the "group configuration" are small. Even though ns orbitals are the highest ones filled in the transition elements, the $(n-1)d$'s are quite close in energy and are also considered to be valence orbitals. Thus, for example, $Cr^{0}[Ar]3d^{5}4s^{1}$ has six valence electrons.

Filling the f orbitals gives rise to the **inner-transition series** of 14 elements. Two known series of inner-transition elements exist: the lanthanides and the actinides. The elements from atomic number 111 up are not currently known. Discovery has been claimed for elements 104–110. Nuclear theory indicates a half-life on the order of years for some of the heavy elements, and therefore both calculations and predictions based on periodic behavior are of current interest.

Groups of elements from the main and transition groups can have the same number of valence electrons and thus display formal similarities in, for example, the formulas of their simple compounds. Thus, P has the electron configuration $\ldots 3s^{2}3p^{3}$ whereas V has $\ldots 3d^{3}4s^{2}$. Both have five valence electrons and form compounds such as P_2O_5 and V_2O_5, PF_3 and VF_3, and so on. This secondary kinship among elements was historically distinguished by naming the groups A and B.[8]

Thus, in the American convention N, P, As, Sb, and Bi were members of Group VA whereas V, Nb, and Ta were in Group VB. This nomenclature emphasized the five valence electrons in both cases. Unfortunately, the European A,B convention was just the opposite. Both conventions were widely used so that the meaning of A and B in the literature was ambiguous. Recently, the IUPAC has proposed numbering the columns of the period table sequentially from 1 to 18, and we shall follow that usage in this book.

► 1.3.5   *The Periodic Table as a Sorting Map*

The periodic table is an example of a **sorting map.** Sorting maps illuminate factors responsible for observed regularities in nature. Figure 1.10*a* shows a sorting map where the question under investigation is what factors influence the occurrence of superconductivity at high temperatures. The hypothesis being tested is that high-T_c superconductivity is influenced by the difference in atomic radii of the elements involved (ΔR) and by the differences in their electron-attracting ability ($\Delta\chi$). A list of high-T_c superconductors was assembled, and ΔR and $\Delta\chi$ values were determined for each. A symbol is placed on the map

---

[8] Mendeleev's table placed both A and B group elements together.

Table 1.5   Electron configuration of the neutral atoms[a]

	H
Z	1
1s	1

	Li	Be
Z	3	4
2s	1	2
2p	—	—

	Na	Mg
Z	11	12
3s	1	2
3p	—	—

	K	Ca	Sc
Z	19	20	21
4s	1	2	2
3d	—	—	1
4p	—	—	—

	Rb	Sr	Y
Z	37	38	39
5s	1	2	2
4d	—	—	1
5p	—	—	—

	Cs	Ba	La		Ce	Pr
Z	55	56	57		58	59
6s	1	2	2		2	2
5d	—	—	1		1	—
4f	—	—	—		1	3
6p	—	—	—		—	—

	Fr	Ra	Ac		Th	Pa
Z	87	88	89		90	91
7s	1	2	2		2	2
6d	—	—	1		2	1
5f	—	—	—		—	2
7p	—	—	—		—	—

Z	119	120	121	122	123	124	125	126	127	128	129	130	131	132	133	134	135	136	137	138	139	140	141	142	143
8s	1	2	2	2	2	2	2	2	2	2	2	2	2	2	2	2	2	2	2	2	2	2	2	2	2
$8p_{1/2}$	—	—	1	1	1	1	1	2	2	2	2	2	2	2	2	2	2	2	2	2	2	2	2	2	2
7d	—	—	—	1	1	—	—	1	—	—	—	—	—	—	—	—	—	—	1	1	2	1	2	2	2
6f	—	—	—	—	1	3	3	2	2	2	2	2	2	2	2	3	4	4	4	3	3	2	3	2	2
5g	—	—	—	—	—	—	1	2	3	4	5	6	7	8	8	8	9	10	11	12	13	14	15	16	17
9s	—	—	—	—	—	—	—	—	—	—	—	—	—	—	—	—	—	—	—	—	—	—	—	—	—

Z	165	166
9s	1	2
$9p_{1/2}$	—	—
$8p_{3/2}$	—	—

[a] After B. Fricke and J. McMinn, *Naturwissenschaften* **1976,** *63,* 162.

Table 1.5   (Continued)

He
2
2

B	C	N	O	F	Ne
5	6	7	8	9	10
2	2	2	2	2	2
1	2	3	4	5	6

Al	Si	P	S	Cl	Ar
13	14	15	16	17	18
2	2	2	2	2	2
1	2	3	4	5	6

Ti	V	Cr	Mn	Fe	Co	Ni	Cu	Zn	Ga	Ge	As	Se	Br	Kr
22	23	24	25	26	27	28	29	30	31	32	33	34	35	36
2	2	1	2	2	2	2	1	2	2	2	2	2	2	2
2	3	5	5	6	7	8	10	10	10	10	10	10	10	10
—	—	—	—	—	—	—	—	—	1	2	3	4	5	6

Zr	Nb	Mo	Tc	Ru	Rh	Pd	Ag	Cd	In	Sn	Sb	Te	I	Xe
40	41	42	43	44	45	46	47	48	49	50	51	52	53	54
2	1	1	2	1	1	—	1	2	2	2	2	2	2	2
2	4	5	5	7	8	10	10	10	10	10	10	10	10	10
—	—	—	—	—	—	—	—	—	1	2	3	4	5	6

Nd	Pm	Sm	Eu	Gd	Tb	Dy	Ho	Er	Tm	Yb	Lu	Hf	Ta	W	Re	Os	Ir	Pt	Au	Hg	Tl	Pb	Bi	Po	At	Rn
60	61	62	63	64	65	66	67	68	69	70	71	72	73	74	75	76	77	78	79	80	81	82	83	84	85	86
2	2	2	2	2	2	2	2	2	2	2	2	2	2	2	2	2	2	1	1	2	2	2	2	2	2	2
—	—	—	—	1	—	—	—	—	—	—	1	2	3	4	5	6	7	9	10	10	10	10	10	10	10	10
4	5	6	7	7	9	10	11	12	13	14	14	14	14	14	14	14	14	14	14	14	14	14	14	14	14	14
																					1	2	3	4	5	6

U	Np	Pu	Am	Cm	Bk	Cf	Es	Fm	Md	No	Lr	104	105	106	107	108	109	110	111	112	113	114	115	116	117	118
92	93	94	95	96	97	98	99	100	101	102	103	104	105	106	107	108	109	110	111	112	113	114	115	116	117	118
2	2	2	2	2	2	2	2	2	2	2	2	2	2	2	2	2	2	2	2	2	2	2	2	2	2	2
1	1	—	—	1	—	—	—	—	—	—	—	2	3	4	5	6	7	8	9	10	10	10	10	10	10	10
3	4	6	7	7	9	10	11	12	13	14	14	14	14	14	14	14	14	14	14	14	14	14	14	14	14	14
—	—										1	—	—	—	—	—	—	—	—	—	1	'2	3	4	5	6

144	145	146	147	148	149	150	151	152	153	154	155	156	157	158	159	160	161	162	163	164
2	2	2	2	2	2	2	2	2	2	2	2	2	2	2	2	2	2	2	2	2
2	2	2	2	2	2	2	2	2	2	2	2	2	2	2	2	2	2	2	2	2
3	2	2	2	2	3	4	3	3	2	2	2	2	3	4	4	5	6	8	9	10
1	3	4	5	6	6	6	8	9	11'	12	13	14	14	14	14	14	14	14	14	14
18	18	18	18	18	18	18	18	18	18	18	18	18	18	18	18	18	18	18	18	18
—	—	—	—	—	—	—	—	—	—	—	—	—	—	1	1	1	—	—	—

167	168	169	170	171	172
2	2	2	2	2	2
1	2	2	2	2	2
—	—	1	2	3	4

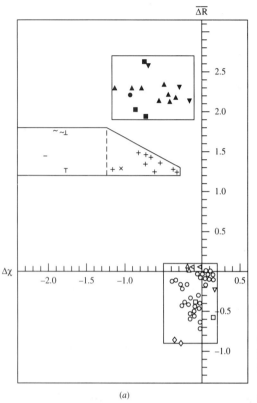

(a)

Figure 1.10 (a) A map of high-T_c supercon-ductors sorted on $\Delta\chi$ and ΔR. (From P. Villars and J. C. Phillips, *Phys. Rev.* **1988**, *B37*, 2345.) (b) A structure map sorted on Mendeleev numbers. (From D. A. Pettifor, *New Scientist*, **1986**, 24 *May*, 52.) The original map includes more structures.

at the point representing the ΔR and $\Delta\chi$ values for that compound. (The different shapes of the symbols represent the various solid-state structures exhibited by the superconductors.) A sorting of high-T_c superconductors into three "islands" is seen. The map is said to sort the superconductors on the variables $\Delta\chi$ and ΔR. This result strongly indicates that high-T_c superconductivity occurs only for the combinations of parameter values represented by the islands. Just why these values lead to superconductivity at high T_c must be the subject of further investigation. The sorting map points to the kind of parameters to be examined.

The periodic table is also a sorting map where one variable that is sorted is atomic number and the other is a less well-defined "similarity in chemical properties." This leads to the familiar row-and-column arrangement of the table. The rows and columns sort the elements according to their reactivity. We presently understand the theoretical reason behind the regularities seen in the periodic table—namely, the periodic recurrence of similar electron configurations.

It is even conceivable that atomic number is not the most fundamental sorting variable for the elements. Pettifor has assigned a "Mendeleev number"[9] to each element which often parallels the atomic number. Figure 1.10*b* depicts a map which sorts binary compounds into various structural types (indicated by the shapes of symbols). Compounds hav-

[9] The Mendeleev number is just the ordinal number which must be assigned to each element in order to achieve the desired sorting and, as yet, has no theoretical basis.

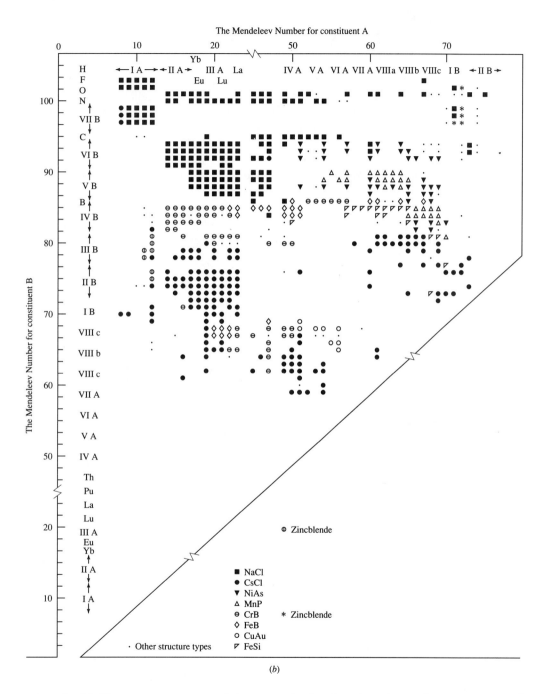

(b)

ing the NaCl structure are sorted into one (rather large) region, whereas compounds having the NiAs structure are sorted into a smaller region. (Just what the NaCl and NiAs structures *are* will be considered in Chapter 5. The point here is that useful sorting can occur with properly chosen sorting variables.) The sorting here is less complete, because islands for compounds of similar structure occur in various parts of the plot.

1.4 SPECTROSCOPIC TERMS AND
THE ZEEMAN EFFECT; ONE-ELECTRON ATOMS[10]

In discussing the Bohr atom, we mentioned the atomic spectrum of H. The experimental "window" to energy levels in atoms is the observation and analysis of light emitted or absorbed by excited atoms as they undergo transitions between allowed energy levels. The energy levels of atoms and ions are referred to as **terms** because of their near-identity with the terms of the spectroscopist; see page 12). We now consider how to determine the number and kinds of terms to be expected in atoms and ions having particular electron configurations. This allows us to determine the allowed energy states of atoms and ions and can lead to an understanding of their atomic spectra. In inorganic chemistry, an important application of terms lies in explaining the fascinating colors exhibited by compounds containing transition elements (Chapter 10). For very simple atoms, the number of possible terms and their designations are easy to deduce. So we look first at simple cases of ground and excited states for H and He.

Spectroscopic terms are classified according to the total values of orbital angular momentum, L, and spin angular momentum, S. The angular momenta of each individual electron, $\vec{l}$ and $\vec{s}$, are vector quantities which add to produce a total value represented by the capital letter $\vec{L}$ or $\vec{S}$. For example, if two electrons are in two different orbitals we might have antiparallel spins ($\vec{s}_1 + \vec{s}_2 = 0 = S$). or parallel spins ($\vec{s}_1 + \vec{s}_2 = 1 = S$).

The term designations are of the form

$$n^a T$$

where n may be the principal quantum number of the valence electrons, or simply a running number, or even the electron configuration; a is a number equal to $(2S + 1)$ and is called the **multiplicity;** T is a capital letter[11] $S, P, D, F, G, H, I, K, \ldots$ according to the value of L (0, 1, 2, 3, 4, 5, 6, 7, . . .). Consider hydrogen; the ground-state electron configuration is $1s^1$. Because there is only one electron, $L = l = 0$ and $S = s = \frac{1}{2}$. Only one term is possible, namely ^2S (read "doublet S"). Investigations of hydrogenlike atoms by high-resolution spectroscopy revealed more energy levels than expected on the basis of the model we have considered.[12] These data led to the conclusion that the $2p^1$ configuration (and higher states beyond the s) could have two energy states which differ in the interaction of the orbital angular momentum and the spin angular momentum of the electron (see Figure 1.11).

Orbital angular momentum[13] is associated with the rotation of the electron cloud. The interaction of the quantized orbital angular momentum $\vec{l}$ with the quantized spin angular momentum $\vec{s}$ produces a resultant total angular momentum $\vec{j}$ which is also quantized.

---

[10] Material on atomic spectra in Sections 1.4 and 1.5 is included to facilitate understanding of the spectra of transition metal complexes. These sections may be omitted here and treated with Chapter 10.

[11] J is omitted in the sequence because it is used as a quantum number. S and L are used as terms and quantum numbers.

[12] If the emission source is cooled sufficiently to reduce Doppler effects, splitting of lines can be found at very high resolution.

[13] Although the wave equation abandons the picture of specific orbits for the electron, the property of angular momentum is retained.

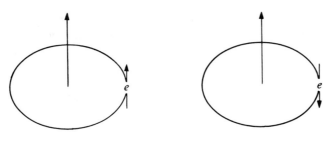

Figure 1.11   Possible alignments of orbit and spin angular momenta for a one-electron atom.

For a $2p$ electron, therefore, the vector sum of $l = 1$ and $s = \frac{1}{2}$ can be $j = \frac{1}{2}$ or $j = \frac{3}{2}$ depending on the relative alignments. The resultant total angular momentum vector has length $\sqrt{j(j + 1)}(h/2\pi)$; the new quantum number j is sometimes referred to as the "inner" quantum number. For hydrogen atoms in the $2p^1$ configuration, the two possible j states have a difference in energy of 0.3561 cm^{-1}. (See footnote 12.)

These two states are designated in spectroscopic notation as $^2P_{1/2}$ and $^2P_{3/2}$, where the J values appear as right-hand subscripts; alternatively, these states can be designated as $2^2P_{1/2}$ and $2^2P_{3/2}$, where the prefix 2 indicates that a second shell orbital is involved.

If an excited H atom is placed in a weak unidirectional magnetic field, the two P_J terms each split into $(2J + 1)$ terms differing in energy by an amount proportional to the applied field strength (Zeeman effect). These states are characterized by a quantum number M_J which may have a value of $J, J - 1, \ldots, -J$. The total angular momentum vector of length $\sqrt{J(J + 1)} (h/2\pi)$ can be oriented only in certain directions with respect to the applied magnetic field—namely, those in which the component of total angular momentum in the direction of the field is $M_J(h/2\pi)$. In a magnetic field, the $^2P_{3/2}$ term yields four closely spaced terms and the $^2P_{1/2}$ term yields two. Figure 1.12 diagrams the splitting for the $J = \frac{3}{2}$ case. (In the one-electron case, $J = j$ and $M_J = m_j$.) The splitting of the ^2P term into six terms in a magnetic field is related to the fact that there are six ways of placing an electron in p orbitals that differ in the values of m_l and/or m_s. In the absence of the magnetic field, some of these terms (all those having the same j) have the same energy (i.e., are **degenerate**).[14]

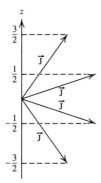

Figure 1.12   Allowed orientations of an angular momentum vector for $J = \frac{3}{2}$ in a magnetic field along z. The length of the vector is
$$|\vec{J}| = \sqrt{\tfrac{3}{2}(\tfrac{3}{2} + 1)} \, (h/2\pi) = \sqrt{\tfrac{15}{4}} (h/2\pi).$$

[14] The four quantum numbers which describe an electron can be chosen as n, l, m_l, and m_s or as n, l, j and m_j. The choice is arbitrary because j values depend on l and m_s. Which set we use depends on the factors which are responsible for the largest energy differences between states. For H, the small energy difference between j states means that the most appropriate choice is n, l, m_l, and m_s. See also Sections 1.5.1 and 1.5.2.

For He, the ground-state configuration is $1s^2$. In this case $L = l_1 + l_2 = 0 + 0 = 0$. Because both electrons occupy the same orbital, they must have opposite spin orientations and $s_1 + s_2 = 0$ to give a ^1S term (read "singlet S"). In the $1s^1 2p^1$ excited state of He we must have $L = l_1 + l_2 = 0 + 1 = 1$. Two possibilities exist for S because the electrons are in different orbitals: $S = 0$ and $S = 1$. Two terms are then possible: ^1P (singlet P) and ^3P (triplet P). For the singlet term, only one value of J is possible: $J = 1 + 0 = 1$. For the triplet term, J may be 0, 1, or 2.[15]

1.5   THE VECTOR MODEL AND TERM STRUCTURE FOR POLYELECTRON ATOMS

As the number of electrons increases beyond two, the number of terms increases even faster. We consider now how to determine the number and relative energies of terms for polyelectron cases.

▶ 1.5.1   *Deriving Spectroscopic Terms from Electron Configurations*

In the $2p^1$ case for H, the total number of terms for the atom in a magnetic field corresponds to the total number of ways of arranging one electron among six p orbital-plus-spin combinations. For polyelectron atoms or ions, we also need to find the number of ways of arranging several electrons among the available orbitals; that is, we determine the number of *microstates*. However, this must be done in accord with the Pauli exclusion principle, which requires that no two electrons have all four quantum numbers the same. We now consider how to do this.

The number of different ways e electrons can be placed in a given set of n orbital sites ($n/2$ orbitals) in accordance with Pauli exclusion principle is expressed as

$$\text{Number of microstates} = \frac{n!}{e!h!} \tag{1.28}$$

where h is the number of "holes" (sites available) and is equal to $(n - e)$. Thus, for a d^5 configuration there are 252 microstates:

$$\frac{10}{1} \times \frac{9}{2} \times \frac{8}{3} \times \frac{7}{4} \times \frac{6}{5}$$

This is equivalent to saying that there are 10 different ways of placing the first electron in the set of orbitals, nine different ways of placing the second, and so on. Because these choices are not independent (i.e., the electrons are indistinguishable), we divide by the

---

[15] Some comment about notation is in order using j as a general symbol for all kinds of angular momentum. The value of j is a quantum number. It is also the label for the angular momentum vector when written as $\vec{j}$. The length of this vector is $|\vec{j}| = \sqrt{j(j + 1)}\,(h/2\pi)$. Thus, we frequently use just j when referring to an angular momentum because the above formula allows calculation of the vector length. It can be confusing when a single symbol is used in several different senses, but, unfortunately, this is the practice in this case.

number of electrons so far added at each stage: that is, $10 \times \frac{9}{2} \times \frac{8}{3} \ldots$. A few of the microstates would be as follows:

$m_l = 2$	1	0	-1	-2
↑	↑	↑	↑	↑
↓	↓	↓	↓	↓
↑	↓	↑	↑	↑

where ↑ represents $m_s = +\frac{1}{2}$ and ↓ represents $m_s = -\frac{1}{2}$.

In a strong magnetic field each microstate has a different energy. If several types of orbitals are partially occupied, the total number of microstates is the product of those possible for each set of orbitals; for d^5s^1, for example, the total is 252×2, or 504.

The individual l values combine vectorially to give a resultant $\vec{L}$, and the individual s values give a resultant $\vec{S}$. As with single electrons, the lengths of the total orbital angular momentum vector and total spin angular momentum vector are $|\vec{L}| = \sqrt{L(L + 1)}(h/2\pi)$ and $|\vec{S}| = \sqrt{S(S + 1)}\,(h/2\pi)$, respectively. These vectors may have only certain orientations with respect to the magnetic field which defines the z direction, and a diagram like that in Figure 1.13 applies (relabeled with the appropriate kind and values of angular momentum). The component of $\vec{L}$ in the z direction, M_L, may take values $-L$, $(-L + 1), \ldots, 0, \ldots, (L - 1), L$; a given L has $(2L + 1)$ possible M_L states. Each M_L value is the sum of the m_l values for the microstate. Likewise, for a given S value there are $(2S + 1)$ possible M_S states, each of which is the sum of the m_s values for the microstate.

Let us consider an atom having a closed shell such as Ne. If any set of orbitals of given n and l is completely filled or completely empty, only one microstate is possible. For every positive value of m_l there is a corresponding negative value, and the sum of the m_l values which gives M_L must be zero. This means that L must be zero. Similarly, the M_S value is zero for a closed shell because there are as many electrons having an "up" as having a "down" spin orientation. Thus S must be zero, and the state resulting from any closed shell of electrons has no resultant orbital angular momentum, spin angular momentum, or total angular momentum. *Closed shells thus give rise to $^1S_0$ spectroscopic terms.* Filled subshells (e.g., ns^2, np^6, etc.) also have a resultant $L = 0$ and $S = 0$ for reasons discussed above.

In general, L may have integer values ranging from the smallest difference of l values to the sum of the l values, whereas S may take on values (differing by whole numbers) from $1/2$ to $n/2$ (for odd n), or from 0 to $n/2$ for even n, where n is the number of un-

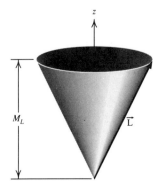

Figure 1.13   An allowed orientation of the angular momentum vector having a well-defined component in the z direction, but of indeterminate position with respect to the x and y axes.

paired electrons. The L and S values, which represent the quantum-mechanically allowed orbital states and the spin-angular momentum states, respectively, combine vectorially to produce J values representing the total angular momentum states. J values may run from $(L - S)$ to $(L + S)$ in integer steps. Thus, the number of J states for given L and S is $(2S + 1)$ if $L \geq S$. This is analogous to the one-electron case treated on page 33.

For nonequivalent electrons (electrons differing in values of the quantum numbers n or l) the restriction of the Pauli principle has already been met within each shell or orbital set. Spectroscopic terms may be derived by obtaining all possible values of L as the vector sums of the individual l values and all possible values of S as the vector sums of the individual s values. Thus for the $2p^1 3p^1$ configuration, the L values are 0, 1, and 2 and the S values are 0 and 1 for each L value, giving six terms: ^1S, ^1P, ^1D, and ^3S, ^3P, and ^3D.

Equivalent electrons (electrons having either n or l, or both, the same) give fewer terms because the Pauli principle forbids certain combinations of values of m_l and m_s. Addition of individual $\vec{l}$ vectors is accomplished by determining the possible M_L values. This is so because we can simultaneously know only the values of l and m_l, the component along z. The components of $\vec{l}$ in the x and y directions are indeterminate. This is tantamount to saying that $\vec{l}$ lies somewhere on the cone making an angle $\Theta = \arccos m_l/l$ with the z direction as shown in Figure 1.13. Hence, we cannot vectorially add $\vec{l}_1$ and $\vec{l}_2$ by laying them head-to-tail because their relative orientations are unknown.

We can, however, determine M_L because this is just the sum of the individual m_l. Thus we can determine the z component of the total angular momentum vector L even though its x and y components remain unknown. Addition of individual $\vec{s}$ vectors to give $\vec{S}$ is subject to the same considerations.

For a $2p^2$ configuration, a value of $M_L = 2$ results only from cases in which $\vec{l}_1$ and $\vec{l}_2$ are oriented so that m_{l_1} and m_{l_2} are both 1. Accordingly, the values of m_{s_1} and m_{s_2} must differ and $M_s = 0$. Obtaining all M_L and M_s values requires writing all allowed microstates and obtaining M_L and M_s for each microstate. For the $2p^2$ configuration, we find:

$m_l = $	$m_s = +\frac{1}{2}$			$m_s = -\frac{1}{2}$			$M_L = \Sigma m_l$	$M_S = \Sigma m_s$
	+1	0	−1	+1	0	−1		
1	↑	↑					1	1
2	↑		↑				0	1
3		↑	↑				−1	1
4				↓	↓		1	−1
5				↓		↓	0	−1
6					↓	↓	−1	−1
7	↑			↓			2	0
8	↑				↓		1	0
9	↑					↓	0	0
10		↑		↓			1	0
11		↑			↓		0	0
12		↑				↓	−1	0
13			↑	↓			0	0
14			↑		↓		−1	0
15			↑			↓	−2	0

These M_L and M_S data are tabulated in the array of Figure 1.14*a*. The array is symmetric about lines through $M_L = 0$ and $M_S = 0$, providing a check of the tabulation.

From the highest M_L value of 2, the array $M_L = 2, 1, 0, -1, -2$ exists corresponding to $L = 2$. These values in the array occur only for $M_S = 0$. So there must be a term with $L = 2$, $S = 0$, or ^1D. We substract the microstates for this ^1D term from the array and are left with numbers of states shown in Figure 1.14*b*.

From the highest remaining $M_L = 1$, the array $M_L = -1, 0, +1$ exists corresponding to $L = 1$. These values occur for $M_S = 1, 0$, and -1. So there must be a term with $L = 1$ and $S = 1$, or ^3P. Subtracting the requisite number (9) of microstates from the array leaves only one microstate having $M_L = 0$ and $M_S = 0$. This corresponds to a ^1S term. Thus, the $2p^2$ configuration gives rise to the terms ^1D, ^3P, and ^1S.[16]

The p^5 configuration gives only a ^2P term as for p^1. If we write all microstates for p^4 we obtain ^1D, ^1S, and ^3P as for p^2. The p^5 configuration has one "hole" in a filled p^6 configuration and p^4 has two holes. Hence, we can determine the microstates of either electrons or holes. The terms are the same for d^1 and d^9, d^2 and d^8, d^3 and d^7, and d^4 and d^6. The p^3 configuration has three electrons or three holes. It is its own hole counterpart as is d^5.

The coupling scheme just discussed is known as the L–S scheme or the Russell–Saunders coupling scheme. The $\vec{L}$ and $\vec{S}$ vectors couple to produce a total angular momentum $\vec{J}$. For light atoms, splitting between terms with different values of J is typically small and occurs only in a magnetic field. Russell–Saunders coupling is a good description of atoms having $Z \leq 30$. Table 1.6 summarizes terms for p^n and d^n configurations in Russell–Saunders coupling.

In heavier atoms, a different coupling scheme is observed in which the orbital and spin angular momenta of individual electrons first couple, giving a resultant $\vec{j}$ for each electron; we say that *spin–orbit coupling* is energetically important. The individual $\vec{j}$'s then couple, producing an overall $\vec{J}$. For most purposes in this book, knowledge of the L–S coupling scheme is sufficient.[17]

(*a*)

↑ M_L	$M_S = -1$	0	$+1$
2		1	
1	1	2	1
0	1	3	1
−1	1	2	1
−2		1	

$M_S \rightarrow$

(*b*)

↑ M_L	$M_S = 1$	0	$+1$
2			
1	1	1	1
0	1	2	1
−1	1	1	1
−2			

$M_S \rightarrow$

Figure 1.14 Arrays for a p^2 configuration showing (*a*) the total number of individual microstates of a given M_L and M_S and (*b*) the array after subtracting out the microstates of ^1D.

[16] For other examples of this approach, see K. E. Hyde, *J. Chem. Educ.* **1975,** *52,* 87. See also D. H. McDaniel, *J. Chem. Educ.* **1977,** *54,* 147 and B. E. Douglas and C. A. Hollingsworth, *Symmetry in Bonding and Spectra,* Academic Press, Orlando, FL, 1985, pp. 128–135.

[17] The term structure in j–j coupling has been treated. See J. Rubio and J. J. Perez, *J. Chem. Educ.* **1986,** *83,* 476.

Table 1.6   Terms for p^n and d^n configurations using Russell–Saunders coupling[a]

p^1, p^5	^2P
p^2, p^4	$^1(S,D)\ ^3P$
p^3	$^2(P,D)\ ^4S$
p^0, p^6, d^0, d^{10}	^1S
d^1, d^9	^2D
d^2, d^8	$^1(S,D,G)\ ^3(P,F)$
d^3, d^7	$^2D\ ^2(P,D,F,G,H)\ ^4(P,F)$
d^4, d^6	$^1(S,D,G)\ ^1(S,D,F,G,I)\ ^3(P,F)\ ^3(P,D,F,G,H)\ ^5D$
d^5	$^2D\ ^2(P,D,F,G,H)\ ^2(S,D,F,G,I)\ ^4(P,D,F,G,)\ ^6S$

[a] In each case the ground-state term is listed last.

▶ 1.5.2   Energy Levels in Polyelectron Atoms

We want to be able to order the energies of atomic terms. To a large extent this is a question that must be answered by experiment. It is, however, possible to select the lowest-energy term (the ground-state term) for any configuration. In order to do this, we briefly consider some interactions in atoms having $L–S$ coupling.

Spin–Orbital Interactions

As mentioned above, in atoms with $Z \leq 30$ the coupling of L and S to produce a resultant J does not produce large energetic effects. A typical example is K having only one electron beyond the Ar configuration. The energy difference between the $4\,^2P_{1/2}$ and $4\,^2P_{3/2}$ states of the $4p^1$ configuration is only 57.72 cm^{-1}. This can be compared with the energy difference of 12,986 cm^{-1} between the $^2S_{1/2}$ ground-state [Ar]$4s^1$ configuration and $^2P_{1/2}$ of [Ar]$4p^1$. Obviously, for atoms obeying the Russell–Saunders coupling scheme, term energies are determined primarily by the values of L and S.

Spin–Spin Interactions

In a partially filled subshell with more than one electron, the effect of the electrons' spin interactions may be large. Thus, in V$^+$, the energy difference between the $3d^4\ ^5D_2$ term and the $3d^4\ ^1D_2$ term is 20,944.87 cm^{-1}. The lower-energy term results from microstates involving four unpaired electrons, whereas the higher-energy term has two pairs of electrons. The primary reason for the differing energy of these terms is the differing **exchange energy** *between pairs of electrons having the same spin*. Four unpaired electrons, ↑ ↑ ↑ ↑ , have six distinct pairs of similar spin orientation, whereas two pairs, ↑ ↓ ↑ ↓ , have only two distinct pairs of similar spin orientation.

Hund's Rules and Ground-State Terms

The size of energy differences due to spin–spin interactions led Hund to formulate the rule that the *ground-state term is the one with maximum S* (maximum **spin multiplicity**

$2S + 1$). Within this restriction the ground state is the one with maximum L. The state with minimum J value lies lowest for subshells less than half filled, and the state with maximum J lies lowest for subshells more than half filled. For example, $^3P_0$ is the ground state for the p^2 configuration, whereas the $^3P_2$ state is the ground state for the p^4 configuration.

Using Hund's rules, we may select from the possible terms the lowest-energy term arising from a given configuration. This can be done easily without having to derive all possible terms: *A microstate is written with the orbitals filled in order of decreasing m_l; each orbital is singly occupied before any is doubly occupied.* The sum of m_l values for this microstate gives the L value of the term; the S value is one-half the number of singly occupied orbitals. For a d^2 case, this gives

$$m_l = \quad +2 \quad +1 \quad 0 \quad -1 \quad -2$$

$$\underline{\uparrow} \quad \underline{\uparrow} \quad \underline{} \quad \underline{} \quad \underline{}$$

$$L = 2 + 1 = 3 \qquad S = \tfrac{1}{2} \times 2 = 1$$

$L = 3$ gives an F term; the $S = 1$ gives a multiplicity of 3. L and S combination gives J values of 2, 3, and 4. The lowest J is selected in this case, because the d orbitals are less than half-filled. The ground-state term for a d^2 configuration is accordingly a $^3F_2$ term. A half-filled subshell gives an S term and, because $L = 0$, $J = S$. Accordingly, the ground-state term for N is $^4S_{3/2}$. See Table 1.6.

EXAMPLE: Determine the ground-state terms for the d^3 and d^8 configurations and for C.

Solution:

$m_l =$	2	1	0	−1	−2	
d^3	$\uparrow$	$\uparrow$	$\uparrow$	—	—	Max. $M_L = 3$, so $L = 3$ and $S = \tfrac{3}{2}$
d^8	$\uparrow\downarrow$	$\uparrow\downarrow$	$\uparrow\downarrow$	$\uparrow$	$\uparrow$	Max. $M_L = 3$, so $L = 3$ and $S = 1$

For d^3, the ground state is $^4F_{3/2}$ (lowest J or $L - S$). For d^8, the ground state is $^3F_4$ (highest J or $L + S$). Although both of these are F terms, the configurations are not hole counterparts like d^3 and d^7, where the number of electrons in one case (3) is equal to the number of holes in the subshell in the other $(10 - 7)$. The d^3 and d^7 hole counterparts have a ^4F ground-state term, but with $J = \tfrac{9}{2}$ for d^7. Note that the configuration written gives the maximum S and L values; this is only one of the microstates of the array for the ground-state term, however. There are $4 \times 7 = 28$ microstates for ^4F. The hole counterpart of d^8 is d^2, giving $^3F_2$ for the ground state.

For C $(2s^2 2p^2)$ we get

$m_l =$	0	1	0	−1	
	$\uparrow\downarrow$	$\uparrow$	$\uparrow$	—	Max. $M_L = 1$, so $L = 1$ and $S = 1$
	$2s$		$2p$		

The ground state is $^3P_0$ (lowest J).

Penetration Effects

We are often interested in the energy of an electron in a particular orbital of a polyelectron atom rather than the energy of a term. *In hydrogenlike species,* all orbitals of a given principal quantum number *n* are degenerate. Thus, for example, the 3*s*, 3*p*, and 3*d* orbitals have the same energy. *In polyelectron atoms,* this *degeneracy is lifted* and we have 3*s* < 3*p* < 3*d*. The reason is difference in **penetration** by electrons in orbitals of different *l*. The radial distribution functions in Figure 1.6 show that *s* electrons of outer shells have some probability of being found at distances close to the nucleus which lie inside the most probable radii for electrons of lower principal quantum number. We say that *s* electrons penetrate these underlying filled shells. Penetration is greater for *s* electrons than for *p* electrons; *p* electrons, in turn, show greater penetration than *d* electrons, and so on.

If there were no penetration effects, the outermost electrons in any atom would be screened from the positive nuclear charge by the negative charges of all electrons of lower *n* which would lie between them and the nucleus. Hence, one would expect the alkali metals with ns^1 configurations to feel an **effective nuclear charge** of +1 and to behave like H. Their spectra are, in fact, quite similar to the atomic spectrum of H. However, the energy differences between n^2S, n^2P, n^2D, and so on, terms are much greater than in H. This must be due to differences in penetration.

An alternative way to view penetration is from a so-called "one-electron" model. We focus on the behavior of a single electron and lump into a single shielding parameter, S, the shielding effects of all the other electrons in the atom. We can then compute the effective nuclear charge Z* felt by an electron as (Z − S), where Z is the atomic number. Slater developed the following set of empirical rules for calculating S.

1. The electrons are grouped in the following order: 1*s*; 2*s* and 2*p*; 3*s* and 3*p*; 3*d*; 4*s* and 4*p*; 4*d*; 4*f*, and so on—*ns* and *np* electrons always being considered as a single group.

2. Electrons in groups lying above that of a particular electron do not shield it at all.

3. A shielding of 0.35 is contributed by each other electron in the same group (except for a 1*s* electron, which contributes 0.30 to the shielding of the other 1*s* electron).

4. For *d* and *f* electrons the shielding is 1.00 for each electron in the underlying groups. For *s* and *p* electrons the shielding from the immediately underlying shell (i.e., *n* − 1) is 0.85 for each electron; the shielding from groups further in is 1.00 for each electron.

Ideally Slater's rules should permit the calculation of the energy of any electron in any atom or ion by the formula

$$E = \frac{-(Z^*)^2}{n^2}(13.6 \text{ eV}) = \frac{-(Z - S)^2}{n^2}(13.6 \text{ eV}) \qquad (1.29)$$

where Z is the nuclear charge, S is the shielding constant, and *n* is the principal quantum number of the electron. Unfortunately, values calculated in this fashion may differ from those obtained from spectroscopic studies by as much as a factor of five. Nevertheless, the effective nuclear charges $Z^* = (Z - S)$ calculated from Slater's rule have been used for such purposes as calculating ionic radii and electronegativities of the elements. More refined calculations can be used when greater accuracy is needed; however, gross trends are nicely reproduced by this admittedly oversimplified model.

EXAMPLE: Predict the energy of the $4s$ electron in K.

Solution: The electron configuration for K is $1s^2 2s^2 2p^6 3s^2 3p^6 4s^1$. The ten electrons with $n = 1$ and 2 contribute 1.00 each to the shielding, whereas the eight $n = 3$ electrons each contribute 0.85. Thus $S = 10(1.00) + 8(0.85) = 16.8$ and $Z^* = 19 - 16.8 = 2.2$. Hence, $E_{4s} = -(2.2)^2/4^2 (13.6 \text{ eV}) = -4.1 \text{ eV}$. Experimental measurements give a value of -4.3 eV.

Wavefunctions for Polyelectron Atoms

Previously, the Pauli principle was introduced by pointing out that it forbids two electrons in the same atom from having the same values for the four quantum numbers n, l, m_l, and m_s. Actually, the original formulation of this principle was more general: The wavefunction of a polyelectron atom must be antisymmetric with respect to pairwise permutation of electrons. That is, for every pairwise exchange of electrons, the wavefunction must change sign.

In writing wavefunctions in analytical form, it is customary to indicate that an electron has $m_s = +\frac{1}{2}$ by assigning it a "spin function" α or that it has $m_s = -\frac{1}{2}$ by assigning it a "spin function" β. α and β are not functions of Cartesian or polar coordinates like $1s$ or $3p_x$, but serve only to indicate m_s values. Hence, for electron 1 with $n = 2$, $l = 0$, $m_l = 0$, and $m_s = +\frac{1}{2}$, we write $2s(1)\alpha(1)$.

Slater recognized that wavefunctions conforming to the Pauli principle could be written as determinants where each column corresponds to a particular wavefunction including spin and where each row corresponds to an electron. Thus for He we have

$$\Psi_{\text{He}} = \begin{vmatrix} 1s(1)\alpha(1) & 1s(1)\beta(1) \\ 1s(2)\alpha(2) & 1s(2)\beta(2) \end{vmatrix} \tag{1.30}$$

which expands into

$$\Psi_{\text{He}} = 1s(1)\alpha(1)1s(2)\beta(2) - 1s(2)\alpha(2)1s(1)\beta(1) \tag{1.31}$$

A property of determinants is that interchanging rows (which corresponds to permuting pairs of electrons) changes the sign.

Wavefunctions resulting from expansion of determinants consist of sums and differences of products of hydrogenlike orbitals. Individual terms of the functions correspond to microstates of the electron configurations. Thus, one can see the sense in which more than one microstate corresponds to a particular energy level. The diagonal term corresponds to the way in which we have written electron configurations if the spin functions are dropped and the information about m_s mentally retained as we do when applying Hund's rule. In fact, a more sophisticated view of the meaning of electron configurations is that they are the diagonal term in the determinant expansion and thus are a shorthand symbol for the entire Slater determinant wavefunction.

For open-shell atoms and for excited states of atoms the wavefunctions are more complicated and consist of linear combinations of Slater determinants. Also, as the number of electrons increases, the determinantal wavefunctions rapidly become quite complicated.

1.6 IONIZATION ENERGIES

The **ionization energy** (IE_1) of an atom is defined as the energy needed to remove an electron from a gaseous atom in its ground state. The second ionization energy (IE_2) is the additional energy needed to remove the second electron, and so forth. Table 1.7 gives the ionization energies of the elements.

Figure 1.15 plots IE_1 against the atomic numbers of the elements. *In general, the IE increases on crossing a period* (e.g., Li through Ne) *and decreases on descending in a group* (He through Rn). These trends, and many of the variations within them, may be rationalized in terms of the electron configurations of elements involved. On crossing a period, the principal quantum number for the electron being removed remains constant but the effective nuclear charge increases, because of incomplete shielding by electrons of the same quantum number. These considerations also rationalize the decrease in atomic radius across a period. On descending in a given group, the principal quantum number increases regularly and relatively more rapidly than the effective nuclear charge changes. Paralleling this trend, atomic radius increases down a group. Variations from or within these trends result primarily from changes in the type of orbital from which the electron is taken, or from changes in the multiplicity of the ground-state term.

Elements with a ground-state $^1S_0$ term have higher IEs than do their neighboring elements (see exceptions below). $^1S_0$ is the ground-state term for any element with an outermost completely filled shell or subshell. The exceptions to this rule are the alkaline earth metals that have a transition metal as a neighbor. In the case of a transition metal, the s electron is removed first. The s-electron orbital of the transition metal may be thought of as penetrating the d orbitals; hence, as the d electrons are filled in to build up the transition metals, the effective nuclear charge increases. Accordingly, IE increases in a fairly regular fashion from the alkali metal through zinc or its congeners. In the case of mercury, not only the d orbitals but also the f orbitals have been filled in, and the penetration effect for the $6s$ electron is much larger. The IE of mercury is almost as great as that of radon. Following the zinc family elements, IE_1 drops sharply for Ga, In, and Tl, because the electron involved is a p electron that occupies a less penetrating orbital. This effect accounts for the near-equality of the first IEs of Ca and Ga and of Sr and Sn. A similar, but smaller, drop in IE following Be and Mg may also be accounted for by the change in the ionizing electron from an s to a p electron.

The higher IEs of nitrogen and its congeners, as compared with those of their neighbors, may be attributed to the stabilizing influence of the spin–spin interactions, which are highest in filled and half-filled subshells.

Successive IEs of an atom increase in magnitude because of the smaller number of electrons left to shield the ionizing electron from the nuclear charge. Variation with atomic number in the second IE is very similar to that in the first IE, except that it is displaced by one atomic number. Some similarity persists in the third and higher ionization energies, again provided that isoelectronic ions are compared. Thus, the highest IEs always occur for ions having the noble gas configurations.

Ahrens has pointed out that the oxidation states of elements in their compounds may be correlated by the difference in IEs for successive states, Δ. If Δ is around 10–11 eV or

Table 1.7   Ionization energies of the elements[a] (in eV; 1 eV/atom = 96.4869 kJ/mol)

Z	Element	I	II	III	IV	V	VI	VII	VIII
1	H	13.598							
2	He	24.587	54.416						
3	Li	5.392	75.638	122.451					
4	Be	9.322	18.211	153.893	217.713				
5	B	8.298	25.154	37.930	259.368	340.217			
6	C	11.260	24.383	47.887	64.492	392.077	489.981		
7	N	14.534	29.601	47.448	77.472	97.888	552.057	667.029	
8	O	13.618	35.116	54.934	77.412	113.896	138.116	739.315	871.387
9	F	17.422	34.970	62.707	87.138	114.240	157.161	185.182	953.886
10	Ne	21.564	40.962	63.45	97.11	126.21	157.93	207.27	239.09
11	Na	5.139	47.286	71.64	98.91	138.39	172.15	208.47	264.18
12	Mg	7.646	15.035	80.143	109.24	141.26	186.50	224.94	265.90
13	Al	5.986	18.828	28.447	119.99	153.71	190.47	241.43	284.59
14	Si	8.151	16.345	33.492	45.141	166.77	205.05	246.52	303.17
15	P	10.486	19.725	30.18	51.37	65.023	220.43	263.22	309.41
16	S	10.360	23.33	34.83	47.30	72.68	88.049	280.93	328.23
17	Cl	12.967	23.81	39.61	53.46	67.8	97.03	114.193	348.28
18	Ar	15.759	27.629	40.74	59.81	75.02	91.007	124.319	143.456
19	K	4.341	31.625	45.72	60.91	82.66	100.0	117.56	154.86
20	Ca	6.113	11.871	50.908	67.10	84.41	108.78	127.7	147.24
21	Sc	6.54	12.80	24.76	73.47	91.66	111.1	138.0	158.7
22	Ti	6.82	13.58	27.491	43.266	99.22	119.36	140.8	168.5
23	V	6.74	14.65	29.310	46.707	65.23	128.12	150.17	173.7
24	Cr	6.766	16.50	30.96	49.1	69.3	90.56	161.1	184.7
25	Mn	7.435	15.640	33.667	51.2	72.4	95	119.27	196.46
26	Fe	7.870	16.18	30.651	54.8	75.0	99	125	151.06
27	Co	7.86	17.06	33.50	51.3	79.5	102	129	157
28	Ni	7.635	18.168	35.17	54.9	75.5	108	133	162
29	Cu	7.726	20.292	36.83	55.2	79.9	103	139	166
30	Zn	9.394	17.964	39.722	59.4	82.6	108	134	174
31	Ga	5.999	20.51	30.71	64				
32	Ge	7.899	15.934	34.22	45.71	93.5			
33	As	9.81	18.633	28.351	50.13	62.63	127.6		
34	Se	9.752	21.19	30.820	42.944	68.3	81.70	154.4	
35	Br	11.814	21.8	36	47.3	59.7	88.6	103.0	192.8
36	Kr	13.999	24.359	36.95	52.5	64.7	78.5	111.0	126
37	Rb	4.177	27.28	40	52.6	71.0	84.4	99.2	136
38	Sr	5.695	11.030	43.6	57	71.6	90.8	106	122.3
39	Y	6.38	12.24	20.52	61.8	77.0	93.0	116	129

[a] From C. E. Moore, "Ionization Potentials and Ionization Limits from Atomic Spectra," NSRDS-NBS 34, 1970.

Table 1.7 *(Continued)*

Z	Element	I	II	III	IV	V	VI	VII	VIII
40	Zr	6.84	13.13	22.99	34.34	81.5			
41	Nb	6.88	14.32	25.04	38.3	50.55	102.6	125	
42	Mo	7.099	16.15	27.16	46.4	61.2	68	126.8	153
43	Tc	7.28	15.26	29.54					
44	Ru	7.37	16.76	28.47					
45	Rh	7.46	18.08	31.06					
46	Pd	8.34	19.43	32.93					
47	Ag	7.576	21.49	34.83					
48	Cd	8.993	16.908	37.48					
49	In	5.786	18.869	28.03	54				
50	Sn	7.344	14.632	30.502	40.734	72.28			
51	Sb	8.641	16.53	25.3	44.2	56	108		
52	Te	9.009	18.6	27.96	37.41	58.75	70.7	137	
53	I	10.451	19.131	33					
54	Xe	12.130	21.21	32.1					
55	Cs	3.894	25.1						
56	Ba	5.212	10.004						
57	La	5.577	11.06	19.175					
58	Ce	5.47	10.85	20.20	36.72				
59	Pr	5.42	10.55	21.62	38.95	57.45			
60	Nd	5.49	10.72						
61	Pm	5.55	10.90						
62	Sm	5.63	11.07						
63	Eu	5.67	11.25						
64	Gd	6.14	12.1						
65	Tb	5.85	11.52						
66	Dy	5.93	11.67						
67	Ho	6.02	11.80						
68	Er	6.10	11.93						
69	Tm	6.18	12.05	23.71					
70	Yb	6.254	12.17	25.2					
71	Lu	5.426	13.9						
72	Hf	7.0	14.9	23.3	33.3				
73	Ta	7.89							
74	W	7.98							
75	Re	7.88							
76	Os	8.7							
77	Ir	9.1							

Table 1.7   *(Continued)*

Z	Element	I	II	III	IV	V	VI	VII	VIII
78	Pt	9.0	18.563						
79	Au	9.225	20.5						
80	Hg	10.437	18.756	34.2					
81	Tl	6.108	20.428	29.83					
82	Pb	7.416	15.032	31.937	42.32	68.8			
83	Bi	7.289	16.69	25.56	45.3	56.0	88.3		
84	Po	8.42							
85	At								
86	Rn	10.748							
87	Fr								
88	Ra	5.279	10.147						
89	Ac	6.9	12.1						
90	Th		11.5	20.0	28.8				
91	Pa								
92	U								
93	Np								
94	Pu	5.8							
95	Am	6.0							

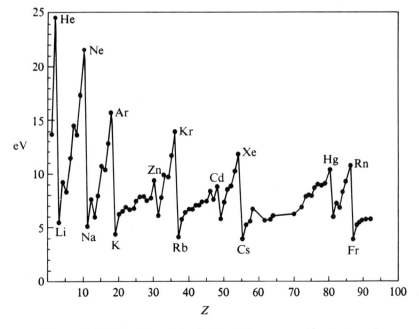

Figure 1.15   Variation of the first ionization energy with atomic number.

less, the lower state is not stable. Consider the values for Al:

$$\Delta_{1,2} = 12.8 \text{ eV}$$

$$\Delta_{2,3} = 9.6 \text{ eV}$$

$$\Delta_{3,4} = 91.5 \text{ eV}$$

Although compounds of Al^I are stable as gaseous species such as $AlCl(g)$, they are unknown as solids. No compound of Al^{II} is known. Al^{III} is the only stable state for Al compounds under usual conditions. Δ values of around 16 or above usually lead to stable states. Because Δ is always high at the noble gas configuration, these configurations frequently are found for the elements in their compounds. The $s^2p^6d^{10}$ configuration is also one in which Δ is high, and it, too, is a common configuration for positive ions.

1.7 ELECTRON AFFINITIES

The **electron affinity** (EA) of an atom is the energy involved in adding an electron to a gaseous atom in its ground state, to form a gaseous anion.

$$X(g) + e^- \rightarrow X(g)^- \tag{1.32}$$

EA sometimes is expressed as the IE of $X^-(g)$. Because for most atoms energy is released when an electron is added, EA in such cases is expressed as a positive number (the amount of energy released), contrary to the thermodynamic convention. Factors affecting EAs are the same as those affecting IEs, and thus the general trend is for EAs to increase (more energy *released*) from left to right in the periodic table, because of the increase in nuclear charge *and* decrease in atomic radius. Table 1.8 displays the EAs of most elements. The trend within a group is less regular than for IEs, but we must remember that the increase in size is expected to decrease the attraction for an electron, whereas an increase in nuclear charge is expected to increase the attraction. For the alkali metals the size effect is more important and EAs decrease from Li to Cs as the IEs do. This is also the general trend for most of the nonmetals with some notable exceptions.

The EAs of N, O, and F are *much* lower than expected from the group trends, which are regular for S → Po and Cl → At. The IEs are regular within each of these groups and highest for N, O, and F of the representative groups. Proceeding across the second period, the sizes[18] decrease and the nuclear charges increase, so that the negative charge density increases, just about reaching the limit for F. Thus, N, O, and F have very high IEs, but the addition of another electron is not so favorable as for P, S, and Cl, respectively. The factors causing low EAs for N, O, and F lead to unusually low bond energies when these elements are bonded to themselves or to one another by single bonds. The amazingly low EA of N and the trends in Groups 15 and 16 are considered in Chapter 16, which treats the group chemistries.

The noble gases have stable ns^2np^6 configurations ($1s^2$ for He), and the EAs are zero or slightly negative (corresponding to endothermic processes), indicating that there is no bound state for the added electron. The Group 2 metals with the ns^2 configuration (beyond

---

[18] A discussion of periodic trends in atomic size is given by J. Mason, *J. Chem. Educ.* **1988**, *65*, 17.

Table 1.8   Electron affinities of the elements[a]

1	2	3	4	5	6	7	8	9
1 H 72.77 0.7542	kJ/(g · atm) eV/atom							
3 Li 59.6 0.618	4 Be −50 −0.5							
11 Na 52.9 0.548	12 Mg −40 −0.4							
19 K 48.4 0.501	20 Ca −30 −0.3	21 Sc −30 −0.3	22 Ti 7.6 0.079	23 V 50.7 0.525	24 Cr 64.3 0.666	25 Mn −50 −0.5	26 Fe 15.7 0.163	27 Co 63.8 0.661
37 Rb 46.89 0.4859	38 Sr −30 −0.3	39 Y ~0 ~0	40 Zr 41.1 0.426	41 Nb 86.2 0.893	42 Mo 72.0 0.746	43 Tc 53 0.55	44 Ru 101 1.05	45 Rh 109.7 1.137
55 Cs 45.50 0.4716	56 Ba −30 −0.3	57 La 50 0.5	72 Hf 10 0.1	73 Ta 31.1 0.322	74 W 78.6 0.815	75 Re 14 0.15	76 Os 110 1.1	77 Ir 151.0 1.565

[a] Data are from H. Hotop and W. C. Lineberger, *J. Phys. Chem. Ref. Data* **1985**, *14*, 731; S. G. Bratsch and J. J. Lagowski, *Polyhedron* **1986**, *5*, 1763 for N, Mn, Hf, Hg, Groups 2 and 13, and the noble gases.

a noble gas configuration) and the Group 12 metals with the $(n - 1)d^{10}ns^2$ configuration have negative EAs. The next higher energy state (^2P) is too high in energy for the favorable addition of an electron. The electron affinity of Hg is ~−50 kJ/mol, but the neighboring elements Pt and Au have very high positive EA values. In fact, only the halogens have higher EAs than do Pt and Au.

EXAMPLE: Suppose a gaseous mixture of He, Ne, Ar, and Kr is irradiated with photons of the frequency appropriate to ionize Ar. What ions will be present in the mixture?

Solution: Obviously Ar^+ will be present. The EA of Ar^+ is 15.76 eV, the same as IE_1 for Ar. IE_1 for Kr is 14.0 eV, and thus electron transfer to Ar^+ will occur exothermically, thereby producing Kr^+. IE_1 for He and Ne are larger than EA of Ar^+, and therefore these atoms will not react.

1.8   ABSOLUTE ELECTRONEGATIVITY AND ABSOLUTE HARDNESS

Ultimately, the goal of chemistry is to make predictions about the way in which atoms, ions, and molecules will react with one another under various sets of conditions and what macroscopic properties the products will have. An important step in doing this involves a

Table 1.8   (Continued)

10	11	12	13	14	15	16	17	18
								2 He −50 −0.5
			5 B 26.7 0.277	6 C 121.9 1.263	7 N −7 −0.07	8 O 141.0 1.461	9 F 328.0 3.399	10 Ne −120 −1.2
			13 Al 42.6 0.441	14 Si 133.6 1.385	15 P 72.0 0.746	16 S 200.43 2.0771	17 Cl 349.0 3.617	18 Ar −96 −1.0
28 Ni 111.6 1.156	29 Cu 118.5 1.228	30 Zn −60 −0.6	31 Ga 30 0.3	32 Ge 120 1.2	33 As 78 0.81	34 Se 194.99 2.0207	35 Br 324.7 3.365	36 Kr −96 −1.0
46 Pd 53.8 0.557	47 Ag 125.6 1.302	48 Cd −70 −0.7	49 In 30 0.3	50 Sn 120 1.2	51 Sb 103 1.07	52 Te 190.16 1.9708	53 I 295.2 3.059	54 Xe −80 −0.8
78 Pt 205.3 2.128	79 Au 222.76 2.3086	80 Hg −50 −0.5	81 Tl 20 0.2	82 Pb 35.1 0.364	83 Bi 91.3 0.946	84 Po 180 1.9	86 At 270 2.8	86 Rn −70 −0.7

model constructed using the ionization energies and electron affinities of gaseous atoms and ions. We attempt to understand the behavior of atoms in molecular environments by considering the properties of the isolated atoms or ions from which we consider the molecules to be composed. Two important fundamental quantities are **absolute electronegativity** (parallelling a suggestion of Mullikan) and **absolute hardness** (recently treated by Pearson and Parr[19]). Absolute electronegativity, χ, is defined as[20]

$$\chi = (IE + EA)/2 \tag{1.33}$$

Absolute hardness, η, is defined as

$$\eta = (IE - EA)/2 \tag{1.34}$$

Because both of these definitions involve the thermodynamic quantities IE and EA, each corresponds to a definite physical process. Consider the two-step process whose overall result is the addition of a pair of electrons to Y^+, the cation of an element Y:

$$
\begin{aligned}
Y^+ + e^- &\rightarrow Y & \Delta H &= -IE \\
Y\ \ + e^- &\rightarrow Y^- & \Delta H &= -EA \\
\hline
Y^+ + 2e^- &\rightarrow Y^- & \Delta H &= -(IE + EA)
\end{aligned}
\tag{1.35}
$$

[19] See R. G. Pearson, *J. Chem. Educ.* **1988**, *64*, 561.

[20] Mullikan applied this definition only to univalent atoms. See R. S. Mullikan, *J. Chem. Phys.* **1934**, *2*, 782.

The more exothermic the reaction, the greater the attraction for the pair of electrons. Hence, absolute electronegativity is a measure of the attraction of Y^+ for an electron pair or, alternatively, the average of the attraction of Y^+ and Y for an electron. To apply this to atoms in molecules, consider the two dissociations of YZ (where Z is another element):

$$Y^+(g) + Z^-(g) \rightleftharpoons YZ \rightleftharpoons Y^-(g) + Z^+(g) \tag{1.36}$$

The energy difference between the products on the right and those on the left is $(\text{IE}_Z - \text{EA}_Y) - (\text{IE}_Y - \text{EA}_Z) = (\text{IE}_Z + \text{EA}_Z) - (\text{IE}_Y + \text{EA}_Y) = 2(\chi_Z - \chi_Y)$—that is, just twice the difference in the absolute electronegativities. For HCl, a calculation gives $(\chi_H - \chi_{Cl}) = (13.598 + 0.754)$ eV $- (12.967 + 3.617)$ eV $= -2.232$ eV, indicating that the energetically favorable dissociation process is to H^+ and Cl^-. Hence, the inference is usually drawn that electron density in the molecule is not spread evenly along the H–Cl axis but is instead distorted (**polarized**) in the sense of H^+Cl^-. χ is indeed a measure of the attraction of atoms for electrons as seen from the result of such a calculation when $Y = Z$; here $\Delta\chi = 0$ because both attract electrons equally.

EXAMPLE: Using values of absolute electronegativity, determine the sense of polarization in the molecules IF, ICl, and BrCl.

Solution: Using data in Tables 1.7 and 1.8, the calculated values of χ (eV) are as follows: F 10.41; Cl 8.30; Br 7.59; I 6.76. $\chi_I - \chi_F = 6.76 - 10.41 = -3.35$ eV, the negative sign indicating that the reaction is exothermic in the direction IF $\rightarrow I^+ + F^-$. Hence, we infer that the molecule is polarized in the sense I^+F^-. Similarly, $\chi_I - \chi_{Cl} = -1.54$ eV and $\chi_{Br} - \chi_{Cl} = -0.71$ eV and the polarizations are I^+Cl^- and Br^+Cl^-.

Absolute hardness corresponds to the energy involved in the disproportionation reaction of Y:

$$Y \rightarrow \tfrac{1}{2}Y^+ + \tfrac{1}{2}Y^- \tag{1.37}$$

The energy for reaction (1.37) is $\tfrac{1}{2}(\text{IE}_Y - \text{EA}_Y) = \eta_Y$. Chemically this quantity is informative as a measure of how easy it is to polarize the electron cloud of Y. The more easily the electron cloud is polarized, the less energy will be required to effect the transfer of an electron. Small species such as F with a high effective nuclear charge hold their electrons tightly (large IE) and tend to have vacant orbitals which could accept electrons at very high energies (small or even negative EA). These hard species thus have large values of η. Conversely, large species such as iodine have rather loosely held electrons (small IE) which can be easily polarized. The many electrons require occupation of orbitals with a high n which are relatively closely spaced. Thus, the unfilled orbitals of lowest energy are not far above the highest filled ones; EA will be large relative to IE, leading to a small value for η. Polarization is to be expected in these cases because it costs little energy to promote electrons into accessible unfilled orbitals which have different shapes and radial extension from those filled in the ground state. Species which have low η values are referred to as *soft*. Alternatively, a softness parameter σ is defined as $1/\eta$.

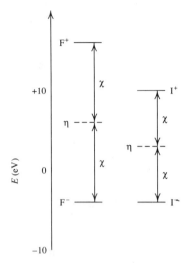

Figure 1.16   The relation between χ and η for hard F and soft I. The zero of energy is the neutral atom.

If we were to change comparable species in the same way, for example, by adding or removing an electron, we expect that the trends in electronegativity and in hardness should be unaltered. Thus we expect F^- to be harder than I^- and expect F^+ to be harder than I^+. Also, O^{2-} should be harder S^{2-}, and so on. Figure 1.16 diagrams the absolute electronegativity and hardness for F and I. Note the relationship between high hardness and high electronegativity.

Although the orbital from which an electron is removed in measuring IE for, say, F is the same one to which an electron is added in measuring EA, IE and EA differ because of the extra electron–electron repulsion energy involved in forming F^- as compared to F^+.

Obviously, Reaction (1.37) involves changes in the number of electrons on both atoms, and the energy would be zero if $\eta = 0$. Large values of EA and small values of IE favor addition and loss of electrons, respectively; moreover, such values also lead to small values of η. Thus, small values of η indicate lack of resistance to change in the number of electrons. Hence, another interpretation of absolute hardness is that it measures the resistance to change in the number of electrons. Because IE is always greater than or equal to EA, the minimum value of η is zero. For bulk metals IE = EA and $\eta = 0$, and these have maximum softness.

In subsequent chapters we shall make use of electronegativities and absolute hardness in understanding molecular properties and reactivity.

GENERAL REFERENCES

P. A. Cox, *The Elements, Their Origin, Abundance, and Distribution*, Oxford University Press, Oxford, 1989.

R. L. DeKock and H. B. Gray, *Chemical Structure and Bonding*, Benjamin/Cummings, Menlo Park, CA, 1980. One of the better treatments.

J. Elmsley, *The Elements,* Oxford University Press, New York, 1989. A handbook of values of the properties of the elements.

G. Herzberg, *Atomic Spectra and Atomic Structure,* Dover, New York, 1944. A classic book that is still useful.

R. M. Hochstrasser, *Behavior of Electrons in Atoms. Structure, Spectra and Photochemistry of Atoms,* Benjamin, New York, 1964. Good treatment at an appropriate level.

L. Pauling, *The Nature of the Chemical Bond,* 3rd ed., Cornell University Press, Ithaca, NY, 1960. This classic book gives much detail of interest to the chemist.

C. J. Suckling, K. E. Suckling, and C. W. Suckling, *Chemistry Through Models,* Cambridge University Press, Cambridge, 1978. A good account of the role of models in chemical thinking.

PROBLEMS

1.1 Use the *Handbook of Chemistry and Physics* as a data source to indicate the nature of the periodic variation of the following properties of the elements in Groups 1, 2, 11, 12, 15, 16, and 18: **(a)** melting points; **(b)** boiling points; **(c)** formulas of oxide; **(d)** colored oxides.

1.2 Bohr postulated that lines in the emission spectrum from hydrogen in highly excited states, such as $n = 20$, would not be observed under ordinary laboratory conditions, because of the large size of such atoms and the much greater probability of atom collision deactivation as compared with radiation emission. Using the Bohr model, calculate the ratio of the cross-sectional area of a hydrogen atom in the $n = 20$ state to that of one in the $n = 1$ state.

1.3 According to the Bohr model of the atom, what would be the size of a Ne^{9+} ion? What would the ionization energy be for this ion? What would the excitation energy be for the first excited state?

1.4 Calculate the energy (eV) released in the transition of a hydrogen atom from the state $n = 3$ to $n = 1$. The wavelength of the radiation emitted in this transition may be found from the relation λ (in Å) $= 12,398/E(eV)$ (often remembered as the approximate $\lambda = 12,345/eV$). Calculate the wavelengths of all spectral lines that could be observed from a collection of hydrogen atoms excited by a potential of 12 eV.

1.5 Assuming a screening of 0.5 for an *s* electron, calculate the wavelength expected for the K_α x-ray line of Tc.

1.6 The Balmer series in the hydrogen spectrum originates from transitions between $n = 2$ states and higher states. Compare the wavelengths for the first three lines in the Balmer series with those expected for similar transitions in Li^{2+}.

1.7 Obtain expressions for $(\partial E/\partial Z)_n$ and $(\partial^2 E/\partial Z^2)_n$ using the Bohr model [Equation (1.9)]. Using the ionization energies for isoelectronic sequences of atoms and ions from Table 1.7, and finite differences to obtain $(\Delta E/\Delta Z)_n$ and $(\Delta^2 E/\Delta Z^2)_n$, show that the $n = 1$ level is filled when the electron occupancy is 2. How would you estimate unknown ionization energies using a finite-difference approximation? Estimate the electron affinities of F and O by this procedure.

1.8   Using Table 1.1 and Equation (1.23), calculate the values of r for which $2p$ and $4d$ hydrogenlike orbitals have radial nodes.

1.9   Using Table 1.3 and Equations (1.27), calculate the values of θ and ϕ for which $2p_z$ and $3d_{x^2-y^2}$ have angular nodes.

1.10   Explain briefly the observation that the energy difference between the $1s^2 2s^1\ ^2S_{1/2}$ state and the $1s^2 2p^1\ ^2P_{1/2}$ state for Li is 14,904 cm^{-1}, whereas for Li^{2+} the $2s^1\ ^2S_{1/2}$ and the $2p^1\ ^2P_{1/2}$ states differ by only 2.4 cm^{-1}.

1.11   Using the p orbitals for an example, distinguish among the angular part of the probability function, the radial part of the probability function, and a probability contour. Draw simple sketches to illustrate. How would each of these be affected by a change in the principal quantum number, n?

1.12   Give the characteristic valence shell configuration for the following periodic groups (e.g., ns^1 for the alkali metals):
    (a) Noble gases   **(d)** Ti family
    (b) Halogens   **(e)** N family
    (c) Coinage metals (Group 11)

1.13   Write the electron configuration beyond a noble gas core (for example, F, [He]$2s^2 2p^5$) for Rb, La, Cr, Fe, Cu, Tl, Po, Gd, and Lu.

1.14   Write the electron configuration beyond a noble gas core and give the number of unpaired electrons (for example, F$^-$, [He]$2s^2 2p^6$) for K$^+$, Ti^{3+}, Cr^{3+}, Fe^{2+}, Cu^{2+}, Sb^{3+}, Se^{2-}, Sn^{4+}, Ce^{4+}, Eu^{2+}, and Lu^{3+}.

1.15   What is the spectroscopic notation for an atom in an energy state in which $L = 3$ and $S = 2$?

1.16   What is the number of microstates for an f^3 configuration? Which of these is unique to the ground-state term arising from this configuration?

1.17   Derive the spectral terms for the d^7 configuration. Identify the ground-state term. Give all J values for the ground-state term and indicate which is lowest in energy.

1.18   Derive the spectral terms for the f^2 configuration. Identify the ground-state term. Give all J values for the ground-state term and indicate which is lowest in energy.

1.19   What spectroscopic terms arise from the configuration $1s^2 2s^2 2p^6 3s^2 3p^5 3d^1$?

1.20   What nodal planes might be expected for the $f_{[x^3-(3/5)xr^2]}$ orbital?

1.21   Use Slater's rules to calculate the effective nuclear charge Z^* for the elements Li through F. How is the trend in Z^* reflected in the ionization energies in this period?

1.22   Use Slater's rules to calculate the effective nuclear charge for the alkaline earth elements. How is the trend in Z^* reflected in the ionization energies in this group?

1.23   Use Slater's rules to calculate the effective nuclear charges for K, Rb, and Cs. Compare these results with those for Cu, Ag, and Au. Comment on the ionization energies of these elements in light of these charges.

1.24   The trend in electron affinities for Group 15 (VB) is opposite to that for Group 16 (VIB). Explain the trend for each family. (See Table 1.8.)

1.25   Rearrange Equation (1.27) to express the "real" set of d orbitals in terms of the set of d orbitals with well-defined m_l values.

1.26   Use the data in Tables 1.7 and 1.8 to predict whether the following hydrides should dissociate to H$^+$ and X$^-$ or H$^-$ and X$^+$ in the gas phase:
    (a) LiH   **(b)** CsH   **(c)** HBr

1.27   Which of the following is softer?

   (a) Cl or Br   **(d)** H or Na

   (b) S^{2-} or Se^{2-} **(e)** Al^{3+} or Tl^{3+}

   (c) S or O   **(f)** K^+ or Cu^+

1.28   Rank the following in the order of increasing absolute electronegativity: F, Li, Ti^{4+}, P, H, C, Mg^{2+}, Li^+.

▸2◂

Molecular Models

••••••••••••••••••••••••••••••••••

Fundamental questions of chemistry include how atoms are bonded together to form compounds and how the formulas and structures of compounds are dictated by bonding forces. The ultimate goal is to be able to predict and understand the properties of compounds on the basis of their composition and structure. In this chapter we treat molecules in which atoms are bonded covalently. We do this from the point of view of a simple electron sharing model which allows also an understanding of why the compounds have the formulas they do. The role of atomic orbitals and their hybrids is emphasized in a discussion of the valence bond model. The shapes of molecules are then treated from the point of view of the hybridization model and the valence shell electron-pair repulsion model (VSEPR theory). Once we know the structures and electron distributions within molecules, we can understand intermolecular attractive forces.

2.1 REVIEW OF COVALENT BONDING

▸ 2.1.1 Introduction

Valence bond (VB) theory has played an important role in advancements in chemistry. Much of the terminology in current bonding applications is derived from VB theory. VB models are often very helpful in visualizing molecules and interactions between molecules. Concepts of VB theory are useful in dealing with electron pairs (or single electrons) as localized bonding, delocalized bonding, and nonbonding pairs. Nonbonding electron pairs often have profound effects on molecular geometry.

Here we use one simple model for covalent molecules. A bond is formed by sharing electrons between adjacent atoms in the region of overlapping orbitals. Modifications are made based on considerations of orbitals used, possible multiple bonding, and resonance. Shapes of molecules are treated in terms of the orientation of atomic or hybrid orbitals and, for finer tuning, repulsions among electron pairs. The varying attraction of atoms in molecules for electrons is addressed by developing the concept of electronegativity. Finally, hydrogen bonding and weak interactions which depend on molecular shape and polarity are presented. Properties of molecular substances that depend on attractions between molecules are examined. You are already familiar with some of the simple models presented here. Early sections of this chapter might be assumed as background from earlier courses. They should provide a convenient review and definitions of terms used throughout the book. The chapter serves as a vehicle to emphasize the differences and similarities of the various models applied to bonding and structure in covalent molecules.

▶ *2.1.2   The Octet Rule*

Halogen atoms have the electron configuration s^2p^5, each atom has only one unpaired electron to share, and we expect one bond. An electron dot picture represents the X_2 molecule as

$$:\!\overset{..}{X}\!:\!\overset{..}{X}\!:$$

In writing electron dot formulas (**Lewis structures**), we place dots representing the shared pair between the symbols for the two bonded atoms. Singly occupied orbitals are indicated by the electron dot picture, as in $:\!\dot{C}\!\cdot$ for the carbon atom. The VB representation replaces a shared pair of dots by a line representing the simultaneous occupancy of two overlapping atomic orbitals (AOs) by a pair of electrons. In cases in which the occupancy of nonbonding orbitals appears obvious, we may show only the bond, as in

$$X - X$$

A pair of electrons occupying a nonbonding AO commonly is referred to as a **lone pair.** In this X_2 molecule there are three lone pairs per atom.

For oxygen the electron configuration of the atom may be represented by

$$:\!\overset{..}{O}\!\cdot \qquad\qquad \underset{1s}{\textcircled{$\cdot\cdot$}} \;\; \underset{2s}{\textcircled{$\cdot\cdot$}} \;\; \underset{2p}{\textcircled{$\cdot\cdot$}\textcircled{$\cdot$}\textcircled{$\cdot$}}$$

We expect oxygen to share two electrons, as it does in almost all of its covalent compounds. The Lewis structures below are typical.

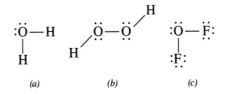

| (a) | (b) | (c) |

The **oxidation number** for atoms in covalent compounds *is obtained by assigning shared electrons to the more electronegative atom (Section 1.8) and then counting the charge on the quasi ion* (see Figure 2.1 for additional examples). Although oxygen forms two bonds in each compound shown, the oxidation number of oxygen is $-II$ in water, $-I$ in hydrogen peroxide, and II in oxygen difluoride.

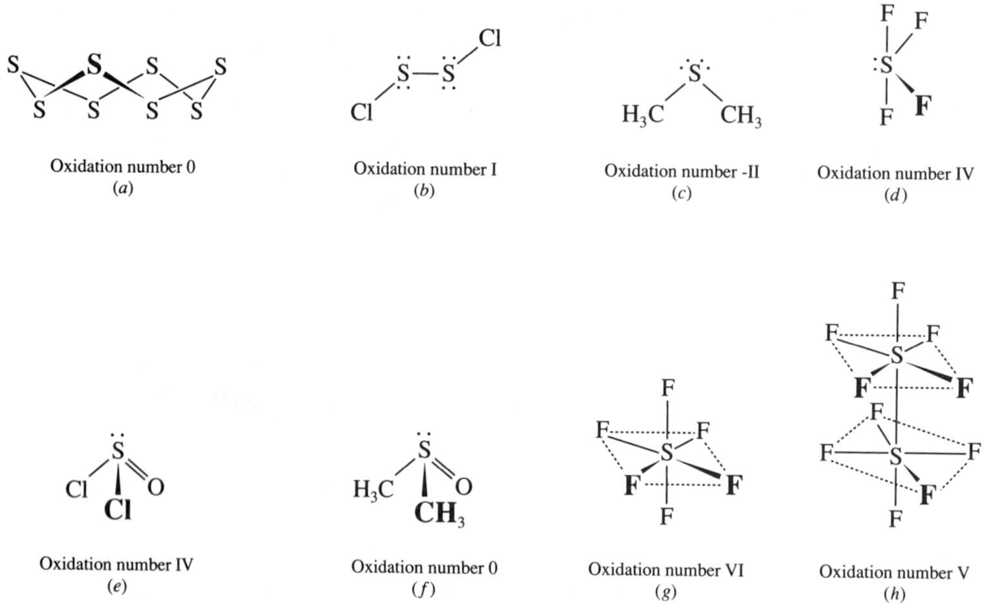

Figure 2.1   Valence bond structures for S compounds.

Commonly we calculate **oxidation numbers** using the known oxidation numbers of partners. Thus Group 1 metal ions (Li through Cs) are always M^+, Group 2 metal ions (Be through Ra) are always M^{2+}, Al is only Al^{3+}, F is always F^-, and oxygen is O^{2-} unless combined with itself (O_2^{2-} and O_2^-) or with F (OF_2). For neutral molecules the sum of oxidation numbers is zero and the sum must equal the charge on an ion.

EXAMPLES: Na_2O_2 $2\,Na^+$ requires O_2^{2-} or -I for O
 OF_2 $2\,F^-$ requires +II for O
 MnO_4^- $4\,O$ (-II) requires VII for Mn + 7 − 8 = −1
 $NaClO_3$ Na^+ and $3\,O$ (-II) requires V for Cl + 1 + 5 − 6 = 0

▶ *2.1.3   Expanded Octet*

As expected, sulfur is chemically similar to oxygen, since both have the same outer electron configuration—and indeed, sulfur forms H_2S, H_2S_2, and SF_2 with structures similar to (a), (b), and (c) above for oxygen. However, sulfur also has empty *d* orbitals of the same principal quantum number as valence electrons (those involved in bonding); that is,

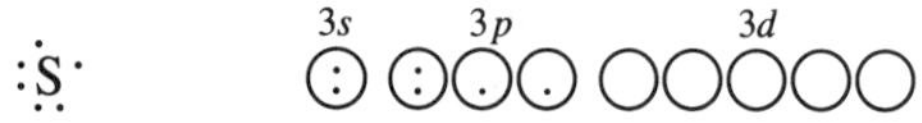

The expenditure of promotional energy produces configurations having a greater number of unpaired electrons. These configurations can form more covalent bonds, that is,

Example Compound

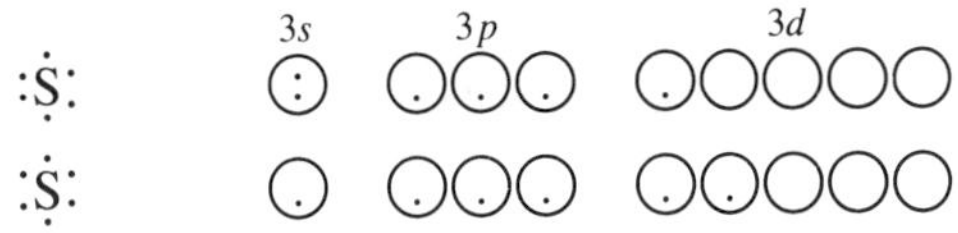

Thus sulfur can form four or even six covalent bonds if another element, combining with sulfur, releases enough energy to pay for promotion of the electrons. In general, such promotion is important only when sulfur combines with the more electronegative elements, F, O, Cl, or Br. Examples are given in Figure 2.1. The assignment of the oxidation number of $-II$ to S for $S(CH_3)_2$ follows the convention of naming CS_2 as a sulfide, even though the electronegativities of C and S are the same (see Table 2.6).

Oxygen is limited to an octet of electrons, because there are only $2s$ and $2p$ orbitals in the second shell. One of the major differences between the second-period elements (C, N, O, and F with $2s\,2p$ orbitals) and those of the following periods is the availability of empty d orbitals (Si, P, S, and Cl with $3s\,3p\,3d$ orbitals and Ge, As, Se, and Br with $4s\,4p\,4d$ orbitals). Chlorine, bromine, iodine, and astatine have outermost electron configurations represented by

Example Compound

HX

and electron promotion could result in the configurations represented by

Example Compound

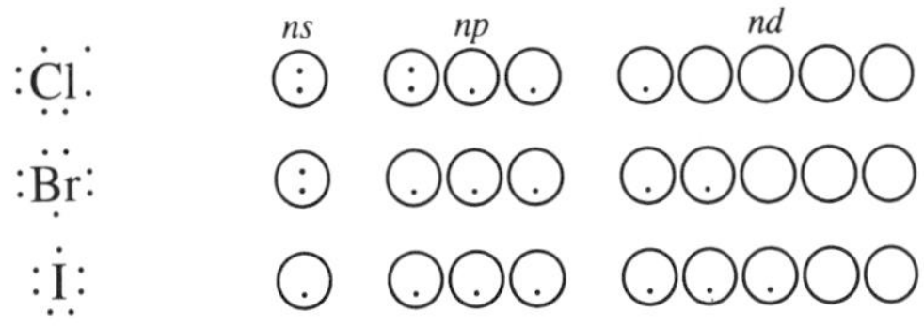

Sharing of the unpaired electrons leads to compounds containing one, three, five, and seven covalent bonds, as illustrated by HX, and so on, and by the highest fluorides of chlorine (ClF_3), bromine (BrF_5), and iodine (IF_7). In this series of XX'_n compounds, more bonds are formed (requiring more electrons to be promoted) with increasing differences in electronegativity. VB structures of these are given in Figure 2.2. In Group 15 (VB) elements, nitrogen, $1s^2 2s^2 2p^3$, forms three bonds, whereas phosphorus, arsenic, and antimony commonly form three or five bonds (Figure 2.3).

In the carbon family the outermost configuration of the isolated ground-state atom is

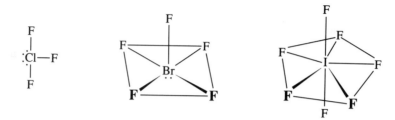

Figure 2.2   Valence bond structures.

The promotional energy necessary to produce the ns^1np^3 configuration is, however, more than compensated for by the formation of the two additional bonds, as in CH_4, CCl_4, $SnCl_4$, and so on:

Each C in ethane forms four bonds and in ethylene, C_2H_4, a $C=C$ double bond is formed to complete the octet for each C. There is a triple bond in N_2, $:N\equiv N:$, to complete octets.

▶ 2.1.4 *Coordinate Covalence (Dative Bonding)*

Ammonia has three covalent bonds and a lone pair. It reacts with H^+ to form NH_4^+. The fourth bond is equivalent to the other three bonds, but it is formed by sharing an electron pair from N with H^+. This bond is called a **coordinate covalent bond** or a **dative bond.** It cannot be distinguished from the other three bonds—all four bonds are equivalent for NH_4^+.

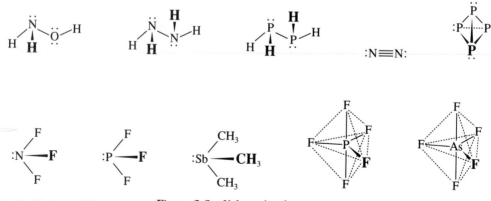

Figure 2.3   Valence bond structures.

Nitrogen can form dative bonds as follows:

$$
\begin{array}{ccc}
\text{H} & \text{CH}_3 & \text{CH}_3 \\
| & | & | \\
\text{N}^{(+)} & \text{N}^{(+)} & \text{N}^{(+)} \\
\text{H} \quad | \quad \text{H} & \text{H}_3\text{C} \quad | \quad \text{CH}_3 & \text{H}_3\text{C} \quad | \quad \text{O}^{(-)} \\
\mathbf{H} & \mathbf{CH_3} & \mathbf{CH_3}
\end{array}
$$

An arrow can be used for the N→B bond in $H_3N{\rightarrow}BF_3$ to identify it as a dative bond, but usually the distinction is not made. Arrows are not appropriate for NH_4^+ or $N(CH_3)_4^+$, because the four bonds are equivalent. The charges shown, called **formal charges**, always arise in dative bonding. *Formal charges are calculated for an electron dot structure by assigning half of the electrons shared between bonded atoms to each atom and calculating the result charge on the quasi ion,* or

$$\text{Formal charge}^1 = \text{No. of valence } e - \tfrac{1}{2}(\text{No. of bonding } e) - \text{No. of nonbonding } e \qquad (2.1)$$

The positive (negative) formal charge is the number of electrons fewer (greater) than the number of electrons for the neutral atom. The sum of the formal charges in a molecule is zero; the sum of the formal charges for an ion is the same as the charge on the ion. Two different atoms do not share electrons equally in an A—X bond. Formal charges are reduced in magnitude by polarization resulting from differences in electronegativities of A and X (see Section 2.4.4).

Carbon monoxide provides a common case of molecule formation in which an *s* electron of carbon is *not* promoted. Without promotion we would expect carbon to form two bonds, giving the configuration

$$
:\text{C}{=}\ddot{\text{O}} \qquad
\begin{array}{cc}
2s & 2p \\
\odot & \text{O}\text{O}\bigcirc
\end{array}
$$

However, the high dissociation energy of CO and the very short C—O bond distance suggest that the Lewis structure is better represented by a triple bond, including a C←O dative bond:

$$^{(-)}:\text{C}{\equiv}\text{O}:^{(+)}$$

Phosphorus can donate its unshared pair in dative bonding, as in

$$\overset{(+)}{\text{Cl}_3\text{P}}{\rightarrow}\overset{(-)}{\text{BBr}_3}$$

or it can accept a pair of electrons into a *d* orbital, as in

$$\overset{(-)}{\text{Cl}_3\ddot{\text{P}}}{\leftarrow}\overset{(+)}{\text{N}(\text{CH}_3)_3}$$

---

[1] The number of valence electrons is the group number in the I → VIII periodic tables.

Phosphorus in PF_5 can add a fluoride ion by accepting a pair of electrons into one of its empty *d* orbitals, as shown below:

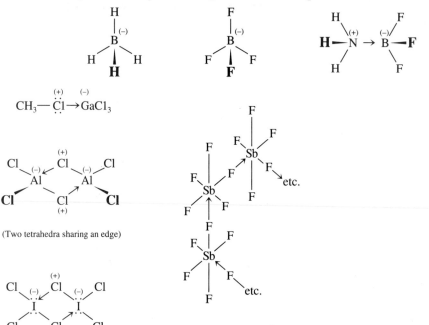

Through dative bonding, phosphorus can form four and six bonds, as well as the usual compounds such as PX_3 and PX_5. This series may be illustrated by the following species. The placement of formal charges is shown for the ions PCl_4^+ and PCl_6^-:

Tne oxidation state of P in PCl_3 is $+III$, and that of P in PCl_4^+, PCl_5, and PCl_6^- is $+V$.

In addition to the compounds XX', XX'_3, XX'_5, and XX'_7, mentioned earlier for the heavier halogens, dative bonding allows the formation of ICl_2^- ($ICl + Cl^-$), ICl_4^-, and IF_6^-. Halogens also can form bridging bonds, as shown in Figure 2.4. In the case of boron and other Group 13 (IIIB) elements, accepting electrons via dative bonds is the rule rather than the exception. Thus the following anions and compound are known for boron:

Figure 2.4 Dative bonds (formal charges shown).

Note that models for the covalent bond used depend only on counting orbitals and electrons. The covalent bond is formed by overlap of two AOs which are occupied by an *e* pair shared by the bonded atoms. The actual identity of the orbitals is not important in predicting formulas. It is necessary to recognize that some orbitals have excessively high energy for bond formation—this always applies to orbitals of higher *n* quantum number than the valence shell. There are serious questions concerning the extent of participation of *d* orbitals because their energy is significantly higher than that of *s* and *p* orbitals of the same quantum shell. Bonding can be explained using more sophisticated models in compounds such as XeF_2 (see Section 4.4), ClF_3, and PF_5 without *d* orbitals. Nevertheless, the simple, useful model using *d* orbitals permits predictions of formulas and geometry of compounds as having expanded octets. This feature of nonmetals of the third (Si, P, S, and Cl) and later periods is a striking difference from C, N, O, and F.

► *2.1.5   Resonance*

Some molecules or ions cannot be represented adequately by a single valence bond formula. The rules of combination lead to the representation $\begin{smallmatrix} (-)O \\ \diagdown \\ C{=}O \\ \diagup \\ (-)O \end{smallmatrix}$ for the carbonate ion, but the oxygen atoms are known to be equivalent and the bond angles are all 120° in the planar ion. The difficulty arises simply because, in reality, electrons are not restricted to particular positions. The "fourth" electron pair, which forms a double bond wherever it is written, is not localized in any one of the three bonds, but rather is delocalized over all three bonds. There are three equivalent bonds, each of which is something between a single and a double bond. Here the "fourth" bond can be written in any one of the three positions, so that each bond is described as having a **bond order** of $1\frac{1}{3}$. The **bond order** (B.O.) of a single bond is 1; of a double bond, 2; of a triple bond, 3.

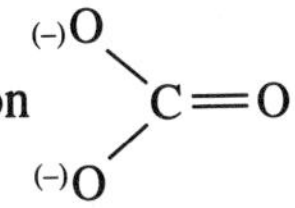

Pauling introduced the concept of **resonance** to adapt the simple valence bond notation to situations in which electrons are delocalized.[2] Three structures are written for the carbonate ion:

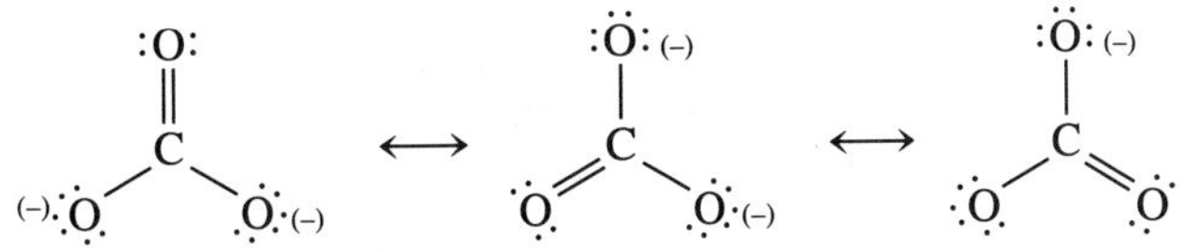

Do not interpret the three **contributing structures** as having any independent existence. The carbonate ion does *not* consist of a mixture of the three structures, nor is there an equilibrium among the three structures. The simple valence bond notation is inadequate to

---

[2] Resonance is needed because of the inadequacy of our notation, which represents electrons by dots and electron pairs by lines. Although the use of a circle inside the benzene ring emphasizes the delocalization of π electrons, it does not take care of the electron bookkeeping.

represent the structure of the carbonate ion, which is neither one nor all of the above structures, but something in between—a **resonance hybrid.** The three bonds and the three oxygen atoms are equivalent. The contributing structures here are separated by double-headed arrows ($\longleftrightarrow$), which should not be confused with the reversible arrows ($\rightleftharpoons$) used in chemical equilibrium.

The B.O. of the C—O bond in carbonates is $1\frac{1}{3}$ because the three most reasonable contributing structures are equivalent. The problem is more complex in the oxoanions of the halogens, the sulfur family, and the phosphorus family. For example, the sulfate ion may be viewed as arising from dative bonding between a sulfide ion and four oxygen atoms (a), or from the sharing of two pairs of electrons between each oxygen and sulfur (b). In addition to these extremes, we may write the intermediate structures (c), (d), and (e):

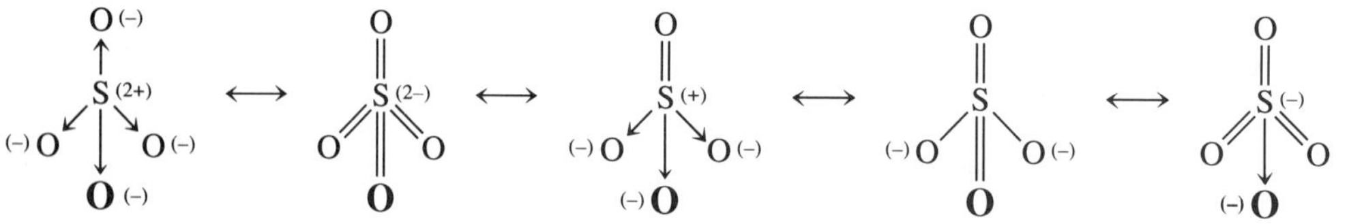

(a)   (b)   Four equivalent structures (c)   Six equivalent structures (d)   Four equivalent structures (e)

The actual structure of the sulfate ion is an average of all structures shown (and even, perhaps, some additional ones) weighted according to the number of equivalent structures and their relative energies.

Pauling proposed an **electroneutrality principle** stating that *electrons are distributed in a molecule in such a way as to make the residual charge on each atom zero or very nearly zero.* Exceptions include hydrogen and the most electropositive metals that can acquire partial positive charge and the most electronegative atoms that acquire partial negative charge. Partial multiple bonding (represented as appropriately weighted structures having double or triple bonds) is a way to achieve electroneutrality. The contributing structure for SO_4^{2-} that gives the best charge distribution is (d) with two double bonds. Considering these six equivalent structures, the formal charges are zero on S and $-\frac{1}{2}$ on each O.

The familiar application of resonance to benzene includes the usual Kekulé structures (a and b); the less important Dewar structures (c–e); and unimportant structures such as (f), in which there is charge separation and one carbon has only six electrons:

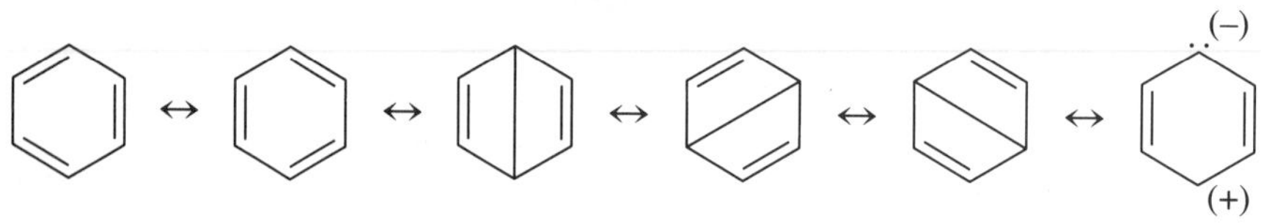

(a)   (b)   (c)   (d)   (e)   (f)

Elimination of unlikely structures and selection of the more important contributing structures—that is, those closer to the true resonance hybrid—is facilitated by the following rules.

1. Contributing structures should have the same or nearly the same atomic positions—they normally differ only in the positions of electrons.

2. All contributing structures of a particular molecule must have the same number of unpaired electrons.

3. Contributing structures should not differ too widely in the positions of electrons—in other words, the contributing structures should not differ greatly in energy. We expect structure (f) for benzene to be a high-energy (or less stable) structure compared with (a) and (b).

4. Like charges should not reside on atoms close together in a contributing structure, but unlike charges should not be greatly separated. In hydroazoic acid, HN_3, structure (c) is relatively unimportant, because of ($+$) charges on adjacent N atoms and also because of the double negative charge on the remaining N atom.

HN$_3$ N$_3^-$

H—N̈=N=N̈:   (a)   :N̈=N=N̈:

H—N̈—N≡N: (b)   :N̈—N≡N:

H—N≡N—N̈:   (c)   :N≡N—N̈:

Because structure (c) for azide ion, N_3^-, does not give ($+$) charges on adjacent N atoms, it should be more important than (c) for HN_3. Still, it is less important than (a) because of the -2 charge on N.

5. Contributing structures in which negative charge resides on an electronegative element and positive charge resides on an electropositive element are more important than those in which the reverse is true. When acetone loses a proton, an anion is obtained. The contributing structure in which the negative charge is on the oxygen, $CH_3\overset{:\overset{..}{O}:^{(-)}}{\underset{|}{C}}=CH_2$, is more important than one with negative charge on carbon.

6. The greater the number of covalent bonds, the greater the importance of a contributing structure. The doubly bonded structures of BF_3, $\overset{F}{\underset{F}{\diagdown}}B\overset{(-)\quad(+)}{=}F$ are important because of the formation of an additional covalent bond, even though a fluorine atom acquires a positive formal charge (see page 59). Here the tendency to utilize all low-energy orbitals in the molecule is more important than the charge distribution[3].

---

[3] The filled F 2_{p_z} orbital overlaps with the empty B 2_{p_z} orbital, and bonding occurs to the extent allowed by the relative energies of the orbitals and the charge distribution determined by the electronegativities (see Sections 1.8 and 2.4). BF is isoelectronic with CO, and the polarity indicates that F is positive relative to B. Two papers by D. K. Straub to be published in *J. Chem. Educ.* in 1994 present much data on: (1) multiple bonding in boron compounds and (2) multiple bonding in oxygen compounds of $3p-5p$ nonmetals.

A molecule for which resonance is important is more stable than predicted from any one of the contributing structures. The discrepancy between calculated and observed heats of formation is called the **resonance energy;** the greater the number of significant contributing structures, the greater the resonance energy. Typical resonance energies (kJ/mol) are benzene, 155; naphthalene, 314; $C \equiv O$, 439; and $O = C = O$, 151. The increased stabilization of molecules by resonance is reflected in the shortening of bond lengths compared to the average expected from the contributing structures.

Planarity is common for molecules of second-period elements in which resonance stabilization is important. In order that a double bond may be written in any one of several positions with good $p-p$ overlap a planar arrangement is favored.

The greater the number of contributing structures, the greater the delocalization of charge, and hence the more stable the molecule. Ions can be stabilized very greatly if the charge can be delocalized. Resonance, in general, is more important in charged species than in neutral ones. The increasing acidity in the series $HClO < HClO_2 < HClO_3 < HClO_4$ is determined largely by the greater stability of the anions containing several oxygen atoms, because of the greater delocalization of the negative charge. For the most part, the negative charge in ClO^- must reside on the oxygen atom. In ClO_4^- the negative charge is spread over four equivalent oxygen atoms.

▶ *2.1.6   Rules for Writing Lewis Octet Structures*

For an AXX′ molecule the least electronegative (Section 1.8) element is expected to be the central atom. A simple rule originally formulated by Langmuir allows octet structures to be written systematically. If each of n atoms achieves an octet, this requires $8n$ electrons. If fewer than $8n$ valence electrons are available, atoms complete their octets by forming covalent bonds in a molecule. Each covalent bond contributes two electrons to the valence shell of each atom bonded. Hence,

$$\text{No. of valence electrons} + 2(\text{No. of covalent bonds}) = 8n \tag{2.2}$$

$$\text{No. of covalent bonds} = \frac{8n - \text{No. of valence electrons}}{2} \tag{2.3}$$

For NO_2^-, for example, the number of covalent bonds is $\{(8 \times 3) - [5 + (2 \times 6) + 1]\}/2 = 3$. We thus have the two resonance structures:

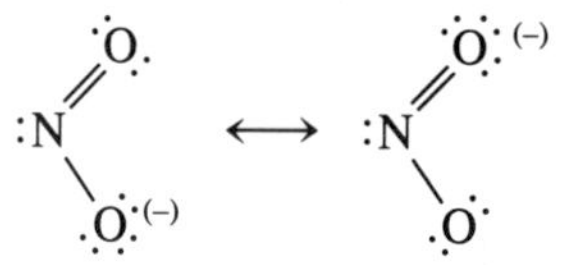

The nonbonding electrons are distributed as lone pairs. For NO_2^+, the number of covalent bonds is $\{(8 \times 3) - [5 + (2 \times 6) - 1]\}/2 = 4$, giving

     I   **II**

The more favorable formal charges for **I** indicates that it is more important than the two equivalent **II** structures.

EXAMPLE: Determine the number of bonds and the octet structure for OF_2.

Solution: OF_2 $\{(8 \times 3) - [6 + (2 \times 7)]\}/2 = 2$ covalent bonds, giving

Equations 2.2 and 2.3 do not apply in cases where expanded octets are possible. However, generating the appropriate octet structure and placing formal charges can lead to expanded octet structures that offer more favorable formal charge distribution. Thus, for example, structure (*a*) for sulfate (page 62) can be generated from Equation 2.3; other structures employing multiple bonding and octet expansion then suggest themselves as a way to remove or reduce the +2 formal charge from S.

If H atoms are present, the equations still hold for the anion without H^+. The H^+ can be added to the framework of "heavy" atoms to give the molecule. There are two electrons in each bond to H which count the same as a lone pair on the heavy atom and which also complete the valence shell of hydrogen. Of course, the H electrons are part of the total pool of valence electrons. Hence for C_6H_6 the number of covalent bonds in the C framework is given by $\{(8 \times 6) - [(6 \times 4) + (6 \times 1)]\}/2 = 18/2 = 9$ bonds. These are three single and three double bonds between pairs of C atoms in the two resonance structures (*a* and *b*) given for benzene on page 62.

EXAMPLE: Use Langmuir's formula (Equation 2.3) to generate valence bond structures for NO_3^-, ClO_3^-, and C_2H_4.

Solution: For NO_3^- we obtain $\{(8 \times 4) - [5 + (3 \times 6) + 1]\}/2 = 4$ covalent bonds, giving

(3 structures)

For ClO_3^- we obtain $\{(8 \times 4) - [7 + (3 \times 6) + 1]\}/2 = 3$ covalent bonds, giving structure (*a*). We can generate the expanded octet structures (*b*) and (*c*) (each of which has two more resonance forms) and the less favorable (*d*).

(*a*)   (*b*)   (*c*)   (*d*)

For ethylene we obtain $\{(8 \times 2) - [(2 \times 4) + (4 \times 1)]\}/2 = 2$ bonds. As pointed out above, this number represents the number of bonds between "heavy" (i.e., nonhydrogen) atoms. Thus, the Lewis octet structure for C_2H_4 is

$$\begin{matrix} H & & & H \\ \diagdown & & & \diagup \\ & C & = & C \\ \diagup & & & \diagdown \\ H & & & H \end{matrix}$$

Langmuir's rule could be extended to include cases where the octet is expanded; however, the extension is not very useful because writing a mathematical expression requires one to know just how many electrons are in the valence shells of atoms with expanded octets. Hence, it is easier to start with the octet structure generated by the octet rule and add multiple bonds to achieve favorable expanded octet structures.

2.2 VALENCE BOND
(PAULING–SLATER) THEORY

The covalent bond was first developed theoretically by Heitler and London for the hydrogen molecule. When two hydrogen atoms are brought together without exchanging electrons (structure I), there is weak attraction at large distance that becomes a strong repulsion at short distances. (The hydrogen nuclei are represented by H_a and H_b, and the electrons are expressed as $\cdot1$ and $\cdot2$.) Because at short distances the alternative structure, in which the electrons exchange positions (structure II), is just as likely, the potential energy curve shows a minimum corresponding to the normal H—H distance. The exchange between two equivalent structures provides the attraction, as in resonance stabilization. For favorable interaction the electrons must have opposite spins.

$$H_a{}^{\cdot1} \; {}^{2\cdot}H_b \qquad H_a{}^{\cdot2} \; {}^{1\cdot}H_b$$
$$\text{I} \qquad\qquad\qquad \text{II}$$

▶ 2.2.1 *Overlap of Atomic Orbitals*

Pauling and Slater extended the Heitler–London theory, making it more general and accounting for the directional character of covalent bonds. The main consideration is that stable compounds result from the tendency to fill all stable orbitals with electron pairs, shared or unshared.

In the VB model, the shapes of orbitals become explicitly important. Now we can consider which orbitals can overlap with which and develop a theory that explains molecular shapes and bond angles. However, in a book of the scope of this one, we cannot do total justice to the VB model because this would require working extensively with the analytical (algebraic) forms of atomic wavefunctions.

The shape of a molecule is determined primarily by the directional characteristics of the orbitals involved. However, molecular shape and the directional character of the orbitals are interrelated. For a given coordination number (the number of atoms bonded to the central atom for a covalent molecule), only a few molecular shapes are reasonable.

The orbitals used for bonding are those with directional characteristics compatible with the symmetry of the molecule as determined experimentally. A covalent bond may be described as resulting from the overlap of orbitals on two atoms so that the combination of orbitals can be occupied by an electron pair. In HF the spherical s orbital of H overlaps with the singly occupied $2p$ orbital of F.

The H_2O molecule results from the overlap of the two singly occupied p orbitals of oxygen with the s orbital of each of two hydrogen atoms, to form an angular molecule (see Figure 2.5*b*). The p orbitals are mutually perpendicular, but the bond angle in water is 104.45°. The ammonia molecule, using the three p orbitals, is pyramidal (Figure 2.5*c*) with a bond angle of 107.2°. (See Section 2.3 for correlations of bond angles.) The bond angles are expected to be greater than 90° because of repulsion between (or among) the H atoms.

▶ 2.2.2 Hybridization

Carbon has the outer configuration $2s^2 2p_x^1 2p_y^1$, but unpairing the $2s$ electrons and promoting one to the vacant $2p_z$ orbital results in $2s^1 2p_x^1 2p_y^1 2p_z^1$, as required for the formation of four bonds as in CH_4. The molecular shape is not immediately apparent from the characteristics of the AOs: The s orbital is spherically symmetrical, and the three p orbitals are mutually perpendicular. Nevertheless, physical and chemical evidence show that the methane molecule is tetrahedral, with four equivalent bonds. The wavefunctions, Ψ, for the tetrahedrally oriented orbitals are linear combinations of the atomic wavefunctions. There are four combinations, differing in the weighting coefficients a, b, c, and d:

$$\Psi = a\Psi_s + b\Psi_{p_x} + c\Psi_{p_y} + d\Psi_{p_z} \tag{2.4}$$

This process of combining atomic orbitals to give mixed orbitals is known as **hybridization**. The combination of the s and three p orbitals gives **tetrahedral hybrid sp^3 orbitals**. Combination of s and two p orbitals gives planar sp^2 hybrid orbitals directed toward the corners of an equilateral triangle. Equivalent sp hybrid orbitals pointing in oppo-

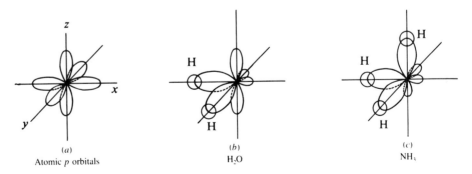

Figure 2.5 Angular characteristics of p orbitals.

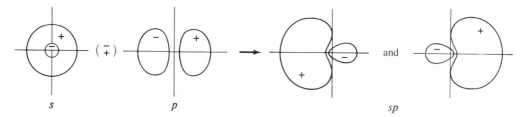

Figure 2.6 Combination of orbitals to give *sp* hybrids.

site directions (linear) result from *s* and one *p* orbital. The hybrid *sp* orbitals are represented in Figure 2.6. The lobe resulting from combination of the *s* orbital with the *p* lobe with the same sign of the Ψ function is enlarged, and the combined lobe of opposite sign is diminished. The enlarged lobe of the resulting orbital gives more favorable overlap with the orbital of another atom and thus forms a stronger bond than can either the *p* or *s* orbital alone. Both equivalent *sp* orbitals are shown separately in Figure 2.6. Of course, they should be superimposed with the same origin.

The shapes of the major lobes for sp^3, sp^2, and *sp* hybrid orbitals differ very little, mainly in the relative sizes of the minor and major lobes. The minor lobe is largest for sp^3 and smallest for *sp*, so the greatest fraction of electron density is in the bonding lobe for the *sp* hybrid orbital. Overlap of bonding orbitals increases in the order $sp^3 < sp^2 < sp$, indicating the greatest overlap for *sp* bonds. The increasing C—H bond energies in the series $CH_4 < C_2H_4 < C_2H_2$ support increasing bond strengths with increasing *s* character for *s–p* hybrids—that is, greatest for *sp*. Hybridization can also include combinations involving *d* orbitals. Figure 2.7 identifies some important hybrids—along with the corresponding geometrical configurations. The particular *d* orbitals involved in hybridization are those directed along bonds: $d_{x^2-y^2}$ for the square plane, d_{z^2} for the trigonal bipyramid, and $d_{x^2-y^2}$ plus d_{z^2} for the octahedron. Hybridization is the optimum combination or mixing of AOs consistent with the observed molecular shape. It is a convenient description based upon the familiar atomic wavefunctions.

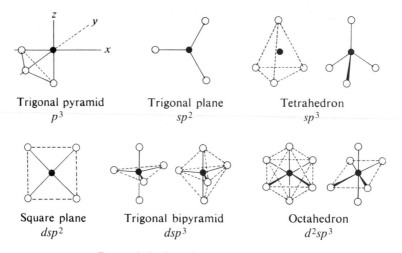

Trigonal pyramid
p^3

Trigonal plane
sp^2

Tetrahedron
sp^3

Square plane
dsp^2

Trigonal bipyramid
dsp^3

Octahedron
d^2sp^3

Figure 2.7 Some bonding configurations.

▶ *2.2.3   Multiple Bonds for Second-Period Elements*

Bonding requirements of each carbon in ethylene, C_2H_4, are satisfied by single bonds to two hydrogen atoms and a double bond between the carbon atoms. A double bond can be represented as two bonds resulting from overlap of sp^3 orbitals (see Figure 2.8a).

The more usual description of multiple bonds is adapted from molecular orbital theory (see page 155). In the ethylene molecule, the carbon orbitals can be considered to be hybridized to give three equivalent sp^2 orbitals (at 120° to each other). Each carbon uses two of the sp^2 hybrid orbitals for overlap with the $1s$ orbitals of two hydrogens, and it uses the remaining sp^2 orbital for the formation of a carbon–carbon bond. These five bonds are known as **sigma (σ) bonds,** *covalent bonds in which the electron density reaches a maximum along the line joining the bonded atoms.* The remaining two electrons from the unmixed p AOs of the carbon atoms pair to form a **pi (π) bond** having a nodal plane containing the bond. The electron pair of the π bond occupies the π orbital obtained by overlap of the two atomic p orbitals. The combination of two atomic p wavefunctions gives the molecular π wavefunction. As shown in Figure 2.8b, the π orbital picture, from the combination of the two p orbitals, shows two regions of high electron density, one above and one below the plane of the molecule, but *only one electron pair occupies the pair of lobes.* (If only *one* electron is in the π orbital, *it occupies both lobes.*) The other electron pair of the double bond occupies the C—C σ bond represented by a line.

Acetylene, C_2H_2, a linear molecule, can be represented by two tetrahedra sharing a face to produce three badly bent bonds joining the carbon atoms. However, the σ and π bond representation is more satisfactory. The σ bond formed by overlap of sp hybrids leaves each carbon with two singly occupied p orbitals. Because only one direction is defined (Figure 2.9a), the unhybridized p orbitals of each carbon atom are doughnut-shaped and they overlap to give a π orbital with cylindrical symmetry extending over both carbon atoms (Figure 2.9b). This result can also be viewed as resulting from the combination of two π orbitals, each having two lobes whose nodal planes intersect at right angles along the internuclear line (Figure 2.9c). The bond distance *shortens* and the bond strength *increases* from single to double to triple bonds (Table 2.1) with more π electrons.

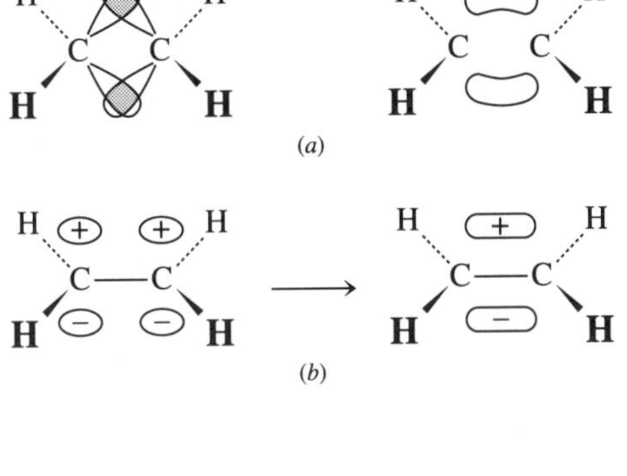

Figure 2.8   Representations of a double bond.

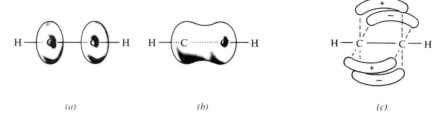

(a)   *(b)*   *(c)*

Figure 2.9   Representation of the orbitals in C_2H_2. Contours of maximum electron density are sketched.

Table 2.1   Bond lengths and bond energies for carbon and nitrogen[a]

	C—C	C=C	C≡C
Bond length, pm:	154	134	120
Bond energy, kJ/mol:	356	682	962

	N—N	N=N	N≡N
Bond length, pm:	145	125	110
Bond energy, kJ/mol:	240	450	942

[a] Data from S. W. Benson, *J. Chem. Educ.* **1965,** *42*, 502, or Table 2.8.

▶ 2.2.4   *Multiple Bonds for Elements Beyond the Second Period*

Double bonding encountered in second-period elements differs in extent and type from that encountered for later elements. Double bonding produced by the formation of $p\pi p$ bonds[4] is common for the second-period elements and diminishes in importance down through a given family. Fluorine and oxygen often form double bonds even at the expense of acquiring a formal positive charge. This serves to delocalize the electrons, diminishing the very high charge density of these small atoms and decreasing the repulsion among the unshared electrons on F or O. The need for such delocalization, or for decreasing the repulsion among lone pairs, is not so great for larger atoms. In these cases $p\pi p$ bonding is less favorable, because the p orbitals are larger and more diffuse, giving less orbital overlap. Increase in atomic radius also favors an increase in coordination number, thereby allowing electrons that might have formed π bonds to form σ bonds.

Table 2.2 displays the significant differences in structure and degree of aggregation between formally similar combinations involving elements of the second and third periods. Elements of the second period tend to form discrete molecules involving $p\pi p$ bonding (N_2, NO_2, O_2, and CO_2). Third-period elements tend to form larger or polymeric molecules. Thus Al_2Cl_6 exists as the dimer in the gas, P_4 is a tetrahedral molecule, in P_4O_6

[4] The notation of R. A. Stairs (*J. Chem. Educ.* **1988,** *65,* 980) shows the atomic or hybrid orbitals on the two atoms and shows the symmetry of the resulting bond—σ, π, and so on. For the C—H bond of acetylene, it is $sp\sigma s$ and S—O π bonding involves $d\pi p$.

Table 2.2   Comparison of melting points and boiling points of some elements and compounds from the second and third periods

$BCl_3(g)$	$CO_2(g)$	$N_2(g)$	$N_2O_3(g)$	$O_2(g)$
bp 12.5°C	subl. $-78.5°$	bp $-195.8°$	dec.	bp $-182.96°$
$Al_2Cl_6(s)$	$SiO_2(s)$	$P_4(s)$	$P_4O_6(s)$	$S_8(s)$
subl. 177.8°	mp 2230°	mp 590°	mp 23.8°	mp 95.5°
	$CS_2(l)$	$HNO_3(l)$	$N_2O_5(l)$	
	bp 46.3°	bp 83°	mp 30°	
	$(SiS_2)_x(s)$	$(HPO_3)_x(s)$	$P_4O_{10}(s)$	
	subl. 109°	dec.	subl. 300°	

the P_4 tetrahedron is expanded to incorporate an O into each tetrahedral edge, S_8 forms rings, and SiO_2 has a three-dimensional network of SiO_4 tetrahedra. In each of these cases for third-period elements there are only single bonds. Carbon disulfide exists as discrete CS_2 molecules (CO_2 structure), whereas SiS_2 forms infinite chains of SiS_4 tetrahedra. In $(HPO_3)_x$ and P_4O_{10} the coordination number (C.N., the number of bonded atoms) of P is four, whereas the C.N. of N is three in the corresponding nitrogen compounds (page 766). Disilenes ($R_2Si=SiR_2$) and digermenes ($R_2Ge=GeR_2$) are known that contain bulky R groups required to prevent polymerization. P and Sn prefer single bonds. The only example of a distannene is $[(Me_3Si)_2CH]_4Sn_2$.[5] Compounds of $Ge=C$[6] and $Sn=C$[7] have been prepared. These compounds also are stable only with bulky substituents.

Although $p\pi p$ bonding is less important for elements of the third and higher periods, multiple bonding may occur in other ways. A coordination number of four leaves no empty p orbitals on the central atom for the formation of $p\pi p$ bonds, but the bond length in an ion such as SO_4^{2-} is shorter than would be expected for single bonds. Sulfur atoms, unlike those of the second-period elements, have available vacant d orbitals of low energy for the formation of $d\pi p$ bonds, which result from the overlap of a filled p orbital on the oxygen with an empty d orbital on sulfur (see Figure 2.10). The d–p overlap may be more favorable than p–p overlap for large atoms, because the d orbitals project out in the general direction of the bond. The second-period elements C in CCl_4 and N in $N(CH_3)_4^+$ show C.N. 4, but in the oxoanions they form CO_3^{2-} and NO_3^- involving $p\pi p$ bonding. Third-period elements form SiO_4^{4-}, PO_4^{3-}, and SO_4^{2-} with the possibility of $d\pi p$ bonding (see footnote 3).

The charge on an atom can be decreased by a change in the amount of multiple bond character of a bond. The formal charge of $+3$ on Cl in the structure $\left[\begin{array}{c} O \\ | \\ O-Cl-O \\ | \\ O \end{array}\right]^-$ is reduced

to $+2$ in $\left[\begin{array}{c} O \\ \| \\ O-Cl-O \\ | \\ O \end{array}\right]^-$ and to zero in $\left[\begin{array}{c} O \\ | \\ O=Cl=O \\ \| \\ O \end{array}\right]^-$. The double-bond character ($d\pi p$) might

[5] A. H. Cowley and N. C. Norman, *Inorg. Chem.* **1986**, *34*, 1.

[6] C. Couret, J. Escudie, J. Satge, and M. Lazraq, *J. Am. Chem. Soc.* **1987**, *109*, 4411.

[7] H. Meyer, G. Baum, W. Massa, S. Berger, and A. Berndt, *Angew. Chem. Int. Ed. Engl.* **1987**, *26*, 546.

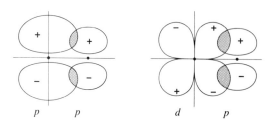

Figure 2.10   Sketches of $p\pi p$ and $d\pi p$ bonding.

be expected to increase in the series $PO_4^{3-} < SO_4^{2-} < ClO_4^-$, because the formal charge on the central atom increases for the single-bonded structures. The decreases in X—O bond lengths $d_{P-O} > d_{S-O} > d_{Cl-O}$ are consistent with an increase in B.O.

Formal charges are calculated assuming that both atoms share the electrons equally or that they have the same attraction for electrons (same electronegativity, see Section 1.8). Unlike atoms have different electronegativities. Because F is more electronegative than H, the electrons are not shared equally in H—F. That is, the bonding electron pair spends more time near F than near H. Electron density of the shared electron pair is shifted or **polarized** toward the F, making F more negative. There is a partial negative charge $(\delta-)$ on F (not -1 as on F^- in the compound Na^+F^-) and a partial positive charge $(\delta+)$ on H. The polarized HF molecule is **polar**. H_2 and F_2 are **nonpolar** because the atoms bonded are identical. Polar molecules can be said to be partially ionic. X—Y molecules have low ionic character if X and Y have similar electronegativities, and they have high ionic character if X and Y differ substantially in electronegativities.

For oxoanions, bond polarity makes the central atom positive because of the high electronegativity of oxygen and thus tends to offset the decrease in positive charge accompanying double bonding. The B.O. is about 1.5 for ClO_4^-. Calculations involving S—O and P—O bonds give reasonable bond lengths even without using d orbitals, but some participation of d orbitals provides better results overall. A B.O. of about 1.25 for SO_4^{2-} (giving S a formal charge of $+1$) and a B.O. only slightly greater than 1 for PO_4^{3-} seem reasonable. The electroneutrality principle, considering polarization as well as formal charge, is a good guide. Oxoanions of fluorine are not stable (but see page 800 for a discussion of HOF), partly because the high electronegativity of fluorine prohibits a formal positive charge. The effect of a positive charge cannot be diminished by double-bond formation because fluorine and oxygen in HOF have only filled valence shell orbitals.[8]

2.3   SHAPES OF MOLECULES

▶ *2.3.1   Molecular Geometry and Hybridization*

In the simplest model, the gross shape of a molecule is determined by its σ bond structure. Because they often occupy hybridized orbitals, unshared electron pairs can be treated the same as bonded groups in determining the hybridization; accordingly, the hybridization in

---

[8] In stable oxoanions (XO_n^{x-}) or oxoacids (H_xXO_n) the electronegativity of X is less than that of O, giving a positive oxidation number of X and a formal charge on X if $n > x$. In HOF the oxidation number of F is -1, as for all compounds of F, and the formal charge is zero. Positive formal charges for F occur in compounds such as BF_3 involving multiple bonding. See Section 2.4.5 for BF.

NH_3 (3σ + LP) and H_2O (2σ + 2LP, where LP stands for lone pair) can be considered to be sp^3. The importance of unshared electron pairs in determining molecular shapes is illustrated in Table 2.3. In writing Lewis structures for AX_n compounds other than those of H = X, A is generally more electropositive than X. Orbitals on X are filled with unshared electrons except as needed for bonding to complete the octet of A.

In many molecules it is difficult to assess *a priori* the relative importance of various contributing structures or the contribution of a particular type of hybridization. Atoms with a greater number of low-lying orbitals than valence electrons have reasonably unambiguous σ bond configurations and hence predictable molecular structures (at least for simple molecules). Atoms with more valence electrons than low-lying orbitals may contribute these electrons in varying degrees in diverse situations (i.e., in different molecules) and, accordingly, may have several distinct and not easily predictable σ bond structures. In trimethoxoboron, $B(OCH_3)_3$ (Figure 2.11), the expected sp^2 hybridization for boron is

Table 2.3 Shapes of some simple molecules

General Formula	Number of unshared electrons on A	σ Bonding orbital hybridization	Shape	Examples
AX_2	None	sp	Linear	$O{=}C{=}O$, $S{=}C{=}S$, $O{=}\overset{(+)}{N}{=}O$ $H{-}C{\equiv}N$ $\quad$ $H{-}C{\equiv}C{-}H^a$ ZnX_2, CdX_2, HgX_2, $AgCl_2^-$
	One or more	sp^2 or sp^3	Angular	
AX_3	None	sp^2	Planar	
	One or two	sp^3	Pyramidal	, H_3O^+, PX_3,
AX_4	None	sp^3	Tetrahedral	CX_4, NH_4^+, BX_4^-, SiX_4, TiX_4, ZnX_4^{2-} SO_4^{2-}, SO_2Cl_2, POX_3, PO_4^{3-}, ClO_4^-

^{a} Each carbon in C_2H_2 and C_2H_4 corresponds to A in AX_2 and AX_3, respectively.

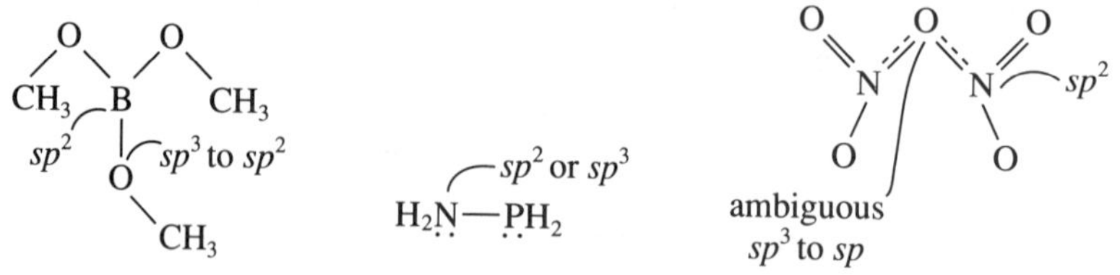

Figure 2.11   Possible assignments of hybridization.

confirmed by 120° bond angles at B. The orbital hybridization of the oxygen atoms could be intermediate between sp^3 and sp^2 because of some π bonding, and the B—O—C bond angle is 113°. Any contribution of resonance structures involving double bonding between boron and oxygen requires a lone pair in an atomic p orbital on oxygen for the formation of a π bond. Hence double bonding favors sp^2 hybridization on O. Similarly, the P—N bond in H_2N—PH_2 might involve an sp^2 or sp^3 N orbital, depending on the involvement of one p orbital used for $p\pi d$ bonding. The hybridization of the σ bonding orbitals of each nitrogen in N_2O_5 must be sp^2 because there is a π bond, and that for the bridging oxygen must be sp^3. Here we are considering expectations for hybridization predicted from VB structures. When structural data are available, the choice of hybridization is likely to be more obvious from experimentally measured bond angles.

The geometry of a molecule in an excited electronic state likely differs drastically from the ground state, because the promotion of an electron can change the hybridization of the orbitals. The first excited state of acetylene has C—C—H bond angles of 120° and a C—C distance similar to that found in benzene. The hybridization of the C orbitals changes from sp to sp^2 in order to accommodate nonbonding electrons.

▶ 2.3.2   *Isoelectronic Molecules*

The species CH_4, NH_4^+, and BH_4^- are **isoelectronic,** since they contain the same number of electrons and the same number of atoms. A molecule (or ion) with a central atom, four hydrogen atoms, and a total of 10 electrons (all shells included) should have the tetrahedral methane structure—that is, be isostructural with methane. Molecules with the same number of *valence* electrons—for example, CH_4 and SiH_4, CO_2 and CS_2, and O_3 and SO_2—are also commonly described as being isoelectronic.

Triatomic molecules (AX_2) with 16 valence electrons (Table 2.3) generally have the linear CO_2 structure, whereas those with more valence electrons have the angular structure—for example, NO_2 (17 e), NO_2^- (18 e) and ClO_2^- (20 e). AX_3 molecules are planar for 24 e (BF_3, NO_3^-) and pyramidal for 26 e (NX_3, PX_3, ClO_3^-). Substances with the same formula types which appear as isoelectronic, but which contain central atoms of different periods, often do not have the same structures: Examples include CO_2 (a gas) and SiO_2 (a solid), and N_2O_5 and P_2O_5 (P_4O_{10}) (see Table 2.2).

The molecules O_3 and SO_2 are isoelectronic based on valence electrons. The same resonance contributing structures can be written for both. The formal charge on S (or central O of O_3) is $+1$, and that for the singly bonded O is -1. The structure gives all atoms zero formal charge. This is a better representation of bonding for SO_2,[9] but the octet limitation prohibits two double bonds for O_3.

The BN unit is isoelectronic with C_2. Borazine, $B_3N_3H_6$, has been described as "inorganic benzene" because its physical properties somewhat resemble those of benzene (see page 846 for the benzenelike structure). Likewise, boron nitride, BN, forms both layer (graphite-type) and diamond structures. We encounter the diamond structure in the elements Si, Ge, and Sn, as well as in Group 13 (IIIB) phosphides, arsenides, and antimonides—except for SbB and these compounds of Tl.

The molecules CH_4 and CF_4 and the ions BH_4^- and BF_4^- differ in the number of valence electrons, but the additional unshared electron pairs on F are nonbonding. These molecules have the same structure. Replacing H by another atom that forms only a single bond should not change the structure. The BF_3 molecule is isoelectronic and isostructural with CO_3^{2-}. BH_3, with no π bonding possible, is the unstable monomer of B_2H_6 (diborane; see page 185).

It is sometimes convenient to view the Ne atom as having four electron pairs arranged tetrahedrally. The molecules HF, H_2O, NH_3, and CH_4 might be considered to differ from Ne in the number of electron pairs that are protonated. For localized nonbonding electron pairs, protonation does not change the geometry greatly. The series of molecules presented in Figure 2.12 are related structurally, although this fact is not immediately evident from the simplest formulas. Basically, there is a linear grouping of three atoms involving sp hybridization of the central atom and two π bonds. The series $CH_3\overset{\overset{\displaystyle O}{\|}}{C}CH_3$ (acetone), $H_2N\overset{\overset{\displaystyle O}{\|}}{C}NH_2$ (urea), $HO\overset{\overset{\displaystyle O}{\|}}{C}OH$ (carbonic acid), and $F\overset{\overset{\displaystyle O}{\|}}{C}F$ (carbonyl fluoride) is related by the replacement of $-CH_3$ by the "isoelectronic" $-NH_2$, $-OH$, and $-F$ groups. (Figure 2.13 shows other of Bent's "isoelectronic" groups.[10]) Do not apply these generalizations blindly, however. Be alert for changes in B.O.—for example, possible $p\pi d$ bonding for S—F where corresponding $p\pi p$ bonding is not possible for O—F (all orbitals filled) or S—H (no p orbitals for H). Also, O_2 is not isoelectronic with C_2H_4, because the O_2 requires a completely different bonding description (page 160).

[9] G. H. Purser, *J. Chem. Educ.* **1989,** *66,* 710.

CO_2 | HNCO (isocyanic acid) | H_2CCO (ketene) | CH_2CCH_2 (allene)

N_2O (nitrous oxide) | HN_3 (hydrazoic acid) | H_2NCN_2 (diazomethane)

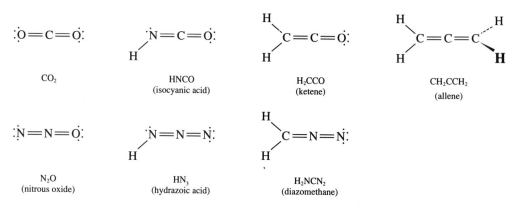

Figure 2.12 Structurally related molecules with three heavy atoms and 16 electrons.

▸ *2.3.3 The Pauli Exclusion Principle and the Prediction of the Shapes of Molecules (Valence Shell Electron-Pair Repulsion Theory)*

Molecules Covered by the Octet Rule

You can predict the approximate shapes of simple molecules by a model in which unshared electron pairs (lone pairs) are considered to be equivalent to bonded groups. A more refined model gives more exact descriptions of the molecular shapes and variations of bond angles, by taking into account differing repulsion among electron pairs of different kinds in the valence shell of the central atom. This approach was introduced by Gillespie and Nyholm and was developed extensively by Gillespie as the **valence shell electron-pair repulsion (VSEPR) theory.** Repulsion among the electrons can be considered to arise from the operation of the Pauli exclusion principle. Although this principle usually is stated

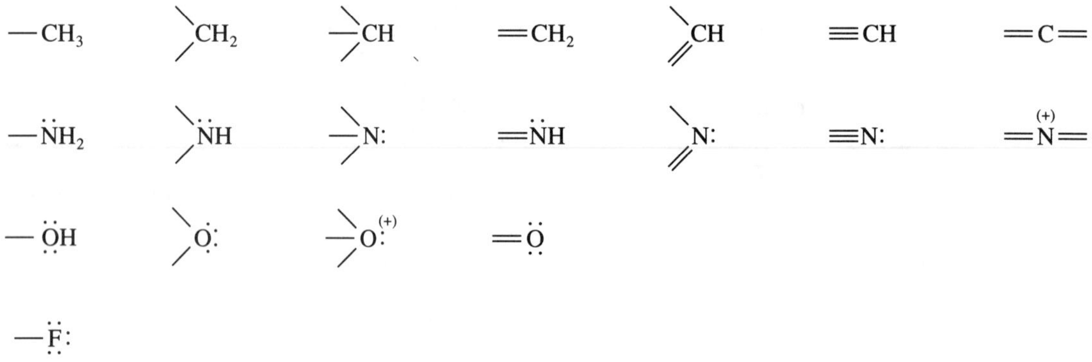

Figure 2.13 Isoelectronic groupings (read vertically).

[10] H. A. Bent, *J. Chem. Educ.* **1966,** *43,* 170.

in terms of the energy or the quantum numbers of the electrons, it effectively rules out the possibility that two electrons might be at the same point at the same time. Two electrons can be confined to the same orbital only if they have opposed spins. Because repulsion is greater for electrons of the same spin, the total repulsion is minimized by arranging the electrons of a complete octet in four localized pairs of electrons of opposed spin, with each pair directed toward the apex of a tetrahedron. This configuration can be visualized for Ne, F^-, and O^{2-}.

The VSEPR model allows us to predict molecular shapes without knowing explicitly the hybridization used, only considering the Pauli principle and Coulomb's law.

If one of the electron pairs is used for bonding, as in HF, the four pairs are no longer equivalent. The bonding pair becomes more localized, and drawn farther from the central atom because its electrons are shared by two positive nuclei. Repulsion among the electrons in the nonbonding pairs causes them to spread out to occupy the space made available by the localization of the bonding pair. If two electron pairs are used for bonding, as in H_2O, the bond angle is affected by repulsions among the four electron pairs. Instead of the normal tetrahedral angle (109.47°) for four equivalent pairs, as found in CH_4 and as expected for O^{2-}, the bond angle in H_2O is 104.45°. Because of the differences in localization, *the repulsion between electron pairs decreases in the order lone pair–lone pair > lone pair–bonding pair > bonding pair–bonding pair*. The stronger repulsion involving lone pairs for H_2O forces the bond angle to decrease.

Figure 2.14 represents the distribution of three electron pairs confined to a plane (*a*) and shows the result of two of the electron pairs shared with H atoms (*b*). The orbitals for the bonding pairs extend out to the H atoms and are more "skinny". The localization of the bonding pairs in H_2O permits the bond angle to close, relative to the tetrahedral angle, in order to decrease repulsion between the lone pairs. In NH_3 the HNH bond angles are 107.2°—an angle greater than that in H_2O because there are no lone pair–lone pair repulsions for NH_3. In addition to the restriction of two nuclei, each bonding pair in NH_3 is subject to the repulsion of two other bonding pairs and one lone pair, whereas in water each bonding pair is subject to the repulsion of one other bonding pair and two lone pairs. We can approach the variation in bond angles by starting with the tetrahedral CH_4 molecule as a reference. The bond angles in NH_3 decrease, relative to CH_4, because of the presence of the lone pair, and in H_2O the bond angle decreases further because of the presence of two lone pairs.

Figure 2.14 (*a*) Most-probable spatial distribution and corresponding trigonal orbitals for three electrons pairs confined to a plane. (*b*) Orbitals for two bonding electron pairs and one lone pair; ● nuclei. (Adapted with permission from R. J. Gillespie, *J. Am. Chem. Soc.* **1960,** *82,* 5978. Copyright © 1960, American Chemical Society.)

Here we are considering four equivalent hybrid orbitals about O and N, with the same amount of s character in each. Hall (*J. Am. Chem. Soc.* **1978**, *100*, 6333; *Inorg. Chem.* **1978**, *17*, 2261) has shown that the bent shape of H_2O can be explained on the basis of the preferential occupancy of the lower-energy s orbital by a lone pair, which gives more p character to the bonding orbitals. Although his treatment of H_2O and other cases raises questions about the theoretical basis for the VSEPR model, in practice it has an exceptionally good record for predicting molecular shapes.

Comparing the bond angles in NO_2 (132°) and in NO_2^- (115°) demonstrates that a single unshared electron in NO_2 causes less repulsion than an unshared electron pair in NO_2^-. The nitryl ion, NO_2^+, with no unshared electron on nitrogen, is linear (see Figure 2.15).

The bond angles decrease in the series $NH_3 > PH_3 > AsH_3 > SbH_3$ and in the series $H_2O > H_2S > H_2Se > H_2Te$ (see Table 2.4). The increasing size and lower electronegativity of the central atom permit the bonding electrons to be drawn out further, thus decreasing repulsion between bonding pairs. The variation in bond angles in a series such as NH_3, PH_3, AsH_3, and SbH_3 is in the direction expected for an increase in the bonds' p character; that is, the bond angles approach the values expected for p^3 bonds, rather than those for sp^3 hybrid bonds. This is just the opposite of what might be expected based upon the energy differences: The energy difference of the s and p orbitals is greatest for nitrogen. The bond angles decrease by large amounts from NH_3 to PH_3 or from H_2O to H_2S, then by very small amounts of the following members of each series.

The most striking argument for the great importance of repulsion among electron pairs, in contrast to steric repulsion between atoms bonded to the central atom, arises from a comparison of bond angles between hydrogen and fluorine compounds of nitrogen and oxygen. Here replacement of H by larger F causes a significant *decrease* in bond angle for N and O compounds. The decrease in the bond angles for NF_3 and F_2O as compared with NH_3 and H_2O can be explained in terms of the decrease in repulsion between bonding pairs resulting from the electrons being drawn out further by the fluorine. The X—F bond is longer than the X—H bond and the electron density is shifted (polarized) toward the highly electronegative F, away from N or O. The FCF bond angles in the C—F compounds shown in Table 2.4 are also smaller than the HCH bond angles in the corresponding C—H compounds.

The bond angles of the fluorine compounds are smaller than those of the corresponding hydrogen compounds only for second-period elements. Multiple bonding is not possible in NH_3, NF_3, or PH_3 but is expected to occur in PF_3, using a filled p orbital on F and an empty d orbital P. The greater B.O. ($\sigma + \pi$ bonding) increases the electron density and increases the repulsion among the bonding electrons. The bond angle of PF_3 is *greater* than that of PH_3.

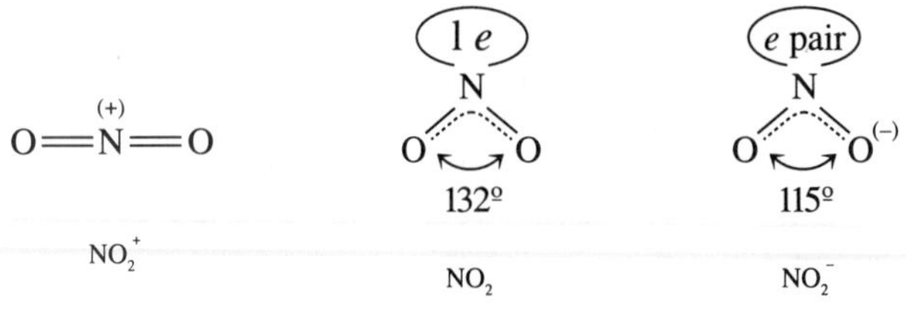

Figure 2.15   Bond angles in NO_2^+, NO_2, and NO_2^-.

Table 2.4   Comparison of bond angles[a]

CH$_4$		109.47°	H$_2$C=CH$_2$	119.9°	H$_2$C=O	116.5°
CH$_3$F	∠HCH	110.6°				
CH$_3$Cl	∠HCH	110.4°				
CHF$_3$	∠FCF	108.6°				
CHCl$_3$	∠ClCCl	111.3°				
			F$_2$C=CF$_2$	114°	Cl$_2$C=O	111.8°
			Cl$_2$C=CCl$_2$	113.5°	F$_2$C=O	107.7°

NH$_3$	107.2°	NF$_3$	102.3°	NCl$_3$	107.1°
PH$_3$	93.2°	PF$_3$	97.7°	PCl$_3$	100.3°
AsH$_3$	92.1°	AsF$_3$	95.8°	AsCl$_3$	98.9°
SbH$_3$	91.6°	SbF$_3$	95.0°	SbCl$_3$	97.1°
				PBr$_3$	101.0°
				PI$_3$	102°

H$_2$O	104.45°	F$_2$O	103.1°	F$_2$SO	∠FSO	106.8°
H$_2$S	92.1°				∠FSF	92.8°
H$_2$Se	90.6°					
H$_2$Te	90.2°					

[a] Data from R. J. Gillespie and I. Hargittai, *The VSEPR Model of Molecular Geometry*, Allyn and Bacon, Boston, 1991; or M. C. Favas and D. L. Kepert, *Prog. Inorg. Chem.* **1980**, 27, 325.

Effects of $d\pi p$ Bonding

Bond angles increase regularly as the size of the halogen increases in each series for the phosphorus and arsenic halides (Table 2.4). The tendency to form multiple bonds is generally greater for small atoms (O, N, F) than for larger atoms. Because the bond angles increase in the order PF$_3$ < PCl$_3$ < PBr$_3$ < PI$_3$, the effect of any decrease in B.O. for the larger halogens might be offset by the increasing size of the bonding orbitals.

The F$_2$SO molecule is isoelectronic with ClO$_3^-$ and, as expected, is pyramidal, with S at the apex. For single-bonded F$_2$SO, the formal charges are 0 for each F, +1 for S, and −1 for O, which forms a dative bond. There is the possibility of $d\pi p$ bonding to S for F and O, but the negative formal charge on O and its lower electronegativity, compared with F, should favor greater π donation by O than by F. Because of the lone-pair repulsion, all bond angles should be less than 109°. The expected higher B.O. for S—O compared to that of S—F results in a larger FSO angle than the FSF angle. (Compare the angles in Table 2.4.) The bond lengths (S—O, 141.2 pm; S—F, 158.5 pm) are consistent with a higher B.O. for S—O than for S—F.

Molecules with More Than Four Electron Pairs—Sigma Bonding Plus Nonbonding Pairs

Table 2.5 gives the overall geometrical configurations usually encountered for different total numbers of σ plus lone pairs. We ignore π bonds here because π bonds are formed along directions of existing σ bonds. Where the total number of σ bonding pairs and lone pairs is five or greater, several arrangements often are possible. The most significant characteristic of lone pairs seems to be their tendency to occupy as much space as available. In structures where axial and equatorial positions are not equivalent, lone pairs preferentially occupy those positions that provide the most space. Thus lone pairs prefer equatorial positions in a trigonal bipyramid, where the bond angles are 120° in the equatorial plane, compared to 90° for the axial bond to those in the equatorial plane. The reverse is true for a pentagonal bipyramid, where the equatorial positions are more crowded. If all positions are equivalent, as in an octahedron, two lone pairs occupy opposite or *trans* positions.

Applying these guidelines to examples shown in Figure 2.16 is simple. Because there are five pairs about I in ICl_2^-, we begin with the trigonal bipyramid. The lone pairs occupy the more spacious equatorial positions, giving structure (*a*), with a linear arrangement of atoms.[11] Given the preference of lone pairs for equatorial positions in a trigonal bipyramid, we predict correctly (*a*) structures for $TeCl_4$ (one lone pair) and ClF_3 (two lone pairs). However, the greater repulsions caused by lone pairs distort the regular geometry shown in Figure 2.16. The lone pair in $TeCl_4$ forces the chlorines closer together to produce a flattened and distorted tetragonal pyramid, and the axial fluorines in ClF_3 bend away from the lone pair to give a bent "T" (giving ⊤). The positions are equivalent for an octahedral arrangement of six pairs, so the two lone pairs of ICl_4^- occupy opposite positions and thus produce a square planar structure (*a*). $(ICl_3)_2$, with two bridging chlorides, is also planar (Figure 2.4). Obviously, only one structure is possible for BrF_5, since the lone pair can occupy any one of the equivalent positions of the octahedron. The influence of the lone pair is evident from the fact that bromine is *below* the plane of the four equivalent fluorine atoms, giving an F(axial)—Br—F(basal) bond angle of 84.5°. In the

Table 2.5   Configurations from the number of electron pairs

Number of σ pairs plus lone pairs	Configuration (considering lone pairs directed as any other group)	Hybridization
2	Linear	sp
3	Trigonal planar	sp^2
4	Tetrahedral	sp^3
5	Trigonal bipyramidal	$d_{z^2}sp^3$
5	Square pyramidal	$d_{x^2-y^2}sp^3$
6	Octahedral	d^2sp^3
7	Pentagonal bipyramidal	d^3sp^3

[11] We use the sum of the (σ + lone pairs) to select the polyhedron and first arrange the lone pairs on this geometrical framework in their preferred positions. *Molecular shape* describes the positions of *atoms*, neglecting lone pairs.

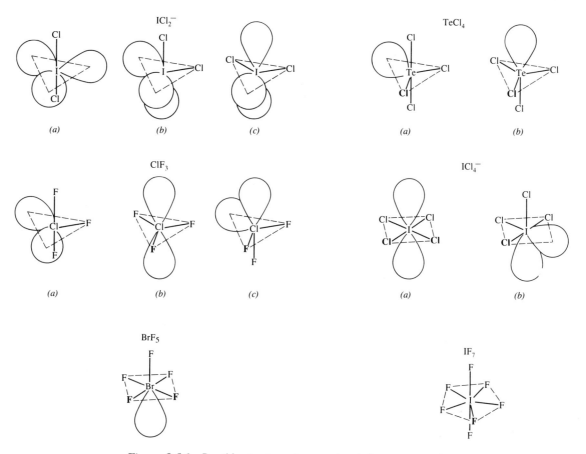

Figure 2.16   Possible structures for some interhalogen compounds.

isoelectronic series TeF_5^-, IF_5, and XeF_5^+, the bond lengths are remarkably constant and the F(apical)—M—F(basal) bond angles are 79°, 81°, and 79°, respectively. You can also predict the pentagonal bipyramidal structure of IF_7 with no lone pairs on I.

Double bonds, with four bonding electrons, are "fatter" than single bonds. Double bonds prefer equatorial positions in a trigonal bipyramid—as observed for SOF_4, where the oxygen occupies an equatorial position because the double-bond character of the S—O bond is greater than that of the S—F bond. The structure of XeO_2F_2 indicates that the two F atoms are in axial positions and both oxygens and the lone pair are in equatorial positions of a trigonal bipyramid.

Stereochemically Inactive Lone Pairs

Unshared electrons in the valence shell are not always stereochemically active; that is, they do not always behave as equivalent to substituents in determining molecular shapes. The ions $TeCl_6^{2-}$, $TeBr_6^{2-}$, and $SbBr_6^{3-}$ have one lone pair on the metal and thus a pentago-nal pyramid or a badly distorted octahedron might be expected, but the ions are octahedral. In these cases ligand–ligand repulsion is of overriding importance. Possibly the lone pair occupies a spherical s orbital; it is described as a **stereochemically inactive pair.** The case of XeF_6, which also contains one lone pair on Xe, is discussed on page 809. However, the ion $[Sb(C_2O_4)_3]^{3-}$ has the pentagonal pyramidal structure expected for a lone pair in an axial position of a pentagonal bipyramid. The Sb is 35 pm below the pentagonal plane, and the Sb—O axial bond is about 20 pm shorter than the average Sb—O equatorial bond length. *For elements to the right of the periodic table with coordination number six plus an unshared electron pair, the electron pair is (a) active if the donor atoms are small (O^{2-} or F^-) and (b) inactive for large donor atoms (Cl^- or Br^-).*

As we shall see later (page 337), the electron clouds of O^{2-} and F^- are not easily deformed (polarized) in an electrostatic field and therefore are considered to be **hard donors (hard bases).** Chloride ion and bromide ion are larger and more polarizable **(soft donors or soft bases).** Here we are using Lewis definitions (page 311) for bases as electron donors and acids as electron acceptors. Because of little polarization, small hard electron donors allow the unshared pair to be stereochemically active. Repulsion among large donors favors the smaller effective coordination number (that is, the unshared pair tends not to act as a seventh substituent). Also, where polarization is important, accommodating the lone pair in the spherically symmetrical s orbital is most favorable.

2.4 ELECTRONEGATIVITY

▶ 2.4.1 *Pauling's Electronegativity Scale*

Absolute electronegativities were introduced in Section 1.8. Within a periodic group the electronegativity decreases with increasing atomic radius and it increases from left to right within a period, with few exceptions. Pauling defined electronegativity as the power of an atom in a molecule to attract electrons to itself.

Pauling established a scale of electronegativities based on "excess" bond energies. The energy of the A—B bond is the sum of a nonpolar contribution (equal sharing electrons) and a polar contribution. The nonpolar contribution is the energy expected from the mean of the nonpolar bond energies of the elements, A—A and B—B. The polar contribution, designated Δ, is a measure of the bond's expected polarity, which results from a displacement of the electronic charge toward one atom. Ordinarily, the polarity of the bond increases with increasing differences in electronegativities (χ) of A and B. Pauling used the relationship

$$\Delta = 96.49(\chi_A - \chi_B)^2 \tag{2.5}$$

or

$$\chi_A - \chi_B = 0.102\sqrt{\Delta} \tag{2.6}$$

The heat of formation (ΔH_f^0) is equal to Δ for a diatomic molecule AB. In the more general case of the compound AB_n, the heat of formation is obtained by summing the expression over all of the bonds in the molecule:

$$-\Delta H_f^0 = n \times 96.49(\chi_A - \chi_B)^2 \tag{2.7}$$

or

$$(\chi_A - \chi_B) = 0.102 \sqrt{\frac{-\Delta H_f^0}{n}} \tag{2.8}$$

These relationships assume that the reactants and products have the same number of covalent bonds.

We must apply a correction for compounds containing N and O, because of the extra stability of N_2 and O_2 in their standard states. The triply bonded N_2 molecule is more stable than a hypothetical molecule for which the bond energy is taken as three times the N—N single-bond energy (160 kJ/mol). This extra stability is $942 - (3 \times 160) = 462$ kJ/mol N_2 or 231 kJ/mol N. The quantity 231 kJ/mol N serves as a correction for the calculation of the heats of formation of nitrogen compounds. The extra stability of the O_2 molecule is 494 kJ/mol $O_2 - 2 \times 142$ kJ/mol O—O = 210 kJ/mol O_2 or 105 kJ/mol O.

Applying the corrections to Equation (2.4) gives

$$-\Delta H_f^0 = n \times 96.49(\chi_A - \chi_B)^2 - 231 n_N - 105 n_O \tag{2.9}$$

where n_N and n_O are numbers of N and O atoms, respectively, in the compound. In this way Pauling was able to extend the table of electronegativities to include most elements.

► 2.4.2   *Allred's Electronegativity Values*

Allred used more recent thermochemical data to calculate the electronegativities of 69 elements. Where the necessary data were available, he calculated electronegativities from single-bond energies as described by Pauling. For the many metals whose M—M bond energies are not known, electronegativities were obtained from the heats of formation of their compounds, using Equation (2.7). Electronegativities were calculated for several compounds of M in most cases, then averaged to give the values in Table 2.6.

Most of Allred's electronegativity values are similar to Pauling's. In several respects, however, Allred's values agree better with the chemical behavior of the elements than do those of Pauling. The alternation in electronegativities of the main Group 13(IIIB) and 14(IVB) elements should be expected from the chemistry of these elements and can be explained by the transition metal contraction and the lanthanide contraction (page 736).

As the oxidation number or formal charge on an atom increases, the attraction for electrons increases, and hence the electronegativity increases. The electronegativity values in Table 2.6 are for the oxidation numbers indicated, and Table 2.7 displays Allred's values for a few elements in different oxidation states. Because the variation is not large, a single value to the nearest 0.1 unit can be used for most purposes.

Table 2.6   The electronegativity scale of the elements[a]

I	II	III	II	II	II	II	II	II	II	I	II	III	IV	III	II	I
H 2.2																
Li 1.0	Be 1.6											B 2.0	C 2.6	N 3.0	O 3.4	F 4.0
Na 0.9	Mg 1.3											Al 1.6	Si 1.9	P 2.2	S 2.6	Cl 3.2
K 0.8	Ca 1.0	Sc 1.4	Ti 1.5	V 1.6	Cr 1.7	Mn 1.6	Fe 1.8	Co 1.9	Ni 1.9	Cu 1.9	Zn 1.7	Ga 1.8	Ge 2.0	As 2.2	Se 2.6	Br 3.0
Rb 0.8	Sr 1.0	Y 1.2	Zr 1.3	Nb 1.6	Mo 2.2	Tc 1.9	Ru 2.2	Rh 2.3	Pd 2.2	Ag 1.9	Cd 1.7	In 1.8	Sn 2.0	Sb 2.1	Te 2.1	I 2.7
Cs 0.8	Ba 0.9	La–Lu 1.1–1.3	Hf 1.3	Ta 1.5	W 2.4	Re 1.9	Os 2.2	Ir 2.2	Pt 2.3	Au 2.5	Hg 2.0	Tl 2.0	Pb 2.3	Bi 2.0	Po 2.0	At 2.2
		Ac 1.1	Th 1.3	Pa 1.5	U 1.4	Np–No 1.3	Pu 1.3									

[a] The oxidation numbers are specified at the top of each group. Values are from A. L. Allred, *J. Inorg. Nucl. Chem.* **1961**, *17*, 215—except those underlined, which are from Pauling.

Pauling used "excess" bond energies of A—B compounds relative to those of A—A and B—B to establish an electronegativity scale. Allred updated the values on the same scale. The "excess" bond energies are related to the polarity of the A—B bond. The atomic properties measuring the attraction for electrons by an isolated atom are the ionization energy (IE) and electron affinity (EA). Mulliken proposed an electronegativity scale based on (IE + EA)/2, where IE and EA are for the *valence state* of an atom. (The valence state is the electron configuration for a combining atom, $2s^1 2p^1 2p^1$ for B.) Trends in Mulliken's values are very similar to those on the Pauling scale and are about 2.8 times as large as those on the Pauling scale. The **absolute electronegativity** (also called the **electronic chemical potential**) (see Section 1.8), (IE + EA)/2, uses the first ionization energy and electron affinity of atom X [EA of X is IE of $X^-(g)$]. Allen[12] proposed the **spectroscopic electronegativity** as the average one-electron energy of the valence shell electrons in ground-state free atoms. Data for calculating spectroscopic electronegativity are available to high accuracy from atomic energy level tables.

Table 2.7   Electronegativities of some elements in different oxidation states

Mo^{II}	2.18	Fe^{II}	1.83	Sn^{II}	1.80
Mo^{III}	2.19	Fe^{III}	1.96	Sn^{IV}	1.96
Mo^{IV}	2.24	Tl^{I}	1.62	Pb^{II}	1.87
Mo^{V}	2.27	Tl^{III}	2.04	Pb^{IV}	2.33
Mo^{VI}	2.35				

[12] L. E. Allen, *J. Am. Chem. Soc.* **1989**, *111*, 9003.

▶ 2.4.3   Group Electronegativities

A single electronegativity value for an element is inadequate, since the value varies with oxidation number, substituents {compare the hydrogen bonding ability (Section 2.6) of $HCCl_3$ with that of CH_4}, and hybridization (compare the acidity and hydrogen bonding ability of HCCH with those of CH_4). Using an extension of Mulliken's approach, Jaffé and others calculated absolute group electronegativities allowing for the charge of a group, effects of substituents, and the hybridization of a bonding orbital. As you might expect, positive groups have higher electronegativities than do neutral groups. CF_3 is much more electronegative than CH_3, and electronegativity increases with an increase in the s character of a hybrid orbital—that is, $sp > sp^2 > sp^3$. These refinements extend the usefulness of the electronegativity concept and make it more generally applicable. The average electronegativity value of an element (Table 2.6) is still useful for qualitative predictions.

▶ 2.4.4   Partial Ionic Character in Covalent Bonds

The electron distribution in covalent bonds joining dissimilar atoms is usually unsymmetrical, with the electron density greater near the more electronegative atom. Polar covalent bonds can be described either in terms of shared electrons distributed unsymmetrically because of polarization or as the resonance hybrid of a covalent structure (electrons shared equally) and an ionic structure (electrons transferred). A **nonpolar covalent bond** involves equal or nearly equal sharing of the bonding electrons; the difference in the electronegativities should be less than 0.5. A **polar covalent bond** is one with an appreciable amount of polarization or ionic character.

The greater the difference in the electronegativities of bonded atoms, the greater the polarity or ionic character of the bond. Thus we should be able to determine the ionic character of a bond from an empirical relationship between ionic character and electronegativity differences. The difficulty has been the lack of a method of giving reliable ionic characters for a wide range of compounds in order to establish the empirical relationship. Pauling used the **dipole moment** as a measure of the polarity and hence of the ionic character of a bond. A diatomic molecule, such as HF, in which the electrons are not shared equally, has a positive and a negative end. Such a molecule acts as a dipole and tends to become aligned in an electrical field. The force that acts on a dipole to align it in an electrical field is a measure of the dipole moment. The greater the ionic character of a bond, the greater the charge separation and the larger the dipole moment.

In discussing the use of dipole moments as a guide to the ionic character of bonds, we must emphasize that it is the bond moment, not the total dipole moment of a molecule, that must be considered in all cases. The C—F bond is distinctly polar, but the tetrahedral arrangement of the four bonds in CF_4 results in a molecule with zero dipole moment. Hybridization of a bonding orbital also affects the nonbonding orbitals, causing the lone pairs of electrons to be distributed dissymmetrically. Coulson reported that the bond moments of the water molecule contribute only about one-quarter of the total dipole moment. The lone pair moment provides the major contribution.

The electric dipole moment, μ, is the product of the charge at one pole, q, and the distance between poles, d, as in

$$\mu = qd \tag{2.10}$$

Dipole moments are expressed commonly in **Debye units.** One Debye unit is 10^{-8} pm · esu (10^{-18} cm · esu). Two charges equal in magnitude to the charge of an electron (4.8×10^{-10} esu) and separated by the distance of 91.7 pm (the interatomic distance for the HF molecule) give a dipole moment of 4.4 Debye units. A value of 4.4 Debye units represents the expected dipole moment for 100% ionic H^+F^-. The percentage of ionic character is obtained from the ratio of the observed dipole moment to that for complete electron transfer. The observed dipole moment for HF is 1.98 Debye units, corresponding to 1.98/4.4 or 45% ionic character. This treatment assumes that the total dipole moment results from the unsymmetrical distribution of charge in the bond.

Pauling derived the relationship

$$\text{Fraction of ionic character} = 1 - e^{-(1/4)(\chi_A - \chi_B)^2} \tag{2.11}$$

from a plot of electronegativity difference versus percentage of ionic character. Percentages of ionic character originally were obtained from the measured dipole moments of HCl, HBr, and HI, which gave 19%, 11%, and 4% ionic character, respectively, and an estimated 60% character for HF. The relationship has been revised by others to take into account the corrected value for the percentage of ionic character of HF from dipole moment measurements. However, Pauling argues that the deviation of HF from the equation is justified and that his relationship is probably as useful as the revised ones. These relationships can be used as qualitative guides only. Rough and easy-to-remember values for electronegativity differences and percentages of ionic character are 1.0, 20%; 1.5, 40%; 2.0, 60%; and 2.5, 80%.

Our approach to describing polar bonding started with the nonpolar covalent model as the limiting case—an approach particularly suited to discrete molecules. You also can start with an ionic model as the limiting case and consider varying degrees of distortion or polarization of the electron cloud of the anion, or even of both ions. Extreme polarization is equivalent to electron sharing. The latter approach is most useful in understanding cases in which the ionic model is a reasonable approximation (page 200).

▶ *2.4.5   Lewis–Langmuir Atomic Charges*

Formal charges are assigned assuming 100% covalent character. Oxidation number assignments assume *no* electron sharing, assigning the bonding electrons to the more electronegative atom. For covalent compounds, Pauling's electroneutrality principle (page 62) is a good guide to reasonable charge distributions. Allen[13] proposed Lewis–Langmuir charges to handle atomic charges as a continuum from formal charges (no differences in electronegativity for atoms bonded considered) to oxidation numbers (very large differences in electronegativities assumed). For a molecule AB:

$$\text{Lewis–Langmuir charge on A} = \text{No. of valence } e - \text{No. of unshared } e \text{ on A} - 2 \sum_{\text{bonds}} \left(\frac{\chi_A}{\chi_A + \chi_B} \right) \tag{2.12}$$

[13] L. E. Allen, *J. Am. Chem. Soc.* **1989,** *111,* 9115.

We sum over two single bonds for AB_2 and sum over two bonds for a double bond in $A{=}B$.

EXAMPLE: Calculate Lewis–Langmuir charges on F in HF, O in H_2O, C in C_2H_4, and F in :B≡F:

Solution: For HF we obtain

$$L{-}L(F) = 7 - 6 - 2\left(\frac{4.0}{4.0 + 2.2}\right) = -0.29$$

For H_2O we obtain

$$L{-}L(O) = 6 - 4 - (2 \times 2)\left(\frac{3.4}{3.4 + 2.2}\right) = -0.43$$

For C_2H_4 we obtain

$$L{-}L(C) = 4 - 0 - (2 \times 2)\left(\frac{2.6}{2.6 + 2.6}\right) - (2 \times 2)\left(\frac{2.6}{2.6 + 2.2}\right) = -0.17$$

For :B≡F: we obtain

$$L{-}L(F) = 7 - 2 - (3 \times 2)\left(\frac{4.0}{4.0 + 2.0}\right) = +1.00$$

BF, isoelectronic with CO, has a calculated[14] dipole moment of 0.9 D (B^-F^+). The formal charge on F is +2 for :B≡F:, and the L–L charge is +1.

EXAMPLE: Calculate the Lewis–Langmuir charges on the terminal nitrogen and oxygen for the two resonance structures given for N_2O.

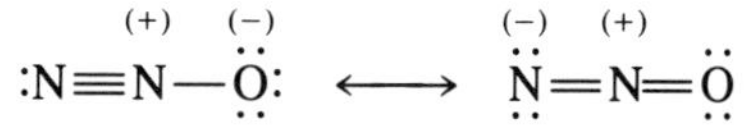

Solution:

Structure I: Terminal N $L{-}L = 5 - 2 - (3 \times 2)\left(\dfrac{3.0}{3.0 + 3.0}\right) = 0$

O $L{-}L = 6 - 6 - (1 \times 2)\left(\dfrac{3.4}{3.4 + 3.0}\right) = -1.06$

Structure II: Terminal N $L{-}L = 5 - 4 - (2 \times 2)\left(\dfrac{3.0}{3.0 + 3.0}\right) = -1$

O $L{-}L = 6 - 4 - (2 \times 2)\left(\dfrac{3.4}{3.4 + 3.0}\right) = -0.12$

[14] H. A. Krutz and K. D. Jordan, *Chem. Phys. Lett.* **1981,** *81,* 104.

Allen[13] showed that the atomic charges agree well with other assignments assuming that structure I is weighted twice that of structure II. The results for L–L charges are:

$$\frac{(2 \times 0) - 1}{3} = -0.33 \text{ terminal N} \quad \text{and} \quad \frac{2(-1.06) - 0.12}{3} = -0.75 \text{ for O}$$

We have no simple basis to assign the weight of each resonance structure for a particular molecule. For N_2O formal charges alone indicate that structure I is favored over structure II.

The L–L charges calculated from Equation (2.12) are pedagogically important in emphasizing that formal charges and oxidation numbers are limits with actual charges usually between these extremes. The L–L charges cannot be taken as accurate charges, nevertheless. The formal charge is -3 for Al in AlF_6^{3-} and the L–L charge is -0.67. This L–L charge does not agree well with the electroneutrality principle. The L–L charge does not agree well in many cases with Mullikan charges from more sophisticated calculations.

2.5 RADII OF ATOMS IN COVALENT COMPOUNDS

▶ 2.5.1 *Covalent Radii*

The bond length—the distance between two bonded atoms—in a covalent molecule such as Cl_2 is the sum of the covalent radii of the atoms. Thus the **covalent radius** of the chlorine atom is one-half of the chlorine–chlorine distance (198.8 pm). The covalent radius for nonmetals is often called the **atomic radius.** The carbon–carbon bond distance in diamond (154.4 pm) is close to that in ethane (153.4 pm) and other saturated hydrocarbons. Carbon has a single-bond covalent radius of 77 pm, so the sum of covalent radii of carbon and chlorine gives 176 pm as the expected C—Cl bond distance; for CCl_4 the distance is 176.6 pm.

The single-bond radii of nitrogen and oxygen are not obtained from the bond lengths in N_2 (110 pm) and O_2 (120.8 pm), since these molecules contain multiple bonds. Instead, we take the single-bond radii from the bond lengths in compounds such as hydrazine (H_2NNH_2) and hydrogen peroxide (HOOH). The radius of nitrogen obtained from N_2 (55 pm) is the triple-bonded radius. A triple-bonded radius for carbon is obtained from the carbon–carbon bond length in acetylene (120 pm). Combining the triple-bonded radii of carbon and nitrogen gives the expected bond length ($55 + 60 = 115$ pm) for a C≡N bond, as compared with the observed bond length of 116 pm for CH_3CN. A double-bonded radius of carbon is obtained from the carbon–carbon bond length in ethylene [134 pm (see Table 2.1)]. The presence of a multiple bond changes the hybridization and affects bond lengths. The C—H bond length for CH_4 (sp^3) is 109.6 pm, for $H_2C=CH_2$ (sp^2) it is 107.1 pm, and for HC≡CH (sp) it is 106.3 pm. Table 2.8 presents bond dissociation energies and bond lengths for calculating covalent radii for molecules of interest.

We usually can assume that covalent radii can be added to give bond lengths, provided that the bond for which the length calculated is similar to the bonds used for the

Table 2.8   Bond energies and bond lengths

Bond	Molecule or crystal	Bond dissociation energy[a] (kJ/mol)	Bond distance[b] (pm)
H—H	H_2	432.08	74.14
H—C	CH_4	**425.1 ± 8**	109.6
H—N	NH_3	**431 ± 8**	101.2
H—O	H_2O	**493.7 ± 0.8**	95.7
H—F	HF	566.3	91.7
H—Si	SiH_4	**398.3**	148.0
H—P	PH_3	~322	143.7
H—S	H_2S	**377 ± 4**	133.5
H—Cl	HCl	427.78	127.4
H—Ge	GeH_4	**365**	153
H—As	AsH_3	~247	151.9
H—Se	H_2Se	276 (?)	146
H—Br	HBr	362.6	141.4
H—Sn	SnH_4	<320	170.1
H—Sb	SbH_3	—	170.7
H—Te	H_2Te	238 (?)	170
H—I	HI	294.68	160.9
B—F	BF_3	**665**	129
C—C	C_2	599	124.3
	Diamond	—	154.45
	C_2H_6	350	153.4
C—N	CH_3NH_2	**331 ± 13** (298 K)	147.4
C—O	CO	1070.2	112.8
	CO^+	804.5	111.5
	CO^-	784	—
C—F	CH_3F	**452 ± 21** (298 K)	138.5
C—Si	SiC	320	—
	$Si(CH_3)_4$	301	187.0
C—P	$P(CH_3)_3$	260	184.1
C—S	CH_3SH	**297 ± 13**	181.9
	$S(CH_3)_2$	270	181.7
C—Cl	CH_3Cl	**335.1**	178.1
	CH_2Cl_2	**314 ± 75**	177.2
	$CHCl_3$	—	176.2
	CCl_4	280	176.6
C—Br	CBr_4	**205 ± 13**	194
C—I	CH_3I	**226 ± 13**	213.9
N—N	N_2	941.66	109.77
	N_2^+	840.67	111.6
	N_2^-	765	119

Table 2.8   (Continued)

Bond	Molecule or crystal	Bond dissociation energy[a] (kJ/mol)	Bond distance[b] (pm)
	N_2H_4	**247 ± 13** (298 K)	145.1
	N_2F_4	88	
N—O	NO	626.86	115.1
	NO^+	1046.9	106.3
	NO^-	487.8	125.8
N—F	NF_3	**238 ± 8**	136
N—Cl	NCl_3	**381** (298 K)	—
O—O	O_2	493.59	120.75
	O_2^+	642.9	111.6
	O_2^-	395.0	(135)
	H_2O_2	**207.1 ± 2.1**	148
O—F	OF_2	**268 ± 13**	142
O—Cl	Cl_2O	**139.3 ± 4**	170
F—F	F_2	154.6	141.2
Si—Si	Si_2	310	224.6
	Si (diamond str.)	—	235.2
Si—F	SiF_4	564	156.1
Si—Cl	$SiCl_4$	380	201.9
Si—Br	$SiBr_4$	309	216
Si—I	SiI_4	234	243.5
P—P	P_2	485.6	189.3
	P (black)	—	222.4
P—F	PF_3	—	153.5
P—Cl	PCl_3	326	203
P—Br	PBr_3	263	223
P—I	PI_3	184	246
S—S	S_2	421.58	188.9
	S_8	226	205
S—F	SF_6	284	156
S—Cl	S_2Cl_2	255	207
S—Br	S_2Br_2	217 (?)	224
Cl—Cl	Cl_2	239.23	198.8
Cl—F	ClF	252.53	162.8
Cl—Br	BrCl	215.46	213.6
Cl—I	ICl	207.75	232.1
Ge—Ge	Ge_2	272	—
	Ge (diamond str.)	188	245
Ge—F	GeF_4	—	168
Ge—Cl	$GeCl_4$	332.5	208
Ge—Br	$GeBr_4$	276	229.7

Table 2.8   (Continued)

Bond	Molecule or crystal	Bond dissociation energy[a] (kJ/mol)	Bond distance[b] (pm)
As—As	As$_2$	382	210.3
	As$_4$	146	243
As—O	As$_4$O$_6$	—	178
As—F	AsF$_3$	464	171.2
As—Cl	AsCl$_3$	**444**	216.1
As—Br	AsBr$_3$	242	233
As—I	AsI$_3$	213	254
Se—Se	Se$_2$	320	216.6
	α-Se	—	237.4
Br—Br	Br$_2$	190.15	228.1
Br—F	BrF	245.8	175.9
Br—I	IBr	175.4	246.9
Sn—Sn	Sn$_2$	190	—
	α-Sn (gray)	—	281.0
Sn—Cl	SnCl$_4$	318	233
Sn—Br	SnBr$_4$	272	246
Sn—I	SnI$_4$	—	269
Sb—Sb	Sb$_2$	298	234.2
	Sb	—	290
Sb—Cl	SbCl$_3$	309	232.5
Sb—Br	SbBr$_3$	—	251
Te—Te	Te$_2$	258.3	255.6
	α-Te	—	283.4
Te—Cl	TeCl$_4$	—	233
Te—Br	TeBr$_2$	—	251
I—I	I$_2$	148.82	266.6
I—F	IF	277.8	191.0

[a] Bond energies and bond lengths for diatomic molecules and ions are for 0 K, from K. P. Huber and G. Herzberg, *Molecular Spectra and Molecular Structure,* Vol. IV, *Constants of Diatomic Molecules,* Van Nostrand Reinhold, New York, 1979. Values for molecules given in boldface type are from B. deB. Darwent, *Bond Dissociation Energies in Simple Molecules,* NSRDS-NBS31, 1970. The values from Darwent are for 0 K except as specified for 298 K. In many cases the differences between the values at 0 K and 298 K are smaller than the uncertainty. Other values are from T. L. Cottrell, *The Strengths of Chemical Bonds,* 2nd ed., Butterworths, London, 1958. The uncertainties in these values likely will be great.

[b] Bond lengths for diatomic molecules and ions are from Huber and Herzberg (see footnote *a*); for most other elements, from J. Donohue, *The Structures of the Elements,* Wiley, New York, 1974. Most of the data for other compounds are from L. E. Sutton, Ed., "Tables of Interatomic Distances and Configuration in Molecules and Ions," Special Publication No. 18, The Chemical Society, London, 1965. In a few cases data are from Sutton's original compilation, Special Publication No. 11, The Chemical Society, London, 1958.

evaluation of the covalent radii in bond order, bond hybridization, and bond strength. Schomaker and Stevenson attributed the discrepancy between calculated and observed bond lengths to variations in bond polarity for most cases where multiple bonding is thought to be unimportant. Thus the bond length, r_{AB} in pm, can be calculated from the nonpolar radii of A and B and the difference in electronegativities of A and B:

$$r_{AB} = r_A + r_B - 9|\chi_B - \chi_B| \tag{2.13}$$

2.6   THE HYDROGEN BOND

Latimer and Rodebush in 1920 introduced the term "hydrogen bond" to describe the nature of association in the liquid state of water, hydrogen fluoride, and so on. Furthermore, they pointed out the necessity for a slightly acidic hydrogen and a nonbonding electron pair in order to form a hydrogen bond, which they formulated as A:H:B. On the basis of this picture they were able to account for the high dielectric constant[15] (ε) of water, the high mobility of the hydrogen ion in water, and the weakness of ammonium hydroxide as a base, as well as melting points and boiling points of compounds capable of hydrogen bonding.

Our modern model takes account of the fact that H has only one electron and is not very electronegative. Hence, in its compounds with electronegative elements such as O, F, and N, H bears a partial positive charge. The attraction of this positive charge for lone pairs on adjacent molecules binds two molecules together with energies ranging from 4 to 50 kJ/mol for neutral molecules. For comparison, typical covalent bond energies range from about 100 to 400 kJ/mol.

▶ *2.6.1   Evidence for Hydrogen Bonding—Influence on the Vaporization Process for Pure Liquids*

We find some of the most striking effects of hydrogen bonding on the physical properties of substances by contrasting the properties of water and methanol with those of dimethyl ether (Table 2.9). There is no hydrogen bonding for $(CH_3)_2O$, and hydrogen bonding is more extensive for H_2O than for CH_3OH. As Figure 2.17 indicates, the dimethyl derivatives of the Group 16 (VIB) elements usually boil about 80–100° higher than the hydrides, expect for H_2O. Accordingly, from the figure it appears that water boils about 200° higher than it would in the absence of hydrogen bonding. Replacing hydroxylic hydrogens with methyl groups in other compounds produces a similar decrease in boiling point.

Hydrogen bonding is not restricted to hydroxylic compounds, although it is most common in this class of compound. Hydrogen fluoride boils at 19.5°C, whereas methyl fluoride boils at −78°C. Hydrazoic acid HN_3 boils at 37°C and methylazide (CH_3N_3) boils

---

[15] The dielectric constant of a medium is the ratio of a condenser's (capacitor's) capacity with the medium between the plates to its capacity with the space between the plates evacuated. The attraction between ions in a medium is reduced by $1/\varepsilon$ compared to that in a vacuum.

**Table 2.9   Some physical properties
of water, methanol, and dimethylether**

	H_2O	CH_3OH	$(CH_3)_2O$
Critical temperature (°C)	374.2	240.6	126.7
Boiling point (°C)	100	64.7	−23.7
Melting point (°C)	0	−97.8	−138.5
Heat of vaporization (kJ/mol)	40.6	35.2	21.5
Dielectric constant	78.54	32.63	5.02
Dipole moment (Debye units)	1.84	1.68	1.30
Viscosity at 20°C (centipoise)	1.005	0.597	(0.2332 for diethylether)

at 20–21°C. The heat of vaporization of hydrogen-bonded substances is also greater than that of their methyl derivatives.

Intermolecular hydrogen bonding in pure substances increases the heat of vaporization in two ways: (**1**) By increasing the attraction between molecules. Either this attraction must be overcome on vaporization (as in the case of H_2O) and/or small polymeric molecules of the liquid must be vaporized—for example, $(HF)_x$. (**2**) By restricting rotation of the molecules in the liquid. Such rotation is possible in the gas, and the energy absorbed in exciting this rotation adds to the heat of vaporization. Hydrogen bonding thus (**1**) increases the boiling point of a liquid beyond that expected from comparison with comparable non-hydrogen-bonded liquids and (**2**) causes a greater entropy of vaporization than that of unassociated liquids.

The boiling point of ammonia, although higher than might be expected from a comparison with phosphine, etc., unlike that of the oxygen compounds, is lower than that of the completely methylated compound trimethylamine (see Figure 2.18). This suggests that the effect of hydrogen bonding is not as great as for H_2O. X-ray studies of solid ammonia indicate that three hydrogen atoms from *different* nitrogen atoms point toward the lone pair of electrons. Thus a normal hydrogen bond is not formed in the solid. From heats of sublimation the energy of interaction for NH_3 with the lone pair is about −5.4 kJ/mol of H, compared to −25 kJ/mol of H bond for H_2O.

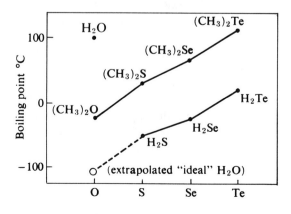

Figure 2.17 Boiling points of hydrogen- and methyl-derivatives of the Group 16 (VIB) elements.

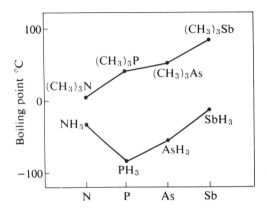

Figure 2.18 Boiling points of hydrogen- and methyl-derivatives of the Group 15 (VB) elements.

▶ 2.6.2 *Consequences of Hydrogen Bonding*

Solubility

When an uncharged organic compound dissolves to an appreciable extent in water, the solubility may be attributed to hydrogen bonding. Thus dimethyl ether (shown below), forming two hydrogen bonds, is completely miscible with water, whereas dimethyl sulfide is only slightly soluble in water.

Benzene is only slightly soluble in water, whereas pyridine is completely miscible in water. The number of carbon atoms in a molecule that is water-soluble depends on the number of hydrogen-bond-forming atoms per molecule. Usually, one to three carbon atoms per oxygen atom gives a highly water-soluble compound. Thus dioxane, $C_4H_8O_2$, is completely miscible in water, whereas diethyl ether, $(C_2H_5)_2O$, is only moderately soluble (7.5 g/100 g H_2O). Sugars are water-soluble because of hydrogen bonding with water. Strong hydrogen bonds formed between the solvent and hydrated ions and oxoanions contribute greatly to the solubility of salts in water.

Hydrogen bonding also plays a role in many nonaqueous solvents. Chloroform is a good solvent for fatty acids, since its polar C—H bond may engage in hydrogen bonding. Ethers can serve as solvents for hydrogen chloride, and acetylene is very soluble in acetone because of hydrogen bond formation. Acetylene under pressure is dangerous to handle, because of its sensitivity to shock. It is handled commercially as a solution in acetone under pressure. The solubility of acetylene in water is low because the solute interferes with the strong hydrogen bonds between water molecules.

► *Infrared Spectra*

Absorption bands in the infrared spectrum of a compound may be associated with particular vibrational modes of the molecule or of functional groups in the molecule. Hydrogen bonding of a group such as the —O—H group affects the frequency, band width, and intensity of the infrared absorption bands assigned to the —O—H stretching and bending vibrations. Dilute solutions of an alcohol (or other substances capable of association through hydrogen bonding) in an inert solvent such as carbon tetrachloride show an absorption band characteristic of the stretching vibration of a free —O—H group. More concentrated solutions, on the other hand, display a new absorption band, at lower frequency and with more intense absorption, resulting from a hydrogen-bonded species (see Figure 2.19). Studies of the effect of temperature on the equilibria between monomeric and polymeric species allow us to evaluate the enthalpy, entropy, and free energy changes associated with the hydrogen bond formation.

► *Crystal Structure*

When hydrogen bonding is possible in molecular crystals, the structure is generally one that yields the maximum number of hydrogen bonds. This gives rise to structures in which the individual molecules are connected through hydrogen bonds to form chains, sheets, or three-dimensional networks. In some cases the crystals contain units consisting of dimeric or trimeric hydrogen-bonded species.

Hydrogen-bonded chains are found in the linear HCN polymer. Zigzag chains are more common, occurring in solid HF,

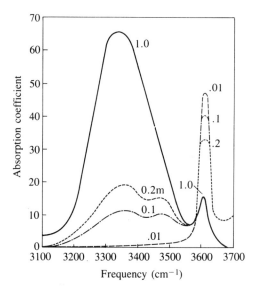

Figure 2.19 Absorption spectra of *tert*-butanol in CCl₄ at various concentrations at 25°C. (From U. Liddel and E. D. Becker, *Spectrochem. Acta* **1957,** *10,* 70, with permission from Pergamon Press, Ltd., Headington Hill Hall, Oxford 6X3 OBW UK.)

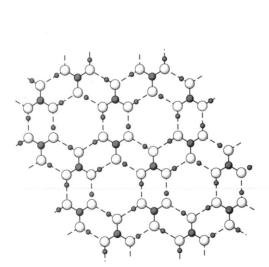

Solid HF

CH₃OH, formic acids, acetanilide (CH₃CONHC₆H₅), and probably *N*-methylamides. Oxalic acid forms a maximum number of hydrogen bonds in either chain or sheets. Both modifications are known. Sheet and three-dimensional structures are typical in cases where each molecule has several hydrogens for hydrogen bonding and also several lone electron pairs. Boric acid [B(OH)₃; see Figure 2.20] provides another example of a compound with a sheet structure. Three-dimensional networks occur in such compounds as water and telluric acid.

In structures such as ice (Figure 2.21), each oxygen is attached tetrahedrally to four hydrogens, because this is required for the maximum number of hydrogen bonds to be formed. Each oxygen has two covalent bonds (short O—H) and two hydrogen bonds to adjacent water molecules. The open channels caused by hydrogen bonding result in the density of ice being less than that of liquid H₂O. A molecular beam study of the water dimer showed that the orientation of a single hydrogen bond is linear within 1° and oriented tetrahedrally with respect to the H—O bonds of the electron-donor molecule. This result is expected for *sp³* hybridization of the oxygen electron donor and suggests a significant covalent contribution to the hydrogen bond.

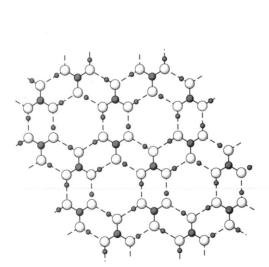

Figure 2.20   Portion of a layer of H₃BO₃. Dashed lines indicate hydrogen bonds. (From A. F. Wells, *Structural Inorganic Chemistry,* 5th ed., Oxford, Oxford, 1984, p. 1067, with permission of Oxford University Press.)

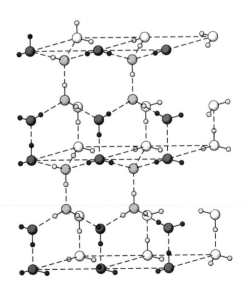

Figure 2.21   The structure of ice showing the hydrogen bonds (dashed lines) and the open structure. (From L. Pauling, *Nature of the Covalent Bond,* 3rd ed., Cornell University Press, Ithaca, NY, 1960, p. 465, with permission of Cornell University Press.)

Hydrogen bonding is very important in polypeptides and cellulose structures. It provides the primary bonding that results in the cross-linking responsible for the α-helix of proteins.

In ionic substances, hydrogen bonding can play a major role in determining the structure. Discrete hydrogen-bonded ions exist in acid salts such as KHF_2 (discussed below). Other acid salts, such as sodium hydrogen carbonate, form polyanion chains or three-dimensional networks, as in KH_2PO_4. The difference in structure between NH_4F (wurtzite structure with 4 F^- arranged tetrahedrally around NH_4^+) and NH_4Cl (CsCl structure with a cubic arrangement of 8 Cl^- around NH_4^+) stems from the greater strength of the N—HF bond than the N—HCl bond.

▶ *2.6.3   Hydrogen Bonding Involving Charged Species*

Table 2.10 lists the range of energies observed for hydrogen bonds between neutral molecules, including weak hydrogen bonding to π clouds of aromatic hydrocarbons. Significant hydrogen bonding occurs for S, Cl, and C in $HCCl_3$ and HCN.

Anion–Molecule Interactions

The combination of hydrogen fluoride with an alkali metal fluoride at room temperature gives an exothermic reaction forming an acid salt. This reaction is used to remove HF

Table 2.10   Energy range of some hydrogen bonds formed between neutral molecules[a]

Bond	Energy range (kJ/mol)	Formed by
F—H ---- O	46	HF with ketones, alcohols, ethers
F—H --- F	29	HF(g)
Cl—H ---- O	25–34	HCl with ketones, alcohols
O—H ---- X	6–15	C_6H_5OH with alkyl halides
O—H --- O	8–34	H_2O, ROH, RCO_2H, C_6H_5OH with themselves and with ketones, ethers
O—H --- N	6–38	H_2O, ROH, C_6H_5OH with amines, ammonia
O—H ---- S	17–21	ROH, C_6H_5OH, R_FOH with thioethers, thioketones
O—H ---- Se	13–17	ROH, C_6H_5OH, R_FOH with selenoethers
O—H ---- π	4–8	ROH, C_6H_5OH with aromatic hydrocarbons
S—H ---- O	4–8	C_6H_5SH with $(BuO)_3PO$, ketones
S—H ---- N	8–12	C_6H_5SH with C_5H_5N
S—H ---- π	1–4	C_6H_5SH with aromatic hydrocarbons
C—H ---- O	6–15	$HCCl_3$ with ketones and ethers
C—H ---- N	6–17	HCN with itself, $HCCl_3$ with amines
N—H ---- N	6–17	NH_3, RNH_2
N—H ---- O	17–21	RNH_2 with ROH, $CH_3CONHCH_3$
N—H ---- S	21–25	*N*-Methylaniline with thiocamphor
N—H ---- π	4–8	RNH_2 with aromatic hydrocarbons

[a] Data largely from S. Singh, A. S. N. Murthy, and C. N. R. Rao, *Trans. Faraday Soc.* **1966**, *62*, 1056.

from gas streams—particularly in the commercial production of F_2, where HF is a contaminant.

$$NaF(s) + (HF)_x(g) \rightarrow NaHF_2(s) \qquad \Delta H = -69 \text{ kJ/mol} \qquad (2.14)$$

Reversal of the reaction—that is, thermal decomposition of $NaHF_2$—serves as a convenient source of anhydrous HF. The heat released in reaction with HF is least with LiF (43 kJ/mol) and increases with the size of the cation: 88 kJ with KF, 93 kJ with RbF, 98 kJ with CsF, and 155 kJ with tetramethylammonium fluoride. X-ray diffraction studies indicate that the alkali metal hydrogen difluorides are salts with a positive metal cation and $(FHF)^-$ as an anion. The F to F distance in $(FHF)^-$ in KHF_2 is only 225 pm, with the hydrogen nucleus equidistant from each fluorine nucleus. The ionic radius of F^- is 119 pm, so the minimum F^-–F^- distance without H bonding is 238 pm.

As HF is absorbed, the crystal lattice of KF must expand to accommodate the $(FHF)^-$ ion. The energy necessary for this expansion is greatest for $LiHF_2$ and least for $CsHF_2$. The weakened lattice on going from the halide to the HF adduct is reflected in a decrease in melting point (mp)—the mp of KF is 846°C, whereas that of KHF_2 is 239°C. The low melting points of these compounds facilitate the production of fluorine by the electrolysis of fused KF–HF baths at lower temperature.

The solubility of many fluorides in liquid HF results from the high degree of solvation of the fluoride ion through hydrogen bonding—that is, $F(HF)_n^-$. The strong acid behavior of liquid HF, in contrast to the behavior of aqueous solutions of HF, indicates that ΔH_{solv} of the fluoride ion is much more negative in HF than in water.

The reactions of water with alkali metal hydroxides are quite similar to those of HF with metal fluroide. Thus solid KOH occasionally is used to remove water vapor from gas streams (especially where it is also desirable simultaneously to remove CO_2)

$$KOH(s) + H_2O(g) \rightarrow KOH \cdot H_2O(s) \qquad \Delta H = -84 \text{ kJ/mol} \qquad (2.15)$$

Cation–Molecule Interactions

The nature of the hydrogen ion in water has been speculated about for the last 80 years. Largely on the basis of the existence of isomorphous crystals of perchloric acid hydrate ($H_3O^+ClO_4^-$) and ammonium perchlorate ($NH_4^+ClO_4^-$), it was assumed that the hydrogen ion in water existed as H_3O^+, termed the **oxonium ion** (commonly called the **hydronium ion**). In 1954 Eigen proposed that the hydrogen ion in water exists as $H_9O_4^+$ with the following structure:

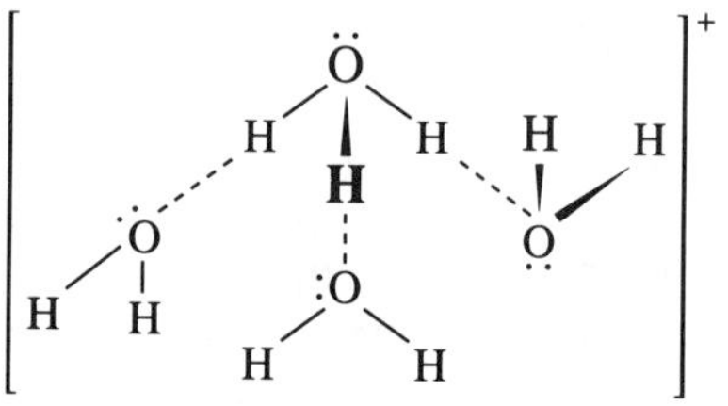

Support for a hydration number of four for the hydrogen ion comes from specific heat studies of aqueous hydrochloric acid solutions, from extraction studies of HCl, HBr, and

$HClO_4$, from activity coefficients of aqueous acids, and from data obtained by field emission mass spectrometry. The most stable gas-phase cluster cation is $H^+(H_2O)_{21}$, an H_3O^+ in a pentagonal dodecahedral cage of 20 water molecules.

The exceptionally high mobility of H^+ in water is unexpected for a species as large as $H_9O_4^+$. It is explained by the ease with which hydrogen bonds realign to form new $H_9O_4^+$ species incorporating water previously in a secondary hydration sphere.

Solids may contain the hydrated hydrogen ion in species varying from H_3O^+ up to $H_9O_4^+$ as illustrated by the hydrates of hydrogen bromide. The increase in melting point from HBr (mp $-86°C$) to $HBr \cdot H_2O$ (oxonium bromide, mp $-4°C$) corresponds to a change from a covalent to an ionic lattice (with hydrogen bonding). The decrease in melting points on going to higher hydrates up to $HBr \cdot 4H_2O$, mp $-56.8°C$, results from expansion of the crystal lattice to accommodate the larger hydrated species accompanied by a loss in lattice energy.[16] Similar hydrates exist for HI, whereas HCl forms lower hydrates, but not $HCl \cdot 4H_2O$.

2.7   WEAK INTERACTIONS IN COVALENT SUBSTANCES

Atoms within a covalent molecule are bound firmly together by covalent bonds, but the molecules usually are attracted very weakly to other molecules by **van der Waals forces**. Molecular substances usually have low boiling points and melting points because the attraction is weak. One contribution to the van der Waals forces can be pictured as the attraction exerted by a positive nucleus on electrons beyond its own radius. There is a finite probability of finding the electrons of an atom at any point out to infinity; the electron density does not fall to zero at the distance used as the radius of the atom. Because a nucleus attracts its own electrons at distances even greater than the atomic radius, it also attracts electrons of other atoms that approach closely. The attractive force increases with the nuclear charge and the number of electrons the two atoms possess. Table 2.11 indicates the magnitude of the energy of various types of molecular interactions in comparison with ionic interaction and covalent bonds. The several interactions between neutral molecules contributing to the van der Waals forces are to be considered next. Ion–molecule interactions are considered also.

▶ *2.7.1   Instantaneous Dipole–Induced Dipole Interaction (Dispersion)*

The attraction between the spherical atoms of the noble gas elements is greater than we might expect for rigidly spherical atoms. Although the distribution of electronic charge is

---

[16] The lattice energy is the energy required to separate ions of a mole of MX(c) as gaseous ions separated by infinite distances (see Section 5.8). Its value decreases as the interionic distances increase, all other things being equal.

Table 2.11   Comparison of intermolecular and intramolecular interactions

Type of interaction	Example	Energy of interaction between molecules or units
Dispersion (Instantaneous dipole–induced dipole)	H_2 (bp 20 K) CH_4 (bp 112 K) CCl_4 (bp 350 K) CF_4 (bp 112 K) $n\text{-}C_{28}H_{58}$ (mp 336 K)	~0.1–5 kJ/mol or ~10 (T_{bp} K) J/mol
Dipole–induced dipole	$Xe(H_2O)_x$ solvation of noble gases or hydrocarbons (see text)	
Ion–induced dipole	Ions in a molecular matrix (see text)	
Dipole–dipole	NF_3—NF_3 (bp 144 K) BrF—BrF (bp 293 K)	5–20 kJ/mol
Ion–Dipole	$[K(OH_2)_6]^+$ Ions in aqueous solution and solid hydrates	67 kJ/mol (energy per bond)
Hydrogen bond	$(H_2O)_x$, $(HF)_x$, $(NH_3)_x$ alcohols, amines HF_2^-	4–50 kJ/mol for neutral molecules
Cation–anion	NaCl, CaO	400–500 kJ/mol of MX "molecules"
Covalent bond	H_2 F_2 Cl_2 Li_2	432.08 kJ/mol 154.6 kJ/mol 239.32 kJ/mol 100.9 kJ/mol

spherically symmetrical on a time average, momentary dipoles can exist. F. London presented a quantum mechanical treatment of the attractive forces between neutral molecules in terms of dispersion effects. These short-lived dipoles polarize adjacent atoms, causing induced dipoles and increasing the force of attraction, which is called the **London force** or **dispersion force.** The dispersion force increases with increasing size and polarizability of the atoms of molecules and varies inversely with the sixth power of the distance between molecules. A repulsive term, caused in the last analysis by the Pauli exclusion principle, drops off more rapidly with distance. Repulsion terms have included sixth to twelfth powers of the intermolecular distance.

For substances such as N_2 and hexane, interaction between molecules arises primarily from the dispersion forces and usually increases with increasing molecular mass. Note, however, that increasing the molecular mass *per se,* without a corresponding increase in dispersion forces, has little or no effect on the boiling point. Thus CD_4 has a slightly *lower* boiling point than CH_4. Similarly, many fluorocarbons' electrons are held so tightly by the highly electronegative fluorine atoms that fluorocarbons boil at temperatures close to, or even less than, the corresponding hydrocarbons. A high degree of symmetry or compactness also favors a low boiling point, by shielding the electrons deep within the molecule and lessening their interaction with neighboring molecules. Thus branching generally lowers boiling points of isomeric organic compounds, e.g., 36°C for $n\text{-}C_5H_{12}$ and 9.5°C for $C(CH_3)_4$.

▶ 2.7.2 Dipole–Induced Dipole and Ion–Induced Dipole Interactions

The attraction between a symmetrical neutral molecule with a molecule having a permanent dipole moment is enhanced greatly because of an induced dipole in the symmetrical molecule. Noble gases form hydrates that could be the result of dipole–induced dipole interactions, since the more stable hydrates are formed by the larger and more polarizable noble gas atoms. The solid hydrates probably are more properly regarded as **clathrate compounds** (inclusion compounds where the host molecule fills voids in the lattice; see Figure 9.16).

There are few clear-cut cases of ion–induced dipole interactions, because more important forces are present in situations where such attractions might occur. Thus interactions between ions and a nonpolar molecule in an ionic medium are much less important than the ionic interactions.

▶ 2.7.3 Dipole–Dipole and Ion–Dipole Interactions

The attraction between molecules with permanent dipole moments represents the strongest van der Waals force. We can see the effect of the attraction between dipoles by comparing the boiling points of $NF_3(-129°C, \mu = 0.234)$ and OF_2 $(-144.8°C, \mu = 0.30)$ with that of $CF_4(-161°C, \mu = 0)$, which has a higher molecular mass but zero dipole moment. The dipole moment of NF_3 (also for OF_2) is small, compared to $\mu = 1.47$ for NH_3 ($\mu = 1.84$ for H_2O), because the polarity of the bonds opposes that of the lone pair on N (and 2 lone pairs on 0).

Ions attract polar water molecules very strongly to form hydrated ions. The hydration energy released provides the energy required for the separation of the ions from the ionic crystal. We shall see later (Section 9.7) that treating hydrated or ammoniated metal ions (water or ammonia metal complexes) in terms of ion–dipole attraction has serious limitations, especially for transition metal ions. The greater stability of ammonia complexes of some metals as compared to the aqua (water) complexes can be explained by the greater polarization of ammonia, which causes the total dipole moment—permanent plus induced—of ammonia in such complexes to be greater than that of water in similar complexes.

▶ 2.7.4 van der Waals Radii

The internuclear separation between nonbonded atoms that are in contact is determined by their **van der Waals radii.** Partitioning the distance between the two atoms may be achieved by using data from systems where both atoms are identical—then, the van der Waals radius is simply half of the nonbonding separation. From a somewhat oversimplified viewpoint, the van der Waals radius for an atom of a nonmetallic element is approximately the same as the radius of the anion formed by the element, since both present to the outside world a completed octet of electrons (see Figure 2.22 and Table 6.8 for ionic radii). Table 2.12 lists van der Waals radii for nonmetals. The van der Waals radius of an atom varies by less than 10 pm in different nonbonding situations. If two atoms are found to be

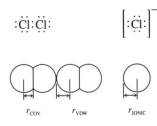

:C̈l:C̈l: [:C̈l:]⁻

r_{COV} r_{VDW} r_{IONIC}

Figure 2.22 Comparisons of covalent, van der Waals, and ionic radii.

Table 2.12   van der Waals radii for nonmetals[a]

H 122				
C 170	N 155	O 152	F 147	Ne 154
Si 210	P 180	S 180	Cl 175	Ar 188
	As 185	Se 190	Br 185	Kr 202
		Te 206	I 198	Xe 216

[a] From A. Bondi, *J. Phys. Chem.* **1964**, *68*, 441. Values are mean van der Waals radii (in pm) for single-bonded forms of the elements.

closer than the sum of their van der Waals radii in a crystal, a hydrogen bond between them usually supplies the explanation. The shortening of the A···B distance in an A—H···B bond compared to that calculated for the van der Waals contact of A and B varies from no observable deviation to as much as 70 pm in FHF^- and $H_2OHOH_2^+$. Typical values for the A···B shortening by hydrogen bonding in oxygen-to oxygen or oxygen-to-nitrogen bonding systems are around 30 pm.

GENERAL REFERENCES

J. K. Burdett, *Molecular Shapes,* Wiley-Interscience, New York, 1980. Excellent coverage of inorganic stereochemistry, including VSEPR, Walsh diagrams, main-group and transition-metal complexes, and cluster compounds.

T. L. Cottrell, *The Strengths of Chemical Bonds,* 2nd ed., Butterworths, London, 1958. Good source of data, even though it is dated.

R. L. DeKock and H. B. Gray, *Chemical Structure and Bonding,* Benjamin/Cummings, Menlo Park, CA, 1980. Good supplement to this text.

R. J. Gillespie and I. Hargittai, *The VSEPR Model of Molecular Geometry,* Allyn and Bacon, Boston, 1991.

R. McWeeny, *Coulson's Valence,* 3rd ed., Oxford, Oxford, 1979. A classic book updated.

L. Pauling, *The Nature of the Chemical Bond,* 3rd ed., Cornell University Press, lthaca, NY, 1960. A classic.

PROBLEMS

2.1 Give the expected possible covalent MX_n compounds (where $X^- = Cl^-$ or Br^-) of Sn, At, and Be. Assuming only σ bonding, predict the geometry of each molecule.

2.2 Assign formal charges and oxidation numbers, and evaluate the relative importance of three Lewis structures of OCN^-, the cyanate ion. Compare these to the Lewis structures of the fulminate ion, CNO^- (encountered in explosive compounds).

2.3 Write a reasonable electron dot structure and assign formal charges and oxidation numbers for each of the following: ClF, ClF_3, ICl_4^-, $HClO_3$ ($HOClO_2$).

2.4 Use Langmuir's formula (Equation 2.3) to generate VB structures for BF_3, SO_3^{2-}, N_2F_2, and H_2CN_2 (consider CN_2^{2-} first).

2.5 Give the oxidation number, formal charge, and hybridization of the central atom in each of the following: NO_3^-, BF_4^-, $S_2O_3^{2-}$, ICl_2^+, ClO_3^-. What are the molecular shapes?

2.6 Select the reasonable electron-dot structures for each of the following compounds. Indicate what is wrong with each incorrect or unlikely structure.

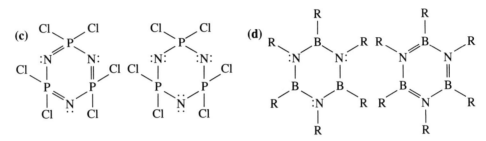

(a) N̈=N̈=Ö   :N̈—N≡O: :N≡N—Ö:

      N̈—N̈≡Ö   :N=N=Ö   :N=N=Ö:

(b) :S̈—C≡N:⁻   :S̈=C=N:⁻   S̈=C=N̈⁻   :S≡C—N̈:⁻

(c)   (d)

2.7 Give the expected hybridization of P, O, and Sb in $Cl_3P—O—SbCl_5$. The P—O—Sb bond angle is 165°.

2.8 Predict both the gross geometry (from the σ orbital hybridization) and the fine geometry (from bond–electron-pair-repulsion, etc.) of the following species: F_2SeO, $SnCl_2$, I_3^-, and $IO_2F_2^-$.

2.9 Which of the following in each pair has the larger bond angle? Why?
   (a) CH_4 and NH_3;   (b) OF_2 and OCl_2;   (c) NH_3 and NF_3;   (d) PH_3 and NH_3.

2.10 Trimethylphosphine was reported by R. Holmes to react with $SbCl_3$ and $SbCl_5$ to form, respectively, $(Me_3P)(SbCl_3)$ or $(Me_3P)_2(SbCl_3)$ and $(Me_3P)(SbCl_5)$ or $(Me_3P)_2(SbCl_5)$. Suggest valence bond structures for each of these and indicate approximate bond angles around the Sb atom.

2.11 Sketch the following hybrid orbitals (indicate the sign of the amplitude of the wavefunction on your sketch):
   (a) an sd hybrid;   (b) a pd hybrid.

2.12 The Ru—O—Ru bond angle in $(Cl_5Ru)_2O$ is 180°. What is the state of hybridization of the oxygen? Explain the reasons for the large bond angle. (See R. J. Gillespie, *J. Am. Chem. Soc.* **1960,** *82,* 5978.)

2.13 Using H. A. Bent's isoelectronic groupings (Figure 2.13), identify the species of list **2** (below) that are "isoelectronic" with each of those of list **1**. Each entry might have more than one "isoelectronic" partner in the other list.
   1. $(H_2N)_2CO$ $HONO_2$ OCO CO $(CN)_2$ ClO_2^+ $Si_3O_9^{6-}$ $TeCl_2$ Bi_9^{5+}
   2. (H_2CCO) $(HO)_2CO$ ONN H_2CNN H_3CNO_2 F_2CO BF B_2O_2 $(CH_3)_2CO$ CH_2CN^- SO_2 cyclic $(SO_3)_3$ ICl_2^+ Pb_9^{4-}

2.14 Consult Pauling's electroneutrality principle (page 62) and describe the reasonable charge distribution in NH_4^+ and in SO_4^{2-}.

2.15 The bromine atom in BrF_5 is below the plane of the base of the tetragonal pyramid. Explain.

2.16 H_2C=SF_4 gives the expected isomer with the double-bonded methylene group in an equatorial position. From the orientation of the π bond, would you expect the H atoms to be in the axial plane or the equatorial plane? (See K. O. Christie and H. Oberhammer, *Inorg. Chem.* **1981,** *20,* 297.)

2.17 The C—CH_3 bond distance in methyl acetylene is unusually short (146 pm) for a single bond. Show how this short bond can be rationalized in more than one way.

2.18 Calculate the dipole moment to be expected for the ionic structure H^+Cl^- using the same internuclear separation as for the HCl molecule. Calculate the dipole moment for HCl assuming 19% ionic character.

2.19 Calculate the heats of formation (from electronegativities) for H_2S, H_2O, SCl_2, NF_3, and NCl_3. Calculate the heats of formation of these compounds from the bond energies for comparison. Compare the results with data from a handbook.

2.20 Calculate the electronegativity differences from the bond energies for H—Cl, H—S, and S—Cl. Using Allred's electronegativity of H, compare the electronegativity values that can be obtained for S and Cl with those tabulated.

2.21 Calculate the Lewis–Langmuir charges on carbon in the following:
(a) CH_4;   (b) C_2H_2;   (c) HCN;   (d) CO_3^{2-}.

2.22 Acetylene, unlike methane, has acidic properties and shows hydrogen bonding. Why do the Lewis–Langmuir charges calculated for CH_4 and C_2H_2 not show a greater negative charge for C in C_2H_2 than for C in CH_4?

2.23 Calculate the formal charges, oxidation numbers, and Lewis–Langmuir charges on S and F in S_2F_2, SF_2, SF_4, and SF_6. Which of these correlate well with the expected changes in bond polarity in the series?

2.24 Although ΔH_{vap} for HF is lower than that for H_2O, HF forms stronger H bonds. Explain.

2.25 The intensity of an infrared absorption is proportional to the change in the dipole moment occurring during the vibration. The asymmetric stretching in IHI^- is far more intense than that found for the stretching vibration of HI. Offer a reasonable explanation. Does your explanation also hold for the intensities shown in Figure 2.19?

2.26 Explain why anhydrous acid $HICl_4$ cannot be isolated, but the crystalline hydrate $HICl_4 \cdot 4H_2O$ may be obtained from ICl_3 in aqueous HCl.

2.27 List the substances in each of the following groups in order of increasing boiling points. (*Hint:* First group the substances according to the type of interaction involved.)
(a) LiF, LiBr, CCl_4, NH_3, CH_4, SiC, CsI
(b) Xe, NaCl, NO, CaO, BrF, Al_2O_3, SiF_4

2.28 When no chemical reaction occurs, the solubility of a gas in a liquid is proportional to the magnitude of the van der Waals interaction energy of the gas molecules (see Table 2.11). Indicate the relative solubility of O_2, N_2, Ar, and He in water. Why do deep-sea divers use a mixture of He and O_2 instead of air?

2.29 Use appropriate radii from pages 89 and 102 to calculate the expected S ···· S distance for SF_6 molecules in contact in the solid.

2.30 Calculate the expected O ···· O separation in a system containing O—H—O if no hydrogen bonding were to occur and the O—H—O atoms are linearly arranged. Compare your results with the nearest O ···· O distance (275 pm) in ice, I_h.

2.31 For each of the following substances, in the liquid state, indicate the major type of interaction between the individual units (atoms, molecules, or ions):
LiF, H_2O, CH_3NH_2, HCl, Xe, CCl_4

2.32 Why is $(CH_3)_2O$ quite soluble in H_2O, and why is C_2H_6 very insoluble in H_2O?

▶ 3 ◀

Symmetry

· · · · · · · · · · · · · · · · · · ·

Many inorganic molecules have high symmetry. Molecular symmetry places significant limitations on electronic and vibrational energy levels and on chemical reactions involving breaking and forming bonds. This chapter serves as an introduction to the application of symmetry concepts to molecules through the use of symmetry point groups. The terminology and notation of symmetry are widely employed in the literature of inorganic chemistry. Familiarity with these is important in understanding the results of molecular orbital treatments and in interpretating various kinds of spectroscopy.

To make full use of symmetry, the mathematical approach of group theory is needed, and this is introduced in Section 3.5. In courses emphasizing a less mathematical approach, this section may be omitted along with the group theoretical approach to molecular orbital theory (Sections 4.5–4.7 and 9.7). Matrix notation is used in introducing symmetry operations, and matrix multiplication is treated in Appendix 3.1. However, use of matrices is not essential for understanding symmetry and the results of molecular orbital treatments.

3.1 INTRODUCTION

Symmetry greatly enhances the beauty of nature, as is very evident in flowers, leaves, and crystals. Molecular structure reveals remarkable ramifications of symmetry, which are developed further in the orderly packing of molecules in crystals. From Chapter 1, you are familiar with the descriptions of atomic orbitals for the hydrogen atom. In a chemical compound orbital shapes are dictated by symmetry, and even energy states can be described in terms of symmetry. Vibrations of molecules can be resolved into characteristic modes that are determined by and described in terms of the molecular symmetry. Spectroscopic selection rules that determine which transitions are allowed can be expressed in terms of molecular symmetry of energy states, whether these are electronic or vibrational levels. Utilizing symmetry greatly diminishes the work necessary in determining a crystal

105

structure or in carrying out molecular orbital calculations. Thus familiarity with symmetry simplifies our task and aids our understanding, even as it greatly enhances our appreciation of the awesome beauty of chemistry and of nature in general.

3.2 SYMMETRY ELEMENTS AND SYMMETRY OPERATIONS

Symmetry is defined here as invariance to transformation; that is, the object appears unchanged after the transformation. Thus an equilateral triangle △ looks the same after rotation by 120° in the plane of the paper. The nature of the transformation defines the type of symmetry. Thus the equation $xy + yz + xz = xyz$ shows **permutational symmetry,** as does the palindrome "Able was I ere I saw Elba." The equation is unchanged by any of the possible permutations of x, y, and z among themselves, whereas the palindrome is unchanged by the specific permutation involving exchange of the first and last letters, the second and penultimate letters, and the . . . x and $n - x$ letters, and thus is invariant to being read forwards or backwards. A **symmetry element** is a feature that permits some transformation to be executed which leaves the appearance of an object (or set of objects) unchanged. We now consider some common symmetry elements and the **symmetry operations** carrying out the implied transformations which are of interest in chemistry.

▶ *3.2.1 Mirror Plane (σ)*

Mirror symmetry denotes invariance to reflection, as illustrated in Figure 3.1. It may be possessed by a single object, or by a *set* of objects (the pair of hands) that need not individually exhibit the particular symmetry. Thus a unit cell of a crystal of a racemate (a 50–50 mixture of optical isomers) may have a mirror plane because of the presence of both d and l molecules in the unit cell.

For invariance under the reflection operation, the reflection must be carried out through a plane that causes the reflected image to be coincident with the object or set of objects. Such a plane is called a **symmetry plane** (or **mirror plane**) and is designated by the Greek letter σ (or by the letter m in crystallography). Humans and most animals have a symmetry plane. The fiddler crab with one large claw has no symmetry plane. For the

planar $\begin{bmatrix} & \text{O} & \\ & \| & \\ & \text{C} & \\ \text{O} & = & \text{O} \end{bmatrix}^{2-}$ ion there is a symmetry plane in the plane of the paper, and a symmetry

plane through each C—O bond perpendicular to the paper. The **reflection operation** that carries out the reflection also is given the designation σ (or m).

Reflection through the yz plane takes an arbitrary point in Cartesian three-dimensional space from the position (x, y, z) to a new position $(-x, y, z)$—only the out-of-plane coordinate changes sign. Alternatively, we can say that the transformation equa-

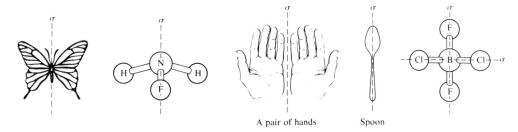

A pair of hands   Spoon

Figure 3.1 Mirror symmetry. The lines labeled σ represent the planes of symmetry perpendicular to the plane of the page.

tions for σ_{yz} are

$$
\begin{aligned}
x' &= (-1)x + (0)y + (0)z \\
y' &= (0)x + (1)y + (0)z \\
z' &= (0)x + (0)y + (1)z
\end{aligned}
\tag{3.1}
$$

We can write these equations more compactly in matrix notation[1] as

$$
\begin{bmatrix} x' \\ y' \\ z' \end{bmatrix} =
\begin{bmatrix} -1 & 0 & 0 \\ 0 & 1 & 0 \\ 0 & 0 & 1 \end{bmatrix}
\begin{bmatrix} x \\ y \\ z \end{bmatrix}
\tag{3.2}
$$

or state simply that the σ_{yz} transformation matrix is $\begin{bmatrix} -1 & 0 & 0 \\ 0 & 1 & 0 \\ 0 & 0 & 1 \end{bmatrix}$. See Appendix 3.1 for a

discussion of matrix multiplication.

EXAMPLE 3.1: Write the transformation matrix for the reflection of a point with coordinates (x, y, z) through the σ_{xy} plane.

Solution: Because σ_{xy} leaves the x and y coordinates unchanged and changes the sign of z, we have

$$
\begin{bmatrix} x' \\ y' \\ z' \end{bmatrix} =
\begin{bmatrix} 1 & 0 & 0 \\ 0 & 1 & 0 \\ 0 & 0 & -1 \end{bmatrix}
\begin{bmatrix} x \\ y \\ z \end{bmatrix}
$$

EXAMPLE 3.2: Write the transformation matrix for the reflection of a point through the plane containing the z axis and at 45° to the x and y axes.

[1] The introduction of matrix notation seems at this point like unnecessary elegance. However, it will later be clear how this notation contributes to identifying the symmetry properties of orbitals and vibrations of inorganic molecules.

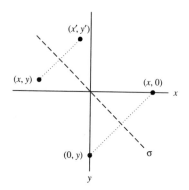

Solution: The z coordinate is unchanged. The point $(x, 0)$ becomes $(0, x)$, and the point $(0, y)$ becomes $(y, 0)$, or $(x, y) \rightarrow (x', y')$, where $x' = y$ and $y' = x$ and $z' = z$. Thus the matrix interchanges x and y:

$$\begin{bmatrix} x' \\ y' \\ z' \end{bmatrix} = \begin{bmatrix} 0 & 1 & 0 \\ 1 & 0 & 0 \\ 0 & 0 & 1 \end{bmatrix} \begin{bmatrix} x \\ y \\ z \end{bmatrix} = \begin{bmatrix} y \\ x \\ z \end{bmatrix}$$

► 3.2.2   *Center of Symmetry (i)*

The **inversion operation** takes a point on a line through the origin (the **inversion center**) to an equal distance on the other side, thus transforming a point with the coordinates (x, y, z) to one with the coordinates $(-x, -y, -z)$. The transformation matrix for the inversion is therefore $\begin{bmatrix} -1 & 0 & 0 \\ 0 & -1 & 0 \\ 0 & 0 & -1 \end{bmatrix}$, the matrix giving the coefficients for an equation of the form of Equation (3.1). An object invariant under the inversion operation is said to possess a **center of symmetry** or an **inversion center** (see Figure 3.2). Both the center of symmetry (the element) and the inversion operation are designated by the symbol i.

EXAMPLE 3.3:   Which of the following molecules have a center of symmetry? Which have mirror planes?

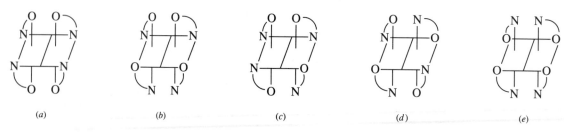

(a)   (b)   (c)   (d)   (e)

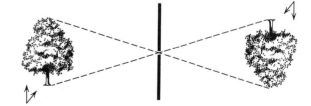

(f)   *(g)*

Isomers of [(gly)$_2$Co(OH)$_2$CO(gly)$_2$]

Solution: Isomers *(a)*, *(d)*, and *(e)* have *i*. Isomers *(a)*, *(b)*, and *(ė)* have σ through the shared edge perpendicular to the drawn plane. The rings formed by glycinate ions represented by O⌢N eliminate possible σ planes perpendicular to the shared edge or in the drawn plane.

▶ 3.2.3   *Axis of Rotation (C$_n$)*

An object has **axial symmetry** (see Figure 3.3) when it is invariant to rotation by some fraction of 360° (or 2π radians). It is said to have an ***n*-fold axis of symmetry,** or a C_n axis, if it is invariant to rotation by $360°/n$ (or $2\pi/n$). In the rotation operation (also designated by C_n) *the object conventionally is rotated clockwise*. The transformation matrix for a clockwise C_n operation about the *z* axis is

$$\begin{bmatrix} \cos 2\pi/n & \sin 2\pi/n & 0 \\ -\sin 2\pi/n & \cos 2\pi/n & 0 \\ 0 & 0 & 1 \end{bmatrix} \tag{3.3}$$

For a $(C_2)_z$ operation (C_2 rotation about *z*), the transformation matrix is thus

$$\begin{bmatrix} -1 & 0 & 0 \\ 0 & -1 & 0 \\ 0 & 0 & 1 \end{bmatrix}$$

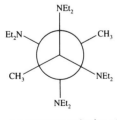

(a) Newman projection of
W$_2$Me$_2$(NEt$_2$)$_4$ in the solid
state, looking along
the W—W bond

(b) Fe(C$_5$H$_5$)$_2$
Ferrocene
(only rings are outlined
to emphasize symmetry)

(c) An object and its pinhole image in equal size

Figure 3.2 Inversion symmetry. The center of symmetry lies at the midpoint of the W—W bond for *(a)* and at the Fe for *(b)*.

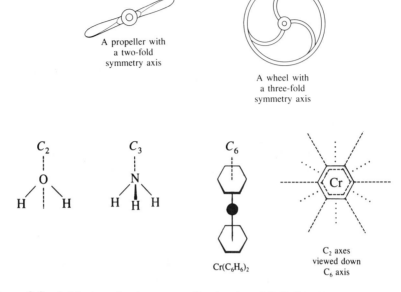

Figure 3.3   Axial or rotational symmetry. For the view of $Cr(C_6H_6)_2$ down the C_6 axis, the dashed hexagon is below the plane of the page.

EXAMPLE 3.4:   Obtain the transformation matrix for rotation by an angle θ about the z axis of a point in the xy plane.

Solution: The position of a point (x, y) can be specified by the polar coordinates r and ϕ as sketched (z axis is perpendicular to the paper). Rotation around the z axis through an angle θ takes the point (x, y) into (x', y'). The new polar coordinates are $[r, 2\pi - (\theta - \phi)] = [r, (\phi - \theta)]$. (Rotations are conventionally measured clockwise, whereas the polar angle is measured counterclockwise.) The relationships between polar and Cartesian coordinates are: for the right triangle $r = \sqrt{x^2 + y^2}$; since $\cos \phi = x/r$, $x = r \cos \phi$; and since $\sin \phi = y/r$, $y = r \sin \phi$. For the new right triangle, $x' = r \cos(\phi - \theta)$ and $y' = r \sin(\phi - \theta)$. From the trigonometric relationships for cos and sin of $(\phi - \theta)$ and $\cos \phi = x/r$ and $\sin \phi = y/r$ we have

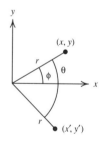

$$x' = r[\cos \phi \cos \theta + \sin \phi \sin \theta)] = x \cos \theta + y \sin \theta$$

$$y' = r[\sin \phi \cos \theta + \cos \phi \sin \theta)] = y \cos \theta - x \sin \theta \text{ or } -x \sin \theta + y \cos \theta$$

Expressing the above in matrix form gives

$$\begin{bmatrix} x' \\ y' \end{bmatrix} = \begin{bmatrix} \cos \theta & \sin \theta \\ -\sin \theta & \cos \theta \end{bmatrix} \begin{bmatrix} x \\ y \end{bmatrix}$$

EXAMPLE 3.5:   What is the transformation matrix for a C_3 operation about the axis $x = y = z$ on a point (x, y, z) (a C_3 axis for a cube, octahedron, or tetrahedron; see Figure 3.13 considering the upper right front C_3 axis)? See Appendix 3.1 for discussion of matrix multiplication.

Solution: Rotation by C_3 about this axis takes y into x, x into z, and z into y, so

$$\begin{bmatrix} 0 & 0 & 1 \\ 1 & 0 & 0 \\ 0 & 1 & 0 \end{bmatrix} \begin{bmatrix} x \\ y \\ z \end{bmatrix} = \begin{bmatrix} x' \\ y' \\ z' \end{bmatrix} = \begin{bmatrix} z \\ x \\ y \end{bmatrix}$$

EXAMPLE 3.6: Find the highest *n*-fold rotation axis for each of the following species. Which of the species has other symmetry axes not coincident with the *n*-fold axis already found?

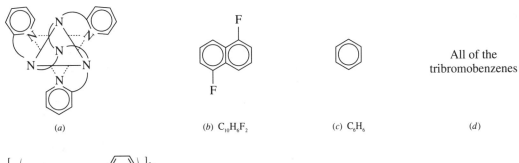

| (*a*) | (*b*) $C_{10}H_6F_2$ | (*c*) C_6H_6 | (*d*) |

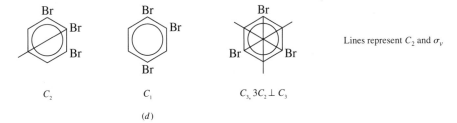

Structure (*a*) is a monocapped octahedron—the Ni at the center of the octahedron is not shown. (From L. J. Wilson and N. J. Rose, *J. Am. Chem. Soc.* **1968**, *90*, 6041. Copyright © 1968, American Chemical Society.)

Solution: (*a*) C_3, (*b*) C_2, (*c*) C_6. C_6H_6 has $6C_2 \perp C_6$.

| C_2 | C_1 | $C_3, 3C_2 \perp C_3$ |

Lines represent C_2 and σ_v

(*d*)

▶ 3.2.4 Identity (E)

The C_1 operation, consisting of a rotation by 360° or an integral multiple of 360°, leaves any object unchanged. This operation, the equivalent of doing nothing, is given the special symbol *E* (German *Einheit*) and is called the **identity** operation. The identity operation seems to be trivial, but it is required mathematically. The transformation matrix for

the identity operation is $\begin{bmatrix} 1 & 0 & 0 \\ 0 & 1 & 0 \\ 0 & 0 & 1 \end{bmatrix}$, the **unit matrix.**

EXAMPLE 3.7: Show that the product of the identity transformation matrix and any other transformation matrix leaves the other matrix unchanged.

Solution:

$$
\overset{E}{\begin{bmatrix} 1 & 0 & 0 \\ 0 & 1 & 0 \\ 0 & 0 & 1 \end{bmatrix}} \begin{bmatrix} a & b & c \\ e & f & g \\ h & i & j \end{bmatrix} = \begin{bmatrix} (a+0+0) & (b+0+0) & (c+0+0) \\ (0+e+0) & (0+f+0) & (0+g+0) \\ (0+0+h) & (0+0+i) & (0+0+j) \end{bmatrix}
$$

▶ 3.2.5 *Improper Rotation (S_n)*

The rotation operations discussed above (C_n) are sometimes termed **proper rotations,** to distinguish them from improper rotations. An **improper rotation** (rotation–reflection) consists of a rotation C_n followed by reflection in the plane perpendicular to the C_n axis. Both the symmetry element and the operation are given the symbol S_n for a rotation–reflection of $360°/n$ ($2\pi/n$ radians). To obtain the product of two symmetry operations $[\sigma_{xy}(C_n)_z]$, multiply the transformation matrices or perform the operations in sequence from right to left—that is, C_n then σ_{xy}. Because the transformation matrix for σ_{xy} is

$$
\begin{bmatrix} 1 & 0 & 0 \\ 0 & 1 & 0 \\ 0 & 0 & -1 \end{bmatrix},
$$

the transformation matrix for the product of σ_{xy} and $(C_n)_z$, $\sigma_{xy}(C_n)_z =$

$$
(S_n)_z, \text{ is } \begin{bmatrix} 1 & 0 & 0 \\ 0 & 1 & 0 \\ 0 & 0 & -1 \end{bmatrix} \begin{bmatrix} \cos 2\pi/n & \sin 2\pi/n & 0 \\ -\sin 2\pi/n & \cos 2\pi/n & 0 \\ 0 & 0 & 1 \end{bmatrix} = \begin{bmatrix} \cos 2\pi/n & \sin 2\pi/n & 0 \\ -\sin 2\pi/n & \cos 2\pi/n & 0 \\ 0 & 0 & -1 \end{bmatrix}
$$

where the rotation is clockwise. Here the sequence does not matter because $[\sigma_{xy}(C_n)_z] = [(C_n)_z\sigma_{xy}]$, but this is not generally true.

Figure 3.4 gives some examples of molecules with S_n axes. Example (*a*) has a mirror plane through the line joining Sn, Mn, and CO, through the upper phenyl ring, and between a pair of phenyls and two pairs of CO. Because σ is unchanged by multiplication by C_1, we have $\sigma C_1 = \sigma = S_1$. We use σ rather than the equivalent S_1. Example (*b*) has a center of symmetry at the center of the P—P bond. Rotating about this point by 180° (C_2) and then reflecting in a plane perpendicular to C_2 reproduces the original positions, so $i = \sigma C_2 = S_2$. All molecules having a C_n axis and perpendicular σ_h always have an S_n axis. The reverse is true when n is odd, but not always true when n is even, so that S_n may exist without the presence of σ. A planar conformer of the $B(OCH_3)_3$ molecule in Figure 3.8 has an S_3 axis (and also C_3 and σ_h because n is odd). Rotating the CH_4 molecule by 90° about a C_2 (Figure 3.4*c*) axis shifts the H atoms to formerly vacant sites of a cube. Reflection through a plane perpendicular to C_2 restores the original arrangement. A tetrahedron has an S_4 axis, but no C_4 axis or σ perpendicular to S_4; the S_4 axis coincides with a C_2 axis. A second example having an S_4 axis is shown (Figure 3.4*d*); the $[Mn(C_5H_5NO)_6]^{2+}$ complex ion has an S_6 axis coincident with the C_3 axis perpendicular to the plane of the paper.

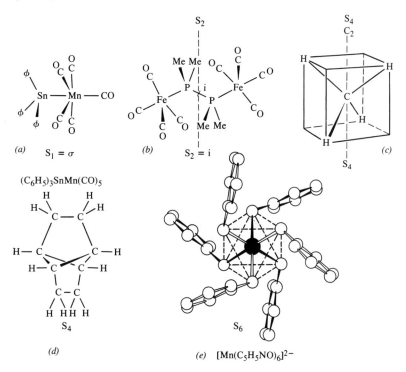

Figure 3.4 Improper rotational axes of symmetry. [Structure (a) is from H. P. Weber and R. F. Bryan, *Acta Cryst.* **1967,** *22,* 822. Structure (d) is after J. F. Chiang and S. H. Bauer, *Trans. Faraday Soc.* **1968,** *64,* 2248. Structure (e) is reprinted with permission from T. J. Bergendahl and J. S. Wood, *Inorg. Chem.* **1975,** *14,* 338; copyright © 1975, American chemical Society.]

EXAMPLE 3.8: Using transformation matrices, show that the S_1 operation is identical to the σ operation, and that the S_2 operation is identical to the i operation.

Solution:

$$S_1 = \sigma_{xy}C_1 = \begin{bmatrix} 1 & 0 & 0 \\ 0 & 1 & 0 \\ 0 & 0 & -1 \end{bmatrix} \begin{bmatrix} 1 & 0 & 0 \\ 0 & 1 & 0 \\ 0 & 0 & 1 \end{bmatrix} = \begin{bmatrix} 1 & 0 & 0 \\ 0 & 1 & 0 \\ 0 & 0 & -1 \end{bmatrix} = \sigma_{xy}$$

$$S_2 = \sigma_{xy}C_2 = \begin{bmatrix} 1 & 0 & 0 \\ 0 & 1 & 0 \\ 0 & 0 & -1 \end{bmatrix} \begin{bmatrix} -1 & 0 & 0 \\ 0 & -1 & 0 \\ 0 & 0 & 1 \end{bmatrix} = \begin{bmatrix} -1 & 0 & 0 \\ 0 & -1 & 0 \\ 0 & 0 & -1 \end{bmatrix} = i$$

It is often said that a molecule is optically active if it has no center of symmetry or plane of symmetry. These are necessary, but not sufficient, conditions. Example 3.8 using matrices and examining of the molecules in Figure 3.4 demonstrate that $S_1 = \sigma$ and $S_2 = i$. The criterion for a molecule to be optically active is that it must not have *any* S_n symmetry axis. A valid test for optical activity is that it should not be possible to superimpose a model of the molecule on its mirror image.

3.3 INTRODUCTION TO GROUPS

We can generate a group and introduce some needed terms using a kaleidoscope. A kaleidoscope has two mirrors arranged at an angle (usually 60°) to each other to produce multiple reflections of bits of colored glass. Consider the reflections of the letter P by the two mirrors σ_1 and σ_2 having a 45° dihedral angle, as illustrated in Figure 3.5. Reflection of P through σ_1 produces the image ь; reflection of P through σ_2 produces the image ᓀ. Each of these images may be reflected in turn, with the following results:

σ_1 on ᓀ → ᓂ $\equiv \sigma_1\sigma_2$ on P $\equiv C_4$ on P

σ_2 on ь → ᓇ $\equiv \sigma_2\sigma_1$ on P $\equiv C_4^3$ on P

Note that the result of reflection through σ_1 and σ_2 *depends on the order* ($\sigma_1\sigma_2$ means σ_2 followed by σ_1) in which the reflection operations are performed; that is, the reflection operations *do not* **commute.** The successive performance of symmetry operations is the **product** of the operations.

In arithmetic and ordinary algebra, multiplication commutes; that is, $2 \times 3 = 3 \times 2$ and $x \times y = y \times x$. Matrix multiplication and the multiplication of symmetry operations do not commute, except in special cases. Thus $E \times \sigma = \sigma \times E$ and the unit matrix $\begin{bmatrix} 1 & 0 & 0 \\ 0 & 1 & 0 \\ 0 & 0 & 1 \end{bmatrix}$ for E commutes with other matrices, but $\sigma_1\sigma_2 \neq \sigma_2\sigma_1$. Note also that the result of successive reflection through two different mirror planes is the same as a single rotation through an angle which is twice that between the mirror planes [for $\sigma_1\sigma_2 = C_4$, $2 \times 45° = 90°$; and for $\sigma_2\sigma_1$, $2 \times 315° = 630° = 270°$ (C_4^3 or C_4 performed three times, $3 \times 90°$)]. The rotational axis coincides with the intersection of the two mirror planes.

Another look at the 45° kaleidoscope shows that the images ᓇ and ᓂ may be reflected further:

σ_1 on ᓇ → ᓄ $\equiv \sigma_1\sigma_2\sigma_1$ on P $\equiv C_4\sigma_1 \equiv \sigma_2'$ on P

σ_2 on ᓂ → q $\equiv \sigma_2\sigma_1\sigma_2$ on P $\equiv C_4^3\sigma_2 \equiv \sigma_1'$ on P

and

σ_1 on q → d $\equiv \sigma_1\sigma_2\sigma_1\sigma_2$ on P $\equiv C_4C_4 \equiv C_2$ on P

σ_2 on ᓄ → d $\equiv \sigma_2\sigma_1\sigma_2\sigma_1$ on P $\equiv C_4^3C_4^3 \equiv C_2$ on P

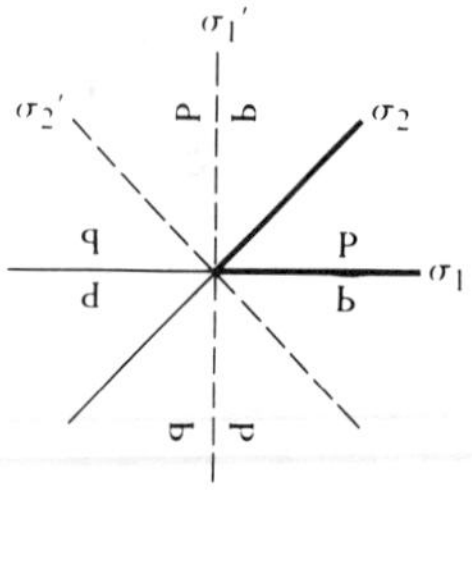

Figure 3.5 Reflections in a kaleidoscope.

where σ_1' and σ_2' are mirror planes perpendicular to σ_1 and σ_2, respectively, and are indicated by dashed lines in Figure 3.5. In the sequence of operations $\sigma_1\sigma_2\sigma_1\sigma_2$, to be performed in order from *right to left,* we may add parentheses in any fashion we choose—for example, $(\sigma_1\sigma_2)(\sigma_1\sigma_2)$ or $\sigma_1(\sigma_2\sigma_1\sigma_2)$ or $(\sigma_1\sigma_2\sigma_1)\sigma_2$, the operations within the parentheses being performed from right to left. From the products evaluated above, note that the groupings can be written as $(\sigma_1\sigma_2)(\sigma_1\sigma_2) = (C_4)(C_4)$, $\sigma_1(\sigma_2\sigma_1\sigma_2) = \sigma_1(\sigma_1')$, and $(\sigma_1\sigma_2\sigma_1)(\sigma_2) = (\sigma_2')\sigma_2$—each of which is equivalent to a C_2 operation. When the result does not depend on the choice of parentheses or the grouping of operations, the operations are said to be **associative** (capable of being joined).

The two mirror operations σ_1 and σ_2 performed in exhaustive combination generate a **group.** In the mathematical sense, a group is a collection of operations having the following properties:

1. The product of any two operations must be an operation of the group (a group is said to be closed under multiplication).

2. Every operation must have an **inverse;** that is, for every operation there must be an operation that will undo the effect of the first operation. The element σ or C_2 is its own inverse because $\sigma^2 = E$ and $C_2^2 = E$ [σ^2 means reflection twice and C_2^2 means rotation by 180° twice (360°) and $C_4^2 = C_2(2 \times 90° = 180°)$].

3. Every group must have an identity operation, E, since the product of an operation and its inverse is the identity. The identity operation has the effect of leaving every group member, or element of the group, unchanged. It commutes with all other operations of the group; that is, $EA = AE$.

4. All operations of the group are associative: $ABC = (AB)C = A(BC)$.

5. The product of any two operations, or elements, is defined. Groups for which all elements commute are called **Abelian** groups.

A group is realized by any object (or set of objects, equations, etc.) that appears to be unchanged after carrying out all of the operations of the group. Such an object, or set, is said to belong to the group.

A group multiplication table can be useful for understanding the properties of a group. In constructing a group multiplication table for the kaleidoscope group, we arbitrarily label one letter (P) with the identity operation and each of the other letters with the single operation that produces it from the letter P (labeled E in Figure 3.6). The product rule, or law of multiplication, is to perform the operations sequentially (from *right to left*) or, in table form, to carry out *the operation at the top of the table (labeling the columns) first and the operation at the left (labeling the rows) second.* The *product* is the single operation that produces the same result. Table 3.1 is our example multiplication table. The product $\sigma_1\sigma_2 = C_4$ as marked in the table. *The sequence is the column operation* C(σ_2), *then the row operation* R(σ_1), written as (RC). The entries were obtained by taking from Figure 3.6 a letter whose label gives the operation at the top of the table, performing one of the operations indicated at the left of the table, and entering the label of the product letter in the table. Thus, performing a C_4 operation on ⌐ (labeled σ_2') gives ⌐ (labeled σ_1'); hence $C_4\sigma_2' = \sigma_1'$. Two operations that have E as their product are each other's *inverse;* an operation and its inverse always commute. Note that each σ is its own inverse—or, $\sigma^2 = E$.

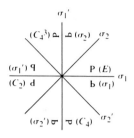

Figure 3.6
Kaleidoscope group with symmetry labels.

Table 3.1   **Multiplication table for the kaleidoscope group**

	E	C_4	$C_4^2(=C_2)$	C_4^3	σ_1	σ_2	σ_1'	σ_2'
E	E	C_4	C_4^2	C_4^3	σ_1	σ_2	σ_1'	σ_2'
C_4	C_4	C_4^2	C_4^3	E	σ_2'	σ_1	σ_2	σ_1'
C_4^2	C_4^2	C_4^3	E	C_4	σ_1'	σ_2'	σ_1	σ_2
C_4^3	C_4^3	E	C_4	C_4^2	σ_2	σ_1'	σ_2'	σ_1
R σ_1	σ_1	σ_2	σ_1'	σ_2'	E	C_4	C_4^2	C_4^3
σ_2	σ_2	σ_1'	σ_2'	σ_1	C_4^3	E	C_4	C_4^2
σ_1'	σ_1'	σ_2'	σ_1	σ_2	C_4^2	C_4^3	E	C_4
σ_2'	σ_2'	σ_1	σ_2	σ_1'	C_4	C_4^2	C_4^3	E

3.4   SYMMETRY POINT GROUPS

The symmetry of a molecule located on symmetry axes, cut by planes of symmetry, or centered at an inversion center is known as **point symmetry.** Crystallographers (along with designers of wallpaper and textiles) are concerned with translational symmetry (space groups), which might involve reorientation of a molecule (or figure) from one lattice site to another. Here we deal only with point symmetry. We are interested in the collection of symmetry operations which arise because of the existence of certain symmetry elements. These collections of symmetry operations constitute groups in the mathematical sense, and each such **point group** has a particular designation. A minimal number of symmetry operations for each group can be regarded as generating the point group; we say that these are the **generators** of the group. We now consider the possible point groups and how they arise from generators. The general approach we use is to begin with high-order C_n operations and add other operations as generators. In the course of the discussion it becomes apparent that the existence of some symmetry operations requires the existence of others as products of operations. The names of point groups are set in **boldface** type in this book, using the Schöenflies symbols. Crystallographers use Hermann–Mauguin notation for symmetry operations: m for σ, 4 for C_4, and so on. These symbols are discussed in Section 6.1.1. Table 6.1 gives this notation along with Schöenflies symbols and international notation for point groups of lattices.

▶ 3.4.1 $\mathbf{C}_n$, $\mathbf{C}_{nh}$, and $\mathbf{C}_{nv}$ Groups

The groups generated by repetition of a C_n operation are termed $\mathbf{C}_n$ **point groups,** containing no other symmetry elements. Because $C_n^n = E$, after n repetitions we will be back where we started; hence the $\mathbf{C}_n$ point group has n operations, or the **order** of the group is n. Some examples of molecules exhibiting $\mathbf{C}_n$ point group symmetry are shown in Figure 3.7. Rotating $As(C_6H_5)_3$ by C_3 (120°) will move ring 1 into the position of ring 2. Two C_3 rotations (240°), written as C_3^2, will move ring 1 into the position of ring 3. Three C_3 rotations (360°) (C_3^3 equivalent to the identity, $C_3^3 = E$) will restore the original positions.

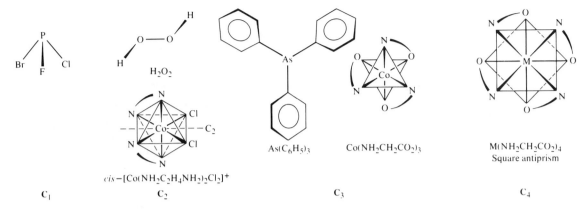

Figure 3.7 Examples of C_n point groups.

The direction of the C_n axis is taken as vertical, so a symmetry plane perpendicular to it is a horizontal plane, σ_h. Adding to the C_n operation (as a new generating operation) a horizontal mirror plane, σ_h, forms new groups termed $\mathbf{C}_{nh}$. Some examples of molecular species belonging to the $\mathbf{C}_{nh}$ point groups are shown in Figure 3.8. The additional operations (besides rotations and reflection in a horizontal plane) generated in a $\mathbf{C}_{nh}$ group, as products of C_n, and σ_h, include an inversion operation, where n is even, and various S_n operations. The $\mathbf{C}_{1h}$ group (having only a mirror plane) is called $\mathbf{C}_s$.

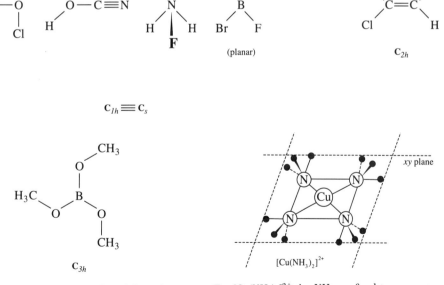

Figure 3.8 Examples of $\mathbf{C}_{nh}$ point groups. For $[Cu(NH_3)_4]^{2+}$ the NH_3 are fixed to preserve $\mathbf{C}_{4h}$ symmetry; actually they rotate freely, giving effective $\mathbf{D}_{4h}$ symmetry.

EXAMPLE 3.9:   Show that C_{nh} groups where n is even have the element i.

Solution: We label the C_n axis as the z axis. $(C_n)^{n/2} \equiv C_2$ transforms x to $-x$ and y to $-y$ for even n, and σ_h transforms z to $-z$, so $\sigma_h(C_n)^{n/2} \equiv (C_n)^{n/2}\,\sigma_h \equiv i$.

If a mirror plane contains the rotational axis, the group is called a C_{nv} group—the v indicating a vertical mirror plane, σ_v. The mirror plane is reproduced n times in a C_{nv} group, examples of which are shown in Figure 3.9. Thus, the existence of C_n and σ_v requires the existence of n σ_v's. The C_{4v} group can be recognized as our 45° kaleidoscope group, which was generated by multiple σ operations. This illustrates that the choice of generators ($C_n + \sigma_v$ or multiple σ planes at $360°/2n$) is arbitrary. Both C_{nv} and C_{nh} groups have $2n$ elements or operations (order $= 2n$).

▶ 3.4.2   *Dihedral Groups (D_n, D_{nh}, and D_{nd})*

Adding a C_2 axis perpendicular to a C_n axis generates one of the **dihedral groups, D_n**. Some examples of molecules belonging to D_n groups are shown in Figure 3.10. Because of the C_n axis, there must be n C_2 axes, (perpendicular to C_n), which, together with n symmetry operations associated with the C_n axis, give a total of $2n$ symmetry operations for a D_n group.

Adding a σ_h to a D_n group generates a D_{nh} group, with $4n$ symmetry operations—the $2n$ operations of the D_n group and $2n$ more that are the products of these operations with

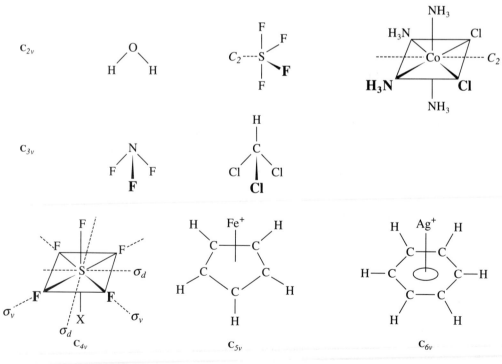

Figure 3.9   Examples of C_{nv} point groups. $C_{1v} \equiv C_s \equiv C_{1h}$ (see Figure 3.8).

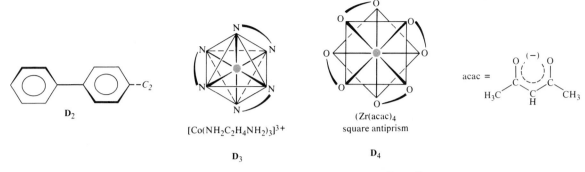

Figure 3.10 Examples of $\mathbf{D}_n$ point groups. $\mathbf{D}_1 \equiv \mathbf{C}_2$.

σ_h. For even values of n, one of these is an inversion center because $(C_n)^{n/2}\sigma_h \equiv i$. Figure 3.11 provides examples of some $\mathbf{D}_{nh}$ groups.

Adding a vertical mirror plane to a $\mathbf{D}_n$ group in such a fashion as to bisect adjacent C_2 axes generates a $\mathbf{D}_{nd}$ group (see Figure 3.12 for examples). See Figure 9.36 (page 429) for the structure of $[\text{Mo(CN)}_8]^{4-}$ that has $\mathbf{D}_{2d}$ symmetry, a dodecahedron with triangular faces. In $\mathbf{D}_{nd}$ groups the mirror planes are referred to as σ_d (**dihedral planes**). The number of planes is n because $(n-1)$ planes are generated by the combination of the first σ_d and C_n. There are $4n$ symmetry operations—the $2n$ operations for $\mathbf{D}_n$ and $2n$ more which are the products of these operations with σ_d. For odd values of n, an inversion center exists generated by combination of C_2 and σ_d operations.

▶ 3.4.3 S_n Groups

The S_n operation serves as a generator for the $\mathbf{S}_n$ groups. For odd n, $(S_n)^n \equiv \sigma_h$ and $(S_n)^{2n} \equiv E$. Hence C_n and σ_h exist as separate symmetry elements and the $\mathbf{S}_n$ group is the *same* as the $\mathbf{C}_{nh}$ group for odd n. The groups are called $\mathbf{C}_{nh}$ (n odd). For even n, $(S_n)^n = E$

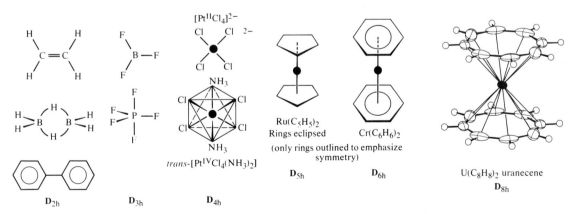

Figure 3.11 Examples of $\mathbf{D}_{nh}$ point groups. (The uranocene is reprinted with permission from A. Zalkin and K. N. Raymond, *J. Am. Chem. Soc.* **1969**, *91*, 5667; copyright © 1969, American Chemical Society.)

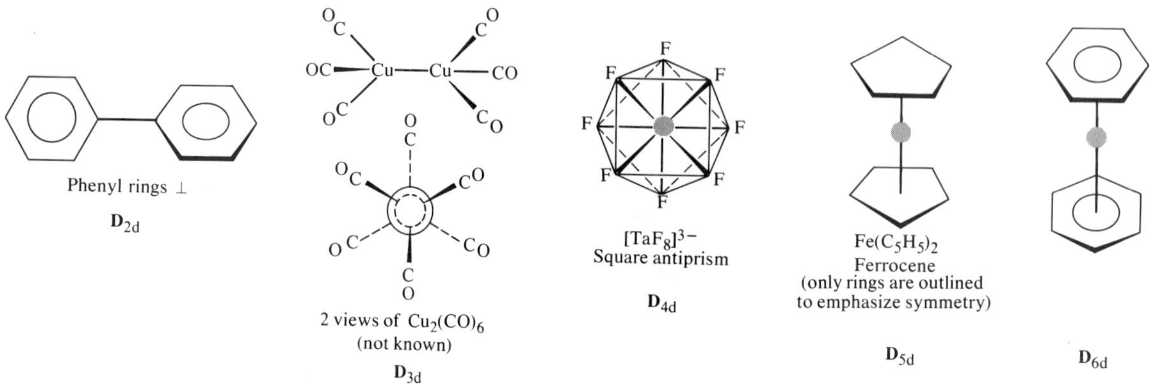

Phenyl rings ⊥

$\mathbf{D}_{2d}$

2 views of $Cu_2(CO)_6$
(not known)

$\mathbf{D}_{3d}$

$[TaF_8]^{3-}$
Square antiprism

$\mathbf{D}_{4d}$

$Fe(C_5H_5)_2$
Ferrocene
(only rings are outlined
to emphasize symmetry)

$\mathbf{D}_{5d}$

$\mathbf{D}_{6d}$

Figure 3.12   Examples of $\mathbf{D}_{nd}$ point groups.

and these groups labeled $\mathbf{S}_n$ have no mirror planes. The absence of mirror planes may be used to distinguish $\mathbf{S}_n$ groups from $\mathbf{D}_{nd}$ groups, which also contain an S_n axis. Each $\mathbf{S}_n$ group contains n operations $(S_n)^1 \cdots (S_n)^n \equiv E$. Examples of $\mathbf{S}_1 \equiv \mathbf{C}_s$ are shown in Figure 3.8 and Figure 3.4a. The group having only an inversion center is called $\mathbf{C}_i$; see Figure 3.4 for an example. Example (d) in Figure 3.4 illustrating the S_4 axis belongs to the $\mathbf{D}_{2d}$ group.

▶ 3.4.4   *Linear Groups* ($\mathbf{C}_{\infty v}$ *and* $\mathbf{D}_{\infty h}$)

Linear molecules possess a C_∞ axis along a line of the nuclei, and there is an infinite number of reflection planes (σ_v) passing through the atoms. If the molecule has no C_2 axis perpendicular to the C_∞ axis, it belongs to the $\mathbf{C}_{\infty v}$ group. Examples of molecules in the $\mathbf{C}_{\infty v}$ group include CO, HCl, HCN, and SCO. If you find a C_2 axis perpendicular to the C_∞ axis (if one is found, an infinite number exist), the molecule or ion belongs to the $\mathbf{D}_{\infty h}$ group. The subscript h indicates the presence of σ_h. Examples belonging to the $\mathbf{D}_{\infty h}$ group are the symmetrical linear molecules: N_2, O_2, HCCH, OCO, and $[NCAgCN]^-$.

▶ 3.4.5   *The Platonic Solids* (*Cubic and Icosahedral Groups*)

The remaining symmetry groups take their names from the **Platonic solids,** which are examples of solid figures belonging to these groups. In a Platonic solid, all vertices, edges, and faces are equivalent. In three-dimensional space there are only five such polyhedra— the tetrahedron, the octahedron, the cube, the regular dodecahedron (with pentagonal faces), and the icosahedron. If the centers of the adjacent faces of a Platonic solid are connected by lines, the solid figure outlined is also a Platonic solid, and the two figures are said to be **conjugates.** Conjugates belong to the same symmetry group. The octahedron and the cube are conjugates, as are the icosahedron and dodecahedron; the tetrahedron is its own conjugate. Figure 3.13 shows an octahedron formed by connecting the centers of the faces of a cube. Platonic solid groups include the icosahedral group and cubic groups

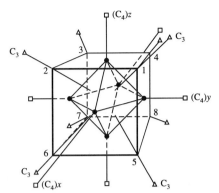

Figure 3.13 An octahedron within a cube, both sharing the same symmetry elements. The C_2 (◊) axes through each edge are not shown.

(the tetrahedral and octahedral groups are called the **cubic groups**). The **rotation subgroup** (eliminating all operations except C_n) of these groups is designated by a single-letter abbreviation (**I, T,** and **O**), and the full symmetry group of the Platonic solids is designated by a two-letter abbreviation (**I_h, T_d,** and **O_h**). The cube and octahedron have the same symmetry elements and belong to the O_h group (see Figure 3.13). The tetrahedron shares many of these elements, but lacks a C_4 axis.

The number of operations belonging to a group is termed the order of the group and denoted by the letter h. The order of the **rotation group** of a Platonic solid is simply the product of the number of faces and the number of edges per face.

EXAMPLE 3.10: Show that the order of the rotation group of a Platonic solid is the product of the number of faces and the number of edges per face.

Solution: The order of a group is the number of distinct operations which leave the object unchanged. Consider flagging one edge of the solid so that the flag extends onto the face. Rotation operations can now be found to move the flag to all edges on all faces because, for a Platonic solid, all edges and all faces are equivalent. For each face, the flag can be moved to e edges and each edge can be moved to any of the f faces. Hence, the total number of rotations is $e \times f$.

The order of the full symmetry group of a Platonic solid is twice the order of the rotation group. Thus a tetrahedron has four triangular faces, so $h = 12$ for **T** or 24 for **T_d**. An octahedron has eight triangular faces, so $h = 24$ for **O** and 48 for **O_h**. A regular dodecahedron has 12 pentagonal faces and a regular icosahedron has 20 triangular faces, so the order of the **I** group is 60 (5×12 or 3×20). The rotational operations arise from the twofold, threefold, and fivefold axes lying along the center of the polyhedra and the centers of the edges, faces, and vertices. The I_h group (Figure 3.14e) has 120 operations; the added 60 improper rotations arise from a center of symmetry, the S_{10} axes are coincident with the C_5 axes, the S_6 axes are coincident with the C_3 axes, and the reflection planes are perpendicular to the C_2 axes.

EXAMPLE 3.11: Identify all symmetry operations associated with the **T_d** group.

Solution: The rotations of the tetrahedron (**T** group) are $4C_3$ (around the C_3 axis through each apex), $4C_3^2$ (equivalent to $4C_3$ involving clockwise rotations from below), and $3C_2$ (around the C_2 axes through opposite edges). The order is 12, including E. The orientation of the tetrahe-

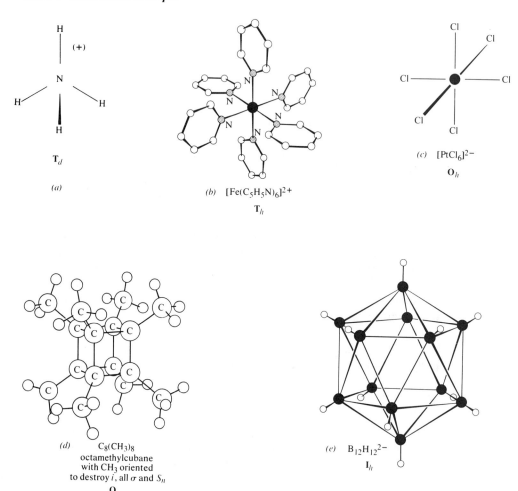

(a) T_d

(b) $[Fe(C_5H_5N)_6]^{2+}$
T_h

(c) $[PtCl_6]^{2-}$
O_h

(d) $C_8(CH_3)_8$
octamethylcubane
with CH_3 oriented
to destroy i, all σ and S_n
O

(e) $B_{12}H_{12}^{2-}$
I_h

Figure 3.14   Examples of cubic and icosahedral groups. [Structure *(b)* is reprinted with permission from R. J. Doedens and L. E. Dahl, *J. Am. Chem. Soc.* **1966**, *88*, 4847; copyright © 1966, American Chemical Society. Structure *(d)* is from *Symmetry: A Stereoscopic Guide for Chemists* by I. Bernal, W. C. Hamilton, and J. S. Ricci, Freeman and Company, New York, copyright © 1972. Structure *(e)* is from E. L. Muetterties, *The Chemistry of Boron and Its Compounds*, Wiley, New York, 1967, p. 233, copyright © 1967.]

dron is chosen for coincidence of the symmetry elements in common with the cube and octahedron (see Figure 3.4c for orientation within a cube). The C_2 axes here coincide with the x, y, and z axes. We can get the T_d group by adding σ_d planes that bisect the edges. On adding a new symmetry element, we must take the product of it with each other symmetry element because the product of any two operations must also be in the group. The product of σ_d and C_2 merely interchanges σ_d planes. We can take the product of σ_d and C_3 by transforming models or using matrices. First, examine the models

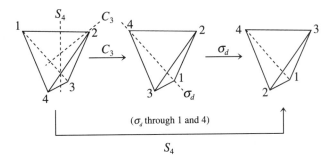

$(\sigma_d$ through 1 and 4)

S_4

The single operation that will accomplish $\sigma_d C_3$, as shown, is S_4. about the axis shown. Using matrices for C_3 about the axis $x = y = z$ (see Example 3.5) and using σ_d containing the x axis (see Example 3.23 for matrix multiplication), we have

$$\overset{\sigma_d}{\begin{bmatrix} 1 & 0 & 0 \\ 0 & 0 & -1 \\ 0 & -1 & 0 \end{bmatrix}} \overset{C_3}{\begin{bmatrix} 0 & 1 & 0 \\ 0 & 0 & 1 \\ 1 & 0 & 0 \end{bmatrix}} = \overset{S_4}{\begin{bmatrix} 0 & 1 & 0 \\ -1 & 0 & 0 \\ 0 & 0 & -1 \end{bmatrix}}$$

There is a σ_d through each of the six edges. There are $3C_2$ axes, and clockwise and anticlockwise rotation (or rotation from above and below) are equivalent for C_2. These operations are not equivalent for S_4, so for the $3S_4$ axes there are six operations, $3 \times (S_4 + S_4^3)$. With $6\sigma_d$ and $6S_4$ operations added to the **T** group, the order of the $\mathbf{T}_d$ group is 24.

The $\mathbf{O}_h$ group is commonly encountered, but the **O** group is rare (see Figure 3.14*d*). Most tetrahedral molecules belong to the $\mathbf{T}_d$ group. The rotation group (**T**) would be achieved for $Si(CH_3)_4$ with the CH_3 staggered to destroy σ_d (and S_4) but to retain the C_3 and C_2 axes. Actually the symmetry of $Si(CH_3)_4$ is $\mathbf{T}_d$ because of free rotation of CH_3 groups.

The tetrahedron does not have a center of symmetry, and if the inversion operation is used to generate a new set of operations from the **T** group, the improper rotations will be i, σ_h (reflections normal to the C_2 axes), and S_6 operations along the C_3 axes. The total of 24 proper and improper rotations comprise the $\mathbf{T}_h$ group. Molecules belonging to this group do not look like tetrahedra (see Figure 3.14*b* for a beautiful example), but they can be distinguished from $\mathbf{O}_h$ molecules, which have C_4 axes.

The sphere belongs to an infinite order group, $\mathbf{O}(3)$, containing all possible proper and improper rotations. The rotation subgroup is labeled $\mathbf{R}(3)$ for all possible proper rotations in three dimensions. All isolated atoms belong to the $\mathbf{O}(3)$ point group.

▶ *3.4.6   Assigning Molecules to Symmetry Groups*

The rotation groups include $\mathbf{C}_n$, $\mathbf{D}_n$, and the Platonic solid rotation groups (**T**, **O**, or **I**). To assign a molecule to a point group, we first assign the rotation group by identifying the highest-order proper rotation axis and then looking for other proper rotations; then we look for other symmetry elements to identify the full symmetry group following Figure 3.15. (The highest-order C_n axis is conventionally taken as the z axis.)

If a molecule has only one C_n axis, the rotation group is $\mathbf{C}_n$ (left of Figure 3.15). If there is a C_2 axis perpendicular to C_n, the rotation group is $\mathbf{D}_n$ and there must be n C_2 axes. If a molecule has more than one C_n axis with $n > 2$, it belongs to one of the Platonic solid rotation groups: $\mathbf{I}$ if $n = 5$, $\mathbf{O}$ if $n = 4$, and $\mathbf{T}$ if $n = 3$. In each case there are several C_5, C_4, or C_3 axes.

After identifying the rotation group, we look for other symmetry properties. If the rotation group is $\mathbf{C}_n$, we look for a mirror plane perpendicular to C_n (σ_h) to identify the full group as $\mathbf{C}_{nh}$. If a molecule belonging to a $\mathbf{C}_n$ rotation group has a vertical plane (slicing through C_n), the group is $\mathbf{C}_{nv}$ and there must be n σ_v.

If the molecule belonging to a $\mathbf{C}_n$ rotation group has no mirror planes, we look for S_n axes, but there are few cases to be examined. For a molecule belonging to the $\mathbf{C}_2$ rotation group, if there is an S_4 axis coincident with the C_2 axis, the full group is $\mathbf{S}_4$. For a molecule belonging to the $\mathbf{C}_3$ rotation group, if there is an S_6 axis coincident with the C_3 axis, the group is $\mathbf{S}_6$. Alternatively, identification of the highest C_n as a C_3 axis and a center of symmetry is adequate to characterize the $\mathbf{S}_6$ full group. For the case of a $\mathbf{C}_4$ rotation group, the presence of an S_8 axis coincident with the C_4 axis identifies the $\mathbf{S}_8$ full group. The $\mathbf{S}_4$ group is rare, and no examples of the $\mathbf{S}_8$ group (or those with higher n) are known. $\mathbf{S}_n$ (n odd) groups are called $\mathbf{C}_{nh}$ ($\mathbf{C}_{3h}$ and $\mathbf{C}_{5h}$). If there are no σ, i, or S_n elements, the full group is $\mathbf{C}_n$.

If a molecule with the rotation group $\mathbf{D}_n$ has a mirror plane (σ_h) perpendicular to the major axis (C_n), the group is $\mathbf{D}_{nh}$. A mirror plane bisecting the angle formed by two C_2 axes, called a dihedral plane (σ_d), gives the $\mathbf{D}_{nd}$ group. There must be n σ_d planes present.

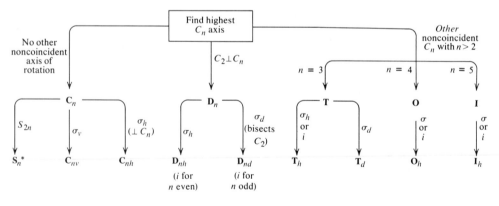

Figure 3.15 Flow chart for assignment of molecules to point groups.

For the **T** rotation group the presence of dihedral planes through the edges (bisecting the angles between the C_3 axes) identifies the $\mathbf{T}_d$ full group. If there are no σ_d planes but we can identify a center of symmetry or mirror plane perpendicular to $C_2(\sigma_h)$, the full group is $\mathbf{T}_h$ (if we identify one of the elements, the other must be there). In the absence of σ_h, σ_d, and i, the full group is **T**. For the **O** rotation group, identification of any σ or the i identifies the $\mathbf{O}_h$ full group. For the **I** rotation group, finding σ or i identifies the $\mathbf{I}_h$ full group. The common groups are $\mathbf{T}_d$, $\mathbf{O}_h$, and $\mathbf{I}_h$.

If a molecule has no symmetry, the group is $\mathbf{C}_1$. If only a mirror plane is present, the group is $\mathbf{C}_s$. The presence of only a center of symmetry identifies the $\mathbf{C}_i$ group.

Linear molecules are easily identified. They have a C_∞ axis, and the rotation group is $\mathbf{C}_\infty$. There is an infinite number of vertical planes (σ_v) giving the full group $\mathbf{C}_{\infty v}$. If there is a mirror plane (σ_h) perpendicular to C_∞, there is also an infinite number of C_2 axes perpendicular to C_∞. The rotation group is $\mathbf{D}_\infty$ and the full group is $\mathbf{D}_{\infty h}$.

EXAMPLE 3.12: Assign the following species to rotation groups and to full symmetry groups: $Mo_6Cl_8^{4+}$ (Figure 17.45), $Ta_6Cl_{12}^{2+}$ (Figure 17.45), $Be_4O(O_2CCH_3)_6$ (Figure 15.5), P_4, P_4O_6, and P_4O_{10}.

Solution:

	$Mo_6Cl_8^{4+}$	$Ta_6Cl_{12}^{2+}$	$Be_4O(O_2CCH_3)_6$	P_4, P_4O_6, and P_4O_{10}
Rotation group:	**O**	**O**	**T**	**T**
Full group:	$\mathbf{O}_h$	$\mathbf{O}_h$	$\mathbf{T}_d$	$\mathbf{T}_d$

EXAMPLE 3.13: Assign the following species to rotation groups and to full symmetry groups:

(a)

(b)

(c)

(*a*) Trigonal bipyramids joined axially. (*c*) Trigonal bipyramids joined by equatorial positions. (Structures for $Co_2(CO)_8$ reprinted with permission from R. L. Sweany and T. L. Brown, *Inorg. Chem.* **1977,** *16,* 415. Copyright © 1977, American Chemical Society.)

Solution:

	P_2Cl_4	B_2F_4	B_2Cl_4	B_4Cl_4	$Co_2(CO)_8$ isomers		
					(a)	*(b)*	*(c)*
Rotation group:	$\mathbf{C}_2$	$\mathbf{D}_2$	$\mathbf{D}_2$	$\mathbf{T}$	$\mathbf{D}_3$	$\mathbf{C}_2$	$\mathbf{D}_2$
Full group:	$\mathbf{C}_{2h}$	$\mathbf{D}_{2h}$	$\mathbf{D}_{2d}$	$\mathbf{T}_d$	$\mathbf{D}_{3d}$	$\mathbf{C}_{2v}$	$\mathbf{D}_{2d}$

EXAMPLE 3.14:   Assign the point group for C_{60}, buckminsterfullerene or buckyball (see Figure 3.16).

Solution: There are C_5 axes through the centers of all pentagons and C_3 axes through the centers of all hexagons giving the **I** rotation group. Identification of any symmetry plane, easily seen as along any C—C bond shared between hexagons, gives the $\mathbf{I}_h$ full group.

▶ 3.4.7 Chemical Properties Determined by Symmetry

CH_4 and NH_4^+ belong to the $\mathbf{T}_d$ group. The H atoms are interchanged by the symmetry operations (except E), and hence the H atoms must be equivalent. All C—H (or N—H) bond lengths must be the same. All H atoms are equivalent in the nuclear magnetic resonance (NMR) spectra, and hence only one signal is seen. The bond angles must be the angles of a regular tetrahedron. PF_5 (trigonal bipyramid) has $\mathbf{D}_{3h}$ symmetry. The two axial F atoms must be equivalent because they are interchanged by C_2 or σ_h. The three equatorial F atoms are equivalent because they are interchanged by C_3 or σ_v. The axial F atoms are equivalent in the NMR spectrum, and these bond lengths are the same. The same is true for the equatorial F atoms. Atoms interchanged by proper rotations are **chemically equivalent.** Those interchanged by improper rotations are **enantiotopically equivalent** and may give rise to separate NMR signals in chiral media. PF_5 contains two sets of chemically equivalent F atoms. The axial and equatorial F atoms are different because they cannot be interchanged by symmetry operations. Symmetry or group theory give no indication whether axial P—F bonds will be longer or shorter than the equatorial bonds. For octahedral SF_6 all F atoms are interchanged by symmetry operations and they are equiva-

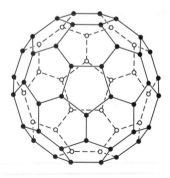

Figure 3.16   Diagram of the C_{60} molecule. (Adapted from K. Hedberg, L. Hedberg, D. S. Bethune, C. A. Brown, H. C. Dorn, R. D. Johnson, and M. deVries, *Science* **1991,** *254,* 410; copyright © 1991 by the AAAS.)

lent. The H atoms are chemically equivalent (and so are the C atoms) for $Fe(C_5H_5)_2$ ($\mathbf{D}_{5d}$, Figure 3.2) and for $Ru(C_5H_5)_2$ ($\mathbf{D}_{5h}$, Figure 3.11) because all are interchanged by proper rotations.

All physical properties having directional character must be invariant to symmetry transformations. Accordingly, only molecules having a single C_n axis of highest order can possess dipole moments. Hence, molecules whose full point group is $\mathbf{C}_n$ or $\mathbf{C}_{nv}$ have a dipole moment (except where accidental cancellation of individual bond moments occurs). Molecules of the $\mathbf{D}_n$ (also $\mathbf{D}_{nh}$ and $\mathbf{D}_{nd}$) groups must have zero dipole moments because adding the C_2 axis perpendicular to C_n ensures that substituents above and below the xy plane are arranged symmetrically, balancing individual bond moments. Molecules with $\mathbf{C}_n$ and $\mathbf{D}_n$ symmetry are optically active since they lack i, σ, or *any* S_n. Figure 9.37 illustrates a Zr complex with $\mathbf{S}_4$ symmetry—an example of a compound that is optically inactive, even though it lacks a center or plane of symmetry.

3.5 CLASS STRUCTURE, REPRESENTATIONS, AND CHARACTER TABLES

▶ 3.5.1 *Symmetry Classes and Character Tables*

Sometimes operations of a group carry one symmetry element into another. Thus an examination of the NH_3 molecule ($\mathbf{C}_{3v}$) shows that all three σ_v mirror planes are converted into one another by a C_3 rotation. An alternative way of thinking about this is that the symmetry *operations* are interconverted by C_3's. Group operations which are interconverted by other operations of the group are said to be in the same **symmetry class.** The σ_v reflections in $\mathbf{C}_{3v}$ are in the same class. In the $\mathbf{C}_{4v}$ group (SXF_4, Figure 3.9) there are two classes of reflection operations involving vertical symmetry planes. The two σ_v through the F—S—F bonds belong to one class, and the two σ_d reflections through dihedral planes bisecting the FSF angles belong to a separate class. No operation carries a σ_v into a σ_d.

As will be discussed below, the number of symmetry operations in each class is given at the top of the **character table** (see Table 3.2 for $\mathbf{C}_{4v}$, see Example 3.18 for $\mathbf{D}_3$, and see Appendix C for each point group). Thus the character table identifies all symmetry operations by class and also describes the way various properties of a molecule behave with respect to each class of operations. For cyclic groups (Abelian groups) in which all operations commute, every operation is in its own class. The cyclic groups are those for which there is a single generator (that is, all operations result from repetition of a single operation); these are the $\mathbf{C}_n$ and $\mathbf{S}_n$ groups.

Rotations have a clockwise sense from above. For each C_n rotation there is a second operation $C_n^{(n-1)}$ that is equivalent to C_n^{-1} (counterclockwise rotation). The $C_n^{(n-1)}$ or C_n^{-1} is also equivalent to clockwise rotation about the axis when viewed from below the molecule. The same symmetry element (the C_n axis) generates (at least) two operations, namely, C_n and C_n^{-1}. For some non-Abelian groups the operations C_n and C_n^{-1} are in the same class, and for some groups they are not in the same class. As an example, for $\mathbf{C}_{4v}$, $\mathbf{D}_4$, $\mathbf{D}_{4h}$, and $\mathbf{D}_{4d}$, operations C_4 and C_4^3 are in the same class, whereas for $\mathbf{C}_{4h}$, operations

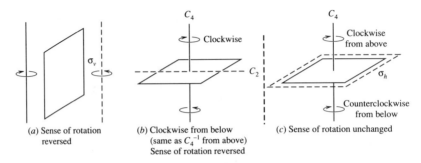

Figure 3.17   Effect of (a) σ_v, (b) C_2, and (c) σ_h on the sense of rotation about an axis.

C_4 and C_4^3 are in separate classes. Examining character tables reveals that C_n and $C_n^{(n-1)}$ are in the same class if there is a vertical plane through the axis ($\mathbf{C}_{nv}$ groups) or if there is a C_2 axis perpendicular to C_n ($\mathbf{D}_n$, $\mathbf{D}_{nh}$, and $\mathbf{D}_{nd}$ groups).

From Figure 3.17 we see that the effect of reflection through σ_v of an arrow representing the sense of rotation reverses the sense of rotation. In effect this interchanges C_4 and C_4^{-1} (or C_4^3), and they are in the same class if there is a vertical plane containing a C_4 axis (as in $\mathbf{C}_{4v}$). The effect of a C_2 rotation about an axis perpendicular to C_4 is to take C_4 (clockwise from above) into C_4 clockwise from below, and that operation is equivalent to C_4^{-1} or C_4^3 from above. The C_n operation about the major axis and the $C_n^{(n-1)}$ operation are in the same class for the dihedral (**D**) groups because these have C_2 axes perpendicular to C_n. Reflecting the arrow representing the clockwise C_4 rotation from above through σ_h leaves the sense of rotation unchanged when viewed from above, and this is counterclockwise C_4^{-1} when viewed from below. These are the same operation because the molecule is rotated in the same direction, or C_4 from above is equivalent to C_4^{-1} from below. Thus σ_h does not interchange C_n and $C_n^{(n-1)}$ so that these operations belong in a different class for $\mathbf{C}_{4h}$ (and all $\mathbf{C}_{nh}$).

Likewise, S_n and $S_n^{(n-1)}$ are in the same class if there is a vertical plane (σ_v or σ_d) through the S_n axis or if there is a C_2 axis perpendicular to S_n.

EXAMPLE 3.15:   Group the C_n and C_n^m operations, and the σ planes by classes for the following groups: $\mathbf{D}_3$, $\mathbf{D}_4$, $\mathbf{D}_{4h}$, and $\mathbf{T}_d$.

Solution: For $\mathbf{D}_3$, a C_2 operation takes the C_3 axis from above into the one below; and because clockwise rotation as viewed from below is equivalent to counterclockwise rotation from above, C_3 and C_3^{-1} (or C_3^2) operations are in the same class.

For $\mathbf{D}_4$, the same situation applies for C_4 from above and from below (equivalent to C_4^{-1} or C_4^3), but $C_4^2 = C_2$ is unique. There are two classes of $C_2 \perp C_4$: one class of the two axes along the x and y axes, and one class of two C_2 axes between the coordinate axes.

For $\mathbf{D}_{4h}$, the C_4, C_4^3, and C_2 operations are grouped in classes as for $\mathbf{D}_4$. The σ_h is unique. There are two classes of vertical reflection planes: two through the coordinate axes (σ_v) and two between the axes (σ_d). The two σ_v are interchanged by C_4, as are the two σ_d.

For $\mathbf{T}_d$, S_4 interchanges C_3 and C_3^2 (or C_3 from below), so they are in the same class. There is also a σ_d through C_3 that interchanges these operations. C_2's are interchanged by C_3's, and thus they are in the same class. C_3 operations interchange σ_d's which thus belong to the same class.

In summary, C_n and C_n^{-1} $(C_n^{(n-1)})$ rotations are interchanged by σ_v and C_2 perpendicular to C_n, but not by σ_h. They belong to the same class for all groups except $\mathbf{C}_n$, $\mathbf{C}_{nh}$, $\mathbf{S}_n$, and $\mathbf{T}$.

▶ *3.5.2   Representations*

A set of matrices (one for each operation of the group) whose multiplication table is the same (under the rules of matrix multiplication—see Appendix 3.1) as that for the symmetry operations of the group is a **representation** of the group.

For the 45° kaleidoscope (or the C_{4v} group), the set of transformation matrices for x, y, and z for each operation form a representation. These may be written (letting the z axis be the vertical axis) as

$$\begin{bmatrix} 1 & 0 & 0 \\ 0 & 1 & 0 \\ 0 & 0 & 1 \end{bmatrix} \begin{bmatrix} 1 & 0 & 0 \\ 0 & -1 & 0 \\ 0 & 0 & 1 \end{bmatrix} \begin{bmatrix} -1 & 0 & 0 \\ 0 & 1 & 0 \\ 0 & 0 & 1 \end{bmatrix} \begin{bmatrix} 0 & 1 & 0 \\ 1 & 0 & 0 \\ 0 & 0 & 1 \end{bmatrix} \begin{bmatrix} 0 & -1 & 0 \\ -1 & 0 & 0 \\ 0 & 0 & 1 \end{bmatrix} \begin{bmatrix} 0 & 1 & 0 \\ -1 & 0 & 0 \\ 0 & 0 & 1 \end{bmatrix} \begin{bmatrix} 0 & -1 & 0 \\ 1 & 0 & 0 \\ 0 & 0 & 1 \end{bmatrix} \begin{bmatrix} -1 & 0 & 0 \\ 0 & -1 & 0 \\ 0 & 0 & 1 \end{bmatrix}$$

$$\quad E \qquad\quad \sigma_1 \qquad\quad \sigma_1' \qquad\quad \sigma_2 \qquad\quad \sigma_2' \qquad\quad C_4 \qquad C_4^3(C_4^{-1}) \qquad C_2(C_4^2)$$

Note that transformations of the z coordinate are completely independent of the x and y coordinates, but that x and y are interchanged by some operations. Accordingly, each of the above representations can be broken into two independent representations consisting of the transformation matrices for x and y (two-dimensional matrices) and the transformation matrices for z (one-dimensional matrices):

$$x, y \quad \begin{bmatrix} 1 & 0 \\ 0 & 1 \end{bmatrix} \begin{bmatrix} 1 & 0 \\ 0 & -1 \end{bmatrix} \begin{bmatrix} -1 & 0 \\ 0 & 1 \end{bmatrix} \begin{bmatrix} 0 & 1 \\ 1 & 0 \end{bmatrix} \begin{bmatrix} 0 & -1 \\ -1 & 0 \end{bmatrix} \begin{bmatrix} 0 & 1 \\ -1 & 0 \end{bmatrix} \begin{bmatrix} 0 & -1 \\ 1 & 0 \end{bmatrix} \begin{bmatrix} -1 & 0 \\ 0 & -1 \end{bmatrix}$$

$$\qquad\quad E \qquad\quad \sigma_1 \qquad\quad \sigma_1' \qquad\quad \sigma_2 \qquad\quad \sigma_2' \qquad\quad C_4 \qquad\quad C_4^3 \qquad\quad C_2$$

$$z \quad\ [1] \qquad [1] \qquad [1] \qquad [1] \qquad [1] \qquad [1] \qquad [1] \qquad [1]$$

As will be seen later, these matrices cannot be reduced further and hence are termed **irreducible representations**.

In general, spotting irreducible representations is not as easy as in the example above. We will not discuss here the techniques for bringing a reducible matrix into block diagonal

form $\begin{array}{|c c|} \hline & 0 \\ \hline & 0 \\ \hline 0\ 0 & \\ \hline \end{array}$ where only zero elements appear outside the blocks; however, when each

representation appears in the same block diagonal form, each set of blocks with similar row and column indices forms a representation.

EXAMPLE 3.16:   Show that the transformation matrices for x and y given above have the same multiplication table as Table 3.1 (Section 3.3).

Solution: C_2^2 and σ^2 give E.

$$\sigma_1^2 = \begin{bmatrix} 1 & 0 \\ 0 & -1 \end{bmatrix}\begin{bmatrix} 1 & 0 \\ 0 & -1 \end{bmatrix} = \begin{bmatrix} 1 & 0 \\ 0 & 1 \end{bmatrix} = E$$

$$\sigma_2^2 = \begin{bmatrix} 0 & 1 \\ 1 & 0 \end{bmatrix}\begin{bmatrix} 0 & 1 \\ 1 & 0 \end{bmatrix} = \begin{bmatrix} 1 & 0 \\ 0 & 1 \end{bmatrix} = E$$

$$C_2^2 = \begin{bmatrix} -1 & 0 \\ 0 & -1 \end{bmatrix}\begin{bmatrix} -1 & 0 \\ 0 & -1 \end{bmatrix} = \begin{bmatrix} 1 & 0 \\ 0 & 1 \end{bmatrix} = E$$

The products RC_4 (giving the column under C_4) are

$$EC_4 = C_4E = C_4$$

$$C_4^2 = \begin{bmatrix} 0 & 1 \\ -1 & 0 \end{bmatrix}\begin{bmatrix} 0 & 1 \\ -1 & 0 \end{bmatrix} = \begin{bmatrix} -1 & 0 \\ 0 & -1 \end{bmatrix} = C_2$$

$$C_4^2 C_4 = \begin{bmatrix} -1 & 0 \\ 0 & -1 \end{bmatrix}\begin{bmatrix} 0 & 1 \\ -1 & 0 \end{bmatrix} = \begin{bmatrix} 0 & -1 \\ 1 & 0 \end{bmatrix} = C_4^3(C_4^{-1})$$

$$C_4^3 C_4 = \begin{bmatrix} 0 & -1 \\ 1 & 0 \end{bmatrix}\begin{bmatrix} 0 & 1 \\ -1 & 0 \end{bmatrix} = \begin{bmatrix} 1 & 0 \\ 0 & 1 \end{bmatrix} = E$$

$$\sigma_1 C_4 = \begin{bmatrix} 1 & 0 \\ 0 & -1 \end{bmatrix}\begin{bmatrix} 0 & 1 \\ -1 & 0 \end{bmatrix} = \begin{bmatrix} 0 & 1 \\ 1 & 0 \end{bmatrix} = \sigma_2$$

$$\sigma_2 C_4 = \begin{bmatrix} 0 & 1 \\ 1 & 0 \end{bmatrix}\begin{bmatrix} 0 & 1 \\ -1 & 0 \end{bmatrix} = \begin{bmatrix} -1 & 0 \\ 0 & 1 \end{bmatrix} = \sigma_1'$$

$$\sigma_1' C_4 = \begin{bmatrix} -1 & 0 \\ 0 & 1 \end{bmatrix}\begin{bmatrix} 0 & 1 \\ -1 & 0 \end{bmatrix} = \begin{bmatrix} 0 & -1 \\ -1 & 0 \end{bmatrix} = \sigma_2'$$

$$\sigma_2' C_4 = \begin{bmatrix} 0 & -1 \\ -1 & 0 \end{bmatrix}\begin{bmatrix} 0 & 1 \\ -1 & 0 \end{bmatrix} = \begin{bmatrix} 1 & 0 \\ 0 & -1 \end{bmatrix} = \sigma_1$$

These commute

Take the other products and check the results against Table 3.1 as practice in multiplying matrices and in reading multiplication tables.

The sum of the diagonal terms of a matrix is called the **trace** (German **spur**) of the matrix. For a representation, the trace of each matrix is called a **character.** Reducible representations furnish **compound characters,** which are sums of the **simple characters** of the irreducible representations contained in the reducible representation. An irreducible representation giving a unique set of characters is said to belong to a particular **symmetry species.** We now give (without proof) and use two important theorems concerning representations of groups and their characters.

The number of symmetry species, or irreducible representations, for any group is the same as the number of classes of operations in the group. For a given symmetry species, the character is the same for all operations in the same class. The character under the identity operation is equal to the dimension of the matrix representation: 1 for [1], 2 for $\begin{bmatrix} 1 & 0 \\ 0 & 1 \end{bmatrix}$, and so on.

If the character under the identity operation is squared, and the sum of the squares is taken over all symmetry species, Γ, the result is equal to the order of the group

$$\sum_{\Gamma}[\chi(E)]^2 = h \qquad (3.4)$$

More generally, for any group operation

$$n_g \sum_{\Gamma} \chi^2 = h \qquad (3.5)$$

where χ is the character under a given operation, n_g is the number of operations in the class, $\sum_{\Gamma}$ indicates the sum is taken over all symmetry species in the group, and h is the order of the group.

The criterion that a set of character be simple (that is, that the representation be irreducible) is that for each irreducible representation Γ we have

$$\sum_{g} n_g\chi^2 = h \qquad (3.6)$$

That is, the sum of each character squared and multiplied by the number of operations in the particular class must be equal to the order of the group. Here the summation is over the g classes of the group.

Returning to the 45° kaleidoscope group, we find that because there are five classes of operations (E, $2C_4$, C_2, $2\sigma_1$, and $2\sigma_2$), there must be five symmetry species. To satisfy Equation (3.4), four of these must be one-dimensional representations and the remaining one must be two-dimensional.

$$1^2 + 1^2 + 1^2 + 1^2 + 2^2 = h = 8$$

By applying Equation (3.5) we see that the characters obtained from the x, y transformation matrices satisfy the criterion for a set of simple characters,

	E	$2C_4$	C_2	$2\sigma_1$	$2\sigma_2$
(x, y)	2	0	-2	0	0
	2^2 +	$2(0)^2$ +	$(-2)^2$ +	$2(0)^2$ +	$2(0)^2$ = $h = 8$

and thus the representation is irreducible. This symmetry species is given the label "E" (not the identity), which denotes a two-dimensional representation. Because the x, y transformation matrices yield the E symmetry species, x and y are said to form a **basis** for this symmetry species or simply to transform as the E species (or E representation). The characters for z are all $+1$. This is the totally symmetric symmetry species, which is labeled A_1. You can show that this set of characters also obeys Equation (3.5).

▶ 3.5.3   *Orthogonality of Representations*

Another important requirement is that the irreducible representations must be orthogonal. The corresponding elements (those in the same row and column) of the matrices for all symmetry operations comprise a vector. Also the characters for all symmetry operations of a representation comprise a vector in h-dimensional space. Two vectors $\mathbf{A}$ and $\mathbf{B}$ are orthogonal if the projection of one on the other is zero, or if $\mathbf{A} \cdot \mathbf{B}$ (the **scalar** or **dot** product) is zero. The scalar product for the vectors $\mathbf{A}$ and $\mathbf{B}$ in two dimensions is

$$\mathbf{A} \cdot \mathbf{B} = AB \cos \Theta \tag{3.7}$$

where Θ is the angle between the vectors and A and B are their lengths. The vectors are orthogonal if $\mathbf{A} \cdot \mathbf{B} = 0$; this is obvious for two dimensions because $\cos 90° = 0$. The irreducible representations of a group comprise a set of mutually orthogonal vectors that must satisfy the requirement that $\Gamma_i \cdot \Gamma_j = 0$ (where Γ_i and Γ_j are two different representations). If we take as the components of each vector (representation) the character (χ) for each symmetry operation (R), orthogonality requires that

$$\sum_R \chi_i(R)\chi_j(R) = 0 \quad \text{if } i \neq j \tag{3.8}$$

Let us check for the orthogonality of the representations A_1 and E for $\mathbf{C}_{4v}$:

$\mathbf{C}_{4v}$	E	$2C_4$	C_2	$2\sigma_1$	$2\sigma_2$
A_1	1	1	1	1	1
E	2	0	-2	0	0

$A_1 \cdot E \quad = \quad 2 \quad + \quad 2(0) \quad -2 \quad + \quad 2(0) \quad + \quad 2(0) \quad = 0 \quad \therefore \text{ orthogonal}$

We have two of the five irreducible representations. We know that a mirror operation carried out twice must give the identity: $\sigma^2 = E$, and $C_2^2 = E$. Because the matrix of a one-dimensional representation for the identity is 1, the matrix (and character) for each σ and for C_2 must be $\sqrt{1}$, or ± 1. The matrices (and the characters) for one-dimensional representations for C_4 also must be ± 1; and because $C_4^2 = C_2$, the matrices (and the characters) for C_2 are actually $+1$ $(\pm 1)^2$. The representations must be different, and we know that the characters for E and C_2 are all $+1$, so the others will have different combintations of ± 1 for C_4, σ_1, and σ_2. Because of the requirement for orthogonality [Equation (3.8)], only one of these can be $+1$, so one of the three additional representations has $+1$ for C_4, one has $+1$ for σ_1, and the third has $+1$ for σ_2. Checking the orthogonality of one of these with A_1, we obtain

$\mathbf{C}_{4v}$	E	$2C_4$	C_2	$2\sigma_1$	$2\sigma_2$
A_1	1	1	1	1	1
Γ_i	1	1	1	-1	-1

$A_1 \cdot \Gamma_i \quad = \quad 1 \quad + \quad 2(1) \quad + \quad 1 \quad + \quad 2(-1) \quad + \quad 2(-1) \quad = 0 \quad \therefore \text{ orthogonal}$

The full character table for the kaleidoscope group, known as the **C₄ᵥ** group, is given in Table 3.2. Looking at all the representations, we see that the characters can differ for σ_1 and σ_2 (they are in different classes) but that both C_4 (C_4^1 and C_4^3) are in the same class and have the same characters. The character table is generated easily for the simple case of **C₄ᵥ**. The generation of more complex character tables is beyond the scope of this book. Appendix C lists character tables for chemically important point groups.

► 3.5.4 Mulliken Labels for Representations

The labels of the symmetry species, which follow a set of rules devised by Mulliken, tell us about the dimensionality of the representation, as well as some of its symmetry behavior. One-dimensional representations are denoted A or B, two-dimensional by E, three-dimensional by T, four-dimensional by G, and five-dimensional by H. The totally symmetric representation (all characters $+1$) is denoted always by A (or by A_1, A_{1g}, or A'). Every group contains the totally symmetric representation. If more than one symmetry species is symmetric with respect to the highest-order rotational axis, the additional species are labeled A_2, A_3, and so on. Such a species will be antisymmetric (character -1) with respect to some other symmetry element(s). Symmetric behavior means that the character is positive, and antisymmetric behavior means that the character is negative under the specified operation. A one-dimensional species that is antisymmetric under the highest-order rotational axis is labeled B, or B_1, B_2, and so on, if there is more than one. For groups with a center of inversion, a subscript g is used to indicate symmetric (German *gerade*) behavior and subscript u is used to indicate antisymmetric (German *ungerade*) behavior under the inversion operation. For **C**$_{nh}$ and **D**$_{nh}$ groups that lack an inversion center (i.e., for odd values of n), the horizontal mirror plane is used to assign $'$ or $''$ superscripts—the $'$ signifying symmetric behavior and the $''$ signifying antisymmetric behavior under the σ_h operation.

Table 3.2   Character table for C₄ᵥ (also the 45° kaleidoscope group)

C₄ᵥ	E	$2C_4$	C_2	$2\sigma_v$	$2\sigma_d$		
A_1	1	1	1	1	1	z	$x^2 + y^2, z^2$
A_2	1	1	1	-1	-1	R_z	
B_1	1	-1	1	1	-1		$x^2 - y^2$
B_2	1	-1	1	-1	1		xy
E	2	0	-2	0	0	$(x, y); (R_x, R_y)$	(xz, yz)

EXAMPLE 3.17:   Verify the orthogonality of A_2 and B_1 for $\mathbf{C}_{4v}$, and show that B_1 obeys Equations (3.6).

Solution:

$\mathbf{C}_{4v}$	E	$2C_4$	C_2	$2\sigma_1$	$2\sigma_2$
A_2	1	1	1	-1	-1
B_1	1	-1	1	1	-1

$$A_2 \cdot B_1 \;=\; (1)(1) \;+\; 2(1)(-1) \;+\; (1)(1) \;+\; 2(-1)(1) \;+\; 2(-1)(-1) \;=\; 0$$

Now verify the orthogonality of the other representations for yourself. Applying Equation (3.6) to B_1, we obtain

$$(1)^2 + 2(-1)^2 + (1)^2 + 2(1)^2 + 2(-1)^2 = 8 = h$$

EXAMPLE 3.18:   Derive the matrices for the transformation of x, y, z for each class of operations of $\mathbf{D}_3$. Determine the characters and derive the character table.

Solution: There are three classes of operations for $\mathbf{D}_3$: E, $2C_3$, and $3C_2$. The matrices for the transformation of x, y, z are obtained using Matrix (3.3).

$$
\begin{array}{ccc}
E & C_3 & (C_2)_x \\[4pt]
\begin{bmatrix} 1 & 0 & 0 \\ 0 & 1 & 0 \\ 0 & 0 & 1 \end{bmatrix} &
\begin{bmatrix} -1/2 & \sqrt{3}/2 & 0 \\ -\sqrt{3}/2 & -1/2 & 0 \\ 0 & 0 & 1 \end{bmatrix} &
\begin{bmatrix} 1 & 0 & 0 \\ 0 & -1 & 0 \\ 0 & 0 & -1 \end{bmatrix} \\[10pt]
3 & 0 & -1
\end{array}
$$

We choose the C_2 axis along x.

Characters

Thus the C_3 operation on x or y gives a function of x and y—or x and y are "mixed" by the C_3 operation—but z is independent. This reducible three-dimensional representation can be reduced using the blocks containing nonzero elements as for $\mathbf{C}_{4v}$.

$$
\begin{array}{cccc}
& E & C_3 & C_2 \\[4pt]
x,\,y & \begin{bmatrix} 1 & 0 \\ 0 & 1 \end{bmatrix} &
\begin{bmatrix} -1/2 & \sqrt{3}/2 \\ -\sqrt{3}/2 & -1/2 \end{bmatrix} &
\begin{bmatrix} 1 & 0 \\ 0 & -1 \end{bmatrix} \\[10pt]
& 2 & -1 & 0 & \text{Characters} \\[6pt]
z & [1] & [1] & [-1] \\[4pt]
& 1 & 1 & -1 & \text{Characters}
\end{array}
$$

Because there are three classes of operations, there are three symmetry species. We have two of them (A_2 and E) and know that there must be a totally symmetric representation, A_1. The character table is

$\mathbf{D}_3$	E	$2C_3$	$3C_2$		
A_1	1	1	1		$x^2 + y^2,\ z^2$
A_2	1	1	-1	$z,\ R_z$	
E	2	-1	0	$(x,\ y);\ (R_x,\ R_y)$	$(x^2 - y^2,\ xy);\ (xz,\ yz)$

Check that these are irreducible representations [Equation (3.6)] and that they are orthogonal [Equation (3.8)].

▶ *3.5.5   Symmetry Species for Functions and Orbitals*

We already have seen how the transformation matrices for x and y give an E symmetry species, and z transforms as the A_1 species in $\mathbf{C}_{4v}$ (and $\mathbf{D}_3$). To see how xy or $x^2 - y^2$ functions transform in $\mathbf{C}_{4v}$, we must first find the behavior of x and y under the group operations and then construct the behavior of the desired functions, as illustrated below.

		E	C_4	C_4^3	C_2	σ_1	σ_1'	σ_2	σ_2'
x	$\rightarrow$	x	y	$-y$	$-x$	x	$-x$	y	$-y$
y	$\rightarrow$	y	$-x$	x	$-y$	$-y$	y	x	$-x$
xy	$\rightarrow$	xy	$-xy$	$-xy$	xy	$-xy$	$-xy$	xy	xy
$x^2 - y^2$	$\rightarrow$	$x^2 - y^2$	$y^2 - x^2$	$y^2 - x^2$	$x^2 - y^2$	$x^2 - y^2$	$x^2 - y^2$	$y^2 - x^2$	$y^2 - x^2$

We find the way xy transforms by taking the product of the transformation of x and y under each operation. For $x^2 - y^2$, we square the element for the transformation of x under each operation, do the same for y, and take the difference. For xy and $x^2 - y^2$, the result of performing each operation is equivalent to multiplying by 1 or -1. Tabulating these characters for each operation gives

$\mathbf{C}_{4v}$	E	$2C_4$	C_2	$2\sigma_1$	$2\sigma_2$	
	1	-1	1	-1	1	for xy
	1	-1	1	1	-1	for $x^2 - y^2$

Examining the $\mathbf{C}_{4v}$ character table (Table 3.2) shows us that xy transforms as the B_2 symmetry species and $x^2 - y^2$ as the B_1 species. Character tables used by chemists, including those given in Appendix C, list polynomials, or combinations of polynomials, that transform as the various symmetry species. These polynomials listed are also those used for the subscripts of various atomic orbitals (p_x, p_y, d_{xy}, $d_{x^2-y^2}$, etc.), and these orbitals also transform as the various symmetry species indicated. Thus we can identify the symmetry species of all of the p and d orbitals from the right side of character tables. The spherical s orbital is always totally symmetric (A_1, A_{1g}, etc.).

The p orbitals transform as x, y, and z, and the representations of the d_{xy}, d_{xz}, and d_{yz} orbitals are identified by those labels (xy, etc.) in character tables. Orbitals belonging to a two- or three-dimensional representation are grouped in parentheses: (xy, yz) for E and (x, y, z) for T. These orbitals are twofold and threefold degenerate, respectively, for the particular point groups. The d_{z^2} orbital is identified as z^2, or occasionally as the full polynomial $2z^2 - x^2 - y^2$. The $d_{x^2-y^2}$ orbital is identified as $x^2 - y^2$, except when x^2, y^2, and z^2 independently belong to A_1 or A_g—as for $\mathbf{C}_{2v}$, $\mathbf{D}_2$, and so on—in which case both d_{z^2} and $d_{x^2-y^2}$ belong to A_1 or A_g. The symbols R_x, R_y, and R_z beside the character tables refer to rotations about x, y, and z, respectively. The corresponding representations are used in applying spectroscopic selection rules (see page 444).

EXAMPLE 3.19: From the character tables in Appendix C, identify the symmetry species for p_x, p_z, d_{z^2}, $d_{x^2 - y^2}$, and d_{xy} for $\mathbf{D_3}$, $\mathbf{C_{4v}}$, $\mathbf{D_{3h}}$, $\mathbf{D_{4h}}$, and $\mathbf{O_h}$.

Solution:

$$\mathbf{D_3} \quad p_z \, (A_2), \, (p_x, p_y) \, (E), \, d_{z^2} \, (A_1), \, (d_{x^2-y^2}, d_{xy}) \, (E).$$

$$\mathbf{C_{4v}} \quad p_z \, (A_1), \, (p_x, p_y) \, (E), \, d_{z^2} \, (A_1), \, d_{x^2-y^2} \, (B_1), \, d_{xy} \, (B_2)$$

$$\mathbf{D_{3h}} \quad p_z \, (A_2''), \, (p_x, p_y) \, (E'), \, d_{z^2} \, (A_1'), \, (d_{x^2-y^2}, d_{xy}) \, (E')$$

$$\mathbf{D_{4h}} \quad p_z \, (A_{2u}), \, (p_x, p_y) \, (E_u), \, d_{z^2} \, (A_{1g}), \, d_{x^2-y^2} \, (B_{1g}), \, d_{xy} \, (B_{2g})$$

$$\mathbf{O_h} \quad (p_x, p_y, p_z) \, (T_{1u}), \, (d_{z^2}, d_{x^2-y^2}) \, (E_g), \, (d_{xz}, d_{yz}, d_{xy}) \, (T_{2g})$$

We note that p_x alone does not transform as a symmetry species in any of the groups listed, but it must be combined with one or more of the remaining p orbitals to give a function that does. This requires that p_x and p_y be degenerate in molecules belonging to the above point groups. In $\mathbf{O_h}$ the orbitals p_x, p_y and p_z are threefold degenerate.

Cyclic Groups

A $\mathbf{C_n}$ (cyclic) group has n elements—C_n^1, C_n^2, and so on—each of which comprises a class by itself. Accordingly, a $\mathbf{C_n}$ group has n one-dimensional representations, for which the matrices, because they are one-dimensional, are the same as the characters. All groups have the totally symmetric representation, A, with all characters $= 1$. For $\mathbf{C_n}$ groups with n even, there is a B representation with the characters of $+1$ or -1, which alternate for E, C_n^1, C_n^2, and so on. All other $\mathbf{C_n}$ groups (except $\mathbf{C_1}$, which has only the A representation; and $\mathbf{C_2}$, which has only A and B representations) have one-dimensional representations that occur in pairs and carry the Mulliken label E. Their characters, except for the identity operation, are imaginary or complex numbers. They can be used as such, but for applications to physical problems they can be combined to give real numbers. The complex characters are expressed in terms of ε (defined for the particular group beside each character table), where

$$\varepsilon = e^{(2\pi i/n)} \tag{3.9}$$

The characters for rotations are evaluated by the relationships

$$e^{i\theta} = \cos \theta + i \sin \theta \tag{3.10}$$

giving for rotations by $\mathbf{C_n}$,

$$\varepsilon_n = \cos 2\pi/n + i \sin 2\pi/n \tag{3.11}$$

and for rotations by $\mathbf{C_n^m}$,

$$\varepsilon_n^m = \cos 2\pi m/n + i \sin 2\pi m/n \tag{3.12}$$

For C_4, $\varepsilon = \cos 90° + i \sin 90° = i$. This is the character for one of the pair of one-dimensional representations; because the paired representations are complex conjugates, the character for C_4 is $-i$ for the other representation. The $\mathbf{C_4}$ character table is given in Table 3.3. For the $\mathbf{C_4}$ group the orbitals s, p_z, and d_{z^2}, belong to A, $d_{x^2-y^2}$ and d_{xy} belong to B (they are not degenerate), and (p_x, p_y) and (d_{xz}, d_{yz}) are doubly degenerate pairs belonging to E.

Table 3.3   Character table for C$_4$

C$_4$	E	C_4	C_2	C_4^3		
A	1	1	1	1	z, R_z	$x^2 + y^2, z^2$
B	1	-1	1	-1		$x^2 - y^2, xy$
E	$\begin{cases} 1 \\ 1 \end{cases}$	$\begin{matrix} i \\ -i \end{matrix}$	$\begin{matrix} -1 \\ -1 \end{matrix}$	$\begin{matrix} -i \\ i \end{matrix}$	$(x, y); (R_x, R_y)$	(xz, yz)

EXAMPLE 3.20:   The E representation for C$_4$ occurs as a pair of one-dimensional representations, with characters 1 and 1 for the identity and i and $-i$ for C_4. From *these values,* derive the characters for C_4^2 and C_4^3.

Solution: Because the characters are the one-dimensional matrices, you can get the characters for new operations as products of other operations.

$$C_4^2 = (i)^2 = -1, \qquad C_4^3 = C_4 C_4^2 = i(-1) = -i$$

and

$$(-i)^2 = -1, \qquad -i(-1) = i$$

For all cyclic groups, if the characters are evaluated for one rotation, the others can be obtained as products. For C$_n$ groups, all operations commute.

▶ *3.5.6   Reducing Reducible Representations*

In the example of the C$_{4v}$ group (the 45° kaleidoscope group—Section 3.3), a reducible representation was reduced, by inspection, into its irreducible representations. Such reductions are carried out more readily in the general case by using the set of characters for the reducible representation and a character table for the group. To find the number of times, n_Γ, a particular irreducible representation is contained in a reducible representation, we use (without proof) the formula

$$n_\Gamma = \frac{1}{h} \sum_g n_g \chi_R \chi_\Gamma \tag{3.13}$$

where h is the order of the group, n_g the number of operations in class g, χ_R is the character of the reducible representation, χ_Γ is the character of the irreducible representation under the operations of class g, and the summation is taken over all classes.

EXAMPLE 3.21:   What symmetry species (irreducible representations) are contained in the representation having the following characters in the T$_d$ group?

T$_d$	E	C_3	C_2	S_4	σ_d
	9	0	1	1	1

Solution: Appendix C supplies the character table for the $\mathbf{T}_d$ group. For each irreducible representation, multiply the number of operations in each class by the corresponding characters for that irreducible representation *and* the reducible representation. These products are summed and divided by 24 (h, the total number of operations), as follows:

$$
\begin{array}{ccccc}
 & E & 8C_3 & 3C_2 & 6S_4 & 6\sigma_d \\
n_{A_1} = 1/24[1\cdot1\cdot9 & + 8\cdot0\cdot1 & + 3\cdot1\cdot1 & + 6\cdot1\cdot1 & + 6\cdot1\cdot1] & = 1\ A_1 \\
n_{A_2} = 1/24[1\cdot1\cdot9 & + 8\cdot0\cdot1 & + 3\cdot1\cdot1 & + 6\cdot1\cdot-1 & + 6\cdot1\cdot-1] & = 0\ A_2 \\
n_E = 1/24[1\cdot2\cdot9 & + 8\cdot0\cdot-1 & + 3\cdot1\cdot2 & + 6\cdot1\cdot0 & + 6\cdot1\cdot0] & = 1\ E \\
n_{T_1} = 1/24[1\cdot3\cdot9 & + 8\cdot0\cdot0 & + 3\cdot1\cdot-1 & + 6\cdot1\cdot1 & + 6\cdot1\cdot-1] & = 1\ T_1 \\
n_{T_2} = 1/24[1\cdot3\cdot9 & + 8\cdot0\cdot0 & + 3\cdot1\cdot-1 & + 6\cdot1\cdot-1 & + 6\cdot1\cdot1] & = 1\ T_2 \\
\end{array}
$$

The reduction must give either zero or a whole number for each irreducible representation. As a check, the dimension of the reducible representation (character for E) must be the sum of the dimensions of the irreducible representations found. Here,

$$A_1 + E + T_1 + T_2$$
$$1 + 2 + 3 + 3 = 9$$

Because large or small characters for some operations of the reducible representation suggest the irreducible representations contained, you might be able to reduce it by inspection, summing the characters of the irreducible representations for each class of operations as a check. In applying the formula, first examine those representations you think are contained, stopping when you can check the sum of the dimensions or representations involved.

3.6   CHEMICAL APPLICATIONS OF SYMMETRY

Is all of this effort worthwhile? Symmetry is extremely important in chemistry and group theory provides the framework for making full use of symmetry. The Greek labels for σ, π, and δ MOs for diatomic molecules were derived from those of the AOs having the same symmetry. The symmetries of some AOs for the $\mathbf{D}_{4h}$ point group are indicated in Figure 3.18a. For AOs to combine forming MOs they must belong to the *same symmetry species*. Because two reacting molecules must come together, break some bonds, and/or form new bonds, it is evident that symmetry restrictions apply to reactions that occur through continuous interaction from initiation to completion. The interacting orbitals of two reacting molecules must belong to the *same symmetry species*. (The Symmetry Rules or Woodward-Hoffmann Rules for chemical reactions apply here.)

MOs are characterized by their symmetry species (representations) which also apply to the mathematical expressions for their wavefunctions. For large molecules with many electrons, high-order secular equations are involved. Symmetry is used to factor these equations into lower-order equations, greatly simplifying their solution.

An animated representation of a vibrating molecule looks as though the motion is random and erratic. Actually the motion can be resolved into **vibrational modes** that must be

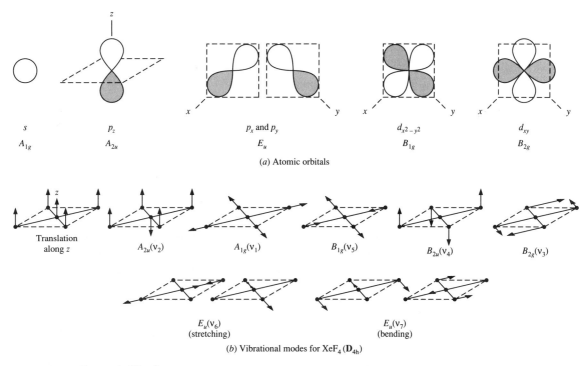

(a) Atomic orbitals

(b) Vibrational modes for XeF₄ (D_{4h})

Figure 3.18 Symmetry species for (a) atomic orbitals and (b) vibrational modes for D_{4h}.

consistent with the symmetry of the molecule itself. That is, these vibrational modes belong to symmetry species for the point group. Vibrational absorption (infrared and Raman spectroscopy) involves transitions from the vibrational ground state to any of various vibrationally excited states. The ground state has the full symmetry of the point group, that is, it is A_{1g} for D_{4h}.

We can represent molecular vibrations by vectors showing the direction of motion of the atoms. If all atoms move along the z direction, then it is translation, not a vibration. The whole molecule moves along z. XeF_4, with D_{4h} symmetry, is used for illustrative purposes in Figure 3.18b. If the Xe atom moves along the $+z$ direction while all F atoms move in the opposite direction so the center of gravity is unmoved, this is a vibration. Performing the D_{4h} operations on this set of vectors shows that this vibration has A_{2u} symmetry. The symmetrical stretching of all four Xe—F bonds (the F atoms move outward and then inward) is totally symmetric, A_{1g}. If two diagonal F atoms move out while the other two move in, the symmetry is B_{1g}. If two diagonal F atoms move upward along z while the other two F move downward this bending vibration is B_{2u}. Another bending vibration involving squeezing adjacent F atoms together on opposite sides of the square is B_{2g}. There is a pair of E_u stretching vibrations and a pair of E_u bending vibrations. A molecule with n atoms has $3n$ degrees of freedom. Subtracting 3 degrees of freedom for translation and 3 degrees of freedom for rotation (2 for linear molecules), *there are $3n - 6$ vibrational degrees of freedom for nonlinear molecules ($3n - 5$ for linear molecules).*

Selection rules determine which energy-absorbing processes are observable for various kinds of absorption spectroscopy (see Section 10.2.2). These are remarkably similar in involving the wavefunction for the ground state, that for the excited state, and an operator

$$\int \Psi_{\text{ground}} \,(\text{operator}) \,\Psi_{\text{excited}} \, d\tau \qquad (3.14)$$

Details of the involvement of this integral depends on the kind of absorption involved. Here we just focus on the symmetry species for these wavefuctions and the operator. The operators are different for various kinds of spectroscopy. The symmetry of the operator is determined by the nature (symmetry) of the displacement of electrons or atoms. We can determine which transitions are "allowed" or "forbidden" by simple examination of the products of the symmetries involved.

An integral of an odd function (u) vanishes when integrated over all space—the positive and negative portions cancel. Thus if the operator is even (g), the integral is nonzero (and the transition is allowed) only if the product of the symmetry species of Ψ_{ground} and Ψ_{excited} are even (g). For odd operators, transitions are allowed only if the product of the symmetry species of Ψ_{ground} and Ψ_{excited} are also odd ($g \times g = g$ and $u \times u = g$, but $g \times u = u$).[2]

Vibrational spectroscopy involves transitions from the ground state to excited vibrations. Absorptions appear in infrared or Raman spectra. The operators differ; the operator for infrared corresponds to translation along x, y, and z. The operator is odd (u) for a centrosymmetric molecule as for XeF_4. For Raman spectroscopy the operator is even (g). Thus we observe only absorption involving vibrations belonging to odd representations in infrared spectra and only vibrations belonging to even representations in Raman spectra. Infrared-active vibrations transform like the Cartesian coordinates x, y, and z and must further result in a change in the molecular dipole moment. The $\mathbf{D}_{4h}$ character table indicates that (x, y) transform like E_u while z transforms like A_{2u}. Hence, the infrared–active vibrations are expected to be A_{2u} and E_u. Note that the combinations of stretches and contractions mirror the signs of lobes of obitals with the same symmetry.

The infrared and Raman spectra[3] for XeF_4 agree well with expectations. The vibrational modes a_{2u} (ν_2 at 291 cm^{-1}) and both e_u (ν_6 at 586 cm^{-1} and ν_7 at 123 cm^{-1}) are observed in the infrared spectrum, as expected because these are the irreducible representations for coordinates x, y, and z. The vibrational modes a_{1g} (ν_1 at 543 cm^{-1}), b_{2g} (ν_3 at 235 cm^{-1}), and b_{1g} (ν_4 at 502 cm^{-1}) are observed in the Raman spectrum. The operator for

---

[2] Selection rules are more restrictive than that just based on parity (odd or even). The expression in the integral must be *totally symmetric* (A_{1g} for $\mathbf{D}_{4h}$). This requires that the product of the two wavefunctions must belong to the *same* representation as the operator (A_{2u} and E_u for infrared under $\mathbf{D}_{4h}$). See Section 10.2.2 for more consideration of selection rules for electronic absorption.

[3] H. H. Claassen, C. L. Chernick, and J. G. Malm, *J. Am. Chem. Soc.* **1963,** *85,* 1927. In this paper the subscripts for b representations are reversed from those used here (b_{1g} for b_{2g}, b_{2g} for b_{1g}, and b_{1u} for b_{2u}). The authors used reversed conventions for C_2' and C_2'', and for σ_v and σ_d. For $\mathbf{D}_{4h}$ we can check the convention used by examining the function $x^2 - y^2$ for $d_{x^2-y^2}$. C_2', here chosen as along the x and y axes, gives character $+1$ for B_{1g}.

Raman spectroscopy transforms as a polarizability tensor. The representations for this operator correspond to quadratic functions of x, y, and z. In character tables, the representations correspond to xz, yz, $x^2 - y^2$, and so on. Thus a_{1g}, b_{1g}, and b_{2g} are expected to be allowed in the Raman spectrum. The b_{1u} mode is inactive in infrared and Raman spectra. It was not observed as a fundamental. (It was questionable as an overtone or combination band.)

3.7   SUMMARY OF TERMINOLOGY AND NOTATION FROM SYMMETRY AND GROUP THEORY

► 3.7.1   *Symmetry Operations*

Identity   (E): Leaving an object unchanged—equivalent to rotation by 360°. This operation is required by group theory, since each operation must have an inverse which undoes it and because the product of two operations (performing them in succession) must be equivalent to another operation of the group.

Rotation   (C_n): Rotation (clockwise) by $360°/n$ about an (n-fold) axis. Also called a *proper* rotation. Performing a C_n operation n times is equivalent to E. (See Figure 3.3.)

$$C_n^n = E \qquad C_n^{n/2} = C_2 \text{ (even } n) \qquad C_n^{-1} \text{ (anticlockwise)} = C_n^{n-1}$$

Reflection   (σ): Reflection through a mirror plane (a plane of symmetry), (See Figure 3.1). A mirror plane perpendicular to C_n is called a **horizontal plane,** σ_h. Planes containing C_n are called **vertical planes,** σ_v, or, if they bisect angles between symmetry axes, **dihedral planes,** σ_d.

$$\sigma^2 = E$$

Improper Rotation   (S_n): Rotation by $360°/n$ followed by reflection through a plane perpendicular to the axis of rotation. Also called a **rotation-reflection operation.** (See Figure 3.4.)

$$S_n = \sigma C_n \quad (C_n \text{ then } \sigma)$$
$$S_1 = \sigma C_1 = \sigma, \text{ since } C_1 = E$$
$$S_n^2 = C_n^2, \quad \text{ since } \sigma^2 = E$$

Inversion   (i): Reflection through a center of symmetry. (See Figure 3.2.)

$$i^2 = E \qquad i C_2 = \sigma$$
$$i = \sigma C_2 = S_2$$

Symmetry elements and symmetry operations are set in *italics* in this book. Labels for point groups are in **boldface.**

▶ 3.7.2 *Symmetry Point Groups*

The presence of some combinations of symmetry elements requires the existence of others. For example, if a molecule has a C_3 axis and σ_v, there must be three σ_v. The natural groupings of symmetry operations which can be performed on account of the existence of the symmetry elements are called **symmetry point groups.** The point groups and the symmetry operations that generate them are given below:

C₁—only E, equivalent to a onefold axis.

Cₛ—only σ (all groups contain the identity E) (see Figure 3.8)

Cᵢ—only i

Axial Groups

Cₙ—C_n axis (see Figure 3.7)

Cₙᵥ—C_n and σ_v through C_n (see Figure 3.9)

Cₙₕ—C_n and a horizontal plane (σ_h) $\perp$ to C_n (see Figure 3.8)

Dₙ—C_n and C_2 axes $\perp$ to C_n (see Figure 3.10)

Dₙₕ—**Dₙ** elements plus σ_h; for even n, there is an inversion center (see Figure 3.11)

Dₙd—**Dₙ** elements plus dihedral planes (σ_d) bisecting the angles between C_2 axes; for odd n, there is an inversion center (see Figure 3.12)

Sₙ—S_n axis (label used for even n)

$$\left. \begin{array}{l} \mathbf{S_1 = C_s} \\ \mathbf{S_2 = C_i} \\ \text{for odd } n, \mathbf{S_n = C_{nh}} \end{array} \right\} \quad \text{The latter symbols are used.}$$

Cubic Groups (see Figure 3.14)

T—$4C_3$ and $3C_2$ (mutually $\perp$)

Tᵈ—**T** elements plus σ (and S_4 and S_4^3).

O—$3C_4$ (mutually $\perp$), C_3, and C_2

Oₕ—**O** elements, i, σ_h, and σ_d

Icosahedral Group

Iₕ—$12C_5$, $20C_3$, C_2, i, and σ

Linear Groups

C∞ᵥ—C_∞ axis and σ_v

D∞ₕ—**C∞ᵥ** elements plus σ_h

Rotation Groups

Only proper rotations (C_n); these include **C_n**, **D_n**, **T**, **O**, and **I**

▶ *3.7.3   Some Terminology of Group Theory (Arranged Alphabetically)*

Characters: Performing a symmetry operation on a point in space described by the coordinates x, y, and z yields new coordinates x', y', and z'—a transformation that can be accomplished mathematically by a matrix. The sum of the diagonal elements of a matrix is its **character.** Characters can be used instead of matrices for many purposes.

Character Table: A character table contains the characters for all of the irreducible representations of a group for all the classes of symmetry operations; in other words, it summarizes the symmetry properties for all representations. Each row in a character table is an irreducible representation of the group. To the right of the character table, x, y, and z identify the representations of the p_x, p_y, and p_z orbitals, respectively. The representations of the d_{xz}, d_{yz}, and d_{xy} orbitals correspond to those of the respective products xz, yz, and xy. The d_{z^2} orbital belongs to the same representation as z^2 and the $d_{x^2-y^2}$ orbital belongs to the same representation as the function $x^2 - y^2$. If $x^2 - y^2$ does not appear, then x^2, y^2, and z^2 all belong to the totally symmetric representation (first row, A, A_1, A_{1g}, etc.), as do the d_{z^2} and $d_{x^2-y^2}$ orbitals. The s orbital, not listed, always belongs to the totally symmetric representation.

Classes of Symmetry Operations: Symmetry operations belong to the same class if one can be transformed into another by some operation of the group. Symmetry operations of the same class have the same characters and form a single column in a character table.

Matrix: A matrix is a rectangular array of numbers. Here we are concerned with the use of matrices for transforming the coordinates of a point in space and products of matrices for symmetry operations. See matrix multiplication (Appendix 3.1).

Mulliken Symbols: See Section 3.5.4.

Orthogonality: All irreducible representations of a point group are mutually orthogonal. A test for orthogonality is that the scalar product of the representations must equal zero. In two or three dimensions, orthogonal vectors are mutually perpendicular.

Products of Representations: The **scalar** (or dot) **product** gives a number (a scalar). The scalar product is used as a test for orthogonality of representations, since $\mathbf{A} \cdot \mathbf{B} = 0$ if A and B are orthogonal. The **direct product** (Appendix 3.2) of two representations gives a representation of the group.

Representation: A set of matrices (or their characters), one for each class of symmetry operations, comprises a **representation.** Properties of molecules, such as wavefunctions and vibrational modes, have symmetry characteristics that are described by one of the representations of the point group for the molecule. The dimension of the representation gives the dimensions of the matrices for the representation. Thus a one-dimensional representation (*A* or *B*) has 1×1 matrices—that is, simple numbers. A two-dimensional representation (*E*) has 2×2 matrices such as $\begin{bmatrix} 1 & 0 \\ 0 & 1 \end{bmatrix}$ for which the **character** is 2. A three-dimensional representation (*T*) has 3×3 matrices with a maximum character of 3. The representations that have matrices of lowest order (or smallest values for their characters) are **irreducible representations** or **symmetry species.** Representations for which the matrices can be reduced to sums of others of lower order are **reducible representations.** For each symmetry operation, the character of the reducible representation is the sum of the characters of the irreducible representations contained.

GENERAL REFERENCES

F. A. Cotton, *Chemical Applications of Group Theory,* 3rd ed., Wiley-Interscience, New York, 1990. The book from which many chemists learned group theory.

B. E. Douglas and C. A. Hollingsworth, *Symmetry in Bonding and Spectra,* Academic Press, Orlando, FL, 1985. An introduction to group theory with inorganic applications.

D. C. Harris and M. D. Bertolucci, *Symmetry and Spectroscopy,* Oxford University Press, New York, 1978. Applications to spectroscopy and molecular orbital theory.

M. Orchin and H. H. Jaffé, "Symmetry, Point Groups, and Character Tables," *J. Chem. Educ.,* **1970,** *47,* 246, 372, 510. Good treatment of topics covered in this text at a suitable level.

J. A. Salthouse and M. J. Ware, *Point Group Character Tables,* Cambridge University Press, Cambridge, England, 1972. Very useful collection of character tables and other tables of interest to chemists.

APPENDIX 3.1 MATRIX REPRESENTATION OF TRANSFORMATION AND MATRIX MULTIPLICATION

► *Equations and Matrix Multiplication*

An $m \times n$ matrix is an array with *m* rows and *n* columns:

$$\mathbf{A} = \begin{bmatrix} a_{11} & a_{12} & \cdots & a_{1n} \\ a_{21} & a_{22} & \cdots & a_{2n} \\ a_{m1} & a_{m2} & \cdots & a_{mn} \end{bmatrix}$$

A matrix does not have a value but rather represents a mathematical operator. It can be viewed as being made up of *n* *m*-dimensional column vectors or of *m* *n*-dimensional row vectors.

Multiplication of matrices involves combinations of the rows of the first matrix and the columns of the second matrix as follows:

$$\mathbf{AB} = \overset{\overset{\text{rows}}{\longrightarrow}}{\begin{bmatrix} a_{11} & a_{12} \\ a_{21} & a_{22} \end{bmatrix}} \begin{bmatrix} b_{11} & b_{12} \\ b_{21} & b_{22} \end{bmatrix} \Big\downarrow \text{columns} = \begin{bmatrix} (a_{11}b_{11} + a_{12}b_{21}) & (a_{11}b_{12} + a_{12}b_{22}) \\ (a_{21}b_{11} + a_{22}b_{21}) & (a_{21}b_{12} + a_{22}b_{22}) \end{bmatrix} \tag{1}$$

Consider **AB** and **BA** for

$$\mathbf{A} = \begin{bmatrix} 2 & 3 \\ 4 & 5 \end{bmatrix} \quad \text{and} \quad \mathbf{B} = \begin{bmatrix} 4 & 3 \\ 2 & 1 \end{bmatrix}$$

$$\mathbf{AB} = \begin{bmatrix} (2 \times 4 + 3 \times 2) & (2 \times 3 + 3 \times 1) \\ (4 \times 4 + 5 \times 2) & (4 \times 3 + 5 \times 1) \end{bmatrix} = \begin{bmatrix} 14 & 9 \\ 26 & 17 \end{bmatrix} \tag{2}$$

$$\mathbf{BA} = \begin{bmatrix} (4 \times 2 + 3 \times 4) & (4 \times 3 + 3 \times 5) \\ (2 \times 2 + 1 \times 4) & (2 \times 3 + 1 \times 5) \end{bmatrix} = \begin{bmatrix} 20 & 27 \\ 8 & 11 \end{bmatrix} \tag{3}$$

Note that unlike arithmetic multiplication, the order of multiplication of matrices is important because **AB** $\neq$ **BA.** Matrices do not commute except in special cases.

If a point with Cartesian coordinates x, y, and z is transformed to a new point x', y', and z', each coordinate depends upon the changes of the point in each direction, giving the equations

$$\begin{aligned} x' &= ax + by + cz \\ y' &= dx + ey + fz \\ z' &= gx + hy + iz \end{aligned} \tag{4}$$

Subsequent transformation of point x', y', and z' to the point x'', y'', and z'', gives the equations

$$\begin{aligned} x'' &= kx' + ly' + mz' \\ y'' &= nx' + oy' + pz' \\ z'' &= qx' + ry' + sz' \end{aligned} \tag{5}$$

Algebraic substituion of (4) into (5) yields the equations for x'', y'', and z'' in terms of x, y, and z as

$$\begin{aligned} x'' &= k(ax + by + cz) + l(dx + ey + fz) + m(gx + hy + iz) \\ y'' &= n(ax + by + cz) + o(dx + ey + fz) + p(gx + hy + iz) \\ z'' &= q(ax + by + cz) + r(dx + ey + fz) + s(gx + hy + iz) \end{aligned} \tag{6}$$

The equations grouped under (6) can be factored to give

$$\begin{aligned} x'' &= (ka + ld + mg)x + (kb + le + mh)y + (kc + lf + mi)z \\ y'' &= (na + od + pg)x + (nb + oe + ph)y + (nc + of + pi)z \\ z'' &= (qa + rd + sg)x + (qb + re + sh)y + (qc + rf + si)z \end{aligned} \tag{7}$$

In matrix notation we have

$$\begin{bmatrix} x'' \\ y'' \\ z'' \end{bmatrix} = \begin{bmatrix} k & l & m \\ n & o & p \\ q & r & s \end{bmatrix}\begin{bmatrix} a & b & c \\ d & e & f \\ g & h & i \end{bmatrix}\begin{bmatrix} x \\ y \\ z \end{bmatrix} \qquad \xrightarrow{\text{rows}} \text{into} \quad \Bigg| \; \text{columns}$$

$$= \begin{bmatrix} (ka + ld + mg) & (kb + le + mh) & (kc + lf + mi) \\ (na + od + pg) & (nb + oe + ph) & (nc + of + pi) \\ (qa + rd + sg) & (qb + re + sh) & (qc + rf + si) \end{bmatrix}\begin{bmatrix} x \\ y \\ z \end{bmatrix} \qquad (8)$$

Each element of the product matrix is made up of the sum of n terms, where n is *the number of columns of the first matrix that must be equal to the number of rows of the second matrix*. The row of the first matrix and the column of the second matrix give the row and column of the element in the product matrix; that is,

$$p_{ik} = \sum_j a_{ij}b_{jk} \qquad (9)$$

where the first subscript denotes the row and the second the column.

EXAMPLE 3.22: Consider the products **AB** and **BA** for

$$\mathbf{A} = \begin{bmatrix} x & 2 & 3 \\ 1 & 2 & y \end{bmatrix} \quad \text{and} \quad \mathbf{B} = \begin{bmatrix} 2x & 0 & 1 & y \\ 2 & 0 & 1 & 0 \\ 1 & 0 & 2 & 1 \end{bmatrix}$$

Solution:

$$\mathbf{AB} = \begin{bmatrix} (2x^2 + 4 + 3) & 0 & (x + 2 + 6) & (xy + 0 + 3) \\ (2x + 4 + y) & 0 & (1 + 2 + 2y) & (y + 0 + y) \end{bmatrix}$$

$$= \begin{bmatrix} (2x^2 + 7) & 0 & (x + 8) & (xy + 3) \\ (2x + 4 + y) & 0 & (2y + 3) & (2y) \end{bmatrix}$$

BA is not defined because the number of columns of **B** (4) is not the same as the number of rows of **A** (2).

EXAMPLE 3.23: Verify the product $\sigma_d C_3 = S_4$ from Example 3.11.

Solution: The σ_d considered contains the x axis changing the coordinates (x, y, z) into $(x, -z, -y)$, and C_3 interchanges (x, y, z) into y, z, x. Matrix multiplication gives

$$\overset{\sigma_d}{\begin{bmatrix} 1 & 0 & 0 \\ 0 & 0 & -1 \\ 0 & -1 & 0 \end{bmatrix}}\overset{C_3}{\begin{bmatrix} 0 & 1 & 0 \\ 0 & 0 & 1 \\ 1 & 0 & 0 \end{bmatrix}} = \begin{bmatrix} 0 & (1 + 0 + 0) & 0 \\ (0 + 0 - 1) & 0 & 0 \\ 0 & 0 & (0 - 1 + 0) \end{bmatrix} = \begin{bmatrix} 0 & 1 & 0 \\ -1 & 0 & 0 \\ 0 & 0 & -1 \end{bmatrix}$$

Here the matrix for x and y coordinates is the matrix for the C_4 operation:

$$\begin{bmatrix} 0 & 1 \\ -1 & 0 \end{bmatrix} = \begin{bmatrix} \cos 90° & \sin 90° \\ -\sin 90° & \cos 90° \end{bmatrix}$$

The matrix for z is $[-1]$, corresponding to reflection through the xy plane.

$$\sigma_h C_4 = S_4$$

Multiplication of the matrix for the S_4 operation with the (x, y, z) coordinates gives the new coordinates:

$$\begin{bmatrix} 0 & 1 & 0 \\ -1 & 0 & 0 \\ 0 & 0 & -1 \end{bmatrix} \begin{bmatrix} x \\ y \\ z \end{bmatrix} = \begin{bmatrix} y \\ -x \\ -z \end{bmatrix}$$

APPENDIX 3.2 THE DIRECT PRODUCT OF REPRESENTATIONS

The **direct product** of two representations gives a representation of the group. For each class of operations, *the product of the characters of two representations gives the character of the resulting representation.* Using the $\mathbf{C}_{4v}$ group, we obtain

$\mathbf{C}_{4v}$	E	$2C_4$	C_2	$2\sigma_1$	$2\sigma_2$
A_1	1	1	1	1	1
A_2	1	1	1	-1	-1
B_1	1	-1	1	1	-1
$A_2 B_1 = B_2$	1	-1	1	-1	1

From the character table we see that this direct product is B_2 or $A_2 B_1 = B_2$. We see that $A_1 \Gamma_i = \Gamma_i$, or the multiplication (direct product) of any representation (Γ_i) by A_1 leaves the representation unchanged. For $\mathbf{C}_{4v}$, the representation E times any other representation gives E. Any one-dimensional representation times itself gives A_1 ($\Gamma_j^2 = A_1$), whereas E^2 gives a four-dimensional (reducible) representation. Although we shall not try to prove it here, it is important for spectroscopic selection rules (page 444) that *the direct product of any representation by itself gives the totally symmetric representation (here, A_1)—or, if the product is reducible, that it contains A_1.*

PROBLEMS

3.1 Use matrices to show that $\sigma_1 \sigma_2 = C_4$ and $\sigma_2 \sigma_1 = C_4^3$ for the kaleidoscope group in Figure 3.5. Take the intersection of σ_1 and σ_2 as the z axis (that is, the axis perpendicular to the plane of the paper) and let the x axis be the intersection of σ_1 and the plane of the paper.

3.2 Select the point group to which each of the species in Example 3.6 belongs.

3.3 Assign the molecules in Example 3.3 to the appropriate point groups.

3.4 Assign the species in Figure 3.4 to the appropriate point groups.

3.5 A scalene triangle has three unequal sides. A regular scalene tetrahedron may be constructed by folding the pattern obtained by joining the midpoints of the sides of a scalene triangle having acute angles. To what point groups does the regular scalene tetrahedron belong?

3.6 Assign point groups to the following figures and their conjugates. What are the shapes of the conjugate figures?

(a) Trigonal prism   *(b)* Hexagonal prism   *(c)* Square pyramid
with alternate sides
of equal area

3.7 Determine the order of the $\mathbf{D}_{3h}$ group, the number of classes of operations, and the dimensions of the irreducible representations.

3.8 Indicate how the slash mark on the "vee bar" shown below would be shifted under each of the operations of the $\mathbf{C}_{2v}$ group. Use your results as a guide to construct the multiplication table for the $\mathbf{C}_{2v}$ group.

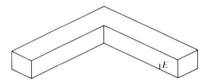

3.9 Identify three symmetry operations that are their own inverses.

3.10 Determine the independent operations of the $\mathbf{C}_{3h}$ group and construct the group multiplication table. Determine whether C_3 and C_3^2 are in the same class.

3.11 Test the mutual orthogonality of the representations of the $\mathbf{D}_{3h}$ group.

3.12 Derive the $\mathbf{C}_{3v}$ character table.

3.13 Derive the character table for the $\mathbf{C}_3$ group using Equations (3.9) and (3.10), expressing the characters as simple and complex numbers.

3.14 Take the following direct products and find the irreducible representations contained.
(a) $E_g \times E_u$ in $\mathbf{D}_{3d}$;   (b) $E_g \times T_{1g}$ in $\mathbf{O}_h$;   (c) $E_g \times T_{2g}$ in $\mathbf{O}_h$;   (d) $E_u \times E_u$ in $\mathbf{O}_h$.

3.15 Give the symmetry labels for the p and d orbitals for $[Co(NH_3)_6]^{3+}$, $[Co(en)_3]^{3+}$, $[Co(edta)]^-$, and *trans*-$[Cr(Cl)_2(NH_3)_4]^+$.

PART II

Bonding and Structure

▶ 4 ◀

Discrete Molecules
Molecular Orbitals
· ·

> Molecular orbital theory has emerged as the most important approach to chemical bonding. Although molecular orbital calculations, even for fairly simple molecules, require computers, we can obtain much of the bonding description from symmetry and group theory. We begin with diatomic molecules and then move to larger molecules. The molecular orbital descriptions for simple molecules can be given pictorially. Symmetry and group theory become more important for larger molecules. The final descriptions obtained are summarized in figures and can be understood even if the group theoretical treatment is omitted. The notation and terminology should be familiar to any chemist who uses the current literature of inorganic chemistry.

4.1 DIATOMIC MOLECULES

▶ 4.1.1 H_2, H_2^+, and He_2^+

According to **molecular orbital (MO) theory,** atomic valence electrons are influenced by all nuclei and other electrons of the molecule.[1] Inner electrons, of course, are localized on a particular atom, and some valence shell electrons are rather highly localized on one atom or on a small group of atoms. We are concerned only with the valence shell electrons—those that participate in chemical bonding. In order to obtain wavefunctions for the molecular orbitals, we assume that these are **linear combinations of atomic orbitals (LCAO).** Atomic orbitals (AOs) that can combine are limited by the symmetry of the molecule and the symmetries of the orbitals. The most effective combinations involve AOs of similar energies as well.

In valence bond theory a chemical bond is considered as resulting from overlap of one AO on each atom. MOs are combinations of all AOs of suitable energy in the molecule consistent with the symmetry of the orbitals.

[1] R. S. Mulliken, *Science* **1967,** *157,* 13.

Consider first the diatomic molecule H_2. The $1s$ AOs give two MOs expressed as linear combinations. *The number of MOs is the same as the number of AOs combined*—we conserve orbitals. The sum of the AOs (actually the wavefunctions are added, $1s_A + 1s_B$) gives a **bonding molecular orbital** (Figure 4.1), because the maximum electron density (the square of the sum of the wavefunctions) occurs in the region between the atoms. The other linear combination, $1s_A - 1s_B$, gives cancellation in the region between atoms, resulting in a nodal plane separating the atoms (Figure 4.1). Because the electron density is depleted between the atoms (zero in the nodal plane), the nuclei are poorly screened from one another and a repulsive interaction results; this is an **antibonding orbital.** MOs with rotational symmetry about the bond axis (C_∞) are called **sigma (σ) orbitals.** The bonding orbital is designated σ and the antibonding orbital is designated σ^*. An occupied σ orbital forms a σ bond.

The wavefunctions for MOs are normalized by multiplying by a normalization constant N such that $\int (N\Psi)^2 d\tau = 1$—or the probability of finding the electron somewhere outside of the nucleus is unity. For our bonding orbital, the normalization equation is

$$\int [N(\psi_{1s_A} + \psi_{1s_B})]^2 \, d\tau = N^2 \int (\psi_{1s_A}^2 + 2\psi_{1s_A}\psi_{1s_B} + \psi_{1s_B}^2) \, d\tau$$

$$= N^2 \left(\int \psi_{1s_A}^2 \, d\tau + 2 \int \psi_{1s_A}\psi_{1s_B} \, d\tau + \int \psi_{1s_B}^2 \, d\tau \right) = 1 \tag{4.1}$$

Because we are using normalized atomic wavefunctions, our expression reduces to

$$1 = N^2 \left(2 + 2\int \psi_{1s_A}\psi_{1s_B} \, d\tau \right) \quad \text{or} \quad N = \pm \frac{1}{\sqrt{2(1 + \int \psi_{1s_A}\psi_{1s_B} \, d\tau)}} \tag{4.2}$$

We choose the positive sign to give a positive wavefunction. The integral $\int \psi_{1s_A}\psi_{1s_B} \, d\tau$, called the **overlap integral,** is usually neglected in the LCAO approximation, giving

$$N = 1/\sqrt{2} \tag{4.3}$$

Our two MOs (Figure 4.2) are

$$\Psi_\sigma = \frac{1}{\sqrt{2}}(\psi_{1s_A} + \psi_{1s_B}) \quad \text{and} \quad \Psi_{\sigma^*} = \frac{1}{\sqrt{2}}(\psi_{1s_A} - \psi_{1s_B}) \tag{4.4}$$

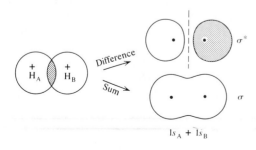

Figure 4.1   The linear combination of atomic orbitals for H_2. The negative lobe for σ^* is shaded.

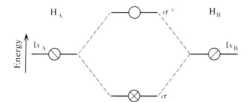

Figure 4.2   The relative energies of the molecular and atomic orbitals for H_2.

Two spin-paired electrons occupy the bonding orbital giving the configuration, σ^2. In the more general case the normalization constant N equals 1 divided by the square root of the sum of the squares of the coefficients in the LCAO for all orbitals:

$$\Psi = a\psi_1 + b\psi_2 - c\psi_3 \cdots \quad \text{and} \quad N = \frac{1}{\sqrt{a^2 + b^2 + c^2 \cdots}} \tag{4.5}$$

In calculating the normalization constant, it seems strange to disregard the overlap integral, which describes the extent of overlap of the bonding orbitals. Neglect of the overlap integral does not affect the symmetry properties of the MO. With this approximation, the antibonding orbitals for H_2 and X_2 molecules are raised in energy by the same amount by which the bonding orbitals are lowered. But if we include the overlap integral, the antibonding orbitals are raised in energy more than the corresponding bonding orbitals are lowered in energy. Consequently, the He_2 molecule discussed below is unstable; it is not just that there is no net gain in its formation.

The H_2^+ ion is stable in gaseous discharge tubes. Its single electron occupies the lower-energy MO giving the configuration σ^1. The resulting one-electron bond is about half as strong as the electron-pair bond encountered for H_2. The bond order (B.O.) is defined as one-half of the number of net bonding electrons (number of bonding electrons minus the number of antibonding electrons), giving 1 for H_2 and $\frac{1}{2}$ for H_2^+. Adding an electron to H_2 produces H_2^-, whose additional electron can be accommodated in the σ^* orbital giving $\sigma^2\sigma^{*1}$. Population of the antibonding orbital weakens the bond, since the *net number of bonding electrons* is reduced to one. The B.O. is $\frac{1}{2}$ for H_2^- and zero for unstable H_2^{2-}. These same MOs apply for possible combinations of the $1s$ orbitals of helium atoms. The B.O. is zero for He_2, an unstable species. However, if an electron is ejected from an He atom in a gaseous discharge tube, the He^+ ion, isoelectronic with H, can form bonds to other species. Two He^+ ions can combine to form He_2^{2+} with B.O. 1, isoelectronic with H_2, but with charge 2+. The He^+ ion can combine with the He atom to form He_2^+. Because there are three electrons, one must occupy the σ^* orbital, giving a B.O. of $\frac{1}{2}$.

▶ 4.1.2   Diatomic Molecules of Second-Period Elements

Sigma Bonding

Sigma bonds also are formed using p orbitals ($s + p_z$ or $p_z + p_z$) or d orbitals ($s + d_{z^2}$, $p_z + d_{z^2}$, or $d_{z^2} + d_{z^2}$) directed along the molecular axis (the z axis is chosen as the unique axis) (Figure 4.3). A hybrid orbital (e.g., sp, sp^2, sp^3) of one atom can form a sigma bond

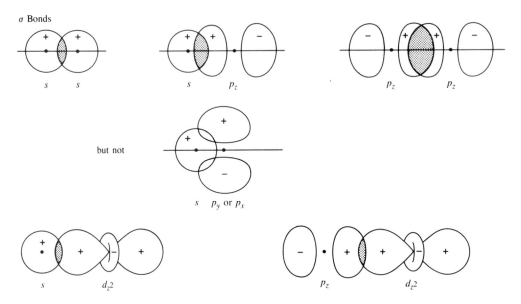

Figure 4.3   Sigma bonding.

to another atom, as long as the bond formed increases electron density along the line join-ing the atoms. An s orbital (or p_z) on one atom cannot form a sigma bond by overlap with a p_x or p_y orbital on another atom (Figure 4.3), because net bonding is precluded by the cancellation caused by the opposite signs of the amplitudes of the wavefunctions in one re-gion of overlap. This combination does not have the rotational symmetry of a σ orbital, nor does it meet the requirements for other types of bonds (π or δ).

Molecular orbitals in a linear (including diatomic) molecule are labeled by the Greek let-ters σ, π, δ, ϕ, and so on, corresponding to the Roman letters s, p, d, f, and so on, for atomic orbitals. The MOs have angular momentum quantum numbers λ 0, 1, 2, 3, . . . , corresponding to the l angular momentum quantum numbers of the atomic orbitals. Or-bital angular momentum is quantized along the molecular axis which is taken as the z di-rection (instead of z being defined as the direction of an externally imposed magnetic field as for atoms). The MO label may be obtained pictorially from the appearance of the cross-section of the MO perpendicular to the line of the bond and the label of the atomic orbital to which the cross-section corresponds. These labels are not strictly appropriate for non-linear molecules, where the labels must correspond to the symmetry species of the point group to which the molecule belongs. Nevertheless, the general labels (σ, π, etc.) are used to describe the local bonding symmetry between any pair of atoms.

Pi Bonding

With the z axis chosen as the bond direction, the p_x (or p_y) orbitals on the two atoms can combine to form **pi** (π) orbitals having $\lambda = 1$ (Figure 4.4). Each of the two equivalent π orbitals ($p_x + p_x$ or $p_y + p_y$) has a nodal plane through the nuclei bonded. The parallel p orbitals overlap, causing an increase in electron density on each side of this plane. Pi or-

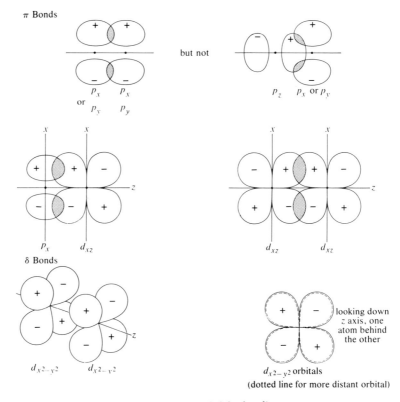

Figure 4.4   Pi and delta bonding.

bitals formed from combinations of d orbitals also result in MOs with a single nodal plane (e.g., $p_x + d_{xz}$ or $d_{xz} + d_{xz}$). Combinations of d orbitals, producing MOs with two nodal planes through the nuclei bonded, are called *delta* (δ) orbitals having $\lambda = 2$ (e.g., $d_{x^2-y^2} + d_{x^2-y^2}$ or $d_{xy} + d_{xy}$; see Figure 4.4). Many examples of compounds, such as $[Re_2Cl_8]^{2-}$, involving M—M quadruple bonds (1σ, 2π, and 1δ) are known (see page 750).

Figures 4.3 and 4.4 show AOs that combine to give bonding MOs. In the corresponding antibonding orbitals, the signs of the wavefunctions[2] are opposed in the region of overlap. The MOs resulting from combinations of s and p AOs for an X_2 molecule are sketched in Figure 4.5 as electron density plots,[3] but the signs of the wavefunctions for individual lobes are shown (by shading negative lobes of MOs) because they are important in determining the proper combinations. In centrosymmetric molecules, orbitals are desig-

---

[2] The signs of the wavefunctions have only mathematical significance—not physical significance. The sine wave pattern has positive and negative amplitudes. In combining waves the amplitudes are combined algebraically. In the region of overlap the amplitudes with the same signs combine to increase the amplitude of the resulting wave, and cancellation for opposite signs decreases the amplitude for the resultant wave. A similar combination of light waves occurs for diffraction experiments, giving (a) regions of increased intensity where amplitudes add and (b) regions of decreased intensity or dark regions where amplitudes cancel. The resultant wave is obtained by adding (algebraically) amplitudes. The intensity of light is the square of the amplitude, so negative amplitude squared gives a positive intensity. Electron density is given by the square of the wavefunction.

[3] A. C. Wahl, *Science* **1966**, *151*, 961.

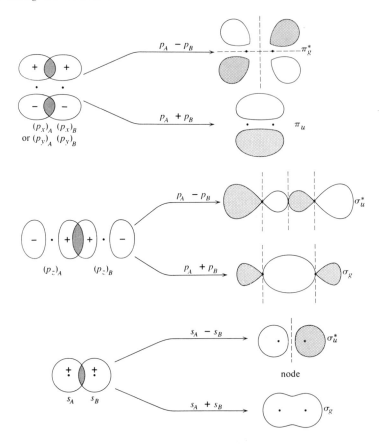

Figure 4.5   Molecular orbitals from combinations of *s* and *p* atomic orbitals. The signs of the wavefunctions (Ψ) are shown even though these are electron density (Ψ^2) contour plots. The negative lobes are shaded for the MOs.

nated as *g* (German *gerade*) if reflection through the center of symmetry does not change sign and *u* (German *ungerade*) if there is a change of sign.

Molecular Orbital Energy Levels

The proper combinations of AOs are obtained unambiguously from symmetry considerations. That is, only AOs which have net overlap can combine to give MOs, and these are AOs having the same symmetry. The second requirement for good MO formation is that the AOs involved be of similar energy. For Li_2, the situation is straightforward. The 2*s* orbitals combine in the same way as the 1*s* orbitals in H_2, giving a bonding and an anti-bonding orbital. The two valence-shell electrons enter the lower-energy σ bonding orbital to produce a B.O. of 1 for Li_2. Inner shells are nonbonding because there is little interaction and as many antibonding electrons as bonding electrons. The B.O. of Be_2 is zero, since there are as many antibonding electrons as bonding electrons (see Table 4.1). Li_2 occurs in the vapor, but there are only Be(*g*) atoms.

Table 4.1 Molecular orbital configurations for some diatomic species[a]

	Configuration	Bond order	Number of unpaired electrons	Bond length (pm)	Bond dissociation energy (kJ/mol)
H_2	σ_{1s}^2	1	0	74.14	432.1
Li_2	σ_{2s}^2	1	0	267.3	101
Be_2	$\sigma_{2s}^2\,\sigma_{2s}^{*2}$	0	0	—	—
B_2	$\sigma_{2s}^2\,\sigma_{2s}^{*2}\,\pi_x^1\,\pi_y^1$	1	2	159	291
C_2	$\sigma_{2s}^2\,\sigma_{2s}^{*2}\,\pi_x^2\,\pi_y^2$	2	0	124.3	599
N_2	$\sigma_{2s}^2\,\sigma_{2s}^{*2}\,\pi_x^2\,\pi_y^2\,\sigma_{pz}^2$	3	0	109.77	942
O_2	$\sigma_{2s}^2\,\sigma_{2s}^{*2}\,\sigma_{pz}^2\,\pi_{x,y}^4\,\pi_x^{*1}\,\pi_y^{*1}$ [b]	2	2	120.75	494
O_2^+	$\sigma_{2s}^2\,\sigma_{2s}^{*2}\,\sigma_{pz}^2\,\pi_{x,y}^4\,\pi^{*1}$	2.5	1	111.6	643
O_2^-	$\sigma_{2s}^2\,\sigma_{2s}^{*2}\,\sigma_{pz}^2\,\pi_{x,y}^4\,\pi_x^{*2}\,\pi_y^{*1}$	1.5	1	135	395
O_2^{2-}	$\sigma_{2s}^2\,\sigma_{2s}^{*2}\,\sigma_{pz}^2\,\pi_{x,y}^4\,\pi_x^{*2}\,\pi_y^{*2}$	1	0	149	126
F_2	$\sigma_{2s}^2\,\sigma_{2s}^{*2}\,\sigma_{pz}^2\,\pi_{x,y}^4\,\pi_x^{*2}\,\pi_y^{*2}$	1	0	141	155
Ne_2	$\sigma_{2s}^2\,\sigma_{2s}^{*2}\,\sigma_{pz}^2\,\pi_{x,y}^4\,\pi_{x,y}^{*4}\,\pi_y^{*2}$	0	0	(315)	0.2

[a] Data (except for O_2^{2-}) from K. P. Huber and G. Herzberg, *Molecular Spectra and Molecular Structure,* Vol. IV, *Constants of Diatomic Molecules,* Van Nostrand–Reinhold, New York, 1979.

[b] See text for discussion of the reversal of the order of energies for σ_{pz} and $\pi_{x,y}$.

We would expect $\sigma_g(p_z)$ to be lower in energy than π_u because of better overlap for p_z orbitals directed along the bond. The B_2 molecule has two unpaired electrons, however, so the energy of the doubly degenerate π orbitals must be lower than that of σ_{2p}, with the order shown in Figure 4.6 (drawn to agree with the magnetic moment) and Table 4.1. The same order of energy levels applies for the molecule C_2, because it is diamagnetic— $\sigma_{2s}^2\sigma_{2s}^{*2}\pi_x^2\pi_y^2$. The N_2 molecule would be diamagnetic with a triple bond regardless of the relative energies of the π and σ_p orbitals. However, since it is known that N_2^+ has a single σ_p electron, the σ_p orbital must be highest in energy, giving $2\sigma_{2s}^2 2\sigma_{2s}^{*2}\pi_x^2\pi_y^2 3\sigma_g^2$ for N_2 (Figure 4.7), or the **highest-energy occupied molecular orbital (HOMO)** is $3\sigma_g^1$ for N_2^+. Two or more orbitals of a given representation are numbered in order of increasing energy ($2\sigma_g$ and $3\sigma_g$, etc.)

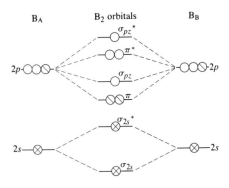

Figure 4.6 Molecular orbital energy-level diagram for B_2.

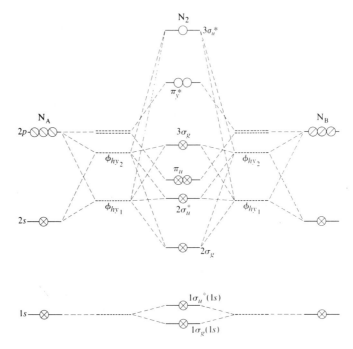

Figure 4.7   Molecular orbital energy-level diagram for N_2 using *sp* hybrid orbitals. Orbitals with the same symmetry are numbered in sequence beginning with $1\sigma_g(1s)$ and $1\sigma_u^*(1s)$. (Adapted with permission from M. Orchin and H. H. Jaffé, *Symmetry, Orbitals, and Spectra*, Wiley-Interscience, New York, 1971, p. 41, copyright © 1971.)

Most stable molecules have only paired electrons and are **diamagnetic.** Diamagnetic substances are repelled by a magnetic field. Molecules containing unpaired electrons are **paramagnetic.** Paramagnetic substances are drawn into a magnetic field. The **magnet moment** (force acting on a molecule in a magnetic field) can be calculated from the number of unpaired electrons (see Section 9.6.1).

The magnetic properties of B_2 and C_2 and the spectroscopic study of N_2^+ show unambiguously the order of the HOMO for these molecules; that is, $\pi_{x,y}(\pi_u)$ is lower in energy than $\sigma_p(3\sigma_g)$. This is the order expected when the energies of the $2s$ and $2p$ orbitals differ only slightly and hybridization occurs. Because the $2s$ and $2p$ orbitals have the same (σ) symmetry, mixing of the *s* and *p* orbitals (hybridization) can occur, so that σ_s has some *p* character and σ_p has some *s* character. As a result, the lower-energy σ orbital (labeled σ_s for B_2 and $2\sigma_g$ for N_2) is lowered still further in energy, and the other σ orbital (labeled σ_p for B_2 and $3\sigma_g$ for N_2) is raised in energy such that its energy exceeds that of the π_u orbitals. The effects can be seen more easily if we consider first formation of *sp* hybrid orbitals (ϕ_{hy1} and ϕ_{hy2}) on each N and then combine these to form MOs (Figure 4.7). Here, in the linear combinations each hybrid contributes to each σ MO, but the greater the contribution the closer in energy an orbital is to the resulting MO. The energy of the more stable bonding orbital is lowered greatly, and the energy of the least stable antibonding orbital is increased greatly. Figure 4.7 shows the effect of hybridization, giving the relative energies of MOs as in Figure 4.6 (where σ_{2s} corresponds to $2\sigma_g$). Hybridization causes the energy observed for B_2 and C_2, so Figure 4.7 is more appropriate than Figure 4.6 for these cases.

The two sp hybrid orbitals considered (page 67) are equivalent and have equal energy, or the weighting coefficient (λ) is 1 in Equations (4.6) and (4.7):

$$\phi_{\text{hy}_1} = \frac{1}{\sqrt{1 + \lambda^2}} (2s + \lambda 2p) \tag{4.6}$$

$$\phi_{\text{hy}_2} = \frac{1}{\sqrt{1 + \lambda^2}} (\lambda 2s - 2p) \tag{4.7}$$

In the case of an X_2 molecule, one hybrid orbital is used for bonding to the other X atom (the combination with greater p character, ϕ_{hy_2} with $\lambda < 1$), whereas the lone pair occupies the sp hybrid orbital with the greater s character. The difference in energies of the two hybrids thus depends on the relative energies of the s and p orbitals. As the nuclear charge increases across a period, the energy separation between $2s$ and $2p$ increases; the more penetrating $2s$ orbital (page 40) decreases in energy as p electrons are added. The greater the energy separation between $2s$ and $2p$, the less mixing occurs.

For diatomic molecules and other linear molecules, the possible terms can be obtained from electron configurations by a procedure analogous to that given in Chapter 1 for atoms. The labels for irreducible representations are Σ for A, Π for E_1, and Φ for E_3 (see character tables for $\mathbf{C}_{\infty v}$ and $\mathbf{D}_{\infty h}$ in Appendix C). A_1 becomes Σ^+ and A_2 becomes Σ^-, where the superscripts are the signs of the characters for reflections in vertical planes. For $\mathbf{D}_{\infty h}$, subscripts g or u apply so A_{1g} is Σ_g^+, A_{2g} is Σ_g^-, and A_{1u} is Σ_u^+.

For molecules having fully occupied orbitals, such as H_2 and N_2, the electron distribution is totally symmetric and the ground-state term is labeled $^1A_{1g}$ or $^1\Sigma_g^+$ (analogous to ^1S for atoms with filled shells or subshells). If there is one unpaired electron, the term representation is that for the singly occupied orbital. For $H_2^+ (1\sigma_g)^1$ and $N_2^+ (3\sigma_g)^1$, the ground-state term is $^2\Sigma_g^+$ (analogous to ^2P for . . . $2s^2 2p^1$ B). For $O_2^+ \cdots (\pi_g^*)^1$, the ground-state term is $^2\Pi_g$.

When two orbitals are singly occupied, as for ground-state $(\pi_g^*)^1 (\pi_g^*)^1$ of O_2, we can proceed to obtain terms from the array of Pauli-allowed microstates. Angular momentum (λ) is quantized along the molecular axis with individual quantum numbers as follows:

Molecular orbital	λ	m_l
σ	0	0
π	1	± 1
δ	2	± 2

For O_2 the possible microstates are

	$m_s = +\frac{1}{2}$		$m_s = -\frac{1}{2}$		$M_L = \Sigma m_l$	$M_S = \Sigma m_s$
$m_l =$	+1	−1	+1	−1		
	↑	↑			0	+1
	↑		↓		2	0
	↑			↓	0	0
		↑	↓		0	0
		↑		↓	−2	0
			↓	↓	0	−1

The numbers of microstates corresponding to individual M_L and M_S values are:

M_L			
2		1	
0	1	2	1
−2		1	
	+1	0	−1
		M_S	

The possible values of the total angular momentum quantum number Λ are 2 and 0. Possible values of S are 1 and 0. The microstates $(1^+, 1^-)$ and $(1^+, 1^-)$ give rise to $^1\Delta_g$, and the four terms with $M_L = 0$ give rise to $^3\Sigma_g$ and $^1\Sigma_g$. Hund's rules allow us to pick $^3\Sigma_g$ as the ground state.

In a similar fashion, the first excited configuration of N_2 is ... $(3\sigma_g)^1(\pi_g^*)^1$, which gives $^3\Pi_g$ and $^1\Pi_g$ states.

Where the energy separation between $2s$ and $2p$ orbitals is great enough, there is little mixing, the s orbitals are primarily nonbonding, and the energy of $\sigma_p(3\sigma_g)$ falls below that of $\pi_{x,y}(\pi_u)$ (Figure 4.8). This appears to be the order of levels for O_2 and F_2. The order of these two filled levels is not very important for O_2 and F_2, because the B.O. and magnetic properties are not affected, and the HOMO is π_g^*. The configuration of the F_2 molecule is identical to that of O_2^{2-}. The bond energy of F_2 (Table 4.1) is very low compared with that of Cl_2 (239 kJ/mol), because of the strong repulsion between lone (nonbonding) pairs for F_2 (see p. 232) with a shorter bond.

The oxygen molecule is paramagnetic with two unpaired electrons. The simple VB model predicts a diamagnetic molecule, $\ddot{O}=\ddot{O}$, which is wrong. The MO set has as the HOMO the doubly degenerate π^* orbitals, $\pi_x^{*1}\pi_y^{*1}$, giving two unpaired electrons and a B.O. of 2 (see Table 4.1). The lowest energy or ground state of O_2 has two unpaired electrons, a triplet (above), described as $^3\Sigma_g$. Excited singlet states[4] (all electrons paired) could occur with both electrons in one π_g^* orbital, described as $^1\Delta_g$, or with the electrons of opposite spins in the two π_g^* orbitals, described as $^1\Sigma_g$. These are both much lower in energy than the configuration with both electrons in the σ_p^* orbital. The $^1\Delta_g$ excited state (7880 cm^{-1} or 94.7 kJ/mol higher than the ground state) has a longer lifetime than $^1\Sigma_g^+$

[4] M. Laing, *J. Chem. Educ.* **1989**, *66*, 453.

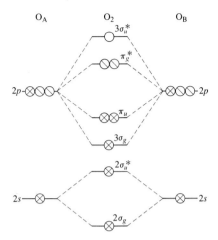

Figure 4.8   Molecular orbital energy-level diagram for O_2.

(158 kJ/mol higher than the ground state). The excited singlet oxygen molecules can be generated photochemically by radiation, chemically, or in an electron discharge tube. Singlet oxygen, a stronger oxidant than triplet O_2, is used in laboratory and industrial selective oxidations.

The significance of the π_g^* orbitals as the HOMO in O_2 is shown by the *shortening* of the bond length on removal of an (antibonding) electron to give O_2^+ (Table 4.1). Adding an electron to O_2, forming the superoxide ion, O_2^-, weakens the bond, and adding another electron to form the peroxide ion, O_2^{2-}, gives a diamagnetic species with B.O. 1.

Heteronuclear Molecules

Heteronuclear diatomic molecules can be described similarly as for X_2 molecules. Consider a molecule containing H bonded to F, a second-period element. First we must judge the energy availability and symmetry of the orbitals on the two atoms. Hydrogen has only a 1s orbital, so symmetry permits σ-bond formation only by overlap with the 2s and 2p_z orbitals of F. The energies of the F orbitals are much lower than those of H (the energy of the highest filled AO can be obtained from the ionization energy: -1.312 MJ/mol for H and -1.681 MJ/mol for F). The 2s orbital is nonbonding, since it is so much lower in energy than the 1s orbital of H (Figure 4.9). The σ bond involves overlap of the H 1s orbital and F 2p_z orbital. With no H orbitals of proper symmetry to combine with them, the other p orbitals on F are nonbonding.

For an XY molecule, where X and Y are second-period elements, we can get the B.O. and number of unpaired electrons from the population of energy levels in Figure 4.6. When the energy difference is small for s and p orbitals, hybridization is important. In the MO energy-level diagram for CO illustrated in Figure 4.10, the orbital energies are lower for the more electronegative oxygen, so that the lower-energy hybrid of carbon is comparable in energy with the higher-energy hybrid of oxygen. These orbitals of comparable energy combine to give bonding and antibonding MOs. The very-low-energy hybrid orbital on oxygen and the high-energy one on carbon are far apart in energy. Hence, even though their symmetries are the same, they are *nonbonding*. Rather than combining to give MOs, they remain as localized orbitals on O and C, respectively. These are the orbitals occupied by the lone pairs on C and O directed away from the CO bond. Because

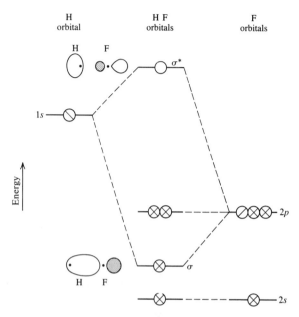

Figure 4.9 Molecular orbital energy-level diagram for HF.

the lone-pair orbital on oxygen is close in energy to the oxygen atomic $2s$, it is said to be mostly s in character; that is, it is much like an oxygen $2s$ orbital. The lone-pair orbital on carbon, on the other hand, is close in energy to the carbon p orbitals, so it is said to have a great deal of carbon p character. With its higher energy, the carbon lone pair is the pair donated in forming dative bonds to other species such as BH_3 in $H_3B \leftarrow CO$ and also to metals in metal carbonyls. The electron cloud for the π orbitals is polarized toward the more electronegative oxygen atom (Figure 4.11). Actually, because only one direction (z)

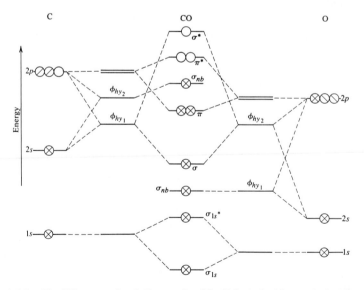

Figure 4.10 The MO energy-level diagram for CO. (Adapted with permission from M. Orchin and H. H. Jaffé, *Symmetry, Orbitals, and Spectra*, Wiley-Interscience, New York, 1971, p. 47, copyright © 1971.)

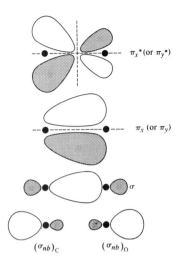

$\pi_x{}^*$ (or $\pi_y{}^*$)

π_x (or π_y)

σ

$(\sigma_{nb})_C$ $(\sigma_{nb})_O$

Figure 4.11 Sketches of MOs for CO. The π and π^* orbitals are shown in cross-section because the π_x and π_y or π_x^* and π_y^* combine, giving cylindrical symmetry along the bond axis.

is defined for the CO molecule, x and y are indistinguishable for the atomic and molecular orbitals. Consequently, π_x and π_y together form a sheath having cylindrical symmetry. Although these bonds have axial symmetry, they are π, not σ, since there is a node instead of a maximum in the electron density along the internuclear axis. The empty π^* orbitals are more localized on (polarized toward) carbon.

4.2   LINNETT'S DOUBLE QUARTET[5]

The MO model gives a good description of paramagnetic O_2. The VB model requires the introduction of one-electron or three-electron bonds to be consistent with O_2 as a diradical. In the case of many diamagnetic molecules, we use resonance to achieve a description consistent with the bond lengths and the equivalence of bonds. Electron pairing is important for many molecules leading to localized bonding and lone electron pairs.

A model by Linnett permits a satisfactory treatment of many paramagnetic molecules and often allows a single acceptable structure to be written instead of a number of contributing resonance structures.

Linnett considered the distribution of electrons to be determined primarily by two factors: spin correlation effects and charge correlation effects. The spin correlation effects arise from the application of the Pauli principle, which leads to the conclusion that electrons with the same spin tend to keep apart, while those with opposed spin tend to come together. The charge correlation effects arise because of the repulsion between particles with same charge. The charge correlation operates in the same direction as the spin correlation to keep electrons of the same spin apart. The spin correlation, however, is opposed to the charge correlation for paired electrons. *Linnett proposed that instead of considering the octet as a group of four localized pairs of electrons, it should be considered as two quartets of opposed spin (Double Quartets).* The most important interaction is the repulsion among a quartet of electrons of the same spin. This quartet is expected to have a dis-

[5] J. W. Linnett, *The Electronic Structure of Molecules,* Wiley, New York, 1964.

position around the nucleus which is approximately tetrahedral. The correlation between the two quartets of an octet is slight in the absence of other restrictions (e.g., bonding or some preferred orientation), because of the opposition of spin and charge correlation effects. The result would be spherical symmetry for a free atom or ion with a complete octet (e.g., Na^+, Ar, F^-, and O^{2-}). The most probable positions for the electrons of a Double Quartet are represented in Figure 4.12*a* by two interpenetrating tetrahedra.

The two tetrahedra of an octet become strongly correlated by the localization of electrons in bonding pairs. A single covalent bond, as in HF, localizes one electron pair, but permits delocalization of the other electrons as shown in Figure 4.12*b*. The presence of two electron-pair bonds, as in H_2O, causes strong correlation for all four electron pairs as seen in Figure 4.12*c*. Here the lone pairs are seen to be highly directional. In molecules such as H_2O, NH_3, or CH_4 where the two tetrahedra are strongly correlated, the representation of the Double Quartet (DQ) description is essentially the same as that for four pairs of electrons.

In the DQ description of multiple bonds, bent bond representations are used. Thus a double bond is represented by two tetrahedra sharing an edge, and a triple bond is represented by two tetrahedra sharing a face.

Where the DQ and four-pair representations are equivalent, the usual (VB) model (four pairs) is used. The DQ model offers no advantage for diamagnetic molecules which can be described by a single VB structure (no resonance). Nevertheless, many molecules exist for which the DQ model extends the usefulness of a VB picture considerably, reducing the necessity for apparently contrived structures or for the use of many contributing structures.

▶ 4.2.1 *Paramagnetic Molecules*

The VB model leads one to expect the oxygen molecule to be diamagnetic. One DQ representation for a diatomic molecule with 12 valence electrons would group the electrons in two sets of 6 each. The two sets of electrons must coincide in space for this structure (**I**):

$$\overset{\circ\times}{\underset{\circ\times}{O}}\overset{\times\circ}{\underset{\times\circ}{O}}\overset{\times\circ}{\underset{\times\circ}{}} \qquad \times\overset{\circ}{\underset{\circ}{O}}\overset{\times}{\underset{\times}{O}}\overset{\circ}{\underset{\circ}{O}}\times$$

$$\text{I} \qquad\qquad \text{II}$$

If we consider the 12 electrons as being of spin sets of 5 and 7 electrons, the set of 5 electrons of one spin is expected to be arranged as shown by the x's in structure **II** (two tetrahedra sharing a face). The set of 7 electrons (circles in **II**) is expected to be arranged to give two tetrahedra sharing one apex. Structure **II** provides for interlocking tetrahedra with the apices of one projecting through the faces of the other. This structure provides for the minimum electron–electron repulsion, both within each group and between groups. Structure **II** contains two one-electron bonds, one two-electron bond, and an octet around each O with the most probable positions of electrons at the corners of a cube.

The DQ model for the paramagnetic NO molecule divides the 11 electrons into a spin set of 5 and a set of 6. Each is arranged so as to maintain a tetrahedral orientation around each atom, as in the structures for O_2, to give structure **III**:

$$°N\!°O° + \,×N\,°O× \longrightarrow °×N×°O×°$$

III

This agrees with the presence of a single unpaired electron. The bond length observed is between that expected for a double and a triple bond. The B.O. for **III** is $2\frac{1}{2}$ because of the presence of a double bond and a one-electron bond.

The failure of NO to dimerize (as is common for odd-electron molecules) can be explained in several ways. There would be no gain in the number of covalent bonds on forming the dimer, $\ddot{O}\!=\!\dot{N}\!-\!\dot{N}\!=\!\ddot{O}$; furthermore, single bonds between very electronegative atoms on which there are unshared electron pairs are surprisingly weak (F_2, H_2O_2, and N_2H_4), whereas π bonds are very strong for N_2 and O_2. Perhaps this new σ bond, which would be formed in N_2O_2, might be weaker than the N—O bond broken in its formation. This DQ model, which is the same as the VB structure, points out the strong electron–electron repulsion existing because the two sets of electrons (of opposed spins) are required to adopt the same spatial arrangement in the dimer, whereas this is not so in the monomer.

The cyanogen monomer, CN, has 9 valence electrons. The set of 5 electrons can be arranged as shown for NO. The only reasonable symmetrical arrangement for the group of 4 electrons about two atoms is that in structure **IV,** where the quartets of the electrons represented by dots are not complete around either atom:

$$×C\,×°N×\qquad\qquad ×C\,×°N°$$

IV **V**

A more favorable distrbution of formal charges (both zero) can be achieved in structure **V.** This structure leads one to expect CN to dimerize to give $(CN)_2$ (this does occur), since an additional bond can be formed in the dimer without any further localization of the electrons.

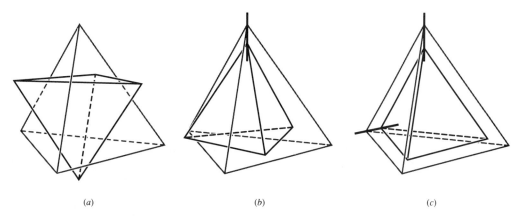

(a) (b) (c)

Figure 4.12 Correlation of two quartets of electrons. (a) No correlation between quartets (b) One bond (c) Two bonds causing the alignment of all apices of the tetrahedra.

▶ *4.2.2 Polyatomic Molecules*

The VB representation for CO_2 (3 atoms and 16 electrons as in OCN^- and CN_2^{2-} also) requires three contributing structures. These same structures and several others can be written using the DQ approach. However, one of these structures (**VI**) describes the molecule very well.

$$°_×^×O ×°_°C°_×^×O_°^°×$$

VI

It shows the delocalization of electrons through the molecule, gives a reasonable B.O., gives each atom a formal charge of zero, and minimizes electron–electron repulsion. The $\ddot{O}{=}C{=}\ddot{O}$ VB description involves each π bond localized over C and one O.

Single structures can be written for O_3 (3 atoms and 18 valence electrons as in SO_2 and NO_2^-) (**VII**) and benzene (**VII**):

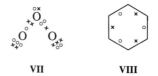

VII   **VIII**

Each DQ description is adequate without writing several contributing structures.

The DQ model of some paramagnetic molecules (or ions) emphasizes the delocalization of electrons, but resonance is still required for an adequate description. NO_2 (3 atoms and 17 electrons) can be represented by one pair of equivalent contributing structures (**IX**):

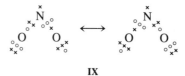

IX

4.3 TRIATOMIC MOLECULES

The formation of stable bonding molecular orbitals requires similar energies and good overlap (same symmetry) of the orbitals combined. If two half-filled orbitals overlap and one is much higher in energy, the electron in the higher-energy orbital will be transferred to the low-energy orbital on the other atom. Only outer orbitals or **"frontier" orbitals** (Section 4.8) overlap significantly for covalent bonding. In the case of diatomic molecules, symmetry restrictions for good overlap are obvious. In general, the possibility of good overlap is determined by symmetry of the molecule and symmetry of the atomic orbitals. Symmetry and group theory are valuable in deducing combinations permitted for larger molecules, where possible combinations might not be obvious. For triatomic molecules, we will use terminology of results from group theory, without having to provide a mathematical treatment.

▶ *4.3.1   Linear Molecules*

The CO_2 molecule is linear, belonging to the $\mathbf{D}_{\infty h}$ point group, whose character table[6] is given in Appendix C. The rotational axis C_∞ (taken as the z direction) is the internuclear axis for σ bonding. The σ bonding orbitals available for carbon and each oxygen are $2s$ and $2p_z$. Because oxygen $2s$ orbitals are much lower in energy than carbon orbitals, they are essentially nonbonding. The $2p_z$ orbital of each oxygen combines with carbon orbitals in the same way, so we may consider them together using the symmetry of the molecule and the symmetry characteristics of the oxygen orbitals. The z axis on each O is defined as "+" in the direction pointing toward C.

The totally symmetric carbon $2s$ orbital is unchanged after carrying out any of the rotation, reflection, or inversion symmetry operations of the group; that is, it belongs to the A_{1g} representation.[7] Only those combinations of oxygen orbitals that also are totally symmetric (belong to the A_{1g} representation) can combine with the carbon $2s$ orbital. The oxygen combination or **group orbitals (or symmetry-adapted orbitals)**—the two linear combinations of the $2p_z$ orbitals—are

$$a_{1g} \qquad \qquad a_{1u}$$

The a_{1g} oxygen group orbital combines with the carbon $2s$ orbital to give a bonding MO ($2\sigma a_{1g}$, the second σa_{1g} orbital in increasing energy); with the signs reversed for both oxygen orbitals (still a_{1g}), it combines with the carbon $2s$ orbital to give an antibonding MO ($3\sigma^*_{a_{1g}}$) (Figure 4.13):

$$2\sigma_{a_{1g}} = (2s)_C + (2p_{z\,O1} + 2p_{z\,O2}) \quad \text{and} \quad 3\sigma^*_{a_{1g}} = (2s)_C - (2p_{z\,O1} + 2p_{z\,O2})$$

The carbon $2p_z$ orbital has the symmetry of the A_{1u} representation; that is, its sign is unchanged by rotation about the z axis or reflection in a plane through the z axis (σ_v), but it changes sign on reflection in the xy plane (σ_h) or upon inversion (i). The symmetry can be identified as A_{1u} in the $\mathbf{D}_{\infty h}$ character table by finding z (for p_z) in the column to the right. It combines with the a_{1u} oxygen group orbital. The MOs (C plus O group orbitals; see Figure 4.13) are

$$2\sigma_{a_{1u}} = (2p_z)_C + (-2p_{z\,O1} + 2p_{z\,O2}) \quad \text{and} \quad 3\sigma^*_{a_{1u}} = (2p_z)_C - (-2p_{z\,O1} + 2p_{z\,O2})$$

These $2a_{1u}$ and $3a^*_{1u}$ orbitals are higher in energy than the respective $2a_{1g}$ and $3a^*_{1g}$ orbitals, since for carbon the energy of $2p$ is higher than that of $2s$.

The π bonds are formed similarly, but there are two equivalent C orbitals (the degenerate $2p_x$ and $2p_y$ orbitals) and corresponding pairs of O group orbitals. These C p_π orbitals belong to the two-dimensional representation E_{1u}. The four O p_π orbitals can com-

[6] M. Orchin and H. H. Jaffé, *J. Chem. Educ.* **1970**, *47*, 510.

[7] The representations themselves and electronic states (terms) are designated by the uppercase italic letters. Lowercase italic letters a_1, b_1, etc. are used for AOs or MOs. The special symbols used to represent linear point groups (Σ, Π, Δ, etc.) are included in the character table, along with the corresponding symbols used for other point groups. Here we use those symbols used also for other groups (A_1, B_1, etc.).

Symmetry	Atomic orbitals C	Group orbitals O	MOs	
A_{1g}	$2s$	$2p_{z_{O1}} + 2p_{z_{O2}}$	$2\sigma_{a_{1g}}$	2 nodes
	$2s$	$-(2p_{z_{O1}} + 2p_{z_{O2}})$	$3\sigma_{a_{1g}}^{*}$	4 nodes
	—	$2s_{O1} + 2s_{O2}$	$1\sigma_{a_{1g}}^{nb}$	0 node
A_{1u}	$2p_z$	$-2p_{z_{O1}} + 2p_{z_{O2}}$	$2\sigma_{a_{1u}}$	3 nodes
	$2p_z$	$-(-2p_{z_{O1}} + 2p_{z_{O2}})$	$3\sigma_{a_{1u}}^{*}$	5 nodes
	—	$2s_{O1} - 2s_{O2}$	$1\sigma_{a_{1u}}^{nb}$	1 node
E_{1u}	$2p_x$ $2p_y$	$2p_{x_{O1}} + 2p_{x_{O2}}$ $2p_{y_{O1}} + 2p_{y_{O2}}$	$\pi_{e_{1u}}$ (one sketched)	1 nodal plane
	$2p_x$ $2p_y$	$-(2p_{x_{O1}} + 2p_{x_{O2}})$ $-(2p_{y_{O1}} + 2p_{y_{O2}})$	$\pi_{e_{1u}}^{*}$ (one sketched)	1 nodal plane and 2 nodes
E_{1g}	— —	$2p_{x_{O1}} - 2p_{x_{O2}}$ $2p_{y_{O1}} - 2p_{y_{O2}}$	$\pi_{e_{1g}}^{nb}$ (one sketched)	1 nodal plane and 1 node

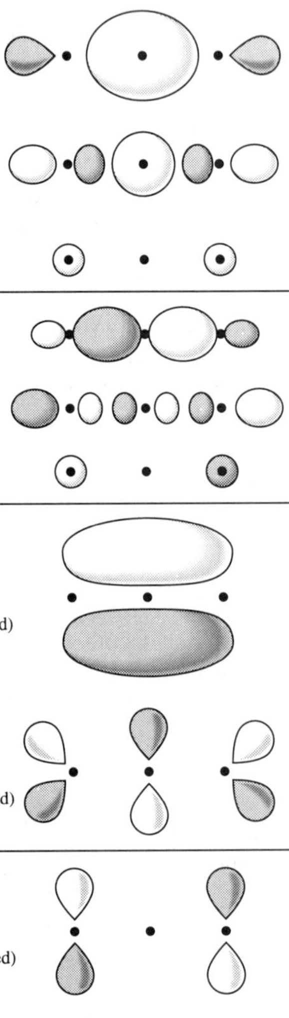

Figure 4.13 Molecular orbitals for CO_2.

bine to give two pairs of group orbitals. One of the pair of degenerate e_{1u} O group orbitals $(p_{xO_1} + p_{xO_2})$ is shown in Figure 4.13 with the corresponding C e_{1u} orbital. The combinations $(2p_x)_C - (2p_{xO_1} + 2p_{xO_2})$ (and with p_y orbitals) gives two $\pi_{e_{1u}}^{*}$ orbitals. The other pair of O group orbitals $(p_{xO_1} - p_{xO_2}$ and $p_{yO_1} - p_{yO_2})$ (Figure 4.13) are centrosymmetric and *together* belong to the E_{1g} representation. Because no C orbitals belong to this representation—that is, have this symmetry–these O group orbitals are *nonbonding*. The e_g orbitals do not combine with C orbitals at all and electrons in these orbitals are localized on the oxygens.

We are now ready to construct a qualitative MO energy-level diagram for CO_2. Recall the two requirements for good bond formation: first, good overlap of the atomic or group orbitals used; second, close match in energy between the orbitals. A consequence of "good" MO formation is a very stable bonding orbital and a correspondingly unstable antibonding orbital, with a large energy gap between. Thus, if the energies of the AOs involved are similar, we can expect the magnitude of energy splittings to decrease in the order $\sigma-\sigma^* > \pi-\pi^* > \delta-\delta^*$ because of decreasing orbital overlap in the same order.

First we consider the energies of MOs involving the $2s_C$, $2p_C$, and $2p_O$ group orbitals, all of which are of similar energies. Note on the right side of the diagram (Figure 4.14) that, although all of the p orbitals of atomic O must be degenerate, the O group orbitals have energies which differ depending on the symmetry. This results from taking into account the overlap integral in the normalization constant when computing the energies of the group orbitals (see page 152). The set of C p orbitals is indicated also as having slightly different energies on the left side of the diagram even though the three p orbitals of C are degenerate. This is for a different reason. The p_z orbital which points directly at the O nuclei is considered to be slightly stabilized by the positive nuclear charge of the

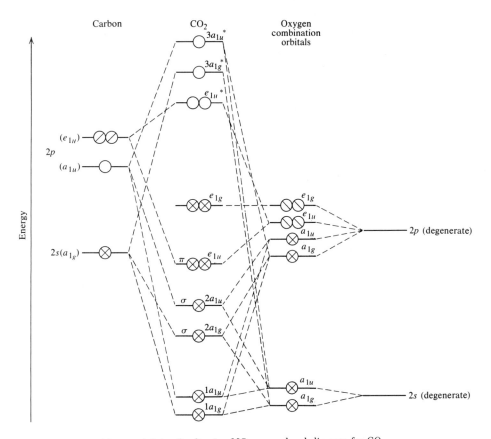

Figure 4.14   Qualitative MO energy-level diagram for CO_2.

oxygens even before any MO formation occurs. The e_{1u} (p_x and p_y) orbitals do not point toward oxygens and consequently experience no such stabilization.

Consider first the σ MOs. In general, the more nodes any MO has, the higher its energy.[8] Hence, the most stable σ orbital is $2a_{1g}$ having two nodes (at the oxygens, but none between carbon and oxygens). Its antibonding counterpart $3a_{1g}^*$ has four nodes (two between carbon and oxygens), and has very high energy. The $2a_{1u}$ orbital has three nodes. Thus, although it is more stable than either of its "parent" orbitals, it has higher energy than $2a_{1g}$. The antibonding counterpart, $3a_{1u}^*$, has five nodes and is the highest-energy orbital in the diagram.

Because of their poorer overlap, the π MOs have smaller splittings and $e_{1u}\pi$ and $e_{1u}\pi^*$ lie inside the energy gap spanned by the σ and σ^* orbitals. The O e_{1g} group orbitals are nonbonding because there are no e_{1g} C orbitals for combination. Their energy in the CO_2 molecule is the same as the energy of the $e_{1g}\pi$ group orbitals.

The $1a_{1g}(2s)$ and $1a_{1u}(2s)$ group orbitals on oxygens are so much lower in energy than the other valence orbitals that they do not effectively form MOs even though their symmetry is correct for overlap with carbon orbitals. In the MO energy-level diagram, they are shown as being very close in energy to the starting group orbitals. This means that these group orbitals are perturbed very little by formation of MOs. The $1a_{1g}(2s)$ and $1a_{1u}(2s)$ oxygen group orbitals are connected to the corresponding and antibonding levels with the same labels by "tie lines".

When the 16 valence electrons are placed in the lowest possible energy levels, the diagram is filled through the e_{1g} orbitals. The nonbonding e_{1g} and (lowest energy) $1a_{1g}$ and $1a_{1u}$ orbitals are localized on oxygens and contain the four lone pairs which appear on the oxygens in the VB structure for CO_2.

We verify that the number of MOs (12) is the same as the number of valence AOs (four orbitals for each of the three atoms). Orbitals can be involved in more than one combination (e.g., $2p_{x,y}$ in e_{1u} and e_{1g}), but the total number of MOs must be 12. A similar bonding scheme applies to other isoelectronic molecules or ions—CS_2, N_2O, N_3^-, OCN^-, and NO_2^+.

Note that the two σ orbitals ($2a_{1g}$ and $2a_{1u}$) have different symmetry and different energy. Both involve all three atoms (three-center bonds). They do not correspond directly to the two equivalent σ bonds in a VB description. A localized MO description, giving two equivalent σ bonds, would result from using two C sp hybrid orbitals for σ bonding. However, in $\mathbf{D}_{\infty h}$ $s(a_{1g})$ and $p(1a_{1u})$, cannot mix because of their differing symmetry.

▶ 4.3.2 Angular Molecules

The NO_2 molecule is angular because of the unshared electron on N. Its symmetry is $\mathbf{C}_{2v}$ (see the character table in Appendix C). We choose the C_2 rotation axis as the z axis and the x axis perpendicular to the molecular plane. We first consider σ orbitals, then π orbitals. Although primarily nonbonding, O $2s$ orbitals are involved in bonding to a greater

---

[8] Just as the energies of ns (or np, nd, etc.) AOs increase with the number of nodes, so do the energies of MOs representing different combinations of the same AOs. The greater the number of nodes, the greater the *localization* of charge. Degenerate orbitals have the same number of nodes.

extent than is true for CO_2, because the energy separation between the O and N orbitals is smaller. The symmetry species (irreducible representations) for the O $2s$ group orbitals in the C_{2v} point group are $a_1(s_1 + s_2)$ and $b_2(s_1 - s_2)$ (see Figure 4.15).

The difference in electronegativities is an indication of the relative energies of the "frontier" orbitals. The electronegativity of O is higher than that of N, so the lowest-energy unoccupied orbital of oxygen is lower in energy than that of N. The ionization energy is a measure of the energy of the highest-energy occupied orbital of an atom or molecule.

The N $2s$ and $2p_z$ orbitals belong to the same representation (a_1), so they both combine with O group orbitals that also have a_1 symmetry. This includes the O $2s$ orbitals, although they participate only slightly because of their relatively low energy. For convenience, we choose to orient the axes for the oxygens so that the $(+)$ z axis for each O is directed toward the N and the x axis is perpendicular to the molecular plane (Figure 4.15). Thus the O $2p_z$ orbitals are oriented ideally for σ bonding. The group orbitals from these

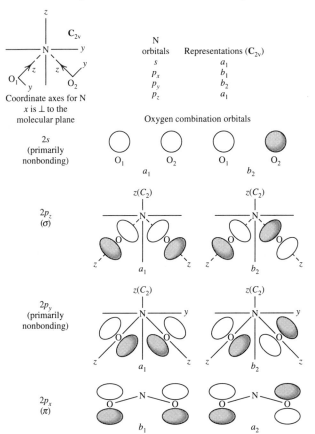

Figure 4.15 Nitrogen and oxygen combination orbitals for NO_2.

orbitals have a_1 and b_2 symmetry, as shown. The a_1 ($2p_z$) O group orbital can combine with the N a_1 orbitals (s and p_z). Because two N orbitals and the O group orbital combine, three MOs must result from the linear combination. We can visualize these best by looking at the combinations of N $2s$ and $2p_z$ orbitals (sp hybrids) and at the way these two hybrids could combine with the a_1 group orbital (Figure 4.16). The lowest energy combination (the one with overlapping lobes having the same signs) is strongly bonding, $2\sigma_{a_1}$, and the highest energy combination (overlapping lobes having opposite signs) is strongly antibonding, $5\sigma a_1^*$. The third combination, $(4a_1)^{nb}$, is mostly nonbonding because of poor overlap, as shown at the top right of Figure 4.16. The b_2 group orbital can combine with the $2p_y$ orbital of N to give a bonding orbital ($2\sigma b_2$) and an antibonding orbital ($4\sigma b_2^*$).

EXAMPLE: Sketch the b_2^* antibonding orbital for NO_2. Why is there no b_2 nonbonding orbital involving N p_y?

Solution: The b_2^* orbital is

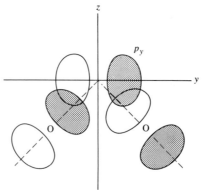

There are two b_2 orbitals to be combined, namely, one b_2 oxygen group orbital and one b_2 orbital on N ($2p_y$). Because we must conserve orbitals, only two MOs are obtained from the combination—the bonding and antibonding orbitals.

The O $2p_y$ orbital combinations of a_1 and b_2 symmetry (Figure 4.15) are primarily nonbonding ($3a_1$ and $3b_2$) because they do not give good overlap with the corresponding N orbitals.

The $2p_x$ orbitals for all three atoms, which are perpendicular to the molecular plane, are available for π bonding. The O group orbitals are b_1 ($p_x + p_x$) and a_2 ($p_x - p_x$) (Figure 4.15). The N p_x orbital gives bonding and antibonding MOs (Figure 4.16) with the b_1 group orbital. The a_2 group orbital is nonbonding, since there is no N orbital with this symmetry.

The σ bonding orbitals are $2\sigma a_1$ and $2\sigma b_2$, and the π bonding orbital, involving all three atoms, is πb_1 as shown in Figure 4.16 and the energy-level diagram (Figure 4.17). The $1a_1$ and $1b_2$ orbitals from combining the O $2s$ orbitals are primarily nonbonding, accommodating O lone pairs. The a_2 orbital is strictly nonbonding, because of symmetry, and accommodates another lone pair localized on the oxygens. The other two O lone pairs

of the five lone pairs in the preferred Lewis structure 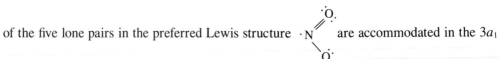 are accommodated in the $3a_1$

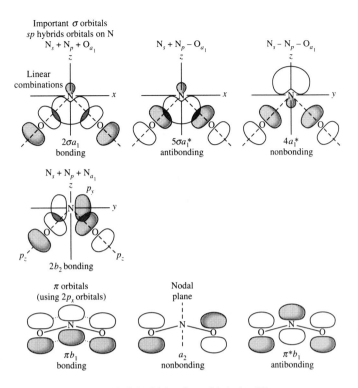

Figure 4.16   Molecular orbitals for NO_2.

$(p_y - p_y)$ and $3b_2$ $(p_y + p_y)$ orbitals. The HOMO $4a_1$ orbital (close in energy to N $2p_z$) accommodating the odd electron is localized primarily on N. Another electron can be added to this orbital to form NO_2^-.

Removing an electron from NO_2 to form NO_2^+ produces a linear ion with the CO_2 structure because another p orbital on N becomes available for π bonding. An NO_2 electron can be promoted from the $3b_2$ level to the $4a_1$ level in the electronic excited state without significant change in bonding or geometry. Promoting an electron in CO_2 to the π^* level can lead to removal of the double degeneracy of this level, with accompanying bending of the molecule. These levels (p_x, p_y) have different energies for a bent molecule because the x and y directions are nonequivalent. The electron goes into the lower-energy orbital in the bent molecule, lowering the energy of the molecule by β, the splitting parameter.

This is an application of the **Jahn-Teller theorem** (see page 407), which states that distortion must accompany splitting that leads to lower energy.

The actual situation is not so simple, since the orbitals from e_{1u}^* that result from the bending can mix with other orbitals belonging to the same representations, raising or low-

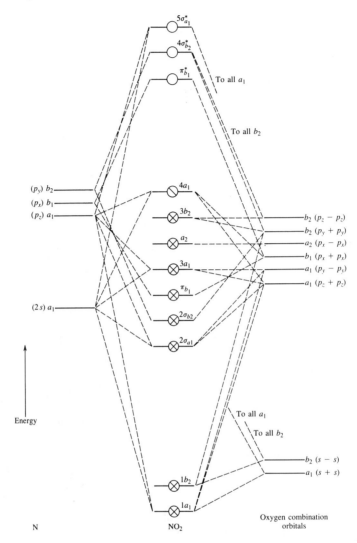

Figure 4.17 Qualitative MO energy-level diagram for NO_2.

ering their energies. Because of this, they do not split symmetrically. Note that electronically excited states need not have the same geometry as the ground state. One of the excited states of CO_2 would be much like NO_2, but with a vacancy in a lower nonbonding orbital.

▶ 4.3.3   Walsh Diagrams

In constructing MO descriptions of bonding in molecules, we assumed that the geometry is known or predicted from valence shell electron-pair repulsion (VSEPR) theory (see Section 2.2.3). **Walsh diagrams** show in the context of the MO model how electron count dictates molecular shape. This is accomplished by considering how MOs change in energy and composition as shape changes. In a series of detailed papers, Walsh[9] treated a range of simple molecules in an empirical and rather qualitative manner. In the past, MO treatments usually started with an assumed geometry. Now, at least for simple molecules, we can use computers to determine the optimum geometry, as well as the ordering of the energy levels. The more complex the molecule, the greater the importance of symmetry in simplifying the problem.

Consider an XH_2 molecule, where X is a second-period element with a maximum of eight electrons in four bonding or nonbonding orbitals. The relative energies of these orbitals and their occupancy determine the molecular shape. For the angular molecule, with the limiting bond angle of 90°, the axes are chosen as for NO_2. For the linear case, the y axis is chosen as the unique σ bond direction, so that when the angle is decreased the axes agree with the choice for the angular molecule. (Otherwise the unique direction would be taken as the z axis.)

Figure 4.18 depicts a Walsh diagram showing the correlation of the energies of the HOMOs only for linear and bent XH_2 molecules. Because the hydrogens have only $1s$ orbitals, the only bonding orbitals available are those with the symmetry of the H group orbitals (a_{1g} and a_{1u} for the linear case). In the bent molecule the a_1 orbital involves both s and p_z orbitals on X, but a_{1g} involves only the s orbital on X in the linear molecule. Note that s and p orbitals have different symmetries (different representations) so there can be no sp hybridization of orbitals of X for the linear case. Because p orbitals have higher energy than s orbitals, mixing in p_z would raise the energy of the a_1 MO for the bent molecule relative to the linear case, but the improved overlap with the hybrid and favor-

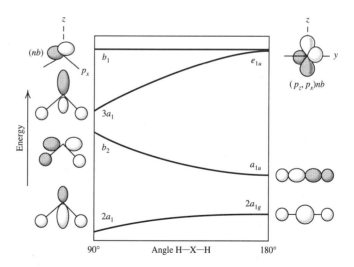

Figure 4.18   Walsh correlation diagram for XH_2.

[9] A. D. Walsh, *J. Chem. Soc.* **1953**, 2260–2331. See also N. C. Baird, *J. Chem Educ.* **1978**, *55*, 412; R. L. DeKock and W. B. Bosma, *J. Chem. Educ.* **1988**, *65*, 194; B. M. Gimarc, *Acc. Chem. Res.* **1974**, *7*, 384; and J. K. Burdett, *Molecular Shapes*, Wiley-Interscience, New York, 1980.

able overlap of the H $1s$ orbitals *decreases* slightly the energy of $2a_1$ on bending. The b_2 (bent)–a_{1u} (linear) pair involves opposite signs of the wavefunctions for the two H atoms. Because it is antibonding with respect to the H atoms themselves, this orbital becomes less stable as the H atoms come closer together with decreasing bond angle. The nonbonding e_{1u} orbitals for the linear molecule give two orbitals (a_1 and b_1) for the bent molecule. The p_x orbital, perpendicular to the plane of the molecule for any bond angle, is affected relatively little by bending because there is no π bonding. This is the b_1 orbital for the bent molecule. The e_{1u} orbitals (linear case) are pure p orbitals, but with increasing bending, p_z mixes more with the s orbital of X, lowering the energy of the a_1 orbital. Orbital $3a_1$ acquires more s character as $2a_1$ loses it. In fact, $2a_{1g}$ (linear) uses a pure s orbital, but $2a_1$ (at 90°) uses the p_z orbital, with sp hybrids involved at intermediate angles.

The BeH_2, BH_2^+, and HgH_2 molecules with four valence-shell electrons prefer the linear arrangement, filling the lower-energy bonding orbitals, $2a_{1g}$ and $1a_{1u}$. Molecules of the XH_2 type with five to eight valence electrons (including H_2O) should be bent. The first excited state of BeH_2 and HgH_2 should be bent. The BH_2 molecule is bent (131°), but has a linear excited state.

4.4 XENON DIFLUORIDE

The electron dot structure for XeF_2 places 10 electrons around Xe, implying the use of Xe $5d$ orbitals, but these orbitals are high in energy for bonding.

$$:\!\ddot{F}\!-\!\ddot{X}\ddot{e}\!-\!\ddot{F}\!:$$

Opposing views have been presented concerning the importance of the d orbitals in bonding. W. Kutzelnigg (*Angew. Chem. Int. Ed. Engl.* **1984,** *23,* 272) concluded that compounds exceeding the octet for elements beyond the second period has little to do with the availability of d orbitals. These compounds are formed because of the larger size of the atoms (thereby reducing steric hindrance between attached atoms), their lower electronegativities, and greater strength of single bonds compared to multiple bonds. See references in footnote 3 on page 63 for bond lengths in favor of multiple bonding for oxoanions.

The symmetry is $\mathbf{D}_{\infty h}$ and the possible combinations of s and p orbitals are the same as for CO_2, except that π bonding is not possible because p_x, p_y orbitals are filled for Xe and F atoms. There are $[(2 \times 7) + 8] = 22$ valence-shell electrons. The average bond energy is 130 kJ/mol and the Xe—F distance is 200 pm. For comparison, the bond energy for IF, which involves an ordinary two-center single bond, is 278 kJ/mol with a bond distance of 191 pm. The weak bonding for XeF_2 suggests a B.O. of about 0.5 for each Xe—F bond.

Rundle (1963) proposed an appealing half-bond model without using d orbitals. There is no net bonding from (p_x, p_y) orbitals on Xe, so they accommodate two of the nonbonding lone pairs. The other electron pair on Xe can be accommodated in the Xe atomic s orbital, or the three pairs can be visualized as being in sp^2 hybrid orbitals (localized MOs).

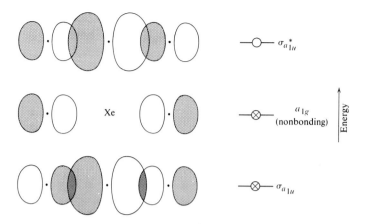

Figure 4.19   The three-center bonding of XeF_2.

The F $2s$, $2p_x$, and $2p_y$ orbitals are also nonbonding, accommodating three lone pairs on each F. The F p_z group orbitals are a_{1g} and a_{1u} (as for CO_2). The XeF_2 a_{1g} orbital is nonbonding because $5s$ is very low in energy. The Xe $p_z(a_{1u})$ combines with the a_{1u} group orbital to give σa_{1u} and σa_{1u}^* MO's. There are 18 nonbonding electrons in AOs (6 on each atom), leaving four electrons for the MOs. One bonding pair occupies σa_{1u} and one pair is in nonbonding a_{1g}, as shown in Figure 4.19. The bonding pair of electrons is delocalized over the three-center bond—or, each Xe—F bond has a B.O. of $\frac{1}{2}$. No antibonding orbitals are populated.

4.5   BORON TRIFLUORIDE—A GROUP THEORETICAL TREATMENT

The results in Figure 4.20 and 4.21 can be appreciated even if we do not work through all of the steps. Group theory is used only for these later sections of this chapter. After studying Section 3.5, you should be able to follow the detailed steps in getting the descriptions of the MOs. If group theory in Section 3.5 was omitted, coverage for BF_3, B_2H_6, and C_5H_5 can be limited to descriptions of the MOs and energy-level diagrams. The results obtained should seem reasonable from the symmetry of the molecules, and they can help your intuitive feel for molecules. For molecules of four or more atoms the group theoretical approach is desirable. The terminology used is summarized in Section 3.6.

The BF_3 molecule is planar with 120° bond angles ($\mathbf{D}_{3h}$ symmetry). The bonding description applies also to the isoelectronic CO_3^{2-} and NO_3^- ions. We choose the boron z axis along C_3, with the z axes of each F parallel to C_3 and the x axes directed toward B. The coordinate axes of the F atoms are brought into coincidence by the C_3 operation (Figure 4.20).

Character table

D_{3h}	E	$2C_3$	$3C_2$	σ_h	$2S_3$	$3\sigma_v$	
A_1'	1	1	1	1	1	1	
A_2'	1	1	-1	1	1	-1	
E'	2	-1	0	2	-1	0	(x,y)
A_1''	1	1	1	-1	-1	-1	
A_2''	1	1	-1	-1	-1	1	z
E''	2	-1	0	-2	1	0	

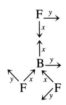

(a) Orientation of axes for BF$_3$.
z is $\perp$ to molecular plane.

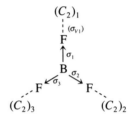

(b) Labels for σ orbitals.
C_2 axes and σ_{v_1}.

Transformation properties of σ orbitals

D_{3h}	E	C_3	C_2	σ_h	S_3	σ_v
σ_1	1	0	1	1	0	1
σ_2	1	0	0	1	0	0
σ_3	1	0	0	1	0	0
Γ_σ	3	0	1	3	0	1

Figure 4.20 Boron trifluoride: character table. (a) Orientation of axes. (b) Sigma bonds.

The choice of axes on B is not arbitrary, but the choices for F atoms are for convenience. If we want to use the character tables available, we must assign axes for the central atom in the same way as for the construction of the character table. The major C_n axis is, conventionally, the z axis.

▶ 4.5.1 Sigma Bonding

Irreducible Representations

We first find the irreducible representations to which the sets of σ bonds belong. The matrices for the transformations of the σ bonds can be written as follows, using the matrices for the symmetry operations (Section 3.2). That is, the matrix for each operation is the one that converts the orbital set into the orbital set resulting from the symmetry operation. Below each matrix the character (χ, the sum of diagonal elements) is given.

$$E \begin{pmatrix} \sigma_1 \\ \sigma_2 \\ \sigma_3 \end{pmatrix} = \begin{pmatrix} 1 & 0 & 0 \\ 0 & 1 & 0 \\ 0 & 0 & 1 \end{pmatrix} \begin{pmatrix} \sigma_1 \\ \sigma_2 \\ \sigma_3 \end{pmatrix} = \begin{pmatrix} \sigma_1 \\ \sigma_2 \\ \sigma_3 \end{pmatrix} \qquad \sigma_h \begin{pmatrix} \sigma_1 \\ \sigma_2 \\ \sigma_3 \end{pmatrix} = \begin{pmatrix} 1 & 0 & 0 \\ 0 & 1 & 0 \\ 0 & 0 & 1 \end{pmatrix} \begin{pmatrix} \sigma_1 \\ \sigma_2 \\ \sigma_3 \end{pmatrix} = \begin{pmatrix} \sigma_1 \\ \sigma_2 \\ \sigma_3 \end{pmatrix}$$
$$x = 3 \qquad\qquad x = 3$$

$$C_3 \begin{pmatrix} \sigma_1 \\ \sigma_2 \\ \sigma_3 \end{pmatrix} = \begin{pmatrix} 0 & 1 & 0 \\ 0 & 0 & 1 \\ 1 & 0 & 0 \end{pmatrix} \begin{pmatrix} \sigma_1 \\ \sigma_2 \\ \sigma_3 \end{pmatrix} = \begin{pmatrix} \sigma_2 \\ \sigma_3 \\ \sigma_1 \end{pmatrix} \qquad S_3 \begin{pmatrix} \sigma_1 \\ \sigma_2 \\ \sigma_3 \end{pmatrix} = \begin{pmatrix} 0 & 1 & 0 \\ 0 & 0 & 1 \\ 1 & 0 & 0 \end{pmatrix} \begin{pmatrix} \sigma_1 \\ \sigma_2 \\ \sigma_3 \end{pmatrix} = \begin{pmatrix} \sigma_2 \\ \sigma_3 \\ \sigma_1 \end{pmatrix}$$
$$x = 0 \qquad\qquad x = 0$$

$$C_2 \begin{pmatrix} \sigma_1 \\ \sigma_2 \\ \sigma_3 \end{pmatrix} = \begin{pmatrix} 1 & 0 & 0 \\ 0 & 0 & 1 \\ 0 & 1 & 0 \end{pmatrix} \begin{pmatrix} \sigma_1 \\ \sigma_2 \\ \sigma_3 \end{pmatrix} = \begin{pmatrix} \sigma_1 \\ \sigma_3 \\ \sigma_2 \end{pmatrix} \qquad \sigma_v \begin{pmatrix} \sigma_1 \\ \sigma_2 \\ \sigma_3 \end{pmatrix} = \begin{pmatrix} 1 & 0 & 0 \\ 0 & 0 & 1 \\ 0 & 1 & 0 \end{pmatrix} \begin{pmatrix} \sigma_1 \\ \sigma_2 \\ \sigma_3 \end{pmatrix} = \begin{pmatrix} \sigma_1 \\ \sigma_3 \\ \sigma_2 \end{pmatrix}$$
$$x = 1 \qquad\qquad x = 1$$

The set of characters obtained comprises a representation for the σ bonds in the group. The characters can be obtained without writing out the matrices by examining the effect of *one symmetry operation of each class* (*one* operation in each column in the character table) on each σ bond of the set treated as vectors (Figure 4.20*b*). If a vector is unchanged by an operation, we write $+1$—as in the case for σ_1 for E, C_2, σ_h, and σ_v, choosing the particular $(C_2)_1$ axis and $(\sigma_v)_1$ plane shown. The choice of the particular operation of each column is arbitrary for the first orbital examined (σ_1), but we use the *same* operations chosen for orbitals σ_1 for examining σ_2 and σ_3. Operations C_3 and S_3 exchange orbital σ_1 with σ_2, so σ_1 "goes into something else." We write 0 unless the vector (or an orbital) is unchanged (giving 1) or merely has its direction (sign) changed (giving -1). None of the operations here reverses the direction of orbital σ_1, so no -1 values are recorded. Using the *same set* of symmetry operations as for σ_1, only E and σ_h leave orbitals σ_2 and σ_3 unchanged. The sum of the numbers in each column is the character of the representation (Γ_σ in Figure 4.20) for the particular symmetry operation for the σ bonds of BF_3; this set of characters does not appear in the character table, so it is a *reducible representation*. The character for each operation is the same as obtained for the matrices above. By inspection (or by using the formula in Section 3.5.7), as we can see that the reducible representation is the sum of A_1' and E'. [Because the character is 3 for σ_h and only single-prime representations are involved, one must be E'. The character $+1$ for C_2 (or σ_v) indicates that the other representation must be A_1'.] The boron orbitals belonging to these representations are s (a_1') and p_x, p_y (e') because s is totally symmetric (here a_1') and p_x and p_y transform as the vectors (x, y) (here e').

Group Orbitals (Symmetry-Adapted Orbitals)

Next we find the combinations for the F group orbitals belonging to a_1' and e'—if such combinations exist—by using the **projection operator method.** Choosing *one* F orbital (σ_1), we perform *every* symmetry operation (not just one of each class) on it and write down the resulting orbital. Thus the E operation leaves orbital σ_1 unchanged, C_3 gives σ_2, C_3^2 (equivalent to C_3^{-1}, the counterclockwise rotation) gives orbital σ_3, and so on, giving

E	C_3	C_3^2	$(C_2)_1$	$(C_2)_2$	$(C_2)_3$	σ_h	S_3	S_3^2	σ_v	σ_v'	σ_v''
σ_1	σ_2	σ_3	σ_1	σ_3	σ_2	σ_1	σ_2	σ_3	σ_1	σ_3	σ_2

To find the linear combinations of F σ orbitals belonging to the A_1' representation, *we multiply each orbital generated by the character of A_1' for the operation used.* Because these characters are all $+1$ for A_1', we get the sum and then divide the result by 4 to simplify the coefficients:

$$\Psi_\sigma a_1' = 4\sigma_1 + 4\sigma_2 + 4\sigma_3 \quad \text{or} \quad \sigma_1 + \sigma_2 + \sigma_3$$

It is always permissible to simplify coefficients at this stage, since next we normalize to obtain the proper coefficients for the normalization condition (page 152).

The wavefunction for this F group orbital is the normalized LCAO (see page 152). (This group orbital is sketched in Figure 4.21.)

What if we try to get a group orbital for A_2', which was not included in Γ_σ? Multiplying the orbitals generated above by the respective characters for A_2' demonstrates that all of the orbitals cancel. There is no $\sigma_{a_2'}$ group orbital for BF_3.

The other group orbital is e', so we multiply the σ orbitals generated above by the respective characters for E', giving the sum

$$4\sigma_1 - 2\sigma_2 - 2\sigma_3 \quad \text{or} \quad 2\sigma_1 - \sigma_2 - \sigma_3$$

and

$$\Psi(\sigma_{e'})_a = \frac{1}{\sqrt{6}}(2\sigma_1 - \sigma_2 - \sigma_3)$$

Note that this orbital (Figure 4.21) has a nodal plane. It is *one* of the LCAOs for e', and we must find the other one of the degenerate pair.[10] Performing a C_3 operation on the LCAO above gives $2\sigma_2 - \sigma_3 - \sigma_1$; a C_3^2 operation gives $2\sigma_3 - \sigma_1 - \sigma_2$. These are not independent orbitals—we have merely interchanged numbers—but subtracting the second of these from the first gives

$$
\begin{array}{l}
 2\sigma_2 - \sigma_3 - \sigma_1 \\
\underline{-(-\sigma_2 + 2\sigma_3 - \sigma_1)} \\
 3\sigma_2 - 3\sigma_3 \quad \text{or} \quad \sigma_2 - \sigma_3
\end{array}
$$

and

$$\Psi(\sigma_{e'})_b = \frac{1}{\sqrt{2}}(\sigma_2 - \sigma_3)$$

---

[10] The systematic way to find the other LCAO of a degenerate set is called the Gram–Schmidt method for orthogonalization.

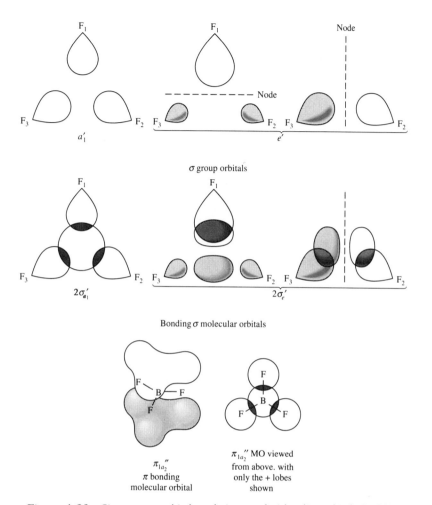

Figure 4.21   Sigma group orbitals and sigma and pi bonding orbitals for BF_3.

It helps to know what we are looking for in trying to arrive at the second LCAO. The a'_1 orbital has no nodal plane, and e'_a has one nodal plane. The other e' orbital (e'_b) also must have a nodal plane, and it will be perpendicular to that for e'_a, slicing through F_1. This means that the coefficient for the σ_1 orbital is zero and the other σ orbitals have opposite signs in this group orbital. Thus we are looking for an orbital combination with zero coefficient for σ_1 and opposite signs for σ_2 and σ_3, or $(\sigma_2 - \sigma_3)$. The group orbitals and the σ-bonding MOs are shown in Figure 4.21.

Orthonormal Wavefunctions

How do we know that these two orbitals are a basis for the E' representation? A basis set of functions must be **orthogonal;** if they are also normalized, they are **orthonormal.** Normalization requires that

$$\int \Psi^2 \, d\tau = 1$$

which is taken care of by the normalization factors (page 152). Orthogonality of two wavefunctions requires that the **scalar product** of the functions must equal zero, as for representations of groups (see Section 3.5.3):

$$\int \Psi_i \cdot \Psi_j \, d\tau = 0$$

Checking our pair of LCAO, we obtain

$$\int \frac{1}{\sqrt{6}}(2\sigma_1 - \sigma_2 - \sigma_3) \cdot \frac{1}{\sqrt{2}}(\sigma_2 - \sigma_3) \, d\tau$$

$$= \frac{1}{\sqrt{12}} \int (2\sigma_1\sigma_2 - 2\sigma_1\sigma_3 - \sigma_2^2 + \sigma_2\sigma_3 - \sigma_3\sigma_2 + \sigma_3^2) \, d\tau$$

Because σ_1, σ_2, and σ_3 are orthonormal atomic wavefunctions—that is, because we start with atomic wavefunctions that are orthonormal—we have

$$\int \sigma_i \cdot \sigma_j \, d\tau = 0 \quad \text{and} \quad \int \sigma_i^2 \, d\tau = 1$$

so

$$\int \Psi_i \Psi_j \, d\tau = \frac{1}{\sqrt{12}}[0 - 0 - 1 + 0 - 0 + 1] = 0$$

This is an orthonormal pair of functions fulfilling the requirements for the E' representation.

▶ 4.5.2   *Pi Bonding*

Pi bonding is handled in the same way. The p_z orbitals on the three F atoms, treated as vectors perpendicular to the molecular plane, give the following characters, considering one symmetry operation of each class:

	E	C_3	C_2	σ_h	S_3	σ_v
Γ_π	3	0	-1	-3	0	1

The identity operation leaves each of the three vectors unchanged giving the character 3. The C_3 operation takes each vector into another—this is recorded as zero for each vector. The $(C_2)_1$ operation interchanges the vectors (representing the p_z orbitals) with reversed signs on F_2 and F_3. These are recorded as zero because they "go into something else". The

C_2 operation causes the p_z orbital on F_1 to reverse it sign, giving the character -1. The σ_h operation reverses the signs of all three vectors representing the orbitals, giving a character -3. The S_3 operation takes each vector into something else—character 0. The σ_v operation leaves one vector unchanged and takes the other two into something else—character $= 1 + 0 + 0$. This set of characters (Γ_π above) does not appear in the character table, so it is a reducible representation. From the -3 character for σ_h we see that the irreducible representations must be double primed. The -1 character for C_2 reveals that $\Gamma_\pi = A_2'' + E''$. Checking the sums of characters for the other columns for A_2'' and E'' verifies this. The result can be obtained also using Equation (3.13).

The boron p_z orbital belongs to A_2'', so it can combine with the A_2'' group orbital for π bonding. The e'' group orbital must be nonbonding, since there is no e'' orbital for boron. We can get the LCAO for the a_2'' group orbital as before. Using all operations on the p_z orbital of one F (p_1) we get

E	C_3	C_3^2	$(C_2)_1$	$(C_2)_2$	$(C_2)_3$	σ_h	S_3	S_3^2	σ_v	σ_v'	σ_v''
$+p_1$	$+p_2$	$+p_3$	$-p_1$	$-p_3$	$-p_2$	$-p_1$	$-p_2$	$-p_3$	$+p_1$	$+p_3$	$+p_2$

and multiplying by the respective characters for A_2'' gives $4p_1 + 4p_2 + 4p_3$ or simplifying $p_1 + p_2 + p_3$ and

$$\Psi_a'' = \frac{1}{\sqrt{3}}(p_1 + p_2 + p_3)$$

Figure 4.21 shows the combination of this π group orbital with the p_z orbital of boron. Because there is only one π bonding group orbital for BF_3, we could sketch the obvious combination and check to see that it has a_2'' symmetry without using the projection operator method. That is, the orbital set is unchanged ($+1$) by the C_3 or σ_v, but reverses sign (-1) for C_2, σ_h, and S_3, corresponding to the characters for A_2'' in the character table; the lobe above the atoms is symmetrical over all atoms and has opposite charge of the equivalent of the lobe below the atoms.

We found that the doubly degenerate e'' group orbitals are nonbonding because B has no e'' orbital. These two group orbitals resemble the e' orbitals in sign patterns, except that the sketches would apply to the lobes of the p_z orbitals on one side of the molecule and the corresponding lobes on the opposite side would have opposite signs. Each group orbital has one nodal plane perpendicular to the plane of the molecule (as for e') and one in the molecular plane.

There are four bonding orbitals ($2a_1'$, $2e'$, $1a_2''$) and four corresponding antibonding orbitals. Adding the nonbonding e'' orbitals gives a total of 10 orbitals. We used all four boron orbitals and two for each fluorine. The σ orbital for F is primarily $2p_x$. The much-lower-energy $2s$ orbitals (in D_{3h} symmetry $1a_1'$ and $1e''$ groups) are nonbonding, and the p_y orbitals (a_2' and $3e'$ groups) are nonbonding because they are not properly oriented for σ or π bonding. A qualitative energy-level diagram is depicted in Figure 4.22. The eight lone pairs on F atoms are in $1a_1'$, $1e'$, e'', a_2', and $3e'$.

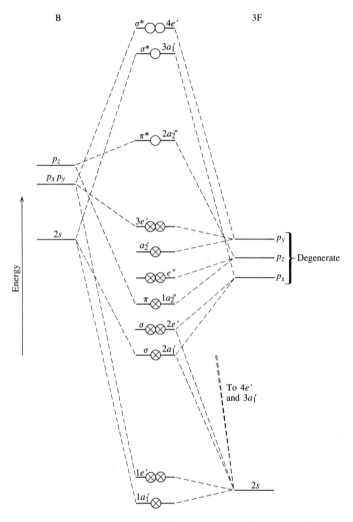

Figure 4.22   Qualitative MO energy-level diagram for BF_3.

Here is a general procedure for obtaining an MO description for a discrete molecule of the AX_n type:

1. Determine the point group for the molecule. (*Ab initio* computer methods can determine the optimum symmetry of a molecule.)

2. Deal separately with σ and π bonding. For BF_3 and SF_6 (octahedral) all σ bonds are equivalent and can be dealt with together. For PCl_5 (trigonal bipyramidal) the axial and equatorial bonds are nonequivalent. No symmetry operations of the $\mathbf{D}_{3h}$ group interchange axial and equatorial bonds. These sets are dealt with separately.

3. Find the irreducible representations for the sets of the σ and π bonds for the central atom from symmetry transformations for each set, which generates a reducible representation.

4. Obtain the group orbitals (symmetry-adapted orbitals) for the σ and π bond sets of the ligands using the projection operator method. In many cases the AOs on the central atom provide templates for the corresponding ligand group orbitals. There might be nonbonding ligand group orbitals or orbitals on the central atom, and these will not have counterparts of the other set.

5. Combine the orbitals from the central atom and the ligand group orbitals to obtain MOs.

6. Draw a qualitative MO energy-level diagram using the relative energies of the AOs and the effectiveness of overlap as a guide to MO energies. The energies of MOs generally increase with greater localization as indicated by the larger number of nodes. This is well illustrated in the case of π orbitals for cyclic molecules (Figures 4.27 and 4.28).

4.6   DIBORANE

Diborane, B_2H_6, presented a challenge in early attempts to describe its bonding. The chemistry and history of the boranes are discussed in some detail in Chapter 17. For now, we focus on describing an unusual molecule's electronic structure in terms of the MO model. Diborane often is described as electron-deficient, because there are only six electron pairs available in the shell although C_2H_6, with the same formula type, requires seven bonds.[11] Actually B_2H_6 does not have the C_2H_6 structure, and the borons and four terminal hydrogens are in a planar ethylene arrangement. The other two hydrogens form bridges between borons in a plane perpendicular to the ethylenelike framework (Figure 4.23). Because this point group (D_{2h}; see Appendix C for the character table) has no axis of highest order, the choice of axes is arbitrary—no atom is at the origin and no choice is advantageous with respect to bond directions. We choose to orient the framework in the xz plane with the z axis through the borons.

Each boron has about it a distorted tetrahedral arrangement of four hydrogens, so for a simple treatment we can use sp^3 hybrid orbitals on each boron. The B—H terminal bonds are ordinary two-center bonds and can be treated as localized MO bonds. We will focus on the bridge bonding. Using two B orbitals for B—H terminal bonds, there remain two sp^3 orbitals per B, one s orbital per bridging H, and four electrons.

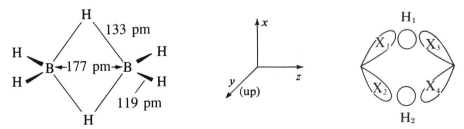

Figure 4.23   Structure and orientation of the orbitals involved in the bridges for B_2H_6.

---

[11] The MO correlation between the C_2H_6 and B_2H_6 structures is instructive. See B. M. Gimarc, *Acc. Chem. Res.* **1974**, *7*, 384.

▶ 4.6.1   *Representations for the Boron and Hydrogen Orbitals*

First we must determine the representations for the B and H bridging orbitals as shown in Figure 4.23. We carry out the symmetry operations on each B and H orbital, recording whether the orbital is left unchanged ($+1$) or changes into another orbital (0). No operation changes the direction (sign) of any of these orbitals. We can write down the totals for the representations for the B orbitals more simply as the number of orbitals of set left unchanged by a particular operation. That is, for the identity and σ_{xz} symmetry operations all 4 boron orbitals are unchanged, and we record 4 for E and σ_{xz}. Each other operation changes any orbital into another orbital, so record 0 for these operations. Check for yourself the values (2 or 0) for the H orbitals tabulated below:

D_{2h}	E	$C_2(z)$	$C_2(y)$	$C_2(x)$	i	σ_{xy}	σ_{xz}	σ_{yz}
Γ_{boron}	4	0	0	0	0	0	4	0
Γ_{H}	2	0	0	2	0	2	2	0

$$\Gamma_{\text{boron}} = A_g + B_{2g} + B_{1u} + B_{3u} \qquad \Gamma_{\text{H}} = A_g + B_{3u}$$

Checking the character table (Appendix C) for D_{2h}, we see that these are reducible representations. For Γ_{boron}, only A_g, B_{2g}, B_{1u}, and B_{3u} have $+1$ characters for σ_{xz}, so these must be irreducible representations contained in Γ_{boron}. The other operations can be used to check this result to ensure that one of these does not occur more than once, eliminating another one. (See Section 3.5.7 for the systematic reduction of reducible representations, although reduction by inspection is simple here.) The reducible representation for the H orbitals can be recognized from the character table as containing A_g and B_{3u} because these are the only representations having $+1$ characters for $C_2(x)$ and σ_{xy}. The a_g and b_{3u} orbitals for boron and hydrogen can combine, and the boron b_{2g} and b_{1u} orbitals must be nonbonding because there are no b_{2g} or b_{1u} H group orbitals.

▶ 4.6.2   *Group Orbitals*

We obtain the LCAO for the group orbitals by using the projection operator. We start with one B orbital (X_1) and one H orbital (H_1) and record the orbital obtained after carrying out *each* operation of the group:

D_{4h}	E	$C_2(z)$	$C_2(y)$	$C_2(x)$	i	σ_{xy}	σ_{xz}	σ_{yz}
X_1	X_1	X_2	X_4	X_3	X_4	X_3	X_1	X_2
H_1	H_1	H_2	H_2	H_1	H_2	H_1	H_1	H_2

Multiplying each orbital obtained by the character of the operation for the representations involved for the B and H orbitals produces the following LCAO and, after normalization, the wavefunctions:

$$a_g \qquad 2X_1 + 2X_2 + 2X_3 + 2X_4, \qquad \Psi_{a_g} = \frac{1}{2}(X_1 + X_2 + X_3 + X_4)$$

$$b_{2g} \qquad 2X_1 - 2X_2 - 2X_3 + 2X_4, \qquad \Psi_{b_{2g}} = \frac{1}{2}(X_1 - X_2 - X_3 + X_4)$$

$$b_{1u} \qquad 2X_1 + 2X_2 - 2X_3 - 2X_4, \qquad \Psi_{b_{1u}} = \frac{1}{2}(X_1 + X_2 - X_3 - X_4)$$

$$b_{3u} \qquad 2X_1 - 2X_2 + 2X_3 - 2X_4, \qquad \Psi_{b_{3u}} = \frac{1}{2}(X_1 - X_2 + X_3 - X_4)$$

$$a_g \qquad 4H_1 + 4H_2, \qquad \Psi_{a_g} = \frac{1}{\sqrt{2}}(H_1 + H_2)$$

$$b_{3u} \qquad 4H_1 - 4H_2, \qquad \Psi_{b_{3u}} = \frac{1}{\sqrt{2}}(H_1 - H_2)$$

Applying the signs obtained to the sketch of the orbitals gives us the bridge MOs shown in Figure 4.24 where shaded lobes have negative signs. A qualitative energy-level diagram (Figure 4.25) can be drawn for bridge bonding. The a_g bonding orbital is lower in energy than the b_{3u} orbital, because the latter has a nodal plane causing more localization of the electron cloud. Similarly, the nonbonding b_{2g} orbital (two nodes) should be higher in energy than b_{1u} (one node). Of the 12 valence shell electrons, four electron pairs are used for the four terminal B—H bonds (localized MOs). The remaining four electrons available for bridging bonding can be accommodated in the two bonding orbitals, leaving the nonbonding and antibonding orbitals empty. A quantitative treatment might change the order of higher-energy orbitals because of differences in the energy separation between B and H AOs. Here they are arranged so that a_g^* is highest because a_g is the more stable bonding orbital.

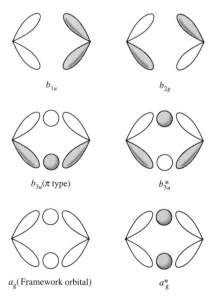

b_{1u}

b_{2g}

$b_{3u}(\pi \text{ type})$

b_{3u}^*

a_g(Framework orbital)

a_g^*

Figure 4.24   Bridge MOs for B_2H_6.

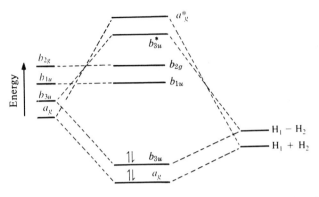

Figure 4.25 Qualitative MO energy-level diagram for the bridge bonding in B_2H_6.

The six atomic or hybrid orbitals are all involved in the bonding a_g and b_{3u} MOs. There are two three-center bridge bonds, each involving one electron pair. In these so-called electron-deficient molecules, only the lowest (bonding) orbitals are filled. Many of these molecules can accept electrons to form anions, but with accompanying structural changes. More commonly, stable molecules have all bonding and nonbonding orbitals filled.

A Double Quartet (Section 4.2) description of B_2H_6 is interesting. The six B electrons can be represented as

$$
\begin{matrix}
\times & & \circ & & \times \\
& B & & B & \\
\circ & & \times & & \circ
\end{matrix}
\qquad \text{giving} \qquad
\begin{matrix}
H & & . & H & . & & H \\
& \searrow & & & & \nearrow & \\
& & B & & & B & \\
& \nearrow & & & & \searrow & \\
H & & . & H & . & & H
\end{matrix}
$$

There are four electrons arranged tetrahedrally around each B (before H atoms are added). Terminal H atoms form localized electron-pair bonds with the four H atoms in a plane. The spiranlike arrangement of the B electrons leads to the two bridging H atoms to be in a plane perpendicular to that of the ethylenelike B_2H_4 plane. The bridging H atoms are involved in three-center bonds. This simple description is consistent with the properties and MO descriptions of B_2H_6. Alternatively, we might consider $B_2H_4^{2-}$, isoelectronic with C_2H_4, representing the double bonds as "bent banana" bonds. The two protons would be embedded in these bent bonds. This describes Pitzer's protonated double-bond structure for B_2H_6:

$$
\begin{matrix}
& & H & & \\
H & & | & & H \\
\searrow & & & & \swarrow \\
& B & == & B & \\
\nearrow & & & & \nwarrow \\
H & & | & & H \\
& & H & &
\end{matrix}
$$

4.7   CYCLIC-PLANAR π MOLECULES

Planar molecules, such as C_6H_6 and C_5H_5 (cyclopentadiene, a radical that forms the stable $C_5H_5^-$ ion), are of interest because of their delocalized π bonding. They form "sandwich"-type metal complexes such as ferrocene, $Fe(C_5H_5)_2$, and dibenzenechromium, $Cr(C_6H_6)_2$.

The bonding description of ferrocene (in Chapter 12) utilizes the C_5H_5 π bonding scheme we shall now examine.

The point group for C_5H_5 is $\mathbf{D}_{5h}$, but it is much easier to identify the π LCAOs using the cyclic subgroup $\mathbf{C}_5$ because of fewer operations involved. When the $\mathbf{C}_5$ representations are identified, their labels in the $\mathbf{D}_{5h}$ group, if desired, can be found by examining the symmetry of the orbitals obtained and adding appropriate subscripts or primes. For a planar C_nH_n molecule, each carbon can be regarded as using sp^2 hybrid orbitals for σ bonding to hydrogen and the neighbor carbons. One p orbital for each C is perpendicular to the plane giving n p orbitals, and in the cyclic $\mathbf{C}_n$ point group, *one* of the resulting π MOs belongs to *each* of the n representations. We do not have to determine the representations of the π MOs as before. For C_5H_5 the five MOs correspond to the representations A, E_1, and E_2. In cyclic groups each doubly degenerate E representation consists of a pair of one-dimensional representations—this solves the problem of finding the linear combination for each of the E pair. Because the operations of the group merely involve the successive application of C_5 rotations, operations on p_1 (the p orbital of carbon atom 1) moves p_1 to p_2, to p_3, and so on, giving

$\mathbf{C}_5$	E	C_5	C_5^2	C_5^3	C_5^4	$(C_5^5 = E)$
p_1	p_1	p_2	p_3	p_4	p_5	

Using the projection operator method (multiplying each orbital obtained by the appropriate character from the character table for each of the five representations), we obtain

$$
\begin{array}{ll}
a & p_1 + p_2 + p_3 + p_4 + p_5 \\
e_1 \begin{cases} e_1(1) & p_1 + \varepsilon p_2 + \varepsilon^2 p_3 + \varepsilon^{2*} p_4 + \varepsilon^* p_5 \\ e_1(2) & p_1 + \varepsilon^* p_2 + \varepsilon^{2*} p_3 + \varepsilon^2 p_4 + \varepsilon p_5 \end{cases} \\
& \hspace{4cm} \varepsilon = e^{-2\pi i/5} \\
e_2 \begin{cases} e_2(1) & p_1 + \varepsilon^2 p_2 + \varepsilon^* p_3 + \varepsilon p_4 + \varepsilon^{2*} p_5 \\ e_2(2) & p_1 + \varepsilon^{2*} p_2 + \varepsilon p_3 + \varepsilon^* p_4 + \varepsilon^2 p_5 \end{cases}
\end{array}
$$

The complex numbers can be eliminated by taking linear combinations of the pairs. Adding $e_1(1)$ and $e_1(2)$, we obtain

$$
e_{1a} = 2p_1 + (\varepsilon + \varepsilon^*)p_2 + (\varepsilon^2 + \varepsilon^{2*})p_3 + (\varepsilon^{2*} + \varepsilon^2)p_4 + (\varepsilon^* + \varepsilon)p_5
$$

and from trigonometric formulas (see Section 3.5.6) for evaluating the conjugate pairs, we obtain

$$
\varepsilon = \cos\frac{2\pi}{5} + i\sin\frac{2\pi}{5}, \qquad \varepsilon^* = \cos\frac{2\pi}{5} - i\sin\frac{2\pi}{5}
$$

$$
\varepsilon + \varepsilon^* = 2\cos\frac{2\pi}{5} = 0.618
$$

$$
\varepsilon^2 = \cos\frac{4\pi}{5} + i\sin\frac{4\pi}{5}, \qquad \varepsilon^{2*} = \cos\frac{4\pi}{5} - i\sin\frac{4\pi}{5}
$$

$$
\varepsilon^2 + \varepsilon^{2*} = 2\cos\frac{4\pi}{5} = -1.618
$$

giving

$$e_{1a} = 2p_1 + 0.618p_2 - 1.618p_3 - 1.618p_4 + 0.618p_5$$

Dividing by two and normalizing gives

$$\Psi(e_{1a}) = \frac{1}{\sqrt{2.5}}(p_1 + 0.314p_2 - 0.809p_3 - 0.809p_4 + 0.314p_5)$$

Subtracting $e_1(2)$ from $e_1(1)$ and normalizing gives

$$\Psi(e_{1b}) = \frac{1}{\sqrt{7.24}}(1.62p_2 + p_3 - p_4 - 1.62p_5)$$

Similarly, for e_2 we get from the sums and difference of $e_2(1)$ and $e_2(2)$

$$\Psi(e_{2a}) = \frac{1}{\sqrt{2.5}}(p_1 - 0.809p_2 + 0.314p_3 + 0.314p_4 - 0.809p_5)$$

$$\Psi(e_{2b}) = \frac{1}{\sqrt{7.24}}(p_2 - 1.62p_3 + 1.62p_4 - p_5)$$

These MOs are sketched in Figure 4.26. Nodal planes are represented by lines separating the lobes of opposite signs. The relative order of the energy levels can be determined easily. The completely delocalized a orbital (a five-center orbital) has lowest energy, and the others increase in energy with increasing numbers of nodes.

We can examine the symmetries of the MOs in Figure 4.26 to identify the representations in D_{5h} using that character table. They are all doubly primed because they change sign for σ_h, giving e_1'' and e_2'' in D_{5h}. The $a(C_5)$ orbital is a_2'' in D_{5h} because of change in sign for σ_h and C_2.

A simple MO description, without the coefficients for the LCAO but with the ordering of the energies of the π orbitals, can be obtained for any planar cyclic C_nH_n system without going through all the steps. The labels for the levels in the cyclic group, C_n, are the representations in the respective character tables: C_3, a and e; C_4, a, e, and b; C_5, a, e_1, and e_2; C_6, a, e_1, e_2, and b; C_7, a, e_1, e_2, and e_3; and so on. The delocalized a orbital is always lowest in energy, with the e's increasing in energy in the order $e_1 < e_2 < e_3$. When n is even, there is a b orbital that is highest in energy with an alternating sign pattern of lobes around the ring. From Figure 4.27, which shows the relative energies of the orbitals for C_nH_n molecules, we readily see why C_5H_5 gives a stable anion $C_5H_5^-$, filling low-energy orbitals, and why C_7H_7 forms the cation $C_7H_7^+$, losing the one high-energy electron. C_5H_5 adds an electron to fill the HOMO, and C_7H_7 loses the single electron in the HOMO. The empirical rule for aromaticity ($4n + 2$ electrons) is also evident, since MOs being filled after the first come in pairs (e). The sequence of energy levels, and their relative energies, correspond to the spacing of vertices of regular polygons standing on

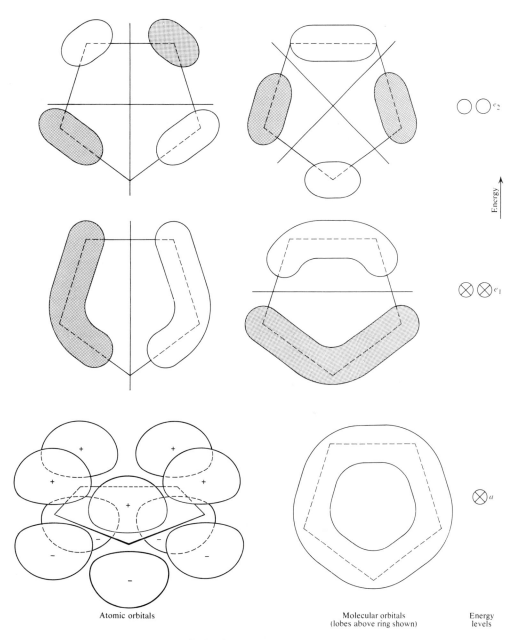

Atomic orbitals

Molecular orbitals
(lobes above ring shown)

Energy
levels

Figure 4.26   Pi MOs for C_5H_5. The labels are for the $\mathbf{C}_5$ point group.

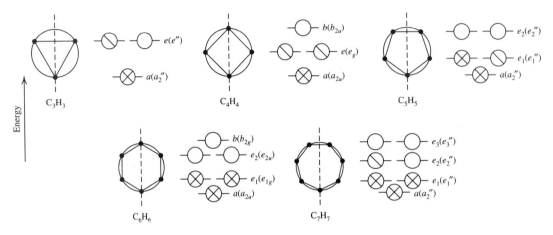

Figure 4.27   Relative energies of the π MOs for planar cyclic C_nH_n molecules.

one vertex as shown in Figure 4.27. Symmetry species are identified for the cyclic group and the full group.

The patterns for the π orbitals for C_nH_n molecules are simple (Figure 4.28). The most stable MO, a, is symmetrical around the ring. Each e_1 MO has a single nodal plane, in addition to the node shared by all of the π orbitals in the plane of C atoms. We superimpose one of these orbitals on the molecule in a manner consistent with its symmetry and arrange the other e_1 MO so that its nodal plane is perpendicular to that of the first choice. For e_2 there are two perpendicular nodal planes. After placing one e_2 MO on the framework of the molecule, the second is placed so that the nodal planes bisect the angles formed in the first e_2 sketch. For C_4H_4 or C_6H_6, superimposing the nodal planes for the highest MO (b) on the framework of carbons such that the nodes—two and three planes, respectively—slice between the carbons, producing a strongly antibonding MO with alternating signs around the ring. If we try the other combination, with the nodal planes bisecting the angles for the first case, the planes go through all of the carbon atoms. This corresponds to zero coefficients for each AO, and the wavefunction vanishes. Consequently, there is only one highest-energy MO (b) for C_4H_4, C_6H_6, or C_nH_n, where n is even.

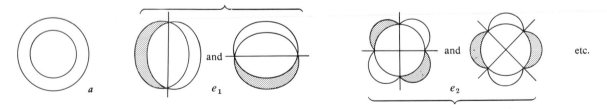

Figure 4.28   Molecular orbital wave patterns for cyclic π systems, showing the lobes on one side of the molecular plane.

4.8   BASIS SETS AND FRONTIER ORBITALS

An advantage of the VB model is the ease of visualizing a simple model of a molecule and its bonding. Hybridization is used extensively, choosing a hybrid combination that "fits" the geometry of the molecule. In the MO models it is not necessary but often convenient to use hybrids as basis sets of orbitals. In B_2H_6 we used sp^3 hybrids on B. Geometry around each B distorts the angles of a regular tetrahedron. Nevertheless, the four bonds around each B can be visualized more readily starting with sp^3 hybrids than starting with s and p AOs. We treated the terminal B—H bonds as localized MOs, and focused on bridge bonding.

The MOs of a molecule are those determined by symmetry. In CH_4 the four bonds are equivalent and the H atoms are equivalent in an NMR spectrum. Each bond can be described as a localized two-center bond resulting from the overlap of a C sp^3 hybrid orbital and the s orbital of H. Such a **localized MO model** does not differ significantly from a VB description. However, there is no MO with four-fold degeneracy. For T_d the highest degeneracy is threefold (T_2). We find the representations of the MOs as LCAO by applying the symmetry operations of the group. In the delocalized MO model of CH_4 there are still four equivalent C—H bonds. Each bond has the same participation of s and p AOs, equivalent to sp^3 hybridization.

Bonding in the H_2O molecule is commonly described as using sp^3 hybrids. The localized MO model for H_2O recognizes that the bond angle (104.5°) more closely matches the tetrahedral angles than the 90° angles for p orbitals. The sp^3 hybrid basis set can form two localized two-center bonds with two localized lone pairs. The essentially tetrahedral distribution of hydrogen bonds about oxygen in ice is consistent with localized lone pairs. The deviation of the bond angles in H_2O from the tetrahedral angle can be explained as the result of repulsions involving the lone pairs or greater oxygen s character for the orbitals occupied by the lone pairs.[12]

The delocalized MO description for H_2O begins with the C_{2v} symmetry:

If we orient the molecule in the xz plane, the representations are A_1 (s and p_z), B_1 (p_x), and B_2 (p_y). The H (1s) combination orbitals are a_1 and b_1 (same as the oxygen s combination orbitals for NO_2 in Figure 4.15). The σ MOs are a_1 (p_z with slight s character) and b_1 (p_x). The a_1 MO involving mostly s character is lowest in energy because of the low energy of the s orbital itself, and it is primarily nonbonding. Note the 14-eV separation in binding energy between $1a_1$ and $2a_1$ in Figure 4.29. The bonding a_1 MO is next higher in energy, and b_1 is still higher in energy because of poorer overlap and no participation by s. The b_2 (p_y) orbital is nonbonding because there is no b_2 H combination orbital.

[12] M. B. Hall, *J. Am. Chem. Soc.* **1978**, *100*, 6333; *Inorg. Chem.* **1978**, *17*, 2261.

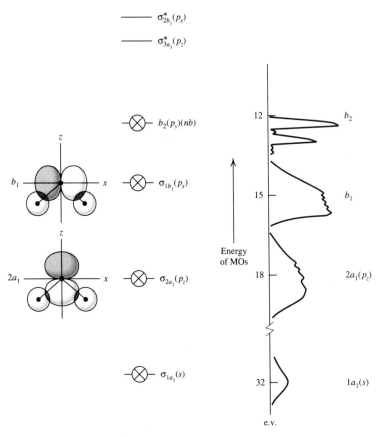

Figure 4.29 Bonding MO's, MO energy-level diagram for H_2O and its photoelectron spectrum. (Adapted from A. W. Potts and W. C. Price, *Proc. R. Soc. London* **1972**, *A326*, 181.)

Photoelectron spectroscopy (PES) involves the ejection of electrons resulting from the collision of photons with atoms or molecules. The kinetic energies of the ionized electrons are monitored. PES yields valence electron binding energies. The PES spectrum of H_2O is shown in Figure 4.29 beside a MO energy-level diagram with sketches of the bonding MOs.

The localized bond description is a satisfactory physical model for H_2O. Generally, the localized bonding approach is adequate for dealing with geometry in simply molecules. However, the PES spectrum matches the energy-level diagram derived from the delocalized MO treatment for H_2O. There are no degenerate MOs. The delocalized MO treatment, using symmetry-adapted MOs, is better for dealing with energy states and spectra.

For electron transitions in spectroscopy and for electron transfer, it is the highest occupied molecular orbitals (HOMOs) and the lowest unoccupied molecular orbitals (LUMOs) that are involved. These are called the frontier orbitals.[13] Because for many pur-

---

[13] K. Fukui, *Angew. Chem.* **1982**, *21*, 801.

poses these are the only orbitals involved, they are very important. Symmetry rules for chemical reactions depend upon the symmetry of these frontier orbitals.[14]

The energies of the frontier orbitals for a molecule in terms of ionization energies and electron affinities from Koopman's theorem are

$$-E_{HOMO} \approx IE, \qquad -E_{LUMO} \approx EA$$

The absolute electronegativity (χ_{abs}), and absolute hardness (η) (Section 1.8) are

$$\chi_{abs} = \frac{(IE + EA)}{2}, \qquad \eta = \frac{(IE - EA)}{2}$$

Hard molecules have a large HOMO–LUMO gap, and soft molecules have a small HOMO–LUMO gap.[15] Figure 4.30 shows the decreases in χ_{abs} and η (or increase in softness) from Cl_2 (IE = 11.6, EA = 2.4), to Br_2 (IE = 10.56, EA = 2.6), to I_2 (IE = 9.4, EA = 2.6). Figure 1.16 is similar for F^+, F^-, I^+, and I^-.

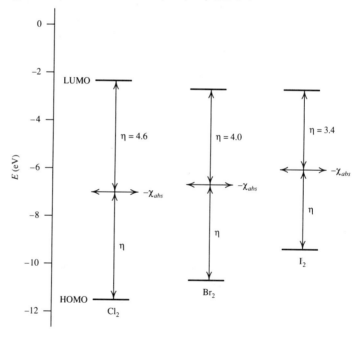

Figure 4.30 Absolute hardness (η) and electronegativity (χ_{abs}) for Cl_2, Br_2, and I_2 from energies of HOMO (IE) and LUMO (EA).

[14] R. B. Woodward and R. Hoffmann, *Angew. Chem.* **1969**, *8*, 781; R. Hoffmann, *Angew. Chem.* **1982**, *21*, 711 (Nobel Lecture); K. Fukui, *Acc. Chem. Res.* **1971**, *4*, 57; R. G. Pearson, *Symmetry Rules for Chemical Reactions*, Wiley-Interscience, New York, 1976.

[15] R. G. Pearson, *Acc. Chem. Res.* **1993**, *26*, 250; R. G. Parr and Z. Zhou, *Acc. Chem. Res.* **1993**, *26*, 256. These are applications of absolute hardness and the Principle of Maximum Hardness.

GENERAL REFERENCES

T. A. Albright, J. K. Burdett, and M.-H. Whangbo, *Orbital Interactions in Chemistry*, Wiley, New York, 1985.

H. Bock and P. D. Mollere, *J. Chem. Educ.* **1974,** *51,* 506. An experimental approach to teaching molecular orbital theory.

J. K. Burdett, *Molecular Shapes,* Wiley-Interscience, New York, 1980. Basically a molecular orbital treatment of inorganic stereochemistry.

F. A. Cotton, *Chemical Applications of Group Theory,* 3rd ed., Wiley-Interscience, New York, 1990.

R. L. DeKock and H. B. Gray, *Chemical Structure and Bonding,* Benjamin/Cummings, Menlo Park, CA, 1980. Good treatment.

B. E. Douglas and C. A. Hollingsworth, *Symmetry in Bonding and Spectra,* Academic Press, New York, 1985. An introduction applying group theory.

W. E. Hatfield and R. A. Palmer, *Problems in Structural Inorganic Chemistry,* Benjamin, New York, 1971. Good problems with solutions.

W. E. Hatfield and W. E. Parker, *Symmetry in Chemical Bonding and Structures,* Merrill, Columbus, OH, 1974. Similar to the level of this text with more quantum mechanics.

S. F. A. Kettle, *Symmetry and Structure,* Wiley, New York, 1985.

J. N. Murrell, S. F. A. Kettle, and J. M. Teddler, *The Chemical Bond,* 2nd ed., Wiley, New York, 1985.

R. McWeeney, *Coulson's Valence,* Oxford Press, Oxford, 1979.

PROBLEMS

4.1 Give the bond order and the number of unpaired electrons for Be_2^+, B_2^+, C_2^+, O_2, O_2^+, O_2^-, O_2^{2-}.

4.2 Sketch σ bonding orbitals that result from combinations of the following orbitals on separate atoms: p_z and d_{z^2}, s and p_z, $d_{x^2-y^2}$ and $d_{x^2-y^2}$ (let the σ bond be along the x axis in the latter case).

4.3 Sketch π bonding orbitals that result from combination of the following orbitals on separate atoms with σ bonding along with the z axis: p_x and p_x, p_x and d_{xz}, d_{xz} and d_{xz}.

4.4 Sketch a δ bond for an X—Y molecule, identifying the atomic orbitals and showing the signs of the amplitudes of the wavefunctions (signs of the lobes).

4.5 Explain why the CO ligand donates the lone pair on carbon in bonding to metals, rather than the pair on oxygen.

4.6 Write the MO configurations (e.g., σ_{1s}^2 for H_2) and give the bond orders of NO^+, NO, and NO^-. Which of these species should be paramagnetic?

4.7 Why must atomic orbitals have very similar energies to form MOs? (Consider the case of combining orbitals of two atoms having much different energies as indicated by ionization energies or electronegativities.)

4.8 (a) Draw an MO energy-level diagram for the MOs that would arise in a diatomic molecule from the combination of unhybridized d orbitals. Label the AOs being combined and the resulting MOs. Let the z axis lie along the σ bond.

    (b) Sketch the shape of these orbitals.

4.9 Account for the differences in dissociation energies and bond lengths in Table 2.8 on addition and removal of an electron from
(a) O_2 (b) N_2 (c) CO (d) NO

4.10 The bond dissociation energy of C_2 (599 kJ/mol) decreases slightly on forming C_2^+ (513 kJ/mol) and increases greatly on forming C_2^- (818 kJ/mol). Why?

4.11 Draw a Double Quartet representation for ClO_2. Is resonance required for the representation?

4.12 The formation of OF_2 from F_2 and O_2 is endothermic. Compare the Double Quartet descriptions of OF_2, F_2, and O_2 with respect to localization of electron pairs.

4.13 The BH^- group is isoelectronic with CH. What would be the effect on the π MOs and energy levels of C_5H_5 if one CH group were substituted by
(a) BH^- giving the planar borole $C_4BH_5^-$?
(b) NH^+ giving the planar pyrrole $C_4NH_5^+$?

4.14 The Lewis structure for CO_2 shows four lone pairs of electrons on the oxygen. There are two nonbonding pairs in the MO description. How do you reconcile the two descriptions?

4.15 The preferred electron dot structure for NO_2 shows five lone pairs on the oxygens and one unshared electron on N. Show how this description is consistent with the MO description.

4.16 In forming NO_2^+ from NO_2, from which MO is the electron lost? On which atom is this electron localized? To which orbital is an electron added in forming NO_2^-? What are the consequences, in terms of molecular shape, of the loss and gain of an electron?

4.17 Why is the first excited state of BeH_2 bent, whereas that of BH_2 is linear? (Consider the Walsh diagram.)

4.18 Apply the group theoretical treatment to obtain the bonding description for NO_2^-.

4.19 Apply the group theoretical treatment to obtain the MO description for σ bonding in PF_5 ($\mathbf{D}_{3h}$). Note that the two axial σ bonds are treated as one set and the three σ bonds in the equatorial plane are in a separate set. They are not geometrically equivalent.

4.20 Apply the group theoretical treatment to obtain the MO bonding description for σ bonding for SF_6 ($\mathbf{O}_h$) using the $3d$ orbitals on S.

▸5◂

Inorganic Solids
Ionic Models
. . . .

Inorganic compounds have a tremendous range of properties, including gases (such as NH_3, SO_2, and SF_6), liquids (such as H_2O, $TiCl_4$, and $SOCl_2$), and solids covering a melting range of more than 2000°. Metal salts are typical of a large and important class of inorganic compounds. An idealized model for ionic compounds treats the structure as an array of ions, regarded as hard spheres for monoatomic ions, arranged according to simple principles of geometry and electrostatics. We begin our study with typical properties of ionic solids resulting from their structures. We examine idealized descriptions and then cases that deviate from the model. In examining structures, we emphasize those features of primary interest to the chemist such as the number and geometry of nearest neighbors about each ion, not minor distortions that are of greater concern to the crystallographer. We can relate a large number of compounds in terms of a few simple structural types and gain a feeling for periodic and structural trends. The structures of metal silicates explain the range of minerals including fibers, thin plates, and three-dimensional networks.

Only three structures are commonly encountered for metals, and the model of metal ions in a sea of electrons make properties associated with the "metallic state" understandable. Bonding in metals, defect structures, and important applications of inorganic solids such as semiconductors and "high temperature" superconductors are treated in Chapter 6.

Some individuals are good at visualizing a drawing as a three-dimensional structure. We all improve with practice in handling models and relating these to drawings. Building models of close-packed arrangements and other structures discussed in detail helps in this process. Ask your instructor if models are available for study.

5.1   SIMPLE IONIC SOLIDS

▶ 5.1.1   *Properties*

Crystals of simple inorganic salts consist of positive and negative ions packed in such a way as to allow a minimum distance between cation and anion and to provide maximum shielding of each ion from other ions of like charge. This arrangement produces maximum attractive forces and minimum repulsive forces. The number of ions of opposite charge around an ion in the crystal is the **coordination number** (C.N.) of that ion. Spherical ions have no directed bonding forces. The arrangement in the crystal is determined by the relative sizes of the ions and coulombic forces among the ions. Many of the inorganic structures commonly encountered can be described most easily in terms of close-packed arrangements of one or more of the ions or atoms. The degree of covalent character varies greatly within a series of compounds of the same formula type (e.g., MX, MX_2, etc.) and even within the same structure type. We will use the terms "atom" and "ion" loosely in describing structures, to avoid lengthy qualifications.

The ions of a salt crystal are held rigidly in place by strong coulombic forces. Because the ions are not free to move about, most ionic crystals are very poor conductors of electricity. The crystal structure is broken down by melting the substance or by dissolving it in a polar solvent, so the ions can migrate freely in an electrical field and hence carry current. The current flows with mass transfer as the ions travel through the melt or solution. The charge transfer processes at the electrodes require oxidation–reduction reactions involving the electrodes and/or the ions in a fused salt. Electrode reactions of a salt in solution might involve oxidation and/or reduction of the solvent. The conductivity of the fused salt or of a solution of the salt is determined by the ions' mobility, which depends on their size and charge. Ions of high charge decrease in mobility when associated with ions of opposite charge (ion pairing) and with solvent molecules (solvation).

▶ 5.1.2   *Electron Configurations of Ions*

Predictions of formula types are simple for ionic compounds. Metals that have one, two, or three electrons in the outer shell and a noble-gas configuration in the next-to-outermost shell commonly combine with nonmetals by losing the outermost electron(s) to form simple cations. The ionic charge is the valence or combining capacity. This leaves an ion with an s^2p^6, **noble-gas** or **octet** configuration, as for K^+, Mg^{2+}, and Al^{3+}. Other common electron configurations for metal ions are $s^2p^6d^{10}$, the **pseudo noble-gas** or **18-electron** configuration, found in such ions as Ag^+, Zn^{2+}, Cd^{2+}, Ga^{3+}; and $(n-1)\,s^2p^6d^{10}ns^2$, the **(18 + 2)-electron** configuration, as in Sn^{2+}, Tl^+, and Sb^{3+} (n is the principal quantum number of the outermost shell). In addition to the (18 + 2)-electron configuration, the post-transition metals (Groups 13–15) also give maximum valence, Sn^{IV}, Tl^{III}, and Sb^V, with 18-electron configurations. Most of the lanthanide metals form tripositive ions, leaving partially filled $4f$ orbitals, whereas many transition elements from di- and tripositive ions having partially filled d orbitals (Au^{3+}, Fe^{2+}, Fe^{3+}, etc.).

Nonmetals that require one or two electrons to complete a noble-gas configuration commonly accept the required number of electrons from metals to form simple anions. In these cases, *valence* refers to an ionic state and has a sign associated with it (e.g., F^- and O^{2-}).

▶ *5.1.3 Nonpolar Character of Ionic Compounds*

Typical ionic compounds, such as NaCl and CaF_2, differ strikingly from covalent compounds such as H_2, CH_4, and H_2O in properties such as melting points, boiling points, and electrical conductance. These properties roughly indicate the degree of nonpolar or covalent character of similar compounds. During melting, solid and liquid phases are in equilibrium, and thus $\Delta G_m = 0$ and, accordingly, $\Delta H_m = T\Delta S_m$. If the C.N. of the atoms/ions does not change on melting (as for CH_4, CO_2, Al_2Cl_6, and $SnCl_4$, for which the molecules retain their identity in the solid, liquid, or vapor state) or vaporizing, the solid is termed a **molecular solid.** For molecular solids the ΔH_m is generally small, since primary bonding interactions within the molecules are unchanged. Because most binary compounds of nonmetals are molecular compounds and the bonding is primarily covalent, low melting points have become associated with covalent bonding character. In view of very high melting compounds such as covalent BN, the relationship between melting points and covalency can be misleading. Like diamond, BN is a covalent solid and bonding extends throughout a crystal. Melting or vaporization requires breaking covalent B—N or C—C bonds, disrupting the molecule. Hence BN sublimes at ~3000°C and diamond melts above 3500°C.

There is a great difference between the melting points of AlF_3 [melting point (mp) 1040°C] and SiF_4 (sublimes at −77°C), but it does not indicate a great difference in the ionic character of the Al—F and Si—F bonds. The silicon in the tetrahedral SiF_4 molecule is shielded effectively from the fluorides of other SiF_4 molecules, and the attractive forces between SiF_4 molecules are weak. Because aluminum in AlF_3 is not shielded on all sides, Al^{3+} interacts very strongly with neighboring F^- ions to give a three-dimensional network and C.N. 6. Thus SiF_4, like CF_4 or CH_4, forms a molecular solid whereas $AlF_3(s)$ is an ionic solid.

Electrostatic forces in a crystal are essentially isotropic (same forces in all directions). Covalent interactions, on the other hand, are anisotropic (directional). Whereas Pauling dealt with variations in polarity in terms of the degree of ionic character in covalent bonds (page 85), a different view was offered by Fajans, who treated the variation in the degree of nonpolar (covalent) character in ionic compounds in terms of polarization effects. **Fajans' rules,** which focus on size and charge relationships and on the electron configuration of the cation, may be summarized as follows:

1. *Nonpolar (or covalent) character increases with decreasing cation size or increasing cation charge.* Very small cations or cations with high charge have high charge density (q/r, charge divided by radius) and consequently tend to distort or polarize the electron cloud around the anions. The greater the polarization of the anions, the more nonpolar or covalent the bond between the atoms. Figure 5.1 shows the effects of increasing polarization. In Table 5.1 for NaCl and $CaCl_2$ the cations have similar radii, but the increased po-

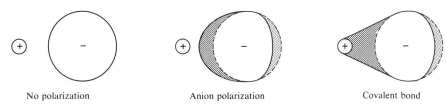

No polarization   Anion polarization   Covalent bond

Figure 5.1   Polarization effects.

larization by the higher charge of Ca^{2+} gives a lower mp for $CaCl_2$. Comparing $MgCl_2$ with $ScCl_3$ shows the effect of higher charge of Sc^{3+}. $ScCl_3$ does not melt, but it sublimes at 800°C. The radius of Al^{3+} is smaller than that of Sc^{3+}, and $AlCl_3$ sublimes at a lower temperature (178°). Comparing the series $BeCl_2$, $MgCl_2$, and $CaCl_2$ shows that for smaller ions for each pair the mp and boiling point (bp) are lower. Comparing $SnCl_2$ with $SnCl_4$ shows that $SnCl_2$ is a solid, but $SnCl_4$, with a higher charge and smaller size, is a covalent liquid (mp −33°C).

$SnCl_4$ is best described as a covalent substance, similar to CCl_4. Sn^{IV} represents the *formal* oxidation state of the hypothetical cation. If you start with the Sn^{4+} cation and four Cl^-, the electron clouds of the Cl^- ions are strongly polarized to envelope the cation. The electrons are shared, with each Cl retaining a partial negative charge and Sn retaining a partial positive charge. The neutral $SnCl_4$ molecules are attracted to one another by weak van der Waals forces, as for CCl_4.

Table 5.1   Effect of cation charge and size upon nonpolar character of anhydrous chlorides

Cation	Cation radius (pm, C.N. 6)	Melting point (°C)	Boiling point (°C)	Equivalent conductance at mp
Na^+	116	800	1470	133
Ca^{2+}	114	772	>1600	51.9
Mg^{2+}	86	712	1412	28.8
Sc^{3+}	88.5	Subl.	Subl. ~800	15
Al^{3+}	67.5	Subl.	Subl. 178	1.5×10^{-5}
Be^{2+}	41 (C.N. 4)	405	~550	0.066
Mg^{2+}	86	712	1412	28.8
Ca^{2+}	114	772	>1600	51.9
Sn^{2+}	112 (C.N. 8)	246	606	21.9
Sn^{IV}	83	−33	114	0

2. *Nonpolar (or covalent) character increases with an increase in anion size or anion charge.* The larger the anion or the higher its negative charge, the more easily it is polarized by the cations, because the electrons are held more loosely.

3. *Nonpolar (or covalent) character is greater for cations with non-noble-gas [18- or (18 + 2)-electron] configurations than for noble-gas type cations of the same size.* Cations with ns^2np^6 configurations cause less distortion, and undergo less distortion themselves, than those with 18-electron configurations. If we compare cations of about the same size (Na^+, Cu^+; see Table 5.2), the 18-e gas cation (Cu^+) has a much higher nuclear charge and the d electrons in the 18-e outer shell do not shield the nuclear charge effectively. Using Slater's rules (page 40) the effective nuclear charge Z^* is 2.2 for Na^+ and 3.7 for Cu^+. An anion near a Cu^+ ion, compared with an Na^+ ion, behaves as though it were under the influence of a greater positive charge and is distorted to a greater extent. The spherical filled d electron cloud is also more easily polarized by anions than is an 8-e shell because the d electrons do not penetrate inner shells as effectively as s and p electrons do. This results in some polarization of an 18-e cation by the anion. Such polarization of the cation results in even greater "exposure" of the cationic nuclear charge, because the electron cloud of the cation is "squeezed" back toward the far side of the cation. The increased effective nuclear charge further enhances the polarization of the anion (see Figure 5.2). The pairs compared in Table 5.2 show the lowering of melting points for the salts of 18-e cations because of greater polarization. This applies even if we compare Na^+ with the larger Ag^+.

Colored Compounds of "Colorless" Ions

One of the manifestations of an increase in nonpolar character of inorganic salts is the appearance or enhancement of color. Although oxides of colorless cations usually are white,

Table 5.2   Effect of electron configuration of the cation upon nonpolar character of anhydrous chlorides[a]

Cation	Configuration	Cation Radius (pm, C.N. 6)	Z^*	Melting point (°C)
Na^+	8 e	116	2.2	800
Cu^+	18 e	91	3.7	430
K^+	8 e	152	2.2	776
Ag^+	18 e	129	3.7	455
Ca^{2+}	8 e	114	3.2	772
Cd^{2+}	18 e	109	4.7	568

[a] Z^*_{eff} are calculated from Slater's rules.

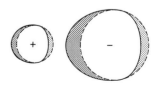

Figure 5.2   Cation and anion polarization.

the corresponding sulfides are likely to be deeply colored if the cation is one that tends to polarize anions. With few exceptions (notably ZnS) the only white sulfides are those of the alkali and alkaline earth metals. In a series of halides of ions such as Ag^+, Hg^{2+}, and Pb^{2+}, the fluorides and chlorides are colorless, the bromides are colorless ($PbBr_2$) or faintly colored (AgBr and $HgBr_2$, pale yellow), and the iodides are more darkly colored (AgI, yellow; HgI_2, yellow or red). The appearance of color in an inorganic salt containing only normally colorless ions usually indicates an appreciable amount of nonpolar character or some unusual structural feature. An appreciable amount of polarization leads to intense visible-ultraviolet absorption bands known as **charge transfer bands**. The absorption involves the transfer of an electron from one atom to another in the excited state (page 463).

Hardness[1]

The strong electrostatic forces between monoatomic ions cause ionic crystals to be relatively hard as well as high melting. The hardness of crystals generally increases with decreasing ionic radius or increasing ionic charge unless the increases are great enough to change the degree of nonpolar character significantly. Salts of cations with 8-e configurations generally are harder (for the same ionic charges and comparable ionic distances) than those with 18-e or (18 + 2)-e configurations. In Table 5.3 note (a) the increase in hardness from LiF to MgO with increasing ionic charge and (b) the decrease from MgO to MgS and CaO to SrO with increasing M—X distance. Observe (c) the decrease in hardness from CaF_2 to CdF_2 (Cd^{2+} has 18 e) and the decrease from SrF_2 to PbF_2 [Pb^{2+} has an (18 + 2)-e configuration].

Table 5.3   Variation of hardness of some crystals with similar structures

NaCl structure	LiF	MgO	MgS	CaO	SrO
M—X distance (pm)	202	210	259	240	257
Hardness (Mohs' scale)	3.3	6.5	4.5	4.5	3.5
CaF_2 structure	CaF_2	CdF_2	SrF_2	PbF_2	
M—X distance (pm)	236	234	250	257	
Hardness (Mohs'scale)	6	4	3.5	3.2	

[1] We considered the chemical hardness or softness earlier (Section 1.8) depending on polarization of electrons of an atom, ion, or molecule. Here we are considering the hardness of solids—macroscopic hardness or resistance to mechanical deformation. The Mohs' scale of hardness is a measure of hardness relative to diamond at 10. Phony diamonds can be scratched by knives or other crystals. A diamond is scratched or cut only by diamond.

5.2 CLOSE PACKING

If we crowd together a layer of spheres of the same radius, such as marbles in a box, the arrangement obtained is close packed with each sphere surrounded by six neighbors arranged hexagonally (Figure 5.3*a*). A sphere rolled onto the layer will come to rest at one of the indentations. An identical second layer can be placed over the first, with each sphere of the second layer in a triangular indentation formed by three spheres of the first layer (Figure 5.3*b*). In order to keep track of the relationships among the layers, the spheres of the first layer are designated as *A* positions while those of the second layer are designated as *B* positions. Two arrangements are possible for a third layer in a close-packed structure. The spheres in the third layer might line up with those in the first layer to give an *ABA* arrangement (repeat the sequence of Figure 5.3*b*). If continued indefinitely, *ABABAB* . . . , this sequence is described as a **hexagonal close-packed** (*hcp*) arrangement. The *only* other possibility is to have the spheres of the third layer in the triangular indentations *not* directly over the spheres of the first layer (Figure 5.3*c*); these are designated as *C* positions. Only three packing positions, *A*, *B*, and *C,* are possible in any close-packed structure. The sequence *ABCABC* . . . describes a **cubic close-packed** (*ccp*) arrangement.[2]

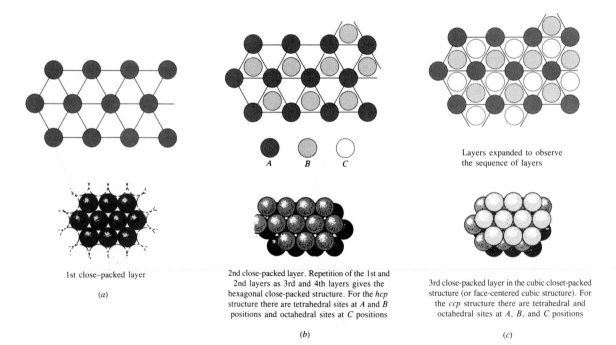

A   *B*   *C*

Layers expanded to observe
the sequence of layers

1st close-packed layer

(*a*)

2nd close-packed layer. Repetition of the 1st and
2nd layers as 3rd and 4th layers gives the
hexagonal close-packed structure. For the *hcp*
structure there are tetrahedral sites at *A* and *B*
positions and octahedral sites at *C* positions

(*b*)

3rd close-packed layer in the cubic closet-packed
structure (or face-centered cubic structure). For
the *ccp* structure there are tetrahedral and
octahedral sites at *A*, *B*, and *C* positions

(*c*)

Figure 5.3   Close-packing layers.

[2] Any sequence without adjacent identical positions (e.g., *ABAC* . . . or *ABCB* . . .) is possible, but the *hcp* and *ccp* sequences are the only ones common.

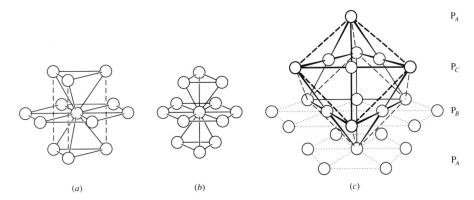

Figure 5.4 (*a*) Hexagonal and (*b*) cubic close-packed structures. (*c*) The *ccp* arrangement along the body-diagonal of the face-centered cubic lattice. Dashed lines outline the face-centered cubic unit cell.

A lattice is an array of points in space each of which has an identical environment. When identical atoms occupy only lattice sites, the lattice label is commonly used as a structural description. For a close-packed structure each sphere has 12 neighbors, with six arranged hexagonally in the same layer, three above and three below (Figure 5.4). The local symmetry of the packing unit consisting of an atom and its 12 neighbors is $\mathbf{D}_{3d}$ for *ccp* (Figure 5.4*b* with three atoms in the layer above staggered relative to the three in the layer below) and $\mathbf{D}_{3h}$ for *hcp* (Figure 5.4*a* with three atoms in the layers above and below are eclipsed). The **face-centered cubic** and *ccp* structures are the *same*—the names emphasize different features. The packing direction—the line perpendicular to the close-packed layers—is along a body diagonal of a face-centered cubic structure, as shown in Figure 5.4*c*; but the packing direction coincides with the crystallographic axis of the *hcp* structure.

Between two close-packed layers (packing layers, *P*) there are interstitial sites or holes with octahedral and tetrahedral symmetry (Figure 5.5). Tetrahedral holes are formed by each sphere of a layer together with the three spheres of the layer above or below. The octahedral holes, which are formed by staggered triangular groups of spheres in each of the two *P* layers, are at $\frac{1}{2}d$, where *d* is the distance between the planes of *P* layers. There are two layers of tetrahedral holes (*T*). One *T* layer (at $\frac{1}{4}d$) is formed by tetrahedra, each of which has a *P* sphere in the upper *P* layer at the tetrahedral apex and three spheres in the lower *P* layer forming the base. The other *T* layer (at $\frac{3}{4}d$) is formed by tetrahedra, each of which has a sphere in the lower *P* layer at the apex and three spheres in the upper *P* layer forming the base. Arrangements of the *T* and *O* sites are *exactly the same between any two adjacent P layers for hcp or ccp. In any close-packed structure the number of O sites is the same as the number of P sites and the number of T sites is twice as great*.

Because *T* sites must line up above or below *P* sites, they are located *only* at *A* and *B* positions for *hcp*, and at *A*, *B*, and *C* positions for *ccp*. The *O* sites are staggered relative to the *P* spheres, so they must occur at *C* positions *only* for *hcp*, but at *A*, *B*, and *C* positions for *ccp*. The *O* positions must be different (staggered) from the positions for adjacent *P* layers, since they are not directly above or below atoms in adjacent *P* layers.

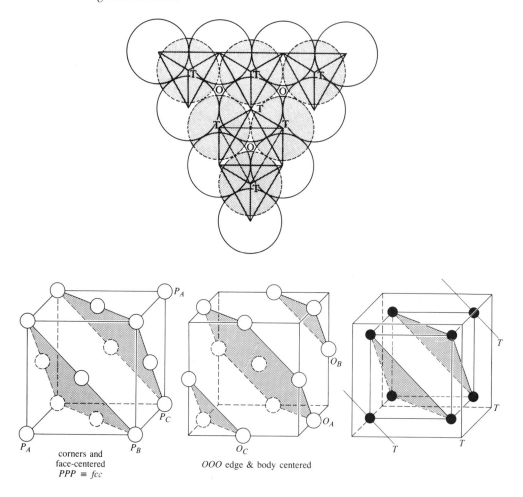

Figure 5.5 (*a*) Tetrahedral (*T*) and octahedral (*O*) holes in a close-packed arrangement of spheres. (*b*) Positions of *P*, *O*, and *T* sites in the usual view of a face-centered cubic (also *ccp*) structure.

5.3   IONIC STRUCTURES BASED UPON CLOSE-PACKED ARRANGEMENTS

Generally, anions are larger than cations. (An atom loses electrons to form a *smaller* cation or gains electrons to form a *larger* anion.) Many ionic crystals can be regarded as close-packed arrays of anions with cations in octahedral and/or tetrahedral sites. There are also structures in which the cations are close-packed with anions in *O* and/or *T* sites. In the ideal cases where the packing spheres touch, the spheres in *O* sites are much smaller than the *P* spheres ($r_O = 0.414 r_P$, where r_O and r_P represent the radii of the spheres filling the respective sites) and those in T sites are still smaller ($r_T = 0.225r_P$). (These results are demonstrated in Section 5.4.) Spheres larger than these ideal cases can be accommodated in T and O sites by pushing the P layers apart, but *without* changing the relationships described for close-packed structures.

▶ *5.3.1 Structures with Cations in Octahedral Sites*

The repeating unit of packing positions for *hcp* is $P_A P_B \ldots$ and that for *ccp* is $P_A P_B P_C \ldots$ with the T and O layers empty. The NaCl structure can be regarded as a *ccp* array of Cl^- with Na^+ in octahedral sites. In the unit cell for NaCl shown in Figure 5.6, *the packing direction is along a body diagonal of the cube.* Packing positions of Cl^- forming layers are shaded in Figure 5.5a. Both Na^+ and Cl^- have C.N. 6 in NaCl. The NiAs structure is similar, in that all octahedral sites are filled by Ni "atoms"; but unlike the Cl^- in NaCl, the As "atoms" are in an *hcp* arrangement (Figure 5.6). Rutile is one of the crys-

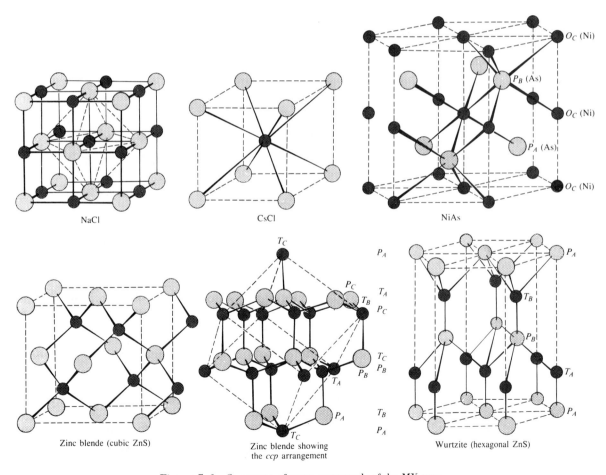

Figure 5.6 Structures of some compounds of the MX type.

tal forms of TiO_2 with oxygens in *hcp* layers. The Ti atoms (these are not Ti atoms, but there are no Ti^{4+} ions in such a structure because of extensive polarization) fill only half of the octahedral sites between the *hcp* layers of oxygens. The unit cell has only two edges of equal length, and so the structure deviates from the ideal hexagonal close-packed model. Ti has C.N. 6 and oxygen has C.N. 3 in rutile (Figure 5.7). Half of the O sites are filled, giving alternate layers of TiO_6 octahedra. In order to achieve effective screening of the ions and the threefold symmetry about oxygen, slight folding or puckering of the packing layers occurs.

The **lattice** is the framework used for arranging the atoms or ions of a structure. Although the choice is arbitrary, the relationships among the atoms or ions are seen more readily if appropriate atoms or ions are located at lattice points (intersections of lines) or edges, reflecting the site symmetry. Some other atoms or ions are commonly within the lattice. The **unit cell** is selected such that the whole crystal structure is reproduced by stacking unit cells in all directions. The unit cell must contain one or several formula units of the substance. A sphere within the cell counts as a whole atom or ion. For a sphere in a face, one-half is within the cell; a sphere on an edge is cut by two faces or counts as $\frac{1}{2} \times \frac{1}{2} = \frac{1}{4}$, and one at a corner is cut by three faces or counts as $\frac{1}{2} \times \frac{1}{2} \times \frac{1}{2} = \frac{1}{8}$. For NaCl:

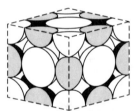

1 Na^+ in center   1 Na^+

12 Na^+ on edges $\frac{1}{4} \times 12 = 3\ Na^+$

6 Cl^- in faces $\frac{1}{2} \times 6$ $= 3\ Cl^-$

8 Cl^- at corners $\frac{1}{8} \times 8$ $= \underline{1\ Cl^-}$

       4 NaCl in unit cell

Figure 5.8   The unit cell of NaCl showing the portions of ions within the cell.

The unit cell of NaCl in Figure 5.8 shows only the portions of spheres in the face centers, edges, and corners within the cell. It is a model using "space-filling spheres" with the ions touching. This is realistic, but figures of the expanded lattices in Figures 5.6 and 5.7 allow us to see the arrangements of ions within the cell.

In some structures, one layer, such as a P layer, is occupied by more than one kind of atom. For the perovskite structure, $CaTiO_3$ (Figure 5.7), calcium and oxygen *together* form the P layers (*ccp*) perpendicular to the cubic-body diagonal. Ti fills only one-fourth of the O sites—those surrounded only by oxygens (Figure 5.7). The ReO_3 structure is re-

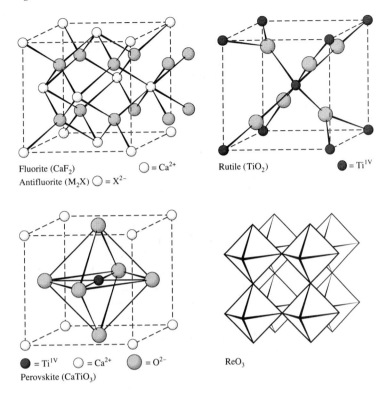

Fluorite (CaF$_2$)   ◯ = Ca^{2+} Rutile (TiO$_2$)   ● = Ti^{IV}
Antifluorite (M$_2$X) ◯ = X^{2-}

● = Ti^{IV} ◯ = Ca^{2+} ◉ = O^{2-} ReO$_3$
Perovskite (CaTiO$_3$)

Figure 5.7   Some common ionic structures. In the rutile structure, one cell dimension (*c*) is shorter than the other two. For ReO$_3$ the connected ReO$_6$ octahedra are outlined.

lated to CaTiO$_3$ with the Ca^{2+} sites missing, giving ReO$_6$ octahedra stacked sharing all apices in three dimensions (Figure 5.7).

The structures encountered for metals are *hcp*, *ccp*, and body-centered cubic (*bcc*; see page 246). If we fill all *O* sites of a *ccp* arrangement by the atoms identical to those in the *P* sites (*T* sites empty), a simple cubic structure results. This is the structure of the element polonium.

▶ 5.3.2   Structures with Cations in One
Layer of Tetrahedral Sites

There are two modifications of ZnS involving close-packed arrangements of S with Zn in half of the *T* sites—that is, *all of the T sites in one of the two T layers*. The S are *hcp* in wurtzite and *ccp* in zinc blende (Figure 5.6). Note that for the *ccp* arrangement the filled *T* layer is very close to the upper *P* layer. The *T* sites near the lower *P* layer are empty. Zn and S have C.N. 4 for both structures. For zinc blende, Zn and S occupy *A*, *B*, and *C* positions. The figure for zinc blende showing the *ccp* arrangement shows the cubic cell outlined for the *fcc* arrangement of Zn^{2+} in *T* sites. If we reverse the interchangeable *P* and *T* sites, we get the packing sequence shown to the right, with top and bottom sites being P_A. For wurtzite (*hcp* S^{2-}) only *A* and *B* positions are used. The empty *C* sites in wurtzite form large open hexagonal channels along the crystallographic *c* axis. This can be seen for ice I_h (Figure 2.21), where the oxygens replace all Zn and S in wurtzite. Diamond has a *ccp* structure corresponding to C in all Zn and S sites of zinc blende.

► 5.3.3 *Structures with Ions Occupying Tetrahedral Sites in Two Layers*

All *T* sites are filled by F^- in a *ccp* array of Ca^{2+} (in *P* sites) in the mineral fluorite, CaF_2 (Figure 5.7). Many MX_2 compounds of large cations have the fluorite structure, and many M_2X compounds have the **antifluorite** structure with reversed roles of cations and anions. MX(1:1) stoichiometry is achieved for NaCl and ZnS structures. In CaF_2 the C.N. is 8 for Ca^{2+} and 4 for F^-. PtS has a *ccp* array of Pt, with S occupying half of the *T* sites in *each* *T* layer. Pt has C.N. 4, but unlike zinc blende with tetrahedral ZnS_4 units, the PtS_4 units are planar (Figure 5.9). This shows us the effect of covalent bonding between Pt and S. For a *ccp* structure each *T* site is at the center of a cube formed by four *P* sites and four *O* sites, each set describing a tetrahedron. For PtS the *P* and *O* sites filled are along parallel face diagonals of the cube, giving planar PtS_4 units. From the lattice, we would expect the bond angles to be 90° for Pt and for S, instead of the normal tetrahedral of 109.5° for S. Both angles bend a little so that the Pt—S—Pt angle is about 97.5° and the S—Pt—S angle is about 82.5°.

We can view the BiF_3 structure as a *ccp* array of Bi (*P* positions), with F in *all* *O* and *T* sites. This is the structure of the intermetallic compound $BiLi_3$ and of cryolite, Na_3AlF_6, treating the entire AlF_6 octahedron as a unit in *P* positions. Similarly, the K_2PtCl_6 structure (Figure 5.10) can be described as the antifluorite structure, a *ccp* array

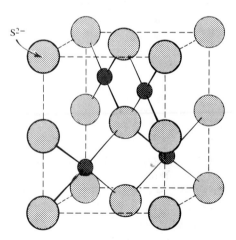

Figure 5.9 The PtS structure showing the planar PtS_4 units. The larger ions are S^{2-}.

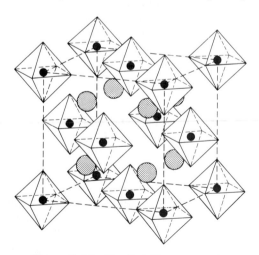

Figure 5.10 The K_2PtCl_6 structure showing the $[PtCl_6]^{2-}$ octahedra in a *ccp* arrangement with K^+ in tetrahedral sites.

of $PtCl_6^{2-}$ octahedra with K^+ ions in all of the T sites. This is a common structure for compounds of the type $M_2M'X_6$. The K_2PtCl_4 structure is similar, replacing the octahedral $PtCl_6^{2-}$ by planar $PtCl_4^{2-}$ units, but the cell length perpendicular to the $PtCl_4^{2-}$ planes is shortened.

5.4   RADIUS RATIOS FOR STABLE ARRANGEMENTS IN IONIC CRYSTALS

The arrangements of ions in a crystal depend to a great extent on the relative sizes of the ions. For a CsCl structure with eight anions touching one another and the cation at the center of the cube, the ratio of $r_M/r_X = 0.732 = \rho$. For a NaCl structure with six anions touching one another and a cation at the center of the octahedron, $\rho = r_M/r_X = 0.414$. For a tetrahedron (ZnS) the ideal value for ρ is 0.225. Starting with the ideal ratio of 0.225, for slightly larger cations at the center with the same anions, the four anions would not touch one another. The C.N. 4 is expected to be stable for cations until the radius ratio becomes 0.414 where there is enough open space to add two anions to give C.N. 6. Table 5.4 provides the radius ratios of geometrical arrangements for the packing of spheres. The consequences of deviations of radius ratios will be considered later (page 223).

EXAMPLE 5.1:   Calculate the ideal r_M/r_X for cation–anion and anion–anion contact for an octahedral arrangement of anions around a cation as sketched in one plane in Figure 5.20a.

Solution: The right triangle sketched here has two sides equal to $(r_M + r_X)$ with the hypotenuse equal to $2r_X$.

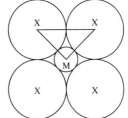

$$(2r_X)^2 = (r_M + r_X)^2 + (r_M + r_X)^2$$

$$4r_X^2 = 2(r_M + r_X)^2$$

$$\sqrt{2}r_X = r_M + r_X$$

$$r_M/r_X = (\sqrt{2} - 1) = 0.414$$

EXAMPLE 5.2:   Show that the ideal r_M/r_X for the cation–anion and anion–anion contact for a tetrahedral arrangement of anions around a cation is 0.225.

Solution: The drawing shows one edge of the tetrahedron with two anions touching and the cation at the center of the tetrahedron. A line perpendicular to the edge bisects the tetrahedral angle. The length of the edge is $2r_X$. The distance from a tetrahedral vertex to the center is $r_M + r_X$. The angle θ is $109°28'/2$, and we have

$$\sin \theta = \frac{r_X}{r_M + r_X} = \sin 54°44'$$

$$0.816 (r_M + r_X) = r_X$$

$$0.816 r_M = 0.184 r_X$$

$$r_M/r_X = 0.225$$

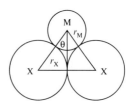

Table 5.4 **Radius ratios for arrangements of rigid spheres**

Coordination number of M	Arrangement of X	Radius ratios		Crystal structure corresponding to cation C.N.
		$\rho = r_M/r_X$	$\rho' = r_X/r_M$	
3	Triangular	0.150–0.225	4.44–5.67	
4	Tetrahedral	0.225–0.414	2.42–4.44	Antifluorite, ZnS
4	Planar	0.414–0.732	1.37–2.42	
6	Octahedral	0.414–0.732	1.37–2.42	NaCl, TiO_2, $CdCl_2$
8	Cubic	0.732–1.00	1.00–1.37	CsCl, CaF_2

5.5   COMMON STRUCTURES OF IONIC CRYSTALS

▶ 5.5.1   *MX Compounds*

Over 700 compounds display the structures represented in Figures 5.6 and 5.7. Most of the alkali halides, alkaline earth oxides, sulfides, and so on—as well as many other 1 : 1 compounds—crystallize in the NaCl structure, Cl^- *ccp* with Na^+ in octahedral holes. In the less common NiAs structure (based on a *hcp* arrangement of As), the As is at the center of a trigonal prismatic arrangement of 6Ni and the Ni fill all octahedral holes. Because all *O* sites, and *only O* sites, occupy *C* positions for an *hcp* structure, the Ni are lined up along the *C* direction. The short Ni—Ni distances account for the semimetallic character of NiAs, a **one-dimensional metal** along the direction of the Ni—Ni—Ni . . . interaction. Because there is no shielding between the Ni atoms, such a structure should occur *only* in appreciably covalent MX compounds. In these cases, polarization of the anions reduces repulsion between the metal centers. The NiAs structure is encountered in salts of highly polarizable anions such as sulfides and selenides of Ti, V, Fe, Co, and Ni, as well as in the tellurides of Ti, V, Cr, Mn, Co, and Ni. The oxides of these metals have the NaCl structure, typical of more ionic compounds. Halides of some of the larger alkali metal ions have the CsCl structure with C.N. 8. Table 5.5 summarizes the structures encountered for some common MX inorganic compounds.

The MX compounds CsCN, TlCN, and CsSH have cubic structures even though the anions are not spherical. Cubic symmetry results from staggered orientations of the anions, whether regular or disordered. Some other salts, containing complex anions, give structures closely related to the NaCl structure. Thus the CaC_2 (called calcium carbide, but it contains the acetylide ion, C_2^{2-}) structure contains face-centered linear C_2^{2-} ions all oriented in the same direction to extend the unit cell along one axis (see Figure 5.11). The CaC_2 structure is encountered for the peroxides CaO_2, SrO_2, and BaO_2 and for the superoxides KO_2, RbO_2, and CsO_2. The calcite ($CaCO_3$) structure occurs for (a) several carbonates of divalent cations, (b) some nitrates, such as $LiNO_3$ and $NaNO_3$, and (c) some borates, such as $ScBO_3$, YBO_3, and $InBO_3$. The calcite structure resembles the NaCl structure, with the lattice expanded and skewed as required by the replacement of the Cl^- by the planar triangular carbonate ions (see Figure 5.11).

Table 5.5 Structures of some compounds of the type MX

Ions	Halides				Ions	Oxygen Family			
	F	Cl	Br	I		O	S	Se	Te
Li	←———— NaCl ————————→				Be	W^a	Z^b	Z	Z
Na	←———— NaCl ————————→				Mg	NaCl	NaCl	NaCl	Z
K	←———— NaCl ————————→				Ca	NaCl	NaCl	NaCl	NaCl
Rb	←———— NaCl ————————→				Sr	NaCl	NaCl	NaCl	NaCl
Cs	NaCl	←———— CsCl ————→			Ba	NaCl	NaCl	NaCl	NaCl
Cu	—	Z	Z	Z	Zn	W	W	W	W
							Z	Z	Z
Ag	NaCl	NaCl	NaCl	W	Cd	NaCl	W	W	Z
							Z		
					Hg	$Other^c$	Z	Z	Z

^{a} Wurtzite.　^{b} Zinc blende.　^{c} See page 731.

We usually consider structures of MX-type solids in terms of the packing of ions, except where covalent bonding is obviously of primary importance (as in the diamond). Using a molecular orbital (MO) approach, Burdett[3] described the zinc blende and wurtzite structures as stacking of puckered sheets (the P and nearest T layers). Because all close-packed layers are identical (differing only in the way they are stacked), both ZnS structures consist of puckered sheets of linked hexagons in chair conformations involving P positions and the T sites in the nearest T layer. You can readily see the chair in the center of the zinc blende structure in Figure 5.6 (showing the *ccp* arrangement) by focusing your attention on a hexagon consisting of three light spheres (P_C sites) above the plane of three dark spheres (T_B sites) connected to the light spheres. Hexagons also can be traced in the vertical stacking, revealing boat conformations of wurtzite but chair conformations for zinc blende. There are three equivalent packing directions for zinc blende because it is cubic, so hexagons traced in any direction have the chair conformation. For wurtzite the packing layers ($P_A P_B P_A P_B$ . . .) are stacked along only one direction, and the chair conformations occur only for the puckered sheets consisting of the P^+ nearest T layers. MO

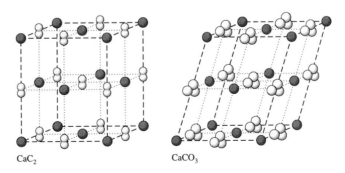

CaC_2　　　　　　$CaCO_3$

Figure 5.11 Structures of CaC_2 and $CaCO_3$. Carbons of CO_3^{2-} are not visible.

[3] J. K. Burdett, *J. Am. Chem. Soc.* **1980**, *102*, 450; *Nature* **1975**, *279*, 121. See also General References.

calculations reveal that the zinc blende structure is favored for small $\Delta\chi$ values (electronegativity differences). Burdett shows how changing the number of valence electrons for MX (eight for ZnS) leads to structural changes for elemental As (10 electrons for As_2, M = X = As), as well as for other cases. Layer structures such as those of $CdCl_2$, CdI_2, and MoS_2 (Figure 5.18) yield to the same approach.

EXERCISE: Build ball-and-stick models of the cage units for the ZnS zinc blende (*ccp*) and wurtzite (*hcp*) structures (Figure 5.6) to see the conformations of the hexagonal rings. For zinc blende in Figure 5.6 showing the *ccp* arrangement, a cage consists of Zn in T_C at the base of the model: three S in one layer (P_A sites), three Zn above these in T_A sites, and three S in the next P_B layer each connected to two of these Zn. The three hexagonal faces on the sides and the one at the top of the cage have the same chair conformations. The simplest wurtzite cage has one Zn in T_A (center of Figure 5.6), three S above these in P_B sites, three Zn in the next layer in T_B sites, and one S in the center of the P_A layer. The three vertical hexagonal faces have boat conformations. The hexagonal faces that can be traced out in the S and nearer Zn layers have chair conformations. If ball-and-stick models are not available, examine models for zinc blende and wurtzite and outline the positions of the cages described above with masking tape.

The expected radius ratios should be used only as rough guidelines in predicting structures. Of the alkali metal halides, only LiF ($\rho = 0.76$), LiCl, LiBr, LiI, NaCl, NaBr, NaI, and KI (0.74) fall within the expected range for r_M/r_X for C.N. 6, yet also NaF, KF, KCl, KBr, and CsF have C.N. 6. An important limitation for MX (1:1) compounds is that the C.N. of the cation is the *same* as that of the anion. Consequently, CsF ($r_M/r_X = 1.53$, using r_M for C.N. 6 from Table 5.8) has C.N. 6, even though this is the *highest* ratio for any of the cesium halides. Here, the C.N. 8 for the small F^- is unfavorable ($r_X/r_M = 0.66$). The requirement of C.N. 8 for the cation *and* anion limit the CsCl structure to MX compounds containing *only* univalent ions. In fact, the CsCl structure is uncommon and found only for CsX (X = Cl^-, Br^-, I^-, CN^-, SH^-, SeH^-, and NH_2^-), TlX (X = Cl^-, Br^-, and CN^-), NH_4X (X = Cl^- and Br^-), and intermetallic compounds where high C.N.s for both metals are tolerable.

Figure 5.12 shows a sorting map, a plot of r_M and r_X for MX compounds with CsCl (C.N. 8), NaCl (C.N. 6), and ZnS (C.N. 4) structures. Because ZnS gives two structures (namely, zinc blende and wurtzite), in the figure the wurtzite modification is identified as the ZnO structure. The solid slanting lines indicate the separations by coordination number based on the radius ratios in Table 5.5. The dashed lines are drawn to separate the structure areas by C.N. more clearly. Because ions are not really like hard billiard balls, the dashed lines are more realistic. A review of the reliability of predictions of structures[4] based on radius ratios (without allowing for polarization—that is, neglecting that ions are not hard spheres) indicated about 67% correct predictions if we consider r_M/r_X *and* r_X/r_M, allowing for the C.N. limitation for either ion.[5] The relative sizes of ions are very important, but the limitations are not as severe as indicated by the geometry for hard spheres.

[4]L. C. Nathan, *J. Chem. Educ.* **1985**, *62*, 215.

[5]Commonly, anions are larger than cations and the ranges of r_M/r_X indicate the number of anions that fit around the cation. For anions with small radius or with high C.N., the number of cations around the anion can be limiting. The r_X/r_M ranges are considered also for this purpose.

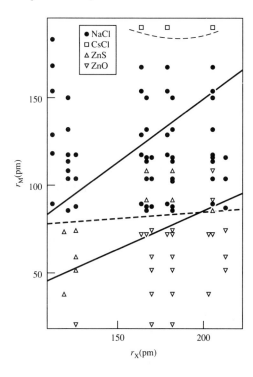

Figure 5.12   Plot of crystal radii r_M and r_X to show the sorting of structures for MX compounds using ratio rules (solid lines) and by areas of predominant structures (dashed lines). The original source included 95 compounds. (Reproduced with permission from J. K. Burdett, S. L. Price, and G. D. Price, *Solid State Commun.* **1981**, *40*, 923. Copyright © 1981, Pergamon Press, Ltd., Headington Hill Hall, Oxford OX3 OBW UK.)

Polarization or partial covalence favors low C.N. If partial negative charge is transferred to the cation from each anion, too much negative charge on the cation would result for high C.N. The ZnS structures (C.N. 4) are encountered for low-ρ compounds (small cations) or for compounds with appreciable covalent character, even with larger ρ values. These structures are found for the copper(I) halides and for the oxides, sulfides, and so on, of Be and the Zn group metals. Of the silver halides, only AgI ($\rho = 0.55$) has C.N. 4.

Several other plots of theoretical radii, empirical ionic size parameters, differences of electronegativities, and covalence parameters separate the MX structures well.[6] Burdett[7] has applied MO theory to examine the role of electron configuration of the cation in determining the structure of compounds having zinc blende, wurtzite, NiAs, and layer structures.

► 5.5.2   *MX₂ and M₂X Compounds*

Compounds of the type MX_2 commonly adopt the fluorite, CaF_2, or rutile, TiO_2, structures (Figure 5.7) for predominantly ionic compounds. Those of large cations usually have the fluorite structure with cations in the *P* positions (C.N. 8). The C.N. is 6 for rutile with cations in *O* sites. These two typical ionic structures are commonly encountered for

[6] J. K. Burdett, G. D. Price, and S. L. Price, *Phys. Rev. B* **1981**, *24*, 2903.

[7] J. K. Burdett, *J. Am. Chem. Soc.* **1980**, *102*, 450.

fluorides and oxides, but not for the more covalent chlorides, bromides, iodides, sulfides, selenides, and tellurides. The antifluorite structure (see page 209) is encountered for alkali metal oxides, and so on, of the type M_2X. Table 5.6 lists some compounds with these structures.

For MX_2 compounds the C.N. is twice as large for the cation (8 for fluorite and 6 for rutile) as for the anion (4 for fluorite and 3 for rutile). If we consider both r_M/r_X and r_X/r_M, because of differences in C.N. for the cations and anions, the ratio rules are somewhat more reliable for MX_2 than for MX in predicting the structures. Most of the ionic compounds of alkali metals of the M_2X type (Table 5.6) have the antifluorite structure with the roles of the cations and anions in CaF_2 reversed. Note that there is no corresponding antirutile structure for M_2X that would give C.N. 3 for the cation.

Large cations that form MX_2 compounds with a high degree of covalency often have the rather complex $PbCl_2$ structure (see references such as Wells). Smaller cations, or those that form even more covalent MX_2 compounds, commonly display one of the layer structures ($CdCl_2$, CdI_2, or MoS_2; see Figures 5.17 and 5.18). The very small Si^{IV} gives one of the SiO_2 structures involving SiO_4 tetrahedra sharing apices. GeS_2 shows a related three-dimensional network of tetrahedra sharing edges; the compound is less ionic than SiO_2.

Figure 5.13 shows a sorting map based on r_M and r_X for common MX_2 structures. The separations are fairly good except for distinguishing the layer structures ($CdCl_2$, CdI_2, and MoS_2 to be discussed later). Better separation is obtained in a plot of pseudopotential radii of the ions in Figure 5.14 (2 errors in predictions out of 112). These quantum mechanical radii are related to experimental radii and inversely related to the valence ionization energy. Essentially equally good separations for MX_2 can be obtained using plots of electronegativities of the elements.[8] Burdett[9] has examined the importance of electron configuration of the cation in rutile and related structures.

▶ 5.5.3 *Perovskite and Spinel*

We encounter the perovskite ($CaTiO_3$) structure (Figure 5.7) in over 250 compounds of the types $M^{II}M^{IV}O_3$ and $M^IM^{II}F_3$. Researchers have prepared many compounds with the idealized or slightly distorted perovskite structure, because of interest in their ferroelectric

Table 5.6   Structures of some compounds of the type MX_2 and M_2X

Fluorite (CaF_2)			*Rutile* (TiO_2)			*Antifluorite*			
CaF_2	CdF_2	ZrO_2	MgF_2	NiF_2	TiO_2	Li_2O	Li_2S	Li_2Se	Li_2Te
SrF_2	HgF_2	ThO_2	MnF_2	ZnF_2	MnO_2	Na_2O	Na_2S	Na_2Se	Na_2Te
BaF_2	PbF_2	CeO_2	FeF_2	PdF_2	MoO_2	K_2O	K_2S	K_2Se	K_2Te
$BaCl_2$		UO_2	CoF_2		GeO_2	Rb_2O	Rb_2S		
					SnO_2				

[8] J. K. Burdett, S. L. Price, and G. D. Price, *Solid State Commun.* **1981**, *40*, 923.

[9] J. K. Burdett, *Inorg. Chem.* **1985**, *24*, 2244.

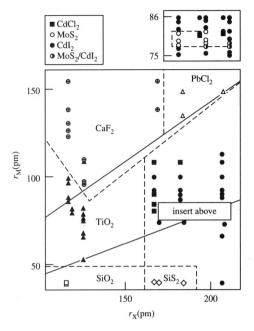

Figure 5.13 Plot for r_M and r_X to show the sorting of structures for MX_2 compounds using ratio rules (solid lines) and by areas of predominant structures (dashed lines). (Adapted with permission from J. K. Burdett, S. L. Price, and G. D. Price, *Solid State Commun.* **1981,** *40,* 923. Copyright © 1981, Pergamon Press, Ltd., Headington Hill Hall, Oxford OX3 OBW UK.)

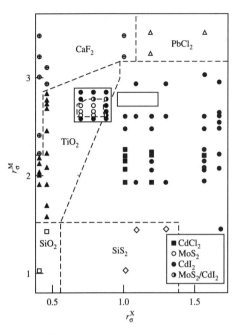

Figure 5.14 Sorting map of structures of MX_2 compounds plotting r_σ^M and r_σ^X (see text). The original reference included 113 compounds. The insert, drawn to a different scale, shows series sulfides, selenides, and tellurides in three vertical columns. (Adapted with permission from J. K. Burdett, S. L. Price, and G. D. Price, *Solid State Commun.* **1981,** *40,* 923. Copyright © 1981, Pergamon Press, Ltd., Headington Hill Hall, Oxford OX3 OBW UK.)

properties. Another common structure for compounds of the type $M^{II}M^{IV}O_3$ is ilmenite, $FeTiO_3$, which is related structurally to α-Al_2O_3 (corundum). For α-Al_2O_3 the oxygens are *hcp* and cations fill two-thirds of the O sites. In $FeTiO_3$ the Fe^{II} and Ti^{IV} are in alternate O layers (each 2/3 filled). The perovskite structure is encountered for large M^{2+} cations, which occupy the P sites; the ilmenite structure is favored if both cations have comparable sizes.

The spinel ($MgAl_2O_4$) structure is found in complex salts of the types $M^{II}M_2^{III}O_4$, $M^{IV}M_2^{II}O_4$, or $M^{VI}M_2^{I}O_4$, as well as in some sulfides, selenides, and fluorides of the same formula types. The atoms in P positions (oxygen) are in a *ccp* pattern, with Mg^{2+} ions in T sites and Al^{3+} ions in O sites for $MgAl_2O_4$, which is regarded as a **normal spinel.** In an **inverse spinel,** such as $MgFe_2O_4$ or written as $Fe(Mg,Fe)O_4$, half of the Fe^{3+} ions are in T sites, and the Mg^{2+} and the remaining Fe^{3+} ions are in the O sites. The formula $Fe(Mg,Fe)O_4$ emphasizes the inverse roles of the cations with half of the Fe^{3+} ions in T sites. We often can predict normal- or inverse-type spinels on the basis of ionic radii (larger ions prefer the octahedral sites) and the preference of some cations for octahedral sites, because of ligand field effects (page 402).

In addition to the spinel structure, AB_2O_4-type compounds display several other structures.[10] The remarkable clustering in domains is shown by plotting the size of one cation against the other (Figure 5.15). Other plots[6] are very successful in differentiating the structures on the basis of theoretical radii or differences in electronegativities.

▶ 5.5.4 Layer Structures

Many salts of highly polarizing cations and easily polarizable anions have structures consisting of layers of cations sandwiched between layers of anions, but without additional cations between the sandwiches. Such salts have relatively low melting points and low solubilities in polar solvents. The effects of polarization are carried further in $CuCl_2$ and $PdCl_2$, which form infinite chains by sharing Cl^- between planar MCl_4 groups (Figure 5.16).

We can view the $CdCl_2$ layer structure as *ccp* layers of Cl^- with Cd^{2+} filling the *O* sites in only alternate *O* layers. The Cl—Cd—Cl "sandwiches" are stacked without cations between them. In effect, the negative charge of the Cl^- layers is polarized inward toward the Cd^{2+} layer to such an extent that the interaction between the Cl layer of one sandwich and the Cl layer of the next sandwich is comparable to the van der Waals interaction between CCl_4 molecules. Other salts with this structure include $MnCl_2$, $FeCl_2$, $CoCl_2$, $NiCl_2$, $MgCl_2$, and $ZnBr_2$. The arrangement for *one* sandwich is the same for $CdCl_2$ and CdI_2, but arrangements between sandwiches differ. The CdI_2 structure (Figure 5.17) has an *hcp* arrangement for I^-. Other compounds with the CdI_2 structure include: $CdBr_2$, $FeBr_2$, $CoBr_2$, $NiBr_2$, MgI_2, CaI_2, ZnI_2, PbI_2, MnI_2, FeI_2, CoI_2; hydroxides of di-

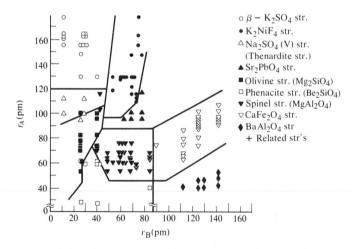

Figure 5.15 Structures of AB_2O_4 compounds. (Reproduced with permission from J. C. Phillips, "Chemical Bonds in Solids," in *Treatise on Solid State Chemistry*, N. B. Hannay, Ed., Plenum, New York, 1973. Copyright © 1973 Bell Telephone Laboratories.)

[10] For details of structures not covered here, see A. F. Wells, *Structural Inorganic Chemistry*, 5th ed., Oxford, Oxford, 1984.

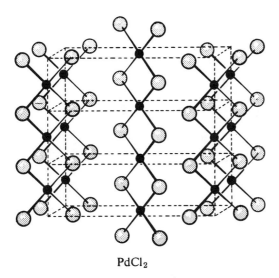

PdCl₂

Figure 5.16   Structure of PdCl₂.

valent Mg, Ca, Cd, Mn, Fe, Co, and Ni; and sulfides of tetravalent Zr, Sn, Ti, and Pt. The $CrCl_3$ layer structure resembles that of $CdCl_2$ with only one-third of all of the octahedral sites filled: that is, with two-thirds of the O sites in alternate layers filled (the other O layers are empty).

The extensive polarization of large cations is demonstrated by Cs_2O that has the anti-$CdCl_2$ structure. The Cs^+ ions are in *ccp P* positions with O^{2-} in only alternate octahedral layers. The Cs—O—Cs sandwiches are stacked with *no shielding by O^{2-} between Cs layers of adjacent sandwiches*. Other alkali metal oxides and sulfides have the antifluorite structure, more typical of ionic compounds.

Molybdenum disulfide and tungsten disulfide have layer structures in which each metal atom has six S neighbors at the apices of a trigonal prism (Figure 5.18), rather than the more common octahedral arrangement seen in $CdCl_2$ and CdI_2. The Mo are *hcp* with S in all tetrahedral sites $|T_B P_A T_B|T_A P_B T_A|$. The S—S distances in eclipsed T sites of each sandwich is 298 pm, indicating S—S bonding. The long S—S distance between sandwich layers are 366 pm, corresponding to van der Waals interaction. The layer structure of

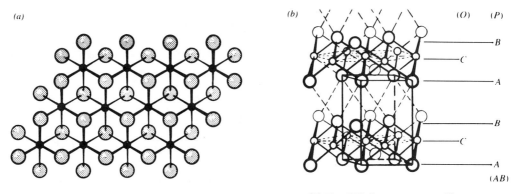

Figure 5.17   (*a*) One layer of the $CdCl_2$ or CdI_2 structure. (*b*) The CdI_2 layer structure. The *P* layers are *hcp*. The hexagonal unit cell is outlined by dark lines.

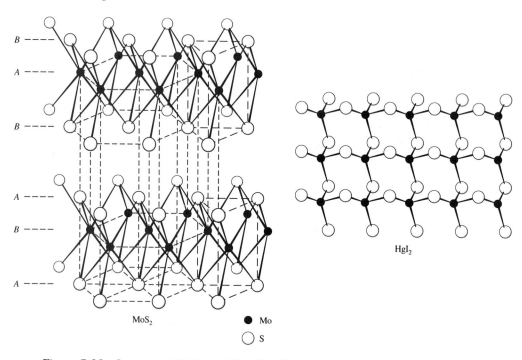

Figure 5.18 Structures of MoS$_2$ and HgI$_2$. The Mo atoms are in an *hcp* arrangement with S in all *T* sites.

MoS$_2$ accounts for the properties that make it a good solid lubricant. The commercial product (Moly-S) is similar to graphite in properties and appearance. As might be expected, the replacement of S by the more polarizable Se or Te improves the properties that make layer-structured sulfides good lubricants.

The mercury(II) halides display a unique structural variation. Mercury(II) fluoride has the typically ionic fluorite structure. The HgCl$_2$ structure consists of discrete HgCl$_2$ molecules. Mercury(II) bromide gives a layer-type structure similar to that of CdCl$_2$, but the octahedral arrangement is distorted, with two Br atoms much closer than the other four. In the layer structure of HgI$_2$, each Hg atom is surrounded by four equivalent iodide ions (Figure 5.18). The HgI$_2$ structure can be regarded as a *ccp* arrangement of I$^-$, with Hg^{2+} ions in one-half of the *T* sites in alternate layers with the other layers of *T* sites empty.

5.6 NOTATION FOR STRUCTURES
BASED UPON CLOSE PACKING

There are two layers of *T* sites and one layer of *O* sites between close-packed *P* layers, or *PTOT*. For *hcp* the sequence is $P_A TOTP_B TOT$ . . . and for *ccp* it is $P_A TOTP_B TOTP_C TOT$ . . . . Each *T* site must line up above or below a *P* site that is the apex of the tetrahedron, but O sites must be staggered relative the P sites above and below. For ccp the full sequence is $P_A T_B O_C T_A P_B T_C O_A T_B P_C T_A O_B T_C P_A$ . . . . A simple notation[11] gives the sequence for a structure. For hcp, $P_A P_B$ is designated $2P$; and for ccp, $P_A P_B P_C$ is designated $3P$. The numerical index is the number of layers in the repeating unit. For NaCl the notation $3 \cdot 2PO$ indicates that only P and O sites are filled with six $(3 \cdot 2)$ layers $(P_A OP_B OP_C O)$ in the repeating units. Because half of the six layers are P layers, they must be ccp, but for simplicity the first number of the index is 3 for ccp and 2 for hcp. For ZnS one P and one T layer are filled, but the two structures are $2 \cdot 2PT$ (hcp) for wurtzite and $3 \cdot 2PT$ (ccp) for zinc blende. In TiO$_2$ the index $2 \cdot 2PO_{1/2}$ indicates that the packing layers are hcp (first index 2) with only one-half of the octahedral sites filled. The number of O sites equals the number of P sites, and the formula TiO$_2$ indicates that the oxides are in filled P sites with Ti in half-filled octahedral sites.

The CaF$_2$ structure is $3 \cdot 3PTT$. The first number indicates that the P layers are ccp, and the total number of layers repeating is $3 \times 3 = 9$, $P_A TTP_B TTP_C TT$ . . . . The number of sites in each T layer is the same as P sites, and here both T layers are filled, so Ca^{2+} ions are in P sites and F$^-$ ions are in T sites, giving C.N. 8 for Ca^{2+} and 4 for F$^-$. In an hcp $2 \cdot 3PTT$ structure the pairs of T sites just above and below a P layer are eclipsed, $P_A T_B [T_A P_B T_A][T_B P_A T_B]$. An hcp structure with all S in T sites is encountered in MoS$_2$. Because the S in adjacent T sites bonded to the same Mo atoms are very close, this is expected only if, as in this case, there is bonding interaction between these atoms.

The ReO$_3$ structure is $3 \cdot 2P_{3/4} O_{1/4}$ (Figure 5.7). Because the number of P sites filled is three times that of O sites, the ReO$_3$ formula indicates that the oxides are in P positions. The first number of the index (3) indicates that these P layers are ccp. Usually anions are in P sites, but antifluorite is a notable exception. The roles of ions in P and O sites are less obvious for perovskite, CaTiO$_3$, designated $3 \cdot 2PO_{1/4}$. Here calcium and oxygen *together* fill the ccp P layers with Ti in one-fourth of the O sites.

Close-packed arrangements provide a model for inorganic crystal structures which aids in recognizing the C.N. and local symmetry of each atom or ion and in seeing relationships among structures. Table 5.7 summarizes the sequence of filling O and T sites for some simple inorganic substances in cubic and hexagonal close-packed arrangements. The description for BiF$_3$ is $3 \cdot 2PTOT$, $T = O$, indicating the same ions (F$^-$) occupy T and O sites. Figure 5.19 shows a face-centered cubic (cubic-close packed) cell and four structures of inorganic compounds derived by filling tetrahedral (T) and octahedral (O) sites. There are compounds corresponding to these based upon an hcp arrangement.

5.7   IONIC RADII

▶ 5.7.1   *Methods of Evaluating Ionic Radii*

The radius of an isolated atom or ion has little meaning. It might be taken as infinite. The "ionic radius" refers to the distance of closest approach by another ion, so the radius is evaluated from the observed distance between centers of nearest neighbors.

[11] S.-M. Ho and B. E. Douglas, *J. Chem. Educ.* **1969**, *46*, 208.

Table 5.7 Some examples illustrating the filling of *P*, *T*, and *O* layers

Layers occupied	*Cubic close packing*	*Hexagonal close packing*
P	Cu, $ccp(3P)$	Mg, $hcp(2P)$
PO	NaCl$(3 \cdot 2PO)$	NiAs$(2 \cdot 2PO)$
PPO	CdCl$_2(3 \cdot 3PPO)$	CdI$_2(2 \cdot \frac{3}{2}PPO)$
	CrCl$_3(PPO_{2/3})$	
PT	ZnS$(3 \cdot 2PT)$	ZnS$(2 \cdot 2PT)$
PTT	CaF$_2(3 \cdot 3PTT)$	MoS$_2(2 \cdot 3PTT)$
	PtS$(3 \cdot 3PT_{1/2}T_{1/2})$	
	SnI$_4(3 \cdot 6PT_{1/8}T_{1/8})$	Al$_2$Br$_6(2 \cdot 6PT_{1/6}T_{1/6})$
PTOT	BiF$_3(3 \cdot 2PTOT, T = 0)$	Mg$_2$SiO$_4(2 \cdot 4PT_{1/8}O_{1/2}T_{1/8})$

Several approaches have been used to evaluate the individual ionic radii. Lande's method for obtaining the radius of the anion assumes that relatively large anions will be in contact with one another when packed around a small cation such as Li$^+$ in LiI (see Figure 5.20c). The radius of the iodide ion is taken as half of the I—I internuclear distance. In crystals such as KI, where the I—I distance is greater than that found in LiI, it is assumed that the anion and cation are in contact with each other, with the I$^-$ ions pushed apart

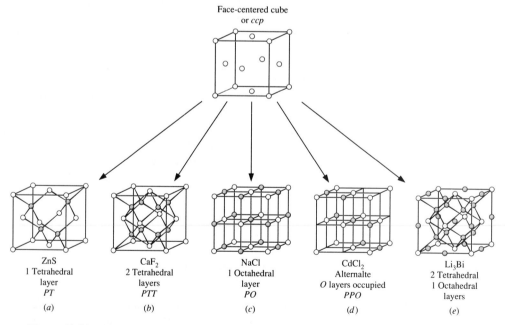

Figure 5.19 Inorganic structures derived from a cubic close-packed arrangement by filling tetrahedral and/or octahedral sites. (Adapted with permission from W. P. Pearson. *The Crystal Chemistry and Physics of Metals and Alloys*, Wiley, New York, 1972, p. 207, copyright © 1972.)

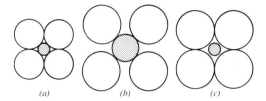

Figure 5.20   Effect of variation of the ratio of radius cation to radius of anion.

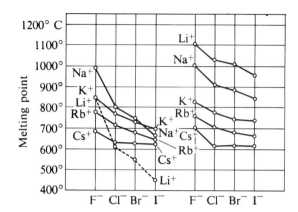

Figure 5.21   The observed melting points of the alkali halides (*left*) and values corrected for the radius-ratio effect (*right*). (Reproduced with permission from Linus Pauling, *The Nature of the Chemical Bond*, 3rd ed., Cornell University, Ithaca, NY. Copyright © 1960 by Cornell University Press.)

(Figure 5.20*b*). Using the radius of I^- from LiI and the internuclear distance in KI, we can calculate the radius of K^+. Similarly the ionic radius of the oxide ion is obtained from the O—O distance in silicates, assuming that the oxides are in contact.

   Ionic radii have been obtained also from electron density plots obtained from x-ray analysis. Taking the radius of each of two adjacent ions as the distance from the nuclear center (electron density maximum) to the electron density minimum between the ions yields radii that are self-consistent, but the radii obtained are larger for cations and smaller for anions than those from other sources. The minimum is broad, making it difficult to assign radii.

► 5.7.2   Radius Ratio Effects

The ratios of (cation radius)/(anion radius) expected for various C.N. are given in Table 5.4. (see also Examples 5.1 and 5.2). Ratios outside of the ranges given are encountered without a change in structure, but in those cases the interionic distances no longer agree with the sum of the ionic radii. We can illustrate this phenomenon by considering a structure with C.N. 6. The spatial relationship can be seen most clearly in a sectional view through any plane of four anions around the central cation. The stable arrangement, where there is cation–anion contact and the anions are just in contact, is shown in Figure 5.20*a*. If the radius ratio is larger (Figure 5.20*b*; larger cation or smaller anions), the decreased anion repulsion would produce an M—X distance shorter than the sum of the ionic radii. Actually, for an exaggerated situation such as that shown in Figure 5.20*b*, we would expect an increase in C.N. If the cation radius is smaller than in Figure 5.20*a* or if the anion radius is larger as in Figure 5.20*c*, the resulting M—X distance is greater than the sum of the ionic radii.

The alkali halides represent a series in which properties dependent on the stability of the ionic lattice should vary in a consistent way. The heats of fusion, melting points, boiling points, and so on, might be expected to vary regularly with the M—X distance. These properties do follow a consistent pattern for salts of K^+, Rb^+, and Cs^+, but Li^+ salts and some Na^+ salts show significant deviations, as illustrated by the melting points in Figure 5.21. We can eliminate these irregularities by applying radius ratio corrections, to allow for the lower stability of the salts for which anion–anion repulsion is important. Similar corrections are needed to remove irregularities in the boiling points.

Some salts crystallize in two or more modifications differing in C.N. Anion–anion repulsion is always greater for the higher C.N., resulting in an increase in the cation–anion distance. Consequently the ionic radii are slightly larger for higher C.N.

▶ 5.7.3 Crystal Radii

Crystal radii are ionic radii evaluated crystallographically as compatible with observed interionic distances in crystal. Several sets of crystal radii have been used. Each set is self-consistent, but because of the different approaches employed, the sets differ sufficiently from each other to prevent combining radii from different sets. The most comprehensive compilation of ionic radii is that of Shannon and Prewitt, as revised and extended by Shannon.

The crystal radii for ions with different C.N.'s are given in Table 5.8, and those for ions of variable oxidation numbers are given in Table 5.9. The paper by Shannon includes some other oxidation states and C.N.'s—in some cases, for C.N. 12 or 14. Table 5.10 lists crystal radii for anions, with van der Waals radii of the noble gases for comparison. Table 5.11 lists radii for some polyatomic ions.

5.8 LATTICE ENERGY

▶ 5.8.1 Born Equation

We can determine the stability of an ionic lattice from the coulombic interactions among the ions. A pair of ions of opposite charge attract one another, and the potential energy varies inversely with the first power of d, the internuclear separation. The force between the ions is $Z_1 Z_2/d^2$, where Z_1 and Z_2 are the charges of the ions. As the ions come very close together, they repel one another because of interpenetration of electron clouds. The repulsion energy is inversely proportional to the nth power of d. The Born exponent, n, increases with an increase in electron density around the ions ($n = 5$ for the He configuration, 7 for Ne, 9 for Ar or Cu^+, 10 for Kr or Ag^+, and 12 for Xe or Au^+; an average value for n is used if the cation and anion have different configurations).

The Born equation [Equation (5.1)] gives the potential energy (PE) for a pair of ions

$$\text{PE} = \frac{Z_1 Z_2 e^2}{d} + \frac{b e^2}{d^n} \tag{5.1}$$

Table 5.8   Crystal radii (pm)[a] (the positive oxidation number is the same as the group number except as noted for actinides)

Main group (s‑ and p‑block)

Coordination number	I	II	III	IV	V	VI	VII
	Li	Be	B	C	N		
3	—	30	15	6	4.4		
4	73	41	25	29	—		
6	90	59	41	30	27		
	Na	Mg	Al	Si	P	S	Cl
4	113	71	53	40	31	26	22
6	116	86.0	67.5	54.0	52	43	41
8	132	103	—	—	—	—	—
	K	Ca	Ga	Ge	As	Se	Br
4	151	—	61	53.0	47.5	42	39
6	152	114	76.0	67.0	60	56	53
8	165	126	—	—	—	—	—
	Rb	Sr	In	Sn	Sb	Te	I
4	—	—	76	69	—	57	56
6	166	132	94.0	83.0	74	70	67
8	175	140	106	95	—	—	—
	Cs	Ba	Tl	Pb	Bi	Po	At
4	—	—	89	79	—	—	—
6	181	149	102.5	91.5	90	81	76
8	188	156	112	—	—	—	—

Transition metals, groups I and II

Coordination number	I	II
	Cu	Zn
2	—	—
4	74	74
6	91	88.0
8	—	104.0
	Ag	Cd
2	81	—
4	114	92
(sq)	116^{Sq}	
6	129	109
8	142	124
	Au	Hg
6	151	110
		116
		128

Transition metals, groups III–VII

Coordination number	III	IV	V	VI	VII
	Sc	Ti	V	Cr	Mn
4	—	56	49.5	40	39
6	88.5	74.5	68	58	60
8	101.0	88	—	—	—
	Y	Zr	Nb	Mo	Tc
4	—	73	62	55	51
6	104.0	86	78	73	70
8	115.9	98	88	—	—
	La	Hf	Ta	W	Re
4	—	72	—	56	52
6	117.2	85	78	74	67
8	130.0	97	88	—	—

Lanthanides

Coordination number	La	Ce	Pr	Nd	Pm	Sm	Eu	Gd	Tb	Dy	Ho	Er	Tm	Yb	Lu
6	117.2	115	113	112.3	111	109.8	108.7	107.8	106.3	105.2	104.1	103.0	102.0	100.8	100.1
7	124	121	—	—	—	116	115	114	112	111	—	108.5	—	106.5	—
8	130.0	128.3	126.6	124.9	123.3	121.9	120.6	119.3	118.0	116.7	115.5	114.4	113.4	112.5	111.7

Actinides

Coordination number	Ac^{III}	Th^{IV}	Pa^{V}	U^{VI}	Np^{IV}	Pu^{IV}	Am^{IV}	Cm^{IV}	Bk^{IV}	Cf^{IV}	No^{II}
6	126	108	92	87	101	100	99	99	97	96.1	124
8	—	119	105	100	112	110	109	109	107	106	

[a] Values from R. D. Shannon, *Acta Cryst.* **1976**, *A32*, 751. The reference includes other oxidation states and coordination numbers.

Table 5.9 Crystal radii (pm)[a] for positive oxidation numbers, C.N. 6 (except as designated in italics)

Oxidation number	Ti	V	Cr	Mn	Fe	Co	Ni	Cu	Zn	Ga	Ge	As	Se
II	100	93	94, 87^{LS}	97.0, 81^{LS}	77(4), 92, 75^{LS}	72(4), 88.5, 79^{LS}	69(4), $63(4)^{LS}$, 83	71(4), $71(4)^{Sq}$, 87	74(4), 88.0		87		
III	81.0	78.0	75.5	78.5, 72^{LS}	78.5, 69^{LS}, 63(4)	75, 68.5^{LS}	74, 70^{LS}	68^{LS}		76.0		72	
IV	74.5	72	69	67	72.5	67	62^{LS}				67.0		64

Oxidation number	Zr	Nb	Mo	Tc	Ru	Rh	Pd	Ag	Cd	In	Sn	Sb	Te
II	—	—	—	—	—	—	$78(4)^{Sq}$	$93(4)^{Sq}$	109	—	122[b](8)	—	—
III	—	86	83	—	82	80.5	90	$81(4)^{Sq}$	—	94.0	—	94(5)	—
IV	86	82	79	78.5	76.0	74	75.5	—	—	—	83.0	—	111

Oxidation number	Hf	Ta	W	Re	Os	Ir	Pt	Au	Hg	Tl	Pb	Bi	Po
I	—	—	—	—	—	—	—	151	133, 111(3)	164	—	—	—
II	—	—	—	—	—	—	$74(4)^{Sq}$	—	116	—	133	—	—
III	—	86	—	—	—	82	—	$82(4)^{Sq}$	—	102.5	—	117	—
IV	85	82	80	77	77.0	76.5	76.5	—	—	—	91.5	—	108

Ce^{IV} 101	Pr^{IV} 99	Sm^{II} 141(8)	Eu^{II} 131	Tb^{IV} 90	Tm^{II} 117	Yb^{II} 116	Cf^{III} 109
Pa^{IV} 104	U^{IV} 103	Np^{VI} 86	Pu^{VI} 85	Am^{III} 111.5	Cm^{III} 111	Bk^{III} 110	

[a] Values from R. D. Shannon, *Acta Cryst.* **1976**, *A32*, 751. Low-spin values (LS) and values for square planar (Sq) coordination are designated by superscripts.

[b] Value for C.N. 8 from R. D. Shannon and C. T. Prewitt, *Acta Cryst.* **1969**, *B25*, 925. The value is probably doubtful, since it was not included in the revised tabulation (footnote a).

Table 5.10 Crystal radii (pm for C.N. 6)[a]

	OH⁻	H⁻	He
	123	153[b]	(93)[c]
N³⁻	O²⁻	F⁻	Ne
132	126	119	(112)
(C.N. 4)	S²⁻	Cl⁻	Ar
	170	167	(154)
	Se²⁻	Br	Kr
	184	182	(169)
	Te²⁻	I⁻	Xe
	207	206	(190)

[a] R. D. Shannon, *Acta Cryst.* **1976,** *A32,* 751.

[b] D. F. C. Morris and G. L. Reed, *J. Inorg. Nucl. Chem.* **1965,** *27,* 1715.

[c] Van der Waals radii of the noble gases are given for comparison.

Table 5.11 Radii of some complex anions[a]

BeF_4^{2-}	CO_3^{2-}	NO_3^-				
245	185	189				
BF_4^-	$GeCl_6^{2-}$	PO_4^{3-}	SO_4^{2-}	ClO_4^-		
228	243	238	230	236		
SiF_6^{2-}	$SnCl_6^{2-}$	AsO_4^{3-}	SeO_4^{2-}		CrO_4^{2-}	MnO_4^-
194	247	248	243		230	240
GeF_6^{2-}	$PbCl_6^{2-}$	SbO_4^{3-}	TeO_4^{2-}	IO_4^-	MoO_4^{2-}	
201	248	260	254	249	254	
	$TiCl_6^{2-}$					
	248					
	$TiBr_6^{2-}$	$IrCl_6^{2-}$				
	261	254				
	$ZrCl_6^{2-}$	$PtCl_6^{2-}$				
	247	259				

[a] Radii for oxoanions, BeF_4^{2-}, and BF_4^- are "thermochemical" radii from T. C. Waddington, *Adv. Inorg. Chem. Radiochem.* **1959** *1,* 157; and A. F. Kapustinskii, *Q. Rev.* **1956,** *10,* 283. Radii for MX_6^{2-} in compounds with the antifluorite structure are from R. H. Prince, *Adv. Inorg. Chem. Radiochem.* **1979,** *22,* 349.

where Z_1 and Z_2 are integral charges with appropriate signs, e is the charge on the electron, d is the internuclear separation, n is the Born exponent, and b is a repulsion coefficient. The potential energy is a minimum at d_0 and becomes positive when d is very small. The repulsion term increases to become dominant for very small values of d.

Equation (5.1) gives the energy released when a cation and anion, separated by an infinite distance, are brought together until they are separated by the distance d. In a crystal of NaCl, however, each Na^+ is surrounded by six Cl^- at a distance d, not by just one. Twelve other Na^+ are located at a distance $\sqrt{2}d$, eight other Cl^- at $\sqrt{3}d$, six more Na^+ at $2d$, 24 Cl^- at $\sqrt{5}d$, 24 Na^+ at $\sqrt{6}d$, and so on (see Figure 5.22). Summing all of the interactions gives the first term in the potential-energy expression for the energy released in bringing a sodium ion from infinity to its stable position in the NaCl lattice:

$$PE = -\frac{6e^2}{d} + \frac{12e^2}{\sqrt{2}d} - \frac{8e^2}{\sqrt{3}d} + \frac{6e^2}{2d} - \frac{24e^2}{\sqrt{5}d} + \frac{24e^2}{\sqrt{6}d} \cdots$$

or

$$= -\frac{e^2}{d}\left(6 - \frac{12}{\sqrt{2}} + \frac{8}{\sqrt{3}} - \frac{6}{2} + \frac{24}{\sqrt{5}} - \frac{24}{\sqrt{6}} \cdots\right)$$

(5.2)

The product $Z_1 Z_2$ is omitted for simplicity, since it is unity in this case; only the appropriate signs are carried. The term in parentheses in Equation (5.2) is the sum of an infinite

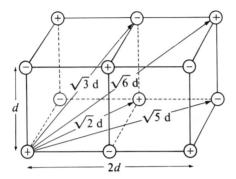

Figure 5.22   Distances to neighboring ions in the NaCl lattice.

series converging toward 1.747558, which is the **Madelung constant**[12] for the NaCl structure. This Madelung constant is used for any salt with the NaCl structure, since it depends only on the geometrical arrangement of the ions. Values of the Madelung constant for other structures are evaluated similarly. Table 5.12 provides the values for some common structures.

The second term in the potential-energy expression for ions in a real crystal allows for the repulsive forces resulting from the interpenetration of electron clouds. This term is simpler to handle than the first term, since the repulsion varies inversely with d^n and hence only nearest neighbors need be considered. Each sodium ion is surrounded by six chloride ions, so the repulsion energy term becomes $6be^2/d^8$ [the Born exponent is the average of that for Na^+ (7) and Cl^- (9)]. The number of nearest neighbors times b is designated by B, the repulsion coefficient.

The potential energy of an ion in a crystal, considering the forces of all neighboring ions, becomes

$$PE = \frac{Ae^2 Z_1 Z_2}{d} + \frac{Be^2}{d^n} \tag{5.3}$$

Table 5.12   Values of Madelung constants[a]

Structure	Madelung constant	Structure	Madelung constant
Sodium chloride	1.74756	Rutile (TiO_2)	2.408
Cesium chloride	1.76267	Anatase (TiO_2)	2.400
Zinc blende (ZnS)	1.63806	Cadmium iodide	2.36
Wurtzite (ZnS)	1.64132	β-Quartz (SiO_2)	2.201
Fluorite (CaF_2)	2.51939	Corundum (Al_2O_3)	4.040

[a] Values are geometrical Madelung constants. "Conventional" values differ in cases where the charges differ for the cations and anions, as for CaF_2, TiO_2, and Al_2O_3.

[12] E. L. Burrows and S. F. Kettle, *J. Chem. Educ.* **1975**, *52*, 58.

where A is the Madelung constant, and B is the repulsion constant. An additional repulsion term must be added where anion–anion repulsion is unusually great, as with many lithium salts.

An ion achieves its most stable equilibrium position when attractive and repulsive forces are balanced; then, the PE is at a minimum and $d = d_0$. Differentiating the PE with respect to d and equating to zero results in

$$d_0 = \left(-\frac{nB}{AZ_1 Z_2}\right)^{1/(n-1)} \tag{5.4}$$

Solving for B, we obtain

$$B = \frac{-d_0^{(n-1)} A Z_1 Z_2}{n} \tag{5.5}$$

By substituting the value of B in Equation (5.3), the potential energy becomes

$$(\mathrm{PE})_0 = \frac{-Ae^2 Z_1 Z_2}{d_0}\left(\frac{1}{n} - 1\right) \tag{5.6}$$

The **lattice energy,** U_0, is the energy required to separate ions as gaseous ions separated from each other by infinite distances from a mole of MX (crystal) ($U_0 = -(\mathrm{PE})_0 N$, where N is Avogadro's number).

$$U_0 = \frac{-NAe^2 Z_1 Z_2}{d_0}\left(1 - \frac{1}{n}\right) \tag{5.7}$$

U_0 is positive; ΔH for the formation of the crystal from gaseous ions is negative. The lattice energy is very useful in correlating properties of ionic substances because the formation or destruction of the crystal is frequently the most important step in reactions involving ionic substances.

The Born treatment employs the model of ions as hard spheres undistorted by the neighboring ions. Because increasing distortion corresponds to increasing covalent character, the lattice energy calculated from Equation (5.7) should not agree well with the experimental value from thermodynamic data, where the covalent character of the bonds is appreciable.

There have been several modifications of Equation (5.7), most of which involve empirical constants or additional terms. The best values of lattice energies for compounds that fit the ionic model well, such as alkali halides, are obtained using equations with additional terms.[13] Equation (5.7) is derived from basic principles and illustrates the important dependencies on charges, sizes, and configurations of ions. The **Born–Mayer equa-**

[13] D. Cubicciotti, *J. Chem. Phys.* **1959,** *31,* 1646.

tion improves the treatment of repulsions:

$$U_0 = \frac{-NAe^2 Z_1 Z_2}{d}\left(1 - \frac{34.5}{d}\right) \tag{5.8}$$

where d is in picometers.

▶ 5.8.2 The Born–Haber Cycle

Lattice energies, which are not obtained easily by direct measurement, can be evaluated using Born–Haber cycles, which relate the lattice energy to other thermochemical quantities. The formation of a solid salt (MX) from the elements by two different paths is formulated as

$$
\begin{array}{ccc}
\mathrm{M}(g) + \mathrm{X}(g) & \xrightarrow{\text{IE} - \text{EA}} & \mathrm{M}^+(g) + \mathrm{X}^-(g) \\
{\scriptstyle S}\uparrow{\scriptstyle \frac{1}{2}D} & & \downarrow{\scriptstyle -U_0} \\
\mathrm{M}(c) + \frac{1}{2}\mathrm{X}_2(g) & \xrightarrow{\Delta H_f^0} & \mathrm{MX}(c)
\end{array}
$$

where U_0 is the lattice energy, IE is the ionization energy of $\mathrm{M}(g)$, EA is the electron affinity of $\mathrm{X}(g)$, D is the heat of dissociation of $\mathrm{X}_2(g)$, S is the sublimation (atomization) energy of the metal, and ΔH_f^0 is the heat of formation of $\mathrm{MX}(c)$ from the elements. Because the change in energy is independent of the path, ΔH_f^0 can be equated to the algebraic sum of the other thermochemical quantities with the appropriate thermodynamic signs.

We can determine the lattice energy from experimental thermochemical quantities. In early applications, the lattice energy was calculated from Equation (5.7) and used with the experimental thermochemical quantities to evaluate EAs, for which experimental values were not available. Using experimental EAs, which are now available from techniques such as electron scattering, an "experimental" U_0 can be obtained from a Born–Haber cycle and compared with the theoretical value from Equation (5.7). Agreement usually is very good, except for salts containing high-charge ions or 18-electron cations, for which bonding has appreciable covalent character. The experimental value of U_0 usually exceeds appreciably the theoretical value for salts such as HgS, HgSe, and PbO_2, for which polarization is important.

By comparing the heats of formation of the alkali halides (Table 5.13), we can see that we must consider all of the thermochemical quantities involved in order to explain the variations in ΔH_f^0. The heats of formation (ΔH_f^0) decrease in magnitude through the series MF, MCl, MBr, and MI, as anion repulsion increases and the coulombic energy decreases because of the increasing interionic distances. The relative contributions of the lattice energy (U_0), dissociation energy (D), and electron affinity (EA) vary through the series. As the radius of M increases, the ΔH_f^0 values increase in magnitude for the chlorides, bromides, and iodides (corresponding to a decrease in anion repulsion), but decrease for the fluorides. Anion repulsion is at a minimum for F^- because of its small size. The lattice en-

Table 5.13   Thermochemical data for the alkali metal halides (kJ/mol)

	$-\Delta H_f{}^a$ 298 K (MX)	S 298 K (M)	$\frac{1}{2}D$ 298 K	IE^b (M)	EA (X)	U_0 298 K	$\Delta H_1{}^c$	U_0 0 K	$U_0{}^d$ (Theor.)
LiF	616.9	160.7	78.9	520.5	328.0	1049.0	−6	1043	966
NaF	573.6	107.8	78.9	495.4	328.0	927.7	−5	923	885
KF	567.4	89.2	78.9	418.4	328.0	825.9	−3	823	786
RbF	553.1	82.0	78.9	402.9	328.0	788.9	−2	787	730
CsF	554.7	77.6	78.9	375.3	328.0	758.5	−1	757	723
LiCl	408.3	160.7	121.3	520.5	348.8	862.0	−5	857	809
NaCl	411.1	107.8	121.3	495.4	348.8	786.8	−2	785	752
KCl	436.7	89.2	121.3	418.4	348.8	716.8	−1	716	677
RbCl	430.5	82.0	121.3	402.9	348.8	687.9	0	688	651
CsCl	442.8	77.6	121.3	375.3	348.8	668.2	+1	669	622
LiBr	350.2	160.7	111.8	520.5	324.6	818.6	−4	815	772
NaBr	361.4	107.8	111.8	495.4	324.6	751.8	−1	751	718
KBr	393.8	89.2	111.8	418.4	324.6	688.6	0	689	650
RbBr	389	82.0	111.8	402.9	324.6	661	+1	662	629
CsBr	395	77.6	111.8	375.3	324.6	635	+2	637	600
LiI	270.1	160.7	106.8	520.5	295.4	762.7	−2	761	723
NaI	288	107.8	106.8	495.4	295.4	703	0	703	674
KI	327.9	89.2	106.8	418.4	295.4	646.9	+1	648	615
RbI	328	82.0	106.8	402.9	295.4	625	+1	626	594
CsI	337	77.6	106.8	375.3	295.4	602	+2	604	569

^{a} I. Barin and O. Knacke, Eds., *Thermochemical Properties of Inorganic Substances*. Springer-Verlag, Berlin, 1973.

^{b} Ionization energies are given in Table 1.7.

^{c} ΔH for changing $M(g)$ and $X(g)$ at 298 K to 0 K (−12 kJ) plus ΔH for changing MX(c) at 298 K to 0 K (6 to 15 kJ). See D. Cubicciotti, *J. Chem. Phys.* **1959**, *31*, 1646 and **1961**, *34*, 2189; as well as D. A. Johnson, *Some Thermodynamic Aspects of Inorganic Chemistry*, Cambridge Press, Cambridge, 1968, Appendix 2, and J. L. Holm, *J. Chem. Educ.* **1974**, *51*, 460.

^{d} Lattice energy calculated using Equation (5.7).

ergy of fluorides is greatest with small cations, because of the short M—X distance. Sublimation energy (S) and ionization energy (IE) of the metal and the lattice energy are the major factors determining the trends.

Considering the individual steps for formation of MX(c) from the elements, the only quantities corresponding to the release of energy are the formation of the lattice and the EA. The EA for the addition of a single electron is greater for each of the halogens than for the member of the oxygen family of the same period, because of the greater nuclear charge, the smaller size, and the formation of a closed electron configuration (see Table 1.8 for EAs). The very small size of the F atom, with its high electron density, makes adding an electron to it slightly less favorable energetically than for Cl. Fluorine is a better oxidizing agent than chlorine partly because of the much lower dissociation energy of F_2, which more than compensates for the difference in EAs. In addition, the very high U_0 of fluorides contributes greatly to the energy released in the formation of solid fluorides.

The fact that the bond energy of F_2 is much lower than that of Cl_2 follows the pattern of the low single-bond energies (in kJ/mol) for O—O (142) and N—N (167) as compared with S—S (268) and P—P (239). The low values for N, O, and F may result from the greater repulsion among the nonbonding electrons of the two small atoms with such high electron densities. The other members of the families, moreover, have vacant d orbitals of the same quantum number, which might be hybridized with the p orbitals to some extent to contribute to the bonding and reduce repulsion. The increased polarizability of the larger atoms also helps decrease repulsion.

Formation of the oxide ion and other anions with a charge greater than one is an endothermic process. Energy is released as the first electron is gained by a gaseous oxygen atom, but adding the second electron requires the expenditure of a greater amount of energy because it is repelled by the negative charge of the O^- ion. Because of the sulfur atom's greater size and lower charge density, the EA for the formation of S^{2-} from S is less endothermic than that for the formation of O^{2-} from O. For oxides and salts of other simple 2− and 3− ions, the formation of the lattice is the only quantity involved in the formation of the solid that releases energy. The only simple 3− ion encountered in ionic crystals is the nitride ion—and then only when combined with the most electropositive metal ions. Forming the N^{3-} ion from N requires 2200 kJ/mol.

Born–Haber cycles are useful in evaluating the contributions of particular steps in determining the relative stability of a compound or in comparisons of compounds. If the lattice energy calculated from data using the Born–Haber cycle is significantly greater than that calculated from Equation (5.7), polarization is probably important. In these cases the hard-sphere model is poor.

▶ *5.8.3 Kapustinskii Equation*

A. F. Kapustinskii[14] noted that the ratio of the Madelung constant to the number of ions per formula is approximately the same for many structures. The ratio increases with C.N., which increases with ionic radius, so A/vd (v is the number of ions per formula unit and $d = r_{M^+} + r_{X^-}$) varies little from one structure to another. If we lack the necessary structural information to obtain a Madelung constant, we can estimate the lattice energy using the **Kaputstinskii equation:**

$$U = -121,000 \frac{\text{kJ pm}}{\text{mol}} \left(\frac{vZ_1 Z_2}{d} \right) \left(1 - \frac{34.5}{d} \right) \tag{5.9}$$

[14] A. F. Kapustinskii, *Q. Rev. Chem. Soc.* **1956,** *10,* 283. See also T. C. Waddington, *Adv. Inorg. Chem. Radiochem.* **1959,** *1,* 157.

Here Z_1 and Z_2 are the actual charges (with signs) of the ions. The radii in Table 5.11 for nonspherical ions are called **thermochemical radii.** These self-consistent radii are obtained as those which give lattice energies in agreement with those obtained from the Born–Haber cycle.

EXAMPLE 5.3:   Estimate the lattice energy of $CaCO_3$ using the Kaputstinskii equation.

Solution: For $CaCO_3$, $v = 2$ and $d = 114 + 185 = 299$ pm.

$$U_0 = -121{,}000 \frac{\text{kJ pm}}{\text{mol}} \left(\frac{2 \times 2 \times (-2)}{299} \right) \left(1 - \frac{34.5}{299} \right) = 2860 \text{ kJ/mol}$$

▶ *5.8.4   Which MX_n Compounds Should Be Stable?*

The steps required to form a gaseous metal ion involve (a) sublimation energy to convert the solid metal in its standard state to the separate gaseous atoms and (b) the ionization energy (IE). The energy needed to remove the second, third, and so on, electrons corresponds to the second, third, and so on, IEs. The periodic variation in IE indicates that the elements most likely to form simple cations (those with low IE values) in ionic compounds will be found at the beginning of each period, and for the members of a family those with larger radii.

The very great increase in successive IEs is the major factor limiting the charge of cations in simple ionic substances to 1+ or 2+, with 3+ uncommon and 4+ encountered only for very large ions, such as Th^{4+} or Ce^{4+}. The other factor involves the increased covalent character of compounds containing "ions" of high charge. As Equation (5.7) makes clear, the U_0 increases greatly with increasing ionic charge. We might expect the U_0 of a salt such as NaF_2 to be very large (the value for MgF_2 can be taken as an estimate), but if we calculate the energy for the reaction

$$NaF_2 \rightarrow NaF + \tfrac{1}{2}F_2 \tag{5.10}$$

we find it to be highly exothermic, primarily because of the very high second IE of Na. Similarly, although the formation of MgF should be an exothermic process (the U_0 of MgF should be about the same as that of NaF), the much greater U_0 of MgF_2 makes the formation of solid MgF_2 much more favorable, and the disproportionation of the solid MgF to give MgF_2 and Mg is exothermic.[15]

EXAMPLE 5.4:   Calculate the lattice energy, the heat of formation, and the heat of disproportionation of CaCl.

Solution: Assume that CaCl has the NaCl structure. The crystal radius of Ca^{2+} is 114 pm, so use 120 pm as the estimate of the radius of Ca^+. The value of e is 4.80×10^{-10} esu, or $\sqrt{\text{dyne}}$ cm.

---

[15] For other examples, see J. L. Holm, *J. Chem. Educ.* **1974,** *51,* 460.

Here the cgs units offer an advantage, and we can convert ergs to kJ easily.

$$U_0 = \frac{(4.80 \times 10^{-10}\sqrt{\text{dyne}}\ \text{cm})^2\ 1^2(6.02 \times 10^{23})1.75}{2.9 \times 10^{-8}\ \text{cm}}\left(1 - \frac{1}{9}\right)$$

$$= \frac{7.44 \times 10^{12}\ \text{erg}}{\text{mol}} \times \frac{1\ \text{kJ}}{10^{10}\ \text{erg}} = 744\ \text{kJ/mol}$$

The Born–Mayer equation gives $U_0 = 737$ kJ/mol.
Using $S = 201$ kJ, $\frac{1}{2}D = 121$ kJ, IE $= 589$ kJ, and EA $= 349$ kJ, we obtain

$$\Delta H_f = 201 + 121 + 589 - 349 - 744 = -182\ \text{kJ/mol CaCl}$$

For the disproportionation reaction,

$$2\text{CaCl} \rightarrow \text{CaCl}_2 + \text{Ca}$$

we have

$$\Delta H = -799 + 0 + 2(182) = -435\ \text{kJ/mol CaCl}_2\ \text{formed}$$

The rate of evaporation of Al from liquid Al can be increased greatly by passing a stream of AlCl_3 over the surface. Upon cooling, the AlCl formed and carried along in the gas stream disproportionates to deposit Al. This process was patented (Gross, 1946) for the purification of Al.

The low negative heat of formation of some noble metal halides, as compared with alkali metal halides, results largely from high IEs, but frequently the differences in sublimation energy also are great. Consider the steps of the Born–Haber cycle that differ for NaCl and AgCl (energies in kJ/mol):

$$\text{Ag}(c) \xrightarrow[S]{284} \text{Ag}(g) \xrightarrow[\text{IE}]{727} \begin{matrix} \text{Ag}^+(g) \\ \text{Cl}^-(g) \end{matrix} \xrightarrow[-U_0]{-910} \text{AgCl}(c)\ \Delta H_f = -127\ \text{kJ/mol} \qquad (5.11)$$

$$\text{Na}(c) \xrightarrow[S]{108} \text{Na}(g) \xrightarrow[\text{IE}]{495} \begin{matrix} \text{Na}^+(g) \\ \text{Cl}^-(g) \end{matrix} \xrightarrow[-U_0]{-787} \text{NaCl}(c)\ \Delta H_f = -411\ \text{kJ/mol} \qquad (5.12)$$

Although the U_0 is greater for AgCl, the heat of formation is much less. Both the IE and sublimation energy contribute significantly to the "nobility" of Ag as compared with Na. (See page 706 for a discussion of periodic variations in sublimation energies.) AgCl's heat of formation would be even lower were it not for the importance of polarization effects. Silver chloride is one of the compounds for which the polarization effects (or increased covalent character) cause the U_0 obtained from the Born–Haber cycle (910 kJ/mol) to be appreciably larger than that calculated using the Born equation and the ionic radii (735).

▶ 5.8.5 Stabilization of Polyatomic Ions by Large Counterions[16]

To isolate a compound such as $\text{M}[\text{IF}_4]$ (page 797) containing the polyatomic or "complex" anion $[\text{IF}_4^-]$, you would not try to make $\text{Li}[\text{IF}_4]$, because the reaction

$$\text{Li}[\text{IF}_4] \rightarrow \text{LiF} + \text{IF}_3 \qquad (5.13)$$

[16] D. H. McDaniel, *Annu. Reports Inorg. Gen. Syntheses* **1972**, 293; F. Basolo, *Coord. Chem. Rev.* **1968**, *3*, 213.

is strongly favored by the high U_0 of LiF and the low U_0 of $Li^+(IF_4)^-$. The U_0 of $LiIF_4$ is low because of the large M—X distance. The IF_4^- ion is more stable in a solid with a large cation such as Cs^+ or $N(C_2H_5)_4^+$. In these cases the U_0 is low also for CsF or $N(C_2H_5)_4F$, the possible products of a reaction such as (5.13). The nitrate ion forms weakly bound complexes such as $[Co(NO_3)_4]^{2-}$ and $[Zr(NO_3)_6]^{2-}$ which can be isolated only with large cations. The value of n in polyhalide complexes $[MX_n]^{x-}$ commonly decreases as the size of the halide ion increases. Hexafluorides are much more common than hexachlorides or hexabromides, with the latter more likely to be isolated as salts of large cations.

The stabilizing effect of the large counterion can take several forms:

Lattice Energy Limiting: The decomposition of $Li[IF_4]$ is favored because of the great gain in U_0 in the formation of LiF. With a cation large enough for cation–cation contact in the $M[IF_4]$ structure, the U_0 would favor $M[IF_4]$ over MF.

Polarization Effects: Small cations may polarize some complex ions so much that they are pulled apart.

Insulating Effects: Ions, such as SbF_4^-, that tend to form polyanions can be isolated and stabilized by large cations in the solid.

5.9   SOLUBILITIES OF IONIC SUBSTANCES

Ionic substances are only very slightly soluble in most common solvents, except for those that are very polar. The strong attractive forces between ions in the crystal must be overcome, and this can be accomplished only if the attractive forces between the ions and solvent molecules are at least comparable to the lattice energy. The energy necessary for the separation of the ions from the crystal comes from the solvation of the ions. The attractive forces between nonbonded neutral molecules are usually very weak, and the forces between an ion and a neutral molecule are not much stronger unless the molecule has a fairly high dipole moment and/or high polarizability. The greater the dipole moment, the stronger the attraction by an ion and, usually, the greater the solvation energy. Solvation energies also increase with the polarizability of the solvent molecules. The energy required to separate ions and keep them apart is diminished by a decrease in the forces between ions, which are dependent on the **dielectric constant** (see page 92) of the medium. Because the dielectric constant of water is 78.54 (25°C), the attractive forces between two ions in water is 1/78.54 of the force between the ions separated by the same distance in a vacuum. Good solvents for ionic substances usually have high dipole moments and high dielectric constants—although few solvents have dielectric constants as high as that of water.

Water is a much better solvent for ionic substances than other solvents with high dielectric constants (e.g., HF). We could view water molecules as donating electron pairs (Lewis bases) to the metal ions (Lewis acids), or the interaction can be interpreted in terms of electrostatic forces. According to either interpretation, a metal ion interacts strongly with a layer of water molecules referred to as the **primary hydration sphere.**

Because the water molecules in the primary hydration sphere, associated with a cation, form stronger hydrogen bonds to other water molecules than they would otherwise, a secondary hydration sphere arises. The water molecules in the secondary hydration sphere, in turn, are hydrogen-bonded to other water molecules with hydrogen bonds weaker than those formed by water in the primary hydration sphere but stronger than in water itself. Anions—particularly oxoanions—also can be strongly hydrogen-bonded to water (see Figure 5.23). HF is a poorer solvent for salts, in spite of its high dipole moment and ability to form strong hydrogen bonds, because it is a much weaker Lewis base (electron donor) than water.

The energy changes involved in the dissolution of a salt can be handled conveniently by a Born–Haber cycle:

$$MX(c) \xrightarrow{U_0} M^+(g) \;+\; X^-(g)$$
$$L \searrow \qquad \downarrow \Delta H_{M^+} \qquad \downarrow \Delta H_{X^-}$$
$$M^+(\text{solv.}) + X^-(\text{solv.})$$

where U_0 is lattice energy of the crystal MX (positive thermodynamic sign as shown), ΔH_{M^+} and ΔH_{X^-} are the energies released (negative sign) as a result of the solvation of the gaseous positive and negative ions, and L is the observed heat of solution at infinite dilution. Because the total energy change in going from MX(c) to the solvated ions is independent of the path, the heat of solution is given by

$$L = \Delta H_{M^+} + \Delta H_{X^-} + U_0 \qquad (5.14)$$

The heat of solution can be positive or negative because the U_0 and the heats of solvation of the ions are comparable. The heat of solution is often, but not always, negative (exothermic) for very soluble substances. The other factor that can account for high solubility—even though the process is endothermic—is the entropy change that accompanies dissolution. NH$_4$Cl is very soluble in H$_2$O, but the solution gets very cold when NH$_4$Cl

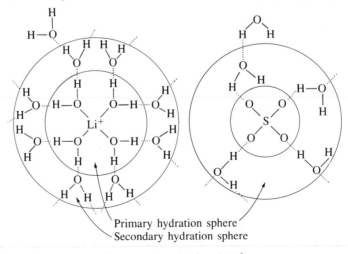

Primary hydration sphere
Secondary hydration sphere

Figure 5.23   Hydrated Li$^+$ and SO$_4^{2-}$ ions.

dissolves. Accompanying the destruction of the well-ordered crystal lattice is a large favorable entropy change, but the ions orient the solvent molecules, causing a decrease in entropy. Using the thermodynamic relationships $\Delta G^0 = \Delta H^0 - T\Delta S^0$ and $\Delta G^0 = -RT \ln K$. The solubility product constant, K_{sp}, can be related to the heat of solution, L, and the change in entropy, ΔS^0, when one mole of solute dissolves to give an ideal one molar solution:

$$RT \ln K_{sp} = -L + T\Delta S^0 \tag{5.15}$$

where R is the gas constant and T is the absolute temperature. Care must be exercised in obtaining thermodynamic quantities from solubility products, because of complications caused by nonideal behavior and competing equilibria–ion association, for example.

Equation (5.7) or (5.8) gives the effects of size and charge on lattice energy. The solvation energy, H, of an ion is given by the Born equation

$$H = -\frac{Z^2}{2r}\left(1 - \frac{1}{\varepsilon}\right) \tag{5.16}$$

where Z is the charge on the ion, r is the radius,[17] and ε is the dielectric constant of the solvent. The solvation energies and solubilities of salts usually increase as the dielectric constant of the solvent increases. Both solvation energy and lattice energy are inversely proportional to the ionic radii (or M—X distance for U_0). The solubilities of salts usually increase as the size of cations or anions increases, presumably because of more favorable entropy changes on solvation—and also, possibly, because of polarization effects in some cases. The charge effect usually is considerably greater than the size effect. With increasing ionic charge, the lattice energy increases more than the solvation energy whenever there is an accompanying increase in the Madelung constant. The entropy change accompanying the dissolution of a salt is usually less favorable for small ions, particularly for ions of high charge. Ions of high charge density cause a great deal of ordering of the solvent molecules, as a result of solvation.

Solubilities of salts are high for low charges of ions—salts of alkali metal and salts of ClO_4^-, NO_3^-, $C_2H_3O_2^-$, and halides. There are fewer soluble salts of anions of higher charge unless paired with univalent cations. Solvation energies and lattice energies are high for small ions. If both the cation and anion are small, the effect of high U_0 is more important in decreasing solubilities. Salts are likely to be soluble if the sizes of the cation and anion differ substantially in size.

Figure 5.24 shows the relation between **charge density** (charge/radius) and the solubilities of salts. The solid lines separate the following: (a) elements with $Z/r < 0.03$/pm which form soluble hydrated cations such as $Cs^+(aq)$ and $Ca^{2+}(aq)$; (b) ones with $Z/r > 0.12$/pm, forming soluble salts of oxoanions such as SO_4^{2-} and PO_4^{3-}; and (c) those with intermediate Z/r, which form insoluble oxides and hydroxides, but are soluble as halides or acetates.

---

[17] Latimer, Pitzer, and Slansky (*J. Chem. Phys.* **1939,** *7,* 108) showed that the Born equation gives poor results unless an empirical adjustment of the radius is made. Reasonable results are obtained if the effective cationic radius is taken to be the crystal radius plus 70 pm and the effective anionic radius is 25 pm larger than the crystal radius. Probably the major factor requiring some adjustment is the expected decrease in ε in the immediate vicinity of an ion.

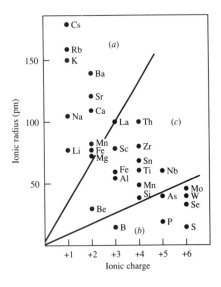

Figure 5.24 Charge density and the aqueous solubility of salts of elements. (Reproduced with permission from P. A. Cox, *The Elements*, Oxford University Press, Oxford, 1989, p. 149).

Equation (5.16) does not take into account polarization effects in evaluating solvation energies, and the greater polarization of cations with 18-electron configurations is allowed for only in the value of the Born exponent used in Equation (5.7) for the calculation of U_0. Solubility depends on the relative polarizability of the anion and of the solvent molecules. Unless the solvent molecules are easily polarizable, the solubility of the salt usually is low if the cation is strongly polarizing and the anion is easily polarizable. The solubilities of salts containing 18- or (18 + 2)-electron cations—$AgCl$, $PbCl_2$, Hg_2Cl_2, $HgCl_2$, for example—usually are lower in water, because of the greater anion polarization, than those containing noble-gas-type cations—alkali and alkaline earth halides, for example. Ammonia has a lower dipole moment than water but is more polarizable; hence, it is a poorer solvent than water for substances of the extreme ionic type, but better than water for salts of strongly polarizing cations and for salts of easily polarized anions. Salts of 18- or (18 + 2)-electron cations are commonly more soluble in ammonia than in water, because of the greater polarizability and basicity of ammonia, giving stable complexes. The solubilities of the silver halides in liquid ammonia increase with increasing anion size (and polarizability), whereas the reverse order is observed in water.

Equation (5.16) also neglects specific solvent interactions such as hydrogen bonding. The hydration energy of F^- is much higher than that of K^+, even though the ionic radii are about the same. The unusually high hydration energy of F^- (and OH^-) results from the formation of strong hydrogen bonds. The high hydration energy of F^- contributes strongly to the great reactivity of F_2 to form solutions of metal fluorides.

5.10 GIANT MOLECULES

Compounds of elements in the same family and of the same formula type sometimes differ strikingly in properties (see page 71). Thus CO_2 is a gas at room temperature and pressure and the solid (Dry Ice) is easily broken into smaller pieces, yet SiO_2 is a very hard, dense, high-melting solid. Within a multiple-bonded CO_2 molecule the valence forces are satisfied, so only weak van der Waals forces hold the molecules to one another in the liquid or solid state. There are no discrete SiO_2 molecules, even at quite high temperatures. The larger size of Si (compared with C) favors a higher C.N., which, together with the decreased tendency of Si to form multiple bonds, leads to the formation of Si—O single bonds in tetrahedral SiO_4 units. The tetrahedra are linked together in a three-dimensional network by the sharing of each O between two Si. In a single crystal of quartz, SiO_2, all SiO_4 units are cross-linked into a single unit—one "molecule". Such substances are called **giant molecules** or **covalent crystals.**

Diamond provides a good example of a giant molecule. Here, each C atom is bonded tetrahedrally to four others. The structure is the same as that described for zinc blende (see Figure 5.6), except that all sites are occupied by carbon atoms. Diamond's extreme hardness stems from this three-dimensional network of covalent atoms. Carborundum, SiC, has the zinc blende structure, and its great hardness makes it useful as an abrasive. Graphite features carbon atoms arranged in two-dimensional layers of hexagonal rings (see Figure 6.28), with conjugated double bonds through each layer; each carbon has one double and two single bonds. Graphite is a solid lubricant because the weak attraction between layers allows them to slide along other layers without breaking any C—C bonds.

5.11 STRUCTURES OF SILICATES

▶ 5.11.1   *Common Silicates*

Silicates provide an interesting array of structural types that show greater variety than usually encountered for compounds of other elements. The silicates are also of great technical importance. The crust of the earth is made up primarily of metal silicates. Silicon and oxygen account for almost 75% by weight of the earth's crust; on a volume basis, oxygen alone accounts for 92% of the earth's crust. Moreover, the eight most abundant elements in the earth's crust[18] usually occur in silicates. The ubiquitous silicates occupy a unique position in mineralogy.

The composition of the silicate minerals used to be given in terms of mole ratios of oxides: For example, forsterite was expressed as $2MgO \cdot SiO_2$ rather than Mg_2SiO_4, and orthoclase was expressed as $K_2O \cdot Al_2O_3 \cdot 6SiO_2$ rather than $KAlSi_3O_8$. The mixed oxide formulations were used partly because, in many cases, the structures or manner of chemical combination were not known. Also, many minerals are not pure chemical compounds that can be represented accurately by simple chemical formulas. The formulas given are idealized; the actual composition might vary considerably.

Variations in the composition of minerals principally result from the isomorphous (having the same structure) replacement of one ion by another. The extent of isomorphous replacement of ions of the same charge is determined by the relative sizes of the ions. If the sizes are very similar (within about 15%), complete isomorphous replacement can oc-

[18] Only eight elements account for 98.5% by weight of the earth's crust: O, 46.6%; Si, 27.7%; Al, 8.1%; Fe, 5.0%; Ca, 3.6%; Mg, 2.1%; Na, 2.8%; and K, 2.6%. The crust is the outer shell to a depth of about 20 km. See Table 18.1.

cur. Mg^{2+} (86 pm) and Fe^{2+} (92 pm) can occupy any proportion of the M^{II} sites in a mineral of the type $M_2^{II}SiO_4$. If little iron is present the mineral is called forsterite, represented as Mg_2SiO_4. Less commonly, where little magnesium is present the mineral is called fayalite, Fe_2SiO_4. The common mineral is olivine containing varying proportions of Mg and Fe, represented as $(Mg,Fe)_2SiO_4$.

Ions of comparable size can replace one another even if the charges differ. The feldspars can be represented by the general formula $M(Al,Si)_4O_8$, where M can be Na, K, Ca, or Ba and where the ratio of Si to Al varies from $3:1$ to $2:2$. For each Ca^{2+} substituting for Na^+ in albite, $NaAlSi_3O_8$, an additional Al^{3+} ion substitutes for Si^{4+}. In the mineral anorthite, $CaAl_2Si_2O_8$, the substitution of Ca^{2+} for Na^+ is complete. Actually, the formulas for albite and anorthite are idealized, and the presence of small amounts of the other cation (Ca^{2+} or Na^+) does not necessitate changing the name of the mineral. Even though there can be a continuous replacement of Ca^{2+} for Na^+, many minerals of intermediate composition ranges are characterized and given their own names. The larger ions K^+ and Ba^{2+} give another series of feldspars.

The silicate minerals have a number of essential features that should be kept in mind. All silicates contain tetrahedral SiO_4 units, which may be linked together by sharing corners, but never by sharing edges or faces. When other cations (alkali or alkaline earth metal ions, Fe^{2+}, etc.) are present in the structure, they usually share oxygens of the SiO_4 groups, to give an octahedral configuration around the cation. Aluminum can replace Si in the SiO_4 tetrahedra, requiring the addition of another cation or the replacement of one by another of higher charge to maintain charge balance. Aluminum also can occupy octahedral sites.

▶ 5.11.2 *Silicates Containing "Discrete" Anions*

In discussing silicates the term "discrete ion" is used for small anions containing 1 to 6 Si in the silicate anions. Silicate ions are not discrete in the sense of independent rotation, as are the perchlorate ions in solid alkali perchlorates at temperatures above $\sim 300°C$. The simplest discrete silicate anion is the orthosilicate ion, SiO_4^{4-}. In the orthosilicates forsterite, $Mg_2[SiO_4]$, and olivine, $(Mg, Fe)_2[SiO_4]$, stacking of the SiO_4 tetrahedra around the divalent cation produces an octahedral configuration. Olivine has oxide ions *hcp* (*P* layers), one-eighth of the *T* sites occupied by Si, and half of the *O* sites occupied by Mg. Orthosilicates (discrete SiO_4^{4-}) are the most compact of the silicate structures.

The uncommon mineral phenacite, $Be_2[SiO_4]$, contains the very small Be^{2+} ions in tetrahedral sites. In zircon, $Zr[SiO_4]$, the large Zr^{4+} ion has C.N. 8. The garnets, $M_3^{II}M_2^{III}[SiO_4]_3$, where M^{II} is Ca, Mg, or Fe and M^{III} is Al, Cr, or Fe, have a more complex structure, in which the M^{II} ions have C.N. 8 and the M^{III} ions have C.N. 6.

We might expect discrete anions containing short chains of SiO_4 tetrahedra, but these are rare. The $Si_2O_7^{6-}$ anion is encountered in thortveitite, $Sc_2[Si_2O_7]$, and in a few other minerals of greater complexity. In condensing two SiO_4^{2-} tetrahedra, one O^{2-} is excised and the two tetrahedra share the O at one vertex, giving $2(SiO_4^{4-}) - O^{2-} = Si_2O_7^{6-}$. There are very few examples of short chains containing more than two SiO_4 groups. Discrete anions consisting of rings of SiO_4 groups are encountered more commonly. Rings of three tetrahedra (to give six-membered rings) containing the anion $[Si_3O_9]^{6-}$ $[3(SiO_4^{4-}) - 3O^{2-}]$ (see Figure 5.25) are encountered in wollastonite, $Ca_3[Si_3O_9]$, and

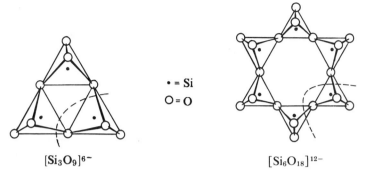

$[Si_3O_9]^{6-}$ $[Si_6O_{18}]^{12-}$

Figure 5.25   Cyclic silicate anions. The repeated SiO_3^{2-} unit is marked by a dashed line.

benitoite, $BaTi[Si_3O_9]$. The anion $[Si_6O_{18}]^{12-}$, a ring of six tetrahedra, is found in the mineral beryl, $Al_2Be_3[Si_6O_{18}]$ (Figure 5.25). In beryl an oxygen is shared by one Si, one Al (C.N. 6), and one Be (C.N. 4). Beryl is the only important Be ore. One Al is substituted for an Si in the ring, giving $[AlSi_5O_{18}]^{13-}$ found in the mineral cordierite, $Mg_2Al_3[AlSi_5O_{18}]$.

Minerals such as alumina, beryl, quartz, and spinel are colorless, but a variety of colors result from the presence of small amounts of transition metal ions. Gem-quality crystals of beryl contain Cr^{3+} in emerald, Fe^{2+} in blue aquamarine, and Mn in morganite (pink). Alumina crystals contain Cr^{3+} in ruby (about 1% Cr_2O_3) and Fe plus Ti in sapphire. With more TiO_2 added, the scattering caused by small crystals of TiO_2 give the star of star sapphires or rubies. Amethyst is quartz containing Fe replacing Si in SiO_4 tetrahedra. Spinels can give a wide range of colors depending on the metal ions present. Blue spinel contains Co^{2+} replacing some Mg^{2+} in $MgAl_2O_4$, and about 1% of Cr_2O_3 added to $MgAl_2O_4$ gives "ruby spinel". Chrysoberyl, $BeAl_2O_4$, a spinel, is encountered as alexandrite (commonly greenish yellow) and the cat's eye.

▶ 5.11.3   Silicates Containing Infinite Chains

Single Chains

Each SiO_4 tetrahedron can share two oxygens to form single chains of indefinite length (Figure 5.26). We can represent the anion by the formula of the repeating unit, SiO_3^{2-}. Minerals of this type are called the pyroxenes, which include enstatite, $Mg[SiO_3]$, diopside, $CaMg[SiO_3]_2$, and spodumene, $LiAl[Si_2O_6]$. Spodumene is an important lithium ore. The nonbridging oxygens are shared with Mg^{2+} (C.N. 6) in enstatite. In diopside the Mg has C.N. 6 and the Ca has C.N. 8. Both Li and Al are six-coordinate in spodumene.

Double Chains

A class of minerals known as the amphiboles contain double chains of SiO_4 tetrahedra (Figure 5.26) joined to form rings of six tetrahedra. The repeating unit is $Si_4O_{11}^{6-}$, with half of the silicon atoms sharing three oxygens with other Si atoms and half sharing only

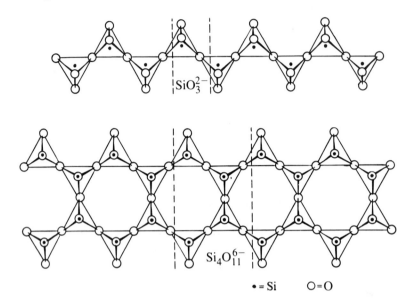

Figure 5.26 Chains of SiO_4 tetrahedra in (*a*) pyroxenes (repeating unit SiO_3^{2-}) and (*b*) amphiboles (repeating unit $Si_4O_{11}^{6-}$). The tetrahedra for *b* are viewed down one O—Si bond, where Si(•) is below the O (o). The repeating units are marked by dashed lines.

two oxygens with other Si atoms. The amphiboles always contain some OH^- groups associated with the metal ion. Tremolite, $Ca_2Mg_5[(OH)_2|(Si_4O_{11})_2]$, is a typical amphibole. The vertical bar used in the formula indicates that the OH^- is not a part of the silicate framework.

The term **asbestos** was originally reserved for fibrous amphiboles such as tremolite, but now the term includes fibrous varieties of layer structures such as chrysotile, $Mg_3[(OH)_4|Si_2O_5]$. The fibers of chrysotile consist of curled ribbons forming cylinders. The detailed structures of the cylinders of chrysotiles are not known. Asbestos was widely used for insulators at high temperature, for partitions and ceilings, for roof shingles, and as a water-suspension to form mats for filtering in the laboratory. Now the uses are limited because of the hazard of lung cancer caused by inhaling the fibers.

▶ **5.11.4 Silicates Containing Sheets**

Each silicon atom can share three oxygen atoms with other Si atoms, to give large sheets. Cross-linking similar to that in the amphiboles, but extending indefinitely in two dimensions, produces rings of six Si in talc, $Mg_3[(OH)_2|Si_4O_{10}]$, and biotite, $K(Mg, Fe)_3[(OH)_2|AlSi_3O_{10}]$ (Figure 5.27 *a*). Biotite is one of the mica minerals. The SiO_4 (and AlO_4) tetrahedra form sheets of interlocking rings with the unshared oxygens pointed in the same direction. Two of these sheets are parallel with the unshared oxygens pointed inward. These oxygens and the OH^- ions are bonded to Mg^{2+} and Fe^{2+} between the sheets. Double sheets weakly bonded together by the K^+ ions account for the characteristic cleav-

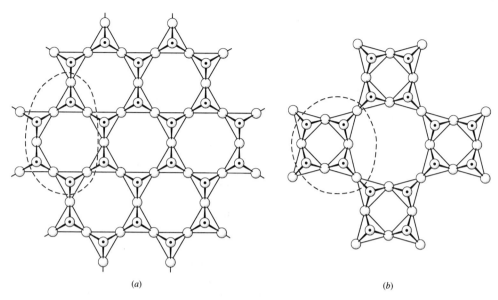

(a)   (b)

Figure 5.27 Sheets of SiO$_4$ tetrahedra. (a) Rings of six SiO$_4^{2-}$ units in talc, Mg$_3$[(OH)$_2$|Si$_4$O$_{10}$], and biotite, K(Mg, Fe)$_3$[(OH)$_2$|AlSi$_3$O$_{10}$]. (b) Alternating rings of four and eight SiO$_4$ units in apophyllite. The repeating Si$_4$O$_{10}^{4-}$ is outlined.

age into thin sheets. The flashes of light reflected off granite often are from flat surfaces of biotite crystals.

Apophyllite, Ca$_4$K[F|(Si$_4$O$_{10}$)$_2$]·8H$_2$O, contains sheets made up of SiO$_4$ tetrahedra linked to form alternating four- and eight-membered rings (Figure 5.27*b*). In this case the sheets are not doubled because the oxygens not shared between Si atoms do not all point in the same direction. The K$^+$ and Ca^{2+} ions lie between the puckered sheets associated with the oxygens uninvolved in the interlocking network of the sheets (Figure 5.28).

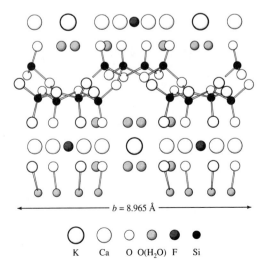

b = 8.965 Å

K   Ca   O   O(H$_2$O)   F   Si

Figure 5.28 Side view showing the ions and water between puckered sheets of apophyllite. (From G. Y. Chao, *Am. Mineral.* **1971**, *56*, 1240. Copyright © 1971 by Mineralogical Society of America.)

▶ 5.11.5   *Framework Silicates*

The partial replacement of Si by Al requires another cation in the framework silicates. The SiO_4, along with randomly distributed AlO_4^-, are linked by the sharing of all four oxygens, as in SiO_2. We encounter three groups of framework silicates: the feldspars, the ultramarines with basketlike frameworks, and the zeolites with open structures.

The **feldspars,** comprising about two-thirds of the igneous rocks, are the most important of the rock-forming minerals. The general formula for the feldspars was given above (page 240) as $M(Al, Si)_4O_8$. When the ratio of Si : Al is 3 : 1, M is Na or K; and when Si : Al is 2 : 2, M is Ca or Ba. The two classes of feldspars are based not on the charge on the cation, but on the crystal symmetry. The **monoclinic** feldspars, including orthoclase, $K[AlSi_3O_8]$, and celsian, $Ba[Al_2Si_2O_8]$, contain the larger cations. Substitution of Na^+ for K^+ or Ca^{2+} for Ba^{2+} does not occur extensively. **Triclinic** feldspars are referred to as the **plagioclase** feldspars. We know of a number of minerals in this series varying in composition from albite, $Na[AlSi_3O_8]$, to anorthite, $Ca[Al_2Si_2O_8]$, corresponding to the isomorphous substitution of Ca^{2+} for Na^+ and the required substitution of Al^{3+} for Si^{4+}.

The **ultramarines** are commonly colored and are used as pigments. They contain negative ions, such as Cl^-, SO_4^{2-}, or S^{2-}, that are not part of the framework. The typical basketlike framework of ultramarines is a truncated octahedron—the polyhedron obtained by cutting off the apices of an octahedron (see Figure 5.29a). Each of the 24 apices of the truncated octahedron is occupied by Si or Al with a bridging O^{2-} along each edge. Sodalite, $Na_8Al_6Si_6O_{24}\cdot Cl_2$, is an ultramarine with a structure formed by sharing all square faces of the truncated octahedron (Figure 5.29b).

The **zeolites**[19] have very open structures formed by joining silicate chains or cages such as the truncated octahedron. Zeolite A (Linde Type A) is a synthetic zeolite formed by bridging (not sharing) the square faces of truncated octahedra. Figure 5.30a shows the open channels of Zeolite A. The size of the open channel is designated by the number of bridging oxygens (along edges), 8 in this case. (This is simpler than saying 8 Si or Al.) Silicalite (ZSM-5) has larger channels formed by 10 oxygens (Figure 5.30b). Faujasite is a

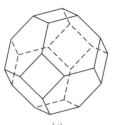

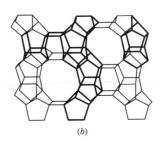

Figure 5.29 (*a*) The truncated octahedron formed from 24(SiO_4 + AlO_4) tetrahedra. (*b*) Sodalite structure. (Reproduced with permission from B. C. Gates, *Catalytic Chemistry.* Wiley, New York, 1992, p. 264.)

Figure 5.30 (*a*) Structure of Zeolite A. (*b*) The framework for silicalite (ZSM-5). (Reproduced with permission from B. C. Gates, *Catalytic Chemistry*, Wiley, New York, 1992, pp. 261 and 265.)

---

[19] B. C. Gates, *Catalytic Chemistry*, Wiley, New York, 1992, pp. 254–272; W. M. Meier and D. H. Olson, *Atlas of Zeolite Structure Types*, 3rd ed., Butterworth-Heinemann, Boston, 1992.

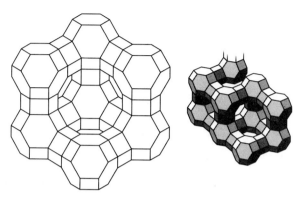

Figure 5.31   Structure of faujasite. (Reproduced with permission from B. C. Gates, *Catalytic Chemistry*, Wiley, New York, 1992, p. 266.)

natural zeolite with 12 oxygens forming the large channels that extend in three dimensions (Figure 5.31). Table 5.14 gives characteristics of some zeolites.

The open structures have made zeolites useful as ion exchangers—softening water by replacing Mg^{2+} or Ca^{2+} by Na^+. The ion exchanger is regenerated by treating it with concentrated NaCl solution. The zeolites are now used in industry in great amounts. Synthetic zeolites can provide channels of optimum size. The size of the channels are of molecular size. They are selective with respect to size, to function as **molecular sieves** or catalysts (see Section 7.5). Molecular sieves can separate gases based on molecular size, as in gas chromatography.

Silica

All crystalline forms of SiO_2 have three-dimensional framework structures with sharing of all four oxygens of the SiO_4 tetrahedra. Of the forms, quartz is stable to 870°, tridymite is stable from 870° to 1470°, and cristobalite is stable from 1470° to 1710° (mp). Each of these forms has a low (α) and high (β) temperature modification. Because transitions among the three are very sluggish, each of these polymorphic forms can be studied at temperatures outside its range of stability.

Table 5.14   Characteristics of some zeolites

Name	Formula	Oxygens in ring	Pore size (Å)
Zeolite A (Linde Type A)	$\{Na_{12}[Al_{12}Si_{12}O_{48}]\cdot 27H_2O\}_8$	8	4.1
Heulandite	$Ca_4[Al_8Si_{28}O_{72}]\cdot 24H_2O$	8 10	2.6 × 4.7 3.0 × 7.6
Ferrierite	$Na_2Mg_2[Al_6Si_{30}O_{72}]\cdot 18H_2O$	8 10	3.5 × 4.8 4.2 × 5.4
Silicalite (ZSM-5)	$Na_n[Al_nSi_{96-n}O_{192}]\cdot \sim 16H_2O$	10 10	5.3 × 5.6 5.1 × 5.5
Faujasite	$(Na_2, Ca, Mg)_{29}[Al_{58}Si_{134}O_{384}]\cdot 240H_2O$	12 (3-dim.)	7.4
Linde Type L	$K_6Na_3[Al_9Si_{27}O_{72}]\cdot 21H_2O$	12	7.1
Mordenite	$Na_8[Al_8Si_{40}O_{96}]\cdot 24H_2O$	12 8	6.7 × 7.0 2.6 × 5.7

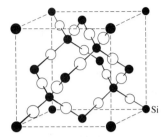

Figure 5.32   The β-cristobalite (SiO$_2$) structure.

The structure of β-cristobalite (Figure 5.32) is related to diamond or zinc blende, with Si in all the C, or Zn and S, positions. The cubic close-packed oxygens lie midway between the Si atoms but slightly shifted away from the line joining the silicons. In the β-tridymite structure, which is related similarly to wurtzite (Figure 5.6), the α modifications involve slight rotations of the SiO$_4$ tetrahedra relative to one another. There are helices of linked tetrahedra in a crystal of β-quartz. The crystal is optically active, since all of the helices will be either right- or left-handed in a particular crystal. Although slight distortions occur among the linked tetrahedra in α-quartz, the crystals are optically active.

5.12   CRYSTAL STRUCTURES OF METALS[20]

Metal atoms in the solid state display three common arrangements: cubic close-packed (also called face-centered cubic), hexagonal close-packed, and body-centered cubic. As their names imply, the first two correspond to the closest packing of similar spheres. These are related to the close-packed layers as shown in Figure 5.33 (see also Figures 5.3 and

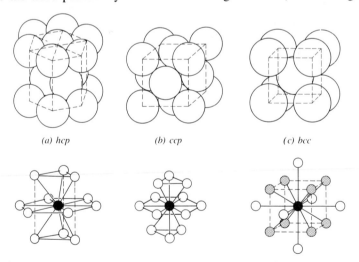

(a) hcp   *(b) ccp*   *(c) bcc*

Figure 5.33   The (*a*) hexagonal close-packed, (*b*) cubic close-packed (face-centered cubic), and (*c*) body-centered cubic structures. The nearest neighbors for each structure are shown below the crystal models. The central atom is black. The closer neighbors for *bcc* are shaded.

[20] J. Donahue, *The Structures of the Elements*, Wiley, New York, 1974; S.-M. Ho and B. E. Douglas, *J. Chem. Educ.* **1972**, *49*, 74.

Table 5.15 Crystal structures of metals[a]

Legend:
- *bcc*: body-centered cubic
- *hcp* (2P): hexagonal close-packed
- *ccp* (3P): cubic close-packed
- 3P′: distorted *ccp*
- 4P: double hexagonal, length of *c* axis is doubled
- 9P: *ABABCBCAC* . . . sequence
- 3·2PT: diamond structure

Main table

1	2	3	4	5	6	7	8	9	10	11	12	13	14
3 Li *bcc*[b] *hcp* *ccp*	4 Be *hcp* *bcc*												
11 Na *bcc* *hcp* *ccp*	12 Mg *hcp*											13 Al *ccp*	14 Si 3·2PT
19 K *bcc*	20 Ca *ccp* *hcp*	21 Sc *hcp* *bcc*	22 Ti *hcp* *bcc*	23 V *bcc*	24 Cr *bcc*	25 Mn 3P′	26 Fe *bcc* *ccp*	27 Co *hcp* *ccp*	28 Ni *ccp*	29 Cu *ccp*	30 Zn *hcp*	31 Ga other	32 Ge 3·2PT
37 Rb *bcc*	38 Sr *ccp* *hcp*	39 Y *hcp* *bcc*	40 Zr *hcp* *bcc*	41 Nb *bcc*	42 Mo *bcc*	43 Tc *hcp*	44 Ru *hcp*	45 Rh *ccp*	46 Pd *ccp*	47 Ag *ccp*	48 Cd *hcp*	49 In 3P′	50 Sn 3·2PT
55 Cs *bcc*	56 Ba *bcc*	71 Lu *hcp* *bcc*	72 Hf *hcp* *bcc*	73 Ta *bcc*	74 W *bcc*	75 Re *hcp*	76 Os *hcp*	77 Ir *ccp*	78 Pt *ccp*	79 Au *ccp*	80 Hg other	81 Tl *hcp* *bcc*	82 Pb *ccp*

Lanthanides

57 La	58 Ce	59 Pr	60 Nd	61 Pm	62 Sm	63 Eu	64 Gd	65 Tb	66 Dy	67 Ho	68 Er	69 Tm	70 Yb
4P *hcp* *bcc*	*ccp* 4P *bcc*	4P *hcp* *ccp*	4P *ccp* *bcc*	4P	9P *bcc*	*bcc*	*hcp* *bcc*	*hcp* *bcc*	*hcp* *bcc*	*hcp* *bcc*	*hcp* *bcc*	*hcp* *bcc*	*ccp* *bcc* *hcp*

Actinides

89 Ac	90 Th	91 Pa	92 U	93 Np	94 Pu	95 Am	96 Cm	97 Bk	98 Cf	99 Es	100 Fm	101 Md	102 No
ccp	*ccp* *bcc*	*bcc* *ccp*	other *bcc*	other *bcc*	other *bcc*	4P *ccp*	4P *ccp*	4P *ccp*	*hcp*				

[a] Adapted from S. -M. Ho and B. E. Douglas, *J. Chem. Educ.* **1972**, *49*, 74.
[b] Room-temperature structures are listed first.

247

5.4). In both close-packed structures, each metal atom has 12 nearest neighbors. In the body-centered cubic structure (see Figure 5.33), each metal atom has eight nearest neighbors and six next-nearest neighbors only slightly (~15%) farther away.

The structures encountered for metals are shown in Table 5.15 on page 247. Many metals adopt more than one structure, depending on temperature and pressure. The relatively small differences in the stabilities of the structures is illustrated by iron: The *bcc* structure is stable at low temperature, the *ccp* structure is more stable in the range 906–1400°C, but the *bcc* structure is again more stable from 1400°C to the mp (1535°C).

The softness of a metal, as noted by its malleability (ability to be beaten or rolled into sheets) and ductility (ability to de drawn into wire), depends on the crystal structure, as well as on other factors such as bond strength. These metallic characteristics are determined by the ability of layers to slide over one another to give an equivalent arrangement of the identical spheres. The highly symmetrical *ccp* structure has slip planes of close-packed layers along four directions (corresponding to the threefold axes, the body diagonals of the cube), as compared with only one direction for the *hcp* structure. Metallic characteristics such as malleability, ductility, and electrical conductivity are important in applications of the *ccp* metals Al, Cu, Ag, Au, Ni, Pd, Pt, Rh, Ir, and Pb. Impurities can create dislocations (local bonding), causing an increase in hardness. Soft metals become "work-hardened" because of dislocations created in the deformation of the metal and disruptions of slip planes. This can be seen in contrasting the ease with which you can bend new copper tubing to the greater difficulty encountered in bending it a second time. A few metals—for example, Sb, Bi, and α-Mn—are brittle. Directional bonding in Sb and Bi forms puckered layers, destroying any slip planes. There is a complex cubic structure for α-Mn with four nonequivalent atoms.

GENERAL REFERENCES

D. M. Adams, *Inorganic Solids,* Wiley, New York, 1974. Good treatment from the chemical viewpoint.

J. K. Burdett, *Molecular Shapes,* Wiley-Interscience, New York, 1980. Chapter 15 deals with a molecular orbital treatment of some solids.

J. K. Burdett, "New Ways to Look at Solids," *Acc. Chem. Res.* **1982,** *15,* 34.

A. K. Cheetham and P. Day, Eds., *Solid State Chemistry,* Clarendon Press, Oxford, 1992.

F. S. Galasso, *Structure and Properties of Inorganic Solids,* Pergamon, Oxford, 1970. Extensive tables of compounds as examples of the common structures—good figures showing relationships among structures, and a good index.

N. B. Hannay, Ed., *Treatise on Solid State Chemistry,* Vol I: *The Chemical Structure of Solids,* Plenum, New York, 1973. Authoritative treatment of selected topics.

B. G. Hyde and S. Andersson, *Inorganic Crystal Structures,* Wiley-Interscience, New York, 1989. A systematic presentation showing how complex structures are related to simple ones.

T. C. W. Mak and G.–D. Zhou, *Crystallography in Modern Chemistry,* Wiley, New York, 1992. A good resource of x-ray structures.

L. Smart and E. Moore, *Solid State Chemistry,* Chapman and Hall, London, 1992.

A. F. Wells, *Structural Inorganic Chemistry,* 5th ed., Oxford, Oxford, 1984. The single most important source for inorganic structures and structural insight for chemists.

R. W. G. Wyckoff, *Crystal Structures,* Vols. I and II, 2nd ed. Wiley-Interscience, New York, 1963 and 1964 (respectively).

PROBLEMS

5.1 Which of each of the following pairs might be expected to be more ionic?
   (a) $CaCl_2$ or $MgCl_2$
   (c) NaCl or CuCl (similar radii)
   (b) NaCl or $CaCl_2$ (similar radii)
   (d) $TiCl_3$ or $TiCl_4$

5.2 For each of the following pairs indicate which substance is expected to be:
   (a) More covalent (Fajans' rules):   **(b)** Harder:

$MgCl_2$ or $BeCl_2$	$SnCl_2$ or $SnCl_4$	NaF or NaBr	MgF_2 or TiO_2
$CaCl_2$ or $ZnCl_2$	$CdCl_2$ or CdI_2	Al_2O_3 or Ga_2O_3	
$CaCl_2$ or $CdCl_2$	ZnO or ZnS		
$TiCl_3$ or $TiCl_4$	NaF or CaO		

5.3 Variations in hardness for ionic substances correlate well with what one thermodynamic property?

5.4 Which of the following are *not* possible close-packing schemes?
   (a) *ABCABC* . . .   **(d)** *ABCBC* . . .
   (b) *ABAC* . . .   **(e)** *ABBA* . . .
   (c) *ABABC* . . .   **(f)** *ABCCAB* . . .

5.5 Calculate the number of formula units in the unit cells for the following:
   (a) CsCl   **(b)** ZnS(zinc blende)   **(c)** CaF_2 **(d)** $CaTiO_3$

5.6 The wurtzite (*hcp* S^{2-}) structure of ZnS has open channels along the packing direction, but the zinc blende (*ccp* S^{2-}) structure does not. Explain.

5.7 Many MX-type compounds with C.N. 6 have the NaCl (Cl^- *ccp*) structure, whereas few have the NiAs (As *hcp*) structure. What features of the NiAs structure limit its occurrence? What characteristics of the compound are necessary for the NiAs structure? What unusual physical property of NiAs results from its structure?

5.8 What do the following pairs or groups of structures have in common, and how do they differ? Give the formula type (MX, MX_2, etc.) for each.
   (a) $3 \cdot 2PO$, $3 \cdot 2\,PT$, and $2 \cdot 2\,PT$.   **(c)** $3 \cdot 3PTT$ and $2 \cdot 2PO_{1/2}$.
   (b) $3 \cdot 3PT_{1/2}T_{1/2}$ and $3 \cdot 2\,PT$.   **(d)** $2(\frac{3}{2})PPO$ and $2 \cdot 2PO_{1/2}$.

5.9 Why are layer structures such as those of $CdCl_2$ and CdI_2 usually not encountered for metal fluorides or compounds of the most active metals?

5.10 Rubidium chloride assumes the CsCl structure at high pressures. Calculate the Rb—Cl distance in the CsCl structure from that for the NaCl structure (from ionic radii for C.N. 6). Compare with the Rb—Cl distance from radii for C.N. 8.

5.11 Calculate the cation/anion radius ratio (by using plane geometry) for a triangular arrangement of anions in which the cation is in contact with the anions but does not push them apart.

5.12 Why are the dashed lines in the sorting maps (Figures 5.12 and 5.13) more successful than radius rules indicated by solid lines in sorting structure types?

5.13 Estimate the density of MgO (NaCl structure) and zinc blende (ZnS) using radii to determine the cell dimensions and the number of formula units per unit cell.

5.14 $NaSbF_6$ has the NaCl structure. The density is 4.37 g/cm^3. Calculate the radius of SbF_6^- using the radius of Na^+, the density, the formula weight, and the number of formula units per unit cell.

5.15 Compare the ionic radii of Na^+ and Mg^{2+}, and those of S^{2-} and Cl^-. Explain the differences within each pair.

5.16 Compare the thermochemical data for the alkali halides. What factors are important in establishing the order of increasing heats of formation within each group of halides (fluorides, chlorides, etc.)?

5.17 Calculate the heat of formation of NaF_2 and the heat of reaction to produce $NaF + \frac{1}{2}F_2$. (Assume the rutile structure.)

5.18 Calculate the heats of formation of Ne^+F^- and Na^+Ne^-, estimating radii of Ne^+ and Ne^-. What factors prohibit the formation of these compounds in spite of favorable lattice energies?

5.19 From spectral data the dissociation energy of ClF has been determined to be 253 kJ/mol. The ΔH_f^0 of $ClF(g)$ is -50.6 kJ/mol. The dissociation energy of Cl_2 is 239 kJ/mol. Calculate the dissociation energy of F_2.

5.20 Although the electron affinity of F is lower than that of Cl, F_2 is much more reactive than Cl_2. Account for the reactivity of F_2: **(a)** with respect to the formation of solid halides MX or MX_2 and **(b)** with respect to the formation of aqueous solutions of MX or MX_2.

5.21 The $[CuCl_5]^{3-}$ is not very stable. What would be a suitable cation to use for the isolation of $[CuCl_5]^{3-}$?

5.22 Draw the net of hexagons of a layer of graphite showing the single and double bonds.

5.23 Suggest a solvent, other than H_2O, for studies of ionic substances. Explain the reasons.

5.24 Synthetic gems, including emerald, sapphire, and ruby, have the same chemical composition as natural gems. Why are the natural gems much more expensive?

5.25 What is the significance of the term "molecular weight" with respect to diamond or SiO_2 (considering a perfect crystal of each)?

5.26 The triclinic feldspars form an isomorphous series involving replacement of Ca^{2+} for Na^+ from albite, $Na[AlSi_3O_8]$, to anorthite, $Ca[Al_2Si_2O_8]$. How can the series be isomorphous with changing ratio of Si/Al? Why do K^+ and Ba^{2+} not occur in this series?

5.27 Demonstrate that the formulas of repeating units can be produced from condensation of SiO_4^{4-} for the following minerals:
(a) SiO_2 **(b)** SiO_3^{2-} in pyroxenes **(c)** $Si_4O_{10}^{4-}$ in talc **(d)** $Si_4O_{11}^{6-}$ in amphiboles

5.28 Metals that are very malleable (can be beaten or rolled into sheets) and ductile (can be drawn into wire) have the *ccp* structure. Why are these characteristics favored for *ccp* rather than for *hcp*?

5.29 Soft metals such as Cu become "work-hardened." Explain this and how the softness is restored by heating the metal.

5.30 Discuss the possible effects of extreme pressure on a metal (assume *bcc* at low P) and on a solid nonmetal.

5.31 Consider the sliding of one layer of a crystal over another until an equivalent arrangement is achieved. What are the consequences with respect to hardness, brittleness, and malleability for **(a)** a metal, **(b)** an ionic crystal, **(c)** a covalent crystal, and **(d)** a molecular crystal. In a crude way, how do these considerations relate to melting point?

▶ 6 ◀

Solid-State
Chemistry
· · · · · · · · · · · · · · · · · · · ·

Many spectacular developments in solid-state materials have occurred within the last few years with the promise of more to come. Melt spinning has produced metal glasses, superionic ceramics have been made from sol-gels, and molecular beam epitaxy has permitted control of the architecture of layered cuprate superconductors. In this chapter we look at some of the models that have been proposed to deal with solid-state phenomena, and we discuss how these models serve as guides in the quest for new materials with specific properties. We deal first with the lattice models of solids originating before x-ray crystallography. These models allowed rapid application as that field developed. The concept of the reciprocal lattice and simple elements of band theory are explored before looking at some recent advances, including superconductors.

6.1 CRYSTAL STRUCTURE

▶ 6.1.1 *The Lattice*

A **crystal lattice** is an array of points, each of which has an identical environment. The space of any lattice can be divided into identical parallelepipeds by suitably connecting the lattice points. The parallelepiped is termed a **unit cell,** and the entire lattice can be reproduced by treating any three intersecting edges as unit vectors along which the cell can be displaced by any integer amount. Alternatively, one can state that the lattice points of a crystal are invariant to any translations of the type

$$\vec{t} = n_1 \vec{a} + n_2 \vec{b} + n_3 \vec{c} \tag{6.1}$$

251

where: $\vec{a}$, $\vec{b}$, and $\vec{c}$ are the unit cell vectors; n_1, n_2, and n_3 are arbitrary integers; and t is the indicated translation vector acting at each lattice point.[1] If *every* lattice point is at a vertex (common to eight such parallelepipeds), the unit cell is said to be **primitive.** An infinite number of primitive unit cells can be drawn for any lattice, all of which have the same volume and contain one lattice point ($8 \times 1/8$). Of the infinite number of possible unit cells, the usual choice is the one of highest symmetry. Any primitive parallelepiped cell belongs to one of seven **crystal systems** (see Figure 6.1 and Table 6.1). Unit cells of higher symmetry than the primitive unit cell can be drawn for the same lattice in some cases by allowing lattice points to reside in the face centers or body center of a unit cell (face-centered or body-centered unit cells).

Auguste Bravais demonstrated (1848) that only 14 different lattices are possible, and these can be constructed of the unit cells shown in Figure 6.1 and further characterized in Table 6.1. Table 6.1 gives the name of the crystal system, the point symmetry of the lattice, and the relationship of the unit cell dimensions and angles resulting as a consequence of the lattice symmetry. The point symmetry of these unit cells displays the symmetry of the corresponding **Bravais lattice** with exception of the hexagonal unit cell, where the dashed lines suggest the sixfold symmetry of the lattice. The symmetry of the Bravais lattice places restrictions on the point symmetry of the collection of atoms around the lattice points; this point symmetry must belong to one of the subgroups of the point symmetry of the lattice itself[2] (Table 6.1).

The notation following the Schöenflies notation in Table 6.1 is called Hermann–Mauguin notation and gives the point group generators. The number denotes an n-fold rotational axis, and m denotes a mirror plane. A bar over the number indicates a combination of rotation and inversion, the order being immaterial because the operations commute. A slash between a number and m indicates that the mirror is normal (perpendicular) to the n-fold axis. The rotation operations are taken along the different crystallographic axes, and the mirror operations are taken normal to the axes. An exception occurs in the cubic system, where 3 and $\bar{3}$ are *always* given as the second symbol referring to the threefold axis along the body diagonal of the cube.

► 6.1.2 *The Unit Cell*

The entire crystal structure is determined by the arrangement of atoms within a **unit cell.** Any atom, or collection of atoms, that does not fall on any symmetry element occurs at h sites within a primitive unit cell, where h is the order of the crystallographic point group. Any atom falling on one or more symmetry elements (special positions) occurs at h/n positions, where n is a factor of h. The International Tables for Crystallography gives detailed information on the symmetry and coordinates of equivalent positions for all possible space groups.

---

[1] Translation is thus a symmetry operation of crystals. In crystals new symmetry elements, **screw axes,** arise from the combination of translation with rotation whereas **glide planes** arise from translation and reflection.

[2] When screw and glide operations are essential operations for the description of the symmetry of the contents of the unit cell, the crystallographic point group is obtained by setting translation equal to zero—that is, by considering an n-fold screw axis as an n-fold rotational axis and considering all glide planes as mirror planes.

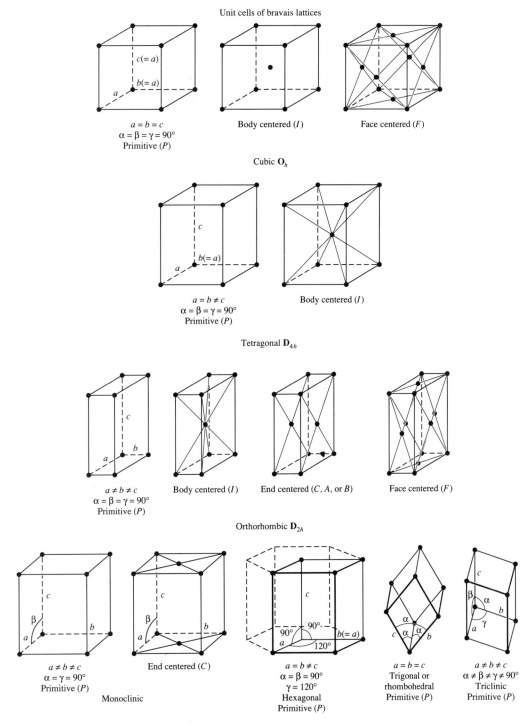

Figure 6.1 Unit cells of Bravais lattices.

Table 6.1 Crystal systems

Name	Point group of lattice		Unit cell	Subgroups		Order
Triclinic	$\mathbf{C}_i(\mathbf{S}_2)$	$\bar{1}$	$a \neq b \neq c$ $\alpha \neq \beta \neq \gamma \neq 90°$	$\mathbf{C}_1$ $\mathbf{C}_i$	1 $\bar{1}$	1 2
Monoclinic	$\mathbf{C}_{2h}$	$2/m$	$a \neq b \neq c$ $\alpha = \gamma = 90° \neq \beta$	$\mathbf{C}_{1h}$ $\mathbf{C}_2$ $\mathbf{C}_{2h}$	m 2 $2/m$	2 2 4
Orthorhombic	$\mathbf{D}_{2h}$	mmm	$a \neq b \neq c$ $\alpha = \beta = \gamma = 90°$	$\mathbf{C}_{2v}$ $\mathbf{D}_2$ $\mathbf{D}_{2h}$	$mm2$ 222 mmm	4 4 8
Tetragonal	$\mathbf{D}_{4h}$	$4/mmm$	$a = b \neq c$ $\alpha = \beta = \gamma = 90°$	$\mathbf{C}_4$ $\mathbf{S}_4$ $\mathbf{C}_{4h}$ $\mathbf{D}_{2d}$ $\mathbf{C}_{4v}$ $\mathbf{D}_4$ $\mathbf{D}_{4h}$	4 $\bar{4}$ $4/m$ $\bar{4}2m$ $4mmm$ 422 $4/mm$	4 4 8 8 8 8 16
Trigonal (rhombohedral)	$\mathbf{D}_{3d}$	$\bar{3}m$	$a = b = c$ $120° > \alpha = \beta = \gamma \neq 90°$	$\mathbf{C}_3$ $\mathbf{S}_6$ $\mathbf{C}_{3v}$ $\mathbf{D}_3$ $\mathbf{D}_{3d}$	3 $\bar{3}$ $3m$ 32 $\bar{3}m$	3 6 6 6 12
Hexagonal	$\mathbf{D}_{6h}$	$6/mmm$	$a = b \neq c$ $\alpha = \beta = 90°$ $\gamma = 120°$	$\mathbf{C}_{3h}$ $\mathbf{C}_6$ $\mathbf{C}_{6h}$ $\mathbf{D}_{3h}$ $\mathbf{C}_{6v}$ $\mathbf{D}_6$ $\mathbf{D}_{6h}$	$\bar{6}$ 6 $6/m$ $\bar{6}m2$ $6mm$ 622 $6/mmm$	6 6 12 12 12 12 24
Cubic	$\mathbf{O}_h$	$m\bar{3}m$	$a = b = c$ $\alpha = \beta = \gamma = 90°$	$\mathbf{T}$ $\mathbf{T}_h$ $\mathbf{T}_d$ $\mathbf{O}$ $\mathbf{O}_h$	23 $m\bar{3}$ $\bar{4}3m$ 432 $m\bar{3}m$	12 24 24 24 48

The location of a point within a unit cell is given by the coordinates x, y, and z that are fractions of a, b, and c and run from 0 to 1. Thus for the primitive unit cell of perovskite shown in Figure 5.7 (page 209) the Ca is located at (000), Ti at $(\frac{1}{2}\frac{1}{2}\frac{1}{2})$, and the O at $(0\frac{1}{2}\frac{1}{2})$. Translational symmetry places Ca at the remaining lattice points shown, and the $m\bar{3}m$ or $\mathbf{O}_h$ symmetry places oxygen atoms in the remaining face centers. The coordinates of the crystal can be said to have a modulus of 1; that is, a point at 2.25, 1.75, 0.50 has the identical environment as a point at 0.25, 0.75, 0.50. Thus the structure of the entire crystalline solid is available from the structure of the unit cell.

As chemists, we are frequently interested in interatomic distances which may be obtained from the general formula

$$d^2 = (x_1 - x_2)^2 a^2 + (y_1 - y_2)^2 b^2 + (z_1 - z_2)^2 c^2 + 2(x_1 - x_2)(y_1 - y_2)ab \cos \gamma$$
$$+ 2(y_1 - y_2)(z_1 - z_2)bc \cos \alpha + 2(z_1 - z_2)(x_1 - x_2)ca \cos \beta \tag{6.2}$$

where: x, y, and z are unit cell coordinates varying between 0 and 1; and a, b, and c are the respective lengths of the edges of the unit cell. α, β, and γ are the angles between unit cell edges. The Greek letter is the counterpart of the English letter omitted from the edge combination. Thus, α is the angle between b and c. For cases in which a desired distance between atoms involves crossing a unit cell boundary, adjustment of the appropriate unit cell coordinates by $+1$ or -1 will be necessary.

Directions in a crystal are indicated by a ray from the origin to a point xyz with the coordinates enclosed in square brackets. Since a ray has an infinite number of points, the smallest whole ratio of numbers is used. Any lattice point may be taken as the origin. Thus the [111] direction indicates a direction along a body diagonal.

Miller Indices

A lattice plane can be designated by selecting an arbitrary lattice point as the origin and indicating the intercepts of the plane with the three crystallographic axes in units of a, b, and c. This is termed the **Weiss index** of a plane. The numbers in a Weiss index can be positive or negative, or, for a plane that is parallel to an axis, infinity. More commonly, the reciprocal of the Weiss index, the **Miller index,** is used to describe planes. The Miller index is usually used to denote a set of equally spaced parallel planes, including one through the origin. Using hkl as a general Miller index for a set of planes, h gives the number of equal divisions of the a edge, k the divisions of the b edge, and l the divisions of the c edge of the unit cell. The unit cell volume is divided into $h + k + l$ volumes. Figure 6.2 gives examples of some sets of planes and their Miller indices. A bar over a num-

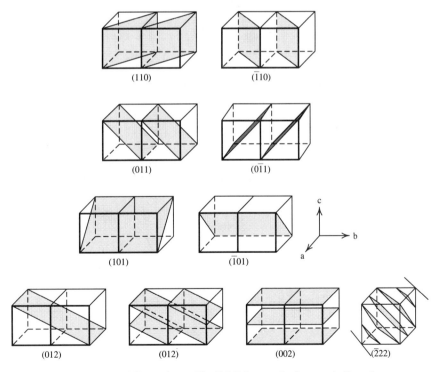

Figure 6.2   Miller indices. The [110] form and others as indicated.

ber in the Miller index indicates a negative value. Crystal growth tends to occur along planes of high surface density of atoms, which coincides with low Miller indices.

EXAMPLE 6.1:   Give the Miller indices for the plane which is the octahedral face cutting the axes at *a*, *b*, and *c*; the opposite (parallel) face; and the joining face sharing the points *a* and *b* with the first face. Take the center of the octahedron as the origin.

Solution: The face cutting the axes at $1a$, $1b$, and $1c$ is 111. The opposite face, which has $-1a$, $-1b$, and $-1c$, is $\overline{1}\,\overline{1}\,\overline{1}$. The face with $1a$, $1b$, and $-1c$ is $11\overline{1}$.

▶ 6.1.3 *Stereographic Projections*

Crystallography was a well-developed science long before x-ray diffraction permitted structure determination of ordered arrays of atoms. The 32 **crystallographic point groups** constitute the possible symmetry of all crystals as determined from examining their external faces. In a perfectly formed crystal the center is located at the intersection of its symmetry elements. More commonly the crystal grows imperfectly, with the symmetry not readily apparent, but the symmetry can be found using the following projection technique. Taking any interior point as the center, normals to each face are constructed. Extending these normals to intersect the surface of a surrounding sphere (having the same center as that selected for the crystal) yields a collection of points whose symmetry belongs to one of the 32 crystallographic point groups. This spherical projection can be reduced to a planar **stereographic projection** that is more convenient for the printed page. This is done by dividing the sphere into hemispheres. Rays connecting the points on the "northern" hemisphere to the "south pole" intersect the equatorial plane to give the projection of these points (represented by solid dots in Figure 6.3). The projections of the "southern" hemisphere using the "north pole" are shown as open circles in Figure 6.3. The "north–south" polar axis is conventionally taken as the highest symmetry axis of the spherical projection. The procedure is illustrated for the octahedron, a common shape for crystals of a number of alums such as $KAl(SO_4)_2 \cdot 12H_2O$.

The faces of crystals can be assigned Miller indices with the origin taken as the center of the crystal, and the smallest whole numbers used for the Miller indices. Crystallographic axes are taken as in Figure 6.1, and the Miller indices are reduced to the smallest whole numbers. Symmetry equivalent faces in a crystal, or planes in a lattice, are termed

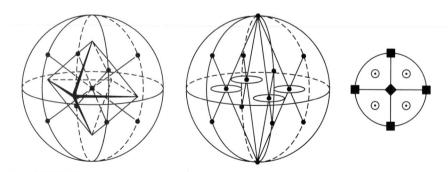

Figure 6.3   Construction of stereographic projection of an octahedron.

a **form** and are indicated by curly brackets. Thus for an octahedral crystal with the crystallographic axes taken through the vertices, the faces are indicated by the {111} form. For any crystal in the cubic system, the Miller indices are identical to the coordinates of the ray normal to the plane and through the origin.

When obtaining a stereographic projection of a crystal to extract point group information, only the faces belonging to the same form needed to be plotted, with the most information coming from the form having the highest number of faces. The stereographic projection of one face of an idealized scheelite crystal ($CaWO_4$) is shown in Figure 6.4. Only the points representing faces of a given form are transformed among themselves by the symmetry operations of the point group of the crystal. Hence, only one set of points is needed to convey the symmetry information, along with the designation of the symmetry operations of the point group. For scheelite these are the C_{4h} (or $4/m$) generators. A point not located on any point group symmetry element is a **general point.**

The stereographic projection for a crystal can show a higher apparent degree of symmetry than is revealed by x-ray crystallography. Thus the {100} faces of a cubic crystal give the same stereographic pattern as the {100}, {010}, and {001} faces of an orthorhombic crystal. This problem arises only if the spherical projection contains points that fall on symmetry elements of the crystal. These are termed **special positions,** and the stereograms indicate such positions by the following symbols:

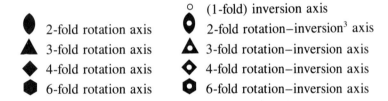

Mirror planes are indicated by solid lines. The stereograms for the 32 crystallographic point groups are given in Figure 6.5. The generators, acting on a single general point, create as many points as there are total operations in the group. Each operation moves an initial general point to a distinct position. The equatorial plane is shown by a dashed circle unless this plane is a reflection plane. Twofold axes in the equatorial plane are shown by a dashed line connecting a pair of ellipses.

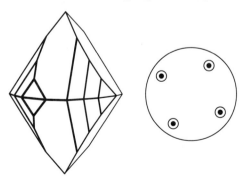

Figure 6.4 Stereographic projection for scheelite (only points for one form are shown).

---

[3] Rotation–inversion combines rotation and inversion through a center of symmetry. S_n is an improper rotation operation involving C_n rotation followed by reflection through σ_h.

Stereograms of the Cubic System

In the crystallographic point groups, only those in the cubic system have symmetry elements that fail to fall on the equator or the north–south pole of a spherical projection. However, we can get a feel for the stereographic projection of the cubic group by labeling the points with the Miller indices of the crystal faces represented. Fortunately, for the cubic system the Miller indices of a face are the same as the *xyz* coordinates of the ray normal to the face and passing through the crystal center. For a cube, all faces belong to the {100} form. In the stereographic projection these fall on the special points marking four-fold rotational axes. There are of course six of these corresponding to all permutations of ±1,0,0 among the three equivalent cubic axes.

The conjugate (see Section 3.4.5) of the cube, the octahedron, has faces belonging to the form {111}, since a ray through the cube vertices would be a normal to the faces of the conjugate octahedron. There are 2^3 permutations of ± among the {111} cubic faces with equivalent crystallographic axes; that is, there are eight faces of the octahedron belonging to the {111} form. These are indicated by solid triangles on the stereographic projection. Although only four such triangles show, there are four more that are directly beneath. These points fall on the three-fold axes, with pairs of the triangles being the poles of three intersecting great circles in a spherical projection. In the stereographic projection, two of these great circles project as ellipses, and the third projects as a straight line.

A third example of a crystal form of the cubic system is given by the rhombododecahedron which can be formed by shaving the edges of a cube to leave only {110} faces. On a stereographic projection these {110} faces give special points falling on two-fold axes. There are 12 such faces, corresponding to the 12 edges of a cube, or to the permutations of ±1 among the {110} faces. The special points in these examples are shown in the stereograms of Figure 6.6.

The {123} faces in the cubic system generate a hexakisoctahedron. In the stereographic projection, these are all general points and are shown in Figure 6.5 on the O_h stereogram as filled and open circles.

EXAMPLE 6.2: Develop a C_{4v} stereogram showing a general point. Mark in the operations that carry the initial point to each of the other points on the stereogram, and use this to generate the multiplication table for the C_{4v} group.

Solution: The symmetry elements of the C_{4v} group are shown in Figure 6.7, with solid straight lines indicating mirror symmetry planes. Points coinciding with the mirror planes or the C_4 axis are special points. We mark a general point at position E, and the effect of operating on this point by C_4, C_4^2, and C_4^3 is to carry the point to the positions indicated by C_4, C_4^2, and C_4^3, respectively. The effect of the mirrors marked σ_1, σ_2, σ_1', and σ_2' on the point marked E is to produce points at σ_1, σ_2, σ_1', and σ_2'. You may recognize this figure as being quite similar to Figure 3.5 for the 45° kaleidoscope group developed earlier. The group multiplication table can be constructed readily from our labeled stereogram. In Table 3.1 (page 116) we perform the operation at the top first, followed by the operation at the left. Note that the product of rotations is a rotation, whereas the product of a rotation and reflection is another reflection. An even number of reflections produces a rotation, while an odd number produces a reflection. As indicated earlier, every representation of a group must have the same multiplication table as the group itself.

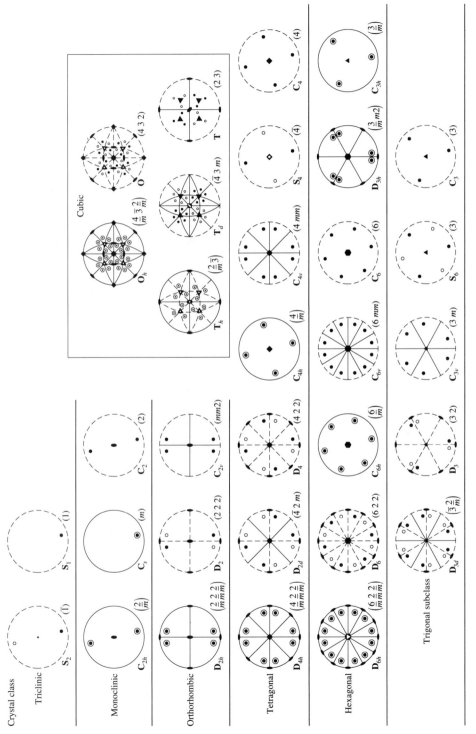

Figure 6.5 Crystallographic point groups.

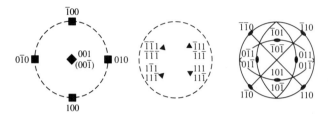

Figure 6.6 Special points in the cubic system (**O** or 432) falling on rotational axes along with the Miller indices of the projected planes.

Space Groups

The combination of lattice translations and the operations of the crystallographic point group gives the **space group,** which is a complete description of the symmetry of a crystal. If the symmetry about a lattice point is given by one of the 32 crystallographic point groups, then the **space lattice** is defined by the point group and the Bravais lattice. In this case the international notation for the point group is preceded by a capital letter that indicates whether the unit cell is primitive (*P*), body-centered (*I*), face-centered (*F*), or end-centered (*A*, *B*, or *C*). The primitive rhombohedral cell is denoted by *R*. There are 73 space groups of this type that are called **symmorphic space groups.**

Wigner–Seitz Cell

The unit cells dealt with so far were made up of parallelepipeds. Some have more than one lattice point per unit cell, or lack the full symmetry of the lattice. We now describe a primitive unit cell that always shows the symmetry of the crystal system. It is called the **Wigner–Seitz cell** and is obtained by constructing planes that are perpendicular bisectors of the lines joining a lattice point to its first and second nearest neighbors, and using these planes as faces, taking the polyhedron of the smallest volume that encloses the lattice point. The Wigner–Seitz cell corresponding to a primitive cubic unit cell is a body-centered cube of the same volume. For a face-centered cubic unit cell, the Wigner–Seitz cell is the rhombic dodecahedron shown in Figure 6.8 along with other unit cells for the cubic system. All the space in a Wigner–Seitz cell is closer to its central lattice point than to any other lattice point. When an atom is located on a lattice point, the vertices of the Wigner–Seitz cell correspond to interstitial sites.

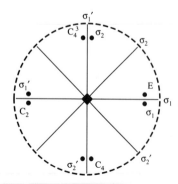

Figure 6.7 Stereogram of the **C**$_{4v}$ group with a general point *E* and the locations to which it is moved by the symmetry operations indicated.

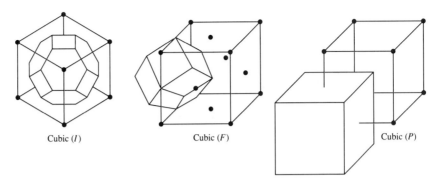

Figure 6.8   Wigner–Seitz unit cells for the cubic system.

▶ 6.1.4   *The Reciprocal Lattice*

In the stereograms we saw the representation of a plane (or a set of parallel planes) by a point. We show how a set of equally spaced parallel planes of a lattice in real space can be treated as a point in **reciprocal space.** To construct a **reciprocal lattice** we take an arbitrary lattice point as the origin and construct a ray normal to the plane of interest, denoted by its usual Miller index. The point in reciprocal space falls on this ray at a distance d^*, where d^*_{hkl} is inversely proportional to the d_{hkl} spacing between hkl planes in the real lattice. (The proportionality constant is usually taken as λ by x-ray crystallographers, and as 2π by theorists.)

We show in Figure 6.9 the construction of a reciprocal lattice for a single layer of graphite represented by tiled hexagons. The direct lattice is shown in heavy lines with the 10 and 01 planes labeled (in a three-dimensional lattice these would be the 100 and 010 planes). We construct perpendicular lines through our arbitrarily chosen lattice point to the 10 and 01 planes. These become the directions of the a^* and b^* vectors in reciprocal space. Units may now be marked out on these in inverse ratio of the distances to the planes above, and a reciprocal lattice grid marked in. Every intersection on the grid just constructed is a reciprocal lattice point, and the distance from the reciprocal lattice origin corresponds to the inverse of the spacing of the set of planes of the same labeling in the direct lattice. The distance from the origin to any point in reciprocal space, d^*_{hkl}, is a simple vector addition of the unit cell vectors, and may be translated back to the direct lattice to give the d_{hkl} spacings. The reciprocal lattice has the same symmetry as the direct lattice.

X-Ray Diffraction

In 1912 a debate raged concerning whether x-rays were particles or waves. No ruled gratings were able to produce diffraction of x-rays, placing an upper limit on their possible wavelength of 10^{-7} m. Unit cell dimensions of crystals had not been measured, but were believed to be on the order of 10^{-10} m. Max von Laue suggested that crystals might be used to diffract x-rays. The experiment, performed by his assistants Friedrich and Knip-

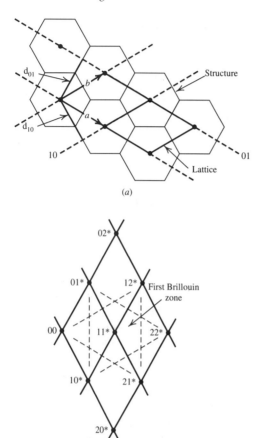

(a)

(b)

Figure 6.9 (a) Planar graphite lattice showing d_{01} and d_{10} planes. (b) Reciprocal lattice for planar graphite.

ping, supported the wave nature of x-rays and launched the field of x-ray crystallography. Laue received the Noble prize in Physics in 1914. Laue photographs are taken with "white" (nonmonochromatic) x-rays with the crystal in the beam between the source and photographic film. A typical Laue photograph in which the crystal symmetry is readily apparent is shown in Figure 6.10.

The Ewald Reflection Sphere

Laue considered that every atom could act as a point of elastic scattering for the incoming x-ray, but *diffraction occurs only when the difference between an incoming and an outgoing x-ray vector is a translation vector of the reciprocal lattice.* P. Ewald developed an elegant construction to relate the diffraction angles to the reciprocal lattice and also to show the agreement with the familiar Bragg equation.

Let a crystal be located at point C (Figure 6.11) with a well-collimated beam of x-rays traversing from A to T. A sphere of unit radius is constructed with C as its center. This unit is taken to be $1/\lambda$, the reciprocal of the x-ray wavelength. The x-rays scattered at point C have the same wavelength as the incoming x-rays; and their vector representation

Figure 6.10   Laue photograph of an Al–Mn alloy showing a picture of the reciprocal lattice with three-fold symmetry. (Reproduced from D. M. Follstaedt and J. A. Knapp, *J. Mater. Res.* **1989,** *4,* 1408.)

starts at C and terminates somewhere on the unit circle. If a reciprocal lattice is constructed for the crystal at C with a lattice point at O* taken as the origin, then only when a lattice point falls on the circle will diffraction occur. Let P* be a point falling on the reflection sphere. By construction, OP* is both the vector difference of the incoming and outgoing x-ray vectors and also a translation vector of the reciprocal lattice. We show agreement with the Bragg condition for refraction by constructing $\overline{BC}$ as the perpendicular bisector of $\overline{OP^*}$ and the bisector of $\angle O^* CP^*$. The length of $\overline{OP^*}$ is then given in our reciprocal space by $(d_{OC})(2 \sin \theta)$. But $\overline{OP^*}$ is $1/d_{hkl}$ and d_{OC} is $1/\lambda$. Therefore this gives the **Bragg equation:**

$$\lambda = 2d_{hkl} \sin \theta \tag{6.3}$$

As the crystal is rotated about C, a corresponding rotation of the reciprocal lattice takes place around O*, bringing other lattice points onto the sphere of reflection. Thus, the formulation by Ewald involving reciprocal space has been shown to be equivalent to the Bragg condition that diffraction occurs from successive parallel planes only when the plane's spacing in the real lattice is such that "extra" distance traveled by x-rays scattered represents some integral number of wavelengths:

$$n\lambda = 2d_{hkl} \sin \theta \tag{6.4}$$

The "order" of diffraction is given by *n*, with first-order diffraction being the most in-

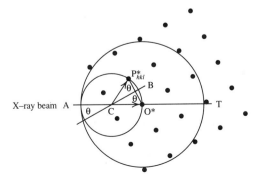

Figure 6.11   Ewald sphere.

tense. The experimental observations are values of θ for which diffraction occurs. Since

$$\sin \theta = \frac{\lambda}{2d_{hkl}} = \frac{\lambda d_{hkl}^*}{2} \tag{6.5}$$

for first-order diffraction, a relationship exists between measured θ values and distances in reciprocal space.

If monochromatic x-rays had been available for the first diffraction experiment with a stationary crystal, the experiment probably would have failed because of the low probability of a reciprocal lattice point falling on the sphere of reflection! Since a continuous range of wavelengths is used in taking Laue photographs, numerous reciprocal lattice points fall on their respective spheres of reflection. This range of wavelength in Laue photographs causes distance information to be lost. For structural determination, monochromatic radiation is employed and the crystal is rotated about an axis, with each reciprocal lattice point causing diffraction as it crosses the sphere of reflection. In powder diffraction, one depends on a finely ground crystalline material having randomly oriented crystallites which allow all possible reflections to be observed. Intensity and 2θ data are obtained that serve as a fingerprint for identifying compounds. The Bragg equation can be applied in a straightforward fashion to obtain d spacings from measured 2θ values. The most accurate unit cell parameters are derived from powder scattering data. The very accurate 2θ data allow correspondingly accurate calculation of cell constants. A method of data analysis due to Rietveld allows full crystal structure determination from powder data.

Diffraction studies may be carried out also with beams of electrons, neutrons, or even charged atoms. Neutron diffraction complements x-ray diffraction in that the scattering factor for x-rays increases with atomic number whereas that for neutrons varies in an irregular fashion. The neutron scattering for O is slightly higher than that for Ba. Thus neutron diffraction allows accurate determination of positions of light elements such as H or O in the presence of heavy elements. Neutron beams also permit the determination of the magnetic substructure of solids by their interaction with unpaired electrons. Most studies of this type are carried out at the National Laboratories (Argonne, Oak Ridge, etc.) because of the necessity of a nuclear reactor as a neutron source. Neutrons suffer very little loss in intensity as they interact weakly with matter; thus bulk samples may be studied. Electrons, by contrast, are not very penetrating, and samples must be thinned to less than a micron for suitable transmission. At high energy, 10–100 keV, the wavelength of the beam will be as short as 0.05 Å, giving an enormous Ewald sphere. Figure 6.10 shows a Laue photograph of an Mn–Al alloy using a high-energy monochromatic electron beam. The three-fold symmetry of the reciprocal lattice is apparent.

6.2 DEFECT STRUCTURES

Thus far, we have treated crystals as perfectly repeating arrays of atoms. But crystals are far less perfect than has been assumed. Imperfections in crystals arise from dislocation of ions, ion vacancies in the lattice, or nonstoichiometric proportion of the ions present—or simply from foreign ions or "impurities" in the lattice.

▶ 6.2.1 *Schottky and Frenkel Defects*

Stoichiometric crystals may display two types of lattice defects. Vacancies of equal numbers of anions and cations are termed **Schottky defects.** Figure 6.12 shows a single-layer sketch representation of an NaCl crystal containing Schottky defects.

    Frenkel defects occur when an ion occupies an interstitial site, leaving its normal site vacant. These defects are most likely when the anion and cation differ greatly in size, so that the smaller ion can fit into the interstitial sites. Frenkel defects in AgBr are illustrated in Figure 6.13.

    The large amount of heat released in the formation of a crystal lattice from gaseous ions is accompanied by a large unfavorable entropy change, resulting from formation of the rigid well-ordered crystal lattice. Introduction of vacancies or dislocations in the crystal lattice certainly decreases a crystal's heat of formation, but the increase in entropy accompanying the decrease in order favors the process. Obviously, a balance is achieved between entropy and enthalpy, and defects are more abundant in crystals where the enthalpy needed to create a defect is low. Defects are less likely for crystals of the extreme ionic type, where the balance of coulombic forces is critical. Frenkel defects, in particular, are more likely to occur in partially covalent substances, where polarization helps decrease the effect of charge dislocations. Pure stoichiometric crystals can exhibit both Frenkel and Schottky defects with a balance in enthalpy and entropy terms. Stoichiometric CrO has 8% vacancies in the anion and cation sites.

▶ 6.2.2 *Solid Electrolytes*

The low-temperature modification (β) of AgI has the ZnS wurtzite structure and is a poor conductor of electricity; this is generally true for solid salts. At 145.8°C there is a sharp transformation to α-AgI, which has a body-centered cubic arrangement of I^-, accompanied by a tremendous increase in ionic conductivity. Such substances are referred to as **solid electrolytes** since their conductivities are closer to those of liquid electrolytes than those of ordinary solid salts. The Ag^+ ions in solid α-AgI (mp 555°C) can move from their normal sites through positions of lower coordination number. These are interstitial sites (Frenkel defects). At the melting point there is a 12% *decrease* in conductivity with melting. Other good solid electrolytes, such as $RbAg_4I_5$ and $[(CH_3)_4N]_2Ag_{13}I_{15}$, have passageways of face-centered tetrahedra, fewer than half of which are occupied by Ag^+. For Ag_2HgI_4 at 50.7°C a phase transition occurs, forming the solid electrolyte α-Ag_2HgI_4,

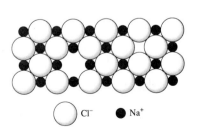

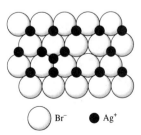

Figure 6.12 Schottky defects in NaCl.   **Figure 6.13.** Frenkel defects in AgBr.

which has a *ccp* arrangement of anions and passageways formed by the face-sharing tetrahedra and octahedra. Figure 6.14*a* shows the zinc blende (ZnS) structure with the cations occupying Zn sites. The high-temperature structure of α-AgHgI$_4$ corresponds to this structure with six atoms of Ag and Hg distributed among the eight cation sites. Figure 6.14*b* shows the structure of β-Ag$_2$HgI$_4$ (low T modification) and shows the distinct sites for Ag$^+$ and Hg^{2+}.

The passageways generally are zigzag for these solid electrolytes—except for (C$_5$H$_5$NH)Ag$_5$I$_6$ (pyridinium salt), which has an essentially *hcp* arrangement of I$^-$ (at *A* and *B* close-packed positions). The pyridinium ions are located at sites centered about *C* positions, and all octahedral holes for Ag$^+$ are lined up along the *C* directions (see page 205), forming straight channels for motion of Ag$^+$. [See Figure 2.21 (page 96) for the structure of ice showing open channels along *C* for a hexagonal structure.] This motion is limited by the number of vacancies, but vacancies can be created by movement of Ag$^+$ ions from octahedral holes into empty tetrahedral holes. The structure is ordered at $-30°$, with Ag$^+$ occupying octahedral sites and only one of the two sets of tetrahedral sites; but the Ag$^+$ positions are disordered at room temperature and above. Solid electrolytes generally display Frenkel defects; and large highly polarizable anions and highly polarizing cations of low charge promote defects and solid electrolyte behavior.

Some new high-energy batteries use nonaqueous systems requiring (a) high temperature, such as liquid Na–liquid S or liquid Li–metal sulfide with a molten electrolyte, (b) solid electrolytes, or (c) nonaqueous solvents for the electrolyte for ambient temperature batteries. One solid electrolyte used is α-alumina,[4] an isomorph of Al$_2$O$_3$. Based on x-ray diffraction studies, the empirical formula of β-alumina is not pure Al$_2$O$_3$ but instead Na$_2$O·11Al$_2$O$_3$. Commonly, the β-alumina used contains ~25% more Na$_2$O than indicated by the formula. Beta-alumina has a layer structure featuring four close-packed layers of oxide ions and Al^{3+} in octahedral and tetrahedral sites. The Na$^+$ occupy planes between the quadruple oxide layers; these planes contain equal numbers of loosely packed Na$^+$ and O^{2-} ions. An excess of this ratio of Na$^+$ ions is accompanied by Al^{3+} vacancies, to maintain charge balance. Adding MgO or Li$_2$O creates a stabilized form referred to as β''-alumina. Both β- and β''-alumina have high ionic conductivity for Na$^+$ and can be used as solid electrolytes for the liquid Na–liquid S battery.

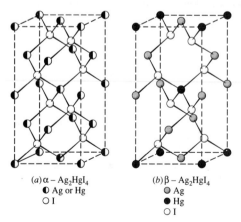

(a) α – Ag$_2$HgI$_4$
◗ Ag or Hg
○ I

(b) β – Ag$_2$HgI$_4$
◕ Ag
● Hg
○ I

Figure 6.14 Structures of α- and β-Ag$_2$HgI$_4$.

[4] G. C. Farrington and J. L. Briant, *Science* **1979**, *204*, 1371; *Inorg, Synth.* **1979**, *19*, 51.

▶ 6.2.3   *Nonstoichiometric Compounds*

Nonstoichiometric compounds often result when a cation can have different oxidation states, as in FeS, FeO, Cu_2O, NiO, CuO, CuI, and so on. These might better be represented by formulas such as $Fe_{1-\delta}S$, where δ represents the fraction of vacant cation sites per formula weight. Electrical neutrality can be maintained by oxidizing the equivalent number of Fe^{2+} to Fe^{3+}, giving the composition $Fe^{2+}_{1-\delta}Fe^{3+}_{2\delta/3}S^{2-}$. The lattice of $Fe_{1-\delta}S$ may be represented as shown in Figure 6.15. The structures of FeO, γ-Fe_2O_3, and Fe_3O_4 all display the *ccp* array of oxide ions. In the case of FeO, which has the NaCl structure, the octahedral sites are up to 95% filled, giving the limiting stoichiometric ratio $Fe_{0.95}O$. The "mixed" oxide Fe_3O_4 is an inverse spinel (page 217), with 32 oxide ions in the unit cell, eight Fe^{2+} in octahedral (*O*) sites, and 16 Fe^{3+} in tetrahedral (*T*) and *O* sites. On average, $21\frac{1}{3}Fe^{3+}$ for the same unit cell (32 oxide ions) are distributed randomly among the *O* and *T* sites. Interconversion between γ-Fe_2O_3 and Fe_3O_4 is accomplished easily. The structure of α-Fe_2O_3 is the same as that of α-Al_2O_3 (corundum, based on an *hcp* scheme), with only *O* sites occupied by Fe^{3+}.

John Dalton interpreted stoichiometric relationships as resulting from the combination of discrete atoms. Berthollet believed that compounds had variable composition. Unfortunately, the beautiful crystals of minerals, considered to represent perfection, were used to test the opposing views. Minerals commonly have variable composition, because of isomorphous substitution of ions. The existence of nonstoichiometric compounds was an obstacle for the acceptance of atomic theory. The term *daltonide* is now used for stoichiometric compounds, and *berthollide* is used for nonstoichiometric compounds.

Nonstoichiometric compounds can occur even if the metal is not present in two or more oxidation states—as in the case of zinc oxide, which loses oxygen on heating to give $Zn_{1+\delta}O$. Sodium chloride reacts with sodium vapor to give $Na_{1+\delta}Cl$. In these cases an electron or an electron pair can occupy a vacant anion site (Figure 6.16).

Square-planar metal complexes can stack in the solid so as to set up direct M–M interaction along the chain.[5] Partially oxidized Pt^{II} complexes such as $K_2Pt(CN)_4X_{0.3}\cdot 3H_2O$ (X = Cl^- or Br^-) stacked in this way provide metallic conduction along the direction of M–M interaction—a **one-dimensional metal.** Square-planar complexes of small coordinated cyanide ions are ideal for close stacking with overlap of the d_{z^2}

Figure 6.15   Defects in FeS.

Figure 6.16   An anion vacancy filled by an electron in $Na_{1+\delta}Cl$.

[5] J. S. Miller and A. J. Epstein, *Prog. Inorg. Chem.* **1976**, *20*, 1; *Inorg. Synth.* **1979**, *19*, 1. J. M. Williams, A. J. Schultz, A. E. Underhill, and K. Carnerio, in *Extended Linear Chain Compounds*, Vol. 1, J. S. Miller, Ed., Plenum, New York, 1982, p. 73.

orbitals. For Pt^{II} (d^8) the d_{z^2} orbital is filled, but partial oxidizing creates vacancies for conduction. Cation-deficient complexes such as $K_{1.75}[Pt(CN)_4] \cdot 1.5H_2O$ behave similarly. The compounds $Hg_{2.86}AsF_6$ and $Hg_{2.91}SbF_6$ are anisotropic metallic conductors that contain infinite linear chains of mercury atoms in two mutually perpendicular directions.

Among other materials of interest as solid-state electrodes are metal oxides with cavities such as that occupied by Ca^{2+} in the perovskite structure (page 209); these include the vanadium oxides V_2O_5, V_6O_{13}, and VO_2. Bronzes with the composition $Li_xV_2O_5$ have been obtained from the reaction of V_2O_5 and LiI in acetonitrile. Acetonitrile serves as a solvent and a mild reducing agent: For each Li^+ incorporated in the lattice, one vanadium(V) is reduced. Such oxides are of interest as cathodes for nonaqueous lithium batteries.

6.3 ELECTRONIC STRUCTURE OF SOLIDS

▶ 6.3.1 *Reciprocal Space Revisited: The First Brillouin Zone*

When atomic orbitals (AOs) are combined to give molecular orbitals (MOs), the orbitals are conserved: The number of MOs formed is exactly the same as the number of AOs combined. In a crystal with N primitive unit cells, each AO per unit cell gives rise to N MOs. These run over the entire crystal and are intimately related to the sets of planes found in crystals. We have seen how these sets of planes in the direct lattice may be represented as points in the reciprocal lattice, and we illustrate this relationship further in Figures 6.17 and 6.18. In these examples the lattice vectors are orthogonal; thus $\vec{a} = d_{10}$ and $\vec{b} = d_{01}$, and $\vec{a}*$ is parallel to $\vec{a}$ and $\vec{b}*$ is parallel to $\vec{b}$. (The small number of points in these examples makes the use of terms such as "unit cell" absurd but may be justified by the simplicity otherwise afforded. A planar lattice is used, so lines in these figures correspond to planes in a three-dimensional lattice and will be referred to as planes.)

We start with four lattice points forming a conventional unit cell (Figure 6.17a). A point is arbitrarily selected as the origin, and the planes through the other points are shown labeled with their Miller indices. In reciprocal space we mark in one point corresponding to the origin and mark the points corresponding to $1/d_{hk}$ for the 10, 01, and 11 planes in the real, or direct, lattice. We enclose the points in square brackets; this now indicates a ray from the origin to the point, as well as indicating the reciprocal lattice point. Each point in reciprocal space may thus be considered to be a vector from the origin to the reciprocal lattice point. The asterisk reminds us that we are referring to reciprocal space.

In Figure 6.17b we have added a lattice point to the direct lattice which gives rise to a new plane denoted by the Miller index 12. Since all unit cells are identical, we can translate the second cell to superimpose it on the first. The new drawing indicates the set of 12 planes passing through any unit cell in the crystal. In reciprocal space the 12* point falls outside the unit cell defined by 00*, 10*, 11*, and 01*, as is true for any set of planes that can be drawn within one unit cell. Each added lattice point generates a new set of parallel planes with unique spacing and Miller indices and a corresponding point in the reciprocal lattice.

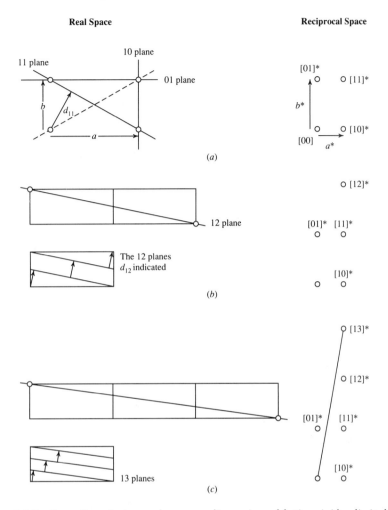

Figure 6.17   Some direct lattices and corresponding reciprocal lattices (with a limited number of lattice points). (*a*) (*Left*) Four lattice points forming a unit cell and their associated planes.) (*Right*) Reciprocal lattice points for the real-space planes. (*b*) (*Left*) An additional lattice point and associated plane. (*Right*) Additional reciprocal lattice point for the 12 planes. (*c*) (*Left*) A third additional lattice point and associated plane. (*Right*) Additional reciprocal space point for the 13 planes.

In Figure 6.18*a* we consider a very small crystal with four-unit cells. The d_{12} and d_{21} spacings are shown below in a single-unit cell with [12]* and [21]* falling outside the reciprocal lattice unit cell as indicated. Let us now look at the new planes in our crystal which fall *outside* the unit cell in the direct lattice; that is, a vector from the origin to the new plane cannot fit within the unit cell (Figure 6.18*b*). These planes are parallel to those drawn within a single-unit cell, and their interplanar separation is simply a multiple of the value for those within a single cell. The reciprocal lattice vectors will therefore have the same direction, but will be a rational fraction of length of the vectors representing parallel planes within a direct lattice cell. These points will fall on, or within, the reciprocal lat-

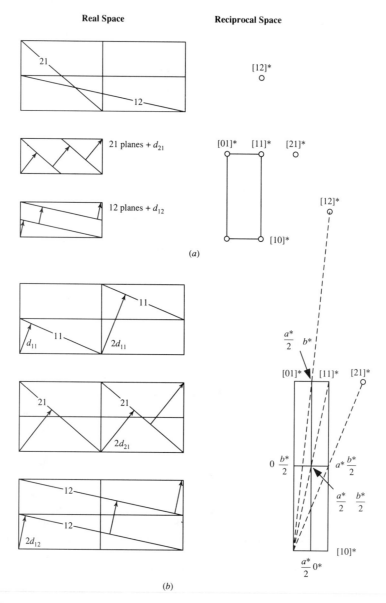

Figure 6.18 A small crystal. (*a*) Four unit cells and some planes within a unit cell. (*b*) Same with interplanar distance on edge or *outside* unit cell. Reciprocal space representation on right.

tice primitive unit cell. Thus the ends of our small crystal have 01 planes separated by $2d_{01}$ and 10 planes separated by $2d_{10}$. These parallel planes are represented in reciprocal space at exactly the midpoint of the [01]* and [10]* vectors, respectively. The spacing between the 11 planes in adjacent unit cells in the direct lattice is $2d_{11}$, and in reciprocal space these parallel planes are represented by a point halfway along the [11]* vector—that is, $\vec{a}*/2, \vec{b}*/2$.

We note that on connecting the points we have added to the unit cell of the reciprocal lattice, we form a grid which resembles the grid of unit cells in the direct lattice. The intersections in the reciprocal-space unit-cell grid correspond to all of the sets of parallel planes with interplanar spacing outside the unit cell of the direct lattice. For a macroscopic crystal the grid within the reciprocal-lattice primitive cell is constructed by dividing $\vec{a}*$, $\vec{b}*$, and $\vec{c}*$ by the number of primitive cells lying along the x, y, and z crystallographic directions in the real crystal, and using these microscopic units along $\vec{a}*$, $\vec{b}*$, and $\vec{c}*$. The grid now has N_a intercepts along $\vec{a}*$, N_b along $\vec{b}*$, and N_c along $\vec{c}*$. There will be $N_a N_b N_c$ intersecting points, corresponding in number to the number of lattice points in either the direct lattice or the reciprocal lattice. The number of these points within the reciprocal-lattice primitive cell also turn out to be the number of MOs formed from a given basis, but to relate these points more directly to the MOs we need to shift the origin to the center of the cell. Such a centered cell is termed **the first Brillouin zone,** and may be formed by constructing a Wigner–Seitz primitive cell for the reciprocal lattice (see Figure 6.9). It will be a polyhedron with sets of parallel faces, with the separation between the faces being $\vec{a}*$, $\vec{b}*$, and $\vec{c}*$. The distance from one of these faces to the origin is $\vec{a}*/2$, $\vec{b}*/2$, or $\vec{c}*/2$, or sums or differences of these. Within the unit cell the lattice vectors will be divided into units of $\vec{a}*/N_a$, $\vec{b}*/N_b$, and $\vec{c}*/N_c$, and these points are numbered as $k_a = 0, \pm1, \pm2, \ldots, \pm N_a/2$. Any point within the first Brillouin zone may be located by a vector $\vec{k}$ given by

$$\vec{k} = \frac{k_a \vec{a}*}{N_a} + \frac{k_b \vec{b}*}{N_b} + \frac{k_c \vec{c}*}{N_c} \qquad (6.6)$$

▶ 6.3.2   The Bloch Theorem

We know that every unit cell in a crystal is identical with regard to the location of atoms of any given type within the cell. An important theorem by Felix Bloch states that wavefunctions can be altered only by a phase factor by any translation by a lattice vector. This means that MOs in solids are formed by imposing a plane wave on an orbital in each primitive cell of the direct lattice. The imposed plane wave will have a wavelength of $1/k$ in the $\vec{k}$ direction, that is, it will be directed normal to the set of planes in the direct lattice which corresponds to the $\vec{k}$ vector in the first Brillouin zone. The k_a, k_b, and k_c values are the quantum numbers designating each MO formed from the contents of the primitive cell, termed a **basis.** If the primitive cell contains a single atom, each orbital type becomes a basis: $1s$, $2s$, $2p$, and so on. If there are several atoms within a primitive cell, each group orbital (or MO within the cell) gives a basis. The $N_a N_b N_c$ MOs formed by a given basis have a nearly continuous set of energies (termed an **energy band** as the $2s$ band, $2p$ band, and so on. If an MO is used as a basis, the band may be called a pi band, and so on. Since there are an astronomical number of individual MOs, wavefunctions and energies are considered for only a few selected k values, or wave vectors, in the Brillouin zone. Special positions in the first Brillouin zone are given particular symbols (see Figure 6.19). The individual psi functions are designated by (a) a superscript for the band index ($1s$ or $2p$, etc.) and (b) a subscript for the k point in the first Brillouin zone.

We first look at a one-dimensional model, a linear lattice where the primitive cell contents may be an atom, a set of atoms, or even an empty lattice! The reciprocal lattice is

One-Dimensional

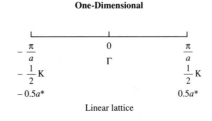

Linear lattice

Two-Dimensional

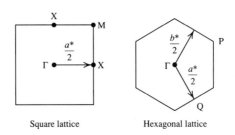

Square lattice   Hexagonal lattice

Three-Dimensional

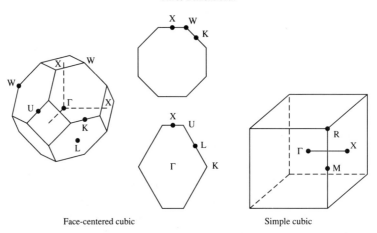

Face-centered cubic   Simple cubic

Figure 6.19 First Brillouin zone for several lattices with special symmetry points labeled. Several notations of linear lattice are indicated. Cross sections for face-centered cubic indicated at right of *fcc* zone.

also linear, with $\vec{a}$ and $\vec{a}^*$ pointing in the same direction and $\vec{a}^* = 1/\vec{a}$. We take the perpendicular bisectors of the reciprocal lattice points and select an origin to obtain the first, second, third, . . . Brillouin zones as indicated in Figure 6.20a. All information about the lattice *outside* the real-space primitive cell is contained *inside* the first Brillouin zone. Figure 6.20a shows the effect of imposing waves corresponding to different $\vec{k}$ values (given by the subscript of Ψ) on an s-orbital located on each atom in the chain. At $k = 0$ (the band center, often labeled Γ) the wave is of infinite wavelength extending over all orbitals with the same amplitude. The MO here is simply the normalized sum of all of the $1s$ orbitals, and is obviously the most bonding and lowest-energy orbital. At the band edge $k = 0.5\vec{a}*$ or $0.5/\vec{a}$ the wavelength is $\vec{a}/0.5$ or $2\vec{a}$. There is now a node between every atom, and this is the most antibonding orbital. The other orbitals for $\vec{k}$ values intermediate between 0 and $0.5\vec{a}*$ have intermediate energies, with the nonbonding orbital occurring at $\vec{k} = 0.25\vec{a}*$. The energy generally increases with an increase in the number of nodes. The actual energy depends on the separation of the atoms and the extension of the particular atomic s orbital used. A plot of energy versus the various points in the Brillouin zone is termed the **band structure.** The band structure for a linear H chain at internuclear separations of 100, 200, and 300 pm as calculated by Roald Hoffmann is given in Figure 6.21 and shows the effect of variation in the orbital overlap on the orbital energies within the band. The maximum range in energy is termed the **band dispersion.** Usually

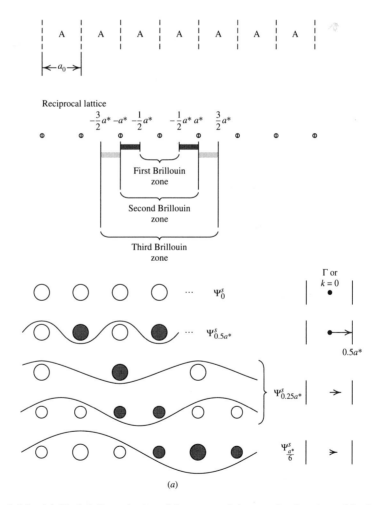

Figure 6.20   (a) (*Top*) A linear lattice of A atoms and the associated reciprocal lattice with several Brillouin zones indicated. (*Bottom*) Points in the first Brillouin zone and the corresponding modulating wave imposed on s orbitals in the lattice at the top. The modulating wave is imposed on the following in each cell: (b) p_z orbital. (c) d_{z^2} orbital. (d) f_{z^3} orbital. (e) Nothing (an empty lattice).

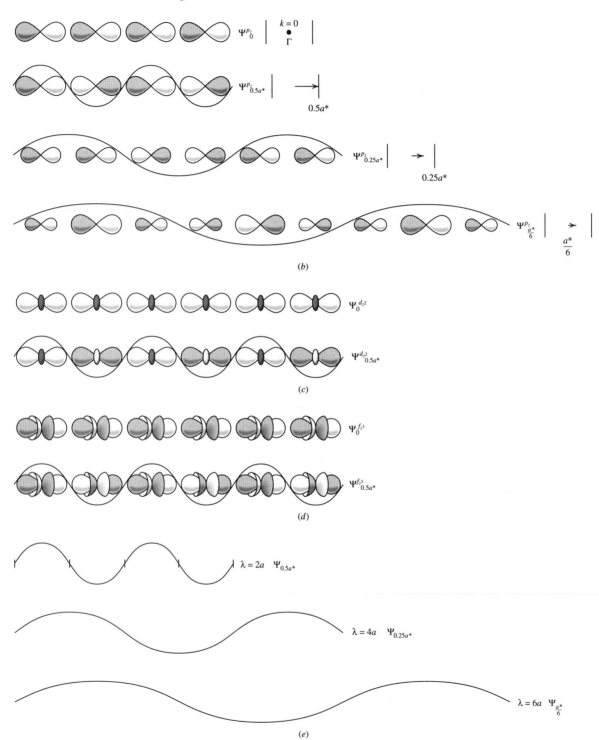

$\Psi^{p_z}_0$ $\quad$ $k = 0$ $\quad$ $\bullet$ Γ

$\Psi^{p_z}_{0.5a*}$ $\quad$ $\longrightarrow$ $\quad$ $0.5a*$

$\Psi^{p_z}_{0.25a*}$ $\quad$ $\longrightarrow$ $\quad$ $0.25a*$

$\Psi^{p_z}_{\frac{a*}{6}}$ $\quad$ $\longrightarrow$ $\quad$ $\frac{a*}{6}$

(b)

$\Psi^{d_{z^2}}_0$

$\Psi^{d_{z^2}}_{0.5a*}$

(c)

$\Psi^{f_{z^3}}_0$

$\Psi^{f_{z^3}}_{0.5a*}$

(d)

$\lambda = 2a$ $\quad$ $\Psi_{0.5a*}$

$\lambda = 4a$ $\quad$ $\Psi_{0.25a*}$

$\lambda = 6a$ $\quad$ $\Psi_{\frac{a*}{6}}$

(e)

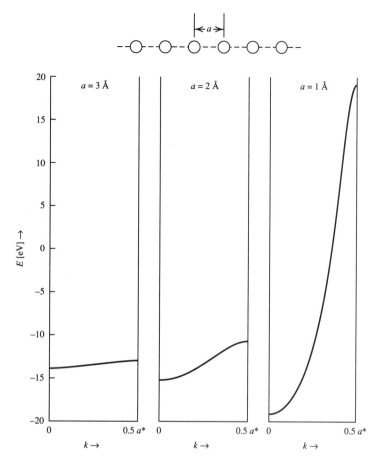

Figure 6.21   The band structure of a chain of H atoms spaced 3, 2, and 1 Å apart. The energy of an isolated H atom is -13.6 eV. (From R. Hoffmann, *Solids and Surfaces*, VCH Publishers, New York, 1988.) Reprinted with Permission by VCH Publishers © 1988.

only positive values of $\vec{k}$ are used in displaying the band structures, as in the figure given, since the $-\vec{k}$ values give a mirrored curve.

We see in Figure 6.20*b–d* what happens when we use p_z, d_{z^2}, or f_{z^3} orbitals as our basis in the one-dimensional structure. This change in basis does not change the modulating plane wave, because the first Brillouin zone is unchanged. Every $\vec{k}$ value in this zone still gives the same wavelength. At the band center the p_z orbitals are identical in every cell obtained by a translation of the unit cell. $\Psi_0^{p_z}$ is now the highest-energy orbital because all interactions between neighboring p_z orbitals are antibonding. In contrast, at the band edge ($\vec{k} = 0.5\vec{a}*$) the alternating phase of the modulating wave reverses the signs of alternate p_z orbitals, so $\Psi_{0.5a*}^{p_z}$ now has only bonding interactions. The p_z-band energies run in the direction opposite to that of the s-band as we move from the Brillouin zone center to the zone edge. Inspection of the several sampled orbitals from the d_{z^2} band and the f_{z^3} band indicates that the d_{z^2} band runs up from the zone center, similar to the s-band behavior, whereas the f_{z^3} band runs down from the zone center, similar to the p_z-band behavior.

The Free Electron in One Dimension

We show in Figure 6.20e the modulating plane wave for an empty lattice with a periodic potential approaching zero. This is essentially the set of Ψ_k for a free electron in a one-dimensional box. Using $1/k = \lambda$ (the wavelength of the electron), from de Broglie's relationship $\lambda = h/p$ (where h is Planck's constant and p is the momentum of the electron) we can obtain the kinetic energy

$$E = \frac{p^2}{2m} = \frac{h^2 k^2}{2m} \tag{6.7}$$

With a completely free electron there would be no restriction at the zone boundary, so values of k in the 2nd, 3rd, ..., nth Brillouin zones become possible. The band structure for a free electron (Figure 6.22) shows this parabolic behavior. Since in reciprocal space, every unit cell is identical, we can show all of these energies as a function of k in the first Brillouin zone. We have to be careful how we translate the cells, since only half of a Brillouin zone is contained between an interval of $a*/2$, with half the zone on one side and half on the other of the first Brillouin zone which is centered about the origin. We obtain a result equivalent to the translation by folding the zones in an accordian pleat as illustrated in Figure 6.22, with the resultant reduced zone picture.

A discontinuity has been introduced at the zone boundaries to reintroduce the effect of the periodic potential of the lattice. Let us look at λ at the zone boundaries. For $\vec{k} = \vec{a}*/2$, $\lambda = 2a$ and there is a node at exactly the unit-cell translational distance. At $\vec{k} = \vec{a}*$, $\lambda = a$, and every translational distance contains two nodes. At $\vec{k} = 3\vec{a}*/2$, $\lambda = 2a/3$, and three nodes occur within every unit-cell translational distance. Since nodes of the modulating wave are not permitted to occur at less than one unit-cell transla-

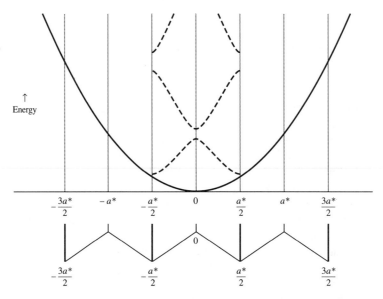

Figure 6.22   The band structure of the free electron under a vanishing periodic potential. Bottom indicates accordian pleat showing the folding of higher Brillouin zones into the first zone to give a reduced zone representation. Dashed lines used for reduced zone representation.

tional distance for any system with a real basis, the *s*-bands, *p*-bands, and so on, live within the first Brillouin zone. However, Altman (see General References) has noted the similarity between the way the band structure runs for the reduced zone picture of the free electron, and the way the bands run for the linear AO chain with the same number of nodes at the band edge if the nodes of the orbital itself are also counted. Thus the *s* band runs up, as does the first free electron band in the first Brillouin zone, both having a maximum of one node/*a*; the p_z band runs down as does the second free electron band in the reduced picture, both having a maximum of two nodes/*a*; and so on. The set of orbitals *s*, p_z, d_{z^2}, f_{z^3}, and so on, look like a folded band representation for a continuous increase in *k* (see Figure 6.23). The nodes increase by one, and the energies run in the direction expected for a folded band. In a one-dimensional case there are theoretical reasons why the bands should not overlap. But real three-dimensional structures apparently do not have this restriction.

The Bloch Equation

While good pictures and qualitative energies associated with any of the Ψ_k orbitals over the crystal can be obtained from the *k* vector, the basis, and the structure, we would like to have explicit coefficients for the linear combination of AOs associated with any $\vec{k}$. The Bloch equation permits us to write the Ψ_k wavefunctions. This equation has the form

$$\Psi_k = \sum_n^N e^{2\pi i \vec{k} \cdot \vec{t}_n} \phi_n \tag{6.8}$$

where: ϕ_n is the basis orbital (an AO or group orbital) residing in the primitive cell; $\vec{t}_n$ is a translation vector of the unit cell; $\vec{k}$ is a particular wave vector in the first Brillouin zone; and the summation is taken over all translations of the lattice (obviously truncated in practice). The dot between $\vec{k}$ and $\vec{t}_n$ indicates that a dot product is to be taken between the two vectors, giving a scalar equal to $k t_n \cos \theta$ where θ is the angle formed between $\vec{k}$ and $\vec{t}_n$. For parallel vectors, $\cos \theta = 1$; for orthogonal vectors, $\cos \theta = 0$. Using the direct

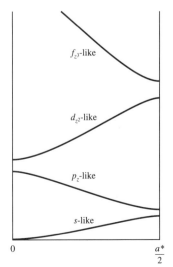

0                                    $\dfrac{a^*}{2}$

Figure 6.23 Comparison of sigma bands of the atomic orbitals with the reduced band structure for the free electron.

lattice vectors and their associated reciprocal lattice vectors causes cross-terms to vanish ($\vec{a}* \cdot \vec{b} = 0$), whereas terms such as $\vec{a}* \cdot \vec{a}$ and $\vec{b}* \cdot \vec{b}$ become simple products. While $\vec{a}*$ will always be perpendicular to $\vec{b}$, it is not always parallel to $\vec{a}$; however, from the nature of its construction $a*a \cos \theta$ will always equal one. (The 2π appearing in the above formula is often omitted by theoreticians who incorporate this term in the reciprocal lattice vectors, setting $\vec{a}* = 2\pi/d_{100}$, etc.)

Equations (6.1) and (6.6) define the $\vec{t}_n$ and $\vec{k}$ vectors. It is convenient to replace the k_i/N_i terms in Equation (6.6) by the fractions, f_i^*, ranging from -0.5 to $+0.5$, representing the position along $a*$, $b*$, and $c*$. Making these substitutions in Equation (6.8), replacing $\vec{a}* \cdot \vec{a}$ with 1, etc., and dropping the cross-terms $\vec{a}* \cdot \vec{b}$, etc., gives

$$\Psi_k = \sum_n^N e^{2\pi i(f_a*n_a + f_b*n_b + f_c*n_c)} \phi_n \tag{6.9}$$

Since $e^{p+q+r} = e^p e^q e^r$, we may write

$$\Psi_k = \sum_n^N e^{2\pi i f_a*n_a} e^{2\pi i f_b*n_b} e^{2\pi i f_c*n_c} \phi_n \tag{6.10}$$

Since Equations (6.8)–(6.10) are all equivalent, we will refer to all of them by the same name, the **Bloch equation.**

Ψ_k takes a particularly simple form at the Brillouin zone center, $k = 0$, and at the zone boundaries.

For $\vec{k} = 0$ we have $f_{a*} = 0$, $f_{b*} = 0$, and $f_{c*} = 0$, so all exponential terms become e^0 or 1. For a one-dimensional lattice we have

$$\Psi_0 = \sum_n^N e^0 \phi_n = \phi_1 + \phi_2 + \phi_3 + \cdots \tag{6.11}$$

or for a three-dimensional lattice we have

$$\Psi_0 = \phi_{111} + \phi_{211} + \cdots + \phi_{121} + \phi_{131} + \cdots + \phi_{112} + \cdots + \phi_{221} + \cdots \tag{6.12}$$

Equations (6.11) and (6.12) tell us that, at the zone center, the basis function orbital is repeated in every unit cell with no change in amplitude or sign.

At the zone boundary in a one-dimensional crystal, $f_{a*} = 0.5$ and the corresponding term in Equation (6.10) becomes $e^{\pi i n_a}$. For even values of n the term is $+1$, whereas for odd powers of n the term becomes -1. Thus in a linear crystal the signs of the basis function alternate as we move from cell to cell:

$$\Psi_{0.5a*} = \phi_0 - \phi_1 + \phi_2 - \phi_3 + \cdots \tag{6.13}$$

For other values of k, the exponential terms in the Bloch equation may be converted to cosine and sine functions (as in Section 4.6), since for every Ψ_k there is a Ψ_{-k}. The sums and differences of these give two wavefunctions of the same energy. In such a sum, for every $+k$ term (or $+f_{a^*}$) there will be a $-k$ term (or $-f_{a^*}$). Thus in Equation (6.10) the $e^{2\pi i f_{a^*} n_a}$ terms become $\cos 2\pi f_a n_a$, giving

$$\Psi_k = \sum (\cos 2\pi f_{a^*} n_a)(\cos 2\pi f_{b^*} n_b)(\cos 2\pi f_{c^*} n_a)\phi_n \qquad (6.14)$$

EXAMPLE 6.3:   Show that Equation (6.14) gives the results presented in Figure 6.17*a*.

Solution: For $f_{a^*} = 0.25$ (halfway to the zone boundary), Equation (6.13) gives

$$\Psi_{0.25a^*} = \sum (\cos 2\pi f_{a^*} n_a)\phi_n = \sum \left(\cos \frac{\pi}{2} n_a\right)\phi_n$$

For $n_a = 0$, $\cos 0 = 1$

For $n_a = 1$, $\cos \dfrac{\pi}{2} = 0$

For $n_a = 2$, $\cos \pi = -1$

For $n_a = 3$, $\cos \dfrac{3\pi}{2} = 0$

For $n_a = 4$, $\cos 2\pi = 1$

One wavelength occurs over an interval of 2π, or $4\vec{a}$.

$$\Psi_{0.25a^*} = 1\phi_0 + (0\phi_1) - \phi_2 + (0\phi_3) + \phi_4 + \cdots$$

With a different choice of origin for the lattice, the second curve of Figure 6.17a is obtained. Only the phase difference between cells is available from the Bloch equation, not the absolute phase.

While it is straightforward to write the form of Ψ_k for crystals using the Bloch equation, evaluating the Ψ's requires much computation effort and high-speed computers. The problem is a multielectron problem such that even with precise AO wavefunctions as beginning basis functions (rarely available at present for transition metals), the change in local electron density as the one-electron wavefunctions are filled requires iterative calculations. The electron density at any point in the unit cell is given by $\Sigma_{\text{electrons}} \ \Psi^2$, the orbitals being filled are the lowest as determined by $\tilde{\mathcal{H}} \Psi = E\Psi$, so the wavefunctions and energies in the many-electron problem are not independent and iterative procedures are required. Despite these hurdles, an increasing number of band stucture calculations are becoming available. The purpose of this section is to provide a background to permit some access to the literature.

Charge Transport

The Bloch equation as represented in Equations (6.8)–(6.10) has a traveling wave as the modulating wave (Ψ_+ and Ψ_-), in contrast to the sine and cosine standing wave solutions of the type represesented by Equation (6.14). If the Ψ_+ and Ψ_- states are equally occupied, no net charge transport is occurring. Under a potential gradient, one of these sets of states will be lowered in energy (say the Ψ_+ states). If both Ψ_- and Ψ_+ bands are filled, with a large energy gap to the next band, nothing happens—the substance is an insulator. If the band is partially filled, the electrons in Ψ_- states will drop into the Ψ_+ states, and there will be more electrons moving in one direction than in the other; that is, an electric current will flow through the metal.

What is the nature of the charge carriers? Here we introduce the physicist's phlogiston. From Equation (6.7) we obtain

$$\frac{d^2E}{dk^2} = \frac{h}{m_{\text{eff}}} \tag{6.15}$$

where we have changed the m to m_{eff}, an effective mass of the electron. If the electron in a crystal were a free electron, the E versus k curve would have the constant increase in curvature shown by the parabola in Figure 6.22. Actual curves behave in this fashion at the bottom of a band, but go through an inflection point, and d^2E/dk^2 becomes negative at the top of the band as the discontinuity at the Brillouin zone boundary is approached. Rather than have an electron with negative mass, we speak of a hole which moves in the direction opposite to that of the electron motion. In an electric field the holes will undergo acceleration, but in the direction opposite to that expected for an electron for the same effective mass. Thus at the top of a band one speaks about holes which behave as positive charge carriers, whereas lower in the band one speaks of electrons as the negative charge carriers.

The E versus k curves for real materials are limited to the first Brillouin zone, and Equations (6.7) and (6.15) apply only to free electrons where the mass of the electron will be constant. However, the picture of electron holes has been a very useful one, particularly in the area of semiconductors.

Some One-Dimensional Solids

While there are no true one-dimensional materials, it is sometimes possible to ignore interactions in all but one direction and obtain at least a qualitative understanding of various properties. The chain of $Pt^{II}(d^8)$ atoms in stacked $[PtX_4]^{2-}$ mentioned in Section 6.2.3 provides such an example. The degeneracy of the d orbitals is lifted in these complexes, the empty $d_{x^2-y^2}$ being the highest in energy with the d_{z^2} orbital next. To a first approximation the MOs running over the chain would resemble those given in Figure 16.20c, with the band energy increasing from the band center at $\vec{k} = 0$ to the band edge at $\vec{k} = 0.5\vec{a}*$. With a filled d band we expect no Pt–Pt bonding since the antibonding character of the wavefunctions at the top of the band are approximately the same as the bonding character of the wavefunctions at the bottom of the band. Since the band is filled, there are as many $+\vec{k}$ states as $-\vec{k}$ states and no conduction occurs. As electrons are removed by partial oxidation from the high-energy antibonding orbitals at the top of the band, the Pt–Pt distance shortens, and we have a Fermi surface in the d_{z^2} band giving rise to conductivity.

In the actual crystal the square-planar $[Pt(CN)_4]^{2-}$ units are staggered as in Figure 6.24, so the unit cell must contain two Pt atoms. The filling of electrons in the picture with one Pt/cell would run the k states into the second Brillouin zone. Just as with the free-electron case, we may fold this zone back into the first zone. However, since the Pt

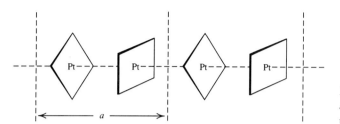

Figure 6.24 The staggered arrangement of stacked $[Pt(CN)_4]^{2-}$ units.

are not identical by translation along the z-axis, we should form a new basis using ϕ_+ as the sum of the d_{z^2} orbitals of the two Pt atoms in the unit cell, and ϕ_- as the difference of the d_{z^2} orbitals. Qualitatively we expect the same bands as the reduced zone structure, since the interaction between d_{xz} and d_{yz} on neighboring Pt is rather weak.

A more complete picture requires a closer look at the effect of the X groups in $[PtX_4]^{2-}$ (good pi acceptors lower the energy levels of the d_{xy}, d_{xz}, and d_{yz}; good pi donors raise these levels; sigma donors raise the $d_{x^2-y^2}$, etc., changing the relative energies of the d_{z^2} with respect to these orbitals) and a closer look at the dispersion of the bands. A band structure calculated by Hoffmann for a closely related $[PtH_4]^{2-}$ stack is shown in Figure 6.25 using several different unit cells. The dispersion of the d bands is as expected from the reduced overlap because the orientation of the orbitals along the z-axis is reduced.

We look next at double boron chains embedded in a transition metal matrix[6]: Here we ignore the transition metal except as a source or sink for electrons. We choose a linear unit cell containing four B atoms and consider only the π orbitals perpendicular to the plane of the chain. The four π MOs of our basis are constructed in C_{2v} symmetry and sketched in Figure 6.26a. The symmetry species of the π_1 and π_3 are the same, as are those of π_2 and π_4.

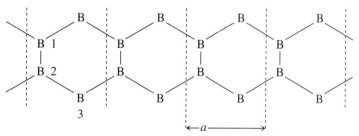

Applying the Bloch equation at $\vec{k} = 0$ gives the Ψ_0 sketched in Figure 6.25b. We have indicated the bonding or antibonding between the B atoms beneath each, from which we can arrange the energy of the $k = 0$ states as (lowest) $\Psi_0^{\pi_1} < \Psi_0^{\pi_2} < \Psi_0^{\pi_3} < \Psi_0^{\pi_4}$ (highest).

Next we apply the Bloch equation at $\vec{k} = 0.5a^*$, where the signs of Φ_n simply alternate in Figure 6.26c. Here π_1 and π_3 give identical solutions to Ψ_k, as do π_2 and π_4. Counting the nodes between adjacent B atoms as antibonding interactions, and similar signs between adjacent B atoms as bonding interactions, we can see that the energy of the $\vec{k} = 0.5a^*$ states should be $\varepsilon_{\pi_1} = \varepsilon_{\pi_3}$; $\varepsilon_{\pi_2} = \varepsilon_{\pi_4}$. However, repulsion of states of same

[6] Adapted from R. M. Minyaev and R. Hoffmann, *Chem. Mater.* **1991**, *3*, 547.

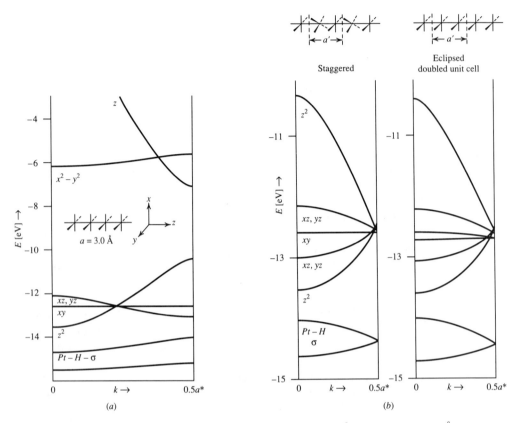

Figure 6.25 (a) Computed band structure of an eclipsed [PtH$_4$]$^{2-}$ stack spaced at 3 Å. The orbital marked *xz, yz* is doubly degenerate. (b) Computed band structure of a staggered [PtH$_4$]$^{2-}$ stack (*left*) compared with the folded-back structure of an eclipsed stack, two [PtH$_4$]$^{2-}$ in a unit cell (*right*). (From R. Hoffmann, *Solids and Surfaces*, VCH Publishers, New York, 1988.) Reprinted with Permission by VCH Publishers © 1988.

symmetry lifts the degeneracy. The resulting band structure is shown in Figure 6.27. Alternatively, physicists would say the symmetry equivalence of B atoms one and two causes a single state to run into the second Brillouin zone, with a discontinuity at the zone boundary. The second zone folded into the first zone is the π_3 band. This is discussed more fully by R. Hoffmann in the general references.

What does our bonding picture tell us? As electrons are added to the π_1 band, the B$_1$—B$_2$ bond distance is expected to shorten since the π_1 interaction is bonding for all values of $\vec{k}$. When this band is about half-filled, the π_2 band will start filling as well. Now the B$_2$—B$_3$ bond distance initially shortens since this interaction is bonding at the band center, but becomes nonbonding at high band filling. The B$_1$—B$_2$ interaction is antibonding throughout the π_2 band, canceling or reversing the effect of bonding from the π_1 band as soon as the π_2 band begins filling.

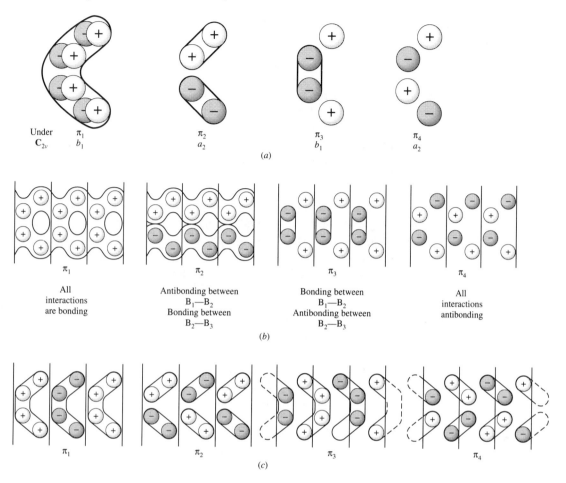

Figure 6.26 (a) Combinations of π orbitals in C_{2v} symmetry. (b) Combinations of π_1, π_2, π_3, and π_4 orbitals for $k = 0$. (c) Combinations for $k = 0.5a^*$.

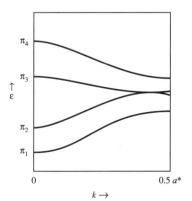

Figure 6.27 Band structure for π electrons of B_4 borides.

6.4 GRAPHITE AS A TWO-DIMENSIONAL SOLID

▶ 6.4.1 *Structure and Bonding in Graphite*

Graphite has the layered structure shown in Figure 6.28. The interlayer spacing of 335 pm is approximately the sum of the van der Waals radii of C. The expected weak interlayer bonding accounts for the softness of graphite, its lubricating properties, and its ease of cleavage. By contrast, the in-plane C—C bond distance is 142 pm, consistent with a bond order of $1\frac{1}{3}$ as predicted from simply counting major valence bond contributions to the resonance hybrid.

Carbon fibers prepared by graphitization of polyacrylonitrile have strong C—C bonds along the fiber axis and thus a high tensile strength. In an epoxy matrix these fibers form extremely strong, lightweight materials that are used for the fabrication of a variety of products ranging from airplane wings to tennis rackets.

Here we look at the band structure of graphite, treating a single layer as an example of a two-dimensional solid. Forming Bloch functions in two dimensions requires the assignment of two f_i^* values, f_a^* and f_b^*, for each Ψ_k. The f_i^* values allow us to identify the components of $\vec{k}$ in terms of $\vec{a}*$ and $\vec{b}*$ and to find the nodes of ψ_k in the direction in the crystal indicated by the k vector in the first Brillouin zone.

The planar unit cell of graphite contains two symmetry-equivalent carbon atoms, and for convenience we select the cell as shown in Figure 6.29. We limit our example to the $2p_z$ orbitals of the π framework that are the frontier orbitals of graphite. We can construct a pair of Φ orbitals as the sum and difference of the orbitals on each C atom. (These would be π and $\pi*$ orbitals for a diatomic species.)

$$\Phi^+ = p_z(C_1) + p_z(C_2)$$
$$\Phi^- = p_z(C_1) - p_z(C_2)$$

$$(6.16)$$

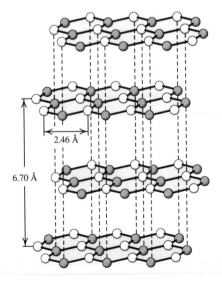

Figure 6.28   Graphite structure. (From R. Steudel, *Chemistry of the Non-Metals*, Walter de Gruyter, Berlin 1977)

2.46 Å

6.70 Å

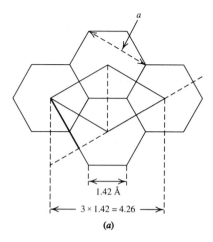

$$d_{10} = d_{01} = d_{11} = 1\tfrac{1}{2}d_{c-c} = 2.13 \text{ Å} = a(\cos 30°) = \frac{\sqrt{3}}{2}a$$

$$a = b = \sqrt{3}(1.42) = 2.46 \text{ Å}$$

$$a^* = \frac{1}{d_{10}} = \frac{2}{\sqrt{3}a} \qquad b^* = \frac{1}{d_{01}} = \frac{2}{\sqrt{3}b}$$

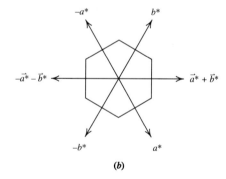

Figure 6.29 (*a*) Planar graphite cell. (*b*) First Brillouin zone for planar graphite. Vectors of first Brillouin zone at boundary are one-half of reciprocal vectors shown (see Figure 6.19 for special positions).

At the zone center, Γ, both f_a^* and f_b^* are zero, so all the coefficients of Φ are $+1$, giving $\Psi_0^{\phi^+} = \phi_0^+ + \phi_1^+ + \phi_2^+ + \phi_3^+ \cdots$. Figure 6.30$a$ shows that Ψ_Γ has only bonding interactions, so this is the lowest energy k state in the ϕ^+ (or simply π) band. With ϕ^- just the opposite holds: There are only antibonding interactions at Γ, and this is the highest energy state in the ϕ^- (or π^*) band.

We can get to the point marked "Q" on the Brillouin zone by letting $f_b^* = 0$ and $\vec{k}_a = 0.5a^* = 0.5/d_{10} = 1/\sqrt{3}a$. Nodes in the planar wave occur at intervals of d_{10} as we move normal to these planes. Applying this wave to Ψ_Γ for ϕ^+ and ϕ^- already obtained gives the pictures shown in Figure 6.30b. The $\Psi_Q^{\phi^+}$ state has twice as many bonding as antibonding interactions. The $\Psi_Q^{\phi^-}$ state is again the reverse with a net antibonding interaction. For other states along the Γ–Q line we choose values of $\vec{k}_a$ between 0 and $0.5\vec{a}^*$, and find intermediate degrees of bonding or antibonding, interactions. For these intermediate cases we need to use the actual coefficients of the ϕ's from the cosine functions to obtain a qualitatively accurate picture.

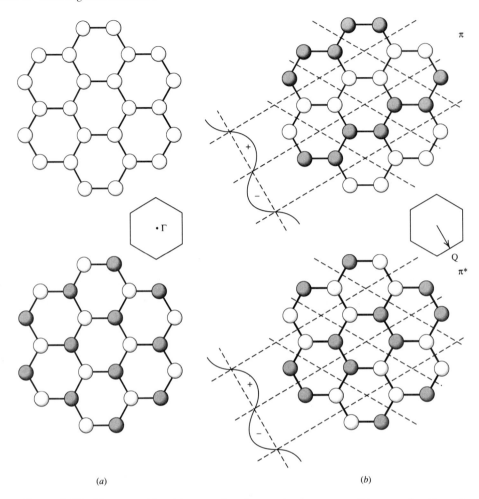

Figure 6.30   Planar graphite. (*a*) π wavefunctions at the zone center. (*b*) π wavefunctions at point Q in the Brillouin zone. (*c*) π wavefunctions at point P in the Brillouin zone (upper curve for ϕ^+, lower curve for ϕ^-). (*d*) Same as (*c*), but with different choice of lattice points and wave phasing. (*e*) Same as (*d*), but with different choice of lattice points ϕ^+ only).

With the single k vector required to describe the modulating wave at point Q, the MOs formed are exactly equivalent to a one-dimensional lattice, with the basis being an extremely long slice of the crystal between the planes designated by the reciprocal-lattice k vector. This is true regardless of the single k vector chosen. Equation (6.14) becomes simple to apply if f_b^* and f_c^* are zero, because the corresponding terms become a factor of one.

As mentioned earlier, the symmetry of the Brillouin zone is the same as that of the real lattice. Q states are found in the center of the hexagonal sides. The pairs of points that differ by a lattice translation must be equivalent, as well as those produced by the three-fold symmetry of the graphite structure.

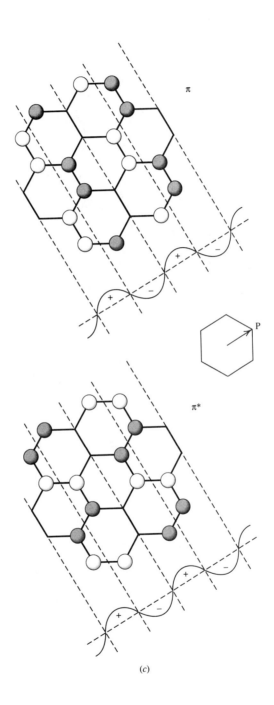

(c)

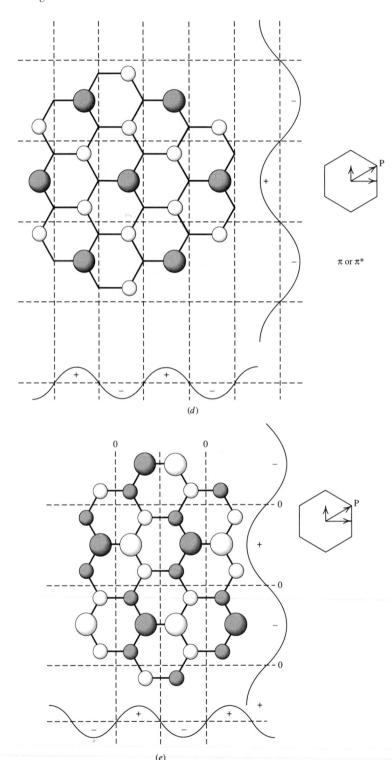

(d)

π or π*

(e)

To get to point P in the Brillouin zone we need coefficients for $\vec{a}$ and $\vec{b}$ such that their vector sum results in the vector $\overline{\Gamma P}$. We can approach this either geometrically or analytically (see Figure 6.31) and find

$$\vec{k} = \tfrac{2}{3}\vec{b}* + \tfrac{1}{3}\vec{a}* \qquad (6.17)$$

But $\tfrac{2}{3}\vec{b}*$ is outside of the first Brillouin zone so we would use $f_b^* = -\tfrac{1}{3}$ (instead of $\tfrac{2}{3}$) and $f_a^* = \tfrac{1}{3}$ in Equation (6.10) to obtain a symmetry equivalent solution. Rather than calculate point by point the phases for individual unit cells, we note that the vector at P is one-third the reciprocal-space distance to point 12*. This means that the modulating plane wave has a wavelength of $3d_{12}$ and is normal to these planes (that is, it is in the direction of the $\overline{\Gamma P}$ vector). We use this wavelength and direction to modulate the solutions found at Γ, the zone center, to produce Figure 6.29c–e which shows that Ψ_P is nonbonding for both the π and $\pi*$ bands. The different pictures result from different locations for the lattice points.

With a knowledge of the relative energies at Γ, Q, and P we sketch the band structure as shown in Figure 6.32.

▶ 6.4.2 Conduction Properties of Graphite

A full treatment of three dimensional (3-D) graphite does not greatly alter the band structure except for opening a small energy gap between the π and $\pi*$ bands and doubling the number of bands since the primitive unit cell now contains four carbon atoms. Of course, the Brillouin zone becomes a 3-D figure with a number of points of special symmetry. We continue with a 2-D solid, but the conclusions are valid for the 3-D case.

The Brillouin zone maps all Ψ_k of a band to a dense set of points within a reciprocal space primitive unit cell. We fill up the zone starting with the lowest energy state and finishing with the state appropriate for the last electron added. For planar graphite at 0 K we expect the Brillouin zone formed from the π_+ to be completely filled since each pair of carbon atoms contributes one orbital and two electrons to the band. As the temperature is raised, some electrons can be excited thermally to the $\pi*$ band. The π band loses its nonbonded and loosely held electrons near the points P in the Brillouin zone and gains these near the P points in the Brillouin zone for the $\pi*$ band.

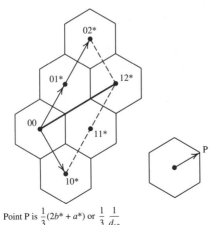

Point P is $\dfrac{1}{3}(2b* + a*)$ or $\dfrac{1}{3}\dfrac{1}{d_{12}}$

Figure 6.31 The relationship of point P to the lattice vectors in reciprocal space.

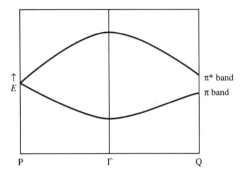

Figure 6.32   Band structure of the π bands of a graphite layer.

To visualize the Bloch functions we have formed standing waves by taking the sum and differences of $\Sigma_n (e^{i\vec{k}\cdot\vec{\tau}_n})\phi_n$ and $\Sigma_n (e^{-i\vec{k}\cdot\vec{\tau}_n})\phi_n$ terms since the Ψ_{+k} and Ψ_{-k} are degenerate. This degeneracy is lifted in an electric field, and the $+k$ and $-k$ states represent traveling waves. In a filled zone there is an equal population of the $+k$ and $-k$ states and no electric current can flow. In a partially filled zone an applied electric field causes an unequal population of the $+k$ and $-k$ states and an electric current flows. The boundary between filled and empty k states in a Brillouin zone is termed the **Fermi surface.** Physicists define a metal as a material with a Fermi surface. At 0 K a substance without a Fermi surface has an infinite resistance to the flow of electricity, while one with a Fermi surface approaches zero resistance at 0 K. The increase in conductivity of metals with a decrease in temperature is related to a decrease in the scattering of the electron waves by interaction with lattice vibrational modes (phonons).

▶ 6.4.3 *Intercalation Compounds*

Compounds with a layer structure offer the possibility of a unique type of reaction in which "guest" molecules can be inserted between the layers of the "host" compound. The reaction is called an **intercalation reaction,** the guest molecule the **intercalate,** and the product an **intercalation compound.** These compounds offer rich possibilities for the fabrication and study of novel materials with tailor-made properties. With graphite as the host, the intercalate behaves as a 2-D solid permitting the study of phenomena such as 2-D melting, 2-D magnetism, and 2-D superconductivity.

Over 200 graphite intercalation compounds are known. Most of these can be formed by direct reaction between graphite and a variety of oxidizing or reducing agents. The alkali metals donate electrons to the empty π^* band, giving rise to metallic conduction. Graphite in contact with molten K, Rb, or Cs under argon, or with the vapor of these metals, swells along the c axis and changes color. Removing excess alkali metal by evaporating under vacuum gives a pyrophoric bronze-colored material with composition MC_8. Further removal of alkali metal produces compounds richer in carbon of the general formula $M(C_{12})_n$ (where $n = 2, 3, 4, 5$), and finally only graphite remains. The carbon layers in contact with the intercalate are shifted to give an eclipsed arrangement, and the other layers retain the staggered arrangement of pristine graphite. The number of carbon layers between the intercalate layers is termed the **stage** of the compound. A cross-sectional view for several stages is shown schematically in Figure 6.33.

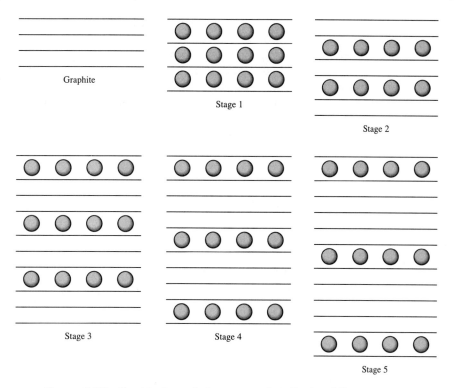

Figure 6.33   Graphite intercalation compounds indicating different staging.

The particular stage of compound which is formed is most readily determined from the prominent 00*l* lines observed in their x-ray patterns. Chlorine promotes the formation of many metal halide graphite intercalation compounds, with the removal of electrons from the top of the π band (Figure 6.31). The intercalated species are chlorometallates, often with chlorine bridging which increases the stoichiometric ratio of neutral metal halide to metallate ion, obscuring the role of the oxidizing agent. With halides such as $SbF\ell_5$, no additional oxidizing agent is needed. The proposed reactions are

$$3SbX_5 + 2e \longrightarrow 2[SbX_6]^- + SbX_3 \tag{6.18}$$

$$[SbF_6]^- + nSbF_5 \longrightarrow [SbF_6 \cdot nSbF_5]^- \tag{6.19}$$

Such complexation reactions greatly reduce the reactivity of the graphite salt compared to that of the free reagent, and use has been made of this in preparative chemistry.[7] The conductivity of some of these compounds approaches that of copper on a weight basis.

The dichalcogenides, MX_2, of the transition-series groups 4, 5, and 6, have layer structures (page 220). Since the bonding is weak between the layers of X^{2-} where the cation sites are vacant, other species can fit between these layers to form intercalation

[7] R. Laali, M. Muller, and J. Sommer, *J. Chem. Soc. Chem. Commun.* **1980**, 1088.

compounds.[8] Layer structures involving octahedral coordination of M (TiS$_2$) and trigonal prismatic coordination of M (Nb, Ta, Mo, and W disulfides) are important for intercalation. Intercalation compounds are formed with NH$_3$, pyridine, various amines and amides, alkali metal hydroxides, and many metals. The metal sulfides capable of forming intercalation compounds reversibly with Na or Li are of interest as solid-state cathodes capable of replacing S in Na–S or Li–S batteries. The reversible reactions are

$$x\,Na^+ + xe + TaS_2 \longrightarrow Na_xTaS_2$$
$$x\,Li^+ + xe + TiS_2 \longrightarrow Li_xTiS_2$$

$$(6.20)$$

6.5 METALS, INSULATORS, AND SEMICONDUCTORS

Although metals generally possess such properties as malleability, ductility, and high tensile strength, the definitive characteristic is their ability to conduct electricity that *increases as the temperature decreases.* In contrast, insulators and semiconductors show an electrical resistivity that approaches infinity as the temperature approaches 0 K. This temperature dependence of conductivity finds ready explanation in the band model of solids. A metal has a partially filled band (Figure 6.34) with immediately accessible higher energy states allowing electrons to be accelerated under the influence of an electric field. Conductivity falls as the temperature increases above 0 K because of electron scattering caused by interaction with lattice vibrations (particularly in the vicinity of impurity atoms and lattice defects).

An insulator has a completely filled band and an appreciable gap (for diamond the next unfilled band is 5.2 eV higher), whereas a semiconductor has a filled band with a small gap to the unfilled band (0.6 eV for Ge) (see Figure 6.34). The filled band is generally referred to as the **valence band,** and the higher unfilled band is called the **conduction band.** At 0 K the conduction band is empty, but as the temperature is increased the conduction band becomes populated following a Boltzmann distribution dependent on the band gap. Charge carriers can be increased by adding a **dopant,** such as a Group 13 or Group 15 element added to a Group 14 semiconductor. Figure 6.25 indicates the effect of these dopants. If doping increases the conduction band electron population, an ***n*-type** semiconductor is produced ("*n*" for negative charge carrier). If doping removes electrons from the valence band, the semiconductor is said to be "hole doped" and is a ***p*-type** semiconductor ("*p*" for positive charge carrier). Density of states distributions for these materials are given in Figure 6.34.

A sample of Ge doped with As adds electrons to the conduction band of Ge and is an *n*-type semiconductor since there are negative carriers. Adding Ga to Ge gives a *p*-type

[8] F. R. Gamble and T. H. Geballe, "Inclusion Compounds," Chapter 2, in *Treatise on Solid State Chemistry,* Vol. 3, N. B. Hannay, Ed., Plenum, New York, 1976. G. C. Farrington and J. L. Briant, *Science* **1979,** *204,* 1371.

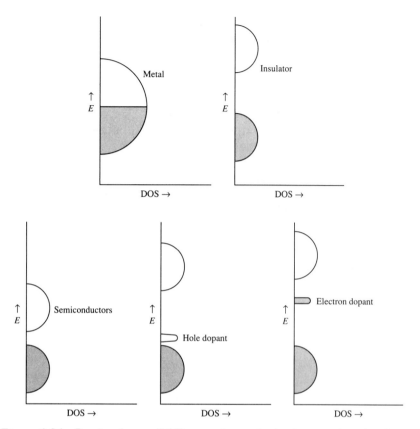

Figure 6.34 Density of states (DOS) curves for metals, insulators, and semiconductors.

semiconductor; the Ga goes into the Ge lattice (diamond structure) but lacks one electron, thus creating an electron hole. Movement of electrons into this hole is the equivalent of movement of the hole about the lattice. When placed in an electric field, such a hole behaves as if it were a positive charge. $Fe_{1-\delta}S$ and $Fe_{1-\delta}O$ are p-type semiconductors, since electron transfer from Fe^{2+} to Fe^{3+} makes the Fe^{3+} appear to move through the lattice. $Na_{1+\delta}Cl$ and $Zn_{1+\delta}O$, on the other hand, are n-type semiconductors, because of the mobility of the "trapped electrons".

Occasionally, impurities are added to reduce rather than promote semiconduction—as in the case of TiO_2 ceramic insulators, where pentavalent impurities are added to reduce the effect of Ti^{3+} ions.

In addition to applications to semiconductors (transistors, etc.), lattice defect compounds are important in catalysis, luminescence (especially color TV), and photography, and they are being studied with regard to radiation damage in materials and corrosion phenomena.

6.6   SOME LIMITATIONS AND EXTENSIONS OF THE SIMPLE BAND MODEL

The band model of metallic behavior just presented does not take into account the **dispersion** of the partially filled band. As the orbital overlap decreases, the band narrows and electrons become bound to individual atoms. The electrons are then termed **localized** electrons in contrast to the **itinerant** or **free** electrons. A transition from a metallic state to an insulating state can occur as a result of changing the distance between atoms or changing the potential "seen" by an electron (by changing the number of electrons in the band through compositional changes). At the point where a metal–insulator transition takes place, the dielectric constant is predicted to be infinity, an event termed the **dielectric catastrophe.**[9] Such a gigantic swing in the dielectric constant has been seen in phosphorus

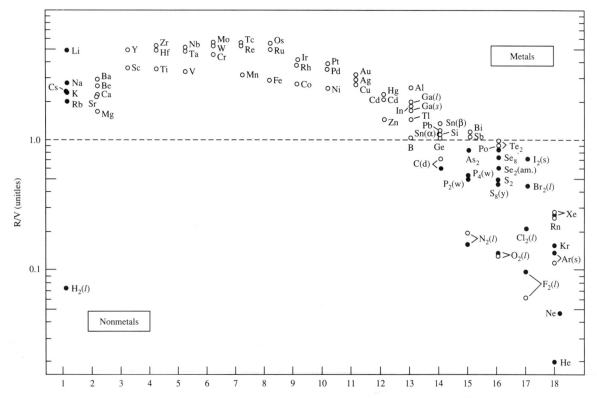

Figure 6.35 The metallization of elements under the ambient conditions imposed on the planet earth. The figure shows the ratio (R/V) for elements of the s, p, and d blocks of the periodic table. Here R is the molar refractivity and V is the molar volume. The closed circles represent elements for which both R and V are known experimentally. The open circles are for elements for which only V is known experimentally and R is calculated. (From P. P. Edwards and M. J. Sienko, Chemistry in Britain, 1983, *19*, 39.)

[9] The dielectric catastrophe occurs when the ratio of the molar polarization to the molar volume for a substance equals 1. Since $R/V = (\varepsilon - 1)/(\varepsilon + 2)$, where R is the molar refractivity, V is the molar volume, and ε is the dielectric constant, this requires $\varepsilon = \infty$ as the only solution.

doped silicon at very low temperatures as the phosphorus concentration is increased. (A search for a dielectric catastrophe in hydrogen at pressures up to 2 megabars has produced no discontinuities, although a number of interesting phase changes were observed.)

It has been proposed that the R/V ratio can be used for the dividing line between metals and insulators. The success of this criterion, under conditions prevailing on the planet earth, is shown in Fig. 6.35.

6.7   SUPERCONDUCTIVITY

▶ 6.7.1   *Discovery*

Superconductivity has intrigued scientists since its discovery by H. Kammerlingh Onnes in 1911. Onnes was the first to liquify helium, and the first to measure the properties of substances at such low temperatures. The electrical resistance of mercury was found to vanish abruptly at 4.2 K. Over the course of the next seventy-five years a number of other metals and alloys were added to the list of superconductors, but the field remained the province of cryogenic researchers. By 1975, superconductivity had been observed at 23 K in Nb_3Ge. In 1986 Bednorz and Mueller discovered superconductivity at 35 K in $La_{1.85}Ba_{0.15}CuO_4$ followed shortly thereafter by the discovery by Chu and Wu of the 93 K superconductor $YBa_2Cu_3O_{7-\delta}$ (termed the 123 compound). These discoveries produced an immediate worldwide flood of research on "high-temperature" superconductors. The temperature, above that of liquid nitrogen, was accessible in almost any laboratory; the superconducting temperature was higher than that which then current theory could explain, the cuprates offered an unexplored field for seeking new superconductors, and the possibility of room temperature superconductivity appeared to exist with a number of technologically important applications.

▶ 6.7.2   *Properties*

In addition to the vanishing electrical resistance below a temperature termed the **critical temperature** (T_C), a number of other phenomena are associated with superconductivity. Magnetic fields below a critical field strength (H_{c1}) are excluded. At field strengths above H_{c1} magnetic flux will penetrate the bulk of the superconductor and at an upper critical field strength (H_{c2}) destroy the superconductivity. The exclusion of a magnetic field by a superconductor is called the Meissner effect and gives rise to the levitation of a superconductor in a magnetic field. When $H_{c1} = H_{c2}$, the superconductor is termed type I. Superconducting elements, other than Nb, are type I superconductors. For a type II superconductor ($H_{c2} > H_{c1}$) magnetic flux will penetrate the bulk sample producing a mixed state with magnetic cores arranged in a regular array in the bulk superconductor. This flux lattice may be seen by a technique known as magnetic decoration. (In this procedure, extremely small magnetic particles suspended in an inert atmosphere are allowed to collect on a single crystal of the superconductor held below T_C in a very small magnetic field. The particles collect on the superconductor around the magnetic field which has penetrated, and may be observed with an electron microscope.) The heat capacity displays a discontinuity at T_C, or more specifically, that part associated with the electron specific heat shows an anomaly. Anomalous effects are also observed in the Debye–Waller factor (a measure of the vibrational stiffness) in the vicinity of T_C.

▶ 6.7.3 *Models*

To explain the heat capacity anomaly of superconductors, Gorter and Casimir proposed (1934) that the conduction electrons were in either a "normal state" or a "superconducting state." Above T_C all conduction electrons are in the normal state, while at 0 K all conduction electrons are proposed to be in the superconducting state. From 0 K to T_C the fraction of electrons in the normal state n_n/n given by this model is

$$n_n/n = (T/T_C)^4 \tag{6.21}$$

This model is called the **two-fluid model,** analogous to the model for superfluidity in ^{4}He where part of the liquid is assumed to be in a single quantum state and part in a normal state.

The Meissner effect required a further extension of the two-fluid model. F. & H. London (1935) proposed that in the presence of a magnetic field, electrons in the superconducting state set up a screening current that completely excludes an external magnet field an internal distance from the surface now called the **London penetration depth, λ.** This penetration depth is inversely proportional to the number of superconducting electrons, leading to the equation

$$\lambda(T) = \frac{\lambda(0)}{\sqrt{1 - (T/T_c)^4}} \tag{6.22}$$

where $\lambda(T)$ is the London penetration depth at temperature T and $\lambda(0)$ is the London penetration depth at 0 K. When $T = T_C$ the penetration depth becomes infinite and the magnetic field completely penetrates the solid destroying the superconductivity.

A coherence length (ξ) for the wavefunction of the superconducting charge carriers was proposed by Ginzburg and Landau. The ratio of the penetration depth, λ, to the coherence length is related to the upper and lower critical magnetic fields by

$$\kappa = \frac{\lambda}{\xi} = \frac{1}{\sqrt{2}}\left(\frac{H_{C_2}}{H_{C_1}}\right)^{1/2} \tag{6.23}$$

Type I superconductors result when the coherence length is much greater than the penetration depth, the dividing line occuring when $\kappa = \dfrac{1}{\sqrt{2}}$.

The above models are essentially phenomenological. A model providing a key to the understanding of superconductivity was developed in 1957 by John Bardeen, Leon N. Cooper, and J. Robert Schrieffer, and is called the **BCS theory.** In the BCS theory the supercurrent is carried by paired electrons called **Cooper pairs.** As we have seen, in metals above 0 K some electrons occupy energy states above the Fermi level and leave holes in the "Fermi sea" of occupied states. Designating the excited electron by a wave vector $\bar{k}$ and the hole state by $\bar{k'}$, Cooper pair states arise from combinations of $\{\phi_k - \phi_{-k}\}$ and $\{\phi_{k'} - \phi_{-k'}\}$, which are coupled though interactions with phonons (lattice vibrations of wave vector $\{\bar{k} - \bar{k'}\}$. The Cooper pair states are bosons (spin 0 and 1) instead of fermions (spin $\frac{1}{2}$) and are no longer subject to the Pauli exclusion principle. Thus the collection of Cooper pairs shows macroscopic occupation of the same quantum state, with the same common momentum. When a phonon interacts with a member of the Cooper pair to change its momentum, an equal and opposite change occurs in the other member of the pair with the reemission of a phonon to the lattice. The interaction of one electron of a Cooper pair with the lattice vibrations leaves a message for the other electron in the pair and allows interaction over a longer distance (the coherence length) and longer time (the vibrational relaxation time) than would be possible for free electrons. The Cooper pair may be considered to be a wave packet with a diameter corresponding to the coherence length ξ. Overlapping wavepackets phase lock the different Cooper pairs. Such overlap would not be possible for single electrons because of the Pauli exclusion principle, but is possible for the Cooper pairs. Because of the phase coherence, we may say that the relationship of superconduction to normal conduction is similar of that of a LASER beam to normal light.

▶ 6.7.4   *Structure and Superconductivity*

The structures of the new high-temperature superconductors have in common CuO_2 planes separated by layers of other metal oxides or metal ions. $La_{2-x}Ba_xCuO_4$ may be described by the layer sequence CuO_2-LaO-LaO-CuO_2-LaO-LaO-CuO (Figure 6.36 left); the 123 compound by CuO-BaO-CuO_2-Y-CuO_2-BaO-CuO (Figure 6.36 middle) and $Nd_{2-x}Ce_xCuO_4$ by CuO_2-(Nd-O_2-Nd)-CuO_2-(Nd-O_2-Nd)-CuO_2. In La_2CuO_4 and $YBa_2Cu_3O_{6.5}$ the CuO_2 layers might be expected to have an exactly half filled σ^* band constructed from the Cu $d_{x^2-y^2}$ orbitals and the appropriate O p_x and p_y orbitals. This half-filled band splits to give a completely empty σ^* band called the upper Hubbard band and a completely filled σ^* band called the lower Hubbard band[10] which falls slightly below a filled π band formed from the Cu d_{xz} and d_{yz} with the O p_z. These compounds may be considered to result from the intergrowth of CuO and other metal oxides with an imperfect matching of lattice constants, which creates internal stresses (see Goodenough under genral references). With La_2CuO_4 the CuO_2 layers are under compression while the LaO layers are under tension. Removing electrons from the top of the π^* band of the CuO_2 plane shortens the CuO in-plane bond distances and relieves the strain. This can be done by replacing some of the La with Ba or Sr which provides one electron fewer per atom than La. The substitution of Ba or Sr for La is termed "hole doping." In the 123 compound, the lower Hubbard band is half-filled at an oxygen stoichiometry of 6.5, which gives the middle structure of Figure 6.36 but with oxygen missing from alternate chains. Doping in the 123 compound is achieved by increasing the oxygen content, which fills in the chain sites but

---

[10] The treatment of the Hubbard bands is beyond the scope of this text. See N. F. Mott, "Metal Insulator Transitions" in *Encyclopedia of Physics*, 2nd ed., R. G. Lerner and G. L. Trigg, eds., VCH Publishers, New York, 1990. This encyclopeida also covers many other topics related to superconductivity.

requires the removal of electrons from the CuO_2 planes. Since the electrons are removed from the top of a previously filled band, the current carriers in the superconducting state appear to be positive Cooper pair states.

When the smaller Nd atom is completely substituted for La to give Nd_2CuO_4, the structure changes to that shown at the right of Figure 6.36. The shift from a rock salt lattice for the LaO layers to a fluorite lattice for the $Nd-O_2-Nd$ layers changes the lattice mismatch with the CuO_2 layers to put the CuO_2 layers under a tensile stress. Doping with Ce, which randomly replaces Nd, relieves the strain by adding electrons to the CuO_2 $\sigma *$ upper Hubbard band, which lengthens the CuO bonds. The electrons at the bottom of the $\sigma *$ upper Hubbard band give negative charge carriers.

It is not possible to dope Nd_2CuO_4 to make a hole superconductor, nor is it possible to dope La_2CuO_4 to make a superconductor with negative charge carriers. Strain is further relieved in the 123 compound by puckering of the CuO_2 layers and in the La compound by tilting the CuO_6 octahedra so the OCuO bond angle is less than 180°. Most of the high T_C superconductors which have been made to date are hole conductors. Some of these are listed in Table 6.2.

▶ 6.7.5 *Processing*

The direct sintering in air or O_2 of precursor metal oxides (or compounds thermally decomposable to the oxides) followed by an "anneal" in O_2 or air is the most commonly used method for preparing the superconducting cuprates.

$$1/2Y_2O_3 + 2BaCO_3 + 3CuO \xrightarrow{950 - 960°C} YBa_2Cu_3O_{6.5} + 2CO_2 \qquad (6.24)$$

$$YBa_2Cu_3O_{6.5} + 0.2O_2 \xrightarrow{400 - 500°C} YBa_2Cu_3O_{6.9} \qquad (6.25)$$

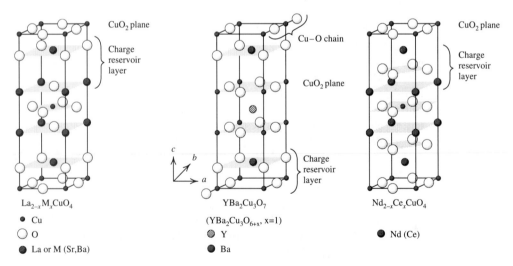

$La_{2-x}M_xCuO_4$	$YBa_2Cu_3O_7$	$Nd_{2-x}Ce_xCuO_4$
● Cu	$(YBa_2Cu_3O_{6+x}, x=1)$	
○ O	◍ Y	● Nd (Ce)
● La or M (Sr,Ba)	● Ba	

Figure 6.36 Structures of three cuprate superconductors. (Reprinted with permission from F. J. Adrian and D. O. Cowan, *Chem. Eng. News*, Dec. 21 1992, p. 26. Copyright 1992 American Chemical Society.)

Table 6.2   Critical Temperatures of Some High-Temperature Superconductors.

Compound[a]	T_C (K)	Compound[a]	T_C (K)
$YBa_2Cu_3O_7$	93	$HgBa_2CuO_4$	95
$YBa_2Cu_4O_8$	80	$HgBa_2Ca_2Cu_3O_8$	135[b]
$Bi_2Sr_2CuO_6$	5 to 20	$TlBa_2CaCu_2O_7$	80
$Bi_2Sr_2CaCu_2O_8$	85	$TlBa_2CaCu_2O_7$	110
$Bi_2Sr_2Ca_2Cu_3O_{10}$	110	$TlBa_2Ca_3Cu_4O_{11}$	125

[a] Nonstoichiometric compounds—oxygen content variable.

[b] 160 K obtained at 280 kbar (*Chemical & Engineering News*, Sept. 27, **1993**, p24).

Since diffusion in solids is a slow process, step (6.24) is usually repeated several times after the sample has been cooled and finely ground. Following the final 960°C heat treatment, the sample is cooled to 500°C in flowing O_2, and after several hours slowly cooled to room temperature.

Some alternative methods to the direct synthesis from solid precursors involve mixing the reactants at a molecular level. In aerosol processing, an aqueous solution of metal nitrates is nebulized, dried by passing over silica gel, and the resulting fine particles are carried in an air stream through a heated reactor tube. During the one to two minutes in the reactor tube the resulting superconductor is formed as a very fine powder that may be caught on a filter. In sol-gel processing, a solution of the precursors may be treated with a complexing agent such as ethylenediaminetetraacetic acid and the resultant solution slowly evaporated to form a gel. This gel is ashed and finally given a high temperature treatment under air or O_2.

Pellets may be pressed from the powders produced by any of the above methods. These are heat-treated and reannealed in oxygen. Since the metal cuprates are not malleable like metal, but brittle like ceramics, forming wire requires special techniques. One approach is to extrude the powder in an epoxy resin, and then thermally decompose the organic material followed by a 960° treatment and O_2 anneal. Another approach to forming wire is to fill silver tubes with 123 powder, extrude and then anneal in an O_2 atmosphere. The Ag is permeable to O_2, and a superconducting strand is produced in a silver sheath.

Fabrication of thin films has been accomplished by a variety of techniques, most of which involve deposition on a substrate having lattice constants closely matched with that of the superconductor (epitaxial deposition). A popular substrate is single crystal $SrTiO_3$ cut along a 100 face. In metal organic chemical vapor deposition (MOCVD), volatile metal complexes such as acetonylacetonates are mixed in the vapor phase and allowed to impinge on a hot substrate where subsequent reaction produces the superconductor.

In molecular beam epitaxy (MBE), beams of neutral atoms from individually heated sources are allowed to sequentially impinge on a target substrate building up the desired structure layer by layer. Beam shutters and computer control permit the individual atomic layers (Bi, Sr, Cu, etc.) to be built up in any desired sequence. An oxidizing atmosphere is provided by a controlled leak (O_3, etc.), with the target temperature closely controlled. This method allows specific architecture of cuprate and metal layers, which may not be achieved from bulk synthesis methods.

▶ 6.7.6   *Applications*

Low-temperature superconductors are already in use in magnets for high field nuclear magnetic resonance spectrometers, magnetic resonance imaging for medical diagnostics, particle accelerators, prototypes of magnetically levitated trains, and a variety of superconducting electronic devices. The high-temperature superconductors are already being used in prototypes for devices, but it will be some time before the high-temperature superconductors will be competitive in high magnetic field applications because of the difficulty of producing large amounts of high quality wire.

A number of devices are based on the Josephson effects, those effects produced by the quantum tunnelling of Cooper pairs when a pair of superconductors is separated by an insulator, or by a normal metal, by a distance that is less than the coherence length of the superconductor. Such an arrangement is called a **Josephson junction.** These Josephson junctions can be used in digital computers as extremely fast switching devices, in detectors for high frequency electromagnetic radiation, and in SQUIDs (superconducting quantum interference devices). A SQUID consists of two Josephson junctions in a loop, and can be used to detect low levels of magnetic field, current or voltage. SQUIDs have recently been used to measure brain waves nonintrusively.

Although the market for high-temperature superconductors is small at present, within the next two decades it is estimated that it will be an annual hundred billion dollar industry.

6.8   BUCKMINSTERFULLERENE

▶ 6.8.1   *Preparation and Properties*

The remarkable buckminsterfullerene was discovered in 1985 using pulsed laser vaporization of graphite, with pulsed He to cool the plasma, and was detected by time-of-flight mass spectrometry.[11] Up to molar mass of approximately 25 carbon atoms, both even and odd numbers of carbon atoms were observed in the clusters, but above C_{40}, only even numbered carbon clusters were observed. C_{60}^+ gave the most intense mass peak observed, and conditions could be adjusted so that it was virtually the only peak. Although the soccer ball structure was proposed in 1985 (Figure 3.16, page 126), it was 5 years before a solid was isolated and its structure could be confirmed.[12] The current method of preparation involves vaporizing carbon by either resistive heating or by an arc discharge between carbon electrodes under helium at 100 torr, and extracting the C_{60}/C_{70} fraction in benzene followed by purification by liquid chromatography. The C_{60} fraction is a magenta color, while C_{70} is orange. By this "bulk" method one can prepare as much as a gram per day.

The name **buckminsterfullerene** was given to C_{60} and the name fullerenes to the general series of carbon cage clusters in honor of Buckminster Fuller whose geodesic dome

[11] H. W. Kroto, J. R. Heath, S. C. O'Brien, R. F. Curl, R. E. Smalley, *Nature* **1985,** *318,* 162.

[12] For personal accounts of the discovery and macroscopic preparation of the fullerenes see: H. W. Kroto, *Angew. Chem. Int. Ed. Engl.* **1992,** *31,* 111; R. F. Curl and R. E. Smalley, *Sci. Am.* **1991,** *Oct.,* 54; D. R. Huffman, *Phys. Today* **1992,** *Nov.,* 22.

structures incorporated patterns of hexagons with sufficient pentagons to give closure. The nickname of "**bucky ball**" for C_{60} was perhaps inevitable. The family of closed carbon polyhedra containing pentagonal faces is termed the **fullerenes**. Some properties of solid C_{60} are given in Table 6.3.

► 6.8.2   *Electronic Structure of C_{60}*

The molecular structure of buckminsterfullerene may be described by a carbon atom at a general point in the **I** group. This carbon atom moves over the surface of a sphere to a unique position for every rotational operation of the **I** group ($h = 60$). In the full $\mathbf{I}_h$ group, four carbon atoms are in each mirror plane, but none of the other improper symmetry elements of $\mathbf{I}_h$ coincide with any carbon atoms. Accordingly we can list the unshifted radial p_z orbitals in C_{60} under the $\mathbf{I}_h$ group operations as

$\mathbf{I}_h$	E	$12C_5$	$12C_5^2$	$20C_3$	$15C_2$	i	$12S_{10}$	$12S_{10}^3$	$20S_6$	15σ
Γp_z	60	0	0	0	0	0	0	0	0	4

Since no operation causes $+p_z$ to go into $-p_z$ in C_{60}, the above is the reducible representation for the π MOs of C_{60}. The energies of these 60π orbitals are given in Figure 6.28a. The ordering of the energy of these states is essentially the order expected for spherical harmonic functions: Under $\mathbf{I}_h$, an s orbital transforms as a_g, the p orbitals as t_{1u}, and so on. Some splitting occurs at the higher levels because of repulsion between orbitals of identical symmetry.

Solid C_{60} has a cubic closest-packed [face-centered cubic (*fcc*)] structure. The large interstitial sites make it extremely difficult to avoid trapping solvent molecules in the solid, but pure material can be obtained by low-pressure sublimation. The pure solid is an insu-

Table 6.3   Physical constants for C_{60}[b]

Quantity	Value	Quantity	Value
Average C—C distance	144 pm	Band gap (HOMO–LUMO)	1.7 eV
fcc Lattice constant	1417 pm	Binding energy per atom	7.40 eV
C_{60} mean ball diameter	683 pm	Debye temperature	185 K
C_{60} ball outer diameter[a]	1018 pm	Thermal conductivity (300 K)	0.4 W/mK
C_{60} ball inner diameter[a]	348 pm	Phonon mean free path	5000 pm
Tetrahedral site radius	112 pm	Static dielectric constant	4.0–4.5
Octahedral site radius	207 pm	Velocity of sound v_t	2.1×10^5 cm/sec
Electron affinity (pristine C_{60})	2.65 eV	Structural phase transitions	255 K, 165 K
Ionization energy (first)	7.58 eV	Mass density	1.72 g/cm³
Ionization energy (second)	11.5 eV		

[a] The inner and outer diameters of the C_{60} molecule are estimated from (683 − 335 pm) and (683 + 335 pm), respectively, where 335 pm is the interlayer spacing in graphite.

[b] From the compilation of M. S. Dresselhaus, G. Dresselhaus, and P. C. Eklund, *J. Mater Res.* **1993**, *8*, 2054. Original references given in source.

lator as we might expect from the band structure shown in Figure 6.37. The band gap is 1.7 eV at X, where X refers to position of the wave vector in the Brillouin zone shown in Figure 6.19a. On exposure to alkali metal vapor, either M_3C_{60}, M_4C_{60}, or M_6C_{60} is formed. No phases having intermediate stoichiometry are found. The M_3C_{60} phase has alkali metal ions occupying all of the octahedral and tetrahedral sites in the *fcc* structure. The alkali metal valence electron is given up to C_{60} and half fills the t_{1u} band, giving rise to metallic conductance at room temperature. The M_6C_{60} goes over to a body-centered cubic lattice with C_{60} spheres centered on the lattice points. The alkali metal now occupies the slightly distorted tetrahedral sites at the vertices of the Wigner–Seitz unit cell. Since the t_{1u} band is entirely filled, this phase is an insulator. M_4C_{60} has a tetragonal unit cell and, perhaps surprisingly, is an insulator. The insulating state can be explained in terms of Jahn–Teller distortion (see Section 9.8.2). Jahn–Teller distortion would lower the symmetry of the C_{60} and split the energy level of the t_{1u} orbitals with a four-electron population.

An even more exciting aspect is that the M_3C_{60} phase becomes superconducting on lowering the temperature. The transition temperature (T_C) increases with the size of the alkali metal. Substituting ^{13}C for ^{12}C decreases the T_C. This is in agreement with predictions from the BCS theory.

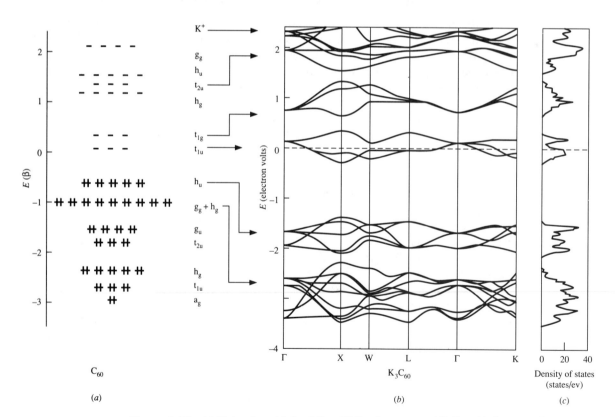

Figure 6.37   (a) Molecular orbitals of C_{60}. (b) Band structure of K_3C_{60}. (c) Corresponding density-of-states curves. (From A. F. Hebard, *Physics Today*, Nov. 1992, 29.)

► 6.8.3 Endohedral Complexes

During the gas-phase synthesis of fullerenes it is possible to trap one or more metal atoms in the carbon cage. These are termed **endohedrally doped fullerenes** and are denoted as $La@C_{60}$, where @ indicates encapsulation. Electron spin resonance spectra show that $La@C_{60}$ contains La^{3+}. It is possible to encapsulate more than one atom in the larger fullerenes to obtain compounds such as $Y_2@C_{82}$. These have not been isolated as solid compounds, so their electrical properties have not yet been investigated.

► 6.8.4 Higher Fullerenes [13]

By varying the laser pulse timing and residence time in the reactor before mass spectrometer analysis, even-numbered carbon clusters containing up to 600 carbon atoms have been observed.

On the more structural side, it is possible to isolate from carbon arc soot a C_{70} fullerene that has a rugby ball shape. Other interesting structures include tubes capped with a hemisphere, chiral tubes, and cages within cages. A chiral fullerene, C_{76}, was resolved using OsO_4 and a chiral alkaloid. [14]

GENERAL REFERENCES

The following books are listed in increasing order of difficulty:

E. A. Wood, *Crystals and Light,* Van Nostrand, Princeton, NJ, 1964 (available from Dover). A classic introduction to optical crystallography.

M. F. C. Ladd, *Symmetry in Molecules and Crystals,* Ellis Horwood Ltd., Chichester, 1989. Stereographic projections, space groups, intro to x-ray structural methods.

G. Burns and A. M. Glazer, *Space Groups for Solid State Scientists,* Academic Press, Orlando, FL, 1990.

R. Hoffmann, *Solids and Surfaces—A Chemist's View of Bonding in Extended Structures,* VCH Publishers, New York, 1988. The Bloch equation and more with numerous examples and figures.

G. Burns, *High-Temperature Superconductivity, An Introduction,* Academic Press, San Diego, 1992.

S. L. Altmann, *Band Theory of Metals—The Elements,* Pergamon Press, Oxford 1970. An excellent tutorial.

C. Kittel, *Introduction to Solid State Physics,* 6th ed., Wiley, New York, 1986. Written at the senior undergraduate physics level; the first edition, published in 1953, defined the area.

G. Burns, *Solid State Physics,* Academic Press, Orlando, FL, 1990. About the same level as Kittel.

---

[13] R. E. Smalley, *Acc. Chem. Res.* **1992,** *25,* 98.

[14] J. M. Hawkins and A. Meyer, *Science* **1993,** *260,* 1918.

The following references are not in sequence of ease of reading:

F. J. Adrian and D. O. Cowan, The New Superconductors, Chem.Eng. News, Dec. 21, **1992**, p24.

W. E. Billups and M. A. Ciufolini, *Buckminsterfullerenes,* VCH Publishers, New York, 1993.

J. A. Duffy, *Bonding, Energy Levels and Bonds in Inorganic Solids,* Longman, Essex, England, 1990.

P. A. Cox, *The Electronic Structure and Chemistry of Solids,* Oxford, Oxford, England, 1987.

J. B. Goodenough, Ch. 16 in *Electron Transfer in Biology and the Solid State,* Ed. M. K. Johnson, *et al.*, American Chemical Society, Washington DC, 1990; MRS Bulletin **1990,** *15,* 23.

S. F. A. Kettle and L. J. Norrby, The Brillouin Zone—An Interface between Spectroscopy and Crystallography, *J. Chem. Educ.* **1990,** *67,* 1022.

H. W. Kroto, C_{60}: Buckminsterfullerene, The Celestial Sphere that Fell to Earth, *Angew. Chem. Int. Ed. Engl.* **1992,** *31,* 111.

M. S. Dresselhaus and G. Dresselhaus, Intercalation Compounds of Graphite, *Adv. Phys.* **1981,** *30,* 139.

M. S. Dresselhaus, G. Dresselhaus, and P. C. Eklund, Fullerenes, *J. Mater. Res.* **1993,** *8,* 2054.

PROBLEMS

6.1 Place a dot anywhere within the set of stick figures below in (**a**) and (**b**) and then form a 2-D lattice by putting dots at all of the sites in an identical environment.

(a)   (b)

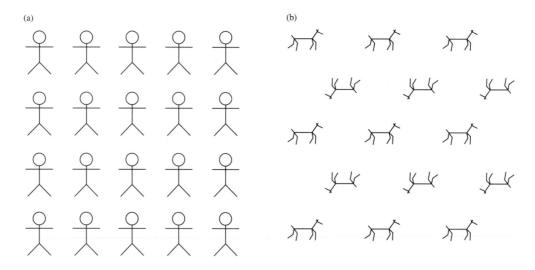

6.2 (**a**) Draw a planar set of lattice points and mark in at least four primitive unit cells.

  (**b**) Demonstrate that these all have the same area.

  (**c**) Mark in a multiply primitive cell (one containing more than one lattice point.) What is its area compared to the primitive cell?

6.3 Give the number of lattice points belonging to a unit cell in each of the following:

  (**a**) Primitive hexagonal

  (**b**) End-centered tetragonal

  (**c**) *I*432

6.4 Show that, for a cubic lattice, the vector $\overrightarrow{1}\,\overrightarrow{1}\,\overrightarrow{1}$ is a perpendicular to the 111 crystal face.

6.5 Consider a tetragonal crystal with $c = 2a = 2b$. Calculate the angle between the 111 vector and the $11\bar{1}$ face.

6.6 A crystal structure belongs to the space group $P\,4_2/n$.
(a) What is the point group of the space group in Schöenflies notation?
(b) To what crystal system does the crystal belong?
(c) How many lattice points may be said to belong to the unit cell?

6.7 With the aid of a stereogram for the $\mathbf{D}_{2h}$ point group, develop the group multiplication for this group.

6.8 List the symmetry elements of the $\mathbf{C}_{3v}$ group and develop a stereogram for the group.

6.9 A unit cell for perovskite is shown at the right.
(a) How many Ca atoms are there per unit cell?
(b) How many Ti/unit cell?
(c) How many O/unit cell?
(d) What is the chemical formula?
(e) Into how many parts does a set of 222 planes divide the unit cell?
(f) Sketch in this set of planes.

6.10 Derive a general formula for the distance between *hkl* planes in a cubic lattice. (*Hint:* What is the distance in reciprocal space from the origin to the point *hkl*?)

6.11 (a) Under what circumstances with regard to relative sizes of ions and degree of nonpolar character are Frenkel and Schottky defects likely?
(b) The phenomenon of "half-melting" of solid electrolytes is related to which type of defect?

6.12 Consider a linear carbyne structure with alternating short and long C—C bond distances as indicated below, with the *z* axis as given:

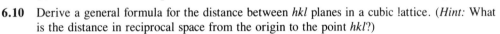

(a) Sketch a set of bonding and antibonding orbitals for the p_x orbitals of a C_2 unit in the above and label the $\pi_{bonding}$ and π^* orbitals.
(b) Sketch a portion of the wavefunction formed from $\pi_{bonding}$ at $k = 0$.
(c) Sketch a portion of the wavefunction formed from $\pi_{bonding}$ at $k = 0.5_a^{-*}$.
(d) Sketch a portion of the wavefunction formed from π^* at $k = 0$.
(e) Sketch a portion of the wavefunction formed from π^* at $k = 0.5_a^{-*}$.
(f) Show band structure (the energy versus *k*) for the bands formed from the $\pi_{bonding}$ and the π^* orbitals for a carbyne chain.

6.13 If Ge is added to GaAs, the Ge is about equally distributed between Ga and As sites. Which sites would Ge prefer if Se is added also? Would GaAs doped with Se be an *n*-type or a *p*-type semiconductor?

6.14 Sketch the curves for the distribution of energy states and their electron populations for a metallic conductor, an insulator, and a semiconductor.

PART III

Chemical Reactions

∙∙

▸7◂

Acids and Bases

••••••••••••••••••••••••••••

> Acid–base interactions are important throughout chemistry. Several models of acids and bases are considered here because we find it advantageous to use one for the reactions of a particular type and another for other reactions, reactions in another solvent, or reactions in the absence of a solvent. Then we examine factors affecting gas-phase proton affinities, since such data eliminate many complications arising from effects such as solvation and characteristics of solids or liquids. We then consider the effect of solvation on acid strength and some chemical implications. The Hard–Soft Acid–Base model gives a good overall view of Lewis acid–base interactions. Chemical reactions in alcohols, liquid ammonia, and sulfuric acid are surveyed. Periodic trends in binary oxides, oxoanions, and hydrated cations are also surveyed. These surveys illustrate acid–base concepts and introduce significant descriptive chemistry.

7.1 MODELS OF ACIDS AND BASES

▸ 7.1.1 Historical Background

Acids as a class of compounds were well known to the alchemists, who noted their sour taste (Latin *acidus,* "sour"), their ability to dissolve many water-insoluble substances, and their action on various vegetable dyes. When Priestly announced his discovery of "dephlogisticated air" in 1775, Lavoisier concluded that (in combination with nonmetals) the newly discovered substance was the common constituent of acids. Accordingly, Lavoisier named the new substance oxygen (French *oxys,* "sharp" or "acid", plus *genesis*). The German name *Sauerstoff* is a translation of the French name. Davy proved that not all acids contain oxygen and proposed that hydrogen was the common constituent of acids. Liebig firmly established the protonic concept of acids and described an acid as a substance of a replaceable hydrogen and an acid radical.

A "radical" is a group of atoms that stays bound together throughout various chemical transformations. An example would be NO_3. Hence, before the nature of chemical bonding was understood, Liebig considered nitric acid to consist of the nitrate "radical" (ions were not known) and hydrogen which could be replaced (for example, by Na to give sodium nitrate). Now we identify the nitrate ion, NO_3^-, as characteristic of nitric acid and nitrate salts. We frequently discuss "free radicals" as species that normally do not exist independently. These are commonly odd-electron molecules, and we show the presence of the unpaired electron using a single dot as in $\cdot OH$ and $\cdot CH_3$.

▶ 7.1.2 *Arrhenius and Ostwald*

Arrhenius and Ostwald's theory of electrolytic dissociation led to the present-day view of acid–base equilibria in water. The acidic hydrogen is recognized as a proton, and the theory focuses on the self-dissociation or autoprotolysis of water.

$$HOH \;\rightleftharpoons\; H^+(aq) + OH^-\,(aq) \tag{7.1}$$

At 25°C about one in every half-billion water molecules is dissociated into hydrogen ions and hydroxide ions. An aqueous solution is said to be neutral when the concentrations of hydrogen ions and hydroxide ions are equal, acidic when there are more hydrogen than hydroxide ions, and basic when there are more hydroxide than hydrogen ions. We can *increase the hydrogen ion concentration either by adding a substance that provides additional hydrogen ions for the system,* such as hydrogen chloride, *or by adding a substance that removes hydroxide ions from the system,* such as boric acid. Such substances are termed **acids.**

$$B(OH)_3 + OH^- \;\rightleftharpoons\; B(OH)_4^- \tag{7.2}$$

Similarly, *the hydroxide ion concentration may be increased by direct dissociation to produce hydroxide ion,* as in the case of sodium hydroxide, *or by combination with hydrogen ions in solution,* as in the case of ammonia to form ammonium ion. Combination processes of the latter type are often referred to as **hydrolysis,** particularly for ions. *Neutralization reactions in water consist of combining hydrogen ions and hydroxide ions to form water.*

Many solvents other than water undergo autoprotolysis, and acid–base equilibria in these solvents may be treated on a conceptual basis similar to the Arrhenius–Ostwald picture for water. Thus in liquid ammonia, a substance that produces ammonium ions is an acid, a substance that produces amide ions is a base, and a neutralization reaction consists of the reaction of ammonium ion and amide ion to produce ammonia:

$$NH_4^+ + NH_2^- \;\rightleftharpoons\; 2\,NH_3 \tag{7.3}$$

A more general **solvent theory** *defines an acid as a substance that produces positive ions derived from the solvent and defines a base as a substance that produces negative ions* *derived from the solvent. Neutralization consists of combining these ions to produce solvent.* Thus in liquid N_2O_4 as a solvent, salts furnishing the nitrosyl ion NO^+ are acids and those furnishing nitrate ion are bases:

$$NO^+ + NO_3^- \longrightarrow N_2O_4(l) \tag{7.4}$$

▶ 7.1.3 Brønsted and Lowry

A protonic picture that does not require solvent participation was presented by Brønsted and Lowry in 1923. According to their views, *an acid is a proton donor and a base is a proton acceptor. In an acid–base reaction, a proton is transferred from an acid to a base to produce another acid, termed the conjugate acid of the original base, and another base, termed the conjugate base of the original acid:*

$$\underset{\text{Acid}}{HCl} + \underset{\text{Base}}{NH_3} \longrightarrow \underset{\substack{\text{Conjugate} \\ \text{acid of } NH_3}}{NH_4^+} + \underset{\substack{\text{Conjugate} \\ \text{base of } HCl}}{Cl^-} \tag{7.5}$$

Water can function as an acid in a reaction with a stronger base:

$$\underset{\text{Acid}}{H_2O} + \underset{\text{Base}}{CN^-} \longrightarrow \underset{\substack{\text{Conjugate} \\ \text{acid}}}{HCN} + \underset{\substack{\text{Conjugate} \\ \text{base}}}{OH^-} \tag{7.6}$$

It can also function as a base in a reaction with a stronger acid:

$$\underset{\text{Base}}{H_2O} + \underset{\text{Acid}}{NH_4^+} \longrightarrow \underset{\substack{\text{Conjugate} \\ \text{acid}}}{H_3O^+} + \underset{\substack{\text{Conjugate} \\ \text{base}}}{NH_3} \tag{7.7}$$

These examples are hydrolysis reactions of CN^- and NH_4^+, but other substances, such as acetic acid, can behave as either acids or bases also:

$$\underset{\text{Acid}}{HC_2H_3O_2} + \underset{\text{Base}}{OH^-} \longrightarrow \underset{\substack{\text{Conjugate} \\ \text{acid}}}{H_2O} + \underset{\substack{\text{Conjugate} \\ \text{base}}}{C_2H_3O_2^-} \tag{7.8}$$

$$\underset{\text{Base}}{HC_2H_3O_2} + \underset{\text{Acid}}{HNO_3} \longrightarrow \underset{\substack{\text{Conjugate} \\ \text{acid}}}{H_2C_2H_3O_2^+} + \underset{\substack{\text{Conjugate} \\ \text{base}}}{NO_3^-} \tag{7.9}$$

▶ 7.1.4 Lewis

A general theory of acids and bases covering all the preceding cases and extending the definition to some substances not included above was set forth by G. N. Lewis, who *defined an acid as an electron-pair acceptor and defined a base as an electron-pair donor.*

Neutralization in the Lewis theory consists of the formation of a new covalent bond between an electron-pair donor and an electron-pair acceptor:

$$BF_3 + (CH_3)_3N: \longrightarrow (CH_3)_3N\!:\!BF_3 \qquad (7.10)$$
$$\text{Acid} \qquad \text{Base}$$

The hydration (or in more general terms to include other solvents, the *solvation*) of metal ions is an acid–base reaction involving the stepwise addition of several water (or *solvent*) molecules:

$$Al^{3+} + 6\,H_2O: \longrightarrow [Al(\!:\!OH_2)_6]^{3+} \qquad (7.11)$$
$$\text{Acid} \qquad \text{Base}$$

The formation of metal complexes or coordination compounds (Chapter 9) involves the metal ion as an acid and the electron-pair donor (or **ligand**) as a base. $[Al(H_2O)_6]^{3+}$ and $[Ag(NH_3)_2]^+$ are examples of metal complexes:

$$Ag^+ + 2:NH_3 \longrightarrow [H_3N\!-\!Ag\!-\!NH_3]^+ \qquad (7.12)$$

▶ *7.1.5 Comparisons of Acid–Base Models*

Usanovitch gave the most general definition of an acid: *An acid is any material that forms salts with bases through neutralization, gives up cations, combines with anions, or accepts electrons. A base is any material that forms salts with acids through neutralization, gives up anions, combines with cations, or gives up electrons.* Thus most reactions, including oxidation–reduction reactions, are classified as acid–base reactions by these definitions.

A classification that includes practically everything may be too broad to be truly useful. In this chapter we concentrate on the other models of acids and bases presented above. All these models are useful, and which one we apply depends on the system under study.

The Arrhenius–Ostwald dissociation picture serves for most situations involving hydrogen ion equilibria in aqueous solution. For the acid HA the dissociation constant, K_a, is defined by the activity quotient, $K_a = \dfrac{a_{H^+}}{(a_{A^-})(a_{HA})}$. The most convenient way to represent the acid strength is by using p$K_a(= -\log K_a)$. The larger the numerical value of the pK_a, the weaker the acid under discussion.

The Brønsted–Lowry model, by emphasizing *proton* transfer as the acid–base reaction, makes acid–base reactions analogous to oxidation–reduction reactions, in which the *electron* transfer is the essential defining process. Just as a total oxidation–reduction reaction can be split into half-reactions involving the oxidized and reduced form of a given reagent, a proton transfer reaction can be broken into half-reactions involving the protonated and unprotonated forms (the conjugate pair) of a given base: $B + H^+ \to BH^+$. The heat released in such a gas-phase reaction is referred to as the **proton affinity** (PA) of B[1]. Through the scale of proton affinities for various bases, the Brønsted–Lowry picture sets limits on the intrinsic protonic acidity or basicity. Dissection of factors influencing the strength of acids and bases without solvation effects is best carried out using PA data.

---

[1] Note that this sign convention is opposite to the thermodynamic one. A positive PA for a protonation reaction corresponds to heat released and a negative ΔH.

Comparison of these gas-phase data with solution data reveals the large degree of attenuation of protonic acidity scales in different solvents and permits us to assess solvation effects.

The many parallelisms between protonic acids and Lewis acids allow us to extend our picture of protonic acids to a wider range of systems, such as coordination complexes and so-called "donor–acceptor" complexes.

7.2   PROTONIC ACIDS

We begin by examining protonic acids because fewer factors are involved in determining their strengths than with Lewis acids. Starting from the Brønsted–Lowry viewpoint, we consider intrinsic acidities and basicities of reagents in the gas phase—how they are measured and what factors determine their strengths. Then we consider the role of solvents in influencing acid and base strengths in order to understand better the behavior of acids and bases as we usually encounter them in the laboratory. We then briefly discuss the properties of a few important nonaqueous protic solvents.

▶ *7.2.1   Proton Affinities*

We can use special types of mass spectrometry to study equilibria of an acid (XH^+ or XH) and a base (B or B^-):

$$XH^+(g) + B(g) \rightleftharpoons BH^+(g) + X(g) \tag{7.13}$$

and

$$XH(g) + B^-(g) \rightleftharpoons BH(g) + X^-(g) \tag{7.14}$$

These are analogous to measuring a complete oxidation–reduction reaction, and we need a reference point for a known reaction of the type

$$X(g) + H^+(g) \rightleftharpoons XH^+(g) \tag{7.15}$$

to establish an absolute PA scale. As is evident from Equation (7.15), a PA is a measure of the gas-phase basicity. Fortunately, sufficient data are available from spectroscopic experiments to establish a number of absolute PAs from the following cycle:

$$
\begin{array}{ccc}
X(g) + H^+(g) & \xrightarrow{\ \Delta H_1 = -PA(X)\ } & XH^+(g) \\[4pt]
{\scriptstyle \Delta H_4 = -IE(X)}\Big\uparrow & & \Big\backslash\;{\scriptstyle \Delta H_2 = D_{X^+-H}} \\[4pt]
X^+(g) + H^+(g) + e & \xleftarrow[\ \Delta H_3\,=\,IE(H)\ ]{} & X^+(g) + H(g)
\end{array}
\tag{7.16}
$$

$$PA(X) = D_{X^+-H} + IE(H) - IE(X) \tag{7.17}$$

Combining absolute values for a few species determined from the above cycle with data from ion cyclotron resonance spectroscopy or high-pressure mass spectrometry (which bracket the PAs of various X atoms or molecules between the PAs of other species) yields results such as the basicities of the binary hydrides shown in Table 7.1. Using Equation (7.17), we can rationalize the trends in basicities observed in these binary hydrides. The ionization energies (IEs) for the hydrides increase smoothly for Groups 15–17 for a period, since the irregularities in spin multiplicity that cause irregularities in the atom IEs disappear for the diamagnetic molecules. The IE for CH_4 is high because there are no non-bonding electrons. Because the IE changes generally are much greater than the changes in the bond dissociation energies, the former dominate the PA behavior. Thus within a given period or a given group, the PA varies conversely with the IE of the binary hydride. For PH_3, NH_3, and AsH_3, however, the IE differences are slight and the decreasing bond energies cause decreases in PAs.

For negative ions, the energy required to remove an electron—corresponding to the electron affinity (EA) of the atom—generally is smaller and changes less in going down a group than the bond energy. Accordingly, within a group the trend in PAs for the anions parallels the trend in bond energies (X—H). On going from one group to another, the EAs change more rapidly than the bond energies and the higher EA (or electronegativity), the lower the PA. Table 7.2 illustrates these trends.

Table 7.1   Ionization energies, proton affinities, and trends in basicities in binary hydrides

	CH_4		NH_3		H_2O		HF
IE^a	12.7		10.2		12.6		16.0
PA^b	552	<	854	>	697	>	399
			V		∧		∧
	SiH_4		PH_3		H_2S		HCl
IE	11.7		10.0		10.5		12.7
PA	~648	<	789	>	712	>	564
			V		∧		∧
			AsH_3		H_2Se		HBr
IE			10.0		9.9		11.7
PA			750	>	717	>	589
							∧
							HI
IE							10.4
PA							628

[a] Ionization energies (eV) from H. M. Rosenstock, K. Draxl, B. W. Steiner, and J. T. Herron, *J. Phys. Chem. Ref. Data* **1977**, *6*, Suppl. No. 1.

[b] Proton affinities (kJ/mol) from S. G. Lias, J. F. Liebman, and R. D. Levin, *J. Phys. Chem. Ref. Data* **1984**, *13*, 695.

**Table 7.2 Acidity order of the binary hydrides
(proton affinities of their anions are in kJ/mol)**[a]

Acidity order					*Proton affinities*		
CH_4	< NH_3	< H_2O	< HF	CH_3^-	NH_2^-	OH^-	F^-
∧	∧	∧	∧	1743	1689	1635	1554
SiH_4	< PH_3	< H_2S	< HCl	SiH_3^-	PH_2^-	SH^-	Cl^-
∧	∧	∧	∧	1554	1550	1478	1395
GeH_4	≈ AsH_3	< H_2Se	< HBr	GeH_3^-	AsH_2^-	SeH^-	Br^-
			∧	1509	1500	1420	1354
			HI				I^-
							1315

[a] Data from J. E. Bartmess and R. T. McIver, Jr., "The Gas-Phase Acidity Scale," in *Gas-Phase Ion Chemistry,* Vol. 2, M. T. Bowers, Ed., Academic Press, New York, 1979.

The EA is defined as the amount of energy involved in adding an electron to an atom. For most atoms, energy is released and the EA is a positive number corresponding to a *negative* ΔH for the process. The IE of the corresponding anion is positive and the ΔH is positive. For example, the EA of F is 328 kJ ($\Delta H = -328$ kJ) and the ΔH for the ionization of F^- is 328 kJ.

Inductive Effects

On going from NH_3 to NF_3, the IE increases by 268 kJ/mol—as expected in view of the electron-withdrawing **inductive effect** of the fluoro groups. On substituting F for H, the more electronegative F atom acquires a greater share of the bonding electrons and induces a partial charge in N relative to its state in NH_3. This not only makes ionizing one of the lone-pair electrons of the N atom more difficult, but also makes nitrogen less willing to share its electrons with a proton, and so the PA drops by 259 kJ/mol on going from NH_3 to NF_3. In these cases, the substituent (H or F) is engaged only in sigma bonding with the reactive center (N). Insertion of a saturated hydrocarbon chain between the substituent and the reactive center generally ensures that the substituent effect is transmitted by sigma-bond polarization—in other words, that the substituent effect is a pure inductive effect. The inductive effect decreases in magnitude as the substituent is moved along the chain from the reactive center.

Resonance Effects

Substituting F for H in PH_3 increases the IE by only 121 kJ/mol. Here we might have expected the difference in electronegativities of P and F to lead to an even greater electron-withdrawing inductive effect than that found with NF_3. In the case of the PF_3, however, the low-lying empty d orbitals permit some of the negative charge withdrawn onto F in σ bonds to be returned via the π system. Hence, resonance structures such as

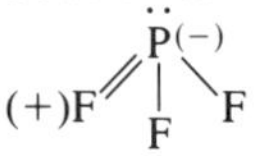

contribute, accounting for the lower-than-expected IE. Hence the PA of PF_3 is only 106 kJ lower than that of PH_3. The positive charge on the HPF_3^+ can be delocalized over the entire molecule as represented in the following resonance structures which rationalize the higher-than-expected PA:

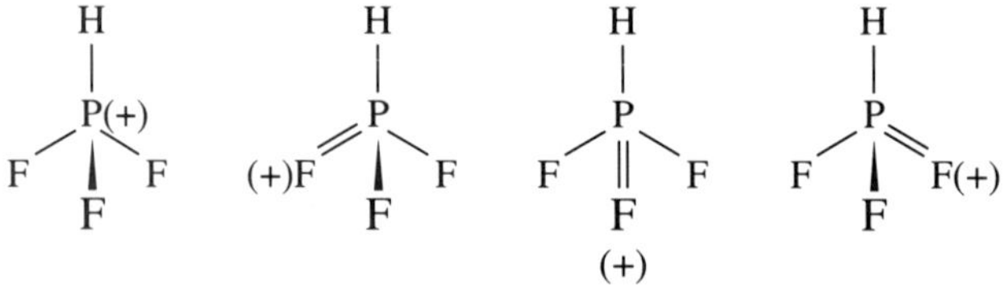

Resonance, or conjugative, effects may be transmitted through carbon chains of alternating (conjugated) double bonds, or from the ortho and para positions of a phenyl group. For the effect to be transmitted in this fashion, direct resonance must be possible when the groups are directly attached (no intervening chain or ring).

Predicting Gas-Phase Proton Transfers

Using the data given in Tables 7.1 and 7.2 (or in other tables of PAs) to predict the direction of proton transfer upon mixing two substances, we find that the proton will go to the species with the highest PA. Thus if we were somehow to mix in the gas phase the hydride ion and H_2O, we would end up with hydroxide ion and elemental hydrogen. On mixing H_2O with the fluoride ion in the gas phase, there is no proton transfer, since the fluoride ion has a lower PA (1554 kJ/mol, compared with 1635 kJ/mol for hydroxide ion), and so the hydroxide ion hangs onto the proton; H_2O and fluoride ion remain in their initial states.

EXAMPLE 7.1: What are the products of PH_3 mixed with NH_4^+? What are the products of $(CH_3)_3P$ (PA = 950 kJ/mol) mixed with NH_4^+?

Solution: Because PA is higher for NH_3 (854) than that of PH_3 (789), NH_4^+ will not lose the proton to PH_3. No reaction occurs. PA is higher for $(CH_3)_3P$ (PA = 950 kJ/mol) than that of NH_3, so in the gas phase $(CH_3)_3P + NH_4^+$ gives $(CH_3)_3PH^+ + NH_3$.

The Leveling Effect of Solvents

Consider the following type of experiment: Into a low-pressure system containing a relatively large number of monomolecular water molecules, we introduce a Brønsted acid or base and then evaluate the acidity or basicity of the system. (For the moment, solvation effects are ignored.) If protonated dinitrogen, N_2H^+ (PA for N_2 = 494 kJ/mol), is introduced, proton transfer will take place immediately to form N_2 and H_3O^+, and the PA of H_2O will determine whether any further proton transfers occur. On the other hand, if H^- is introduced, H_2 and OH^- will be formed, and the PA of OH^- determines the basicity of the system for further reactions. A third situation arises when a species such as NH_3 or HCl is added (separately) to the $H_2O(g)$ system: No reaction takes place because $PA(OH^-) > PA(NH_3)$ and $PA(Cl^-) > PA(H_2O)(g)$.

$$H_2O(g) + NH_3(g) \rightleftharpoons NH_4^+ + OH^- \qquad (7.18)$$
PA 697 854 1635

$$HCl(g) + H_2O(g) \rightleftharpoons H_3O^+ + Cl^- \qquad (7.19)$$
PA 564 697 1395

The NH_3 controls the basicity of the $H_2O(g)/NH_3(g)$ system because it has greater PA than H_2O, while HCl controls the acidity of the $HCl(g)/H_2O(g)$ system because it is the better proton donor. The PA of species capable of proton transfer in a gaseous system with excess monomolecular H_2O molecules must lie between the PAs of H_2O (697 kJ/mol) and OH^- (1635 kJ/mol). If the PA of X^- is greater than 1635, X^- will take H^+ from H_2O; and if the PA of X^- is less than 697, HX will lose H^+ to H_2O. Thus in gas-phase H_2O, nothing can be more acidic than H_3O^+ nor more basic than OH^-, and acids or bases outside this range are leveled to these values upon being added to the gaseous H_2O system by reacting to form these species. The energy released on hydrating a proton increases with the number of water molecules added to the cluster, as in high-pressure mass spectrometry, approaching a limit of 1130 kJ/mol. This is the effective "PA of liquid water."

What happens to the PA of OH^- as it becomes solvated? Here, hydrogen bonds formed in the $OH^- \cdot nH_2O$ cluster are broken when the cluster disintegrates upon protonation (charge neutralization).

$$H^+(g) + OH^- \rightleftharpoons H_2O(g) \qquad -\Delta H = PA_{OH^-} = 1635 \text{ kJ/mol} \quad (7.20)$$

$$H^+(g) + OH^-(aq) \rightleftharpoons H_2O(l) \qquad -\Delta H = 1188 \text{ kJ/mol} \qquad (7.21)$$

Accordingly, we may conclude that in aqueous solution, the effective "PA" range is $1130 - 1188 = |58|$ kJ/mol. *No species more acidic than $H^+(aq)$ and no base stronger than $OH^-(aq)$ can exist in aqueous solution.*

▶ 7.2.2 Solvent Effects

If no solvation were to take place for any species other than H^+ and OH^-, all negative ions would form $OH^-(aq)$ when introduced into water, since the lowest PA for a negative ion is above 1250 kJ/mol, and all positive ions bearing a hydrogen atom would form $H^+(aq)$, since the highest PA for the conjugate bases of these cations is below 1050 kJ/mol. However, this does not occur because the other ions themselves are solvated to some degree. The solvation of negative ions releases much more energy than the solvation of their neutral conjugate acids, thereby lowering the effective PAs of the negative ions in aqueous solution. Because solvation of the positive ions is much more effective than that of the neutral species produced when the proton is given up, the effective PAs of the neutral molecules increase as they go into aqueous solution. The anions of strong acids have sufficient solvation energy that their effective PA is well below 1130 kJ/mol; and thus in aqueous solution these acids are completely dissociated, to give $H^+(aq)$ and the solvated anion. On the other hand, solvation of anions such as H^- and CH_3^- is insufficient to lower the effective PA of these substances enough to permit hydrogen (H_2) or methane to show acidic properties in aqueous solution.

Summarizing, we see that *acid species stronger than the* **oxonium** *ion* (H_3O^+, *commonly called the* **hydronium ion**) *in water transfer their proton to water and the acidity is* **leveled** *to the acidity of* H_3O^+. Likewise, *proton acceptors (bases) that are stronger than the aquated hydroxide ion all form the hydroxide ion in water*. Thus the range of acidities available in water must fall between aquated oxonium ion–neutral water couple and the neutral water–aquated hydroxide ion couple. We have just noted the enormous compression of the acidity scale that takes place on going from the gas phase to the aqueous solution phase. This compression stems from the large degree of solvation of the protonated solvent species, the **lyonium** ion, and of the deprotonated solvent species, the **lyate** ion. *The acidity range available in any solvent falls between that of the lyonium ion and that of the lyate ion*. The more solvation stabilizes these ions, the greater the solution acidity scale is compressed compared with the PA differences found in the gas phase. The most important specific solvation effect is hydrogen bonding (see Section 2.6), and the extent of stabilization of the solvent ions depends primarily on the number and strength of the hydrogen bonds that may form. In addition to the compression of the available range of acidities caused by solvation, inversion in the order of acidities of individual pairs of acids is common on going from the gas phase to solution, or from one solvent to another.

In referring to the gas phase, we have used a difference in PAs—that is, an enthalpy change—as the criterion for the direction of proton transfer. We assume negligible entropy contributions for the gas-phase proton transfer (and indeed the entropy effects are small) and thus $\Delta G^0 \simeq \Delta H^0$. In solution, entropy effects are far less likely to cancel, and hence we must use the ΔG^0 values. The acidity range available in a solvent may be obtained from $\Delta G^0 = -RT \ln K_S$, where K_S refers to the autoprotolysis constant of the solvent.

▶ 7.2.3 Some Protonic Solvents

Alcohols

Whereas the (gas-phase) PAs of alcohols increase with their degree of alkylation [H_2O (697 kJ/mol), CH_3OH (761), C_2H_5OH (778), *i*-PrOH (791), *t*-BuOH (799)], the PAs of alkoxides decrease with their degree of alkylation [OH^- (1635), CH_3O^- (1586), $C_2H_5O^-$ (1573), *i*-PrO$^-$ (1565), *t*-BuO$^-$ (1561)]. Protonation of an alcohol produces a charged species (ROH_2^+), whereas protonation of an alkoxide destroys a charged species (RO^-). The stabilization of the charged species by alkylation may be accounted for primarily by the large polarizability of the alkyl group relative to hydrogen[2]. In solution, the lyonium ion of the alcohols can form only two primary hydrogen bonds (compared with three for that of water), and both the primary H-bonds and those in the secondary solvation sphere are weakened by the steric strain caused by interference between the alkyl groups. Hence, the

---

[2] For a discussion of the separation of the inductive effect and the polarization effect from data such as those given here, see R. W. Taft et al., *J. Am. Chem. Soc.* **1978,** *100,* 7765.

lyonium ions are less stabilized by solvation for alcohols than for water. Furthermore, the stabilization decreases with increasing size of the alkyl group. Although the alkoxide ions, like hydroxide ions, can form three primary hydrogen bonds, steric effects weaken these bonds. So although the PA range in the gas phase between alcohols and their alkoxide ions is less than that of water and the hydroxide ion, *the acidity range in solution between the lyonium and lyate ions is greater for alcohols than for water*. This is reflected in the ability to titrate both (a) ammonium salts in ethanol with alcoholic KOH (which is in equilibrium with the alkoxide ion) and (b) carboxylate salts with alcoholic HCl. In part, the greater acidity range of alcohols arises from the smaller value of the dielectric constant (see page 92) of alcohols compared with that of water (32.6 for methanol, 24.3 for ethanol, and 78.5 for water—all at 25°C).

Liquid Ammonia

To establish the available acid–base range for a solvent accurately, we must know two facts: (1) the absolute enthalpy (or better, the free energy) of solvation of the proton in the solvent and (2) the autoprotolysis constant of the solvent. Although in the last decade much progress has been made toward ascertaining the absolute enthalpies of solvation of the proton and other ionic species, extrathermodynamic assumptions are still required to arrive at such values. Even so, estimates are available for relatively few solvents. The difference in the proton's free energy of solvation in water and in liquid ammonia has been estimated as 100 kJ/mol H^+, based on (a) the electrode potentials for Rb/Rb^+ and H_2/H^+ in the two solvents and (b) the assumption that the solvation energy of Rb^+ is the same in both solvents. This greater free energy solvation for H^+ means that ammonia is a more basic solvent than water. The value of the autoprotolysis constant of NH_3 is 10^{-33} at $-50°C$, at which temperature the dielectric constant is 25.

EXAMPLE 7.2: Calculate the PA range for $NH_3(l)$ at $-50°C$ using the estimate of 1130 [PA for $H_2O(l)$] + 100 = 1230 kJ/mol for the effective PA of liquid ammonia. Calculate the effective PA of NH_2^-.

Solution: $\Delta G° = -RT \ln 10^{-33} = -8.31$ J mol^{-1} $K^{-1} \times 223$ K $\times (-76) = 141$ kJ, the PA range for $NH_3(l)$ at $-50°C$. The effective PA of NH_2^- is $1230 + 141 = 1371$ kJ.

Compared to water with an effective PA range of 58 kJ, the effective PA range for ammonia is more than twice as great. Figure 7.1 shows the acid–base range for several solvents. Water has the greatest effect of compressing the range compared to the gas phase.

The greater basicity of liquid ammonia relative to water, coupled with the greater range of basicity stemming from its lower autoprotolysis constant, permits acidic behavior to be observed for many compounds that display virtually no acidic properties in water. This is advantageous for synthesis, as the following examples show.

$$PH_3 + NaNH_2 \xrightarrow{NH_3(l)} NaPH_2 + NH_3$$
$$C_2H_2 + NaNH_2 \longrightarrow NaC_2H + NH_3 \tag{7.22}$$

$$NaPH_2 + CH_3Cl \longrightarrow CH_3PH_2 + NaCl$$
$$NaC_2H + BuCl \longrightarrow Bu-C\equiv CH + NaCl \tag{7.23}$$

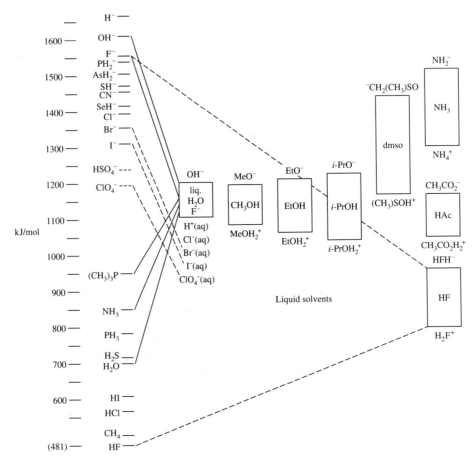

Figure 7.1   Acid–base range available in the gas phase and in various solvents (vertical range of boxes).

In Equations (7.23), the negatively charged ion produced by deprotonation undergoes a displacement reaction with the alkyl halide. Further alkylation may be carried out by repeating the process.

Negative ions—particularly multiply charged monoatomic ions—need a basic environment in order to be stable in protonic solvents. If the solvent is too acidic, protons are transferred to the negative species. In liquid ammonia we can have appreciable concentrations of ions such as S^{2-}, Se^{2-}, As^{3-}, and PH^{2-}. We might consider the ultimate in negative ions to be solvated electrons or electron pairs. When hydrogen is introduced at high pressure into an amide solution in liquid ammonia, solvated electrons and possibly even hydride ions are produced:

$$\tfrac{1}{2}H_2 + NH_2^- \longrightarrow NH_3 + e(NH_3)$$
$$H_2 + NH_2^- \longrightarrow NH_3 + H^-$$

$$(7.24)$$

Solvated electrons are produced more readily by dissolving alkali metals in liquid ammonia (see Section 15.2.5). Dilute solutions containing the alkali metal ions and the solvated electron are deep blue in color, whereas concentrated solutions are bronze. The volume of the solution is greater than the sum of its components: Expansion occurs because of cavities in the liquid, which contain electrons. Transitions from blue to bronze solutions occur above a concentration of approximately 0.5 M and are interpreted as a transition to a metallic state (bronze color). Although solutions of alkali metal in ammonia are thermodynamically unstable with respect to hydrogen evolution and amide formation, the reaction is extremely slow in the absence of a catalyst, such as Fe_2O_3. Accordingly, one of the best ways to obtain anhydrous ammonia is to distill it from its sodium solution.

As might be expected, solutions of alkali metals in liquid ammonia are very good reducing agents. Among the reactions observed in liquid ammonia are:

$$
\begin{aligned}
K_2Ni(CN)_4 + 2\,K &\longrightarrow K_4Ni(CN)_4 \\
Fe(CO)_5 + 2\,Na &\longrightarrow Na_2Fe(CO)_4 + CO \\
[Pt(NH_3)_4]Br_2 + 2\,K &\longrightarrow Pt(NH_3)_4 + 2\,KBr \\
9\,Pb + 4\,Na &\longrightarrow Na_4Pb_9
\end{aligned}
\tag{7.25}
$$

The plumbide ion in the last reaction is a "naked" metal cluster ion whose geometry is that of a C_{3v} tricapped trigonal prism. Adding lead iodide to the ammonia solution of Pb_9^{4-} precipitates metallic lead:

$$
Pb_9^{4-} + 2\,Pb^{2+} \longrightarrow 11\,Pb
\tag{7.26}
$$

A number of polynuclear anions of Group 14, 15, and 16 elements have been prepared using alkali metal solutions in liquid ammonia.

E. C. Franklin[3] pointed out the closely analogous behavior of certain nitrogen-containing species in liquid ammonia and the corresponding oxygen species in water:

$$
\begin{array}{ccccc}
NH_4^+ & NH_3 & NH_2^- & NH^{2-} & N^{3-} \\
\searrow & \searrow\searrow & \searrow\nearrow & \searrow\nearrow & \nearrow \\
H_3O^+ & H_2O & OH^- & O^{2-} &
\end{array}
$$

Thus nitrides solvolyze in liquid ammonia to produce basic solutions of amides, just as oxides in water produce basic solutions of hydroxides. Ammonium ion reacts with active metals, thereby liberating H_2. One of the properties of aqueous acids is the ability to dissolve active metals with the evolution of hydrogen. Aluminum in the form of an amalgam demonstrates amphoterism by dissolving in liquid ammonia solution containing an ammonium salt *or* an amide.

Amphoteric character is found also with zinc ions, as indicated by the following sequence of reactions:

[3] E. C. Franklin, *The Nitrogen System of Compounds*, Reinhold, New York, 1935.

$$ZnI_2(\text{solvated}) + 2\,KNH_2 \longrightarrow 2\,Zn\,(NH_2)_2(s) + 2\,KI$$
$$Zn(NH_2)_2(s) + 2\,KNH_2 \longrightarrow K_2[Zn(NH_2)_4]$$
(7.27)

The tetraamidozincate ion ($[Zn(NH_2)_4]^{2-}$) formed corresponds to the $[Zn(OH)_4]^{2-}$ ion formed in aqueous solution by addition of excess hydroxide ion. Adding ammonium ion to $[Zn(NH_2)_4]^{2-}$ reprecipitates zinc amide, and further addition of ammonium ion redissolves the precipitate, corresponding to the addition of acid to an aqueous solution of $[Zn(OH)_4]^{2-}$.

The ammono acid corresponding to acetic acid is acetamidine, $CH_3C{\overset{\displaystyle NH}{\underset{\displaystyle NH_2}{\diagup\!\!\!\diagdown}}}$, which behaves as a weak acid in liquid ammonia. Acetonitrile, CH_3CN, is the ammono equivalent of acetic anhydride. The following equations show the utility of the water–ammonia analogy:

$$2\,H_2O + CO_2 \longrightarrow H_3O^+ + HOCO_2^-$$

$$2\,NH_3 + CO_2 \longrightarrow \underset{\text{Ammonium}}{NH_4^+} + \underset{\text{Carbamate}}{H_2NCO_2^-}$$

$$Cl_2SO_2 + 2\,H_2O \longrightarrow H_2SO_4 + 2\,HCl$$

$$\underset{\substack{\text{Sulfuryl}\\\text{chloride}}}{Cl_2SO_2} + 4\,NH_3 \longrightarrow \underset{\text{Sulfamide}}{(NH_2)_2SO_2} + 2\,NH_4Cl \qquad (7.28)$$

$$Cl_2 + H_2O \rightarrow HOCl + HCl(aq)$$

$$Cl_2 + 2\,NH_3 \rightarrow H_2NCl + NH_4Cl$$

$$CH_3CO_2H \xrightarrow[\text{electrolysis}]{} CO_2 + CH_4$$

$$CH_3C(NH)NH_2 \xrightarrow[\text{electrolysis}]{} \underset{\text{Cyanamide}}{NH_2CN} + CH_4$$

Organic amines serve as even more basic solvents than liquid ammonia. These include primary, secondary, and tertiary amines, since the chemistry of amines is dominated by the electron-pair donor role, a characteristic common to all amines. The tertiary amines, of course, cannot serve as proton donors.

Sulfuric Acid

In dilute solutions, the pH may be determined spectrophotometrically, using indicators of known pK_a, by means of the relationship

$$pH = pK_a + \log \frac{[\text{Ind}]}{[\text{HInd}]} + \log \frac{\gamma_{\text{Ind}}}{\gamma_{\text{HInd}}} \qquad (7.29)$$

where [Ind] is the concentration of the unprotonated form of the indicator and [HInd] is the concentration of the protonated form. The γ terms are activity coefficients. [Ind] and [HInd] can be measured from absorption spectra, and the ratio of activity coefficients approaches 1 at sufficient dilution. The charge on the indicator (neutral or negative) or on the protonated indicator is temporarily ignored here.

In concentrated acidic or alkaline solutions and in nonaqueous media the activity coefficient ratio is uncertain. Thus, the pH is not meaningful, but an "effective pH scale" can be defined by the Hammett acidity function[4] H. For a neutral unprotonated indicator (or a base in general),

$$B + H^+ \rightleftharpoons BH^+$$

and

$$H_0 = pK_{BH^+} + \log \frac{[B]}{[BH^+]} \tag{7.30}$$

For an acidic indicator HInd, the Hammett function is H_-:

$$HInd \rightleftharpoons H^+ + Ind^-$$

$$H_- = pK_a + \log \frac{[Ind^-]}{[HInd]} \tag{7.31}$$

A given indicator may be used to measure H functions over a range of approximately two units—that is, from $[Ind^-]/[HInd] = 0.1$ to $[Ind^-]/[HInd] = 10$. By using structurally related indicators, it was felt that the activity coefficient ratios would cancel effectively and the H functions could be referenced to dilute aqueous systems. The H functions provide a guide, in the absence of information about the free energy of proton solvation, for setting the acidity or basicity of a solvent, or of a solvent containing a known concentration of the anion or cation derived from autoprotolysis. As you can see from the definition of H, the more negative the H value, the higher the protonating power (effective acidity) of the solution.

The H_0 function for pure sulfuric acid has a value of -12.1; the dielectric constant of H_2SO_4 is 100 and its autoprotolysis constant is 2.7×10^{-4} at 25°C. Sulfuric acid is thus a strongly acidic solvent with a very small acid–base range.

We can prepare aqueous sulfuric acid solutions with H_0 values that range from aqueous pH values up to the H_0 of pure sulfuric acid of -12.1, permitting us to determine pK_a values (for protonated bases) of very weak bases. For example, since phosphoric acid is 50% protonated in approximately 80% H_2SO_4, which has an H_0 value of -7.4, the pK_a of $P(OH)_4^+$ may be taken as approximately -7.4.

Sulfuric acid is a good dehydrating agent. It has a freezing-point depression constant of 6.12°C/kg/mol. When nitric acid is dissolved in sulfuric acid, the freezing point depression indicates that each mole of nitric acid produces four moles of solute species. Coupled with conductance data, this is interpreted by the equation

$$HNO_3 + 2\,H_2SO_4 \longrightarrow NO_2^+ + H_3O^+ + 2\,HSO_4^- \tag{7.32}$$

---

[4] For a complete treatment of these Hammett acidity functions, consult C. H. Rochester, *Acidity Functions*, Academic Press, New York, 1970.

This reaction can be thought of as resulting from protonation to form $H_2NO_3^+$, which subsequently dehydrates to NO_2^+. There is strong evidence that NO_2^+ is the active species in nitration reactions with aromatic compounds. In similar fashion, the species SO_3H^+ has been proposed as playing a role in sulfonation reactions.

The behavior of mesitoic acid in lowering the freezing point of sulfuric acid is very similar to that of nitric acid; that is,

$$2,4,6\text{-}(CH_3)_3C_6H_2CO_2H + 2H_2SO_4 \longrightarrow (CH_3)_3C_6H_2CO^+ + H_3O^+ + 2HSO_4^- \quad (7.33)$$

Pouring a solution of mesitoic acid in sulfuric acid into an alcohol produces esterification in 100% yield—a reaction difficult to accomplish even in low yield by direct reaction of an alcohol with this highly hindered acid. Benzoic acid shows a lowering of the freezing point of sulfuric acid consistent with the production of two moles of solute ions per mole of benzoic acid:

$$C_6H_5CO_2H + H_2SO_4 \longrightarrow C_6H_5CO_2H_2^+ + HSO_4^- \quad (7.34)$$

The differing behavior of these aromatic acids may be explained by the resonance stabilization of the protonated benzoic acid, whereas steric inhibition of resonance in the mesitoic acid makes the protonated form much higher in energy.

Formic acid dehydrates in 95–100% sulfuric acid to give carbon monoxide, providing a convenient laboratory source of CO. The HCO^+ species is a probable intermediate in this reaction and would transfer rapidly a proton to H_2SO_4 or HSO_4^-. Oxalic acid also decomposes in sulfuric acid, yielding in this case an equimolar mixture of CO and CO_2.

Most organic compounds, except for the saturated hydrocarbons, are protonated by and dissolve in sulfuric acid. The insolubility of saturated hydrocarbons in sulfuric acid can be used to separate mixtures of saturated and unsaturated hydrocarbons to determine the fraction of saturated hydrocarbon present.

Sulfuric acid itself is a mild oxidizing agent, and, particularly when hot, its oxidizing strength increases with its concentration. Up to about 12 M, the hydrogen ion is the oxidizing agent; above 12 M, the sulfuric acid is reduced to SO_2 during the oxidation of solutes. Just as basic media permit the existence of a variety of reduced species that are not stable in water (even the solvated electron), acidic media allow the existence of a variety of oxidized or positive species such as S_8^{2+}, Se_8^{2+}, Se_4^{2+}, Te_6^{4+}, and Te_4^{2+} (see Section 16.3.7). It was observed more than 150 years ago that Se and Te give brightly colored solutions in H_2SO_4. These contain polyatomic cations that can be obtained also in liquid SO_2 containing AsF_5 or SbF_5 and in molten $AlCl_3$. I_3^+ is apparently produced by the reaction

$$HIO_3 + 7I_2 + 8H_2SO_4 \longrightarrow 5I_3^+ + 3H_3O^+ + 8HSO_4^- \quad (7.35)$$

Dilution of the sulfuric acid media results in the destruction of these species by disproportionation reactions such as

$$16H_2O + 16Se_8^{2+} \longrightarrow 120Se + 8SeO_2 + 32H^+ \quad (7.36)$$

Few substances behave as acids in sulfuric acid. Perchloric acid, chlorosulfuric acid ($HClSO_3$), and fluorosulfuric acid (HSO_3F) are among the protonic acids that show weak acidic character in sulfuric acid. The Lewis acids sulfur trioxide and boron tris(hydrogensulfate) behave as moderately strong acids in sulfuric acid, increasing the lyonium ion by reducing the concentration of the lyate ion.

$$SO_3 + 2\,H_2SO_4 \longrightarrow H_3SO_4^+ + HS_2O_7^-$$
$$B(HSO_4)_3 + 2\,H_2SO_4 \longrightarrow H_3SO_4^+ + B(HSO_4)_4^- \tag{7.37}$$

Alternatively, we may think of these substances as forming protonic acids by reaction with sulfuric acid, with the lyonium ion subsequently being increased by the dissociation of these acids.

$$SO_3 + H_2SO_4 \longrightarrow H_2S_2O_7$$
$$H_2S_2O_7 + H_2SO_4 \longrightarrow H_3SO_4^+ + HS_2O_7^-$$
$$B(HSO_4)_3 + H_2SO_4 \longrightarrow HB(HSO_4)_4$$
$$HB(HSO_4)_4 + H_2SO_4 \longrightarrow H_3SO_4^+ + B(HSO_4)_4^- \tag{7.38}$$

Solutions of SO_3 in H_2SO_4 are called "fuming sulfuric acid" or "oleum". As the SO_3 concentration increases, higher polysulfuric acid species—$HO(SO_3)_nH$—are formed, with the acidity increasing slightly as n increases. The hydrogen tetrakis(hydrogen sulfato)borate is formed *in situ* by dissolving either boric acid or boric oxide in sulfuric acid and adding sufficient fuming sulfuric acid to react with the oxonium ion formed.

$$H_3BO_3 + 6\,H_2SO_4 \longrightarrow B(HSO_4)_4^- + 3\,H_3O^+ + 2\,HSO_4^-$$
$$\searrow 3\,H_2S_2O_7 \tag{7.39}$$
$$HB(HSO_4)_4 + 8\,H_2SO_4$$

Antimony pentafluoride is a very strong Lewis acid. In a very acidic protonic solvent such as HF, $H_2SO_4\cdot SO_3$, or $HSO_3F\cdot nSO_3$, SbF_5 produces extremely strong proton donors known as **"superacids."**[5] The superacids are protonated solvent molecules of the already strong acids.

$$SbF_5 + 2\,HF \longrightarrow H_2F^+ + SbF_6^- \tag{7.40}$$

$$SbF_5 + HSO_3F \longrightarrow HO-\overset{\overset{\displaystyle O}{\|}}{\underset{\underset{\displaystyle F}{|}}{S}}-O-SbF_5 \xrightarrow{\ HSO_3F\ } H_2SO_3F^+ + FSO_3SbF_5^-$$

In SbF_5/HSO_3F the acidity is even greater on addition of several moles of SO_3. In such media even protonated alkanes are obtained.

[5] R. J. Gillespie, *Acc. Chem. Res.* **1968,** *1,* 202; G. A. Olah, G. K. S. Prakash, and J. Sommer, *Superacids,* Wiley, New York, 1985.

7.3   ACID–BASE BEHAVIOR OF THE BINARY OXIDES AND AQUA CATIONS

Aqueous acid–base chemistry plays important roles in laboratory chemistry, industrial chemistry, geology, and biology, among other areas. Thus, we discuss the strengths of common reagents in water as acids and bases: binary oxides, oxoacids, and cations. Our viewpoint is that the Arrhenius–Ostwald or Brønsted–Lowry model seems most appropriate.

▶ 7.3.1   *Relative Acid or Base Strength of Binary Oxides and Oxoanions*

Most binary oxides dissolve either in acids or in alkali and thus may be classified as basic or acidic oxides, respectively. Alternatively, if in an aqueous solution the element in question forms oxoanions, its oxide is classified as **acidic;** if the element forms cations (with or without oxygen content), its oxide is classified as **basic.** Upon dissolving in water the central atom coordinates with water and a proton transfer takes place from the bonded water to a bonded oxide, producing an oxoacid $XO_n(OH)_m$. It is customary to write the formulas of oxoacids as $H_mXO_{(m+n)}$ (e.g., $HClO_4$) even though the acidic protons are attached to the O atoms. The final coordination number of X is determined primarily by the size of X, the number of lone electron pairs on X, and, to a lesser extent, the pH. In only a few known ions do oxo and aqua ligands coexist (ligands are groups donating electron pairs to the central atom). Examples include the hydrated cations (a) the uranyl ion, UO_2^{2+} [*trans*-tetraaquadioxouranium(VI) ion], (b) the vanadyl ion, VO^{2+} [pentaaquaoxovanadium(IV) ion], and (c) the dioxovanadium(V) ion, VO_2^+ [*cis*-tetraaquadioxovanadium(V) ion].

The periodic trends in acid–base behavior may be noted in the following sequence of oxides, where B indicates basic, A indicates acidic, Am indicates amphoteric, and S and W are strong and weak. Oxidation numbers of the positive partner of the oxides are given below:

I	II	III	IV	V	VI	VII
Na_2O	MgO	Al_2O_3	SiO_2	P_4O_{10}	SO_3	Cl_2O_7
SB	WB	Am	WA	A	SA	VSA

We can rationalize this trend readily on an electrostatic model by assuming that the charge on the central atom parallels the oxidation number. Similarly, the variation of basicity or acidity on descending a given periodic group may be rationalized on the basis of an electrostatic model and the size variation of the central atom.

Let us consider the general formula of an hydroxide or oxoacid as X—O—H. *If X has low positive charge density* (as for a metal of Group IA with the largest radius of the period and low charge or oxidation number, +I), *XOH is basic, giving X^+ and OH^-. If X has high positive charge density* (as for a nonmetal with small radius and high oxidation number—because of additional oxo ligands), *XOH is acidic, giving XO^- (or XO_nO^- from XO_nOH) and H^+*. Thus we see the decrease in basicity of hydroxides from metals of Groups 1 to 2 to 3 (IA → IIIA). The basicity increases with increasing radii going down each of these groups. Similarly the acidity of oxoacids increases with increasing oxidation state of X or a decrease in radius. Thus $HClO_4$ is a very strong acid and $HClO$ is a very weak acid; acidity decreases from $HClO$ to $HBrO$ to HIO.

It is clear that the distinction between bases and acids is not just a metal/nonmetal distinction. For chromium, CrO (large radius and low charge on Cr^{2+}) is basic, Cr_2O_3 is amphoteric, and CrO_3 is acidic. Thus CrO_3, with chromium in its highest oxidation state and with the smallest effective radius, dissolves in water to produce H_2CrO_4 (chromic acid), whose acid strength is comparable to that of H_2SO_4.

Table 7.3 lists the dissociation constants (as pK) of some of the acids formed by some nonmetal oxides. (Base constants are given in the following section on hydrolysis.) Many of the acids for which dissociation constants are given cannot be isolated as anhydrous compounds. The K_1 constants for "H_2CO_3" and for "H_2SO_3" apply to the following equilibria:

$$CO_2(aq) + H_2O \; \rightleftharpoons \; H^+ + HCO_3^-$$
$$SO_2(aq) + H_2O \; \rightleftharpoons \; H^+ + HSO_3^- \tag{7.41}$$

The greater acidity of H_2SO_3, in part results from the greater solubility of SO_2 compared to that of CO_2.

Pauling, among others, formulated an empirical rule for estimating the strength of oxoacids: For an acid of the general formula $XO_n(OH)_m$, the value of pK_1 varies in approximate steps of five units (differences of 10^5 in K_1) and is related to n, the number of oxo ligands (O atoms not bonded to H), in the following approximate manner:

$$pK_1 = 7 - 5n \tag{7.42}$$

Pauling's rule may be interpreted as indicating that the resonance stabilization of the anion is increased by each oxo ligand, but not by an OH. Also each oxo ligand increases the positive oxidation number of X by two. Thus H_3BO_3 and $HOCl$ are very weak acids (both $n = 0$), but H_2SO_4 ($n = 2$) is stronger than H_2SO_3 ($n = 1$) and HNO_3 ($n = 2$) is stronger

Table 7.3 pK_a values of some oxoacids

	H_2CO_3	H_2SO_4	H_2SeO_4	H_6TeO_6	H_3BO_3	
pK_1	6.73	≈−3	≈−3	7.68	9.14	
pK_2	10.33	1.92	1.92	11.29		
	HNO_3	H_3PO_4	H_3AsO_4	H_2SO_3	H_2SeO_3	H_2TeO_3
pK_1	−1.4	2.16	2.25	1.81	2.46	2.48
pK_2		7.21	6.77	6.91	7.31	7.70
	$HClO$	$HClO_2$	$HClO_3$	$HClO_4$	$HBrO$	HIO
pK_1	7.4	2.0	−1	≈−10	8.7	11

than HNO_2 or H_3PO_4 (both $n = 1$). Because of the electrostatic effect, pK_2 of a polyprotic acid is approximately five units greater than pK_1; that is, $K_1 \simeq K_2 \times 10^5$.

Phosphorus Oxoacids

The oxoacids of phosphorus provide constants (Table 7.4) that reflect interesting structural changes. The pK_1 value for phosphoric acid is consistent with the predicted value of 2. The value of 2.00 for phosphorous acid, H_3PO_3, suggests that there is also one oxo ligand as in

The structure was confirmed by analysis of the nuclear magnetic resonance (NMR) spectrum of aqueous phosphorous acid.

EXAMPLE 7.3: The pK value for arsenious acid, H_3AsO_3, is 9.2 and that for hypophosphorus acid, H_3PO_2, is ~1. What structures are consistent with these values?

Solution: The high pK for H_3AsO_3 indicates no oxo ligand, as below. The pK for H_3PO_2 indicates $n = 1$, with one oxo ligand giving a monoprotic acid.

Table 7.4   Strength of the oxoacids of phosphorus[a]

Acid	Formula	pK_1	pK_2	pK_3	pK_4	pK_5	pK_6
Phosphoric[b]	H_3PO_4	2.161	7.207	12.325			
Phosphorous[b]	H_3PO_3	*2.00*	6.58				
Hypophosphorus[b]	H_3PO_2	~1					
Pyrophosphoric[b]	$H_4P_2O_7$	1.52	2.36	6.60	9.25		
Triphosphoric[c]	$H_5P_3O_{10}$		*1.06*	2.30	6.50	9.24	
Tetraphosphoric[c]	$H_6P_4O_{13}$			*1.36*	*2.23*	7.38	9.11
Trimetaphosphoric[b]	$H_3P_3O_9$	2.05					
Tetrametaphosphoric[b]	$H_4P_4O_{12}$	2.74					
Hypophosphoric[b]	$H_4P_2O_6$	*<2.2*	*2.81*	7.27	*10.03*		

[a] At 25°C and zero ionic strength—except for italicized values, which are for other ionic strengths and/or temperatures.

[b] From J. Bjerrum, G. Schwarzenbach, and L. G. Sillén, "Stability Constants, Part II: Inorganic Ligands," *Chem. Soc. (London) Spec. Publ.,* **1958,** 7.

[c] From J. I. Watters, P. E. Sturrock, and R. E. Simonaitis, *Inorg. Chem.* **1963,** 2, 765.

The pK_2 and pK_3 values of H_3PO_4 show the expected decrease in acidity with increasing negative charge on the anion which the proton is leaving. It is somewhat surprising that the effect of increasing charge is so slight for the successive dissociation constants of the polyphosphoric acids. The phosphorus units in a polyphosphoric acid appear to behave almost independently. Thus there is one fairly acidic hydrogen for each phosphorus atom in a polyphosphoric acid. The terminal units carry an additional weakly acidic hydrogen, with a pK comparable to the pK_2 of phosphoric acid. The difference in the last two pK values for polyphosphoric acids indicates some interaction between the ends of the chain, arising from its flexibility.

We can determine the ratio of the terminal units to total units in a polyphosphoric acid from the ratio of weakly acidic hydrogen atoms to moderately acidic hydrogen atoms (i.e., from the titration curves). Largely from this type of data, Van Wazer and Holst[6] concluded that no branched polyphosphates exist in aqueous solution.

▶ 7.3.2 Hydrolysis and Aqua Acids

Metal Ions

We can relate the dissociation constant (K_b) of the metal hydroxides in water to the dissociation constants (K_a) of the cationic aqua acids (or hydrolysis constants). Hydrolysis of a cation is often considered as $M^+ + H_2O \leftrightarrows MOH + H^+$, but for hydrated ions $M(H_2O)_n^+ \leftrightarrows M(H_2O)_{(n-1)}OH + H^+$ is more realistic since H_2O is bonded to the cation. K_{hyd} is the same as K_a for the aquated ion.

$$
\begin{array}{lll}
M(H_2O)_n^+ & \rightleftharpoons & M(H_2O)_{n-1}OH + H^+(aq) \quad K_a \text{ (or } K_{hyd}) \\
MOH(aq) & \rightleftharpoons & M^+(aq) + OH^-(aq) \quad\quad\quad K_b \\
\hline
\text{Adding, we get} \quad H_2O & \rightleftharpoons & H^+(aq) + OH^-(aq) \quad\quad\quad K_w
\end{array}
$$

and $K_a K_b = K_w$, or $pK_a + pK_b = pK_w$. Hydrolysis is slight for cations forming strong bases (e.g., Na^+).

For cations with a noble-gas configuration, the interaction between the metal ion and the hydroxide ion is essentially electrostatic—that is, the pK_b varies in a linear fashion with q^2/r, where q is the charge on the ion and r is the ionic radius, and pK_b refers to the loss of the last hydroxide group. Thus the basicity of LiOH is lower than that of all alkali metal hydroxides, and the hydrolysis of lithium salts is greater than that of the salts of other alkali metals.

$$\text{LiOH}(aq) \quad \rightleftharpoons \quad \text{Li}^+(aq) + \text{OH}^-(aq) \qquad pK_b = 0.18 \tag{7.43}$$

[6] J. G. Van Wazer and K. A. Holst, *J. Am. Chem. Soc.* **1950**, *72*, 639.

or

$$Li(H_2O)_3OH(aq) + H_2O \rightleftharpoons Li(H_2O)_4^+(aq) + OH^-(aq)$$

Hence Li^+ ions in Li salts behave as weak acids:

$$Li^+(aq) + H_2O \rightleftharpoons LiOH(aq) + H^+(aq)$$

or $\quad Li(H_2O)_4^+ \rightleftharpoons Li(H_2O)_3(OH)(aq) + H^+(aq) \quad pK_a = 13.8$ (7.44)

In each pair, the first equation in no way implies the degree of solvation occurring, and the second equation indicates speculation regarding the primary hydration of the lithium ion and the lithium hydroxide ion pair. The increase in ionic radius for Na^+ compared to Li^+ results in NaOH being a stronger base ($pK_b = -0.8$) than LiOH. KOH, RbOH, and CsOH are even stronger bases approaching the limiting pK_b value of -1.7; in other words, K^+, Rb^+, and Cs^+ are not measurably hydrolyzed in aqueous solution.

The ionic radii of Be^{2+}, Mg^{2+}, Ca^{2+}, Sr^{2+}, and Ba^{2+} are about 30 pm less than the corresponding isoelectronic alkali metal ions, and the ionic charge is, of course, twice as great. Accordingly, all these ions are hydrolyzed more highly in solution than the lithium ion, with the pK_a values varying from about 11.4 for Mg^{2+} to 13.5 for Ba^{2+} for the reaction

$$M^{2+}(aq) + H_2O \rightleftharpoons MOH^+(aq) + H^+(aq)$$ (7.45)

The Be^{2+} ion is even more extensively hydrolyzed, forming polynuclear species in solution. Appreciable concentrations of the unhydrolyzed Be^{2+} ion occur in dilute solutions under fairly acidic conditions.

Extending the electrostatic approach beyond the alkaline earth metal ions gives less satisfactory results, as Figure 7.2 indicates. Thus although we might expect a decrease in hydrolysis of the ions in the series Mg^{2+}, Zn^{2+}, Cu^{2+}, Cd^{2+}, and Hg^{2+}, because of the increasing size of the respective ions, we actually find the reverse: Mg^{2+} is least hydrolyzed ($pK_a = 11.4$) and Hg^{2+} is most hydrolyzed ($pK_a = 3.4$). This inversion of order results from the increasing polarizability of the metal ions in the series given, along with their increasing electronegativity (Mg^{2+} 1.3 to Hg^{2+} 2.0). Both factors lead to an increase in the covalence of the M—O bond.

The second dissociation constant of the aqua mercury(II) cationic acid is *greater* than the first.

$$Hg(H_2O)_2^{2+} \rightleftharpoons Hg(OH)(H_2O)_y^+ + H^+ \quad pK_{a_1} = 3.4$$
$$Hg(OH)(H_2O)_y^+ \rightleftharpoons Hg(OH)_2 + H^+ \quad pK_{a_2} = 2.6$$ (7.46)

Probably a change in the coordination number of mercury during hydrolysis accounts for these results. The difference between pK_{a_1} and pK_{a_2} of other cationic aqua acids is smaller than the five units generally found for oxoacids (for example, for $Zr^{4+}(aq)$ $pK_1 = -0.3$, $pK_2 = 1.7$, $pK_3 = 5.1$). This indicates that the charge at the metal ion center is not greatly affected by the loss of a proton from a coordinated water molecule. In contrast,

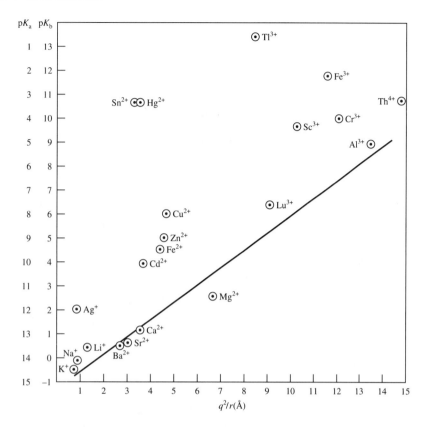

Figure 7.2   Relationship between acidity and aqua ions and the ionic charge squared to radius ratio.

changing a unit charge on the metal ion brings about a large change in hydrolysis [for $Fe^{2+}(aq)$, $pK_a = 9.5$; for $Fe^{3+}(aq)$, $pK_a = 2.2$]. We can conclude that the loss of a proton by a water molecule is not transmitted, but a change in charge of M affects all M—O bonds and profoundly affects the acidity. Al^{3+} salts are often applied to the soil around acid-loving plants because the q^2/r ratio leads to a relatively large K_a. Aluminum sulfate is added to the soil to alter the color of hydrangea from pink (high pH) to blue (low pH).

Metal Oxides and Amphoterism

Weakly basic oxides begin to polymerize and eventually precipitate from solution as the solution is made more alkaline—that is, as the pH increases. Weakly acidic oxides begin to polymerize and eventually precipitate from solution as the solution is made more acidic. Amphoteric oxides precipitate from solution and then redissolve as the pH is increased, starting with the cationic species, or as pH is decreased, starting with the anionic species. The predominance diagram for vanadium(V) species as a function of pH and log concentration is shown in Figure 7.3. At very high pH, simple vanadates occur and the cation

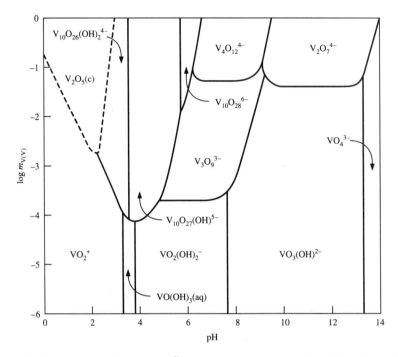

Figure 7.3 Predominance diagram for V^V species as a function of pH at 25°C. The solid lines represent conditions under which the species in adjacent regions contain equal amounts of V^V. The dashed lines represent the solubility of V_2O_5 in terms of V^V concentration. (Reproduced with permission from C. F. Baes, Jr. and R. E. Mesmer, *The Hydrolysis of Cations*, Wiley–Interscience, New York, 1976, p. 210, copyright 1976.)

VO_2^+ exists in strongly acidic solution. Polyvanadates are important at intermediate pH values, particularly in slightly acidic solutions.

Separating one element from another frequently requires judicious use of acid–base and/or redox properties. In the processing of vanadium ore, after roasting to convert the sulfoxide to the pentaoxide the V_2O_5 is purified by heating with Na_2CO_3 to form $NaVO_3$, which is water-soluble at high pH but precipitates as V_2O_5 on acidifying. In the production of vanadium steels, V_2O_5 added is reduced along with iron oxide.

Cerium can be separated from the other rare earth elements by oxidizing it to Ce^{IV} while leaving the other rare earths in the +III state. Increasing the pH precipitates the very weak base $Ce(OH)_4$ before any of the $RE^{III}(OH)_3$ starts to separate. Manganese occurs in igneous rocks as Mn^{2+} accompanying the more abundant ions with comparable radii, Fe^{2+} and Mg^{2+}. During the weathering process Mn^{2+} is oxidized and precipitates as MnO_2 (see Section 8.6.2 for pH/potential diagrams for Mn). The bulk of metals as aluminosilicates are dispersed in the original igneous rocks, but important ores commonly occur in the sedimentary rocks resulting from oxidation and/or hydrolysis during the weathering process. We take advantage of separations effected by nature. Thus most of the Al occurs in the aluminosilicates, accounting for about two-thirds of the earth's crust, but the Al ore is bauxite, $Al_2O_3 \cdot n\, H_2O$, deposited by hydrolysis of aluminosilicates.

It is likely that most oxides could exhibit amphoteric character; however, normal conditions in aqueous solution give an insufficient range of acidity to show both acidic and basic properties. Thus N_2O_3 forms NO_2^-, nitrite ion, in basic solution, whereas in highly acidic media NO^+, the nitrosyl cation, is formed. Likewise, N_2O_5 forms NO_3^- in most media but forms NO_2^+, the nitryl cation, in concentrated sulfuric acid. In 1933–1941, Scholder and co-workers isolated from basic solution a series of salts containing $Zn(OH)_4^{2-}$, $Zn(OH)_6^{4-}$, $Cu(OH)_4^{2-}$, $Cu(OH)_6^{4-}$, $Pb(OH)_4^{2-}$, $Pb(OH)_6^{4-}$, and so on. In a medium in which bases of greater strength than OH^- are available, it should be possible to remove the remaining protons in these complexes. The oxide is much more basic than the hydroxide ion, but in water it is converted to the hydroxide ion. In fused alkali or alkaline earth metal oxides, however, we can prepare compounds[7] such as Li_8PbO_6 or Ba_3MoO_6.

7.4 NONPROTONIC CONCEPTS OF ACID–BASE REACTIONS

▶ 7.4.1 Lux Concept

In the reaction sequence

$$BaO + H_2O \longrightarrow Ba(OH)_2(aq)$$
$$CO_2 + H_2O \rightleftharpoons H_2CO_3(aq)$$
$$\searrow BaCO_3(s) + 2\,H_2O \qquad (7.47)$$

the second stage is clearly an acid–base reaction. The reaction can be carried out directly with BaO and CO_2; that is,

$$BaO + CO_2 \longrightarrow BaCO_3 \qquad (7.48)$$

We know of many other examples of the direct reaction between acidic and basic anhydrides in the absence of water or of hydrogen ions:

$$3\,Na_2O + P_2O_5 \longrightarrow 2\,Na_3PO_4$$
$$CaO + SiO_2 \longrightarrow CaSiO_3 \qquad (7.49)$$

The Brønsted definition of an acid by the equation

$$\underset{\text{Acid}}{A\text{H}} \rightleftharpoons \underset{\text{Base}}{A^-} + \underset{\text{Proton}}{H^+} \qquad (7.50)$$

is not applicable. Lux proposed for oxide systems the defining equation

$$\underset{\text{Base}}{B} \rightleftharpoons \underset{\text{Oxide ion}}{O^{2-}} + \underset{\text{Acid}}{A} \qquad (7.51)$$

[7] R. Scholder, *Angew. Chem.* **1958**, *70*, 583.

According to this definition, *a base is an oxide ion donor and an acid is an oxide ion acceptor*. This is obviously a special case of the Usanovitch definition. The Lux view applies in particular to high-temperature chemistry, as in the fields of ceramics and metallurgy. Thus the ores of Ti, Ta, and Nb can be brought into solution at around 800°C in sodium disulfate ($Na_2S_2O_7$, also known as sodium pyrosulfate) or in potassium disulfate:

$$\underset{\text{Base}}{TiO_2} + \underset{\text{Acid}}{Na_2S_2O_7} \longrightarrow Na_2SO_4 + (TiO)SO_4 \tag{7.52}$$

Displacement of the more volatile acid from solution is seen in the following reactions of sand or clay (the reactant is shown as SiO_2):

$$\begin{aligned}
3\,SiO_2 + Ca_3(PO_4)_2 &\xrightarrow{\Delta} 3\,CaSiO_3 + P_2O_5(g) \\
SiO_2 + CaSO_4(\text{gypsum}) &\xrightarrow{\Delta} CaSiO_3 + SO_3(g)
\end{aligned} \tag{7.53}$$

By this same scheme, we may classify substances as amphoteric if they show a tendency both to take up or to give up oxide ions depending on the circumstances—that is,

$$\underset{\text{Base}}{ZnO} + S_2O_7^{2-} \longrightarrow Zn^{2+} + 2SO_4^{2-} \tag{7.54}$$

$$Na_2O + \underset{\text{Acid}}{ZnO} \longrightarrow 2\,Na^+ + ZnO_2^{2-}$$

The order of decomposition temperature of metal sulfates and carbonates with evolution of SO_3 or CO_2 is $Ba^{2+} > Li^+ > Ca^{2+} > Mg^{2+} > Mn^{2+} > Cd^{2+} > Pd^{2+} > Co^{2+} > Ag^+ > Fe^{2+} > Ni^{2+} > Cu^{2+} > Fe^{3+} > Be^{2+}$. This order of stability is approximately the order of base strength of the hydroxides. This is just what we would expect because the decompositions are the reverse of acid–base reactions such as (7.48).

The Lux oxide transfer picture of acid–base reactions can be extended to any negative ion—the halides, sulfides, or even carbanions! The following illustrate these reactions.

$$\begin{array}{ccccc}
\textit{Negative ion} & & \textit{Negative ion} & & \\
\textit{donor} & & \textit{acceptor} & & \\
3\,NaF & + & AlF_3 & \xrightarrow{\text{high temp.}} & 3\,Na^+ + AlF_6^{3-} \\
Na_2S & + & CS_2 & \longrightarrow & 2\,Na^+ + CS_3^{2-} \\
EtNa & + & Et_2Zn & \longrightarrow & Na^+ + ZnEt_3^- \\
\text{Base} & & \text{Acid} & & \text{Acid} \quad\ \text{Base}
\end{array} \tag{7.55}$$

▶ 7.4.2 Lewis Acids and Bases

In 1938 G. N. Lewis proposed the following operational definition of acids and bases in terms of the reactions they could undergo.

a. Neutralization—acids and bases react rapidly to neutralize each other.

b. Displacement—a strong acid will displace a weaker acid from its compounds; a strong base will displace a weaker base from its compounds.

    c. Titration—an indicator may be used to determine the neutralization endpoint.

    d. Catalysis—acids (and bases) may catalyze reactions.

These properties already had been widely associated with protonic acid–base reactions. Lewis pointed out that a number of nonprotonic compounds could exhibit these properties of acids—SO_3, $AlCl_3$, $SnCl_4$, and BF_3, among others. He proposed that the fundamental quality shared by these acids is a *vacant orbital* that could accept an electron pair with covalent bond formation. A base is a compound with an unshared pair of electrons. The Lewis acid picture covers (with occasional stretching) the ion-transfer process as a special case and extends the concept of acids and bases to nonionic compounds, while focusing attention on the strength of the new covalent bond formed.

Strength of Lewis Acids and Bases

Comparing the strengths of Lewis acids and bases is somewhat difficult because of the many different methods used to establish orders. These methods include gas-phase dissociation data, calorimetric heats of reaction, competition experiments among several competing acids (or bases) for an insufficient amount of base (or acid), displacement methods, and studies of the volatility of addition compounds formed (the more volatile, the less stable the adduct within a series). Nevertheless, we can draw conclusions within a given series, and overlap of several series permits some generalizations, although caution should be exercised.

Position in the Periodic Table. With regard to position in the periodic table, the strength of Lewis bases toward Lewis acids parallels their basicity toward the proton, provided that only sigma bonds are formed and steric factors are not great. The following examples illustrate this.

	Molar heat of reaction with BMe_3	*Properties of the product of reaction with HCl*
Me_3N	-74 kJ	Sublimes at 250°C
Me_3P	-67 kJ	Sublimes at 125°C
Me_3As	Exists only below -80°C	Unstable at room temperature
Me_3Sb	No compound formed	Unstable above -80°C

Likewise, the base strength decreases in going from Me_3N to Me_2O to MeF. Trimethylamine readily displaces dimethyl ether from boron trifluoride methyl etherate. The variation in strength of Lewis acids according to position in the periodic table is somewhat more complex, but for oxygen and nitrogen bases the trends in base strength parallel the tendency toward hydrolysis discussed earlier (i.e., B > Be > Li > Be > Mg > Ca, etc.). Variations within the transition metals will be discussed more fully in Section 9.4, which deals with stability of coordination compounds.

Effect of Substituents. Often we may predict the effect of substituents on the strength of Lewis acids and bases from a study of the inductive, resonance, and steric effects of the substituents. Thus an electron-withdrawing inductive effect makes an acid stronger (more willing to accept an electron pair) and a base weaker (less willing to donate an electron pair). Hence the base strength decreases in the series $Me_3N > NH_3 > NF_3$, whereas the acid strength increases in the series $Me_3B < BH_3{}^8 < BF_3$. This order is precisely that expected based on the inductive effect of these groups. The order of acid strength $BF_3 < BCl_3 < BBr_3$ is not the order expected from the inductive effects, but this can be explained by the greater resonance stabilization of BF_3 (as compared with that of BCl_3) resulting from the decreasing tendency to form double bonds as the size of the halogen atom increases.

3 equivalent forms

The weakness as an acid of methylborate, $(MeO)_3B$, compared with trimethylboron may be attributed to resonance.

Steric Effects. In general, steric effects only slightly influence ion-transfer equilibria, especially when the ions differ by only a proton. In Lewis acid–base reactions, steric effects may be quite large. Thus whereas 2-methylpyridine is a stronger base toward a proton than pyridine (as would be expected from the inductive effect), it is a much weaker base than pyridine toward Lewis acids such as BMe_3.

ΔH		
Reaction with $B(CH_3)_3$	-71 kJ stable	-74 kJ stable

Trimethylboron does not react at all with 2,6-dimethylpyridine (2,6-lutidine), even though this base has a higher pK_a than the 2-methylpyridine or pyridine itself. We can attribute the effect of these *ortho*-methyl groups in reducing the stability of the Lewis acid adduct to the strain between nonbonded groups attached to different atoms that have conflicting steric requirements.

A change in hybridization during the formation of a molecular addition compound may also bring about a change in steric requirement. Thus boron changes from sp^2 to sp^3 hybridization during acid–base reactions of trialkylboron compounds. Branching of the alkyl groups greatly reduces the acid strength of trialkylboron. Trimesitylboron (see below) is inert even to such strong bases as methoxide ion, since there would be too much crowding if the mesityl groups were forced from a trigonal planar to a tetrahedral bonding arrangement.

---

[8] After correction for bridge breaking of B_2H_6.

Inert due to strain in sp^3 configuration

Some Inverted Orders.   Although BF_3 is "normally" a stronger acid than diborane, B_2H_6, the latter forms a compound with CO—that is, H_3BCO—but the former does not. Likewise, the compound dimethyl sulfide is a stronger base than dimethyl ether toward BH_3, although the reverse is true with BF_3. Both of these may be explained on the basis of hyperconjunction of the BH_3 group, stabilizing the addition compound; that is,

3 equivalent structures

The latter involves the $3d$ orbitals of S and hence would be of no importance for oxygen derivatives.

▶ 7.4.3   The Hard–Soft Acid–Base Model

G. N. Lewis's criterion of the strength of a base is that *a stronger base should be able to displace a weaker base from its compounds* (the criterion for acid strength is similar).

$$:B_{strong} + A_{acid}:B_{weak} \longrightarrow A_{acid}:B_{strong} + :B_{weak} \qquad (7.56)$$

For aqueous solution, using hydrogen ion as the reference acid, the order of base strength is the same as the order of pK_a of the BH^+ species; that is,

$$NH_3(aq) + HF(aq) \rightleftharpoons NH_4^+(aq) + F^-(aq) \qquad (7.57)$$

pK_a $\qquad$ 3.45 $\qquad$ 9.25

The log K_{eq} for the displacement reaction is 5.80, simply the difference in the pK_a values of the ammonium ion and of hydrofluoric acid. Hence we conclude that the equilibrium lies far to the right and that NH_3 is a better base than F^- toward the reference acid H^+.

For a metal ion combining with a ligand (a Lewis base bonded to the metal ion) the equilibrium constant is called the **formation constant** or **stability constant.** Addition of ligands occurs stepwise, and the overall formation constant is designated as β (see Section 9.4).

$$Ag^+ + NH_3 \;\rightleftharpoons\; Ag(NH_3)^+ \qquad K_1 \tag{7.58}$$

$$Ag(NH_3)^+ + NH_3 \;\rightleftharpoons\; [Ag(NH_3)_2]^+ \qquad K_2 \tag{7.59}$$

$$Ag^+ + 2\,NH_3 \;\rightleftharpoons\; [Ag(NH_3)_2]^+ \qquad \beta = K_1 K_2 \tag{7.60}$$

The formation constants for one metal ion can give the order of base strength for a series of ligands, and with a fixed ligand for a series of metal ions the formation constants can give the order of acid strength of the metal ions with respect to the particular ligand and solvent system. Lewis noted that the order of base strengths obtained using metal ions as the reference is not necessarily the same as that obtained using hydrogen ion as the reference acid. With ammonia and the fluoride ion as ligands, Ag^+ gives the same order of base strengths as $H^+(aq)$, whereas Ca^{2+} ion gives the opposite order.

$$\underset{\log K_1}{NH_3(aq)} + \underset{0.51}{CaF^+(aq)} \;\rightleftharpoons\; \underset{-0.2}{Ca(NH_3)^{2+}(aq)} + F^-(aq) \tag{7.61}$$

$$\log K_{\text{displacement}} \qquad = -0.2 - 0.51 = -0.7$$

$$\underset{\log K_1}{NH_3(aq)} + \underset{0.36}{AgF(aq)} \;\rightleftharpoons\; \underset{3.32}{Ag(NH_3)^+(aq)} + F^-(aq) \tag{7.62}$$

$$\log K_{\text{displacement}} \qquad = 3.32 - 0.36 = 2.96$$

Thus, in aqueous solution, ammonia is a stronger base than F^- toward silver ion and is a weaker base than F^- toward calcium ion. In the gas phase, or a suitably nonsolvating nonaqueous medium, F^- ion would be more basic toward both metal ions, because of strong electrostatic attraction for the anion.

Using the equilibrium constants in reactions (7.57), (7.61), and (7.62), we obtain the following orders of acid strength in aqueous solution.

$$H^+ > Ag^+ > Ca^{2+} \qquad \text{toward } NH_3(aq) \tag{7.63}$$

$$H^+ > Ca^{2+} > Ag^+ \qquad \text{toward } F^-(aq) \tag{7.64}$$

The ligand $I^-(aq)$ gives us a still different order, $Ag^+ > Ca^{2+} > H^+$. Acidity orders using different reference bases should become much more uniform, but still not identical, if all reactions were carried out in the gas phase.

Let us now consider a metathesis, or interchange, reaction involving two pairs of acid–base adducts. The equilibrium constant for the interchange reaction may be obtained in a straightforward fashion from the equilibrium constants for the displacement reactions, or from the appropriate formation constants. Thus, for the interchange between AgF and $Ca(NH_3)^{2+}$, we have

$$AgF + Ca(NH_3)^{2+} \rightleftharpoons Ag(NH_3)^+ + CaF^+$$

$$\log K_{interchange} = \log K(7.62) - \log K(7.61) = 3.66$$

(7.65)

where $K(7.61)$ and $K(7.62)$ refer to the displacement reactions (7.61) and (7.62). The direction of this metathesis reaction could also be predicted qualitatively from the order of acidity toward various bases obtained above. As with protonic acids, we should like to develop some rationalization of these Lewis acidities and basicities based on the structure of the reactants.

In order to explain the differences in relative base strength toward various Lewis acids, Pearson proposed the classification of hard and soft bases. *Hard acids* (e.g., H^+, Na^+, Mg^{2+}) *and hard bases* (e.g., F^-, OH^-) *hold electrons tightly, and thus they are not readily polarized. Soft acids* (e.g., Ag^+, Hg^{2+}) *and soft bases* (e.g., I^-, H^-, S^{2-}) *do not bind electrons tightly, and thus the electron clouds are easily distorted or polarized.* Hard acid–base interactions are primarily electrostatic, while soft acid–base interactions involve significant polarization (covalence). Pearson proposed that *a hard acid prefers to combine with a hard base, and a soft acid prefers to combine with a soft base.* We will use this rule [now known as the **hard–soft acid–base (HSAB) principle**] as an operational definition of hardness or softness when the assignment of hard and soft to a reference pair of acids (or bases) seems relatively unambiguous. Although classification of a reagent as hard or soft is fairly straightforward as a function of charge and position in the periodic table, the relative ranking on the hardness or softness scale is often not obvious from these considerations. The orderings of softness and hardness are established purely empirically, and we now discuss some ways of obtaining these orderings.

Comparisons of Softness and Hardness

Although we readily can calculate $K_{interchange}$ from the stability of the acid–base pairs, an affinity diagram is useful when a large amount of data must be inspected. Here the affinity diagram is a plot in which the abscissa is the affinity of a base for various acids [as measured by $(1/n) \log \beta_n$] and the ordinate is a similar scale for a second base. Placing unit slope lines through points on such a diagram makes the lateral (or vertical) distance between the lines a measure of the $\log K$ (or, depending on the data used, ΔH or ΔG) for the interchange reaction. The $\log K$ for the interchange can be obtained from the difference in the intercepts on either axis. The $\log K$ is positive when the product acid–base adduct is represented by the abscissa and the line with the greater intercept with the abscissa (the other product acid–base adduct, of course, is represented by the ordinate and the line with the greater intercept with the ordinate). Figure 7.4 depicts the affinity diagram for the Ag^+ and Ca^{2+} ions with NH_3 and F^-. You can see here that the equilibrium represented by Equation (7.61) lies to the right from the fact that the intercept of the Ag^+ line with

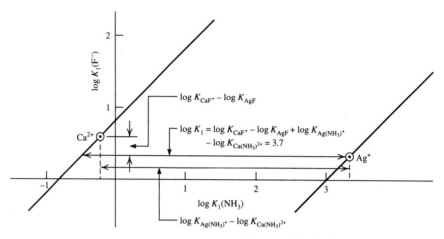

Figure 7.4 An affinity diagram for Ca^{2+} and Ag^+ with F^- and NH_3.

the abscissa lies farther to the right (is more positive) than the intercept of the Ca^{2+} line. Of course, similar diagrams could be constructed using the affinities for two different acids as ordinates and abscissa. The unit slope lines would then each correspond to a different base.

G. Schwarzenbach[9] proposed that the methyl mercury(II) ion would provide an ideal soft ion to compare with the hard hydrogen ion, since both form 1:1 acid:base complexes as their primary product in aqueous solution. An affinity plot of pK_a versus log $K_{CH_3Hg^+}$ gives the following order of softness for the bases:

$$I^- > Br^- > Cl^- > S_2O_3^{2-} > S^{2-} > (C_6H_5)_3P > CN^- >$$
$$SCN^- > SO_3^{2-} > NH_3 > F^- > OH^- \qquad (7.66)$$

Although the use of an interchange reaction removes much of the effect of solvation, the position of NH_3 in the above order undoubtedly stems from the high solvation energy of the NH_4^+ ion. Likewise, the ability of OH^- to fit into the water structure places OH^- as a harder base than it would otherwise fall.

Avoiding solvation entirely, Phillips and Williams[10] used heats of formation data for pairs of metal ions of comparable size to construct affinity plots. Using Ca^{2+} compounds as hard acid complexes, and Cd^{2+} compounds as soft acid complexes, the order of decreasing softness of ligands found is

$$Te^{2-} > Se^{2-} > S^{2-} > I^- > Br^- > O^{2-} > Cl^- > OH^- > CO_3^{2-} >$$
$$\text{hydrated ion} > NO_3^- > SO_4^{2-} > F^- \qquad (7.67)$$

[9] In *Chemical and Engineering News*, **1965**, 90–103. This article is a condensation of 12 symposium papers on the HSAB concept.

[10] C. S. G. Phillips and R. J. P. Williams, *Inorganic Chemistry*, Vol. II, Chapter 31, Oxford University Press, Oxford, 1966. Their approach is not actually the affinity-plot approach, but is mathematically equivalent to it.

In the gas phase, all metal ions, hard or soft, prefer to bond to F^- rather than to I^-. Absolute hardness[11] applies to interactions without effects of solvation and possible rehybridization. The hardness or softness of acids and bases are evaluated from acid–base interactions. In an interchange reaction between fluorides and metal iodides in aqueous solution, soft metal ions prefer I^- and hard metal ions prefer F^-. Assigning F^- as a hard base and I^- as a soft base, Williams and Hale[12] arrived at the following order of hardness for cations from interchange reactions:

$$\text{SOFT } Au^+ < Ag^+ < Hg^{2+} < Cu^+ < Cd^{2+} < Cs^+ < Rb^+ < K^+ < Na^+ < Cu^{2+} <$$
$$Pb^{2+}, Ca^{2+} < Tl^+, Mn^{2+} < Zn^{2+} < In^+ < Fe^{2+} < Co^{2+}, Ni^{2+}, Sn^{4+} <$$
$$Cr^{2+}, Sn^{2+}, Ge^{2+}, Ba^{2+}, Sr^{2+} < In^{3+} < Ni^{3+}, Fe^{3+} < Co^{3+}, Bi^{3+} < Li^+ < \qquad (7.68)$$
$$Mn^{3+} < Ga^{3+} < Ti^{2+} < Cr^{3+} < Sc^{2+} < Sb^{3+} < Ge^{4+} < Mg^{2+} < V^{3+} < V^{2+} <$$
$$Ga^+ < Y^{3+}, Ti^{3+} < Sc^{3+} < As^{3+} < Ti^{4+}, La^{3+} < Nb^{5+} < Zr^{4+} < Al^{3+} <$$
$$Be^{2+} \text{ HARD}$$

Pearson's classification of some Lewis acids and bases is given in Table 7.5.

Charge/Size Effects and Hardness in Exchange Reactions

Emphasis on polarization and polarizing power resulting in soft acids and bases may have implied that increased covalent character in soft–soft interactions is *the* driving force for exchange reactions. We can show readily that in the products in an exchange reaction between ion pairs, the highly charged anions will pair with the highly charged cations, or, if charge is kept constant, the smallest cation will pair with the smallest anion.

Although not a necessary consequence of the above, the empirical generalization has been made that metal complex ions that are difficult to isolate often can be isolated as salts of large ions having an equal but opposite charge.

Huheey and Evans[13] have shown that the preference of small ions for other small ions does not necessarily support an electrostatic model for the acid–base pair, since covalent interactions also favor the matching of small with small. They made a convincing argument that the matching of small with small (i.e., hard with hard) is a major factor in the HSAB principle. Increasing the charge on the cation serves to intensify the $q_1 q_2/r$ term, as well as to decrease the ionic radius. This helps justify classifying I^{7+}, Mn^{7+}, Cr^{6+}, and so on,

[11] R. G. Pearson, *Inorg. Chem.* **1988**, *27*, 734; *Coord. Chem. Rev.* **1990**, *100*, 403.

[12] R. J. P. Williams and J. D. Hale, in *Structure and Bonding*, Vol. 1, C. J. Jørgensen, Ed., Springer-Verlag, Berlin, 1966, pp. 255ff.

[13] J. E. Huheey and R. S. Evans, *J. Inorg. Nucl. Chem.* **1970**, *32*, 383; *Chem. Commun.* **1969**, 968.

Table 7.5   Classification of hard and soft acids and bases

Hard Acids	Borderline Acids	Soft Acids
H^+, Li^+, Na^+, K^+, Be^{2+}, Mg^{2+}, Ca^{2+}, Sr^{2+}	$B(CH_3)_3$	$(BH_3)_2$
BF_3, Al^{3+}, $AlCl_3$, $Al(CH_3)_3$	Fe^{2+}, Co^{2+}, Ni^{2+}, Cu^{2+}, Zn^{2+}	$GaCl_3$, $GaBr_3$, GaI_3
Mn^{2+}, Cr^{3+}, Cr^{VI}, Mn^{VII}, Mo^{VI}, W^{VI}	Ru^{2+}, Rh^{2+}, Sn^{2+}, Sb^{3+}	Cu^+, $Co(CN)_5^{3-}$, Ag^+, Cd^{2+}
Sc^{3+}, La^{3+}, Ce^{3+}, Lu^{3+}, Ti^{4+}, Zr^{4+}, Hf^{4+}	Rh^{3+}, Ir^{3+}, Pb^{2+}, Bi^{3+}	Pt^{2+}, Pt^{4+}, Au^+, Hg_2^{2+}, Hg^{2+}, Tl^+
VO^{2+}, UO_2^+, Th^{4+}, Pu^{4+}		M^0
CO_2, SO_3		

Hard Bases	Borderline Bases	Soft Bases
O^{2-}, OH^-, F^-, Cl^-, CO_3^{2-}, NO_3^-, $CH_3CO_2^-$	$C_5H_5^-$, N_2, $:NO_2^-$	H^-, R^-, CN^-, I^-
PO_4^{3-}, SO_4^{2-}, ClO_4^-	$C_6H_5NH_2$, N_3^-, $:SO_3^{2-}$, Br^-	C_2H_4, RNC, CO
H_2O, ROH, RO^-, R_2O	$SCN:^-$ (N donor)	R_3P, $(RO)_3P$, R_3As, RSH, R_2S
NH_3, RNH_2, N_2H_4		RS^-, $S_2O_3^{2-}$, $:SCN^-$ (S donor)

as hard acids. Unfortunately, there appear to be insufficient thermodynamic data to settle this classification because most soft bases undergo oxidation in the presence of these acids. They have been classified as hard presumably on the basis of electrostatic arguments and on the nonexistence of complexes with soft bases!

Qualitative impressions of relative hardness are sometimes misleading. For compounds absolute hardness (η) is given by $\eta = (IP - EA)/2$. Very stable compounds commonly have wide separations of the highest occupied molecular orbital (HOMO) and the lowest unoccupied molecular orbital (LUMO) and have high hardness. Compounds with small separations of HOMO and LUMO are softer and more reactive. In these terms hardness is not necessarily an indication of polarizability. Datta[14] tabulated hardness of many compounds and examined exchange reactions as follows:

$$\Delta H^0 \text{ (kcal/mol)}$$

$$H^+Cl^- + Li^+H^- \longrightarrow Li^+Cl^- + H^+H^- \qquad -56.1 \qquad (7.69)$$
$$\eta \quad 8.0 \qquad 4.08 \qquad\qquad 4.75 \qquad 8.7$$

$$HO^+F^- + Li^+H^- \longrightarrow Li^+F^- + HO^+H^- \qquad -144.1 \qquad (7.70)$$
$$\eta \quad 7.82 \qquad 4.08 \qquad\qquad 5.87 \qquad 9.5$$

$$Li^+F^- + H^+Br^- \longrightarrow Li^+Br^- + H^+F^- \qquad -10.8 \qquad (7.71)$$
$$\eta \quad 5.87 \qquad 6.83 \qquad\qquad 4.04 \qquad 11.0$$

$$SiH_3^+I^- + Li^+H^- \longrightarrow Li^+I^- + SiH_3^+H^- \qquad -43.9 \qquad (7.72)$$
$$\eta \quad 4.54 \qquad 4.08 \qquad\qquad 3.2 \qquad 6.8$$

$$CH_3^+F^- + H^+I^- \longrightarrow CH_3^+I^- + H^+F^- \qquad -12.3 \qquad (7.73)$$
$$\eta \quad 9.4 \qquad 5.3 \qquad\qquad 4.7 \qquad 11.0$$

$$Li^+H^- + H^+F^- \longrightarrow Li^+F^- + H^+H^- \qquad -49.0 \qquad (7.74)$$
$$\eta \quad 4.08 \qquad 11.0 \qquad\qquad 5.87 \qquad 8.7$$

[14] D. Datta, *Inorg. Chem.* **1992**, *31*, 2797. See also reference 15 page 195.

In reactions (7.69–7.73) the hardest of the four possible combinations is formed. Note that H_2O is harder (HOMO − LUMO) than LiF. For reaction (7.71) our expectation in terms of polarizabilities might be that the reverse reaction would occur, thereby forming LiF, but it is the great hardness of HF that is important. *The driving force is usually the production of the single hardest species.* In the last reaction, and a few other examples considered by Datta, the hardest species (HF here) is not formed. In almost all of these cases *the average hardness of the products is greater than that of the reactants.*

7.5   THE DRAGO-WAYLAND EQUATION

In 1965 Drago and Wayland[15] introduced an empirical four-parameter equation to describe and predict the enthalpy change accompanying the reaction of weak neutral acids with weak neutral bases in poorly solvating solvents or in the gas phase. This equation takes the form

$$-\Delta H = E_A E_B + C_A C_B \qquad (7.75)$$

where E_A and C_A are parameters for the acid (arbitrarily set as $E_A = 0.50$ and $C_A = 2.00$ for I_2) and where E_B and C_B are parameters for the base. It is felt that the $E_A E_B$ term reflects the electrostatic part of the acid–base interaction and that the $C_A C_B$ term reflects the covalent part. The equation is similar in form to a number of other equations in the literature, but differs in that the others correlate independently observable quantities such as stability constants with the Brønsted acidity and oxidative coupling constants. The beauty of the Drago–Wayland equation lies in its minimum of constraints, which in turn allows the best possible fitting of the enthalpy data by an equation of its form. Accordingly, it can perform equally the task of fitting a given set of data as an equation of similar form that uses observable properties as parameters. It thus sets a limit to the degree to which any equation of its form can fit a set of data.

With the constants listed in Table 7.6, we can predict the enthalpy change for 900 acid–base reactions. If an interchange reaction between two acid–base adducts is considered, we can predict approximately one million such enthalpies by the equation

$$\Delta H = \Delta E_A \Delta E_B + \Delta C_A \Delta C_B \quad \text{for} \quad A_1 B_1 + A_2 B_2 \longrightarrow A_1 B_2 + A_2 B_1 \quad (7.76)$$

where the Δ terms are all consistent (either all A and B subscripted 1 subtracted from those subscripted 2, or vice versa). We can use such data to construct affinity diagrams, and it can be shown that a consistent ordering of hardness or softness cannot be achieved with the set of acids and bases for which we have E and C parameters.

Along with predicting enthalpy data, the Drago–Wayland equation can be used to select solvents that have nearly the same extent of acid–base interactions with solutes matching the E and C parameters. Thus dimethyl sulfoxide (dmso) and dimethylformamide (dmf) would show similar acid–base behavior toward solutes, since they have almost identical E and C values.

[15] R. S. Drago and B. B. Wayland, *J. Am. Chem. Soc.* **1965,** *87,* 3571.

Table 7.6   *E* and *C* parameters [Equation (7.78)] for some acids and bases[a-c]

Acid	E_A	C_A	R_A	Acid	E_A	C_A	R_A
I_2	0.50	2.00	—	H^+	45.00	13.03	130.21
H_2O	1.54	0.13	0.20	Li^+	11.72	1.45	24.21
H_2S	0.77	1.46	0.56	K^+	3.78	0.10	20.79
HF	2.03	0.30	0.47	NH_4^+	4.31	4.31	18.52
HCl	3.69	0.74	0.55	$(CH_3)_2NH_2^+$	3.21	0.70	20.72
HCN	1.77	0.50	0.54	$(CH_3)_3NH^+$	2.60	1.33	15.95
CH_3OH	1.25	0.75	0.39	$(CH_3)_4N^+$	1.96	2.36	8.33
C_2H_5OH	1.34	0.69	0.41	$C_5H_5NH^+$	1.81	1.33	21.72
C_6H_5OH	2.27	1.07	0.39	H_3O^+	13.27	7.89	20.01
$(CH_3)_3COH$	1.36	0.51	0.48	$(H_2O)_2H^+$	11.39	6.03	7.36
$HCCl_3$	1.49	0.46	0.45	$(H_2O)_3H^+$	11.21	4.66	2.34
HCF_3	1.32	0.91	0.27	$(H_2O)_4H^+$	10.68	4.11	−3.25
CH_3CO_2H	1.72	0.86	0.63	CH_3^+	19.70	12.61	55.09
$B(OCH_3)_3$	0.54	1.22	0.84				
$B(C_2H_5)_3$	1.70	2.71	0.61				
PF_3	0.61	0.36	0.87				
AsF_3	1.48	1.14	0.78				
SO_2	0.56	1.52	0.86				

EXAMPLE 7.4:   Would pyridine (C_5H_5N) be closest in acid–base behavior toward solutes to a primary, secondary, or tertiary alkyl amine?

Solution: The *E* and *C* values for pyridine fall between those for CH_3NH_2 and $(CH_3)_2NH$, so pyridine should be intermediate in acid–base behavior between primary and secondary amines.

A modification of Equation (7.75) by Drago et al.[16] adds a term, *W*, a constant contribution to the enthalpies of reaction for a particular acid (or base) that is independent of the bases (or acids) reacting.

$$-\Delta H = E_A E_B + C_A C_B + W \tag{7.77}$$

The term *W* is usually zero for neutral acids reacting with neutral bases. As an exception, this term would be the enthalpy required for the dissociation of a dimeric acid prior to combination with a base.

For reactions of a cationic acid with a neutral base, a neutral acid with an anionic base, or a cationic acid with an anionic acid base, the amount of charge transfer involved requires an equation of the form of Equation (7.77). Reactions of ions can be handled by replacement of *W* with two additional parameters: R_A for the acid as the receptor and T_B for the base as the transmitter. Values for R_A and T_B are given in Table 7.6.

$$-\Delta H = E_A E_B + C_A C_B + R_A T_B \tag{7.78}$$

[16] R. S. Drago, N. Wong, C. Bilgrien, and G. C. Vogel, *Inorg. Chem.* **1987**, *26*, 9.

**Table 7.6   E and C parameters [Equation (7.78)]
for some acids and bases^{a-c} (Continued)**

Base	E_B	C_B	T_B	Base	E_B	C_B	T_B
NH_3	2.31	2.04	0.56	CH_3OH	1.80	0.65	0.70
CH_3NH_2	2.16	3.12	0.59	C_2H_5OH	1.85	1.09	0.70
$(CH_3)_2NH$	1.80	4.21	0.64	C_6H_6	0.70	0.45	0.81
$(CH_3)_3N$	1.21	5.61	0.75	H_2S	0.04	1.56	1.13
$C_2H_5NH_2$	2.35	3.30	0.54	HCN	1.19	0.10	0.90
$HC(C_2H_4)_3N$	0.80	6.72	0.83	H_2O	2.28	0.10	0.43
C_5H_5N	1.78	3.54	0.73				
CH_3CN	1.64	0.71	0.83	F^-	9.73	4.28	37.40
$HC(O)N(CH_3)_2$				Cl^-	7.50	3.76	12.30
(dmf)	2.19	1.31	0.74	Br^-	6.74	3.21	5.86
$(C_2H_5)_2O$	1.80	1.63	0.76	I^-	5.48	2.97	6.26
$O(C_2H_4)_2O$	1.86	1.29	0.71	CN^-	7.23	6.52	9.20
$(CH_3)_2SO$				OH^-	10.43	4.60	50.73
(dmso)	2.40	1.47	0.65	CH_3O^-	10.03	4.42	33.77
$(CH_3)_2O$	1.68	1.50	0.73				
$(CH_3)_2S$	0.25	3.75	1.07				
$(CH_3)_3P$	1.46	3.44	0.90				

^{a} Data reproduced with permission from R. S. Drago, D. C. Ferris, and N. Wong, *J. Am. Chem. Soc.* **1990,** *112,* 8953 and R. S. Drago, N. Wong, and D. C. Ferris, *J. Am. Chem. Soc.* **1991,** *113,* 1970.

^{b} These parameters yield enthalpy values in kcal/mol adduct.

^{c} The original sources give parameters for additional acids and bases and give limitations of data for some values.

Gas-phase acid–base reactions of ions involve very large charge transfer. Note the very large change in parameters for H^+ compared to the $H^+(H_2O)_n$ species given in Table 7.6. These effects show that proton affinities are not good guides to the interpretation of acid–base reactions in solution.

EXAMPLE 7.5:   Calculate $-\Delta H$ for the combination of H^+ with H_2O successively to produce H_3O^+, $H^+(H_2O)_2$, $H^+(H_2O)_3$, and $H^+(H_2O)_4$. Compare the results.

Solution: For $H^+ + H_2O \longrightarrow H_3O^+$, $-\Delta H = (45.00)(2.28) + (13.03)(0.10) + (130.21)(0.43) = 159.89$ kcal/mol (669 kJ/mol) Experimental value: $-\Delta H = 166.5$ kcal/mol.

For $H_3O^+ + H_2O \longrightarrow H^+(H_2O)_2$,

$$-\Delta H = 39.65 \text{ kcal/mol (166 kJ/mol)}$$

For $H^+(H_2O)_2 + H_2O \longrightarrow H^+(H_2O)_3$,

$$-\Delta H = 29.7 \text{ kcal/mol (124 kJ/mol)}$$

For $H^+(H_2O)_3 + H_2O \longrightarrow H^+(H_2O)_4$,

$$-\Delta H = 27.0 \text{ kcal/mol (113 kJ/mol)}$$

Obviously the enthalpy change is very large for formation of H_3O^+, with smaller enthalpy changes for each additional H_2O.

EXAMPLE 7.6:   Calculate $-\Delta H$ for the reactions of H^+ with NH_3 and H_3O^+ with NH_3.

Solution: For $H^+ + NH_3 \longrightarrow NH_4^+$,

$$-\Delta H = (45.00 \times 2.31) + (13.03 \times 2.04) + (130.21 \times 0.56)$$
$$= 203.45 \text{ kcal/mol (851 kJ/mol)}$$

For $H_3O^+ + NH_3 \longrightarrow H_2O + NH_4^+$, $-\Delta H = (13.27 \times 2.31) + (7.89 \times 2.04) +$
$(20.01 \times 0.56) = 57.96$ kcal/mol (242 kJ/mol)

EXAMPLE 7.7:   Calculate $-\Delta H$ for $HCl(g) + NH_3(g)$.

Solution: For $HCl(g) + NH_3(g) \longrightarrow NH_4^+ + Cl^-$,

$$-\Delta H = (3.69 \times 2.31) + (0.74 \times 2.04) + (0.55 \times 0.56) = 10.34 \text{ kcal/mol (43.26 kJ/mol)}$$

Note that the $R_A T_B$ term is small for the reaction of neutral molecules.

7.6 SOLID ACIDS[17]

Many organic compounds are produced using acid catalysts, for example, a strong acid such as H_2SO_4 or H_3PO_4, or a Lewis acid such as $AlCl_3$. Now there are large scale industrial processes using **solid acids** as catalysts for preparing ethyl acetate in one step from ethylene and acetic acid, isooctane from lower hydrocarbons, and methyl *t*-butyl ether (a gasoline additive to replace tetraethyllead) from methanol and 2-methylpropene, and many others. The reactions are much cleaner and more environmentally "friendly" than older processes. Important solid acid catalysts are zeolites (page 244) and clay. Both of these are aluminosilicates with open channels. The pores of zeolites can be as much as 50% of the total volume. Natrolite $[Na_{16}(Al_2Si_3O_{10})_8 \cdot 16H_2O]$ and mordenite $[Na_8(AlSi_5O_{12})_8 \cdot 24H_2O]$ are zeolites that serve as cation exchangers. Minerals in clays commonly have layer structures similar to that of apophyllite (Figure 5.28) with replaceable cations between layers. In zeolites and clays the mobile cations can be replaced by H^+, thereby forming solid acids.

The open channels of zeolites vary in size. Table 5.14 gives the formulas and pore diameters for some zeolites. Synthetic zeolites can be produced to obtain the desired size of channels (molecular sieves). A synthetic zeolite ZSM-5 $[Na_3(AlSi_{31}O_{64})_3 \cdot 16H_2O]$ is used to produce gasoline from methanol on a large scale in countries without petroleum resources. The shape-size selectivity is illustrated by the synthesis of *p*-xylene in preference to the unwanted usual by-product *o*-xylene. The rod-shaped *p*-xylene molecules fit into the pores of ZSM-5, but *o*-xylene molecules do not.

Isopoly acids (Section 15.8.4) are solid acids that show promise as catalysts for organic reactions. They lack the pores of zeolites.

[17] J. M. Thomas, *Sci. Am.* **1992**, *266*(4), 112; B. C. Gates, *Catalytic Chemistry*, Wiley, New York, 1992.

GENERAL REFERENCES

C. F. Baes, Jr. and R. E. Messner, *The Hydrolysis of Cations,* Wiley-Interscience, New York, 1976.

J. E. Bartmess and R. T. McIver, Jr., "The Gas Phase Acidity Scale" in *Gas-Phase Ion Chemistry,* M. T. Bowers, Ed., Academic Press, New York, 1978.

R. P. Bell, *The Proton in Chemistry,* 2nd ed., Cornell University Press, Ithaca, NY, 1973.

R. S. Drago and N. A. Matwiyoff, *Acids and Bases,* Heath, Boston, 1968.

W. B. Jensen, *The Lewis Acid–Base Concepts,* Wiley, New York, 1980.

R. G. Pearson, *Survey of Progress in Chemistry,* Vol. 5, Chap. 1, A. Scott, Ed., Academic Press, New York, 1969.

W. Stumm and J. J. Morgan, *Aquatic Chemistry,* Wiley-Interscience, New York, 1981. Coverage of the chemistry of natural waters.

PROBLEMS

7.1 Which has broader applications in terms of definitions of acids and bases, the Arrhenius–Ostwald or the Brønsted–Lowry definitions?

7.2 What is the great departure in terms of acid–base definitions in the Lewis theory compared to those of Arrhenius–Ostwald and Brønsted–Lowry?

7.3 **(a)** Indicate the conjugate bases of the following: NH_3, NH_2^-, NH^{2-}, H_2O, HI.
 (b) Indicate the conjugate acids of the above species.
 (c) What relationship exists between the strength of a conjugate acid and a conjugate base for a neutral substance, such as NH_3 in parts **a** and **b** above?

7.4 The heat of neutralization of $H^+(aq)$ and $OH^-(aq)$ is considerably higher in concentrated salt solution ($\Delta H_{neut.} = -85.4$ kJ/mol at $\mu = 16$) than in dilute aqueous solution ($\Delta H_{neut.} = -56.5$ kJ/mol at $\mu \rightarrow 0$). Explain what might cause the increase in the heat of neutralization when a high concentration of NaCl is present.

7.5 Write a thermochemical cycle from which you can obtain the hydride affinity of a positive ion (E^+) from the bond dissociation energy of EH and other suitable data.

7.6 Why is there a smaller difference in the PAs of PH_3 and PF_3 compared to those of NH_3 and NF_3?

7.7 Why is $H^+(aq)$ the strongest acid and $OH^-(aq)$ the strongest base possible in water?

7.8 Are the solution acidity scales compressed more for solvents that stabilize lyonium and lyate ions greatly or for those that stabilize them minimally?

7.9 What would be a good solvent for studying
 (a) a series of very weak bases?
 (b) a series of very weak acids?
 (c) a wide range of strengths of acids and bases?

7.10 In what media might one obtain protonated species of very weakly basic substances such as alkanes?

7.11 Give the approximate pK_a values for the following acids.
   (a) H_3PO_3 pK_1 =   pK_2 =
   (b) HNO_3 pK_1 =
   (c) $HClO_4$ pK_1 =
   (d) H_5IO_6 pK_1 =   pK_2 =

7.12 Select the best answer and give the basis for your selection.
   (a) Thermally most stable: PH_4Cl PH_4Br PH_4I
   (b) Strongest acid: H_2O H_2S H_2Se H_2Te
   (c) Acidic oxide: Ag_2O V_2O_5 CO Ce_2O_3
   (d) Strongest acid: MgF_2 $MgCl_2$ $MgBr_2$
   (e) Stronger base (toward a proton): PH_2^- NH_2^-

7.13 Give equations to explain why adding ammonium acetate to either zinc amide(s) in liquid ammonia or zinc acetate(s) in acetic acid causes the solid to dissolve.

7.14 What are the acid–base properties expected for the following?
   (a) Mn^{2+} **(b)** Mn^{IV} **(c)** Mn^{VII}

7.15 Transition metal ions usually are encountered in igneous rocks in low oxidation states. Why are many of their ores found in sedimentary rocks formed by the weathering of igneous rocks?

7.16 Would the following *increase, decrease,* or have *no effect* on acidity of the solution? Why?
   (a) Addition of Li_3N to liquid NH_3
   (b) Addition of HgO to an aqueous KI solution.
   (c) Addition of SiO_2 to molten $Fe + FeO$
   (d) Addition of $CuSO_4$ to aqueous $(NH_4)_2SO_4$
   (e) Addition of $Al(OH)_3$ to $NaOH$
   (f) Addition of $KHSO_4$ to H_2SO_4
   (g) Addition of CH_3CO_2K to liquid NH_3

7.17 Some transition metals such as vanadium and molybdenum in their highest oxidation state form aggregated poly acids. Are the poly acids formed to the greater extent in **(a)** strongly acidic solution **(b)** strongly basic solution, or **(c)** intermediate pH values?

7.18 In general, anhydrous nitrites of a given metal ion have lower thermal stability than the nitrates, and similarly for the sulfites and sulfates. How may this be explained?

7.19 Select the best response within each horizontal group and indicate the major factor governing your choice.

   Strongest protonic acid
   (a) SnH_4 SbH_3 H_2Te
   (b) NH_3 PH_3 SbH_3
   (c) H_5IO_6 H_6TeO_6 HIO
   (d) $[Fe(H_2O)_6]^{3+}$ $[Fe(H_2O)_6]^{2+}$ H_2O
   (e) $Na(H_2O)_x^+$ $K(H_2O)_x^+$

   Strongest Lewis acid
   (f) BF_3 BCl_3 BI_3
   (g) $BeCl_2$ BCl_3
   (h) $B(n\text{-}Bu)_3$ $B(t\text{-}Bu)_3$

   More basic toward BMe$_3$
   (i) Me_3N Et_3N
   (j) 2-MePy   4-MePy   Py (=pyridine)
   (k) $2\text{-}MeC_6H_4CN$ C_6H_5CN

7.20 Tetrahydrofuran (thf) has an ionization energy of 11.1 eV, and diethyl ether has an ionization energy of 9.6 eV. How can we account for this difference? What difference is expected in their PAs? Which is higher? thf is much more basic toward $MgCl_2$ and Grignard reagents than Et_2O. How might this observation be rationalized?

7.21 Geochemical classifications distinguish metals as *lithophile* if they tend to occur as oxides and *chalcophile* if they tend to occur as sulfides. What properties of the metals determine this distinction? Give two examples of each classification.

7.22 Indicate which of the following Lewis bases (ligands) should be able to coordinate through more than one donor atom at a time (ambidentate character), and indicate for such bases the hard and soft ends: NO_3^-, SO_3^{2-}, $S_2O_3^{2-}$, $S_2O_7^{2-}$, NO_2^-, Me_2SO, CN^-, $SeCN^-$.

7.23 Use Table 7.6 and the Drago–Wayland Equation to decide which of the following in each pair is the stronger base:

(a) Acetone or dimethyl sulfoxide (dmso)

(b) Dimethyl sulfide or dmso

Comment on your conclusions regarding possible ambiguity.

▶8◀

Oxidation–Reduction Reactions

·················

> Thermodynamic data can summarize a great deal of information concerning chemical stability and properties. Many inorganic reactions involve reduction and oxidation. For these "redox" reactions, standard potentials of electromotive forces are useful in calculating cell voltages and predicting which possible reactions occur. The stability of species in water and the products of reactions can be detemined from standard potentials. We must also be able to deal with effects of concentrations of the species of an element as well the acidity of the solution on standard potentials. Potential–pH diagrams are useful for summarizing the regions of stability for possible species of various oxidation states for an element.

8.1 CONVENTIONS COVERING STANDARD STATES

At equilibrium the free-energy change accompanying a reaction at a fixed temperature and pressure is zero. Thus, from the equilibrium partial pressures at 1000 K of 0.564 atm SO_2, 0.102 atm O_2, and 0.333 atm SO_3, we may write

$$2\,SO_2(0.564\text{ atm}) + O_2(0.102\text{ atm}) \rightleftharpoons 2\,SO_3(0.333\text{ atm}) \qquad \Delta G_{1000K} = 0$$

Assuming ideal gas behavior for the above system, we then may write

$$K_p = \frac{(P_{SO_3})^2}{(P_{SO_2})^2(P_{O_2})} = 3.42\text{ atm}^{-1} \quad \text{or} \quad 4.5 \times 10^{-3}\text{ torr}^{-1}$$

We can discover the ΔG value for the above reaction under conditions other than the equilibrium value by means of the following equation, known as the van't Hoff reaction isotherm:

$$\Delta G_T = RT \ln \frac{Q}{K} \qquad (8.1)$$

where Q is an activity quotient similar in form to the equilibrium constant but with activity values for some beginning combination of reactants and products. Thus at 1000 K and with $P_{SO_2} = 5$ atm, $P_{O_2} = 2$ atm, and $P_{SO_3} = 10$ atm, we have

$$Q = \frac{(10)^2}{(5)^2(2)} = 2 \text{ atm}^{-1}$$

and since

$$K = 3.42 \text{ atm}^{-1}$$

then

$$\Delta G_{1000 \text{ K}} = 2.303 \times 8.314 \text{ J} \times 1000(\log 2 - \log 3.42) = -4461 \text{ J}$$

Under the stated conditions, there is a considerable driving force and equilibrium is approached by the formation of SO_3.

The units used in expressing Q and K must be the same, of course. When the arbitrary activity of unity is selected for both the reactants and the products, the free-energy change for the reaction is called the "standard" free-energy change, designated by the superscript zero: that is ΔG^0.

From the relationship $\Delta G^0 = -RT \ln K$, we calculate a "standard" free-energy change for the SO_2 oxidation as follows:

$$2 SO_2(1 \text{ atm}) + O_2(1 \text{ atm}) \; \rightleftarrows \; 2 SO_3(1 \text{ atm}) \qquad \Delta G^0_{1000 \text{ K}} = -9372 \text{ J}$$

or[1]

$$2 SO_2(1 \text{ torr}) + O_2(1 \text{ torr}) \; \rightleftarrows \; 2 SO_3(1 \text{ torr}) \qquad \Delta G^0_{1000 \text{ K}} = +44980 \text{ J}$$

This example reveals that the equilibrium constant and the standard free-energy change depend upon the standard states selected for the reactants and the products.

For thermodynamic equilibrium constants, the quantities used in the equilibrium product should be activities rather than concentrations. We define the standard state of any substance as a state of unit activity. According to convention, the standard state for *gases* is unit fugacity—that is, the gas behaving ideally at atmospheric pressure. For most gases

---

[1] The standard state of 1 torr assumed here is not the conventional one, and this ΔG^0 would normally not be designated as ΔG^0 (see below).

the standard state may thus be taken as 1 atm. For *pure liquids* and *solids,* which occur as separate phases in reactions, the standard state is defined conventionally as the pure solid or pure liquid at 1 atm pressure. Conventions covering *solutions* are not as uniform as those for gases and pure liquids or solids. We usually define the standard state for the solvent as that of the pure liquid. For the solute the standard state may be taken as an **activity,** *a* (thermodynamic concentration), of 1 molal. For approximate calculations, *concentrations* (*m*) in units of molality may be used instead of activities. These two quantities are related to each other through the use of activity coefficients (γ):

$$a = \gamma m \tag{8.2}$$

At high dilution, γ approaches 1. Note that using activities in terms of molarities (or mole fractions) involves different standard states from those selected by the use of molalities.

EXAMPLE 8.1:   A standard free-energy change of 45.6 kJ is calculated from the K_{sp} for $PbSO_4$. Write the chemical equation for which this ΔG^0 applies.

Solution: $PbSO_4(s) \longrightarrow Pb^{2+}(\gamma^{\pm}m = 1) + SO_4^{2-}(\gamma^{\pm}m = 1)$

EXAMPLE 8.2:   At 20°C, K_N is 4 for the esterification reaction of acetic acid with ethanol. Mole fractions are used as concentrations in the expression K_N. Calculate ΔG^0 and write the hypothetical reaction to which it applies. What data would be needed to calculate K_p?

Solution: $\Delta G^0 = -RT \ln K_N = -3430$ J for the reaction

$$CH_3CO_2H(l) + C_2H_5OH(l) \longrightarrow CH_3CO_2C_2H_5(l) + H_2O(l)$$

This must also be the free-energy change for the above reaction when each of the substances is in the gas phase with a partial pressure equal to the vapor pressure of the pure liquid. We thus can find K_p by using the ΔG^0 as ΔG in Equation (8.1) and evaluating Q from the vapor pressures.

8.2   METHODS OF DETERMINING CHANGES IN FREE ENERGY

Changes in free energy, enthalpy, and entropy at constant temperature are related through the equation

$$\Delta G = \Delta H - T \, \Delta S \tag{8.3}$$

Knowing any two of these quantities allows us to calculate the other, and knowing all three enables us to check the assumptions involved in the calculation and/or in the experimental methods. The most common methods used to obtain one or more of these pieces of information are:

1. Direct calorimetric measurement of heats of reaction (ΔH).
2. Heat capacity measurements (ΔS).
3. Direct determination of equilibrium constants (ΔG).
4. Galvanic cell potentials (ΔG).

5. Statistical mechanics and appropriate spectral data (ΔG).

6. Approximation methods.

We shall discuss only galvanic cell potentials in this chapter.

8.3 SCHEMATIC REPRESENTATION OF GALVANIC CELLS

An **electrode** consists of a metallic conductor in contact with (or part of) a phase boundary across which a difference in electrical potential occurs. The phase boundary is represented by a single vertical line, $|$, separating the components of the phases. The schematic representation of the electrode signifies a half-reaction in which positive ions move from left to right and negative ions move from right to left. The representations below thus imply the half-reactions that follow.

Electrode	*Half-Reaction*
$Ag(s) \mid Ag^+(aq)$	$Ag(s) \longrightarrow Ag^+(aq) + e$
$Ag^+(aq) \mid Ag(s)$	$Ag^+(aq) + e \longrightarrow Ag(s)$
$Pt(s) \mid Fe^{2+}(aq), Fe^{3+}(aq)$	$Fe^{2+}(aq) \longrightarrow Fe^{3+}(aq) + e$
$Ag(s), AgCl(s) \mid Cl^-(aq)$	$Ag(s) + Cl^-(aq) \longrightarrow AgCl(s) + e$
$Pt(s), H_2(g) \mid H^+(aq)$	$\frac{1}{2}H_2(g) \longrightarrow H^+(aq) + e$

The last electrode listed above is called the **hydrogen electrode**, which has a defined electrode potential of zero under standard conditions. It consists of a platinum electrode coated with platinum black in contact with a solution saturated with hydrogen. An early design by Hildebrand, still widely used, is shown in Figure 8.1.

Figure 8.1   The hydrogen electrode. (Reproduced with permission from J. H. Hildebrand, *J. Am. Chem. Soc.* **1913,** *35,* 847. Copyright 1913, American Chemical Society.)

8.4 CONVENTIONS REGARDING CELLS

To construct a galvanic cell we must be able to carry out an oxidation half-reaction and a reduction half-reaction in separate places (i.e., at the electrodes). Electrolytic conduction (current carried by ions) must be possible internally, and metallic conduction (current carried by electrons) must be possible externally. For some systems the electrolyte may be the same throughout the cell; the cell may be represented by the appropriate combination of electrodes, as in the following example:

Cell: $Pt, H_2(g) \mid HCl(aq) \mid AgCl(s), Ag$

Reaction: $AgCl(s) + \frac{1}{2}H_2 \longrightarrow Ag(s) + HCl(aq)$

To construct a cell in which the reaction $Zn + 2HCl \rightarrow H_2 + ZnCl_2$ may be carried out reversibly, we must prevent H^+ from coming into direct contact with Zn. One way of doing this is to insert a salt bridge (designated by two parallel vertical lines, $\parallel$) that prevents the direct mixing of the electrolytes of the two electrode half-cells. (The salt bridge may introduce liquid junction potentials, which will not be discussed here.) The reaction of hydrogen ion with Zn could thus be carried out reversibly in the following cell:

$$Zn(s) \mid ZnCl_2(aq) \parallel HCl(aq) \mid H_2(g), Pt(s) \tag{8.4}$$

▶ 8.4.1 Electrode Potentials

When two electrodes are coupled to make a cell, the voltage developed by the cell is simply the difference in the electrical potential of the two electrodes. If we assign the hydrogen electrode under standard conditions a potential of zero, then we can assign the potential of any electrode relative to hydrogen. By this convention the potential of the Zn electrode in contact with Zn ions under standard conditions is -0.763 V. Note that the sign of the **potential** of the Zn electrode is independent of the cell in which the Zn electrode occurs and of whether oxidation or reduction takes place at the Zn electrode.

▶ 8.4.2 Relationship Between Cell Voltage and the Free-Energy Change for the Cell Reaction

To obtain the free-energy change for a reaction occurring in a galvanic cell, we simply multiply the potential difference of the cell electrodes by the number of coulombs passing through the circuit, or

$$\Delta G = -n\mathscr{F}E \tag{8.5}$$

where

ΔG is the free-energy change for a redox reaction
n is the number of equivalents oxidized or reduced (i.e., the number of electrons exchanged per formula unit)
$\mathscr{F}$ is the Faraday (96,500 coulombs/Faraday)
E is the voltage produced by the galvanic cell under reversible reactions

By convention, the voltage for a spontaneous reaction is positive whereas the free-energy change for a spontaneous reaction is negative. If the reactants and products are all present at unit activity, then E is E^0 and ΔG is ΔG^0.

▶ *8.4.3 Half-Cell Electromotive Force (emf) Values*

Under the conventions stated earlier, the emf for the reversible operation of the cell

$$\text{Zn}(s) \mid \text{Zn}^{2+}(aq) \parallel \text{H}^+(aq) \mid \text{H}_2(g), \text{Pt}(s) \tag{8.6}$$

is related to the free-energy change for the reaction

$$\text{Zn}(s) + 2\,\text{H}^+(aq) \longrightarrow \text{Zn}^{2+}(aq) + \text{H}_2(g) \tag{8.7}$$

through the relationship

$$\Delta G^0 = -n\mathscr{F}E^0 \tag{8.8}$$

Because the conventions assign ΔG_f^0 values (standard free energies of formation) of zero for hydrogen and hydrogen ion, the ΔG_f^0 value obtained may be assigned entirely to the free energy change for the oxidation half-reaction

$$\text{Zn} \longrightarrow \text{Zn}^{2+} + 2e \qquad E^0 = +0.763 \text{ V} \tag{8.9}$$

Because reaction (8.7) occurs spontaneously under standard conditions, ΔG^0 must be negative and, accordingly, E^0 must be positive. For the reverse of reaction (8.7),

$$\text{Zn}^{2+}(aq) + \text{H}_2(g) \longrightarrow \text{Zn}(s) + 2\,\text{H}^+(aq) \tag{8.10}$$

the signs of ΔG^0 and E^0 are reversed and, accordingly, E^0 for the reduction half-reaction

$$\text{Zn}^{2+}(aq) + 2e \longrightarrow \text{Zn}(s) \qquad E^0 = -0.763 \text{ V} \tag{8.11}$$

is reversed. In tabulations which compare half-reactions, it is conventional to represent the half-reactions as reductions. Note that the **standard emf** for a reduction half-reaction is identical in sign and magnitude to the value of the **standard potential for the electrode at which the half-reaction is carried out.** As pointed out above, the **electrode potential** is independent of whether an oxidation or reduction reaction is taking place at the electrode. Accordingly, terms such as oxidation potentials or reduction potentials are inappropriate in talking properly about emf values for half-cell oxidation or reduction reactions. Nevertheless, the International Union of Pure and Applied Chemistry has agreed to the use of the word "potential" as synonymous with the emf value for the **reduction half-cell reaction.** Because of lack of adherence to a single convention in many sources of emf data, you should remember the sign of a half-cell emf reaction such as that given in reaction (8.11) and check the sign given this half-reaction in any tabulation you use.

8.5   CALCULATIONS USING emf DATA

An obvious use of measured emf values is in predicting which redox reactions will be spontaneous.

► *8.5.1*   *emf Diagrams*

Given below is an emf diagram for the standard half-cell reduction reactions of Mn in acidic solution at 25°C:

$$E_A^0 \quad MnO_4^- \xrightarrow{0.564} MnO_4^{2-} \xrightarrow{2.27} MnO_2 \xrightarrow{0.95} Mn^{3+} \xrightarrow{1.51} Mn^{2+} \xrightarrow{-1.18} Mn \tag{8.12}$$

The arrows indicate the direction of the half-reactions. Except for writing the half-reactions as reductions, the above diagram is identical to those given in Latimer's[2] extensive treatment. Our diagram abbreviates the half-reactions by showing only the substance that changes in oxidation number. Thus E^0 of 2.27 V is for the half-reaction[3]

$$MnO_4^{2-} + 4H^+ + 2e \longrightarrow MnO_2 + 2H_2O \qquad E^0 = 2.27 \text{ V} \tag{8.13}$$

The pair of species related by oxidation–reduction is often referred to as a **couple**.

     The superscript zero implies that all substances shown in the equation are in their standard states; E_A^0 specifically indicates that the hydrogen ion activity is unity, and E_B^0 indicates that the hydroxide ion activity is unity.

► *8.5.2*   *Use of Half-Cell emf Data to Predict Chemical Reactions*

Because $\Delta G = -n \mathscr{F} E$, a positive value of E means that the reaction will be "spontaneous" in a thermodynamic sense—that is to say, that the reactants and products are not in equilibrium and equilibrium will be approached by the formation of more product. The major advantage of using emf data rather than the free-energy data is that with emf data the stoichiometries need not be taken into account, because the half-cell emf data are given in terms of volts per electron transferred. Thus the sum of the emf values for any oxidation half-reaction and any reduction half-reaction supplies an answer as to whether such a reaction may occur "spontaneously."

---

[2] W. M. Latimer, *The Oxidation States of the Elements and Their Potentials in Aqueous Solution*, 2nd ed., Prentice–Hall, New York, 1952.

[3] We can write the balanced half-reactions readily by the following sequence: (1) Balance the equation with respect to the atoms undergoing change in oxidation number. (2) Obtain a balance of oxygen atoms by adding H_2O to the appropriate side of the equation. (3) Balance the equation with respect to hydrogen atoms by adding H^+ to the appropriate side of the equation. (4) Balance the equation with respect to charge by adding electrons to the appropriate side of the equation.

When a substance is added to water, it may be unstable for one of the following reasons: (1) It may undergo disproportionation (auto-oxidation–reduction) in which species of higher and lower oxidation states are produced; (2) It may react with water to evolve H_2; or (3) It may react with water to evolve O_2.

EXAMPLE 8.3:   Is Mn^{2+} stable in aqueous solution?

Solution: Referring to the emf diagram for manganese [Equation (8.12)], we can obtain a sum for the half-reactions as follows:

$$
\begin{array}{lll}
 & & E^0 \\
Mn^2 + 2e \longrightarrow \quad Mn & & -1.18 \text{ V} \\
\underline{2(Mn^{2+} \longrightarrow \quad Mn^{3+} + e)} & & \underline{-1.51 \text{ V}} \\
3\,Mn^{2+} \longrightarrow \quad Mn + 2\,Mn^{3+} & & -2.69 \text{ V}
\end{array}
$$

Because E^0 is negative, the reaction does not occur and Mn^{2+} does not disproportionate. Note that the emf for the oxidation half-reaction

$$Mn^{2+} \longrightarrow \quad Mn^{3+} + e$$

is simply the negative of the emf for the reduction half-reaction given in the emf diagram.

We can readily identify species which undergo disproportionation in an emf diagram. Such a species has a more positive emf value to its right (for reduction) than the emf to its left. Here only Mn^{3+} and MnO_4^{2-} would disproportionate in acidic solution. We can verify these predictions from the method used above for Mn^{2+}.

Appendix E gives emf diagrams for the elements.

For many practical purposes, we can predict a reaction directly, without evaluating activity coefficients, using **formal emf values.** The formal emf, E^F, is the emf of a half-reaction with the *concentrations* of oxidant and reductant equal to unity, and with arbitrarily chosen concentrations of other electrolytes, including acids. Unfortunately, however, formal emf values vary with the ionic strength of the solutions, and a given list applies to one ionic strength only.

If formal emf values are available for a particular medium, we can make reaction predictions based on them conveniently. In many practical situations the necessary activity coefficients needed to permit the use of the standard emf are not easily available. Formal emf values in 1 M perchloric acid solution often are used as close approximations of E_A^0 values.

▶ *8.5.3   Combination of Half-Cell emf Values to Obtain Other Half-Cell emf Values: The Use of Volt-Equivalents*

From the emf diagram (8.12) given for manganese, we can find the half-cell emf values for a half-reaction that occurs between any pair of the species given in the diagram. Because E^0 refers to the voltage per electron, evaluating E^0 for the new half-reaction corresponds to calculating a weighted average. **Volt-equivalents**—that is, the product of the number of electrons involved in the half-reaction and the E^0 of the half-reaction—are used. That this gives the same E^0 as the use of the more familiar handling of ΔG^0 values can be seen from the following example.

EXAMPLE 8.4:   Calculate E^0 for $Mn^{3+} + 3e \rightarrow Mn$ from diagram (8.12) using ΔG^0 and using volt-equivalents.

Solution:

Reaction	E^0	$\Delta G^0 = -n\mathscr{F}E^0$	nE^0(*volt-equivalent*)
$Mn^{2+} + 2e \rightarrow Mn$	-1.18	$2\mathscr{F}(1.18)$	-2.36
$Mn^{3+} + \ e \rightarrow Mn^{2+}$	1.51	$-\mathscr{F}(1.51)$	1.51
$Mn^{3+} + 3e \rightarrow Mn$		$0.85\mathscr{F}$	-0.85

From ΔG^0 we have $E^0 = -(0.85\mathscr{F})/3\mathscr{F} = -0.28$ V; from nE^0 we have $3E^0 = -0.85$ and $E^0 = -0.28$ V.

In summary, we can obtain E^0 for a couple by multiplying each of the E^0 values of intervening couples by the number of electrons involved, adding the products, and dividing the sum by the total number of electrons involved in the new half-reaction. Thus the E_A^0 value for $MnO_4^- \rightarrow Mn^{2+}$ is given by

$$\frac{(1)(1.51) + (1)(0.95) + (2)(2.27) + (1)(0.564)}{5} = 1.51 \text{ V}$$

The free energies are algebraically additive in combining equations. The volt-equivalents are additive also because $\mathscr{F}$ cancels in converting E^0 values to ΔG^0 and back to E^0 for the final equation.

Hence in the example concerning Mn^{2+}, we need not consider disproportionation of Mn^{2+} to oxidation states higher than Mn^{3+}, since the half-cell emf for oxidation of Mn^{2+} to *any* higher oxidation state is negative.

▶ 8.5.4 *Effect of Concentration on Half-Cell emf Values*

Appropriately combining Equations (8.1) and (8.5) leads to the well-known **Nernst equation,** which gives the effect of variation of activity (thermodynamic concentration) on the emf of a cell or half-cell; that is,

$$E = E^0 - \frac{RT}{n\mathscr{F}} \ln Q \tag{8.14}$$

where $R = 0.314$ J or 8.314 volt-coulombs and $\mathscr{F} = 96{,}500$ coulombs. Q is an activity product identical in form to the equilibrium constant, but in which the activities need not represent equilibrium values. If there are no changes in the solution species involved as pH changes, the Nernst equation readily permits calculation of the effect of pH on the emf values.

EXAMPLE 8.5: Given E_A^0 for the half-reaction $MnO_4^{2-} \rightarrow MnO_2(s)$, express the E value as a function of pH.

Solution: From the Nernst equation and the balanced half-reaction, we may write

$$E = E_A^0 - \frac{0.059}{2} \log \frac{1}{[H^+]^4[MnO_4^{2-}]}$$

where the constant 0.059 incorporates R and $\mathcal{F}$, a temperature of 298 K, and the conversion factor from ln to log; 1 is used for the activity of the solid phase. If the manganate activity is kept at 1, the pH dependence is

$$E = E_A^0 - \frac{(0.059)(4)}{2} \text{pH} = 2.27 - 0.118 \text{ pH}$$

E_B^0 is used to indicate that the emf's (potentials) are standard potentials for base solution—all reactants and products, including hydroxide ion, are at unit activity. In the above example, we find the E_B^0 value by using a pH value of 14 (since pH + pOH = pK_w = 14), giving E_B^0 = 0.62 V.

EXAMPLE 8.6: Given E_A^0 for the couple $MnO_4^- \rightarrow MnO_4^{2-}$, what is E_B^0?

Solution: Because hydrogen is not involved in the half-reaction, E_B^0 and E_A^0 are identical, but E_A^0 is not significant because MnO_4^{2-} is stable only in strongly basic solution.

We can use the methods developed above to determine whether Mn^{3+} and MnO_4^{2-} would be stable in neutral water. A new emf diagram should be calculated for pH = 7. For

$$MnO_2(s) + 4H^+ + e \longrightarrow 2H_2O + Mn^{3+}$$

we have

$$E = E_A^0 - \frac{0.059}{1} \log \frac{1}{(10^{-7})^4} = -0.70 \text{ V}$$

Similar calculations should be made for the $MnO_4^{2-} \rightarrow MnO_2$ couple. These emf values are the emf values in neutral water, E_N. The E_N diagram for Mn is

$$MnO_4^- \xrightarrow{0.564} MnO_4^{2-} \xrightarrow{1.44} MnO_2 \xrightarrow{-0.70} Mn^{3+} \xrightarrow{1.51} Mn^{2+} \xrightarrow{-1.18} Mn \qquad (8.15)$$

with the lower path under MnO_4^{2-} to MnO_2 labeled $(2.88 + 0.56)/3$.

Inspection of this diagram indicates that MnO_4^{2-} will disproportionate in neutral water (E_N = 1.44 − 0.564) and that Mn^{3+} will disproportionate (E_N = 1.51 + 0.70). All other species are stable with respect to disproportionation in neutral water. Disproportionation cannot occur for terminal species (Mn, MnO_4^-), since such species cannot undergo both oxidation and reduction.

EXAMPLE 8.7:   CALCULATE E_B^0 FOR THE $CLO^- \to CL_2$ COUPLE FROM E_A^0.

Solution: $E_A^0 = 1.63$ V for $HClO + H^+ + e \to \frac{1}{2} Cl_2 + H_2O$

HClO is a weak acid. We calculate [HOCl] from K_a for use in the Nernst equation:

$$K_a = 4.0 \times 10^{-8} = \frac{[H^+][OCl^-]}{[HOCl]}$$

At pH 14: $[H^+] = 1 \times 10^{-14} m$, $[OCl^-] = 1 m$

$$[HOCl] = \frac{1 \times 10^{-14} \times 1}{4.0 \times 10^{-8}} = 2.5 \times 10^{-7} m \text{ ([HOCl] is very small, but not zero.)}$$

$$E_B^0 = 1.63 - \frac{0.0591}{1} \log \frac{1}{1 \times 10^{-14}(2.5 \times 10^{-7})} = 0.41 \text{ V}$$

Overvoltage

Many redox reactions involving breaking or making bonds or rearrangements of atoms are slow relative to simple ionic reactions such as $H^+ + OH^- \to H_2O$. In such cases a reaction might not occur even though E is positive and ΔG is negative. In order to carry out an electrolysis reaction resulting in gas evolution, a potential difference, which exceeds that calculated for a reversible galvanic cell, must be applied to the electrodes. This excess potential, called **overvoltage,** increases as the current density at the electrodes increases. The overvoltage depends on the gas evolved and on the material used for the electrode. Thus at a current density of 0.01 amp/cm², shiny platinum has an H_2 overvoltage of 0.07 V, whereas the O_2 overvoltage is 0.40 V. Under the same conditions, nickel has an H_2 overvoltage of 0.56 V, whereas the O_2 overvoltage is 0.35 V. Metal deposition does not show an overvoltage effect. Hence, by working at high current density and low temperatures (where overvoltages are highest), we can electroplate from aqueous solution metals, such as nickel, that lie above hydrogen in the activity series. These are metals with negative E^0 values for reduction of M^{n+} in the lowest oxidation state.

For gas evolution reactions to occur as reasonable rates, the emf for the reaction generally must be on the order of $+0.4$ V or more. This is only an estimate of overvoltage because the effects are specific for both the metal (and even its physical state) and the gas evolved.

The overvoltage effect is not limited to reactions carried out in an electrolytic cell, but occurs in many reactions producing a gas in solution—particularly those producing H_2 and O_2, as noted above. Overvoltage affects the stability of species in solution by suppressing gas evolution in some cases where it is thermodynamically feasible.

Hydrogen Evolution

E_A^0 for the half-reaction

$$2 H^+ + 2e \longrightarrow H_2 \qquad E_A^0 = 0.00 \text{ V} \tag{8.16}$$

by convention is given a value of zero. Using the Nernst equation, we obtain $E_N = -0.414$ and $E_B^0 = -0.828$. For the reaction of Mn with acid to form Mn^{2+} and H_2, an E_A^0

of $+1.18$ is obtained. Because E_A^0 for the reaction of manganese with acid exceeds the expected H_2 evolution overvoltage (~ 0.4 V), reaction should take place. Referring to the example above, E_N for H_2 evolution by Mn is $+0.77$ and H_2 evolution should occur in neutral water. Mn^{2+} (with oxidation to Mn^{3+}) does not evolve H_2 in acidic or neutral solution.

Oxygen Evolution

We write the following half-cell emf value for O_2 evolution:

$$2\,H_2O \longrightarrow O_2 + 4\,H^+ + 4e \qquad E = -E_A^0 = -1.229 \text{ V} \tag{8.17}$$

From this, we may obtain $E_N = -0.815$ V and $E_B = -0.401$ V.

EXAMPLE 8.8:   Will permanganate ion oxidize H_2O to evolve O_2 in acidic solution?

Solution: For the reaction

$$2\,MnO_4^- + H_2O \longrightarrow 2\,MnO_4^{2-} + \tfrac{1}{2}O_2 + 2\,H^+$$

we have

$$E = +0.564 - 1.229 = -0.665 \text{ V}$$

Hence this reaction does not occur. However, for the reaction

$$2\,MnO_4^- + 2\,H^+ \longrightarrow 2\,MnO_2 + \tfrac{3}{2}O_2 + H_2O$$

we have

$$E_A^0 = 1.70 - 1.23 = 0.47 \text{ V}$$

This value (0.47 V) is barely in excess of the expected overvoltage needed for O_2 evolution, and we find that O_2 evolution in acidic permanganate solutions take places very slowly. The E_N value shows that O_2 evolution does not take place in neutral solution unless a suitable catalyst, such as platinum, is present to reduce the overvoltage. Actually if a standardized solution of $KMnO_4$ is stored for weeks, a dark coating of MnO_2 can be observed on the inside of the bottle. Restandardization of the solution is required.

Formation of Insoluble Hydroxides

When a change of pH results in the formation of an insoluble hydroxide, we can calculate the emf value for the half-cell at the new pH (if the value of K_{sp} is known) by application of the Nernst equation, as the following free-energy cycle demonstrates.

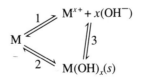

If M^{x+} exists in a solution in contact with $M(OH)_x$ (that is to say, in equilibrium), ΔG for step 3 must be zero and ΔG for steps 1 and 2 must be identical. K_{sp} can be evaluated from emf measurements for known OH^- activity.

EXAMPLE 8.9: Given $E_A^0 = -1.18$ for $Mn^{2+} + 2e \rightarrow Mn$ and $K_{sp} = 2 \times 10^{-13}$ for $Mn(OH)_2$, what is E_B^0 for the $Mn(OH)_2(s) \rightarrow Mn$ couple?

Solution: K_{sp} is the equilibrium constant for the reaction

$$Mn(OH)_2(s) \rightleftharpoons Mn^{2+} + 2OH^-$$

$$K_{sp} = [Mn^{2+}][OH^-]^2 \text{ [the activity is 1 for } Mn(OH)_2(s)]$$

In a 1 M base, $[OH^-] = 1$ and thus $[Mn^{2+}] = K_{sp} = 2 \times 10^{-13}$. From the Nernst equation,

$$E = E^0 - \frac{0.059}{2} \log \frac{1}{[Mn^{2+}]}$$

$$= -1.18 - \frac{0.059}{2} \log \frac{1}{2 \times 10^{-13}}$$

$$= -1.55 \text{ V} = E_B^0$$

The calculated result in the example is that expected qualitatively on the basis of LeChatelier's principle. As the metal ion concentration decreases, it becomes easier to form the ion from the metal or the reduction of Mn^{2+} is more difficult.

Formation of a Metal Complex

Similar considerations apply to the effect of complexing of metal ions by ligands such as cyanide, chloride, ammonia, and so on, that form complex ions. The structure and chemistry of complex ions forms the subject of Chapters 9 and 10. We have looked at the formation of complexes as Lewis acid–base reactions (Chapter 7). The relative effectiveness of such ligands in decreasing the free-metal ion concentration is revealed by the following emf values:

$$[Zn(H_2O)_w]^{2+} + 2e \longrightarrow Zn + w\,H_2O \qquad E^0 = -0.76 \text{ V} \qquad (8.18)$$

$$[Zn(NH_3)_4]^{2+}(aq) + 2e \longrightarrow Zn + 4\,NH_3 \qquad E^0 = -1.03 \text{ V} \qquad (8.19)$$

$$[Zn(CN)_4]^{2-}(aq) + 2e \longrightarrow Zn + 4\,CN^- \qquad E^0 = -1.26 \text{ V} \qquad (8.20)$$

The $[Zn(CN)_4]^{2-}$ complex is more stable than $[Zn(NH_3)_4]^{2+}$ because it is more difficult to reduce than $[Zn(NH_3)_4]^{2+}$.

For the case of two oxidation states of a metal (e.g., $Mn^{3+} \rightarrow Mn^{2+}$), the emf value can become more positive or more negative with the formation of complexes, depending on the relative stabilization of Mn^{3+} and Mn^{2+} in their complexes. The emf value becomes less positive (or more negative) if Mn^{3+} is stabilized (the Mn^{3+} complex is more stable than that of Mn^{2+}). Greater stabilization of Mn^{2+} makes the emf for the reduction of Mn^{3+} more positive or more favorable.

We examine in more detail the use of emf data to obtain both stoichiometries and stability constants of complex ions on page 385.

8.6 POURBAIX (OR PREDOMINANCE AREA) DIAGRAMS

▶ 8.6.1   emf–pH Diagrams

Figure 8.2 shows the potentials of some manganese couples as a function of pH. The activity of all solution species containing manganese, and involved in the couples whose potentials are plotted, is taken as unity. The potentials of couples unaffected by pH, such as Mn/Mn^{2+}, appear as horizontal lines. To the right of the intersection of the Mn/Mn^{2+} line with the $Mn/Mn(OH)_2$ line, the Mn/Mn^{2+} line is dashed, since a concentration of 1 m Mn^{2+} ion cannot be maintained at high pH values because of precipitation of $Mn(OH)_2$. To the left of this intersection, the $Mn/Mn(OH)_2$ line is dashed, since the solubility of $Mn(OH)_2$ is greater than 1 m at low pH values. For similar reasons the Mn^{2+}/Mn_3O_4 line is dashed to the right of the intersection with the $Mn(OH)_2/Mn_3O_4$ line, and the latter is dashed to the left of the intersection. Species undergoing disproportionation are either omitted entirely (Mn^{3+}, for example) or, in the case of couples involving such species, are shown as dotted lines in the region of instability of one of the species involved (MnO_4^{2-}, for example).

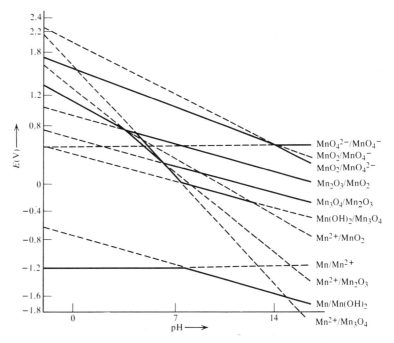

Figure 8.2   Variation with pH of the potentials of some Mn couples.

▶ *8.6.2 Pourbaix Diagram for Manganese*

Figure 8.3 presents the same data as Figure 8.2, except that only the solid lines are shown for Mn couples and several vertical lines have been added to represent acid–base equilibria not involving electron transfer. In addition, the potentials for H_2O/H_2 and O_2/H_2O at 1 atm pressure are shown by dashed lines; outside of this area, oxidizing agents produce O_2 and reducing agents produce H_2, allowing for overvoltage. This type of figure is known as a **Pourbaix diagram**, a **potential–pH diagram**, or a **predominance area diagram.** The latter name reflects the fact that the species listed in any area is the predominant species of that element within the *E–pH* range given by its boundaries. The oxidation state always increases with increasing *E* on crossing a boundary. For solution species, the boundaries are concentration-dependent, in accordance with the Nernst equation. Lowering the concentration of a solution species has the effect of increasing the predominance area of the solution species at the expense of solid species. Geologists and corrosion engineers frequently use diagrams with solution concentrations of 10^{-6} molal.[4] Note that infor-

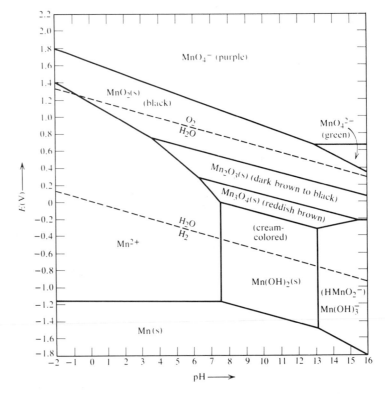

Figure 8.3 Pourbaix diagram for Mn species. (Reproduced with permission from M. Pourbaix, *Atlas of Electrochemical Equilibria in Aqueous Solution*, 2nd ed., translated by J. A. Franklin, National Association of Corrosion Engineers, Houston, TX, 1974, p. 290.)

---

[4] Some predominance area diagrams use dotted lines to separate solution species, such as MnO_4^{2-} and MnO_4^-, and solid lines to indicate the existence of phase boundaries.

mation not included in the construction of the diagram (species for which data may not be available, etc.) cannot be obtained from the diagram. For temperatures other than 25°C, you should construct a new Pourbaix diagram using the temperature coefficients of the potentials of the appropriate couples. Finally, remember that the Pourbaix diagrams give the results expected *at equilibrium*. Frequently, equilibrium is achieved too slowly to make the conclusions valid—particularly in cases involving gas evolution, as pointed out earlier. The area between the dashed lines in Figure 8.3 delineates the region of stability of substances in contact with water, but as noted already, overvoltage effects permit the existence of species within a range of about 0.4 V above or below the indicated area for periods of time sufficient to carry out experiments in the laboratory.

▶ 8.6.3 Redox Chemistry from Pourbaix Diagrams

Mn^{II} is stable over an extremely wide range of E values. On increasing the pH of a deaerated 1 m solution of Mn^{II} ion (very pale pink), white $Mn(OH)_2$ begins to precipitate at pH $\simeq 8$. If air is allowed to contact the solution, the E gradient varies from about 600 mV at the air interface to about 0 at the solution–precipitate interface, and in several hours the precipitate develops striations that are cream-colored on the bottom, through reddish-brown, dark brown, or black where the surface is in contact with the solution. After standing several days, only the black MnO_2 will be present. The reddish-brown Mn_3O_4, a mixed valence oxide containing both Mn^{2+} and Mn^{3+} ions in a slightly distorted normal spinel structure (see page 217 for spinels), may be made directly by roasting any oxide of manganese in air at 1000°C.

If we add I_2 crystals to an Mn^{II} solution, the solution is "buffered" at an E around 600 mV. Very slowly adding base precipitates $Mn(OH)_3$ ($K_{sp} = 10^{-36}$), which, in the absence of air goes over to a hydrated Mn_2O_3 that is brown in color. The Mn^{2+}/Mn^{3+} couple may be added to the Pourbaix diagram to give a horizontal line at 1.51 V. In the usual pH range of 0 to 14, and with concentrations (activities) of solution species set at one, the Mn^{2+}/Mn^{3+} couple has a higher potential than the Mn^{2+}/MnO_2 couple—thus Mn^{3+} is not stable under the conditions depicted in Figure 8.3. As shown in Section 8.5.2, it disproportionates. If the range of acidity is extended, however, the Mn^{2+}/Mn^{3+} line eventually falls below the Mn^{2+}/MnO_2 line, and the Mn^{3+} species exhibits a predominance domain. One hundred percent H_2SO_4 has an effective pH of -12 (or more precisely, an H_0 value; see page 323). When MnO_2 is heated in concentrated sulfuric acid, $Mn_2(SO_4)_3$ forms and O_2 is evolved. We also can produce Mn^{III} by dissolving Mn_2O_3 in concentrated acids, or by reducing MnO_2 or MnO_4^- in concentrated acid. (**Caution:** Permanganates are dehydrated in concentrated sulfuric acid and produce the very explosive Mn_2O_7, a dark greenish-brown oil.) Decreasing the solution concentrations on the Pourbaix diagram increases the Mn^{3+} domain at the expense of the MnO_2 domain. Complexing agents, such as F^- and CN^-, reduce the concentration of free Mn^{3+} and can stabilize this oxidation state even at moderate total concentration of Mn^{3+} and moderate acidities.

Mn^I does not appear in the diagram, and the Mn^I/Mn^{II} couple must have a potential less than that of Mn/Mn^{II}, but if the free manganese solution concentrations are low enough, the Mn^I species has a domain between Mn and Mn^{II}. This occurs with the cyano complexes, and reduction of Mn^{II} in the presence of cyanide gives the Mn^I complex.

As might be surmised from our discussion of air oxidation of $Mn(OH)_2$, manganese ions in air-saturated natural waters are oxidized to Mn^{IV} and precipitated as MnO_2, pyrolusite. The name of this mineral (Greek *pyro*, "fire"; and *louein*, "to wash") derives from its use in glassmaking, in which it decolorizes green glass by oxidizing the Fe^{2+} impurity to Fe^{3+} while itself being reduced to Mn^{2+}. Both of the product ions are virtually colorless. High concentration of Fe^{2+} ion in glass can reduce solar transmission by as much as 10% of the total energy.

MnO_2 can coexist with MnO_4^- except in very basic solution. Because the coexistence boundary lies above the H_2O/O_2 boundary over the entire pH range of coexistence, permanganate ion cannot be produced directly from MnO_2 by air oxidation. It can be produced, however, by an indirect air-oxidation route. At very high pH, MnO_2 has a coexistence boundary with the manganate ion (MnO_4^{2-}) that crosses the H_2O/O_2 boundary at a pH = 17.4 for 1 *m* solution concentrations; for 10^{-7} *m* MnO_4^{2-}, the boundaries cross at pH = 14. MnO_4^{2-} salts are isomorphous with sulfates, and air oxidation of MnO_2 in $Ba(OH)_2$ solution results in the precipitation of $BaMnO_4$. Acidifying manganates results in disproportionation to MnO_2 and MnO_4^-. To complete the conversion of the manganate to permanganate industrially, KOH is used instead of $Ba(OH)_2$, and the last step is carried out using chlorine or ozone, or electrolytically.

The effective pH of 10 *M* KOH (actually the H_- value) is 19. In this pH range the blue MnO_4^{3-} tetraoxomanganate(V) ion has a predominance domain with an *E* between that of MnO_4^{2-} and MnO_2. It can be prepared by reducing the permanganate ion with Na_2SO_3 in 25–30% NaOH.

Permanganate ion is a strong oxidizing agent that is thermodynamically unstable in water at all pH values. Because of overvoltage effects, only in acid solutions is the oxidation of water to oxygen of any consequence. The reaction is catalyzed by MnO_2, particularly in the presence of light. For this reason, solutions of $KMnO_4$ used as analytical reagents should be completely free of dust or organic matter and stored in dark glass bottles. The reduction product in acid solution is the almost colorless Mn^{2+} ion, and the endpoint of permanganate titrations in acid may be detected readily by the disappearance of the red-violet color of MnO_4^- ion. In neutral or alkaline solutions, the reduction product of MnO_4^- is the insoluble brown to black MnO_2.

8.7   GEOCHEMISTRY OF MANGANESE[5]

We have seen already several examples of the increase in a domain area as the solution activity of an oxidation state is decreased. From solubility products or free energies of formation, geochemists have been able to construct Pourbaix-type diagrams showing the *E* and pH domains of minerals for a set of conditions for other dissolved substances. Thus, setting the Mn^{2+} solution concentration at 10^{-6} *m* and the concentration of total dissolved carbonate species at $10^{-1.4}$ *m* corresponding to a system first saturated with CO_2 under 1 atm pressure and then closed gives us Figure 8.4. The small field for pyrochroite $[Mn(OH)_2]$ indicates the very limited conditions under which this mineral forms and reflects its rarity. Assuming the concentrations of S and CO_2 corresponding to an open

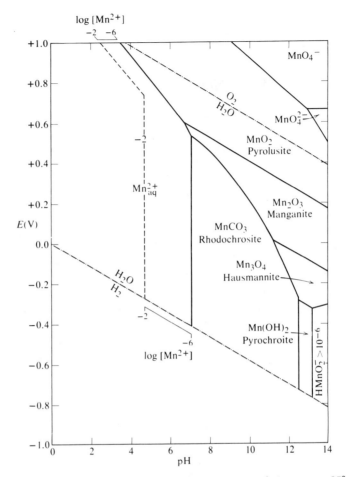

Figure 8.4 Stability relations among some manganese compounds in water at 25°C and 1 atmosphere total pressure, total dissolved carbonate species $= 10^{-1.4}$. (Adapted with permission from R. M. Garrels and C. L. Christ, *Solutions, Minerals, and Equilibria*, 2nd ed., Jones and Bartlett, Boston, 1990, p. 389.)

system in marine sediment gives a similar diagram.[5] The ratio of manganese to other transition metals occurring in ocean-floor nodules has been explained in terms of both the ease of mobility by ionic diffusion of the lower oxidation states of pairs below and the E boundary between these states at pH $= 8$.

Element:	Ni	Co	Mn	Cr	Cu	V	Fe	U
	III/II	III/II	IV/II	VI/III	II/I	V/IV	III/II	VI/IV
E (volts):	+0.8	+0.6	+0.5	+0.3	+0.1	0.0	−0.2	−0.3

[5] R. M. Garrels and C. L. Christ, *Solutions, Minerals, and Equilibria*, Freeman, Cooper, San Francisco, 1965.

8.8   PERIODIC TRENDS AMONG THE TRANSITION ELEMENTS[6]

▶ *8.8.1   Technetium Chemistry*

The Pourbaix diagrams shown in Figures 8.3 and 8.5 indicate the general trends in redox behavior among transition element species in aqueous media. Going from Mn to Tc, we find an increase in nobility of the metal—Tc does not liberate H_2 from acidic, neutral, or basic solution, but it may be oxidized by dissolved oxygen in water. The important species are Tc, TcO_2, and TcO_4^- with Tc^{2+} stable at low concentrations in acidic solution (below pH 1). The crosshatched area in Figure 8.5 is the region of stability of Tc^{2+}. The horizontal lines in this area represent Tc/Tc^{2+}, and the sloping lines represent Tc^{2+}/TcO_2. The nobility of the heavier transition metals extends to Re and through the transition elements to the right of Tc and Re: Ru, Rh, Pd, and Ag in the second transition series and Os, Ir, Pt, and Au in the third. The technetate ion (TcO_4^-) is stable in aqueous solution, occurring as $HTcO_4$ in strong acid. This trend in oxidation states for Mn and Tc is general for transition metals: The higher oxidation states are relatively more stable for the heavier elements in a given group, and the lower states are either unstable or exhibit relatively narrow stability domains. Thus the higher states are much milder oxidizing agents for the heavier elements, and the lower states may behave as reducing agents.

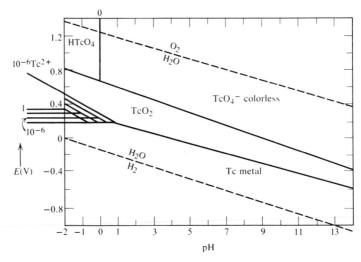

Figure 8.5   Pourbaix diagram for Tc species. Concentrations are shown for the Tc^{2+} − Tc and TcO_2 − Tc^{2+} regions. (Reproduced with permission from M. Pourbaix, *Atlas of Electrochemical Equilibria in Aqueous Solution*, 2nd. ed., translated by J. A. Franklin, National Association of Corrosion Engineers, Houston, TX, 1974, p. 290.)

---

[6] Campbell and Whiteker constructed a large number of Pourbaix diagrams for the stability domain range in water, which you are urged to examine—particularly if access to Pourbaix's work is not available. J. A. Campbell and R. A. Whiteker, *J. Chem. Educ.* **1969,** *46,* 90.

The chemistry of Tc has excited much interest, because of the use of ^{99m}Tc in radio-pharmaceuticals as heart- and bone-scanning agents. ^{99m}Tc is a short-lived gamma emitter produced from the β-decay of ^{99}Mo and is separated from it by column chromatography.

▶ 8.8.2 Vanadium Chemistry

From the Pourbaix diagram in Figure 8.6 we see that, except in very strong acid, the V^V species are relatively weak oxidizing agents. As the atomic number increases on crossing a transition metal series, the oxidizing ability of the highest oxide or oxoanion increases. Thus chromates are better oxidizing agents than vanadate, permanganate is still better, and FeO_4^{2-} is such a strong oxidizing agent that it oxidizes water. On the other hand, V^{2+} is a reducing agent, thermodynamically unstable with respect to reducing water, but because of overvoltage effects it can be prepared in aqueous solution. In solution the stable species are V^{3+}, VO^{2+}, and vanadates. The pH-dependent polymerization equilibria of the highest oxidation state are perhaps most pronounced with V, but are seen to a lesser extent with chromates and molybdenates.

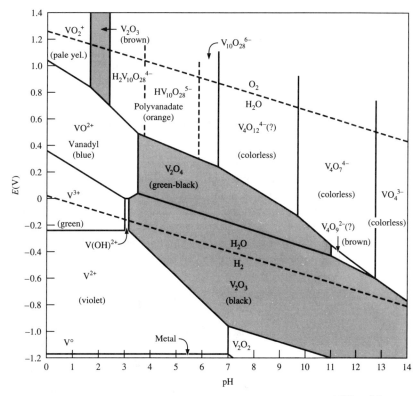

Figure 8.6 Stability of some vanadium compounds and ions in water at 25°C and 1 atm total pressure. Solids exist in the shaded areas. (Adapted with permission from H. T. Evans, Jr. and R. M. Garrels, *Geochim. et Cosmochim. Acta* **1958,** *15,* 131.)

EXAMPLE 8.10: **(a)** To examine the significance of a predominance area, calculate the concentration of V^{2+} in the presence of $1\ m\ V^{3+}$ at $E = -0.26$ V and pH = 0. What is the concentration of V^{2+} at $E = -0.26$ V and pH = 2?

$$V^{3+} + e \longrightarrow V^{2+} \qquad E^0 = -0.26 \text{ V}$$

(b) Calculate the concentration of V^{2+} in the presence of $1\ m\ V^{3+}$ at $E = -0.20$ V and at $E = 0$ V.

Solution: **(a)** $E = E^0 - \dfrac{0.059}{1} \log \dfrac{[V^{2+}]}{[V^{3+}]}$

At $E = -0.26$ V we have

$$0 = 0.059 \log [V^{2+}]$$

$$[V^{2+}] = 1\ m$$

The concentrations of V^{2+} and V^{3+} do not change with changing pH because the half-cell reaction does not involve H^+. The line in Figure 8.6 is horizontal.

(b) At -0.20V we have

$$-0.20 = -0.26 - 0.059 \log [V^{2+}]$$

$$0.06 = -0.059 \log [V^{2+}]$$

$$[V^{2+}] = 0.10\ m$$

At 0 V we have

$$0 = -0.26 - 0.059 \log [V^{2+}]$$

$$[V^{2+}] = 4.0 \times 10^{-5} m$$

8.9   SOME REDOX REACTIONS OF NONMETALS

If two Pourbaix diagrams are superimposed, species can coexist only if their predominance areas overlap; otherwise, a reaction will take place and the resulting products will have overlapping predominance areas. The emf line diagrams connecting species with a value given for the potential of the couple under standard conditions gives the boundary points on a Pourbaix diagram at a pH of 0 for E_A^0 values, or a pH of 14 for E_B^0 values. The chemistry of aqueous solutions at the appropriate pH often can be understood by studying such line potential diagrams. Consider the following data for chlorine species:

$$E_A^0 \ \text{ClO}_4^- \xrightarrow{1.20} \text{ClO}_3^- \xrightarrow{1.18} \text{ClO}_2 \xrightarrow{1.19} \text{HClO}_2 \xrightarrow{1.70} \text{HClO} \xrightarrow{1.63} \text{Cl}_2 \xrightarrow{1.36} \text{Cl}^- \tag{8.21}$$

$$E_B^0 \ \text{ClO}_4^- \xrightarrow{0.37} \text{ClO}_3^- \xrightarrow{-0.48} \text{ClO}_2 \xrightarrow{1.07} \text{ClO}_2^- \xrightarrow{0.68} \text{ClO}^- \xrightarrow{0.42} \text{Cl}_2 \xrightarrow{1.36} \text{Cl}^- \tag{8.22}$$

In the laboratory, we can produce elemental chlorine from chloride ion sources using a variety of oxidizing agents in acidic solution—$KMnO_4$, $K_2Cr_2O_7$, $KClO_3$, MnO_2, PbO_2, and so on. Pertinent E_A^0 values are

$$MnO_4^- \xrightarrow{1.51} Mn^{2+} \qquad ClO_3^- \xrightarrow{1.48} Cl_2$$
$$MnO_2 \xrightarrow{1.23} Mn^{2+} \qquad PbO_2 \xrightarrow{1.46} Pb^{2+}$$
$$Cr_2O_7^{2-} \xrightarrow{1.38} Cr^{3+}$$

MnO_2 and $K_2Cr_2O_7$ require a strong acid with a hydrogen ion activity greater than 1; in actual practice, sulfuric acid usually is used because sulfuric acid itself is too weak an oxidizing agent to oxidize the Cl^- ion. Sulfuric acid reacts with Cl^- simply to form HCl.

$$E_A^0 \qquad SO_4^{2-} \xrightarrow{0.16} H_2SO_3 \xrightarrow{0.45} S \qquad (8.23)$$

Although Cl_2 is produced by the action of concentrated nitric acid on chlorides, lower oxides of nitrogen are produced also, giving Cl_2 + N oxides.

A mixture of concentrated nitric and hydrochloric acids (*aqua regia*) does serve as a strong oxidizing medium for even noble metals. Complexing action of the chloride ion stabilizes the metal ion, lowering the emf for oxidation of the metal.

Commercially, the electrolysis of brine solutions is used to produce chlorine. Although the evolution of O_2 should take place at lower voltages than those needed for Cl_2, the overvoltage for O_2 at high current densities is much greater than that for Cl_2.

$$E_N \qquad H_2O \xrightarrow{-0.815} O_2$$

Chlorine reacts slowly with water to liberate O_2:

$$Cl_2 + H_2O \longrightarrow 2\,HCl + O_2 \qquad E_N = 0.543 \text{ V} \qquad (8.24)$$

More immediate, however, is a disproportionation reaction that is self-repressed by the hydrogen ion formed.

$$Cl_2 + H_2O \longrightarrow HOCl + H^+ + Cl^- \qquad E_N = +0.14 \text{ V} \qquad (8.25)$$
$$E_{pH4.5} = 0$$

In basic solution, hydrolysis is complete ($E_B^0 = +0.94$) to give hypochlorite (OCl^-) and chloride ions (household bleach). In one commercial procedure a brine solution is stirred during electrolysis, so that the anode and cathode products mix and OCl^- is formed. More commonly, Cl_2 is allowed to react directly with "slaked lime" to form bleaching powder of "chloride of lime":

$$Cl_2 + Ca(OH)_2 \longrightarrow CaCl(OCl) + H_2O \qquad (8.26)$$

Although the couples $Cl_2 \xrightarrow{1.36} Cl^-$ and $O_2 \xrightarrow{1.23} H_2O$ indicate that in acidic solution, Cl_2 and O_2 should have about the same ability as oxidizing agents, Cl_2 actually is much stronger. This is because hydrogen peroxide is an intermediate when O_2 is reduced to H_2O and the $O_2 \rightarrow H_2O_2$ couple is the effective emf available in O_2 oxidations.

$$E_A^0 \qquad O_2 \xrightarrow{0.682} H_2O_2 \xrightarrow{1.776} H_2O \qquad (8.27)$$
$$\underset{1.229}{\vdash\!\!-\!\!-\!\!-\!\!-\!\!-\!\!-\!\!-\!\!-\!\!\dashv}$$

Accordingly, O_2 cannot be used to displace Br_2 from bromides. Instead, Cl_2 is used to displace Br_2 to recover Br^- from sea water with the pH adjusted to 3.5. The pH adjustment prevents disproportionation of Cl_2 or Br_2 to the halide and hypohalite (see Problem 8.18).

As the emf data indicate, chlorous acid is unstable with respect to disproportionation in acidic solution. In base, a mixture of chlorites and chlorates is obtained from the disproportionation of ClO_2:

$$2\,ClO_2 + 2\,OH^- \longrightarrow ClO_2^- + ClO_3^- + H_2O \qquad E_B^0 = 1.55 \text{ V} \qquad (8.28)$$

Reaction of ClO_2 with sodium peroxide in aqueous solution produces pure chlorites:

$$OH^- + 2\,ClO_2 + O_2H^- \longrightarrow 2\,ClO_2^- + O_2 + H_2O \qquad E_B^0 = 1.00 \text{ V} \quad (8.29)$$

In hot basic solution the disproportionation of Cl_2 yields chlorates and chlorides:

$$3\,Cl_2 + 6\,OH^- \longrightarrow 5\,Cl^- + ClO_3^- + 3\,H_2O \qquad E_B^0 = +0.89 \text{ at } 25°C \ (8.30)$$

We also can carry out this reaction by allowing mixing of the anode and cathode compartments during electrolysis of a hot brine solution.

Hot concentrated perchloric acid is an extremely explosive oxidizing agent. Cold dilute $HClO_4$ does not even oxidize I^-. The E_{cell}^0 is very favorable, but the reaction is slow. Reactions involving atom transfer, commonly involved for oxoanions, are often slow.

8.10 EQUILIBRIUM CONSTANTS FROM emf DATA

In addition to qualitative predictions about the spontaneity of a reaction, the E^0 value for a reaction permits us to calculate the equilibrium constant for the reaction. At equilibrium, $E = 0$ and $Q = K_{eq}$, the Nernst equation then gives

$$E^0 = \frac{0.059}{n} \log K_{eq} \tag{8.31}$$

EXAMPLE 8.11: Calculate the equilibrium constant for the reaction

$$Pb + Sn^{2+} \longrightarrow Pb^{2+} + Sn$$

Solution: From the emf data, we calculate $E^0 = 0.125 - 0.137 = -0.012$ and hence $K_{eq} = 10^{-2(0.012)/0.059} = 0.40$.

EXAMPLE 8.12: Utilizing standard emf data, calculate K_p for the reaction

$$2\,PbO(s) \longrightarrow 2\,Pb(s) + O_2$$

Solution:

$$
\begin{array}{lll}
4e + 2\,PbO + 2\,H_2O \longrightarrow 2\,Pb + 4\,OH^- & & E_B^0 = -0.58 \\
4\,OH^- \longrightarrow O_2 + 2\,H_2O + 4e & & E_B^0 = -0.40 \\
\hline
2\,PbO(s) \longrightarrow 2\,Pb(s) + O_2 & & E_B^0 = -0.98 \text{ V}
\end{array}
$$

$$K_p = 10^{4(-0.98)/0.059} = 10^{-66.4} = 2.8 \times 10^{-66} \text{ atm at } 25°C$$

We often can obtain solubility products of slightly soluble substances using cell measurements, reversing the calculation procedure from emf data by procedures closely resembling those used to obtain solubility products.

GENERAL REFERENCES

See Appendix E for a summary of half-cell emf data.

A. J. Bard, R. Parsons, and J. Jordan, Eds., *Standard Potentials in Aqueous Solution*, Marcel Dekker, New York, 1985.

G. Milazzo and S. Caroli, *Tables of Standard Electrode Potentials*, Wiley, New York, 1978.

W. M. Latimer, *The Oxidation States of the Elements and Their Potentials in Aqueous Solution*, 2nd ed., Prentice–Hall, New York, 1952.

PROBLEMS

8.1 Should the value of the heat of a reaction calculated from the variation of K_p with temperature be affected by the units in which K_p is expressed? Under what circumstances would ΔH be unaffected and under what circumstances affected by the selection of units for the K_{eq}?

8.2 Write balanced ionic half-reactions for

(a) $HgO \xrightarrow{\text{base}} Hg$

(b) $VO^+ \xrightarrow{\text{acid}} HV_2O_5^-$

(c) $HGeO_3^- \xrightarrow{\text{base}} Ge$

(d) $S_2O_3^{2-} \xrightarrow{\text{acid}} S$

(e) $F_2O \xrightarrow{\text{acid}} HF$

(f) $C_{12}H_{22}O_{11} \xrightarrow{\text{acid}} CO_2$

8.3 The following are somewhat more challenging equations to balance.

(a) $H_2O + P_2I_4 + P_4 \rightarrow PH_4I + H_3PO_2$

(b) $ReCl_5 + H_2O \rightarrow Re_2Cl_9^{2-} + ReO_4^- + Cl^- + H^+$

(c) $B_{10}H_{12}CNH_3 + NiCl_2 + NaOH \rightarrow Na_4[B_{10}H_{10}CNH_2]_2Ni + NaCl + H_2O$

(d) $ICl + H_2S_2O_7 \rightarrow I_2^+ + I(HSO_4)_3 + HS_3O_{10}^- + HSO_3Cl + H_2SO_4$

8.4 Calculate E^0 values for the following cells and write the reactions for which these apply.

(a) $Pt, H_2 \mid HCl \parallel KCl \mid Hg_2Cl_2(s), Hg$ $(E^0 \; Hg_2Cl_2(s) \xrightarrow{0.268} Hg + Cl^-)$

(b) $Cu \mid Cu^{2+} \parallel I^- \mid CuI(s), Cu$ $(E^0 CuI(s) \xrightarrow{-0.185} Cu + I^-)$

(c) $Pt, CuI(s) \mid Cu^{2+} \parallel I^- \mid CuI(s), Cu$

From parts **b** and **c**, comment on the necessity of knowing the half-cells involved in making predictions of (1) the spontaneity of a reaction and (2) the free-energy change of a reaction.

8.5 Calculate the emf values for the following cells.

(a) $Pt, H_2 \mid H^+(a = 0.1) \parallel H^+(a = 10^{-7}) \mid H_2, Pt$

(b) $Zn \mid Zn^{2+}(a = 1) \parallel Cu^{2+}(a = 10^{-4}) \mid Cu$

(c) $Fe, Fe(OH)_2 \mid OH^-(a = 0.1) \parallel Fe^{2+}(a = 1) \mid Fe$

8.6 Describe the conditions of acidity most appropriate for the following processes.

(a) $Mn^{2+} \rightarrow MnO_4^-$

(b) $CrO_4^{2-} \rightarrow Cr_2O_7^{2-}$

(c) $Fe^{3+} \rightarrow FeO_4^{2-}$

(d) $ClO_4^- \rightarrow ClO_3^-$

(e) $C_2O_4^{2-} \rightarrow 2\,CO_2$

(f) $H_2O_2 \rightarrow H_2O$

(g) $H_2O_2 \rightarrow O_2$

8.7 Given the following emf diagram in acidic solution:

$$E_A^0 \qquad Sb_2O_5(s) \xrightarrow{0.48} Sb_2O_4(s) \xrightarrow{0.68} Sb^{III} \xrightarrow{0.20} Sb \xrightarrow{-0.51} SbH_3$$

$$O_2 \xrightarrow{1.23} H_2O$$

(a) At a pH of 4 the $Sb^{III} \rightarrow Sb$ emf is $= 0.04$ V. From these data determine whether the predominant species for Sb^{III} in the pH range of 0 to 4 is Sb_2O_3, SbO_2^-, SbO^+, or Sb^{3+}.

(b) Which of the oxidation states of Sb are unstable in acidic solution? Write balanced equations for all reactions of the unstable species occurring spontaneously in acid.

(c) What is the emf for $Sb_2O_5 \rightarrow SbH_3$ in acidic solution?

8.8 The emf is more positive for the $Fe^{III} \rightarrow Fe^{II}$ couple in the presence of *o*-phenanthroline and more negative in the presence of F^- as compared to the value in the absence of these ligands forming complex ions. What do these values tell us about the tendency of these ligands to stabilize Fe^{2+} or Fe^{3+}?

8.9 K_{sp} for $Cu(OH)_2$ is 1.6×10^{-19}. (a) Calculate the emf for $Cu(OH)_2 \rightarrow Cu$ in 1 *M* base. (b) What information other than the emf values or $Fe^{III} \rightarrow Fe^{II}$ in acid and in base is needed to calculate the K_{sp} values for both $Fe(OH)_2$ and $Fe(OH)_3$?

8.10 By extrapolation from known data for Cr, W, and Mo, Seaborg estimated the following emf values for the couples involving possible species for the currently unknown element 106:

$$E_A^0 \qquad MO_3 \xrightarrow{-0.5} M_2O_5 \xrightarrow{-0.2} MO_2 \xrightarrow{-0.7} M^{3+} \xrightarrow{0.0} M$$

Predict the results of the following:

(a) M is placed in 1 *m* HCl.

(b) MCl_3 is placed in an acidic solution containing $FeSO_4$.

(c) MO_2 and MO_3 are in contact in acidic solution.

(d) M^{3+} and MO_3 are in contact in acidic solution.

(e) M is treated with excess concentrated HNO_3.

8.11 Given the following half-cell emf diagram for osmium:

$$E_A^0 \qquad OsO_4(s) \xrightarrow{1.0} OsCl_6^{2-} \xrightarrow{0.85} OsCl_6^{3-} \xrightarrow{0.4} Os^{2+} \xrightarrow{0.85} Os$$

(a) Which of the above species, if any, would be unstable in 1 *m* HCl? Give balanced equations for any reactions that occur.

(b) Which couple(s) would remain unchanged in their emf value on altering the pH?

(c) Which couple(s) would remain unchanged in their emf value on altering the chloride ion concentration?

(d) Calculate the value of E_A^0 for $OsO_4(s) \rightarrow Os$.

(e) Predict the results of mixing osmium with solid OsO_4 in contact with 1 *m* HCl.

8.12 (a) What E is required to reduce the V^{3+} concentration to 10^{-6} *m* at pH 1 in the presence of 1 *m* VO^{2+}?

(b) What E is required to reduce the VO^{2+} concentration to 10^{-6} *m* at pH 1 in the presence of 1 *m* V^{3+}?

$$VO^{2+} + 2H^+ + e \longrightarrow V^{3+} + H_2O \qquad E^0 = 0.34 \text{ V}$$

8.13 (a) Use data from Appendix E to construct an emf diagram for Cl in acidic and basic solutions, showing only species stable with respect to disproportionation.

(b) Construct a Pourbaix diagram for Cl with unit activity for predominant species.

8.14 The ionization energies and sublimation energies of Li and Be are regular for their families. The emf for reduction of Li^+ is very much out of line for the family, but Be^{2+} is not. Explain.

8.15 The oxidation of $C_2O_4^{2-}$ by MnO_4^- in acidic solution is usually very slow initially, but it can be speeded up by the addition of Mn^{2+}. Explain how the Mn^{2+} might speed the process.

8.16 From the potential diagrams for Np (Appendix E):

  (a) Write equations for the reactions which would occur for unstable oxidation states in acidic solution and in basic solution.

  (b) Give a suitable preparation for each stable oxidation state starting with Np.

8.17 Give the products expected for the following reactions:

  (a) Cl_2 (excess) + HI (1 *m*)

  (b) $HClO_4$ (1 *M*) + NaI (1 *m*)

  (c) HClO (excess) + NaI (small amount)

  (d) HClO (small amount) + NaI (excess)

  (e) Which of the above reactions might be too slow to observe?

8.18 Calculate E_{cell} at pH = 0, 14, and 3.5 for

  (a) $Cl_2 \rightarrow HOCl + Cl^-$ **(b)** $Br_2 \rightarrow HOBr + Br^-$

PART IV

Coordination Chemistry

· · · · · · · · · · · · · · · · · · · ·

▶9◀

Models and Stereochemistry of Coordination Compounds

·····················

We can think of coordination chemistry either as a branch of inorganic chemistry or as an approach to—an interpretation of—many aspects of chemistry. Certainly the chemistry of many metal compounds in solution, in the melt, or in the solid is best understood in terms of coordination chemistry. To a great extent, coordination interactions dictate crystal structures and site preferences of metal ions. Here we begin consideration of bonding with valence bond theory and then use an electrostatic model leading to the development of ligand field theory. The stereochemistry of coordination compounds is surveyed illustrating the great variety of structures for various coordination numbers.

9.1 INTRODUCTION

A **coordination compound** or **metal complex** might be defined as a central atom or ion attached to a sheath of ions or molecules. Perhaps your first obvious encounter with a coordination compound or complex ion was seeing the deep blue $[Cu(NH_3)_4]^{2+}$ ion formed by adding aqueous ammonia solution to a solution of a copper(II) salt. Although this reaction is usually described by Equation (9.1), Equation (9.2) furnishes a more complete representation.[1] The copper ion in solution is already a complex ion—an aqua or water complex.

[1] Actually, these representations for both the aqua (hydrated) and the ammonia complexes are also incomplete. The Cu^{2+} ion usually has four molecules held rather strongly and two more held more loosely. Attached to the ammonia complex might be two molecules of water or one or two additional ammonia molecules, depending on the concentrations. More H_2O molecules are held in a second sphere.

$$Cu^{2+} + 4NH_3 \quad \rightleftarrows \quad [Cu(NH_3)_4]^{2+} \tag{9.1}$$

$$[Cu(H_2O)_4]^{2+} + 4NH_3 \quad \rightleftarrows \quad [Cu(NH_3)_4]^{2+} + 4H_2O \tag{9.2}$$

This reaction is a Lewis acid–base substitution reaction in which ammonia displaces the weaker base water from the **coordination sphere** of the metal ion—a Lewis acid. The coordination sphere includes those molecules or ions bonded directly to the metal ion. These groups bonded to the metal ion are called **ligands.**

If acid is added to the solution of $[Cu(NH_3)_4]^{2+}$, the protons combine with NH_3 to give NH_4^+ and the ammonia complex is destroyed. The ammonia molecules are removed because H^+ is a stronger acid than Cu^{2+}. The less basic water molecules replace them to give $[Cu(H_2O)_4]^{2+}$. When a minimum amount of hydrochloric acid is added to neutralize the ammonia, the solution takes on the pale blue color of the $[Cu(H_2O)_4]^{2+}$ ion. Adding an excess of concentrated hydrochloric acid turns the solution green because some of the yellow $[CuCl_4]^{2-}$ ion is formed while some of the blue $[Cu(H_2O)_4]^{2+}$ remains. The intermediate species $[CuCl_3(H_2O)]^-$, $[CuCl_2(H_2O)_2]$, and $[CuCl(H_2O)_3]^+$ will also be present in varying amounts, depending on the concentration of Cl^-. The $[Cu(H_2O)_4]^{2+}$ ion is never converted completely to $[CuCl_4]^{2-}$, even in concentrated hydrochloric acid solution, because the concentration of water is still greater than that of Cl^-. Upon dilution, the green solution in concentrated HCl again becomes blue, because the Cl^- loses out in the competition when the concentration of chloride ion is low. Solid brownish-yellow $CuCl_2$ contains chains of $CuCl_4$ units. Adding an excess of HBr or NaBr to a solution of $[Cu(H_2O)_4]^{2+}$ turns the solution green and finally, in the presence of a high Br^- concentration, dark brownish-green, because of extensive, but incomplete, conversion to brown $[CuBr_4]^{2-}$. This complex ion also is destroyed by dilution reforming $[Cu(H_2O)_4]^{2+}$.

The chemistry of metals in solution is essentially the chemistry of their complexes. A metal ion in solution is coordinated to water molecules or to other ligands. The transition metal ions are fairly good Lewis acids, and their complexes are quite stable. The cations of the more electropositive metals such as the alkali metals and alkaline earth metals are weaker Lewis acids that form fewer and less stable complexes. Although they are hydrated in solution, the interaction with water is much weaker than in the case of transition metal ions.

Metal ions are almost never encountered without effective shielding from one another. In solution the shielding is provided by the solvent or some other ligand. In the solid state, cations are surrounded by anions. For an ionic substance such as NaCl, the interactions between the Na^+ and the six surrounding Cl^- ions are primarily electrostatic. As mentioned previously (see page 218), the more acidic metal ions, such as Cd^{2+} and Pd^{2+}, tend to form lattices in which the interactions are not the same in all directions.

CuCl$_2$ crystallizes as the dihydrate, which contains planar $H_2O{-}\overset{\displaystyle Cl}{\underset{\displaystyle Cl}{|}}\overset{|}{Cu}{-}OH_2$ units. The solid containing $[CuCl_2(H_2O)_2]$ dissolves to form $[Cu(H_2O)_4]^{2+}$ in dilute solution. In many cases the complexes found in well-ordered crystals do not persist in solution, because of solvent in competition with the ligands for metal coordination positions.

The role of coordination compounds in nature is discussed in Chapter 18. Coordinated metal ions are at the active centers of many enzymes.

9.2   ANALYTICAL APPLICATIONS

Equilibria involving complex formation are encountered frequently in analytical chemistry. Ag^+ is separated from Hg_2Cl_2 and $PbCl_2$ in the qualitative analysis scheme by dissolving AgCl as $[Ag(NH_3)_2]^+$. We separate antimony, arsenic, and tin sulfides from the remaining Qual Group II sulfides by dissolving them in the presence of an excess of sulfide ion as the complexes SbS_4^{3-}, AsS_4^{3-}, and SnS_3^{2-}. Organic reagents that give insoluble complexes are used widely for separation, identification, and quantitative determination of metals. Two of the most familiar examples are dimethylglyoxime, which is a fairly specific reagent for nickel, and 8-hydroxyquinoline (oxine), which precipitates many metals.

Bis(8-hydroxyquinolinato)zinc(II)

These compounds usually are insoluble, because the metal is incorporated into a large organic molecule and the resulting complex has no net charge. Some uncharged complexes are fairly volatile and can be sublimed; for example, $Fe(acetylacetonate)_3$ can be purified by sublimation at reduced pressure.

Because many uncharged complexes are insoluble in water but soluble in organic solvents, they also are used for separations based on solvent extraction procedures. At an appropriate pH (i.e., the pH at which formation of the uncharged complex is favored), we can get reasonable separations of a metal such as Ga^{III} from others by extracting the 8-hydroxyquinoline complex with chloroform. The equilibria involved are

$$HQ \;\rightleftharpoons\; H^- + Q^- \qquad K_a = \frac{[H^+][Q^-]}{[HQ]} \tag{9.3}$$

$$3Q^- + M^{3+} \;\rightleftharpoons\; MQ_3(aq) \qquad K_f = \frac{[MQ_3(aq)]}{[M^{3+}][Q^-]^3} \tag{9.4}$$

$$MQ_3(aq) \;\rightleftharpoons\; MQ_3(org) \qquad K_d = \frac{[MQ_3(org)]}{[MQ_3(aq)]} \tag{9.5}$$

K_a depends only on the complexing agent. K_f depends on the specific metal ion and the complexing agent. K_d, the partition coefficient, depends on the solvation of the complex in the two solvents; it is particularly dependent on the choice of the organic solvent. At low pH (low concentration of Q^-), only the more stable complexes (high K_f) are formed to be extracted. Complexes of the same ligand with different metal ions but the same stoichiometry usually have partition coefficients of the same magnitude, because of the similar external appearance to the solvent.

Acetylacetone also forms complexes used in solvent extraction procedures. Acetylacetone exists in the keto and enol forms, which are in equilibrium:

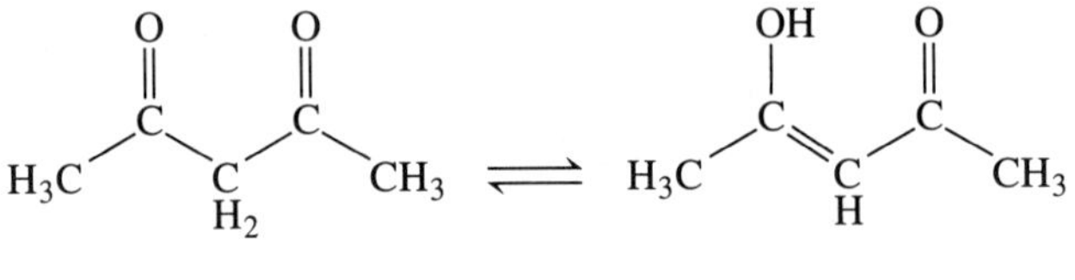

Keto   Enol

The enol form loses a proton and coordinates through both oxygen atoms to form a **chelate ring** (see Figure 9.1). The term "chelate" (Greek "crab's claw") was first used by Morgan to describe the formation of similar complexes. The term **metal chelate** is now used to refer to the compounds formed as a result of chelation. Chelating ligands must have two or more points of attachment. Ammonia, with only one unshared pair of electrons, is a monodentate (one-toothed) ligand. Chelating ligands such as acetylacetone and ethylenediamine ($NH_2C_2H_4NH_2$) are didentate. Table 9.1 lists some common polydentate ligands and their abbreviations. For the rules of nomenclature as applied to ligands and coordination compounds, see Section 9.3 and Appendix B.

Many titration procedures now use the hexadentate ligand ethylenediaminetetraacetate ion (edta; see Table 9.1) in the determination of metals. These procedures usually involve competition for the metal ion between edta and a dye that also can serve as a chelating ligand. One of the common oxidation–reduction indicators, ferroin, is a complex containing the $[Fe(phen)_3]^{2+}$ ion (see Table 9.1 for phen), which is deep red, whereas the Fe^{3+} compound is pale blue. In a suitable oxidation–reduction titration the removal of the color of the ferroin indicates that the endpoint has been reached. The ferroin is oxidized only after the reducing agent being titrated is oxidized completely.

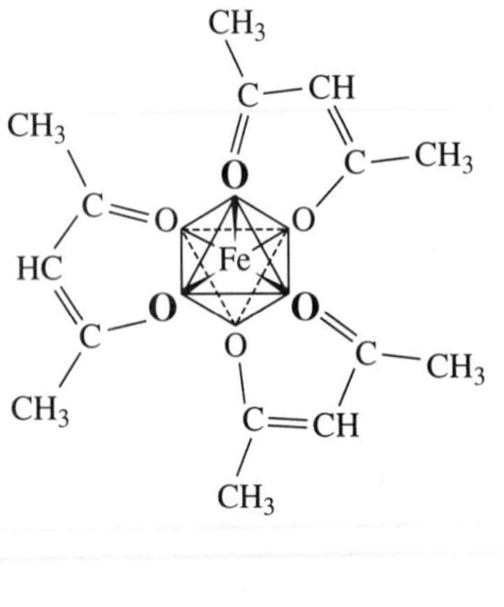

Figure 9.1   Chelate rings in the tris(acetylacetonato)iron(III) compound.

Table 9.1 Some common polydentate ligands

Name	Formula	Abbreviation	Classification[a]
Carbonato	CO_3^{2-}		Didentate
Oxalato	$C_2O_4^{2-}$	ox	Didentate
Ethylenediamine	$NH_2C_2H_4NH_2$	en	Didentate
1,2-Propanediamine	$NH_2CH(CH_3)CH_2NH_2$	pn	Didentate
Acetylacetonato	$CH_3\!-\!\overset{\displaystyle O^-}{\underset{\displaystyle \vert}{C}}\!=\!CH\overset{\displaystyle O}{\overset{\displaystyle \Vert}{C}}\!-\!CH_3$	acac	Didentate
8-Hydroxyquinolinato		oxine	Didentate
2,2'-Bipyridine		bipy	Didentate
1,10-Phenanthroline		phen	Didentate
Glycinato	$NH_2CH_2CO_2^-$	gly	Didentate
Diethylenetriamine	$NH(C_2H_4NH_2)_2$	dien	Tridentate
Triethylenetetraamine	$H_2NC_2H_4NHC_2H_4NHC_2H_4NH_2$	trien	Tetradentate
Nitrilotriacetato	$N(CH_2CO_2)_3^{3-}$	nta	Tetradentate
Tetraethylenepentaamine	$NH(C_2H_4NHC_2H_4NH_2)_2$	tetraen	Pentadentate
Ethylenediamine-tetraacetato	$[(O_2CCH_2)_2NC_2H_4N(CH_2CO_2)_2]^{4-}$	edta[b]	Hexadentate

[a] Previously, Latin prefixes were used with Latin stems (e.g., bidentate and quadridentate), and Greek prefixes were used with Greek stems (e.g., mononuclear, trinuclear, and tetranuclear). Now one set of numerical prefixes (Greek) is used: mono-, di-, tri-, tetra-, penta-, hexa-, hepta-, octa-, . . . , poly-.

[b] Capitals usually have been used for edta and other amino polyacid anions. The IUPAC rules recommend lower case letters for abbreviations of all ligands.

9.3 BRIEF SUMMARY OF NOMENCLATURE OF COORDINATION COMPOUNDS

Early names of coordination compounds used the names of discoverers (e.g., Magnus' green salt, $[Pt(NH_3)_4][PtCl_4]$, and Zeise's salt, $K[PtCl_3(C_2H_4)]$) or color (e.g., luteo for the yellow $[Co(NH_3)_6]^{3+}$ ion and roseo for the rose-red $[Co(H_2O)(NH_3)_5]^{3+}$ ion). Werner developed a system of nomenclature that is the basis of the current IUPAC rules. A very brief introduction to the IUPAC rules is presented here. For more complete rules see Appendix B.

The symbol for *the central atom is placed first in the formula of coordination compounds followed by anionic ligands in alphabetical order of the symbols and then neutral ligands in alphabetical order*. Abbreviations of complicated organic formulas may be used in formulas. The formula for the complex molecule or ion is enclosed in square brackets []. In names the central atom is placed after the ligands. The neutral ligand H_2O is called *aqua,* and the neutral ligand NH_3 is called *ammine*. For names *the ligands are listed in alphabetical order regardless of the charge on the ligand or the number of each*. Thus diammine (for two NH_3 ligands) is listed under "a" and dimethylamine is listed under "d".

The names of anionic ligands ends in -o (-ido, -ito, and -ato commonly). The number of ligands of each kind is indicated by the prefixes di-, tri-, tetra-, penta-, hexa-, and so on, unless the prefix could be misinterpreted as part of the name of the ligand. In such cases the prefixes bis, tris, tetrakis, and so on, are used—for example, diammine for $2NH_3$ but bis(methylamine) for $2CH_3NH_2$, since dimethylamine is $(CH_3)_2NH$. Exceptions for names of anionic ligands (omit -id of -ido): F^-, fluoro; Cl^-, chloro; Br^-, bromo; I^-, iodo; O^{2-}, oxo; OH^-, hydroxo; O_2^{2-}, peroxo; S^{2-}, thio; CN^-, cyano.

The oxidation number of the central atom is indicated by the Stock notation (Roman number) in parentheses after its name. Formulas and names may be supplemented by italicized prefixes *cis, trans, fac, mer*, and so on. Names of metal complex anions end in -ate. Complex cations and neutral molecules are given no distinguishing ending. See Appendix B for examples.

9.4 STABILITIES OF COORDINATION COMPOUNDS

The coordination process is an acid–base reaction. In general, an increase in basicity of the ligand and/or an increase in the acidity of the metal enhances the stability of the complex formed. Now we will consider some factors that contribute to thermodynamic stability and some specific effects that do not conform to the general pattern.

Saying that a compound such as ethanol is stable might refer to thermodynamic stability with respect to the formation from the elements carbon, hydrogen, and oxygen, or it might refer to kinetic stability—the fact that ethanol can be handled under ordinary circumstances without decomposition. Ethanol is kinetically stable in air, even though it burns in air (once ignited) and the products of the reaction with O_2 (CO_2 and H_2O) are more stable thermodynamically.

The stability of a metal complex, MX_m^{n+}, is usually expressed in terms of the formation constant, β_m, for the reaction

$$M^{n+} + mX = MX_m^{n+} \qquad \beta_m = \frac{[MX_m^{n+}]}{[M^{n+}][X]^m} \qquad (9.6)$$

The β_m expresses the thermodynamic stability, since it is an equilibrium constant from which we can calculate a free energy change for the formation of the complex. Considering the constants for each step in the formation of the complex often is worthwhile:

$$M^{n+} + X = MX^{n+} \qquad K_1 = \frac{[MX^{n+}]}{[M^{n+}][X]} \qquad (9.7)$$

$$MX^{n+} + X = MX_2^{n+} \qquad K_2 = \frac{[MX_2^{n+}]}{[MX^{n+}][X]} \tag{9.8}$$

$$MX_{m-1}^{n+} + X = MX_m^{n+} \qquad K_m = \frac{[MX_m^{n+}]}{[MX_{m-1}^{n+}][X]} \tag{9.9}$$

The overall constant β_m is the product of individual K values or expressions:

$$\beta_2 = K_1 K_2$$
$$\beta_m = K_1 K_2 \cdots K_m \tag{9.10}$$

Ligands are Lewis bases (and usually Brønsted bases)—for example, $NH_2C_2H_4NH_2$ or $C_2O_4^{2-}$ (oxalate ion, the anion of weak oxalic acid). For consistency, we express the basicity of the ligand in terms of pK ($-\log K$) of the acid dissociation constant of its conjugate acid. The symbols HL, HL^+, H_2L, and so on, are used to indicate the species to which the pK refers.

$$H_2C_2O_4 \;\rightleftharpoons\; H^+ + HC_2O_4^- \qquad K(H_2L) = \frac{[H^+][HC_2O_4^-]}{[H_2C_2O_4]} \tag{9.11}$$

$$HC_2O_4^- \;\rightleftharpoons\; H^+ + C_2O_4^{2-} \qquad K(HL^-) = \frac{[H^+][C_2O_4^{2-}]}{[HC_2O_4^-]} \tag{9.12}$$

$$^+NH_3C_2H_4NH_3^+ \;\rightleftharpoons\; H^+ + NH_2C_2H_4NH_3^+ \qquad K(H_2L^{2+}) = \frac{[H^+][NH_2C_2H_4NH_3^+]}{[NH_3C_2H_4NH_3^{2+}]} \tag{9.13}$$

$$NH_2C_2H_4NH_3^+ \;\rightleftharpoons\; H^+ + NH_2C_2H_4NH_2 \qquad K(HL^+) = \frac{[H^+][NH_2C_2H_4NH_2]}{[NH_2C_2H_4NH_3^+]} \tag{9.14}$$

The formation constants are evaluated by the usual methods for determining equilibrium constants: potentiometric titrations, spectrophotometric methods, polarography, electrode potential measurements, solubility measurements, ion exchange procedures, and so on—any procedure for determining the activities (or concentrations) of the species present at equilibrium.

EXAMPLE 9.1:   Calculate the equilibrium constant β_6 (see Section 8.5.4) for $Fe^{2+}(aq) + 6CN^- \rightarrow [Fe(CN)_6]^{4-}(aq)$ from the E^0 values.

$$Fe \longrightarrow Fe^{2+} \qquad E^0 = 0.44 \text{ V}$$

and

$$Fe \longrightarrow [Fe(CN)_6]^{4-} \qquad E_c^0 = 1.5 \text{ V}$$

Solution: The E_c^0 value is the *standard* emf for

$$Fe(s) + 6CN^- (aq) \longrightarrow [Fe(CN)_6]^{4-} + 2e$$

that is, for a solution containing CN^- and $[Fe(CN)_6]^{4-}$ at unit activity. Accordingly, the Fe^{2+} activity is given by

$$\beta_6 = \frac{[Fe(CN)_6^{4-}]}{[Fe^{2+}][CN^-]^6}$$

$$[Fe^{2+}] = \frac{1[Fe(CN)_6^{4-}]}{\beta_6[CN^-]^6} = \frac{1(1)}{\beta_6(1)^6}$$

Iron in equilibrium with $[Fe(CN)_6]^{4-}$ and CN^- ion at unit activity is thus in equilibrium with Fe^{2+} at an activity of $1/\beta_6$; and from the Nernst equation we obtain

$$E_c^0 = E^0 - \frac{0.059}{n} \log [Fe^{2+}]$$

$$1.5 = +0.44 - \frac{0.059}{2} \log \frac{1}{\beta_6}$$

$$\beta_6 = 10^{36}$$

If stoichiometry of the complex had not been known, we could have found it from the slope of E_c plotted against log $[CN^-]$. If the complexes of other stoichiometries are present over the ligand concentration studied, a plot of E_o versus log (ligand concentration) will not be linear.

▶9.4.1 Effect of Metal Ion

Thermodynamic stabilities of the complexes formed by various metals follow some regular trends, such as those involving size and charge effects, factors that determine the Lewis acid strength of a metal ion. Although metals display pronounced differences in their tendencies to form complexes with various ligand atoms, groups of similar metals show some helpful trends.

Generally, the stabilities of complexes decrease with increasing atomic number for the electropositive metals—for example, Group IIA or hard Lewis acids—and increase with increasing atomic number for the more noble metals (soft acids), following the general trend for the ionization energies. The ionic charge density (q/r) is the determining factor for electropositive metal ions (d^0); for example, F^- complexes of Be^{2+} are more stable than those of Mg^{2+}. Other factors must be considered for d^1–d^9 transition metal ions (see Section 9.8.1).

Classification of Metal Acceptor Properties

As previously mentioned, for the more electropositive metals (hard acids) the order of stability of the halide complexes is $F > Cl > Br > I$, but for highly polarizing (and also polarizable) soft acid metal ions such as Hg^{2+}, we see the reverse order. The most electropositive metals (hardest acids) show a great preference for forming complexes with hard ligands such as F^- or oxygen-containing ligands. As the electropositive character of the metals decreases (or the metal becomes softer), the complexes of N donors increase in stability with respect to those of O donors. Still more-electronegative (softer or more noble) metals form more stable complexes with soft donor atoms such as S relative to O, and P relative to N. The tendency to form stable olefin (alkene) complexes is greatest for very soft noble metals.

Table 9.2 shows metals classified according to their acceptor properties. The metals of class (*a*) are hard acids and show affinities for ligands that are roughly proportional to the basicities of the ligands toward the proton. The class (*b*) acceptors are soft acids, and they form stable olefin complexes (see Section 12.3.1). The border regions are not well-defined in all cases and, of course, the classification depends on the oxidation state of the metal ion. Copper(I) is a class (*b*) acceptor, but copper(II) is in the border region. Class (*a*) or hard acceptors form most stable complexes with hard ligand atoms in the second period (N, O, or F), whereas soft class (*b*) acceptors form their most stable complexes with ligand atoms from the third or a later period. Thus, trimethyl gallium shows the following tendencies toward complex formation involving alkyls of the ligand atoms from Groups V(15) and VI(16): N > P > As > Sb and O > S < Se > Te. On the other hand, for the soft Pt^{II} the order is apparently N << P > As > Sb; O <<< S >> Se < Te and F < Cl < Br < I.

▶ 9.4.2 *Basicity and Structure of Ligand*[2]

Although class (*a*) metal atoms display a preference for different ligand atoms than do class (*b*) metals, with a given donor atom the stability constants generally increase as the basicity of the ligand increases. Thus the complexes of simple amines are generally more stable than those of NH_3, and the stability increases with the basicity of the amine. Generally the complexes of chelate ligands—for example, $NH_2C_2H_4NH_2$—are much more

Table 9.2 Classification of metals according to acceptor properties[a]

H																
Li	Be											B	C			
Na	Mg											Al	Si			
K	Ca	Sc	Ti	V	Cr	Mn	Fe	Co	Ni	Cu	Zn	Ga	Ge	As		
Rb	Sr	Y	Zr	Nb	Mo	Tc	Ru	Rh	Pd	Ag	Cd	In	Sn	Sb		
Cs	Ba	La[b]	Hf	Ta	W	Re	Os	Ir	Pt	Au	Hg	Tl	Pb	Bi	Po	
Fr	Ra	Ac[c]														

☐ Class (*a*)   ■ Class (*b*)   ⬚ Border region

[a] S. Ahrland, J. Chatt, and N. R. Davies, *Q. Rev. Chem. Soc.* **1958**, *12*, 265.

[b] Lanthanides.

[c] Actinides.

[2] See D. P. N. Satchell and R. S. Satchell, *Q. Rev.* **1971**, *25*, 171; *Chem. Rev.* **1969**, *69*, 251. Also see J. R. Chipperfield, Chapter 7 in *Advances in Linear Free Energy Relationships*, N. V. Chapman and J. Shorter, Eds., Plenum Press, New York, 1972; and J. Hine, *Structural Effects on Equilibria in Organic Chemistry*, Wiley-Interscience, New York, 1975.

stable than those of monodentate ligands such as NH_3. This enhanced stability is referred to as the **chelate effect.** The stability increases with a greater number of chelate rings. The log of the formation constant for the addition of four ammonia molecules to Cu^{2+} is 11.9; for the addition of two en molecules, 20.0; and for triethylenetetraamine (four coordinated nitrogen atoms with three chelate rings), 20.4.

The chelate rings of greatest stability are generally five-membered, because the bond angles at the metal ion are 90° for planar and octahedral complexes. Six-membered rings are most often encountered in organic compounds, and six-membered chelate rings involving conjugation in the chelate ring are more stable than five-membered rings. Rings of more than six members are rare, and only a few ligands—CO_3^{2-}, for example—give four-membered rings.

The chelate effect can be largely an entropy effect for nontransition metal ions, but there may be a significant enthalpy effect for transition metal ions.[3] After one end of an ethylenediamine molecule has attached to a metal ion, the "effective concentration" of the other —NH_2 group is increased, because its motion is restricted to a small volume in the vicinity of the metal ion. Also each ethylenediamine molecule replaces two water molecules, increasing the total number of particles in the system, and hence the entropy. Although most alkaline earth metal complexes with uncharged ligands are not very stable, ethylenediaminetetraacetate ion (edta) forms quite stable complexes with metals such as Ca^{2+}. The edta can wrap around a metal ion, displacing as many as six coordinated water molecules. These complexes of the alkaline earth metal ions form primarily because of entropy effects resulting from the increase in the number of particles in the system and the decrease of charge on M^{2+} and edta^{4-}. Charge neutralization brings about a very favorable entropy change in a polar solvent, because it removes a large part of the ordering of the solvent molecules caused by the ions. The ligand is often added to foods and pharmaceuticals to **sequester** (tie up) metal ions, because some metal ions accelerate deterioration of organic compounds.

Lower-than-expected stabilities of complexes with chelating ligands often reveal structural differences in the complexes being formed. The stability of complexes of Ag^+ with didentate ligands is generally low compared to that of metals such as Cu^{2+} and Ni^{2+}. The normal *sp* hybridization of Ag^+ prevents it from participating in chelate formation.

Sepulchrate Ligands

The stable $[Co(en)_3]^{3+}$ complex ion can be encaged by condensation with formaldehyde and ammonia in a basic medium. The octahedron is capped above and below (along the C_3 axis), thereby forming the coordinated ligand sep:

$$N \underset{CH_2-NH-C_2H_4-NH-CH_2}{\overset{CH_2-NH-C_2H_4-NH-CH_2}{-CH_2-NH-C_2H_4-NH-CH_2-}} N$$

The condensation of each CH_2O involves one H of ammonia and one of NH_2 of en with elimination of H_2O. The caged complex is known as a **sepulchrate.**[4] One optical isomer

[3] J. J. R. Frausto da Silva, *J. Chem. Educ.* **1983,** *60,* 390.

[4] I. I. Creaser, J. M. Harrowfield, A. J. Herlt, A. M. Sargeson, J. Springborg, R. J. Geue, and M. R. Snow, *J. Am. Chem. Soc.,* **1977,** *99,* 3181; A. Hammershøi and A. M. Sargeson, *Inorg. Chem.,* **1983,** *22,* 3554.

of the very stable $[Co(sep)]^{3+}$ can be reduced to the stable $[Co(sep)]^{2+}$ and reoxidized to $[Co(sep)]^{3+}$ without racemization. Most complexes of Co^{II} exchange ligands rapidly and cannot be resolved because they racemize.

Forced Configurations

A ligand such as porphyrin or phthalocyanine (see Figure 9.2), which has a completely fused planar ring system, forms extraordinarily stable complexes with metal ions that tend to give planar complexes—for example, Cu^{2+}. These ligands impose planar configurations on metal ions that show no tendency to form planar complexes with monodentate ligands—for example, Be^{2+} and Zn^{2+}.

The ligands tris(2-aminoethyl)amine, $N(CH_2CH_2NH_2)_3$ (referred to as **tren**), and triethylenetetraamine, $(NH_2C_2H_4NHC_2H_4NHC_2H_4NH_2)$ (referred to as **trien**), are both tetradentate, but only trien can assume a planar or nearly planar configuration. From the pK values of the ligands, we would expect tren to form more stable complexes—as indeed is the case for Zn^{2+}, which can form tetrahedral complexes (Table 9.3). The Ni^{2+} complex of tren is only slightly more stable than that of trien, and Cu^{2+}, which tends to form planar complexes or octahedral complexes with two long bonds, gives a slightly more stable complex with trien in spite of its lower basicity.

▶ 9.4.3 *Symbiosis and Innocent Ligands*

The F^- ligand always has oxidation number $-I$. Such a ligand is described as **innocent.** For complex ions, assigning oxidation numbers is not always easy. Considerable ambiguity of oxidation numbers surrounds the complexes in which NO may function as NO^-, NO, or NO^+. In balancing redox equations, we normally assume that the addition of a hydrogen ion does not affect the oxidation state of atoms in a species (SO_4^{2-}, HSO_4^-, H_2SO_4, etc.). Even a hydrogen ion is not always "innocent", however. In the following reaction, which may be considered as an oxidative-addition reaction, the Pt goes from Pt^0 to Pt^{II}:

$$Pt[P(C_2H_5)_3]_3 + H^+ \rightleftharpoons HPt[P(C_2H_5)_3]_3^+$$

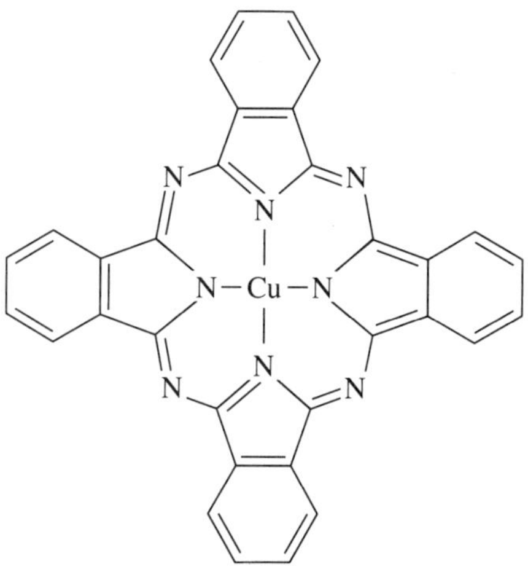

Figure 9.2 The planar copper(II) phthalocyanine complex.

Table 9.3   Some formation constants

	Tren	*Trien*
Ni^{2+} (log K_1)	14.8	14.0
Cu^{2+} (log K_1)	19.1	20.4
Zn^{2+} (log K_1)	14.65	12.1
pK (HL$^+$)	10.29	9.92

Jørgensen introduced the terms *innocent* and *noninnocent* to describe ligands forming complexes in which the assignment of oxidation states is *rigorous* and *ambiguous,* respectively.[5] The probability of noninnocent behavior is related to the softness of a ligand.

The same metal sometimes can function as a class (*a*) (hard) or a class (*b*) (soft) acid. Because softness (or hardness) is related to the highest occupied molecular orbital (HOMO)–lowest unoccupied molecular orbital (LUMO) gap (see Figure 4.30), it is not surprising that the greater the number of soft ligands surrounding a metal ion, the smaller the gap. Also, the greater number of soft ligands on a metal ion, the softer the metal ion becomes as an acid. Conversely, the greater number of hard ligands around a metal ion, the harder the metal ion becomes as an acid. Jørgensen called this phenomenon **symbiosis**[6]—a term used in biology to describe organisms that live together in harmony. The complex ion $[CoX(NH_3)_5]^{2+}$ can be prepared with X = F^-, Cl^-, Br^-, or I^-. With the soft CN^- ligand replacing NH_3, $[Co(CN)_5I]^{3-}$ is stable, but the complex containing F^- is unknown. Other examples of symbiosis include the ability of metal carbonyls to form carbonyl hydrides but not usually carbonyl fluorides and the disproportionation of HBF_3^- into BH_4^- and BF_4^- in ether solution (the HBF_3^- ion being an intermediate in the reaction of BF_3 with NaH or with the BH_4^- ion). The concept has been used extensively in the synthesis of complexes of a ligand such as a SCN^- which can bond through the soft S end or the hard N end. Farona and Wojcicki proposed that when good π-bonding ligands are present, the high oxidation states of the metal prefer the S end of SCN^- and the low oxidation states prefer the N end. This proposal was extended to the CNS^- complexes by Sloan and Wojcicki.[7]

9.5   WERNER'S COORDINATION THEORY

Compounds now considered to be coordination compounds have been known for more than 150 years, but no satisfactory explanation for their formation was available until the imaginative work of Alfred Werner.[8] In 1893, Werner formulated his coordination theory,

---

[5] Initially, the term "innocent" was used to describe ligands in colorless complexes—color being defined as a transition below 20,000 cm^{-1}. C. K. Jørgensen, *Struct. Bonding* **1966**, *1*, 234. Note that H bonded to a metal is treated as H$^-$.

[6] C. K. Jørgensen, *Inorg. Chem.* **1964**, *3*, 1201.

[7] See T. E. Sloan and A. Wojcicki, *Inorg. Chem.* **1968**, *7*, 1268, and references therein.

[8] Kauffman reviewed the early work in *J. Chem. Educ.* **1959**, *36*, 521. See also his translations of early papers under General References.

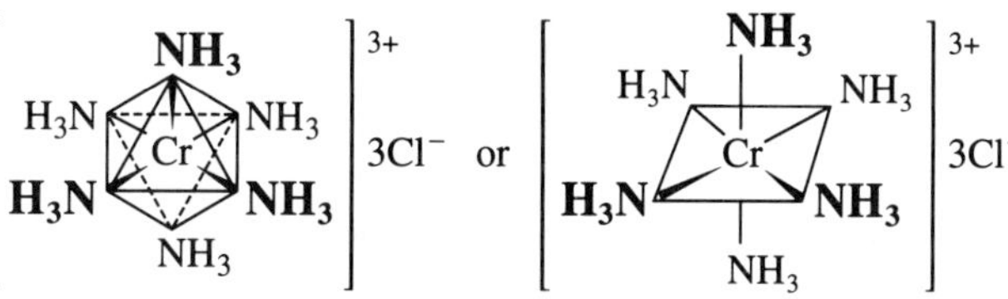

Figure 9.3 Octahedral structure of $[Cr(NH_3)_6]Cl_3$.

which provided the basis for modern theories. In order to appreciate Werner's insight, you must remember that the electron, the basis for all modern theories of chemical bonding, was unknown at the time. Werner suggested that each metal has two kinds of valence: **primary** (or ionizable) **valence,** which can be satisfied only by negative ions, as in simple salts such as $CrCl_3$; and **secondary valence,** which can be satisfied by negative ions or neutral molecules. The secondary valences are responsible for the addition of ammonia to produce compounds such as $CrCl_3 \cdot 6NH_3$. Whereas each metal has a characteristic number of secondary valences directed in space to give a definite geometrical arrangement, the primary valences are nondirectional. Werner assumed that the six secondary valences were arranged octahedrally (see Figure 9.3). When a $1-$ ion is in the inner coordination sphere, it satisfies one of the primary, as well as one of the secondary, valences of the metal (see $[CrCl(NH_3)_5]Cl_2$ in Figure 9.4). When two anions are in the coordination sphere, the directional character of the secondary valences gives rise to *cis* and *trans* isomers—for example, $CrCl_3 \cdot 4NH_3$ or $[CrCl_2(NH_3)_4]Cl$ (see Figure 9.4). Only negative ions outside of the coordination sphere are readily ionizable. Thus, the molecule $CrCl_3 \cdot 3NH_3$ or $[CrCl_3(NH_3)_3]$ is a nonelectrolyte and the chloride ion is not precipitated by addition of silver nitrate, except as it is displaced from the coordination sphere by another ligand. The species bonded to the metal ion in the coordination sphere are said to be in the **inner sphere.** The next sheath of ions or solvent molecules, held more loosely, makes up the **outer sphere.**

Werner spent his productive life placing his intuitive coordination theory on a firm experimental basis. He compiled such an impressive amount of evidence for the octahedral configuration from the isomerism and reactions of six-coordinate complexes that his configuration was accepted generally long before it had been confirmed by modern structural investigations.

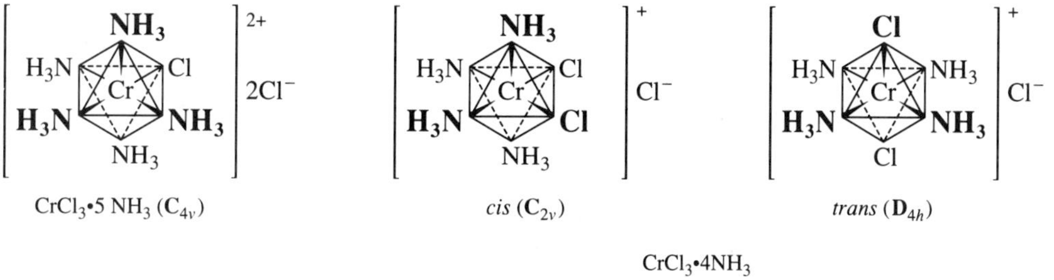

$CrCl_3 \cdot 5 NH_3 (C_{4v})$ *cis* (C_{2v}) *trans* (D_{4h})

$CrCl_3 \cdot 4NH_3$

Figure 9.4 Octahedral structures of chromium ammines.

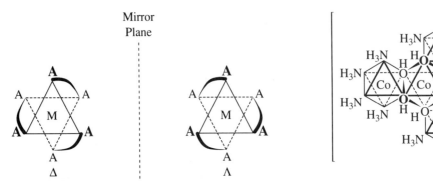

Figure 9.5   Representation of the mirror images of a complex of the type $M(\widehat{AA})_3$ as right- and left-handed spirals ($\mathbf{D}_3$).

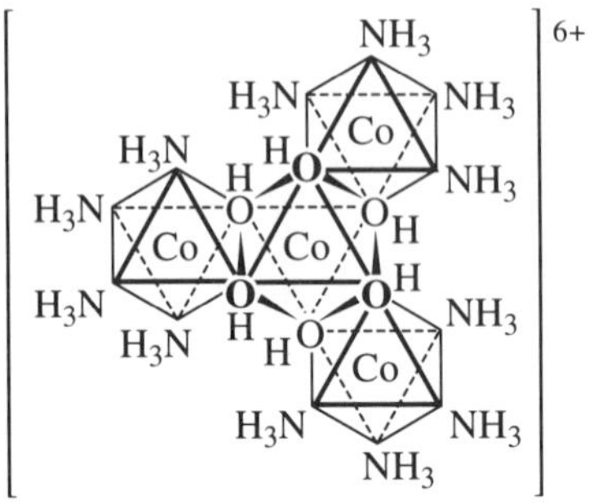

Figure 9.6   A completely inorganic optically active complex ion.

Some critics were skeptical of the optical activity attributed to the octahedral arrangement around a central ion in a complex such as $[Co(en)_3]Cl_3$. Figure 9.5 shows the optical isomers looking down the threefold axis in order to illustrate that the Δ and Λ isomers are really right- and left-handed spirals and are not superimposable (see page 419 for

notation)[9]. Werner prepared and resolved $\left[Co \left< \begin{Bmatrix} OH \\ \\ OH \end{Bmatrix}_3 Co(NH_3)_4 \right> \right]^{6+}$, which contains no

carbon (see Figure 9.6), to answer the criticism that the optical activity of $[Co(en)_3]Cl_3$ in some way might be caused by carbon. He was awarded the Nobel prize in 1913.

9.6   VALENCE BOND MODELS

▶ 9.6.1   *Effective Atomic Number*

Counting electrons is an important way of rationalizing the formulas of chemical compounds. For light elements, the octet rule provides a useful guide to understanding why the chloride of Mg (for example) has the formula $MgCl_2$ (rather than $MgCl$ or $MgCl_3$). The formulas of coordination compounds are "complex"—for example, $CoCl_3 \cdot 6NH_3$, which, in Werner's terms, has the primary valence satisfied by Cl^- and the secondary valence satisfied by NH_3. The nature of secondary valence follows from G. N. Lewis' treatment of the coordinate covalent bond in which the ligand atom furnishes an electron pair that is shared with the central metal ion. Sidgwick argued that the coordination process provided the opportunity for a transition metal ion to reach a noble gas configuration, or the same number of electrons as the atomic number of a noble gas. The **effective atomic number** (EAN) of a metal ion is calculated by adding the electrons of the metal ion to those shared with it through coordination. In $[Co(NH_3)_6]^{3+}$ the cobalt(III) ion has 24 elec-

---

[9] X-ray methods have proven that the Λ isomer (left-handed spiral) is the isomer of $[Co(en)_3]^{3+}$ formerly designated as D (for *dextro*, with a positive rotation at the Na–D line).

trons plus the six electron pairs from the ammonia molecules, for a total of 36 electrons—the configuration of krypton. Just as the octet is useful in formulating the bonding in compounds of the light elements, the notion of an EAN provides a rough guide for bonding in coordination compounds. Quite a few, but not all, metals achieve the EAN of a noble gas through coordination (see Table 9.4). The EAN concept has been particularly successful for complexes of low-valent metals (oxidation number $\leq$II). For complexes having high oxidation numbers ($\geq$II), EAN can be taken only as a rough guide. Most of the complexes of Ni^{II}, for example, give EAN values which are 34 [coordination number (C.N.) 4] or 38 (C.N. 6) rather than 36, but the EAN is 36 for $Ni(CO)_4$ containing Ni^0.

This electron-counting scheme is also called the **eighteen-electron rule,** in contrast to the usual octet rule for "simple" compounds. The octet rule applies to main-group elements for which s and p are the only low-energy orbitals to be filled. For transition metals the five d orbitals are also included in the valence shell, since they are comparable in energy to the filled s and p orbitals of the same quantum number. The energy of the d orbitals is lowered with increasing nuclear charge through a transition series, so that after several d electrons have been added, the addition of ligand electrons can complete a shell of 18 (a pseudo-noble-gas configuration). The octet rule for representative elements and the eighteen-electron rule for transition metals recognize the tendency to use all low-energy atomic orbitals either in forming chemical bonds or for containing unshared electron pairs. Nickel(II) and other d^8 ions form stable square-planar complexes such as $[Ni(CN)_4]^{2-}$. These and some other complexes correspond to the **sixteen-electron rule.**

▶ 9.6.2   *Hybridization and Orbital Occupancy*

Much of the valence bond (VB) model of coordination compounds was developed by Pauling. Qualitative VB theory was very important in fostering the development of this field to its present status, although we shall see that later treatments offer some important advantages.

Table 9.4   Valence electron count for metals

Complex	Number of electrons on M^{n+}	Number of electrons from ligands	EAN
$Pt(NH_3)_6^{4+}$	$78 - 4 = 74$	$6 \times 2 = 12$	86 (Rn)
$Co(NH_3)_6^{3+}$	$27 - 3 = 24$	$6 \times 2 = 12$	36 (Kr)
$Fe(CN)_6^{4-}$	$26 - 2 = 24$	$6 \times 2 = 12$	36
$Fe(CO)_5$	$26 - 0 = 26$	$5 \times 2 = 10$	36
$Cr(CO)_6$	$24 - 0 = 24$	$6 \times 2 = 12$	36
$Ni(CO)_4$	$28 - 0 = 28$	$4 \times 2 = 8$	36
$Ni(NH_3)_6^{2+}$	$28 - 2 = 26$	$6 \times 2 = 12$	38
$Ni(CN)_4^{2-}$	$28 - 2 = 26$	$4 \times 2 = 8$	34
$Cr(NH_3)_6^{3+}$	$24 - 3 = 21$	$6 \times 2 = 12$	33

According to this interpretation, coordination compounds result from the use of available empty bonding orbitals on the metal for the formation of coordinate covalent bonds. The C.N. and geometry are determined in part by size and charge effects, but also to a great extent by the orbitals available for bonding. The common hybridized orbitals encountered in coordination compounds are shown in Table 9.5, which also lists the orbitals suitable for forming strong π bonds. The d orbitals used are those available with the lowest quantum number. For octahedral complexes of first transition series metals, these are $3d$.

Magnetic susceptibility measurements have been used widely to determine the number of unpaired electrons in complexes, and from this information the number of d orbitals used for bond formation usually can be inferred for ions containing four to eight d electrons. Table 9.6 shows that for many Co^{3+} and Fe^{2+} complexes, for example, the d electrons pair off to make available d orbitals which form d^2sp^3 hybrids used for bond formation. Such complexes were referred to as "covalent" complexes by Pauling. Those complexes in which there were no low-energy d orbitals available for bond formation, as in $[Ga(C_2O_4)_3]^{3-}$, $[Ni(NH_3)_6]^{2+}$, or $[Fe(C_2O_4)_3]^{3-}$ (five unpaired electrons), were referred to as "ionic" complexes. The terms **low-spin** (or spin-paired) and **high-spin** (or spin-free) are preferred, since they describe the electron population of the d orbitals, as determined from magnetic properties, without any assumptions concerning the nature of the bonding. In Table 9.6, "covalent" octahedral complexes are shown using the d^2sp^3 orbitals, and the sp^3 orbitals are used for tetrahedral $[ZnCl_4]^{2-}$ for ligand electrons.

Table 9.5   Hybrid orbitals for metal complexes[a]

Coord. number	Hybridized orbitals (σ)	Configuration	Strong π orbitals	Examples
2	sp	Linear	p^2, d^2	$Ag(NH_3)_2^+$
3	sp^2	Trigonal planar	p, d^2	BF_3, NO_3^-, $Ag(PR_3)_3^+$
4	sp^3	Tetrahedral		$Ni(CO)_4$, MnO_4^-, $Zn(NH_3)_4^{2+}$
4	dsp^2	Planar	d^3, p	$Ni(CN)_4^{2-}$, $Pt(NH_3)_4^{2+}$
5	$d_{z^2}sp^3$ or d^3sp	Trigonal bipyramidal	d^2	TaF_5, $CuCl_5^{3-}$, $[Ni(PEt_3)_2Br_3]$
5	$d_{x^2-y^2}sp^3$, d^2sp^2, d^4s, or d^4p	Tetragonal pyramidal	d	IF_5, $[VO(acac)_2]$
6	d^2sp^3	Octahedral	d^3	$Co(NH_3)_6^{3+}$, $PtCl_6^{2-}$
7	d^5sp or d^3sp^3	Pentagonal bipyramidal	—	ZrF_7^{3-}
7	d^4sp^2 or d^5p^2	Trigonal prism with an extra atom in one tetragonal face	—	TaF_7^{2-}, NbF_7^{2-}
8	d^4sp^3	Dodecahedral	d	$Mo(CN)_8^{4-}$, $Zr(C_2O_4)_4^{4-}$
8	d^5p^3	Antiprismatic	—	TaF_8^{3-}, $Zr(acac)_4$
8	d^3fsp^3 or d^3f^4s	Cubic	—	$U(NCS)_8^{4-}$

[a] G. E. Kimball, *J. Chem. Phys.* **1940**, *8*, 188.

Table 9.6 Electron configurations for some metal complexes in valence bond (VB) terms and classified as high or low spin

Ion or complex	Configuration									High or low spin
	3d					4s	4p			
Cr^{3+}	↑	↑	↑	—	—	—	—	—	—	
$[Cr(NH_3)_6]^{3+}$	↑	↑	↑	[↑↓	↑↓	↑↓	↑↓	↑↓	↑↓]	[a]
Co^{3+}, Fe^{2+}	↑↓	↑	↑	↑	↑	—	—	—	—	
$[Co(NH_3)_6]^{3+}$	↑↓	↑↓	↑↓	[↑↓	↑↓	↑↓	↑↓	↑↓	↑↓]	Low spin
$Fe(CN)_6^{4-}$										
Co^{2+}	↑↓	↑↓	↑	↑	↑	—	—	—	—	
$[Co(NH_3)_6]^{2+}$	↑↓	↑↓	↑	↑	↑	"Ionic" (VB)				High spin
Zn^{2+}	↑↓	↑↓	↑↓	↑↓	↑↓	—	—	—	—	
$[ZnCl_4]^{2-}$ (tetrahedral)	↑↓	↑↓	↑↓	↑↓	↑↓	[↑↓	↑↓	↑↓	↑↓]	[b]
$[Zn(NH_3)_6]^{2+}$	↑↓	↑↓	↑↓	↑↓	↑↓	"Ionic" (VB)				[b]
Fe^{3+}	↑	↑	↑	↑	↑	—	—	—	—	
$[Fe(C_2O_4)_3]^{3-}$	↑	↑	↑	↑	↑	"Ionic" (VB)				High spin
$[Ni(NH_3)_6]^{2+}$	↑↓	↑↓	↑↓	↑	↑					

[a] All octahedral complexes of Cr^{III} have three unpaired electrons.

[b] All complexes of Zn^{II} are diamagnetic.

For transition metal ions, the total magnetic moment is given by the Landé expression:

$$\mu_M = \left[1 + \frac{S(S+1) - L(L+1) + J(J+1)}{2J(J+1)}\right]\sqrt{J(J+1)}$$

When the metal ions are surrounded by ligands in complexes, the orbital contribution to the magnetic moment is often quenched ($L = 0$ and $J = S$, so the term in brackets becomes 2) so that the magnetic moment can be calculated from the spin contribution alone using the spin-only formula, $S = n(\frac{1}{2})$:

$$\mu_M = \sqrt{n(n+2)} \quad \text{Bohr magnetons}$$

where n is the number of unpaired electrons. See inside the back cover for the Bohr magneton value.

Square-planar complexes of Ni^{II} are diamagnetic ($n = 0$). For octahedral Ni^{II} complexes, $n = 2$ and $\mu_M = \sqrt{2(2+2)} = 2.82$ Bohr magnetons. Both octahedral and tetrahedral complexes of Ni^{II} have $n = 2$, but they can be distinguished in some cases because the orbital contribution is not quenched to the same extent in the two cases. The magnetic moment for Ni^{II} is closer to the spin-only value for octahedral complexes.

Nickel(II) forms octahedral complexes with ligands such as H_2O and NH_3. However, with ligands such as CN^-, which display a greater tendency to form strong covalent bonds, Ni^{II} forms diamagnetic planar complexes. Planar complexes are also common for other d^8 ions: Pd^{II}, Pt^{II}, Au^{III}, Rh^I, and Ir^I. The *empty d* orbital necessary for forming dsp^2 hybrid orbitals is available or can be made available in the d^8 ions by pairing electrons in the four other d orbitals. A d orbital cannot be vacated by pairing for Cu^{II} (d^9). If the ninth d electron in $[Cu(NH_3)_4]^{2+}$ were promoted to an outer ($4d$) orbital, it might be lost easily, accompanying oxidation to a Cu^{III} complex. But in fact, the Cu^{II} complexes are not oxidized easily. The interpretation of the planar Cu^{II} complexes as "ionic" or as "covalent" using $4d$ orbitals is unsatisfactory because of the great stability of the complexes and because the bonding does not seem to be particularly different from that in planar nickel complexes. The lack of a really satisfactory account of the bonding in the very stable complexes labeled here as "ionic" is a major weakness of the VB treatment. These include $[Cu(NH_3)_4]^{2+}$, $[Ni(NH_3)_6]^{2+}$, $[Zn(NH_3)_4]^{2+}$, and other complexes of d^9 and d^{10} ions. Also, absorption spectra and other properties involving excited states are not treated adequately by simple VB theory.

▶ ## 9.6.3 *Stabilization of Oxidation States and Pi Bonding*

The aqua or hydrated ion $[Co(H_2O)_6]^{2+}$ is difficult to oxidize to $[Co(H_2O)_6]^{3+}$, but $[Co(NH_3)_6]^{2+}$ is oxidized to diamagnetic $[Co(NH_3)_6]^{3+}$ by O_2 in air. The difference in reactivity with ligand type is not so great for other first transition series ions, but in general, coordination with ligand atoms less electronegative than oxygen stabilizes the oxidation state next higher than that of the most stable aqua ion, unless other factors, such as π bonding, are involved. This is because less electronegative ligand atoms are more effective donors toward the more highly charged ions. Compounds of the highest oxidation state are strongly oxidizing and are stable only with the most electronegative ligands, F and O, as in fluoride complexes and oxoanions. For example, $[FeF_6]^{3-}$ is stable, but $[FeI_6]^{3-}$ is not. I^- is such a good donor of electrons that it reduces Fe^{3+} to Fe^{2+}.

The general electron configuration for the first transition series neutral atoms is $3d^n4s^2$, with Cr ($3d^54s^1$) and Cu ($3d^{10}4s^1$) as exceptions. All transition elements give M^{2+} ions, corresponding to removal of the $4s$ electrons (or one s and one d electron for Cr and Cu). For early members of the transition series, however, higher oxidation states are more stable. Sc^{III} is the only important oxidation state for Sc. For Ti, Ti^{IV} is most important, with Ti^{III} and especially Ti^{II} obtainable only under strongly reducing conditions. V^{II} and Cr^{II} also are good reducing agents (easily oxidized). Manganese is the first member of the series for which II is the most important oxidation state.

The generalization that higher oxidation states are stabilized in compounds containing ligand atoms of low electronegativity has some notable exceptions. The most striking common examples are compounds of CO, CN^-, and 1, 10-phenanthroline; some examples are given in Table 9.7. A ligand atom such as C (CO and CN^- coordinate through C nor-

Table 9.7   Stabilization of low oxidation states through coordination

Complex	Oxidation number of M	Comment
$Ni(CO)_4$, $Fe(CO)_5$, $Cr(CO)_6$	0	Stable
$[Ni(CN)_4]^{4-}$, $[Pd(CN)_4]^{4-}$	0	Easily oxidized
$[Ni(CN)_4]^{3-}$, $[Mn(CN)_6]^{5-}$	I	Easily oxidized
$[Fe(phen)_3]^{2+}$	II	Fe^{II} more stable than Fe^{III}

mally) could give a predominantly covalent coordinate bond with a transition metal atom, leading to the accumulation of negative charge on the metal atom. The Ni—C bond distance in $Ni(CO)_4$, however, is shorter than expected for a single bond, suggesting that the bond has an appreciable amount of double-bond character. The VB representation of a contributing structure containing a metal–carbon double bond (Figure 9.7a) reveals *that* particular contributing structure to have a carbon–oxygen double bond instead of the triple bond in the other main contributing structure. Figure 9.7b shows the VB orbital representation. The sketch of orbitals available for ligand–metal π-bonding in a molecular orbital (MO) description (Figure 9.7c) reveals that if a metal d orbital of appropriate symmetry is filled, electrons can be donated into the CO $\pi*$ orbital. This reduces the CO bond order and is represented in the VB model as double bond character (since only pairs of electrons can be moved in the VB theory). In either model, the C—O bond is weakened and lengthened while the M—C π-bonding strengthens the M—C bond. Bonding descriptions are similar for cyano complexes. Both CO and CN^- stabilize low oxidation numbers (see Table 9.7) where metal electrons are available for M—C π bonding.

Several transition metal complexes of bipyridine or 1, 10-phenanthroline (which also contain empty orbitals of π symmetry) can be reduced to unusually low (even zero and negative) oxidation numbers. Thus $[Fe(bipy)_3]^{2+}$ can be reduced polarographically in three one-electron steps to $[Fe(bipy)_3]^-$, and the $[Ni(bipy)_2]^{2+}$ ion can be reduced in two steps to $[Ni(bipy)_2]^0$. The tris(bipy) complexes of Ti^0, Ti^{-1}, Cr^I, Cr^0, and V^0 can be obtained by chemical reduction using the lithium salt of $bipy^{2-}$ as the reducing agent in tetrahydrofuran:

$$MX_y + yLi_2(bipy) + nbipy \xrightarrow{\text{thf}} M(bipy)_n + yLiX + yLi(bipy) \qquad (9.15)$$

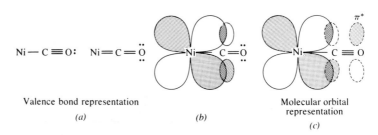

Valence bond representation

(a)   (b)

Molecular orbital representation

(c)

Figure 9.7   Metal–carbon double bonding.

The compounds $M(bipy)_2$ of Be, Mg, Ca, and Sr appear as formally M^0 compounds, but various experimental techniques lead to conflicting conclusions as to whether they should be represented as $M^0(bipy)_2$ or $M^{2+}(bipy^-)_2$. Because of extensive electron delocalization, bipy is a noninnocent ligand (Section 9.4.3) here, and the assignment of oxidation states is not clear for the reduced complexes.

9.7 SIMPLE ELECTROSTATIC MODELS OF BONDING

Here we consider a simple electrostatic model of metal ion–ligand interaction, modify it for polarization of the ligand, and then modify it for polarization of the metal ion and the ligand. In the next section we examine the additional consideration of transition metal ions with partially filled *d* orbitals.

▶ 9.7.1 Ion–Dipole Interaction

The electrostatic interpretation of the interaction between metal ions and ligands in complexes developed concurrently with the covalent bond approach. In the early 1920s the formation of complexes such as AlF_6^{3-} was discussed in terms of the attraction between oppositely charged ions. Water (aqua) and ammonia complexes were considered in terms of ion–dipole interaction. This model leads to the correct prediction that large cations of low charge (e.g., weak Lewis acids such as Na^+ and K^+) form few complexes. In Group 2 (IIA) the larger cations show little tendency to form stable complexes, but Be^{2+} does form some very stable complexes (e.g., BeF_4^{2-}). The charge density (q/r) of a cation is useful in comparing complexing ability of ions of different charges in the electrostatic model, since the energy of interaction varies with the charge (q) and inversely with the radius (r) of an ion. We expect ions with high charge densities (see Table 9.8), such as Be^{2+}, Al^{3+}, and

Table 9.8   Charge densities [charge/radius(Å)] for cations[a]

Ion	q/r	Ion	q/r
Be^{2+}(tet)	4.88	Al^{3+}	4.44
Mg^{2+}	3.32	Sc^{3+}	3.39
Ca^{2+}	1.75	Cr^{3+}	3.97
Fe^{2+}	2.17	Fe^{3+}	3.82
Co^{2+}	2.26	Co^{3+}	4.00
Ni^{2+}	2.40	Rh^{3+}	3.72
Cu^{2+}	2.20		
Zn^{2+}	2.27	Zn^{2+}(tet)	2.70
Hg^{2+}	1.72	Hg^{2+}(tet)	1.82

[a] Radii in Ångstroms for C.N. 6 except as noted for tetrahedral configurations (C.N. 4).

many of the transition metal ions, to form stable complexes. The simple electrostatic interpretation, however, does not account for the great stability of complexes of some cations (such as Hg^{2+}) whose charge density is small compared with Mg^{2+}, which has a high charge density but forms few stable complexes.

Because the permanent dipole moment of the water molecule is greater than that of ammonia, aqua complexes should be considerably more stable than those of ammonia. Fulfilling this expectation, ammonia complexes of the alkaline earth metals are unstable in aqueous solution. Nevertheless, the ammonia complexes are much more stable than the aqua complexes for many of the transition metal ions, such as Ni^{2+}. The charge density is smaller for Ni^{2+} than that of Mg^{2+}, so the difference in relative stabilities of the aqua and ammonia complexes of these ions cannot be explained by simple electrostatic theory.

▶ 9.7.2   *Ligand Polarization*

The simple electrostatic treatment views metal ions and ligands as rigid and undistorted. The picture can be improved by taking into account polarization of the ligands to explain the reversal in the stabilities of water and ammonia complexes.

The stability of complex ions containing neutral ligands depends on the charge density of the cation and the total dipole moment of the ligand—the sum of the permanent (μ) and induced (μ_i) dipole moments:

$$\text{Total dipole moment} = \mu + \mu_i \qquad (9.16)$$

The induced dipole moment is given by the relationship $\mu_i = \alpha q/d$, where q/d (cation charge/distance between cation and ligand) represents the strength of the cation's electrical field and α is the electronic polarizability of the molecule. The polarizability of an atom is greater for low absolute hardness (see Section 1.9). The energy of interaction, E (polarization of the cation is neglected), is given by

$$E = (q/d^2)(\mu + \mu_i)$$
$$= (q/d^2)\mu + (q^2/d^3)\alpha \qquad (9.17)$$

The permanent dipole moment of the water molecule is greater than that of ammonia, but the ammonia molecule is more polarizable and the induced dipole moment in a strong field can be quite large. Thus the total dipole moment of NH_3 is greater than that of H_2O in a strong electric field. Cations that are hard acids form more stable complexes with H_2O. Further refinement, to include polarization of the cation as well as the ligand atom, is required to explain the relative stabilities of some of the halide complexes and the preference for NH_3 over H_2O by some cations with low charge densities.

▶ 9.7.3   *Mutual Polarization of Cation and Ligand*

Polarization of the ligand molecule does not fully explain the differences in the stabilities of the complexes of the metals in Groups 1(IA) and 2(IIA) compared with those of Groups 11(IB) and 12(IIB). Even for metal ions with about the same charge density (e.g., Na^+

and Cu^+, Ca^{2+} and Hg^{2+}), soft B subgroup (post-transition) metals form much more stable complexes than do hard A group metals (see Section 7.4.3). The polarization of soft B subgroup cations themselves must be considered in order to account for differences in the coordinating ability of ions such as Na^+ and Cu^+. Noble-gas cations with an outer shell of eight electrons are hard and not easily deformed, but polarization effects are quite important for cations that have the soft pseudo-noble-gas-type configuration (18 electrons in the outer shell) or those with similar configurations $[(n - 1)s^2p^6d^{10}ns^2$, or an almost complete group of d orbitals in the outer shell; see page 202]. The greater polarization of electrons in the d subshell results from the low penetration of inner shells by the d electrons (see page 40). Cation polarization is particularly important, because it decreases the screening of the nuclear charge of the cation and decreases the metal–ligand distance. Both factors favor increased polarization of the ligand. Mutual polarization of the cation and of the ligand accounts for the greater stability of the complexes of ammonia, compared with aqua complexes, for the metals near the end of, and just following, the transition series.

Halide complexes of hard cations with noble-gas-type configurations decrease in stability in the order

$$F^- > Cl^- > Br^- > I^-$$

This is the order expected on the basis of the anion sizes: The smaller anions release more energy (q_1q_2/d) because of the closer approach to the cation. The reverse order of stability is observed for soft cations such as Hg^{2+}, for which mutual polarization of the cation and anion favors combination with larger, more polarizable anions.

Alkyl substitution in the hard ligands H_2O and NH_3 (an alkyl group is represented by R) decreases the tendency toward complex formation in the order

$$H_2O > ROH > R_2O$$
$$NH_3 > RNH_2 > R_2NH > R_3N$$

These orders correspond to decreases in dipole moments of the molecules, but hydrogen-bonding stabilization of aqua and ammine complexes also play a great—perhaps predominant—role in aqueous solution. Steric effects can decrease the stability of complexes with R groups substituting for H, particularly for large R groups. However, alkyl substitution *increases* the tendency toward complex formation for soft ligands H_2S and PH_3, corresponding to the order of *increasing* dipole moment in the series

$$H_2S < RSH < R_2S$$
$$PH_3 < RPH_2 < R_2PH < R_3P$$

The electrostatic approach, including polarization effects, permits us to predict the "bond" energies for many complexes. Table 9.9 compares calculated energies per bond and those obtained from thermochemical data for the reaction

$$M^{m+}(g) + nL(g) \longrightarrow ML_n^{m+}(g) \qquad (9.18)$$

Table 9.9   Some metal–ligand bond energies[a]

Complex	Energy per bond (calculated)	Energy per bond (experimental)[b]	Energy per bond corrected for ligand-field stabilization
$[Fe(H_2O)_6]^{2+}$	209 kJ	243 kJ	218 kJ
$[Fe(H_2O)_6]^{3+}$	456	485	456[c]
$[K(H_2O)_6]^{+}$	54	67	54[d]
$[Cr(H_2O)_6]^{3+}$	464	510	502
$[Co(NH_3)_6]^{3+}$	490	560	523
$[AlF_6]^{3-}$	887	975	887[d]
$[Zn(NH_3)_4]^{2+}$	360	372	360[e]

[a] Data from F. Basolo and R. Pearson, *Mechanisms of Inorganic Reactions,* 2nd ed., Wiley, New York, 1967, p. 63.
[b] From thermochemical data.
[c] No correction, high-spin d^5.
[d] No correction, d^0.
[e] No correction, d^{10}.

Because metals of the first transition series have nearly the same radii, the electrostatic treatment predicts that the complexes they form in the same oxidation state should have approximately the same stability. Failure to account for striking stability differences among the complexes of the transition metals is a great weakness of the simple electrostatic approach. The $[Cr(NH_3)_6]^{3+}$ and $[Co(NH_3)_6]^{3+}$ complexes are very stable, but manganese and iron do not form stable ammonia complexes in aqueous solution. Such differences among the transition metal complexes can be explained by applying the ligand field theory.

9.8   THE LIGAND-FIELD THEORY

The five d orbitals are degenerate (of equal energy) in gaseous metal ions. The simple electrostatic model fails to predict differences of bond energies among complexes of different transition metals because it ignores the fact that the degeneracy of the d orbitals may be removed in the presence of ligands. For a d^1 metal ion in an octahedral field the electron can go into one of the lower-energy d orbitals, thereby stabilizing the complex. The field caused by the ligands resembles the electric field around ions in ionic crystals. This similarity accounts for the name **crystal-field theory.** Although crystal-field theory was first proposed by Bethe in 1929, its application to coordination compounds is more recent. In the pure crystal-field theory, we consider that only electrons belonging to the metal occupy the metal's d orbitals. The ligand lone pairs are considered to belong entirely to the ligand and to provide an electric field of the symmetry of the ligand arrangement around the metal. The name **ligand-field theory** is used to refer to the approach in present use; it is basically the same as the pure crystal field approach, except that covalent interaction is taken into account as necessary.

▶ *9.8.1 Octahedral Complexes*

Ligand-Field Stabilization Energy

In an octahedral complex oriented as shown in Figure 9.8, two of the d orbitals, d_{z^2} and $d_{z^2-y^2}$, are directed toward the ligands. The repulsion caused by the electron pairs of the ligands raises the energy of any electrons occupying these orbitals[10] (e_g) more than that of the other three orbitals, d_{xy}, d_{xz}, and d_{yz} (t_{2g}), which are directed at 45° to the axes (see Figure 9.9). The stronger the electric field caused by the ligands, the greater the splitting (energy separation) of the energy levels. The energy separation for an octahedral field is $10Dq$ (or Δ). This splitting causes no net change in the energy of the system if all five orbitals are occupied equally because a filled or half-filled orbital set is totally symmetrical with $L = 0$. The t_{2g} level is lowered by $4Dq(\frac{4}{10}\Delta)$, and the e_g level is raised by $6Dq(\frac{6}{10}\Delta)$ relative to the average energy of the d orbitals. Hence the net change is zero for a d^{10} ion with 6 t_{2g} and 4 e_g electrons.

$$6(-4Dq) + 4(+6Dq) = -24Dq + 24Dq = 0$$

Similarly, no stabilization results from the splitting for a d^5 configuration if each orbital is singly occupied. For all configurations other than d^0, d^5 (high-spin), and d^{10}, splitting

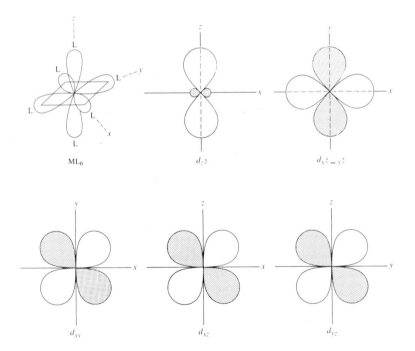

Figure 9.8 The d orbitals.

[10] See Section 3.5.4 for an explanation of the symmetry MO notation. The e_g^* orbitals carry the asterisk to indicate that they are really antibonding MOs; bonding orbitals $e_g(d)$, $a_{1g}(s)$, and $t_{1u}(p)$ are the d^2sp^3 orbitals accommodating electron pairs from the six ligands. The asterisk for e_g^* usually is omitted because the "pure" crystal-field theory treatment neglects covalent bonding. We will omit it except for the MO treatment.

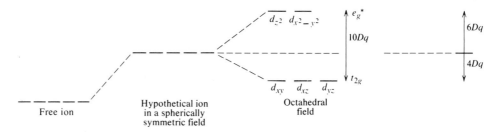

Figure 9.9   Splitting of the d energy levels in an octahedral complex.

lowers the total energy of the system as compared to the spherically symmetric case because electrons preferentially occupy the lower-energy orbitals. The decrease in energy caused by the splitting of the energy levels is the **ligand-field stabilization energy** (LFSE).

Electrons enter the t_{2g} orbitals in accordance with Hund's rule for d^1, d^2, and d^3 configurations. Thus $[Cr(NH_3)_6]^{3+}$ has three unpaired electrons with LFSE of $3(-4Dq) = -12Dq$. The LFSE for any octahedral chromium(III) complex is $-12Dq$, but the value of Dq varies with the ligand. Two configurations, $t_{2g}^3 e_g^1$ and t_{2g}^4, are possible for a d^4 ion in the ground state (see Figure 9.10). For $t_{2g}^3 e_g^1$, the fourth electron occupies the e_g level, giving four unpaired electrons. The LFSE is $3(-4Dq) + 1(6Dq) = -6Dq$. For t_{2g}^4 the LFSE is $4(-4Dq) = -16Dq$. Which one of these configurations a given complex adopts depends on the relative magnitudes of the ligand field (Dq) and the unfavorable potential energy of repulsion of two electrons occupying the same orbital with spins paired (the pairing energy, PE). If Dq (and therefore $10Dq$) is relatively small compared to the PE, it will be energetically more favorable to expend $10Dq$ in energy to place the fourth electron in an e_g orbital instead of pairing it with another electron in a t_{2g} orbital. The electron configuration will be $t_{2g}^3 e_g^1$. This is referred to as the **weak-field case,** since the splitting is small with respect to PE, or the **high-spin case,** since there is a maximum number of unpaired spins. If $10Dq$ is very large compared to the PE, placing the fourth electron in a t_{2g} orbital produces the t_{2g}^4 configuration, the lower energy state. This is referred to as the **strong-field case,** or the **low-spin (or spin-paired)** case. The net energy

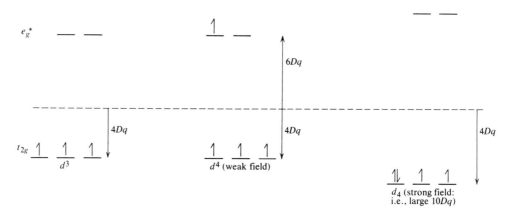

Figure 9.10   Occupation of d orbitals in octahedral complexes for d^3 and d^4 configurations.

gain in the strong-field case is of smaller magnitude than $-16Dq$ by the amount of energy required for pairing the electrons. Maximum LFSE is obtained for the d^6 configuration in a strong field, since the lower-energy t_{2g} orbitals are occupied completely and the high-energy orbitals are empty.

Table 9.10 lists the LFSE and the numbers of unpaired electrons for strong and weak fields for all d^n configurations. Of course, an LFSE of $-4Dq$ for the high-spin case is smaller than $-4Dq$ for the low-spin case, because of the smaller Dq for the weak field.

Table 9.9 compares metal–ligand bond energies calculated using the simple electrostatic approach, experimental values, and the bond energies corrected for LFSE. Even though the ligand-field corrections are important in explaining differences among the transition-metal complexes, note that simple electrostatic attraction accounts for *most* of the bond energy.

Spectrochemical Series

The order of ligand-field strength (representing a *decrease* in Dq) for common ligands is approximately

$$CN^- > phen \sim NO_2^- > en > NH_3 \sim py > H_2O > C_2O_4^{2-} > OH^-$$
$$> F^- > S^{2-} > Cl^- > Br^- > I^- \qquad (9.19)$$

This is called the **spectrochemical series**, since the original order for ligands X was determined from spectral shifts for complexes such as $[Co^{III}(NH_3)_5X]^{n+}$. The halides, at least, are in the order expected from electrostatic effects. In many other cases, we must consider covalent bonding to explain the order. The high ligand-field effect of CN^- and 1,10-phenanthroline is attributed to π bonding as discussed in Section 9.6.3 and in Section

Table 9.10   Ligand-field stabilization energy of octahedral complexes

Configu-ration	Examples	Strong field				Weak field			
		t_{2g}	e_g	Number of unpaired e	LFSE	t_{2g}	e_g	Number of unpaired e	LFSE[a]
d^0	Ca^{2+}, Sc^{3+}	0	0	0	0	0	0	0	0
d^1	Ti^{3+}	1	0	1	$-4Dq$	1	0	1	$-4Dq$
d^2	V^{3+}	2	0	2	-8	2	0	2	-8
d^3	Cr^{3+}, V^{2+}	3	0	3	-12	3	0	3	-12
d^4	Cr^{2+}, Mn^{3+}	4	0	2	-16	3	1	4	-6
d^5	Mn^{2+}, Fe^{3+}	5	0	1	-20	3	2	5	0
d^6	Fe^{2+}, Co^{3+}	6	0	0	-24	4	2	4	-4
d^7	Co^{2+}	6	1	1	-18	5	2	3	-8
d^8	Ni^{2+}	6	2	2	-12	6	2	2	-12
d^9	Cu^{2+}	6	3	1	-6	6	3	1	-6
d^{10}	Cu^+, Zn^{2+}	6	4	0	0	6	4	0	0

[a] Dq is smaller for weak-field ligands.

10.8 in terms of MO theory. Perhaps the greatest surprise in the order is the fact that the ligand field strength is greater for H_2O than for OH^-. We expect charged ligands to produce greater splitting that neutral ligands using the electrostatic model because of the stronger interaction between a positive metal ion and a negative ligand. There are several sequences in the spectrochemical series that cannot be explained based upon electrostatic interaction. We will examine these cases in terms of MO theory later.

Complexes of the very-strong-field ligands CN^- and 1,10-phenanthroline have low spin, corresponding to a large splitting of the energy levels and maximum occupancy of the t_{2g} orbitals. Water and halides are weak-field ligands and form complexes which have high spin. As might be expected, the splitting ($10Dq$) increases as the charge on the metal ion increases, because the cation radius decreases with increasing charge and the ligands would be more strongly attracted by the higher charge (see Fe^{2+} and Fe^{3+} in Table 9.11). The splitting, for the same ligands, increases and the PE decreases for the second and third transition series compared with the first transition series (compare Co^{3+} and Rh^{3+}). Both trends, which favor low-spin complexes, stem from the larger size of the $4d$ and $5d$ orbitals compared with the $3d$ orbitals. The larger orbitals, which extend farther from the central atom, are affected to a greater extent by the ligands. Because $4d$ and $5d$ orbitals are larger than $3d$ orbitals, it is less energetically unfavorable to pair two electrons in the same orbital; PEs are smaller. PEs (Table 9.12) for the same number of d electrons also increase with increasing oxidation number, because of increased electron repulsion resulting from the contraction of d orbitals with increasing positive charge.

The PEs are higher for the d^5 configuration than for the d^6 configuration, because there is a greater loss in exchange energy (see page 38) for the d^5 case. Consequently, there are more low-spin d^6 complexes than d^5 complexes (compare Co^{3+} and Fe^{3+} in Table 9.11). For Fe^{3+} (d^5) the low-spin complexes are stable when the stabilization for *two* electrons moved from e_g to t_{2g}, $2(10Dq)$, is greater than the PE (60,000 cm^{-1}) going from 5 unpaired electrons to 1 unpaired electron. Therefore low-spin complexes are expected if $10Dq > 30,000$ cm^{-1}. For Co^{3+} (d^6) the low-spin complexes are expected if $10Dq > 21,000$ cm^{-1} ($\frac{1}{2}$PE). From Table 9.11 we see that $[Co(H_2O)_6]^{3+}$ has low spin even though

Table 9.11   Ligand-field splitting for some octahedral complexes (10 *Dq* values in italics indicate low-spin complexes)[a]

	Ions	*Ligands*				
		6F$^-$	6H$_2$O	6NH$_3$	3 en	6CN$^-$
d^3	Cr^{3+}	15.06 kK	17.4 kK	21.5 kK	21.9 kK	26.6 kK
d^5	Mn^{2+}	7.75	8.5	—	10.1	~30
	Fe^{3+}	13.35	14.0	—	—	*35.0*
d^6	Fe^{2+}	—	10.4	—	—	*32.2*
	Co^{3+}	13.1	*20.7*	*22.87*	*23.6*	*32.2*
	Rh^{3+}	*22.6*	*27.2*	*34.1*	*34.6*	*44.9*
d^8	Ni^{2+}	7.25	8.50	11.0	11.85	—

[a] The values of 10 *Dq* are in kilokaysers (1 kK = 1000 cm^{-1}); 10 kK, or 1000 nm, corresponds to 119.7 kJ/mol (28.6 kcal/mol); 20 kK, or 500 nm, corresponds to 239.4 kJ/mol (57.2 kcal/mol).

Table 9.12   Pairing energies[a]

Configuration	Ion	PE	Configuration	Ion	PE
$d^4(1e)$	Cr^{2+}	23.5 kK	$d^6(2e)$	Fe^{2+}	35.2 kK
	Mn^{3+}	28.0		Co^{3+}	42.0
$d^5(2e)$	Mn^{2+}	51.0	$d^7(1e)$	Co^{2+}	19.5
	Fe^{3+}	60.0			

[a] Pairing energies in kK (1 kK = 1000 cm^{-1}) for free ions for 1 or 2 electrons as noted.

$10Dq = 20,700$ cm^{-1}. We expect low-spin complexes at $10Dq$ values about 10–30% less than the PE because of decreased interelectron repulsion for the complex compared to the free ion—there is some delocalization (covalence) involving the ligand orbitals. This lowered interelectron repulsion in complexes is called the **nephelauxetic effect.** This effect is greatest for soft (highly polarizable) ligands—for example, I^- and those containing S as the donor—and for ligands that favor π-bonding to the metal.

We can evaluate the magnitude of the ligand-field splitting from spectra. For a d^1 ion, such as $[Ti(H_2O)_6]^{3+}$, only a one-electron transition, from t_{2g} to e_g, is expected. The difference in energy between t_{2g} and e_g is $10Dq$. One broad absorption band is seen in Figure 9.11 for $[Ti(H_2O)_6]^{3+}$ with a maximum at 20.1 kK (490 nm), which we take as the value of $10Dq$. Although theoretical (calculated) values of $10Dq$ can be used as approximations, the spectral values are obviously measures of the real values. The empirical (experimental) $10Dq$ values from spectra reflect the *actual* splitting of energy levels resulting from the combination of electrostatic and covalent bonding effects. Spectra of complexes are discussed in Section 10.2.

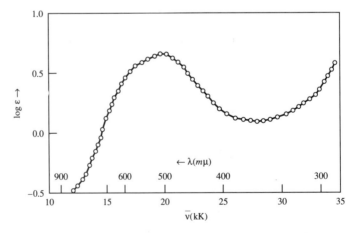

Figure 9.11   Absorption spectrum of $[Ti(H_2O)_6]^{3+}$ (1 kK = 1000 cm^{-1}). (From H. Hartman, H. L. Schlafer, and K. H. Hansen, *Z. Anorg. Chem.* **1957**, *289*, 40.)

▶ *9.8.2   Distorted Octahedral Complexes*

The model of ligand-field splitting of the *d* orbitals into two levels applies to a regular octahedral arrangement with six identical ligands. In many cases some distortion of the octahedron can increase the stability of the complexes. Just as removing the degeneracy of the five *d* orbitals can result in ligand-field stabilization, further removal of the degeneracy of the *d* orbitals can sometimes result in additional stabilization. The **Jahn–Teller Theorem** *predicts that distortion will occur whenever the resulting splitting of energy levels yields additional stabilization.* We have seen this result for a linear molecule with one electron in a pair of degenerate orbitals (page 173). Bending the molecule removes the orbital degeneracy with the electron going into the lower energy orbital. The molecule assumes the bent shape, thus lowering the energy.

For an example of a metal complex, if an octahedron is elongated along the *z* axis by stretching the metal–ligand distances along this axis, the degeneracy of the t_{2g} (3-fold) and e_g (2-fold) orbitals is lowered. Because the ligand field decreases rapidly as the distance is increased, the ligand field is lowered along *z* and for the e_g orbitals, d_{z^2} is lowered in energy, while the $d_{x^2-y^2}$ orbital is raised in energy (see Figure 9.12). Withdrawing the ligands along the *z* axis decreases the ligand–ligand repulsion and tends to shorten the metal–ligand distances along the *x* and *y* axes, thereby, increasing their interaction with the metal. Because only two orbitals (for e_g) are involved, the increase in energy of one orbital is the same as the decrease of the other. The t_{2g} orbitals are affected to a much smaller extent by the distortion, because they are not directed toward the ligands. For the t_{2g} orbitals, d_{xy} (contained in the *xy* plane) is raised in energy and the doubly degenerate d_{xz} and d_{yz} orbitals are lowered in energy.

Copper(II) complexes (d^9 ion) generally are tetragonal ($\mathbf{D}_{4h}$ with four short M—X distances in a plane and two long M—X distances perpendicular to the plane), as ex-

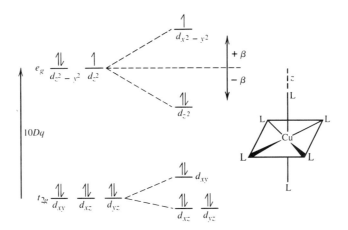

Figure 9.12 Splitting of the energy levels in an octahedral field elongated along the *z* direction (tetragonal).

pected from the Jahn–Teller effect. With the high-energy $d_{x^2-y^2}$ orbital occupied singly (destabilization $+\beta$), the net additional stabilization over the regular octahedron is $+\beta - 2\beta = -\beta$, as shown in Figure 9.12. The t_{2g} orbitals are filled, so no net stabilization results from their splitting. No additional stabilization by distortion is possible for a high-spin d^8 complex ion ($[NiL_6]^{2+}$) because each e_g orbital is occupied singly. Of course, no ligand field stabilization occurs for d^0, d^5 (high-spin), or d^{10} ions for regular or distorted octahedra because each d orbital is empty or equally populated. We expect distorted octahedral complexes for low-spin d^7 ions ($t_{2g}^6 e_g^1$) and for high-spin d^4 ions ($t_{2g}^3 e_g^1$) (one electron in d_{z^2} in each case). Co^{2+} (d^7) forms a low-spin complex, $[Co(ttcn)_2]^{2+}$, with the cyclic ligand 1,4,7-trithiacyclononane $\left[ttcn, \; S \underset{(CH_2)_2-S}{\overset{(CH_2)_2-S}{<}} >(CH_2)_2 \right]$. The tridentate ligand coordinates on opposite trigonal faces of a slightly distorted octahedron.[11] The constraints of the rings probably limit the expected Jahn–Teller tetragonal distortion. Slight distortion is expected for complexes involving unequal occupancy of the t_{2g} orbitals, but the effect is small because of the smaller splitting of the energies of these orbitals (Figure 9.12).

▶ *9.8.3 Planar Complexes*

If the ligands along the z axis are removed completely, forming a planar complex (still D_{4h}), the orbital energy splitting is increased greatly. The most favorable electron configuration for planar complexes is d^8—for example, Ni^{2+}, Pt^{2+}, and Au^{3+}—in which splitting can be great enough to bring about pairing of all electrons. The very-high-energy $d_{x^2-y^2}$ orbital is left vacant (see Figure 9.13). Planar complexes are expected also for d^7 and d^9 ions. In a low-spin planar complex of a d^7 ion such as Co^{2+}, the highest energy orbital is vacant and the next-to-highest energy orbital is occupied singly. A planar copper(II) (d^9) complex has a single electron in the highest energy orbital.

The ideal radius ratio ($r_+/r_- \geq 0.414$) for a square-planar arrangement is identical to that for an octahedral complex. Because the additional energy released from the interaction with the two extra ligands favors the octahedral complex, the higher C.N. should be expected. The square-planar arrangement is encountered only in cases where the additional splitting of the levels is most advantageous or where the planar configuration is imposed by the geometry of the ligand. Additional splitting will be most advantageous with strong-field ligands. Thus, Ni^{2+} forms octahedral $[Ni(H_2O)_6]^{2+}$ with weak-field H_2O, but square-planar $[Ni(CN)_4]^{2-}$ with strong-field CN^-. The copper phthalocyanine complex (see Figure 9.2) has the square-planar geometry imposed by the ligand; the very stable complex can be sublimed *in vacuo* at about 580°C without decomposition and is not decomposed by strong mineral acids. The divalent metal ions that produce planar complexes with this ligand (e.g., Be, Mn, Fe, Co, Ni, and Pt) include several that normally form tetrahedral complexes. The planar fused-ring system of phthalocyanine is similar to that of the porphyrins (see Section 18.1.5).

[11] W. N. Setzer, C. A. Ogle, G. S. Wilson, and R. S. Glass, *Inorg. Chem.* **1983**, *22*, 266; G. S. Wilson, D. D. Swanson, and R. S. Glass, *Inorg. Chem.* **1986**, *25*, 3827.

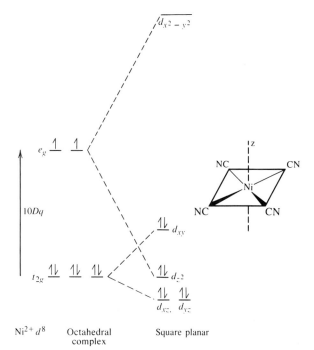

Figure 9.13   Energy levels for square-planar nickel(II) complexes such as $[Ni(CN)_4]^{2-}$.

► 9.8.4   Tetrahedral Complexes

A tetrahedral arrangement is the structure of least ligand–ligand repulsion for a four-coordinated metal ion and is the structure expected for a complex with a radius ratio between 0.225 and 0.414. The ligand field splitting is *smaller* in tetrahedral complexes because there are only four ligands and they do not approach along the direction of any of the d orbitals. The tetrahedral splitting of ML_4 ($10Dq_{tet}$) is calculated to be only four-ninths that of a regular octahedral complex ML_6 ($10Dq_{oct}$, same M and the same ligands). The coordinate axes of the tetrahedron are chosen as the twofold symmetry axes that bisect the six edges (see Figure 9.14). These C_2 (or S_4) axes are coincident with the C_4 axes of the octahedron or cube. The high-energy orbitals for the tetrahedron are the d_{xy}, d_{xz}, and d_{yz} orbitals, those directed most nearly toward the ligands (see Figure 9.14). A cube is formed by adding four ligands in the vacant apices of the cube in Figure 9.14, so $10Dq_{cubic} = 2(10Dq_{tet}) = \frac{8}{9}(10Dq_{oct})$. (Figure 3.12 shows the octahedron within a cube, and a tetrahedron can be outlined by connecting corners 1, 3, 6, and 8.)

Because of the smaller ligand-field splitting, low-spin complexes generally are not encountered and tetrahedral complexes are not expected in cases where the ligand field stabilization is great for octahedral or planar complexes. Tetrahedral complexes are more likely for nontransition metals and transition metals with no ligand field stabilization (d^0, high-spin d^5, and d^{10}; see Table 9.13). Tetrahedral complexes of d^2 and d^7 ions demonstrate high ligand-field stabilization. Maximum LFSE occurs in the low-spin d^4 case, and

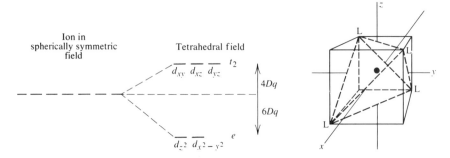

Figure 9.14   Ligand-field splitting for tetrahedral complexes. The orientation of the tetrahedron relative to the axes is shown.

$[Cr\{N(Si(CH_3)_3)_2\}_3NO]$ is a low-spin complex with a distorted tetrahedral ($\mathbf{C}_{3v}$) structure (Figure 9.15). It is formulated as a Cr^{II} (d^4) complex containing three bulky $N(Si(CH_3)_3)_2^-$ and NO^+. Tetrakis(1-norbornyl)cobalt $\left(\text{1-norbornyl anion} = :C \begin{smallmatrix} CH_2CH_2 \\ -CH_2- \\ CH_2CH_2 \end{smallmatrix} CH \right)^-$ has been established as a low-spin tetrahedral complex of Co^{IV}.[12] Large ligands or those with high negative charge favor the formation of tetrahedral complexes over planar or octahedral complexes, in order to minimize ligand–ligand repulsion. The oxoanions and chloro complexes, such as those shown in Table 9.13, are common examples. Identification of d^n configurations leading to no Jahn–Teller distortions of the tetrahedron is left as an exercise (Problem 9.13).

Ligand-field stabilization greatly favors planar complexes for Cu^{II}, Ni^{II}, and Co^{II}, but $CuCl_4^{2-}$, $NiCl_4^{2-}$, and a few other complexes of these metals are tetrahedral.[13] The importance of covalent bonding—particularly π bonding—is difficult to assess in many cases. Planar nickel complexes usually are those with an appreciable amount of π bonding (e.g., $[Ni(CN)_4]^{2-}$), and it seems likely that the necessary splitting for spin pairing requires covalent bonding. Although the pure crystal field treatment does not take covalent character into account, the effect of increased covalent character or π bonding can be handled empirically as an increase in $10Dq$. The ligand P,P-di-*tert*-butylphosphinic-N-isopropyl-

Table 9.13   Examples of regular tetrahedral complexes

d^0	d^2	d^5 (high-spin)	d^7 (high-spin)	d^{10}
$AlCl_4^-$				
$TiCl_4$	FeO_4^{2-}	$FeCl_4^-$	$CoCl_4^{2-}$	$ZnCl_4^{2-}$
MnO_4^-				$GaCl_4^-$
CrO_4^{2-}				

[12] E. K. Byrne, D. S. Richeson, and K. H. Theopold, *Chem. Commun.* **1986**, 1491.

[13] $[CuCl_4]^{2-}$ exists as a flattened tetrahedron in Cs_2CuCl_4, but at high pressure the structure changes to square planar. Some hydrogen-bonded solids contain both square-planar and tetrahedrally distorted square-planar $[CuCl_4]^{2-}$. See J. R. Ferraro and G. J. Long, *Acc. Chem. Res.* **1975**, *8*, 171.

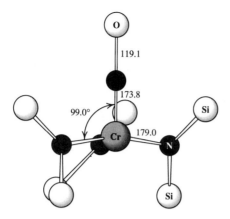

Figure 9.15 Structure of the distorted tetrahedral (C_{3v}) complex of $Cr[N(Si(CH_3)_3)_2]_3NO$. CH_3 are not shown. (From D. C. Bradley, *Chem. Br.* **1979**, *11*, 393.)

amidato ion was found to form a paramagnetic *trans*-planar complex[14] with Ni^{II}. The LFSE is great enough for a planar structure, but does not give electron pairing.

9.9 ISOMERISM OF COORDINATION COMPOUNDS

▶ 9.9.1 Simple Types of Isomerism

Reagent chromium chloride, labeled $CrCl_3 \cdot 6H_2O$, is usually a green crystalline compound. When dissolved in water it produces a green solution, which, upon standing for a week or more, turns violet. Violet crystals, which have the composition $CrCl_3 \cdot 6H_2O$, can be obtained from the solution. Actually, there are three isomers with this composition. They differ as follows:

$[Cr(H_2O)_6]Cl_3$ $[CrCl(H_2O)_5]Cl_2 \cdot H_2O$ $[CrCl_2(H_2O)_4]Cl \cdot 2H_2O$
     Violet   Blue-green   Green

In the violet isomer the six water molecules are coordinated to the chromium and the three chloride ions are separate ions in the solid and in solution. In the second isomer, a chloride ion displaces from the coordination sphere one water molecule, which is held as a molecule of water of crystallization. In the third isomer there are two coordinated chloride ions and two molecules of water of crystallization. Commercial chromium chloride is usu-

[14] T. Frömmel, W. Peters, H. Wunderlich, and W. Kuchen, *Angew. Chem. Int. Ed. Engl.* **1992**, *31*, 612.

ally a mixture of the two green isomers. This type of isomerism is known as **hydrate isomerism.**

Several other kinds of isomerism are illustrated in Table 9.14. **Coordination isomers** involve the exchange of the metal ions between the cation and anion. **Ionization isomers** differ only in the exchange of a coordinated anion for one present to maintain charge balance. Conceivably, many ligands might form **linkage isomers** similar to the nitro complex (coordination through N) and nitrito complex (coordination through O) shown. Ligands that form linkage isomers are called **ambidentate ligands.**[15] Most ligands normally coordinate through the same atom. Cyanide ion usually coordinates through C, except when it serves as a bridging group. In the clathrate compound, $Ni(CN)_2(NH_3)C_6H_6$ (Figure 9.16), the C_6H_6 is enclosed in a cage. The six-coordinate Ni^{2+} ions are bonded to two NH_3 and the N atoms of four bridging CN^- ions; the four-coordinate Ni^{2+} ions are bonded to the C of CN^- ions. When bonded to the less electronegative carbon, Ni should have the lower coordination number, because more negative charge is transferred to Ni in the Ni—C bond. The thiocyanate ion coordinates through the nitrogen atom in most cases—for example, $[Co(NCS)_4]^{2-}$—but it can coordinate through S, as in $[Hg(SCN)_4]^{2-}$. These are the expected isomers, since Co^{II}, a relatively hard Lewis acid, is bonded through N (hard Lewis base) and Hg^{II}, a soft Lewis acid, is bonded through S (soft Lewis base). Two isomers have been obtained for $[Pd(SCN)_2(bipy)]$. Although the *S*-bonded isomer is stable in the solid at room temperature, it rearranges to give the *N*-bonded isomer at elevated temperature or in solution. The structures were assigned from infrared data.

▶ 9.9.2 *Stereoisomerism and Geometric Isomerism*

Stereoisomers are isomers that differ only in the spatial arrangement of bonded groups. Stereoisomers bearing a mirror-image relationship to each other are termed **enantiomers.** Enantiomers have identical bond lengths and bond angles, except for the sense of the angles. These mirror-image isomers are nonsuperimposable only if the molecule has no S_n axis (see page 127). Geometric isomers differ in the spatial arrangement of the connected atoms, other than the handedness—a pair of enantiomers is classified as the *same* geometrical isomer. The words *cis* and *trans* (Latin "on the same side" and "across", respectively) were first used to describe isomers in van't Hoff's classic 1874 paper, in which he not only developed the molecular model for optical activity, but also *predicted* the exis-

Table 9.14 Some types of isomerism among coordination complexes

Examples	Type of Isomerism
$[Co(NH_3)_6][Cr(CN)_6]$ and $[Cr(NH_3)_6][Co(CN)_6]$	Coordination
$[PtCl_2(NH_3)_4]Br_2$ and $[PtBr_2(NH_3)_4]Cl_2$	Ionization
$[Co(NO_2)(NH_3)_5]Cl_2$ and $[Co(ONO)(NH_3)_5]Cl_2$	Linkage
$[Pd(SCN)_2(bipy)]$ and $[Pd(NCS)_2(bipy)]$	

[15] A. H. Norbury and A. I. P. Sinha, *Q. Rev.* **1970**, *5*, 69; J. L. Burmeister, *Coord. Chem. Rev.* **1968**, *3*, 225.

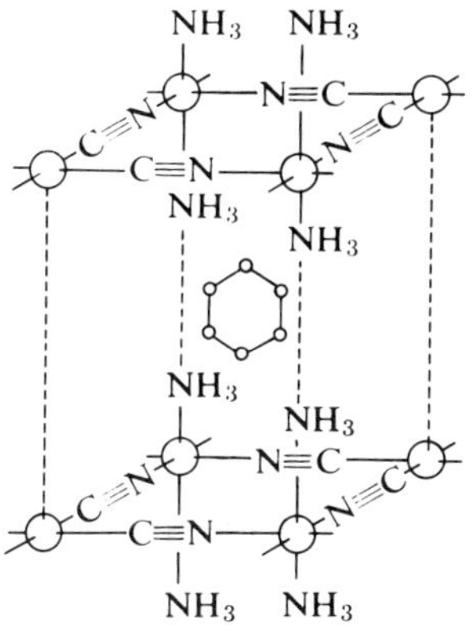

Figure 9.16   The $Ni(CN)_2(NH_3)C_6H_6$ clathrate compound. (From J. H. Rayner and H. M. Powell, *J. Chem. Soc.* **1952**, 319.)

tence of geometrical isomers of alkenes. The *cis, trans* nomenclature applies also to square-planar and octahedral complexes. We shall consider stereoisomerism of octahedral complexes in more detail in Sections 9.11.1 and 9.11.2. See also Figures 9.4 and 9.5.

9.10   COORDINATION NUMBER FOUR

Coordination compounds with C.N. 4 are encountered with tetrahedral and square-planar configurations.

▶ 9.10.1   Tetrahedral Complexes

The only isomerism for tetrahedral complexes is optical isomerism. Because of lability (rapid exchange of ligands), few optical isomers in which four different groups are coordinated to the metal have been prepared. Most of these are organometallic compounds of Fe, Mn, and Ti of the type shown in Figure 9.17. In this case, when the acetyl group is reduced, the configuration at the Fe center is retained. When the Fe center is involved in the

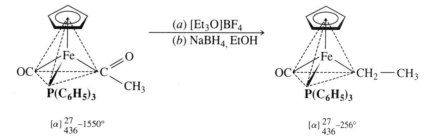

Figure 9.17   Optically active compounds of iron. The angles of rotation of plane-polarized light at 27°C and 436 nm are given. (From H. Brunner, "Optical Activity," in *The Organic Chemistry of Iron*, E. A. Koerner Von Gustof, F.-W. Grevels, and I. Fishler, Eds., Academic Press, New York, 1978, p. 299.)

reaction, inversion sometimes occurs.[16] Most of the tetrahedral complexes that have been resolved contain unsymmetrical chelating agents, giving isomerism similar to that of the organic spiran compounds. Thus bis(benzoylpyruvato)beryllium(II) can be separated into enantiomers (see Figure 9.18). Other tetrahedral complexes of Be, B, Zn, and Cu^{II} have been resolved. In many cases we cannot rule out the possibility that the Cu^{II} and Zn^{II} complexes are really six-coordinate (water- or anion-coordinated).

▶ 9.10.2 Square-Planar Complexes

Isolation of two isomers of [PtCl$_2$(NH$_3$)$_2$] led Werner (1893) to conclude that these are *cis* and *trans* isomers of a square-planar complex. The planar structures of K$_2$[PtCl$_4$] and K$_2$[PdCl$_4$] were not confirmed by x-ray studies until 1922. Pauling (1932) predicted that Ni^{II} also would form planar complexes. Many planar complexes are now known for d^7 (Co^{II}), d^8 (Rh^I, Ir^I, Ni^{II}, Pd^{II}, and Pt^{II}, and Au^{III}), and d^9 (Cu^{II} and Ag^{II}). The compounds *cis*-[PtCl$_2$(NH$_3$)$_2$] and *cis*-[PtCl$_2$(en)] are effective antitumor drugs; *trans*-[PtCl$_2$(NH$_3$)$_2$] is not very effective.

Three isomers have been reported for [Pt(NO$_2$)(py)(NH$_3$)(NH$_2$OH)]NO$_2$, corresponding to the ammonia *trans* to each of the other three groups. The isomers, represented as [Pt(*abcd*)], are as follows:

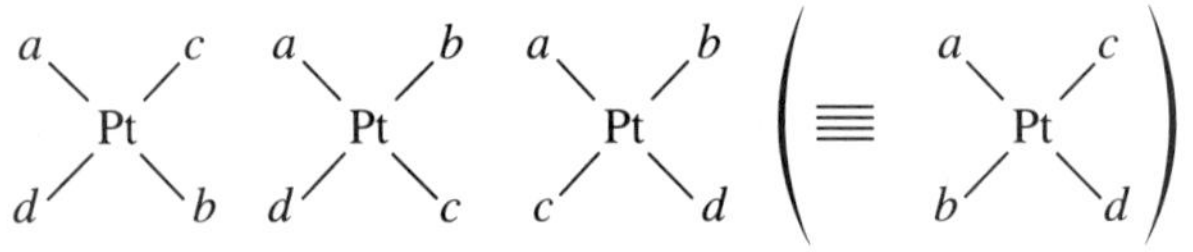

Even coordination of four different groups to the metal ion does not lead to optical isomerism, because each isomer has a plane of symmetry (the molecular plane) and hence mirror images are superimposable. Nevertheless, the ingenuity of Mills and Quibell (1935) provided an example of an optically active planar complex. The compound [Pt(NH$_2$CHC$_6$H$_5$CHC$_6$H$_5$NH$_2$)(NH$_2$C(CH$_3$)$_2$CH$_2$NH$_2$)]Cl$_2$ should be optically active if planar, because the plane of symmetry is eliminated; if it were tetrahedral, on the other hand, it would possess a plane of symmetry and hence could not be optically active (Figure 9.19). In the representation of the tetrahedral configuration, the ring to the right is in the plane of the paper, whereas the one to the left is perpendicular to it. The plane of the paper represents a plane of symmetry.

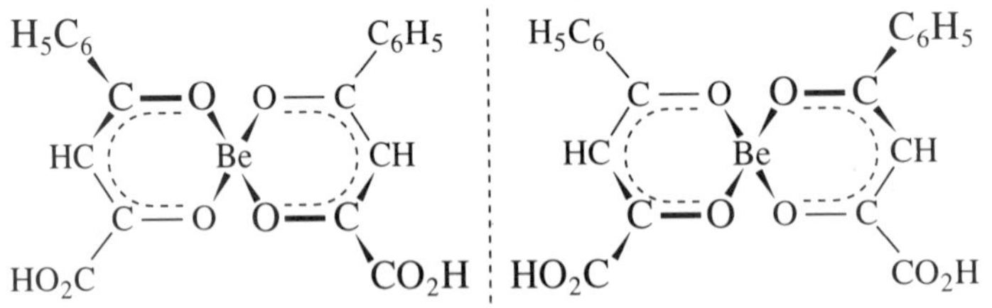

Figure 9.18 Enantiomers of bis(benzoylpyruvato)beryllium.

[16] H. Brunner, *Adv. Organomet. Chem.* **1980,** *18,* 151.

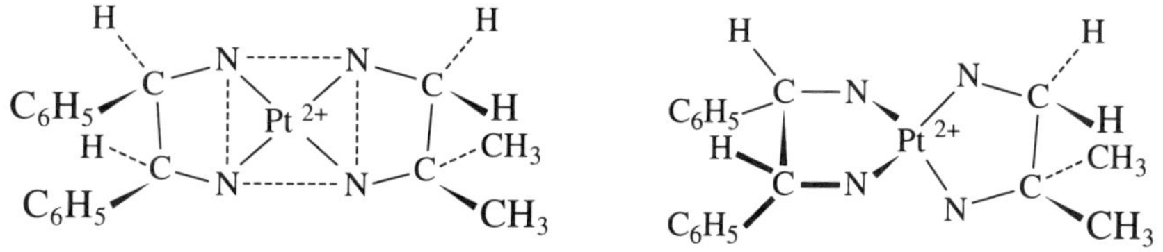

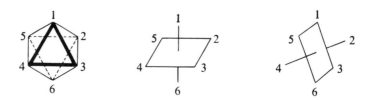

Figure 9.19   Possible configurations of a Pt^{II} complex. (H atoms on N are omitted.)

9.11 COORDINATION NUMBER SIX

The most common and certainly the most thoroughly studied complexes are those with C.N.6. Several possible geometries have been considered, but the weight of physical and chemical evidence strongly favors the octahedron, although several trigonal prismatic complexes are now known (Section 9.11.3). The octahedron is sometimes distorted by the structure of the ligands or because of the Jahn–Teller effect (Section 9.8.2).

▶ 9.11.1 *Enumeration of Isomers*

That the octahedron can be represented in many ways is illustrated in Figure 9.20, which emphasizes both the threefold axis of the octahedron and the equivalence of the substitution sites. Here we shall use the first representation unless the symmetry being discussed is discerned more easily in another representation.

In drawing isomers of a complex, do the following: (1) Identify the unique structural features, (2) draw the structure in a fashion that emphasizes the symmetry present, and (3) use enumeration equations to check the results.

It is useful to draw a *cis* pair of ligands to emphasize the C_{2v} parent geometry and a *trans* pair to emphasize the D_{4h} parent geometry of the remaining positions (see Figure 9.21). For the octahedral complex M($abcde_2$), the *trans*-e_2 arrangement gives rise to the three isomers having *trans ab*, *ac*, and *ad*. All of these are optically inactive, because of the mirror plane through M($abcd$).

For cases involving just a few isomers, we can generate these from drawings, taking advantage of symmetry. The situation becomes more complex if there are several different ligands. For *n* different ligands the total number of stereoisomers possible is $n!/\alpha$,

Figure 9.20   Representations of the octahedron.

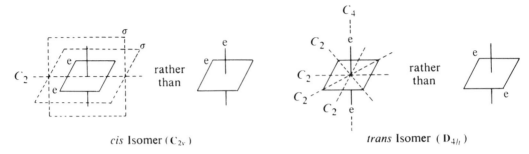

Figure 9.21   Convenient representations of *cis* and *trans* isomers of an octahedral complex.

where α is the **symmetry number**[17] of the complex. The symmetry number is identical to the *order of the rotational group* (see inset paragraph below) except for linear molecules where it is 2 for $\mathbf{D}_{\infty h}$ and 1 for $\mathbf{C}_{\infty v}$. For the M($abcde_2$) case just considered, we deal with two frameworks, *trans*-[M(e)$_2$] and *cis*-[M(e)$_2$], separately. Thus the *trans*-[M(e)$_2$] arrangement pictured in Figure 9.21 belongs to the $\mathbf{D}_{4h}$ point group and the rotational group is $\mathbf{D}_4$, a group of order 8. The number of stereoisomers is 4!/8 = 3, with *a trans* to *b*, *c*, and *d*. For the *cis*-[M(e)$_2$] arrangement as pictured in Figure 9.21, the point group is $\mathbf{C}_{2v}$ and the rotational group is $\mathbf{C}_2$. For [M($abcde_2$)] with the *cis*-e_2 arrangement, the number of stereoisomers would be 4!/2 = 12 (where 2 is the order of the $\mathbf{C}_2$ group). Because neither of the mirror planes of the $\mathbf{C}_{2v}$ *cis*-e_2 structure remains after adding the other ligands, the 12 isomers exist as 6 pairs of enantiomers. Using the axial *trans* positions, we can immediately make the *trans* pairings

(i.e., is *ab*, *ac*, *ad*, *bc*, *bd*, and *cd*) and draw these and their mirror images.

The rotation group is the group or subgroup with only proper rotations—no center of symmetry, planes of symmetry, or S_n axes. Thus the rotation groups lack any letter subscripts—$\mathbf{O}$, $\mathbf{D}_4$, $\mathbf{D}_3$, $\mathbf{C}_4$, $\mathbf{C}_3$, and so on.

EXAMPLE 9.2:   Draw all the isomers of M($abcd\widehat{ee}$), where $\widehat{ee}$ is a didentate ligand.

Solution: The presence of a didentate ligand limits the number of isomers possible, because normally a chelated ligand can span only *cis* (adjacent) positions. In this case we introduce the didentate ligand first and use it to establish our basic framework. For the configuration M($\widehat{ee}$), see the *cis* isomer in Figure 9.21 that shows the C_2 axis and σ planes. The point group is $\mathbf{C}_{2v}$ and the rotation group is $\mathbf{C}_2$, giving 4!/2 = 12 stereoisomers (6 pairs of enantiomers) for

[17] For a group theoretical approach to determining the number of isomers, see D. H. McDaniel, *Inorg. Chem.* **1972**, *11*, 2678.

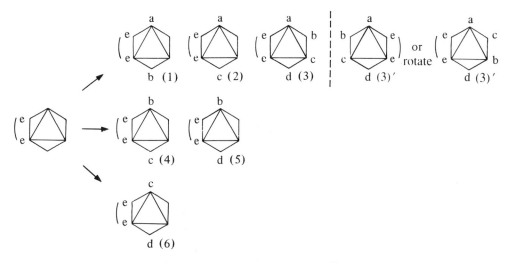

Figure 9.22   Isomers of [M($abcd\widehat{ee}$)].

[M($abcd\widehat{ee}$)]. Placing the a ligand in the apical position in Figure 9.22 (structure 1) leads to three *trans* pairs, namely, ab, ac, and ad. Starting with b (structure 4) and then c (structure 6) in the apical position, we get bc and bd and then cd as the unique *trans* pairs. Each of the six geometrical isomers can exist as a pair of enantiomers, for a total of 12 stereoisomers; one pair of enantiomers is shown. The enantiomer of (3) is (3)′, and this is rotated by 180° to orient it as (3).

A set of three similar ligands may be arranged on an octahedron in an all-*cis* fashion, giving the facial or *fac* isomer, or with one pair *trans,* giving the meridional or *mer* isomer (Figure 9.23). The *fac* and *mer* isomers of an Ma_3b_3 octahedral complex are optically inactive.

The more general nomenclature scheme uses the numbering system shown in Figure 9.20. The *cis*- and *trans*-[M(a_2b_4)] isomers are 1,2- and 1,6-isomers, respectively. For M(a_3b_3), the facial isomer is 1,2,3 and the meridional isomer is 1,2,6.

The known examples of optically active octahedral complexes have at least one chelate ring. Three of the five possible geometrical isomers of [PtCl$_2$(NH$_3$)$_2$(py)$_2$]$^{2+}$ (see Figure 9.24) have been prepared. The geometrical isomers with all like groups *cis* to one another should be optically active.

1,2,3 Isomer
Facial C$_{3v}$

1,2,6 Isomer
Meridional C$_{2v}$ **Figure 9.23**   Isomers of M(a_3b_3).

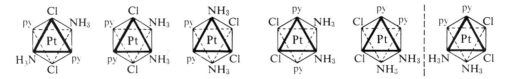

Figure 9.24   Isomers of $[PtCl_2(NH_3)_2(py)_2]^{2+}$.

▶ 9.11.2 *Optical Activity*

Enantiomers Based Upon the Chirality of Chelate Rings

The compounds $[Co(NH_3)_6]Cl_3$ ($\mathbf{O}_h$) and $[Co(en)_3]Cl_3$ ($\mathbf{D}_3$) have similar properties, but only the latter can be resolved into enantiomers (Figure 9.25). We can prove the non-superimposability by examining models, or, with experience, by examining sketches as shown. The rotation of the Λ-isomer by 180° demonstrates that this orients one chelate ring as for the Δ-isomer, whereas the other two rings are oriented differently. The other method of proving nonsuperimposability is to examine the symmetry properties. Enantiomers can exist only if there is no S_n axis (including $\sigma = S_1$ and $i = S_2$). The $\mathbf{O}_h$ point group has several of these symmetry elements, but there are none for $\mathbf{D}_3$ (or any rotation group). In this case, the **chirality** (Greek word for "handedness") arises from the spiral configuration of the chelate rings, providing the basis for the Δ (right) and Λ (left) notation (see page 420). The backbones of the chelating ligands form a right-handed spiral (Δ) or a left-handed spiral (Λ).

The complexes *cis*-$[CoCl_2(NH_3)_4]Cl$ ($\mathbf{C}_{2v}$, violet) and *trans*-$[CoCl_2(NH_3)_4]Cl$ ($\mathbf{D}_{4h}$, bright green) have the same colors as the corresponding complexes *cis*-$[CoCl_2(en)_2]Cl$ ($\mathbf{C}_2$) and *trans*-$[CoCl_2(en)_2]Cl$ ($\mathbf{D}_{2h}$). Of these, only *cis*-$[CoCl_2(en)_2]Cl$ has no S_n and is optically active. The isomers of $M(a_3b_3)$ (see Figure 9.23) are not optically active, but the complex $M(ab)_3$, with three unsymmetrical didentate ligands—for example,

$$\left[Co\left(\overset{\diagup NH_2 \diagdown}{\underset{O—C}{\,}} CH_2 \right)_3 \right] ,$$ tris(glycinato)cobalt(III)—exists as facial ($\mathbf{C}_3$) and meridional

($\mathbf{C}_1$) isomers, each of which exists as enantiomeric pairs.

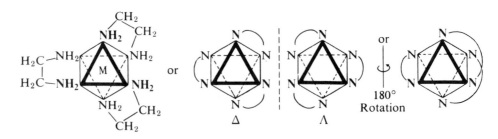

Figure 9.25   Enantiomers of $[M(en)_3]^{n+}$ ($\mathbf{D}_3$).

Table 9.15 lists metals for which optically active octahedral complexes have been resolved. Some of the gaps in the table doubtless stem from the small amount of work done with metals, other than Group VIII, beyond the first transition series.

Resolution Procedures.   The physical properties of the enantiomers of a complex such as $[Co(en)_3]Cl_3$ are identical except for the interaction with polarized light. However, if the chloride ions are replaced by optically active anions, the resulting salts might differ in properties such as solubility. $[Co(en)_3]^{3+}$ is easily resolved by replacing two chloride ions by the d-tartrate ion, followed by fractional crystallization. The diastereomers[18] (+)-$[Co(en)_3]Cl$ d-tart and (−)-$[Co(en)_3]Cl$ d-tart, are not mirror images, because the tartrate ion has the same configuration in each salt. Upon crystallization, the (+)-$[Co(en)_3]Cl$ d-tart separates in the form of large crystals. The (−) complex is much more soluble, and the solution becomes a thick gelatinous mass before fine needles begin to crystallize. We can convert the separated diastereoisomers containing d-tart to the active chloride complexes by adding concentrated hydrochloric acid and crystallizing the chlorides (tartrate ion becomes tartaric acid). The resolved complex is stable in solution for many weeks without racemization.

The antimony tartrate[19] and the d-α-bromocamphor-π-sulfonate ion have been useful resolving agents for cationic complexes. Anionic complexes—for example, $[M(C_2O_4)_3]^{3-}$—have been resolved using an optically active cation such as those formed by the bases strychnine or brucine. Resolved complex ions such as $[Co(edta)]^-$, $[Co(en)_3]^{3+}$, or *cis*-$[Co(NO_2)_2(en)_2]^+$ are often used as resolving agents.

Designation of Absolute Configurations.   Various symbols (*D* or *L*, *R* or *S*, *P* or *M*, and Δ or Λ) have been used to designate absolute configurations of octahedral complexes. Piper suggested Δ for a right-handed spiral (and Λ for a left spiral) of chelate rings about the C_3 axis of a tris(didentate) complex such as $[Co(en)_3]^{3+}$. Such a complex ($\mathbf{D}_3$ symmetry) also has three C_2 axes (perpendicular to C_3). For the $\Delta(C_3)$ enantiomer, two chelate rings define the opposite (Λ) chirality about each of the C_2 axes. There are many more optically active complexes with C_2 axes than with C_3 axes, and for a while the symbols Δ and Λ were used to define chirality with reference to either C_2 or C_3 axes. The (+)-$[Co(en)_3]^{3+}$ ion (positive optical rotation at the sodium D line, unless otherwise designated) is $\Lambda(C_3)$, but $\Delta(C_2)$.

Table 9.15   Metals for which octahedral complexes have been resolved

								Al	Si	
	Ti	Cr	Fe	Co	Ni	Cu	Zn	Ga	Ge	As
Y			Ru	Rh			Cd			
Gd,Nd			Os	Ir	Pt					

[18] Whereas enantiomers are mirror-image isomers, diastereoisomers (also called diastereomers) are not. The optical isomers (+)-$[Co(L\text{-alanine})_3]$ and (−)-$[Co(L\text{-alanine})_3]$ are diastereomers because both contain L-alanine, not one L-alanine and the other the mirror image D-alanine.

[19] The antimony tartrate ion often is represented as $SbO(d\text{-tartrate})^-$, but two tartrate ions bridge two antimonate ions, (+)-bis[μ-tartrato(4-)]diantimonate (2-). (See Appendix B on Nomenclature for μ notation for bridging.)

The IUPAC rules[20] avoid the use of a reference axis. Two skewed lines not orthogonal to each other define a helical system; either of them can be considered the axis of the cylinder, with the other line tangent to a helix on the surface of the cylinder. Helices are designated as Δ (right) or Λ (left). For $(+)$-$[Co(en)_3]^{3+}$, using any pair of octahedral edges spanned by chelate rings (Figure 9.26) as skewed lines, the designation is Δ—the same as relative to the C_3 axis. For $(-)$-*cis*-$[CoCl_2(en)_2]^+$, the only pair of octahedral edges spanned by chelate rings defines a right (Δ) helix—the same as relative to the "pseudo" C_3 axis. In each case, orient the octahedron so that one octahedral edge is spanned on the back side and horizontal (A---A in Figure 9.27); the other edge spanned (B—B in front) should be tipped down to the right for Δ and to the left for Λ.

The same system applies to the designation of enantiomeric chelate-ring conformations. An ethylenediamine chelate ring (*gauche* configuration) is viewed with the octahedral edge spanned horizontal (A---A). The C—C bond (closer to the viewer) is skewed down to the right for a δ conformation and down to the left for a λ conformation (see Figure 9.27). The direction of viewing is not important, but the more distant line should be horizontal. If you view the ring from the side of the metal and tip the ring so that the more distant line (the C—C bond) is horizontal, the nearer line (the octahedral edge) is tipped down to the right for the δ isomer.

The sign of optical rotation (at the sodium D line, unless otherwise specified) of an optically active ligand is designated. The absolute configuration of the ligand (R for *rectus;* and S, for *sinstra*) can be given if known—for example, S-$(+)$-pn. The symbols D and L are still in limited use, particularly for naturally occurring amino acids, all of which have the L (or S) configuration.

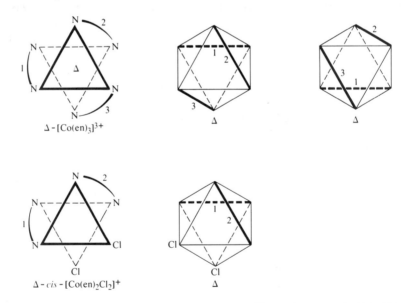

Figure 9.26   Absolute configurations of Δ-$(-)$-$[Co(en)_3]^{3+}$ and Δ-$(-)$-*cis*-$[CoCl_2(en)_2]^+$. Octahedra are oriented for each complex ion to examine the relative orientations of pairs of heavy edges spanned by en.

[20] See *Inorg. Chem.* **1970,** *9,* 1; or the references cited in Appendix B.

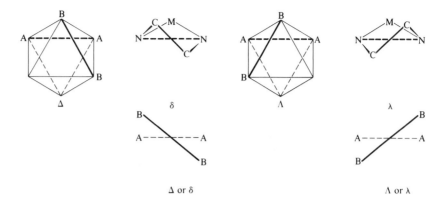

Figure 9.27 Projection of skewed lines with A — A below the plane of the paper and B — B above the paper. The corresponding elements of the Δ and Λ octahedra are shown as well as the δ and λ conformations of a chelate ring formed by ethylenediamine.

Stereospecific and Stereoselective Effects. The complex ion [Co(edta)]$^-$ can be resolved into enantiomers, but if we use optically active (−)-propanediaminetetraacetic acid, only $(+)_{546}$-[Co(−)-pdta]$^-$ results. The reaction appears to be completely stereospecific, and the complex ion does not racemize under conditions for the racemization of [Co(edta)]$^-$. At first it was believed that racemic propanediamine formed only two enantiomeric tris complexes, (+)-[Co(+)-(pn)$_3$]$^{3+}$ and (−)-[Co(−)-(pn)$_3$]$^{3+}$. Others of the eight combinations are formed, even though these two isomers are favored. Such a reaction is stereoselective.

In a classic paper, Corey and Bailar[21] provided the first explanation of stereoselective effects in metal complexes. The *gauche* configuration of the chelate ring in an ethylenediamine complex can be compared to the chair form of cyclohexane. The substituents on the puckered ring can be classified as approximately axial or equatorial. In a square-planar complex of the type [M(en)$_2$]$^{n+}$, the two puckered rings can be arranged as shown in Figure 9.28. In the $\lambda\delta$ form the two en rings are mirror images. The clockwise sequence of substituents, *a* or *e*, nearer to you viewed from above in the chelate ring for λ is M*eaea*, for δ it is M*aeae*. The $\lambda\lambda$ form, with the hydrogens staggered around both rings, should be more stable than the $\lambda\delta$ form, in which the hydrogen atoms on adjacent nitrogen atoms are directly opposed. In octahedral complexes the interactions are not so apparent, but the situation is similar. There are four different forms possible for an octahedral complex of the type [M(en)$_3$]$^{n+}$; $\lambda\lambda\lambda$, $\lambda\lambda\delta$, $\lambda\delta\delta$, and $\delta\delta\delta$. The four forms differ in energy, with the energies of $\lambda\lambda\delta$ and $\lambda\delta\delta$ intermediate between those of $\lambda\lambda\lambda$ and $\delta\delta\delta$. The hydrogen–hydrogen and carbon–hydrogen interactions were evaluated for the $\Delta(\lambda\lambda\lambda)$ form [or the enantiomeric $\Lambda(\delta\delta\delta)$ form] and the $\Delta(\delta\delta\delta)$ form [or the enantiomeric $\Lambda(\lambda\lambda\lambda)$ form]. In Figure 9.29 these are designated *lel* (the C — C bonds are approximately paral*lel* to the C_3

[21] E. J. Corey and J. C. Bailar, Jr., *J. Am. Chem. Soc.* **1959**, *81*, 2620. The ring conformations were designated *k* and *k'* as shown in Figure 9.29, but an error in another figure reversed these designations and this error was carried forth in some papers, leading to confusion. The error apparently was caused by the reversal of a photographic print in the paper. Because the original print supplied by Professor Bailar was used in the first edition of this book, it was correct. The symbols λ and δ are used now instead of *k* and *k'*.

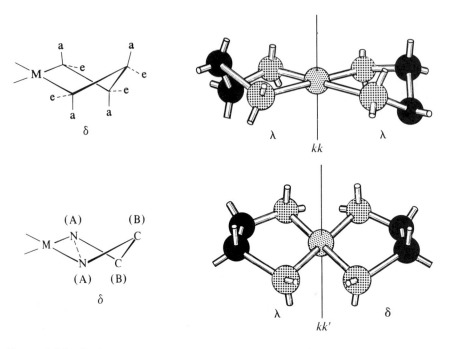

Figure 9.28   Conformation of the chelate ring in a planar ethylenediamine complex. A side view of the skewed lines (AA and BB) considered in Figure 9.27 are shown here for comparison. These δ chelate rings are oriented as the δ ring to their right. (Adapted from E. J. Corey and J. C. Bailar, Jr., *J. Am. Chem. Soc.* **1959,** *81,* 2620. Copyright 1959, American Chemical Society.)

axis) and *ob* (the C—C bonds are *ob*lique to the C_3 axis). The *lel* isomer was found to be more stable that the *ob* isomer by about 7.5 kJ/mol (or 2.5 kJ/ligand). For an optically active ligand such as (+)-propanediamine, one conformation (δ) with the methyl group equatorial (Figure 9.30) is more stable than the other by about 8 kJ/mol of pn. Thus (+)-pn has a strong preference for the δ conformation, so the preferred isomers are $\Lambda(\delta\delta\delta)$ (the more stable *lel* configuration with all methyls equatorial) and $\Delta(\delta\delta\delta)$ (the less stable *ob* configuration, but with all methyls equatorial). Because of the opposing effects, *lel* is more stable than *ob* by only 6.7 kJ/mol (~2 kJ/pn). Both isomers as well as mixed isomers containing (+)-pn and (−)-pn have been isolated.

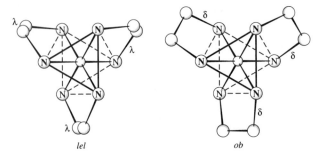

Figure 9.29   The *lel* and *ob* conformations of Δ-[M(en)$_3$]$^{n+}$ viewed along the C_3 axis. (Adapted with permission from Y. Saito in K. Nakamoto and P. J. McCarthy, *Spectroscopy and Structure of Metal Chelate Compounds,* Wiley, New York, 1968, p. 29, copyright 1968.)

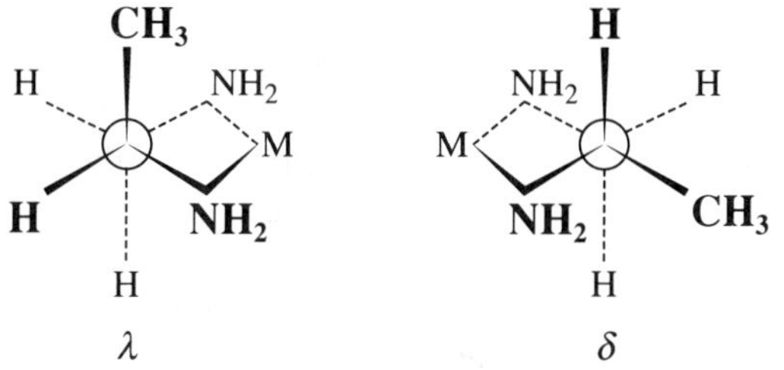

Figure 9.30   Enantiomeric conformations of a *gauche* chelate ring of *S*-(+)-propylenediamine placing the methyl group in axial (λ) and equatorial (δ) positions. Viewed along the C—C bond.

Stereoselectivity can arise because of differences in thermodynamic stability (as noted), or because of kinetic effects. Metal ions might play important roles in the remarkable stereospecificity encountered in biological systems.

▶ 9.11.3   *Trigonal Prismatic Coordination*[22]

General acceptance of the existence of square-planar complexes (C.N. 4) was delayed by the remarkable success of the tetrahedron in organic chemistry. Ligand–ligand repulsion is minimized for tetrahedral coordination. Similarly, octahedral coordination minimizes ligand–ligand repulsion for C.N. 6, and exceptions came as surprises.[23] Trigonal prismatic coordination was first observed for the layer lattices of MoS_2 and WS_2 (Figure 5.18), but these could be rationalized as peculiarities of the solid state. In 1965 the complex

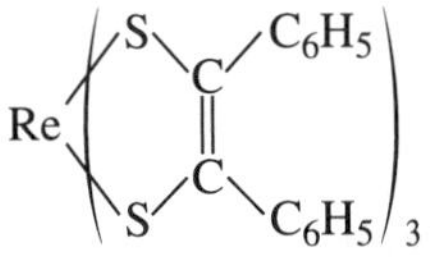

was reported to have trigonal prismatic coordination of Re. However, similar complexes of Mo, Re, and V have remarkably similar M—S distances, varying from 232.5 pm for Re to 233.8 pm for V—even though V has the smallest radius! The tetragonal faces are almost square (as is true also for MoS_2) with very short S—S distances (~306 pm). The S—S distance along the edges spanned by the chelate ring are shorter than for the other edges but are within 1 pm of the S—S distance between ligands. These short S—S distances suggest that S—S bonding might be important in establishing the trigonal prismatic stereochemistry. The square faces are reminiscent of square-planar S_4^{2+} rings observed by Gillespie.

Other 1,2-dithiolene and 1,2-ethanedithiolate complexes of these metals, as well as Fe and W, have been obtained, and more examples should be forthcoming. Researchers found that two-electron reduction of the neutral complex to produce $M[S_2C_2(CN)_2]_3^{2-}$ causes twisting about the C_3 axis, to give stereochemistry intermediate between a trigonal prism and an octahedron. Presumably, the electrons occupy orbitals that are antibonding with respect to S—S, thereby increasing repulsion. An octahedron would result from a twist angle of 60°.

[22] R. A. D. Wentworth, *Coord. Chem. Rev.* **1972**, *9*, 171.

[23] Various geometries for C.N. 6 were considered theoretically by R. Hoffmann, J. M. Howell, and A. R. Rossi, *J. Am. Chem. Soc.* **1976**, *98*, 2484.

Of the few reported trigonal pyramidal complexes of ligands without sulfur, most involve three relatively rigid didentate ligands that can coordinate along the tetragonal edges of a prism, with three ligands joined to cap one or both ends of the trigonal prism. Such a ligand is *cis,cis*-1,3,5-tris(pyridine-2-carboxaldiimino)cyclohexane [the Schiff base from *cis,cis*-1,3,5-triamino-cyclohexane (tach) with pyridine-2-carboxaldehyde]. Figure 9.31 shows the Zn^{2+} complex ion $[Zn\{(py)_3tach\}]^{2+}$. The nitrogens form a slightly tapered trigonal prism only slightly twisted ($4°$). The Ni^{2+} complex with high octahedral ligand-field stabilization energy shows a twist angle of about $32°$, about halfway between an octahedron and a trigonal prism. Less rigid encapsulating ligands give varying twist angles, depending on the preference of the metal ion for octahedral coordination.

The most favorable cases for trigonal prismatic coordination are d^0, d^1, d^5, d^{10}, and high-spin d^7. These are cases that do not involve strong preference for octahedral coordination because of ligand-field stabilization.

9.12 COORDINATION NUMBER FIVE[24]

Stoichiometry often is a poor indication of the actual C.N. Thus PCl_5 exists in the solid as $[PCl_4]^+[PCl_6]^-$, and commonly C.N. 6 is achieved by halide bridging—$MoCl_5$, $TaCl_5$, SbF_5, and $Tl_2[AlF_5]$ are examples.

▶ 9.12.1 *Square-Pyramidal Complexes*

Antimony(III) forms the square-pyramidal (SP) complexes $[SbF_5]^{2-}$ (Figure 9.32) and $[SbCl_5]^{2-}$ because of the steric requirements of the lone pair of electrons. In $[SbF_5]^{2-}$ the Sb is slightly below the base of the pyramid because of the influence of the unshared electron pair. For transition metal complexes the metal ion commonly is above the square

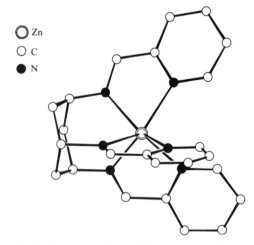

○ Zn
○ C
● N

Figure 9.31 The structure of the $[Zn(py)_3tach]^{2+}$ ion in the anhydrous perchlorate salt. (From W. O. Gillum, J. C, Huffman, W. E. Streib, and R. A. D. Wentworth, *Chem. Commun.* **1969**, 843.)

[24] B. F. Hoskins and F. D. Williams, *Coord. Chem. Rev.* **1973**, *9,* 365; R. Morassi, I. Bernal, and L. Sacconi, *Coord. Chem. Rev.* **1974**, *11,* 343; A. R. Rossi and R. Hoffmann, *Inorg. Chem* **1975**, *14,* 365; M. C. Favas and D. L. Kepert, *Prog. Inorg. Chem.* **1980**, *27,* 325; R. R. Holmes, *Prog. Inorg. Chem.* **1984**, *32,* 119.

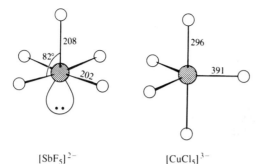

$[SbF_5]^{2-}$ $[CuCl_5]^{3-}$

Figure 9.32   Structures of $[SbF_5]^{2-}$ and $[CuCl_5]^{3-}$. Bond lengths are in pm.

base of the pyramid (by 34 pm for $[Ni(CN)_5]^{3-}$). The apical position is not equivalent to the basal positions for the SP (C_{4v}). The vanadyl group, VO^{2+}, retains its identity in many compounds and the acetylacetonate complex, $VO(acac)_2$, is square-pyramidal. The didentate ligand o-$C_6H_4[As(CH_3)_2]_2$ (diars) forms SP complexes of divalent Ni, Pd, and Pt with an anion in the apical position, $[M^{II}(diars)_2X]^+$.

Five-coordinate transition metal complexes without constraints of chelating ligands occur with the trigonal-bipyramidal (TBP) structure—for example, $[CuCl_5]^{3-}$ (Figure 9.32), $[Co(NCCH_3)_5]^+$, and $Fe(CO)_5$—and with the square-pyramidal structure—for example, $[Ni(CN)_5]^{3-}$ (Figure 9.33). These two structures differ less than the usual representations suggest. In general, chelated complexes or those with two or more different ligands coordinated, display versions of these structures so distorted that it often is difficult to say whether the structure is closer to TBP or to SP. The stabilities of the two regular structures are very similar; in fact, $[Cr(en)_3][Ni(CN)_5] \cdot 1.5H_2O$ contains *both* SP and TBP anions. The corresponding compound containing the $[Cr(1,3\text{-propanediamine})_3]^{3+}$ cation contains only the SP $[Ni(CN)_5]^{3-}$.

Reaction of $(C_5H_5)Re(CO)_3$ with Br_2 gives $(C_5H_5)Re(CO)_2Br_2$ (C.N. 5 considering C_5H_5 as one ligand, see Section 12.3 for bonding for C_5H_5) that gave by column chromatography two isomers.[25] The two CO and Br groups form the base of a square pyramid

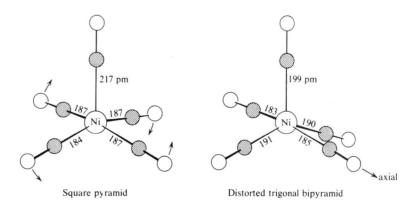

Square pyramid   Distorted trigonal bipyramid

Figure 9.33   Structures of $[Ni(CN)_5]^{3-}$ in $[Cr(NH_2C_2H_4NH_2)_3][Ni(CN)_5] \cdot 1.5H_2O$. Only slight shifts in the directions of the arrows are required to convert from SP–TBP.

[25] R. B. King and R. H. Reimann, *Inorg. Chem.* **1976**, *15*, 179.

with C_5H_5 at the apex. One isomer is described as diagonal (*trans*), and one is described as lateral (*cis*). Mild heating converts the lateral isomer to the diagonal isomer. The SP Mo complex[26] *cis*-[(C$_5$H$_5$)Mo(CO)$_2$Cl(*S*-L)] [where *S*-L = (*S*)-(+)- P —N—C⟨Me, with P the donor] was separated into the optical isomers, and the absolute configuration was determined by x-ray crystallography for the less soluble diastereoisomer. It racemizes quickly in solution at room temperature.

▶ 9.12.2   *Trigonal-Bipyramidal Complexes*

Five-coordinate compounds of nonmetals—PF$_5$, for example—generally have the trigonal bipyramidal structure. The TBP, (**D**$_{3h}$) has nonequivalent axial and equatorial positions. Whereas in PX$_5$ molecules the axial bonds are longer, in [CuCl$_5$]$^{3-}$ (Figure 9.32) the equatorial bonds are slightly longer than axial bonds.

That only slight displacement of ligands is required to go from SP to TBP, or *vice versa,* is illustrated in Figure 9.33, in which the regular SP is shown for [Ni(CN)$_5$]$^{3-}$ alongside the distorted TBP that occurs in the same crystal structure. Two diagonally related ligands in the base of the SP are displaced upward to become the axial ligands for the TBP, and the other two basal ligands are displaced downward to equatorial positions for the TBP. In this distorted TBP, one C—Ni—C bond angle in the equatorial plane is much larger (142°) than the other two (107.3° and 111.5°). Further displacement to close the large angle would give a regular TBP.

Several tripod ligands, such as N[C$_2$H$_4$N(CH$_3$)$_2$]$_3$ and P[C$_3$H$_6$As(CH$_3$)$_2$]$_3$, form TBP complexes with Ni^{II}, Co^{II}, and some other metals. These tetradentate umbrella-like ligands coordinate symmetrically, with another ligand—for example, CN$^-$, Cl$^-$, or Br$^-$—in the remaining axial position.

Ligand–ligand repulsion alone favors the TBP arrangement, but bonding interactions, and especially π bonding, involving d orbitals can favor the SP structure. Both arrangements give a pair of d orbitals of lowest energy and one d orbital of very high energy. Low-spin five-coordinate complexes with d^5, d^6, d^7, and d^8 configurations leave this highest-energy orbital empty.

Although diffraction studies and infrared spectroscopy show the axial and equatorial groups in Fe(CO)$_5$ and PF$_5$ to be nonequivalent, on the nuclear magnetic resonance (NMR) time scale all five groups are equivalent. The easy TBP $\rightleftarrows$ SP interconversion can scramble all positions. These are examples of fluxional molecules[27] (see Section 9.15).

Co^{2+}(d^7) forms a low-spin complex [Co(dpe)$_2$Cl]$^+$ [dpe = 1,2-bis(diphenylphosphino)ethane, Ph$_2$PC$_2$H$_4$PPh$_2$] that gives red crystals containing SP ions (Cl axial) and

[26] G. M. Reisner, I. Bernal, H. Brunner, M. Muschiol, and B. Siebrecht, *J. Chem. Soc. Chem. Commun.* **1978,** 691.

[27] J. R. Shapley and J. A. Osborn, *Acc. Chem. Res.* **1973,** *6,* 305.

green crystals containing TBP ions.[28] In the TBP ion, two P atoms are axial with the other two P atoms and Cl^- in equatorial positions.

9.13   LOWER COORDINATION NUMBERS

▶ *9.13.1   Coordination Number Two*

We encounter C.N. 2 for Cu^+, Ag^+, Au^+, and Hg^{2+}, d^{10} ions. Cu^+ forms linear $[Cu(NH_3)_2]^+$ and $CuCl_2^-$, but most complexes of Cu^+ are tetrahedral. In solid $K[Cu(CN)_2]$ the C.N. of Cu^+ is 3 and solid halide complexes usually involve tetrahedral coordination. $[Ag(CN)_2]^-$, $[Ag(NH_3)_2]^+$, $Hg(CN)_2$, and $HgCl_2$ involve linear coordination in the solid state. The Ag^+ and Hg^{2+} ions add more ligands to form tetrahedral complexes, but two ligands are bound weakly. In solution the MX_2 species might be solvated strongly to give an effective C.N. of 4. Only for Au^+ is C.N. 2 characteristic, with little tendency to add additional ligands.

Although ethylenediamine complexes generally are much more stable than those of ammonia, this is not true for $[Ag(en)]^+$ compared with $[Ag(NH_3)_2]^+$. We attribute the unexpectedly low stability of $[Ag(en)]^+$ to the fact that the en molecule cannot span the Ag^+ to permit the desired linear arrangement. In chelate compounds, five-membered rings generally are most stable. In the case of silver, larger rings permit less strained bonding and compounds containing six-, seven-, and eight-membered rings are all more stable than $[Ag(en)]^+$.

We have encountered linear molecules such as CO_2, N_2O, and BeH_2 involving sp hybridization. More favorable bonding can be achieved for linear metal complexes by participation of the d_{z^2} orbital (Problem 4.2). Infrared studies of linear $HgCl_2$, $HgBr_2$, and HgI_2 in the vapor state indicate that the bending force constants are extremely low, suggesting that the bonding orbitals have only weak directional properties.

▶ *9.13.2   Coordination Number Three*[29]

Apparently, no metal ions show a preference for C.N. 3, which is encountered primarily for metal ions of low oxidation number with bulky ligands or when imposed in the solid. The C.N. of Cu^+ is 3 in $K[Cu(CN)_2]$ (through bridging CN^-) but 4 in $K_2[Cu(CN)_3]$. Planar trigonal ($\mathbf{D}_{3h}$) coordination occurs for HgI_3^-, $Pt[P(C_6H_5)_3]_3$, $Cu[SP(CH_3)_3]_3$, and $Cu[SC(NH_2)_2]_3^+$ in the solid. The bulky ligand $N[Si(CH_3)_3]_2^-$ gives three-coordinate ML_3 molecules with Fe^{3+}, Cr^{3+}, and even Mn^{3+} (d^4), Co^{3+} (d^6)[30] and some rare earth ions.

[28] J. K. Stalick, P. W. R. Corfield, and D. W. Meek, *Inorg. Chem.* **1973**, *2*, 1668.

[29] P. G. Eller, D. C. Bradley, M. B. Hursthouse, and D. W. Meek, *Coord. Chem. Rev.* **1977**, *24*, 1; D. C. Bradley, *Chem. Br.* **1975**, *11*, 393.

[30] J. J. Ellison, P. P. Power, and S. C. Shoner, *J. Am. Chem. Soc.* **1989**, *11*, 8044.

9.14 HIGHER COORDINATION NUMBERS

▶ 9.14.1 *Coordination Number Seven*[31]

The molecule IF_7 has a pentagonal-bipyramidal structure ($\mathbf{D_{5h}}$), as do $[UO_2F_5]^{3-}$, $[UF_7]^{3-}$, $[ZrF_7]^{3-}$, $[HfF_7]^{3-}$, and $[V(CN)_7]^{4-}$ (Figure 9.34). The structure of $[NbF_7]^{2-}$ and $[TaF_7]^{2-}$ ions derives from a trigonal prism, with a seventh F^- added to one of the tetragonal faces ($\mathbf{C_{2v}}$), a **capped** trigonal prism. Subsequent distortion to diminish anion–anion repulsion results in a configuration closer to that shown in Figure 9.34, with a tetragonal face opposite a trigonal face. The La_2O_3 structure (A-type M_2O_3) involves layers of LaO_6 octahedra with an oxygen of a neighboring octahedron directly over one triangular face, to give C.N. 7. The compound $(NH_4)_3SiF_7$ contains $[SiF_6]^{2-}$ and F^- anions rather than a seven-coordinate species.

Studies have shown that pentadentate macrocyclic and related ligands (Figure 9.35) form seven-coordinate complexes with several metal ions. The five ligand atoms are nearly in a plane, allowing plenty of room for coordination of anions (e.g., NCS^-, Cl^-, or Br^-) or water molecules above and below the plane, to give C.N. 7. Complexes such as $[Co(edta)]^-$ are octahedral with some strain in the glycinate rings in the plane of the ethylene ring. With the larger Fe^{3+} ion, edta cannot span and screen the cation well, so the complex is seven-coordinate, $[Fe(edta)H_2O]^-$. The octahedral LFSE for d^5 Fe^{3+} is zero. The complex is approximately pentagonal bipyramidal with H_2O in the plane of 2N and 2O of acetate arms.

M^{2+} ions (M = Mn, Fe, Co, Ni, Cu, and Zn) form complexes with py_3tren

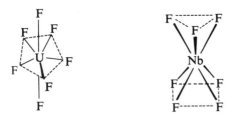

the M^{2+}, forming a trigonal antiprism (a distorted octahedron) with the bridging N above one triangular face. A C.N. 7 is imposed by the ligand, but M—N distances to the bridging N are too long to indicate bonding except for weak bonds to Mn^{2+} (d^5) and Co^{2+} (d^7), cases without strong octahedral LFSE.

Figure 9.34 Structures of $[UF_7]^{3-}$ and $[NbF_7]^{2-}$.

[31] D. L. Kepert, *Prog. Inorg. Chem.* **1979**, *25*, 41; R. K. Boggess and W. D. Wiegele, *J. Chem. Educ.* **1978**, *55*, 156.

[32] R. M. Kirchner, C. Mealli, M. Bailey, N. Howe, L. P. Torre, L. J. Wilson, L. C. Andrews, N. J. Rose, and E. C. Lingafelter, *Coord. Chem. Rev.* **1987**, *77*, 89.

Figure 9.35   A pentadentate macrocyclic ligand (L) capable of forming seven-coordinate complexes of the type $[FeCl_2(L)]ClO_4$.

▶ 9.14.2   *Coordination Number Eight*[33]

After C.N. 6 and 4 the most common C.N. is 8, which is encountered in solids for large cations in the CsCl and CaF_2 structures, $ZrSiO_4$, and garnets, $M_3^{II}M_2^{III}[SiO_4]_3$. Eight-coordination is cubic in solids with close-packed anions. A cubic coordination polyhedron has been found for discrete complex ions in $(Et_4N)_4[U(NCS)_8]$ and $Na_3[PaF_8]$. The [2.2.2]cryptate, $N(C_2H_4OC_2H_4OC_2H_4)_3N$, gives a cubic arrangement[34] with high-spin Mn^{II}. More commonly found is the dodecahedron with triangular faces (D_{2d})[35] (Figure 9.36). Whereas the dodecahedron occurs for $[Mo(CN)_8]^{4-}$, $[Mo(CN)_8]^{3-}$, and $[Zr(C_2O_4)_4]^{4-}$, the square antiprism (D_{4d}) occurs for $[TaF_8]^{3-}$, $Cs_4[U(NCS)_8]$, and $Zr(acac)_4$ (D_4); in the last-named case, the acetylacetonate chelate rings span edges on the square faces. It is interesting that $[U(NCS)_8]^{4-}$ gives both the antiprismatic and cubic structures: The Et_4N^+ cations fit into the cubic faces to stabilize the cube, while the Cs^+ ions are above the triangular faces of the antiprism.

Both the dodecahedron and antiprism give one d orbital of lowest energy, and many examples have d^0, d^1, or d^2 configurations. The $[Mo(CN)_8]^{4-}$ ion is diamagnetic (d^2). Having similar energies, the two stereochemistries provide no firm basis for choice be-

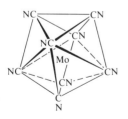

$[Mo(CN)_8]^{4-}$ (D_{2d})
Doderahedron

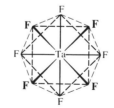

$[TaF_8]^{3-}$ (D_{4d})
Square antiprism

Figure 9.36   Complexes with C.N. 8 (Mo—C bonds are omitted).

[33] M. G. B. Drew, *Coord. Chem. Rev.* **1977,** *24,* 179; J. K. Burdett, R. Hoffmann, and R. C. Fay, *Inorg. Chem.* **1978,** *17,* 2553; D. L. Keppert, *Prog. Inorg. Chem.* **1978,** *24,* 179.

[34] K. S. Hagen, *Angew. Chem. Int. Ed. Engl.* **1992,** *31,* 764.

[35] There is also a dodecahedron with pentagonal faces (I_h), sometimes used for calendars. The D_{2d} dodecahedron with triangular faces has two equivalent sets of four vertices, with each set forming a trapezoid. The planes of the intersecting trapezoids are perpendicular.

tween them. As the "**bite**" (closeness of the donor atoms determined by the size of the chelate ring) is increased, however, the square antiprism (D_{4d}) should become more favored than the dodecahedron (although ligands with large "bites" are likely to form bridged complexes). Physical measurements on $[Mo(CN)_8]^{4-}$ in solution are inconclusive with respect to effective D_{2d} or D_{4d} symmetry.

C.N. 8 is favored for large cations and small ligands—F^- and CN^-, for example. Chelate ligands with a small "bite" show C.N. 8 with smaller cations—

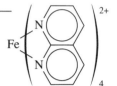

and $[Co(NO_3)_4]^{2-}$, for example; both have approximately dodecahedral coordination, with the average positions of each ligand approximately tetrahedral. The spectral characteristics of $[Co(NO_3)_4]^{2-}$ resemble those of tetrahedral complexes. A similar dodecahedral structure (Figure 9.37) was found for tetrakis(*N*-ethylsalicyldiiminato)zirconium(IV) (the didentate ligand is the Schiff base from ethylamine and salicylaldehyde). The actual symmetry of the complex is S_4. This is an interesting example of a compound with no center or plane of symmetry, but it is optically inactive because of an S_4 axis. No substantiated cases of geometric isomerism are known for C.N. 8.

C.N. 8 is achieved for compounds of dioxo cations—for example, uranyl ion, UO_2^{2+}—through hexagonal bipyramidal coordination. The $[UO_2(NO_3)_3]^-$ ion has three didentate NO_3^- slightly twisted from the hexagonal equatorial plane. The axial positions of the uranyl oxygens and the average positions of the nitrate ions describe a trigonal bipyramid.

▶ 9.14.3 Coordination Numbers Nine, Ten, Twelve, and Fourteen[36]

The $[Nd(H_2O)_9]^{3+}$ ion in $Nd(BrO_3)_3 \cdot 9H_2O$ has the Nd^{3+} ion at the center of a trigonal prism with an additional water molecule over each tetragonal face; the polyhedron is described as a tricapped trigonal prism (see Figure 15.6). Other hydrated rare earth ions with this structure have been reported. The structures of $[ReH_9]^{2-}$ and $[TcH_9]^{2-}$ are similar. C.N. 9 is achieved through sharing of anions in $PbCl_2$, UCl_3, $La(OH)_3$, and other

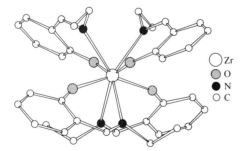

○ Zr
● O
● N
○ C

Figure 9.37 Structure of the eight-coordinate complex $Zr(OC_6H_4CH{=}NC_2H_5)_4$ (S_4 symmetry). (From D. C. Bradley, M. B. Hursthouse, and I. F. Rendall, *Chem. Commun.* **1970**, 368.

[36] M. G. B. Drew, *Coord. Chem. Rev.* **1977**, *24*, 179; R. E. Robertson, *Inorg. Chem.* **1977**, *16*, 2735; D. L. Kepert, *Prog. Inorg. Chem.* **1981**, *28*, 309.

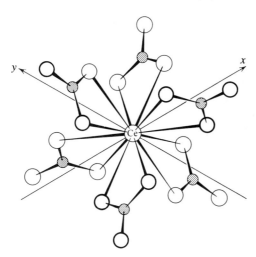

Figure 9.38 Twelve-coordination of $[Ce(NO_3)_6]^{2-}$ in $(NH_4)_2[Ce(NO_3)_6]$. (From T. A. Beineke and J. Delgaudio, *Inorg. Chem.* **1968**, *7*, 715. Copyright 1968, American Chemical Society.)

compounds having these structures. In the solid, $[ThF_8]^{2-}$ achieves C.N. 9 in a capped (shared F^-) square antiprism.

Cerium(IV) shows C.N. 10 in $[Ce(NO_3)_4\{OP(C_6H_5)_3\}_2]$, where the four nitrate ions are didentate. If we view the phosphine oxygens as occupying axial positions of an octahedron, the nitrate ions would be in the four equatorial positions, but with the coordinated oxygens tipped out of the equatorial plane. The C.N. is also 10 for $[Ce(NO_3)_5]^{2-}$. Here we can visualize the structure by viewing the nitrate ions as arranged about a trigonal bipyramid. The coordinated oxygens are twisted about the Ce—N lines, to minimize repulsion. The structure of $[Ce(CO_3)_5]^{6-}$ is a distorted bicapped square antiprism ($\mathbf{D}_{4d}$ for the regular polyhedron).

C.N. 12 is achieved in the compound $Mg_3Ce_2(NO_3)_{12} \cdot 24H_2O$ (containing discrete $[Ce^{III}(NO_3)_6]^{3-}$ ions) and in $(NH_4)_2Ce(NO_3)_6$, both involving didentate nitrate ligands. The

Table 9.16 Simplified shapes of some complexes based on the mean positions of didentate ligands

Coordination number	Figure described by positions of ligand atoms	Figure described by mean position of ligands	Examples
4	Tetrahedron ($\mathbf{D}_{2d}$)	Straight line	$Be(acac)_2$, $Cu(NO_3)_2{}^a$
6	Octahedron[b] ($\mathbf{D}_3$)	Triangle	$Co(NO_3)_3$
8	Dodecahedron ($\mathbf{D}_{2d}$)	Tetrahedron	$Ti(NO_3)_4$, $Fe(NO_3)_4^-$
8	Square antiprism ($\mathbf{D}_{4d}$)	Square plane	$Zr(acac)_4$
10	"Bicapped" trigonal antiprism ($\mathbf{C}_1$)[c]	Trigonal bipyramid	$Ce(NO_3)_5^{2-}$
12	Icosahedron ($\mathbf{T}_h$)	Octahedron	$Ce(NO_3)_6^{3-}$, $Ce(NO_3)_6^{2-}$

[a] The structure of $Cu(NO_3)_2$ in the vapor is distorted tetrahedral; the O—Cu—O angle is 70°.

[b] The octahedron is distorted toward the trigonal prism because of the small "bite" of the ligand. The simplified shape would still be trigonal planar even if the structure were the trigonal prism.

[c] Each cap is a didentate nitrate ligand.

approximately octahedral positions of the nitrate ions result in a figure with a center of symmetry. Opposite pairs of nitrate ligands are twisted slightly from the xy, xz, and yz planes. The coordinated oxygens are at the 12 corners of an irregular icosahedron (Figure 9.38). The actual symmetry is nearly T_h for the Ce^{IV} complex. Icosahedral complexes are interesting, because there is no splitting of the energies of the d orbitals in an icosahedral field; they are still fivefold degenerate. Chelated nitrate complexes have high C.N.s because of the small "bite" of the didentate nitrate ligand, and their mean positions—the positions of the N of the nitrate ion—usually correspond to a lower polyhedron (C.N./2) (see Table 9.16).

9.15 STEREOCHEMICALLY NONRIGID AND FLUXIONAL MOLECULES[37]

For most molecules, one stereochemical arrangement is much more stable than others. These molecules usually have large HOMO–LUMO gaps (see Section 4.8). The molecule displays the stable structure unless it is in an excited vibrational or electronic state or a transition state for a reaction or stereochemical transformation. For some molecules, two or more structures have comparable stability, and at ordinary temperatures the molecules change between or among these continually. As noted above, $[Ni(CN)_5]^{2-}$ is found as a square pyramid or trigonal bipyramid, and both even exist in the same crystal (Figure 9.33). Such complex ions with two structures of comparable stability are said to be **stereochemically nonrigid**. If the two (or more) configurations are chemically equivalent, the molecule is **fluxional**. Molecular vibrations offer a path for permuting nuclear positions and traversing the collection of possible structures.

Whether or not a molecule is judged stereochemically nonrigid depends on the time period required to make experimental observations on its structure. Structural techniques (for example, NMR) that have a slower time scale than the vibrations permuting nuclear positions "see" a time-averaged structure.

The time required to make an observation (its **time scale**) is related (by the Uncertainty Principle) to the frequency of exciting electromagnetic radiation employed:

$$\Delta E \Delta t \sim h$$

$$\Delta(h\nu)\Delta t = h\Delta\nu\Delta t \sim h \qquad (9.20)$$

$$\Delta\nu\Delta t \sim 1$$

Because we adopt the most pessimistic view in uncertainty calculations, $\Delta\nu$ can be set $\simeq\nu$. Then $\Delta t \simeq 1/\nu$. All other things being equal, the time scale is inversely proportional to the frequency of the exciting radiation.[38] Other considerations connected with the sample (e.g., relaxation times) may effectively lengthen Δt. Table 9.17 gives time scales for several common structural techniques.

---

[37] F. A. Cotton, *Acc. Chem. Res.* **1968,** *1,* 257; E. L. Muetterties, *Inorg. Chem.* **1965,** *4,* 769; E. L. Muetterties, *Acc. Chem. Res.* **1970,** *3,* 266; J. R. Shapley and J. R. Osborn, *Acc. Chem. Res.* **1973,** *6,* 305.

[38] Spectroscopists commonly use the width of the signal at half-height as a measure of uncertainty.

Table 9.17   Time scale for structural techniques[a]

Technique	Approximate time scale (s)
Electron diffraction	10^{-20}
Neutron diffraction	10^{-18}
X-ray diffraction[b]	10^{-18}
Ultraviolet	10^{-15}
Visible	10^{-14}
Infrared-Raman	10^{-13}
Electron spin resonance[c]	10^{-4}–10^{-8}
Nuclear magnetic resonance[c]	10^{-1}–10^{-9}
Quadrupole resonance[c]	10^{-1}–10^{-8}
Mössbauer (iron)	10^{-7}
Molecular beam	10^{-6}
Experimental separation of isomers	$>10^2$

[a] From E. L. Muetterties, *Inorg. Chem.* **1965**, *4*, 769.

[b] Individual measurements of this duration are taken over a long time span. Hence a time-averaged structure is obtained.

[c] Time scale sensitivity defined by chemical systems under investigation.

NMR spectroscopy is suited to investigating stereochemical nonrigidity because, at low temperature, time scales for rotations and vibrations that permute nuclei often are much greater than the scale for nuclear spin transitions. At high temperatures they are often smaller. At 273 K (Figure 9.39a), sufficient thermal energy is available so that free rotation of $BrCl_2CCBrF_2$ occurs so rapidly relative to the NMR time scale that the F's "see" an averaged environment over the time required for observation. On lowering the temperature to 193 K (Figure 9.39b), the available thermal energy is less than the internal

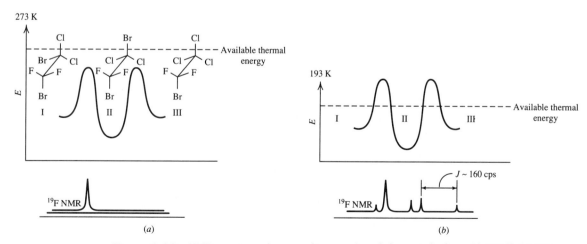

Figure 9.39   NMR spectra and energy diagrams for a halogenated ethane (a) 273 K (b) 193 K. (Spectra from J. D. Roberts. *Nuclear Magnetic Resonance.* Copyright 1959, McGraw–Hill book Company, used with permission.)

rotational barrier. Hence, internal rotation is slowed sufficiently to detect the presence of rotational isomers in solution. (Since I and III are enantiomers, their NMR spectra are the same in an achiral solvent.) Rotational barriers rarely are so high. More commonly, vibrational barriers result in detection of rigid structures at low temperatures.

For the trigonal bipyramid, the axial positions are not equivalent geometrically to the equatorial positions; this generally results in different bond lengths. However, minor deformations passing through a square-pyramidal intermediate can interchange the axial and equatorial positions (see Figure 9.33). The axial and equatorial F's in PF_5 are expected to give different signals in the ^{19}F NMR because no symmetry operation in $\mathbf{D}_{3h}$ exchanges the two positions. (This is the necessary condition for **chemical inequivalence** of the two sets of F's.) However, the NMR spectrum down to the lowest accessible temperatures displays only one signal, suggesting that NMR "sees" the five F's as equivalent. Some processes on a time scale faster than NMR observations permute the F's.[39] PF_5 is a fluxional molecule.

In any spectroscopic measurement, absorbance of the irradiating electromagnetic radiation occurs when the frequency corresponds to a difference in energy levels of the compound being observed—the so-called resonance condition.[40] F has a nuclear spin $I = \frac{1}{2}$; thus, in a magnetic field there are two possible orientations of the nuclear spin: parallel to the field ($m_I = +\frac{1}{2}$) and antiparallel ($m_I = -\frac{1}{2}$). Thus, a transition with absorbance of radiofrequency radiation is expected to occur for each chemically different type of F as it realigns its nuclear spin moment from lower energy ($+\frac{1}{2}$) to higher energy ($-\frac{1}{2}$). P also has a nuclear spin of $\frac{1}{2}$. Hence each equivalent type of F has the possibility of being bound to a P with spin orientation $+\frac{1}{2}$ or $-\frac{1}{2}$. This means that the lower-energy $+\frac{1}{2}$ state of F will be split into two states according to the nuclear spin orientation of P. In the simplest approximation, the signals for equatorial and axial F should be split into two. P and F are said to display **spin–spin coupling.** The situation may be diagrammed as follows for the F resonances:

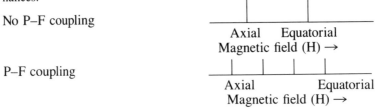

No P–F coupling

Axial   Equatorial
Magnetic field (H) →

P–F coupling

Axial   Equatorial
Magnetic field (H) →

By employing a magnetic field H of different magnitude, the NMR spectrum of P could also be observed. In this case the single signal expected from the one kind of P present is split by coupling to the two sets of F's. ^1H, ^{31}P, and ^{13}C all have $I = \frac{1}{2}$ and are expected to show one NMR signal per chemically equivalent type of atom as long as nuclear permuta-

---

[39] The exact nature of the atomic motions which permute nuclei can often be elucidated by analyzing the shapes of the NMR signals. Such an analysis is beyond the scope of this book. Sometimes, however, we do present accepted mechanisms for permutations, but these particular mechanisms are compatible with experimental data rather than being rigorously derived therefrom.

[40] For a clear and brief account of NMR spectra and their interpretation, see R. S. Macomber, *NMR Spectroscopy, Basic Principles and Applications,* Harcourt Brace Jovanovich, San Diego, CA, 1988; R. S. Drago, *Physical Methods for Chemists,* 2nd ed., W. B. Saunders, Orlando, FL, 1992, Chapters 7 and 8.

tion is slow. If any of these atoms is attached to n identical atoms of different elements with $I = \frac{1}{2}$, the signal is split into $(n + 1)$ components corresponding to the $(n + 1)$ possible orientations of I. The relative intensities of the components are given by the terms of Pascal's triangle in which the members of each succeeding row are obtained from adding adjacent numbers in the row above:

$$
\begin{array}{ccccccc}
 & & & 1 & & & \\
 & & 1 & & 1 & & \\
 & 1 & & 2 & & 1 & \\
1 & & 3 & & 3 & & 1, \quad \text{etc.}
\end{array}
$$

These intensity ratios reflect the number of ways of achieving (and therefore the probability of attaining) various combinations of nuclear-spin orientations for the coupling nuclei. For example, when $n = 1$, there is only one way of achieving the $+\frac{1}{2}$ and one of achieving $-\frac{1}{2}$ orientation. Hence, both are equally probable and the nucleus undergoing transition has an equal probability of "seeing" either, leading to the $1:1$ intensity ratio for the split components. On the other hand, if $n = 2$ we have the possible orientation combinations $++$, $+-$, $-+$, and $--$. The probability of having one attached atomic nucleus with spin $+\frac{1}{2}$ and one with spin $-\frac{1}{2}$ is twice as large as for the other values, and so on.

If we were to observe the ^{31}P NMR spectrum of PF_5, the one expected signal from the $+\frac{1}{2} \rightarrow -\frac{1}{2}$ nuclear transition would be split into a sextet of relative intensities $1:5:10:10:5:1$ because all F's are equivalent on the time scale of the measurement. If the trigonal bipyramid were rigid, we would expect P to be coupled separately to two equivalent axial and three equivalent equatorial F's, giving either a triplet of quartets (if axial coupling were larger than equatorial) or a quartet of triplets (if equatorial coupling were larger).

Figure 9.40 depicts the ^{19}F NMR spectrum of PCl_2F_3 at several temperatures. Three possible geometries can be entertained for this compound on the basis of the valence shell electron-pair repulsion (VSEPR) model, which predicts a trigonal bipyramidal geometry:

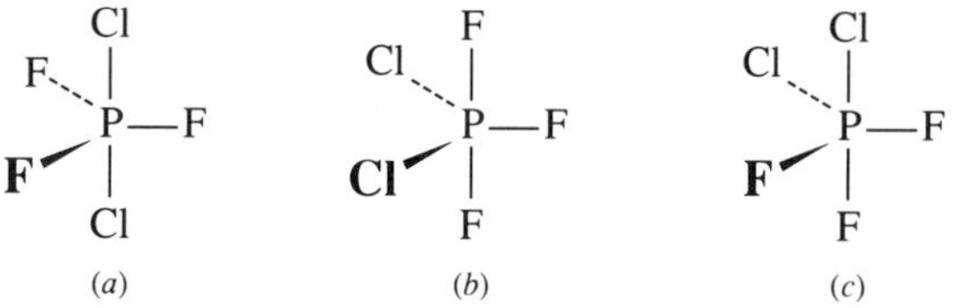

At the highest temperature, one F signal split by P is seen. As the temperature is lowered, the signal collapses into the baseline and finally, in the low-temperature limit, signals for two chemically different F sets are seen, each one split into a doublet by P–F coupling. That this is P–F coupling is evident from the similarity of the splitting between each component to that in the high-temperature limiting spectrum. Without P–F splitting, the two signals would be a triplet and a doublet. The relative intensities reflect the numbers of different kinds of F's. The triplet is half as intense as the doublet. Thus, the spectrum indicates two sets of chemically equivalent F's with a $1:2$ population ratio. One kind of F is split by two equivalent F's (the triplet) and, for the other set, two chemically equivalent F's are split by one different F (the doublet). This rules out structure (a) as the low-temperature isomer. Either (b) or (c) is consistent with the number and splitting of signals.

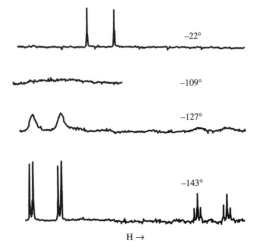

Figure 9.40 ^{19}F NMR spectra of PCl_2F_3 in isopentane at several temperatures. (Reprinted with permission from R. R. Holmes, R. P. Carter, and G. E. Peterson, *Inorg. Chem.* **1964**, *3*, 1750. Copyright 1964, American Chemical Society.)

The position of one signal at higher magnetic field (with respect to some agreed-upon signal as 0) is called the **chemical shift,** and its value is indicative of the chemical environment. For example, the study of a number of P–F compounds has shown that equatorial F resonates at higher field than axial F.[41] Thus the low-intensity triplet is an equatorial F; it is split by two axial F. The lower-field signal is due to two axial F's split by the single equatorial F [Structure (*b*)]. As the temperature is raised, the signals broaden as the exchange rate becomes faster relative to the time scale of observation. When the observation time and exchange rate for F's are comparable, the F's have no defined environment over the time required to measure, and the signal is reduced to almost zero. Finally, as the temperature is raised sufficiently so that the exchange rate is far above the time scale for measurement, only a single F environment is seen (the **fast-exchange limit**). The chemical shift for the high-temperature spectrum is the weighted average of the separate low-temperature chemical shifts. The fact that P–F coupling persists at this temperature indicates that the F's remain bonded to P at all temperatures. [If we had not narrowed the possible geometries through VSEPR considerations, we could entertain the square-pyramidal structure (*d*).]

EXAMPLE 9.3: Figure 9.41 depicts the ^{19}F NMR spectrum of ClF_3. What can you say about the geometry of this molecule? Is the molecule fluxional?

Solution: Two different types of F are present as indicated by the two different signals: a triplet and a lower-field doublet in a 1 : 2 intensity ratio. Thus, at this temperature, no process inter-

[41] Tables of expected chemical shifts for various nuclei in different chemical environments are available in the references cited in footnote 40 and in E. D. Becker, *High Resolution NMR*, 2nd ed., Academic Press, New York, 1980.

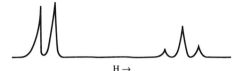

H →

Figure 9.41 ^{19}F NMR spectrum of neat ClF$_3$ at −60°C. (Reproduced with permission from E. L. Muetterties and W. D. Phillips, *J. Am. Chem. Soc.* **1957**, *79*, 322. Copyright 1957, American Chemical Society.)

changes F's faster than the NMR observation; thus the molecule is not fluxional. The so-called **static structure** involves one F split by two equivalent F's because the low-intensity signal is a triplet; also two equivalent F's are split by one F because the higher intensity signal is a doublet. (Cl has $I = 0$ so that it does not split the F signals.) The appearance of the spectrum is consistent with a static structure containing one equatorial and two axial F's. This is to be expected as the T-shaped geometry predicted from the VSEPR model.

Ni^{II} complexes such as [NiX$_2$(PR$_3$)$_2$] are tetrahedral when R = C$_6$H$_5$; when R = cyclohexyl the geometry is square-planar. For other phosphines, an equilibrium mixture—corresponding to a tetrahedral $\rightleftarrows$ square-planar transformation—is formed in solution. Such solutions exhibit magnetic moments and visible spectra which vary with temperature depending on the relative proportions of the two isomers. Because the planar and tetrahedral structures are not equivalent chemically, these complexes are *stereochemically nonrigid, but not fluxional*. Both green paramagnetic tetrahedral and brown diamagnetic square-planar [NiBr$_2$(PEtPh$_2$)$_2$] can be isolated.

The structure of [Mo(CN)$_8$]$^{4-}$ in the solid is the dodecahedron; some studies have favored a square antiprism in solution. This case is inconclusive, with stereochemically nonrigid behavior suggested. Stereochemically nonrigid or fluxional behavior is rather common among organometallic molecules and cluster compounds (see pages 603 and 661) and accounts for the conflicting structures obtained from structural techniques with different time scales.

GENERAL REFERENCES

C. J. Hawkins, *Absolute Configuration of Metal Complexes,* Wiley-Interscience, New York, 1971. Covers more than implied by the title.

G. B. Kaufman, Ed., *Classics in Coordination Chemistry,* Part I, *The Selected Papers of Alfred Werner,* 1968; Part II, *Selected Papers (1798–1899),* 1976; Part III, *Twentieth Century Papers (1904–1935),* 1978, Dover, New York.

A. E. Martell, Ed., *Coordination Chemistry,* Vol. I, Van Nostrand-Reinhold, New York, 1971; Vol. II, ACS Monograph 174, SIS/American Chemical Society, Washington, D.C., 1978.

K. Nakamoto and P. J. McCarthy, Eds., *Spectroscopy and Structure of Metal Chelate Compounds,* Wiley, New York, 1968.

Y. Saito, "Absolute Stereochemistry of Chelate Compounds," *Topics in Stereochemistry* **1978**, *10*, 95.

G. Wilkinson, Ed., *Comprehensive Coordination Chemistry,* Pergamon Press, Oxford, 1987. Seven volumes; a definitive reference work.

PROBLEMS

9.1 Name the following compounds according to the IUPAC rules (see Appendix B):

K_2FeO_4 $K_2[Co(N_3)_4]$

$[Cr(NH_3)_6]Cl_3$ $K[Co(edta)]$

$[CrCl_2(NH_3)_4]Cl$

$K[PtCl_3(C_2H_4)]$ $[(NH_3)_4Co\diagdown^{OH}_{NH_2}\diagup Co(en)_2]Cl_4$

$K_3[Al(C_2O_4)_3]$ $[Cr(OH)(H_2O)_3(NH_3)_2](NO_3)_2$

9.2 The values for formation constants for each step in the formation of $[Ni(en)_3]^{2+}$ are log $K_1 = 7.52$, log $K_2 = 6.28$, and log $K_3 = 4.26$ at 30°C in 1.0 M KCl. **(a)** What is log β_3 for the overall formation from Ni^{2+} and 3 en? **(b)** Why do values of K decrease with $K_1 > K_2 > K_3$?

9.3 Considering the charge densities (Table 9.8), what is the expected order of decreasing stabilities of F^- complexes of Be^{2+}, Mg^{2+}, and Ca^{2+}?

9.4 The K_1 for HgI^+ is greater than that for MgI^+ even though Mg^{2+} has a considerably higher charge density. How can this be explained?

9.5 $K_1 = 10.7$ for $[Cu(en)]^{2+}$ and $\beta_2 = 7.8$ for $[Cu(NH_3)_2]^{2+}$. $K_1 = 4.7$ for $[Ag(en)]^+$ and $\beta_2 = 7.2$ for $[Ag(NH_3)_2]^+$. **(a)** Why is the formation constant for adding one en to Cu^{2+} greater than that for adding two NH_3? **(b)** Why is the order the reverse for Ag^+?

9.6 For complexes of Fe^{3+} and Ag^+ with SCN^-, would you expect coordination of SCN^- through S or N to these cations? (*Hint:* Consider hardness/softness.)

9.7 Which of the following complexes obey the 18-electron rule (EAN rule)?

(a) $[Cu(NH_3)_4]^{2+}$, $[Cu(en)_3]^{2+}$, $[Cu(CN)_4]^{3-}$ **(d)** $[Fe(CN)_6]^{3-}$, $[Fe(CN)_6]^{4-}$, $[Fe(CO)_5]$

(b) $[Ni(NH_3)_6]^{2+}$, $[Ni(CN)_4]^{2-}$, $[Ni(CO)_4]$ **(e)** $[Cr(NH_3)_6]^{3+}$, $[Cr(CO)_6]$

(c) $[Co(NH_3)_6]^{3+}$, $[CoCl_4]^{2-}$

9.8 Calculate the magnetic moment for the spin-only contribution for:

(a) $[Ni(CN)_4]^{2-}$ (planar)   **(e)** $[Co(NH_3)_6]^{3+}$

(b) $[Ni(NH_3)_6]^{2+}$ **(f)** $[CoCl_4]^{2-}$ (tetrahedral)

(c) $[Cu(NH_3)_4]^{2+}$ (planar)   **(g)** $[Cr(NH_3)_6]^{3+}$

(d) $[Ag(NH_3)_2]^+$ (linear)   **(h)** $Ni(CO)_4$ (tetrahedral)

9.9 Determine the number of unpaired electrons and the LFSE for each of the following:

(a) $[Fe(CN)_6]^{4-}$ **(e)** $[Ru(NH_3)_6]^{3+}$

(b) $[Fe(H_2O)_6]^{3+}$ **(f)** $[PtCl_6]^{2-}$

(c) $[Co(NH_3)_6]^{3+}$ **(g)** $[CoCl_4]^{2-}$ (tetrahedral)

(d) $[Cr(NH_3)_6]^{3+}$

9.10 Discuss briefly the factors working for and against the maximum spin state of d electrons in transition metal complexes.

9.11 Explain why square-planar complexes of transition metals are limited (other than those of planar ligands such as porphyrins) to those of **(a)** d^7, d^8, and d^9 ions and **(b)** d^8 configurations for very strong field ligands which can serve as π acceptors.

9.12 Construct a table of ligand-field stabilization energies (LFSE) similar to Table 9.10 for all d^n configurations in tetrahedral complexes with weak fields. Which d^n configurations would be diamagnetic for a strong field?

9.13 For which d^n configurations would no Jahn–Teller splitting be expected for the tetrahedral case (ignore possible low-spin cases)?

9.14 Give the orbital occupancy (identify the orbitals) for the Jahn–Teller splitting expected for tetrahedral complexes with high-spin d^3 and d^4 configurations. Indicate the nature of distortions expected.

9.15 **(a)** Why are low-spin complexes usually not encountered for tetrahedral coordination? **(b)** Octahedral splitting is expressed as $10Dq$. What would be the splitting for ML_8 with cubic coordination? Assume the same ligands at the same distance as for octahedral and tetrahedral cases.

9.16 Sketch all possible geometrical isomers for the following complexes and indicate which of these would exhibit optical activity.

    (a) $[CoBrCl(en)(NH_3)_2]^+$ **(d)** $[CoCl(NH_2CH_2CO_2)_2(NH_3)]$

    (b) $[PtBrCl(NO_2)(NH_3)]^-$ **(e)** $[CoCl_2(trien)]^+$ (consider the different ways of linking trien to Co)

    (c)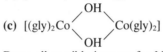

9.17 Draw *all* possible isomers for Ma_2bcd assuming the complex forms a square pyramid.

9.18 How might one distinguish between the following isomers?

    (a) $[CoBr(NH_3)_5]SO_4$ and $[Co(SO_4)(NH_3)_5]Br$

    (b) $[Co(NO_2)_3(NH_3)_3]$ and $[Co(NH_3)_6][Co(NO_2)_6]$

    (c) *cis*- and *trans*-$[CoCl_2(en)_2]Cl$

    (d) NH_4^+ *cis*- and *trans*-$[Co(NO_2)_4(NH_3)_2]^-$

    (e) *cis*- and *trans*-$[Pt(gly)_2]$

9.19 Give examples of the following types of isomerism:

    (a) Hydrate isomerism   **(d)** Ionization isomerism

    (b) Coordination isomerism   **(e)** Geometrical isomerism

    (c) Linkage isomerism

9.20 Draw all isomers possible for octahedral $M(abcdef)$. (*Hint:* Calculate the number of stereoisomers of $[M(abcdef)]$. Write unique pairings *trans* to *a* and calculate the number of possible isomers for each of these. Then consider unique *trans* pairings for the remaining groups.)

9.21 Draw all possible isomers for

    (a) $M(abcde\overparen{f})$ ($e\overparen{f}$ is didentate).

    (b) $M(\overparen{aa}bcd_2)$ (*Hint:* leave $\overparen{bc}$ in a fixed position in all drawings except for enantiomers; $\overparen{aa}$ is a symmetrical didentate ligand and $\overparen{bc}$ is an unsymmetrical didentate ligand.)

9.22 Sketch the isomers possible for a trigonal prismatic complex $M(\overparen{aa})_3$, where $\overparen{aa}$ is a planar didentate ligand. Could any of these be optically active? Assign the point groups.

9.23 Sketch octahedral and trigonal prismatic $[M(NO_3)_3]$ complexes to show that the N atoms are in a trigonal-planar arrangement in each case. (*Hint:* Sketch the complexes looking down the threefold axes.) Assign the point groups.

9.24 Determine the number of Δ and Λ pairs of chelate rings for the isomer of $[Co(edta)]^-$ shown. The pairs of chelate rings attached to a common point and those spanning parallel edges are omitted.

9.25   Which chelate ring conformation is favored for coordinated S-(+)-propylenediamine, and why?

9.26   For *trans*-[CoCl$_2$(trien)]$^+$ (trien = NH$_2$C$_2$H$_4$NHC$_2$H$_4$NHC$_2$H$_4$NH$_2$) there are three isomers possible, depending on the chelate ring conformations. Two are optically active and one is *meso* with the central diamine ring in an eclipsed "envelope" conformation. Sketch the two optically active *trans* isomers (all rings *gauche*) and assign the absolute configurations (δ or λ) to the chelate rings. (See D. A. Buckingham, P. A. Marzilli, and A. M. Sargeson, *Inorg. Chem.* **1967,** *6,* 1032.)

9.27   What is (are) the number(s) of d electrons for metals usually encountered with the following stereochemistries?
   (a) Linear   **(c)** Trigonal prismatic
   (b) Square planar   **(d)** Dodecahedral

9.28 ^{63}Cu has $I = \frac{1}{2}$. When CuI is dissolved in P(OMe)$_3$, the ^{63}Cu NMR spectrum shows a five-line pattern with relative intensities $1:4:6:4:1$. What inference can be made about the environment of Cu in this solution? (See J. K. M. Sanders and B. K. Hunter, *Modern NMR Spectroscopy, A Guide for Chemists,* Oxford University Press, Oxford, 1987, p. 48.)

9.29   Figure 9.42 shows the low- and high-temperature limiting ^{1}H NMR spectra of (H)$_2$Fe[P(Ph)(OPh)$_2$]$_4$ in the hydride region. Account for the appearance of the spectra. (*Hint:* In this case, coupling between two H's and between H and P in the same plane in the *cis*-isomer is too small to produce noticeable splittings.)

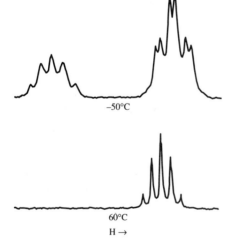

−50°C

60°C

H →

Figure 9.42 ^{1}H NMR spectra of (H)$_2$Fe[P(Ph)(OPh)$_2$]$_4$ in toluene. (Reprinted with permission from J. P. Jesson and E. L. Muetterties, in *Dynamic Nuclear Magnetic Resonance Spectroscopy,* L. M. Jackman and F. A. Cotton, Eds., Academic Press, New York, 1975, p. 289.)

▶10◀

Spectra and Bonding of Coordination Compounds

· ·

> The variety of colors of transition metal compounds, and particularly their complexes, is striking. We can interpret their electronic absorption spectra using ligand-field theory (Chapter 9) considering electron configurations and the strengths and symmetry of the ligand field. Molecular orbital theory is applied to metal complexes using a pictorial approach and then using group theory. An overall view shows that the valence bond and ligand-field results are contained in the molecular orbital treatment.

10.1 INTRODUCTION

The colors of chemical compounds have long been of interest to chemists. The bright greens, violets, yellows, and so on, led to the names praseo, violeo, luteo, and so on, which were used over a century ago for the appropriate transition metal complexes. However, our understanding of the origin of these colors has come relatively recently and is still unfolding. Although it was recognized early that compounds containing transition metal ions having partially filled d orbitals are usually colored, the role of the d electrons was not clear.

We now begin to consider the reasons why transition metal complexes absorb light in the visible region and hence are colored. That is, we will study their electronic absorption spectra. To do this, we adopt the point of view of ligand-field theory and deal first with the energy states of metal atoms or ions in complexes. Optical rotatory dispersion and circular dichroism spectra are discussed briefly as sources of information about stereochemistry. After considering spectra we will examine the broader treatment of bonding in coordination chemistry using molecular orbital theory.

10.2 LIGAND-FIELD SPECTRA
OF OCTAHEDRAL COMPLEXES[1]

▶ 10.2.1 *Energy States from Spectral Terms*

Spectral terms in atoms or ions arise from coupling of the angular momentum (l) of individual electrons in ways limited by the Pauli exclusion principle to afford states of differing total angular momentum (L) and different energy (arising from differences in electron repulsion; see page 34): The values of l associated with the various hydrogenlike orbitals arise from the solution of the Schrödinger equation for H, a problem having spherical symmetry. On placing a transition metal ion in an electric field of less-than-spherical symmetry such as O_h, l ceases to be a valid quantum number and orbitals are described by their symmetry labels (irreducible representations) for O_h, or the appropriate symmetry, instead. A physical manifestation of this is sometimes the splitting of a set of orbitals which were degenerate in the free ion.

In spherical symmetry the character of the representation for a state having an orbital angular momentum quantum number of l (or L) under an operation of the spherical group $R_h(3)$ (Douglas and Hollingsworth, pp. 87–90, see General References) is given by

$$\chi[C(\phi)] = \frac{\sin(l + \frac{1}{2})\phi}{\sin \phi/2} \quad \text{for rotation by an angle of } \phi \text{ radians} \tag{10.1}$$

$$\chi[S(\phi)] = \pm \frac{\sin(l + \frac{1}{2})(\phi + \pi)}{\sin \frac{1}{2}(\phi + \pi)} \quad \begin{array}{l}\text{for an improper rotation by an angle of } \phi \\ \text{radians (minus sign for } \textit{ungerade} \text{ states)}\end{array} \tag{10.2}$$

from which the more specific equations for symmetry operations E, σ, and i below are easily obtained:

$$\chi(E) = 2l + 1 \qquad \begin{array}{l}\text{(the character under the identity operation} \\ \text{is the degeneracy of the orbital set)}\end{array} \tag{10.3}$$

$$\chi(\sigma) = \pm \sin(l + \tfrac{1}{2})\pi \qquad \begin{array}{l}\text{(minus sign for } \textit{ungerade} \text{ states, } \sigma \\ \text{represents a symmetry plane)}\end{array} \tag{10.4}$$

$$\chi(i) = \pm(2l + 1) \qquad \text{(minus sign for } \textit{ungerade} \text{ states)} \tag{10.5}$$

To obtain the symmetry species into which a set of orbitals split in a symmetry lower than spherical, one obtains the appropriate characters by Equations (10.1)–(10.5) and reduces the representation by means of Equation (3.13). Thus, for a set of d orbitals ($l = 2$) or a D spectroscopic term ($L = 2$) in O_h symmetry, one obtains the following representation which reduces to $e_g + t_{2g}$ for d and $E_g + T_{2g}$ for D:

---

[1] This section uses the notation and terminology of group theory but does not require the detailed manipulations of Section 3.5. The material of Sections 1.4 and 1.5 is useful as background.

O_h	E	$8C_3$	$6C_2$	$6C_4$	$3C_2$	i	$6S_4$	$8S_6$	$3\sigma_h$	$6\sigma_d$	
Γ_d	5	-1	1	-1	1	5	-1	-1	1	1	$\longrightarrow e_g + t_{2g}$ or $D \rightarrow E_g + T_{2g}$

The e_g orbitals (d_{z^2} and $d_{z^2-y^2}$) and t_{2g} orbitals (d_{xy}, d_{xz}, d_{yz}) are shown in Figure 9.8. As you can verify for yourself, by using the operations on p orbitals or using the equations above, an octahedral field does not split the p orbitals. They remain triply degenerate and are characterized by their symmetry properties as t_{1u}. Likewise, the nondegenerate, totally symmetric s orbital becomes a_{1g}. See R. L. DeKock, A. L. Kromminga, and T. S. Zwier, *J. Chem. Educ.* **1979**, *56*, 510.

Ions in symmetric ligand fields have different energy states which result not only from differences in electron repulsion energy (as they do in free ions) but also from the splitting of orbital degeneracies by the ligand field. The d^1 case is quite simple because there is no electron repulsion. The electron is in the t_{2g} orbital in the ground state in an octahedral field. The d^1 ion $[Ti(H_2O)_6]^{3+}$ shows a single absorption band in the visible region of the spectrum corresponding to the transition $t_{2g}^1 \rightarrow e_g^1$ (see Figure 9.11). The energy of the band maximum is the splitting parameter $10Dq$. *The energy states of the complex follow by analogy to the behavior of orbitals in going from the free ion to the ion in the complex.* Just as the $d(l = 2)$ orbitals split into $t_{2g} + e_g$, so the ^2D term ($L = 2$) of the d^1 configuration splits into $^2T_{2g}$ and ^2E_g with the same spin multiplicity in an O_h field.

In a similar fashion, $S(L = 0)$ terms become A_{1g} and P terms become T_{1g} in an octahedral field. In a centrosymmetric point group (e.g., O_h), *all* spectral terms arising from d^n configurations give g states. Terms arising from p^n or f^n configurations give states which are u if n is odd and g if n is even ($u \times u = g$). If the point group is noncentrosymmetric (e.g., T_d), the g or u subscript is omitted. Table 10.1 lists the states arising from various spectral terms for d^n configurations in an O_h field. Note that the degeneracy (5-fold for D, 7-fold for F, etc.; the same as for the corresponding orbital sets) is conserved.

EXAMPLE 10.1:   What states are expected in an octahedral field from a ^1G free ion term? Verify the result as the same degeneracy as ^1G.

Solution: From Table 10.1 the singlet states in O_h are $^1A_{1g}$, ^1E_g, $^1T_{1g}$, and $^1T_{2g}$. The degeneracy is $1 + 2 + 3 + 3 = 9$.

Table 10.1   Splitting of spectral terms for d^n configurations in an octahedral field

Term	Degeneracy	States in an octahedral field
S	1	A_{1g}
P	3	T_{1g}
D	5	$E_g + T_{2g}$
F	7	$A_{2g} + T_{1g} + T_{2g}$
G	9	$A_{1g} + E_g + T_{1g} + T_{2g}$
H	11	$E_g + T_{1g} + T_{1g} + T_{2g}$
I	13	$A_{1g} + A_{2g} + E_g + T_{1g} + T_{2g} + T_{2g}$

▶ *10.2.2   Selection Rules*

Not every possible transition is seen in a spectrum. In studying any type of spectrum, one object is to learn what the characteristics of the ground and excited states must be, in order that transitions between them may be observed. The statements of required characteristics are called **selection rules.** If ground and excited states for a possible transition possess the required characteristics, the transition is said to be **allowed.** If they do not, the transition is **forbidden.**

The selection rules are derived from specific idealized theoretical models of the absorption process. It often happens that the ground- and excited-state wavefunctions are not the simple ones assumed in theoretical treatments. In such cases, "forbidden" transitions are often observed. However, forbidden transitions have low probability and, consequently, their observed intensities are generally low.

The model for absorption of electromagnetic radiation which is most often applicable in spectroscopy is the **electric dipole model.** In this model, radiation absorption is accompanied by a change in the electric dipole moment of the molecule, and the intensity of an electronic transition is proportional to the dipole strength, D. The dipole strength is the square of an integral involving the wavefunctions for the ground and excited states and an operator called the **dipole moment vector;** $d\tau$ is a volume element.

$$D = \left[\int \Psi_{\text{ground}} \, (\text{operator}) \Psi_{\text{excited}} \, d\tau \right]^2 \tag{10.6}$$

The operator (the dipole moment vector) transforms as a vector having components in the x, y, and z directions and thus has the form $a\vec{x} + b\vec{y} + c\vec{z}$.

LaPorte's Rule

A couple of generally applicable selection rules can be derived from Equation (10.6). **LaPorte's Rule** states that *in centrosymmetric environments, transitions can occur only between states of opposite parity* ($u \to g$ or $g \to u$). This means that $d \to p$, $s \to p$, and so on, are allowed, but not $d \to d$, $s \to d$, and so on. Transitions allowed by LaPorte's rule are quite intense, having $\varepsilon \cong 10{,}000$.

That Equation (10.6) leads to LaPorte's rule can be demonstrated readily. The dipole moment operator transforms as x, y, and z which in any centrosymmetric point group must be odd (u). The product of the operator and the wavefunctions in the integral must be g if the integral is not to equal zero. [The integral of an odd (u) function vanishes when integrated over all space because the positive and negative portions cancel one another.] Thus, even products result from $g \times g = g$ and $u \times u = g$, but an odd product results from $g \times u = u$. These dictate that because the operator is odd (or u), the integrand is g only if the product of the wavefunctions is also u. This is the basis for LaPorte's rule.

Actually, the selection rules are even more restrictive than this. The function involved in the integral must not only be even, it must be *totally symmetric*—it must belong to the totally symmetric representation (A_{1g} for the **O**$_h$ case). The product of two representations is A_{1g} (or contains A_{1g} for a reducible representation) *only* if the two representations are identical. Therefore, the *product of the representations* for the ground and excited states (this product is called the **symmetry of the transition**) must belong to the *same representation* as that of one of the vectors **x**, **y**, and **z**. This provides a basis for detailed assignments of transitions.

Transitions involving only d orbitals ($d \rightarrow d$ transitions) are LaPorte-forbidden in centrosymmetric point groups; yet, as we shall soon see, they do appear—however, with intensities ($\varepsilon \cong 5-100$) that are only about 10^{-2} of allowed transitions. The intensity-giving mechanism is the so-called **vibronic mechanism** which relaxes partially LaPorte's rule for complexes. A vibration of odd parity (e.g., T_{1u}) distorts an octahedron, for example, destroying the center of symmetry so that the g and u designations are not applicable during the time required for the transition and allows mixing of d and p orbitals. The vibrational and electronic transitions are coupled (**vibronic**). Because the energy contribution of the vibrational part of the transition is so much less than that of the electronic transition, the energies of the absorptions may be related to differences in electronic energy levels alone.

If a complex lacks a center of inversion, then the *gerade* and *ungerade* labels no longer have meaning. LaPorte's rule might not be expected to apply to tetrahedral complexes or to trigonal bipyramidal complexes that are noncentrosymmetric. Indeed, such complexes show more intense $d \rightarrow d$ absorptions ($\varepsilon \cong 100-200$) than do octahedral complexes, but they are still much weaker than LaPorte-allowed transitions.

Spin Selection Rule

A second selection rule is the **spin selection rule,** which states that *transitions may occur only between energy states with the same spin multiplicity*. That this should be so is not obvious from the general form of the integral in Equation (10.6). However, it can be shown that the integral should vanish unless the spin function of both Ψ_{ground} and $\Psi_{excited}$ are the same. Allowed transitions involve promotion of electrons without change of spin. Spin-forbidden transitions are less intense ($\varepsilon \cong 10^{-2} - 1$) than those of spin-allowed transitions. The fact that they are seen at all is because the wavefunctions are often not as simple as those corresponding to Russell–Saunders coupling which have well-defined values of S. In particular, the j–j coupling scheme (page 37), which applies especially to the heavier elements, mixes the orbital and spin-angular momentum through so-called spin–orbit coupling to yield states of the same *total* angular momentum made up partly of orbital and partly of spin contributions. Thus, the spin function is no longer separable and well-defined; that is, the experimental situation does not fully correspond to the assumption of definite values for S. This effect becomes more important for the heavier elements of the second and third transition series. Hence, **spin-forbidden** transitions become more common for complexes of these complexes. Spin–orbit coupling makes the interpretation using Russell–Saunders coupling for the free-ion states (which we assume in examples used here) less reliable and thus introduces greater uncertainty in the prediction of spectra for these elements.[2]

▶ *10.2.3*   *Splitting Diagrams*

Having discussed some of the requirements for transitions between energy states to occur, we now turn to an examination of what energy states are actually possible for transition metal ions in complexes. In Section 10.2.1 we discussed the number of states into which each free-ion term splits for an ion in a ligand field. We also need to understand how the energy separations between states depend on the identity of ligands through the value of $10Dq$.

Splittings for d^1, d^9, and High-Spin d^4 and d^6

The d^1 case, without electron repulsion, is simplest to treat because there is a one-to-one correspondence between electron configurations and states. The t_{2g}^1 configuration in $\mathbf{O}_h$ gives rise to a $^2T_{2g}$ state of energy, $-4Dq$. Promotion of the electron gives the e_g^1 configuration corresponding to the excited ^2E_g state of energy, $+6Dq$.

The right side of Figure 10.1 shows how the energy of each state changes as the field Dq increases. (This could be accomplished in principle by replacing the ligands with others lying successively higher in the spectrochemical series.) As the splitting between the states increases with the value of Dq, so does the energy of the transition between them. Thus, while the $^2T_{2g} \rightarrow {}^2E_g$ transition[3] occurs at 20,100 cm^{-1} for $[\mathrm{Ti}(\mathrm{H_2O})_6]^{3+}$, it lies at 22,300 cm^{-1} for $[\mathrm{Ti}(\mathrm{CN})_6]^{3-}$.

The absorption maximum for $[\mathrm{Ti}(\mathrm{H_2O})_6]^{3+}$ in Figure 9.11 has a shoulder at about 17,400 cm^{-1} because of Jahn–Teller distortion (Section 9.8.2). This distortion splits the ^2E_g excited state to give $^2A_{1g}$ and $^2B_{1g}$. The distortion causes small splitting of the ground state ($^2T_{2g}$).

---

[2] The **magnetic dipole model** provides another mechanism for absorption of light by molecules in which the magnetic dipole oscillates when light is absorbed. Intensities for transitions in this model are orders of magnitude lower than those of electric-dipole-allowed transitions and have opposite selection rules—*they occur only with retention of parity* (i.e., $d \rightarrow d$ and $f \rightarrow f$ are allowed, but $p \rightarrow d$ is forbidden). Elements of the *f*-block absorb light according to this model, which we will not discuss further. Circular dichroism and optical rotatory dispersion involve the product of the integrals for electric dipole and magnetic dipole changes.

[3] Electronic spectroscopy began with atomic spectroscopy concerned with emissions from excited states to the ground state ex $\rightarrow$ gr. Early absorption spectroscopy used ex $\leftarrow$ gr notation, but here we use gr $\rightarrow$ ex, as for $^2T_{2g} \rightarrow {}^2E_g$.

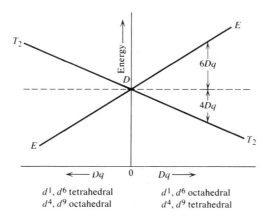

Figure 10.1   Orgel D term splitting diagram.

Because the splitting of the d orbitals is reversed for a tetrahedral field, so is the splitting diagram as shown on the left of Figure 10.1 with ^2E as the ground state.

We have seen that d^9 gives only one spectral term, ^2D—the electron can be missing from any one of the 10 possibilities. The transition in an octahedral field is $t_{2g}^6 e_g^3 (^2E_g) \rightarrow t_{2g}^5 e_g^4 (^2T_{2g})$—the "hole" has moved from e_g to the t_{2g} orbital set. A positive hole would be affected by the ligand field in just the way as would a negative electron. This corresponds to a reversal of the sign of Dq and a ^2E_g ground state with a $^2T_{2g}$ excited state. The splitting diagram for octahedral d^9 is the reverse of d^1 (the left side of Figure 10.1). As before, the d^9 case for a tetrahedral complex is the reverse of that for an octahedron, or it corresponds to the d^1 octahedral case. In all these cases the transition energy is $10Dq$. The subscripts g and u are omitted from Figure 10.1 so that it applies to both $\mathbf{O}_h$ *and* $\mathbf{T}_d$ complexes.

The high-spin d^6 case gives only one quintet term ^5D, and this behaves like the d^1 case. We can consider that the half-filled configuration is unaltered and only the one electron of unique spin can move. Similarly, the high-spin d^4 case (also ^5D) corresponds to d^9—there is one "hole" in the half-filled d^5 configuration.

$$d^6 \quad \frac{\uparrow}{\uparrow\downarrow}\frac{\uparrow}{\uparrow}\frac{\ }{\uparrow}\, \begin{matrix} e_g \\ t_{2g} \end{matrix} \qquad d^4 \quad \frac{\uparrow}{\uparrow}\frac{\ }{\uparrow}\frac{\ }{\uparrow}\, \begin{matrix} e_g \\ t_{2g} \end{matrix}$$

Thus, a single diagram (called an **Orgel diagram**) in Figure 10.1 shows the change in energy with Dq for four different configurations, all corresponding to one electron or one hole. It is qualitative because even for the same ligand $Dq_{\text{oct}} > Dq_{\text{tet}}$; besides, Dq depends on the particular metal ion. No spin multiplicities are shown for the states in Figure 10.1, so that the diagram applies to the four configurations discussed.

Configurations d^4 and d^6 give rise to many spectral terms other than D, but they differ in spin multiplicity. If we concentrate only on transitions which obey the spin-selection rule (the most intense ones), then also for high-spin d^4 and d^6 a single $d \rightarrow d$ band of appreciable intensity is expected. The spectrum of the d^4 complex $[CrCl_6]^{4-}$ (present in $CrCl_2$ in molten $AlCl_3$ at 227°C and 5.6 atm) in Figure 10.2a shows the expected single absorption band with a maximum at 10,200 cm^{-1} with no apparent splitting. The spectrum of $[Fe(H_2O)_6]^{2+}$ (high-spin d^6) in Figure 10.2b shows a maximum at 10,400 cm^{-1} with a pronounced shoulder at 8300 cm^{-1}. The splitting results from Jahn–Teller distortion as for $[Ti(H_2O)_6]^{3+}$.

The spectra of Figures 9.11 and 10.2 show that light absorption occurs over a range of wavelengths. That is, complex ions do not give line spectra as are seen for gaseous atoms; instead we see *broad absorption bands*. This is so, even though the choice of a particular ligand would be expected to dictate a particular value of Dq and an exact energy of the transition corresponding to the distance between the state energies in the Orgel diagram. A major reason for the broad bands is that the value of Dq is very sensitive to the metal–ligand distance. Vibrations of the M—L bond change this distance and, in effect, sweep out a range of Dq values. These broad bands cover weak spin-forbidden transitions unless their peaks are isolated from other transitions.

One absorption band with the energy $10Dq$ is expected for d^9 with possible splitting of the band because of Jahn–Teller distortion. The only common d^9 ion is Cu^{2+}. The CuL_6^{2+} complexes such as $[Cu(NH_3)_6]^{2+}$ have large distortion with two long M—L bonds along z. This *tetragonal* distortion, $\mathbf{D}_{4h}$, is great enough to require the use of the lower ($\mathbf{D}_{4h}$) symmetry rather than $\mathbf{O}_h$ for interpretation of the spectra. In addition, spin–orbit coupling is significant for copper, and this causes additional splitting of energies.

Splittings for d^2, d^3, d^8, and High-Spin d^7

The d^2 case is the simplest one where we deal with both (a) the splitting of orbital degeneracy and (b) the effect of electron repulsion.[4] The d^2 configuration has two terms of maximum spin-multiplicity for the free ion: ^3F (the ground state) and ^3P. These differ in energy by $15B$.[5] As Table 10.1 shows, ^3P gives $^3T_{1g}$ and the ground state ^3F term splits in an octahedral field into $^3T_{1g} + ^3T_{2g} + ^3A_{2g}$. The relative energies of these states can be calculated by techniques beyond the scope of this book and are given in Figure 10.3a. Note that the zero of energy is the ground-state energy of the ^3F term of the free ion obtained from atomic spectroscopy. The ^3P term (as for p orbitals) does not split and becomes $^3T_{1g}$ in $\mathbf{O}_h$. The diagram shows how the state energies change with Dq.

The two $^3T_{1g}$ states in Figure 10.3 are distinguished by their "parentage" in parentheses: $^3T_{1g}(F)$ and $^3T_{1g}(P)$. Because they have the same symmetry properties in an octahedral field, these two states can interact with each other. As Dq_{oct} increases, the wavefunc-

---

[4] To be sure, electron repulsion is present for the 4-, 6-, and 9-d-electron cases. However, the high-spin cases all correspond to the motion of just one electron among the manifold of split d orbitals while the rest of the electrons remain in a fixed configuration. Thus, as long as we confine ourselves to cases where no spin changes occur (the spin-allowed transitions), the electron repulsion does not change in any of the transitions and thus does not contribute to their energies. In contrast, two d electrons can adopt a variety of orbital arrangements, all corresponding to triplet spin states.

[5] The energy separations of spectroscopic terms are expressed in terms of Racah parameters B and C. For terms of maximum spin multiplicity, this separation is only a function of B. B and C are different for each atom or ion and are evaluated from atomic spectra. The value of B, labeled as B', is lower in complexes because of electron delocalization.

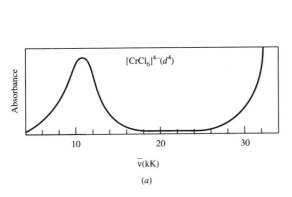

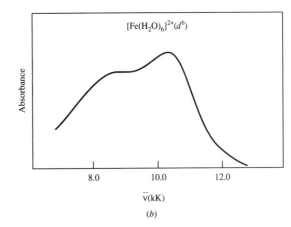

Figure 10.2 (*a*) Absorption spectrum of $[CrCl_6]^{4-}$ using $CrCl_2$ in $AlCl_3$ at 227°C and 5.6 atm. (Adapted with permission from H. A. Øye and D. M. Gruen, *Inorg. Chem.* **1964**, *3*, 836. Copyright 1964 American Chemical Society.) (*b*) Absorption spectrum of $[Fe(H_2O)_6]^{2+}$.

tion of the $^3T_{1g}(F)$ ground state mixes in increasing amounts of the $^3T_{1g}(P)$ wavefunction, and vice versa. (We say that mixing of the states, or configuration interaction, occurs.) The result is that the energies of these two states deviate from what their values would have been in the absence of interaction, bending away from each other instead of displaying the linear behavior (dashed lines) otherwise expected as shown in Figure 10.3*b*. This is referred to as the **noncrossing rule.**[6]

In an extremely weak O_h field (so weak that no curvature of the $^3T_{1g}$ states yet occurs), a d^2 complex is expected to display three transitions from the $^3T_{1g}(F)$ ground state to: $^3T_{2g}(\nu_1 = 6Dq + 2Dq = 8Dq)$, $^3A_{2g}(\nu_2 = 18Dq)$, and $^3T_{1g}(P)(\nu_3 = 6Dq + 15B)$, as shown in Figure 10.3*a*. For a stronger field (after curvature has set in) the energies of all the transitions will be increased by the amount that the ground-state energy has been lowered due to curvature. Band ν_3 will increase by the additional amount corresponding to the raising of the $^3T_{1g}(P)$ energy. At still higher values of Dq, the $^3T_{1g}(F) \rightarrow {}^3T_{1g}(P)$ transition is at lower energy than $^3T_{1g}(F) \rightarrow {}^3A_{2g}$.

For d^8, the two-hole case, holes are affected in the opposite way when compared to two electrons (Dq is reversed in sign), and we can extrapolate the straight lines to the left in Figure 10.3*b*. Here the ground state is $^3A_{2g}$. The noncrossing rule still applies, and the lines for the $^3T_1$ states bend away from each other. The resulting curvature is greater than for d^2 because the energies of the unperturbed states are closer and must not cross. Now because of bending, the energy for $^3A_2 \rightarrow {}^3T_1(F)$ is *less* than $18Dq$ and that of $^3A_2 \rightarrow {}^3T_1(P)$ is *greater* than $12Dq + 15B$. The hole formalism relates the configurations d^8 to d^2

[6] We have already seen in Chapter 4 that only orbitals with the same symmetry properties can interact to form molecular orbitals (MOs). This is also true of the many-electron wavefunctions (states) which are required by quantum mechanics to interact if their symmetry behavior is the same. Just as mixing of orbitals to form MOs is greater when the orbitals are of similar energies, so is the mixing of states more extensive as the unperturbed states approach more closely in energy; thus, the curvature of the plot increases for both the lower and upper states. See R. W. Jotham, *J. Chem. Educ.* **1975**, *52*, 377.

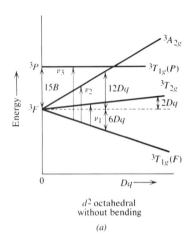

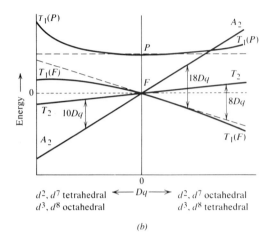

Figure 10.3   Orgel term splitting diagram for d^2, d^3, d^7, and d^8 cases in octahedral and tetrahedral fields. (Adapted with permission from L. E. Orgel, *J. Chem. Phys.* **1955**, *23*, 1004.)

and d^3 to high-spin d^7 as shown in the diagram. Figure 10.4 shows the effect of splitting and curving from state mixing for transitions for d^3 and d^8 in an octahedral field. The energies of the transitions for d^3 and d^8 ($\mathbf{O}_h$) with c for the bending are

$$\nu_1 = 10Dq$$
$$\nu_2 = 18Dq - c \tag{10.7}$$
$$\nu_3 = 15B' + 12Dq + c$$

EXAMPLE 10.2: $[Cr(H_2O)_6]^{3+}$ shows three absorption bands at 17.4, 24.6, and 37.9 kK. Assign the three bands and calculate Dq, c, and B'.

Solution:

$$\nu_1 = 10Dq = 17.4 \text{ kK} \quad \text{or} \quad Dq = 1.74 \text{ kK}$$
$$\nu_2 = 18Dq - c = 18(1.74) - c = 24.6 \quad \text{and} \quad c = 6.7 \text{ kK}$$
$$\nu_3 = 15B' + 12Dq + c = 37.9$$
$$15B' = 37.9 - 20.9 - 6.7 = 10.3, \qquad B' = 0.69 \text{ kK}$$

B for Cr^{III} is 1.06 kK for the free ion.

Summary for High-Spin d^n Ions

Only two diagrams (Figures 10.1 and 10.3) are needed to describe the splitting of terms of maximum spin multiplicity for all configurations, except d^5, because they all have D or F ground-state terms. The d^5 configuration is unique. It is its own hole counterpart and there are no spin-allowed transitions; there is only one sextet, the ground-state $^6A_{1g}$. Octahedral high-spin d^5 ions (e.g., $[Mn(H_2O)_6]^{2+}$) are faintly colored because all $d \rightarrow d$ transitions are spin-forbidden. However, because there are several ways to arrange four electrons of one spin set and one of the other spin set, there are four free-ion quartet terms and many quartet states arise in an octahedral field, leading to many peaks of very low intensity. This spectrum will be discussed on page 455.

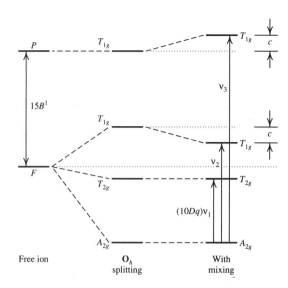

Figure 10.4 Splitting and mixing of states for d^3 and d^8 ions in an octahedral field, for a single value of Dq.

A single absorption band is expected for octahedral complexes of d^1 and d^9, as well as high-spin d^4 and d^6 ions. The band might be unsymmetrical or split because of Jahn–Teller splitting of the e_g level whether these orbitals are occupied unequally in the ground state (d^4 and d^9) or in the excited state (d^1 and d^6). The splitting is greater for e_g than for t_{2g} occupancy (see Section 9.8.2). Octahedral complexes of ions with the other high-spin configurations (except d^5) are expected to show three ligand-field absorption bands.[7] The energies of two of the transitions [to $T_{1g}(P)$ and A_{2g}] are nearly the same in fields of intermediate strength (near where these lines in Figure 10.3 cross), and they might appear as a single broad band.

Orgel diagrams apply only to high-spin complexes. They describe what happens to energies of states starting from the free ions (where only electron repulsion energies are significant) as orbital energies split because the ligand field increases from zero to a magnitude comparable to that of the electron-repulsion energy. In the range of applicability of Orgel diagrams, both these energy contributions determine the placement of electrons. If Dq were increased sufficiently, the orbital-splitting energies would ultimately become much greater than the electron repulsion energies and would be the major factor determining electron placement. As we saw in Section 9.8.1, this leads to low-spin complexes with change in spin multiplicity of the ground state. In Section 10.2.5 we will examine how to deal with this case where $10Dq \gg$ electron repulsion energy.

EXAMPLE 10.3: What are the lowest energy transitions for high-spin octahedral ML_6^{n+} complexes for V^{3+}, Cr^{3+}, Mn^{2+}, Ni^{2+}, and Cu^{2+}?

Solution: Using Figures 10.1 and 10.3, we obtain

$$V^{3+}\ d^2:\quad {}^3T_{1g} \longrightarrow {}^3T_{2g} \qquad Ni^{2+}\ d^8:\quad {}^3A_{2g} \longrightarrow {}^3T_{2g}$$
$$Cr^{3+}\ d^3:\quad {}^4A_{2g} \longrightarrow {}^4T_{2g} \qquad Cu^{2+}\ d^9:\quad {}^2E_g \longrightarrow {}^2T_{2g}$$
$$Mn^{3+}\ d^4:\quad {}^5E_g \longrightarrow {}^5T_{2g}$$

[7] Ligand-field bands are those involving $d \rightarrow d$ transitions.

▶ *10.2.4 The Spectrum of $[CoF_6]^{4-}$*

An example of the splitting of states described above is seen in the absorption spectrum of crystals of $KCoF_3$ in which $Co^{2+}(d^7)$ is surrounded octahedrally by $6F^-$. Absorption bands centered at 7.15, 15.2, and 19.2 kK (1kK = 1000 cm^{-1}) are seen in Figure 10.5. For a weak-field ligand such as F^- we expect from Figure 10.3 that these bands should correspond to $^4T_{1g} \rightarrow {}^4T_{2g}$, $^4T_{1g} \rightarrow {}^4A_{2g}$, and $^4T_{1g} \rightarrow {}^4T_{1g}(P)$, respectively. We could calculate Dq from ν_1, which is $8Dq$, but this does not allow for configuration interaction (bending) between the T_{1g} states. The $(\nu_2 - \nu_1)$ difference is $10Dq$ $(18Dq - 8Dq)$ giving $10Dq = 15.2 - 7.2 = 8.0$ kK. From the expressions

$$\nu_1 = 8Dq + c \tag{10.8}$$

$$\nu_2 = 18Dq + c \tag{10.9}$$

(where c is the bending from configuration interaction), we get $c = 0.8$ kK in each case. We see from Equations (10.8) and (10.9) that c cancels for $\nu_2 - \nu_1$.

The Racah parameter B for the complex can be evaluated from ν_3; the value of B in complexes (B') is always less than the value for the free ion.

$$\nu_3 = 15B' + 6Dq + 2c \tag{10.10}$$

This gives $B' = 0.85$ kK, compared with the free-ion value for Co^{2+} of 0.97 kK. Typically, $B' = 0.7B$ to $0.9B$ because of the decreased electron repulsion caused by greater delocalization of the d electrons in the complex. Jørgensen calls this decreased repulsion the **nephelauxetic** ("cloud expanding") **effect.** It arises because the actual bonding in the complexes departs from the crystal field model in which the only function of ligands is to provide an electric field of particular symmetry. Actually, ligand electrons interact with the central metal atom in two ways: First, the electrons of the ligand partly invade the space of the metal electrons, thus partially screening the metal nuclear charge from its

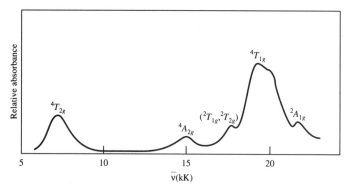

Figure 10.5 The absorption spectrum of $[CoF_6]^{4-}$ in a crystal of $KCoF_3$ at 150 K. (Adapted from J. Ferguson, D. L. Wood, and K. Knox, *J. Chem. Phys.* **1963,** *39,* 881.)

own electrons. Because the electrons are then less tightly held, the electron cloud expands and interelectronic repulsion is reduced. Second, MO formation between metal and ligand orbitals occurs. (This is treated in detail in Section 10.6.) For now, we note that the contribution of ligand orbitals to occupied MOs can lead to the delocalization of metal electrons onto the ligand, further reducing electron repulsion. We obtain values of B' empirically from spectra, and for a given metal the ratio B'/B decreases roughly as follows[8]:

$$F^- > H_2O > NH_3 > en > NCS^- > Cl^- \sim CN^- > Br^- > S^{2-} > I^-$$

This ranking is referred to as the **nephelauxetic series.** The value of B' for $[CoF_6]^{4-}$ is $0.85B$, indicating the small amount of interaction between F^- and Co^{2+} electrons. If we do not know the value of B', we can use $\sim 0.8B$ as a reasonable approximation.

The example of $[CoF_6]^{4-}$ is for a crystal of $KCoF_3$, and the spectrum of $[CrCl_6]^{4-}$ (Figure 10.2) is for $CrCl_2$ in molten $AlCl_3$. These examples are used because well-defined species are present. In solution, many complexes of Co^{III} and Cr^{III} are stable with respect to substitution by other ligands—the complexes are called **inert** (substitution-inert). We can get stable geometrical and optical isomers of these complexes. For complexes of many other metal ions, substitution of one ligand by another occurs readily—these complexes are called **labile.** In these cases there can be equilibria among several species in solution, depending upon concentrations. Species such as $[CoF_6]^{4-}$ and $[CrCl_6]^{4-}$ cannot be identified in aqueous solution.

▶ 10.2.5   Tanabe–Sugano Diagrams

The simple Orgel diagrams are useful for spin-allowed transitions when the number of peaks observed is not fewer than the number of empirical parameters (Dq, B', and bending). Other cases require the use of Tanabe–Sugano (T–S) diagrams, which differ from Orgel diagrams in having the ground state as the constant reference (it becomes the horizontal axis). Configuration interaction is included as bending of lines for the excited states. The energies of states with spin multiplicity different from that of the ground state are also represented. These states require Racah parameters B and C in the analytical expressions for their energies. Thus, state energies in T–S diagrams are functions of Dq, B', and C' and would require a four-dimensional plot. Because this is obviously not feasible, some simplification must be made. A common procedure is to assume that $C' = 4.5B'$ and to express both the state energies and Dq as multiples of B'. Hence, the ordinate is in units of E/B' and the abscissa is in units of Dq/B'. This reduces the dimensionality to two. Because it may not be true in every case that $C' = 4.5B'$, T–S diagrams have only qualitative significance; but they are valuable aids in interpretating spectra and, in many cases, in finding approximate values of Dq. For high-spin complexes the T–S diagrams are similar to Orgel diagrams except for including states of lower spin multiplicity.

---

[8] C. K. Jørgensen, *Absorption Spectra and Chemical Bonding,* Pergamon Press, New York, 1962, pp. 134ff.

EXAMPLE 10.4: For $[Cr(H_2O)_6]^{3+}$ there is a very weak absorption peak at 15.0 kK in addition to those in Example 10.2. What is this transition and why is it weak?

Solution: The weak peak is for a spin-forbidden transition. Because it occurs at lower energy than the spin-allowed transitions, from the T–S diagram (Appendix D) it must be the unresolved $^4A_{2g} \rightarrow (^2T_{2g}, 2E_g)$ transition.

Spectra of d^7 Ions

Consider the T–S diagram for d^7 (Figure 10.6a). At low field (Dq/B', left side of the figure), the $^4T_1$ ground state and other quartets are represented by solid lines. Because F^- is low in the spectrochemical series, we expect that the complex $[CoF_6]^{4-}$ should have a rather small value of Dq and Dq/B'. The absorption bands for $[CoF_6]^{4-}$ are identified in Figure 10.5. The field strength obtained in Section 10.2.4 was $Dq/B' = 0.8/0.85 =$

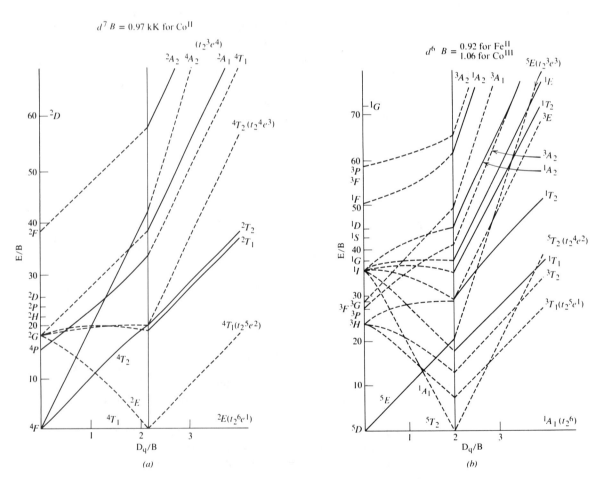

Figure 10.6 Semiquantitative energy-level diagrams for (a) d^7 and (b) d^6 in octahedral symmetry. (After T. Tanabe and S. Sugano, *J. Phys. Soc. Japan* **1954**, *9*, 753.)

0.94. Using this value for the left side (low field) of the T–S diagram for d^7 (Figure 10.6a), we see that E/B' for $^4T_2$ is about 9, or $E \cong 7.6$ (observed at 7.1 kK). The spin-forbidden transition to ^2E_g, expected from the T–S diagram at about $E/B' = 14$, does not appear in the spectrum. Energies of the other spin-allowed bands agree reasonably with the spectrum: $^4A_{2g}$ at $E/B' \cong 18$, $E \cong 15$ and $^4T_{1g}$ at $E/B' \cong 23$, $E \cong 19.5$. The weak peaks are assigned to spin-forbidden transitions as $[^4T_{1g} \rightarrow (^2T_{1g}, {}^2T_{2g})]$ unresolved at about 17.4 kK and $^2A_{1g}$ at about 21.5 kK. In the spectrum the intensity[9] of the $^4A_{2g}$ band is very low compared to the other spin-allowed bands. Transitions from $^4T_{1g} \, (t^5_{2g} e^2_g)$ to $^4T_{2g}$ and $^4T_{1g} \, (t^4_{2g} e^3_g)$ are one-electron transitions, but $^4T_{1g} \rightarrow {}^4A_{2g} \, (t^3_{2g} e^4_g)$ is a two-electron transition. Simultaneous promotion of two electrons is less probable, and often this transition has low intensity.

In the d^7 T–S diagram the ^2E state derived from ^2G decreases in energy as Dq increases. Discontinuities occur at $Dq/B' = 2.2$, which represents the ligand-field strength required to overcome the pairing energy and enter the low-spin regime. Above this field strength (right side of Figure 10.6a), the ground state becomes ^2E_g, and the diagram is tipped to make the energy of this state the reference (a horizontal line). The energy of $^4T_1$ increases with increasing field as an excited state. Spin-allowed transitions, involving states represented by solid lines, are now $^2E_g \rightarrow (^2T_{1g}, {}^2T_{2g})$ (nearly the same energy) and $^2E_g \rightarrow {}^2A_{1g}$ at much higher energy, as well as others too high in energy to be observed ordinarily. At very high Dq values, electron-repulsion terms become negligible compared to orbital splittings and the states are specified by giving the orbital occupancy of t_{2g} and e_g orbitals as seen (in parentheses) in the high-field limit for Figure 10.6a. These are referred to as **strong-field configurations.**

Spectra of d^5 Ions

Compare the T–S diagram[10] for d^5 with the assignments for the $[Mn(H_2O)_6]^{2+}$ spectrum in Figure 10.7. Note that these spin-forbidden bands have very low intensities and that all involve $^6A_{1g}$ as ground state and quartet excited states. Any transition for the high-spin d^5 case must change the number of unpaired electrons, but a two-electron transition (to pair four electrons giving a doublet excited state) is most improbable. The $^4A_{1g}$ and ^4E_g peaks are unresolved in the spectrum, and their energies are the same in the T–S diagram. Note the sharp peaks for the transitions to the $^4A_{1g}$, $^4E_g(^4G)$, and $^4E_g(^4D)$ states. The sharpness stems from their energies' independence of ligand-field strength (horizontal lines in the T–S diagram with freedom from vibrational broadening) which comes about because they involve pairing electrons *within* the t_{2g} ($^4A_{1g}$, ^4E_g) or e_{2g} (states derived from ^4D) rather than promotion of electrons from t_{2g} to e_g. These field-independent bands are useful in making assignments not only because they stand out, but also because their energies do not change with a change in ligands. If a sharp peak occurs, try to identify it with a horizontal line and, if possible, check the energy of the peak for another complex of the same metal ion.

---

[9] The intensity of a band is given by the area under the band, subtracting the background, not just by the maximum (height) of the band.

[10] Doublets at low field are omitted for simplification except for $^2T_2$ that becomes the ground state for high-field (low-spin) cases.

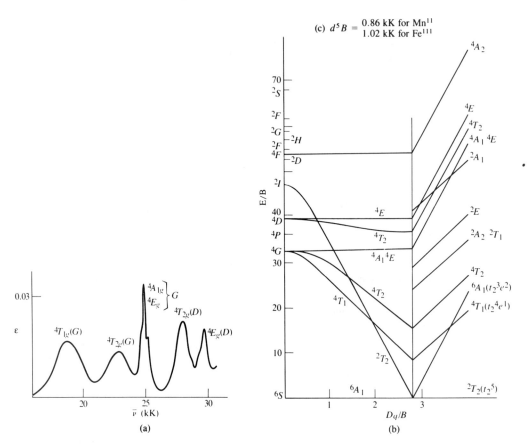

Figure 10.7 (*a*) Absorption spectrum of $[Mn(H_2O)_6]^{2+}$. (From C. K. Jørgensen, *Acta Chem. Scand.* **1954**, *8*, 1502.) (*b*) Tanabe–Sugano diagram for d^5.

Low-Spin d^6 Complexes

Many well-characterized complexes of d^6 ions are known, and many of these have stable geometrical or optical isomers. There are many more low-spin complexes for d^6 than for d^5 because of the higher ligand-field stabilization energy (LFSE) and lower pairing energy for d^6 (Table 9.12). The low-spin case is achieved at lower field strength ($Dq/B' = 2.0$) than with d^5 (pairing occurs only for $Dq/B' \geq 2.8$). For Co^{III}, only $[CoF_6]^{3-}$ is a well-characterized high-spin complex with $^5T_{2g} \rightarrow {}^5E_g$ as the only spin-allowed transition. The ground state for low-spin d^6 is $^1A_{1g}$ (the t_{2g} orbitals are filled). The spin-allowed transitions (see the right side of the T–S diagram[11] for d^6, Figure 10.6*b*) within the usual visible

---

[11] A. L. Hormann and C. F. Shaw III (*J. Chem. Educ.* **1987**, *64*, 918) called attention to the common omission of one $^1T_2$ state in the T–S diagram for d^6. The states from the ^1I spectral term in $\mathbf{O}_h$ are identified in Table 10.1. The original T–S diagram for d^6 showed only the lower-energy $^1T_2$ state. The diagram for d^6 in Figure 10.6 includes both $^1T_2$ states. Omission of the higher-energy $^1T_2$ state does not create a problem. The transition to this state would normally be too high in energy to be observed.

and near-ultraviolet range are $^1A_{1g} \rightarrow {}^1T_{1g}$ and $^1A_{1g} \rightarrow {}^1T_{2g}$, so two absorption bands are observed for $[Co(NH_3)_6]^{3+}$, a spectrum very similar to that of $[Co(en)_3]^{3+}$ shown in Figure 10.8. The actual symmetry of $[Co(en)_3]^{3+}$ is $\mathbf{D_3}$, but the absorption spectrum shows no $\mathbf{D_3}$ splitting, so its *effective symmetry* is $\mathbf{O}_h$. The chelate rings do not show an effect here, so the *chromophore* is CoN_6. Metal complexes generally absorb strongly in the ultraviolet region (below ~300 nm) because LaPorte-allowed transitions or charge-transfer transitions occur in this region (see Section 10.4). T–S diagrams for $d^n(n = 2$–$8)$ configurations are given in Appendix D.

EXAMPLE 10.5:   What are the lowest-energy spin-allowed transitions for low-spin octahedral $[ML_6]^{n+}$ complexes for Mn^{2+} and Mn^{3+}?

Solution: Using T–S diagrams (Appendix D), we obtain

$$Mn^{2+} \ d^5: \quad {}^2T_{2g} \quad \longrightarrow \quad {}^2A_{2g}, {}^2T_{1g} \text{ (very close in energy)}$$

$$Mn^{3+} \ d^4: \quad {}^3T_{1g} \quad \longrightarrow \quad {}^3E_g \text{ (three others at slightly higher energies)}$$

A **correlation diagram** uses experimentally determined energies of spectral terms and the energies of strong-field configurations (from orbital energies) to obain the change in energies of states with increasing field strength (Dq). The result is an Orgel diagram for a particular d^n configuration and symmetry. All spin multiplicities can be included, but we are usually most interested in spin-allowed transitions. We can draw a d^2 correlation diagram by plotting the energies of the free-ion spectral terms (^3F, ^1D, ^3P, etc.) obtained from atomic spectroscopy as given in the d^2 T–S diagram (Appendix D), on the vertical axis at the left of Figure 10.9a (zero applied field). On the right vertical axis we plot the relative energies of the infinitely strong-field configurations in $\mathbf{O}_h$ symmetry (all possible arrangements of two electrons in t_{2g} and e_g). Because we know that t_{2g} is lower in energy than e_g, the order of energies for the configurations in $\mathbf{O}_h$ is $t_{2g}^2 < t_{2g}^1 e_g^1 < e_g^2$. In $\mathbf{O}_h$ symmetry the orbital energies are $-4Dq$ for $t_{2g}(d_{xy}, d_{xz}, d_{yz})$ and $6Dq$ for $e_g(d_{z^2}$ and $d_{x^2-y^2})$. The energy of the t_{2g}^2 configuration is 2 (for two electrons) $\times$ $(-4Dq) = -8Dq$. For $t_{2g}^1 e_g^1$ it is the sum of the orbital energies $-4Dq + 6Dq = 2Dq$, and for e_g^2 it is $2 \times (6 \ Dq) = 12 \ Dq$. These are plotted on the right vertical axis (Figure 10.9d). Relaxing the infinitely strong field produces strong-field states from each configuration. The states derived from these strong-field configurations are the direct products (see Appendix 3.2) of the orbitals occupied: $T_{2g} \times T_{2g}$ (for t_{2g}^2) gives $^3T_{1g}$, $^1T_{2g}$, ^1E_g, $^1A_{1g}$; $T_{2g} \times E_g$ (for $t_{2g}^1 e_g^1$) gives $^3T_{2g}$, $^3T_{1g}$, $^1T_{2g}$, $^1T_{1g}$; and $E_g \times E_g$ (for e_g^2) gives $^3A_{2g}$, ^1E_g, and $^1A_{1g}$ [all states are shown in Figure 10.9c; the triplets (with higher spin multiplicity) are expected to have energy lower than that of

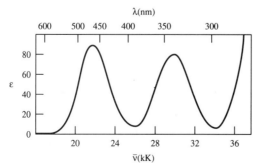

Figure 10.8   Absorption spectrum of $[Co(en)_3]^{3+}$.

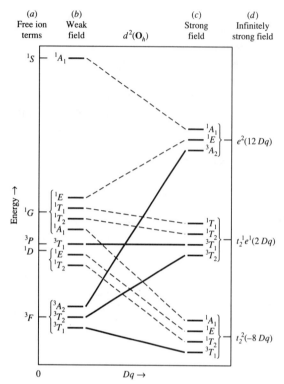

Figure 10.9   Correlation diagram for a d^2 ion in O_h symmetry (the g subscripts are omitted).

the singlets].[12] Next we identify the weak-field O_h states (Figure 10.9*b*) derived from the free-ion spectral terms from Table 10.1. These are the states produced from the free-ion spectral terms by a weak octahedral field. The weak-field states on the left are connected by lines with the corresponding states from the strong-field configurations on the right, observing the noncrossing rule: *Tie lines for states of the same designation (including multiplicity) cannot cross.* The correlation diagram includes triplets and singlets. Focusing on triplet states (solid lines), we see that $^3T_{1g}$ from $t_{2g}^1 e_g^1$ correlates with $^3T_{1g}$ from ^3P) and that the energies of the states from ^3F increase in the order $^3T_{1g} < ^3T_{2g} < ^3A_{2g}$. The correlation diagram for the triplets is an energy-level splitting diagram corresponding to the Orgel diagram in Figure 10.3*a* for d^2. Strong-field configuration interactions giving bending is included on the right side of Figure 10.3*b* and in the T–S diagram (Appendix D).

	Configuration degeneracy	Total degeneracy of terms
t_{2g}^2	$15 = (6 \times 5)/2$	$3 \times 3\ (^3T_{1g}) + 1 \times 3\ (^1T_{2g}) + 1 \times 2\ (^1E_g) + 1\ (^1A_{1g}) = 15$
$t_{2g}^1 e_g^1$	$24 = (6 \times 4)$	$3 \times 3\ (^3T_{2g}) + 3 \times 3\ (^3T_{1g}) + 3\ (^1T_{2g}) + 3\ (^3T_{1g}) = 24$
e_g^2	$6 = (4 \times 3)/2$	$3\ (^3A_{2g}) + 2\ (^1E_g) + 1\ (^1A_{1g}) = 6$

[12] Methods for assigning triplet and singlet designations are beyond the scope of this book. See General References on spectroscopy at the end of this chapter. Also see D. I. Ford, *J. Chem. Educ.* **1972,** *49,* 336; and D. H. McDaniel, *J. Chem. Educ.* **1977,** *54,* 147. Here we can check the results based on total degeneracies as above.

10.3 COMPLEXES OF LOWER SYMMETRY

► *10.3.1 Energies of d Orbitals in Fields of Various Symmetries*[13]

Krishnamurthy and Schaap's useful approach to determine the relative energies of the d orbitals in fields of various symmetries is based on the additivity of effects of combined groupings of ligands. They calculated crystal-field energies for three primary arrangements of ligands. Geometries usually encountered can be generated from combinations of these three arrangements, ML, ML_2, and ML_4—as shown in Figure 10.10. Table 10.2 gives the relative energies of the d orbitals for these cases. The field of a linear ML_2 complex, with 2L along the z axis, is twice that of ML, so the energies of each orbital are doubled. The square-planar field in the xy plane is twice that of the ML_2 primary ligand group. The field for octahedral geometry of ML_6 in terms of the primary ligand groups is thus $V_{O_h} = 2V_{ML} + 2V_{ML_2}$. Adding the energies of the d orbitals for a square-planar arrangement of four ligands (xy plane, $2V_{ML_2}$) and for a linear arrangement (along the z axis, $2V_{ML}$) of two ligands thus gives the relative d orbital energies for an octahedral field:

	d_{z^2}	$d_{x^2-y^2}$	d_{xy}	d_{xz}	d_{yz}
Square planar (xy) $\mathbf{D}_{4h}$:	$-4.28Dq$	$12.28Dq$	$2.28Dq$	$-5.14Dq$	$-5.14Dq$
Linear (z) $\mathbf{D}_{\infty h}$:	10.28	-6.28	-6.28	1.14	1.14
Octahedral $\mathbf{O}_h$:	$6.00Dq$	$6.00Dq$	$-4.00Dq$	$-4.00Dq$	$-4.00Dq$

For a square pyramid the energies are given by $V_{C_{4v}} = V_{ML} + 2V_{ML_2}$ (1L along z and 4L in the xy plane) or $V_{C_{4v}} = V_{O_h} - V_{ML}$ (remove one ligand along z from an octahedron).

The *total* potential energy for six unit charges in the xy plane is the same whether there are -3 units of charge along x and -3 units along y, or $-1\frac{1}{2}$ units each at $\pm x$ and $\pm y$, or unit charges arranged in a planar hexagon. The energy for charges on a sphere is determined only by the number of units of charge and the radius of the sphere, not by the arrangement of the charges. For three unit charges in a planar-trigonal (triangular) arrangement ML_3, the effects would be half as great as for six units of charge in a plane. Multiplying the energies of the d orbitals for the ML_2 case by $\frac{3}{2}$ gives us the energies for three units of charge—either $-1\frac{1}{2}$ at $+x$ and $-1\frac{1}{2}$ at $+y$, or $-\frac{3}{4}$ each at $\pm x$ and $\pm y$:

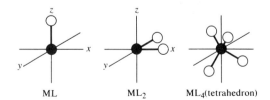

ML ML_2 ML_4(tetrahedron)

Figure 10.10 The three primary ligand groups. (From R. Krishnamurthy and W. B. Schaap, *J. Chem. Educ.* **1969**, *46*, 799.)

[13] R. Krishnamurthy and W. B. Schaap, *J. Chem. Educ.* **1969**, *46*, 799.

Table 10.2   **Relative *d* orbital energies for three primary geometric configurations**

Configuration	Relative energies in units of Dq				
	d_{z^2}	$d_{x^2-y^2}$	d_{xy}	d_{xz}	d_{yz}
M-L (along z axis)	5.14	−3.14	−3.14	0.57	0.57
ML_2 (two ligands at right angles, along x and y axes)	−2.14	6.14	1.14	−2.57	−2.57
ML_4 (regular tetrahedron)[a]	−2.67	−2.67	1.78	1.78	1.78

[a] The original reference also gives values for distorted tetrahedra.

	d_{z^2}	$d_{x^2-y^2}$	d_{xy}	d_{xz}	d_{yz}
ML_3(in xy plane) $[\frac{3}{2} \times ML_2(90°)]$:	−3.21	9.21	1.71	−3.85	−3.85

The *total potential energy* is the same as this for three unit charges in a trigonal planar ML_3 arrangement, but the $\mathbf{D}_{3h}$ character table (Appendix C) indicates two doubly degenerate pairs of *d* orbitals: d_{xz} and d_{yz}, and $d_{x^2-y^2}$ and d_{xy}. The first pair is already degenerate (same energies), and we can make the other pair equivalent by averaging the energies to give the values shown below for planar ML_3, $\mathbf{D}_{3h}$.

Adding the energies of the *d* orbitals for a linear ML_2 arrangement (along z) to those for trigonal-planar ML_3 (xy plane) gives the energies of the *d* orbitals for a trigonal-bipyramidal complex, ML_5, with five identical ligands:

	d_{z^2}	$d_{x^2-y^2}$	d_{xy}	d_{xz}	d_{yz}
Planar ML_3, $\mathbf{D}_{3h}$:	−3.21	5.46	5.46	−3.85	−3.85
Linear ML_2:	10.28	−6.28	−6.28	1.14	1.14
Trigonal-bipyramidal ML_5, $\mathbf{D}_{3h}$:	7.07	−0.82	−0.82	−2.71	−2.71

EXAMPLE 10.6:   What are the energies of the *d* orbitals in a pentagonal antiprismatic ($\mathbf{D}_{5d}$) field?

Solution: The solution is simple, but not obvious. The *d* orbitals are fivefold degenerate in an icosahedral, $\mathbf{I}_h$ (see Figure 3.14e), field; there is no splitting of the energies. Removing two ligands along an axis (choose z, or the linear ML_2 case) of the icosahedron leaves us with a pentagonal antiprism, such as the staggered arrangement of carbons in ferrocene, $Fe(C_5H_5)_2$. The corresponding treatment of the relative energies of the *d* orbitals gives the energies of the *d* orbitals for a pentagonal antiprism ($\mathbf{D}_{5d}$):

	d_{z^2}	$d_{x^2-y^2}$	d_{xy}	d_{xz}	d_{yz}
ML_{12}, $\mathbf{I}_h$:	0	0	0	0	0
ML_2 (linear):	10.28	−6.28	−6.28	1.14	1.14
ML_{10}, $\mathbf{D}_{5d}$ (pentagonal antiprism):	−10.28	+6.28	+6.28	−1.14	−1.14

The splitting is exactly opposite of that for linear ML_2.

We can use this approach to determine the relative energies of d orbitals for various geometries and for the further splittings of energy levels of complexes from some parent symmetry (here O_h) for *cis*- and *trans*-$[MX_4Y_2]$ (see Problem 10.10). The original paper discusses limitations of this approach, which applies specifically to the d^1 case (no interelectron repulsion) or, using the hole formalism, to the d^9 case (one hole in the filled d^{10} set).

▶ *10.3.2 Effects of Lowering Symmetry*

The complex *cis*-$[CoCl_2(en)_2]Cl$ is violet, whereas *trans*-$[CoCl_2(en)_2]Cl$ is bright green. Obviously with the same ligands a change in symmetry at the metal center (the chromophore) can change the color and absorption spectrum greatly. Lowering the symmetry of an ML_6 complex by substituting other ligands, as in the case of *cis*- and *trans*-$[CoCl_2(en)_2]^+$, partially removes the degeneracy of the T states. The states that arise can be obtained from group theory, or we can use the results from a correlation table for O_h and several important subgroups in Table 10.3. The actual symmetry of the *cis* isomer is C_2 {C_{2v} for *cis*-$[CoCl_2(NH_3)_4]^+$}, and that of the *trans* isomer is D_{2h} {D_{4h} for *trans*-$[CoCl_2(NH_3)_4]^+$}. We can ignore the chelate rings of the ethylenediamine complexes for interpretating absorption spectra and use the higher symmetry of the corresponding ammonia complexes, corresponding to the CoX_2N_4 chromophore. For $[Co(en)_3]^{3+}$, the effective symmetry is very similar to that of $[Co(NH_3)_6]^{3+}$.

The spectra for *cis*- and *trans*-$[CoF_2(en)_2]^+$ are shown in Figure 10.11. The low-energy band (corresponding to T_{1g} in O_h) is split into two well-resolved components for *trans*-$[CoF_2(en)_2]^+$. Note the higher intensity for the noncentrosymmetric *cis* isomer; the greater intensity for noncentrosymmetric complex is usually observed, and was noted above for tetrahedral complexes. *Trans* isomers often show distinct splitting of the first absorption band [parentage T_{1g} (O_h)] if the two different ligands (N and F) differ appreciably in field strength. The *cis* isomers commonly show no distinct splitting; perhaps, they only

Table 10.3   Correlation table for some groups derived from O_h

O_h	T_d	D_{4h}	D_3	C_{4v}	C_{2h}	C_{2v}
A_{1g}	A_1	A_{1g}	A_1	A_1	A_g	A_1
A_{2g}	A_2	B_{1g}	A_2	B_1	B_g	A_2
E_g	E	$A_{1g} + B_{1g}$	E	$A_1 + B_1$	$A_g + B_g$	$A_1 + A_2$
T_{1g}	T_1	$A_{2g} + E_g$	$A_2 + E$	$A_2 + E$	$A_g + 2B_g$	$A_2 + B_1 + B_2$
T_{2g}	T_2	$B_{2g} + E_g$	$A_1 + E$	$B_2 + E$	$2A_g + B_g$	$A_1 + B_1 + B_2$
A_{1u}	A_2	A_{1u}	A_1	A_2	A_u	A_2
A_{2u}	A_1	B_{1u}	A_2	B_2	B_u	A_1
E_u	E	$A_{1u} + B_{1u}$	E	$A_2 + B_2$	$A_u + B_u$	$A_1 + A_2$
T_{1u}	T_2	$A_{2u} + E_u$	$A_2 + E$	$A_1 + E$	$A_u + 2B_u$	$A_1 + B_1 + B_2$
T_{2u}	T_1	$B_{2u} + E_u$	$A_1 + E$	$B_1 + E$	$2A_u + B_u$	$A_2 + B_1 + B_2$

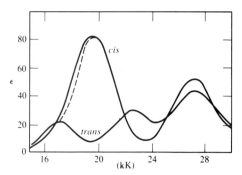

Figure 10.11   The absorption spectra of *cis*- and *trans*-[CoF₂(en)₂]NO₃. The dashed line outlines the main Gaussian band. (After F. Basolo, C. J. Ballhausen, and J. Bjerrum, *Acta Chem. Scand.*

show a shoulder on one side of the main band. In Figure 10.11 the *cis* isomer shows slight broadening on the low-energy side. The higher energy band [parentage $T_{2g}(O_h)$] usually shows no splitting for either isomer.

For absorption spectra the effective symmetry—the symmetry consistent with the splitting observed—is determined by the identity of the ligand atoms directly bonded to the metal rather than by the overall symmetry. Ligands that are *trans* to one another interact with the same *d* orbital, and the *effective field along this axial direction is the average of the field strengths of the two ligands*. The simplified diagram of the *trans* isomer, considering only the fields along each axis, has $\mathbf{D}_{4h}$ (tetragonal) symmetry with a weak field (2F) along *z* (Figure 10.12). The *cis* isomer also can be treated as a tetragonal case, with a strong field (2N) along the unique *z* axis and the weaker field along *x* and *y* (the average of the fields for NH₃ and F⁻ for each axis). The splitting between the levels derived from $T_{1g}(O_h)$ depends on the differences between the axial and in-plane field strengths and is *twice as great* for the *trans* isomer as for the *cis* isomer (see Problem 10.11 for verification), and in the opposite direction. In the *trans* case, the unique d_{z^2} orbital is lower in energy than $d_{x^2-y^2}$ because it interacts with the weaker field (2F); in the *cis* case, d_{z^2} interacts with the stronger field (2N).

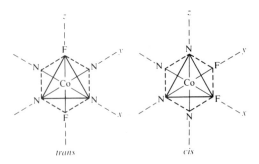

Figure 10.12   Diagrams for the *trans*- and *cis*-[CoF₂N₄] isomers.

10.4 CHARGE TRANSFER BANDS

At energy higher than the ligand-field absorption bands, we commonly see one or more very intense bands that go off scale unless log ε is plotted. These normally are **charge-transfer bands,** corresponding to electron transfer processes that might be ligand → metal (L → M) or metal → ligand (M → L). M → L transitions occur for metal–ion complexes that have filled, or nearly filled, t_{2g} orbitals with ligands that have low-lying empty orbitals. These empty orbitals are ligand π^* orbitals in complexes such as those of pyridine, bipyridine, 1,10-phenanthroline, CN^-, CO, and NO. See Figure 9.7 for overlap of a metal t_{2g} metal orbital and π^* of CO.

The L → M charge-transfer (C-T) spectra have been studied more thoroughly. The intense bands are for LaPorte-allowed transitions commonly of the (ligand) $p \rightarrow d$ (metal) type. The ionization energy for the ligand (or its ease of oxidation) as well as the oxidation state and electron configuration of the metal (or its ease of reduction) determine the energy of the transition. No net oxidation–reduction usually occurs, because of the short lifetime of the excited state. However, many complexes are decomposed photochemically by this process, so they are not stored in strong light. Figure 10.13 shows the spectra of $[CoX(NH_3)_5]^{2+}$ ions, where X is a halide ion. The spectra are plotted on a log ε scale so that we can see the weak ligand-field and intense C-T bands. Two ligand-field ($d \rightarrow d$) bands are expected for Co^{III} (low-spin d^6). The first (lowest energy, related to T_{1g} in $\mathbf{O}_h$) band occurs at about 18,000 cm^{-1} (18,200, 18,700, 19,500 cm^{-1}, respectively, for the

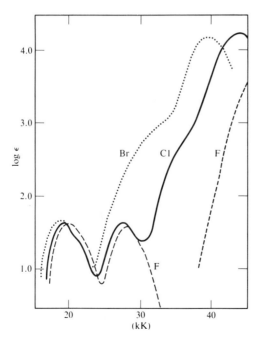

Figure 10.13 The spectra of the $[CoX(NH_3)_5]^{2+}$ ions, where X is a halide ion. (After M. Linhard and M. Weigel, *Z. Anorg. Chem.* **1951,** *266,* 49.)

complexes where X = Br^-, Cl^-, and F^-). The second $d \rightarrow d$ band (related to T_{2g} in O_h) occurs at 27,500 cm^{-1} for $[CoCl(NH_3)_5]^{2+}$ and 28,300 cm^{-1} for $[CoF(NH_3)_5]^{2+}$. This second $d \rightarrow d$ band is covered by a much more intense C-T band for the bromo complex. The $d \rightarrow d$ bands shift slightly toward lower energy as the field strength of X decreases (F^- highest field and highest energy of bands and Br^- lowest). Little splitting[14] of the first band is observed for $[CoF(NH_3)_5]^{2+}$, because the field strength of F^- does not differ greatly from that of NH_3. A shoulder appears on the higher-energy side of this band for the Cl^- and Br^- complexes, because of the significantly lower field strength along the unique (N—Co—X) direction. As noted for $[CoF_2(en)_2]^+$, little splitting of the second band (~28 kK) occurs for any of the $[CoX(NH_3)_5]^{2+}$ ions.

C-T bands ($X^- \rightarrow M$) are shifted with changing X^- much more than the ligand-field bands. The lowest-energy C-T band appears as a shoulder for $[CoBr(NH_3)_5]^{2+}$ (32,000 cm^{-1}) and $[CoCl(NH_3)_5]^{2+}$ (37,100 cm^{-1}). A second C-T band is a distinct peak with log ε above 4.0 for the bromo and chloro complexes. We see only the shoulder of the first C-T band for $[CoF(NH_3)_5]^{2+}$. The large shift (compared to ligand-field shifts) toward lower energy for each C-T band corresponds to the change in ease of oxidation, $Br^- > Cl^- > F^-$ (highest energy).

EXAMPLE 10.7:   What is the nature of the electron transition causing light absorption for intensely colored MnO_4^- (purple-red), MnO_4^{2-} (green), and CrO_4^{2-} (yellow) ions?

Solution: These are d^0 metal "ions"; thus there are no $d \rightarrow d$ transitions. Colors arise from C-T ($O \rightarrow M$). The source of color in metal sulfides, selenides, and tellurides also results from C-T transitions (see Section 5.1.3).

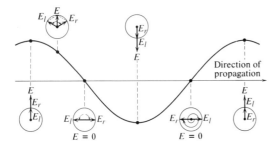

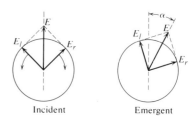

Figure 10.14   A beam of plane polarized light viewed from the side (sine wave) and along the direction at specific times (circles), where the resultant electric vector E and the circularly polarized components E_l and E_r are shown.

Figure 10.15   Plane polarized light before entering and after emerging from an optically active substance.

---

[14] The splitting for $[CoF(NH_3)_5]^{2+}$ is expected to be one-half of that for *trans*-$[CoF_2(NH_3)_4]^+$. The field strength along the unique axis of $[CoF(NH_3)_5]^{2+}$ is the average of the field strengths of N and F.

► 10.4.1 Optical Electronegativities

C. K. Jørgensen used C-T bands of metal complexes to assign **optical electronegativities** to both metal ions and the ligands.[15] The energy of these C-T bands is essentially equal to the difference between the energies of the highest occupied molecular orbital (HOMO) of the ligand and the lowest unoccupied molecular orbital (LUMO) of the metal ion as perturbed by the formation of the MO of the complex. Hard bases such as F^- and H_2O have high optical electronegativity, whereas soft bases such as I^- and $(C_2H_5O)_2PS_2^-$ have low optical electronegativity. The optical electronegativities of the metals increase with charge and, generally, with d-orbital occupancy. Compounds of metals that exhibit color not caused by $d \rightarrow d$ transitions are generally of the soft acid–soft base type, in which the low energy of the C-T band places it in the visible region.

In MnO_4^-, a complex of Mn^{VII}, absorption occurs in the green region of the visible spectrum (~500 nm), so that red and blue light are transmitted. In Mn^{VI}, MnO_4^{2-}, the lower oxidation state and hence lower charge on Mn makes the metal harder to reduce. Hence, the CT moves to the violet and blue regions (~400 nm) so that green light is transmitted. In CrO_4^{2-}, the Cr^{VI} ion has smaller effective nuclear charge than Mn in MnO_4^{2-} or MnO_4^-. Hence CT moves into the ultraviolet, and a tail in the blue region allows transmission of yellow light, the complementary color to blue.

10.5 CIRCULAR DICHROISM AND OPTICAL ROTATORY DISPERSION

► 10.5.1 Optical Rotatory Dispersion

Optical rotation, the rotation of the plane of polarization of plane-polarized light, is the usual test for optical activity. Fresnel (1823) explained this phenomenon as the result of different interactions of the optically active substance with the two circularly polarized components of plane-polarized light. The right circularly polarized component's electric vector spirals to the right (E_r) along the direction of propagation (with the viewer looking toward the light source) and thus sweeps out a right spiral on the surface of a cylinder of radius equal to E_r, whereas the left circularly polarized component spirals to the left. These two components of equal magnitude add vectorially to produce the sine wave of plane-polarized light as shown in Figure 10.14. They travel with the same velocity in an optically inactive medium, but in an optically active medium one component is slowed relative to the other; that is, the indices of refraction differ, $n_l \neq n_r$. Because one component is slowed more than the other, the resultant vector is rotated through an angle α after passing through the optically active medium (Figure 10.15). A plot of angle of rotation (or

[15] C. K. Jørgensen, *Orbitals in Atoms and Molecules,* Academic Press, New York, 1962.

a plot of $n_l - n_r$) versus wavelength (or frequency) is an **optical rotatory dispersion** (ORD) curve (Figure 10.16). The relationship of the ORD curve with the absorption band for the same electronic transition is shown.

▶ 10.5.2   *Circular Dichroism*

An optically active substance absorbs the two circularly polarized components to different extents. Differential absorption $\Delta\varepsilon$ is known as **circular dichroism** (CD).

$$\Delta\varepsilon = \varepsilon_l - \varepsilon_r \qquad (10.11)$$

Optical rotation and CD together are known as the **Cotton effect.** We need not measure both ORD and CD, since the curves are described by the same set of three parameters: (1) the position of the electronic transition in wavelength or frequency, $\Delta\varepsilon_{max}$, for CD and the inversion point for ORD; (2) intensity, $\Delta\varepsilon_{max}$, for CD and the trough-to-peak height for ORD; and (3) width or half-width (half-width at $\frac{1}{2}\Delta\varepsilon_{max}$) for CD and trough-to-peak spread for ORD (Figure 10.16).

CD curves have the advantage of simple shape (Gaussian), with $\Delta\varepsilon$ vanishing not far from $\Delta\varepsilon_{max}$. When adjacent CD peaks have different signs, it can be helpful in separating overlapping peaks in the absorption spectrum. In ORD studies you always have to contend with an appreciable background (the tails) from other transitions.

The tails of ORD curves extend indefinitely outside of absorption bands. The ORD tails extending far from the inversion point are responsible for the observed optical rotation of a colorless substance such as sugar or tartaric acid in the visible region. The observed rotation, in general, is a composite of the tails of ORD curves for all optically active transitions of the molecule throughout the visible and ultraviolet regions. The smooth tails are referred to as "normal" dispersion curves. The ORD curve within the absorption band reverses sign and is referred to as an "anomalous" dispersion curve, but it is typical of all optically active substances.

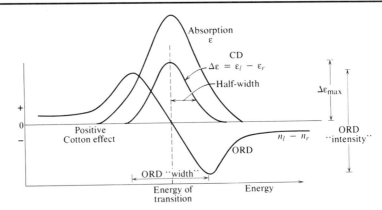

Figure 10.16   Theoretical curves for absorption, circular dichroism, and optical rotatory dispersion for a single electronic transition of one optical isomer. For the other optical isomer the Cotton effect is negative for this transition, the CD curve is negative, and the ORD curve is the mirror image of the one shown, with the trough at lower energy than the peak.

▶ *10.5.3 Stereochemical Information from CD*

$[Co(en)_3]^{3+}$ demonstrates the value of CD studies. The absorption spectrum (Figure 10.17) is very similar to that of $[Co(NH_3)_6]^{3+}$, showing two symmetrical bands in the visible region. One could conclude that the effective field symmetry is O_h, because there is no splitting observed, and that these are the $A_{1g} \rightarrow T_{1g}$ and $A_{1g} \rightarrow T_{2g}$ bands. In the region of the first (lower energy) absorption band, the ORD curve shows a hint of a second Cotton effect near 24 kK, but two CD peaks of opposite sign are unmistakable within this band region. The actual symmetry of $[Co(en)_3]^{3+}$ is D_3 and the triply degenerate states (for O_h symmetry) should split to give $T_{1g}(O_h) \rightarrow (A_2 + E)(D_3)$ and $T_{2g}(O_h) \rightarrow (A_1 + E)(D_3)$ (see the correlation table, Table 10.3). We can observe both A_2 and E CD peaks for the first absorption band, but only a weak E CD peak for the second absorption band, since A_1 is magnetic-dipole-forbidden.[16] The absorption spectrum does not reveal the presence of two components, because the peaks are broad and merge into one for small splittings. Mason[17] used CD studies of single crystals of $[Co(en)_3]^{3+}$ to identify the major CD peak as E. This is an important case, since it was the first complex for which the absolute configuration was determined by x-ray crystallography. First resolved by Werner, it is used widely as a

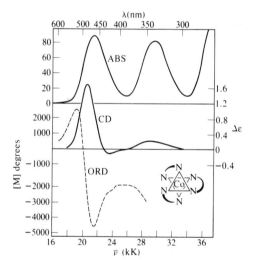

Figure 10.17 Absorption, ORD, and CD curves of Λ-(+)-$[Co(en)_3]Cl_3$.

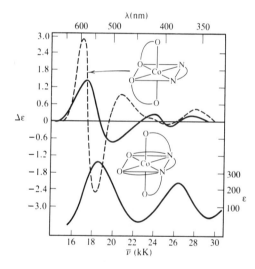

Figure 10.18 Absorption and CD curves of $K\Lambda(-)_{546}$-$[Co(edta)] \cdot 2H_2O$(——) and CD curve of $K\Lambda(-)_{546}$-$[Co(mal)_2(en)] \cdot 2H_2O$ (---).

---

[16] The absorption bands for transition-metal complexes involve electric dipole transitions. For ORD and CD the intensities depend on the product of integrals for the electric dipole transition *and* the magnetic dipole transition. If either of these is zero, the transition is forbidden. One of these is zero if the molecule possesses a plane or center of symmetry of *any* S_n axis (page 127). In the case of $[Co(en)_3]^{3+}$ the E and A_2 transitions are electric- and magnetic-dipole-allowed, but the A_1 transition is magnetic-dipole-forbidden (see Problem 10.14). "Optically active transitions" are those that are electric- and magnetic-dipole-allowed.

[17] R. E. Ballard, A. J. McCaffery, and S. F. Mason, *Proc. Roy. Soc.,* **1962**, 331.

reference complex for assigning absolute configurations of closely related complexes on the basis of the signs of CD peaks for transitions that can be related to those of $[Co(en)_3]^{3+}$.

The absorption spectrum of $[Co(edta)]^-$ shows two absorption bands in the visible region with no apparent splitting (Figure 10.18). This was interpreted to mean that the "effective" symmetry about Co^{III} was cubic—that is, the field was averaged out to become effectively $\mathbf{O}_h$ instead of the actual $\mathbf{C}_2$ symmetry. The CD spectrum shows two peaks in the region of the first absorption band and three weak peaks in the second-band region. We probably see only two of the three peaks expected in the first band for $\mathbf{C}_2$ symmetry because of small splittings; that is, only two peaks are resolved. The expected three peaks appear for a ligand-field model compound, which also has $\mathbf{C}_2$ symmetry, $[Co(mal)_2(en)]^-$ (mal = malonate ion). The absolute configurations of both complexes are known from x-ray crystallography. CD studies have proven useful in distinguishing among geometrical isomers formed by a number of complexes, even in cases where the isomers differ only in the chelate ring conformations, such as for triethylenetetraamine. Absolute configurations can be assigned from CD spectra for closely related complexes if one of the series has known configuration. Uncertainty arises whenever there is doubt about assignments of the electronic transitions.

10.6 PICTORIAL MOLECULAR-ORBITAL
DESCRIPTION OF BONDING

We follow the same approach for a molecular-orbital (MO) description of a metal complex, ML_n, that we used for other molecules, AX_n (Sections 4.5 and 4.6). Thus bonding is very much the same in $AlCl_4^-$ as in $SiCl_4$, and in AlF_6^{3-} as in SF_6. Even though metal complexes are not special cases, they often exhibit features that justify separate discussion. In transition metal complexes, the participation of the $(n-1)$ d orbitals is particularly important.

Consider σ bonding in an octahedral complex, ML_6^{n+}, where M is a transition metal ion, such as Ni^{2+}, and L is a Lewis base, such as :NH_3. In the complex each NH_3 molecule has a filled orbital directed toward the metal ion. We identify all orbitals with reasonable energies and proper symmetry for bonding. Group theory helps us obtain all proper combinations of metal and ligand orbitals. In simple cases, such as this one, we can use a pictorial approach.

We recognize that the energy of the $4s$ electrons is only slightly higher than that of the $3d$ electrons, since the transition metals give stable positive oxidation numbers corresponding to the removal of $3d$ electrons after the $4s$ electrons are removed. Also, at atomic numbers 19 (K) and 20 (Ca), the $4s$ energy level is *lower* than the $3d$ level because $3d$ is not occupied. The energies of the $4p$ orbitals are slightly higher than those of the $4s$ and $3d$ orbitals. Thus the metal orbitals to consider are $4s$, $3d$ and $4p$, for a total of nine. Now we must choose the ligand orbitals of proper symmetry.

Considering σ bonding only, each ligand has a directed hybrid or atomic orbital for bonding. The metal $4s$ orbital is spherically symmetrical and can interact with all six ligands. One bonding MO corresponds to the combination of the $4s$ orbital with the σ orbital of the same sign on each of the six ligands (Figure 10.19). For convenience these six ligand orbitals are treated together as a **ligand-group orbital** (LGO) (or a symmetry-adapted

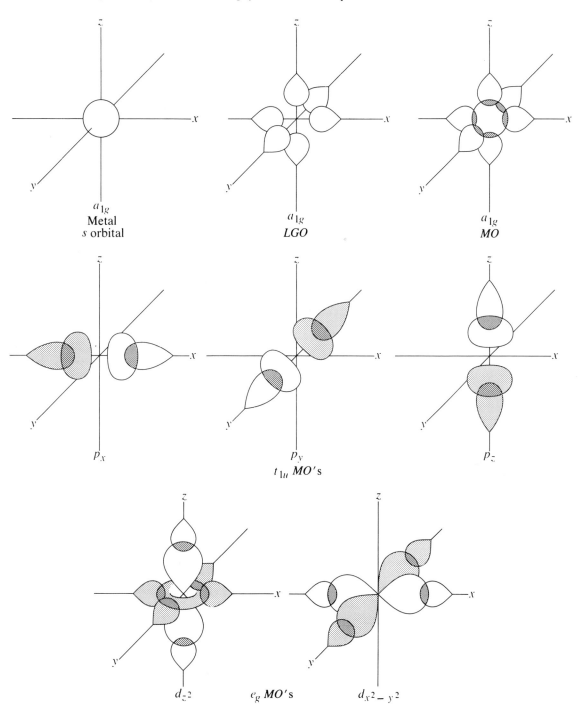

Figure 10.19   Combination of atomic orbitals and LGOs to form sigma bonds in an octahedral complex.

orbital). The LGOs for a particular symmetry species result from taking linear combinations of ligand σ orbitals whose signs match those of the lobes of the metal orbital of that symmetry. Thus, the a_{1g} LGO is the combination of ligand σ's with all positive (+) signs which matches the everywhere positive sign of $4s$. The combination of the $4s$ orbital with this LGO of opposite phase (sign) is antibonding. The $4s$ orbital and both combinations with this LGO are totally symmetric; that is, they are symmetric with respect to every operation of the point group. For the $\mathbf{O}_h$ group, they belong to the A_{1g} representation. The bonding MO is designated a_{1g}, and the antibonding combination is designated a_{1g}^*.

Each metal $4p$ orbital is directed along one of the coordinate axes and can combine with the LGO constructed from orbitals of the two ligands lying along the axis (Figure 10.19) with + and − signs which match those of p lobes. The three LGOs together belong to the triply degenerate representation T_{1u}. They can combine with the metal $4p$ orbitals, giving threefold degenerate bonding (t_{1u}) and antibonding (t_{1u}^*) combinations.

Two of the d orbitals $(d_{x^2-y^2}$ and $d_{z^2})$ are directed along the coordinate axes and hence are suitable for σ bonding. The two bonding LGOs are those that overlap the lobes of the appropriate metal orbitals with the signs matching (Figure 10.19). These two d orbitals belong to the E_g representation, as do the two corresponding LGOs. For the other three d orbitals (t_{2g}) directed at 45° to the coordinate axes, no σ LGOs can be constructed because the t_{2g}'s have a nodal plane along the bond. Hence the t_{2g}'s are not suitable for σ bonding; they are nonbonding.

We have a total of six bonding MOs, six antibonding MOs, and three nonbonding MOs. This total (15) must agree with the total number of orbitals combined—nine from

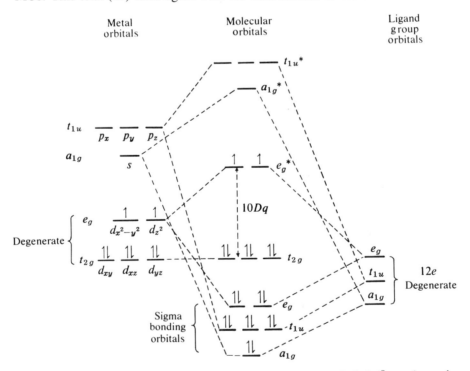

Figure 10.20 Qualitative diagram for the MOs of an octahedral d^8 complex such as $[Ni(NH_3)_6]^{2+}$ (without π bonding).

M and six from the ligands.

A qualitative MO energy level diagram for $[Ni(NH_3)_6]^{2+}$ (d^8) considering σ bonding only is shown in Figure 10.20. Without π bonding, the filled six Ni^{2+} t_{2g} orbitals are non-bonding. Two d electrons of Ni^{2+} occupy antibonding e_g^* orbitals, giving two unpaired electrons. The ligand-field splitting $10Dq$ is the result of the electrostatic effect of six ligands in an $\mathbf{O}_h$ field *plus* the increase in energy of the antibonding e_g^* orbitals determined by covalent bonding.

EXAMPLE 10.8:   What are the metal orbitals and irreducible representations involved in σ bonding for square-planar $[ML_4]^{n+}$ ($\mathbf{D}_{4h}$)?

Solution: The $s(a_{1g})$ orbitals can bond with all four ligands; and for p_x and p_y (e_u), each p orbital can bond two ligands. The $d_{x^2-y^2}$ orbital (b_{1g}) can bond all four ligands. These are the orbitals used to construct dsp^2 hybrids in the VB model. The ligand group orbitals for σ bonding must belong to A_{1g}, E_u, and B_{1g}.

10.7   EFFECTS OF π BONDING

In Figure 10.20 for only σ bonding, the metal t_{2g} orbitals are nonbonding. The metal t_{2g} and e_g orbitals for a d^3 ion involved only in σ bonding are shown in the center of Figure 10.21. Two different cases involving π bonding for metal complexes affect the charge on M in opposite fashions. If the ligands have filled t_{2g} orbitals whose energies are close to those of the empty or only partially filled metal t_{2g} orbitals, then L → M donation can occur. This is the situation for metal ions with high oxidation numbers and few d electrons. The consequences of L → M π donation are shown in Figure 10.21a. The metal d electrons go into the antibonding t_{2g}^* orbitals, since the shared ligand electrons occupy the π-bonding t_{2g} orbitals. For this case, $10Dq$ is *smaller* for L → M donation than without π bonding (compare the center and the left side of Figure 10.21). This probably is the case for complexes of F^-, Cl^-, or OH^-. The lower $10Dq$ value for OH^- (weaker-field ligand) compared with that for H_2O comes about because OH^- is a better π donor than H_2O. If there are as many as six t_{2g} metal electrons, the π bonding does not increase the stability of the complex because there are as many antibonding as bonding electrons.

The other type of π bonding involves M → L donation, which is important for (a) metals with filled or nearly filled t_{2g} orbitals and (b) ligands with empty low-energy orbitals, such as the d orbitals of P or S or the π^* orbitals of CO, CN^-, NO^+, 1, 10-phenanthroline, and so on (Section 9.6.3). Of course, the ligands with empty π^* orbitals also have filled π orbitals and could function as π donors. The acceptor role, however, is more important, especially when there are several metal t_{2g} electrons. The metal (t_{2g}) → π^* (ligand) bonding strengthens the metal–ligand bond but weakens the C—N bond in CN^-, for example, because of population of the antibonding (with respect to the C—N bond) π^* orbital. For M → L π donation the metal t_{2g} orbitals become bonding and their energy decreases, thereby increasing the value of $10Dq$ (Figure 10.21b). Although M → L donation is more important for metals with more d electrons, the d^3 case is chosen in Figure 10.21 so that the role of ligand and metal electrons is obvious. Most ligands near the top of the spectrochemical series (Section 9.8.1) are π acceptors, whereas π donors are near the bottom of the series.

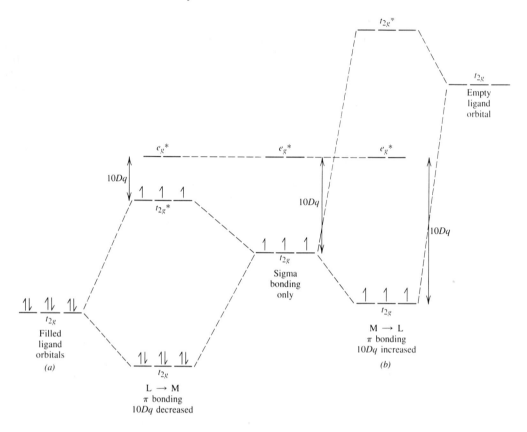

Figure 10.21 Comparison of the effects of π bonding using (a) filled low-energy ligand π orbitals for L $\rightarrow$ M donation and (b) empty ligand orbitals of π symmetry for M $\rightarrow$ L donation.

Figure 10.20 is the energy diagram for σ bonding only. Replacing these t_{2g} orbitals with the t_{2g} and t_{2g}^* orbitals of Figure 10.21a or 10.21b gives a diagram including σ and π bonding.

10.8   GROUP THEORETICAL TREATMENT[18]

▶ 10.8.1   *Sigma Bonding in Octahedral Complexes*

The systematic approach is to apply group theory (see Sections 3.5, 4.5, and 4.6). Transformation properties of the metal orbitals identify their representations, but this case does not require that. We know that the s orbital is totally symmetric (A_{1g}), the p orbitals transform as vectors along x, y, and z (T_{1u}) (also identified in the character table), and the representations for the d orbitals (E_g and T_{2g}) can be identified from the right column of the

[18] This section makes use of the methods presented in Chapter 3 and applied in Chapter 4. See also S. F. A. Kettle, *J. Chem. Educ.* **1966**, *43*, 21.

Table 10.4   Character table for the *O* point group

O	*E*	$6C_4$	$3C_2 (=C_4^2)$	$8C_3$	$6C_2$		
A_1	1	1	1	1	1		
A_2	1	−1	1	1	−1		
E	2	0	2	−1	0		$(z^2, x^2 - y^2)$
T_1	3	1	−1	0	−1	(x, y, z)	
T_2	3	−1	−1	0	1		(xy, xz, yz)

O$_h$ character table (Appendix C). To obtain the representations to which the LGOs belong, we examine the effects of *each class* of symmetry operations on the individual ligand orbitals. Of importance here is not whether the ligand orbitals are atomic or hybrid orbitals, but only that there is a component along the M—L bond for σ bonding. We can simplify the procedure by using the **O** character table (Table 10.4 for **O,** the pure rotation subgroup with fewer symmetry operations) instead of the larger **O**$_h$ table.

The ligand σ orbitals in Figure 10.22 are represented as vectors and are numbered in accordance with nomenclature rules. We tabulate the effect of *one* symmetry operation (the *same* one for all ligands) of *each class* on each ligand orbital. Each symmetry operation either leaves an orbital unchanged (enter character 1) or transforms it into another orbital (enter character 0). (If we were examining the metal orbitals, with the nucleus at the origin, it would also be possible to reverse the sign of an orbital to give a character −1— for example, to reflect a p orbital through the plane of symmetry perpendicular to it.) The results, as tabulated in Table 10.5, do not identify the representations as g or u required for **O**$_h$, but that can be done later by examining the effects on LGOs of the center of symmetry, i, and/or planes of symmetry.

The sums of characters (χ_σ) for the orbital set correspond to a reducible representation. Inspection (or the method discussed in Section 3.5.6) gives us the irreducible representations contained. We can obtain the results more directly by inspection, recognizing that there will be an A_1 representation (to match the s orbital) and keeping other metal orbitals in mind. The result is

$$\Gamma_\sigma = A_1 + E + T_1 \tag{10.12}$$

The A_1 combination with all ligand orbitals having the same sign must be A_{1g} (totally symmetric) for **O**$_h$. Because the character for the reducible representation (χ_σ) is zero for the center of symmetry (each orbital goes into another), the characters must sum to

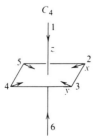

Figure 10.22   Orientation of the ligand σ orbitals as vectors in an octahedral complex.

Table 10.5   **Effects of the symmetry operations of the O point group on the ligand orbitals of an octahedral complex**

O	E	C_4^z	$C_2^z(C_4^2)$	C_3	C_2	i (*for* $\mathbf{O}_h$)
L_1	1	1	1	0	0	0
L_2	1	0	0	0	0	0
L_3	1	0	0	0	0	0
L_4	1	0	0	0	0	0
L_5	1	0	0	0	0	0
L_6	1	1	1	0	0	0
Γ_σ	6	2	2	0	0	0

$+3 - 3 = 0$. For i in $\mathbf{O}_h$, A_{1g} has character $+1$, E is ± 2, and T_1 is ± 3, so E must be $E_g(+2)$ and T_1 is $T_{1u}(-3)$ or

$$\Gamma_\sigma = A_{1g} + E_g + T_{1u} \tag{10.13}$$

Superimposing these LGOs on their metal counterparts in Figure 10.19, we see that there is no LGO belonging to T_{2g}. Hence, the metal t_{2g} orbitals must be nonbonding.

LGOs Using the Projection Operator

The systematic way to obtain the detailed description of the LGOs and their LCAO (linear combinations of atomic orbitals) wavefunctions is to apply the **projection operator** approach (page 179). Here we choose one ligand orbital, tabulate *all* of the symmetry operations, and write down the orbital obtained as a result of performing each operation on this orbital. Choosing L_1, for example: It is still L_1 after the identity operation or $C_4(1)$, $C_4(6)$, or $C_2(1)$ (along z), but it becomes L_5 after the $C_4(2)$ operation, L_2 after $C_3(1)$, and so on,[19] as tabulated in Table 10.6. Next we multiply each orbital in Table 10.6 by the character for the corresponding symmetry operation from Table 10.4. For the A_1 representation, characters are all $+1$. In this case the sum is $a_1 = 4L_1 + 4L_2 + 4L_3 + 4L_4 + 4L_5 + 4L_6$; or, simplifying, $a_1 = L_1 + L_2 + L_3 + L_4 + L_5 + L_6$. This is normalized to give the LCAO

Table 10.6   **Results of applying all symmetry operations to L_1**

E	$C_4(1)$	$C_4(2)$	$C_4(3)$	$C_4(4)$	$C_4(5)$	$C_4(6)$	$C_2(1)$	$C_2(2)$	$C_2(3)$
L_1	L_1	L_5	L_2	L_3	L_4	L_1	L_1	L_6	L_6
$C_3(1)$	$C_3(2)$	$C_3(3)$	$C_3(4)$	$C_3(5)$	$C_3(6)$	$C_3(7)$	$C_3(8)$		
L_2	L_3	L_4	L_5	L_5	L_2	L_3	L_4		
$C_2'(1)$	$C_2'(2)$	$C_2'(3)$	$C_2'(4)$	$C_2'(5)$	$C_2'(6)$				
L_6	L_6	L_2	L_3	L_5	L_4				

[19] The numbering of C_4 and C_2 axes correspond to the numbers in Figure 10.22. The C_2' axes are through edges (2 each in xy, xz, and yz planes). The C_3 axes are through the center of the eight triangular faces. Any consistent choice of numbering gives the same final result.

wavefunction by dividing by the square root of the sum of the squares of the coefficients:

$$\Psi_{a_1} = \frac{1}{\sqrt{6}}(L_1 + L_2 + L_3 + L_4 + L_5 + L_6) \tag{10.14}$$

Multiplying the orbitals in Table 10.6 by the corresponding characters for the representation E and summing gives

$$e(1) = 4L_1 + 4L_6 - 2L_2 - 2L_3 - 2L_4 - 2L_5$$

or

$$2L_1 + 2L_6 - L_2 - L_3 - L_4 - L_5 \tag{10.15}$$

and

$$\Psi_{e(1)} = \frac{1}{\sqrt{12}}(2L_1 + 2L_6 - L_2 - L_3 - L_4 - L_5) \tag{10.16}$$

This gives us one of the pair of e LGOs—the one that matches perfectly with the metal d_{z^2} orbital (see Figure 10.19). To get the other one, we could start with a different ligand orbital (e.g., L_2) and construct the table analogous to Table 10.6 or we can try performing an operation that interchanges ligands along z with those in the xy plane. Using a C_2 axis such that L_1 goes into L_2 and L_6 goes into L_4, we get

$$2L_2 + 2L_4 - L_1 - L_3 - L_5 - L_6 \tag{10.17}$$

This is equivalent to the LGO already obtained (10.15) with a different numbering scheme. It is not the other e LGO. Adding it to the one obtained just gives still another equivalent combination. Keeping in mind that the other metal e orbital is $d_{x^2-y^2}$, we recognize that we want an LGO entirely contained in the xy plane, with L_1 and L_6 not participating (coefficients zero). Multiplying Equation (10.17) by two and adding it to the first LGO [Equation(10.15)], L_1 and L_6 cancel, thereby giving us the second e orbital:

$$3L_2 + 3L_4 - 3L_3 - 3L_5 \quad \text{and} \quad \Psi_{e(2)} = \tfrac{1}{2}(L_2 - L_3 + L_4 - L_5) \tag{10.18}$$

We can obtain the t_1 LGOs in similar fashion, recognizing that for each we need a pair of orbitals of opposite sign along one axis. The metal orbitals of the same symmetry provide templates indicating the proper combinations of ligand orbitals for LGOs:

$$t_1(1) = \frac{1}{\sqrt{2}}(L_1 - L_6) \tag{10.19}$$

$$t_1(2) = \frac{1}{\sqrt{2}}(L_3 - L_5) \tag{10.20}$$

$$t_1(3) = \frac{1}{\sqrt{2}}(L_2 - L_4) \tag{10.21}$$

Symmetry and group theory do not help in arriving at the energies of the MOs. We can get some qualitative ordering of the energy levels from the relative energies of the metal and ligand orbitals and the extent of orbital overlap. Such a qualitative energy level is shown in Figure 10.20.

EXAMPLE 10.9: What are the LGOs for σ bonding in a square-planar $[ML_4]^{n+}$ complex examined in Example 10.8?

Solution: a_{1g} is totally symmetric, so it must be $a_{1g} = \frac{1}{2}(L_1 + L_2 + L_3 + L_4)$. e_{1u} corresponds to the sign patterns of the p_x and p_y orbitals.

$$(e_{1u})(1) = (1/\sqrt{2})(L_1 - L_3) \quad \text{and} \quad (e_{1u})(2) = (1/\sqrt{2})(L_2 - L_4)$$

b_{1g} corresponds to the sign patterns of $d_{x^2-y^2}$, or $b_{1g} = \frac{1}{2}(L_1 - L_2 + L_3 - L_4)$.

▶ *10.8.2 Pi Bonding in Octahedral Complexes*

Pi bonding can be treated independently of σ bonding because no group operation interchanges σ and π orbitals. With the representations for the metal orbitals already identified, we find the reducible representations for the ligand orbitals oriented properly for π bonding as before. Figure 10.23 depicts these orbitals as vectors oriented so that they are interchanged by some symmetry operation. Vectors that transform in this way belong to the same set. (If we were dealing with a trigonal bipyramid, there would be two independent sets: those on ligands in the equatorial plane and those in axial positions.)

We examine the effects on each vector by one operation of each class (each column in the character table). Using the **O** point group, we see that each vector is unchanged by the identity operation, giving a total character of 12 for E; this is always the number of vectors in the set. For the C_2 axis along z, C_2^z, the four vectors for ligands 1 and 6 have their directions reversed—giving a character of -4 for the four vectors (all other vectors are changed into others by C_2^z). Each of the other vectors is changed into another vector by any operation (other than E and C_2^z), so the characters are zero.

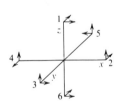

Figure 10.23
Orientation of the ligand π orbitals as vectors in an octahedral complex.

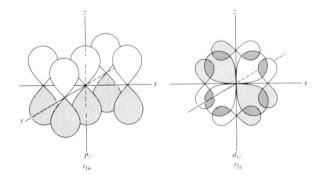

Figure 10.24 The two symmetry types of π bonds in octahedral complexes. (a) One of the t_{1u} orbitals (there are identical orbitals along the x and y directions). (b) One of the t_{2g} π orbitals (there are identical orbitals in the xy and yz planes).

O	E	C_4	$C_2^z(C_4^2)$	C_3	C_2'	$i(\mathbf{O}_h)$
Γ_π	12	0	-4	0	0	0

The reducible representation obtained can be identified by inspection as $2T_1$ and $2T_2$ because only these have negative characters for C_4^2 and their sums of characters give zero for other rotations, giving the characters for Γ_π. Because the character for i is zero for all 12 vectors, in the $\mathbf{O}_h$ point group we have

$$\Gamma_\pi = T_{1g} + T_{2g} + T_{1u} + T_{2u} \tag{10.22}$$

Because there are no metal t_{1g} and t_{2u} orbitals, these LGOs must be nonbonding. We can sketch the bonding LGOs using the metal orbitals as guides or templates (see Figure 10.24). The metal t_{2g} orbitals are important for π bonding unless both ligand and metal orbitals are empty or both filled. The t_{1u} metal orbitals are used effectively for σ bonding, where they provide better overlap, so they generally are not involved much in π bonding.

EXAMPLE 10.10: Find the irreducible representations for LGOs for π bonding in square-planar $[ML_4]^{n+}$.

Solution: The orbitals are those in the xy plane in Figures 10.23 and 10.24. These eight vectors are changed into something else except for E and C_2' for $\mathbf{D}_4$, giving

D$_4$	E	C_4	C_2^z	C_2'	C_2''	$i(\mathbf{D}_{4h})$
Γ_π	8	0	0	-4	0	0

The metal orbitals that can participate in π bonding are $p_z(a_{2u})$, $(p_x, p_y)(e_u)$, (d_{xz}, d_{yz}) (e_g), and $d_{xy}(b_{2g})$. There must be corresponding bonding LGOs with the characters from the character table:

D$_4$	E	C_4	C_2^z	C_2'	C_2''	$i(\mathbf{D}_{4h})$
a_{2u}	1	1	1	-1	-1	-1
b_{2g}	1	-1	1	-1	1	1
e_g	2	0	-2	0	0	2
e_u	2	0	-2	0	0	-2
$\Gamma_\pi(b)$	6	0	-2	-2	0	0 (sum)
$\Gamma_\pi(nb)$	2	0	2	-2	0	0 [difference, $\Gamma_\pi - \Gamma_\pi(b)$]
a_{2g}	1	1	1	-1	-1	1
b_{2u}	1	-1	1	-1	1	-1

Characters for i distinguish g and u subscripts for $\mathbf{D}_{4h}$. The characters for a_{2g} and b_{2u} sum to those of $\Gamma_\pi(nb)$, so these are the nonbonding LGOs. They are nonbonding because there are no metal orbitals with their symmetries.

10.9   COMPARISON OF THE DIFFERENT APPROACHES TO BONDING IN COORDINATION COMPOUNDS

Research in the field of coordination compounds made rapid progress utilizing the valence bond (VB) approach for about 25 years, starting in the early 1930s. As more results were obtained, the qualitative nature of this approach became a serious limitation. Even some of the qualitative predictions of the relative stabilities of complexes based on the availability of low-energy metal orbitals are incorrect; for example, Cu^{2+} and Zn^{2+}, with no vacant $3d$ orbitals, form more stable complexes than do some of the metals with vacant $3d$ orbitals.

Crystal-field theory offered the advantages of permitting the interpretation of spectra of complexes and more detailed interpretation and explanation of the magnetic behavior, stability, stereochemistry, and reaction rates of complexes. The pure crystal-field approach, which ignores covalent bonding, might seem as limited as the extreme valence bond theory approach, which considers only covalent bonding. The effect of σ bonding, however, can be treated empirically as though it were the result of a very strong ligand field. Without π bonding, the results of the MO treatment are the same as the ligand-field representation for the LUMO and HOMO, so ligand-field theory is adequate for most applications. The important spectral transitions and the differences that result in ligand-field stabilization involve the t_2 (nonbonding) and e_g^* (antibonding) levels.

For the present, the ligand-field theory seems to offer the most practical approach to bonding in coordination compounds. It is essentially the crystal-field theory modified to take covalent character into account when necessary. In applying ligand-field theory, we can take advantage of simple pictorial models similar to those that proved so useful in the valence bond approach. Ligand-field theory lends itself to simple qualitative applications and predictions, but it also permits quantitative applications.

MO theory encompasses the ligand-field and VB approaches as special cases. VB theory focuses on the metal bonding and nonbonding t_{2g} orbitals, and ligand-field theory focuses on the t_{2g} and e_g^* orbitals in Figure 10.20. Applications of MO theory to metal complexes have become increasingly important with the availability of large computers. For spectroscopy involving higher-energy levels, such as C-T spectra, the MO approach is appropriate.

GENERAL REFERENCES

F. Basolo and R. G. Pearson, *Mechanisms of Inorganic Reactions*, 2nd ed., Wiley, New York, 1967.

J. K. Burdett, *Molecular Shapes*, Wiley-Interscience, New York, 1980. A good treatment of crystal-field theory MO, and angular-overlap approaches.

B. E. Douglas and C. A. Hollingsworth, *Symmetry in Bonding and Spectra*, Academic Press, Orlando, FL, 1985. More detailed treatment of bonding and spectra.

J. A. Duffy, *Bonding, Energy Levels and Bonds in Inorganic Solids*, Longman, Essex, England, 1990.

A. B. P. Lever, *Inorganic Electronic Spectroscopy*, 2nd ed., Elsevier, Amsterdam, 1984. Excellent for theory and spectral data.

G. Wilkinson, R. D. Gillard, and J. A. McCleverty, Eds., *Comprehensive Coordination Chemistry*, Vols. I–VII, Pergamon Press, Oxford, 1987. The most complete compilation available.

PROBLEMS

10.1 Identify the ground-state terms with the spin multiplicity for the following cases: **(a)** octahedral complexes and **(b)** tetrahedral complexes:

$$Cu^{2+}, \ V^{3+}, \ Cr^{3+}, \ Mn^{2+}, \ Fe^{2+}, \quad and \quad Ni^{2+}$$

10.2 For $[V(H_2O)_6]^{2+}$ absorption bands are observed at 12,300, 18,500, and 27,900 cm^{-1}. Assign the bands and calculate or estimate Dq, B', and the bending. Why would transitions to $(^2E_g, {}^2T_{2g})$ and $^2A_{1g}$ not be expected to be observed?

10.3 For $[V(H_2O)_6]^{3+}$ (in a hydrated crystal), absorption bands are observed at 17,800 and 25,700 cm^{-1}. Assign the bands. Calculate or estimate the values of Dq, B', and bending. What difficulty is encountered in calculating the three parameters in this case?

10.4 Common glass used for windows and many bottles is green because of Fe^{2+}. It is decolorized by adding MnO_2 to form Fe^{3+} and Mn^{2+}. Why is the glass decolorized? Would you expect broad, very weak absorption peaks for Fe^{3+} and Mn^{2+} or many weak peaks?

10.5 Identify the lowest-energy spin-allowed transitions for high-spin $[Co(Hi)_6]^{2+}$ and low-spin $[CoLo_6]^{2+}$.

10.6 For $[Cr(NH_3)_6]^{3+}$ there are two absorption bands observed at 21,500 cm^{-1} and 28,500 cm^{-1} and a very weak peak at 15,300 cm^{-1}. Assign the bands and account for any missing spin-allowed bands. Calculate Dq for NH_3 using the Orgel diagram. Account for any discrepancy between the observed position of any of the spin-allowed bands and that expected from the Orgel diagram.

10.7 The following data are available for the complexes identified:

$[Ni(H_2O)_6]^{2+}$	$[Ni(NH_3)_6]^{2+}$	
8,600 cm^{-1}	10,700 cm^{-1}	
13,500	17,500	
25,300	28,300	
15,400 ⎫	15,400 ⎫	Very weak peaks for both
18,400 ⎭	18,400 ⎭	complexes

Assign the bands. Calculate $10Dq$ and the expected positions of the spin-allowed bands. Account for any discrepancy between the experimental and calculated energies of the bands. Account for the relative positions of corresponding bands for the two complexes.

10.8 Interpret the following comparisons of intensities of absorption bands for transition metal complexes.

(a) Two isomers of a Co^{III} complex believed to be *cis* and *trans* isomers give the following spectral features:

Both give two absorption bands in the visible region. Complex A has two symmetrical bands with $\varepsilon \cong 60$–80. The lower-energy band for complex B is broad with a shoulder and has lower intensity. Assign the isomers. Explain.

(b) An octahedral complex of Co^{III}, with an amine and Cl^- coordinated, gives two bands with $\varepsilon = 60$–80, one very weak peak with $\varepsilon = 2$, and a high-energy band with $\varepsilon = 20{,}000$. What is the presumed nature of these transitions? Explain.

10.9 Calculate the relative energies of the d orbitals for an ML_6 complex with trigonal prismatic coordination ($\mathbf{D}_{3h}$), assuming that the ligands are at the same angle relative to the xy plane as for a regular tetrahedron. [*Hint:* Start with the tetrahedral case, but allow for three ligands up and down, instead of two. The degeneracy of the d orbitals is the same as for other $\mathbf{D}_{3h}$ complexes, such as trigonal (ML_3) and trigonal bipyramidal (ML_5) cases.]

10.10 Calculate the relative energies of the d orbitals for the following complexes, assuming $Dq(X) = 1.40Dq(Y)$ and that where X and Y are along the axis (*trans* to one another), the field strength is the same as for two equivalent ligands with the average field strength of X and Y. Use z as the unique axis.

(a) MX_5Y (c) *cis*-$[MX_4Y_2]$ (both Y's in the xy plane)

(b) *trans*-$[MX_4Y_2]$ (d) *facial*-$[MX_3Y_3]$ (see Figure 9.23)

10.11 Compare the splitting of the $e_g(\mathbf{O}_h)$ orbitals (difference in energies of d_{z^2} and $d_{x^2-y^2}$) for the cases in Problem 10.10 with the spectral results for *cis*- and *trans*-$[CoF_2(en)_2]^+$ (Figure 10.11).

10.12 Calculate the relative energies of the d orbitals (see Section 10.3.1) for ML_5 in the TBP($\mathbf{D}_{3h}$) and SP($\mathbf{C}_{4v}$) geometries. For high-spin d^6–d^9 metal ions, for which cases is there a significant preference for one of the two geometries?

10.13 Explain why the ligand field (d–d) bands are shifted only slightly for $[CoX(NH_3)_5]^{2+}$ ions ($X^- = F^-$, Cl^-, Br^-, and I^-), but charge-transfer bands are shifted greatly for the series.

10.14 Selection rules for electric dipole transitions (those usually observed for transition metals) require that the symmetry of the transition (the direct product of the representations for the ground and excited states) be the same as the representation for one of the electric dipole moment operators, and these correspond to the representations for x, y, and z. Selection rules for magnetic dipole transitions (observed for lanthanide metals) require that the symmetry of the transition be the same as the representation of one of the magnetic dipole moment operators, and these correspond to rotations about x, y, and z (R_x, R_y, and R_z in the character tables). For CD and ORD, the transitions must obey the selection rules for electric *and* magnetic dipole transitions. For $[Cr(NH_3)_6]^{3+}$ ($\mathbf{O}_h$) the $d \rightarrow d$ transitions are $^4A_{2g} \rightarrow {}^4T_{2g}$ and $^4A_{2g} \rightarrow {}^4T_{1g}$. For $[Cr(en)_3]^{3+}$ ($\mathbf{D}_3$) the transitions are $^4A_2 \rightarrow {}^4E$ and $^4A_2 \rightarrow {}^4A_1$ (both derived from $^4A_{2g} \rightarrow {}^4T_{2g}$ by lowering symmetry). Which of these are electric-dipole-allowed? Which are *both* electric- and magnetic-dipole-allowed?

10.15 For a linear ML_2 complex ($\mathbf{D}_{\infty h}$) with the ligands lying along the z axis, indicate the metal orbitals that may participate in σ bonds. Sketch the ligand-group orbitals that could enter into σ MOs. Identify the metal orbitals for π bonds.

10.16 (a) Identify the metal orbitals for σ bonding for a trigonal-planar complex, ML_3 ($\mathbf{D}_{3h}$).
(b) Sketch the σ MOs involving s and p orbitals.

10.17 Use the group theoretical approach to obtain the representations and LGO for σ bonding in trigonal-planar complexes, ML_3.

10.18 Negative ions might be expected to create stronger ligand fields than neutral molecules on the basis of the pure crystal-field model. Explain why OH^- has a weaker ligand field than does H_2O. Why does CO have such a strong ligand field?

10.19 For the MO diagram for $[Ni(NH_3)_6]^{2+}$ (Figure 10.20): **(a)** Identify the portion of the diagram considered by crystal-field theory. **(b)** Designate the portion of the diagram considered by valence bond theory. Why does the d^8 case present a problem for valence bond theory for an octahedral complex? **(c)** Sketch the effect on the e_g^* levels in an octahedral complex with a π-donor ligand.

▶11◀

Reaction Mechanisms of Coordination Compounds

. . . .

> The chemistry of transition metal complexes involves ligand coordination to metals, exchange of ligands, coordination-sphere rearrangement, oxidation and reduction, and alteration of ligand reactivity by coordination. A thorough appreciation of complex chemistry requires a knowledge of reaction mechanisms. The relatively simple complexes discussed here serve as models displaying the main mechanistic features for coordination compounds.
>
> In this chapter we look first at the model for understanding reaction kinetics and mechanisms. Then we discuss substitution reactions of octahedral Werner complexes, developing the concepts needed for relating the model to experiment. After dealing briefly with recemization reactions of optically active octahedral complexes, we treat ligand substitution reactions of square-planar complexes. Next, the mechanisms of redox reactions are explored. Finally, a brief look is taken at photochemical reactions. Reaction mechanisms for complexes containing π-acid ligands are discussed in Chapter 14.

11.1   THE KINETIC MODEL

The model that chemists currently use to rationalize and predict what goes on during chemical reactions was developed in the 1930s from transition-state theory. The course of reactions is seen as involving collision of reactants sufficiently energetic to form the **activated complex** (or **transition state**). Along the path from reactants to products, kinetic energy is converted to potential energy by bond stretching, partial bond formation, angular distortions, and so on. The exact processes that occur depend on the particular reaction. The transition state is the species of maximum potential energy resulting from these motions. All energetically unfavorable processes have occurred by the time a transition state is reached. The energy required to reach the transition state from the reactant ground state is called the **activation energy**. After the transition state, the collection of atoms rearranges toward more stable species: old bonds are fully broken, new ones are fully made, and geometric rearrangement toward a more stable configuration (**relaxation**) occurs. In multistep reactions a new transition state is reached for each distinguishable step.

Plots can be made (at least in principle) of the energy of a reacting system as a function of relative atomic positions. As an example, consider the reaction[1] $H_2 + F \rightarrow$ $HF + H$ when all reactants are confined to a straight line. The distances between atoms r_{H-H} and r_{H-F} describe relative atomic positions. A plot of energy versus two variables is three-dimensional. Figure 11.1*a* shows a two-dimensional **contour plot** representation. Lines represent the shape of the intersection of a plane having the indicated energy value with the three-dimensional energy surface. (The energy zero is the energy of $H_2 + F$.) Such a contour plot is like a terrain map showing contours of constant altitude. At very large r_{H-F} (bottom right), a cut through the plot (A–B) is just the H–H energy versus distance curve. At very large r_{H-H}, a cut through the plot (C–D) gives the H–F energy versus distance curve. Figure 11.1*b* shows computer-generated three-dimensional plots for the two different sections of the potential surface. At very large interatomic distances (top right of the plot), the energy is essentially independent of the distance and the surface is a plateau. Any linear collection of HHF atoms takes the lowest-energy path along the contour plot's valleys. This path (indicated by a dashed line in Figure 11.1*a*) involves changes in both r_{H-H} and r_{H-F}; it is called the **reaction coordinate**.

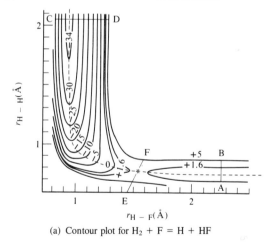

(a) Contour plot for $H_2 + F = H + HF$

Figure 11.1 Plots for the reaction $H_2 + F \rightleftarrows H + HF$. (Parts *a* and *b* are from The University of California, Lawrence Livermore National Laboratory and the Department of Energy, under whose auspices the work was performed when these figures were used; and C. F. Bender, S. V. O'Neill, P. K. Pearson, and H. F. Schaeffer, III, *Science* **1972**, *176*, 1412. Copyright 1972, American Association for the Advancement of Science.)

[1] C. F. Bender, S. V. O'Neill, P. K. Pearson, and H. F. Schaeffer, III, *Science* **1972**, *176*, 1412.

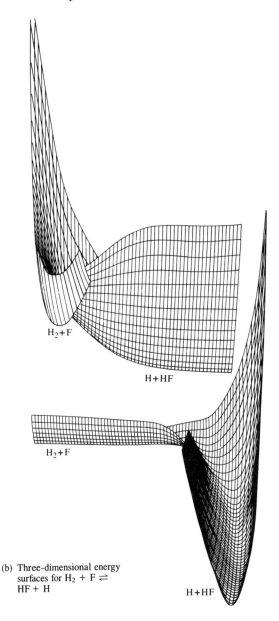

H$_2$+F

H+HF

H$_2$+F

(b) Three-dimensional energy
     surfaces for H$_2$ + F $\rightleftharpoons$
     HF + H

H+HF

The intersection of the two parts of the potential surface in Figure 11.1*b* involves a saddle point (marked * in Figure 11.1*a*), which is an energy maximum as the reaction travels along the valleys of the plot (reaction coordinate), but a minimum along a path such as E–F. The surface around the point is saddle-shaped. Figure 11.1*c* is a two-dimensional plot of the system energy as it moves along the reaction coordinate, a so-called **reaction profile.**

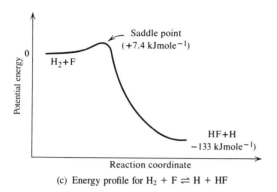

(c) Energy profile for $H_2 + F \rightleftharpoons H + HF$

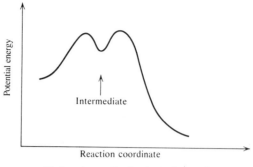

(d) Reaction profile showing an intermediate

Whether the $H_2 + F \rightarrow HF + H$ reaction proceeds forward or backward, it follows the reaction coordinate. Atomic positions simply reverse themselves as the reaction is reversed, like a film shown backwards. This is a statement of the **principle of microscopic reversibility.**

The principle of microscopic reversibility is extremely important in mechanistic studies, because it requires that *at equilibrium* both forward and reverse reactions proceed at equal rates along the reaction coordinate. Hence, the study of exchange reactions where leaving and entering groups X are the same is quite important, because the route of entry for the new ligand must be just the reverse of that for loss of the leaving ligand. *Insofar as some other entering ligand Y resembles the leaving one X,* the principle of microscopic reversibility places limitations on the possible mechanisms for replacement of X by Y; that is, we expect the mechanism to resemble that for replacement of X by X (exchange). This is true for each separate step of a multistep reaction. Instead of a saddle point, some potential surfaces (although not the HHF one) have a potential well or depression corresponding to a slightly stable configuration of reacting atoms, an **intermediate.** Slight stability means that the intermediate can persist for some small length of time. However, it eventually follows the reaction coordinate down into a valley. Figure 11.1d shows a reaction profile for a system with one intermediate. (Several successive intermediates are also possible in multistep reactions.)

The width of the low-energy valley is related to entropy requirements for the reaction. A steeply sloping and narrow energy valley is associated with very negative entropy of activation involving constriction of the reactant atomic motions. Conversely, a wide valley is associated with relative freedom in the motion and orientation of the reactants (more positive entropy of activation).

More complex reactions are hard to depict in plots of energy as a function of all the relative atomic positions. A nonlinear N-atom molecule has $(3N - 6)$ interatomic distances and angles that could be varied to change the energy [a linear molecule has $(3N - 5)$]. Thus, a plot of energy versus each coordinate would require a $(3N - 5)$-dimensional hyperspace [or $(3N - 4)$ for a linear system]. Also, calculating such a multidimensional energy surface is tremendously difficult and time-consuming—especially when our interest lies along the path of minimum energy from reactants to products.

Hence, we use a two-dimensional reaction profile. All the complexities of many variables are lumped into the definition of the reaction coordinate, a parameter that collects all the bond distances and angles that must change as the reactants rearrange to products. The reaction profile supplies a qualitative model, and experiments are designed to show what reacting species are present in the transition state and what factors are important in determining the energy required to reach this state. Also of interest is whether intermediates exist along the reaction path.

The reaction rate is determined by the energy required to surmount the activation barrier—that is, to reach the transition state. The higher the barrier, the smaller the fraction of reactant molecules that will have this much energy at a given temperature, and thus, the slower the reaction. Fast reactions are those with low activation energies.

It would be a good idea to review the essential features of transition-state theory.[2] A major theme of this chapter is how features of the model provided by transition state theory can be verified experimentally. After all, only those concepts that can be investigated by experiment have scientific meaning. These considerations are quite important in elucidating the nature of the transition state, which is (by definition) an energy maximum and hence too short-lived for direct observation. Even intermediates (which are relative minima on the energy surface) seldom last long enough for spectroscopic detection. Hence, designing and interpreting experiments bearing on these points is a nontrivial and subtle endeavor.

11.2 INTRODUCTION TO LIGAND SUBSTITUTION REACTIONS

In ligand substitution reactions the composition of the metal's first coordination sphere changes. The rates of such reactions vary widely, ranging from completion within the time for reactant mixing (such as the formation of blue-purple $[Cu(NH_3)_4]^{2+}$) to years (such as substitution by H_2O on $[Co(NH_3)_6]^{3+}$). Taube called complexes having substitution half-life $t_{1/2} < 30$ sec **labile** and called those with longer $t_{1/2}$ **inert.** Studying reactions of labile complexes requires techniques such as stopped-flow, P-jump, or T-jump[3] (in which a system at equilibrium is perturbed by a sudden change in pressure or temperature and its relaxation to a new equilibrium monitored).

Slower reactions can be monitored by conventional techniques [including nuclear magnetic resonance (NMR), ultraviolet–visible (UV–Vis) spectroscopy, and polarimetry]. Hence, considerably more information is available on reactions of inert species.

Lability and **inertness** are kinetic terms referring to how quickly a system reaches equilibrium. They are unrelated to thermodynamic stability. Table 11.1 gives exchange rates and formation constants (see Section 9.4) for several cyano complexes. There is no necessary connection between the magnitude of the formation constant β_n and the rate of

---

[2] See K. J. Laidler and J. H. Meiser, *Physical Chemistry,* Benjamin-Cummings Publishing Co., Menlo Park, CA, 1982, or see any undergraduate physical chemistry text.

[3] See R. G. Wilkins, *Kinetics and Mechanism of Reactions of Transition Metal Complexes,* 2nd ed., VCH Publishers, New York, 1991 for a discussion of these techniques.

Table 11.1 Formation constants and exchange rates of some cyano complexes[a]

Complex	Formation constant[b] (β_n)	Exchange rate
$[Ni(CN)_4]^{2-}$	10^{30}	Very fast
$[Hg(CN)_4]^{2-}$	10^{42}	Very fast
$[Fe(CN)_6]^{3-}$	10^{44}	Very slow
$[Fe(CN)_6]^{4-}$	10^{37}	Very slow
$[Pt(CN)_4]^{2-}$	$\sim 10^{40}$	$t_{1/2} = 1$ min

[a] From F. Basolo and R. Pearson, *Mechanisms of Inorganic Reactions,* 2nd ed., Wiley, New York, 1967.

[b] $\beta_n = \dfrac{[M(CN)_n]^{y-}}{[M^{(n-y)+}][CN^-]^n}$ for the reaction $M^{(n-y)+} + nCN \rightleftharpoons [M(CN)_n]^{y-}$.

exchange with labeled CN^-. Thus, $[Hg(CN)_4]^{2-}$ is both thermodynamically stable and kinetically labile. In solution, it exchanges ligands many times each second, but the entering ligand is always cyanide instead of more abundant water. Inertness is sometimes called **kinetic stability**—as differentiated from thermodynamic stability which is given by the value of β_n.

We shall study the ligand substitution reactions of octahedral and of square-planar complexes in turn. Many concepts developed for octahedral complexes also apply to square-planar ones.

11.3 OCTAHEDRAL COMPLEXES: LIGAND SUBSTITUTION REACTIONS

▶ 11.3.1 *Kinetics of Water Exchange*

The simplest substitution reaction is exchange of coordinated water with bulk solvent (in the absence of other ligands). Figure 11.2 displays values of k for the exchange reaction.[4] Relative k values show how such features as ionic charge and size affect exchange rates. Langford and Gray divided metal ions into four classes, based on exchange rates.

Class I. Very fast (diffusion-controlled) exchange of water occurs; $k \geq 10^8$ s^{-1}. Ions in this class are those of the alkali metals and alkaline earths (except for Be^{2+} and Mg^{2+}), Cd^{2+}, Hg^{2+}, Cr^{2+}, and Cu^{2+}, and several trivalent lanthanides.

Class II. Exchange-rate constants are between 10^4 and 10^8 s^{-1}. The divalent first-row transition metal ions (except for V^{2+}, Cr^{2+}, and Cu^{2+}) and Ti^{3+}, as well as Mg^{2+} and the other trivalent lanthanide ions, are members of this class.

---

[4] In this chapter, k is used for rate constants and K is used for equilibrium constants.

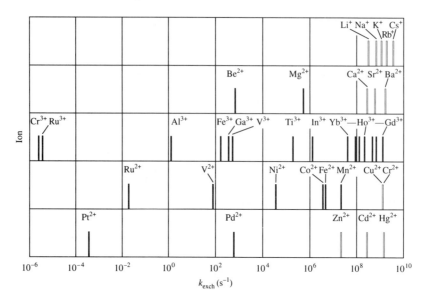

Figure 11.2 Exchange rate constants for metal aqua complexes. ⎯⎯, estimated from complex formation. (Reprinted with permission from Y. Ducommum and A. E. Merbach, in *Inorganic High Pressure Chemistry*, R. van Eldik, Ed., Elsevier, Amsterdam, 1986.)

Class III. Exchange-rate constants are between 1 and 10^4 s^{-1}. This class includes Be^{2+}, V^{2+}, Al^{3+}, Ga^{3+} and several trivalent first-row transition metal ions.

Class IV. Ions in this class are inert in Taube's sense; their rate constants for exchange fall between 10^{-6} and 10^{-2} s^{-1}. Members of the set are Cr^{3+}, Co^{3+}, Rh^{3+}, Ru^{2+}, Ir^{3+}, and Pt^{2+}.

Obviously, one factor determining exchange rates for ions with the noble gas and pseudo-noble-gas configurations is the charge density q/r (see Section 9.7.1). Ions with high q/r exchange relatively slowly, suggesting that a principal contribution to the activation energy is breaking of the bond to the leaving molecule. With transition metals the behavior is more complex. One factor involved is still q/r, since trivalent ions tend to exchange more slowly than divalent ones. However, the d-electron configuration also plays an important role. Rapid exchange by Cr^{2+} and Cu^{2+} (which places them in Class I) is related to Jahn–Teller distortion in the ground state (see Section 9.8.2), which makes axial bonds longer than equatorial ones. The ground-state geometry thus is close to that of the transition state for water exchange, and axial water ligands can exchange more rapidly than for undistorted octahedral complexes.

▶ 11.3.2 *Lability and d-Electron Configuration*

One way of viewing the relation of electron configuration to lability is to note that inert complexes are those with very high ligand-field stabilization energy (LFSE): d^3 and low-spin d^4, d^5, and d^6. The d^8 configuration is borderline. The octahedral d^8 complexes encountered are the weak-field complexes of Ni^{II}. Strong-field complexes of d^8 ions—for example Rh^I, Ir^I, Ni^{II}, Pd^{II}, Pt^{II}, and Au^{III}—are square-planar and low-spin. The weak-field (high-spin) complexes of Ni^{II} usually react much more rapidly than those of d^3 or d^6 ions, but slower than those of labile Cu^{II}, Co^{II}, and Zn^{II} complexes.

Two extreme possibilities for substitution mechanisms in octahedral complexes come to mind. A dissociative path involves prior departure of the leaving ligand followed by attack of the entering ligand on the resulting five-coordinate species. An associative path involving prior attack of the entering ligand and giving a seven-coordinate species also is possible. Taube accounted for the lability of ions having an empty t_{2g} orbital or at least one electron in an e_g orbital by postulating that the unoccupied low-energy t_{2g} bonded the entering ligand followed by departure of the leaving ligand.

The ligand-field model (see Section 9.8.1) indicates a large loss in LFSE for d^3, low-spin d^4, low-spin d^5, low-spin d^6, and d^8 ions in going from six- to either five- or seven-coordinate species (see problem 11.3). For both dissociative and associative pathways, these configurations experience a ligand-field contribution to the activation energy. Hence, these configurations are kinetically inert. Other configurations lose little or no LFSE and are labile.[5]

Note that for a dissociative mechanism the activation energy must be equal to or greater than the bond dissociation energy. The LFSE is only one factor in determining this energy and, when the change in LFSE is small, other factors might be more important. This should apply particularly to complexes involving π bonding.

▶ *11.3.3   Mechanisms for Ligand Substitution Reactions*

We now consider in more detail dissociative and associative pathways for ligand substitution. Henceforth, for consistency in notation, we call the *leaving group* X and the *entering group* Y. For ease in reading, the leaving ligand is generally represented last in formulas for emphasis—sometimes in violation of IUPAC rules. In particular cases, X and/or Y may be identical with other ligands in the complex. Any of these ligands may be water or other solvent. Thus, the general reaction (omitting charge) may be written

$$L_5MX + Y \longrightarrow L_5MY + X \tag{11.1}$$

Two issues are important: first, whether the main factor controlling the activation energy (and hence the rate) is the breaking of the bond to the leaving group (dissociative activation) or the making of the bond to the entering group (associative activation); and second, what the sequence of elementary steps leading from reactants to products is. Langford and Gray labeled these the *intimate* and *stoichiometric* mechanisms, respectively.

---

[5] F. Basolo and R. G. Pearson, *Mechanisms of Inorganic Reactions*, 2nd ed., John Wiley, New York, 1967, pp. 141–158.

Dissociative (d) Mechanisms

To understand this distinction better, consider first the case of a ligand substitution in which the activation energy is determined primarily by the energy required to break the bond to the leaving group, a so-called **dissociative** or **d** reaction. At the start, the complex with its coordination sphere of L and X is surrounded by an outer sphere containing loosely held solvent molecules, Y, and other species present in solution. In one possible **d** mechanism, two elementary steps are detectable: The complex accumulates enough energy to break completely the M—X bond, leaving a five-coordinate intermediate; then, this intermediate reacts with Y (which could be solvent) from the second coordination sphere. This is the so-called **D** mechanism.

$$L_5MX \;\; \underset{k_{-1}}{\overset{k_1}{\rightleftharpoons}} \;\; L_5M + X$$

$$L_5M + Y \;\; \overset{k_2}{\longrightarrow} \;\; L_5MY \tag{11.2}$$

The rate law for this mechanism is

$$\frac{d}{dt}[L_5MY] = k_2[L_5M][Y] \tag{11.3}$$

Applying the steady-state approximation to $[L_5M]$, we obtain

$$\frac{d}{dt}[L_5M] = 0 = k_1[L_5MX] - k_{-1}[L_5M][X] - k_2[L_5M][Y] \tag{11.4}$$

$$[L_5M] = \frac{k_1[L_5MX]}{k_{-1}[X] + k_2[Y]} \tag{11.5}$$

And in terms of experimentally measurable concentrations the rate becomes

$$\frac{d}{dt}[L_5MY] = \frac{k_1 k_2 [L_5MX][Y]}{k_{-1}[X] + k_2[Y]}$$

$$= \frac{k_1 k_2 [L_5MX][Y]/k_{-1}[X]}{1 + k_2[Y]/k_{-1}[X]} \tag{11.6}$$

If we are to show that any reaction does in fact proceed via this mechanism, L_5M must have a sufficiently long lifetime to be experimentally detectable. (Later, we see some ways to do this.)

In carrying out kinetics experiments, reactant solutions are made up to particular concentrations and the rate is measured for several sets of concentrations. The proportionality between rate and concentration is expressed in terms for the "observed" rate constant k_{obs}. For the expression in Equation (11.6), the experimental data would be expressed as rate $k_{obs}[ML_5X]$. The interpretation of k_{obs} as $k_1 k_2 [Y]/\{1 + (k_2[Y]/k_1[X])\}$ involves selecting a model mechanism and showing that an appropriate combination of the constants defined by the separate steps corresponds to k_{obs}.[6]

Consider now the second case in which L_5M does *not* have an appreciable lifetime, the $\mathbf{I}_d$ mechanism. L_5MX accumulates sufficient thermal energy into the M—X vibration to begin to break the bond. Before it can be broken fully, M begins to form a bond with whatever species Y happens to be in a suitable geometric position. This is the limiting case as the intermediate's lifetime becomes progressively shorter, and it finally does not survive long enough to be detectable—corresponding to the relative energy minimum in Figure 11.1*d*, becoming progressively shallower until it disappears. The likelihood of any Y being present in the outer sphere is proportional to [Y]. Also involved is the equilibrium constant for outer-sphere complex formation. The magnitude of K depends on the ionic charge, being larger when Y and the complex are oppositely charged. This possible mechanism can be written as follows, where the presence of Y in the outer sphere (formation of an outer-sphere complex) is indicated explicitly in the first step.

$$
\begin{array}{lll}
L_5MX + Y & \rightleftharpoons & (L_5MX, Y) \qquad K \\[4pt]
(L_5MX, Y) & \longrightarrow & (L_5MY, X) \qquad k \\[4pt]
(L_5MY, X) & \xrightarrow{\text{fast}} & L_5MY + X
\end{array}
\tag{11.7}
$$

The accessible experimental parameters are the initial concentrations of Y, [Y], and of the reactant complex, $[L_5MX]_0$, which (in a solution containing Y) exists partly in the form of the outer-sphere complex.

$$
\begin{aligned}
[L_5MX]_0 &= [L_5MX] + [(L_5MX, Y)] \\[4pt]
&= [L_5MX] + K[L_5MX][Y]
\end{aligned}
\tag{11.8}
$$

Hence

$$
[L_5MX] = \frac{[L_5MX]_0}{1 + K[Y]}
\tag{11.9}
$$

and the rate is

$$
\begin{aligned}
\frac{d}{dt}[L_5MY] &= k[(L_5MX,Y)] = \frac{kK[L_5MX]_0[Y]}{1 + K[Y]} \\[6pt]
&= k_{obs}[L_5MX]_0
\end{aligned}
\tag{11.10}
$$

For each mechanistic possibility just discussed, the energy necessary for bond breaking mainly determines the activation energy and hence the rate. The intimate mechanism in both cases is **d** (dissociative). The two cases differ in that an intermediate of reduced coordination number may or may not be detectable experimentally. These two possibilities lead to differing sequences of elementary steps (different stoichiometric mechanisms).

---

[6] The rather complicated dependence of k_{obs} on [X] and [Y] in Equation (11.6) can be sorted out by plotting values of $1/k_{obs}$ versus $1/[Y]$ for a series of runs in which [X] is kept constant. If X = solvent, this condition is automatically met. Such a plot is called a **double reciprocal plot.** You should verify that it gives a straight line and note the values of the slope and intercept in terms of the parameters of the **D** mechanism.

When the five-coordinate intermediate is detectable, the stoichiometric mechanism is labeled **D** (dissociative); the other case is labeled $\mathbf{I}_d$ (dissociative interchange). Table 11.2 shows the relationship between intimate and stoichiometric mechanism (associative mechanisms are discussed below) and presents notation relating the stoichiometric mechanisms to the S_N1 and S_N2 labels of organic chemistry.

Associative (a) Mechanisms

Bond-making to an entering group also might be the main factor determining the size of the activation energy. Such reactions are designated **a** (associative). Again, two types of stoichiometric mechanism are possible. In the first, an intermediate of higher coordination number survives long enough for detection; this is the so-called **A** (or S_N2 limiting) stoichiometric mechanism:

$$
\begin{aligned}
L_5MX + Y &\underset{k_{-1}}{\overset{k_1}{\rightleftharpoons}} L_5MXY \\
L_5MXY &\overset{k_2}{\longrightarrow} L_5MY + X
\end{aligned}
$$

(11.11)

The **A** mechanism of Equation (11.11) leads to a rate law of the form

$$
\text{rate} = \frac{k_1[L_5MX][Y]}{k_{-1} + k_2} = k_{\text{obs}}[L_5MX][Y]
$$

(11.12)

At present, no firmly established examples of **A** mechanisms have been observed for octahedral substitution reactions. The other possibility is the $\mathbf{I}_a$ mechanism, the associative counterpart of $\mathbf{I}_d$. The outer-sphere complex between the reactants is formed first. The bond between metal and the entering group begins to form before the bond to the leaving group begins to break. In the transition state, both the entering and leaving group are bonded to the metal, but the extent of bond making exceeds that of bond breaking. For $\mathbf{I}_a$ and $\mathbf{I}_d$, the two transition states could be diagrammed as follows where " $\cdots$ " denotes the *less* important bond energy contribution.

$$
\text{Y} - \text{M} \cdots \text{X} \qquad \text{Y} \cdots \text{M} - \text{X}
$$
$$
\mathbf{I_a} \qquad\qquad\qquad \mathbf{I_d}
$$

The steps in the $\mathbf{I}_a$ stoichiometric mechanism and the form of the $\mathbf{I}_a$ rate law resemble those in Equations (11.7) and (11.10). The difference is that in the $\mathbf{I}_a$ mechanism, the magnitude of k is controlled by the energy of bond formation. Another way of viewing the $\mathbf{I}_a$ mechanism is as the limiting case of an **A** mechanism in which bond breaking begins too soon for detection of a seven-coordinate intermediate.

Table 11.2   Relation between intimate and stoichiometric reaction mechanisms

Intimate mechanism		*d*		*a*	
Stoichiometric mechanism		**D** (S_N1 lim)	$\mathbf{I}_d$ (S_N1)	$\mathbf{I}_a$ (S_N2)	**A** (S_N2 lim)

The relationship among these four models is shown in Table 11.2. As with any models, a crucial question is what experimental evidence can be used to distinguish among them. (Analysis of experimental results also leads to refinement of models themselves in favorable cases. Such considerations led to the notion that the S_N1 and S_N2 models of reactivity did not suffice for inorganic complexes.) Plainly, the form of the rate law will not serve to distinguish: **D**, $\mathbf{I_d}$, and $\mathbf{I_a}$ all lead to a rate law of the form

$$\text{rate} = \frac{a[\text{L}_5\text{MX}][\text{Y}]}{1 + b[\text{Y}]} \tag{11.13}$$

Under certain circumstances ($b[\text{Y}] \gg 1$), the reaction could be first-order for any of the mechanisms. If $b[\text{Y}] \ll 1$, the reaction would be second-order. We see mixed-order kinetics when $b[\text{Y}] \simeq 1$. As long as the final step is not reversible, second-order kinetics would result for **A** reactions. Thus, the preeminence of bond breaking in **d** intimate mechanism does not necessarily lead to first-order kinetics for inorganic complexes, and so molecularity is not necessarily indicative of mechanism.

▶ *11.3.4   Experimental Tests of Mechanism*

No reaction mechanism ever can be really "proven". Evidence in favor of one mechanism can be advanced and experiments designed to refute or confirm the proposed reaction path. The accepted mechanism turns out to be the one with the least evidence against it.

Intimate Mechanism

The most straightforward distinction is between **d** and **a** activation (the intimate mechanism). Most data involve reactions of inert ions such as Co^{III} (especially), Cr^{III}, Rh^{III}, Ir^{III}, Pt^{II}, and the borderline Ni^{II}. Two frequently studied reactions (along with exchange), are **aquation** (sometimes called **acid hydrolysis**),

$$\text{L}_5\text{MX}^{n+} + \text{H}_2\text{O} \longrightarrow \text{L}_5\text{M(OH}_2)^{(n+1)+} + \text{X}^- \tag{11.14}$$

and **anation,**

$$\text{L}_5\text{M(OH}_2)^{(n+1)+} + \text{Y}^- \longrightarrow \text{L}_5\text{MY}^{n+} + \text{H}_2\text{O} \tag{11.15}$$

Sensitivity to the Nature of the Entering or Leaving Group.   Comparing reaction rates for a series of complexes with different leaving groups and the same entering group (or *vice versa*) affords evidence about intimate mechanisms. Table 11.3 reports rates of aquation for $[\text{Co(NH}_3)_5\text{X}]^{n+}$. The rates depend heavily on the nature of X (the leaving group) and vary over some six orders of magnitude. In contrast, the rates of anation of $[\text{Co(NH}_3)_5(\text{H}_2\text{O})]^{3+}$ (Table 11.4) display relatively little sensitivity to the nature of Y^-. Clearly, these reactions are much more sensitive to the kind of bond being broken than to the kind being formed. This is what is meant by **d** activation. Further support comes from the aqueous chemistry of $[\text{Co(NH}_3)_5\text{X}]^{2+}$ complexes where no direct replacement of X^- by

Table 11.3   Rate constants for acid aquation of some octahedral complexes[a] of Co^{IIIb} at 25°C

Complex	$k\ (s^{-1})$	Complex	$k\ (s^{-1})$
$[Co(NH_3)_5(OP(OMe)_3)]^{3+}$	2.5×10^{-4}	$[Co(NH_3)_5Cl]^{2+}$	1.8×10^{-6}
$[Co(NH_3)_5(NO_3)]^{2+}$	2.4×10^{-5}	$[Co(NH_3)_5(SO_4)]^+$	8.9×10^{-7}
$[Co(NH_3)_5I]^{2+}$	8.3×10^{-6}	$[Co(NH_3)_5F]^{2+}$	8.6×10^{-8}
$[Co(NH_3)_5(H_2O)]^{3+}$	5.8×10^{-6}	$[Co(NH_3)_5N_3]^{2+}$	2.1×10^{-9}
		$[Co(NH_3)_5(NCS)]^{2+}$	3.7×10^{-10}

[a] Leaving ligand written last.

[b] M. L. Tobe, *Adv. Inorg. Bioinorg. Mech.* **1984,** *2,* 1.

Y^- is observed. *Instead, aquation occurs, followed by anation with Y^-.* Such behavior indicates that the energetically significant process is Co—X bond breaking and that bond making has so little energetic significance that whatever species is present in greatest concentration (in this case, the solvent water) serves as the initial entering group independent of its identity.

Substitution rate constants for $[Ti(H_2O)_6]^{3+}$ reported in Table 11.5 stand in striking contrast, indicating a rate variation with the nature of the entering groups over a factor of $\sim 10^4$ and hence an **a** intimate mechanism.

Steric Effects of Inert (Nonleaving) Ligands.   Crowding around the metal ion is expected to retard the rates of reactions that occur by an **a** mechanism and to speed up those occurring via a **d** mechanism. Data in Table 11.6 show that aquation reactions of *trans*-$[Co(N—N)_2Cl_2]^+$ (where N—N is a didentate amine ligand) exhibit steric acceleration. This behavior is indicative of **d** activation, as is the rate increase on going from cyclam complexes to those of its hexamethyl analogue tet-*b* (see Figure 11.3 for ligand structures).

Table 11.4   Limiting rate constants for anation by Y^{n-} and water exchange (k_e) at 45°C[a] of $[Co(NH_3)_5(H_2O)]^{3+}$

Y^{n-}	$k\ (s^{-1})$	k/k_e
H_2O	100×10^{-6}	—
N_3^-	100×10^{-6}	1.0
SO_4^{2-}	24×10^{-6}	0.24
Cl^-	21×10^{-6}	0.21
NCS^-	16×10^{-6}	0.16
H_2O	$5.8 \times 10^{-6\ b}$	—
$H_2PO_4^-$	$7.7 \times 10^{-7\ b}$	0.13

[a] From R. G. Wilkins, *The Study of Kinetics and Mechanism of Reactions of Transition Metal Complexes,* Allyn and Bacon, Boston, 1974.

[b] At 25°C.

**Table 11.5 Rates of anation of
$[Ti(H_2O)_6]^{3+}$ by Y^{n-} at 13°C.**

Y^{n-} $(n = 0)$	k $(M^{-1} s^{-1})$	Y^{n-} $(n = 1)$	k $(M^{-1} s^{-1})$
$ClCH_2CO_2H$	6.7×10^2	NCS^- (8–9°C)	8.0×10^3
CH_3CO_2H	9.7×10^2	$ClCH_2CO_2^-$	2.1×10^5
H_2O	8.6×10^3	$CH_3CO_2^-$	1.8×10^6

[a] H. Diebler, *Z. Phys. Chem.* **1969**, *68*, 64.

Electronic Effects of Inert Ligands.   Along with steric effects, inert ligands can also exert electronic effects on reaction rates. Data on acid aquation of $[Co(en)_2LX]^{n+}$ complexes (Table 11.6c) are explained nicely by a **d** mechanism. Consider the *cis*-$[Co(en)_2LX]^{n+}$ series. When the leaving group X departs, the orbital of the d^2sp^3 hybrid set that bound it to Co is empty. If the *cis* ligand is a good π-donor, it can supply electrons to electron-deficient Co, thereby stabilizing the transition state and lowering the activation energy (Figure 11.4a). As a consequence, *cis* complexes where L is a good π-donor react more rapidly than *cis* complexes of π-acceptor ligands (such as CN^-) or σ-donor-only ligands (such as NH_3). When L is *trans* to the leaving group X, no π-donation into the vacant Co orbital can occur without rearrangement to a trigonal bipyramid (Figure 11.4b,c). The energy of rearrangement raises the activation energy for aquation of *trans*-$[Co(en)_2LX]^{n+}$ and lowers the rates relative to those of the corresponding *cis* complexes. The reactivity pattern of *cis*- and *trans*-$[Co(en)_2LX]^{n+}$ complexes would be difficult to explain assuming an **a** intimate mechanism.

Comparison of Rates of Anation and Water Exchange.   Water exchange is an important reaction. Because the concentration of water (or any solvent) cannot be varied, its presence in the transition state (or lack thereof) cannot be determined kinetically. The form of the rate law for Reaction (11.1) (X = Y = H_2O) is always rate = $k_{obs}[L_5MX]$ because $[H_2O] = 55.5$ *M*. We might consider a complex having X = $H_2^{18}O$ placed in H_2O = Y so that solvent exchange could be detected by entry of H_2O of natural isotopic abundance into the coordination sphere. For a **D** mechanism, $k_2[H_2O] \gg k_{-1}[H_2^{18}O]$ in Equation (11.6) and the rate becomes $k_1[L_5MX]$. In this case, k_{obs} can be interpreted as the rate constant for breaking of the M—O bond. In the $\mathbf{I}_d$ mechanism, $K[H_2O] \gg 1$ and k_{obs} [Equation (11.10)] is identified with k, the rate constant for dissociative interchange of a water molecule between the outer and inner coordination spheres. For the $\mathbf{I}_a$ mechanism, k_{obs} is the rate of associative interchange of Y = H_2O between the outer and inner coordination spheres.

For ligands Y other than water, a similar limiting rate is often reached if [Y] is made large enough. The reaction then is just first-order in L_5MX and k_{obs} is in units of s^{-1}. When [Y] is very small (impossible when Y = solvent), the rate laws have the limiting forms summarized in Table 11.7. In the most general case, second-order kinetics are observed at low [Y]; at high [Y], we see the limiting case of first-order behavior (saturation). Figure 11.5 shows a plot of k_{obs} versus [Y] for two reactions. In one, the limiting first-order behavior is not seen at high [Y], because either k_2 or K is too small.

Table 11.6 Effect of nonleaving ligands on acid hydrolysis rates of some Co^{III} complexes

a. *trans*-[Co(N—N)$_2$Cl$_2$]$^+$ + H$_2$O → [CoCl(N—N)$_2$(H$_2$O)]$^{2+}$ + Cl^{-} [a,b]

N—N	k (s^{-1})
NH$_2$CH$_2$CH$_2$NH$_2$ (en)	3.2×10^{-5}
NH$_2$CH$_2$CH(CH$_3$)NH$_2$	6.2×10^{-5}
d, l-NH$_2$CH(CH$_3$)CH(CH$_3$)NH$_2$	1.5×10^{-4}
meso-NH$_2$CH(CH$_3$)CH(CH$_3$)NH$_2$	4.2×10^{-4}
NH$_2$CH$_2$C(CH$_3$)$_2$NH$_2$	2.2×10^{-4}
NH$_2$C(CH$_3$)$_2$C(CH$_3$)$_2$NH$_2$	Instantaneous
NH$_2$CH$_2$CH$_2$NH(CH$_3$)	1.7×10^{-5}

b. *trans*-[Co(N$_4$)LCl]$^{n+}$ + H$_2$O → *trans*-[Co(N$_4$)L(H$_2$O)]$^{(n+1)+}$ + Cl^{-} [a,c,d]

Complex	k (s^{-1})
trans-[Co(cyclam)Cl$_2$]$^+$	1.1×10^{-6}
trans-[Co(cyclam)(NCS)Cl]$^+$	1.1×10^{-9}
trans-[Co(cyclam)(CN)Cl]$^+$	4.8×10^{-7}
trans-[Co(tet-*b*)Cl$_2$]$^+$	9.3×10^{-4}
trans-[Co(tet-*b*)(NCS)Cl]$^+$	7.0×10^{-7}
trans-[Co(tet-*b*)(CN)Cl]$^+$	3.4×10^{-4}

c. [Co(en)$_2$LCl]$^{n+}$ + H$_2$O → [Co(en)$_2$L(H$_2$O)]$^{(n+1)+}$ + Cl^{-} [a,e]

Complex	k (s^{-1})
cis-[Co(en)$_2$(OH)Cl]$^+$	0.012
trans-[Co(en)$_2$(OH)Cl]$^+$	1.60×10^{-3}
cis-[Co(en)$_2$Cl$_2$]$^+$	2.4×10^{-4}
trans-[Co(en)$_2$Cl$_2$]$^+$	3.5×10^{-5}
cis-[Co(en)$_2$(NH$_3$)Cl]$^{2+}$	5×10^{-7}
trans-[Co(en)$_2$(NH$_3$)Cl]$^{2+}$	3.4×10^{-7}
cis-[Co(en)$_2$(H$_2$O)Cl]$^{2+}$	1.6×10^{-6}
trans-[Co(en)$_2$(H$_2$O)Cl]$^{2+}$	2.5×10^{-6}
cis-[Co(en)$_2$(CN)Cl]$^+$	6.2×10^{-7}
trans-[Co(en)$_2$(CN)Cl]$^+$	8.2×10^{-5}

[a] At 25°C.

[b] R. G. Pearson, C. R. Boston and F. Basolo, *J. Am. Chem. Soc.* **1953,** *75,* 3089.

[c] N$_4$-tetradentate amine ligand; see Figure 11.3 for structures.

[d] From the work of Poon, Tobe, and others.

[e] From the work of Chan, Tobe, and others.

Comparing values of limiting first-order rate constants for anation and for solvent exchange allows us to choose between **a** and **d** intimate mechanisms. In the case of **d** activation, the main contribution to the activation energy is the breaking of the M—X bond. Bond formation to the entering group tends to take place rather unselectively with any Y in correct position in the second coordination sphere. Because of its tremendous concentration advantage, water is the most likely entering group in aqueous solution, and so we

$$NH_2CH_2CH_2NHCH_2CH_2NH_2$$

(*a*) dien (1, 4, 7– triazaheptane)

$$NH_2CH_2CH_2NHCH_2CH_2NHCH_2CH_2NH_2$$

(*b*) trien (1, 4, 7, 10– tetraazadecane)

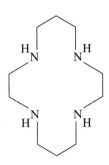

(*c*) cyclam (1, 4, 8, 11 – tetraazacyclotetradecane)

(*d*) tet–*b* (*d, l*–1, 4, 8, 11– tetraaza–5, 5, 7, 12, 12, 14 – hexamethylcyclotetradecane)

tet–*a* (*meso*–1, 4, 8, 11– tetraaza–5, 5, 7, 12, 12, 14 – hexamethylcyclotetradecane)

Figure 11.3   Structures of some chelating ligands. (*a*) dien (1,4,7–triazaheptane). (*b*) trien (1,4,7,10–tetraazadecane). (*c*) cyclam (1,4,8,11–tetraazacyclotetradecane. (*d*) tet–*b* (*d, l*-1,4,8,11–tetraaza–5,5,7,12,12,14–hexamethylcyclotetradecane); tet–*a* (*meso*–1,4,8,11–tetraaza–5,5,7,12,12,14–hexamethylcyclotetradecane).

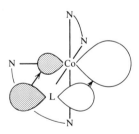

(*a*) π-Donation by a *cis* ligand into the *p* component of an empty d^2sp^3 hybrid.

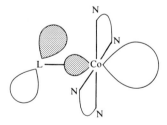

(*b*) π-Orbital of *trans* L is orthogonal to d^2sp^3 and no donation occurs.

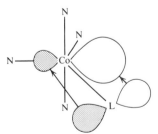

(*c*) π-Donation from *trans* L can occur upon rearrangment to trigonal bi pyramid.

Figure 11.4   (*a*) π-Donation by a *cis* ligand into the *p* component of an empty d^2sp^3 hybrid. (*b*) π-Orbital of *trans* L is orthogonal to d^2sp^3 and no donation occurs. (*c*) π-Donation from *trans* L can occur upon rearrangement to trigonal bipyramid. (After F. Basolo and R. Pearson, *Mechanisms of Inorganic Reactions*, 2nd ed., Wiley, New York, 1967.)

Table 11.7   **Limiting forms of rate laws and significance of rate constants**[a]

Stoichiometric mechanism	Conditions	Rate law	k_{obs}	Comments
D	$k_2[Y]$ very large	$k_1[L_5MX]$	k_1	k_{obs} represents rate of M—X dissociation.
D	$k_2[Y]$ very small	$\dfrac{k_1 k_2[L_5MX][Y]}{k_{-1}[X]}$	$\dfrac{k_1 k_2[Y]}{k_{-1}[X]}$	If X = solvent, [X] is constant; otherwise, can observe mass law retardation.
I$_d$	$K[Y]$ very large	$k[L_5MX]$	k	k_{obs} represents interchange rate.
I$_d$	$K[Y]$ very small	$kK[L_5MX][Y]$	$kK[Y]$	k_{obs} is composite.
I$_a$	$K[Y]$ very large	$k[L_5MX]$	k	k_{obs} represents rate of ligand interchange.
I$_a$	$K[Y]$ very small	$kK[L_5MX][Y]$	$kK[Y]$	k_{obs} is composite.
A **A**	[Y] very large [Y] very small	$\dfrac{k_1[L_5MX][Y]}{k_{-1}+k_2}$	$\dfrac{k_1[Y]}{k_{-1}+k_2}$	Kinetics always second-order if second step is irreversible $(k_{-2}=0)$.

[a] Results from Equations (11.6), (11.10), and (11.12).

expect that the anation rate should never be greater than the water exchange rate for **d** activation. Data presented in Table 11.4 show that anation of $[Co(NH_3)_5(H_2O)]^{3+}$ is **d**. In cases of **a** activation, we expect to see selectivity toward the entering ligand, and it is entirely possible that some ligands will be better entering groups than water. The data in Tables 11.5 and 11.8 indicate an **a** mechanism. The fact that limiting first-order rates are reached indicates the mechanism to be **I$_a$** rather than **A**. Although a limiting first-order rate constant for anation larger than that for exchange is a sufficient criterion for establishing

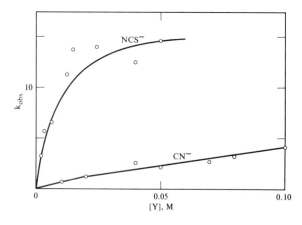

Figure 11.5   Plots of k_{obs} versus concentration of entering ligand for the anation reactions $[Co^{III}(hematoporphyrin(IX))(H_2O)_2] + Y^-$ showing almost linear behavior for $Y^- = CN^-$ and transition from second- to first-order behavior for $Y^- = NCS^-$. (Reprinted with permission from E. B. Fleischer, S. Jacobs, and L Mestichelli, *J. Am. Chem. Soc.* **1968,** *90,* 2527. Copyright 1968, American Chemical Society.)

an I_a mechanism, it is not a necessary one. Even though the anation reaction reactions of $[Cr(H_2O)_6]^{3+}$ are thought to proceed via the I_a mechanism, no soft base is as effective as H_2O for attacking the hard acid Cr^{3+}.[7,8]

Effect of Charge on Reaction Rate.   All other things being equal, increased positive charge should make bond breaking between ligands and metal more difficult. Hence, we expect the rate to decrease with increasing positive charge for **d** activation. The fact that the water exchange reactions of a number of main-group metals (Section 11.3.1) follow this trend, coupled with the similarity of substitution to aquation rates, provides evidence for **d** activation. With transition metal complexes, LFSE considerations (Section 11.3.2) are superimposed on those of charge. Moreover, for a metal ion in a given oxidation state, there is no way to vary the overall charge on the complex without introducing differently charged ligands, which leads to change in σ- and π-bonding. Even so, an increase in rate as charge decreases often has been accepted as evidence of **d** activation. Table 11.9 shows rate constants for anation of $[Fe(H_2O)_6]^{3+}$ and $[Fe(OH)(H_2O)_5]^{2+}$. The rate increase for less positively charged hydroxo complexes implies a **d** mechanism. It may be that electron density can be supplied to the metal by OH^-, thus compensating for loss of the electron pair during the bond-breaking step.

The expectations for **a** mechanisms as the positive charge on the complex increases are not clear. Although bond making should be enhanced (thus increasing rates for **A** reactions), bond breaking should be retarded, leading to opposing effects in I_a pathways. The above considerations do not lead to any conclusions about substitution on the reference system $[Fe(H_2O)_6]^{3+}$. However, the rather pronounced dependence of rate on entering ligand identity and the fact that some ligands enter as fast as water point to an I_a mechanism for these substitutions.

Table 11.8   Limiting rate constant[a] for anation by Y^{n-} and water exchange (k_e)

Complex	Y^{n-}	k (s^{-1})	k/k_e
$[Rh(NH_3)_5(OH_2)]^{3+b,d}$	Br^-	7.9×10^{-3}	4.9
	Cl^-	4.2×10^{-3}	2.6
	SO_4^{2-}	1.7×10^{-3}	1.0
	H_2O	1.6×10^{-3}	—
$[Ir(NH_3)_5(OH_2)]^{3+c,e}$	Cl^-	9.2×10^{-4}	~4
	H_2O	2.2×10^{-4}	—

[a] See Table 11.7 for explanation.

[b] F. Monacelli, *Inorg. Chim. Acta* **1968**, *2*, 263.

[c] E. Borghi, F. Monacelli, and T. Prosperi, *Inorg. Nucl. Chem. Lett.* **1970**, *6*, 667.

[d] At 65°C.

[e] At 85°C.

[7] J. H. Espenson, *Inorg. Chem.* **1969**, *8*, 1554.

[8] A metal ion with its normal complement of water ligands is sometimes written as M^{n+}.

Table 11.9   Rate constants for anation by Y^{n-} for $[Fe(H_2O)_6]^{3+}$ and $[Fe(OH)(H_2O)_5]^{2+}$ at 25°C^a

Y^{n-}	k (M^{-1} s^{-1}) for $[Fe(H_2O)_6]^{3+}$	k (M^{-1} s^{-1}) for $[Fe(H_2O)_5(OH)]^{2+}$
SO_4^{2-}	1.1×10^5	2.3×10^3
Cl^-	5.5×10^3	4.8
Br^-	2.6×10^3	1.6
NCS^-	5.1×10^3	90
$ClCH_2CO_2^-$	4.1×10^4	1.5×10^2
H_2O	1.2×10^5 (s^{-1})	1.6×10^2 (s^{-1})

^{a} From M. Grant and R. B. Jordan, *Inorg. Chem.* **1981**, *20*, 55 and references therein.

Activation Parameters.   The activation energy defined in Figure 11.1 ($E_a = RT \ln k_{obs}$) can be shown to be related to $\Delta G^{\ddagger} = \Delta H^{\ddagger} - T\Delta S^{\ddagger}$. Thus, by studying a reaction at different temperatures, values of $\Delta H^{\ddagger}$ (which relates to the energy requirements for reaching the transition state) and $\Delta S^{\ddagger}$ (which relates to the change in ordering on reaching the transition state) may be extracted. Because of the definition of a transition state, $\Delta G^{\ddagger}$ and $\Delta H^{\ddagger}$ will be > 0; since the transition state may be more or less ordered than the reactants, $\Delta S^{\ddagger}$ may be either positive or negative.

Examining activation parameters provides added information on mechanism for a series of related complexes because opposite trends in $\Delta H^{\ddagger}$ and $\Delta S^{\ddagger}$ are masked by similar values of k_{obs}. Similarities in trends of activation parameters for related reactions constitutes evidence for similarity in mechanism.

An interesting case is aquation of ammonia and methylamine complexes of Cr^{3+} and Co^{3+}; see Table 11.10. As predicted on steric grounds for a **d** mechanism, $[Co(NH_2Me)_5Cl]^{2+}$ aquates faster than $[Co(NH_3)_5Cl]^{2+}$. The opposite behavior for the analogous Cr^{3+} complexes has been taken as evidence of an **a** mechanism. However, activation parameters show that faster aquation of $[Co(NH_2Me)_5Cl]^{2+}$ is not due to any weakness of the Co—Cl bond because $\Delta H^{\ddagger}$ is about the same as for $[Co(NH_3)_5Cl]^{2+}$. It is rather a result of less negative $\Delta S^{\ddagger}$ which is probably a result of poorer solvation of the methylamine complex leading to less order in the second coordination sphere. (Much other evidence indicates that Co^{3+} complexes generally are $\mathbf{I_d}$.) Turning to the Cr com-

Table 11.10   Rate constants and activation parameters for aquation of Co^{III} and Cr^{III} ammine complexes at 25°C$^{a, b}$

Complex	$10^6 k$ (s^{-1})	$\Delta H^{\ddagger}$ (kJ/mol)	$\Delta S^{\ddagger}$ (J/mol K)
$[Co(NH_3)_5Cl]^{2+}$	1.72	93	-44
$[Co(NH_2Me)_5Cl]^{2+}$	39.6	95	-10
$[Cr(NH_3)_5Cl]^{2+}$	8.70	93	-29
$[Cr(NH_2Me)_5Cl]^{2+}$	0.26	110	-2

^{a} From the work of Lawrance, Swaddle, van Eldik, Lay, and others.

^{b} Cl^- is the leaving ligand.

plexes, the slower rate for $[Cr(NH_2Me)_5Cl]^{2+}$ results from a considerably larger $\Delta H^{\ddagger}$, suggesting a stronger Cr—Cl bond (which must be weakened to reach the transition state) than in $[Cr(NH_3)_5Cl]^{3+}$. The conclusion is that the Cr reactions are also I_d (although many Cr^{3+} substitutions are **a**). This illustrates the necessity of comparing groups of related reactions and doing it in several ways in order to establish mechanism.

Recently, measurements at various pressures have been employed to obtain volumes of activation $\Delta V^{\ddagger}$.[9] k_{obs} is proportional to $\exp(-P\Delta V^{\ddagger}/RT)$ so that a plot of $\ln k_{obs}$ versus P gives a straight line of slope $-\Delta V^{\ddagger}/RT$. Values of $\Delta V^{\ddagger}$ are not always straightforward to interpret. The naive expectation is that **d** mechanisms should lead to $\Delta V^{\ddagger} > 0$ because a ligand is at least partially released into the bulk solvent in the transition state. In contrast, an **a** mechanism partially bonds a species formerly present in the bulk solvent and thus might be expected to lead to $\Delta V^{\ddagger} < 0$. This is probably true insofar as only the bond angles and distances intrinsic to the reacting species are concerned. Solvent effects are also important, however, and any process which creates charge leads to solvent electrostriction and a negative contribution to $\Delta V^{\ddagger}$. Conversely, charge neutralization gives a positive contribution. The volume of activation is thus partitioned into two contributions: $\Delta V^{\ddagger} = \Delta V^{\ddagger}_{intrinsic} + \Delta V^{\ddagger}_{solvent}$. $\Delta V^{\ddagger}$ is often large and negative when anionic ligands leave cationic complexes in **d** mechanisms. Volumes of activation for aquation of $[Co(NH_3)_5SO_4]^+$, $[Co(NH_3)_5Cl]^{2+}$, and $[Co(NH_3)_5(H_2O)]^{3+}$ of -18.3, -10.6, and $+1.2$ cm^3/mol, respectively, are seen even though all these are **d** reactions. $\Delta V^{\ddagger}$ is large and positive when anionic ligands leave anionic complexes (total charge remains the same, but two species of lesser charge are less tightly solvated). This suggests domination of $\Delta V^{\ddagger}$ by $\Delta V^{\ddagger}_{solvent}$ when charged ligands are involved. Another factor not yet assessed is the contribution by the nonlabile ligands whose bond strengths and bond distances from M will surely change on going to the transition state.[10, 11]

The data in Table 11.11 show different patterns in $\Delta V^{\ddagger}$ for aquation of a series of Co^{3+} and Cr^{3+} complexes with neutral leaving groups. This is evidence that the Cr^{3+} reactions are **a** because much other evidence points to a **d** mechanism for the Co^{3+} reactions.

Summary of Results on Intimate Mechanism.   In this section we have developed models for mechanisms of ligand substitution reactions within the transition-state theory. Equally importantly, we have discussed what kinds of experimental data can be used to distinguish among possible pathways. The rate law for any reaction gives only the chemical composition of the transition state. The energetic importance of various structural features and of reactant identity must be assessed by examining series of similar complexes, a point of view exemplified in the preceding material on octahedral complexes. We have focused on identifying reactions as **d** or **a**.

In many cases, evidence is not sufficiently definitive to label an interchange (**I**) as either **a** or **d** in a series of related reactions. One can only say that some members of the series are more or less dissociative than others.

[9] R. van Eldik, T. Asano and W. J. LeNoble, *Chem. Rev.* **1989**, *89*, 549; A. E. Merbach, *Pure Appl. Chem.* **1987**, *59,* 161.

[10] C. H. Langford, *Inorg. Chem.* **1979**, *18*, 3288.

[11] K. E. Newman and A. E. Merbach, *Inorg. Chem.* **1980**, *18*, 2481.

Table 11.11 Rate constants (25°C) and activation parameters for aquation of $[M(NH_3)_5X]^{3+}$, M = Cr, Co

X	$k \times 10^5$ (s^{-1})	$\Delta H^{\ddagger}$ (kJ/mol)	$\Delta S^{\ddagger}$ (J/mol K)	$\Delta V^{\ddagger}$ (cm³/mol)
M = Cr				
H_2O	5.2	97.0	0.0	−5.8
$OSMe_2$	1.95	95.3	−15	−3.2
$OCHNH_2$	5.1	94.0	−12	−4.8
$OC(NH_2)_2$	2.0	93.5	−22	−8.2
$OP(OMe)_3$	6.0	89.7	−23	−8.7
M = Co				
H_2O	0.59	111	+28	+1.2
$OSMe_2$	1.8	103	+10	+2.0
$OCH(NH_2)$	0.58	107	+12	+1.1
$OC(NH_2)_2$	5.5	94	−10	+1.3
CH_3OH	6.5	98	+5	+2.2

[a] N. J. Curtis, G. A. Lawrance, and R. van Eldik, *Inorg. Chem.* **1989**, *28*, 329.

The choice of **a** or **d** activation as a function of metal is not understood completely. On the basis of progressively more positive volumes of activation across the first transition series, water exchange is considered to progress from **a** to **d** activation. All other things being equal, trivalent metals are more likely to exhibit **a** mechanisms than divalent ones. However, all complexes of Co[III] investigated so far have been found to react via a **d** mechanism. Aside from Co[III], the most-studied metal ion has been Cr[III]. It now seems clear that Cr[III] complexes can react with either **d** or **a** activation, although most are **a**.[12] A considerable number of Ru[III] complexes react via **a** pathways, and a claim to the discovery of an **A** mechanism has been advanced[13] on the basis of the very negative $\Delta V^{\ddagger}$ for both aquation of $[Ru(NH_3)_5Cl]^{2+}$ and Cl⁻ anation of $[Ru(NH_3)_5(H_2O)]^{3+}$ ($\Delta V^{\ddagger}$ = −30 and −20 cm³/mol, respectively).

Table 11.12 summarizes available information on intimate mechanism as a function of metal ion. As more experiments are done, we can anticipate that evidence for **a** activation will be found for more of the species now in the **d** column (especially those with high positive charge).

As compared with Co[III] complexes, Ru[III], Rh[III], and Ir[III] complexes react very slowly, presumably because of the large LFSE. In addition, bond making by these large cations is likely to be of greater energetic importance than in Co[III], even when the intimate mechanism is still **d** (Table 11.13).

[12] Studies of Cr[III] complexes are complicated by oligomerization of Cr[III] aqua species except under the most acidic conditions and the tendency for Cr—N bond rupture.

[13] M. T. Fairhurst and T. W. Swaddle, *Inorg. Chem.* **1979**, *18*, 3241.

[14] For an **A** reaction the second step in Equation (11.11) could be reversible; $k_{-2} \neq 0$. A departure from strict second-order kinetics would be seen. This kinetic detection could occur even if the intermediate were too short-lived for spectroscopic observation.

Table 11.12   Intimate mechanisms for substitutions of octahedral complexes

a	a *and* d	d
Ti^{3+}, V^{2+}, V^{3+}, Mo^{3+}, Cd^{2+}, In^{3+}	Cr^{3+} (mostly **a**), Mn^{2+}, Fe^{3+}, Co^{2+} (mostly **d**), Ru^{2+} (mostly **d**), Ru^{3+}, Rh^{3+}, Ir^{3+}	Mg^{2+}, Al^{3+}, TiO^{2+}, VO^{2+}, Fe^{2+}, Co^{3+}, Ni^{2+}, Ni^{3+}, Cu^{2+}, Zn^{2+}, Ga^{3+}

Stoichiometric Mechanisms

The previous section outlined methods for distinguishing between dissociative (**d**) and associative (**a**) intimate substitution mechanisms. We now address the problem of how to differentiate between the stoichiometric mechanisms **D** [Equation (11.2)] and I_d [Equation (11.7)] and between **A** (Equation (11.11)) and I_a [Equation (11.7)] mechanisms. The distinction in each case rests on the ability to detect a five-coordinate (for **D**) or seven-coordinate (for **A**) intermediate. Detection of such intermediates is a rigorous requirement that can be fulfilled only if the intermediate's lifetime is sufficiently long for us to observe it. If not, the reaction must be regarded as concerted (I_d or I_a).

A Versus I_a.   Although a seven-coordinate intermediate ML_5XY could survive long enough for spectroscopic detection if X were a very poor leaving group,[14] no such intermediates have been detected so far. A claim for an **A** mechanism for anation of $[Ru(NH_3)_5(H_2O)]^{3+}$ relies on $\Delta V^{\ddagger}$ measurements (*vide supra*) that do not depend on activation energies. The extremely wide variation of rate with entering ligand has been taken as evidence for an **A** mechanism in substitutions on $[Ti(H_2O)_6]^{3+}$ (Table 11.5). Pending further evidence, most presently known **a** reactions all must be regarded provisionally as I_a.

D Versus I_d.   Very large positive volumes of activation ($\Delta V^{\ddagger} \sim +22 \text{ cm}^3$) have been taken as evidence of **D** mechanisms as in the aquation of $[Fe(CN)_5L]^{3-}$ complexes.[15]

Table 11.13   Acid hydrolysis rate constants for some inert octahedral complexes at 50°C[a,b]

Complex	$k \times 10^6$ (s^{-1})	Complex	$k \times 10^6$ (s^{-1})
$[Ru(NH_3)_5Cl]^{2+}$	328 (80.1°)	$[Rh(NH_3)_5I]^{2+}$	0.24
$[Ru(NH_3)_5Br]^{2+}$	309 (80.1°)	*trans*-$[Ir(en)_2Cl_2]^+$	5.9
$[Ru(NH_3)_5I]^{2+}$	164 (80.1°)	*trans*-$[Ir(en)_2BrCl]^+$	9.0
$[Rh(NH_3)_5Cl]^{2+}$	1.13	*trans*-$[Ir(en)_2ICl]^+$	159
$[Rh(NH_3)_5Br]^{2+}$	1.00		

[a] Leaving group written last in formula.

[b] From the work of Kane-Maguire, Pöe, Basolo, and others.

[15] K. B. Reddy and R. van Eldik, *Inorg. Chem.* **1991**, *30*, 596.

Second-Order Limiting Rate Constant. An intermediate of reduced coordination number need not be detected spectroscopically, but can be recognized by so-called kinetic tests involving discrimination in reaction with various entering groups or by product stereochemistry. Just as the relative values of rate constants under first-order limiting conditions enable us to recognize **a** and **d** mechanisms, so values of rate constants under second-order limiting conditions (low [Y]) can permit a choice between **D** and **I**$_d$ stoichiometric mechanisms. We can think of ligand interchange between outer and inner sphere in the **I**$_d$ path as occurring so rapidly that the outer coordination sphere lacks sufficient time to reorganize (relax) before the reaction is done. Consequently, whatever ligand happens to be in the proper geometric orientation in the outer sphere enters ("accidental bimolecularity"). In the **D** mechanism there is sufficient time for rearrangement in the outer sphere before a new group enters the inner sphere, so the coordinatively unsaturated intermediate can discriminate among incoming ligands.

Under second-order limiting conditions for an **I**$_d$ reaction, k_{obs} can be identified with kK [Equation (11.10)]. K often can be measured from spectroscopic experiments on nonreacting systems or calculated from theoretical expressions.[16] If we compare a series of reactions in which a variety of entering groups displace the same leaving group, we expect any differences in k_{obs} to stem from differences in K if the mechanism is **I**$_d$. Table 11.14 presents data for ligand replacement reactions of $[Ni(H_2O)_6]^{2+}$. Values of k obtained by dividing k_{obs} by a calculated K are essentially the same, as expected for an **I**$_d$ mechanism, even though the charge on the entering group varies from -2 to $+1$.[17]

For any reaction proceeding via a **D** mechanism, the second-order limiting value of k_{obs} can be identified as $k_1 k_2 [Y]/k_{-1}[X]$ (Table 11.7). When X = solvent, [X] will be a constant. Moreover, in comparing a series of reactions featuring a solvent leaving group and different entering groups Y, k_1/k_{-1} will be constant and differences in k_2 will control the changes in k_{obs} on varying Y. Variation in k_{obs} with Y reflects the ability of a five-coordinate intermediate to discriminate among Y. Indeed, a truly five-coordinate intermediate has no "memory" of the leaving group and always displays the same selectivity toward Y, no matter what its origin. For a **D** pathway, any difference in k_{obs} is attributable to the identity of the entering group.

We might expect such behavior to be apparent if outer-sphere complexation with the entering reagent (required for the **I**$_d$ path) could be minimized—a situation likely to obtain in reactions of two anionic species such as

$$[RhCl_5(H_2O)]^{2-} + Y^- \longrightarrow [RhCl_5Y]^{3-} + H_2O \qquad (11.16)$$

A plot of k_{obs}^{-1} versus $[Y^-]^{-1}$ [Equation (11.6)] corrected for the possibility of reversibility in the second step of the mechanism of Equation (11.2) permits extraction of the ratio k_2/k_{-1}, the competition ratio between Y^- and H_2O for a five-coordinate intermediate (Table 11.15). Note that the values quoted represent evidence for a **d** intimate mech-

---

[16] D. B. Rorabacher, *Inorg. Chem.* **1966,** *5,* 1891.

[17] Although it may be possible to demonstrate the presence of an outer-sphere complex, such a demonstration does not necessarily mean that it is an intermediate along the reaction path. See J. Halpern, *J. Chem. Educ.* **1968,** *45,* 372.

Table 11.14 Observed rate constants (k_{obs}), calculated outer-sphere complexation equilibrium constants (K), and resulting values of $k\left(= \dfrac{k_{obs}}{K} \right)$ for $[Ni(H_2O)_6]^{2+}$ reactions with ligands Y^{n-} at 25°C[a]

Y^{n-}	$k_{obs}(M^{-1}\,s^{-1})$	$K(M^{-1})$[b]	$k = \dfrac{k_{obs}}{K}\,(s^{-1})$
CH_3COO^-	1×10^5	3	3×10^4
SCN^-	6×10^3	1	6×10^3
F^-	8×10^3	1	8×10^3
HF	3×10^3	0.15	2×10^4
H_2O			3×10^3
NH_3	5×10^3	0.15	30×10^3
$NH_2(CH_2)_2N(CH_3)_3^+$	4×10^2	0.02	20×10^3

[a] From R. G. Wilkins, *Acc. Chem. Res.* **1970**, *3*, 408.
[b] Calculated from the expression

$$K = \frac{4\pi N r^3}{3000}\, e^{-U(r)/kT}$$

where

$$U(r) = \frac{Z_1 Z_2 e^2}{rD} - \frac{Z_1 Z_2 e^2 \kappa}{D(1 + \kappa r)}$$

(representing the sum of the attractive and repulsive potentials between two ions of charge $Z_1 e$ and $Z_2 e$ at a distance r in a medium of dielectric constant D)

$$\kappa^2 = \frac{8\pi^2 N e \mu}{1000 D k T}$$

μ is the ionic strength, and r is taken as 500 pm.

Table 11.15 k_2/k_{-1} ratios for anation of $[RhCl_5(H_2O)]^{2-}$[a]

Y^{n-}	k_2/k_{-1}	Y^{n-}	k_2/k_{-1}
H_2O (exchange)	1.0	SCN^-	0.079
Cl^-	0.021	NO_2^-	0.10
Br^-	0.016	N_3^-	0.14
I^-	0.018		

[a] D. Robb, M. M. Steyn, and H. Krueger, *Inorg. Chim. Acta* **1969**, *3*, 383.

anism because the rates are all smaller than that for H_2O exchange. The 60-fold range indicates discrimination among the various possible entering groups.

Other techniques for detecting five-coordinated intermediates are discussed in Section 11.3.5.

So far, rather few **d** reactions can be characterized as **D.** Lack of definitive evidence for the unsaturated intermediate forces us to label them as I_d.[18]

EXAMPLE 11.1:   Anation of $[Cr(NH_3)_5(H_2O)]^{3+}$ is thought to proceed by an interchange mechanism. The table below presents rate constants and activation parameters for interchange of the entering and leaving groups in the Cr coordination sphere at 50°C.

Entering ligand	$10^4\ k(s^{-1})$	$\Delta H^{\ddagger}$ (kJ/mol)	$\Delta S^{\ddagger}$ (J/mol K)
NCS^-	6.12	102	12
$HC_2O_4^-$	6.2	112	39
$C_2O_4^{2-}$	6.2	104	33
H_3PO_4	1.45		
$H_2PO_4^-$	1.45		
$[Co(CN)_6]^{3-}$	2.5	103	26
H_2O (exchange)	13.7	97	0

What do these data suggest about the intimate mechanism of these reactions? (See D. C. Gaswick and S. M. Malinak, *Inorg. Chem.* **1993**, *32*, 175 and references therein.)

Solution: All the anions and neutral molecules display activation parameters which are virtually the same, indicating a common mechanism. If the mechanism were **a,** we might expect the anionic entering groups to have different *k* values than comparable neutrals. But this is not the case. Furthermore, one might expect that a bulky group such as the complex $[Co(CN)_6]^{3-}$ (which reacts to give $[(NH_3)_5CrNCCo(CN)_5]$) to react more slowly than other ligands; but, over a wide variety of types of entering groups the rate varies by a factor of only about 3. Moreover, all the anation rates are smaller than those of water exchange. This suggests that bond breaking is the important mechanistic step and that the mechanisms are I_d. (Other substitutions on Cr have been interpreted as being **a,** and the question is not yet settled.)

▶ 11.3.5 *Base Hydrolysis*

Replacement of a ligand by OH^- is called **base hydrolysis** [Equation (11.1), $Y = OH^-$]. In Co^{III} complexes containing amine ligands, the rate of base hydrolysis is very much faster than acid hydrolysis (aquation). Compare the second-order rate constants for base hydrolysis in Table 11.16 with aquation rate constants (Tables 11.4 and 11.6). Base hydrolysis is ordinarily second-order and never seems to reach the mixed-order behavior (rate saturation) of the rate laws for the **D** and I_d mechanisms (Section 11.3.3). Base hydrolysis of $[Co(NH_3)_5Cl]^{2+}$ obeys the rate expression

---

[18] One seemingly simple piece of evidence of a **D** mechanism would be the observation of the mass-law retardation by X predicted in Equation (11.6). However, it has been shown that outer-sphere complexation by X released in the reaction leads to a rate law of the same form. See J. A. Ewen and D. J. Darensbourg, *J. Am. Chem. Soc.* **1976**, *98*, 4317.

Table 11.16   Rate constants for base hydrolysis of some octahedral complexes[a,b]

Complex	k (M^{-1} s^{-1}) (25°C)	Complex	k (M^{-1} s^{-1}) (0°C)
$[Co(NH_3)_5OP(OMe)_3]^{3+}$	79	cis-$[Co(OH)(en)_2Cl]^+$	0.37
$[Co(NH_3)_5NO_3]^{2+}$	5.5	trans-$[Co(OH)(en)_2Cl]^+$	1.7×10^{-3}
$[Co(NH_3)_5I]^{2+}$	3.2	cis-$[Co(en)_2Cl_2]^+$	15.1
$[Co(NH_3)_5Cl]^{2+}$	0.23	trans-$[Co(en)_2Cl_2]^+$	85
$[Co(NH_3)_5SO_4]^+$	4.9×10^{-2}	cis-$[Co(en)_2(NH_3)Cl]^{2+}$	0.50
$[Co(NH_3)_5F]^{2+}$	1.3×10^{-2}	trans-$[Co(en)_2(NH_3)Cl]^{2+}$	1.25
$[Co(NH_3)_5(NCS)]^{2+}$	5.0×10^{-4}	cis-$[Co(CN)(en)_2Cl]^+$	8.9×10^{-3}
$[Co(NH_3)_5N_3]^{2+}$	3.0×10^{-4}	trans-$[Co(en)_2(CN)Cl]^+$	0.13

[a] Leaving ligand written last.

[b] From the work of Lalor, Chan, Bailar, Taube, Basolo, Pearson, Wallace, Kane-Maguire, and others.

$$\text{rate} = k[Co(NH_3)_5Cl^{2+}][OH^-] \tag{11.17}$$

up to $[OH^-] = 1.0$ M. These facts point to a special mechanism for base hydrolysis. In fact, it was argued for some time that base hydrolysis provided an example of an **A** reaction involving rate-determining attack by OH^-. However, it would be extremely surprising if OH^- were the *only* nucleophile in aqueous solution capable of attack. Currently, the mechanism first proposed by Garrick in 1937 is considered to operate:

$$[Co(NH_3)_5Cl]^{2+} + OH^- \underset{K_h}{\rightleftharpoons} [Co(NH_2)(NH_3)_4Cl]^+ + H_2O$$
$$[Co(NH_2)(NH_3)_4Cl]^+ \xrightarrow{k} [Co(NH_2)(NH_3)_4]^{2+} + Cl^- \tag{11.18}$$
$$[Co(NH_2)(NH_3)_4]^{2+} + H_2O \xrightarrow{\text{fast}} [Co(NH_3)_5(OH)]^{2+}$$

The first step involves the proton removal by OH^- in a rapid acid–base equilibrium, giving a complex ion of lower charge that loses Cl^- more rapidly than the starting complex. The last step is relatively fast, so the second step is rate-determining. This is referred to as a **D–CB** mechanism, indicating that it involves a **D** reaction of the conjugate base (**CB**) of the starting complex[19] (Figure 11.6).

The rate law [Equation (11.17)] is easily shown to be consistent with the **D–CB** mechanism if the first step of Equation (11.18) is a rapidly established equilibrium. The rate would be first-order in conjugate base [$CoCl(NH_2)(NH_3)_4^+$], but the concentration of conjugate base can be related to the concentrations of the initial complex and OH^-. Consider the hydrolysis reaction

$$\text{Complex} + OH^- \rightleftharpoons \text{Conjugate base} + H_2O \qquad K_h$$

$$K_h = \frac{K_a}{K_w} = \frac{[\text{Conjugate base}]}{[\text{Complex}][OH^-]} \tag{11.19}$$

$$[\text{Conjugate base}] = \frac{K_a[\text{Complex}][OH^-]}{K_w} \tag{11.20}$$

[19] In the older literature, this is referred to as the S_{N1}**CB** mechanism.

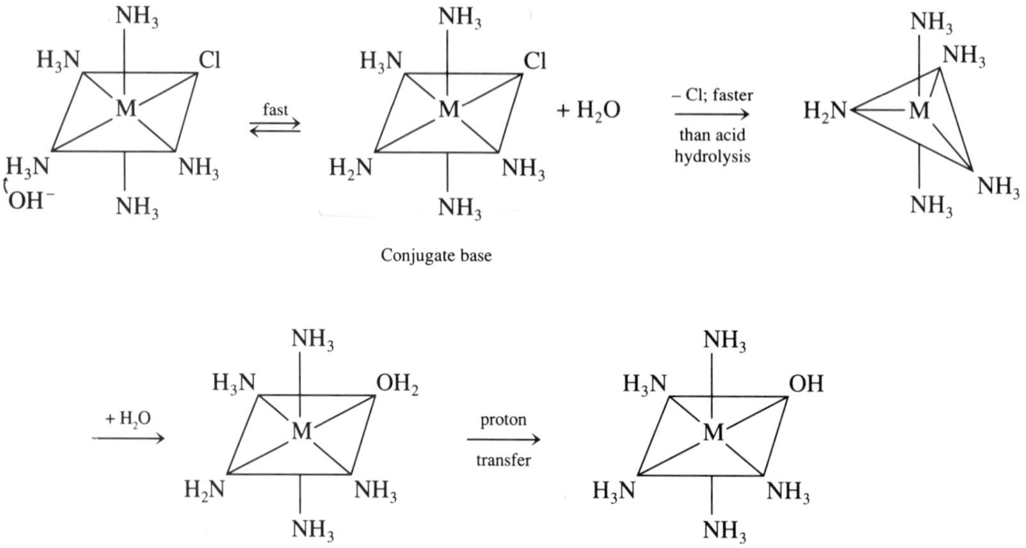

Figure 11.6　The **D–CB** mechanism.

Because

$$\text{rate} = k'[\text{Conjugate base}] = k'\frac{K_a}{K_w}[\text{Complex}][\text{OH}^-] \qquad (11.21)$$

Equation (11.17) is the observed rate law where $k = k'K_a/K_w$. The **D–CB** mechanism requires a moderately acidic proton in the starting complex. A complex without an acidic proton should react with OH^- much more slowly, and the rate would be expected to be independent of $[\text{OH}^-]$. This, in fact, is observed for the base hydrolysis of $[\text{Co(CN)}_5\text{Br}]^{3-}$ and *trans*-$[\text{Co(py)}_4\text{Cl}_2]^+$. However, $[\text{Pt(CN)}_4\text{Br}_2]^{2-}$, *cis*-$[\text{CoL}_2\text{Cl}_2]^+$ (L = bipy, phen), and *cis*-$[\text{CrL}_2\text{Cl}_2]^+$ (L = bipy, phen)[20] all undergo base hydrolysis in spite of a lack of acidic protons for reasons that are not well understood.

　　The important features of the Garrick mechanism [Equation (11.18)] have been well documented in the literature.[21] The conjugate base is believed to be about 10^6 more reactive to substitution than the parent. A true five-coordinate intermediate is produced having trigonal-bipyramidal geometry (unless the ligands are too rigid to rearrange). The existence of the intermediate is inferred by several lines of evidence.

Effects of Nonleaving Ligands

The trigonal bipyramid is stabilized by the π-donor NH_2^-. (See Figure 11.4, where L = NH_2^-.) On rearrangement to a trigonal bipyramid, ligands both *cis* and *trans* to the leaving group occupy equivalent positions. Hence, the reactivities of *trans*-$[\text{CoL(en)}_2\text{Cl}]^+$ reflect π-donor abilities of L. The low reactivity for X = CN^- is related to its π-acceptor

---

[20] J. Josephson and C. E. Schaeffer, *J. Chem. Soc., Chem. Commun.* **1970**, 61.

[21] An excellent overall review is provided by M. L. Tobe, *Acc. Chem. Res.* **1970**, *3*, 377.

properties. Although a good π-donor *cis* to the leaving ligand can donate electrons without the necessity for rearrangement to a trigonal bipyramid, it seems likely that the five-coordinate intermediates of base hydrolysis survive long enough to rearrange. Note that both *cis*- and *trans*-[Co(en)$_2$LX]$^{n+}$ often display similar reactivities in base hydrolysis—in contrast to acid hydrolysis, where the *cis* complexes are more reactive. Further evidence supporting this view lies in the stereochemical course of base hydrolysis discussed in Section 11.3.6.

Although steric effects of nonleaving ligands have not been investigated as intensively as for aquation, base hydrolysis rates for [M(NH$_2$Me)$_5$(OSO$_2$CF$_3$)]$^{2+}$ (M^{III} = Co, Rh, Ir, Cr) were enhanced by a factor of >1000 for Co, 150 for Rh, and 800 for Cr compared to the NH$_3$ complexes. In contrast, only small enhancements were seen for acid hydrolysis for Co and Rh, and a small retardation was seen for Cr.[22]

Product Analysis

If a five-coordinate intermediate exists, it must be able to discriminate among various entering groups (Section 11.3.4). In aqueous solution in the presence of Y$^-$ the intermediate could be captured either by Y$^-$ or by H$_2$O.

$$[\text{Co(NH}_2)(\text{NH}_3)_4]^{2+} \xrightarrow{\;+\text{H}_2\text{O}\;} [\text{Co(OH)(NH}_3)_5]^{2+} \qquad (11.22)$$

$$[\text{Co(NH}_2)(\text{NH}_3)_4]^{2+} \xrightarrow{\;+\text{Y}^-\;} [\text{CoY(NH}_2)(\text{NH}_3)_4]^{+} \xrightarrow{\;\text{H}_2\text{O}\;} [\text{CoY(NH}_3)_5]^{2+} + \text{OH}^- \qquad (11.23)$$

The percent of the intermediate captured by Y$^-$ (obtained from separation and isolation of products) should depend on the nature of Y$^-$ and should be the same, no matter how [Co(NH$_3$)$_5$(NH$_2$)]$^{2+}$ is generated. Data in Table 11.17 give percents of *S*- and *N*-bonded thiocyanate products for base hydrolysis of several Co^{III} complexes in the presence of

Table 11.17   Thiocyanate ion competition results for base hydrolysis of [Co(NH$_3$)$_5$X]$^{n+}$ at 25°C[a]

$n+$	X	% S-isomer	% N-isomer	Total % thiocyanate
3+	OP(OMe)$_3$	12.3	5.19	17.5
3+	OSMe$_2$	12.0	5.93	17.9
2+	I$^-$	8.90	4.69	13.6
2+	OSO$_2$CF$_3^-$	8.88	4.62	13.4
2+	OSO$_2$CH$_3^-$	8.68	4.70	13.4
1+	OSO$_3^{2-}$	3.7	3.1	6.8

[a] W. G. Jackson and C. N. Hookey, *Inorg. Chem.* **1984**, *23*, 668.

[22] N. J. Curtis, G. A. Lawrance, P. A. Lay, and A. M. Sargeson, *Inorg. Chem.* **1986**, *25*, 484.

NaSCN. All complexes studied were found to give a ratio of 2.0 ± 0.1 S-isomer : N-isomer.[23] The total percent captured by NCS^- is constant for complexes of a particular charge and decreases with increasing negative charge on X. This is consistent with formation of a five-coordinate intermediate in which more-negative ligands remain in the second coordination sphere for a longer time, thus preventing capture by another negative ligand (NCS^-). *The course of the base hydrolysis in the presence of competing ions differs from reactions in acid solution where $[Co(NH_3)_5X]^{2+}$ first undergo aquation, then anation.*

Proton Exchange

In the Garrick **D–CB** mechanism, hydroxide simply acts to deprotonate the starting complex to the more labile conjugate base (see problem 11.10). Hydroxide as the lyate ion of the solvent is the strongest base that can exist in aqueous solution and thus is uniquely effective. Base-catalyzed proton exchange between ammine ligands and solvent water has been amply demonstrated with the protons *trans* to X in $[Co(NH_3)_5X]^{2+}$ exchanging more rapidly than those *cis* to it. The other possibility, that OH^- is the best entering group and nucleophile (which would support interpretation of the result via an **A** mechanism), has been ruled out in a classic labeling experiment.[24] Other evidence against OH^- as an entering group includes competition experiments[25] with HO_2^- and reactions in nonaqueous solvents containing added OH^-.[26]

Activation Parameters

As might be expected if a five-coordinate intermediate is formed, $\Delta V^{\ddagger}$ and $\Delta S^{\ddagger}$ have been found to be large and positive for base hydrolysis. For a large number of Co^{III} complexes, there is also a linear relationship between $\ln k_{OH}$ and $\ln k_H$, the rate constant for acid hydrolysis. This also means, of course, that $\Delta G^{\ddagger}_{OH}$ is proportional to $\Delta G^{\ddagger}_H$, suggesting a common (**d**) mechanism. In contrast, no such relationship exists for Cr ammine complexes. This is interpreted to mean that for Cr^{III}, acid hydrolysis is I_a whereas base hydrolysis is **D**. Cr^{III} complexes undergo base hydrolysis around 3300 times slower than their Co^{III} counterparts. This may result from lower acidity of the ligand protons or reduced reactivity of the five-coordinate intermediate.

In summary, base hydrolysis in complexes with acidic protons occurs via a **D–CB** mechanism involving a five-coordinate trigonal-bipyramidal intermediate. Water, rather than OH^-, is the entering group.

EXAMPLE 11.2: Some volumes of activation for base hydrolysis are given below. Rationalize the trends observed. (See Y. Kitamura, G. A. Lawrance, and R. van Eldik, *Inorg. Chem.* **1989**, *28*, 333.)

[23] W. G. Jackson and C. N. Hookey, *Inorg. Chem.* **1984**, *23*, 668.

[24] M. Breen and H. Taube, *Inorg. Chem.*, **1963**, *2*, 948.

[25] R. G. Pearson and D. N. Edgington, *J. Am. Chem. Soc.* **1962**, *84*, 4607.

[26] R. G. Pearson, H. H. Schmidtke, and F. Basolo, *J. Am. Chem. Soc.* **1960**, *82*, 4434.

Complex	$\Delta V^\ddagger$ (cm^3/mol)
$[\mathrm{Co(NH_3)_5(O\!=\!C(NMe_2)H]^{3+}}$	$+43.2$
$[\mathrm{Co(NH_2Me)_5Cl}]^{2+}$	$+32.7$
$[\mathrm{Co(NH_2Et)_5Cl}]^{2+}$	$+31.1$
trans-$[\mathrm{Co(en_2)Cl_2}]^+$	$+24.8$
cis-$[\mathrm{Co(en_2)Cl_2}]^+$	$+27.9$

Solution: The volumes of activation are all large and positive as we would expect a **D** mechanism
in which a departing ligand is lost and a five-coordinate intermediate formed. However, the
values become less positive as the charge on the starting complex declines. This is because dissociation of an anionic ligand creates charge and therefore creates structure in the solvent, giving a negative contribution to $\Delta V^\ddagger$ as a result of solvent electrostriction. Intermediates with
bulky NH_2R are less tightly solvated than the others.

▶ *11.3.6   Stereochemistry of Octahedral Substitution Reactions*

General

Any satisfactory theory of substitution reactions must explain their stereochemistry.
Werner believed that substitution stereochemistry was determined by the geometry of the
second coordination sphere. According to Werner, if the incoming ligand were located on
an octahedral face opposite the departing one in a *trans* complex, a *cis* product resulted.
On the other hand, if the incoming group were on a face adjacent to the departing ligand,
a *trans* product was formed.

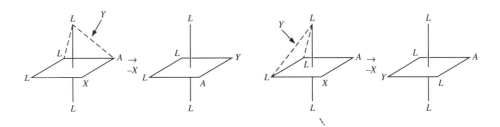

As more data were accumulated, it became apparent that additonal factors must be involved. In particular, variable stereochemistry when solvent (which occupies all positions
in the second coordination sphere) is the entering group provides evidence that other factors, including the geometry of the transition state, must be significant. **A** reactions could
be rationalized as proceeding through pentagonal-bipyramidal intermediates, and stereochemical predictions could be made. Hughes, Ingold, and Nyholm viewed stereochemistry
as resulting from edge displacement of the departing ligand by the incoming one.

These models for the stereochemical outcome of substitutions do not take into account the possibility for rearrangements of reaction intermediates which we now believe exists. In a **D** substitution reaction a square pyramid is initially generated by loss of X. A square pyramid and a trigonal bipyramid can be interconverted readily by a vibrational motion, if the species is sufficiently long-lived. The angle between two equatorial ligands E opens and the angle between two axial ligands A closes, giving a square pyramid with E_2 axial.

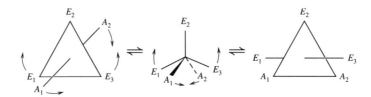

If angle A_1–M–A_2 closes further and angle E_1–M–E_3 opens further, a trigonal bipyramid forms again. Two formerly equatorial ligands are now axial, and vice versa. This motion, providing a mechanism for exchanging all equatorial and axial ligands, is called Berry **pseudorotation** because, if all ligands are equivalent, the net effect appears to be rotation of the molecule to give new spatial orientation.[27] (See Section 9.15.)

Figure 11.7 depicts possible geometries of five-coordinate intermediates arising in **D** mechanisms for $[M(N\!-\!N)_2LX]$, where $N\!-\!N$ is didentate and L is a stereochemical marker. The existence of two coordinated didentate $N\!-\!N$ ligands permits the existence of optical isomers. Analyzing product stereochemistry provides insight into the geometry of intermediates.

We can see from Figure 11.7 that reactions occurring through dissociation to a square pyramid (SP) such as **A** lead to retention of both geometric and optical configuration if **A** is stable, whereas if **A** goes on to form a trigonal bipyramid (TBP), rearrangement occurs. Collapse of an initially formed **A** can give **B** or **C**. The relative energies of SP and TBP isomers depend on the identities of their ligands. If the SPs are much more stable than the TBPs (Figure 11.8*a*), only the SPs will be accessible; on the other hand, if the energy difference is small, TBP intermediates are also formed, and these are accessed via other SPs (Figure 11.8*b*) if we restrict consideration to Berry pseudorotation. Three kinds of TBP are possible: mirror images having equatorial L, and *meso* having a plane of symmetry, and axial L. Each TBP with equatorial L can pseudorotate via SPs to three others while each *meso* TBP goes to two others. The products of attack by Y on each kind of TBP can be ascertained. The result is the same whether we consider attack on each accessible trigonal edge or whether we view attack and exit from the energy surface as occurring on the SPs that can be accessed from each TBP. Each *meso* TBP gives equal amounts of Λ-

---

[27] Another possible motion that interconverts axial and equatorial ligands is the turnstile rotation, in which two ligands (say A_1 and E_2) are held constant while the other three undergo a threefold rotation around an axis that is not a symmetry element.

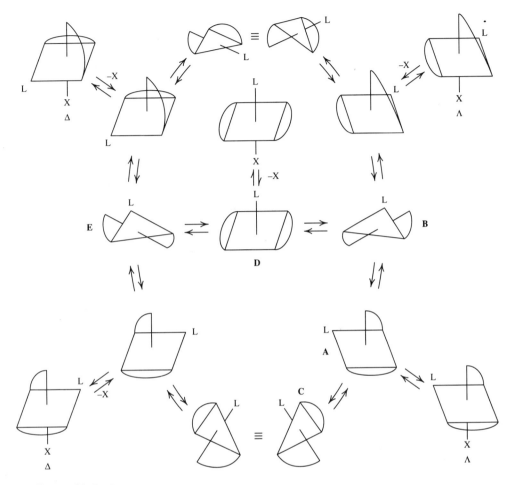

Figure 11.7 Rearrangement of [M(N—N)₂L] five-coordinate intermediates via Berry pseudorotation.

and Δ-*cis*-[M(en)₂LY]. Each mirror image isomer (equatorial L) gives 66.7% Δ- or Λ-*cis* and 33.3% *trans* product. Thus, starting with Λ-*cis*-[M(en)₂LX], the product would consist of 80% racemic *cis*- and 20% *trans*-[M(en)₂LY] if all possible TBPs were energetically accessible.

Starting with *trans*-[M(N—N)₂LX], the initial SP isomer **D** can collapse to two TBPs **B** and **E**, each having equatorial L. Immediate exit from these leads to a 2:1 *cis*:*trans* ratio of products. On the other hand, the accessibility of **C** having axial L from *cis*-[M(N—N₂LX] gives more chances for *cis* product (presuming that not all TBPs will be energetically accessible). Thus the amount of *cis* product from the *cis* isomer is always greater than the amount of *cis* product from the *trans* isomer of [M(N—N)₂LX] (Table 11.18).

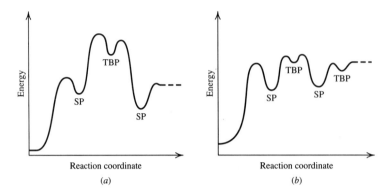

Figure 11.8   Energy profiles for **D** substitution of $[M(N-N)_2LX]$ when TBP intermediates are much less stable then SP intermediates (*a*) and when SP and TBP energies are similar (*b*).

Figure 11.7 is a simplified case of a mapping among all possible TBP isomers. In the most general case, ligands 1–5 are monodentate and all different. A total of 20 isomers is possible (five choices for the first axial ligand, four choices for the second). The possible interconversions among TBP isomers by way of Berry pseudorotation can be worked out on paper and are summarized in the Desargues–Levi map shown in Figure 11.9.[28] Such maps are also useful in monitoring interconversions of P and Si intermediates. For isomers of $[M(N-N)_2L]$, the small bite of the chelate prevents the existence of 12, $\overline{12}$, 34 and $\overline{34}$. Also, since 1 = 2 = 3 = 4, most isomers are not distinct; for example, 12 is the same as 34, and so on. Making such deletions and substitutions on the full map results in the simplified version.

Stereochemistry of Acid Hydrolysis and Anation

Table 11.18 records stereochemical consequences of acid aquation of some complexes. The results for several *trans*-$[Co(en)_2LX]^{n+}$ are consistent with a TBP species along the reaction coordinate. (See Section 11.3.4.) Retention of configuration is the rule for *cis*-$[Co(en)_2LX]^{n+}$ and for complexes containing stereochemically rigid tetradentate chelating ligands. Presumably, an SP species is involved here. A correlation exists between rearrangement and positive $\Delta S^{\ddagger}$.

Stereochemistry of Base Hydrolysis

The extensive rearrangement observed in base hydrolysis of *cis*-$[Co(NH_3)(en)_2X]^{n+}$ complexes (Table 11.19) is consistent with production of a five-coordinate intermediate that survives long enough to convert to a TBP and perhaps to undergo pseudorotation. The

[28] See K. Mislow, *Acc. Chem. Res.* **1970**, *3*, 321; K. F. Purcell and J. C. Kotz, *Inorganic Chemistry*, W. B. Saunders, Philadelphia, 1977, pp. 404ff.

Table 11.18 Stereochemical results of acid hydrolysis of some octahedral complexes[a,b]

Complex	$\Delta S^{\ddagger}$ (J/K mol)	Percent retention
trans-[Co(NH$_3$)$_4$Cl$_2$]$^+$	+36	45 ± 10
cis-[Co(CN)(en)$_2$Cl]$^+$	−20	100
trans-[Co(CN)(en)$_2$Cl]$^+$	−8	100
trans-[Co(NH$_3$)(en)$_2$Cl]$^{2+}$	−46	100
cis-[Co(NCS)(en)$_2$Cl]$^+$	−59	100
trans-[Co(NCS)(en)$_2$Cl]$^+$	+38	40 ± 10
cis-[Co(OH)(en)$_2$Cl]$^+$	−42	100
trans-[Co(OH)(en)$_2$Cl]$^+$	+84	25
cis-[Co(en)$_2$Cl$_2$]$^+$	−20	100
trans-[Co(en)$_2$Cl$_2$]$^+$	+59	65
trans-[Co(OH)(cyclam)Cl]$^+$	−29	100
cis-[Co(cyclam)Cl$_2$]$^+$	−25	100
trans-[Co(cyclam)Cl$_2$]$^+$	−13	100
cis-[Ru(NH$_3$)$_4$Cl$_2$]$^+$	—	100
trans-[Ru(NH$_3$)$_4$Cl$_2$]$^+$	—	100

[a] Leaving ligand written last in formula.

[b] From the work of Linck, Tobe, Kane-Maguire, and others.

cis : *trans* ratio of the [Co(en)$_2$(NH$_3$)(OH)]$^{2+}$ product is constant, as would be predicted if the same five-coordinate intermediate is produced from all the complexes. Greater retention occurs for some X, suggesting that a "partly bonded" leaving group influences stereochemistry as expected for an **I$_d$** path rather than **D − CB.** Identity of inert ligands is important; Co complexes of the tetradentate cyclam ligand (Figure 11.3) react with complete retention. For Rh^{III}, Ir^{III}, and Ru^{III}, as well as for Cr^{III}, configuration generally is retained. With the Rh, Ir, and Ru complexes, the identity of the *trans* ligand is quite important in determining the rate of base hydrolysis, suggesting a **d** reaction proceeding via an SP.

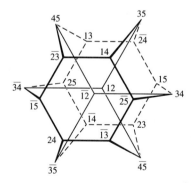

Figure 11.9 Desargues–Levi map. Vertices represent isomers indexed by their axial ligands; 13 has ligands 1 and 3 axial with 2,3,4 in clockwise order as viewed from the lower-numbered axis. $\overline{13}$ is the mirror image. Lines connecting the vertices represent pseudorotations. (Reprinted with permission from K. Mislow, *Acc. Chem. Res.* **1970,** *3,* 321. Copyright 1970, American Chemical Society.)

Table 11.19   Stereochemical results of base hydrolysis of *cis*-[Co(NH₃)(en)₂X]ⁿ⁺ at 25°C[a]

n^+	X	% trans	% cis	% cis R	% cis I[b]
3+	OP(OMe)₃	22.5	77.5	84.5	15.5
3+	OSMe₂	23.1	76.7	84.4	15.6
2+	Br⁻	22.0	78.0	78.0	23.0
2+	Cl⁻	22	78	81	19
2+	NO₃⁻	23	77	80	20

[a] D. A. Buckingham, C. R. Clark, and T. W. Lewis, *Inorg. Chem.* **1979**, *18*, 1985; D. A. Buckingham, I. I. Olson, and A. M. Sargeson, *J. Am. Chem. Soc.* **1968**, *90*, 6654.

[b] R = retention; I = inversion.

▶ 11.3.7   Substitution Without Breaking the Metal–Ligand Bond

In acid solution, carbonato complexes such as $[Co(NH_3)_5(CO_3)]^+$ are converted to aqua complexes with release of CO_2. When the reaction is carried out in the presence of $H_2^{18}O$, no ^{18}O is found in the resulting aqua complex. Hence, the Co—O bond must be retained during the reaction. Presumably, the Co-bonded O is protonated and CO_2 is then removed (Figure 11.10). $[Co(NH_3)_4(CO_3)]^+$, having didentate CO_3^{2-}, is converted to *cis*-$[Co(NH_3)_4(H_2O)_2]^{3+}$ in acid solution. In $H_2^{18}O$, half the oxygen in the product is found to have come from solvent. The first step[29] involves breaking the chelate ring, with an H_2O

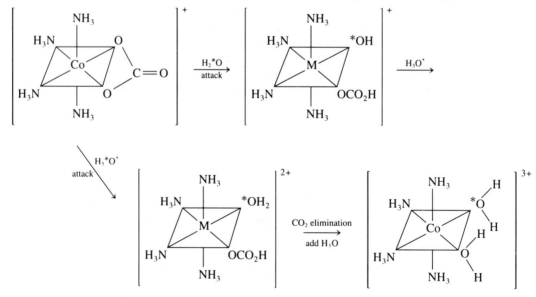

Figure 11.10   Aquation of $[Co(CO_3)(NH_3)_4]^+$.

[29] T. P. Dasgupta and G. M. Harris, *J. Am. Chem. Soc.* **1969**, *91*, 3207.

or H_3O^+ substituting for one carbonato O. The second step involves removal of CO_2 without rupture of the Co—O bond, as in the case of $[Co(NH_3)_5(CO_3)]^+$. Another example is the aquation of nitrito complexes $[Cr(NH_3)_5(ONO)]^{2+}$ and $[Co(NH_3)_5(ONO)]^{2+}$ catalyzed by Cl^- and Br^- ions. The mechanism is thought to proceed through species such as

$$M-O-N\underset{\displaystyle \overset{\Vert}{\ddot{O}}}{\overset{\displaystyle \overset{H}{|}}{\diagup}}X$$

and to involve elimination of XNO (which is rapidly hydrolyzed).

11.4 RACEMIZATION REACTIONS

Octahedral complexes can undergo substitutions involving geometrical isomerization and/or racemization.[30] We will not discuss isomerization here, but instead treat racemization of tris-chelate complexes such as $[Co(en)_3]^{3+}$, $[Fe(phen)_3]^{3+}$, $[Cr(C_2O_4)_3]^{3-}$, and so on.[31, 32] Because chelating ligands are more difficult to remove than monodentate ones, we might expect intramolecular racemization pathways to be important for chelate complexes. And, although this generally is true, the complexes $[Ni(phen)_3]^{2+}$ and $[Ni(bipy)_3]^{2+}$ provide striking exceptions. The rates of racemization and ligand exchange (as well as the activation parameters) are the same within experimental error for both complexes. Hence the mechanism must be

$$\Lambda-[Ni(phen)_3]^{2+} \;\rightleftharpoons\; [Ni(phen)_2]^{2+} + \text{phen} \;\rightleftharpoons\; \Delta-[Ni(phen)_3]^{2+} \qquad (11.24)$$

The bis-chelate intermediate (which may contain coordinated solvent molecules) either is symmetric or loses optical activity faster than it recombines with phen.

A number of tris-chelate complexes of both inert and labile metals racemize faster than they exchange ligands. These racemizations must be intramolecular. Most classical studies have been on inert complexes ($k_{rac} < 10^{-2}$ s^{-1}) requiring isomer separation and polarimetric measurements. NMR techniques are employed to investigate racemizations of labile complexes ($k_{rac} \sim 10^{-2}$ to 10^6 s^{-1}) in which rearrangement often occurs too fast to permit isolation of optical isomers. Figure 11.11 shows intramolecular pathways for racemization of $[M(chel)_3]$ complexes. Pathways *a* and *b* involve the dissociation of one end of a chelate to produce a TBP and an SP, respectively. Attack by the asterisked (dissociated) end at the position indicated by arrows leads to optical inversion. Attack at the equivalent position not designated by the arrow simply leads back to the starting isomer. Pathway *c* is the Bailar (trigonal) twist. The figure depicts an alternative view of the

[30] See F. Basolo and R. G. Pearson, *Mechanisms of Inorganic Reactions,* 2nd ed., Wiley, New York, 1967.

[31] J. J. Fortman and R. E. Sievers, *Coord. Chem. Rev.* **1971,** *6,* 331; N. Serpone and D. G. Bickley, *Prog. Inorg. Chem.* **1972,** *17,* 391.

[32] L. H. Pignolet, *Top. Curr. Chem.* **1975,** *56,* 91.

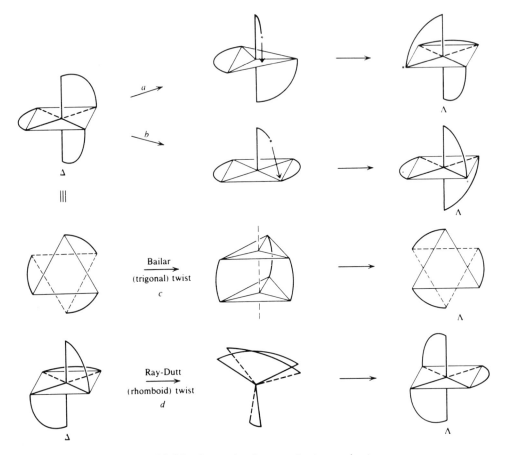

Figure 11.11   Intramolecular racemization mechanisms.

chelate complex looking down the C_3 axis. Clockwise twisting of the triangle of chelate atoms in the upper plane (dashed lines indicate a plane below the page) leads to a trigonal prism after 60°. If the rotation is continued yet another 60°, the Λ configuration results. The bottom left view shows the chelate rotated around an "imaginary" C_3 axis. A 60° rotation around this axis (Ray–Dutt twist) leads to the intermediate shown, and a further 60° produces a complex of Λ configuration. We must now consider how to decide whether one (or both) of the dissociative mechanisms (*a* or *b*) or one (or both) of the twist mechanisms (*c* or *d*) is operative for a particular intramolecular racemization.

A dissociative pathway has been postulated[33] in the racemization of optically active

$$\text{Co}\left\{\begin{matrix} O=C\diagdown^{CH_3} \\ \qquad\qquad CCD_3 \\ O=C\diagup_{\overset{\|}{\underset{CH_3}{O}}} \end{matrix}\right\}_3$$

[33] A. Y. Girgis and R. C. Fay, *J. Am. Chem. Soc.* **1970**, *92*, 7061.

The rate constant for linkage isomerism to give

$$
Co\left\{
\begin{array}{l}
O=C \overset{CH_3}{\underset{CCH_3}{\diagdown}} \\
\quad\quad\quad \overset{\|}{O} \\
O=C \diagdown \\
\quad\quad CD_3
\end{array}
\right\}_3
$$

is reported to be 3.0×10^{-5} s^{-1} at 90° in chlorobenzene. Under the same conditions the rate constant for racemization is 4.4×10^{-4} s^{-1}. It was argued that both processes involve dissociation of one end of the chelate to give a five-coordinate species. Linkage isomerism is slower because it also requires rotation around a C—C bond. In another investigation, the ^{18}O exchange for oxygens not bonded to the metal in $[Co(C_2O_4)_3]^{3-}$ was found to occur much faster than racemization, whereas six more O's exchanged much more slowly and at a rate similar to that for racemization.[34] Presumably these last two processes go through a common intermediate that must involve dissociation of one O for exchange.

Direct evidence for twist mechanisms is difficult to amass. Some criteria have included a racemization rate far in excess for that of ligand exchange and a low $\Delta S^{\ddagger}$. The transition state for the Bailar twist has $\mathbf{D}_{3h}$ symmetry so that the vertical distances of the trigonal prism which represent the bite distance of the chelate are all required to be equal as are the distances between the ligands bonded on a face. The required transition-state symmetry for the Ray–Dutt twist is lower; in particular, the edge of one triangular face is required to have the bite of the chelate while the length of the other two is dictated by the distance required to exist between ligand atoms bonded to the metal. That is, the triangular faces may be isosceles. Calculations[35] indicate that the Bailar twist should be favored for chelating ligands with a small bite distance compared to the L—L distance of bonded chelate atoms. To distinguish between the Bailar (c) and the Ray–Dutt (d) twists, we must investigate the behavior of complexes of unsymmetric chelates such as $C_6H_5C(O)CHC(O)CH_3^-$. In these situations, both optical and geometric isomerism are possible, giving a greater number of parameters for tracing the stereochemical course.[36]

11.5   SQUARE-PLANAR COMPLEXES: LIGAND SUBSTITUTION REACTIONS

▶ 11.5.1   General Features

Metal ions with the d^8 configuration [Au^{III}, Ni^{II}, Pd^{II}, Pt^{II}, Rh^{I}, Ir^{I}] usually form four-coordinate square-planar complexes, especially with strong-field ligands. We might anticipate (as turns out to be true) that lack of steric crowding and availability of an empty p

[34] L. Damrauer and R. M. Milburn, *J. Am. Chem. Soc.* **1968**, *90*, 3884.

[35] A. Rodger and B. F. G. Johnson, *Inorg. Chem.* **1988**, *27*, 3062.

[36] J. G. Gordon and R. H. Holm, *J. Am. Chem. Soc.* **1970**, *92*, 5319; D. G. Bickley and N. Serpone, *Inorg. Chem.* **1979**, *18*, 2200.

orbital perpendicular to the molecular plane would lead to an **a** mechanism. Two general characteristics have emerged: The mechanisms are associative and the rate law has two terms. For the reaction

$$ML_2AX + Y \longrightarrow ML_2AY + X \qquad (11.25)$$

rate $= (k_1 + k_2[Y])[ML_2AX]$. Moreover, substitutions are stereospecific: A *trans* reactant gives a *trans* product ML_2AY, and a *cis* reactant gives a *cis* product.

The **a** character is shown by the importance of entering group identity in determining reaction rate. We can visualize the course as shown in Figure 11.12, proceeding through a TBP. Obviously, this accounts nicely for the stereospecificity. In some cases, for example in the reactions of $[AuCl_4]^-$ with NCS^-,[37] five-coordinate intermediates actually have been detected. Also, many stable five-coordinate d^8 species are known to exist. For example, $[Ni(CN)_4]^{2-}$ with excess CN^- gives $[Ni(CN)_5]^{3-}$, isolated by crystallization of its $[Cr(en)_3]^{3+}$ salt (Figure 9.33).

During substitution both entering and leaving groups are coordinated and give a five-coordinate species. This species may be an intermediate (**A** mechanism) or a transition state ($\mathbf{I}_a$ mechanism), depending on the particular reaction. It is convenient to model substitution reactions of square-planar complexes as proceeding by the **A** mechanism of Figure 11.12 involving the TBP intermediate **C**. Much of the rest of this section consists of rationalizing kinetic data in terms of this model. The vast majority of data have been accumulated on inert complexes of Pt^{II}. The reactivity order is $Ni^{II} > Pd^{II} > Pt^{II}$. Complications in studying reactions include the tendency of Ni^{II} to form octahedral complexes with weak-field ligands and the ease with which Pd^{II} is reduced to Pd^0. Relatively few kinetic studies have been done on complexes of Rh^I, Ir^I, and Au^{III}.

▶ 11.5.2   *Significance of the Rate Law*

The two-term rate law indicates two parallel reaction paths. Reactions are studied conveniently under pseudo-first-order conditions. A large excess of Y is introduced so that [Y] remains effectively constant throughout the reaction. The rate is measured at several values of [Y], and the rate constant k_{obs} is plotted as a function of [Y]. Figure 11.13 shows a

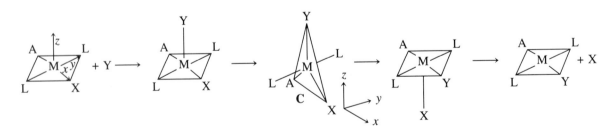

Figure 11.12   Steric course of square-planar substitution.

[37] A. J. Hall and D. P. N. Satchell, *J. Chem. Soc. Dalton Trans.* **1977**, 1403.

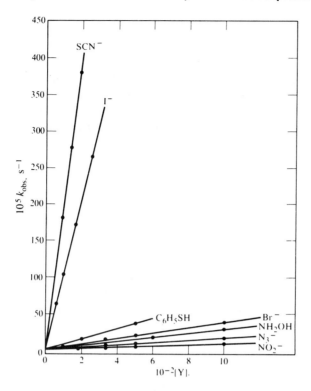

Figure 11.13 Rates of reaction of *trans*-[PtCl$_2$(py)$_2$] at 30°C in methanol as a function of concentration of incoming nucleophile. (Reprinted with permission from U. Belluco, L. Cattalini, F. Basolo, R. Pearson, and A. Turco, *J. Am. Chem. Soc.* **1965**, *87*, 24. Copyright 1965, American Chemical Society.)

plot of k_{obs} versus the concentration of several different entering nucleophiles replacing Cl^- in *trans*-[PtCl$_2$(py)$_2$]. The straight-line plot shows that $k_{obs} = k_1 + k_2[Y]$. Values of k_2 are found from the slope and differ for different nucleophiles. The k_2 values reflect the different nucleophilicities of entering ligands. A particularly noteworthy feature is that all lines have the same intercept—that is, k_1 is the same for all the nucleophiles. This implies attack by the same species in all these different reactions. The only other nucleophile present in all cases is the solvent methanol. Indeed, we generally find that the value of k_1 is the same for all nucleophiles and depends only on the solvent. Hence, the two terms in the rate law represent attack by the solvent and by the entering group Y as parallel paths to the product.[38]

$$
\begin{array}{c}
ML_2AX \\[-2pt]
\Big\downarrow S \qquad Y \\
ML_2AS + X
\end{array}
\;\xrightarrow{\;Y\;}\; ML_2AY
\tag{11.26}
$$

Mechanism (11.26) leads to the observed rate law if displacement of solvent S by Y is much faster than displacement of X by S.

---

[38] In some reactions, one of the rate constants is large enough to swamp the other and only a one-term rate law is found.

▶ 11.5.3   Effect on Rates of Entering and Leaving Ligands

If the intimate mechanism of square-planar substitution reactions is indeed **a,** we might expect the nature of both entering and leaving ligands to affect rates substantially. Data in Table 11.20 show that this is so. Values of k_2 for various ligands provide a measure of their nucleophilicities toward *trans*-[Pt(py)$_2$Cl$_2$]. The identity of Y affects the rate over a range of ~10^9 in this series of reactions.

A nucleophilic reactivity parameter n_{Pt} can be defined as

$$n_{Pt} = \log \frac{k_2(Y)}{k(CH_3OH)} = \log k_2(Y) - \log k(CH_3OH) \qquad (11.27)$$

n_{Pt} measures the nucleophilicity of a particular Y relative to that of methanol. The factors making for good nucleophilicity toward Pt complexes differ from those making for activity in nucleophilic displacements on CH_3I and for basicity. These include both basicity toward Pt^{II} and reduction potential of Y.[39] Values of n_{Pt} are given in Table 11.20*a*. The same nucleophilicity order holds for other Pt complexes in other solvents.

The results of classic studies by Gray on the effect of leaving groups involve py as the common entering group with [Pt(dien)X]$^+$ complexes. The effect of the nature of X is also large, spanning a range of ~10^6 (Table 11.20*b*).

The diagrams of Figure 11.14 show that in a particular case, either the entering or the leaving group could make the overriding contribution to activation energy. Hence for separation of the two effects in the [Pt(py)$_2$Cl$_2$] studies, Cl$^-$ was chosen because it is a relatively good leaving ligand. That is, the energy profile of Figure 11.14*b* applies, and the reaction rate reflects the height of the first activation barrier.

▶ 11.5.4   The Trans Effect

An extremely significant aspect of square-planar substitution is the *trans effect,* the effect on the rate of replacement of some ligand by the identity of the *trans* ligand.[40] This effect was recognized rather early on in the preparation of the geometric isomers of $PtCl_2(NH_3)_2$.

---

[39] For many organic and inorganic reactions, nucleophilicities can be related explicitly to basicity and reduction potential by the Edwards equation

$$\log \frac{k_2(Y)}{k_1} = -\alpha E^0 + \beta H$$

where E^0 is the reduction potential for Y and H is related to its proton basicity (J. O. Edwards, *J. Am. Chem. Soc.* **1954,** *76,* 1540; J. O. Edwards, *Inorganic Reaction Mechanisms,* W. A. Benjamin, New York, 1964). However, this seems not to be the case for substitution reactions of Pt^{II} (R. G. Pearson, H. Sobel and J. Songstad, *J. Am. Chem. Soc.* **1968.** *90,* 319).

[40] The effects of the identity of *cis* ligands are smaller by a factor of 10^2.

Table 11.20   Rate constants for ligand displacement in some square-planar Pt^II complexes

a. *trans*-[Ptpy$_2$Cl$_2$] + Y → *trans*-[PtClpy$_2$Y] + Cl^{-a}			b. [Pt (dien)X]$^+$ + py → [Pt (dien)py]$^{2+}$ + X^{-b}	
Y	$k_2(M^{-1} s^{-1})$	n_{Pt}	X	$k_{obs}(M^{-1} s^{-1})$
CH$_3$OH (25°)	2.7×10^{-7}	0.00	CN$^-$	1.7×10^{-8}
CH$_3$O$^-$ (25°)	Very slow	<2.4	SCN$^-$	3.0×10^{-7}
N(CN)$_2^-$	3.03×10^{-4}	2.87	I$^-$	1.0×10^{-5}
Cl$^-$	4.5×10^{-4}	3.04	Cl$^-$	3.5×10^{-5}
NH$_3$	4.7×10^{-4}	3.07	H$_2$O	1.9×10^{-3}
N$_3^-$	1.55×10^{-3}	3.58	NO$_3^-$	Very fast
I$^-$	1.07×10^{-1}	5.46		
(Me$_2$N)$_2$CS	0.30	5.87		
(MeNH)$_2$CS	2.5	6.79		
(PhNH)$_2$CS	4.13	7.01		
CN$^-$ (25°)	4.00	7.14		
PPh$_3$ (25°)	249	8.93		

[a] In CH$_3$OH, at 30°C except as noted. From U. Belluco, L. Cattalini, F. Basolo, and A. Turco, *J. Am. Chem. Soc.* **1965**, *87*, 241; R. G. Pearson, H. Sobel, and J. Songstad, *J. Am. Chem. Soc.* **1968**, *90*, 319; M. Becker and H. Elias, *Inorg. Chim. Acta* **1986**, *116*, 47.

[b] At 25°C in water. From F. Basolo, H. B. Gray, and R. G. Pearson, *J. Am. Chem. Soc.* **1960**, *82*, 4200; H. B. Gray and R. J. Olcott, *Inorg. Chem.* **1962**, *1*, 481.

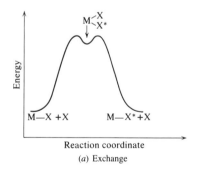

(a) Exchange

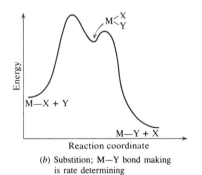

(b) Substition; M—Y bond making is rate determining

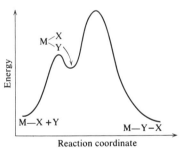

(c) Substition; M—Y bond breaking is rate determining

Figure 11.14   Reaction profiles for square-planar substitution. (a) Exchange. (b) Substitution; M—Y bond making is rate-determining. (c) Substitution; M—X bond breaking is rate-determining. (Reprinted from C. H. Langford and H. B. Gray, *Ligand Substitution Processes*, 1966, with permission of the publisher, Benjamin-Cummings, Reading, MA.)

Data in Table 11.21 give some idea of the magnitude of the *trans* effect. Allowing for the difference in temperature, varying the *trans* ligand can change the speed of replacement over a range of 10^6! The ordering of *trans* labilization is not precisely the same for every substrate. However, a sort of "average" order is

$$CN^-, CO, C_2H_4 > PR_3, H^- > CH_3^-, SC(NH_2)_2 \gg C_6H_5^-, NO_2^-, I^-, \tag{11.28}$$
$$SCN^- > Br^- > Cl^- > py > NH_3 > H_2O \gg OH^-$$

The *trans* effect is very useful in synthesis of square-planar complexes. In preparing the three isomers of [PtBrI(NH₃)(py)], the *trans* effect and the bond-strength order (Pt-NH₃ > Pt-py > Pt-Br > Pt-Cl) govern the results.

$$\begin{bmatrix} Cl \\ | \\ Cl-Pt-Cl \\ | \\ Cl \end{bmatrix}^{2-} \xrightarrow[-Cl^-]{+NH_3} \begin{bmatrix} NH_3 \\ | \\ Cl-Pt-Cl \\ | \\ Cl \end{bmatrix}^- \xrightarrow[-Cl^-]{+Br^-} \begin{bmatrix} NH_3 \\ | \\ Cl-Pt-Br \\ | \\ Cl \end{bmatrix}^- \xrightarrow[-Cl^-]{+py} \begin{bmatrix} NH_3 \\ | \\ py-Pt-Br \\ | \\ Cl \end{bmatrix} \tag{11.29}$$

When Br⁻ is added, one of the ligands *trans* to Cl⁻ is replaced faster. The product turns out to be [PtBrCl₂(NH₃)]⁻, rather than [PtBrCl₃]²⁻, because of the greater Pt—NH₃ bond strength.

$$\begin{bmatrix} Cl \\ | \\ Cl-Pt-Cl \\ | \\ Cl \end{bmatrix}^{2-} \xrightarrow[-Cl^-]{+py} \begin{bmatrix} py \\ | \\ Cl-Pt-Cl \\ | \\ Cl \end{bmatrix}^- \xrightarrow[-Cl^-]{+Br^-} \begin{bmatrix} py \\ | \\ Cl-Pt-Br \\ | \\ Cl \end{bmatrix}^- \xrightarrow[-Cl^-]{+NH_3} \begin{bmatrix} py \\ | \\ H_3N-Pt-Br \\ | \\ Cl \end{bmatrix} \tag{11.30}$$

Table 11.21 Effect of *trans* ligand L on the rate of the reactions *trans*-[Pt(PEt₃)₂LCl] + py → *trans*-[Pt(PEt₃)₂L(py)]⁺ + Cl⁻[a,b]

L	$k_1(s^{-1})$	$k_2(M^{-1}s^{-1})$	$T(°C)$
H⁻	1.8×10^{-2}	4.2	0
CH₃⁻	2×10^{-4}	7×10^{-2}	25
C₆H₅⁻	2×10^{-5}	2×10^{-2}	25
Cl⁻	1.0×10^{-6}	4.0×10^{-4}	25

[a] In ethanol.

[b] From F. Basolo, J. Chatt, H. B. Gray, R. G. Pearson, and B. L. Shaw, *J. Chem. Soc.* **1961**, 2207.

The reversal in the order of py and NH_3 addition reverses their positions in the product in (11.30).

$$
\begin{bmatrix} & \text{Cl} & \\ \text{Cl} & | & \\ & \text{Pt} & \text{Cl} \\ & | & \\ & \text{Cl} & \end{bmatrix}^{2-}
\xrightarrow[-\text{Cl}^-]{+2\text{py}}
\begin{bmatrix} & \text{py} & \\ \text{Cl} & | & \\ & \text{Pt} & \text{py} \\ & | & \\ & \text{Cl} & \end{bmatrix}^{+}
\xrightarrow[-\text{Cl}^-]{\text{NH}_3}
\begin{bmatrix} & \text{py} & \\ \text{Cl} & | & \\ & \text{Pt} & \text{py} \\ & | & \\ & \text{NH}_3 & \end{bmatrix}^{-}
\xrightarrow[-\text{py}]{+\text{Br}^-}
\begin{bmatrix} & \text{py} & \\ \text{Cl} & | & \\ & \text{Pt} & \text{Br} \\ & | & \\ & \text{NH}_3 & \end{bmatrix}
$$

$$(11.31)$$

The second step in Equation (11.31) may seem to contradict the notion of the *trans* effect, but this is not so. The *trans* effect does *not* predict that *any* ligand *trans* to Cl^- should be replaced faster than *any* ligand *trans* to py. Rather, it leads us to expect that a Cl^- *trans* to a Cl^- should be replaced faster than a Cl^- *trans* to py if there were a choice (here there is not). The lesser bond strength of Pt—Cl can serve as a rationale for Cl^- replacement. Events in the final step are not rationalized easily. This reminds us that the *trans* effect is an empirical correlation that cannot necessarily predict what reactions will occur. Given the reaction stoichiometry (that is, what the entering and leaving groups are), the *trans* effect permits us to assign product geometry. In the final step of Equation (11.31), if we know that Br^- is the entering and py the leaving group, the *trans* effect leads us to expect replacement of the py *trans* to Cl^-.

▶ *11.5.5 Theories of the Trans Effect*

The *trans* effect is essentially an empirical observation on reaction rates. However, numerous efforts have been made to explain the origin of this effect on some model of electronic structure. The presence of certain ligands *trans* to a leaving ligand in substitution lowers the activation energy relative to that when some other ligand is in the *trans* position. The lowering can be accomplished in either of two ways: by raising the energy of the reactants' ground state or by lowering the energy of the transition state. Good *trans*-directing ligands must act in one or both of these ways.

Ground-State Effects

A good *trans*-directing ligand could weaken the bond between the metal and the *trans* ligand. This is a thermodynamic effect that has been called the **trans influence**[41] to distinguish it from the kinetic *trans* effect. Several theories exist as to how a ligand might weaken the bond *trans* to itself. One of the earliest was the polarization theory of Grinberg, in which a good *trans*-directing ligand was visualized as being very polarizable and forming an ionic bond to metal. A dipole then serves to repel the *trans* ligand, thus weakening the bond. This could account for the presence of H^- and CH_3^- high in the *trans* effect series.

---

[41] See T. G. Appleton, H. C. Clark, and L. E. Manzer, *Coord. Chem. Rev.* **1973,** *10,* 335.

A second view emphasizes the effectiveness of good σ-donor ligands at weakening the *trans* bond. With dsp^2 hybrids, *trans* ligands share $d_{x^2-y^2}$, s, and one p. A good σ donor claims a larger share of the metal σ orbitals, leaving a lesser share to bond the *trans* ligand.

Structural studies indicate that the nature of a ligand *can* have a substantial effect on the strength of the bond to the *trans* group. For example, the Pt—Cl distance in *trans*-[PtCl$_2$(PEt$_3$)$_2$] is 229.4 pm. In *cis*-[PtCl$_2$(PMe$_3$)$_2$] where Cl$^-$ is *trans* to PMe$_3$, the Pt—Cl distances are lengthened to 236.4 and 238.8 pm. Thus (presuming a relation between bond length and bond strength) a *trans*-phosphine weakens a Pt—Cl bond as compared with a *trans*-chloride. Not every ligand that is an effective *trans* director displays a large *trans* influence. In particular, π-acid ligands have low *trans* influence. Several other techniques have been employed to demonstrate the dependence of bond strengths on the identity of the *trans* ligand, including infrared, ^{1}H, ^{31}P, and ^{19}F NMR (examination of both chemical shifts and coupling constants).[42]

Transition-State Effects

Any feature that stabilizes the TBP (Figure 11.12) lowers the activation energy barrier and speeds the reaction. If the mechanism is **A,** this TBP will represent an intermediate. However, Hammond's postulate tells us that the geometry of the transition state (which is not far away from **C** along the reaction coordinate) will be quite similar. Thus we expect features that stabilize a TBP to lower the activation energy barrier. Ligands capable of a high degree of σ donation or π acceptance can confer stability and are good *trans* directors. In brief, π-acceptor ligands can help to drain off some of the extra electron density provided by the addition of a fifth ligand while good σ-donors claim a large share of metal orbitals for bonding.[43]

The role of good *trans* directing ligands in stabilizing the transition state is evident from data in Tables 11.20 and 11.21. Ligands high in the *trans* effect series are both good entering groups and poor leaving groups.

▶ 11.5.6 *Steric Effects of Nonleaving Ligands*[44, 45]

We can make no clear-cut prediction about the effect of bulky ligands on the rates of square-planar substitution by **a** mechanisms: Whereas attack by an entering ligand should be retarded, departure of the leaving group should be assisted. Nevertheless, steric retardation provides evidence in favor of an **a** mechanism, since there is no way to rationalize such retardation on the assumption of a **d** mechanism. Table 11.22 shows data for substitution of Cl$^-$ by I$^-$ in a series of progressively more hindered complexes

---

[42] There also is evidence for a *trans* effect in octahedral complexes. Sulfite is an especially effective *trans* labilizer.

[43] Recent calculations support the views given here. See Z. Lin and M. B. Hall, *Inorg. Chem.* **1991**, *30*, 646.

[44] G. Faraone, V. Ricevuto, R. Romeo, and M. Trozzi, *Inorg. Chem.* **1969**, *8*, 2207; *Inorg. Chem.* **1970**, *9*, 1525.

[45] R. Romeo, D. Minniti, and M. Trozzi, *Inorg. Chem.* **1976**, *15*, 1134.

Table 11.22   Kinetic parameters for the k_2 path of the reaction $[Pd(N—N—N)Cl]^+ + I^- \rightarrow [Pd(N—N—N)I]^+ + Cl^-$ at 25°C^a

N—N—N	k_2 (M^{-1} s^{-1})	$\Delta V^{\ddagger}$ (cm^3/mol)
dien	4446	−10.3
1,4,7-Me$_3$dien	3542	−18.9
1,4,7-Et$_3$dien	932	−11.1
1,1,4-Et$_3$dien	21.2	−11.3
1,1,7,7-Me$_4$dien	0.28	—
1,1,7,7-Et$_4$dien	8.0×10^{-4}	—

^{a} From the work of van Eldik, Paimer, Kelm, and others.

$[Pd(N—N—N)Cl]^+$, where N—N—N is dien (Figure 11.3) or substituted dien. The values of k_2 span a range of 10^7, much larger than the range for sterically hindered octahedral complexes (Table 11.6a). Such data were once regarded as evidence that severe steric hindrance can block associative pathways completely, leading to a much slower **d** substitution. However, the negative values of $\Delta V^{\ddagger}$ point to an **a** mechanism.[46]

EXAMPLE 11.3: Exchange of NH$_3$ in $[Pt(NH_3)_4]^{2+}$ has recently been studied (B. Brønnum, H. Saaby Johansen, and L. H. Skibsted, *Inorg. Chem.* **1992**, *31*, 3023). The reaction

$$[Pt(NH_3)_4]^{2+} + {}^{15}NH_3 \longrightarrow [Pt({}^{15}NH_3)(NH_3)_3]^{2+} + NH_3$$

was followed in buffered aqueous solutions having $0.050 < [NH_3] < 0.50$. The rate law was found to be rate $= (k_1 + k_2[{}^{15}NH_3])[Pt(NH_3)_4]^{2+}$. $k_1 = 3.9 \times 10^{-10}$ s^{-1} and $k_2 = 9.5 \times 10^{-10}$ M^{-1} s^{-1}. **(a)** What do these data suggest about the exchange mechanism? **(b)** Calculate the half-life for ammonia exchange in 1.0 M NH$_3$. **(c)** $[Au(NH_3)_4]^{3+}$ was found to undergo similar exchange with ^{15}NH$_3$; the rate law is of the same form with $k_1 = 3 \times 10^{-6}$ s^{-1} and $k_2 = 1.69 \times 10^{-1}$ M^{-1} s^{-1}. What can you say about the mechanism of exchange for Au^{III}?

Solution: **(a)** The two-term rate law is that typically seen for square-planar substitution. The first-order term represents **a** attack by solvent water, and the second represents direct attack by ^{15}NH$_3$.

(b) The reaction is extremely slow. If we maintain [NH$_3$] at 1.0 M, the conditions are pseudo-first-order and $k' = k_1 + k_2[NH_3] = 13.4 \times 10^{-10}$ s^{-1}. The half-life is the time τ required for the concentration of reactant to drop to half its initial value. The integrated form of the rate law is $[Pt(NH_3)_4^{2+}]_t = [Pt(NH_3)_4^{2+}]_0 e^{-k't}$. When $t = \tau$, $[Pt(NH_3)_4^{2+}]_t/[Pt(NH_3)_4^{2+}]_0 = 1/2 = \ln(-k't)$, or $\tau = 0.693/k'$. Here $\tau = 0.693/13.4 \times 10^{-10}$ s$^{-1} = 5.1 \times 10^8$ s $= 17$ years!

(c) The rate law is that expected for associative square-planar substitution. That the reaction is **a** is reflected in the fact that substitution is much faster on Au^{III} than on Pt^{II}, as would be expected for nucleophilic attack. If we compare k_2/k_1 we have a measure of the discriminating power of each metal toward attack by NH$_3$ and H$_2$O. The values of 2 for Pt and 6×10^4 for Au indicate that the nucleophilic discrimination factor is far larger for Au^{III}.

[46] Dissociative mechanisms are known for some reactions of square-planar complexes. For a recent review, see R. Romeo, *Comments Inorg. Chem.* **1990**, *XI*, 21.

11.6 CATALYSIS OF SUBSTITUTION BY REDOX PROCESSES

Substitutions on octahedral low-spin d^6 complexes proceed quite slowly (see Table 11.13). Pt^{IV} also forms octahedral complexes that undergo slow substitution. In the presence of catalytic amounts of Pt^{II} complexes, however, substitution rates are accelerated by factors of 10^4–10^5. The mechanism of these catalytic reactions is thought to be[47]

$$Pt^{II}L_4 + Y \xrightarrow{\text{fast}} Pt^{II}L_4Y$$

$$Pt^{II}L_4Y + X^*Pt^{IV}L_4Z \longrightarrow X^*Pt^{IV}L_4Z\!-\!Pt^{II}L_4Y$$

$$X^*Pt^{IV}L_4Z\!-\!Pt^{II}L_4Y \longrightarrow X^*Pt^{II}L_4\!-\!ZPt^{IV}L_4Y \qquad (11.32)$$

$$X^*Pt^{II}L_4\!-\!ZPt^{IV}L_4Y \longrightarrow X^*Pt^{II}L_4 + ZPt^{IV}L_4Y$$

$$X^*Pt^{II}L_4 \longrightarrow X + {}^*Pt^{II}L_4$$

In the third step, Pt^{II} is oxidized to Pt^{IV} while the $^*Pt^{IV}$ is reduced to $^*Pt^{II}$. The two metals exchange roles via a redox process. The relatively fast substitution on Pt^{II} enables the reaction to occur rapidly, and the product $ZPt^{IV}L_4Y$ contains a metal that originally was Pt^{II}. We also find redox catalysis in Pd^{II}/Pd^{IV}, Cr^{II}/Cr^{III}, and Ru^{II}/Ru^{III} substitutions where one oxidation state is more labile than the other and substitution on the labile center can be followed by rapid oxidation or reduction to an inert product.

11.7 REDOX REACTIONS

Redox reactions are of wide importance in chemistry. Many classical analytical methods are based on rapid redox reactions, including Fe^{II} determination by titration with $HCrO_4^-$, as well as oxidations of substances such as $C_2O_4^{2-}$ by MnO_4^-. The role of transition metal ions in life processes (see Chapter 18) depends on their ability to participate selectively in electron-transfer reactions. Redox reactions involving two transition-metal complexes generally occur fairly rapidly. Thus, values of E^0 are a rather good guide to the chemistry that actually occurs on a convenient time scale.

Below we consider models for redox reactions of two transition-metal-containing complexes. Oxidations of nonmetallic species by metallic ions have been reviewed.[48]

▶ 11.7.1 Inner- and Outer-Sphere Reactions

Two models for redox mechanisms of metal complexes currently are considered operative. In the **inner-sphere mechanism,** the coordination spheres of the two metals interpenetrate in the transition state. A bridging ligand is coordinated to both the oxidant and the

[47] W. R. Mason, *Coord. Chem. Rev.* **1972**, *7*, 241.

[48] A. McAuley, *Coord. Chem. Rev.* **1970**, *5*, 245; J. K. Beattie and G. P. Haight, *Prog. Inorg. Chem.* **1972**, *17*, 93; D. Benson, "Mechanisms of Oxidation by Metal Ions," in *Reaction Mechanisms in Organic Chemistry*, Vol. 10, Elsevier, New York, 1976.

reductant and forms part of the first coordination sphere of each. In the **outer-sphere mechanism,** each separate coordination sphere remains intact.

The first definitive evidence for an inner-sphere process was provided by Taube and Myers, who investigated the Cr^{2+} reduction of $[Co(NH_3)_5Cl]^{2+}$.[49]

The final products in acid solution are Co^{2+}, Cr^{3+}, $CrCl^{2+}$, Cl^-, and NH_4^+. It is possible to imagine both inner- and outer-sphere mechanisms that would lead to these products.

Inner Sphere

$$[Co(NH_3)_5Cl]^{2+} + Cr^{2+} \rightleftharpoons [Co(NH_3)_5ClCr]^{4+}$$

$$[Co(NH_3)_5ClCr]^{4+} \longrightarrow CrCl^{2+} + [Co(NH_3)_5]^{2+} \qquad (11.33)$$

$$[Co(NH_3)_5]^{2+} + 5H^+ \longrightarrow Co^{2+} + 5NH_4^+$$

Outer Sphere

$$[Co^{III}Cl(NH_3)_5]^{2+} + Cr^{2+} \longrightarrow [Co^{II}Cl(NH_3)_5]^+ + Cr^{3+}$$

$$[CoCl(NH_3)_5]^+ + 5H^+ \longrightarrow Co^{2+} + 5NH_4^+ + Cl^- \qquad (11.34)$$

$$Cr^3 + Cl^- \rightleftharpoons CrCl^{2+}$$

However, the inner-sphere pathway must be operative, because Cr^{3+} is inert to substitution on the time scale for the reaction. At 25°C, the second-order rate constant for Cl^- anation of Cr^{3+} is $2.9 \times 10^{-8} M^{-1} s^{-1}$, whereas that for reduction is $6 \times 10^5 M^{-1} s^{-1}$. Hence, the $CrCl^{2+}$ could not have arisen by substitution with free Cl^- on the time scale required to separate the products. The electron-transfer step converts labile Cr^{2+} into substitution-inert Cr^{III}, which retains the bridging chloride in its coordination sphere when the bridged species breaks up. Notice also that Cr^{2+} is sufficiently labile ($k_{exch} \sim 10^8 s^{-1}$ at 25°C) that substitution of the bridging Cl^- into its coordination sphere is not rate-limiting.

A sufficient condition for establishing that a reaction is inner-sphere is that one product be substitution-inert and retain the bridging ligand originally coordinated to the other reactant. This means that the oxidant and reductant must be chosen so that one is inert while the other is labile and the products are labile and inert in the opposite sense. Rather few inner-sphere reactions fulfill this requirement (that is, the condition is not a necessary one). Later, we shall devote attention to other ways of deciding whether a reaction is inner-sphere.

The above discussion reveals that a sufficient condition for the outer-sphere mechanism is that the redox rate be much faster than substitution on either metal center. An example is the reduction of $[RuBr(NH_3)_5]^{2+}$ by V^{2+}. The second-order rate constant for the reduction is $5.1 \times 10^3 M^{-1} s^{-1}$, whereas aquation reactions on Ru^{III} and V^{II} centers have first-order rate constants of 20 and 40 s^{-1}, respectively.

Before discussing other ways of assigning mechanism, we review some general features of outer- and inner-sphere reactions.

---

[49] In discussing redox reactions, M^{n+} ordinarily is taken to represent the metal ion with its first coordination sphere occupied by aqua ligands. Thus $Cr^{2+} \equiv [Cr(H_2O)_6]^{2+}$. ML^{n+} represents the complex in which one aqua ligand is replaced by L.

▶ *11.7.2   Marcus Theory*

Marcus[50] has calculated from theory the contributions to $\Delta G^{\ddagger}$ for outer-sphere reactions; this work won him the 1992 Nobel prize in chemistry. In his model, outer-sphere reactions are regarded as involving five steps:

1. Reactants diffuse together to form an outer-sphere complex in which both metal coordination spheres remain intact.
2. Bond distances around each metal change to become more "productlike".
3. The solvent shell around the outer-sphere complex reorganizes.
4. The electron is transferred.
5. The products diffuse away; this step is generally fast.

Calculation of contributions by the various steps to $\Delta G^{\ddagger}$ can be achieved because no bond breaking is involved. Marcus[50] showed that

$$\Delta G^{\ddagger} = \Delta G^{\ddagger}_{\text{inner sphere}} + \Delta G^{\ddagger}_{\text{solvation}} + \frac{\Delta G^0_r}{2} + \frac{(\Delta G^0_r)^2}{16(\Delta G^{\ddagger}_{\text{inner sphere}} + \Delta G^{\ddagger}_{\text{solvation}})} \quad (11.35)$$

where ΔG^0_r is the overall free-energy change for the reaction corrected for the work required to bring the reactants to the distance r where redox occurs, $\Delta G^{\ddagger}_{\text{inner sphere}}$ is the energy required for changing bond lengths, and $\Delta G^{\ddagger}_{\text{solvation}}$ is the solvent reorganization energy. We now discuss each of these contributions in more detail with a view toward understanding the factors involved in each rather than with the goal of actually calculating these terms ourselves.

ΔG^0, the overall free-energy change, is easily calculated from reduction potentials: $-\Delta G^0 = RT \ln K_{eq} = nFE^0$. The work involved in bringing reactants together is just $U(r)$, where $U(r)$ is the sum of attractive and repulsive potentials for complex ions as they approach. $U(r)$ is involved in the calculation of K for outer-sphere complex formation and is defined in the footnote to Table 11.14. $\Delta G^0_r = \Delta G^0 + NU(r)$.

$\Delta G^{\ddagger}_{\text{inner sphere}}$ results from bond length changes occurring in the separate coordination spheres of the outer-sphere complex before electron transfer can happen. We first consider why such changes are necessary. A well-known outer-sphere reaction is the so-called **self-exchange;** for example,

$$[Fe(H_2O)_6]^{3+} + [Fe^*(H_2O)_6]^{2+} \rightleftharpoons [Fe^*(H_2O)_6]^{3+} + [Fe(H_2O)_6]^{2+} \quad (11.36)$$

The equilibrium constant for this reaction is 1 and $\Delta G^0 = 0$; hence, $\Delta G^{\ddagger} \approx \Delta G^{\ddagger}_{\text{inner sphere}} + \Delta G^{\ddagger}_{\text{solvation}}$. The measured value of k is $4\ M^{-1}\ s^{-1}$ at 25°C. Before an electron can be transferred, the Fe—O bond lengths must distort. Those of the Fe^{3+} complex lengthen to a distance halfway between the Fe^{2+} and Fe^{3+} distances. Those of the Fe^{2+} complex contract to the same distance. This requires energy expenditure. This model for the outer-sphere process is in accord with the **Franck–Condon principle,** which states

[50] R. A. Marcus, *Annu. Rev. Phys. Chem.*, **1964**, *15*, 155; T. W. Newton, *J. Chem. Educ.*, **1968**, *45*, 571.

that electron transfer is much faster than nuclear motion. As far as electrons are concerned, nuclear positions are frozen during the time period required for electron transfer.

If this reorganization energy (Franck–Condon energy) were not expended, initial products of self-exchange would be (a) $[*Fe(H_2O)_6]^{3+}$ with bond distances characteristic of $[*Fe(H_2O)_6]^{2+}$ and (b) $[Fe(H_2O)_6]^{2+}$ with bond distances characteristic of $[Fe(H_2O)_6]^{3+}$. The energy released when these species relax to their stable geometries would be created from nothing, in violation of the first law of thermodynamics.

The reaction profile[51] for self-exchange is shown in Figure 11.15a. The energy is the total energy for a pair of ions in the outer-sphere complex. The reaction coordinate represents all changes in bond lengths and angles in the coordination spheres for the *pair* of complex ions. The curve is symmetric, since products and reactants are identical. In this picture, electron transfer occurs where the energy curve for products intersects that for reactants; this occurs at some value of the coordinate for which energy is not a minimum. As the reaction coordinate approaches its transition-state value, the wavefunctions for reactants and products mix, creating two separate energy states instead of the parabolas crossing. Dashed lines show the course of the parabolas if no mixing occurred; and the distance between the two solid curves ($2H_{AB}$) is a measure of the degree of mixing. The path of the self-exchange is the motion of a point from the reactant well up the Franck–Condon barrier and on down into the product well.

$\Delta G^{\ddagger}_{\text{inner sphere}}$ is usually calculated by using the expression

$$\Delta G^{\ddagger}_{\text{inner sphere}} = \frac{3Nf_{av}(\Delta d)^2}{8} \tag{11.37}$$

where Δd is the change in M—L distance on going from the ground to the transition state and f_{av} is the average value of the force constant for the symmetric breathing vibrational mode by which bond-length changes are considered to be achieved.

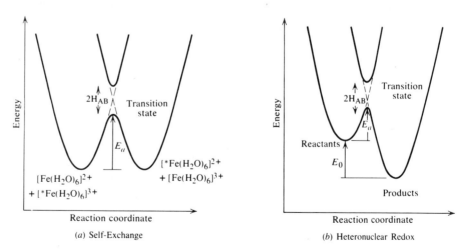

Figure 11.15   Reaction profiles for outer-sphere redox reactions.

[51] N. A. Lewis, *J. Chem. Educ.* **1980,** *57,* 478.

$\Delta G_{\text{solvation}}^{\ddagger}$ can be calculated from the expression

$$\Delta G_{\text{solvation}}^{\ddagger} = N \frac{\Delta e^2}{4}\left(\frac{1}{2a_1} + \frac{1}{2a_2} - \frac{1}{r}\right)\left(\frac{1}{n^2} - \frac{1}{D_s}\right) \tag{11.38}$$

where Δe is the charge transferred from one metal the other, a_1 and a_2 are the ground-state radii of the complexes, r is the separation between the metals in the transition state (often taken as $a_1 + a_2$), n is the refractive index of the medium and D is the dielectric constant of the medium. This calculation assumes that the solvent is a dielectric continuum instead of consisting of individual molecules.

Values of k are calculated from the formula[52]

$$k \sim \frac{4\pi N r^3}{3000} \kappa_{\text{el}} \nu_n e^{-\Delta G^{\ddagger}/RT} \tag{11.39}$$

$$\sim K_{\text{os}} \kappa_{\text{el}} \nu_n e^{-(\Delta^{\ddagger}\,\text{inner sphere} \,+\, \Delta G^{\ddagger}\,\text{solvation} \,+\, 0.5\Delta G^0)/RT}$$

where κ_{el} is the so-called **electron transmission coefficient** and represents the probability of electron transfer at the transition-state geometry. For reactions in which there is strong interaction between the reactant and product wavefunctions, a large energy gap ($2H_{\text{AB}}$) exists between the solid curves of Figure 11.15a; the system stays on the lower-energy curve and passes smoothly from reactants to products. Such systems are called **adiabatic** and $\kappa_{\text{el}} = 1$. When coupling is small, $2H_{\text{AB}}$ is small and the reactant system may reach the transition state which is the intersection of reactant and product energy curves and continue on up the reactant parabola (and roll back down) without executing the transition to products. In such nonadiabatic systems, $\kappa_{\text{el}} \ll 1$. In calculations, κ_{el} is taken as $= 1$. ν_n is a nuclear frequency factor and represents the frequency of a vibrational mode between the two metal atoms which removes the system from its transition-state geometry; it is often taken to be 1×10^{13} s^{-1}. K_{os} is the outer-sphere complex formation constant.

Table 11.23 compares some calculated and observed values for self-exchange rate constants. The agreement is good for Fe and Co, but not for Ru$^{3+/2+}$. In cases where k_{calc} is much greater than k_{obs}, the difference is usually attributed to nonadiabaticity. A number of approximations are involved in calculations beyond those made in the theory. For instance, it is not clear what the bond distances in the transition state are; Δd is usually estimated as the difference between bond distances in the reduced and oxidized forms of the couple, but this may not be correct. Also, force constants for symmetric breathing vibrations are often not available, and so on. However, we can see some important trends. Complexes with larger estimated values of Δd have higher $\Delta G_{\text{inner sphere}}^{\ddagger}$ and lower k. Electron transfer from $e_g\sigma^* \rightarrow t_{2g}\pi$ leads to larger Δd (compare Co complexes with others). Bond-length changes are especially important because $\Delta G_{\text{inner sphere}}^{\ddagger}$ depends on $(\Delta d)^2$; this effect contributes to the small values for self-exchange of Co complexes. Larger ligands lead to smaller $\Delta G_{\text{solvation}}^{\ddagger}$ and larger k, all other things being equal. (Compare NH$_3$, en, and bipy complexes.)

---

[52] This formula is approximate and holds when

$$\Delta G_r^{02} < (\Delta G_{\text{inner sphere}}^{\ddagger} + \Delta G_{\text{solvation}}^{\ddagger})$$

This will certainly be the case for self-exchange when $\Delta G^0 = 0$ and for reactions where ΔG^0 is small.

Table 11.23 Calculation of self-exchange constants for redox couples[a]

	$Fe^{3+,2+}$	$Ru^{3+,2+}$	$Ru(NH_3)_6^{3+,2+}$	$Ru(bipy)_3^{3+,2+}$	$Co(NH_3)_6^{3+,2+}$	$Co(en)_3^{3+,2+}$	$Co(bipy)_3^{3+,2+}$
E^0 (volts):	+0.74	+0.045	+0.051	+1.26	+0.058	-0.24	—
r (pm):	65	65	67	136	66	84	136
$K_{os}/10^{-1}\ M$:	0.05	0.33	1.0	3.3	0.33	1.2	3.3
$\Delta G^{\ddagger}_{\text{inner sphere}}$ (kJ/mol):	35.2	15.9	3.3	0	73.7	—	57.3
$\Delta G^{\ddagger}_{\text{solvation}}$ (kJ/mol):	28.9	28.9	28.0	13.8	28.5	5.3	13.8
Δd (pm):	1.3	0.9	0.2	0.0	2.2	2.1	1.9
k_{calc} ($M^{-1}\ s^{-1}$):	3	4×10^3	1×10^5	1×10^9	$4.8 \times 10^{-6\ b}$	$2 \times 10^{-5\ b}$	20
k_{obs} ($M^{-1}\ s^{-1}$):	4	50	2.8×10^4	4.2×10^8	8×10^{-6}	7.7×10^{-5}	18

[a] Data from B. S. Braunschweig, C. Creutz, D. H. Macartney, T. -K. Sham, and N. Sutin, *Disc. Faraday Soc.* **1982**, *74*, 113; R. A. Marcus and N. Sutin, *Biochem. Biophys. Acta* **1985**, *811*, 265.

[b] A. Hammershor, D. Geselowitz, and H. Taube, *Inorg. Chem.* **1984**, *23*, 979.

▶ *11.7.3   Heteronuclear Redox Reactions and Simplified Marcus Theory*

Most redox reactions of interest are heteronuclear—they involve two different metals. Nevertheless, each metal complex must undergo separately the kind of rearrangement described above. Thus, we might expect that $\Delta G^{\ddagger}_{\text{inner sphere}}$ for the heteronuclear reaction will be related to $\Delta G^{\ddagger}_{\text{inner sphere}}$ for self-exchange of each of the reactants. Figure 11.15*b* shows a reaction profile for a heteronuclear redox reaction that is thermodynamically favorable. The activation barrier height depends on the relative placement of reactant and product potential energy curves—which is equivalent to saying that the value of the rate constant also depends on the overall free-energy change ΔG^0.

A simplified version of the Marcus equation which incorporates these considerations is

$$k_{12} = \sqrt{k_{11}k_{22}K_{12}f_{12}} \tag{11.40}$$

where

$$\log f_{12} = \frac{(\log K_{12})^2}{4 \log \dfrac{k_{11}k_{22}}{Z^2}} \tag{11.41}$$

and Z is the number of collisions per second between particles in solution ($\sim 10^{11}\ M^{-1}\ s^{-1}$ at 25°C). The Marcus equation shows that rates of redox reactions depend on an intrinsic factor (through k_{11} and k_{22}) and a thermodynamic factor (through K_{12}). The rate is related to the driving force, and more thermodynamically favorable reactions are faster. This equation is an example of a linear free-energy relationship.

One use of the simplified Marcus equation is to predict the outer-sphere rate constant for a cross-reaction. To predict k_{12} for reduction of $[Co(bipy)_3]^{3+}$ by $[Co(terpy)_2]^{2+}$, we require self-exchange constants.

$$[Co(bipy)_3]^{2+} + [Co^*(bipy)_3]^{3+} \longrightarrow [Co(bipy)_3]^{3+} + [Co^*(bipy)_3]^{2+}$$
$$k_{11} = 9.0\ M^{-1}\ s^{-1}\ \text{at}\ 0°C \tag{11.42}$$

$$[Co(terpy)_2]^{2+} + [Co^*(terpy)_2]^{3+} \longrightarrow [(Co(terpy)_2]^{3+} + [Co^*(terpy)_2]^{2+}$$
$$k_{22} = 48\ M^{-1}\ s^{-1}\ \text{at}\ 0°C \tag{11.43}$$

Reduction potentials for $[Co(terpy)_2]^{3+}$ and $[Co(bipy)_3]^{3+}$ are $+0.31$ and $+0.34$ V, respectively. Hence, $\log K_{12} = 0.553$ and $K_{12} = 3.57$.

$$\log f_{12} = \frac{(0.553)^2}{4 \log \dfrac{9.0 \times 48.0}{10^{22}}} = -3.95 \times 10^{-3}, \qquad f_{12} = 0.99 \tag{11.44}$$

$$k_{12} = \sqrt{(9.0\ M^{-1}\ s^{-1})(48.0\ M^{-1}\ s^{-1})(3.57)(0.99)} = 39\ M^{-1}\ s^{-1}$$

This calculated value compares favorably with the measured value of 64 $M^{-1}s^{-1}$.[53] Agreement between calculated and measured rate constants generally is good as long as the driving force is not too great ($K \leq 10^6; f \geq 0.2$); it has been estimated[54] that rate constants can be calculated to within a factor of ~25. Data in Table 11.24 are fairly typical. In the case of an unknown mechanism, agreement between the observed rate constant and that calculated from the Marcus equation is regarded as evidence for an outer-sphere pathway.[54]

Self-exchange rates can also be calculated from the simplified Marcus equation. Some values calculated for $Ru^{2+/3+}$ self-exchange are shown in Table 11.25; the agreement with the measured value of 50 M^{-1} s^{-1} is within a factor of ten for all cases, and the average is ~40 M^{-1} s^{-1}. Calculations of this sort have been employed to estimate self-exchange rates which cannot be easily measured directly—for example, in metalloproteins.

Self-exchange constants calculated for $Fe^{3+/2+}$ in this way usually turn out to be orders of magnitude smaller than the measured one (Table 11.26). It has been suggested that a value of ~10^{-3} for the Fe self-exchange would lead to better agreement with experiment. One interpretation is that the measured exchange rate is really that for inner-sphere electron transfer. Some evidence in favor of this view is that the calculated $\Delta G^{\ddagger}_{\text{inner sphere}}$ is much larger for Fe than for Ru, which leads to the expectation that k_{11} for Fe should be orders of magnitude lower than that for Ru. Also, $V^{2+/3+}$, where $\Delta d \sim 1.5$ pm compared to 1.3 for Fe, has $k_{11} = 1.1 \times 10^{-2}$ M^{-1} s^{-1}. However, recent measurement[55] of $\Delta V^{\ddagger}$ suggests that Fe self-exchange is indeed outer-sphere. It may be that heterometallic electron transfers involving Fe exhibit nonadiabaticity which is absent in the self-exchange.

▶ 11.7.4 Inner-Sphere Reactions

A generalized mechanism for inner-sphere electron-transfer reactions involves three separate steps:

$$M^{II}L_6 + XM'^{III}L_5' \underset{k_{-1}}{\overset{k_1}{\rightleftharpoons}} L_5M^{II}-X-M'^{III}L_5' + L$$

$$L_5M^{II}-X-M'^{III}L_5' \longrightarrow L_5M^{III}-X-M'^{II}L_5' \qquad k_t \qquad (11.45)$$

$$L_5M^{III}-X-M'^{II}L_5' \longrightarrow \text{products} \qquad k$$

The first involves substitution by the bridging group X into the coordination sphere of the labile reactant (usually the reductant[56]) to form the **precursor complex.** The precursor complex then undergoes the same kind of reorganization described for outer-sphere pathways, followed by electron transfer to give the **successor complex.** Thus, events in this step are related to those in outer-sphere reactions. In the last step the successor complex

[53] R. Farina and R. G. Wilkins, *Inorg. Chem.* **1968,** *7,* 514.

[54] For a critical review of the applicability of the Marcus equation, see M. Chou, C. Creutz, and N. Sutin, *J. Am. Chem. Soc.* **1977,** *99,* 5615.

[55] W. H. Jolly, D. Stranks, and T. W. Swaddle, *Inorg. Chem.* **1990,** *29,* 1948.

[56] The bridging ligand may be carried by the reductant, as in the $[Fe(CN)_6]^{4-}$ reduction of $HCrO_4^-$.

Table 11.24 Comparison between calculated and observed rate constants for outer-sphere reactions[a]

Reaction	k_{12} obsd. $(M^{-1} s^{-1})$	k_{12} calcd. $(M^{-1} s^{-1})$	k_{11} $(M^{-1} s^{-1})$	k_{22} $(M^{-1} s^{-1})$	E^0 $V(K_{12})$	f
$[Fe(CN)_4]^{4-}-[IrCl_6]^{2-}$	3.8×10^5	1×10^6	7.4×10^2	2.3×10^5	$0.24(1.2 \times 10^4)$	5.1×10^{-1}
$[Fe(CN)_6]^{4-}-[Fe(phen)_3]^{3+}$	$>10^8$	$>1 \times 10^8$	7.4×10^2	$>3 \times 10^7$	$0.39(3.5 \times 10^6)$	1.2×10^{-1}
$[Fe(CN)_6]^{4-}-MnO_4^-$	1.7×10^5	6×10^4	7.4×10^2	3×10^3	$0.20(2.5 \times 10^3)$	6.6×10^{-1}
$[Fe(CN)_6]^{4-}-Ce^{IV}$	1.9×10^6	6×10^6	7.4×10^2	4.4	$0.76(5.8 \times 10^{12})$	4.7×10^{-3}
$[Fe(phen)_3]^{2+}-Ce^{IV}$	1.4×10^5	$>7 \times 10^6$	$>3 \times 10^7$	4.4	$0.36(1.1 \times 10^6)$	2.1×10^{-1}
$[Fe(phen)_3]^{2+}-MnO_4^-$	6.1×10^3	>3	$>3 \times 10^7$	3×10^3	$-0.50(3.4 \times 10^{-9})$	2.4×10^{-2}
$Fe^{2+}-[Fe(phen)_3]^{3+}$	3.7×10^4	$>5 \times 10^6$	4.0	$>3 \times 10^7$	$0.35(7.6 \times 10^5)$	2.4×10^{-1}
$Fe^{2+}-Ce^{IV}$	1.3×10^6	5×10^5	4.0	4.4	$0.71(8.3 \times 10^{11})$	2.0×10^{-2}
$Fe^{2+}-[IrCl_6]^{2-}$	3.0×10^6	2×10^4	4.0	2.3×10^5	$0.16(5.0 \times 10^2)$	8.9×10^{-1}
$Cr^{2+}-Fe^{3+}$	2.3×10^3	$<1 \times 10^6$	$\leq 2 \times 10^{-5}$	4.0	$1.18(6.6 \times 10^{19})$	1.7×10^{-4}

[a] From D. A. Pennington, in *Coordination Chemistry*, Vol. 2, A. E. Martell, Ed., American Chemical Society, Washington, D.C., 1978. k_{11} and k_{22} are the rate constants for self-exchange for reductant and oxidant, respectively, for the given redox pair.

Table 11.25   Calculated self-exchange rate constants for $Ru^{3+/2+}$ using Equation (11.40)[a]

Cross-reaction	K_{12}	k_{12}^{obs}	k_{22}^{obs}	k_{11}^{calc}
$Ru^{3+} + V^{2+}$	5.8×10^7	2.5×10^2	1.5×10^{-2}	0.43
$Ru^{3+} + [Ru(NH_3)_6]^{2+}$	6×10^2	1.4×10^4	$\sim 3 \times 10^3$	140
$[Co(phen)_3]^{3+} + Ru^{2+}$	4.2×10^2	53	~ 40	0.20
$[Ru(NH_3)_5py]^{3+} + Ru^{2+}$	35	1.1×10^4	4.7×10^5	8.0

[a] Data from J. T. Hupp and M. J. Weaver, *Inorg. Chem.* **1983**, *22*, 2557 and M. J. Weaver and E. L. Yee, *Inorg. Chem.* **1980**, *19*, 1936 and references therein.

breaks up to give the products. If M^{III} is inert to substitution and M'^{II} is labile, transfer of the bridging ligand occurs, affording $M^{III}XL_5$ as one product—which provides evidence of the inner-sphere mechanism. This is not always the case, however. For example, in Reaction (11.46) no ligand transfer occurs, since both Fe complexes are inert.

$$[Co(edta)]^{2-} + [Fe(CN)_6]^{3-} \longrightarrow [Co(edta)]^- + [Fe(CN)_6]^{4-} \qquad (11.46)$$

In principle, any one of the three steps could be rate-determining. Figure 11.16 shows potential-energy diagrams corresponding to each possibility.

Type I.   Most inner-sphere redox reactions correspond to the case in which the electron-transfer step (with attendant bond-length and solvation changes) that transforms the precursor into the successor complex is slowest. ($k_{ct} \ll k_1$, k_{-1} and k) All the inner-sphere reactions mentioned so far are examples of this type. Inner-sphere redox reactions are placed in this class if no evidence exists for assigning them to the other types. For these reactions $k_{obs} = Kk_{et}$, where $K = k_1/k_{-1}$ is the equilibrium constant for precursor complex formation and k_{et} is the rate constant for e transfer defined in Equations (11.45).

Because the e-transfer step involves a reorganization barrier similar to the outer-sphere case, a Marcus-type relationship applies k_{et}. The study of dinuclear complexes such as $\left[(bipy)_2ClRu^{II}-N\bigcirc N-Ru^{III}(bipy)_2Cl\right]^{3+}$ as model systems for inner-sphere precursor complexes has enabled evaluation of Franck–Condon barriers, solvation ener-

Table 11.26   Calculated[a] self-exchange rate constants for $Fe^{3+/2+}$ using Equation (11.40)

Cross-reaction	K_{12}	k_{12}^{obs}	k_{22}^{obs}	k_{11}^{calc}
$Fe^{3+} + Ru^{2+}$	5×10^8	2.3×10^3	50	1.4×10^{-3}
$Fe^{3+} + [Ru(NH_3)_5py]^{2+}$	1.5×10^7	5.8×10^4	4.7×10^5	2.3×10^{-3}
$Fe^{3+} + [Ru(NH_3)_4bipy]^{2+}$	4.5×10^3	7.2×10^3	$\sim 1 \times 10^7$	2×10^{-3}
$[Fe(bipy)_3]^{3+} + Fe^{2+}$	1.2×10^5	3.7×10^4	$\sim 1 \times 10^9$	4×10^{-6}

[a] Data from J. T. Hupp and M. J. Weaver, *Inorg. Chem.* **1983**, *22*, 2557 and M. J. Weaver and E. L. Yee, *Inorg. Chem.* **1980**, *19*, 1936 and references therein.

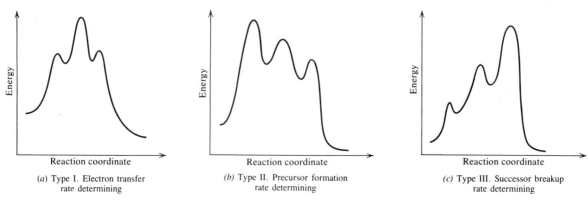

(*a*) Type I. Electron transfer
rate determining

(*b*) Type II. Precursor formation
rate determining

(*c*) Type III. Successor breakup
rate determining

Figure 11.16   Reaction profiles for inner-sphere redox reactions. (From R. G. Linck, in *M. T. P. International Review of Science, Chemistry*, Vol. 9, series one, M. L. Tobe, Ed., University Park Press, Baltimore, 1972.)

gies, and so on.[57] Reaction profiles such as those shown in Figure 11.15 are applicable to the *e*-transfer step in inner-sphere reactions.

Type II.   When formation of the precursor complex is rate-determining (Figure 11.16*b*), the reaction rate is controlled by the substitution rate into the coordination sphere of the labile reactant ($k_1 \ll k_{et}$, k). Table 11.27 gives kinetic parameters for some reductions by V^{2+} which are substitution-controlled. The reduction rates are quite similar, and the kinetic parameters closely match those for substitution on V^{2+}. For water exchange on V^{2+}, $k = 100$ s^{-1}, $\Delta H^{\ddagger} = 68.6$ kJ/mol, and $\Delta S^{\ddagger} = -23$ J/mol K. For NCS$^-$ substitution, $k = 28$ M^{-1} s^{-1}, $\Delta H^{\ddagger} = 56.5$ kJ/mol, and $\Delta S^{\ddagger} = -29$ J/mol K. That the rates should be so similar for a variety of possible bridging ligands could be interpreted on an outer-sphere model. However, the similarity of the activation parameters to those for substitution is evidence that the process is controlled by the dissociative activation of the substitution (see Section 11.3.4).

Table 11.27   Rate parameters for some reductions by V^{2+} at 25°C[a]

Oxidant	k (M^{-1} s^{-1})	$\Delta H^{\ddagger}$ (kJ/mol)	$\Delta S^{\ddagger}$ (J/mol K)
$[Co(NH_3)_5C_2O_4H]^{2+}$	12.5	51.0	−54
$[Co(NH_3)_5C_2O_4]^+$	45.3	—	—
cis-$[Co(NH_3)(en)_2(N_3)]^{2+}$	10.3	52.7	−50
cis-$[Co(H_2O)(en)_2(N_3)]^{2+}$	16.6	50.6	−50
trans-$[Co(en)_2(N_3)_2]^+$	26.6	51.0	−46
trans-$[Co(H_2O)(en)_2(N_3)]^{2+}$	18.1	46.0	−67
Cu^{2+}	26.6	46.7	−57.7

[a] From the work of Taube, Espenson, Sutin, Linck, and others.

[57] T. J. Meyer, *Acc. Chem. Res.* **1978,** *11,* 94.

A corollary of these considerations is that any V^{2+} reduction with a rate constant substantially in excess of the substitution rate constant must be outer-sphere (e.g., the Fe^{3+} reduction for which $k_{12} = 1.8 \times 10^4 \, M^{-1} \, s^{-1}$).

Substitution-controlled reductions by Cr^{2+}, Fe^{2+}, Cu^+, and Eu^{2+} are known.

Type III.   You might expect the breakup of the successor complex to be rate-determining ($k \ll k_{et}$) when the electron configuration of both metals in the successor complex leads to substitution inertness. Usually, the existence of a binuclear complex in equilibrium with reactants is indicated by the form of the rate law—as, for example, in the Cr^{2+} reduction of $[Ru(NH_3)_5Cl]^{2+}$. A complex having both Ru and Cr is kinetically detectable. Because the successor complex would contain inert Ru^{II} and Cr^{III}, we assume that this must be the species detected.[58] The mechanism involved is

$$[Ru^{III}(NH_3)_5Cl]^{2+} + Cr^{2+} \rightleftharpoons [Ru^{III}(NH_3)_5ClCr^{II}]^{4+} \quad K_{eq}$$
$$[Ru^{III}(NH_3)_5ClCr^{II}]^{4+} \xrightarrow{fast} [Ru^{II}(NH_3)_5ClCr^{III}]^{4+} \xrightarrow{k} Ru^{II} + CrCl^{2+} \quad (11.47)$$

This is consistent with the observed rate law

$$-\frac{d}{dt}[Ru^{III}] = \frac{kK_{eq}}{1 + K_{eq}[Cr^{2+}]}[Ru^{III}]_0[Cr^{2+}] \quad (11.48)$$

which you should verify as an exercise.[59]

Sometimes successor complexes are sufficiently stable that they can be prepared independently and their breakup followed. An example is the successor complex from the reduction of VO^{2+} by V^{2+}: $[V^{III}(OH)_2V^{III}]^{4+}$ [60]; this species can also be prepared by hydrolysis of V^{III} solutions. Comparatively few redox reactions have been shown to belong to Type III.

The Bridging Ligand

Even in reactions in which precursor complex formation is not rate-determining, the nature of the bridging ligand can be quite important because the activation barrier for precursor formation may represent a substantial fraction of the overall activation energy and because electron transfer is mediated by the bridging ligand. This section points out some aspects of the role of the bridging ligand and their mechanistic consequences.

If a ligand is to function in a bridging capacity, an unshared pair of sufficiently basic electrons is required. Thus, complexes such as $[Co(NH_3)_6]^{3+}$ and $[Co(NH_3)_5(py)]^{3+}$ react only via outer-sphere paths. Current evidence also indicates that unshared pairs on water

[58] It is true that the mere detection of a bridged species does not mean that it is involved in the redox reaction. Other evidence, including comparison with related systems, is needed to establish this point. See, for example, L. Rosenheim, D. Speiser, and A. Haim, *Inorg. Chem.* **1974**, *13*, 1571; D. H. Huchital and J. Lepore, *Inorg. Chem.* **1978**, *17*, 1134.

[59] *Hint:* Rate = k_{obs} $[Cr^{2+}][Ru^{III}]_0$, where $[Ru^{III}]_0$ represents the concentration of $[RuCl(NH_3)_5]^{2+}$ used in preparing the reacting solution. However, this is now partitioned between two solution species, $[RuCl(NH_3)_5]^{2+}$ and the Ru—Cr dimer. Cr^{2+} is assumed present in large excess.

[60] T. W. Newton and F. B. Baker, *Inorg. Chem.* **1964**, *3*, 569; L. Pajdowski and B. Jezowska-Trzebiatowska, *J. Inorg. Nucl. Chem.* **1966**, *28*, 443.

are not sufficiently basic to allow aqua complexes an inner-sphere path.

Evidence for an inner-sphere mechanism follows in selected cases from changes in rate as a function of bridging ability of ligands. Table 11.28 contains rate data for some reductions as the possible bridging ligand is changed from H_2O to OH^- and from N_3^- to NCS^-. The $[Ru(NH_3)_6]^{2+}$ reductions must be outer-sphere (*o.s.*) because the reductant is both inert to substitution on the time scale of the redox reaction and without electron pairs for bridging. The rates of reaction with aqua and hydroxo complexes are quite similar. The large rate enhancement on going from H_2O to OH^- complexes when Cr^{2+} is the reductant supplies strong evidence that OH^- is a good bridging ligand and that Cr^{2+} reduction of the hydroxo complexes is inner-sphere (*i.s.*).

In monatomic bridges, the function of the bridging ligand can only be to increase the probability of *e* transfer by tunneling.

Remote Attack.   Table 11.28's information on reduction of thiocyanato and azido complexes raises several interesting points. First, notice that all the products are consistent with transfer of the bridging group in an *i.s.* mechanism. Another feature is the fact that lone pairs for bridging are not always available on the atom adjacent to the oxidant. Thus, for example, the only site available for bridging in $[Co(NH_3)_5NCS]^{2+}$ is on S remote from the metal center. The product, $CrSCN^{2+}$, indicates that remote attack occurs on S. Of course, the symmetry of the azido ligand makes such a distinction experimentally impossible without isotopic labeling. However, drawing the Lewis structure will show that only the remote N can act as a bridge. In $[Co(SCN)(NH_3)_5]^{2+}$, both adjacent and remote attack occur. On standing, the $CrSCN^{2+}$ product rearranges to the more stable $CrNCS^{2+}$ isomer. Because Cr^{2+} is a hard acid, it prefers to bond with hard N. This preference is reflected in a lesser stability of the $Cr—SCN—Co$ precursor and accounts for the rate differences between *S*- and *N*-bonded complexes.[61] In general, N_3^- is a better bridging ligand than $M—NCS$, and a difference in rates of $\sim 10^4$ for inner-sphere reactions is common. If comparable azido and *N*-bonded thiocyanato complexes are reduced by the same reagent at comparable rates, the reaction is probably outer-sphere. Finally, these data show that electron transfer can occur through a multiatom bridge.

Table 11.28   Rate constants for some redox reactions at 25°C

Oxidant	Reductant	$k(M^{-1}\ sec^{-1})$	Mechanism
$[Co(NH_3)_5(H_2O)]^{3+}$	Cr^{2+}	≤ 0.1	Probably *o.s.*
$[Co(NH_3)_5(OH)]^{2+}$	Cr^{2+}	1.5×10^6	*i.s.*
$[Co(NH_3)_5(H_2O)]^{3+}$	$[Ru(NH_3)_6]^{2+}$	3.0	*o.s.*
$[Co(NH_3)_5(OH)]^{2+}$	$[Ru(NH_3)_6]^{2+}$	0.04	*o.s.*

Reaction	Product	$k\ (M^{-1}\ s^{-1})$
$[Co(NH_3)_5NCS]^{2+} + Cr^{2+}$	$CrSCN^{2-}$	19
$[Co(NH_3)_5N_3]^{2+} + Cr^{2+}$	CrN_3^{2+}	3×10^5
$[Co(NH_3)_5SCN]^{2+} + Cr^{2+}$	71% $CrNCS^{2+}$ + 29% $CrSCN^{2+}$	1.9×10^5

a From the work of Haim, Linck, Taube, Endicott, and others.

[61] Sulfur seems to be an especially good mediator of electron transfer. Thus, attack at S sites often occurs kinetically, even when the product is not the most thermodynamically stable one.

One such electron transfer invokes remote attack by Cr^{2+} on O in

$$(NH_3)_5Co-N\langle\bigcirc\rangle-C\overset{\ddot{O}}{\underset{NH_2}{\diagdown}}{}^{3+}$$

The electron pair on N apparently is not sufficiently basic to provide a site for attack. This seems to be a general feature of $-NH_2$ groups. Whereas $\left[(NH_3)_5Co \leftarrow NH_2\underset{\overset{\parallel}{O}}{C}H\right]^{3+}$ un-

dergoes reduction with Cr^{2+} via remote attack on the carbonyl O, its linkage isomer,

$$\left[(NH_3)_5Co\longleftarrow O=C\overset{H}{\underset{NH_2}{\diagup}}\right]^{3+}$$ follows an *o.s.* path. This has been interpreted as evi-

dence that, in remote attack, the oxidant and reductant must belong to a conjugated system involving the bridging ligand. The function of the bridging ligand is to couple the metal orbitals. This is termed the **resonance mechanism** for electron transfer.

The Chemical Mechanism.[62] Organic bridging ligands having conjugated π-systems display further interesting effects on redox chemistry. A study of the Cr^{2+} reduction of $\left[(NH_3)_5Co-C(O)-\langle\bigcirc\rangle-C(O)H\right]^{2+}$ shows that remote attack occurs on the carbonyl O and gives Co^{2+} as the product. However, $\left[(NH_3)_5Co-C(O)-\langle\bigcirc\rangle-NO_2\right]^{3+}$ gives no Co^{2+} on reduction. Instead, the nitro group on the ligand is reduced. The Cr^{2+} reduction of $\left[(NH_3)_5Co-OC(O)-\langle\overset{N^-}{\underset{N}{\bigcirc}}\rangle\right]^{2+}$ has been found to involve two detectable intermediates. The first decays to give the successor complex, which then affords Co^{2+} as the product.[63] Apparently, when the organic ligand is reducible to a fairly stable radical, electron transfer can first reduce the bridging ligand, which then may or may not in turn reduce the other metal. This mode of electron transfer [Mechanism (11.49)] is called the **chemical** or **radical-ion mechanism.**

$$\begin{aligned}
M^{III}-L + Cr^2 &\longrightarrow M^{III}-L-Cr^{2+}\\
M^{III}-L-Cr^{2+} &\longrightarrow M^{III}-L^{\cdot-}-Cr^{III}\\
M^{III}-L^{\cdot-}-Cr^{III} &\longrightarrow M^{II}-L-Cr^{III}\\
M^{II}-L-Cr^{III} &\longrightarrow M^{II} + Cr^{III}L
\end{aligned}\qquad(11.49)$$

[62] H. Taube and E. S. Gould, *Acc. Chem. Res.* **1969**, *2*, 321.
[63] E. S. Gould , *J. Am. Chem. Soc.* **1972**, *94*, 4360.

We also can recognize the occurrence of the radical-ion mechanism by the insensitivity of reaction rates to the identity of the metal in the oxidant. For example, Co^{III} complexes typically are reduced in inner-sphere reactions about 10^5–10^7 times as fast as Cr^{III} complexes with the same ligands. However, for $\left[ML_5N\bigcirc-C(O)NH_2\right]^{3+}$ complexes, the reaction is only about 10 times as fast for $ML_5 = Co(NH_3)_5$ as for $ML_5 = Cr(H_2O)_5$, implying that the electron is transferred to the ligand rather than the metal.

Orbital Symmetry in Electron Transfer. In our model of electronic structures, electrons must always exist in orbitals. Electrons transferred to unsaturated organic ligands are presumed to occupy π^* orbitals. They are transferred into e_g orbitals of σ symmetry when Cr^{III} and Co^{III} are reduced. The inner-sphere reorganization energy required to achieve overlap of the ligand π and oxidant σ orbitals slows the reduction of the metal center. The lifetime of the radical-ion intermediate is prolonged. However, when the metal orbital is t_{2g} of π-symmetry, resonance electron transfer can occur immediately.

The complex $\left[Ru^{III}(NH_3)_5-N\bigcirc-C(O)NH_2\right]^{3+}$ has the t_{2g}^5 configuration and is reduced by Cr^{2+} 30,000 times as fast as the Co^{III} complex. The resonance transfer mechanism is not possible for saturated bridging ligands. This is consistent with the bridging inefficiency of multiatom saturated ligands. Such complexes react by outer-sphere pathways.

Bridged Outer-Sphere Electron Transfer

A rather surprising role for organic bridging ligands has been found[64]; in some cases, the bridging group simply holds oxidant and reductant in proximity while the actual *e* transfer occurs via an outer-sphere process. At least this seems the most plausible explanation for relative rates of electron transfer in a series of long-lived precursor complexes

$[(NC)_5Fe-L-Co(NH_3)_5]$, where $L = N\bigcirc-X-\bigcirc N$ and $X = (CH_2)_n$

($n = 1$–3) and $C{=}O$. Because the complex with $X = C{=}O$ has a conjugated π-system, we would expect it to undergo resonance electron transfer faster than the complexes with saturated X. Instead, it reacts an order of magnitude slower. This implies that the flexible, saturated X group simply hold the two reactants together and permits them to achieve a suitable conformation for outer-sphere *e* transfer!

EXAMPLE 11.4: The ratios of rate constants for reduction of $[Co(edta)Cl]^{2-}$ and $[Co(edta)(H_2O)]^-$ by various reductants at 25°C are given below. What can you say about the inner- or outer-sphere nature of these reactions? (See H. Ogino, E. Kikkawa, M. Shimata, and N. Tanaka, *J. Chem. Soc. Dalton Trans.* **1981**, 894.)

[64] J. J. Jwo, P. L. Gaus, and A. Haim, *J. Am. Chem. Soc.* **1979**, *101*, 6189.

Reductant	k_{Cl}/k_{aqua}
$[Fe(CN)_6]^{4-}$	33
Ti^{3+}	31
Cr^{2+}	2×10^3
Fe^{2+}	$>3 \times 10^2$

Solution: Because d^6 Co^{III} complexes are inert and $[Fe(CN)_6]^{4-}$ (d^6) is inert to substitution itself, the ferrocyanide reductions must be *o.s.*; the rates are quite similar for the chloro and aqua complexes. The ratio for Ti^{3+} is the same as that for ferrocyanide, and this is reasonable if these reductions are also *o.s.* The nature of the possible bridging ligand makes a big difference for the other reactions, suggesting that the chloro complexes, at least, must undergo *i.s.* reduction. Chloride is known to be a good bridging ligand in the *i.s.* reduction of $[Co(NH_3)_5Cl]^{3+}$ by Cr^{2+}. These data by themselves do not reveal whether or not the aqua complexes are reduced by an *i.s.* mechanism.

▶ *11.7.5   Other Issues in Inner-Sphere Reactions*

Nonbridging ligands play a role in rates of redox reactions because they affect the relative energies of donor and acceptor orbitals. Also, ionic strength and medium effects greatly influence the determined form of the rate law.[65]

Table 11.29 summarizes methods for distinguishing between inner- and outer-sphere models for the transition state.

11.8   PHOTOCHEMICAL REACTIONS

In **photochemistry,** an area of rapidly growing interest in inorganic chemistry, energy is imparted to molecules by irradiating them with visible and ultraviolet light. Energy of light quanta ranges from ~170 to 590 kJ/mol—up to 3.5 times that available thermally at ordinary temperatures. Only certain wavelengths of light are absorbed by transition metal complexes—namely, those corresponding to differences between electronic energy states (See Sections 10.2 and 10.4.).

Excited complexes can undergo several kinds of energy transfer. The most straightforward is luminescence, in which light is reemitted as the excited state returns to the ground state (a radiative process). Alternatively, energy may be converted to vibrational energy and dissipated thermally to the environment as the complex returns to the ground state (internal conversion). Of more interest to chemists is the fact that relatively long-lived excited states of complexes may display very different reaction chemistry from the thermally activated ground state. Figure 11.17 diagrams the possible fates of a photochemically activated complex and compares them with thermally activated processes. The photochemical

---

[65] See, for example, D. E. Pennington in *Coordination Chemistry,* Vol. 2, A. E. Martell, Ed., American Chemical Society, Washington, D. C., 1978. See also references at the end of the chapter.

Table 11.29 Methods for distinguishing between inner- and outer-sphere redox reactions

Method	Comments
Form of rate law	Usually, no distinction possible; most have rate = k[oxidant][reductant]. Can sometimes detect stable dinuclear complex.
Analysis of products	Detection of product with transfer of bridging ligand is evidence of *i.s.* mechanism. Product must be substitution-inert on experiment time scale.
Rate prediction from Marcus equation	Agreement indicates *o.s.* May be applicable in *i.s.* cases.
Rate ratio	Agreement with prediction from ratio for known *o.s.* mechanism implies *o.s.* mechanism. May also apply to *i.s.* mechanism.
Rate enhancement by bridging ligand	Applicable to limited class of ligands; shows *i.s.* behavior. N_3^- vs. NCS^-; OH^- vs. H_2O; not applicable to halides.
Insensitivity of rate to possible bridging ligand	Comparison to reactions of known mechanism must be made carefully. Could represent *o.s.* or substitution-controlled or radical-ion-type *i.s.*
Value of rate constant and activation parameters	If these are similar to values for substitution, *i.s.* substitution-controlled mechanism applies. If rate is $\gg$ substitution rate for either reactant, mechanism is *o.s.*
Detection of binuclear species	If this is a kinetically detected intermediate, *i.s.* mechanism applies. If spectroscopically detected, most likely *i.s.*
Sensitivity to metal ions	Probably results from thermodynamic factors in *o.s.* or *i.s.* case. Main significance is in cases where separate evidence for *i.s.* or *o.s.* behavior exists.
Ligand identity	If no ligand has a lone pair for bridging, must be *o.s.* Even H_2O and NH_2 probably do not allow *i.s.* path.

excited state may luminesce with a rate constant k_r, undergo internal conversion with a rate constant k_q, or react (perhaps with other species present) with rate constants k_a, k_b, . . . , and so on. The figure also indicates schematically how products thermodynamically (*A*) or kinetically inaccessible (*C*) in thermal reactions might result from photochemical excitation. A further complicating possibility is so-called **intersystem crossing,** in which a transition occurs to a different excited state, which can then undergo any of the fates depicted in the figure.

Four types of photochemically excited states can be achieved for transition metal complexes. Ligand-field excited states, which result from *d–d* transitions and generally place electron density in $e_g\sigma^*$ orbitals, are expected to lead to weakened L—M bonds and to favor ligand labilization. Charge-transfer (CT) excited states may be L → M or M → L. In the latter, more common, case, the reactivity of the excited state should parallel that of L^- or involve electron transfer to the oxidized metal center. Transitions localized on the ligands result in intraligand excited states. Insofar as anything is known about the reactivity to be expected, ligand rearrangement and ligand reaction with other substrates may be involved. Finally, a charge transfer to solvent state involving an oxidized metal ion and a solvated e_s could lead to electron transfer to the oxidized metal and reduction of solvent or other substrate by e_s. As one example, photoaquation of $[Co(NH_3)_6]^{3+}$ to give

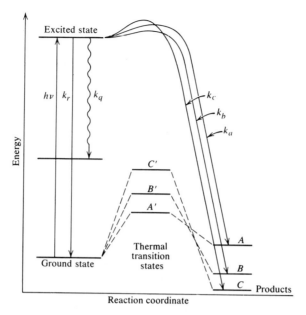

Figure 11.17 Comparison of thermal and photochemical activation. (After Balzani and Carassiti, in *Photochemistry of Coordination Compounds*, Academic Press, London, 1970.)

$[Co(NH_3)_5(H_2O)]^{3+}$ occurs readily upon irradiating the lower-wavelength ligand-field band of the hexaammine. In contrast, an acid solution of $[Co(NH_3)_6]^{3+}$ can be boiled for days without a reaction occurring. The photochemically excited state is $\sim 10^{13}$ as reactive as the ground state.

Irradiation of $[Ru(bipy)_3]^{2+}$ and related complexes produces a CT excited species, $[Ru^*(bipy)_3]^{2+}$, with an oxidized Ru and an electron localized on bipy. This excited molecule is both a better oxidizing agent and a better reducing agent than $[Ru(bipy)_3]^{2+}$, as the reduction potentials in Table 11.30 show. $[Ru^*(bipy)_3]^{2+}$ is luminescent; quenching of the luminescence by other species in solution Q is evidence of a reaction that may be oxidation, reduction, or energy transfer.

$$[Ru^*(bipy)_3]^{2+} + Q \longrightarrow [Ru(bipy)_3]^{3+} + Q^- \qquad (11.50)$$

$$[Ru^*(bipy)_3]^{2+} + Q \longrightarrow [Ru(bipy)_3]^{1+} + Q^+ \qquad (11.51)$$

$$[Ru^*(bipy)_3]^{2+} + Q \longrightarrow [Ru(bipy)_3]^{2+} + Q^* \qquad (11.52)$$

Table 11.30   Reduction potentials for some Ru complexes at 25°C[a]

Reaction	E^0 (volts)
$[Ru(bipy)_3]^{3+} + e \rightarrow [Ru(bipy)_3]^{2+}$	1.26
$[Ru^*(bipy)_3]^{2+} + e \rightarrow [Ru(bipy)_3]^{1+}$	0.84
$[Ru(bipy)_3]^{3+} + e \rightarrow [Ru^*(bipy)_3]^{2+}$	−0.84
$[Ru(bipy)_3]^{2+} + e \rightarrow [Ru(bipy)_3]^{1+}$	−1.28

[a] From the work of Creutz and Sutin.

The oxidation and reduction reactions represent ways of converting light energy into chemical energy. Energy storage could also be achieved if a suitable quencher could convert the excited species into complexes such as the excellent oxidizing agent $[Ru(bipy)_3]^{3+}$ and the excellent reducing agent $[Ru(bipy)_3]^+$. (The first of these has been shown to oxidize OH^- to O_2—representing a possible step in the decomposition of water to H_2 and O_2 by solar energy.) One such set of thermodynamically feasible quenching reactions is

$$[Ru^*(bipy)_3]^{2+} + [Ru(NH_3)_6]^{3+} \longrightarrow [Ru(bipy)_3]^{3+} + [Ru(NH_3)_6]^{2+} \qquad (11.53)$$

$$[Ru^*(bipy)_3]^{2+} + [Ru(NH_3)_6]^{2+} \longrightarrow [Ru(bipy)_3]^{1+} + [Ru(NH_3)_6]^{3+} \qquad (11.54)$$

Unfortunately, both thermodynamics and kinetics favor the back-reaction of the products to regenerate $[Ru(NH_3)_6]^{2+}$, dissipating the stored energy to the solution. A possible solution for this problem is to use an organic quencher such that Q^- or Q^+ reacts further to produce stable products, leaving energy stored as either $[Ru(bipy)_3]^{3+}$ or $[Ru(bipy)_3]^+$. Gray and co-workers have prepared isocyanide complexes such as $[Rh_2(\mu_2\text{-}NCCH_2CH_2CN)_4]^{2+}$, which on irradiation reduces H^+ to H_2. Both kinds of complexes have potential application to solar energy storage and splitting of water. However, our main focus here is on the fact that photolytic activation can lead to chemistry quite different from that caused by thermal activation.

An interesting connection exists between photochemistry and the redox reactions just discussed. Taube, Creutz, Meyer, and others have prepared a series of mixed-valence compounds of which $\left[(bipy)_2ClRu^{II}-N\bigcirc N-Ru^{III}(bipy)_2Cl \right]^{3+}$ is an example. These contain the same metal in two different oxidation states and serve as models for the precursor complex in inner-sphere e transfer reactions when the bridging ligand contains some unsaturation.[66] We can study the rates of electron transfer apart from considerations relating to the precursor stability.

Mixed-valence complexes usually are intensely colored because of an absorption in the low-energy region. This absorption is not present in the spectrum of $[Ru(bipy)_2Cl(py)]^{n+}$ ($n = 2, 3$), and results from electron transfer from one metal to the other—the so-called **intervalence transfer** (IT). Figure 11.18a makes clear the origin of this band. This figure is like Figure 11.15a in that the reactants and products are identical. However, E_{op} (the IT energy) differs from E_{th} (the thermal energy for e transfer) because of the Franck–Condon principle. Most of the complex ions in solution are at the equilibrium position for the ground state when the photon energy is absorbed. Because nuclear motions are slow compared with electronic transition, the energy absorbed in going to the upper state is E_{op}, corresponding to a vertical transition. For symmetrical complexes, we can show that $E_{op} = 4E_{th}$.[67]

EXAMPLE 11.5:   Show that $E_{op} = 4E_{th}$ for a symmetrical mixed-valence complex.

Solution: Suppose a is the distance between energy minimum and the thermal transition state. Both curves are parabolas. Hence, at the thermal transition state, $E_{th} = k(a)^2$, where k is the proportionality constant. At the value of reaction coordinate corresponding to the reactants, the distance from the product curve minimum is $2a$. Thus the energy on the upper parabola is $E_{op} = k(2a)^2$. Hence $E_{op} = 4E_{th}$.

[66] See. T. J. Meyer, *Acc. Chem. Res.* **1978**, *11*, 94.

[67] N. S. Hush, *Prog. Inorg. Chem.* **1967**, *8*, 391; *Chem. Phys.* **1975**, *10*, 361; J. J. Hopfield, *Proc. Natl. Acad. Sci. USA* **1974**, *71*, 3640.

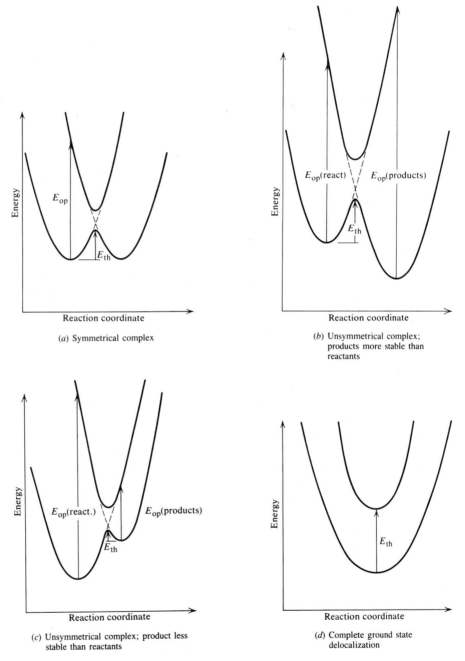

Figure 11.18 Reaction profiles for intervalence transfer. (*a*) Symmetrical complex. (*b*) Unsymmetrical complex; products more stable than reactants. (*c*) Unsymmetrical complex; products less stable than reactants. (*d*) Complete ground-state delocalization.

At least theoretically, measurements of E_{op} should lead to estimates of self-exchange in symmetrical complexes. Figures 11.18b and 11.18c show alternative energy possibilities for unsymmetrical mixed-valence complexes that might contain different metals or differ-

ent ligands—for example, $\left[Cl(NH_3)_4Ru^{III}-N \overset{\frown}{\bigcirc} N-Ru^{II}(bipy)_2Cl \right]^{3+}$.

Unsaturated bridging ligands such as pyrazine give mixed-valence complexes that are good models for inner-sphere precursor complexes, because of orbital overlap between π^* ligand orbitals and each separate metal orbital. If the bridging ligand is saturated and sufficiently long, electronic coupling between redox sites is essentially from outer-sphere orbital overlap. Such a complex is $[Cl(bipy)_2Ru^{II}Ph_2PCH_2CH_2PPh_2Ru^{III}Cl(bipy)_2]^{3+}$, which also exhibits an IT band, but one of much lower intensity, because of unfavorable orbital overlap.[68] Thus, complexes with low-intensity IT bands are models for outer-sphere e-transfer reactions.

Redox chemistry can often be initiated by irradiating the IT band.

$$[(NH_3)_5Co^{III}N\equiv C-Ru^{II}(CN)_5]^- \xrightarrow{h\nu} 5NH_3 + Co^{2+} + [Ru^{III}(CN)_6]^{3-} \qquad (11.55)$$

A third group of complexes also may be prepared from metals in two different oxidation states. Where the bridging ligand has delocalized orbitals and the metal orbitals can overlap with particular effectiveness (because of a short physical separation, low oxidation state, etc.), the two metal centers may "communicate" electronically in the ground state. In these cases, delocalized molecular orbitals exist and the electrons are distributed equivalently over both metals in the ground state. An example is $[(NH_3)_5Ru-O-Ru(NH_3)_5]^{5+}$,[69] where each Ru is assigned 5.5 d electrons. This corresponds to the more-or-less complete coalescence of the curves in Figures 11.8a–c to give a curve like that in Figure 11.18d. Notice that a band corresponding to a transition between two delocalized molecular orbitals now is expected. The delocalized case can be recognized from electrochemical, UV–Vis spectral, and magnetic measurements.

Inorganic photochemistry is a comparatively new research area. Important questions demand investigation: What is the exact electronic state that undergoes photoreaction? (This is the photochemists' equivalent of stoichiometric mechanism.) What are the reaction products (the equivalent of intimate mechanism)? What fraction of the light quanta absorbed leads to a particular reaction? (This is the quantum yield for the reactions and corresponds to the thermal rate constant.) Besides photoreaction, what other processes do excited molecules undergo? Further information about this exciting field can be found in the references.[70]

[68] B. P. Sullivan and T. J. Meyer, *Inorg. Chem.* **1980,** *19,* 752; S. A. Stein and H. Taube, *J. Am. Chem. Soc.* **1978,** *100,* 1635.

[69] J. A. Baumann and T. J. Meyer, *Inorg. Chem.* **1980,** *19,* 345.

[70] V. Balzani and V. Carassiti, *Photochemistry of Coordination Compounds,* Academic Press, London, 1970; A. W. Adamson and P. D. Fleischauer, *Concepts of Inorganic Photochemistry,* Wiley, New York, 1975; M. S. Wrighton, Ed., *Inorganic and Organometallic Photochemistry,* Academic Press, New York, 1979.

GENERAL REFERENCES

F. Basolo and R. Pearson, *Mechanisms of Inorganic Reactions,* 2nd ed., Wiley, New York, 1967. The classic text in the field.

R. B. Jordan, *Reaction Mechanisms of Inorganic and Organometallic Systems,* Oxford University Press, New York, 1991. An excellent introduction.

R. G. Wilkins, *The Study of Kinetics and Mechanisms of Reactions of Transition Metal Complexes,* 2nd ed., VCH Publishers, New York, 1991. Excellent discussions of rate laws, their interpretation, and experimental establishment.

Substitution Reactions

C. H. Langford and H. B. Gray, *Ligand Substitution Processes,* Benjamin-Cummings, Reading, MA, 1966. A very readable account detailing the development of the **D, A, I$_d$** and **I$_a$** nomenclature.

D. W. Margerum, G. R. Caley, D. E. Weatherburn, and G. K. Pagenkopf, in *Coordination Chemistry,* Vol 2, A. E. Martell, Ed., ACS monograph 174, American Chemical Society, Washington, D.C., 1978. A discussion of ligand-exchange and substitution reactions, focusing on labile ions and multidentate ligands.

M. L. Tobe, in *Comprehensive Coordination Chemistry,* Vol. 1, G. Wilkinson, Ed., Pergamon Press, Oxford, U.K., 1987.

Redox Reactions

A. Haim, *Progress in Inorg. Chem.* **1983,** *30,* 273. An authoritative account of inner-sphere reactions.

D. E. Pennington, in *Coordination Chemistry,* Vol. 2, ACS Monograph 174, American Chemical Society, Washington, D.C., 1978. An excellent account of mechanistic investigations in redox chemistry.

W. L. Reynolds and R. W. Lumry, *Mechanisms of Electron Transfer,* Ronald Press, New York, 1966.

N. Sutin, *Prog. Inorg. Chem.* **1983,** *30,* 441. An authoritative account of outer-sphere reactions.

H. Taube, *Electron Transfer Reactions of Complex Ions in Solution,* Academic Press, New York, 1970.

PROBLEMS

11.1   Explain how the data on water exchange given in this chapter (see page 488) suggest a dissociative mechanism.

11.2   Account for the difference in rate constants for the following two reactions:

$$[Fe(H_2O)_6]^{2+} + Cl^- \rightarrow [FeCl(H_2O)_5]^+ + H_2O \qquad k = 10^6 \, M^{-1} \, s^{-1}$$
$$[Ru(H_2O)_6]^{2+} + Cl^- \rightarrow [RuCl(H_2O)_5]^+ + H_2O \qquad k = 10^{-2} \, M^{-1} \, s^{-1}$$

11.3 Octahedral complexes in which the loss of ligand-field stabilization energy (LFSE) is large on going to the activated complex were said to substitute slowly. A model of the transition state for **D** substitution is a five-coordinate square pyramid. Using the approach of Krishnamurthy and Schaap (page 460), calculate the LFSE for strong-field d^n configurations. The difference LFSE(oct.) − LFSE(sq. py.) is the ligand-field contribution to the activation energy for substitution, the so-called **ligand-field activation energy** (LFAE). Compute this quantity for each configuration.

11.4 Distinguish between the intimate and stoichiometric mechanism of a reaction.

11.5 What is the significance of the following facts taken together for the mechanism of substitution at Co^{III} in aqueous solution?
(a) The rates of aquation are always given by the following expression: rate = k_{aq} [$Co(NH_3)_5X^{2+}$].
(b) No direct replacement of X^- by Y^- is ever observed. Instead, water enters first and is subsequently replaced by Y^-.

11.6 The following data have been reported for aquation at 298.1 K of complexes *trans*-[$Co(N_4)LCl$]$^{n+}$, where N_4 is a tetradentate chelate.

| | $k \times 10^4$ (s^{-1}) | |
L	(N_4) = cyclam	(N_4) = tet-*b*
Cl$^-$	0.011	9.3
NCS$^-$	0.000011	0.0070
NO$_2^-$	0.43	410
N$_3^-$	0.036	210
CN$^-$	0.0048	3.4
NH$_3$	0.00073	—

What do these data suggest about the intimate mechanism for aquation? (See W.-K. Chau, W.-K. Lee, and C. K. Poon, *J. Chem. Soc. Dalton Trans.* **1974**, 2419.)

11.7 The reactions [$Cr(NCS)_6$]$^{3-}$ + solv → [$Cr(NCS)_5(solv)$]$^{2-}$ + NCS^- have been investigated and found to have the following rate constants near 70°C.

Solvent	k (s^{-1})
Dimethylacetamide	9.5×10^{-5}
Dimethylformamide	12.4×10^{-5}
Dimethylsulfoxide	6.2×10^{-5}

What do these values suggest about the intimate mechanism of these reactions? (See S. T. D. Lo and D. W. Watts, *Aust. J. Chem.* **1975**, *28*, 1907.)

11.8 Figure 11.19 shows plots of k_{obs} versus [Y^-] for the anation reactions in dimethylsulfoxide (dmso) solvent:

$$[Co(NO_2)(en)_2(dmso)]^{2+} + Y^- \xrightarrow{\text{dmso}} [Co(NO_2)(en)_2 Y]^+ + dmso$$

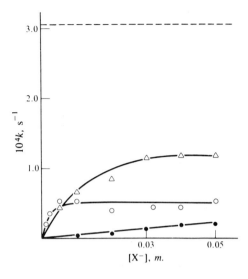

Figure 11.19 Rate of anation of *cis*-[Co(NO₂)(en)₂dmso]²⁺ as a function of concentration of the entering anion Y⁻. △, Y⁻ = NO₂⁻; ○, Y⁻ = Cl⁻; ●, Y⁻ = NCS⁻. The dashed line shows the rate of the dmso exchange reaction. (Reprinted with permission from W. R. Muir and C. H. Langford, *Inorg. Chem.* **1968**, *7*, 1132. Copyright 1968, American Chemical Society.)

All these reactions are presumed to have the same mechanism.

(a) What is the significance of the shapes of the curves?

(b) What is the significance of the fact that the first-order limiting rate constants are smaller than that for dmso exchange?

(c) If the mechanism were **D**, to what would the limiting rate constants correspond?

(d) If the mechanism were I_d, to what would the limiting rate constants correspond?

(e) The limiting rate constants are 5.0×10^{-5} s⁻¹ and 1.2×10^{-4} s⁻¹ for Cl⁻ and NO₂⁻, respectively. For NCS⁻, the limiting rate constant can be estimated as 1×10^{-4} s⁻¹. Do these values constitute evidence for a **D** or an I_d mechanism? (See W. R. Muir and C. H. Langford, *Inorg. Chem.* **1968**, *7*, 1032.)

11.9 The following data refer to water exchange on $[Ln(H_2O)_8]^{3+}$:

Ln^{3+}	r (pm)	$10^{-8}\, k_{exch}$ (s⁻¹)	$\Delta H^{\ddagger}$ (kJ/mol)	$\Delta S^{\ddagger}$ (J/mol K)	$\Delta V^{\ddagger}$ (cm³/mol)
Tb^{3+}	109.5	5.58	12.08	−36.8	−5.7
Dy^{3+}	108.3	4.34	16.57	−23.9	−6.0
Ho^{3+}	107.2	2.14	16.36	−30.4	−6.6
Er^{3+}	106.2	1.33	18.37	−27.7	−6.9
Tm^{3+}	105.2	0.91	22.68	−16.3	−6.0
Yb^{3+}	104.2	0.47	23.39	−29.9	—

Comment on the significance of these data for the exchange mechanism. (See C. Cossy, L. Helm, and A. E. Merbach, *Inorg. Chem.* **1989**, *28*, 2699.)

11.10 In 1.0 *M* OH⁻ solution, at least 95% of $[Co(NH_3)_5Cl]^{2+}$ that undergoes base hydrolysis must be present in the pentaammine form. If 5% or more were present as the conjugate base $[Co(NH_2)(NH_3)_4Cl]^+$, it could be detected by departure from second-order kinetics which is *not* observed to happen.

(a) Using this fact together with expressions in Equations (11.19)–(11.21), show that for $[Co(NH_3)_5Cl]^{2+}$, $K_a \leq 5 \times 10^{-16}$.

(b) Show the form of the rate law that would result if the conjugate base were present in appreciable amount. (*Hint:* The concentration of $[Co(NH_3)_5Cl]^{2+}$ used to make up the solution will be partitioned between the complex and its conjugate base.)

11.11 The complexes *trans*-$[RhL(en)_2X]^+$ react with various Y^-, giving $[RhL(en)_2Y]^+$. The rate law for the appearance of the product is as follows: rate $= (k_1 + k_2 [Y^-])[RhL(en)_2X^+]$. Do these data constitute evidence for an **a** mechanism? Why or why not? (See A. J. Pöe and C. P. J. Vuik, *Inorg. Chem.* **1980,** *19,* 1771.)

11.12 Anation of *trans*-$[Rh(en)_2(H_2O)_2]^{3+}$ with Cl^- has been studied. Two plots of k_{obs} versus concentration of species in solution are reproduced in Figure 11.20.

(a) Account for the linearity of the plots in Figure 11.20*a*.

(b) What does the nonzero intercept for each of the two lines in Figure 11.20*a* indicate about the reaction mechanism?

(c) Account for the shape of the two curves in Figure 11.20*b*.

(d) Do these curves alone allow you to distinguish what the stoichiometric mechanism is?

11.13 (a) A linear relationship between log k's for a series of related complexes is taken to imply similarity in intimate mechanism. Plot log k for aquation versus log k for base hydrolysis for the $[Co(NH_3)_5X]^{n+}$ complexes given in Tables 11.3 and 11.16. Use linear least-

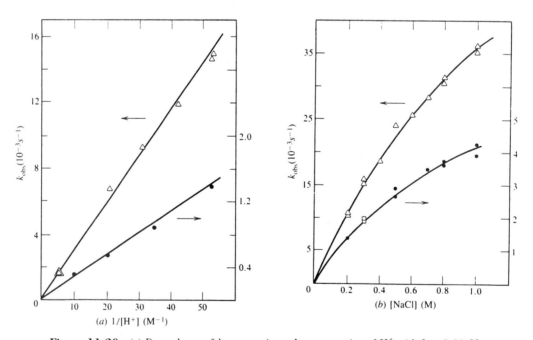

Figure 11.20   (*a*) Dependence of k_{obs} on reciprocal concentration of H^+ with I = 1.00 *M*, [complex] = 0.001 *M*; △, 65°C, [NaCl] = 0.801 *M*; ●, 50°C, [NaCl] = 0.641 *M*. Arrows indicate corresponding ordinate. (*b*) Dependence of k_{obs} on concentration of chloride with T = 40°C, I = 1.00 *M*, and [complex] = 0.0190 *M*; △, [H$^+$] = 0.0190*M*; ●, [H$^+$] = 0.199 *M*. Arrows indicate corresponding ordinate. (Reprinted with permission from M. J. Pavelich, *Inorg. Chem.* **1975,** *14,* 982. Copyright 1975, American Chemical Society.)

squares regression to fit the best straight line. How does this relationship accord with what you know about substitution on Co^{III} complexes?

(b) The table below gives base hydrolysis constants for $[ML_5Cl]^{2+}$ complexes at 25°C, where M = Cr and Co.

L_5	k_{Cr} $(M^{-1} s^{-1})$	k_{Co} $(M^- s^{-1})$
$(NH_3)_5$	1.8×10^{-3}	0.86
fac-(en)(dien)	1.56×10^{-2}	4.71
fac-(NMetn)(dien)	2.21×10^{-1}	7.57
$(NH_2Me)_5$	$>4.3 \times 10^{-1}$	8×10^2
mer-(en)(dpt)	1.05×10^{-1}	2.2×10^3
mer-(tn)(dpt)	6.06×10^{-1}	2.8×10^3
mer-(en)(2, 3-tri)	7.35×10^{-1}	1.24×10^3

NMetn = $MeNH(CH_2)_3NH_2$; dpt = $NH_2(CH_2)_3NH(CH_2)_3NH_2$;
tn = $NH_2(CH_2)_3NH_2$; 2, 3-tri = $NH_2(CH_2)_2NH(CH_2)_3NH_2$

What is the relationship between the two sets of data, and what does this suggest about the mechanism of base hydrolysis for Cr^{III} complexes?

(c) Acid hydrolysis rate constants for the Cr complexes in **b** are given in the same order: $10^7 k_H$ = 95, 224, 37.0, 2.48, 5.34, 5.04, 2.87. Plot log k_{OH} versus log k_H for the series of Cr complexes. What does the plot suggest about the mechanism of acid hydrolysis? (See D. A. House, *Inorg. Chem.* **1988,** *27,* 2587.)

11.14 Given below are overall volume changes ($\Delta \overline{V}_{BH}$) and volumes of activation ($\Delta V^{\ddagger}$) for base hydrolysis of some Co complexes:

Complex	$\Delta \overline{V}_{BH}$ (cm³/mol)	$\Delta V^{\ddagger}$ (cm³/mol)
$[Co(NH_3)_5(Me_2SO)]^{3+}$	21.2	42.0
$[Co(NH_3)_5(OC(Me)NMe_2)]^{3+}$	23.9	43.2
$[Co(NH_3)_5Cl]^{2+}$	9.9	33.0
$[Co(NH_3)_5Br]^{2+}$	11.1	32.5
$[Co(NH_3)_5I]^{2+}$	15.5	33.6
$[Co(NH_3)_5(SO_4)]^+$	−3.9	22.7

(See Y. Kitamura, G. A. Lawrance, and R. van Eldik, *Inorg. Chem.* **1989,** *28,* 333.

(a) Explain why the values of $\Delta V^{\ddagger}$ are qualitatively reasonable given your knowledge of the mechanism of base hydrolysis.

(b) Rationalize the trend in $\Delta V^{\ddagger}$ as a function of charge on the starting complex and on the leaving ligand.

(c) Rationalize the trend in $\Delta \overline{V}_{BH}$ as a function of charge on the starting complex and on the leaving ligand.

(d) Construct a volume profile for base hydrolysis of $[Co(NH_3)_5I]^{2+}$ analogous to the energy profile depicted in Figure 11.1c.

11.15 (a) By drawing the products obtained from Berry pseudorotations of the TBP isomers of (12345)M where the ligands are monodentate, verify the Desargues–Levi map of Figure 11.9.

(b) Show that simplifying the above diagram by making $N_1N_2 = N_3N_4$ with two didentate ligands with all N's equivalent gives the map shown in Figure 11.7.

(c) Modify the diagram of Figure 11.9 for the case of A_2B_3M TBP complexes.

11.16 Photolysis of a square-planar complex M(ABCD) produces distortion to a tetrahedron. Relaxation back to the planar geometry leads to all possible isomers of M(ABCD). Construct a Desargues–Levi map connecting the possible isomers. (*Hint:* There are few enough isomers that their full structures can appear on the map.)

11.17 Predict the isomer distribution resulting from associative attack of Y on the faces of *trans*-MN_4LX using the model shown in Section 11.3.6.

11.18 A plot of k_{obs} versus $[Y^-]$ is shown in Figure 11.21 for the reactions $[Pd(Et_4dien)Br]^+ + Y^- \rightarrow [PdY(Et_4dien)]^+ + Br^-$. Account for the shape of the plot. In particular, what mechanism can you propose to account for the zero slope when $Y^- = N_3^-$, I^-, NO_2^-, SCN^-?

11.19 Base hydrolysis, a well-known reaction of octahedral complexes, is generally about 10^6 times faster than acid hydrolysis. Given the following facts, what can you say about the existence and importance, if any, of a special mechanism for base hydrolysis of square planar-complexes?

(a) Au^{III} complexes, which are also d^8, typically react about 10^4 times as fast as Pt^{II} complexes. The following data are representative:

$$[M(dien)Cl]^{n+} + Br^- \rightarrow [M(dien)Br]^{n+} + Cl^-$$

M	n	k_1 (s^{-1})	k_2 $(M^{-1}\,s^{-1})$
Pt	1	8.0×10^{-5}	3.3×10^{-3}
Au	2	0.5	154

(b) The reactivity of $[Au(dien)Cl]^{2+}$ and $[Pt(dien)Cl]^+$ toward anions increases in the order $OH^- \ll Br^- < SCN^- < I^-$.

(c) The replacement of chloride in $[Au(Et_4dien)Cl]^{2+}$ by anions Y^- is independent of $[Y^-]$. The rate constant (25°C) increases with pH from $k = 1.9 \times 10^{-6}$ to a maximum of $1.3 \times 10^{-4}\,s^{-1}$.

(d) The anation rate of $[(R_5dien)Pd(H_2O)]^{2+}$ by I^- decreases from pH 5 to 9, but increases at higher pH.

11.20 Rate constants for the reaction *trans*-$[Pt(py)_2Cl_2] + XC_6H_4SC_6H_4Y \rightarrow$ *trans*-$[PtCl(py)_2(XC_6H_4SC_6H_4Y)]^+ + Cl^-$ were measured in methanol at 30°C. From these data calculate n_{Pt} for the following ligands.

X	Y	$k_2 \times 10^3$ $(M^{-1}\,s^{-1})$
NH_2	NH_2	7.25
OH	OH	2.72
CH_3	CH_3	1.84
NH_2	NO_2	0.78
NO_2	NO_2	0.096

See J. R. Gaylor and C. V. Senoff, *Can. J. Chem.* **1971**, *49*, 2390.

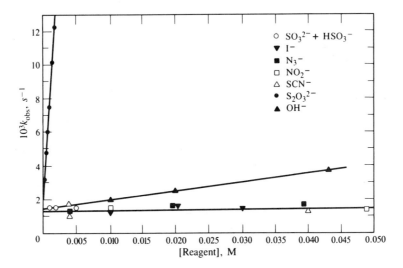

Figure 11.21 Plot of k_{obs} versus concentration of entering nucleophile for anation of $[PdBr(Et_4dien)]^+$ in water at 25°C. (Reprinted with permission from J. B. Goddard and F. Basolo, *Inorg. Chem.* **1968**, *7*, 936. Copyright 1968, American Chemical Society.)

11.21 The following activation parameters have been measured for the reactions

$$[PtClL_2X] + py \longrightarrow [PtClL_2py]^+ + X^-$$

in methanol at 30°C.

Complex	$k_2 \times 10^3$ (M^{-1} s^{-1})	$\Delta H^\ddagger$ (kJ/mol)	$\Delta S^\ddagger$ (J/mol K)	$\Delta V^\ddagger$ (cm³/mol)
trans-[Pt(NO₂)py₂Cl]	7.35	49.3	−94	−8.8
cis-[Pt(NO₂)py₂Cl]	0.150	55.2	−110	−19.8
trans-Pt(PEt₃)₂Cl₂	0.53	53.9	−100	−13.6

What mechanistic information can be extracted from these values? (See M. Kosikowski, D. A. Palmer, and H. Kelm, *Inorg. Chem.* **1979**, *18*, 2555.)

11.22 Activation volumes for acid hydrolysis of *cis*-[PtCl₂(NH₃)₂] and [PtCl₂en] are −9.5 (at 45°C) and −9.2 (at 41°C) cm³/mol, respectively. How are these values consistent with what you know about the mechanism for square-planar substitution?

11.23 Explain how the following data for the reaction

$$cis\text{-Pt(N---N)Cl}_2 + Me_2SO \longrightarrow cis\text{-[PtCl(N---N)(Me}_2SO)]^+ + Cl^-$$

are in accord with an **a** mechanism for substitution in square-planar Pt complexes.

N—N	$10^4 k_{meso}$	$10^5 k_{d,1}$
NH₂CHMeCHMeNH₂	1.39	9.3
C₆H₁₀(CH₂CH₂NH₂)₂	1.06	9.1
MeNHCH₂CH₂NHMe	0.92	7.7
i-PrNHCH₂CH₂NH*i*-Pr	0.47	3.0

See F. P. Fanizzi, F. P. Intini, L. Mareaca, G. Natile, and G. Ucello-Barretta, *Inorg. Chem.* **1990**, *29*, 33.

11.24 Show that for a redox reaction of the type $A^+ + B \overset{K}{\rightleftarrows}$ intermediate $\overset{k}{\to} A + B^+$, where B is present in excess, the rate law will be of the form

$$\text{rate} = \frac{a[A^+]_0[B]}{1 + b[B]}$$

whether the intermediate is a precursor or successor complex. Electron transfer is fast. (*Hint:* $[A]_0$ will be partitioned between free A and the intermediate.)

11.25 Calculate predicted rate constants for the following outer-sphere redox reactions from the information provided. Measured values of k_{12} are given for comparison.

Reaction	k_{11} (M^{-1} s^{-1})	k_{22} (M^{-1} s^{-1})	E^0 (volts)	k_{12}^{obs} (M^{-1} s^{-1})
$Cr^{2+} + Fe^{3+}$	$\leq 2 \times 10^5$	4.0	1.18	2.3×10^3
$[W(CN)_8]^{4-} + Ce^{IV}$	$> 4 \times 10^4$	4.4	0.90	$> 10^8$
$[Fe(CN)_6]^{4-} + MnO_4^-$	7.4×10^2	3×10^3	0.20	1.7×10^5
$[Fe(phen)_3]^{2+} + Ce^{IV}$	$> 3 \times 10^7$	4.4	0.36	1.4×10^5

11.26 Use the value of 1.0×10^{-3} M^{-1} s^{-1} for $Fe^{3+/2+}$ self-exchange to estimate k_{12} for the Fe^{2+}–$[Fe(phen)_3]^{3+}$ cross-reaction in Table 11.24.

11.27 Assign an outer- or inner-sphere mechanism for each of the following:
  (a) The main product of the reaction between $[CrF(NCS)(H_2O)_4]^+$ and Cr^{2+} is $[CrF(H_2O)_5]^{2+}$. (See F. N. Welch and D. E. Pennington, *Inorg. Chem.* **1976**, *15*, 1515.)
  (b) When $[(VO)(edta)]^{2-}$ reacts with $[V(edta)]^{2-}$, a transient red color is observed. (See F. J. Kristine, D. R. Gard, and R. E. Shepherd, *Chem. Commun.* **1976**, 944.)
  (c) The rates of reduction of $[Co(NH_3)_5(py)]^{3+}$ by $[Fe(CN)_6]^{4-}$ are insensitive to substitution on py. (See A. J. Miralles, R. E. Armstrong, and A. Haim, *J. Am. Chem. Soc.* **1977**, *99*, 1416.)
  (d) The rate of reduction of $[Co(NCS)(NH_3)_5]^{2+}$ by Ti^{3+} is 36,000 times smaller than the rate of $[Co(N_3)(NH_3)_5]^{2+}$ reduction. (See J. P. Birk, *Inorg. Chem.* **1975**, *14*, 1724.)
  (e) Activation parameters for some reductions by V^{2+} are as follows:

Complex	$\Delta H^{\ddagger}$ (kJ/mol)	$\Delta S^{\ddagger}$ (J/mol K)
$[CoF(NH_3)_5]^{2+}$	46.4	-77.4
$[CoCl(NH_3)_5]^{2+}$	31.4	-120
$[CoBr(NH_3)_5]^{2+}$	30.1	-115
$[CoI(NH_3)_5]^{2+}$	30.5	-103
$[Co(N_3)(NH_3)_5]^{2+}$	48.9	-58.5
$[Co(SO_4)(NH_3)_5]^+$	48.5	-54.8

See M. R. Hyde, R. S. Taylor, and A. G. Sykes, *J. Chem. Soc. Dalton Trans.* **1973**, 2730.

  (f) A series of Co^{III} carboxylate complexes is known to be reduced by Cr^{2+} in an inner-sphere mechanism. The rate constants for Eu^{2+} give the log–log plot versus $k_{Cr^{2+}}$, shown in Figure 11.22. (See F. R. Fan and E. S. Gould, *Inorg. Chem.* **1974**, *13*. 2639.)

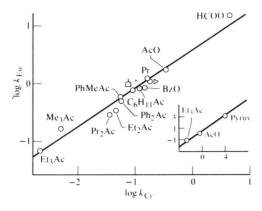

Figure 11.22   Log–log plot comparing specific rates of reductions of carboxylatopentaammine Co^{III} complexes $[R(NH_3)_5Co]^{2+}$ by Eu^{2+} and Cr^{2+} at 25°C. Reprinted with permission from F.-R. Fan and E. S. Gould, *Inorg. Chem.* **1974,** *13,* 2639. Copyright 1974, American Chemical Society.)

(g) Reduction of **(a)** by Cr^{2+} is much slower than reduction of **(b)**. (See R. J. Balahura and N. A. Lewis, *Can. J. Chem.* **1975,** *53,* 1154.)

$$\left[(en)_2Co \begin{array}{c} O = \overset{Me}{} \\ \\ O = \underset{Me}{} \end{array} \right]^{2+} \qquad \left[(en)_2Co \begin{array}{c} O = \overset{Me}{} \\ \\ O = \underset{Me}{} \end{array} \overset{O}{\underset{H}{C}} \right]^{2+}$$

$$\textbf{(a)} \qquad\qquad\qquad\qquad \textbf{(b)}$$

(h) The reduction rates of $[(NH_3)_5CoOC(O)R]^{2+}$ (R = Me, Et) by Eu^{2+}, V^{2+}, and Cr^{2+} decrease as the pH decreases. (See J. C. Thomas, J. W. Reed, and E. S. Gould, *Inorg. Chem.* **1975,** *14,* 1696.)

(i) On mixing $\left[(NH_3)_5CoOC(O) - \left\langle \overset{N-}{\underset{-N}{\bigcirc}} \right\rangle \right]^{2+}$ and Cr^{2+}, a transient ESR signal could be observed. The *g*-value indicated that the odd electron resided mainly in the aromatic ring. (See H. Spiecker and K. Wieghardt, *Inorg. Chem.* **1977,** *16,* 1290.)

11.28 The following rate constants (in $M^{-1}\ s^{-1}$) were measured for reduction of Co complexes:

Complex	$[Ru(NH_3)_6]^{2+}$	$[Ru(NH_3)_5(H_2O)]^{2+}$
$[Co(N_3)(CN)_5]^{3-}$	1.71	2.06
$[Co(NCS)(CN)_5]^{3-}$	0.81	0.99

Comment on the inner- or outer-sphere mechanism of these four reactions. (See A. B. Ageboro, J. F. Ojo, O. Olubuyide, and O. T. Sheyin, *Inorg. Chim. Acta* **1987,** *131,* 247.)

11.29 Using data in Tables 11.23 and 11.30, calculate equilibrium constants for Reaction (11.53).

11.30 The complex $\left[(bipy)_2ClRu - N \bigcirc N - Ru(bipy)_2Cl \right]^{5+}$ displays an intervalence transfer absorption band at $7.79 \times 10^3\ cm^{-1}$. Estimate the energy barrier to thermal electron transfer. If E_{th} can be very approximately equated to $\Delta G^{\ddagger}$ for thermal electron transfer, estimate the rate constant. The thermal rate constant has been reported to be $\leq 3 \times 10^{10}\ s^{-1}$.

► **PART V**

Organometallic Chemistry

· ·

▸12◂

General Principles of Organometallic Chemistry

·················

This chapter deals with the chemistry of species containing metal–carbon bonds, including compounds between metal and CO (carbonyls) but excluding metal cyanides, metal carbides, and graphite intercalation compounds. Carbonyls exist only for transition metals of low oxidation number. Their stability has been interpreted in terms of back-bonding between an electron-rich metal and empty CO orbitals of low enough energy to accept electrons—so-called **π-acid behavior.** Exploring this bonding model provides some insight into the applicability of the 18-electron rule to transition metal compounds. We will find that simple electron counting provides a guide to possible reactions, particularly when coupled with other concepts such as oxidation state and Lewis acidity.

Bonding of organic groups to metals via their π systems is introduced through a treatment of olefin complexes, which are typical of π-donor complexes. We then briefly consider the application of the 18-*e* rule to complexes of more complicated π-donor ligands, without dealing explicitly with their chemistry. This allows us to explain and predict the stoichiometry of a wide variety of organometallic complexes. Compounds with metal–carbon σ bonds are then presented.

Experimental evidence bearing on the evaluation of π acidity in ligands is then introduced through consideration of the difficulties of defining accurately the oxidation number in organometallic complexes and a discussion of the infrared (IR) spectra of carbonyl complexes. Techniques for characterizing organometallic complexes are briefly treated. An important analogy showing the relationship between electronic structures of octet species and 18-*e* compounds is briefly presented. Finally, dihydrogen complexes are discussed.

12.1   INTRODUCTION

Organometallic chemistry is the chemistry of compounds containing metal–carbon bonds. Main-group metals and transition metals, as well as lanthanides and actinides, form bonds to C.

Organometallic chemistry has undergone a renaissance in the last 40 years. However, Zeise prepared the first organometallic compound in 1827, and Frankland developed the chemistry of alkyl zinc compounds in the mid- and late nineteenth century.[1] Much interest centers on organometallic chemistry because of its importance in catalysis. Transformations in organic molecules on laboratory and industrial scales often involve catalysis by metals. Metals serve as reaction templates which bond organic species (at least transiently), providing a low-energy reaction pathway for their combination with other bonded species, and release the weakly bonded products. For example, Reppe found over 40 years ago that $Ni(CN)_2$ catalyzes tetramerization of acetylene to cyclooctatetraene, which presumably involves assembling the reactants by coordination to Ni:

$$(12.1)$$

Synthetic, spectroscopic, and kinetic techniques have been used to probe processes that occur with reactants in the same phase (homogeneously). Investigation of the fundamental chemistry of organometallic complexes has uncovered a tremendous number of intriguing reactions and modes of ligand attachment. The results have been applied to improving product yield and selectivity of catalyzed reactions. A variety of novel and synthetically useful stoichiometric reactions also have been developed.

12.2   CARBONYL COMPLEXES

▶ *12.2.1   CO—The Most Important π-Acid Ligand*

CO provides a paradigm for bonding of π-acid ligands to metals. The valence-bond (VB) structure of CO shows two nonbonding electron pairs:

$$:C{\equiv}O:$$

The molecular orbital (MO) energy diagram of Figure 4.10 shows that the electron pair in the MO localized on C is more loosely bound and is the one available for electron donation

---

[1] An excellent presentation of the history of organometallic chemistry is given in J. S. Thayer, *Adv. Organomet. Chem.*, **1975**, *13*, 1.

to a metal. Figure 12.1 shows the bonding resulting from overlap of the filled C σ orbital and an empty metal σ orbital. CO also has a pair of empty, mutually perpendicular π^* orbitals that overlap with filled metal orbitals of π symmetry and help to drain excess negative charge from the metal onto the ligands. Because ligands containing empty π^* orbitals accept electrons, they are Lewis acids and are called **π acids.** Metal-to-ligand electron donation is referred to as **back-bonding.**

Energetically, the most important bonding component is σ L $\rightarrow$ M donation. Back-bonding (the π component) assumes greater relative importance when the metal has many electrons to dissipate; thus, low oxidation states are stabilized by π-acid ligands.

σ- and π-bonding reinforce each other: The greater the electron donation from a filled ligand σ orbital, the greater the partial positive charge on the ligand and the more stable the π^* orbitals become, making them better acceptors. This mutual reinforcement is called **synergism.** The importance of synergic bonding is indicated by the fact that CO forms a very large number of complexes with transition metals in low oxidation states, even though it is an extremely poor Lewis base toward other species. (The description of back-bonding given here is the MO equivalent of the VB formulation in Figure 9.7.)

Other π-acid ligands, such as CN$^-$, also exhibit synergism. Organic ligands that also act as two-electron σ-donors and have empty π^* orbitals include isocyanides (:CNR) and carbenes (:C(X)(Y)). Still other ligands are capable of back-bonding but do not contain C, including NO$^+$ (isoelectronic with CO), phosphines R_3P, arsines R_3As, stibines R_3Sb, bipyridine (bipy), and 1,10-phenanthroline (phen). Both CNR and NO$^+$ have two mutually perpendicular π^* orbitals, like CO. For bipy and phen (Table 9.1), only one delocalized π^* orbital perpendicular to the molecular plane is properly oriented for back-bonding. For Group 15 ligands just one empty d orbital, and for carbenes a single empty C $2p_z$ of π symmetry, can back-bond. Accordingly, complexes containing these ligands are treated with organometals, even though, strictly speaking, not all of them are organic.

▶ *12.2.2   Binary Carbonyl Complexes*

Binary carbonyls are the simplest class of π-acid complexes. Table 12.1 lists neutral carbonyls, which often are used as reactants in the preparation of other compounds. Most are available commercially, enabling one to avoid syntheses involving high pressures of CO.

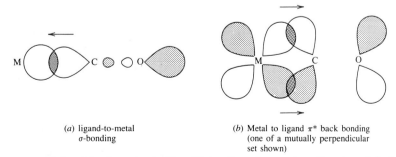

(*a*) ligand-to-metal   (*b*) Metal to ligand π^* back bonding
σ-bonding   (one of a mutually perpendicular
                                set shown)

Figure 12.1   Orbital overlap in M—CO bonding. Arrows show direction of electron flow. (*a*) Ligand-to-metal σ bonding. (*b*) Metal-to-ligand π^* back-bonding (one of a mutually perpendicular set shown).

Table 12.1   Binary metal carbonyls

$Ti(CO)_6$	$V(CO)_6$	$Cr(CO)_6$	$Mn_2(CO)_{10}$	$Fe(CO)_5$	$Co_2(CO)_8$	$Ni(CO)_4$
Green; dec. 40–50 K	Blue-green; O_h; dec. 60–70°C; air-stable	Colorless; O_h; mp 150–152°C (*in vacuo*); air-stable	Golden yellow; D_{4d}; mp 154–155°C; heat-/air-sensitive	Light yellow; D_{3h}; mp −20°C; bp 103°C	Dark orange D_{3d} (sol'n); C_{2v} (solid); dec. 51–52°C; air-sensitive	Colorless; T_d; mp −17°C; bp 42°C; dec. to Ni + CO 180°C
		$Mo(CO)_6$	$Tc_2(CO)_{10}$	$Ru(CO)_5$	$Rh_2(CO)_8$	
		Colorless; O_h; mp 146°C (*in vacuo*); air stable	Colorless; D_{4d}	Colorless; D_{3h} (sol'n); mp −22°C; difficult to purify	Stable only at low T and high CO pressure	
		$W(CO)_6$	$Re_2(CO)_{10}$	$Os(CO)_5$	$Ir_2(CO)_8$	
		Colorless; O_h; mp 166°C (*in vacuo*); air-stable	Colorless; D_{4d}; dec. 170°C; air-stable	Colorless; D_{3h} (sol'n);	Stable only at low T and high CO pressure	
				$Fe_2(CO)_9$		
				Yellow-orange; D_{3h}; dec. 100°C; more reactive than $Fe(CO)_5$		
				$Ru_2(CO)_9$		
				$Os_2(CO)_9$		
				$Fe_3(CO)_{12}$	$Co_4(CO)_{12}$	
				Green-black; C_{2v} (solid); dec. 140°C; air-sensitive	Black; C_{3v}; dec. 60°C; air-sensitive	
				$Ru_3(CO)_{12}$	$Rh_4(CO)_{12}$	
				Orange; D_{3h}; dec. 155°C; air-stable	Red; C_{3v}; dec. 150°C; air-stable	
				$Os_3(CO)_{12}$	$Ir_4(CO)_{12}$	
				Yellow; D_{3h}; mp 224°C; air-stable	Yellow; T_d; dec. 230°C; air-stable	

Synthesis of Carbonyls

Simple transition-metal carbonyls are made in two ways: **direct reaction** of a metal with CO, and **reductive carbonylation** in which a metal salt reacts with CO in the presence of a reducing agent (which may also be CO).

Only Fe and Ni react directly with CO under mild conditions to give $Fe(CO)_5$ and $Ni(CO)_4$, both of which are very toxic because they decompose thermally to release very toxic CO. The extreme ease of thermal decomposition for $Ni(CO)_4$ is the basis of the Mond process for Ni purification. Impure Ni reacts with CO at 100°C to give gaseous $Ni(CO)_4$, which is separated easily and decomposed back to Ni and CO (which can be recycled) by heating.

Particularly common is synthesis from a metal halide; a reducing agent that is also a halide acceptor is employed in a CO atomsphere, often under pressure:

$$CrCl_3 + Al + 6\,CO \xrightarrow[C_6H_6]{AlCl_3} Cr(CO)_6 + AlCl_3 \qquad (12.2)$$

When halide is not present, CO can serve as the reducing agent:

$$Re_2O_7 + 17\,CO \longrightarrow Re_2(CO)_{10} + 7\,CO_2 \qquad (12.3)$$

Higher nuclearity carbonyls (those containing more metals) result from thermolysis of lower ones. Metal–CO bond cleavage produces unsaturated fragments which combine. In some cases, such as $Os(CO)_5$, the lower carbonyl is unstable even at ambient temperature:

$$3\,Os(CO)_5 \xrightarrow{\Delta} Os_3(CO)_{12} + 3\,CO \qquad (12.4)$$

Photochemical bond cleavage also occurs, as in the synthesis of diiron enneacarbonyl:

$$2\,Fe(CO)_5 \xrightarrow{h\nu} Fe_2(CO)_9 + CO \qquad (12.5)$$

**Molecular and Electronic Structures
of Carbonyls—The 18-Electron Rule**

The first carbonyls discovered were $Ni(CO)_4$, $Fe(CO)_5$, and $Co_2(CO)_8$. Their formulas, as well as the formulas of carbonyls discovered later, were suggestive to Sidgwick of a stable 18-electron (pseudo-noble-gas) valence shell configuration around the metal, comparable to the stable octet for the lighter elements suggested by Lewis. A very large number of metal carbonyls—including many anionic and cationic species, nitrosyl-, hydrogen-, and halogen-substituted metal carbonyls, and small metal cluster carbonyls—conform to the **18-*e* rule.** This rule can also be formulated in terms of the total number of electrons around the metal—in which case this number is usually found to be 36, 54, or 86, corresponding to the atomic numbers of the noble gases Kr, Xe, and Rn. The metals are then said to have the *effective atomic number* of the noble gases, or to obey the **EAN rule.** Because the pseudo-noble-gas valence shells of Kr, Xe, and Rn contain 18 *e*'s, it is simpler just to count valence shell electrons.

Counting rules help to predict and understand the stoichiometry and structures of the binary metal carbonyls. In counting the electrons around a single metal atom, follow these simple (but arbitrary) rules:

1. Count two electrons for each CO.
2. Count one electron for each metal–metal bond.
3. Find the number of electrons that formally belong to the metal atom alone by (a) adding up the charges on the ligands and changing the sign, (b) finding the metal oxidation number by adding this number to the total charge on the complex, and (c) subtracting the oxidation number from the valence electron count of the neutral metal.
4. Add together the counts from steps **1–3**.

EXAMPLE 12.1: Count electrons in $Mo(CO)_6$ (its O_h structure is shown in Figure 12.2).

Solution: In neutral binary carbonyls, the metal oxidation number is always 0.

$$Mo^0 + 6\,CO$$

$$6e + (6 \times 2e) = 18e$$

The counting rules merely bookkeep electrons; they provide no information about the actual density distribution.

EXAMPLE 12.2: Count electrons around each Re in $Re_2(CO)_{10}$ (the D_{4d} structure is shown in Figure 12.2).

Solution: We consider a single Re atom:

$$Re^0 + 5\,CO \quad + Re{-}Re$$

$$7e + (5 \times 2e) + \quad 1e \quad = 18e$$

The molecular structure of $Fe_3(CO)_{12}$ (Figure 12.2) reveals a new feature—CO's that bridge two metals. In line formulas, doubly bridging carbonyls are written as μ-CO or μ_2-CO to emphasize that two metals are bridged. Hence, a formula for triirondodecacarbonyl that conveys structural information is $Fe_3(\mu_2\text{-CO})_2(CO)_{10}$.[2]

Each bridging CO is considered to be sp^2-hybridized and to contribute one hybrid orbital and one electron to each metal.

EXAMPLE 12.3: Count electrons around each Fe in $Fe_3(CO)_{12}$.

Solution:

$$\mu_2\text{-Fe's:} \quad Fe^0 \quad + 3\,CO \quad + 2\,\mu\text{-CO} \quad + 2\,Fe{-}Fe$$

$$8e \quad + (3 \times 2e) + (2 \times 1e) + (2 \times 1e) = 18e$$

$$\text{Unique Fe:} \quad Fe^0 + 4\,CO \quad + 2\,Fe{-}Fe$$

$$8e \quad + (4 \times 2e) + (2 \times 1e) = 18e$$

[2]C. H. Wei and L. F. Dahl, *J. Am. Chem. Soc.* **1969**, *91*, 1351.

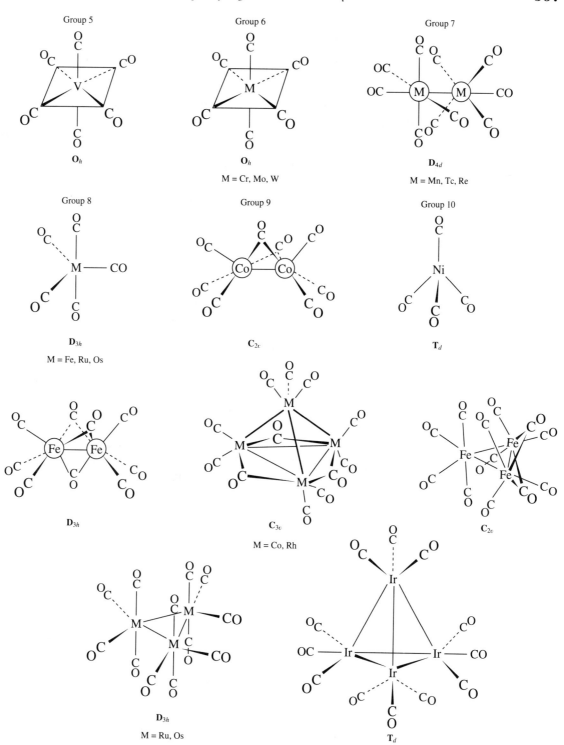

Figure 12.2 Solid-state structures of some neutral binary metal carbonyls.

We cannot predict via the EAN rule the presence of μ_2-CO's. Without any structural information we still calculate the number of e's/Fe as $[(3 \times 8e) + (12 \times 2e) + (3 \times 2e)]/3\,\mathrm{Fe} = 54e/3\,\mathrm{Fe} = 18e/\mathrm{Fe}$. The number of e's per M is the total number around all metal atoms divided by the number of metal atoms which share them. Both $Ru_3(CO)_{12}$ and $Os_3(CO)_{12}$ adopt structures containing only terminal (nonbridging) carbonyls. Presumably, this occurs because distances between larger metal atoms are too great for effective bridge-bond formation. Similar considerations apply to $M_4(CO)_{12}$ (M = Co, Rh, Ir), where Co and Rh give a $\mathbf{C}_{3v}$ structure whereas Ir forms a $\mathbf{T}_d$ one (Figure 12.2).

Different structures consistent with the 18-e rule can occur in different environments. For example, $Co_2(CO)_8$ has a solid-state structure with two μ_2-CO's. In solution, however, no fewer than three isomers exist: the bridged form, a form containing all terminal

CO's, and a third form of unknown structure. The structures 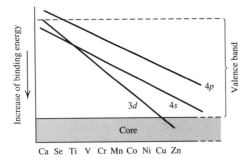 are equivalent in electron count, and bridge-terminal tautomerism occurs frequently.

CO is also known to bridge metals through both C and O. An example of this relatively uncommon coordination made is $(PPh_3)W(CO)_4(CO{\rightarrow}AlBr_3)$.

Some known exceptions to the 18-e rule among binary carbonyls are $V(CO)_6$ (17e) and $Rh_6(CO)_{16}$ (discussed in Chapter 17).

Conformity of transition-metal carbonyls to the 18-e rule is a result of the fact that the nine nd, $(n + 1)\,s$, and $(n + 1)\,p$ orbitals are all valence orbitals in the transition series, and that all their bonding capacity is used when the 18-e configuration is reached.

Figure 12.3 shows the changes in energy with atomic number for the 3d, 4s, and 4p orbitals of the first-transition-series atoms. All the orbitals become more stable as the effective nuclear charge increases across the series. Around the middle (V—Co), all the orbitals are of fairly similar energy, and thus all are available for bonding and the 18-e rule is expected to apply. As the end of the series is approached, the 3d and 4s orbitals drop in energy faster than the 4p, until at Ni the 3d's are part of the atomic core and too stable to

Figure 12.3 The change in energy of the 3d, 4s, and 4p orbitals of the first transition series. (After C. S. G. Phillips and R. J. P. Williams, *Inorganic Chemistry*, Vol. II, Oxford University Press, Oxford, 1955.)

[3] For a discussion of the structures and reactions of various types of bridging carbonyls, see C. P. Horowitz and D. F. Shriver, *Adv. Organomet. Chem.* **1984**, *23*, 219.

participate in bonding. Hence, we expect that Cu, Zn, and the main-group metals beyond should not obey the 18-*e* rule. (For Ni^0 it makes no difference whether we consider the *d*'s to be part of the core or not, since they are filled completely whereas the $4s$ and $4p$ orbitals are still available for bonding, leading to a prediction of 18 electrons around Ni.) As some orbitals disappear into the core toward the end of the series, 16- and even 14-electron species become more stable. Early in the series (before V^0) *d* orbitals are of relatively high energy. Also, atoms early in the series, for steric reasons, cannot always accommodate the number of ligands required to supply 18 electrons. Hence, the 18-*e* rule often fails here. Trends in the second and third series are similar. In short, for metals with low or negative oxidation state, we expect the 18-*e* rule to hold through the middle part of the transition series. Sixteen-electron complexes and intermediates are common enough that we sometimes speak of the **16-, 18-*e* rule.**

An MO perspective on the 18-*e* rule is also useful. A transition metal with nine valence orbitals binds *n* ligands. Then (Figure 12.4) *n* bonding orbitals and *n* antibonding orbitals form, leaving $(9 - n)$ nonbonding orbitals of relatively low energy. Filling all the low-lying bonding and nonbonding orbitals with electrons requires $2 \times [n + (9 - n)] = 18$ electrons. This configuration results in a large energy gap between the highest occupied molecular orbitals (HOMOs) and lowest-unoccupied molecular orbitals (LUMOs), and thus it is stable. The reasons for the failure of the 18-*e* rule at the extremes of the transition series are also intelligible in this model. Early transition metals have few electrons. Steric crowding may prevent enough ligands from coordinating to supply the complement of 18 electrons. Late transition metals are electron-rich, and their electrons may occupy too many nonbonding metal orbitals to leave sufficient vacant orbitals for bonding with all the ligands needed to supply 18 electrons.

As metal oxidation number increases, the slopes of the binding energy versus atomic number curves decrease more steeply and their separation widens. Thus, at sufficiently positive metal oxidation number ($\geq$ II), some orbitals enter the core and the 18-*e* rule no

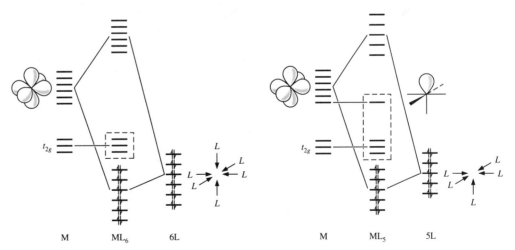

Figure 12.4   Schematic MO energy diagrams for ML_n ($n = 6, 5$). Nonbonding orbitals (surrounded by a dashed box) contain metal *d* electrons (After R. Hoffmann, *Angew. Chem., Int. Ed. Engl.* **1982**, *21*, 711.)

longer applies. For example, $[Cr(NH_3)_6]^{3+}$ does not obey the EAN rule whereas $Cr(CO)_6$ does. Within its region of applicability, however, the 18-*e* rule is a rather reliable guide to molecular structure and stoichiometry for stable organometallics.

The wide applicability of the 16-,18-*e* rule undoubtedly results from the fact that most hitherto developed organometallic chemistry has relied on nuclear magnetic resonance (NMR) and *x*-ray crystallography to characterize compounds. The 16- and 18-*e* species are usually diamagnetic, permitting NMR measurements, and most are sufficiently thermally stable for *x*-ray crystallography. Modern techniques are currently making possible the preparation and characterization of very thermally and kinetically reactive species as well as their detection when they are transient species, and we can expect that more and more compounds will be known which do not conform to the 16-,18-*e* rule. A number of radical reactions discussed in Chapter 14 involve 17-*e* and 19-*e* intermediates.

► *12.2.3 Substituted Carbonyls, Other π-Acid Complexes, and Carbonylate Anions*

CO replacement by another 2-*e* donor ligand or by electron addition produces 18-*e* species. Thus, replacing CO by Br^- in $Mo(CO)_6$ gives $[Mo(CO)_5Br]^-$. Likewise, replacing Mo by isoelectronic Mn^+ leads to $[Mn(CO)_6]^+$. The anions $[Co(CO)_4]^-$ and $[Fe(CO)_4]^{2-}$ also are known 18-*e* species. Table 12.2 gives electron counts for selected ligands.

Phosphine-Containing Complexes

Neutral Lewis bases such as PR_3 (phosphines), $P(OR)_3$ (phosphites), and CNR (isocyanides) often replace CO's. P^{III} ligands have a P lone pair for σ-donation, filled π orbitals (depending on the identity of R) as well as empty orbitals of π symmetry (which could be the P *d* orbitals or antibonding P—C or P—O orbitals). The donor and acceptor ability of these ligands is influenced by the identity of R. The σ-donating ability is expected to parallel the pK_a's of HPR_3^+ or $HP(OR)_3^+$, and π-acidity is promoted by electron-withdrawing groups such as F, Cl, and OR on P.

Table 12.2 Electron counts for some ligands

Ligand	Number of valence electrons contributed
H^-, F^-, Cl^-, Br^-, I^-, CN^-, NCS^-, CO, CNR, NO^+, NO^-, PR_3, $P(OR)_3$, AsR_3, NR_3, SbR_3, SR_2, R^-, $C^-(O)R$, $Ar^-(Ar = aromatic)$, $:C(X)(Y)$, CR^+	2
NO	3
$:\overset{..}{C}(X)(Y)^{2-}$	4
$:\overset{..}{C}R^{3-}$	6

One can sort out whether a ligand is acting as a σ donor only or also as a π acid by drawing a graph such as that shown in Figure 12.5a. In brief, a plot of $-\Delta H°$ (or some other measured property) versus pK_a is made for a particular reaction. Some points lie on a straight line, and these are considered to belong to the class of σ-donor ligands. Those which lie above the line (more-negative-than-expected $\Delta H°$) are considered to exhibit enhanced stability due to π acidity. Those lying below the line (less-negative-than-expected $\Delta H°$) belong to the class of σ and π donors. Three classes of ligands were distinguished: Class I, σ donor and π donor; Class II, σ donor only; Class III, σ donor and π acceptor. Class membership changes somewhat with the metal, but, in general, trialkyl phosphines are Class I whereas phosphites and $P(CH_2CH_2CN)_3$ are Class III. Based on these results and IR and NMR studies, the generally accepted order of π acidity for P^{III} ligands is

$$PF_3 > PCl_3 > P(OAr)_3 > P(OR)_3 > PAr_3 > PR_3$$

Also affecting bonding ability are steric factors measured by the cone angle Θ. Tolman obtained values for the apex angle Θ (Figure 12.6) of cones centered 228 pm from the P atom and tangent to the van der Waal's radii of R groups by constructing molecular models to scale and measuring the angle. Some typical values are reported in Table 12.3. All other things being equal, small cone angles should lead to better bonding by permitting closer ligand approach.

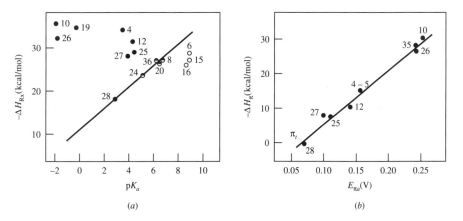

(a)   (b)

Figure 12.5   (a) Plot of ΔH^0 versus pK_a for the reactions $[(C_5H_9)Ni(\mu\text{-Me})]_2 + L \rightarrow 2(C_5H_9)Ni(\mu\text{-Me})L$. ◗, Class II ligands; ○, Class I ligands; ●, Class III ligands. 12.5(b) Plot of deviation from straight line in Figure 12.5a for class III ligands versus $E_{\pi a}$. $E_{\pi a}$ measures the π-bonding ability of L. A plot of pK_a for various LH^+ versus reduction potential of some ML_n complexes is made. A straight-line correlation results when L's are only σ-bonding. The positive deviation from this line of the reduction potential for some L is $E_{\pi a}$, which is taken as a measure of L's π-back-bonding ability. See M. N. Golovin, M. M. Rahman, J. E. Belmonte, and W. P. Giering, *Organometallics* **1985**, *4*, 1981. No significant steric effects are observable for this reaction. (Reprinted with permission from Md. M. Rahman, H. Y. Liu, A. Prock, and W. P. Giering, *Organometallics* **1987**, *6, 650*. Copyright 1987, *American Chemical Society.*)

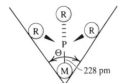

Figure 12.6 Tolman cone angle for PR$_3$.

Although it is difficult to disentangle steric and electronic factors, the approach given in Figure 12.5 has helped to sort them out.[4] If we take those points in Figure 12.5a that fall above the line and plot their "extra" enthalpy contribution (ΔH_π), measured by their deviations from the line, against an electrochemical parameter measuring π acidity, any negative deviation from the expected straight line is attributed to a steric effect which again is measured by the distance from the line. This is done in Figure 12.5b. No deviations are seen, suggesting that steric effects are unimportant in the reaction examined. Cone angles of 140–160° are usually required for significant steric effects.

Isocyanide-Containing Complexes

Isocyanides, :C≡NR, are isoelectronic with :C≡O:. The lower electronegativity of N as compared to that of O makes them slightly better σ donors and slightly poorer π acceptors than CO. Thus, isocyanides stabilize higher oxidation states as in $[V(CNR)_6]^+$, $[Cr(CNR)_7]^{2+}$, and $[Pt(CNR)_4]^{2+}$. Only recently has a **homoleptic** (containing only one kind of ligand) isocyanide *anion*, $[Co(2,6-Me_2C_6H_3NC)_4]^-$, been synthesized.[5] Electronic properties can be fine-tuned by the choice of R, and aryl isocyanides are better π acceptors than alkyl. Several homoleptic isocyanides correspond to stable carbonyls. Examples are $Ni(CNR)_4$, $Co_2(CNR)_8$, and $Mo(CNR)_6$. Among the VB models for the electronic structures of isocyanide complexes is structure **b** below, indicative of an important contribution by metal–ligand back-bonding. If back-bonding is important enough, **b** might actually contribute substantially to the overall complex structure.

Table 12.3 Tolman[a] cone angle (Θ) for selected P^{III} ligands

Ligand	(Θ) (deg)	Ligand	(Θ) (deg)
P(OCH$_2$)$_3$Et	101	PEt$_3$	132(137[b])
P(OEt)$_3$	109 (134[b])	P(*n*-Bu)$_3$	132
P(OMe)$_3$	107 (128[b])	PPh$_3$	145
PMe$_3$	118	PCy$_3$ (Cy = *c*-C$_6$H$_{11}$)	170
PCl$_3$	124	P(*o*-MeC$_6$H$_4$)$_3$	194

[a] C. A. Tolman, *Chem. Rev.* **1977**, *77*, 313.

[b] Values in parentheses are estimated assuming folding which was not taken into account in the original Tolman measurements. See L. Stahl and R. G. Ernst, *J. Am. Chem. Soc.* **1987**, *109*, 5673.

[4] Md. M. Rahman, H. Y. Liu, A. Prock, and W. P. Giering, *Organometallics* **1987**, *6*, 650. See also Md. M. Rahman, H. Y. Liu, K. Eriks, A. Prock, and W. P. Giering, *Organometallics* **1989**, *8*, 1.

[5] G. F. Warnock and N. J. Cooper, *Organometallics* **1989**, *8*, 1826.

$$\overset{(-)}{\underset{\cdot\cdot}{M}}\!\leftarrow\!C\!\equiv\!\overset{(+)}{N}\!-\!R \qquad M\!=\!C\!=\!\overset{R}{N}: \qquad \overset{(+)}{M}\!\leftarrow\!\overset{(-)}{\underset{\cdot\cdot}{C}}\!=\!\overset{R}{N}:$$

(a)   (b)   (c)

One would then expect appreciable bending of the $M-C-N$ angle. Equatorial iso-cyanide ligands in $Fe(CNt\text{-}Bu)_5$ have been found to have angle $Fe-C-N \sim 137°$ (although this much bending is unusual). Isocyanides are not confined to substituting for carbonyls, and their complexes have a rich chemistry.[6]

Nitrosyl Complexes

NO contains one more electron than CO (in an $N-O$ π^* orbital) and thus acts as three-electron donor. In electron-counting terms, $3CO = 2NO$. Thus, $Fe(NO)_2(CO)_2$ is isoelec-tronic with $Fe(CO)_5$, and $Cr(NO)_4$ and $V(CO)_5NO$ are isoelectronic with $Cr(CO)_6$. Three VB structures can be written for nitrosyl complexes:

$$\overset{\cdot\cdot}{\underset{(2-)}{M}}\!\leftarrow\!\underset{(+)}{N}\!\equiv\!\underset{(+)}{O}: \qquad \underset{(-)}{M}\!=\!\underset{(+)}{N}\!=\!O: \qquad M\!-\!\overset{\cdot\cdot}{N}\!\diagdown\!\underset{\cdot\cdot}{O}:$$

(a)   (b)   (c)

Structure **a** implies that NO places its π^* electron in a singly occupied metal π orbital giving a nonbonding pair on M and donates two electrons in a σ bond. This is the so-called **NO$^+$ coordination mode.** The formal charges in **a** are larger than the ones in the analogous isocyanide structure. In structure **b**, the nonbonding pair is back-donated into the π^* orbital of NO, thereby lowering the NO bond order. In **c**, an electron pair is local-ized on the N in an sp^2 orbital, leading to a bent structure. In a formal sense, one lone-pair electron must have come from the metal, giving the so-called **NO$^-$ coordination mode.** Nitrosyl complexes are known with both linear and bent NO.[7] Transition metals late in the series with relatively many electrons tend to adopt the bent mode. For example, treating d^8 $IrCl(CO)(PPh_3)_2$ with $NO^+PF_6^-$ gives square-pyramidal $[IrCl(CO)(PPh_3)_2NO]^+$ with bent apical NO (Figure 12.7a). This could be considered a compound of Ir^{III} in which two electrons from Ir^I have converted NO^+ to NO^- with bent geometry. On the other hand, isoelectronic $[RuCl(PPh_3)_2(NO)_2]^+$ (Figure 12.7b) has a linear equatorial and a bent apical NO. Likewise, $[Co(diars)_2(NO)]^{2+}$ (diars = $o\text{-}(Me_2As)_2C_6H_4$) has a trigonal-bipyramidal structure with a linear equatorial NO. When an additional electron pair is introduced into the Co coordination sphere, the $[Co(NCS)(diars)_2(NO)]^+$ is octahedral with bent NO. In contrast, complexes of the electron-poor early transition metals, such as $[Cr(CN)_5NO]^{3-}$, display linear NO. The energetic factors involved in bending are corre-lated with other factors, so that it is not always possible to predict which mode of coordi-nation will be observed.

Routes to Substitution

Three preparative routes are common for substituted carbonyls:

    1. Direct replacement [Equations (12.6), (12.7), (12.9), and (12.10)]

---

[6] See E. Singleton and H. E. Oosthuizen, *Adv. Organomet. Chem.* **1983,** *22,* 209.

[7] For a discussion of factors influencing the mode of NO bonding, see R. Eisenberg and C. D. Meyer, *Acc. Chem. Res.* **1975,** *8,* 26; J. H. Enemark and R. D. Feltham, *Coord. Chem. Rev.* **1974,** *13,* 339.

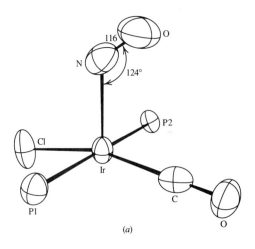

(a)

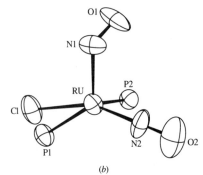

(b)

Figure 12.7   (a) The inner coordination sphere of [IrCl(CO)(PPh₃)₂(NO)]⁺ (BF₄⁻ salt). (Reprinted with permission from D. J. Hodgeson and J. A. Ibers, *Inorg. Chem.* **1968,** 7, 2345. Copyright 1968, American Chemical Society.) (b) The inner coordination sphere of [RuCl(NO)₂(PPh₃)₂]⁺(PF₆⁻ salt). (Reprinted with permission from C. G. Pierpont, D. G. Van Deveer, W. Durland, and R. Eisenberg, *J. Am. Chem. Soc.* **1970,** 92, 4760. Copyright 1970, American Chemical Society.)

2. **Oxidation** by halogens [Equations (12.11) and (12.12)]
3. **Reactions between metal halides and CO** [Equation (12.14)]

$$Ir_4(CO)_{12} + 2\,PPh_3 \longrightarrow Ir_4(CO)_{10}(PPh_3)_2 + 2\,CO \tag{12.6}$$

$$Mo(CO)_6 + Br^- \longrightarrow [Mo(CO)_5Br]^- + CO \tag{12.7}$$

Some monomers form bridged dimers on thermally induced CO loss:

$$2Mn(CO_5)Br \xrightarrow{\Delta} Mn_2(\mu\text{-}Br)_2(CO)_8 + 2CO \tag{12.8}$$

Photochemical activation sometimes can aid in breaking the M—CO bond, but many reactions are carried out thermally. Except when the substituting ligands are very good π acids, substitution usually stops after two or three carbonyls have been replaced, because the remaining CO's become saturated in their ability to withdraw electron density from the metal. With good π acids, however, all CO's can be replaced, giving compounds such as Ni(PF₃)₄, Cr(CNPh)₆, and so on. For reasons that are not well understood, Pt(CO)₄ is unstable, but Pt(PPh₃)₄ can be prepared. The relatively large cone angle (145°) for PPh₃ promotes ligand dissociation in solution to Pt(PPh₃)₃ and even Pt(PPh₃)₂.

Direct displacement with NO is a route to nitrosyl complexes as is treatment with acidified nitrites:

$$Fe(CO)_5 + 2\,NO \longrightarrow Fe(CO)_2(NO)_2 + 3\,CO \tag{12.9}$$

$$[Co(CO)_4]^- + H^+ + HNO_2 \longrightarrow Co(CO)_3(NO) + H_2O + CO \tag{12.10}$$

Carbonyl halides result from halogen oxidation of carbonyls:

$$Mn_2(CO)_{10} + Cl_2 \longrightarrow 2\,Mn(CO)_5Cl \tag{12.11}$$

$$Fe(CO)_5 + Br_2 \longrightarrow Fe(CO)_4Br_2 + CO \tag{12.12}$$

A formally related reaction is

$$Co_2(CO)_8 + H_2 \longrightarrow 2\,HCo(CO)_4 \tag{12.13}$$

It seems strange to think of an "oxidation" by H_2! However, it is an arbitrary convention to obtain oxidation states of transition metals in complexes by "removing" ligands in closed-shell configuration. (The exception to this is μ_2 $2e$-donor ligands such as CO and CNR.) Thus, M—H compounds are treated as containing H^-. In fact, the chemical behavior of $HCo(CO)_4$ is consistent with the presence of protonated $[Co(CO)_4]^-$ containing Co^{-I}. (See Section 12.4.1.)

Typical of reactions between metal halides and CO is

$$2\,RhCl_3 + 4\,CO \longrightarrow \tag{12.14}$$

Note that a noncarbonyl ligand bridges in the product, as in most carbonyl halides including $Cu_2(CO)_2(\mu\text{-Cl})_2(H_2O)_2$. Bridged species can be cleaved by Lewis bases; for example,

$$[Rh(CO)_2(\mu\text{-Cl})]_2 + PPh_3 \longrightarrow 2\ \textit{trans-}[Rh(PPh_3)_2(CO)Cl] \tag{12.15}$$

Although metal carbonyls undergo many different kinds of reactions, substitutions are among the most important. In later sections we shall encounter examples in which carbonyls are substituted by organic groups.

Carbonylate Anions

We know of a very large number of carbonyl anions but very few cations. This is reasonable in view of the π acidity of CO, which stabilizes low oxidation states.

Carbonylate anions often are made *in situ* and used for reaction without isolation. Three common preparations are:

1. Base-induced redox reactions. Examples include:

$$13\,Mn_2(CO)_{10} + 40\,OH^- \longrightarrow 24[Mn(CO)_5]^- + 2\,Mn^{2+} + 10\,CO_3^{2-} + 20\,H_2O \tag{12.16}$$

$$Co_2(CO)_8 + 5\,CNR \longrightarrow [Co^I(CNR)_5]^+ + [Co^{-I}(CO)_4]^- + 4\,CO \qquad (12.17)$$

Reactions (12.16) and (12.17) amount to base-induced disproportionation which, although convenient, involves conversion of part of the metal to a cationic product.

2. Reaction of metal carbonyls with reducing agents. Alkali metal amalgams, hydride reagents, and Na/K alloy in basic solvents, including liquid ammonia, have been used. Examples include:

$$Fe(CO)_5 + 2\,Na \xrightarrow{NH_3(l)} Na_2[Fe(CO)_4] + CO \qquad (12.18)$$

$$Cr(CO)_6 + Na/Hg \xrightarrow{thf,\ 65°} Na_2[Cr(CO)_5] + CO \qquad (12.19)$$

$$Co_2(CO)_8 + 2\,Li[HBEt_3] \xrightarrow{thf,\ 25°} 2\,Li[Co(CO)_4] + 2\,BEt_3 + H_2 \qquad (12.20)$$

$$Mn_2(CO)_{10} + 2\,KH \xrightarrow{thf} 2\,K[Mn(CO)_5] + H_2 \qquad (12.21)$$

3. Carbonyl substitution by anions. Substitutions such as Reactions (12.7) and (12.22) lead to carbonylate anions:

$$Re(CO)_5Cl + 2\,KCN \xrightarrow{MeOH} K[Re(CO)_4(CN)_2] + CO + KCl \qquad (12.22)$$

Two especially important reactions of carbonylate anions are those with alkyl or acyl halides to give organic derivatives [Equations (12.23) and (12.24)] and protonation to afford metal hydrides [Equation (12.26)]:

$$[Re(CO)_5]^- + MeI \longrightarrow MeRe(CO)_5 + I^- \qquad (12.23)$$

$$[Mn(CO)_5]^- + Me\overset{\overset{\displaystyle O}{\displaystyle \|}}{C}Cl \longrightarrow Me\overset{\overset{\displaystyle O}{\displaystyle \|}}{C}Mn(CO)_5 + Cl^- \qquad (12.24)$$

When carbonylate anions displace halide from carbonyl halides, heteronuclear carbonyls result. An example is

$$[Mn(CO)_5]^- + Re(CO)_5Br \longrightarrow MnRe(CO)_{10} + Br^- \qquad (12.25)$$

The above reactions are analogous to S_N2 displacements on organic halides. Protonation of carbonylate anions is a route to hydride complexes:

$$[Mn(CO)_5]^- + H^+ \longrightarrow HMn(CO)_5 \qquad (12.26)$$

EXAMPLE 12.4: What will be the product(s) of mixing the following reactants?

(a) $NbCl_5 + CO + Na \xrightarrow[\text{high pressure}]{\text{diglyme}}$

(Diglyme is an ether $(CH_3OCH_2CH_2)_2O$).

(b) $[Co(CO)_4]^- + C_6H_5CH_2Cl \xrightarrow{thf}$

(c) $Mo(CO)_6$ + CNMe $\xrightarrow{\text{thf}}$

Solution: (a) Because CO is one of the reactants, a carbonyl is a likely product. Nb is in the +V oxidation state, and carbonyls are stable for low oxidation states. However, Na is a reducing agent. We need to know the stable carbonyl(s) of Nb. Because Nb is in the same group as V, it is a reasonable guess that $Nb(CO)_6$ could be formed because the analogous carbonyl is known for V. However, this is a 17-*e* compound. Thus, like $V(CO)_6$, it should be easily reduced to $[Nb(CO)_6]^-$ by Na. Thus, we predict that $Na[Nb(CO)_6]$ and NaCl will be formed.

(b) $[Co(CO)_4]^-$ is a carbonylate anion and a nucleophile. It displaces Cl^- from the alkyl halide to give $C_6H_5CH_2Co(CO)_4$.

(c) CNMe is a two-electron donor ligand isoelectronic with CO. Depending on the stoichiometric proportions, it will displace one or more CO's, giving $Mo(CO)_{6-n}(CNMe)_n$ ($n = 1-3$) and CO.

12.3 BONDING OF ORGANIC LIGANDS TO METALS

▶ *12.3.1   Olefin Complexes—The Paradigmatic π-Donor Complexes*

The first organometallic compound was prepared in 1827 by Zeise, who heated a $PtCl_2$ + $PtCl_4$ mixture in ethanol, evaporated the solvent, and treated the residue with aqueous KCl. The product was Zeise's salt $K[PtCl_3(C_2H_4)]$, whose structure was not understood until x-ray studies in the early 1950s showed that ethylene was coordinated approximately perpendicular to the $PtCl_3$ plane with the H's bent away from the metal.

An MO approach to the bonding, first developed by Dewar, Chatt, and Duncanson, serves as the paradigm for bonding in π complexes (Figure 12.8). The VB structure of ethylene (unlike that of CO) has no lone pair. However, the perpendicular orientation of C_2H_4 situates the filled π and empty π* orbitals properly for overlap with metal orbitals. The ethylene–metal bond has a σ component in which electrons are donated from the filled π orbital into an empty metal $5d6s6p^2$ hybrid orbital. In the π component, Pt electrons are shared with the olefin through overlap of a filled hybridized $5d6p$ orbital (a hybridized orbital is used to provide better overlap than occurs with d_{xz} or d_{yz}) with the π* orbitals of the olefin. The π* orbital is antibonding with respect to the olefin molecule. Its use weakens the C–C bond, but its overlap with a Pt orbital strengthens the Pt–olefin bond.

The two olefin–metal bond components reinforce one another synergistically. Like CO, olefins behave as two-electron donors, the two electrons coming from an orbital of π symmetry in the olefin. Polyolefins with isolated double bonds also form complexes, acting as ($n \times 2$)-electron donors to one or several metals. One such ligand is 1,5-cyclooctadiene (1,5-cod), which functions as a 2×2-electron donor. Another is 1,3,5,7-cyclooctatetraene (cot), which may function as a 2×2-electron donor to one or two different metals (see Figure 12.9) or a 3×2-electron donor to a single metal (Table 12.4). Metal–olefin bonds often are represented by arrows pointing from olefin to metal, since the lig-

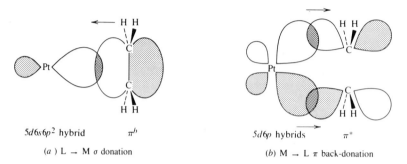

Figure 12.8 The Chatt–Dewar–Duncanson picture of bonding in a Pt–olefin complex. Arrows show direction of electron flow. (*a*) L → M σ-bonding; (*b*) M → L π* back-bonding.

ands are formally π donors. Silene complexes such as $[(i\text{-Pr})_2\text{Si}=\text{Si}(i\text{-Pr})_2]\text{Pt}(\text{Ph}_2\text{PCH}_2\text{CH}_2\text{PPh}_2)$ have recently been prepared[8]; the Pt–silene bonding can be understood on the model of olefin–metal bonding. Isoelectronic with $CH_2=CH_2$ is diazene, $HN=NH$. $[\text{W}(HN=NH)(CO)_2(NO)(PPh_3)_2]^+$ containing a *trans*-diazene has also been made.[9]

▶ 12.3.2 The 18-e Rule for π-Donor Complexes

Besides olefins, other unsaturated organic species can act as π donors. The chemistry of some of these compounds, along with special features of their bonding, is discussed in the following chapter. All complexes with π-donor ligands feature electron donation from filled ligand π-orbitals to the metal and donation from the metal into empty π* ligand orbitals. The most stable obey the 18-*e* rule. As the ligands grow more complicated, so does the number and symmetry of the orbitals. For now, we focus simply on electron-counting

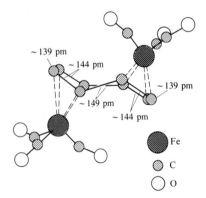

Fe
C
O

Figure 12.9 Structure of $(\eta^4, \eta^{4'}\text{-cot})\text{Fe}_2(\text{CO})_6$. (Reprinted with permission from B. Dickens and W. N. Lipscomb, *J. Am. Chem. Soc.* **1961**, *83*, 489. Copyright 1961, American Chemical Society.)

[8] E. K. Pham and R. West, *J. Am. Chem. Soc.* **1989**, *111*, 7667.

[9] M. R. Smith, III and G. L. Hillhouse, *Abstracts of Papers*, 199th ACS Meeting, Boston, 1990, Paper INOR135.

Table 12.4   Electron counting for π-donor ligands

Electrons contributed	Ligand	Structure	Example
2	η^2-C_2H_4	$H_2C{=}CH_2$	$[PtCl_3(C_2H_4)]^-$ trichloro(ethylene)platinate(1−)
4	η^3-$C_3H_5^-$ η^3-allyl		bis(η^3-ally)di-μ-bromodipalladium
4	η^4-C_4H_8 η^4-butadiene		η^4-butadienetricarbonyliron
4	η^4-C_5H_6 η^4-cyclopentadiene		dicarbonylbis(η^4-cyclopentadiene)-molybdenum
4	η^4-C_8H_8 η^4-cyclooctatetraene (cot)		$(1, 2, 5, 6$-η^4-cyclooctatetraene)(η^5-cyclopentadienyl)cobalt
6	η^4-$C_4H_4^{2-}$ η^4 cyclobutadiene		tricarbonyl(η^4-cyclobutadiene)iron
6	η^5-$C_5H_5^-$ η^5-cyclopentadienyl (Cp)		tricarbonylchloro(η^5-cyclopentadienyl)molybdenum
	η^5-$Me_5C_5^-$ pentamethylcyclopentadienyl (Cp*)		

(Continued)

Table 12.4   Electron counting for π-donor ligands *(Continued)*

Electrons contributed	Ligand	Structure	Example
6	η^5-C$_5$H$_6^-$ η^5-pentadienyl		Fe(CO)$_3$ tricarbonyl(η^5-pentadienyl)iron (1+)
6	η^6-C$_6$H$_6$ η^6-benzene		bis(benzene)chromium
6	η^7-C$_7$H$_7^+$ η^7-tropylium		Mo(CO)$_3$ tricarbonyl(η^7-tropylium)- molybdenum(1+)
6	η^6-C$_7$H$_8$ η^6-cycloheptatriene		Mo(CO)$_3$ tricarbonyl(η^6- cycloheptatriene)- molybdenum
6	η^6-C$_8$H$_8$ η^6-cyclooctatetraene (cot)		Cr(CO)$_3$ tricarbonyl(η^6-cyclooctatetraene)- chromium

rules for π-donor ligands, as given in Table 12.4. Note that some ligands can bond using varying numbers of electrons; for example, cyclooctatetraene can donate electrons from two, three, or all four double bonds. This is reflected by the η^n nomenclature, where n denotes the number of atoms bonded to the metal. η is from the Greek prefix *hapto* (derived from *haptein*, "to fasten"). Thus η^3-C_3H_5 is read "trihaptoallyl" and denotes that all three carbons are bonded to the metal by π orbitals delocalized over all three C's; η^5-C_5H_5 (abbreviated Cp) is read "pentahaptocyclopentadienyl". Index numbers of the C atoms bound to the metal must be specified whenever confusion could arise—as, for example, in the $(\eta^4-1,2,5,6)-1,3,5,7$-cyclooctatetraene (cot) complex shown in Table 12.4, in which alternant (instead of adjacent) double bonds are attached. (See also Figure 12.9.) In structural formulas, the totality of π bonds usually is represented by a single line instead of an arrow from each bond under the assumption that the hapticity is that needed to satisfy the 18-e rule.

Ligands existing in the free state as neutral molecules which have an even number of metal-bonded C's are named as such. Examples are olefins (ethylene, butadiene, etc.) and arenes. A more complex and arbitrary nomenclature is used for ligands having an odd number of C's (and delocalized π systems); these are named as if they were odd-electron radicals. For example, $\cdot C_5H_5$ (cyclopentadienyl) has the contributing structures

Because it has both olefin (−*ene*) and radical (−*yl*) functionality, it is named as an −*enyl* as are other ligands bonded through an odd number of carbons. *For electron counting purposes*, however, we treat π-donor ligands as having a closed-shell configuration. We assume odd-C cyclic π systems to have the $(4n + 2)$ π electrons required for aromaticity. Although this convention is completely arbitrary, it has some connection to π-donor chemistry. For example, $Na^+C_5H_5^-$ is a source of Cp^- for preparing cyclopentadienyl compounds. [However, the electron count is the same for $Fe^0(C_5H_5)_2$ or $Fe^{II}(C_5H_5^-)_2$.] Noncyclic π donors are also treated as closed-shell species for electron counting because their compounds are not fundamentally different from those of cyclic ligands. Thus, the allyl (propenyl) ligand is considered to be $C_3H_5^-$. Finally, we note that some ligands are related to those in Table 12.4 by closing a chain compound with CH_2 groups. For example, η^3-$C_5H_7^-$ is just an allyl closed by two CH_2 groups; it is named η^3-cyclopentenyl.

Electron counts for π-donor ligands appear in both Tables 12.2 and 12.4.

EXAMPLE 12.5:   Count electrons in

Solution: We use the rules on page 566. For each Pd, the oxidation number is $-(-1$ [for Br^-]$+$ -1 [for allyl]) $= +2$; each Br forms one dative bond that does not contribute to the charge.

$$Pd^{II} + \eta^3\text{-allyl} + Br^- + \mu\text{-Br}$$

$$8e\ \ + 4e\ \ \ \ \ \ + 2e\ + 2e\ \ = 16e$$

EXAMPLE 12.6: Count electrons in

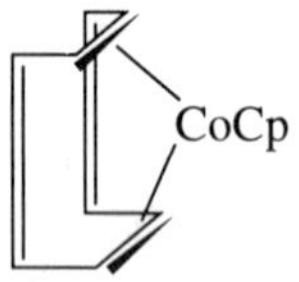

Solution:

$$Co^I + Cp^- + 2 \times \eta^2\text{-cot}$$
$$8e \ + 6e \ \ + 2 \times 2e \ \ \ \ = 18e$$

EXAMPLE 12.7: Count electrons in

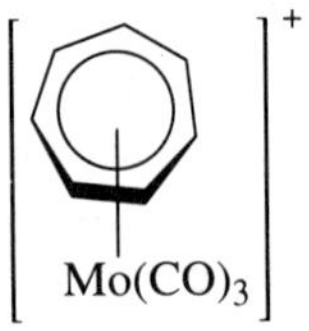

Solution: $C_7H_7^+$ is tropylium; thus, the oxidation number of Mo is $-(+1) + 1 = 0$.

$$Mo^0 + \eta^7\text{-}C_7H_7^+ + 3\,CO$$
$$6e \ + 6e \ \ \ \ \ \ + 3 \times 2e = 18e$$

Familiarity with applying the 18-e rule to several kinds of ligands now allows you to predict the formulas and understand some structural features of organometallic compounds. Hence, we can discuss compounds countaining several different kinds of ligands, even though we may wish to focus on the chemical and structural features associated with only one.

▶ *12.3.3 Cyclopentadienyl Complexes*

Current interest in organometallic chemistry began in the early 1950s, when two research groups, working independently, prepared the astonishing Fe hydrocarbon derivative (η^5-$C_5H_5)_2$Fe or Cp_2Fe, by reaction of $FeCl_2$ with C_5H_5MgBr and by reaction of reduced Fe with cyclopentadiene under nitrogen at 300°C in the presence of K_2O. The product, now known as ferrocene, is an orange solid melting around 170°C with sublimation. Ferrocene is diamagnetic, but is easily oxidized to paramagnetic blue $[Cp_2Fe]^+$ having one unpaired electron. All the first-row transition elements except Ti form Cp_2M complexes (metallocenes); a few second- and third-row metals do so (Table 12.5), but those which do not obey the 18-e rule are extremely unstable and difficult to isolate. In the solid state, ferrocene (Figure 12.10) has planar Cp rings, with all C—C distances equal; these are arranged in the staggered sandwich configuration (D_{5d}), rather than with eclipsed carbons (D_{5h}). Both ruthenocene and osmocene have the eclipsed configuration in the solid—as does ferrocene itself in the gas phase. This indicates that the energy barrier to ring free rotation is quite small—only ~5kJ/mol.

Table 12.5   Some bis(cyclopentadienyl) complexes of transition metals

Compound	Color	mp (°C)	Number of unpaired electrons	Structure and comments
"Cp$_2$Ti"	Dark green	Dec. 200	0	Dimer with μ-H; structure **A**
	Gray-black		>0	Dimer containing M—M bond; structure **B**
	Purple		0	Dimer with antiferromagnetic spin coupling; structure **C**
(Me$_5$C$_5$)$_2$Ti	Yellow-orange	Dec. 60	2	Monomer, tilted rings
"Cp$_2$Zr"	Purple-black	Dec. 300	0	Str. similar to Ti μ-H dimer **A** or **C**
Cp$_2$V	Purple	167–168	3	Air-sensitive; reacts with CO, X$_2$, RX
Cp$_2$Nb			>0	Thermally unstable
"Cp$_2$Nb"	Yellow		0	Dimer; structure **C**
"Cp$_2$Ta"			0	Isomorphous with "Cp$_2$Nb" **C**
Cp$_2$V$^-$	Red		2?	Prepared by K reduction of Cp$_2$V
Cp$_2$Cr	Scarlet crystals	172–173	2	Air-sensitive;
Cp$_2$Mo	Yellow		2	Thermally unstable; very reactive; observed in lo-temperature matrix
"Cp$_2$Mo"	Yellow (**E**); green (**G**)		0	Dimers; structures **E**, **F**, and **G**
Cp$_2$W			2	Thermally unstable; very reactive intermediate; observed in lo-temperature matrix
"Cp$_2$W"	Yellow			Dimer; same structure as **F**
Cp$_2$Cr$^+$	Brown-black (CpCr(CO)$_3^-$ salt)		3	Prepared by oxidation of Cp$_2$Cr (Cp$_2$Cr$^+$ + $e \rightarrow$ Cp$_2$Cr, $E° = -0.67$ V vs. S.C.E.)
Cp$_2$Cr$^-$				Very reactive. Cp$_2$Cr + $e^- \rightarrow$ Cp$_2$Cr$^-$, $E° = -2.30$ V vs. S.C.E.
Cp$_2$Mn	Amber		5	Stable form $\leq$ 159°C, chain structure
	Pale pink	172–173	5	Stable form > 159°C
(Me$_5$C$_5$)$_2$Mn	Orange		1	Monomer; Jahn–Teller distorted in solid
Cp$_2$Re		Dec. $\geq$ 20 K	1	Thermally unstable
(Me$_5$C$_5$)$_2$Re	Deep-purple		1	Planar, eclipsed rings in solid
"Cp$_2$Re"	Purple-black	Dec. 60	0	Dimer; structure **H**
(Me$_5$C$_5$)$_2$Re$^-$	Orange (K$^+$ salt)		0	
Cp$_2$Fe	Orange	173	0	Staggered configuration in solid, eclipsed in gas phase; air-stable, stable > 500°C
Cp$_2$Ru	Light yellow	199–201	0	Eclipsed configuration in solid; most thermally stable metallocene (>600°C)
Cp$_2$Os	Colorless	229–230	0	Eclipsed configuration in solid
Cp$_2$Fe$^+$	Blue (PF$_6^-$ salt) Dichroic in aq. sol'n. (blue-green to blood red)		1	Prepared by oxidation of Cp$_2$Fe (Cp$_2$Fe$^+$ + $e \rightarrow$ Cp$_2$Fe, $E° = -0.34$ V vs. S.C.E.)
"Cp$_2$Os"$^+$	Green-black		0	Product of oxidation of Cp$_2$Os; structure **H** with halves rotated by 90°
Cp$_2$Co	Purple-black	173–174	1	19-e species; easily oxidized by air to Cp$_2$Co$^+$
Cp$_2$Rh	Brown-black		>0 ·	19-e species; monomeric $\leq -196°$

Table 12.5   **Some bis(cyclopentadienyl) complexes of transition metals (*Continued*)**

Compound	Color	mp (°C)	Number of unpaired electrons	Structure and comments
"Cp$_2$Rh"	Yellow-orange	Dec. 140	0	Dimer linked through rings; structure **I**
Cp$_2$Ir	Colorless		>0	19-e species; monomeric $\leq -196°$
"Cp$_2$Ir"	Yellow	Dec. 230	0	Same structure as Rh dimer **I**
Cp$_2$Co$^+$	Yellow (colorless anion)		0	Prepared by oxidation of Cp$_2$Co or directly from Co halide + Cp$^-$
Cp$_2$Co$^-$	Brown-orange		0?	Prepared by electrochemical reduction in solution
Cp$_2$Rh$^+$	Colorless		0	Prepared directly from RhCl$_3$ + Cp$^-$
Cp$_2$Ir$^+$	Yellow (PF$_6^-$ salt)		0	Prepared directly from IrCl$_3$ + Cp$^-$
Cp$_2$Ni	Dark green	Dec. 173–174	2	20-e species; toxic
Cp$_2$Ni$^+$	Yellow-orange (colorless anion)		1	(Cp$_2$Ni$^+$ + $e \rightarrow$ Cp$_2$Ni, $E° = -0.09$ V vs. S.C.E.)
Cp$_2$Ni^{2+}			0	(Cp$_2$Ni^{2+} + $e^- \rightarrow$ Cp$_2$Ni$^+$, $E° = +0.74$ V vs. S.C.E.)

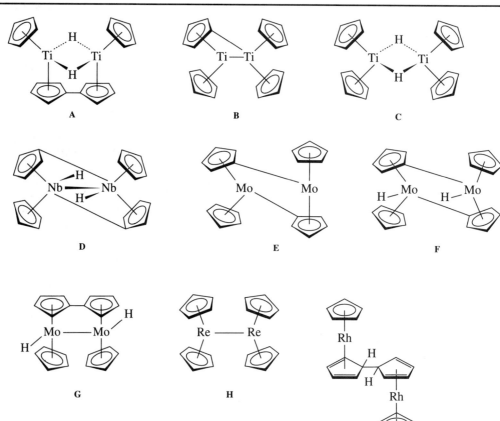

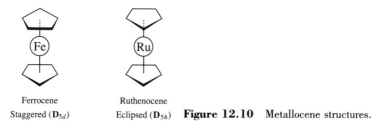

Ferrocene | Ruthenocene
Staggered (D_{5d}) | Eclipsed (D_{5h})   **Figure 12.10**   Metallocene structures.

The sandwich structure of the metallocenes gave the first clue that organic ligands could be bonded to metals via their π systems. Although the structure of ferrocene was known to involve the ligand π system, formulation of its electronic structure presented a considerable challenge. Because of the number of VB structures that can be written for the Cp radical or the Cp^- anion, a VB description of ferrocene requires 560 contributing structures![10] In this situation the MO description is more appealing.

MOs for Cp are depicted in Figure 4.25. We consider each ligand to be Cp^- having the closed-shell configuration $a^2 e_1^4$ and acting as a six-electron donor. In describing the electronic structure of ferrocene, we construct ligand-group orbitals (LGOs) conforming to the D_{5d} molecular symmetry from MOs on the two separate rings. Figure 12.11 depicts these symmetry orbitals along with the metal orbitals with which they can overlap. Figure 12.12 shows an energy diagram for ferrocene and other metallocene complexes. The most energetically significant bonding interaction is between the e_{1g} (d_{xz}, d_{yz}) metal orbitals and the ligand e_{1g} pair. Although metal e_{1u} (p_x, p_y) orbitals overlap favorably with ligand e_{1u}, the energy difference is too large for effective MO formation. Hence, the $1e_{1u}$ orbitals are localized largely on the ligands. Because the metal e_{2g}'s ($d_{x^2-y^2}$, d_{xy}) and a_{1g} (d_{z^2}) do not overlap very effectively with ring orbitals, they are nonbonding and localized on the metal. The 18 electrons of ferrocene exactly fill the energy levels through $2a_{1g}$, above which a substantial HOMO–LUMO energy gap exists. This gap, together with effective occupation of all the coordination positions by Cp rings, accounts for the remarkable kinetic and thermal stability of ferrocene.

The energy-level diagram of Figure 12.12 applies to other metallocenes of Table 12.5. It accounts for their magnetic properties if the reasonable assumption is made that electron-pairing energy sometimes is greater than the $1e_{2g}$–$2a_{1g}$ energy separation (see Problem 12.15). Cp_2Co, Cp_2Rh, Cp_2Ir, and Cp_2Ni contain electrons in the antibonding $2e_{1g}$ orbital. Hence, it is not surprising that these compounds undergo reactions leading to 18-electron species such as

$$2\,Cp_2Rh + 3\,Br_2 \longrightarrow 2[Cp_2Rh]^+\,Br_3^- \tag{12.27}$$

$$Cp_2Ni + C_2F_4 \longrightarrow CpNi \qquad \tag{12.28}$$

Manganocene is unique in exhibiting the maximum possible number of unpaired electrons (five). Its bonding (in the high-temperature form) has been described as being more ionic than that in other metallocenes, with a large contribution from the structure

---

[10] This number can be reduced by considering only d orbitals of proper symmetry for metal-ring bonding.

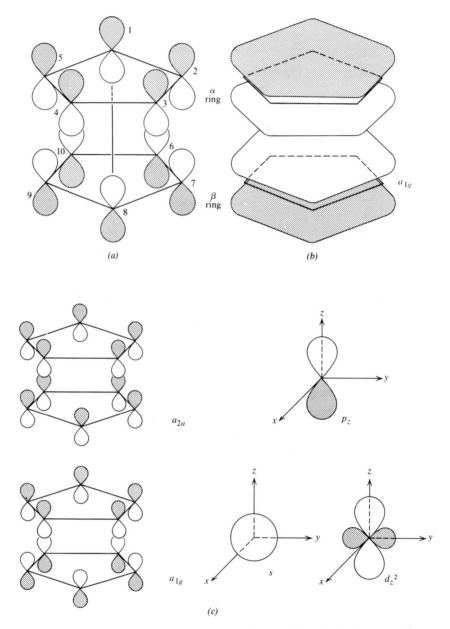

Figure 12.11 (*a*) Orientation of orbitals for consideration of bonding in ferrocene. (*b*) a_{1g} orbitals for the two rings. (*c*) Symmetry combinations of Cp orbitals and their metal counterparts in the $\mathbf{D}_{5d}$ symmetry of ferrocene.

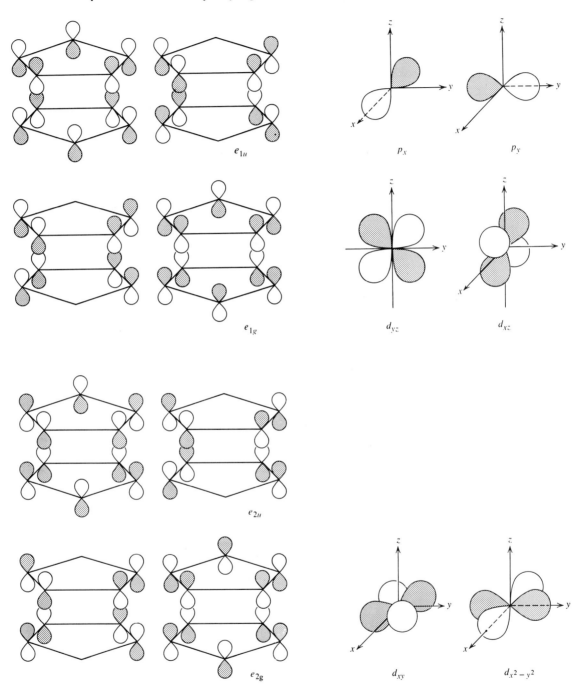

e_{1u}

p_x

p_y

e_{1g}

d_{yz}

d_{xz}

e_{2u}

e_{2g}

d_{xy}

$d_{x^2-y^2}$

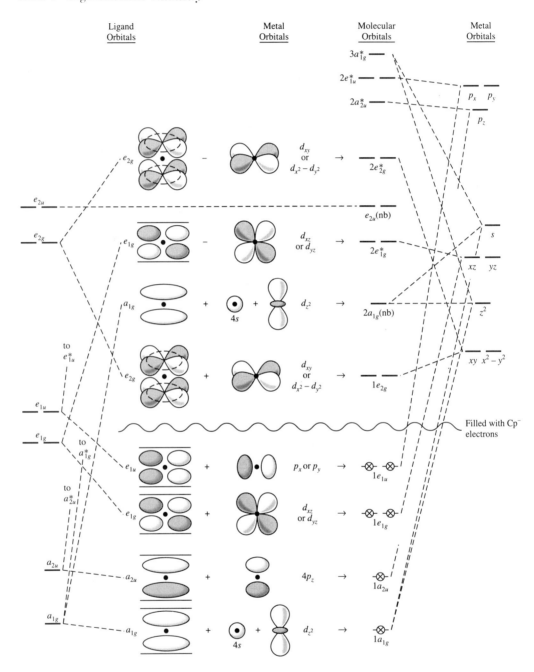

Figure 12.12 Molecular orbital energy level diagram for metallocenes. (From B. E. Douglas and C. A. Hollingsworth, *Symmetry in Bonding and Spectra*, Academic Press, New York, 1985.)

$(Cp^-)_2Mn^{2+}$. This is borne out in the chemistry of Cp_2Mn, which, in contrast to other metallocenes, is hydrolyzed easily and reacts with CO_2 to give carboxylic acid salts. However, manganocene is soluble in organic solvents, displays the same structure with planar rings as other metallocenes, and gives solutions of low conductivity. Probably the best picture involves MOs that are close together, permitting the large stabilization gained from exchange energy of five unpaired electrons to outweigh the smaller orbital energy separations.

The H atoms on the Cp ring can, of course, be substituted by organic groups. Many η^5-Me_5C_5 complexes are known, and this ligand is common enough to be abbreviated as Cp^*.

► 12.3.4   Some Chemistry of Ferrocene

Cp ligands in ferrocene display the chemistry of aromatic rings (lending credence to our formulation of the ligand as $C_5H_5^-$). Indeed, ferrocene is more reactive than benzene toward electrophilic reagents. It undergoes Friedel–Crafts acylation, giving (η^5-$C_5H_4C(O)CH_3$)(η^5-C_5H_5)Fe with equimolar acetyl chloride. With excess $CH_3C(O)Cl$, two disubstituted isomers are produced:

$$Cp_2Fe + xs\ CH_3C(O)Cl \xrightarrow{AlCl_3}$$

60   :   1   (12.29)

Entry of electron-withdrawing acetyl deactivates the ring to further substitution. Hence the major product is 1,1′-diacetylferrocene. When ferrocene is alkylated under Friedel–Crafts conditions, dialkylation occurs homoannularly because electron-donating alkyl groups activate each ring to further electrophilic attack.

Because ferrocene is oxidized easily to $[Cp_2Fe]^+$, direct halogenation and nitration cannot be carried out on the ring.

► 12.3.5   Cyclopentadienyl Complexes
Containing Other Ligands

Many organometallic complexes, including Cp_2TiCl_2, Cp_2ReH, and Cp_2NbOCl, contain two cyclopentadienyl rings along with other ligands. An important feature of their structures is the bending back of the planar Cp rings to make orbitals sterically available for bonding to other ligands. Many other complexes contain only one Cp ring acting as a planar six-electron donor occupying three coordination positions. Examples of these half-sandwiches include $CpTi(CO)_2$, $[CpCr(NO)_2]_2$, $[CpFe(CO)_2]_2$, and $CpMo(O)_2Cl$.

In general, Cp rings are less reactive than other ligands and can be considered inert ligands that occupy three coordination positions. When no other ligands are present, the chemistry is forced to take place at Cp; a few examples were discussed in the last section.

▶ 12.3.6   σ Bonds Between C and Metals

Table 12.2 includes electron counts for some organic ligands that are not π acids. Alkyl, aryl, and acyl ligands are capable of σ donation to metals. Notice that we again consider ligands as having closed-shell configurations. Thus, even though alkyl, aryl, and acyl ligands are named as radicals, we consider them negatively charged carbanions for electron-counting purposes. They are like halides or alkoxides in terms of electron count. For example, $CH_3Mn(CO)_5$, $PhMn(CO)_5$, $CH_3C(O)Mn(CO)_5$, and $ClMn(CO)_5$ are all Mn^I complexes that obey the 18-e rule. However, halides also have filled π orbitals and can act as π donors when metals have empty d orbitals, a role which cannot be taken by alkyl and acyl complexes. Aryl rings having π-donor substituents could formally act as π donors as the following structures show:

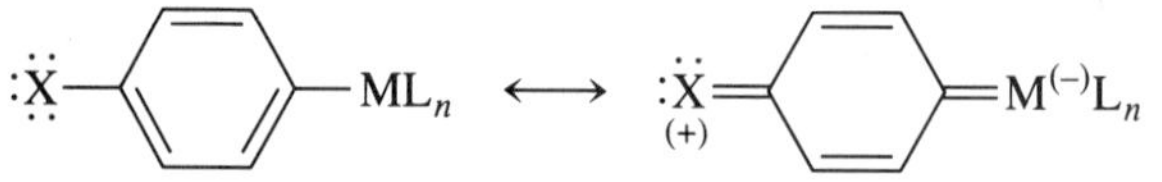

$$\ddot{X}\!-\!\!\langle\!\!\!\!\bigcirc\!\!\!\!\rangle\!\!-\!ML_n \longleftrightarrow \underset{(+)}{\ddot{X}}\!=\!\!\langle\!\!\!\!\bigcirc\!\!\!\!\rangle\!\!=\!M^{(-)}L_n$$

Canonical structures can also be written showing that acyl and aryl ligands are capable of π^* back-donation. However, they are much less effective than CO and other neutral π-acid ligands.

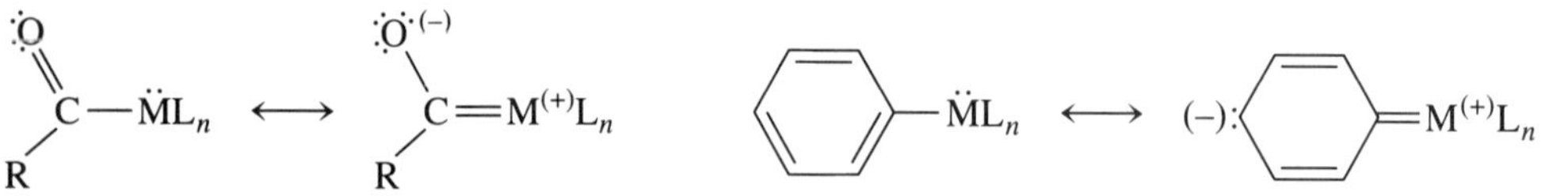

$$\underset{R}{\overset{\ddot{O}}{\underset{|}{C}}}\!\!-\!\ddot{M}L_n \longleftrightarrow \underset{R}{\overset{\ddot{O}^{\cdot(-)}}{\underset{|}{C}}}\!\!=\!M^{(+)}L_n \qquad \langle\!\!\!\!\bigcirc\!\!\!\!\rangle\!\!-\!\ddot{M}L_n \longleftrightarrow (-)\!\!:\!\langle\!\!\!\!\bigcirc\!\!\!\!\rangle\!\!=\!M^{(+)}L_n$$

Of course, the electron count is not changed by π bonding.

A very significant reaction of alkyls containing β-H is the so-called **β-elimination**:

$$M\!-\!CH_2CH_2R \rightleftharpoons M\!-\!\!\overset{\overset{\displaystyle H}{|}}{\underset{\underset{\displaystyle CHR}{\|}}{CH_2}} \longrightarrow M\!-\!H + CH_2\!=\!CHR \qquad (12.30)$$

Olefin is eliminated and H remains bonded to the metal. This reaction affords a facile decomposition path for metals which have a vacant metal orbital to interact with the β-C—H bond and led to the supposition that transition-metal alkyls were thermodynamically unstable. However, organometals which lack β-H, such as methyls, benzyls, neopentyls (CH_2CMe_3), and so on, are stabilized because β-elimination is unavailable.

▶ *12.3.7   Carbenes and Alkylidenes*

Divalent carbon ligands (**carbenes**), $:C(X)(Y)$, where X and Y may be OR, NR, alkyl, aryl, H, X, and so on, can be stabilized by coordination to transition metals. Fischer and Maasböl in 1964 were the first to prepare a carbene complex in a planned synthesis:

$$W(CO)_6 + PhLi \longrightarrow \left[(CO)_5W\overset{\overset{O}{\|}}{C}Ph \right]^- Li^+ \xrightarrow{Me_4N^+Cl^-}$$

$$\left[(CO)_5W\overset{\overset{O}{\|}}{C}Ph \right]^- Me_4N^+ \xrightarrow{H^+} (CO)_5W{=}C\underset{Ph}{\overset{OH}{<}} \qquad (12.31)$$

$$\xrightarrow{CH_2N_2} (CO)_5W{=}C\underset{Ph}{\overset{OCH_3}{<}}$$

(It is more usual now to alkylate with trialkyloxonium salts or with methyl fluorosulfonate.) Nucleophilic attack on a coordinated unsaturated ligand followed by proton transfer also leads to carbene ligands:

$$\textit{trans-}[Cl(PEt_3)_2Pt(C{\equiv}NR)]^+ + HY \longrightarrow \textit{trans-}\left[Cl(PEt_3)_2Pt{=}C\underset{Y}{\overset{NHR}{<}} \right]^+$$

$$(12.32)$$

$$HY = ROH, NH_2R, NHR_2$$

Three canonical structures describe the metal–carbene bond. (In Fischer-type carbenes, X is a heteroatom.)

$$M{=}C\underset{R}{\overset{\ddot{X}:}{<}} \longleftrightarrow \overset{(-)}{M}{\leftarrow}\overset{(+)}{C}\underset{R}{\overset{\ddot{X}:}{<}} \longleftrightarrow \overset{(-)}{M}{\leftarrow}C\underset{R}{\overset{\overset{(+)}{\ddot{X}:}}{<}} \equiv M{-}C\underset{R}{\overset{X}{⫶}}$$

$$\quad\text{(a)} \qquad\qquad\quad\text{(b)} \qquad\qquad\quad\text{(c)}$$

Their relative importance depends on the identity of X and R. M$=$C distances are shorter than those of M$-$C single bonds (but longer than M$-$CO bonds), indicating the importance of structure **a.** MO calculations suggest that the carbene C carries a positive charge corresponding to structure **b**; however, carbonyl ligands are calculated to have greater positive charge. Structure **c** contributes because NMR studies indicate a barrier to rotation around the C—X bond which is smaller for X = OMe than for isoelectronic NMe$_2$ (a better electron donor). The importance of **c** is also indicated by the low stability of carbene complexes which do not contain a heteroatom.

Low-oxidation-state d^6–d^{10} metals are a feature of Fischer-type carbene complexes. :C(X)(Y) is considered a neutral two-electron donor; an empty p_π orbital is available on C for back-donation from the metal (Figure 12.13). The net result is a partial positive charge on C, and in Equation (12.33) a carbanion attacks the carbene C; methanol elimination follows, leading to a new carbene.

$$(CO)_5W{=}C\overset{OMe}{\underset{Ph}{\diagup}} \xrightarrow[-780°C]{PhLi} (CO)_5W^-{\leftarrow}C\overset{OMe}{\underset{Ph}{\diagdown}}{-}Ph \xrightarrow[-78°C]{HCl} (CO)_5W{=}CPh_2 + MeOH + LiCl$$

$$(12.33)$$

Substitution of a heteroatom group by amine (aminolysis) also proceeds via initial nucleophilic attack at C followed by alcohol elimination:

$$(CO)_5Cr{=}C\overset{OMe}{\underset{Ph}{\diagup}} + Me_2NH \longrightarrow (CO)_5Cr^-{\leftarrow}C\overset{OMe}{\underset{Ph}{\diagdown}}{-}NHMe_2 \longrightarrow$$

$$(CO)_5Cr{=}C\overset{NMe_2}{\underset{Ph}{\diagup}} + MeOH$$

$$(12.34)$$

OR can be substituted by OR$'$ or enolate ions by a route similar to that in Equation (12.34), but amino groups are resistant to replacement because of the good donor ability of N leading to stabilization via structure **c**. The electrophilicity of the carbene C enhances the acidity of methyl protons in $(CO)_5Cr{=}C(OMe)Me$ by stabilizing the intermediate resulting from proton abstraction as shown in Equation (12.35):

$$(CO)_5Cr{=}C\overset{OMe}{\underset{Me}{\diagup}} \xrightarrow[MeONa]{MeOD} (CO)_5Cr{=}C\overset{OMe}{\underset{CH_2^{(-)}}{\diagup}} \longleftrightarrow (CO)_5\overset{(-)}{Cr}{\leftarrow}C\overset{OMe}{\underset{CH_2}{\diagdown\diagdown}}$$

$$+ MeOH \longrightarrow (CO)_5Cr{=}C\overset{OMe}{\underset{CH_2D}{\diagup}} \longrightarrow (CO)_5Cr{=}C\overset{OMe}{\underset{CD_3}{\diagup}}$$

$$(12.35)$$

Diaminocarbenes such as $(CO)_5Cr{=}C(NR_2)_2$ show virtually no electrophilic behavior at the carbene C because N is a good π donor. Carbenes also serve as excellent transfer agents for :C(X)(Y) as in the following cyclopropanation reaction:

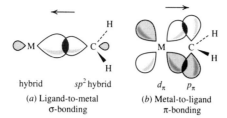

hybrid sp^2 hybrid d_π p_π

(a) Ligand-to-metal   (b) Metal-to-ligand
σ-bonding   π-bonding

Figure 12.13   Orbital overlap in metal-carbene bonding. Arrows show direction of electron flow. (*a*) L → M σ-bonding; (*b*) M → L π back-bonding.

$$\left[Cp^*Fe(CO)_2 = C\begin{smallmatrix} H \\ \\ Me \end{smallmatrix} \right]^+ + PhCH = CH_2 \longrightarrow >99:1 \; cis:trans \quad \triangle_{Me \quad Ph} \qquad (12.36)$$

$$Cp_2Ti \underset{Cl}{\overset{CH_2}{\diamondsuit}} AlMe_2 \;(Tebbe's\ reagent)\ is\ the\ functional\ equivalent\ of\ Cp_2Ti = CH_2$$

and, when treated with a base such as pyridine, transfers CH_2.

$$RCR \overset{O}{\underset{\|}{}} + "Cp_2Ti = CH_2" \longrightarrow \overset{CH_2}{\underset{\|}{RCR}} \atop +[Cp_2TiO]_x \xrightarrow{"Cp_2Ti=CH_2"} Cp_2Ti\diamondsuit_R^R \qquad (12.37)$$

Formally similar complexes of early transition metals in high oxidation states have been prepared by Schrock[11] as shown in Equation (12.38). The route would be expected to result in $Ta(CH_2CMe_3)_5$, since $TaMe_5$ is prepared similarly.

$$M(CH_2CMe_3)_3Cl_2 + 2\,Me_3CCH_2Li \longrightarrow (Me_3CCH_2)_3M = C\begin{smallmatrix} H \\ \\ CMe_3 \end{smallmatrix} + CMe_4 + 2\,LiCl \qquad (12.38)$$

$$M = Nb,\ Ta$$

The absence of β-hydrogens and the presence of bulky ligands apparently leads to intramolecular abstraction of an α-hydrogen and elimination of Me_4C from an $M(CH_2CMe_3)_4Cl$ or $M(CH_2CMe_3)_5$ intermediate.

The crystal structure of a related complex is shown in Figure 12.14. In $Cp_2Ta(CH_2)CH_3$, the ~20-pm shortening of the M—C distance for the methylene compared to that for the methyl indicates the importance of the $M=CH_2$ formulation. Moreover, NMR studies indicate a rather substantial (60–80 kJ/mol) barrier to rotation around the Ta=C bond—that is, a substantial contribution from resonance structure **a** on page 591. Besides this, compounds without heteroatoms are quite stable. These features offer a striking contrast to the behavior of Fischer-type carbene complexes, and we distinguish

[11] R. R. Schrock, *Acc. Chem. Res.* **1979,** *12,* 98.

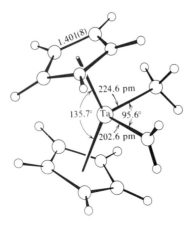

Figure 12.14 Structure of the alkylidene complex Cp_2TaMe-($=CH_2$). (Reprinted with permission from L. H. Guggenberger and R. R. Schrock, *J. Am. Chem. Soc.* **1975**, *97*, 6578. Copyright 1975, American Chemical Society.)

Schrock-type complexes by calling them **alkylidenes**.[12] Alkylidene ligands are considered as dianions $:CRR'^{2-}$ which donate four electrons, namely, the lone-pair C electrons and a pair from the filled p_π orbital on C. Many alkylidene complexes contain early transition metals in high oxidation states and have too few electrons to obey the 18-*e* rule. CRR'^{2-} ligands help to stabilize these species because they are good π donors into the empty stable orbitals on metals.

The chemistry of alkylidenes is also quite different, involving behavior of the alkylidene C as a *nucleophile*. That is, the polarity of the $M^{\delta+}=C^{\delta-}$ bond is opposite to that of carbenes. Electrophiles such as electron-deficient $AlMe_3$ or ketones (whose organic

$$R-\overset{\displaystyle :\overset{..}{O}:^{(-)}}{\underset{|}{C^{(+)}}}-R'$$

reactivity involves the resonance form attack alkylidene C while nucleophiles attack the metal:

$$Cp_2Ta\overset{\displaystyle Me}{\underset{\displaystyle CH_2}{\diagdown\diagup}} + AlMe_3 \longrightarrow Cp_2Ta\overset{\displaystyle (+)\diagup Me}{\underset{\displaystyle (-)\diagdown CH_2AlMe_3}{}} \qquad (12.39)$$

$$(Me_3CCH_2)_3Ta(=CHCMe_3) + R\overset{\displaystyle O}{\overset{\|}{C}}R' \xrightarrow[25°C]{pentane} \begin{array}{c} (-):\overset{..}{\underset{..}{O}}-\overset{(+)}{C}RR' \\ (Me_3CCH_2)_3Ta=CHCMe_3 \end{array}$$

$$(12.40)$$

[12] Although it is generally true that carbene complexes can be recognized because they contain other good π-acid ligands and metals in low oxidation states, it is not always possible to classify complexes just from their formulas. In case of uncertainty the term **alkylidene** is applied because the $=CRR'$ ligand is formally derived from an alkyl by H abstraction; thus, the name describes the connectivity of the ligand without necessarily imputing any particular reactivity trend.

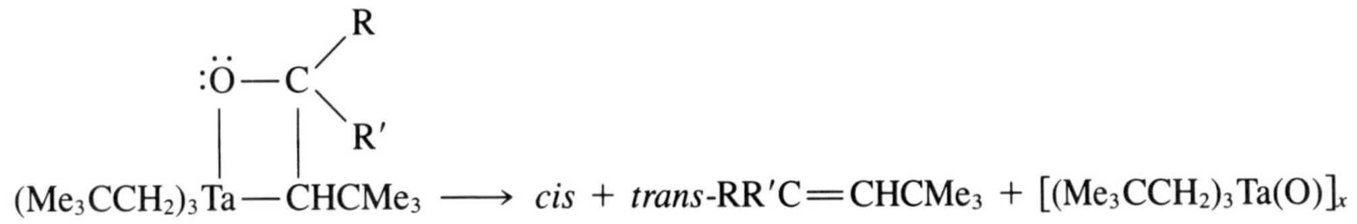

$$:\ddot{O}-C\overset{\Large R}{\underset{\Large R'}{\diagup}}$$

$$(Me_3CCH_2)_3Ta-CHCMe_3 \longrightarrow cis + trans\text{-}RR'C{=}CHCMe_3 + [(Me_3CCH_2)_3Ta(O)]_x$$

Alkylidenes display analogies with phosphorus ylids such as $Ph_3P{=}CH_2$ [Equation 12.40)]. Compare reaction (12.37) using Tebbe's reagent.

▶ 12.3.8 Carbynes and Alkylidynes

$M{\equiv}CR$ bonds are present in **carbynes** and **alkylidynes** which are related to carbenes and alkylidenes, respectively. Electrophilic BX_3 (X = Cl, Br, I) removes the heteroatom substituent of Fischer-type carbenes to give a metal carbyne cation which is then attacked by X^- which displaces CO and gives a neutral product[13]:

$$(CO)_5W{=}C\overset{\Large OMe}{\underset{\Large Ph}{\diagup}} + BBr_3 \longrightarrow [(CO)_4W{\equiv}CPh]^+ + (MeO)BBr_2 + Br^- \longrightarrow$$

$$trans\text{-}Br(CO)_4W{\equiv}CPh + CO \tag{12.41}$$

The $M{\equiv}C$ bond length is quite short, as we expect for a triple bond as shown in Figure 12.15.[14]

An alkylidene α-hydrogen is removed to prepare Schrock-type alkylidynes. The function of PMe_3 in Equation (12.42) is apparently to increase steric crowding around the metal which leads to α-H abstraction by an alkyl ligand. Alternatively, a proton can be abstracted by an external reagent as in Equation (12.43).

$$CpTaCl(CH_2CMe_3)({=}CHCMe_3) + 2\ PMe_3 \longrightarrow$$
$$CpTa(PMe_3)_2Cl({\equiv}CCMe_3) + CMe_4 \tag{12.42}$$

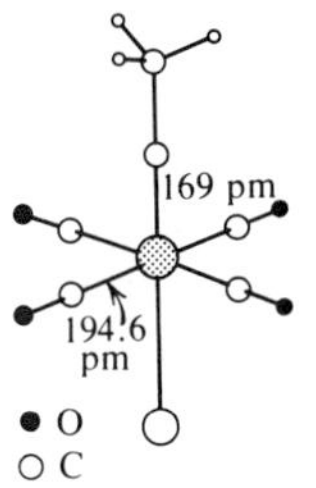

Figure 12.15 Structure of the carbyne complex $(CO)_4ICr{\equiv}CMe$. (From G. Huttner, H. Lorenz, and W. Gartzke, *Angew. Chem. Int. Ed. Engl.* **1974**, *13*, 609.)

[13] E. O. Fischer, *Adv. Organomet. Chem.* **1976**, *14*, 1.
[14] See H. P. Kim and R. J. Angelici, *Adv. Organomet. Chem.* **1987**, *27*, 51.

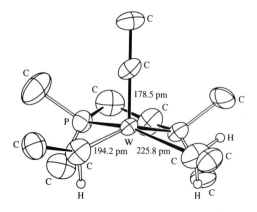

Figure 12.16 Structure of $W(\equiv CCMe_3)$-$(\!=\!CHCMe_3)(CH_2CMe_3)$ (dmpe). Methyl groups on terminal C's omitted; H on dmpe omitted for clarity. (Reprinted with permission from M. R. Churchill and W. J. Youngs, *Inorg. Chem.* **1979**, *18*, 2454. Copyright 1979, American Chemical Society.)

$$Ta(\!=\!CHCMe_3)(CH_2CMe_3)_3 + LiBu + dmp \longrightarrow$$
$$[Li(dmp)][Ta(\equiv CCMe_3)(CH_2CMe_3)_3 + C_4H_{10} \quad (12.43)$$

$$dmp = N, N'\text{-dimethylpiperazine}$$

Alkylidyne complexes are known for Nb, Ta, Mo, W and Re. Figure 12.16 shows the structure of $W(\equiv CCMe_3)(\!=\!CHCMe_3)(CH_2CCMe_3)[Me_2PCH_2CH_2PMe_2]$. The linear $\equiv CCMe_3$ ligand forms a very short $W\equiv C$ bond of 178.5 pm as compared with 194.2 pm for $W\!=\!C$ and 225.8 pm for $W\!-\!C$.

For counting electrons we (arbitrarily) consider a carbyne to be CR^+ in complexes derived from Fischer-type carbenes. Thus, the carbyne σ orbital is filled and acts as a two-electron donor. This leaves a pair of empty perpendicular p_π orbitals which can accept electrons from filled metal orbitals (Figure 12.17). (The situation is formally analogous to bonding with CO or NO^+. Calculations place negative charge on the carbyne C.) Thus, $Cl(CO)_4Cr\equiv CMe$ and related compounds obey the 18-*e* rule. In Schrock-type alkylidynes, we (arbitrarily) assume CR^{3-}, a six-electron donor (the C lone pair plus pairs from the two filled p_π orbitals on C). Counting in this way is in accord with the convention for CR_2 in carbenes and alkylidenes. In many alkylidyne complexes, metals display the group oxidation number such as $(t\text{-BuO})_3W\equiv CCMe_3$ and $[Re(\equiv CCMe_3)(NHPh)Cl_4]^-$. Schrock-type alkylidynes in general do not obey the 18-*e* rule, but are stabilized by the presence of other π-donating ligands such as halides, oxides, alkoxides, or amides.

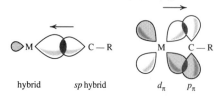

Figure 12.17 Orbital overlap in metal–carbyne bonding. Arrows show direction of electron flow. (*a*) L → M σ-bonding; (*b*) M → L π back-bonding (one of a mutually perpendicular set shown).

12.4   EXPERIMENTAL EVIDENCE
FOR BACK-DONATION

▶ 12.4.1   *The Problem of Oxidation Numbers*

Computing metal oxidation number is straightforward and is a very important way of classifying complexes. However, the concept of oxidation number has little *physical* meaning for π-acid and organometallic complexes, since many ligands both donate and accept electrons. The charge on each ligand is assumed arbitrarily as that necessary to give a closed-shell electron configuration (except for μ-CO), and it may bear little resemblance to the actual electron distribution. Indeed, it may not be the same for all complexes, depending on the ability of the metal to back-donate electrons. Consider olefin complexes. The formal effect of back-donation into π^* orbitals is to destroy the π^b component of the C=C bond. If back-donation from the metal were as great as one electron pair, the ligand would be represented better as the carbanion $H_2C^- - CH_2^-$ capable (formally) of bonding to the metal as two σ-bonded alkyl groups. The two extreme situations can be represented as

$$M^{n+} \longleftarrow \parallel^{CH_2}_{CH_2} \qquad \text{versus} \qquad M^{(n+2)+} \underset{^-CH_2}{\overset{^-CH_2}{\diagdown}}$$

(a) Metal–olefin   (b) Metallacyclopropane

The oxidation number of the metal changes by two units when comparing structure **a** to structure **b**. The real situation likely falls somewhere between the two and corresponds to the electroneutrality principle (see Section 2.1.5).

We can estimate the contribution of structure **b** by noting that a carbanion would have sp^3 hybridization around each C (109°48″ bond angles), whereas an olefin would have sp^2 hybridization (120° bond angles). Also, the C—C bond distance would be longer in a metallacyclopropane. Hence, x-ray structural and NMR evidence can be brought to bear. On coordination, the best π-acid olefins should exhibit distortions toward a metallacyclopropane structure. A good example is the structure of $Ir(Br)(CO)(PPh_3)_2(tcne)$, where tcne is $(NC)_2C=C(CN)_2$ (Figure 12.18). The substituents on the olefinic carbons are bent away from the metal, and the coordinated molecule is no longer planar. The C—C bond length is ~17 pm longer than that in free tcne. The extent of bending in a series of olefin complexes has been observed to increase with the length of the C—C bond,[15] and is thus related to the π-acid capacity. ^{19}F NMR studies[16] on compounds such as $Pt(PPh_3)_2(C_2F_4)$ show that chemical shifts typically are more characteristic of F bonded to sp^3 than of sp^2-hybridized C. These experimental results confirm the importance of back-donation and indicate the difficulties it introduces in employing the concept of oxidation number.

An example from the metal carbonyl hydrides also illustrates the difficulty of defining a meaningful oxidation number in organometals. The series of compounds $Mn(CO)_5Cl$, $Mn(CO)_5Br$, and $Mn(CO)_5I$ can be viewed as containing halide ligands and Mn^I. Hence,

[15] J. K. Stalick and J. A. Ibers, *J. Am. Chem. Soc.* **1970**, *92*, 5333.
[16] F. G. A. Stone, *Pure Appl. Chem.* **1972**, *30*, 551.

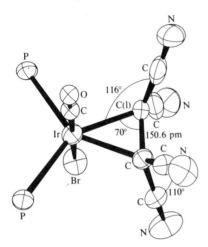

Figure 12.18   The molecular structure of IrBr(CO)-(PPh₃)₃(tcne). (From L. Manojovic-Muir, K. W. Muir, and J. A. Ibers, *Disc. Faraday Soc.* **1969**, *47*, 84.)

we can consider Mn(CO)₅H as a hydride complex by analogy. The series of hydride complexes HMn(CO)₅, H₂Fe(CO)₄, and HCo(CO)₄ (Table 12.6) show NMR signals in the high-field region typical of hydrides. However, these compounds become progressively stronger acids until HCo(CO)₄ behaves as a strong acid in water, dissociating into H⁺ and [Co(CO)₄]⁻!

Plainly, oxidation number is only a device for counting electrons according to arbitrary, if plausible and consistent, rules. (It is also possible to use a convention that considers ligands as uncharged radicals[17] and considers the metals as possessing the charge on the complex.)

▶ 12.4.2   *Infrared Spectra of Carbonyl Complexes*[18]

CO Stretching Frequencies and Electron Density

Back-donation in carbonyl complexes is monitored conveniently by IR spectroscopy. CO stretching vibrations usually are quite intense and occur in a region (2100–1500 cm⁻¹) fairly well isolated from other types of vibrations.

Table 12.6   Some properties of metal carbonyl hydrides[a]

	$HV(CO)_6$	$HMn(CO)_5$	$H_2Fe(CO)_4$	$HCo(CO)_4$	$HRe(CO)_5$
Color:	—	Colorless	Colorless solid, yellow liquid	Yellow	Colorless
mp(°C):	—	−24.6	−70	−26.2	12.5
δ(M—H):	—	−17.50[b]	−20.8[b]	−20.7 ± 2[b]	−15.66[c]
K_a:	Strong	8 × 10⁻⁸ (20°C)	4 × 10⁻⁵ (17.5°C) = K_1 4 × 10⁻¹⁴ (17.5°C) = K_2	1	Very weak

[a] From A. P. Ginsberg, *Transition Met. Chem.* **1965**, *1*, 112.

[b] Neat.

[c] In thf.

[17] J. E. Ellis, *J. Chem. Educ.* **1976**, *53*, 2.

[18] See P. S. Braterman, *Struct. Bonding* **1976**, *26*, 1; a more complete and rigorous treatment is given in P. S. Braterman, *Metal Carbonyl Spectra*, Academic Press, New York, 1975.

One use for IR spectra is in structural diagnosis. Terminal $C\equiv O$ ligands in neutral molecules have stretches in the region $2100-1850$ cm^{-1}, whereas μ_2-CO's absorb between 1860 and 1700 cm^{-1} and μ_3-CO's absorb below 1750 cm^{-1}. The energies involved reflect a progressively decreasing C—O bond order. For example, two different structures for Mn_2CO_{10} are consistent with the 18-e rule:

$$(OC)_5Mn—Mn(CO)_5 \quad \text{and} \quad (OC)_4Mn \diagup\diagdown Mn(CO)_4$$

(a)   (b)

The IR spectrum in the carbonyl stretching region shows only bands in the region for terminal CO, ruling out **b.** In contrast, $Cp_2Fe_2(CO)_4$ was shown to contain μ-CO's and to exist in several geometries in solution, depending on solvent.[19]

In compounds containing terminal carbonyls, the electron-donor abilities of the metal and other ligands are reflected in the CO stretching vibration frequencies ν_{CO}. The greater the electron density supplied by the metal and other ligands, the greater the back-donation into π^* orbitals of CO, and hence the lower will be the CO bond order and the stretching frequency, ν_{CO}. (See Figure 12.1.) Data in Table 12.7 demonstrate that decreasing the positive charge on the metal in a series of isoelectronic and isostructural hexacarbonyls lowers ν_{CO} and indicates increased back-bonding. In fact, ν_{CO} for $[Ag(CO)]^+$[20] is actually *higher* than that for free CO, indicating that the C—O bond is strengthened on coordina-

Table 12.7 ν_{CO} **for some metal carbonyl compounds**

Compound	ν_{CO} (cm^{-1})	Compound	ν_{CO} (cm^{-1})
$[Ti(CO)_6]^{2-}$	1748	$Ir(PPh_3)_2(CO)(Cl)_2H$	2046[b]
$[V(CO)_6]^-$	1858[a]	$Ir(PPh_3)_2(CO)Cl(I)_2$	2067[b]
$[Cr(CO)_6]$	1984[a]	$Ir(PPh_3)_2(CO)Cl(Br)_2$	2072[b]
$[Mn(CO)_6]^+$	2094[a]	$Ir(PPh_3)_2(CO)(Cl)_2I$	2074[b]
CO	2143[a]	$Ir(PPh_3)_2(CO)Cl_3$	2075[b]
$[Ag(CO)][B(OTeF_5)_4]$	2204[c]	$Ir(PPh_3)_2(CO)Cl(O_2)$	2015[b]
$H_3B\leftarrow CO$	2165	$Ir(PPh_3)_2Cl(CO)[CNHC\equiv CHCN]$	2029[b]
		$Ir(PPh_3)_2Cl(CO)(C_2F_4)$	2049[b]
		$Ir(PPh_3)_2Cl(CO)(tcne)$	2054[b]

[a] From P. S. Braterman, *Metal Carbonyl Spectra*, Academic Press, New York, 1975.

[b] From the work of Vaska and Shriver.

[c] P. K. Hurlburt, O. P. Anderson, and S. H. Strauss, *J. Am. Chem. Soc.* **1991**, *113*, 6277.

[19] A. R. Manning, *J. Chem. Soc. A* **1968**, 1319; P. A. McArdle and A. R. Manning, *J. Chem. Soc. A* **1969**, 1498; J. G. Bullitt, F. A. Cotton and T. J. Marks, *J. Am. Chem. Soc.* **1979**, *92*, 2155.

[20] P. K. Hurlburt, Q. P. Anderson and S. H. Strauss, *J. Am. Chem. Soc.* **1991**, *113*, 6277.

tion to the cation. This is in accord with the MO diagram in Figure 4.9 which shows that the HOMO which acts as the electron-pair donor to a metal is slightly antibonding between C and O. Thus, donation of electron density from this orbital should increase the C—O bond strength and raise ν_{CO}. Consistent with this, the C—O bond length in $[Ag(CO)]^+$ is only 107.7 pm compared with 112.8 pm in free CO. The carbonyl ligand is apparently acting only as a σ donor here because ν_{CO} is higher than that in $H_3B \leftarrow CO$ where no π-back-donation is possible.

Electron-donating abilities of other ligands are revealed in data on the two sets of Ir complexes.[21] In general, the more electronegative the other ligands, the less electron density available for back-donation and the higher the ν_{CO}. In the first group of $[Ir(PPh_3)_2(CO)Cl(X)(Y)]$ compounds (where X and Y are σ and π donors only), the most electronegative (poorest donor) X and Y ligands lead to the highest ν_{CO}'s. In the second group of $Ir(PPh_3)_2(CO)ClL$ compounds (whose structures resemble that shown in Figure 12.18), the ligands L have empty π^* orbitals which compete with those of CO for back-donated electrons. The best competitors (best π acids) lead to the highest values for ν_{CO}.

Another point worth noting is that ν_{CO} is lower for coordinated than for free CO (except in $[Ag(CO)]^+$), indicating that back-bonding populating the π^* orbitals actually does occur.

Normal Modes

Compounds with more than one carbonyl ligand usually have more than one IR-active CO stretch. Each stretching frequency corresponds to the symmetry of the complex; that is, each ν_{CO} corresponds to a linear combination of stretching and compression of CO bonds that transforms like some irreducible representation of the molecular point group. Such a linear combination is a **normal vibrational mode** (and also is important, of course, for M—C, etc. vibrations). The C—O stretching modes belong to the same symmetry species as the M—C bonds. $M(CO)_6$ has stretching modes A_{1g}, E_g, and T_{1u}. Only normal modes that have the symmetry properties of (transform like) Cartesian coordinates x, y, and z are active in the IR. Thus, the IR spectrum of $M(CO)_6$ displays a single T_{1u} carbonyl stretch.

Table 12.8 lists expected IR-active bands in the carbonyl region for several common geometries. Note that the varying number of IR-allowed bands permits us to distinguish among geometric isomers. More symmetric species have fewer IR-active bands. However, distortions from the idealized point group may lower the symmetry sufficiently to permit additional bands to appear (i.e., if L is less than cylindrically symmetrical around the M—L bond). A case in point is that of *cis*- and *trans*-$Mo(CO)_4[P(OPh)_3]_2$ whose IR spectra appear in Figure 12.19. The more symmetric *trans* compound has the simpler spectrum. However, the A_{1g} and B_{1u} modes (nonallowed for $\mathbf{D}_{4h}$) appear as weak bands, because of distortions from $\mathbf{D}_{4h}$ symmetry brought about by bulky $P(OPh)_3$ ligands[22] whose threefold axial symmetry is lower than the idealized fourfold symmetry of the complex.

---

[21] For an application to phosphine ligands, see Md. M. Rahman, H.-Y. Liu, K. Eriks, A. Prock, and W. P. Giering, *Organometallics*, **1989**, *8*, 1.

[22] Values of CO force constants can be extracted from normal stretching frequencies; see F. A. Cotton and C. S. Kraihanzel, *J. Am. Chem. Soc.* **1962**, *84*, 4432. These can be used to evaluate quantitatively the donor properties of other ligands present; see W. A. G. Graham, *Inorg. Chem.* **1968**, *7*, 315.

Table 12.8   Infrared-active modes for some carbonyl complexes

Complex			Point group	Symmetry of IR-active CO normal modes	Number ν_{CO} expected
1. $M(CO)_6$			$\mathbf{O}_h$	T_{1u}	1
2. $M(CO)_5L$			$\mathbf{C}_{4v}$	$A_1 + E\,(+A_1)$	2 or 3
3. $M(CO)_4L_2$	*trans:*		$\mathbf{D}_{4h}$	E_u	1
	cis:		$\mathbf{C}_{2v}$	$A_1 + B_1 + B_2 + (+A_1)$	3 or 4
4. $M(CO)_3L_3$	*mer:*		$\mathbf{C}_{2v}$	$2A_1 + B_2$	3
	fac:		$\mathbf{C}_{3v}$	$A_1 + E$	2
5. $M(CO)_5$			$\mathbf{D}_{3h}$	$A_2'' + E'$	2
6. $M(CO)_4L$	*ax:*		$\mathbf{C}_{3v}$	$2A_1 + E$	3
	eq:.		$\mathbf{C}_{2v}$	$2A_1 + B_1 + B_2$	4

(Continued)

Table 12.8   **Infrared-active modes for some carbonyl complexes** *(Continued)*

Complex		Point group	Symmetry of IR-active CO normal modes	Number ν_{CO} expected
7. $M(CO)_3L_2$		$\mathbf{D}_{3h}$	E'	1
		$\mathbf{C}_s$	$2A' + A''$	3
		$\mathbf{D}_{3h}$		
8. $M(CO)_2L_3$			A_2''	1
		$\mathbf{C}_{2v}$		
			$A_1 + B_1$	2
		$\mathbf{T}_d$		
9. $M(CO)_4$			T_2	1

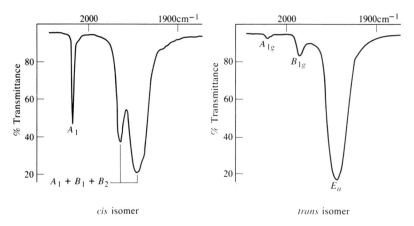

Figure 12.19   Infrared spectra of the geometric isomers of $Mo(CO)_4[P(OPh_3)]_2$ in the CO stretching region. From M. Y. Darensbourg and D. J. Darensbourg, *J. Chem. Educ.* **1970**, *47*, 33.)

12.5   STRUCTURAL CHARACTERIZATION OF ORGANOMETALLIC COMPOUNDS

The combination of IR spectroscopy in the carbonyl region and ^{13}C NMR provides a valuable means of determining the structures of organometallic compounds. (^1H NMR spectroscopy is also useful, but is not treated here because it is generally covered in the organic course.) As mentioned in Chapter 9, the value of the chemical shift of a resonating atom provides information on its chemical environment. Table 12.9 gives chemical shifts for several types of C found in organometallic species. ^{13}C has a natural isotopic abundance of only 1.1%, and the quality of NMR spectra can be enhanced by preparing isotopically enriched samples. Because of complications arising from relaxation times which are beyond the scope of this treatment, ^{13}C NMR signals are not in the same intensity ratios as the number of C's in each chemically equivalent set, and C's with no H's attached give very-low-intensity (or sometimes zero-intensity) signals.

EXAMPLE 12.8:   A complex with the formula $(CO)_5CrC_5H_7N$ was recently prepared. In the carbonyl region there are stretches at 2057, 1980, and 1908 cm^{-1}. The NMR spectra are as follows:

^1H NMR δ (relative intensity): 3.65 (3H), 3.89 (3H), 5.91 (1H)

^{13}C NMR δ (splitting): 47.97 (quartet), 49.71 (quartet), 84.16 (doublet), 116.1 (doublet), 216.98, 223.82, 248.85 (singlets)

What is the structure of the compound? (See A. Rahm and W. D. Wulff, *Organometallics*, **1993**, *12*, 597.)

Table 12.9 ^{13}C NMR chemical shifts for organometallic compounds$^{a, b}$

Ligand	^{13}C Chemical shift range
CO	177 to 275
M—CH$_3$	−28.9 to 53
M—C(=O)—X	240 to 300
M=C	192 to 360
M≡C	200 to 400
M—olefin	3 to 110
M—η^1-C$_3$H$_5$	C$_1$, −18 to 33; C$_2$, 20 to 155; C$_3$, 100 to 135
M—η^3-C$_3$H$_5$	C$_1$ and C$_3$, 40 to 78; C$_2$, 90 to 140
M—η^5-C$_5$H$_5$	75 to 122
M—C$_6$H$_5$	M—C, 131 to 193; *o*-C, 132 to 142; *m*-C, 127 to 131; *p*-C, 121 to 131

^{a} For diamagnetic complexes in ppm downfield from SiMe$_4$.

^{b} From B. E. Mann, *Adv. Organomet. Chem.* **1974**, *12*, 135; and P. W. Jolly and R. Mynott, *Adv. Organomet. Chem.* **1981**, *19*, 257.

Solution: The compound has five carbonyls, so the structure is probably derived from an octahedron with an organic group occupying the sixth position. The idealized symmetry is C_{4v}, and Table 12.8 indicates that the expected number of CO stretches is two or three. This is in accord with the observed IR spectrum. In the ^{1}H NMR, the H's fall into two sets of three and a single proton. The chemical shifts for the sets of three are in the range for methyl protons. These must be attached to the C's which give quartet signals. Thus, there are two inequivalent methyl groups. There are two types of CO's: equatorial and axial which can be assigned the ^{13}C signals at 216.98 and 223.82. The singlet at 248.85 could represent either M=C or M≡C. The two ^{13}C doublets remain unassigned; the doublet structure indicates that each is split by one proton. Because only one proton signal is left unassigned, we must have the same proton splitting both C's. This can happen in unsaturated molecules. Only one proton attached to a C is indicative of a —C≡C—H grouping with H splitting both C's. The following structure is consistent with the spectra, the 18-*e* rule, and the octet rule:

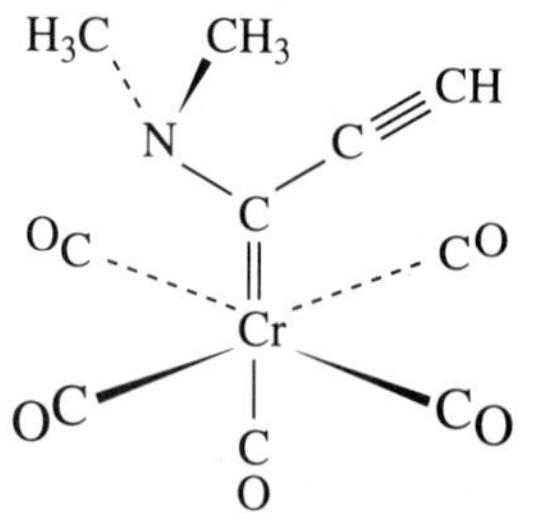

EXAMPLE 12.9:   Square-planar Ir(CO)(Ph$_2$PCH$_2$CH$_2$PPh$_2$)(2-mesityl)(mesityl = 1,3,5-trimethylphenyl) is constrained by the bite of the didentate ligand to be *cis*. When it is allowed to react with CO, the product is Ir(CO)$_3$(Ph$_2$PCH$_2$CH$_2$PPh$_2$)(2-mesityl). Given the following spectroscopic data, what is the structure of the product? (See B. P. Cleary and R. Eisenberg, *Organometallics* **1992,** *11,* 2335.)

IR: three strong bands at 1986, 1935, and 1601 cm^{-1}

^{1}H NMR δ **(relative intensity):** 2.17 (3H), 1.75 (6H) (signals for the phosphine ligand omitted)

^{31}P NMR δ **(splitting):** 25.2 (doublet), 20.3 (doublet)

Solution: The proton NMR data are not complete, but the two signals in the methyl region in a 1:2 ratio indicate the presence of the *p*- and two equivalent *o*-methyls of the mesityl ring. Because P has a nuclear spin of $\frac{1}{2}$ (see page 434), the two ^{31}P signals suggest two chemically inequivalent P atoms each split by the other. This requires that no plane of symmetry pass through the center of the didentate P ligand. The IR band at 1601 cm^{-1} indicates a C=O bond. A structure compatible with the above data and with the 18-*e* rule is a trigonal bipyramid with equatorial CO's.

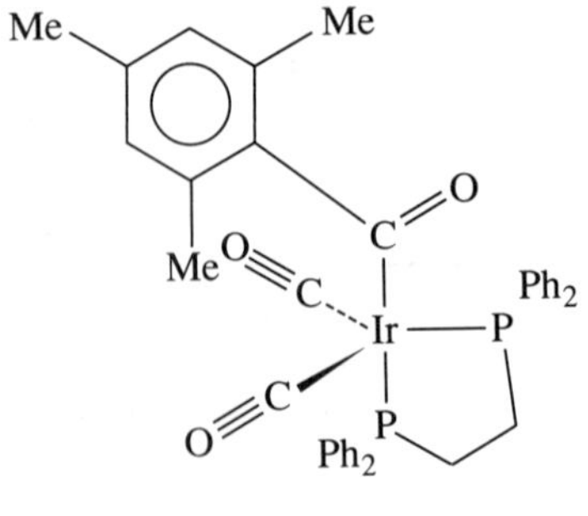

12.6   THE ISOLOBAL ANALOGY

A useful analogy between the bonding properties of organic and inorganic fragments has been pointed out by Hoffmann and co-workers[23] as a model for relating the structures of organometallic compounds to organic ones. If we consider the electronic structure of the 17-e fragment $Mn(CO)_5^{\cdot}$, a simple model views the carbonyls as donating electron pairs to five d^2sp^3 hybrids on Mn. Six of the Mn electrons are in "t_{2g}" orbitals, and the seventh is in the hybrid pointing away from $Mn(CO)_5$. This is analogous to the situation in 7-e $CH_3^{\cdot}$ where three sp^3 hybrids on C form C—H bonds and the fourth hybrid points away from the CH_3 group and contains the seventh electron. Figure 12.20 shows calculated contour diagrams for the half-filled orbitals in MnH_5^{5-} (isoelectronic with $Mn(CO)_5$) and CH_3. The shape and extent of these frontier orbitals resemble one another, and so we expect that their chemistry may also. For example, both fragments can dimerize to give C_2H_6, $Mn_2(CO)_{10}$ or the cross-product $CH_3Mn(CO)_5$. These two fragments are said to be **isolobal,** and this relationship is indicated by the notation $CH_3 \overset{\curvearrowright}{} Mn(CO)_5$. In general, we call two fragments isolobal *"if the number, symmetry properties, approximate energy, and shape of the frontier orbitals and the number of electrons in them are similar."* [23]

Hoffmann and co-workers have calculated MO energy and contour diagrams for many organometallic fragments. It turns out that an isolobal relation exists between organometallic fragments having $(18 - n)$ electrons and organic fragments having $(8 - n)$ electrons. Thus, 16-e $Fe(CO)_4$ is isolobal with 6-e CH_2. Table 12.10 lists isolobally equivalent fragments for first-row transition metal species. Extension to later rows is obvious; that is, $Mn(CO)_5$ is isolobal with $Re(CO)_5$, and so on. Thus the following compounds are all "cyclopropanes": C_3H_6; $(CO)_4Fe(\eta^2-C_2H_4)$ when viewed as a metallacyclopropane; $[(CO)_4Fe]_2(\mu_2-CH_2)$; and $Os_3(CO)_{12}$. Note that the frontier orbitals are not changed for hydrocarbon fragments by addition or removal of a proton. It may seem strange to find both $Ni(CO)_3$ and $Ni(CO)_2$ listed as isolobal with the same fragment CH_2. This results from the breakdown of the 18-e rule at the end of the transition series. $Ni(CO)_3$ has $(18 - 2)e$ while $Ni(CO)_2$ has $(16 - 2)e$.

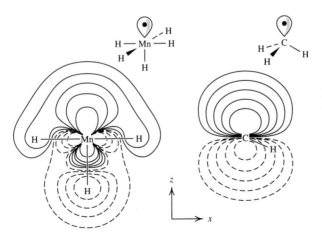

Figure 12.20   Contour diagram of the isolobal a_1 orbitals of MnH_5^{5-} (left) and $CH_3^{\cdot}$ (right). The contours of Ψ, plotted in a plane passing through Mn and three H's (left) and C and one H (right) are ± 0.2, ± 0.1, ± 0.055, ± 0.025, and ± 0.01. (Reprinted from R. Hoffmann, *Angew. Chem. Intl. Ed. Engl.* **1982**, *21*, 711.)

[23] R. Hoffmann, *Angew. Chem. Int. Ed. Engl.* **1982**, *21*, 711.

Table 12.10   Isolobal fragments[a]

Organic Species

CH_3	CH_2	CH	C
CH_2^-	CH^-	CH_2^+	CH^+
	CH_3^+		

Organometallic Fragment

$Co(CO)_4$	$Fe(CO)_4$	$Co(CO)_3$	$Fe(CO)_3$
$Mn(CO)_5$	$Cr(CO)_5$	$Mn(CO)_4$	$Cr(CO)_4$
$CpFe(CO)_2$	$CpCo(CO)$	$CpNi$	$CpCo$
$CpCr(CO)_3$	$CpMn(CO)_2$	$CpFe(CO)$	$CpMn(CO)$
	$Ni(CO)_2$	$CpCr(CO)_2$	
	$Ni(CO)_3$		

[a] Read vertically.

Isoelectronic metal fragments are isolobal with one another and with appropriate organic fragments. For example, in 14-*e* $Pt(PMe_3)_2$, we can replace Pt with Ni and PMe_3 with the two-electron donor CO to arrive at $Pt(PMe_3)_2$ ⌒ $Ni(CO)_2$ ⌒ CH_2. Similarly, CH_3 ⌒ $CpFe(CO)_2$; we consider that six-electron Cp^- is isolobal with three CO's occupying three coordination positions; $CpFe(CO)_2$ ⌒ $[Fe(CO)_5]^{.+}$ ⌒ $Mn(CO)_5$ ⌒ CH_3. Also, $CpFe(CO)_2^-$ ⌒ $Mn(CO)_5^-$ ⌒ CH_3^-, and so on.

EXAMPLE: With what organic compound is $Ir_4(CO)_{12}$ isolobal?

Solution: The structure of $Ir_4(CO)_{12}$ (Figure 12.2) shows that it consists of four $Ir(CO)_3$ fragments. Table 12.10 shows that 15-*e* $Ir(CO)_3$ is isolobal with CH. Hence $Ir_4(CO)_{12}$ ⌒ tetrahedrane C_4H_4.

EXAMPLE 12.10: What organic compound is isolobal with

$$Me_3P \diagdown \underset{Pt}{} \diagup PMe_3$$
$$HC \equiv\!\!\equiv W(CO)_2Cp \quad ?$$

Solution: According to Table 12.10, $CpW(CO)_2$ ⌒ CH, and $Pt(PMe_3)_2$ ⌒ CH_2. Thus, cyclopropene is the isolobal compound.

12.7 DIHYDROGEN COMPLEXES[24]

Although H_2 is not organic, the similarity of its bonding to metals to that in metal–olefin complexes makes discussion of H_2 complexes appropriate here. The first H_2 complexes, including $W(H_2)(CO)_3(Pi-Pr_3)_2$ (Figure 12.21), were prepared by Kubas in 1985. Figure

[24] See (a) G. J. Kubas, *Acc. Chem. Res.* **1988**, *21*, 120; (b) R. H. Crabtree and D. G. Hamilton, *Adv. Organomet. Chem.* **1988**, *28*, 299; (c) R. H. Crabtree and D. G. Hamilton, *Acc. Chem. Res.* **1990**, *23*, 97; (d) D. M. Heinekey and W. J. Oldham, *Chem. Rev.* **1993**, *93*, 913.

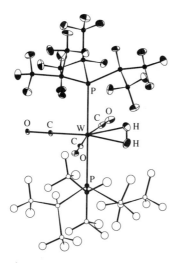

Figure 12.21   Molecular structure of W(CO)₃(P*i*-Pr₃)₂(H₂). (Reprinted with permission from G. J. Kubas, R. R. Ryan, B. I. Swanson, P. J. Vergamini, and H. J. Wasserman, *J. Am. Chem. Soc.* **1984,** *106,* 451.)

12.22 shows orbital interactions in dihydrogen–metal bonding featuring (a) electron donation from the filled H_2 σ orbital into a vacant metal orbital and (b) back-donation from a filled metal π orbital into the σ* on dihydrogen. Back-donation of an entire electron pair into σ* would rupture the H—H bond and form a dihydride complex. In terms of a VB model, the extremes are the dihydrogen (η^2-H_2) complex (**a**) and the dihydride complex (**b**).

$$M^{n+} \longleftarrow \overset{\text{H}}{\underset{\text{H}}{|}} \qquad \text{versus} \qquad M^{(n+2)+} \overset{\diagup \text{H}^-}{\underset{\diagdown \text{H}^-}{}}$$

(**a**) dihydrogen complex   (**b**) dihydride

The most straightforward experimental technique for distinguishing the two types of dihydrogen bonding is x-ray or neutron diffraction which locates the hydrogens. The structure shown in Figure 12.21 reveals an H_2 ligned with its axis parallel to the *trans* phosphines. This arrangement accords with the notion that back-donation to H_2 occurs from the filled W d_π orbital which also interacts with the phosphines (and which is therefore less stabilized than the filled π orbitals which interact with CO). Interestingly, W(H₂)(CO)₃(P*i*-Pr₃)₂ exists in solution as a mixture of tautomers consisting of 85% dihydrogen W⁰ complex and 15% Wᴵᴵ dihydride. NMR can distinguish between coordinated H_2 and hydride in

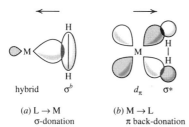

(*a*) L → M
σ-donation

(*b*) M → L
π back-donation

Figure 12.22   Bonding in metal dihydrogen complexes.

solution. Especially diagnostic is the magnitude of coupling in HD complexes; large values of the coupling constant indicate η^2 complexes. IR measurements are also employed. M—H stretches occur at around 2100–2200 cm^{-1}, and η^2-H_2 has an H—H stretch at 2400–2600 cm^{-1} (which, unfortunately, is often buried in C—H stretches). Because structure **b** has directional properties whereas structure **a** is cylindrically symmetric about the metal–H_2 bond, back-donation is expected to be associated with a larger barrier to rotation of dihydrogen, and this has been found to be so in several cases.[25] Several long-known "polyhydrides" which seemed to have unusually large oxidation numbers have recently been reformulated as dihydrogen complexes. For example, "$M^{IV}(PR_3)_3(H)_4$" (M = Fe, Ru) have been shown to be $M^{II}(H_2)(H)_2(PR_3)_3$.

Although some back-bonding is undoubtedly needed to stabilize dihydrogen species, strongly electron-donating metals form dihydride complexes. For example, $[Re(H)_2(H_2)(CO)(PMe_2Ph)_3]^+$ contains coordinated dihydrogen, but replacing the CO by phosphine (a poorer π acid) gives the tetrahydride $[Re(H)_4(PMe_2Ph)_4]^+$. Most well-established dihydrogen complexes involve d^6 metals. Neutral d^8 fragments such as $Fe(CO)_4$ tend to form dihydrides such as $H_2Fe(CO)_4$. $Fe(CO)(NO)_2(H_2)$ and $Co(CO)_2(NO)(H_2)$ both contain dihydrogen presumably due to the π acidity of NO ligands. Donation from the H_2 σ orbitals seems to be the more important bonding component, so dihydrogen is often said to form complexes using a σ-bond pair as contrasted with π complexes. An alternative view is of a three-center, two-electron bond such as in H_3^+. The net donation leaves dihydrogen positively polarized so that $[IrH(H_2)(PPh_3)_2(bq)]^+$ (bq = 7, 8-benzoquinolinate) which contains both hydride and dihydrogen can be deprotonated, the H^+ being lost from the dihydrogen. In contrast to CO and olefins where varying degrees of back-donation can be experimentally detected, only the extremes of η^2-H_2 and dihydride are accessible.

Dihydrogen complexes are often prepared by addition of H_2 to coordinatively unsaturated complexes. In fact, their discovery resulted from attempts to maximize the yield of $M(CO)_3(PR_3)_2$ complexes by conducting the synthesis under H_2 as an "inert" gas. Protonation of hydrides such as $Fe(Ph_2PCH_2CH_2PPh_2)_2(H)_2$ can also lead to dihydrogen complexes $[Fe(Ph_2PCH_2CH_2PPh_2)_2(H)(H_2)]^+$. H_2 is quite easily displaced by other ligands.

Agostic Interactions

Complexes which would otherwise be coordinatively unsaturated can be stabilized by the interaction of a C—H bonding pair of electrons with the metal in the same way as the H—H pair interacts in dihydrogen complexes. One example of this **agostic**[26] (from the Greek "to clasp") interaction is in the structure of $W(CO)_3(PCy_3)_2$ (Cy = cyclohexyl) (Figure 12.23) where one ligand distorts to permit close approach of the C—H bond to W and thus increases the electron count from 16 to 18. In solution, agostic interactions can be detected by unusual positions for ^1H signals and values of coupling constants $J(^{13}C–H)$. A vacant metal orbital is the necessary but not the sufficient condition for agostic interactions to occur. Many compounds are known whose structures and/or spectra give no evidence of agostic interactions even though there is an empty metal orbital.[27]

---

[25] J. Eckert, A. Albinati, R. P. White, C. Bianchini, and M. Perruzini, *Inorg. Chem.* **1992**, *31*, 4241.

[26] M. Brookhart and M. L. H. Green, *J. Organomet. Chem.* **1983**, *250*, 395.

[27] Si—H bonds can also be coordinated in η^2 fashion. See U. Schubert, *Adv. Organomet. Chem.* **1990**, *30*, 151.

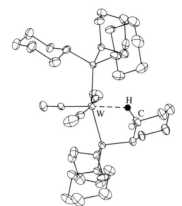

Figure 12.23   Molecular structure of $W(CO)_3(PCy_3)_2$ showing agostic interaction of a C—H bond with W. (Reprinted with permission from H. J. Wasserman, G. J. Kubas, and R. R. Ryan, *J. Am. Chem. Soc.* **1986**, *108*, 2294. Copyright 1986, American Chemical Society.)

GENERAL REFERENCES

J. P. Collman, L. S. Hegedus, J. R. Norton, and R. G. Finke, *Principles and Applications of Organotransition Metal Chemistry,* University Science Books, Mill Valley, CA, 1987. Authoritative and extensive treatment of the field.

Ch. Elschenbroich and A. Salzer, *Organometallics: A Concise Introduction,* 2nd ed., VCH Publishers, New York, 1992. A good introduction presented in outline format.

P. Powell, *Principles of Organometallic Chemistry,* 2nd ed., Chapman and Hall, London, 1988. A very readable update of the classic text by Coates, Green, and Wade.

G. B. Richter-Addo and P. L. Legzdins, *Metal Nitrosyls,* Oxford University Press, New York, 1992. A very readable and complete treatment of the subject.

G. Wilkinson, F. G. A. Stone and E. W. Abel, Eds., *Comprehensive Organometallic Chemistry,* Pergamon Press, New York, 1982. A multivolume treatise covering all aspects of organometallic chemistry.

PROBLEMS

12.1   Give the valence electron counts for the following species. Which ones obey the 18-*e* rule?

(a) $W(CO)_6$
(b) $Cr(CNMe)_6$
(c) $Co_2(\mu\text{-}CO)_2(CO)_6$
(d) $[Fe(CN)_6]^{4-}$
(e) $Mn(CO)_5CH_2C_6H_5$
(f) $Fe(CO)_4Br_2$
(g) $[Mn(CO)_5]^-$
(h) $HRh(CO)_5$
(i) $Ru_3(CO)_{12}$
(j) $[Co(CN)_5]^{3-}$

12.2   Name each of the species in Problem 12.1.

12.3   The shapes of the binary carbonyls can be rationalized using a VB model in which each CO donates an electron pair to the metal in a dative σ bond and the 18-*e* rule applies. Metal lone pairs are stereochemically inactive, and the geometry is fixed by the hybridization of sigma bonds alone. Use this model to predict the structure of (a) $Ni(CO)_4$, (b) $Fe(CO)_5$, and (c) the simplest Co carbonyl.

12.4   (a) Explain why the 18-*e* rule applies to transition metal complexes and particularly to those in the middle of the series.

(b) Why does the 16-*e* configuration become more stable at the end of the transition series?

12.5 Give the valence electron count for the following species. Which ones obey the 18-*e* rule?

(a) $Ni(PPh_3)_4$ (e) $Mo(CO)_5(Pn\text{-}Bu_3)$ (i) $[W(CO)_4]^{4-}$

(b) $H_2Fe(CO)_4$ (f) $[Ta(CO)_6]^-$ (j) $V(CO)_5NO$

(c) $[Cr(CO)_5]^{2-}$ (g) $[FeRu(CO)_8]^{2-}$ (k) $[Ir(PPh_3)_2(CO)Cl(NO)]^+$

(d) $[Co(CNMe)_5]^{2+}$ (h) $[Fe(CN)_5NO]^{3-}$ (l) $Hf(CO)_2(Me_2PCH_2CH_2PMe_2)_2I_2$

12.6 Name each species in Problem 12.5.

12.7 Transition metal compounds differ from those of the *s* and *p* block elements in that non-bonding *d* electrons do not participate in the σ-bond hybridization in the VB bonding model. Compare changes in the σ-bond hybridization and changes in geometry of the molecules in the following sets of reactions:

(a) $H^+ + PH_3 \rightarrow PH_4^+$

    $H^+ + [Co(CO)_4]^- \rightarrow HCo(CO)_4$

(b) $SO_3 + 2e \rightarrow SO_3^{2-}$

    $[Ni(CN)_4]^{2-} + 2e \rightarrow [Ni(CN)_4]^{4-}$

12.8 Figure 12.24 shows a plot of ν_{CO} versus E^0 for CpFeL(CO)Me, where L is a phosphine. E^0 is the reduction potential; the more positive the E^0 values, the more favorable the reaction $CpFeL(CO)Me + e \rightarrow [CpFeL(CO)Me]^-$.

(a) What is the significance of the linear relationship displayed for the ligands represented by filled circles?

(b) What is the significance of the position of the unfilled-circle ligands above the line?

12.9 Explain in your own words how the MO and VB descriptions of $C_3H_5^-$ are equivalent.

12.10 Give the valence electron count for the following species. Which ones obey the 18-*e* rule?

(a) $Cp_2Ru_2(CO)_4$ (e) $[(\eta^4\text{-cod})Ru(\mu\text{-Cl})]_2$ (h) $Rh(C_2H_4)(PPh_3)_2Cl$

(b) $(\eta^3\text{-CH}_2CMeCH_2)_2Ni$ (f) $Ru(Me_2PCH_2CH_2PMe_2)_2$ (i) $(\eta^6\text{-C}_6H_6)_2Mo$

(c) $(\eta^4\text{-cot})Fe(CO)_3$ (g) $Cp_2^*ZrCl_2$ (j) $MeC(O)Re(CO)_4(CNPh)$

(d) $CpCr(NO)_2Me$

12.11 Name each species in Problem 12.10.

12.12 Give the formula of the most stable compound of the type $M(olefin)(CO)_x$ expected for each of the following metals with each olefin listed.

Metals	Olefins
Cr, Mn, Fe	$C_3H_5^-$, Cp^-, C_6H_6, $C_7H_7^+$

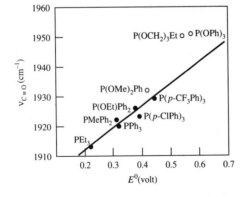

Figure 12.24 ν_{CO} versus E^0 for CpFeL(CO)Me, where L = phosphine or phosphite. (Data from M. M. Rahman, *et al.*, *Organometallics*, **1989**, *8*, 1.)

12.13 Given the following pairs of π-donor rings, choose the proper d^6 ion from the following list to form a neutral mixed sandwich compound: Cr^0, Mn^I, Fe^{II}, Co^{III}.

12.14 The complex ion $[(\eta^3\text{-}C_3H_5)Ir(PPh_3)_2(NO)]^+$ can be crystallized with either linear or bent NO depending on the counterion. Give the electron count for each isomer.

12.15 Using Figure 12.12, write electron configurations that account for the magnetic properties (given in Table 12.5) of Cp_2V, Cp_2V^-, Cp_2Cr, Cp_2Mn, Cp_2Fe, Cp_2Fe^+, Cp_2Co, Cp_2Ni, and Cp_2Ni^+.

12.16 Give the valence electron count for the following species. Which ones obey the 18-e rule?
(a) $[CpCo(NO)]_2$
(b) $Cp*Re(\eta^2\text{-}EtC\equiv CEt)Cl_2$
(c) $Cp_2Co_2(\mu\text{-}NO)_2$
(d) $[CpTi(CO)_4]^-$
(e) $Cr(CH_2CMe_3)_2[i\text{-}Pr_2PCH_2CH_2Pi\text{-}Pr_2]_2$
(f) $Cp*(CO)Ir(\mu\text{-}CO)_2Re(CO)Cp$

12.17 Name each of the species in Problem 12.16.

12.18 Give the valence electron count for the following species. Which ones obey the 18-e rule?
(a) $(CO)_5CrC(OEt)(NMe_2)$
(b) $CpMn(CO)_2CPh_2$
(c) $cis\text{-}Cl_2Pt[C(Oi\text{-}Pr)(Me)]_2$
(d) $[CpFe(CO)(PPh_3)(CF_2)]\,[BF_4]$
(e) $TaCl_3(PEt_3)_2(CHCMe_3)$
(f) $(Me_3CO)_3W(CCMe_3)$
(g) $CpMo[P(OMe)_3]_2(CCH_2CMe_3)$
(h) $Br(CO)_4Cr(CPh)$
(i) $CpIr(PMe_3)(CH_2)$
(j) $Cp*Ta(PMe_3)_2ClCPh$

12.19 Name each species in Problem 12.18.

12.20 The first carbene complex was prepared in 1915 by Tschugaev, although it was not recognized as such until 1970. The reaction involved $[(MeNC)_4Pt]^{2+}$ and hydrazine N_2H_4 followed by deprotonation of the product by excess hydrazine. Draw the (several) canonical forms for the product. *Hint:* see Equation (12.32).

12.21 Rationalize the trends in the following sets of IR-active CO stretching frequencies (in cm^{-1}):
(a) $(\eta^6\text{-}C_6H_6)Cr(CO)_3$ 1980, 1908
     $CpMn(CO)_3$ 2027, 1942
(b) $CpV(CO)_4$ 2030, 1930
     $CpMn(CO)_3$ 2027, 1942
     $[CpFe(CO)_3]^+$ 2120, 2070
(c) $W(CO)_5(Pn\text{-}Bu_3)$ 2068, 1936, 1943
     $W(CO)_5(PPh_3)$ 2075, 1944, 1944
     $W(CO)_5[P(OPh)_3]$ 2079, 1947, 1957
(d) $Ni(CO)_4$ 2040
     $[Co(CO)_4]^-$ 1890
     $[Fe(CO)_4]^{2-}$ 1730
     $[Mn(CO)_4]^{3-}$ 1670

12.22 (a) Draw at least two possible structures of $Os_3(CO)_9(PPh_3)_3$.
(b) The IR spectrum of this compound in CH_2Cl_2 has CO stretches at 1962 and 1917 cm^{-1}. How does this knowledge help to narrow the possible structures?

12.23 The IR spectrum of $Rh_2I_2(CO)_2(PPh_3)_2$ has CO stretches at 2061 and 2005 cm^{-1}. Suggest a structure consistent with this.

12.24 The IR spectrum of $(CO)_5CrC(OMe)Ph$ shows a band belonging to the CO stretch for the CO *trans* to the carbene at 1953 cm^{-1}. For comparison, the totally symmetric A_{1g} mode for $Cr(CO)_6$ is seen in the Raman at 2108 cm^{-1}. Compare the π-acid strengths of CO and the carbene ligand.

12.25 Treating $W(CO)_6$ with $LiC_6H_4$4-CMe_3 gives an anion of composition $[(CO)_6WC_{10}H_{13}]^-$ which can be crystallized as the NMe_4^+ salt. What is the structure of anion from its spectroscopic data? (See K. A. Belsky, M. F. Asaro, S. Y. Chan, and A. Mayr, *Organometallics* **1992**, *11*, 1926.)

IR: 2042 cm^{-1} (weak), 1901 cm^{-1} (strong), ~1600 cm^{-1}

^{1}H NMR δ (relative intensity): 7.54 (2H), 7.35 (2H), 1.31 (9H)

^{13}C NMR δ: 279.3, 209.0, 204.7, 155.5, 152.2, 126.4, 125.2, 31.7

12.26 In the ^{13}C NMR spectrum of $CH_3Mn(CO)_5$, the CO's *cis* to Me absorb at 213.8 ppm and the *trans* CO absorbs at 211.3 ppm. A sample of $CH_3\overset{\displaystyle O}{\overset{\displaystyle \|}{C}}Mn(CO)_5$ was prepared by reaction of $CH_3^{13}C(O)Cl$ with $[Mn(CO)_5]^-$ [Equation (12.24)]. When this labeled sample was heated, $CH_3Mn(CO)_5$ was produced. The ^{13}C NMR spectrum of the product showed dramatic signal enhancement at only the 213.8-ppm position. What conclusions can you draw about the mechanism of the CO loss? (See T. C. Flood, J. E. Jensen, and J. A. Statler, *J. Am. Chem. Soc.* **1981**, *103*, 4410.)

12.27 With what organic compound is each of the following isolobal?

(a)

$$
\begin{array}{ccc}
\text{CO} & & \text{CO} \\
| & & | \\
\text{CpRh} & \!\!-\!\! & \text{RhCp} \\
\diagdown & & \diagup \\
& \text{C} & \\
& \text{H}_2 &
\end{array}
$$

(b) $[CpFe(CO)_2]_2$

(c)

$$
\begin{array}{ccc}
(CO)_3Co & \!\!-\!\!\!-\!\!\!-\!\! & Co(CO)_3 \\
\diagdown & \text{CH} & \diagup \\
& \diagdown | \diagup & \\
& Co(CO)_3 &
\end{array}
$$

(d)

$$
\begin{array}{ccc}
\text{H}_2\text{C} & \!\!-\!\!\!-\!\! & \text{CH}_2 \\
| & & | \\
(CO)_4Os & \!\!-\!\!\!-\!\! & Os(CO)_4
\end{array}
$$

(e)

$$
\begin{array}{c}
\text{Me}_2 \\
\text{As} \\
\diagup \quad \diagdown \\
(CO)_4Fe \!\!-\!\!\!-\!\! Co(CO)_3
\end{array}
$$

12.28 Dihydrogen complexes of Mo and W can be prepared via the following route: $M(CO)_3(PCy_3)_2 + H_2 \rightarrow M(CO)_3(PCy_3)_2(\eta^2\text{-}H_2)$, where M = Mo or W and where Cy = cyclohexyl. Why is the above route better for preparing dihydrogen complexes than, for example, thermal CO replacement as in $W(CO)_6 + H_2 \rightarrow (CO)_5W(\eta^2\text{-}H_2) + CO$?

12.29 $[Rh(PP_3)(H_2)]^+$ where $PP_3 = (Ph_2PCH_2CH_2)_3P$ reacts with D_2, giving $[(PP_3)Rh(D_2)]^+$ but no $[(PPh_3)Rh(HD)]^+$. On the other hand, $W(CO)_3(Pi\text{-}Pr_3)_2(H_2) + D_2 \rightarrow W(CO)_3(Pi\text{-}Pr_3)_2(HD)$ even in the solid state. What significance do these results have for the formulation of the Rh and W complexes as dihydrides or dihydrogen species? (See G. J. Kubas, C. J. Unkefar, B. I. Swanson, and E. Fukushima, *J. Am. Chem. Soc.* **1986**, *108*, 7000; C. Bianchini, C. Mealli, M. Peruzzini, and F. Zanobini, *J. Am. Chem. Soc.* **1987**, *109*, 5548).

12.30 The x-ray structure of $W(CO)_3(PCy_3)_2$ shows an agostic interaction between W and cyclohexyl C—H. Give electron counts for $W(CO)_3(PCy_3)_2$ both with and without the agostic interaction.

12.31 Predict the products of the following substitution reactions:
(a) $W(CO)_6 + PPh_3 \rightarrow$
(b) $Fe(CO)_5 + CF_2{=}CF_2 \xrightarrow{h\nu}$
(c) $CpFe(CO)_2Cl + PPh_3 \rightarrow$
(d) $Cr(CO)_6 + xs\ NO \rightarrow$
(e) $Mn(CO)_5Br + CN\text{-}t\text{-}Bu \rightarrow$

12.32 Predict the products of the following oxidations:
(a) $Re_2(CO)_{10} + Br_2 \rightarrow$
(b) $[CpFe(CO)_2]_2 + I_2 \rightarrow$
(c) $Pt(PPh_3)_4 + Cl_2 \rightarrow$

12.33 Predict the products of the following reductions:
(a) $[CpFe(CO)_2]_2 + Na/Hg \rightarrow$
(b) $Re_2(CO)_{10} + Li[HBEt_3] \rightarrow$
(c) $[CpMo(CO)_3]_2 + KH \rightarrow$

12.34 Identify the following reactions by type and predict the products; for example, $Mo(CO)_6 + PPh_3$ is CO substitution by a nucleophile, giving $Mo(CO)_5(PPh_3) + CO$.
(a) $Re_2(CO)_{10} + Na/Hg \rightarrow$
(b) $W(CO)_6 + (n\text{-}Bu_4N)\ I \rightarrow$
(c) $CpCo(CO)_2 + PPh_3 \rightarrow$
(d) $Cp_2Co_2(NO)_2 + Na/Hg \rightarrow$
(e) $Fe(CO)_5 + EtCO_2CH{=}CHCO_2Et \xrightarrow{h\nu}$
(f) $Mo(CO)_3(PCy_3)_2(\eta^2\text{-}H_2) + CO \rightarrow$
(g) $[CpMo(CO)_3]^- + C_3H_5Br \rightarrow \mathbf{1} \xrightarrow{h\nu} \mathbf{2}$
(h) $[Re(CO)_5]^- + MeI \rightarrow$

▸ 13 ◂

Survey of Organometallic Compounds

· · · · · · · · · · · · · · · · · · · ·

A vast number of complexes are known which contain bonds between metals and organic ligands. These can involve orbitals of π symmetry within the organic ligand forming σ bonds to the metal and/or σ orbitals of the organic forming bonds to metals. Chapter 12 treated the general model of bonding and electron counting for these ligand types as well as a few important types of compounds. This chapter contains a systematic survey of bonding, structures, methods of syntheses, and some typical reactions for complexes of two- through eight-carbon π-donor ligands. Because of the truly enormous number of π complexes known, only some of the main features can be emphasized. Following this, the chemistry of complexes with M—C σ bonds is discussed. This chapter is included for courses in which a particularly complete treatment of organometallic chemistry is desired; Chapter 14 on organometallic reactions does not depend on the material here.

13.1 SOME CHEMISTRY OF OLEFIN COMPLEXES

Synthetic methods for olefin complexes include thermal or photochemical displacement of CO or of more weakly coordinated ligands:

$$Cr(CO)_5[P(OMe)_3] + 1, 5\text{-cod} \xrightarrow{h\nu} \text{Cr(CO)}_3[P(OMe)_3] + 2CO \qquad (13.1)$$

(cod = cyclooctadiene)

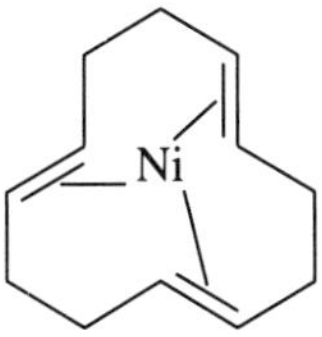

$$\text{(structure)} + 3\,C_2H_4 \longrightarrow Ni(\eta^2\text{-}C_2H_4)_3 + C_{12}H_{18} \tag{13.2}$$

Halide removal can be assisted by $AlCl_3$ or Ag^+ salts:

$$Re(CO)_5Cl + C_2H_4 + AlCl_3 \longrightarrow [Re(CO)_5(\eta^2\text{-}C_2H_4)]^+[AlCl_4]^- \tag{13.3}$$

As with carbonyls, reduction of a metal salt in the presence of the ligand is possible. In the following reaction, ethylene is the reductant:

$$RhCl_3 + xs\ C_2H_4 \longrightarrow \text{(structure)} \tag{13.4}$$

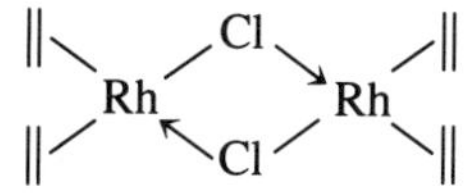

A recently introduced technique involves co-condensation of metal vapor and ligand at very low temperature on the cooled walls of a reactor. Reaction occurs as the matrix warms; re-formation of solid metal competes. Isolated metal atoms are extremely reactive because the molar enthalpy is increased by ΔH_{vap}, so many otherwise inaccessible compounds (not just olefin complexes) can be prepared by this route:

$$Pd(g) + 3\,CF_3CF{=}CFCF_3(g) \xrightarrow{-78°C} Pd(\eta^2\text{-}CF_3CF{=}CFCF_3)_3(s) \tag{13.5}$$

$$\xrightarrow{-30°C}$$

1. $+2py$, $-78°C$
2. warm to $25°C$

$$Pd(s) + 3\,CF_3CF{=}CFCF_3 \qquad (py)_2Pd(\eta^2\text{-}CF_3C{=}CCF_3)$$
$$+ 2\,CF_3CF{=}CFCF_3$$

Coordinated olefins (like coordinated CO) are activated toward attack by nucleophiles, because electron density is withdrawn by donation to the metal—especially if the complex is positively charged and/or the metal has a high positive oxidation state.

$$Cl_2(PPh_3)Pt{-}\overset{CH_2}{\underset{CH_2}{\|}} + Et_2NH \longrightarrow Cl_2(PPh_3)Pt^{(-)}CH_2CH_2N^{(+)}HEt_2 \tag{13.6}$$

This contrasts with the behavior of free olefins, which, on account of the electron-rich double bond, ordinarily are attacked by electrophiles.

13.2   ALKYNE COMPLEXES

Alkynes, $RC{\equiv}CR$, have *two* mutually perpendicular sets of π and π^* orbitals. In mononuclear complexes,[1] alkynes often function like olefins. An example is *trans-*

[1] S. Otsuka and A. Nakamura, *Adv. Organomet. Chem.* **1976**, *14*, 245.

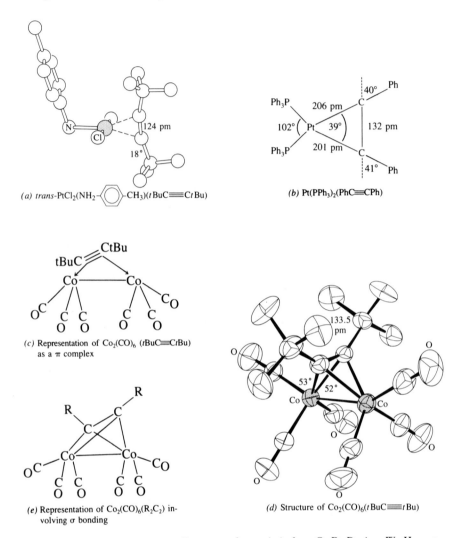

*(a) trans-*PtCl₂(NH₂—⟨benzene ring⟩—CH₃)(*t*BuC≡C*t*Bu)

(b) Pt(PPh₃)₂(PhC≡CPh)

(c) Representation of Co₂(CO)₆ (*t*BuC≡C*t*Bu) as a π complex

(e) Representation of Co₂(CO)₆(R₂C₂) involving σ bonding

(d) Structure of Co₂(CO)₆(*t*BuC≡*t*Bu)

Figure 13.1   Structures of some alkyne complexes. (*a* is from G. R. Davies, W. Hewertson, R. H. B. Mais, P. G. Owston, and C. G. Patel, *J. Chem. Soc. A*, **1970**, 1873; *b* is from J. O. Glanville, J. M. Stewart, and S. O. Grim, *J. Organomet. Chem*, **1967**, *7*, P9; *d* is reprinted with permission from F. A. Cotton, J. D. Jamerson, and B. R. Stults, *J. Am. Chem. Soc.* **1976**, *98*, 1774. Copyright 1976, American Chemical Society.)

[Pt^{II}Cl₂(*p*-toluidine)(*t*-BuC≡C*t*-Bu)] (Figure 13.1*a*). The alkyne is perpendicular to the molecular plane (like ethylene in Zeise's salt), and the *t*-Bu groups are bent back by ~20°. The C—C bond is lengthened from 120 pm for a typical C≡C bond to 124 ± 2 pm (a barely significant change), and $\nu_{C≡C}$ is lowered from that of free *t*-BuC≡C*t*-Bu. Although caused by back-bonding, these effects are sufficiently small that we can better represent the bonding as a π-bonded alkyne (**a**), rather than a metallacyclopropene (**b**).

(a) Alkyne complex   **(b)** Metallacyclopropene

However, in $Pt^0(PPh_3)_2(PhC\equiv CPh)$ (Figure 13.1*b*) the C—C bond is lengthened to 132 pm and the phenyl groups are bent back by ~40°. The $C\equiv C$ stretching frequency is lowered to 1750 cm^{-1} from ~2200 cm^{-1} in free $PhC\equiv CPh$. The increased back-bonding by Pt^0 over Pt^{II} indicates that the structure is better represented by **b**.

Alkynes can also behave as four-electron donors.[2] Figure 13.2 (in which the π system analogous to that of olefins is labeled $\pi_{\parallel}$ and the other is labeled $\pi_{\perp}$) shows orbital overlap possibilities. Calculations indicate that back-donation into $\pi_{\perp}^*$ is negligible. However, σ-donation from $\pi_{\parallel}$ and π-donation from $\pi_{\perp}$ lead to classification as a four-electron donor. The four-electron-donor role is expected to be important for metals having relatively few d electrons. This is generally true, with a large number of Mo^{II} and W^{II} (d^4) complexes being known. (However, d^8 $[MeSi(CH_2PMe_2)_3]Fe(PhC\equiv CPh)$ also is known.) Besides the sometimes ambiguous criterion of electron count, 4-e-donor alkynes can be recognized by the low-field ^{13}C nuclear magnetic resonance (NMR) shift of the acetylenic carbons as well as by x-ray structural data. For example, the average Co—C distance in $[Co(PhC\equiv CPh)(PMe_3)_3(MeCN)]^+$ is 198 pm whereas it is 185 pm in $[Co(PhC\equiv CPh)-(PMe_3)_3]^+$, where donation of two more alkyne electrons compensates for MeCN loss. An extensive series of complexes $MX_2L_2(CO)(RC\equiv CR)$ (M = Mo, W) display structures like that of Figure 13.3 with alkyne parallel to the M—CO axis. This orientation is un-

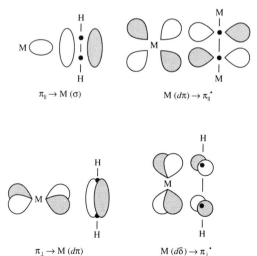

$\pi_{\parallel} \rightarrow M\ (\sigma)$ $M\ (d\pi) \rightarrow \pi_{\parallel}^*$

$\pi_{\perp} \rightarrow M\ (d\pi)$ $M\ (d\delta) \rightarrow \pi_{\perp}^*$

Figure 13.2 Metal–alkyne overlap interactions. (After J. L. Templeton, *Adv. Organomet, Chem.* **1989,** *29,* 1.)

[2] J. L. Templeton, *Adv. Organomet. Chem.* **1989,** *29,* 1.

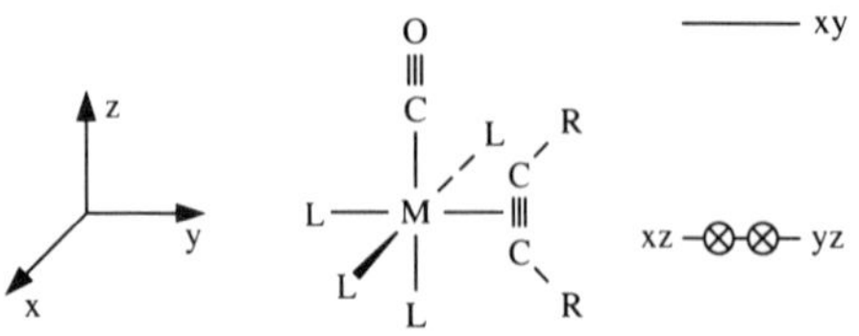

Figure 13.3 Geometry and *d*-orbital splitting diagram for d^4 $L_4M(CO)(RC\equiv CR)$ complexes. (After J. L. Templeton, *Adv. Organomet. Chem.* **1989**, *29*, 1.)

derstandable on the bonding model of Figure 13.2. Ignoring σ-bonding and the differences between X and L in π-bonding, we see that the π^* orbitals of CO interact with d_{xz} and d_{yz} to stabilize them. These stable metal-localized orbitals contain the four *d* electrons leaving the empty d_{xy} oriented for electron donation from the alkyne $\pi_\perp$.

Alkynes form a number of dinuclear complexes,[3] in which each perpendicular π system may be considered formally as bonding to a different metal. $(t\text{-BuC}\equiv Ct\text{-Bu})Co_2(CO)_6$ (Figure 13.1*d*) is fairly typical. The C—C bond length and bond angles indicate rehybridization to give some σ character to the M—C bonds (see Figure 13.1*c* and 13.1*e*).

Alkyne complexes usually are prepared by addition or by displacement of other ligands, as in the following reactions.

$$IrCl(CO)(PPh_3)_2 + CF_3C\equiv CCF_3 \longrightarrow IrCl(CO)(PPh_3)_2(\eta^2\text{-}CF_3C\equiv CCF_3) \qquad (13.7)$$

$$Co_2(CO)_8 + PhC\equiv CPh \longrightarrow Co_2(CO)_6(Ph_2C_2) + 2\,CO \qquad (13.8)$$

$$(\text{olefin})Pt(PPh_3)_2 + PhC\equiv CPh \longrightarrow (\eta^2\text{-}PhC\equiv CPh)Pt(PPh_3)_2 + \text{olefin} \qquad (13.9)$$

Acetylenes with at least one substituted C are most likely to form isolable complexes. Especially with late transition metals, acetylenes tend to dimerize to form metallacyclopentadienes, to trimerize to form substituted benzenes,[4] or to incorporate CO to give cyclic dienones:

$$Fe(CO)_5 + xs\ RC\equiv CR \longrightarrow \qquad (R = H, Ph, CF_3) \qquad (13.10)$$

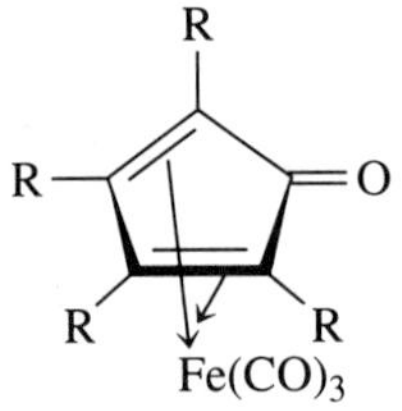

$$Fe(CO)_5 + 2\,CH_3C\equiv CCH_3 \xrightarrow{h\nu} \qquad (13.11)$$

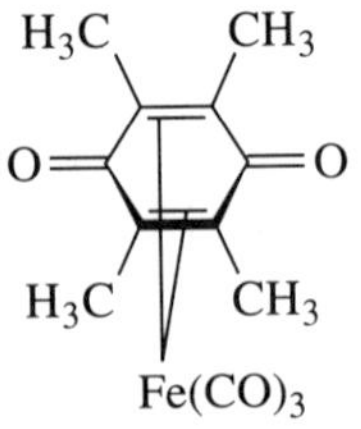

[3] E. L. Muetterties *et al.*, *J. Am. Chem. Soc.* **1978**, *100*, 2090.
[4] K. P. C. Vollhardt, *Acc. Chem. Res.* **1977**, *10*, 1.

$$CpZr(Me_2PCH_2CH_2PMe_2)_2Cl + 2MeC\equiv CMe \longrightarrow CpCl(Me_2PCH_2CH_2PMe_2)_2Zr \qquad (13.12)$$

$$Ph_3Cr(thf)_3 + xs\ CH_3C\equiv CCH_3 \longrightarrow \qquad + 3\ thf \qquad (13.13)$$

The coupling reactions obviously are preparative routes to alkene and arene complexes, many of which also can be made directly from the free ligands.

With terminal acetylenes, H loss or transfer is common [although Reaction (13.12) also occurs with MeC≡CH].

$$CpZr(Me_2PCH_2CH_2PMe_2)_2Cl + 2\,t\text{-}BuC\equiv CH \longrightarrow$$

$$Me_2PCH_2CH_2PMe_2 + CpZr\underset{\diagdown}{\overset{\diagup}{}}(Me_2PCH_2CH_2PMe_2)Cl \qquad (13.14)$$

with $CH=CHt-Bu$ and $C\equiv Ct-Bu$ groups on Zr.

Like coordinated olefins, coordinated alkynes are subject to attack by nucleophiles, especially in cationic complexes:

$$\left[CpFe(CO)[P(OPh)_3]-\underset{Ph}{\overset{Me}{\underset{\|}{\underset{C}{\overset{C}{\|}}}}} \right]^{+} + MeLi \longrightarrow Cp(CO)[P(OPh)_3]Fe-\underset{Me}{\overset{Me}{C}}=C\underset{Me}{\overset{Ph}{<}} + Li^{+} \qquad (13.15)$$

13.3 ALLYL COMPLEXES[5]

The allyl group $^{(-)}CH_2-CH=CH_2$ can bond to metals in a $\sigma(\eta^1\text{-})$ fashion or as a 4-e π donor with η^3 coordination. The valence bond (VB) model indicates a delocalized π system:

(a)   (b)   (c)

[5] For a review of allyl complexes and their role in catalysis, see H. Kurosawa, *J. Organomet. Chem.* **1987**, *334*, 243.

Figure 13.4 shows the allyl$^-$ π molecular orbitals (MOs). Ψ_1 and Ψ_2 are filled with four electrons, whereas Ψ_3 is empty and available for back-donation of metal electrons. Although allyl complexes exist for many transition metals, the most important are those of Ni, Pd, and Pt, which have an extensive organic chemistry and are useful catalysts.

A typical structure is that of $(\eta^3\text{-}C_3H_5PdCl)_2$ (Figure 13.5). The plane of the allyl group is tilted by 111.5° with respect to the Pd_2Cl_2 plane. Moreover, the Pd_2Cl_2 plane is closer to the end carbons than to the central one. As expected from the MO description, both C—C distances are the same (136 pm). A structure with the same electron count is **d,** in which an M—C single bond and coordinated C=C are present. It is sometimes convenient to think of allyl complexes in this way, and there are even some known complexes which display the different C—C bond lengths implicit in this structure. For example, $CpMo(C_3H_5)(NO)$ has one C—C bond of 142 and one of 137 pm.

$$\underset{\textbf{(d)}}{\text{M}}$$

Four preparative methods are commonly employed to synthesize allyl complexes: **(1)** metathesis (partner-exchange) with metal halides using allyl$^-$, **(2)** reaction of allyl halides with carbonylate anions, **(3)** proton abstraction from olefin complexes, and **(4)** nucleophilic addition to diene complexes.

Bis(allyl) complexes are prepared conveniently via route **1**:

$$2\,C_3H_5MgBr + NiCl_2 \xrightarrow[-10°]{\substack{\text{diethyl} \\ \text{ether}}} \text{Ni} + 2\,MgBrCl \tag{13.16}$$

Allyl halides with carbonylate anions usually give the η^1-allyl, which can be converted to an η^3-allyl by CO loss (route **2**):

$$[Mn(CO)_5]^- + CH_2\!=\!CHCH_2Cl \longrightarrow (CO)_5MnCH_2CH\!=\!CH_2 \xrightarrow{\Delta} {-Mn(CO)_4} + CO \tag{13.17}$$

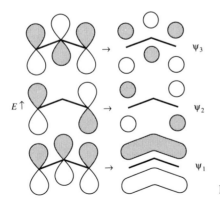

Figure 13.4   Molecular orbitals for $C_3H_5^-$.

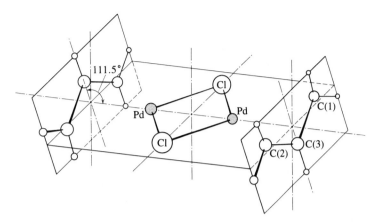

Figure 13.5 Structure of $[(\eta^5\text{-}C_3H_5)PdCl]_2$. (From A. F. Smith, *Acta Cryst.* **1965**, *18*, 331.)

In Reaction (13.18), elimination of an allylic proton from propene (as HCl) brings the third C into conjugation, giving a $C_3H_5^-$ (propenyl = allyl) ligand (monoolefin $-$ H$^+$ $\rightarrow$ enyl$^-$; route **3**).

$$2\,CH_2{=}CHCH_3 + 2\,PdCl_2 \longrightarrow \left[\text{...} \right] \longrightarrow \left\langle\!\!\left\langle \text{...} \right\rangle\!\!\right\rangle + 2\,HCl \qquad (13.18)$$

Reaction (13.19) reverses Reaction (13.18); allyl ligands can be converted to olefin ligands on treatment with nucleophiles[6]:

$$\left[\underset{CpRe(CO)_2}{\overbrace{\qquad}} \right]^+ + Li(CuMe_2) \longrightarrow \underset{CpRe(CO)_2}{\underline{\qquad}}{-}Me \qquad (13.19)$$

For examples of route **4**, see Reactions (13.23) and (13.25) in Section 13.4.

Complexes of terminally substituted allyls are capable of isomerism; the substituent may be on the same side as the center C (*syn* isomer), or on the opposite side (*anti* isomer). This same feature leads to nonequivalence of terminal protons. In solution, inter-

[6] Diene complexes are converted to allyl complexes by removing one C from conjugation. Reaction (14.43), provides another example.

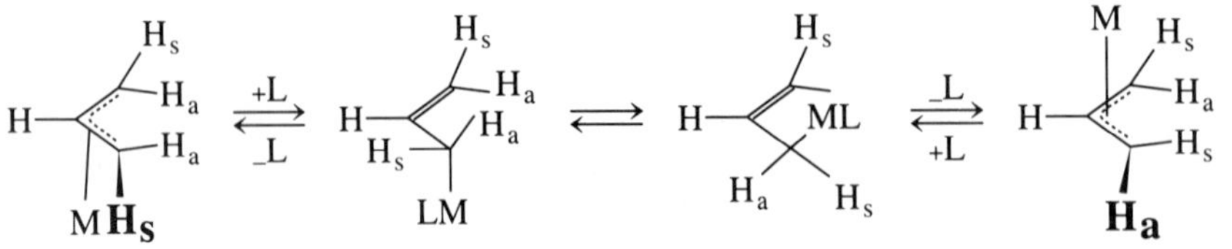

Figure 13.6   Exchange mechanism for *syn* and *anti* protons in allyl complexes.

conversion of terminal groups sometimes occurs when free basic ligands are added; the mechanism is shown in Figure 13.6. The free ligand displaces an olefinic bond, converting the η^3-allyl to an η^1-allyl. Free rotation around the M—C single bond interconverts *syn* and *anti* protons. The process can often be followed by NMR. (See Section 9.15 and Problem 13.6).

13.4   BUTADIENE COMPLEXES

Butadiene and substituted butadienes, as four-electron donor ligands, form an especially large number of complexes with the $Fe(CO)_3$ group. Butadienes are conjugated (have alternating double and single bonds), and the free ligand exists almost entirely in the *s-trans* form (**a**) with the *s-cis* geometry (**b**) lying about 12 kJ/mol higher in free energy. The *s-cis* form is most common as a ligand and is described in terms of the contributing VB structures **b, c,** and **d** (with analogous ones for *s-trans*):

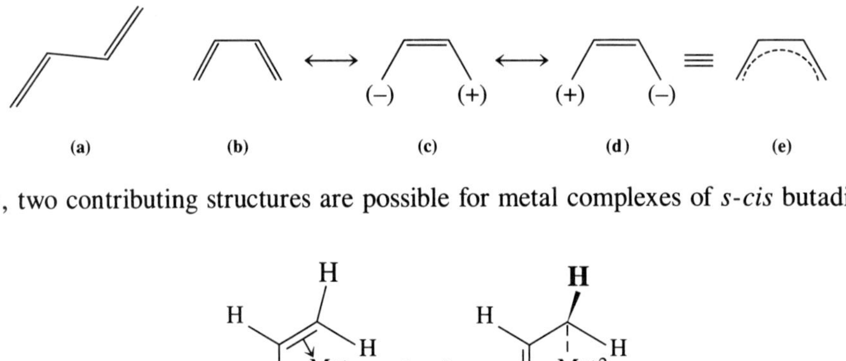

Hence, two contributing structures are possible for metal complexes of *s-cis* butadiene:

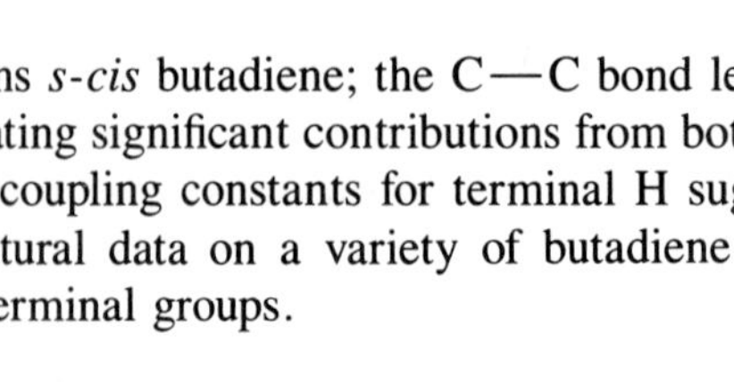

$(\eta^4\text{-}C_4H_6)Fe(CO)_3$ contains *s-cis* butadiene; the C—C bond lengths are identical within experimental error, indicating significant contributions from both structures **f** and **g**. However, the ^{13}C—H NMR coupling constants for terminal H suggest sp^2, rather than sp^3, hybridization at C. Structural data on a variety of butadiene complexes indicate some bending out of plane by terminal groups.

Complexes of Zr and Hf such as $Cp_2Zr(\eta^4\text{-}2, 3\text{-}Me_2C_4H_4)$ (Figure 13.7a) are known[7] in which structure **g** above is the main contributor. In this complex, the C_1—C_2 and C_3—C_4 distances are longer than the C_2—C_3 distance; the Zr—C_1 and Zr—C_4 distances are typical of Zr—C σ bonds. Mooveover, Zr is farther away from C_2 and C_3 than from C_1 and C_4 as might be expected from **g**. In contrast, late-transition-metal butadiene complexes display shorter bonds to C_2 and C_3 than to C_1 and C_4 as if the π system were slightly tipped toward the metal.

A few complexes such as $Cp_2Zr(\eta^4\text{-}1, 4\text{-}Ph_2C_4H_4)$ have coordinated *s-trans* butadiene (Figure 13.7b). This form easily converts to an η^2 coordinatively unsaturated isomer, thus enhancing its reactivity.

The MO π-bonding model resembles that for allyl complexes; Figure 13.8 shows π-MOs for *cis*-butadienes. Equalization of C—C bond lengths in this model results from ligand-to-metal π-donation from filled Ψ_1 and Ψ_2 and metal-to-ligand back-donation into empty Ψ_3 (bonding between C_2 and C_3). Even when the best description of the bonding is structure **e** on page 622, emphasizing delocalization for the C_4H_6 ligand, it is usual to draw structure **b** to emphasize its 4-*e* donor ability.

Like olefin and acetylene complexes, butadiene complexes generally are prepared by direct reaction between the ligand and a metal complex. Often, CO is displaced.

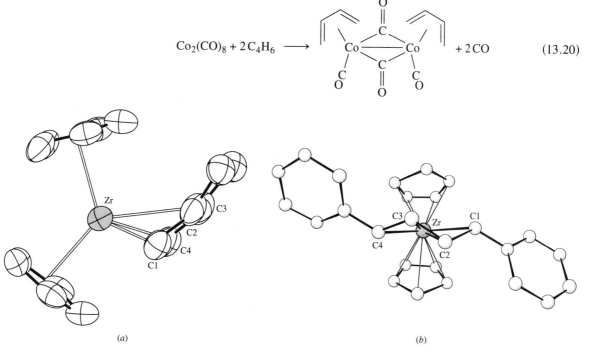

$$Co_2(CO)_8 + 2\,C_4H_6 \longrightarrow \quad\quad\quad + 2\,CO \quad\quad (13.20)$$

Figure 13.7   (a) Molecular structure of $Cp_2Zr(\eta^4\text{-}2, 3\text{-}Me_2C_4H_4)$. (G. R. Erker, K. Engel, C. Krüger, and A. -P. Chiang, *Chem. Ber.* **1982**, *115*, 3311. Reprinted from R. Erker and C. Krüger, *Adv. Organomet. Chem.* **1985**, *24*, 1.) (b) Molecular structure of $Cp_2Zr(\eta^4\text{-}1,4\text{-}Ph_2C_4H_4)$. (From Y. Kai, N. Kanehisa, K. Miki, N. Kasai, K. Mashima, K. Nagasuna, H. Yasuda, and A. Nakamura, *J. Chem. Soc., Chem. Commun.* **1982**, 191.)

[7]G. Erker, C. Krüger, and G. Müller, *Adv. Organomet. Chem.* **1985**, *24*, 1.

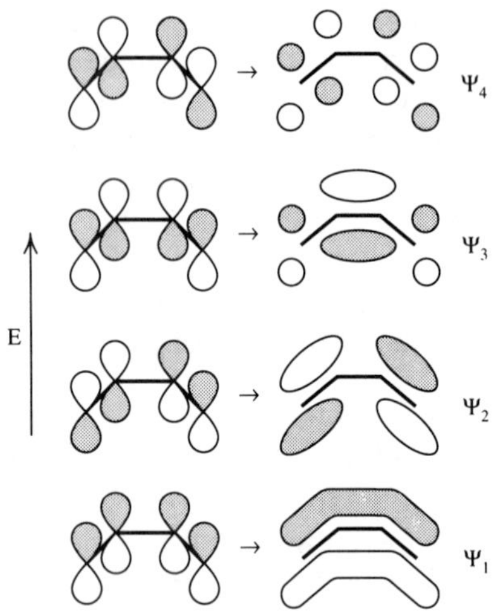

Figure 13.8   Molecular orbitals for *s*-*cis*-C_4H_6.

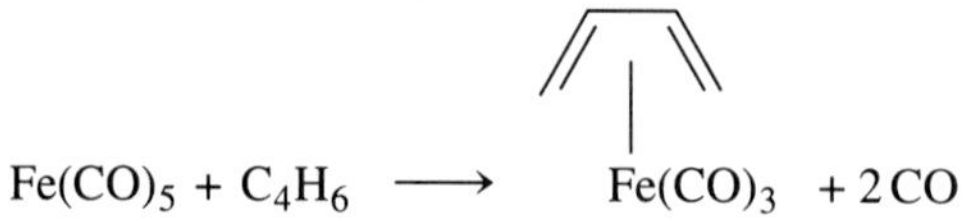

$$Fe(CO)_5 + C_4H_6 \longrightarrow Fe(CO)_3 + 2\,CO \tag{13.21}$$

Early-transition-metal complexes are prepared by reduction of halides, and so on, in the presence of the ligand:

$$Cp_2ZrCl_2 + 2\,Li + C_4H_6 \longrightarrow Cp_2Zr(\eta^4\text{-}C_4H_6) + 2\,LiCl \tag{13.22}$$

Coordinated butadiene does not undergo the hydrogenation and Diels–Alder reactions of the free molecule. Diene and methallyl complexes can be interconverted; depending on the metal and charge, nucleophiles or electrophiles can be used to abstract or add H. Formal hydride (H^-) addition to butadiene saturates an olefinic carbon, thereby removing it from conjugation, and giving a methallyl$^-$. (Diene + H^- → enyl$^-$.) H comes from the solvent (thf) in Reaction (13.23) and from a metal hydride in Reaction (13.24):

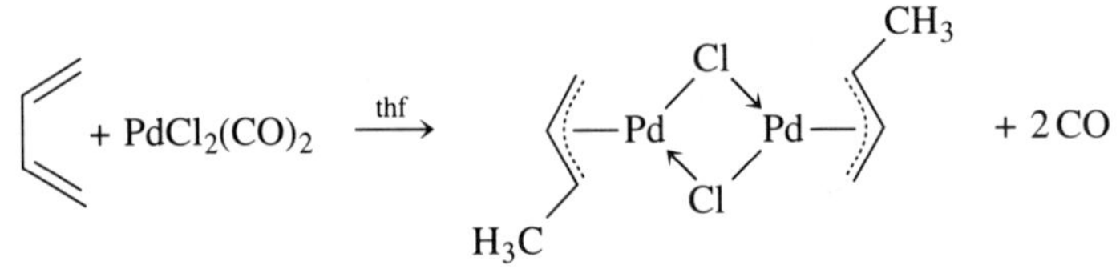

$$\text{(13.23)}$$

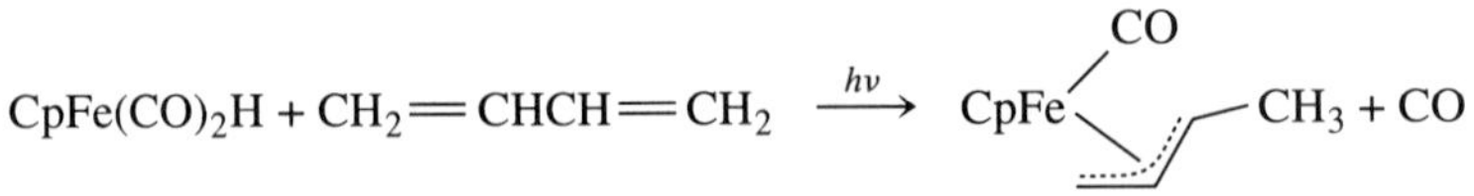

$$\text{(13.24)}$$

H^+ probably adds first to the metal in Equation (13.25), making this also a case of oxidation of Fe which formally converts H^+ to H^- followed by hydride addition to butadiene:

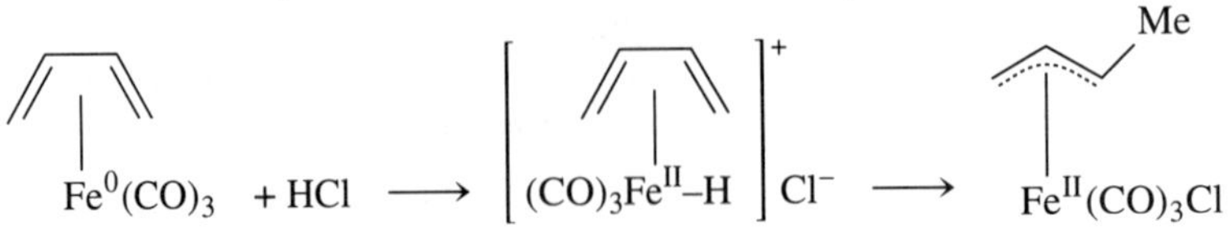

$$\text{(13.25)}$$

H^- abstraction by Ph_3C^+ on cyclic dienes places a fifth C in conjugation, affording di-enyl$^-$ complexes; the metal ion is formally oxidized.

$$+ Ph_3C^+ \longrightarrow \qquad\qquad + Ph_3CH \qquad (13.26)$$

Fragment Analysis

Bonding in organometallic species can often be understood from the point of view of **fragment analysis**, an approach developed largely by Hoffmann and co-workers.[8] Instead of analyzing molecular composition into metal and ligands, we break it up into a fragment such as $Fe(CO)_3$ and other ligands, such as butadiene. The MOs of the metal-containing fragment are constructed. Interactions between the frontier orbitals of the metal fragment and those of the organic ligands can then be analyzed to understand the bonding. This approach is particularly useful for molecules of low symmetry where delocalized MOs may make the bonding difficult to understand pictorially.

To construct the MOs for ML_3 fragment with C_{3v} symmetry, we start with octahedral ML_6 where only σ-bonding is considered. We orient the octahedron with the unconventional choice of a C_3 axis as the z-axis and classify orbitals in the C_{3v} subgroup of O_h (Figure 13.9). d_{xz} and d_{yz} are directed closest to the ligands and thus are used to construct d^2sp^3 hybrids for σ-bonding to L, leaving $d_{x^2-y^2}$, d_{xy} (an e set), and d_{z^2} (a_1) as nonbonding. Three d^2sp^3 hybrids point down, and three point up. The "down" set consists of $(s - p_z)$ (a_1) and the e set $(p_x - d_{xz})$, $(p_y - d_{yz})$ which are used to bond the three ligands and are omitted from Figure 13.9 because they are not frontier orbitals. The "up" orbitals are $(s + p_z)$ (a_1) and the e set $(p_x + d_{xz})$, $(p_y + d_{yz})$ and remain empty hybrids if we remove three L to give ML_3. For $ML_3 = Fe(CO)_3$, eight metal electrons fill the nonbonding metal orbitals and half fill each e of the "up" set. If we now bring up a butadiene, we note that the frontier orbitals which can interact are a_1 and e on Fe and Ψ_1, Ψ_2, and Ψ_3 on butadiene. MOs can be constructed from these and the 8 Fe + 4 butadiene electrons allowed to occupy energy levels. Figure 13.9 shows the energy-level diagram. Bonding and nonbonding levels are exactly filled, leaving a large HOMO–LUMO gap. Bonding between the $Fe(CO)_3$ fragment and conjugated dienes is so favorable that 1, 5-cyclooctadiene is isomerized to conjugated 1, 3-cyclooctadiene on complexation with $Fe(CO)_3$.[9]

---

[8] M. Elian and R. Hoffmann, *Inorg. Chem.* **1975**, *14*, 1058; M. Elian, M. M. L. Chen, D. M. P. Mingos, and R. Hoffmann, *Inorg. Chem.* **1976**, *15*, 1148; R. Hoffmann, *Angew. Chem. Int. Ed. Engl.* **1982**, *21*, 711.

[9] Alternatively, we recall that the Jahn–Teller theorem (Section 9.8.2) requires that $Fe(CO)_3$ distort so as to remove degeneracy. Suppose the direction of the distortion is to make the d_{yz} orbital more stable; then it will contain both electrons. If we now bring up a butadiene, the filled Ψ_1 of butadiene has proper symmetry to donate electrons to the metal a_1; Ψ_2 can donate electrons into the empty d_{xz}. Interaction with empty Ψ_3 stabilizes filled d_{yz}.

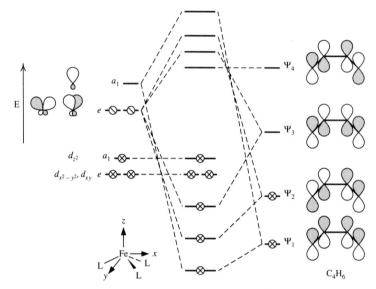

Figure 13.9   Frontier orbital interactions in $(\eta^4\text{-}C_4H_6)Fe(CO)_3$.

13.5   CYCLIC π COMPLEXES

Several cyclic organic species form complexes involving transition metals. Complete electron delocalization around the ring ordinarily occurs. We consider that the cyclobutadiene, cyclopentadienyl, and arene ligands all behave as 6-e donors, possessing the $(4n + 2)$ π electrons necessary for aromaticity (Table 12.4). All these ligands are bonded by donation of electrons from their filled π orbitals to the metal and by back-donation from the metal into empty ligand π* orbitals.[10] Bonding in Cp complexes was discussed in Chapter 12.

▶ 13.5.1   *Cyclobutadiene Complexes*

Unlike straight-chain C_4H_6, cyclobutadiene (C_4H_4) is not stable in the free state, since its four π-electrons render it antiaromatic. An early consideration (1956) of orbital symmetries led Orgel and Longuet-Higgens to predict that cyclobutadiene could be stabilized by coordination to a metal.[11] Shortly thereafter, a tetramethylcyclobutadiene complex was

---

[10] For orbital pictures see P. Powell, *Principles of Organometallic Chemistry*, Chapman and Hall, London, 1988.

[11] Bonding in (η^4-cyclobutadiene)Fe(CO)$_3$ can be easily understood in terms of fragment analysis involving the interaction of an empty a_1 orbital on [Fe(CO)$_3$]$^{2+}$ with the filled cyclobutadiene^{2-} a_1 and between the empty e's on [Fe(CO)$_3$]$^{2+}$ and the filled cyclobutadiene^{2-} e's.

prepared by Criegee. Petit and co-workers prepared an unsubstituted analogue by a similar reaction:

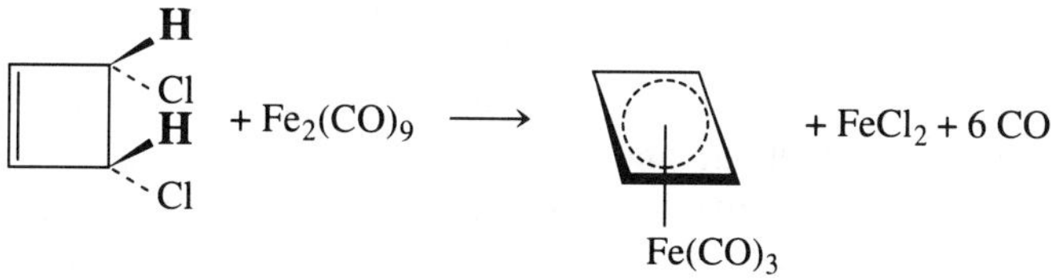

$$\text{(structure)} + \text{Fe}_2(\text{CO})_9 \longrightarrow \text{(structure)} + \text{FeCl}_2 + 6\,\text{CO} \qquad (13.27)$$

Cyclobutadiene complexes often result from coupling reactions of substituted acetylenes. As shown in Equation (13.28), several coupling products often result:

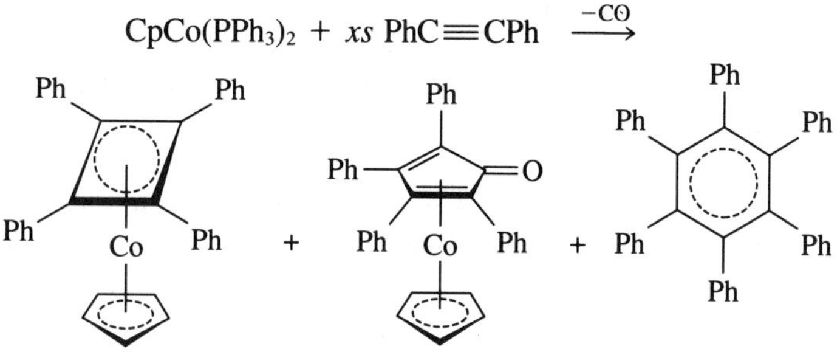

$$\text{CpCo(PPh}_3)_2 + xs\,\text{PhC}\equiv\text{CPh} \xrightarrow{-\text{CO}}$$

$$\text{(structures)} \qquad (13.28)$$

Justification for considering the cyclobutadiene ring as an aromatic dianion (Table 12.4) arises from the variety of electrophilic substitution reactions that occur on the ring, such as

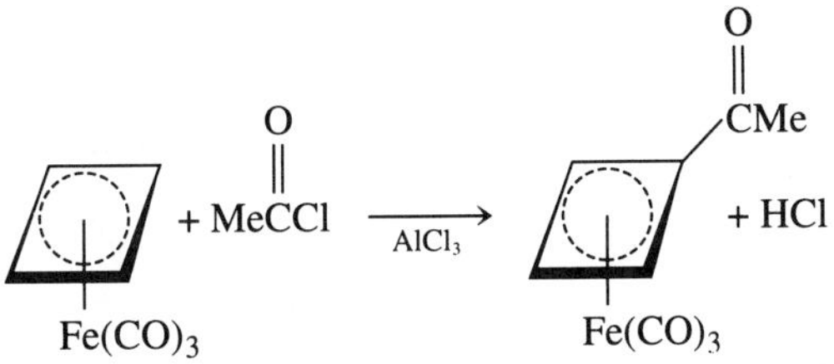

$$\text{(structure)} + \text{MeCCl} \xrightarrow{\text{AlCl}_3} \text{(structure)} + \text{HCl} \qquad (13.29)$$

$$\text{(structure)} + \text{NaCl} \xrightarrow{\text{Hg(OAc)}_2} \text{(structure)} \qquad (13.30)$$

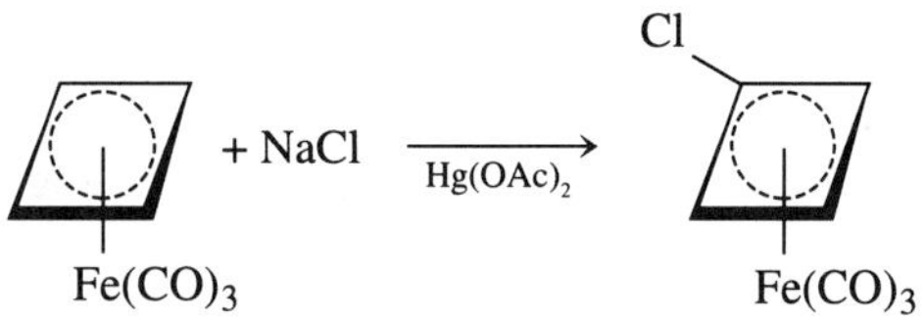

Oxidation of Fe in a cyclobutadiene complex with Ce^{4+} releases the free ligand; this is a common technique for releasing organic ligands. Free cyclobutadiene can be trapped by a Diels–Alder reaction with an alkyne to give the bicyclic product shown below:

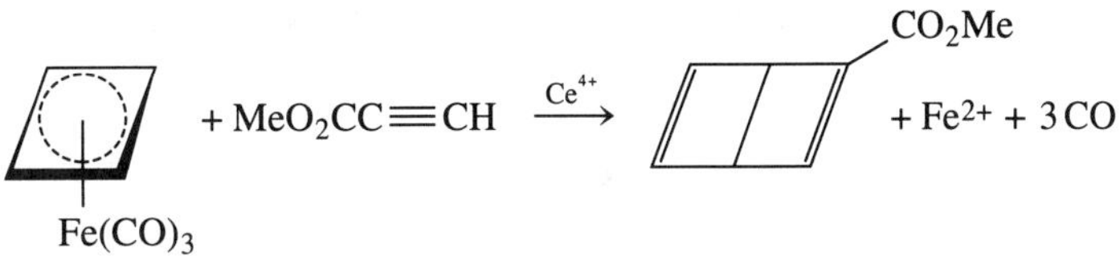

$$\text{(structure)} + \text{MeO}_2\text{CC}\equiv\text{CH} \xrightarrow{\text{Ce}^{4+}} \text{(structure)} + \text{Fe}^{2+} + 3\,\text{CO} \qquad (13.31)$$

▶ *13.5.2 Pentamethylcyclopentadienyl Complexes*

Substituting the Cp ring with methyls gives Me_5C_5, pentamethylcyclopentadienyl ligand (often abbreviated Cp*). The bonding scheme is formally similar to that of Cp; however, Cp* differs from Cp in three important respects: It is (1) more sterically bulky, (2) more electron-donating, and (3) more "organic". Property **3** increases the solubility of Cp* complexes in organic solvents. Property **1** permits preparation of monomeric complexes whose Cp analogues are di- or polymeric. For example, Cp_3Sc is a polymer having two η^5-C_5H_5 and one μ-η^1, η^1-C_5H_5 per Sc, but Cp_3^*Sc is monomeric. Cp* ligand bulk also promotes kinetic stability by blocking attacking reagents. The electron-donating nature of Me as compared to that of H makes Cp* a stronger-field ligand than Cp. Thus Cp_2^*Mn is low-spin in contrast to Cp_2Mn. The visible spectra of Cp_2^*M reveal the greater splitting between the $2a_{1g}$ and $2e_{1g}$ orbitals; values of B' (see page 452) are also higher than those in Cp_2M, reflecting greater electron density on the metal. The electron density buildup on M makes Cp_2^*M much easier to oxidize than Cp_2M, with reduction potentials typically being shifted by -0.5 V.[12] The combination of steric blockage and a substantial energy gap between HOMOs and LUMOs leads to enhanced kinetic and thermal stability for Cp* (as compared to Cp) compounds.

▶ *13.5.3 η^3-Cyclopentadienyls*

$C_5H_5^-$ is almost always an η^5 ligand donating six electrons. Removing one double bond from coordination gives η^3-Cp^- which donates four electrons (isoelectronic with allyl). Ring slippage could remove two C's from bonding, permitting a metal to coordinate an additional two-electron donor ligand. The indenyl (Ind) ligand presents a favorable opportunity for η^3 bonding because the "free" double bond is conjugated with the phenyl ring fused to Cp:

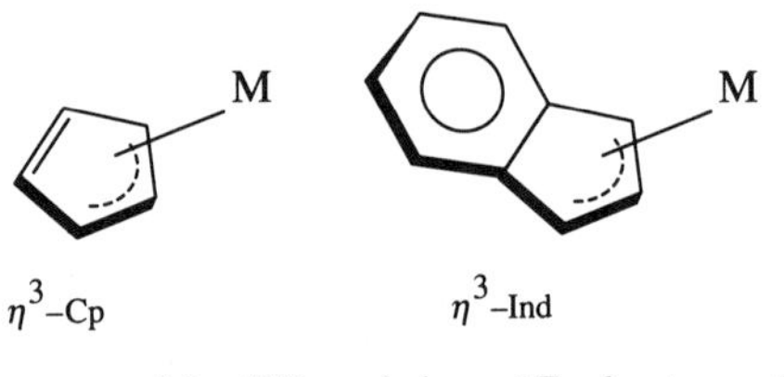

$(\eta^5$-Ind$)_2$V was found to react with CO, giving 17-electron $(\eta^5$-Ind$)(\eta^3$-Ind$)$V(CO)$_2$ (Figure 13.10).

▶ *13.5.4 Synthesis of Cyclopentadienyl Complexes*

The usual way to add Cp is to allow alkali metal cyclopentadienides to react with metal complexes in ether solution. The cyclopentadienides are prepared *in situ* from cyclopentadiene:

$$2\,Na + 2\,C_5H_6 \longrightarrow 2\,C_5H_5^- + 2\,Na^+ + H_2 \tag{13.32}$$

[12] J. L. Robbins, N. Edelstein, B. Spencer, and J. C. Smart, *J. Am. Chem. Soc.* **1982,** *104,* 1882.

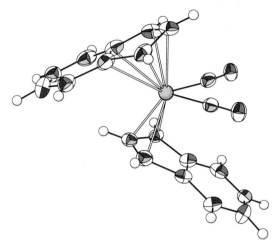

Figure 13.10 Molecular structure of $(\eta^3\text{-Ind})$-$(\eta^5\text{-Ind})V(CO)_2$. (From R. M. Kowalski, A. L. Rheingold, W. C. Trogler, and F. Basolo, *J. Am. Chem. Soc.* **1986**, *108*, 2460. Copyright 1986, American Chemical Society.)

$$VCl_3 + 3\,NaC_5H_5 \longrightarrow Cp_2V \qquad (13.33)$$

In Reaction (13.33), Na^+Cp^- also reduces V^{III} to V^{II}.

$$W(CO)_6 + NaC_5H_5 \longrightarrow Na^+[CpW(CO)_3]^- + 3\,CO \qquad (13.34)$$

$Tl(C_5H_5)$ is a convenient source of Cp, because it can be prepared separately and stored. Similarly, $(Me_5C_5)MgCl(thf)$, which is crystalline and easily stored and weighed, as well as $n\text{-Bu}_3Sn(C_5Me_5)$ are important sources of Cp*; NaCp* and KCp* are also employed.

$$TiCl_4 + 2\,Tl(C_5H_5) \longrightarrow Cp_2TiCl_2 + 2\,TlCl(s) \qquad (13.35)$$

Direct reaction between cyclopentadiene and metal complexes, often in the presence of a base acting as a proton acceptor, produces Cp complexes:

$$2\,C_5H_6 + 2\,Fe(CO)_5 \xrightarrow[\text{reflux}]{CO} [2\,CpFe(CO)_2H] \longrightarrow [CpFe(CO)_2]_2 + H_2 \qquad (13.36)$$

$$NiCl_2 + 2\,C_5H_6 + 2\,Et_2NH \longrightarrow Cp_2Ni + 2\,[Et_2NH_2]^+Cl^- \qquad (13.37)$$

▶ *13.5.5 Arene Complexes* [13]

Arene complexes were first prepared in 1919 by Hein, who treated $CrCl_3$ with C_6H_5MgBr in ether. However, these were not recognized as π complexes until 1954. Upon hydrolysis, Hein's reaction mixtures afforded $(\eta^6\text{-}C_6H_6)_2Cr$, $(\eta^6\text{-}C_6H_5\text{-}C_6H_5)(\eta^6\text{-}C_6H_6)Cr$, and $(\eta^6\text{-}C_6H_5\text{-}C_6H_5)_2Cr$.

The x-ray structure of bis(benzene) chromium (Figure 3.11) shows eclipsed rings and $\mathbf{D}_{6h}$ symmetry. Arene ligands behave as six-electron donors, and each occupies three coor-

[13] E. L. Muetterties, J. R. Bleeke, J. Wucherer, and T. A. Albright, *Chem. Rev.* **1982**, *82*, 499.

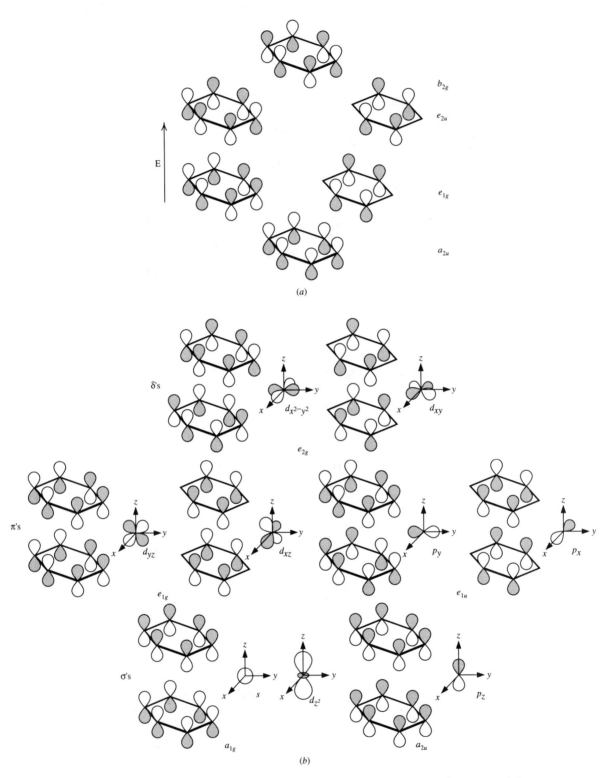

Figure 13.11 (a) Benzene π-MOs. (b) Ligand-group orbitals and their metal counterparts for $\mathbf{D}_{6h}$ bis(benzene)chromium.

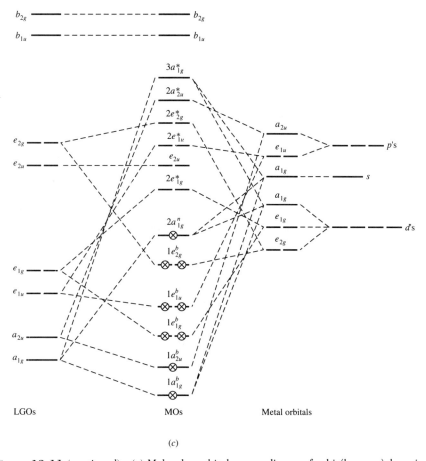

LGOs   MOs   Metal orbitals

(c)

Figure 13.11 (*continued*)   (c) Molecular-orbital energy diagram for bis(benzene)chromium.

dination positions around the metal. Figure 13.11 presents the π orbitals of benzene, the ligand-group orbitals (LGO's) with their metal counterparts, and a qualitative MO diagram for $(\eta^6\text{-}C_6H_6)_2Cr$. The most important bonding interaction is between d_{xz}, d_{yz}, and the e_{1g} ligand combination. The 18 electrons fill the levels through a_{1g} (d_{z^2})—which accounts for the diamagnetism of $(\eta^6\text{-}C_6H_6)_2Cr$. This energy-level scheme probably also applies to other bis(arene) complexes.

Not all bis(arene) complexes whose formulas seem to violate the 18-e rule actually do so. For example, $(Me_6C_6)_2Ru$ would be a 20-e species with two unpaired electrons if both arene rings were planar. Instead, the complex is diamagnetic and has a structure[14] in which one hexamethylbenzene behaves as an η^4-four-electron donor and is nonplanar. Several electron-rich metals form complexes featuring bending away of the uncoordinated portion of the arene ring or the destruction of ring aromaticity, as shown by alternating C—C bond lengths.

[14] G. Huttner and S. Lange, *Acta Cryst. Sect. B* **1972,** *28,* 2049.

Most neutral bis(arene) complexes are soluble in organic solvents, can be sublimed, and are oxidized by air to mono cations. The metal–ring bond strength is less than that in ferrocene.

Coordination of an arene ligand usually is achieved by (**1**) the Fischer–Hafner synthesis, (**2**) cyclic condensation of alkynes, or (**3**) ligand replacement. The Fischer–Hafner method is applicable to most transition metals [except for Ti^0] and generally leads to bis(arene) products. A metal halide is allowed to react with the arene solvent in the presence of $AlCl_3$ or $AlBr_3$, which scavenges halide, and Al metal, which acts as a reducing agent:

$$3\,CrCl_3 + 2\,Al + AlCl_3 + 6\,\text{arene} \longrightarrow 3\,[(\text{arene})_2Cr]^+[AlCl_4]^- \quad (13.38)$$

Cationic compounds are reduced to neutral ones by aqueous dithionite ($S_2O_4^{2-}$). If no reduction is required, Al is omitted. Reaction conditions are Friedel–Crafts conditions. The method of Equation (13.38) is not applicable to aryl halides because the reaction conditions lead to dehalogenation. Cyclic condensation of alkynes to form arenes has been mentioned previously (Section 13.2). Chemical vapor deposition is often used; $(C_6H_6)_2Fe$ has been made in 10-gram quantities in this way.

Carbonyls can be replaced by arene donors:

$$M(CO)_6 + \text{arene} \longrightarrow (\text{arene})M(CO)_3 + 3\,CO \qquad M = Cr,\ Mo,\ W \quad (13.39)$$

The yields in Reaction (13.39) are low. Starting materials with more easily replaced ligands improve yields.

$$(MeCN)_3Cr(CO)_3 + C_6H_6 \longrightarrow (\eta^6\text{-}C_6H_6)Cr(CO)_3 + 3\,MeCN \quad (13.40)$$

Halides can be removed by $AlCl_3$ along with CO's:

$$Mn(CO)_5Cl + \text{arene} + AlCl_3 \longrightarrow [(\text{arene})Mn(CO)_3]^+[AlCl_4]^- + 2\,CO \quad (13.41)$$

Carbonyl displacement is a good way to make mixed complexes:

$$(\eta^4\text{-}Ph_4C_4)Co(CO)_2Br + \text{arene} + AlBr_3 \longrightarrow$$
$$[(\eta^6\text{-arene})(\eta^4\text{-}Ph_4C_4)Co]^+[AlBr_4]^- + 2\,CO \quad (13.42)$$

In discussing the reactivity of coordinated arenes, we concentrate on complexes in which the ligand is coordinated through its entire π system and is planar. A major organic reaction of free arenes is attack by electrophiles, leading to substitution. Coordination to a metal withdraws charge from the arene ring and deactivates the ring to electrophilic attack. Although most bis(arene) complexes do not survive Friedel–Crafts reaction conditions, $(\eta^6\text{-}C_6H_6)Cr(CO)_3$ can be acetylated, although not very readily. Other evidence of the "electron-poor" state of coordinated arenes includes the fact that benzoic acid is a weaker acid than $(\eta^6\text{-}C_6H_5COOH)Cr(CO)_3$ and that $(\eta^6\text{-}C_6H_5NH_2)Cr(CO)_3$ is a weaker base than aniline.

In contrast to free arenes, coordinated arenes are activated toward attack by nucleophiles, as the following reactions demonstrate:

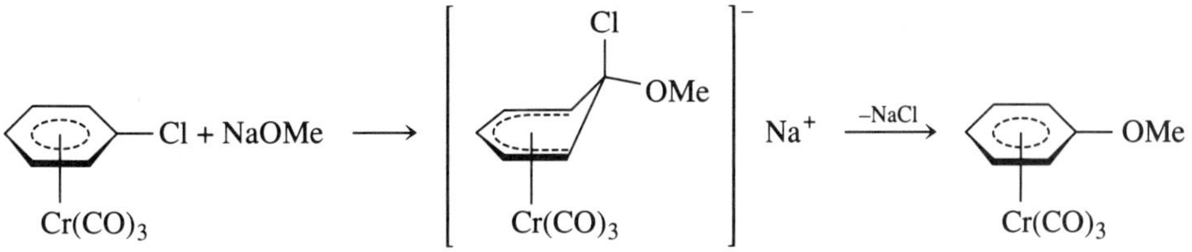

$$\text{(structure: arene-Cr(CO)}_3\text{)}\text{—Cl} + \text{NaOMe} \longrightarrow \left[\text{(structure with Cl, OMe on arene-Cr(CO)}_3\text{)} \right]^{-} \text{Na}^{+} \xrightarrow{-\text{NaCl}} \text{(arene-Cr(CO)}_3\text{)}\text{—OMe} \qquad (13.43)$$

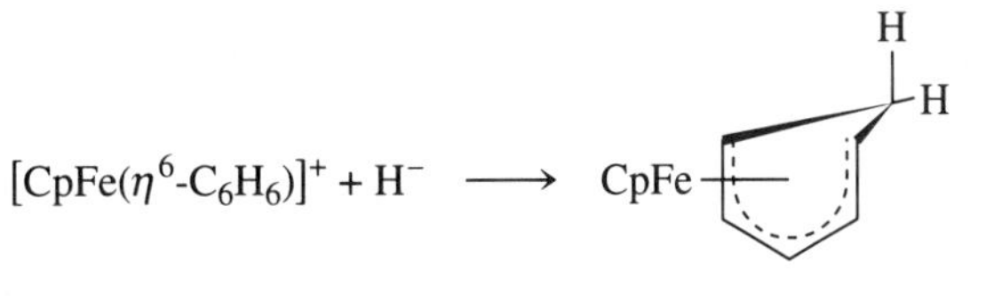

$$[\text{CpFe}(\eta^6\text{-C}_6\text{H}_6)]^{+} + \text{H}^{-} \longrightarrow \text{CpFe} \text{(structure with H, H)} \qquad (13.44)$$

The other common reaction is arene displacement by other arenes or by donor ligands. For example, arene ligands may be exchanged in the presence of AlCl$_3$:

$$[(\eta^6\text{-1,3,5-Me}_3\text{C}_6\text{H}_3)_2\text{Cr}]^{+} + 2\,\text{C}_6\text{H}_6 \xrightarrow{\text{AlCl}_3}$$
$$[(\eta^6\text{-C}_6\text{H}_6)_2\text{Cr}]^{+} + 2\,(1,3,5,\text{-Me}_3\text{C}_6\text{H}_3) \qquad (13.45)$$

Arene exchange occurs in donor solvents such as ketones, ethers, or nitriles and involves an η^6- to η^4-**haptotropic shift** (i.e., change in the number of C's coordinated) of the leaving arene resulting from solvent displacement of an aromatic electron pair. As might be expected from this picture, $[(\eta^4\text{-C}_6\text{H}_6)\text{Cr(CO)}_3]^{2-}$ (which can be generated by K reduction of the neutral compound) exchanges benzene for naphthalene easily.[15]

As a rule, bis(arene) complexes are somewhat less reactive than mono(arene) complexes.

► *13.5.6 Cycloheptatrienyl and Cyclooctatetraene Complexes*

Seven-carbon C$_7$H$_8$ (cycloheptatriene) can be a 3×2-e donor toward metals. The coordination geometry shown in Figure 13.12, in which the saturated C is bent away from the metal, is typical. Note the longer C—C bonds to the C bent away. Such complexes can undergo hydride abstraction (using carbocations such as Ph$_3$C$^+$) to bring the remaining C into conjugation, producing planar delocalized cycloheptatrienyl (tropylium) ligand C$_7$H$_7^+$ (-ene − H$^-$ → -enyl$^+$).

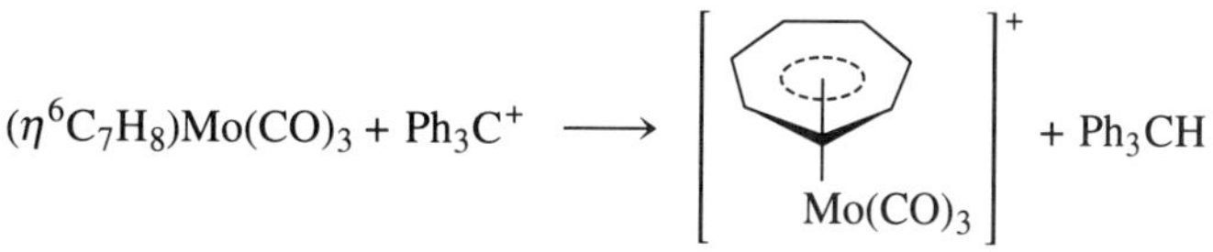

$$(\eta^6\text{C}_7\text{H}_8)\text{Mo(CO)}_3 + \text{Ph}_3\text{C}^+ \longrightarrow \left[\text{(C}_7\text{H}_7\text{-Mo(CO)}_3\text{ structure)} \right]^{+} + \text{Ph}_3\text{CH} \qquad (13.46)$$

[15] V. S. Leong and N. J. Cooper, *Organometallics* **1988**, *7*, 2058.

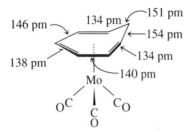

Figure 13.12 Molecular structure of $(\eta^6\text{-}C_7H_8)Mo(CO)_3$. (From J. D. Dunitz and P. Pauling, *Helv. Chem. Acta* **1960,** *43*, 2188.)

The planar delocalized ring of $\eta^7\text{-}C_7H_7^+$ shown in Figure 13.13a has very similar C—C distances as expected; they range from 137 to 143 pm. The boron-containing ring shown in Figure 13.13b is isoelectronic with $C_7H_7^+$, but displays alternating bond distances. The two B—C ring distances are ~150 pm, whereas the C—C distances in the seven-membered ring range from 137 to 145 pm and alternate in the same fashion as those in Figure 13.12. Interestingly, heating the complex does *not* result in migration of Cr to the fused benzene ring. Nucleophilic attack on tropylium complexes occurs on the ring and converts them back to cycloheptatriene species.

Most cycloheptatrienyl complexes of transition metals are mixed-sandwich complexes. The fact that they do not contain ligands such as CO, Cl, or NO may be because such 18-*e* complexes would have formulas such as $(\eta^7\text{-}C_7H_7)Co(CO)$, and have one side open for attack, thus reducing kinetic stability.

Cyclooctatetraene (cot) is *not* a $(4n + 2)$ aromatic system. Cot generally behaves toward transition metals as a ligand with isolated double bonds. The example in Table 12.4 shows its function as a $2 \times 2\text{-}e$ donor with nonbonded carbons bent away from the metal. With $Fe(CO)_3$ groups, cot acts as a $1 \times 4\text{-}e$ or $2 \times 4\text{-}e$ donor, behaving as one or two butadiene ligands (see Figure 12.9). Presumably d orbitals are not sufficiently diffuse and lack proper symmetry for effective overlap with the MOs of planar aromatic cot^{2-}. Hence, transition metal complexes of cot seldom display planar cot^{2-} rings.

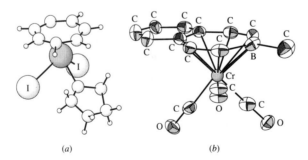

(a)　　　　　　　(b)

Figure 13.13 (a) Molecular structure of $(\eta^7\text{-}C_7H_7)MoI_2(thf)$. (From A. Gourdon and K. Prout, *Acta Cryst. B* **1982,** *38*, 1596.) (b) Molecular structure of $(\eta^6\text{-}C_{10}H_{10}BMe)Cr(CO)_3$. (Reprinted with permission from A. J. Ashe, J. W. Kempf, C. M. Kausch, H. Konishi, M. O. Kristen, and J. Kroker, *Organometallics,* **1990,** *9*, 2944. Copyright 1990, American Chemical Society.)

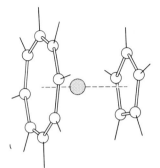

Figure 13.14   Molecular structure of $(\eta^8\text{-}C_8H_8)\text{-}$ $(\eta^5\text{-}C_5H_5)\text{Ti}$. (From P. A. Kroon and R. B. Helmholdt, *J. Organomet. Chem.* **1970**, *25*, 451.)

$[C_8H_8]^{2-}$ does, however, act as a planar aromatic (10 π electrons) with nearly equal C—C distances (138–141 pm) in $\text{Ti}(C_8H_8)\text{Cp}$ (Figure 13.14). (C—C distances range from 139 to 141 pm in the Cp ring.) It is probable that the Ti-ring bonding has considerable ionic character. The f orbitals in actinides are suitable for overlap with cot^{2-}, and uranocene $(\eta^8\text{-}C_8H_8)_2\text{U}$ (Figure 3.11) has been prepared from UCl_4 and cot^{2-} in thf. The green product is pyrophoric in air and not very soluble in organic solvents. Uranocene does not undergo electrophilic substitutions or metallation. Thorocene, plutonocene, neptunocene, $[\text{Pu(cot)}]_2^-$, and $[\text{Np(cot)}]_2^-$ are also known. Structural evidence indicates that bonds in cot^{2-} complexes are ionic.

Main-group elements also form π complexes, although they are not nearly as numerous as those of the transition metals.[16] Most of the well-characterized compounds involve Cp (or substituted Cp) or aromatic ligands. Only s and p orbitals on metals are available for bonding, and the large energy differences between filled d levels (if any) and ligand π^* orbitals preclude synergistic effects.

13.6   COMPOUNDS WITH METAL–CARBON σ BONDS

► ## 13.6.1   *Metal Alkyls and Aryls*

Main-Group Alkyls and Aryls

Both main- and transition-group elements form alkyls and aryls; but those of the main group are more numerous and more industrially and synthetically useful. Their utility springs primarily from the weakness of the main-group-metal–carbon bond (relative to M—O or M—X bonds, which are typically 1.5 to 2 times as strong). This feature enables ready transfer of alkyl and aryl groups to other atoms. M—C bond strength decreases down main-group families; some typical figures for the average M—C bond dissociation energy in MMe_4 are: C, 358 kJ/mol; Si, 311; Ge, 249; Sn, 215; Pb, 152. Compounds of the type R_nM and $\text{R}_{n-y}\text{MX}_y$ are of importance.

[16] P. Jutzi, *Adv. Organomet. Chem.* **1986**, *26*, 217.

Synthesis.   Of the several synthetic routes available, the most straightforward is the so-called **direct method.**[17] The elemental metals make good starting materials because most of their compounds have exothermic heats of formation. Thus, extraction of metals from their ores consumes energy which is "stored" in the metals and tends to make reaction enthalpies more negative. Metals themselves react with alkyl or aryl halides. The best-known example is the Grignard reagent:

$$Mg(s) + RX \xrightarrow[\text{or thf}]{\text{ether}} RMgX \tag{13.47}$$

The reaction occurs by a free-radical process at the Mg surface. Si reacts similarly:

$$Si(s) + 2\,MeCl \xrightarrow{\text{catalyst}} 2\,Me_2SiCl_2(g) \tag{13.48}$$

The reactivity order for metals parallels their electropositive character, and more active metals displace less active ones from their compounds (see *transmetallation* below):

$$
\begin{array}{c|ccc}
\text{Li} & \multicolumn{3}{c}{\longleftarrow \qquad \text{Si}} \\
 & \text{Cu} & \text{Zn} & \text{Ge} \\
 & \text{Ag} & \text{Cd} & \text{Sn} \\
\text{Cs} \downarrow & \text{Au} & \text{Hg} & \text{Pb} \downarrow
\end{array}
$$

For alkyl or aryl halide, the reactivity order is RF $<<$ RCl $<$ RBr $<$ RI and Ar $<$ R $<$ allyl or benzyl, reflecting the order of C—X bond strengths. However, the high reactivity of iodides (except MeI) and, sometimes, of the bromides is too high and leads to coupling of organic groups as the major path (termed the **Wurtz reaction** when the product is an alkane or the **Fittig reaction** when the product is aromatic). For example:

$$
\begin{aligned}
RI + 2\,Na(s) &\longrightarrow RNa + NaI \\
RI + RNa &\longrightarrow R\!-\!R + NaI
\end{aligned}
\tag{13.49}
$$

The direct method is often limited by positive ΔH_f^0 for the products which results from the large enthalpy contribution required to separate individual metal atoms from M(s) (ΔH_{vap} for the metal). Two ways of circumventing this difficulty have recently been developed. One involves vaporizing the more metallic reactant and condensing the vapor with RX at low temperature; reaction occurs on warming. For example, ΔH_f^0 for B(g) is ~540 kJ/mol and it reacts with C_6H_5Br, thereby producing $PhBBr_2$, whereas elemental B(s) is too electronegative to be reactive in direct synthesis. The other involves preparation of "highly reactive" metal powders[18] by reducing metal halides with Li, Na, or K (sometimes in the presence of an electron carrier such as naphthalene). Frankland discovered around 1850 that Zn(s) reacts with RBr and RI, giving RZnX. However, Zn powder produced by reduction of anhydrous Zn salts also reacts with RCl, ArBr, and ArI. Highly reactive Mg forms Grignards from fluoro compounds.

---

[17] E. G. Rochow, *J. Chem. Educ.* **1966**, *43*, 58.
[18] R. D. Riecke, *Science* **1989**, *246*, 1260.

Direct synthesis is also applicable to nonmetallic elements of Groups 15 and 16.

Although a large number of organometallic compounds can be prepared by the direct method, laboratory syntheses usually start with a Grignard reagent, RMgX, or an organolithium, RLi. More active metals react with organometals of less active metals, displacing them from their compounds (**transmetallation**):

$$2\,Na + R_2Hg \longrightarrow 2\,NaR + Hg(l) \tag{13.50}$$

On the other hand, organometallics of the less active metals can be formed by **metathesis** reactions between their halides (or alkoxides) and alkyls of the more active metal. These reactions are driven by the large lattice energies of the active-metal halides or alkoxides which often precipitate from the organic solvent:

$$2\,RMgX + HgX_2 \xrightarrow{\text{hexane}} R_2Hg + MgX_2(s) \tag{13.51}$$

$$3\,PhLi + AlCl_3 \cdot OEt_2 \xrightarrow{Et_2O} Ph_3Al \cdot OEt_2 + 3LiCl(s) \tag{13.52}$$

Whereas RLi alklyates all metal–halogen bonds, alkylation is often incomplete with alkyls of the less electropositive metals, leading to loss of efficiency.

$$3\,Et_3Al \cdot OEt_2 + 2\,BF_3 \cdot OEt_2 \xrightarrow{Et_2O} 2\,Et_3B + 3\,Et_2AlF + 5\,Et_2O \tag{13.53}$$

$$Ph_4Sn + 2\,BCl_3 \longrightarrow 2\,PhBCl_2 + Ph_2SnCl_2 \tag{13.54}$$

Metals may replace acidic hydrogens of organics in the **metallation** reaction. The equilibrium lies on the side of the weaker organic acid. Acidic hydrocarbons such as acetylenes and cyclopentadiene (Reaction (13.32)) react directly with metals:

$$CH_3Li + Ph_3CH \longrightarrow Ph_3CLi + CH_4 \tag{13.55}$$

$$Hg[N(SiMe_3)_2] + 2\,Me\overset{\overset{\displaystyle O}{\displaystyle \|}}{C}Me \longrightarrow [MeC(O)CH_2]_2Hg + 2\,NH(SiMe_3)_2 \tag{13.56}$$

$$HC\equiv CH + Na \xrightarrow{NH_3(l)} Na^+\ ^-C\equiv CH \tag{13.57}$$

Na also reacts in organic solvents with conjugated organics such as naphthalene and benzophenone ($Ph_2C{=}O$) to transfer an electron, forming deeply colored paramagnetic solutions of green sodium naphthalide or blue-black sodium benzophenone ketyl. These compounds are strong reducing agents. They reduce water easily so that sodium metal and a catalytic amount of benzophenone are often used to dry nonhalogenated organic solvents.

In **hydrometallation,** a metal hydride adds to an olefin. This reaction occurs with hydrides of the less electropositive metals (B, Al, Si, Ge, Sn, Pb); addition ease (Si—H < Ge—H < Sn—H < Pb—H) parallels decreasing M—H bond strength. Although aluminum does not react directly with H_2 to give a hydride, this reaction does occur in the presence of aluminum alkyls; the Al—H bonds formed react readily with terminal olefins (hydroalumination). Synthesis of Al alkyls is very significant industrially, with hundreds of thousands of metric tons being produced each year in spite of their flammability in air. Because they are liquids, they serve as reaction solvents as well as alkylating agents in Ziegler–Natta polymerization (see Section 14.1.2). B—H bonds also add to olefins in the hydroboration reaction which is discussed in Section 17.3.3.

$$Me_3SnH + H_2C{=}CHCO_2Me \longrightarrow Me_3SnCH_2CH_2CO_2Me \qquad (13.58)$$

$$Al + 3H_2 + 3RCH{=}CH_2 \xrightarrow{AlR_3} (RCH_2CH_2)_3Al \qquad (13.59)$$

Metal–halogen exchange is important with Group 1 metals and occurs cleanly only when the product carbanion stabilizes negative charge better than the reactant carbanion. Wurtz coupling is often a side reaction here:

$$n\text{-BuLi} + PhBr \longrightarrow PhLi + n\text{-BuBr} \qquad (13.60)$$

EXAMPLE: Predict the products of the following reactions:
  (a) $HgCl_2$ + MeNa →
  (b) Me_2Cd + $MgBr_2$ →
  (c) PhI + Ca →

Solution: (a) Na is a more active metal than Hg; thus, we expect that metathesis occurs to give $HgMe_2$ driven by the large lattice energy of NaCl.
   (b) Mg is the more active metal; thus, we do not expect metathesis, which is disfavored by the lattice energy contribution.
   (c) Ca is an active metal. The product of this direct synthesis is PhCaI.

Reactivity.   Main-group organometallics are thermodynamically unstable with respect to oxidation to CO_2, H_2O, and metal oxide. Nevertheless, their kinetic stability varies widely. Group 1, 2, and 13 alkyls are quite reactive to O_2, and most inflame spontaneously in air. Except for BR_3 compounds, hydrolysis to give R—H and M—OH occurs, often explosively. Reactivity toward CO_2 is confined to the most electropositive metal compounds, but these often inflame spontaneously in a carbon dioxide atmosphere. Organo compounds of the less electropositive metals are less reactive in general, although several, including $CdMe_2$ and $ZnMe_2$, still inflame spontaneously in air. In contrast, $HgMe_2$ is stable to air and water. Group 14 organometals are the least reactive.

These trends have at least two causes. First, high bond polarity promotes attack by reagents. The very electropositive elements of Groups 1 and 2 form alkyls with extremely polar $M^{\delta+}$—$C^{\delta-}$ bonds. Second, compounds with empty valence orbitals or lone pairs (which can serve as sites of attack) are more reactive than those with central atoms bonded to four or more groups; thus, $AlMe_3$ and $:BiMe_3$ are pyrophoric, but $SiMe_4$ and $SnMe_4$ are inert to O_2. However, even coordinatively saturated species can be quite reactive if the bonds are very polar; for example, $SiCl_4$ is readily hydrolyzed. Even nonpolar species can be reactive if groups surrounding the central atom are too small to prevent attack on empty d orbitals; for example, SiH_4 hydrolyzes easily and inflames spontaneously in air.[19]

---

[19] Surprisingly, GeH_4 is much less reactive, being air-stable and unaffected by aqueous acid or 30% NaOH, thus illustrating our current inability to rationalize every fact of reactivity.

The tendencies to hydrolyze and to inflame in air which we have used as a general index of reactivity do not always parallel one another. Thus, $AlMe_3$ readily hydrolyzes, whereas $:BiMe_3$ and other Group 15 methyls do not. Even the relatively inert Group 14 tetraalkyls react with very active reagents such as HX. Aryl compounds are less reactive than alkyl. For a given metal, the moisture- and oxygen-sensitivity of an organometallic halide $R_{n-y}MX_y$ parallels the sensitivity of the corresponding homoleptic alkyl R_nM.

A very significant reaction of alkyls containing β-H is the so-called **β-elimination**:

$$M{-}CH_2CH_2R \iff M{-}\underset{\underset{\displaystyle CHR}{\|}}{\overset{\overset{\displaystyle H}{|}}{CH_2}} \longrightarrow M{-}H + CH_2{=}CHR \quad (13.61)$$

Olefin is eliminated and H remains bonded to the metal. This reaction affords a facile decomposition path for M = Li, Na, Mg, Be, B, and Al, all of which have a vacant metal orbital to interact with the β-C—H bond and, ultimately, to bond the eliminated H^-. A common way of increasing the stability of organometals is to tie up vacant metal orbitals by complexation, thus blocking β-elimination. N, N, N', N'-tetramethylethylenediamine (tmeda) is added to solutions of RLi in order to block β-elimination as well as to increase the carbanionic character of R^- by reducing its interaction with Li^+. Organometals which lack β-H are stabilized because β-elimination is unavailable.

Electron-Deficient Organometals. Organometallic compounds of the smallest and most electropositive main-group metals (Li, Be, Mg, Al, and, to some extent, B and Zn) have a structural chemistry and reactivity strongly influenced by their having too few electrons and too many valence orbitals to obey the octet rule as monomers. They form aggregates in the solid state in which the "sigma" electrons are delocalized in multicentered bonds. The tendency toward aggregation increases with metal electropositivity; it decreases as the steric requirements of the alkyl group increase. LiMe(s) (Figure 13.15) is a tetramer whose unit cell contains a body-centered packing of Li_4 tetrahedra. A methyl caps each triangular face. Each C in the $[LiMe]_4$ unit interacts with an adjacent Li_4, providing bond-

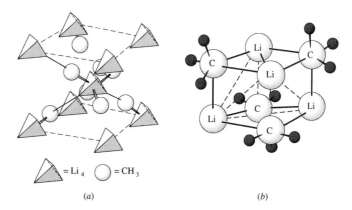

= Li$_4$ = CH$_3$

(a)   (b)

Figure 13.15 (a) Unit cell of LiMe(s). (After E. Weiss and E. A. C. Lucken, *J. Organomet. Chem.* **1964**, 2, 197. (b) Schematic drawing of (LiMe)$_4$ unit.

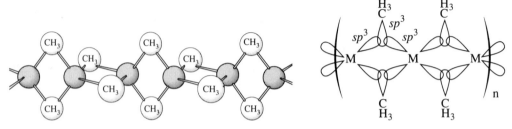

Figure 13.16   Chain structure of MMe₂ (M = Be, Mg). (The chlorides also have this structure.)

Figure 13.17   Three-center bonding with bridging methyls.

ing throughout the solid lattice which accounts for the low solubility of LiMe in poorly solvating media. NaMe(s) has a similar structure. Alkyls of other alkali metals are best considered as ionic compounds. [KMe has the NiAs structure (Figure 5.6) with methyl carbanions.] They are intractable white solids that decompose on heating and are insoluble in virtually all solvents. Extremely bulky R groups can lead to monomers; for example, $Li(2,4,6\text{-}Ph_3C_6H_2)\cdot OEt_2$ is a monomer with three-coordinate Li.

BeMe₂(s) and MgMe₂(s) (Figure 13.16) are long-chain polymers having metals tetrahedrally coordinated by μ-Me. The bridging entails three-center bonding in which each metal contributes two empty sp^3 orbitals while each Me contributes one sp^3 hybrid filled with two electrons (Figure 13.17). In contrast, more electronegative Zn, Cd, and Hg form alkyls R₂M which are linear volatile monomers. However, Ph₂Zn crystallizes as a $PhZn(\mu\text{-}Ph)_2ZnPh$ dimer.

(AlMe₃)₂ (Figure 13.18) is a dimer having terminal and bridging methyl groups both in the solid and vapor phase. "Simple" organoaluminum compounds such as (Ph₃Al)₂ are dimers or, in some cases, such as (PhCH₂)₃Al, chain oligomers. A dimer–monomer equilibrium exists for straight-chain Al alkyls in solution in inert solvents. The equilibrium lies progressively farther toward the monomer with increasing chain length and steric bulk. (2,4,6-Me₃C₆H₂)₃Al is a monomer in solution. Ready formation of bridging alkyls provides a facile mechanism for alkyl exchange. Hence, mixtures of Al alkyls equilibrate with a statistical distribution of different alkyl groups around each Al. Bridging alkyls are involved in the ability of organometals to behave as good alkylating agents. The lower dialkyl aluminum halides R₂AlX form structures with μ-halides (Figure 13.19); the halide bridges involve only two-center bonds.

In contrast to Al, the trialkyls of less electropositive B, Ga, In, and Tl are all monomeric trigonal-planar molecules. B forms a wide range of compounds with μ-H which are treated in Chapter 17.

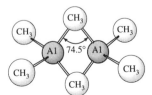

Figure 13.18   Structure of (AlMe₃)₂.

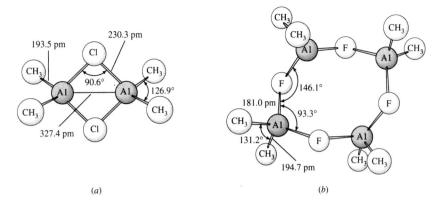

Figure 13.19 Molecular structure of (*a*) (Me₂AlCl)₄. (From K. Brendhaugen, A. Haaland, and D. P. Novak, *Acta Chem. Scand.* **1974,** *28A,* 45.) (*b*) Molecular structure of (Me₂AlF)₄. (From G. Gundersen, N. T. Haugen and A. Haaland, *J. Organomet Chem.* **1973,** *54,* 77.)

Availability of empty metal orbitals leads to Lewis acid behavior by Groups 1, 2, 12, and 13 compounds. Complexes such as Me₃B←NMe₃, Et₃Al←OEt₂, Li[ZnEt₄], and Ph₂Mg←OEt₂ are formed. Complexation of Li⁺ in LiR by tmeda which <u>enhances</u> the reactivity has been noted above. In contrast, trialkylboranes and trialkylalanes form strong complexes with ethers which <u>reduce</u> the reactivity by blocking the one remaining coordination site.

Solid-state structures of organo halides, alkoxides, and amides of the most electropositive elements are often polymeric as a result of Lewis acid–base behavior. Figure 13.20*a* shows the solid-state structure of EtZnI which consists of sheetlike layers in which each

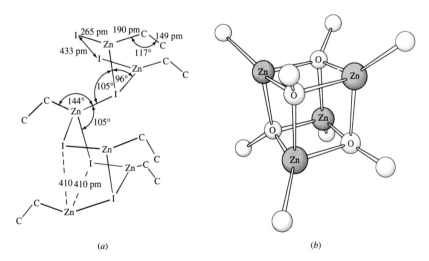

Figure 13.20 Structure of (*a*) EtZnI. (From P. T. Moseley and H. M. M. Shearer, *J. Chem. Soc. Chem. Commun.*. **1966,** 876.) (*b*) Structure of (EtZn(OMe))₄, with only Zn-attached C's shown. (From H. M. M. Shearer and C. B. Spencer, *J. Chem. Soc., Chem. Commun.* **1966,** 194.)

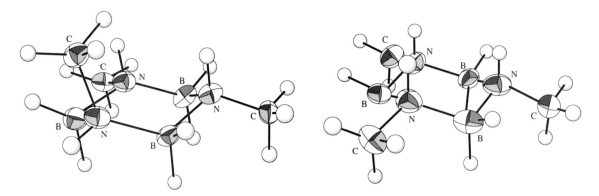

Figure 13.21   The structures of two isomers of (H₂BNHMe)₃ featuring six-membered B₃N₃ rings and differing in the axial versus equatorial orientation of one methyl group. (Reprinted with permission from C. K. Narula, J. K. Janik, E. N. Duesler, R. T. Paine, and R. Schaefer, *Inorg. Chem.* **1986**, *25*, 3346. Copyright 1986, American Chemical Society.)

Zn bonds to two additional I's, giving a four-coordinate environment. Reaction of ZnEt₂ with half an equivalent of alcohol eliminates CH₄ to give [EtZn(OR)]₄ in which Zn is again four-coordinate (Figure 13.20b). (EtZnCl)₄ and (EtZnBr)₄ have similar structures. In a reaction typical of Lewis acid–base addition products containing acidic H, H₃B←NH₂Me eliminates H₂, forming a borazane [H₂BNHMe]₃ (Figure 13.21; see also Section 17.9.1). The dimeric structure of RMgX is discussed next.

▶ 13.6.2 A Survey of Some Important Main-Group Organometals

Grignard Reagents

Grignard reagents are the best-known organometal halides. The product that crystallizes from an ether solution of EtMgBr displays distorted tetrahedral geometry, with two moles of diethyl ether completing the octet around Mg (Figure 13.22). Other oligomeric species are obtained from other Grignard reagents and solvents. The species present in solution

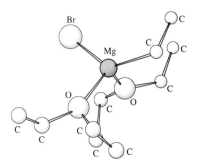

Figure 13.22   Structure of EtMgBr·OEt₂.

correspond to the Schlenck equilibrium:

$$R-Mg \overset{X}{\underset{X}{\diamond}} Mg-R \;\rlap{\rightleftharpoons}{}\; 2\,RMgX \;\rlap{\rightleftharpoons}{}\; R_2Mg + MgX_2 \qquad (13.62)$$

An equimolar solution of R_2Mg and MgX_2 is identical with a Grignard solution upon attainment of equilibrium. The equilibration rate falls in the order $X = Cl > Br > I$ and $R = 1° > 2° > 3°$. Equilibrium position is a function of R, X, and solvent. Basic thf displaces bridging X, shifting the equilibrium to the right while less basic Et_2O competes less effectively with the halide bridges. Colligative property measurements show that Grignard reagents are associated in Et_2O, with the degree of association increasing with concentration.

Alkylaluminums

Karl Ziegler discovered in 1960 that in the presence of some pre-formed $AlEt_3$, Al powder reacts with H_2, thereby forming Et_2AlH. If ethylene is introduced at 110°C under pressure, hydroalumination occurs followed by ethylene insertion into the Al—C bond, leading to alkyl chains which can contain up to 200 C atoms. (Insertions are a general class of reactions; they are discussed in Section 14.1.2.)

$$Et_2Al-H + CH_2{=}CH_2 \longrightarrow Et_2Al-CH_2CH_3 \qquad (13.63)$$

$$(13.64)$$

$$R_2Al-R + CH_2{=}CH_2 \longrightarrow R_2AlCH_2CH_2R \overset{CH_2=CH_2}{\longrightarrow} R_2Al(CH_2CH_2)_nR$$

In the Alfol process the chain length is allowed to grow to the desired number of carbons, and the alkyls are then oxidized and hydrolyzed to give long-chain alcohols. A distribution of chain lengths is obtained, and C_6–C_{10} alcohols are utilized in making plasticizers such as dialkylphthalates; C_{12}–C_{16} alcohols are utilized in making detergents and surfactants.

$$Al{\overset{(CH_2CH_2)_nEt}{\underset{(CH_2CH_2)_pEt}{-(CH_2CH_2)_oEt}}} + \tfrac{3}{2}O_2 \longrightarrow Al{\overset{O(CH_2CH_2)_nEt}{\underset{O(CH_2CH_2)_pEt}{-O(CH_2CH_2)_oEt}}} \qquad (13.65)$$

$$Al{\overset{O(CH_2CH_2)_nEt}{\underset{O(CH_2CH_2)_pEt}{-O(CH_2CH_2)_pEt}}} + 3\,H_2O \longrightarrow 3\,Et(CH_2CH_2)_{n,o,p}OH + Al(OH)_3 \qquad (13.66)$$

At higher temperatures the reverse of hydroalumination, namely β-elimination, competes with chain growth, and the Al—H bonds formed insert more ethylene, giving $AlEt_3$:

$$Al{\overset{(CH_2CH_2)_nEt}{\underset{(CH_2CH_2)_pEt}{-(CH_2CH_2)_oEt}}} \overset{CH_2=CH_2}{\longrightarrow} 3\,CH_2{=}CH(CH_2CH_2)_{n,o,p}H + AlEt_3 \qquad (13.67)$$

$AlEt_3$ then re-enters the cycle, thus achieving ethylene polymerization catalytically. With bulkier 1-propene, insertion stops at $n = 2$ and elimination occurs to give 2-methyl-pent-1-ene, a precursor to isoprene which is polymerized to yield synthetic rubber. (See Problem 13.8.)

Al alkyls are employed along with Ti as catalysts in the industrially important polymerization of ethylene to polyethylene (See Section 14.1.2.)

Silicones

Silicones are polymers $(-\overset{\overset{\displaystyle R}{|}}{\underset{\underset{\displaystyle R}{|}}{OSi}}-)_n$; the name was coined by analogy with ketones whose formulas are similar, but whose properties differ drastically. The C—O single bond energy (358 kJ/mol) is less than half the C=O energy (803 kJ/mol), leading to a preference for monomeric ketones $R_2C=O$. The Si—O single bond energy is 464 kJ/mol compared to only 640 kJ/mol for Si=O, leading to a polymeric structure for silicones.

Silicones are valuable for their extreme chemical inertness, water repellency, constancy of viscosity over a wide temperature range, (presumed) physiological inertness, and resistance to ultraviolet irradiation.

Production of silicones involves hydroylsis of organosilicon chlorides made by the direct process [Equation (13.48)] followed by thermal treatment with catalytic quantities of H_2SO_4. The reaction products of direct synthesis are only about 75% Me_2SiCl_2. Also present are $MeSiCl_3$ (10%), Me_3SiCl (4%), and $MeSiHCl_2$ (6%) as well as small amounts of $SiMe_4$, $SiCl_4$, $HSiCl_3$, and disilanes. Most of these play a role in polymerization. Hydrolysis of Me_2SiCl_2 gives silanols $Me_2Si(OH)_2$ which are more acidic than alcohols and dehydrate to give both cyclic and linear polymers:

$$2n\ Me_2Si(OH)_2 \xrightarrow{H^+} \underset{\text{cyclic}}{(Me_2SiO)_n} + \underset{\text{linear}}{H(OSiMe_2)_nOH} + (2n - 1)H_2O \qquad (13.68)$$

On heating with sulfuric acid, cyclic polymers are converted to linear chains, presumably by cleaving Si—O bonds to form sulfate esters and silanols which then recondense in a linear fashion. Me_3SiCl hydrolyzes and condenses to form $Me_3SiOSiMe_3$; these siloxanes are also cleaved by H_2SO_4 and when the sulfate esters recondense with linear chains, they provide end groups which terminate the chains:

$$Me_3Si-OSO_3H + H(OSiMe_2)_nOH \longrightarrow Me_3Si(OSiMe_2)_nOH + H_2SO_4 \qquad (13.69)$$

Chain lengths are controlled by the relative proportion of chain-forming groups ($-OSiMe_2-$) and chain-terminating groups (Me_3Si-) in the reactant mixture. Depending on chain length, silicone fluids, oils, greases, rubbers, or resins can be produced.

Chain-branching groups $(Me\overset{\overset{\displaystyle O}{|}}{Si}O-)$ are formed from $MeSiCl_3$, and cross-linking groups

$(-O\overset{\overset{\displaystyle O}{|}}{\underset{\underset{\displaystyle O}{|}}{Si}}O-)$ are formed from $SiCl_4$. These help increase polymer viscosity and are thus essential in making silicone rubbers and resins. Other cross-linking groups are also introduced. Addition of vinyl silicon halides provides vinyl groups which can condense with Si—H bonds (hydrosilation) from $MeSiHCl_2$ [Equation (13.70)]. Benzoyl peroxide reacts with Me groups, providing radicals which can condense to cross-link chains [Equation (13.71)]:

$$\underset{\overset{|}{O}}{\overset{\overset{O}{|}}{MeSi}}-CH{=}CH_2 + H-\underset{\overset{|}{O}}{\overset{\overset{O}{|}}{SiMe}} \longrightarrow \underset{\overset{|}{O}}{\overset{\overset{O}{|}}{MeSi}}-CH_2CH_2-\underset{\overset{|}{O}}{\overset{\overset{O}{|}}{SiMe}} \qquad (13.70)$$

$$ROOR \longrightarrow 2\,RO^{\cdot}$$

$$2\,RO^{\cdot} + 2\,\underset{\overset{|}{O}}{\overset{\overset{O}{|}}{MeSiMe}} \longrightarrow 2^{\cdot}\underset{\overset{|}{O}}{\overset{\overset{O}{|}}{CH_2SiMe}} \longrightarrow \underset{\overset{|}{O}}{\overset{\overset{O}{|}}{MeSi}}-CH_2CH_2-\underset{\overset{|}{O}}{\overset{\overset{O}{|}}{SiMe}} \qquad (13.71)$$

Chains of molar mass 5×10^5–10^7 are found in silicone rubbers; a cross-link occurs every 100–1000 Si atoms. SiO_2 is used as a reinforcing agent. These rubbers retain elasticity at extremely low temperatures, presumably because of the very low energy required for O—Si—O bending. Also, they are useful at higher temperatures than are possible with carbon-based rubbers. Silicone resins have a three-dimensional structure with cross-links provided by Me_3SiCl and $SiCl_4$ in the hydrolysis mixture. Varying the nature of the organic group can fine tune the properties of the resulting silicones.

▶ 13.6.3   Transition-Metal Alkyls

Thermochemical data indicate[20] that transition-metal–carbon bonds are considerably stronger (100–200 kJ/mol) than had been realized earlier, though still somewhat weaker than M—F, M—OR, or M—Cl bonds (300–400 kJ/mol). The instability of transition metal alkyls thus is of kinetic rather than thermodynamic origin, and so these species can be stabilized by blocking reaction pathways. β-elimination [Equation (13.61)] is a principal decomposition pathway and requires a vacant metal orbital and a vacant coordination site to form a hydride–olefin complex from an alkyl. Hence, ligands that are strongly bonded and occupy all coordination sites stabilize alkyls. Examples include Cp, Cp*, CO, and PR_3, as well as NH_3, in substitution-inert $[Rh(NH_3)_5C_2H_5]^{2+}$. Another strategy is to employ alkyls containing no β-H's such as CH_3, $CH_2C_6H_5$, 1-norbornyl, CH_2CMe_3 (neopentyl), and CH_2SiMe_3 (trimethylsilylmethyl). The last three groups are quite bulky and also tend to block attack at all coordination positions. Thus, stable homoleptic alkyls such as $Cr(CH_2SiMe_3)_4$, WMe_6, $[Li(thf)_4][Co(1\text{-nor})_4]$ (nor = norbornyl), $Co(1\text{-nor})_4$, and $[Co(1\text{-nor})_4]BF_4$ have been prepared. Because alkyls are excellent σ donors, they are capable of stabilizing high oxidation states such as Co^{IV} and Co^{V}. Aryls tend to be stable because they also contain no β-H, and complexes such as $CrPh_3(thf)_3$ and $Pt(PPh_3)_2(Ph)I$ are known.

[20] J. A. Connor, *Top. Curr. Chem.* **1977**, *71*, 71.

Synthesis and Structure

Methods applicable to main group alkyls and aryls are also important in making transition-metal alkyls and aryls. Metathesis using main-group compounds is widely employed as in the following examples. Li reagents and Grignard reagents are most reactive whereas alkyl aluminums and alkyls of more electronegative metals give incomplete reaction.

$$CrCl_3(thf)_3 + 3\,MeMgBr \longrightarrow CrMe_3(thf)_3 + 3\,MgClBr(s) \qquad (13.72)$$

$$TiCl_4 + Me_2AlCl \longrightarrow MeTiCl_3 + MeAlCl_2 \qquad (13.73)$$

$$TiCl_4 + 4\,MeLi \xrightarrow{\text{ether}} Me_4Ti + 4\,LiCl(s) \qquad (13.74)$$

The stability of carbonylate anions permits a different type of metathesis resulting from $S_N 2$ displacement. See also Equations (12.23) and (13.17).

$$Na[Cp(CO)_2Fe] + C_6H_5CH_2Cl \longrightarrow CpFe(CO)_2CH_2C_6H_5 + NaCl(s) \qquad (13.75)$$

Addition of a transition-metal hydride to an alkene (hydrometallation) is also possible:

$$Pt(H)(PEt_3)_2Cl + C_2H_4 \longrightarrow Pt(C_2H_5)(PEt_3)_2Cl \qquad (13.76)$$

Reaction (13.76) is often referred to as an **insertion reaction.** Such reactions will be discussed more fully in Section 14.1.2.

Because transition metals have more than one stable oxidation state, they sometimes undergo oxidative coupling. In Equation (13.77) a diolefin complex is converted to a dialkyl with a change in oxidation state from Fe^0 to Fe^{II}:

$$Fe^0(CO)_3(\eta^2\text{-}C_2F_4)_2 \xrightarrow{\text{CO}} (CO)_4\overset{\displaystyle CF_2CF_2}{\underset{\displaystyle CF_2CF_2}{Fe^{II}}} \qquad (13.77)$$

Two other preparative methods (oxidative-addition and extrusion reactions) will be discussed in Section 14.1.3.

The structures of transition metal alkyls and aryls generally involve discrete molecules, mostly monomers.

The most important reactivity mode for transition metal alkyls and aryls is their involvement in insertion reactions which will be discussed in Section 14.1.2.

► *13.6.4 Metal Acyl Complexes*

Metal acyls (which are unknown for all main-group metals except Si) are related to alkyl complexes. They are prepared from carbonylate anions and acyl halides [see also Equation (12.24)]:

$$[CpFe(CO)_2]^- + C_6H_5\overset{O}{\overset{\|}{C}}Cl \longrightarrow CpFe(CO)_2\overset{O}{\overset{\|}{C}}C_6H_5 + Cl^- \qquad (13.78)$$

and by insertion reactions (see Section 14.1.2). Their chemistry is discussed further in Chapter 14.

13.7   COMPOUNDS WITH MULTIPLE METAL–CARBON BONDS

Carbenes, alkylidenes, carbynes, and alkylidynes were discussed in Chapter 12. Here we give fragment analyses of bonding in carbenes and alkylidenes and mention a few added aspects of their chemistry. Figure 13.23 shows a fragment analysis of the bonding in a simple carbene complex. The filled b_1 orbital on CH_2 is one of the two C—H bonding MOs. The net result of σ donation and π acceptance by the carbene is that C has a positive charge and characteristically behaves as an electrophile. Even though carbonyls have higher positive charge, the LUMO ($3b_2$ in Figure 13.23) is localized on the carbene C, leading to orbital-controlled reactivity.

Figure 13.24 shows a fragment analysis of bonding in a carbyne complex. Carbyne and alkylidyne chemistry is much less well understood than that of carbenes and alkylidenes. Instances are known of both nucleophiles and electrophiles which attack at C and at metals. Both charge and orbital control of reactions occur, and it is currently not possible to predict which will be favored. This means that several HOMOs and several LUMOs are rather close together in energy. A few patterns have emerged, though. In cationic com-

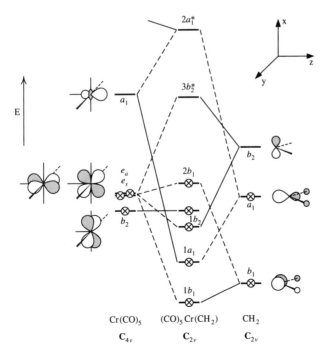

Figure 13.23   Bonding in (hypothetical) $(CO)_5Cr=CH_2$ showing HOMOs and LUMOs of the fragments and MOs produced from their combination. (After P. Hofmann, in *Transition Metal Carbene Complexes,* Verlag Chemie, Deerfield Beach, FL, 1983.)

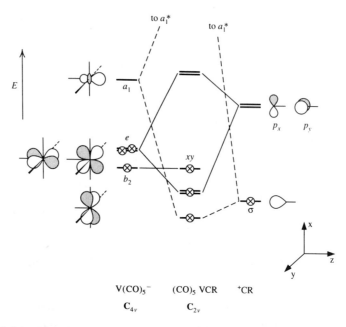

Figure 13.24 Bonding in (hypothetical) (CO)₅V≡CR showing HOMOs and LUMOs of the fragments and MOs produced from their combination. R is assumed to have twofold or higher symmetry around the V–C axis. (After P. Hofmann in *Transition Metal Carbyne Complexes*, VCH Publishers, New York, 1988.)

Table 13.1 Summary of M—C σ-bond types

Bond type	Structure	Ligand charge	C ligand donates
Alkyl	M—C—	−1	2e
Acyl	M—C—C—	−1	2e
Vinyl	M—C=C<	−1	2e
Acetylide	M—C≡C—	−1	2e
Carbene	M=C<	0	2e
Alklylidene	M=C<	−2	4e
Vinylidene	M=C=C<	0	2e
Carbyne	M≡C—	+1	2e
Alkylidyne	M≡C—	−3	6e

plexes the carbyne C is electrophilic and reversal of the preparative route yields novel carbene complexes, especially for anionic nucleophiles:

$$[(CO)_5M\equiv CNR_2]^+ + Nu^- \longrightarrow (CO)_5M=C(Nu)NR_2$$

$$M = Cr, Mo, W; Nu = F, Cl, Br, I, CN, NCO, NCS, SPh, NPh_2, \text{ and so on.}$$

(13.79)

In neutral complexes, the electrophilicity of the carbyne C is lowered so that CO substitution sometimes occurs preferentially to attack on the carbyne C:

$$\text{trans-}Br(CO)_4Cr\equiv CPh + PPh_3 \longrightarrow \text{mer-}Br(PPh_3)(CO)_3Cr\equiv CPh + CO \quad (13.80)$$

With the better nucleophile PMe_3 addition to C is seen:

$$\text{trans-}Br(CO)_4Cr\equiv CPh + PMe_3 \longrightarrow \text{trans-}Br(CO)_4Cr=C(PMe_3)Ph \quad (13.81)$$

Electrophiles can add to the $M\equiv C$ bond in neutral species, and this is the usual pattern for protonation:

$$Cl(PMe_3)_4W\equiv CH + CF_3SO_3H \longrightarrow [Cl(PMe_3)_4W=CH_2]^+[CF_3SO_3]^- \quad (13.82)$$

Table 13.1 summarizes M—C bonded species, including several with metal–carbon multiple bonds. Figure 13.25 shows how ligands with M—C bonds can be interconverted by treatment with electrophiles and nucleophiles; nonbonding pairs on the metal are involved in these reactions.

Another way of thinking about the differences between Fischer and Schrock carbenes is that, in cases of metals having electron-withdrawing ligands (Fischer-type carbenes), interaction between filled metal orbitals and the empty carbene p_π is small, leaving a positive charge on C and localizing the LUMO there. In complexes with very electron-donating ancillary ligands (Schrock-type carbenes), enhanced metal basicity results in good back-donation to C, giving it a negative charge and delocalizing the LUMO over both metal and C. This argument rationalizes the behavior of some recently prepared carbenes of Ru, Os, and Ir which have a chemistry intermediate between that of Fischer- and Schrock-type carbenes. In particular, d^6 $Cl_2(CO)(PPh_3)_2Ru=CF_2$ reacts with MeOH to give $Cl_2(CO)(PPh_3)_2Ru=CF(OMe)$, but not with electrophiles. However, d^8 $(CO)_2(PPh_3)_2Ru=CF_2$ reacts with HCl to add H^+ to C and coordinate the Cl^- to Ru, giving $Cl(CO)_2(PPh_3)_2RuCF_2H$.[21]

---

[21] M. A. Gallop and W. R. Roper, *Adv. Organomet. Chem.* **1986,** 25, 121.

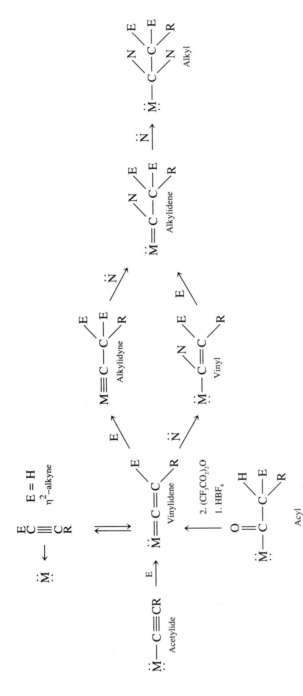

Figure 13.25 Interconversions among metal–carbon bonded species. E's are electrophiles such as H^+ and Me^+; N's are nucleophiles such as H^-. Charges are not shown. M represents a metal with ancillary ligands. (After K. R. Birdwhistle, T. L. Tonker, and J. L. Templeton, *J. Am. Chem. Soc.* **1985,** *107,* 4474.)

GENERAL REFERENCES

J. P. Collman, L. S. Hegedus, J. R. Norton, and R. G. Finke, *Principles and Applications of Organotransition Metal Chemistry*, University Science Books, Mill Valley, CA, 1987. Authoritative and extensive treatment of the field.

Ch. Elschenbroich and A. Salzer, *Organometallics: A Concise Introduction*, 2nd ed., VCH Publishers, New York, 1992. A good introduction presented in outline format.

F. R. Hartley and S. Patai, Eds., *The Chemistry of the Metal-Carbon Bond*, Wiley, Chichester, 1985–87. Four volumes of excellent reviews.

P. Powell, *Principles of Organometallic Chemistry*, 2nd ed., Chapman and Hall, London, 1988. A very readable update of the classic text by Coates, Green, and Wade.

F. G. A. Stone and R. West, Eds., *Advances in Organometallic Chemistry*, Academic Press, San Diego. A series of volumes published yearly containing authoritative reviews on organometallic chemistry.

G. Wilkinson, F. G. A. Stone, and E. W. Abel, Eds., *Comprehensive Organometallic Chemistry*, Pergamon Press, Oxford, 1982. A nine-volume treatise with extensive treatment of all aspects of organometallic chemistry.

PROBLEMS

13.1 Give the valence electron count for the following species. Which ones obey the 18-*e* rule?

(a) $CpMo(CO)_3(\eta^1\text{-}C_3H_5)$

(b) $Ir(PPh_3)_2(CO)Cl(\eta^2\text{-}C_3H_6)$

(c) $Cp_2Ti(CO)(\eta^2\text{-}C_2Ph_2)$

(d) $[(\eta^4\text{-}C_4Ph_4)PdCl_2]_2$

(e) $[(\eta^5\text{-}C_7H_9)Ru(\eta^6\text{-}C_7H_8)]^+$

(f) $CpMn(\eta^6\text{-}C_6H_6)$

(g) $Pb[CH(SiMe_3)_2]_2$

(h) $[Me_2Al(\mu\text{-}NMe_2)]_2$

(i) Cp_2^*Sn

(j) $[Li(thf)_4][Mn_2(\mu\text{-}Ph)_2Ph_4]$

(k) $Li[CuMe_2]$

(l) $[(\eta^3\text{-}C_3H_5)_2(\mu\text{-}Cl)Rh]_2$

13.2 Name each species in Problem 13.1.

13.3 $PhC{\equiv}CPh$ has a $C{\equiv}C$ distance of 119.8 pm and $\nu_{C{\equiv}C}$ = 2223 cm^{-1}. Formulate the best VB structure for each of the following compounds in light of the structural and spectroscopic data given:

(a) $CpRe(CO)_2(C_2Ph_2)$: d_{C-C} = 123.2 pm; $< C{-}C{-}Ph$ = 151.6°; $\nu_{C{\equiv}C}$ = 1848 cm^{-1}.

(b) $[CpMo(CO)(PPh_3)(C_2Ph_2)]^+$: d_{C-C} = 129 pm; $< C{-}C{-}Ph$ = 135° and 141°; $\nu_{C{\equiv}C}$ = 1645 cm^{-1}.

13.4 Give an MO analysis of the bonding between the allyl$^-$ and $Co(CO)_3^+$ fragments in $(\eta^3\text{-}C_3H_5)Co(CO)_3$. Draw pictorial representations of interacting frontier orbitals. Give a qualitative MO diagram.

13.5 (a) The allylic H in propene is easily lost to give an allyl. A reaction thought to be important in catalysis is the so-called 1,3-hydride shift. An allylic H of an η^2-propene adds to the metal, giving an η^3-allyl hydride product with the metal's oxidation number increased by two. Show how the reversibility of this process can afford a mechanism for shifting H from C_1 to C_3—a so-called **1,3-hydride shift.**

(b) Account for the following observation by a mechanism involving the 1,3-hydride shift:

In the complex [structure with D] $Fe(CO)_4$, olefin is isomerized to give [structure with (1/3D) labels] $Fe(CO)_4$, and D is scrambled to all methyl groups. (See C. P. Casey and C. R. Cyr, *J. Am. Chem. Soc.* **1973**, *95*, 2248.)

13.6 $[(\eta^5\text{-}C_5Me_5)Ir(acetone)_3]^{2+}$ reacts with $CH_2=CH_2$ in acetone at room temperature to produce $[(\eta^5\text{-}C_5Me_5)IrC_5H_9]^+$, where C_5H_9 represents more than one organic ligand. Given the following NMR data, what is the structure? (See J. B. Wakefield and J. M. Stryker, *Organometallics* **1990**, *9*, 2428.)

^{1}H NMR δ (splitting, relative intensity): 4.11 (triplet of triplets, 1H), 3.79 (doublet of doublets, 2H), 3.11 (singlet, 2H), 2.60 (singlet, 2H), 2.04 (doublet of doublets, 2H), 1.91 (singlet, 15H)

^{13}C NMR δ (splitting): 100.6 (quartet), 86.3 (doublet), 47.6 (triplet), 47.0 (broad triplet), 8.5 (quartet)

13.7 Explain how the facts that $(\eta^6\text{-}C_6H_5CO_2H)Cr(CO)_3$ is a stronger acid than benzoic acid and that $(\eta^6\text{-}C_6H_5NH_2)Cr(CO)_3$ is a weaker base than aniline show that the $Cr(CO)_3$ group withdraws electrons from the aromatic rings.

13.8 Write a mechanism for dimerization of propene to 2-methylpent-1-ene using Al, H_2, and a catalytic quantity of $AlEt_3$.

13.9 **(a)** The butene product from the reaction

$$CpFe(CO)(PPh_3)(n\text{-}C_4H_9) \longrightarrow CpFe(CO)(PPh_3)H + butene$$

consists of 1-butene as well as *cis*- and *trans*-2-butene. Keeping in mind the reversibility of the β-hydride elimination, write a mechanism to account for this.
(b) When the reaction in **a** is run with $CpFe(CO)(PPh_3)(CD_2CH_2Et)$, complete scrambling of D occurs. Write a mechanism to account for this observation.

13.10 Coordination of ligands to transition metals often makes them subject to nucleophilic attack. Make a list of reactions given in this chapter involving nucleophilic attack on coordinated ligands.

13.11 Identify the following reactions by type and predict the products; for example, $Mo(CO)_6 + PPh_3$ is CO substitution by a nucleophile, giving $Mo(CO)_5(PPh_3) + CO$.

   (a) $Fe(CO)_5 + EtCO_2CH=CHCO_2Et \xrightarrow{h\nu}$
   (b) $[CpMo(CO)_3]^- + C_3H_5Br \rightarrow \mathbf{1} \xrightarrow{h\nu} \mathbf{2}$
   (c) $[\{Os(\eta^4\text{-}C_8H_{12})Cl_2\}_x] + Sn(n\text{-}Bu_2)_2(C_5Me_5)_2$
   (d) $[1\text{-}4\text{-}\eta^4\text{-}C_6H_8]Fe(CO)_3 + Ph_3C^+$
   (e) $EtMgI + SnCl_4$
   (f) $[CpFe(CO)_2]^- + C_6H_5CH_2Cl$
   (g) $MeLi$ (excess) $+ CuCl_2$
   (h) $Cp_2Fe + n\text{-}BuLi$
   (i) $(CO)_5CrC(OMe)Ph + BI_3$
   (j) $Li + C_6H_5Cl \xrightarrow{ether}$
   (k) $(CO)_5MoC(OMe)Ph + CH_2=CHCO_2Et$
   (l) $(Me_3CCH_2)_3NbCHCMe_3 + PMe_3$
   (m) $Cp(CO)_2WCMe + HI$
   (n) $[(CO)_5WCNPh_2]^+ + CN^-$
   (o) $[CpRe(CO)_2(\eta^3\text{-}C_3H_5)]^+ + CH(CO_2Et)_2^-$

·14·

Organometallic Reactions,
Mechanisms, and Catalysis

· · · · · · · · · · · · · · · · · · · ·

> The theme of this chapter is the remarkable transformations, both stoichiometric and catalytic, which can be carried out using organometallic compounds. Although organic chemistry has long been characterized by organization into "type" reactions (nucleophilic substitutions, electrocyclic ring closures, etc.), inorganic chemistry is just now becoming amenable to such organizing principles. Hence, we begin by considering four fundamental reaction types: ligand substitutions, insertions, oxidative additions, and reactions of coordinated ligands. Some features of the first and last classes have already been mentioned in Chapter 12. Here, we also concentrate on what is known about reaction mechanisms. We then apply this knowledge to industrially important chemistry of transition-metal organometallics.

14.1 FUNDAMENTAL REACTIONS

▶ *14.1.1 Substitution in Carbonyl Complexes*[1]

Numerous examples of substituted carbonyls are discussed in Section 12.2.3. Here we concentrate on mechanistic aspects of ligand substitution reactions.

Rate Laws

Mechanistic studies on organometallic compounds are considerably fewer than those on coordination compounds (see Chapter 11). However, interesting and similar reactivity

[1] For a review, see J. A. S. Howell and P. M. Burkinshaw, *Chem. Rev.* **1983,** *83,* 557.

patterns have been revealed. Typical substitution behavior is exhibited by d^6 metal carbonyls with Lewis bases. For the reaction

$$LM(CO)_5 + L' \longrightarrow LL'M(CO)_4 + CO \qquad (14.1)$$

(where L may also be CO), a two-term rate law is observed:

$$\text{rate} = k_1[LM(CO)_5] + k_2[LM(CO)_5][L'] \qquad (14.2)$$

As usual, the two terms reflect two parallel substitution paths. For some reactions the second-order path is undetectably slow ($k_2 \ll k_1$).

Significance of the Rate Law

d *Mechanisms.* It is generally agreed that the first-order path is **D** (Section 11.3) in character and involves rate-determining CO dissociation. For substitution reactions of Group 6 hexacarbonyls, relative values of k_1 are 100 : 1 : 0.01 for Mo : Cr : W; these do not parallel the mean M—CO bond dissociation energies [Cr (109 kJ/mol) < Mo(150) < W(176)]. The same reactivity trend with the second-transition-series metal forming the most reactive compound is observed in other substitution reactions also. In compounds containing noncarbonyl ligands, the identity of L is important in determining the rate and site for CO dissociation. A study of ^{13}CO exchange with $Mn(CO)_5Br$[2] showed that equatorial carbonyls exchange much faster than axial ones. This would be understandable on the basis of the ground-state properties of the molecule. In a simple bonding model,[3] both the d_{xz} and d_{yz} orbitals on Mn are shared by Br^- and the *trans* carbonyl; hence, Br^- (which has filled π orbitals) is able to act as a π donor toward *trans* CO, thus strengthening the Mn—C_{trans} bond. The d_{xz} orbital is also shared among Br^- and a pair of equatorial CO's that are *trans* to one another (likewise for d_{yz}). Hence, bond-strengthening to each equatorial CO is (to a first approximation) half as much as to axial CO, and equatorial carbonyls are more labile. (See Figure 14.1.)

The argument given above might lead you to expect that *cis*-carbonyl ligands in $Mn(CO)_5Br$ should be less labile to substitution than those in $[Mn(CO)_6]^+$, which has no π-donor ligands. Instead, the bromide complex is many orders of magnitude more reactive, much beyond the difference accounted for by its smaller charge. As another example, $Cr(CO)_5(PPh_3)$ substitutes about 100 times as fast as $Cr(CO)_6$, whereas $[Cr(CO)_5Br]^-$ is 10^7 as reactive. Thus, ground-state effects cannot account for the faster CO displacement in $M(CO)_nL$ (where L is a poorer π acid) than in $M(CO)_{n+1}$. L exerts its influence by stabilizing a five-coordinate transition state or intermediate resulting from dissociation of a *cis* CO. Calculations indicate that the initially formed square pyramid relaxes to a trigonal bipyramid with equatorial L; in this geometry, an empty metal d orbital is of proper symmetry to be stabilized by donation from a filled π orbital of L.[4]

---

[2] J. D. Atwood and T. L. Brown, *J. Am. Chem. Soc.* **1975**, *97*, 3380.

[3] A more complete model would use *p–d* hybrids, but this would not change the symmetry argument on which the conclusion is based.

[4] J. D. Atwood and T. L. Brown, *J. Am. Chem. Soc.* **1976**, *98*, 3160; D. L. Lichtenberger and T. L. Brown, *J. Am. Chem. Soc.* **1978**, *100*, 366; W. G. Jackson, *Inorg. Chem.* **1987**, *26*, 3004; T. L. Brown, *Inorg. Chem.* **1989**, *28*, 3229.

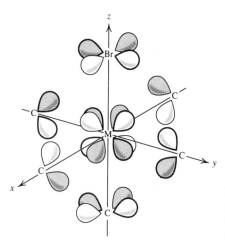

Figure 14.1 Orbital overlap in $Mn(CO)_5Br$. The d_{yz} orbital and the ligand orbitals that overlap with it are outlined in black. d_{xz} and its overlapping orbitals are not outlined.

$$\text{(diagram)} \longrightarrow \text{(diagram)} \longrightarrow OC\!-\!M\!-\!CO \xrightarrow{+L'} cis\text{-}MLL'(CO)_4 \cdot \qquad (14.3)$$

Hence L = I⁻, py, Br⁻, Cl⁻, NO₃⁻, and so on, enhance the rate of *cis* ligand dissociation. The final products are usually the *cis* complexes indicated in Equation (14.3)—but not always, since the intermediate may last long enough to rearrange before reacting with L'. This is why CO exchange studies are so important in establishing the pattern of *cis* CO activation: The principle of microscopic reversibility requires entry of ¹³CO into the position vacated by the leaving carbonyl, whereas no such requirement exists for L' ≠ CO. Even when *cis*-$M(CO)_4LL'$ is the kinetic product, rearrangement often occurs to the *trans* product when it is favored on steric grounds. Electronic considerations favor the *cis* isomer where each noncarbonyl ligand is *trans* to the excellent π-acid CO.

Complexes derived from trigonal-bipyramidal $M(CO)_5$ and tetrahedral $M(CO)_4$ ordinarily substitute by CO dissociation. Steric effects can also come into play in dissociative substitutions. For example, the rate of CO substitution in $Ru(CO)_4(PR_3)$ yielding *trans*-$Ru(CO)_3(PR_3)_2$ follows the order of decreasing cone angle $PPh_3 > PPh_2Me > PBu_3$.

The second-order term in Equation (14.2) most often represents the operation of an I_d mechanism with considerable bond-making in the transition state. Activation parameters bear out this interpretation. For the k_1 pathway, large positive values of both $\Delta H_1^\ddagger$ and $\Delta S_1^\ddagger$ are seen (Table 14.1), whereas the k_2 pathway typically displays smaller positive $\Delta H_2^\ddagger$ and negative $\Delta S_2^\ddagger$.

a *Mechanisms.*   In particular cases, the k_2 path may represent associative (**a**) activation. The first common case is attack on a carbonyl ligand by basic reagents.[5] CO is polarized

[5] T. L. Brown and P. A. Bellus, *Inorg. Chem.* **1978**, *17*, 3727; P. A. Bellus and T. L. Brown, *J. Am. Chem. Soc.* **1980**, *102*, 6020.

Table 14.1   Activation parameters for some carbonyl substitutions[a]

Reaction		$\Delta H_1^{\ddagger}$, kJ/mol	$\Delta S_1^{\ddagger}$, J/mol K	$\Delta H_2^{\ddagger}$, kJ/mol	$\Delta S_1^{\ddagger}$, J/mol K
$Cr(CO)_6 + {}^*CO \rightarrow Cr(CO)_5({}^*CO) + CO$		161.9	77.4		
$Mo(CO)_6 + {}^*CO \rightarrow Mo(CO)_5({}^*CO) + CO$		126.4	-1.7		
$W(CO)_6 + {}^*CO \rightarrow W(CO)_5({}^*CO) + CO$	gas	166.5	46		
$Cr(CO)_6 + PBu_3 \rightarrow Cr(CO)_5(PBu_3) + CO$		168.0	94	106.6	-61
$Mo(CO)_6 + PBu_3 \rightarrow Mo(CO)_5(PBu_3) + CO$		132.5	28	90.7	-62
$W(CO)_6 + PBu_3 \rightarrow W(CO)_5(PBu_3) + CO$		166.8	58	122.0	-29
$Mn(CO)_5X + AsPh_3 \rightarrow cis\text{-}Mn(CO)_4(AsPh_3)X + CO$					
X = Cl		114.9	66		
X = Br in CHCl$_3$		124.6	79		
X = I		134.6	87		

[a] From J. A. S. Howell and P. M. Burkinshaw, *Chem. Rev.* **1983,** *83,* 557 and references therein.

in the sense shown below when coordinated to cationic or neutral metals (see Section 12.2.1):

$$M^{(\delta-)} \leftarrow C^{(\delta+)} \equiv O$$

Sufficiently nucleophilic reactants can attack the positive C:

$$[Mn(CO)_6]^+ + MeLi \longrightarrow Me\overset{\overset{\displaystyle O}{\displaystyle \|}}{C}Mn(CO)_5 \qquad (14.4)$$

In Reaction (14.5), excess amine deprotonates N, leading to a coordinated carbamoyl:

$$[Mn(PPh_3)(CO)_5]^+ + MeNH_2 \longrightarrow cis\text{-}[Mn(PPh_3)(CO)_4\overset{\overset{\displaystyle O}{\displaystyle \|}}{C}NH_2Me]^+ \qquad (14.5)$$

$$cis\text{-}[Mn(PPh_3)(CO)_4\overset{\overset{\displaystyle O}{\displaystyle \|}}{C}NH_2Me]^+ + MeNH_2 \longrightarrow cis\text{-}Mn(PPh_3)(CO)_4\overset{\overset{\displaystyle O}{\displaystyle \|}}{C}NHMe + NH_3Me^+$$

By employing a base that could not undergo deprotonation, Bellus and Brown demonstrated that the second-order term in the rate law for py substitution in $[Mn(CO)_5(MeCN)]^+$ represents attack on CO. The cation is known to be stable in CH_3NO_2 for a week or more and not to undergo exchange with ${}^{13}CO$ in $C_2H_5NO_2$ over at least 72 hr at 30°C. On addition of excess py, however, the following reaction occurs rapidly:

$$[Mn(CO)_5(MeCN)]^+ + MeNO_2 + 3\,py \longrightarrow$$
$$fac\text{-}[Mn(CO)_3(py)_3]^+ + MeCN + CO + CO_2 + CH_2NOH \qquad (14.6)$$

The mechanism proposed is shown below. **d** substitution lability at the *cis* position is en-

hanced by the $\overset{\text{O}}{\overset{\|}{\text{C}}}CH_2NO_2^-$ ligand. This pathway represents the equivalent of base hydrolysis for carbonyl complexes. It is also a redox reaction in which CO is oxidized and CH$_3$NO$_2$ is reduced to CH$_2$NOH.

$$\text{py} + \text{CH}_3\text{NO}_2 \rightleftharpoons {}^-\text{CH}_2\text{NO}_2 + \text{pyH}^+$$

$$[\text{Mn(CO)}_5(\text{MeCN})]^+ + {}^-\text{CH}_2\text{NO}_2 \rightleftharpoons cis\text{-}[\text{Mn(CO)}_4(\text{MeCN})(\overset{\text{O}}{\overset{\|}{\text{C}}}\text{CH}_2\text{NO}_2)]$$

$$cis\text{-}[\text{Mn(CO)}_4(\text{MeCN})(\overset{\text{O}}{\overset{\|}{\text{C}}}\text{CH}_2\text{NO}_2)] \rightleftharpoons [\text{Mn(CO)}_4(\overset{\text{O}}{\overset{\|}{\text{C}}}\text{CH}_2\text{NO}_2)] + \text{MeCN}$$

$$[\text{Mn(CO)}_4(\overset{\text{O}}{\overset{\|}{\text{C}}}\text{CH}_2\text{NO}_2)] + \text{py} \longrightarrow cis\text{-}[\text{Mn(CO)}_4(\text{py})(\overset{\text{O}}{\overset{\|}{\text{C}}}\text{CH}_2\text{NO}_2)]$$

$$cis\text{-}[\text{Mn(CO)}_4(\text{py})(\overset{\text{O}}{\overset{\|}{\text{C}}}\text{CH}_2\text{NO}_2)] \longrightarrow [\text{Mn(CO)}_3(\text{py})(\overset{\text{O}}{\overset{\|}{\text{C}}}\text{CH}_2\text{NO}_2)] + \text{CO} \quad (14.7)$$

$$[\text{Mn(CO)}_3(\text{py})(\overset{\text{O}}{\overset{\|}{\text{C}}}\text{CH}_2\text{NO}_2)] + \text{py} \longrightarrow fac\text{-}[\text{Mn(CO)}_3(\text{py})_2(\overset{\text{O}}{\overset{\|}{\text{C}}}\text{CH}_2\text{NO}_2)]$$

$$fac\text{-}[\text{Mn(CO)}_3(\text{py})_2(\overset{\text{O}}{\overset{\|}{\text{C}}}\text{CH}_2\text{NO}_2)] \longrightarrow [\text{Mn(CO)}_3(\text{py})_2]^+ + \text{CO}_2 + \text{CH}_2\text{NO}^-$$

$$[\text{Mn(CO)}_3(\text{py})_2]^+ + \text{py} \xrightarrow{\text{fast}} fac\text{-}[\text{Mn(CO)}_3(\text{py})_3]^+$$

$$\text{pyH}^+ + \text{CH}_2\text{NO}^- \longrightarrow \text{CH}_2\text{NOH} + \text{py}$$

By way of contrast, reaction of $[\text{Mn(CO)}_5(\text{MeCN})]^+$ with nucleophilic (but weakly basic) PPh$_3$ does not involve attack on CO and is 10^2 slower than Reaction (14.6).

The second common instance of **a** mechanisms involves direct metal attack. This is most common for 16- or 14-electron complexes or for 18-e complexes when a ligand is capable of removing an electron pair from coordination or of delocalizing a pair of electrons to avoid a 20-e intermediate. V(CO)$_6$ (a stable 17-e compound) reacts with phosphines displaying only the second-order term in the rate law, and the rate varies with the basicity of the phosphine as expected for an **a** mechanism (Table 14.2). Steric factors also come into play; for example, bulky PPh$_3$ reacts more slowly than might be predicted from its basicity. Activation parameters for the associative reaction with PMePh$_2$ are $\Delta H_2^{\ddagger} = 37$ kJ/mol and $\Delta S_2^{\ddagger} = -108$ J/mol K. Seventeen- and nineteen-electron species are not usually thermodynamically stable and are most often transition states or intermediates rather than isolable complexes. However, they need not be present in large concentration to be kinetically significant, and the high associative reactivity of 17-e radicals involving 19-e transition states is an important theme in organometallic chemistry (*vide infra*).

Indenyl and cyclopentadienyl ligands can convert from η^5- to η^3-coordination to accommodate the electron pair of the attacking reagent. The "indenyl effect" is apparent in substitution of $[(\eta^5\text{-C}_9\text{H}_7)\text{Mn(CO)}_3]$ by P(OEt)$_3$, which has a second-order rate constant of 4.0×10^{-5} M^{-1} s^{-1} at 120°C in decalin. The analogous Cp complex is completely inert to

Table 14.2 Data for the reaction
$V(CO)_6 + L \rightarrow V(CO)_5L + CO^a$

L	Cone angle (*deg*)	ΔHNP^b	log k_2
PMe$_3$	118	114	2.12
P(*n*-Bu)$_3$	132	131	1.70
PMePh$_2$	136	424	0.60
P(O*i*-Pr)$_3$	135	520	−0.03
P(OMe)$_3$	107	580	−0.16
PPh$_3$	143	573	−0.61
P(*i*-Pr)$_3$	160	33	−0.98

[a] From Q. Z. Shi, T. G. Richmond, W. C. Trogler, and F. Basolo, *J. Am. Chem. Soc.* **1984,** *106,* 71.

[b] ΔHNP is the half-neutralization potential measured with respect to guanidine as the zero. The lower the value, the more basic toward H$^+$ is the phosphine.

substitution under these conditions. η^5- to η^3-ring slippage is promoted by the fused benzene ring which conjugates with the electron pair removed from coordination. Figure 13.10 depicts such ring slippage in an isolable compound. The fluorenyl analogue of the Mn complex has two benzene rings fused to the five-membered ring; it substitutes faster yet, having a $k_2 = 1.1 \times 10^{-2} M^{-1}s^{-1}$. Allyl ligands can rearrange from η^3- to η^1-coordination.

Nitrosyls can rearrange from the "NO$^+$" to the "NO$^-$" mode of coordination (see Section 12.2.3) with delocalization of an electron pair.[6] Thus V(CO)$_5$NO reacts with phosphines and amines exclusively via a second-order path below −15°C. (At higher temperatures and with weak nucleophiles a dissociative mechanism begins to operate.) The great π-acidity of NO weakens the *trans* V—CO bond, and the substitution products are *trans*-V(CO)$_4$(NO)L.

Free-Radical Substitution. Free-radical mechanisms are beginning to be increasingly apparent in organometallic chemistry.[7] Often these are initiated by traces of adventitious substances, so that the kinetics frequently are not reproducible. An early example was substitution of HRe(CO)$_5$ involving labile 17-*e* radicals and a chain reaction:

$$\text{Int}^{\cdot} + \text{HRe(CO)}_5 \longrightarrow \text{IntH} + \text{Re(CO)}_5^{\cdot} \quad (\text{Int} = \text{initiator}) \quad \text{Initiation}$$
$$\text{Re(CO)}_5^{\cdot} + \text{L} \longrightarrow \text{Re(CO)}_4\text{L}^{\cdot} + \text{CO} \qquad\qquad (14.8)$$
$$\text{Re(CO)}_4\text{L}^{\cdot} + \text{HRe(CO)}_5 \longrightarrow \text{HRe(CO)}_4\text{L} + \text{Re(CO)}_5^{\cdot} \quad \text{Propagation}$$

[6] In fact, a number of 19-*e* radicals (in which the unpaired electron is delocalized onto an NO ligand) are known to be fairly stable in solution. Examples include Fe(NO)$_3$(CO), Co(NO)$_2$(CO)$_2$, and CpW(NO)$_2$[P(OPh)$_3$].

[7] D. R. Tyler, *Prog. Inorg. Chem.* **1988,** *36,* 125.

Initiation can also be accomplished photochemically when a catalytic amount of $Re_2(CO)_{10}$ is added, since irradiation produces $Re(CO)_5{}^{\cdot}$. Quantum yields of >1 are common for photochemical initiation. That is, for every mole of photons injected, more than one mole of product is obtained due to the chain propagation step which requires no photon energy.

A second class of free-radical substitutions is the so-called **electron-transfer chain** (ETC) reactions. A generalized example starting with 18-e $M(CO)_n$ is

$$
\begin{array}{lll}
M(CO)_n + e & \longrightarrow & [M(CO)_n]^{\overline{\cdot}} \quad \text{Initiation} \\
[M(CO)_n]^{\overline{\cdot}} & \longrightarrow & [M(CO)_{n-1}]^{\overline{\cdot}} + CO \\
[M(CO)_{n-1}]^{\overline{\cdot}} + L & \longrightarrow & [M(CO)_{n-1}L]^{\overline{\cdot}} \qquad\qquad (14.9) \\
[M(CO)_{n-1}L]^{\overline{\cdot}} + M(CO)_n & \longrightarrow & M(CO)_{n-1}L + [M(CO)_n]^{\overline{\cdot}} \quad \text{Propagation} \\
[M(CO)_n]^{\overline{\cdot}} & \longrightarrow & M(CO)_n + e \quad \text{Termination}
\end{array}
$$

The initiating radical can be produced by chemical, electrochemical, or photochemical reduction. This 19-e species dissociatively substitutes CO, and the chain is propagated by electron transfer from the 19-e substituted radical to the starting material. A necessary condition for operation of this path is that L make $M(CO)_{n-1}L$ a poorer electron acceptor than $M(CO)_n$. This is likely to happen when L is a poorer π acceptor than CO. An example of an ETC substitution is

$$
Mo(CO)_4(bipy) + PPh_3 \longrightarrow Mo(CO)_3(PPh_3)(bipy) + CO \qquad (14.10)
$$

Oxidative initiation can also occur. In this case, reaction proceeds through 17-e radicals and electron transfer must take place from the starting complex to $[M(CO)_nL]^{\overset{+}{\cdot}}$. A reaction which proceeds via oxidative ETC substitution is

$$
Mo(CO)_4(py)_2 + 2\,PPh_3 \longrightarrow Mo(CO)_4(PPh_3)_2 + 2\,py \qquad (14.11)
$$

in which a weaker π-acceptor is replaced by a stronger one.

An ETC mechanism is likely in the base-induced disproportionations of metal carbonyls to produce carbonylate anions [Equations (12.16) and (12.17)]. For Reaction (12.17) the following mechanism applies:

$$
\begin{array}{lll}
Co_2(CO)_8 + Co(CNR)(CO)_4{}^{\cdot} & \longrightarrow & [Co_2(CO)_8]^{\overline{\cdot}} + [Co(CNR)(CO)_4]^+ \\
[Co_2(CO)_8]^{\overline{\cdot}} & \longrightarrow & [Co(CO_4)]^- + Co(CO)_4{}^{\cdot} \\
Co(CO)_4{}^{\cdot} + CNR & \longrightarrow & Co(CNR)(CO)_4{}^{\cdot} \qquad\qquad (14.12)\\
[Co(CNR)(CO)_4]^+ + 4\,CNR & \xrightarrow{\text{fast}} & [Co(CNR)_5]^+ + CO
\end{array}
$$

A number of reactions go by parallel free-radical and conventional 16- and 18-e paths.

EXAMPLE: For the reaction $Ru(CO)_4(PR_3) + PR_3 \rightarrow Ru(CO)_3(PR_3)_2 + CO$, activation parameters are as follows. What do they suggest about the substitution mechanism?

R_3	$\Delta H_1^{\ddagger}$ (kJ/mol)	$\Delta S_1^{\ddagger}$ (J/mol K)
Ph$_3$	124.0	59
MePh$_2$	132	77

Solution: Only a first-order term is observed in the rate law. $\Delta H_1^{\ddagger}$ and $\Delta S_1^{\ddagger}$ are quite large and positive. This is consistent with a **D** mechanism.

EXAMPLE: $W(CO)_4(NO)X$ undergoes carbonyl substitution with phosphines in toluene. With PPh$_3$ a one-term rate law is observed, whereas a two-term rate law is seen with $P(n$-Bu$)_3$. What can you say about the mechanism given the following activation parameters? (See Y. Sulfab, F. Basolo, and A. L. Rheingold, *Organometallics* **1989**, *8*, 2139.)

X	$\Delta H_1^{\ddagger}$ (kJ/mol)	$\Delta S_1^{\ddagger}$ (J/mol K)	$\Delta H_2^{\ddagger}$ (kJ/mol)	$\Delta S_2^{\ddagger}$ (J/mol K)
PPh$_3$				
Cl	27.8	18.4		
Br	28.1	15.8		
I	30.7	21.3		
$P(n$-Bu$_3)$				
Cl	Not measured		9.6	−39.0
Br	40.8	56.2	15.3	−20.8
I	Not measured		16.3	−20.3

Solution: The independence from [PPh$_3$] suggests only a dissociative pathway. In agreement with this are (a) large positive enthalpies of activation as would be required for W—CO bond-breaking and (b) positive entropies of activation as expected for production of two particles from one in the transition state. The enthalpies of activation increase in the order Cl < Br < I. As the halogens become larger and more polarizable, they become better donors to the metal, thereby increasing back-bonding to the CO π^* orbitals and making the W—CO bond stronger and harder to dissociate. For $P(n$-Bu$)_3$, the first-order term in the rate law represents a dissociative mechanism whereas the second-order term represents an associative one. This can be seen from the small enthalpies of activation and the negative entropies. $P(n$-Bu$)_3$ is expected to be more nucleophilic than PPh$_3$ and can attack the large metal atom, giving a seven-coordinate 18-e intermediate with coordinated NO$^-$ draining off the "extra" electron pair.

Summary

Several reactivity patterns for substitution are now established: (1) Production by ligand dissociation of intermediates with reduced coordination number; such coordinately unsaturated species are encountered more often in organometallic chemistry than with coordination compounds. (2) Activation of ligands *cis* to non-π-acid ligands toward dissociation. The group that dissociates is the one with the weakest bond to the metal, and the π- donor stabilizes the transition state. (3) Nucleophilic attack on carbonyl (and other) ligands polarized by coordination to give stable compounds or reactive intermediates. (4) Direct nucleophilic attack on metals. (5) Free-radical mechanisms, sometimes involving one-electron transfer processes.[8]

Redox (electron-transfer) reactions of metal complexes can be investigated by **cyclic voltammetry.** An electrode is immersed in a solution of the complex and voltage is swept while current flow is monitored. No current flows until oxidation or reduction occurs. After the voltage is swept over a set range in one direction, the direction is reversed and swept back to the original potential. The cycle may be repeated as often as desired. Figure 14.2 shows the cyclic voltammogram (CV) for a reversible one-electron redox reaction such as

$$CpFe(CO)LMe \rightleftharpoons CpFe(CO)LMe^+ + e^- \tag{14.13}$$

Sweeping the potential in an increasing direction oxidizes the complex as the anodic current i_a flows; reversible reduction of $CpFe(CO)LMe^+$ generates cathodic current i_c on the reverse sweep. The magnitude of the current is proportional to the concentration of the species being oxidized or reduced.

Values of the voltages required for electron transfer (the half-wave potentials) reflect the thermodynamic tendency of species to be reduced or oxidized. Such measurements demonstrated[9] the following values of the potentials E^0 (measured in CH_2Cl_2 relative to the potential of acetylferrocene) for oxidation of $CpFe(CO)LMe$: -486 mV for L = PMe_3, -205 mV for L = $P(p\text{-}CF_3C_6H_4)_3$, and -602 mV for L = PCy_3. This shows that $CpFe(CO)(PCy_3)Me$ is easiest to oxidize (i.e., the least positive potential is required) and that PCy_3 is the most effective net donor L. (This is in line with general results quoted in Chapter 12 on σ-donor and π-acceptor properties of phosphines and results from both a high pK_a and a large cone angle of 170° which prevents close approach for optimum π back-donation.)

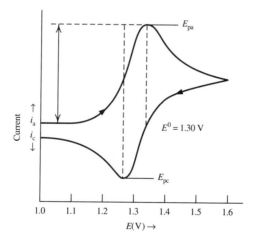

Figure 14.2   A reversible one-electron reduction and oxidation.

[8] For an extensive treatment of free-radical processes, see J. K. Kochi, *Organometallic Mechanisms and Catalysis,* Academic Press, New York, 1978. See also P. R. Jones, *Adv. Organomet. Chem.* **1977**, *15*, 273.

[9] A. Prock, W. P. Giering, J. E. Greene, R. E. Meirowitz, S. L. Hoffman, D. C. Woeska, M. Wilson, R. Chang, J. Chen, R. Magnuson, and K. Erics, *Organometallics* **1991**, *10*, 3479.

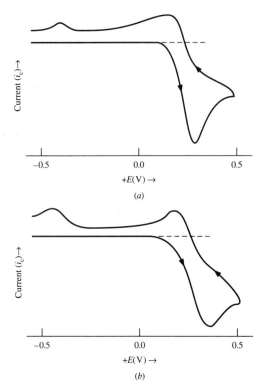

Figure 14.3   Cyclic voltammogram for CpFe(PPh$_3$)(CO)Me in CH$_2$Cl$_2$ containing MeCN. (*a*) Sweep rate is 0.1 V/s. (*b*) Sweep rate is 1.6 V/s. (Reprinted with permission from D. C. Woeska, M. Wilson, J. Bartholomew, K. Eriks, A. Prock, and W. P. Giering, *Organometallics* **1992**, *11*, 3343.)

In contrast, CVs in CH$_2$Cl$_2$ containing MeCN (Figure 14.3) show a decrease in i_c at +0.2 V on the return sweep relative to i_a, indicating that the concentration of CpFe(CO)LMe$^+$ has been depleted before reduction could occur. This could indicate decomposition and/or reaction to form a different product which is reduced at another potential. The appearance of a second cathodic wave at −0.5 V indicates production of a new species. At a constant sweep rate, the ratio of i_c for the −0.5-V peak to i_c for the +0.2-V peak was found to increase linearly with [MeCN], showing that MeCN is involved in the production of the species reduced at −0.5 V. This was identified in separate experiments as CpFeL(NCMe)C(O)Me$^+$. The faster sweep rate used for Figure 14.3*b* shows a greater total i_c/i_a ratio, indicating that more cationic products are detected before they have a chance to decompose. Kinetic information can be obtained by monitoring changes in i as the sweep rate is varied.

Depending on the chemistry expected, the initial sweep direction can be to higher or lower voltages.

▶ 14.1.2   *Insertion Reactions*[10]

A large class of reactions involves **insertion** of small molecules X—Y into metal–ligand bonds, especially M—C and M—H bonds. As shown in Equations (14.14) and (14.15), both 1,1 and 1,2 insertions are possible.

[10] For a review, see J. J. Alexander in *The Chemistry of the Metal-Carbon Bond*, Vol. 2, F. R. Hartley and S. Patai, Eds., Wiley, Chichester, 1985. p 339.

$$M—L + X—Y \longrightarrow M—\overset{\overset{\displaystyle Y}{|}}{X}—L \qquad (14.14)$$

$$M—L + X—Y \longrightarrow M—X—Y—L \qquad (14.15)$$

The term "insertion" *describes only the result of the reaction;* it has no mechanistic significance. Many insertion reactions are reversible, and the reverse reaction is called **extrusion** or **elimination.** Table 14.3 shows a representative sampling of insertion reactions involving transition metals. We will discuss only insertions of CO, SO_2, and olefins.

CO Insertion

A good deal of what is known about CO "insertion" (or **carbonylation**)[11] comes from studies on $RMn(CO)_5$ as a model. $CH_3Mn(CO)_5$ reacts with ^{14}CO to give $CH_3\overset{\overset{\displaystyle O}{\|}}{C}Mn(CO)_4(^{14}CO)$; that is, the inserted CO is one previously coordinated to the metal, not the labeled one added to the complex. This suggests that other Lewis bases L can also bring about insertion. The generally accepted mechanism is

$$RMn(CO)_5 + Sol(= solvent) \; \rightleftharpoons \; R\overset{\overset{\displaystyle O}{\|}}{C}Mn(CO)_4(Sol)$$

$$R\overset{\overset{\displaystyle O}{\|}}{C}Mn(CO)_4(Sol) + L \; \longrightarrow \; cis\text{-}R\overset{\overset{\displaystyle O}{\|}}{C}Mn(CO)_4L + Sol \qquad (14.16)$$

$$cis\text{-}R\overset{\overset{\displaystyle O}{\|}}{C}Mn(CO)_4L \; \rightleftharpoons \; trans\text{-}R\overset{\overset{\displaystyle O}{\|}}{C}Mn(CO)_4L$$

An acyl intermediate is formed with solvent occupying one coordination position (if the solvent is basic enough). Increasing the size of solvent molecules when the basicity is similar hinders formation of the solvated acyl and slows down carbonylation. The Lewis base L enters to give the *cis* complex, which then equilibrates with the *trans* isomer. Both basicity and size of L influence its ability to capture the intermediate before it reverts back to the alkyl. Alkyl complexes of Mo, Fe, Co, Rh, and Ir behave similarly.

The labeling experiment depicted in Figure 14.4 shows that the Mn alkyl group moves during the carbonylation because the product distribution observed is just that predicted by the "alkyl migration" model. (See also Problems 12.26 and 14.9.) Electron-withdrawing R groups migrate more slowly than electron-donating ones, presumably because of the greater strength of the Mn—R bond which must be seriously weakened in the transition state. For example, $CF_3Mn(CO)_5$ does not insert even under 333 atm of CO and temperatures up to 200°C even though $CH_3Mn(CO)_5$ inserts at 1 atm at room tempera-

[11] A. A. Wojcicki, *Adv. Organomet. Chem.* **1973**, *11*, 87; F. Calderazzo, *Angew. Chem. Int. Ed. Engl.* **1977**, *16*, 299; E. J. Kuhlmann and J. J. Alexander, *Coord. Chem. Rev.* **1980**, *33*, 195.

Table 14.3   **Some representative insertion reactions**[a]

"Inserted" molecule	Bond	Product
CO	M—CR$_3$	MC(O)CR$_3$
	M—OH	MC(O)OH
	M—NR$_2$	MC(O)NR$_2$
CO$_2$	M—H	MO$_2$CH
	M—R	MC(O)OR
	M—NR$_2$	MOC(O)NR$_2$
	M—OH, M—OR	MOCO$_2$H(R)
CS$_2$	M—M	MSC(S)M
	M—H	MS$_2$CH and MSC(S)H
C$_2$H$_4$	M—H	MC$_2$H$_5$
C$_2$F$_4$	M—H	MCF$_2$CF$_2$H
RC≡CR′	M—H	MC(R)=CH(R′) *cis* or *trans*
CH$_2$=C=CH$_2$	M—R	M(η^3-allyl)
SnCl$_2$	M—M	MSn(Cl)$_2$M
RNC	M—H	MCH=NR
	M—R′	MCR′=NR
	M—η^3-C$_3$H$_5$	MC(=NR)(CH$_2$CH=CH$_2$)
RNCS	M—H	M(RNCHS)
RN=C=NR	M—H	M(RNCHNR)
SO$_2$	M—C	MS(R)(O)$_2$ or MOS(O)R
	M—M	MOS(O)M
	M—(η^3-C$_3$H$_5$)	MS(O)$_2$CH$_2$CH=CH$_2$
SO$_3$	M—CH$_3$	M—OS(O)$_2$CH$_3$
O$_2$	M—H	M—OOH
	M—CR$_3$	M—OOCR$_3$, MOCR$_3$
S$_8$	M—CR$_3$	MSCR$_3$
CH$_2$N$_2$	M—Cl	MCH$_2$Cl

[a] From F. A. Cotton and G. Wilkinson, *Advanced Inorganic Chemistry* 5th ed., Wiley, New York, 1988.

ture. Configuration is retained at the alkyl C. Stereochemistry at the metal can be quite variable. Reaction of optically active CpFe*(CO)(PR$_3$)R with CO leads to inversion or retention, depending on the solvent.

Many acyl complexes may be decarbonylated either thermally or photochemically. As required by the principle of microscopic reversibility, a terminal CO is lost followed by migration of R to fill the vacant coordination position. For example,[12]

$$\tag{14.17}$$

[12] J. J. Alexander, *J. Am. Chem. Soc.* **1975**, *97*, 1729.

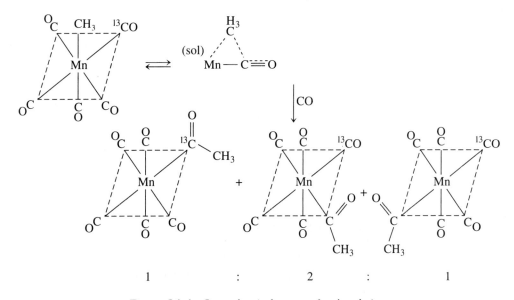

Figure 14.4   Stereochemical course of carbonylation.

SO$_2$ Insertion

In contrast to CO, SO$_2$ inserts directly into M—C bonds, often producing an *S*-sulfinate.[13] For example, CpFe(CO)$_2$CH$_3$ gives CpFe(CO)$_2\overset{\overset{\text{O}}{\|}}{\underset{\underset{\text{O}}{\|}}{\text{S}}}CH_3$ when refluxed in SO$_2(l)$.

Kinetic studies have shown that SO$_2$ behaves as a Lewis acid attacking the alkyl ligand rather than the metal (Figure 14.5) so that more electron-donating alkyls react faster. As indicated in the figure, backside attack leads to inversion of configuration at C. Also in contrast to CO, extrusion of SO$_2$ occurs much less readily and its insertion is seldom reversible.

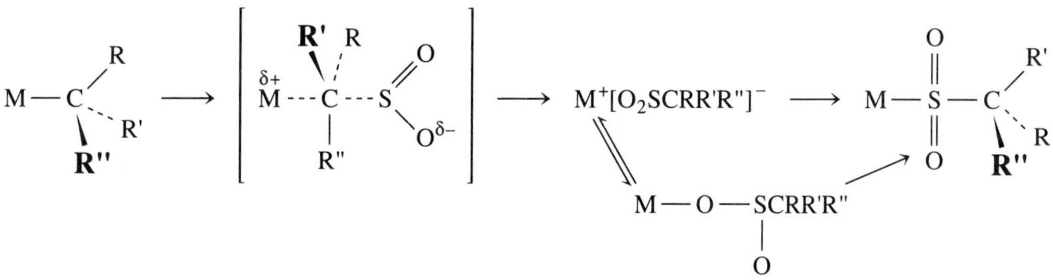

Figure 14.5   Mechanism of SO$_2$ insertion. (See A. Wojcicki, *Adv. Organomet. Chem.* **1974**, *12*, 31.)

[13] Products with different structures can be formed with other metals, as Table 14.3 indicates.

Olefin Insertion

Olefins can insert into M—C and M—H bonds. When the metal has sufficient d electrons for back-bonding, the olefin is often coordinated before insertion. This may require ligand dissociation to vacate a coordination position. For example,

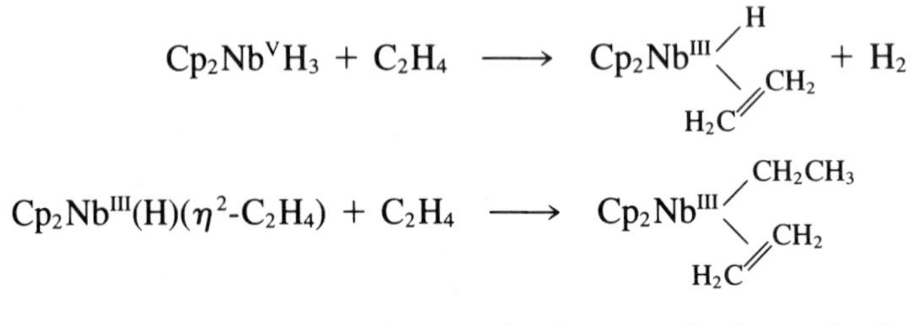

$$Cp_2Nb^VH_3 + C_2H_4 \longrightarrow Cp_2Nb^{III}\overset{H}{\underset{H_2C}{<}}_{CH_2} + H_2 \tag{14.18}$$

$$Cp_2Nb^{III}(H)(\eta^2\text{-}C_2H_4) + C_2H_4 \longrightarrow Cp_2Nb^{III}\overset{CH_2CH_3}{\underset{H_2C}{<}}_{CH_2}$$

Insertion converts a hydrido- or alkyl-olefin complex into an alkyl complex by addition of H or C to the β-carbon of the olefin. Note that the metal oxidation state does not change.[14] In contrast to CO, olefins often undergo multiple insertions providing a mechanism for their polymerization (*vide infra*). [However, the final product shown in Reaction (14.18) resists further insertion.]

The reverse of olefin insertion into an M—H bond is the β-elimination reaction of metal alkyl complexes, which can occur only if a vacant orbital and a vacant coordination position are available and β-H is present (Figure 14.6).[15] If these conditions are fulfilled, a facile route exists for the decomposition of metal alkyl complexes (see also Section 13.6.1). The initial product is a hydride/olefin complex which may then go on to dissociate the olefin.

A study[16] of the thermal decomposition of CpFe(CO)(PPh$_3$)(n-C$_4$H$_9$) [Equation (14.19)] revealed that the reaction

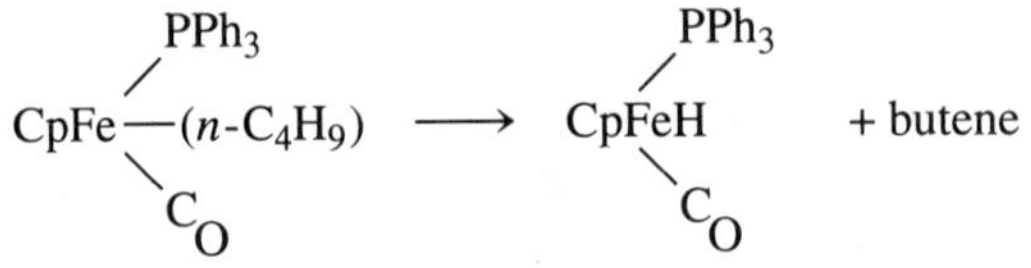

$$\overset{PPh_3}{\underset{CO}{CpFe}}\text{—}(n\text{-}C_4H_9) \longrightarrow \overset{PPh_3}{\underset{CO}{CpFeH}} + \text{butene} \tag{14.19}$$

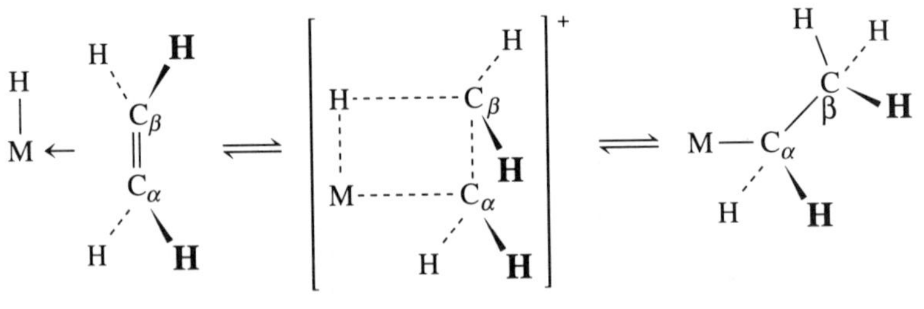

Figure 14.6   Olefin insertion and β-elimination.

---

[14] This insertion is related to attack on coordinated olefins by nucleophiles (Section 13.1), except that here the hydride or alkyl nucleophile is precoordinated to the metal.

[15] α-Eliminations are also known [see Equation (12.38)]. These are much rarer than β-eliminations and are most often seen with early transition metals. α-Eliminations are promoted by steric crowding around the metal, addition of phosphines (which may abstract the α-H or increase steric crowding), and lack of β-H.

[16] D. L. Reger and E. C. Culbertson, *J. Am. Chem. Soc.* **1976**, *98*, 2789.

was severely retarded by the addition of PPh_3. This suggests that dissociation of PPh_3 to create a vacant coordination position on Fe is a necessary step in the decomposition via β-elimination for this 18-*e* complex.

Fundamental features of olefin insertion were brought out in an investigation of some Sc complexes[17] ($Cp^* = \eta^5\text{-}Me_5C_5$).

$$Cp_2^*ScR + CH_2{=}CH_2 \longrightarrow Cp_2^*ScCH_2CH_2R$$
$$R = H, CH_3, CH_2CH_3, CH_2CH_2CH_3$$
(14.20)

Cp_2^*ScR are coordinatively unsaturated d^0, 14-*e* compounds stabilized by bulky electron-donating Cp^* ligands. They represent ideal models for studying insertion and β-elimination because no complications arise from the need to dissociate other ligands to provide vacant coordination sites nor from the kinetics of olefin coordination because this does not happen for d^0 complexes which are incapable of back donation.

The rate of ethylene insertion by Cp_2^*ScR decreases in the order $R = H \gg CH_2(CH_2)_nCH_3$ ($n \geq 2$) $\geq CH_2CH_2CH_3 > CH_3 > CH_2CH_3$. This ordering can be understood in terms of a cyclic, planar, four-center transition state involving incipient formation of a bond between the metal and an olefin carbon. This creates a partial positive charge on the β-carbon, and H or alkyl C migrates as an anion to β-carbon. Metal and R add in *cis* fashion to the olefin.

$$L_nM \overset{CH_2}{\underset{R}{\diagdown}} CHR \rightleftarrows L_nM \overset{\overset{H_2}{C}}{\underset{R^{\delta-}}{\cdots}} C^{\delta+}HR \rightleftarrows L_nMCH_2CHR'R$$
(14.21)

H inserts fastest because its nondirectional *s* orbital can overlap more effectively with the β—C orbital than can alkyl sp^3 hybrids. Among alkyls, rates decrease with increasing strength of the M–alkyl bond except for R = Et. The ground state of Cp_2^*ScEt is apparently stabilized by agostic interaction of the unsaturated metal with the electrons in a C—H bond, which is impossible for larger alkyls due to steric hindrance of the Cp^* rings and for Me because of its size and geometry.

The rates of β—H elimination from a series of substituted complexes $Cp_2^*ScCH_2CH_2C_6H_4p\text{-}X$ showed that the rate decreased with the electron-withdrawing power of X, indicating development of positive charge on the β-carbon in the transition state.

Ethylene polymerization catalyzed heterogeneously by Ti^{III} salts and Al alkyls (Ziegler–Natta catalysts[18]) surely involves insertion. R_3Al could alkylate the surface of a $TiCl_3$ crystal. Coordination of ethylene to a vacant Ti site may be followed by insertion, coordination of another ethylene, and so on:

$$R_3Al + TiCl_3 \longrightarrow R{-}Ti \xrightarrow{C_2H_4} R\underset{\overset{|}{CH_2{=}CH_2}}{-}Ti \longrightarrow RCH_2CH_2Ti$$
$$\xrightarrow{C_2H_4} RCH_2CH_2\underset{\overset{|}{CH_2{=}CH_2}}{-}Ti \longrightarrow R(CH_2CH_2)_2Ti \xrightarrow{C_2H_4} \text{etc.}$$
(14.22)

[17] B. J. Burger, M. E. Thompson, W. D. Cotter, and J. E. Bercaw, *J. Am. Chem. Soc.* **1990**, *112*, 1566.

[18] See H. Sinn and W. Kaminsky, *Adv. Organomet. Chem.* **1980**, *18*, 99.

In spite of the reactivity of M—H bonds toward olefin insertion, only two examples of CO insertion to give an isolable $M\overset{\overset{\displaystyle O}{\|}}{C}H$ formyl complex have yet been found—namely, with HRh(octaethylprophyrin) and Cp$_2$ThH(OR). This likely results from the greater M—H than M—C bond strength which makes the reactions thermodynamically unfavorable.

▶ 14.1.3 Oxidative Addition[19]

The Reaction and Its Scope

A ubiquitous reaction in organometallic chemistry is **oxidative addition,** in which a low-valent transition-metal complex reacts with a molecule XY to yield a product in which both the oxidation number and coordination number of the metal are increased. An example is

$$\begin{array}{c}\text{Cl}\diagdown\underset{\diagup}{\overset{\diagup}{\text{Ir}}}\diagup\text{PPh}_3 \\ \text{Ph}_3\text{P}\diagup \quad \diagdown\text{C}_\text{O}\end{array} + \text{Cl}_2 \longrightarrow \begin{array}{c}\text{Cl} \\ \text{Cl}\diagdown \; | \diagup\text{PPh}_3 \\ \text{Ir} \\ \textbf{Ph}_3\textbf{P}\diagup | \diagdown\text{C}_\text{O} \\ \text{Cl}\end{array} \tag{14.23}$$

The d^8, four-coordinate, 16-e starting complex is converted into a d^6, six-coordinate, 18-e product by addition of Cl$_2$. The oxidation number of Ir increases from I to III. Consistent with this change, $\nu_{C\equiv O}$ increases from 1967 cm^{-1} to 2075 cm^{-1}. The Cl—Cl bond is broken and two Ir—Cl bonds are formed. Various molecules (see Table 14.4) react in this fashion, including XY = Cl$_2$, Br$_2$, I$_2$, HX, RX, RC(O)X, H$_2$, R$_3$SiH, and R$_3$GeH.[20] As Table 12.7 shows, oxidative-addition products of carbonyl complexes exhibit increased values of $\nu_{C\equiv O}$ compared to the starting complexes.

The high C—H bond energy (~470 kJ/mol) and their low polarity make C—H bonds relatively unreactive in general. Oxidative addition of C—H forms M—C bonds and opens the way to further reaction of the organic group on the metal (insertion, for example). Addition to a metal is said to "activate" the C—H bond (see Section 14.1.4). Oxidative addition of H$_2$ affords a way of "activating" the rather strong H—H bond (430 kJ/mol) to reaction by first coordinating it to a metal as two hydride ligands which could then undergo further reaction. The first step in oxidative addition of H$_2$ might be regarded as coordination of the intact molecule in an η^2 fashion. (Complexes containing η^2-H$_2$ ligands were discussed in Section 12.7.) That η^2-H$_2$ complexes represent the first step in H—H bond-breaking is seen by the fact that they are sometimes in equilibrium in solution with dihydride complexes as shown by variable-temperature nuclear magnetic resonance (NMR) for W(CO)$_3$(Pi-Pr$_3$)$_2$(H$_2$), for example. The water molecule (average bond

---

[19] For reviews, see J. P. Collman and W. R. Roper, *Adv. Organomet. Chem.* **1968,** 7, 53 and J. K. Stille in *The Chemistry of the Metal-Carbon Bond,* Vol. 2, F. R. Hartley and S. Patai, Eds., Wiley, Chichester, 1985, p. 625.

[20] Oxidative addition reactions could be thought of as insertions into the X—Y bond. However, the term *insertion* conventionally is reserved for reactions involving no change in metal oxidation number.

Table 14.4 Substances that can be added oxidatively[a]

Atoms separate	Atoms remain attached
$X-X$	O_2
H_2,	SO_2
Cl_2, Br_2, I_2, $(SCN)_2$, RSSR	
	$CF_2=CF_2$, $(CN)_2C=C(CN)_2$
$C-C$	
	$RC\equiv CR'$
Ph_3C-CPh_3, $(CN)_2$ C_6H_5CN,	RNCS
$MeC(CN)_3$	
	RNCO
$H-X$	
	$RN=C=NR'$
HCl, HBr, HI, $HClO_4$, C_6F_5OH, C_6H_5SH, H_2S, H_2O,	
CH_3OH	
$C_6F_5NH_2$, ⬡NH, $HC\equiv CR$, C_5H_6, CH_3CN, HCN,	
HCO_2R, C_6H_6, C_6F_5H, HR, $HSiR_3$, $HSiCl_3$	
H-$B_{10}C_2HPMe_2$, H-B_5H_8	
	$RCON_3$
$C-X$	
	$R_2C=C=O$
CH_3I, C_6H_5I; CH_2Cl_2, CCl_4	
CH_3COCl, $C_6H_5CH_2COCl$	
C_6H_5COCl, CF_3COCl	CS_2
	$(CF_3)_2CO$, $(CF_3)_2CS$, CF_3CN
$M-X$	
Ph_3PAuCl, $HgCl_2$, $MeHgCl$, R_3SnCl, R_3SiCl, $RGeCl_3$	
H_8B_5Br, Ph_2BX	
Ionic	
H^+, PhN_2^+, BF_4^-, Ph_3C^+, BF_4^-	

[a] From F. A. Cotton and G. Wilkinson *Advanced Inorganic Chemistry, 5th ed.*, Wiley, New York, 1988.

energy 493 kJ/mol) has recently been activated by oxidative addition to $[Ir(PMe_3)_4]^+$ to give *cis*-$Ir(H)(OH)(PMe_3)_4]^+$.[21]

Other Group 8, 9, and 10 complexes having d^8 metals also undergo oxidative addition; the tendency to become oxidized to d^6 increases on going down a triad or to the left, as shown in Figure 14.7.

We know of cases where early transition metals undergo oxidative addition also: d^6 $W^0 \rightarrow d^4$ W^{II}, for example.

$$W(CO)_4(bipy) + SnCl_4 \longrightarrow W(CO)_3(bipy)(SnCl_3)Cl \qquad (14.24)$$

We can view reactions of the main-group elements as oxidative additions—for example, the addition of F_2 to BrF_3 and PF_3, yielding BrF_5 and PF_5.

[21] D. Milstein, J. C. Calabrese, and I. D. Williams, *J. Am. Chem. Soc.* **1986**, *108*, 6387.

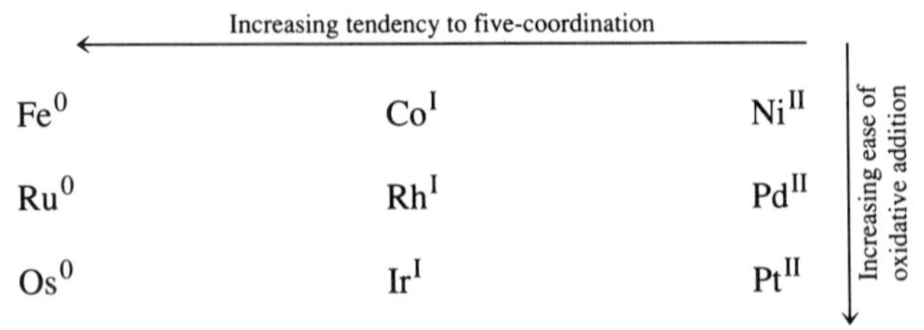

← Increasing tendency to five-coordination

Fe^0	Co^I	Ni^{II}
Ru^0	Rh^I	Pd^{II}
Os^0	Ir^I	Pt^{II}

Increasing ease of oxidative addition ↓

Figure 14.7 Reactivity of Group 8, 9, and 10 d^8 metal atoms and ions. (From J. P. Collman and W. R. Roper, *Adv. Organomet. Chem.* **1968**, *7*, 54.)

One-Electron Oxidative Addition. $d^7 \rightarrow d^6$. $[Co(CN)_5]^{2-}$ undergoes free-radical oxidative addition in one-electron steps:

$$[\cdot Co(CN)_5]^{2-} + RX \longrightarrow [Co(CN)_5X]^{2-} + R\cdot$$
$$R\cdot + [\cdot Co(CN)_5]^{2-} \longrightarrow [RCo(CN)_5]^{2-}$$
$$\overline{}$$
$$2[Co(CN_5]^{2-} + RX \longrightarrow [RCo(CN)_5]^{2-} + [Co(CN)_5X]^{2-} \tag{14.25}$$

$d^8 \rightarrow d^7$. Some dinuclear complexes undergo oxidative addition at two centers:

$$[Rh(\mu_2\text{-}CN(CH_2)_3NC)_4Rh]^{2+} + 4Br_2 \longrightarrow [BrRh\mu_2\text{-}(CN(CH_2)_3NC)_4RhBr](Br_3)_2 \tag{14.26}$$

Addition of Oxygen—An Intermediate Case. Some molecules $X=Y$ undergo oxidative addition without cleavage of the $X-Y$ bond. A good example is addition of O_2 to $Ir(PPh_3)_2(CO)X$. The adduct could be described formally as either of the two valence bond (VB) structures shown below:

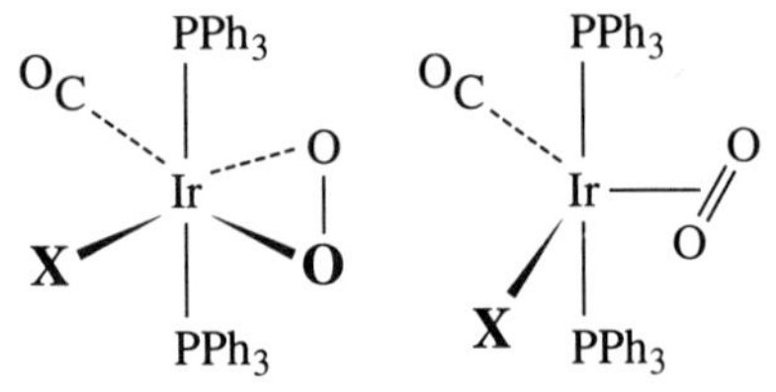

(a)   (b)

Structure **a** involves a six-coordinate Ir^{III} complex of O_2^{2-}, whereas structure **b** involves a five-coordinate Ir^I complex of neutral O_2. As usual, structural data can help make the distinction. For $X = Cl$, the $O-O$ distance is 130 pm—comparable to that in the superoxide ion O_2^- (128 pm). This indicates transfer of electron density from the metal into the π^* orbitals of O_2. When $X = I$, the $O-O$ distance is 151 pm, very near that in organic peroxides $RO-OR$ (149 pm). Obviously, as the metal complex becomes a better base, an entire pair of electrons is donated to the dioxygen, giving an Ir^{III} complex. These results show that all degrees of electron donation are possible. Table 12.7 presents several adducts of $Ir(PPh_3)_2(CO)Cl$ with neutral molecules in which electron density is transferred from the metal, as shown in $\nu_{C\equiv O}$ values. Because all X—Y bonds are not broken in the adduct, we would *formally* label these complexes of Ir^I.[22] The actual extent of electron transfer can be determined only by structural and spectroscopic data. This is the same consideration alluded to in the discussion of the π-acid properties of olefins (see Section 12.4.1).

Mechanisms

The mechanisms for oxidative additions are of three general types: concerted, $S_N 2$, and radical. We discuss each of these in turn. Figure 14.8 diagrams possible mechanisms of the first two kinds. (The "pathways" of Figure 14.8 and 14.10 constitute individual steps of stoichiometric reaction mechanisms, so that each diagram represents several possible stoichiometric mechanisms.) The emphasis here is on what *types* of mechanisms there are and how to recognize them. Subtle changes in ligands and in oxidizing substrate can lead to changes in mechanism; apparently, energetic differences are not great and some reactions proceed by more than one route simultaneously.

Five-Coordinate Eighteen-Electron Reactants (Figure 14.8*a*).   Lower-valent Group 8, 9, and 10 metals tend to form five-coordinate complexes that conform to the 18-*e* rule (coordinatively saturated). Hence their oxidative additions may involve $S_N 2$ displacement which increases coordination by one and oxidation state by two units (pathway *a*). For example,

$$Os(CO)_3(PPh_3)_2 + Br_2 \longrightarrow \textit{cis-}[Os(CO)_2(PPh_3)_2(Br)_2] + CO \qquad (14.27)$$

Reaction (14.27) proceeds via pathway *a* with XY = Br$_2$. This involves $S_N 2$ attack by the metal on XY to afford [ML$_5$X]$^+$ that may (pathway *e*) or may not undergo subsequent attack by Y$^-$ to displace L (in either case, oxidative addition has occurred). In the above reaction, [Os(CO)$_3$(PPh$_3$)$_2$Br]$^+$Br$^-$ can be isolated; on heating to 60°C, Br$^-$ displaces CO to form *cis*-[Os(CO)$_2$(PPh$_3$)$_2$Br$_2$].

$S_N 2$ displacement by carbonylate anions is common and is expected to result in inversion of configuration at C when XY is an optically active halide such as Ph(Me)HC*Cl. Table 14.5 shows relative nucleophilicities of several carbonylate anions. [CpFe(CO)$_2$]$^-$ (and other carbonylate anions) may also react with H$^+$ to produce hydrides such as CpFe(CO)$_2$H; because we conventionally count the H ligand as hydride (H$^-$) and alkyls as R$^-$, the metal oxidation number is increased by two in both cases.

Alternatively, prior ligand loss (pathway *b* of Figure 14.8) may occur to give a 16-*e* species that could go on react via $S_N 2$ displacement (pathway *c*) or addition (pathway *d*). A probable case of pathway *b* + *d* is Equation (14.28), which occurs only under conditions vigorous enough to promote ligand dissociation—in contrast to H$_2$ addition to 16-*e* square-planar complexes, which takes place at room temperature and atmospheric pressure:

$$H_2 + Os(CO)_5 \xrightarrow[\text{80 atm}]{100°C} H_2Os(CO)_4 + CO \qquad (14.28)$$

---

[22] Another way to think of these reactions is as the addition of neutral Lewis acid ligands.

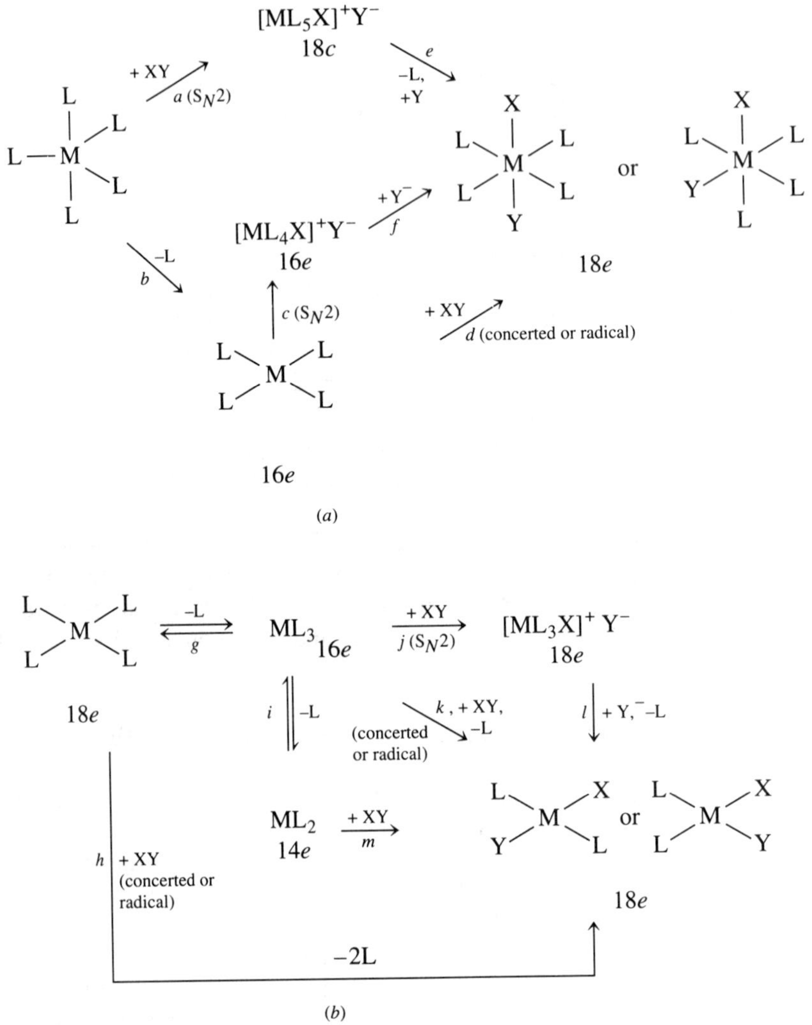

Figure 14.8 Oxidative addition mechanisms. (L represents a neutral ligand.)

Four-Coordinate Sixteen-Electron Reactants. Sixteen-electron square-planar complexes (which may be intermediates produced from ligand dissociation from ML_5) can undergo both S_N2 (c) and concerted (d) additions. Addition of MeI to Vaska's compound is an S_N2 process characterized by a large negative value of $\Delta S^{\ddagger}$ in nonpolar solvents; $\Delta S^{\ddagger}$ becomes less negative in polar solvents, thereby increasing the rate. The rates of S_N2 reactions (which involve ionic intermediates) generally increase with solvent polarity. S_N2 additions are promoted by "polar" substrates such as acyl halides, alkyl halides, hydrogen halides, halogens, and pseudohalogens which contain good leaving groups. Concerted additions to square-planar complexes (d) involve less polar XY and include those of H_2 and R_3SiH to $Ir(CO)X(Ph_2PCH_2CH_2PPh_2)$.

Table 14.5   Relative nucleophilicities of some carbonylate anions[a]

Anion	Relative nucleophilicity
$[CpFe(CO)_2]^-$	7×10^7
$[CpRu(CO)_2]^-$	7.5×10^6
$[CpNi(CO)]^-$	5.5×10^6
$[Re(CO)_5]^-$	2.5×10^4
$[CpW(CO)_3]^-$	5×10^2
$[Mn(CO)_5]^-$	77
$[CpMo(CO)_3]^-$	67
$[CpCr(CO)_3]^-$	4
$[Co(CO)_4]^-$	1

[a] R. B. King, *Acc. Chem. Res.* **1970**, *3*, 417.

Pearson[23] has pointed out that reactions proceeding with reasonably low activation energies involve electron flow between orbitals on each reactant with the same symmetry properties. For oxidative additions, electrons must flow from a filled metal orbital into an antibonding $X-Y$ orbital; this allows the $X-Y$ bond to be broken and new bonds to the metal to be formed. As shown in Figure 14.9, the $X-Y$ antibonding orbital must overlap in phase with a filled metal orbital. The figure shows three possibilities for such overlap, leading to three different mechanisms for oxidative addition of $X-Y$ to square-planar ML_4. Both *cis* and *trans* addition have been observed. However, these stereochemical results at the metal are not mechanistically diagnostic, since there is no proof that the geometries are controlled kinetically.

Four-Coordinate Eighteen-Electron Reactants (Figure 14.8*b*).   With d^{10} four-coordinate complexes (18*e*), ligand loss must also occur on oxidative addition as the product formula shows. This can be prior to (pathway *g*, etc.) or concurrent with (pathway *h*) oxidative addition. For example, solutions of $Pt(PPh_3)_4$ contain $Pt(PPh_3)_3$ as the major species as well as some $Pt(PPh_3)_2$. It is likely that oxidative addition proceeds via *g*:

$$Pt(PPh_3)_3 + RBr \longrightarrow \textit{trans-}RPt(PPh_3)_2Br + PPh_3 \tag{14.29}$$

On the other hand, $Ni[P(OEt_3)]_4$ is barely dissociated and may react via *h* unless kinetically significant quantities of unsaturated species too small to detect are present. Timing of dissociation and addition is still unclear for most oxidative additions to d^{10} complexes. The reactivity order for d^{10} complexes in general parallels ease of oxidation by two units:

$$Pt^0 \approx Pd^0 \approx Ni^0 > Au^I > Cu^I \gg Ag^I$$

---

[23] R. G. Pearson, *Symmetry Rules for Chemical Reactions*, Wiley, New York, 1976.

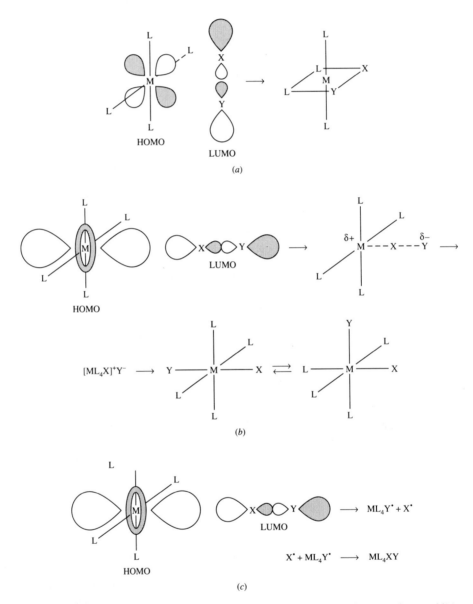

Figure 14.9 Symmetry-allowed overlap of orbitals for electron transfer in oxidative addition to 16-*e* ML$_4$. (*a*) concerted addition. (*b*) S_N2-type addition. (*c*) free-radical addition.

Concerted Versus Free-Radical Mechanisms

The pathways labeled d, h, k, and m in Figure 14.8 may be either concerted or radical in nature. Radical mechanisms occur more often in oxidative additions of RBr and RI to Pt^0 and Ir^I than to Pd^0 and Rh^I. Figure 14.10 amplifies the situation for radical addition. Electron transfer produces the radical pair $L_xM^{(n+1)+}X^{\cdot}$, $Y^{\cdot}$ which may collapse to form the product or diffuse apart to give free radicals. Rate laws do not necessarily distinguish; first-order dependence on both complex and RX is expected for S_N2 (a, c, or j in Figure 14.8) as well as for the formation of a radical pair or concerted mechanism. A free-radical path can often be detected. Racemization of optically active alkyl in additions of $XY = R^*X$, inversion in reactivity order of RX ($3° > 2° > 1°$ for radical, $3° < 2° < 1°$ for S_N2), and a very fast radical-chain process provide evidence of a free-radical addition.

Reductive Elimination

Many species can undergo **reductive elimination,** the reverse of oxidative addition. Reductive elimination decreases both the coordination number and the oxidation state of the metal. Sometimes the molecule eliminated is different from the one added oxidatively. For example, the Pd^{IV} species resulting from oxidative addition in Reaction (14.30) is not isolable, but its intermediacy explains the observed products:

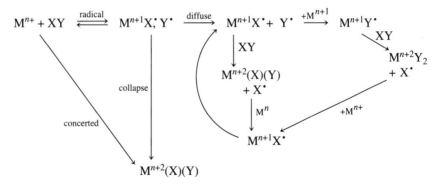

$$\text{(14.30)}$$

Reaction (14.30) scheme:

Ph₃P / Me Pd^II Ph₃P / Me + MeI ⟶ [Me, Ph₃P, Pd^IV, Ph₃P, I, Me] ⟶ I / PPh₃ Pd^II Ph₃P / Me + C₂H₆

Figure 14.10 Concerted and radical mechanisms for oxidative addition. Free-radical chain processes are represented on the right. M represents the metal complex with its complement of ancillary ligands. n is the metal oxidation state in the starting complex. Y is the more electropositive group. MY_2 is a side product.

A requirement for reductive elimination seems to be that eliminated groups be adjacent. This is nicely exemplified in the product distributions for some gold complexes where L = PPh_3:

$$\begin{array}{c} L \diagdown \\ \diagup Au \diagdown \\ Me \diagup \quad \diagup Et \quad Me \end{array} \quad \underset{L}{\rightleftarrows} \quad \begin{array}{c} Me \\ | \\ Au-Et \\ | \\ Me \end{array} \quad \overset{+L}{\longrightarrow} \quad MeEt + LAuMe$$

(14.31)

$$\begin{array}{c} L \diagdown \quad \diagup Me \\ Au \\ Et \diagup \quad \diagdown Me \end{array} \quad \underset{L}{\rightleftarrows} \quad \begin{array}{c} Me \\ | \\ Au-Me \\ | \\ Et \end{array} \quad \overset{+L}{\longrightarrow} \quad MeEt + MeMe + LAuMe$$

The fact that the reaction is retarded by added phosphine suggests that reductive elimination proceeds with ligand loss. This is reasonable because elimination increases electron density on the metal. In contrast, some reductive eliminations are promoted by added ligands. For example cis-$Pt(PPh_3)_2Ph_2$ eliminates biphenyl at a lower temperature in the presence of added phosphine. Presumably, the trigonal-bipyramidal intermediate formed places the Ph groups in a more favorable geometry for elimination. A different role for an added ligand seems plausible for the reductive elimination of alkane from (bipy)NiR_2 where dissociation of one end of the chelate is not favored. Added olefins promote this reaction; the more electron-withdrawing the olefin, the faster the elimination of R_2. This can be understood if the function of the olefin is to coordinate to Ni and drain off excess electron density, thereby stabilizing a transition state in which added electron density is placed on the metal.

Reductive elimination involving two metal centers is also known to occur. An example is

$$HCo(CO)_4 + Me\overset{\overset{\displaystyle O}{\|}}{C}Co(CO)_4 \longrightarrow Me\overset{\overset{\displaystyle O}{\|}}{C}H + Co_2(CO)_8$$

(14.32)

▶ *14.1.4   C—H Activation, An Application of Oxidative Addition*

Hydrocarbons are plentiful raw materials, being found in natural gas and petroleum (see Section 14.2.1). Unfortunately, they are rather inert so that converting them to organic chemicals containing reactive functional groups involves breaking C—H bonds with expenditure of a considerable amount of energy. It would obviously be of interest to find a low-energy catalytic pathway that would accomplish this.

Recently, oxidative addition of C—H bonds to electron-rich unsaturated organometallic fragments has been achieved. Jones and Feher[24] have studied Reaction (14.33):

$$Cp^*Rh(PMe_3) + R—H \longrightarrow Cp^*Rh(PMe_3)(R)(H) \qquad R = Ph, C_3H_7 \quad (14.33)$$

[24] W. D. Jones and F. J. Feher, *Acc. Chem. Res.* **1989**, *22*, 91.

ΔH^0_{rxn} will be favorable only when the sum of M—C and M—H bond energies in the product exceeds the C—H bond energy. Typical values of C—H bond energies are 460 kJ/mol for Ph—H and 427 kJ/mol for alkyl—H.

The required coordinatively unsaturated fragments can be generated *in situ* by either thermally or photochemically induced reductive elimination. For example, Cp*Rh(PMe₃)(H)₂ reductively eliminates H₂ on irradiation. This is the method of choice when low temperatures are required. Generation of Cp*Rh(PMe₃) at −55°C in a mixed benzene/propane solvent gives Cp*Rh(PMe₃)(Ph)(H) and Cp*Rh(PMe₃)(C₃H₇)(H) in a 4:1 ratio. This represents the kinetic product distribution because the available thermal energy at −55°C is not sufficient to permit rearrangement to the thermodynamic product mixture. This shows that both arenes and alkanes can be activated and that the kinetic preference is slightly in favor of arene addition, the difference in $\Delta G^{\neq}$ being ~2.5 kJ/mol/.

At higher temperatures, available thermal energy can provide the activation energy necessary for reductive elimination from Cp*Rh(PMe₃)(R)(H); this process can be followed by the disappearance of NMR signals belonging to starting material as a function of time. Cp*Rh(PMe₃)(Ph)(H) is relatively stable, undergoing reductive elimination only on heating to +60°C. Measuring the rate at several temperatures permits evaluation of the activation parameters $\Delta H^{\ddagger} = 128$ kJ/mol and $\Delta S^{\ddagger} = 62.3$ J/mol K. These parameters can be used to calculate $\Delta G^{\ddagger}(-17°C) = 111.7$ kJ/mol for benzene elimination. On the other hand, Cp*Rh(PMe₃)(C₃H₇)(H) eliminates propane at −17°C. This corresponds to $\Delta G^{\ddagger}(-17°C) = 77.8$ kJ/mol. The energy profile in Figure 14.11 shows the much greater thermal stability (36.4 kJ/mol) for the products of arene addition. The equilibrium constant is

$$K_{eq} = \frac{[Cp*Rh(PMe_3)(Ph)(H)][propane]}{[Cp*Rh(PMe_3)(C_3H_7)(H)][benzene]} = 2.7 \times 10^7 \tag{14.34}$$

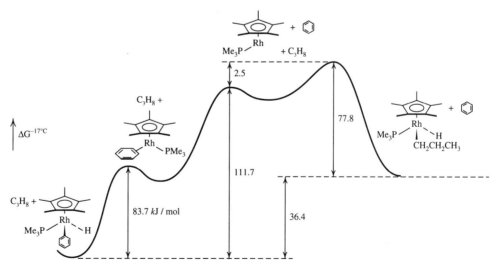

Figure 14.11 Energy profile for oxidative addition to Cp*Rh(PMe₃) and reductive elimination from Cp*Rh(PMe₃)(R)(H). (Reprinted with permission from W. D. Jones and F. J. Feher, *Acc. Chem. Res.* **1989**, *22*, 91. Copyright American Chemical Society, 1989.)

Such great thermodynamic preference for activation of an arene is interesting in view of the fact that alkanes have weaker C—H bonds. Further studies have shown that $Cp^*Rh(PMe_3)(\eta^2\text{-arene})$ is the initial product when benzene reacts with $Cp^*Rh(PMe_3)$; addition of the C—H bond to Rh follows. The stability of this intermediate plays an important role in the preference for arene activation. Besides this, the Rh—aryl bond is apparently much stronger ($\sim$75 kJ/mol) than an Rh—alkyl bond.

The isoelectronic $Cp^*Ir(PMe_3)$ system investigated by Bergman and co-workers[25] exhibits some interesting differences. As expected, the system is kinetically more inert and the chemistry requires higher temperatures. In contrast to the Rh system, a variety of alkyl complexes can be prepared by thermally induced reductive elimination of cyclohexane from $Cp^*Ir(PPMe_3)(Cy)(H)$ in the presence of alkanes. Again, aryl addition products are more thermodynamically stable. The phenyl complex does not thermally eliminate benzene below 200°C, so that the importance of η^2-arene intermediates cannot be assessed.[26]

The relative strengths of various types of Ir—C bonds can be established through equilibria such as (14.35) in mixed hydrocarbon solvents:

$$Cp^*Ir(PPMe_3)(Cy)(H) + C_5H_{12} \;\rightleftharpoons\; Cp^*Ir(PPMe_3)(C_5H_{11})(H) + C_6H_{12} \qquad (14.35)$$

Adjusting the solvent composition to 91.5 mol % cyclohexane and 8.5 mol % pentane[27] produced a 1.0 ± 0.1 ratio of *n*-pentyl:cyclohexyl complex at 140°C as determined by NMR. This corresponds to $K_{eq} = 10.8$ and $\Delta G^0 = -8.4$ kJ/mol. Assuming that $T\Delta S^0 \approx 0$ and that the bond dissociation energies (BDEs) are the same in solution and in the gas phase, we obtain

$$
\begin{aligned}
\Delta H^0_{rxn} \approx \Delta G^0_{rxn} &= -8.4 \text{ kJ/mol} \\[4pt]
&= [- \text{ BDE(cyclohexane secondary C—H)} - \text{BDE(Ir}-n\text{-pentyl)} \\[4pt]
&\quad + \text{BDE(}n\text{-pentane primary C—H)} + \text{BDE(Ir—cyclohexyl)}] \\[4pt]
&= -400 \text{ kJ/mol} - \text{BDE(Ir}-n\text{-pentyl)} + 410 \text{ kJ/mol} \\[4pt]
&\quad + \text{BDE(Ir—cyclohexyl)} \\[4pt]
&= +10 \text{ kJ/mol} + [\text{BDE(Ir—cyclohexyl)} - \text{BDE(Ir}-n\text{-pentyl)}]
\end{aligned}
\qquad (14.36)
$$

From this expression we obtain

$$[\text{BDE(Ir}-n\text{-pentyl)}] - [\text{BDE(Ir—cyclohexyl)}] = 18 \text{ kJ/mol} \qquad (14.37)$$

[25] J. M. Buchanan, J. M. Stryker, and R. G. Bergman, *J. Am. Chem. Soc.* **1986**, *108*, 1537.

[26] The possibility of π-complex formation does not always facilitate C—H activation. For example, $Cp^*Ir(PMe_3)$ was found to give two products when allowed to react with $CH_2{=}CH_2$: $Cp^*Ir(PMe_3)(CH{=}CH_2)(H)$ and $Cp^*Ir(PMe_3)(\eta^2\text{-}C_2H_4)$. On heating, the former rearranged to the latter indicating that the π complex cannot be an intermediate in the activation of the ethylene C—H bond. (P. O. Sutherland and R. G. Bergman, *J. Am. Chem. Soc.* **1985**, *107*, 4581.)

[27] The solvent composition is adjusted to produce a $\sim$1:1 ratio of products so that both can be accurately detected by NMR.

The Ir—n-pentyl bond is estimated to be 18 kJ/mol stronger than the Ir—cyclohexyl bond. Setting up a series of such equilibria has led to the following qualitative trend in Ir—C bond strengths: neopentyl < cyclohexyl ~ cyclopentyl < primary 2,3-dimethylbutyl < primary pentyl << phenyl. The Ir—Ph bond is estimated to be ~102 kJ/mol stronger than Ir—cyclohexyl.[28]

Thus far, it has proven difficult to use this chemistry to functionalize hydrocarbons in a one-pot reaction because the presence of reagents such as CO or olefins which might react with the M—C bond tends to prevent formation of the coordinatively unsaturated species required for C—H activation.

Intramolecular C—H activation has been known for some time and is referred to as **cyclometallation** [Equation (14.38)] or, when the *ortho* proton of an aromatic ring is involved, as **orthometallation** [Equation (14.39)]:

$$(14.38)$$

$$(14.39)$$

In the presence of D_2, reductive elimination of a C—D bond can lead to complete *ortho* deuteration in reaction (14.39), because both oxidative addition and reductive elimination of H_2 (D_2) and C—H are reversible. In fact, observation of H—D exchange on C is often a good way of detecting C—H activation. (See also Problem 13.5.)

C—H bond activation has been achieved with d^0 complexes as well. Obviously, this cannot involve oxidative addition. Instead, a four-center transition state is proposed. The reaction is often termed **σ-bond metathesis.** Examples are

$$(14.40)^{29}$$

---

[28] Agostic C—H bonds (see page 608) to metals are probably a good model of the transition state for oxidative addition to unsaturated fragments.

[29] P. L. Watson, *J. Am. Chem. Soc.* **1983,** *105,* 6491.

$$(14.41)^{30}$$

► 14.1.5   *Attack on Coordinated Ligands*

Electron distribution in ligands is changed when they bond to metals. This was a main theme in Chapter 12, where it was amplified via molecular orbital (MO) model and VB model descriptions of the bonding (see also p. 397). We say that the molecules are polarized. This changes their reactivity. Hence, coordinated ligands are often activated to attack by nucleophiles or electrophiles, which does not occur in the free molecules. A number of such reactions have been given in the discussions of various classes of ligands. This section treats the subject from a more descriptive point of view. Some reactions of Chapter 13 are cited as examples. Even if Chapter 13 was not covered, the chemistry in the reactions cited here can be recognized and appreciated as bearing on the theme at hand. However, this is a large subject, and our treatment is not complete.

[30] L. R. Chamberlain, J. L. Kershner, A. P. Rothwell, I. P. Rothwell, and J. C. Huffman, *J. Am. Chem. Soc.* **1987,** *109,* 6471.

Attack by Nucleophiles[31]

Reactivity to nucleophilic attack on ligands is increased by (a) overall positive charge on the complex, (b) the presence of other π-acid ligands, (c) coordinative saturation of the metal (which prevents attack on the metal), and (d) high nucleophilicity of the attacking reagent. Ligand displacement competes, and it is not always possible to predict which will occur.

Examples of nucleophilic attack on coordinated CO [Equations (12.31), (14.4), (14.5), and (14.7)], CNR [Equation (12.32)], η^2-olefins [Equation (13.6)], and η^2-alkynes [Equation (13.15)] have been given.

Attack on π-Donor Ligands. η^3-allyls constitute an especially important class of ligands which undergo nucleophilic attack. Equation (13.19) is one example. Pd—allyl complexes play an important role in organic synthesis of substituted olefins. For example,

$$\longrightarrow CH_2=CHCMe_2\{CH(CO_2Me)_2\} + Me_2C=CHCH_2\{CH(CO_2Me)_2\} + Pd^0 + 2PPh_3$$

Dissociation of the olefin allows Pd to be recycled, and the process can be made catalytic by using 3-methyl-2-butenyl acetate which can add oxidatively to Pd^0 species to generate an allyl complex. Good control of regiochemistry can be achieved by the proper choice of ancillary ligands on Pd and the steric demands of the nucleophile in this process pioneered by Trost and co-workers.

Nucleophilic addition to an η^4-diene is exemplified in the reaction

$$Nuc^- = H^-, OMe^-, SMe^-, CN^-$$

[31] For a review see L. S. Hegedus in *The Chemistry of the Metal-Carbon Bond*, Vol. 2, F. R. Hartley and S. Patai, Eds., Wiley, Chichester, 1985, p. 401.

Addition saturates terminal C, removing it from conjugation and leaving an enyl (allyl). The starting cation in Reaction (14.43) has three π-donor ligands which could be attacked by nucleophiles. The susceptibility to nucleophilic attack *on cationic 18-e species* has been found generally to decrease in the order

That is, ligands with an even number of C's are attacked in preference to those with an odd number. This is reasonable because we count *even* ligands (except for cyclobutadiene) as neutral and *odd* ligands (except for tropylium, $C_7H_7^+$) as anionic. Open ligands are attacked in preference to cyclic (closed) ones. Except when the metal fragment is very electron-donating, attack occurs on the end carbon. These rules (formulated by Green, Davies, and Mingos) predict the structure of the kinetic product. Thus, even, open butadiene is attacked in Equation (14.43).

For Nuc$^-$ = OMe$^-$, Reaction (14.43) is reversible:

$$\text{(14.44)}$$

Cp$^-$ is the most common η^5 ligand, and it is usually quite inert to nucleophilic attack except when the complex is cationic and the nucleophile very strong:

$$\text{(14.45)}$$

Cyclohexadienyl complexes are more reactive:

$$\text{(14.46)}$$

Nucleophilic attack on coordinated η^6-arenes is shown in Equations (13.43) and (13.44). Another example (which does not obey the "rules" given above) is

$$[\text{Ru}]^{2+} + xs\ \text{LiAlH}_4 \longrightarrow \left[\text{Ru} + \text{Ru}\right] \quad (14.47)$$

The neutral product of two hydride additions "should" be $(\eta^5\text{-C}_6\text{H}_7)_2\text{Ru}$, since attack at an *even* ligand is expected to be favored. The expected pattern *is* observed for the analogous Fe complex. As is the case with other π-donor ligands, the free molecules are unreactive toward nucleophiles because their electron-rich multiple bonds generally react with electrophiles.

There is no guarantee that attack will occur at just one site. A recent case in point[32] is that of $[(\eta^6\text{-C}_6\text{Me}_6)\text{Mn(CO)}_3]^+$ which reacts with halide ions giving the product of carbonyl displacement, $(\eta^6\text{-C}_6\text{Me}_6)\text{Mn(CO)}_2\text{X}$. With hydride,

$(\eta^6\text{-C}_6\text{Me}_6)\text{Mn(CO)}_2\overset{\overset{\textstyle O}{\|}}{\text{C}}\text{H}$, $(\eta^5\text{-C}_6\text{Me}_6\text{H})\text{Mn(CO)}_3$ or $(\eta^5\text{-C}_6\text{Me}_5\text{=CH}_2)\text{Mn(CO)}_3$ can result, depending on the reaction conditions and the hydride source.

Fischer-type carbenes are attacked at the carbene C. The product may be either the adduct [Equation (13.81)] with very strong nucleophiles or the product of elimination of a molecule containing the original heteroatom [Equation (12.34)] for protic nucleophiles. Alkylidenes having α—H may undergo H^+ abstraction to give alkylidynes on attack by an external [Equation (12.43)] or internal [Equation (12.42)] nucleophile. Because carbynes are synthesized by abstraction of nucleophilic groups, nucleophiles add (especially to cationic) carbynes to regenerate carbenes [Equation (13.79)].

Attack by Electrophiles[33]

Factors which promote attack by electrophiles include (a) zero or negative charge on the complex, (b) low oxidation state, and (c) late transition metals (high d-electron count).

The discussion will be mostly limited to attack by H^+ and Ph_3C^+. Electrophilic substitutions on aromatic Cp^- [Reaction (12.29)] and cyclobutadiene^{2-} [Reactions (13.29) and (13.30)] ligands are known to occur. Because transition metals generally contain some

[32] D. M. LaBrush, D. P. Eyman, N. C. Baenziger, and L. M. Maillis, *Organometallics* **1991**, *10*, 1026.

[33] For a review see M. D. Johnson in *The Chemistry of the Metal-Carbon Bond*, Vol. 2, F. R. Hartley and S. Patai, Eds., Wiley, Chichester, 1985., p. 513.

nonbonding electrons, the metal may be the initial site of electrophilic attack followed usually by transfer to a π-donor ligand. For example, ferrocene can be protonated by strong acids to afford $[Cp_2FeH]^+$. See also Equation (13.25), where initial protonation (and oxidation) of Fe is followed by transfer of H^- to the η^4-butadiene, converting it to an enyl$^-$ (allyl$^-$). A similar reaction occurs with a hexamethylbenzene complex converting a "triene" ligand into a dienyl$^-$ ligand:

$$\text{(structures)} + HCl + NH_4PF_6 \xrightarrow{\text{acetone}} \text{(structures)}\ PF_6^- + NH_4Cl \qquad (14.48)$$

The net result of H^+ addition to a π-donor polyene ligand is to convert it to a polyenyl ligand containing one fewer C in the π system.

Enyl complexes are generally not subject to protonation. An exception is the 20-*e* complex which is converted to an 18-*e* species:

$$\text{(structures)} + H^+ \longrightarrow \text{(structures)} \qquad (14.49)$$

The Ph_3C^+ cation is commonly employed to abstract H^-. Hydride abstraction at a CH_2 group frees a π orbital on C and converts polyene ligands into polyenyls containing one more C. Thus, η^4-cyclohexadiene is converted to η^5-cyclohexadienyl [Equation (13.26)], and η^6-cycloheptatriene is converted to η^7-cycloheptatrienyl (tropylium) [Equation (13.46)]. Also η^6-cyclooctatriene is converted to η^7-cyclooctatrienyl:

$$\text{(structures)} + Ph_3C^+BF_4^- \longrightarrow \text{(structures)}\ BF_4^- + Ph_3CH \qquad (14.50)$$

For all cases the metal oxidation state is dictated by the even electron count for the ligand π-system required by our formalism.

EXAMPLE: When $CpCo(PPh_3)Me_2$ is allowed to react with excess $CH_2{=}CH_2$, the products are $CpCo(PPh_3)$, $CH_3CH{=}CH_2$ (propylene), and CH_4. Propose a mechanism for this reaction. (See E. R. Evitt and R. G. Bergman, *J. Am. Chem. Soc.* **1979**, *101*, 3973.)

Solution: Because this problem is posed here, it is likely that the steps involved in the mechanism correspond to reactions discussed in this chapter. The starting complex is one of Co^{III}, and the product is Co^{I}. Hence, reductive elimination is probably involved. One organic product (CH_4) must be derived from a methyl ligand. Propylene is likely derived from ethylene; but an additional C must have been added, presumably from methyl. The role of transition metals is to act as a template on which the reactants are assembled. These then react with one another in the metal coordination sphere, and the products dissociate. Thus, the first step is to coordinate $CH_2{=}CH_2$ to Co. An electron count reveals that $CpCo(PPh_3)Me_2$ has 18-e's. Hence, it is not likely to add ethylene to give a 20-e species. One ligand coordinated to Co must dissociate. PPh_3 is a good candidate because it is a stable molecule which can exist in solution. Thus, the first step is

$$CpCo(PPh_3)Me_2 \;\rightleftharpoons\; CpCoMe_2 + PPh_3$$

Coordination of ethylene can now occur to produce an 18-e species:

$$CpCoMe_2 + CH_2{=}CH_2 \;\rightleftharpoons\; CpCoMe_2(\eta^2\text{-}CH_2{=}CH_2)$$

An insertion reaction transfers methyl to coordinated olefin:

$$CpCoMe_2(\eta^2\text{-}CH_2{=}CH_2) \;\longrightarrow\; CpCoMe(CH_2CH_2CH_3)$$

$CpCoMe(CH_2CH_2CH_3)$ is a coordinatively unsaturated 16-e species; it could form 18-e $CpCo(PPh_3)Me(CH_2CH_2CH_3)$ by adding dissociated PPh_3 present in solution. This may indeed happen, but this species will not yield the products experimentally observed. (It could produce C_4H_{10} by reductive elmination.) If phosphine addition occurs, it must be reversible. In order to obtain the desired organic products, we must transfer H from the propyl ligand to methyl. A β-hydride elimination can occur because $CpCoMe(CH_2CH_2CH_3)$ has a vacant orbital and a vacant coordination position. Irreversible reductive elimination of CH_4 follows:

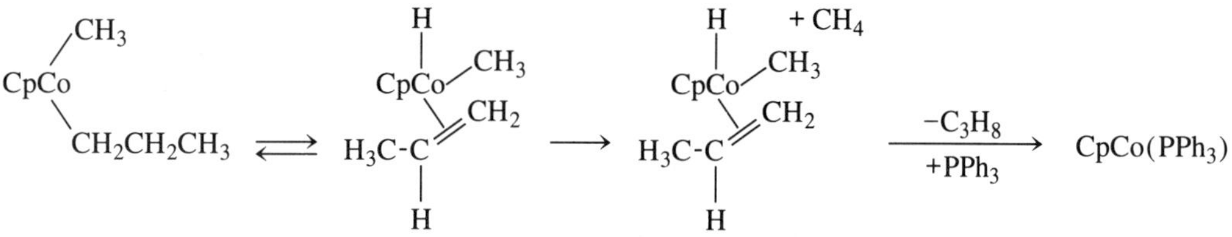

PPh$_3$ could then displace propylene to give the other two products. Note that we cannot without further information distinguish among the possible mechanisms for ligand displacement or reductive elimination. This may or may not be the actual mechanism, but it does allow us to rationalize the observed products in terms of reaction types which are known to occur.

14.2   HOMOGENEOUS CATALYSIS BY SOLUBLE TRANSITION METAL COMPLEXES

Much of the impetus for the development of organometallic chemistry over the past 50 or so years has arisen from the ability of transition metals to catalyze various kinds of organic transformations. The function of a catalyst is to provide an alternative, low-activation-energy path by which a reaction can occur. The catalytic species is regenerated and used repeatedly, rather than consumed stoichiometrically. The best catalysts are both effective (producing high product yields rapidly) and selective (producing mainly the desired product). Catalysis by organometallic compounds is thought to involve several of the stoichiometric reactions that have been discussed in this chapter: coordination of ligands to metals, insertion of olefins into M—C and M—H bonds, insertion of CO into M—C bonds, and attack on coordinated ligands and oxidative addition, as well as the reverse of each of these reactions (dissociation, β-elimination, decarbonylation, and reductive elimination, respectively). Under reaction conditions as practiced industrially (usually involving high temperatures and pressures), organometallic species often are too labile to be detected spectroscopically. (After all, if a catalyst is to be useful, any reaction intermediates ought to react rather rapidly to form later intermediates and products.) Hence, many of the steps in catalytic sequences are postulated to occur by analogy with reactions known to take place with related but more stable systems and/or under mild reaction conditions. Such reactions perform as models for the catalytic processes.

Catalysis may occur homogeneously (catalyst, reactants, and products all in the same phase) or heterogeneously (usually with the catalyst in a solid phase and the reactants and products in a solution phase).[34] This section focuses on homogeneously catalyzed reactions (which also serve as models for possible mechanisms in heterogeneous catalysis), because organometallic species thought to be present resemble those treated in this chapter. Evidence for their presence can be amassed by solution spectroscopic and kinetic studies on model systems and, sometimes, even under actual industrial conditions. Such studies have revealed considerable information about the mechanisms of homogeneous catalytic processes and thus have contributed to the design of better catalysts.

► *14.2.1 Feedstocks for Chemical Industry*

Substances available in large quantity directly from natural sources or involving only minimal treatment of naturally obtained mixtures are the starting materials or **feedstocks** for industrial synthesis. Over the past 40 years, the heavy organic chemicals industry has come to rely almost entirely on feedstocks derived from petroleum. Crude petroleum is a mixture consisting largely of saturated hydrocarbons (paraffins) with a few percent aromatics. The exact composition varies with the source, but up to 2% may consist of oxygenated compounds such as phenols, cresols, and carboxylic acids. Small amounts of N-containing compounds (up to 0.8%, about half of which is pyridine and quinoline) and S-containing species (up to 5% consisting of elemental S, thiols, and aliphatic and cyclic sulfides) may also be present. Initial distillation of crude petroleum gives a fraction volatile below 400°C and a nonvolatile residual oil. The volatile fraction is then redistilled and gives, in order of increasing boiling temperature: (a) C_1–C_4 hydrocarbons, (b) light gasoline, (c) naphtha (or heavy gasoline), (d) kerosene, and (e) light gas oil. The residual oil is distilled under reduced pressure for fuel oil. Heavy oils undergo **cracking;** that is, they are broken up into

---

[34] Heterogeneous catalysis is also used extensively in the chemical industry. See Section 17.10.1.

lower-molar mass hydrocarbons suitable for fuel use. A dual-function heterogeneous catalyst permits concurrent **reforming**—that is, the conversion of straight-chain to branched-chain hydrocarbons for smoother fuel burning.

Thermal (noncatalytic) as well as catalytic cracking of the naphtha fraction yields olefins and acetylene, with the exact product mixture depending on the temperature. This method is used in Europe for olefin production. In the United States, natural gas (which consists mostly of CH_4 and very little ethane, propane, and butane with small amounts of hydrogen, ethylene, CO, and CO_2) is used to produce olefins by cracking the heavier fraction. These olefins (ethylene, propylene, butenes, butadiene, and C_5 unsaturated hydrocarbons) along with methane and synthesis gas (CO + H_2, *vide infra*) constitute the principal building blocks for the heavy organic chemicals industry. Also used are higher alkanes and arenes, most of which are oxidized catalytically by O_2.

Synthesis gas is prepared from the combustion of CH_4 or other hydrocarbons to give a mixture of CO and H_2. In the presence of steam, the **water gas equilibrium** [Equation (14.52)] is also established.

$$-CH_2- + 0.5\,O_2 \longrightarrow CO + H_2 \qquad \Delta H = -92 \text{ kJ/mol} \qquad (14.51)$$

$$-CH_2- + H_2O(g) \rightleftharpoons CO + 2H_2 \qquad \Delta H = +151 \text{ kJ/mol} \qquad (14.52)$$

Alternatively, coke (C) prepared by heating coal in the absence of air can be allowed to react with hot steam in the **water gas reaction:**

$$C + H_2O(g) \longrightarrow CO + H_2 \qquad \Delta H = +130 \text{ kJ/mol} \qquad (14.53)$$

The requisite heat is supplied by admitting air to burn coke to CO_2 ($\Delta H° = -394$ kJ/mol); after CO_2 is vented, steam is admitted to the hot coke bed and the endothermic water gas reaction cools the reactants. Another cycle begins with admission of air which converts coke to CO_2 and heat. When petroleum became the basis for the organic chemical industry, this method of making synthesis gas vastly decreased in importance. However, the large reserves of coal as compared to oil may make such a route more important again in the future.

The composition of synthesis gas can be varied by the **water gas shift:**

$$CO + H_2O \rightleftharpoons CO_2 + H_2 \qquad \Delta H = -42 \text{ kJ/mol} \qquad (14.54)$$

The task of the organic chemicals industry is to convert the available feedstocks—olefins, synthesis gas and arenes—to compounds possessing additional functional groups that confer desired properties of reactivity. Among the industrially important compounds produced are alcohols, aldehydes, acids, and polymers. This chapter briefly discusses only a few of the many industrial reactions that are catalyzed homogeneously.[35]

---

[35] For a more complete treatment and many excellent references, see G. W. Parshall and S. D. Ittel, *Homogeneous Catalysis*, 2nd ed., Wiley–Interscience, New York, 1992; and C. Masters, *Homogeneous Transition-Metal Catalysis*, Chapman and Hall, New York, 1981.

As petroleum resources grow more scarce and expensive, it becomes increasingly desirable to optimize petroleum use and to replace petroleum feedstocks. Synthesis gas can be obtained from cheaper coal, as well as from combustion of petroleum hydrocarbons. Direct combination of CO and H_2 over a catalyst can produce CH_3OH and other alcohols. At least in theory, hydrocarbons could be produced by hydrogenation of CO—making oil from coal! CO hydrogenation can be catalyzed heterogeneously by solid oxides of transition metals (the Fischer–Tropsch synthesis), but the process is not yet very economically feasible except in places which lack petroleum but have abundant coal reserves. Heterogeneous catalysis and its relation to inorganic chemistry are discussed more thoroughly in Section 17.10.1.

▶ 14.2.2 *Hydroformylation (The Oxo Reaction)*

Olefins add the elements of H_2 and CO across the double bond to give aldehydes in the presence of a Co (or Rh) catalyst, the reaction being

$$RCH{=}CH_2 + H_2 + CO \xrightarrow{Co} RCH_2CH_2\overset{\overset{\displaystyle O}{\|}}{C}H + \underset{\underset{\displaystyle HC{=}O}{|}}{RCHCH_3} \qquad (14.55)$$

The relative amounts of straight- and branched-chain aldehydes produced depend on the identity of R and other constituents of the reaction mixture. Straight-chain aldehydes, the more desirable products, usually are hydrogenated (affording straight-chain alcohols) or self-condensed (affording more complex aldehydes).

The process is run at high pressure (~200 atm) and temperature (~90°C to 140°C) in the presence of a solvent. Finely divided Co metal or any Co salt can be used as the catalyst precursor, which is converted to $Co_2(CO)_8$ by reduction and CO coordination. The active catalyst is $HCo(CO)_4$, formed via the reaction

$$Co_2(CO)_8 + H_2 \longrightarrow 2\,HCo(CO)_4 \qquad (14.56)$$

High pressures of CO are required to prevent decomposition of $HCo(CO)_4$. The first step in the catalytic process involves CO dissociation followed by olefin coordination.

$$HCo(CO)_4 \rightleftharpoons HCo(CO)_3 + CO \qquad (14.57)$$

$$HCo(CO)_3 + RCH{=}CH_2 \rightleftharpoons H(CO)_3Co{-}\|\overset{\overset{R\diagup\diagdown H}{C}}{\underset{\underset{H\diagup\diagdown H}{C}}{}} \qquad (14.58)$$

Subsequently, coordinated olefin is inserted into the Co—H bond [Equation (14.59)] and the resulting alkyls are stabilized by CO coordination [Equations (14.60) and (14.61)], giving 18-*e* species:

$$\text{(structure)} \longleftrightarrow RCH_2CH_2Co(CO)_3 + CH_3CHCo(CO)_3 \quad (14.59)$$

$$RCH_2CH_2Co(CO)_3 + CO \rightleftharpoons RCH_2CH_2Co(CO)_4 \quad (14.60)$$

$$CH_3CHCo(CO)_3 + CO \rightleftharpoons CH_3CHCo(CO)_4 \quad (14.61)$$

Note that hydride can migrate to either olefin carbon, yielding two possible alkyl products. Now CO can be inserted to give acyl complexes:

$$RCH_2CH_2Co(CO)_4 \rightleftharpoons RCH_2CH_2\overset{\overset{\displaystyle O}{\|}}{C}Co(CO)_3 \quad (14.62)$$

$$CH_3CHCo(CO)_4 \rightleftharpoons CH_3CH\overset{\overset{\displaystyle O}{\|}}{C}Co(CO)_3 \quad (14.63)$$

These 16-*e* Co^I species now can undergo oxidative addition of H_2, giving 18-*e* Co^{III} dihydrides.

$$RCH_2CH_2\overset{\overset{\displaystyle O}{\|}}{C}Co(CO)_3 + H_2 \rightleftharpoons RCH_2CH_2\overset{\overset{\displaystyle O}{\|}}{C}Co(CO)_3 \quad (14.64)$$

$$CH_3CHCCo(CO)_3 + H_2 \rightleftharpoons CH_3CHCCo(CO)_3 \quad (14.65)$$

Reductive elimination of aldehyde leaves $HCo(CO)_3$, and the cycle can be reentered with a new olefin molecule at Equation (14.58):

$$RCH_2CH_2\overset{\overset{\displaystyle O}{\|}}{C}Co(CO)_3 \longrightarrow RCH_2CH_2\overset{\overset{\displaystyle O}{\|}}{C}H + HCo(CO)_3 \quad (14.66)$$

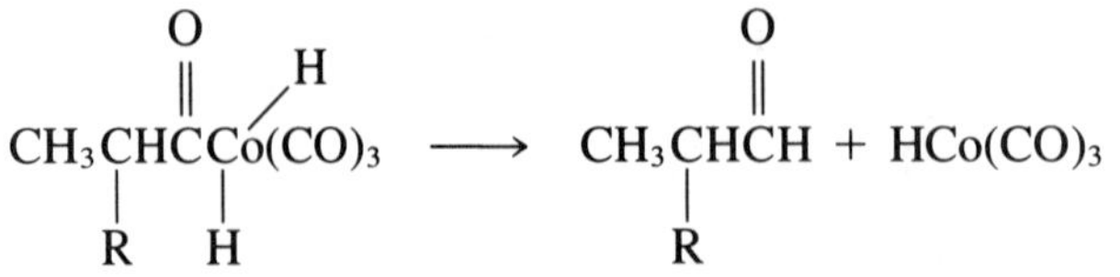

$$CH_3CHCC_o(CO)_3 \longrightarrow CH_3CHCH + HCo(CO)_3 \qquad (14.67)$$

Alternatively, cleavage may occur via the two-center reductive elimination of Equation (14.32).

The information in Equations (14.56)–(14.67) is summarized conveniently in the diagram of Figure 14.12, which also emphasizes the cyclic nature of the process.

The yield of straight-chain product can be increased by adding (n-Bu)$_3$P to the reaction mixtures, presumably because steric interactions destabilize the intermediate $CH_3CH(R)Co(CO)_3[P(n\text{-}Bu_3)]$ relative to the straight-chain alkyl complex.

▶ 14.2.3 *Hydrogenation of Olefins*

At present, the homogeneous catalytic hydrogenation of olefins is not of much industrial importance. However, a relatively small-scale operation is conducted by the Monsanto Company in order to produce *L*-DOPA for treatment of Parkinson's disease from a substituted cinnamic acid using an Rh catalyst. We mention it here because the selectivity for just one optical isomer from an optically inactive precursor provides a good example of the kinds of advantages offered by homogeneous catalysis. The catalytic mechanism probably is related to the one operative for Rh(PPh$_3$)$_3$Cl (Wilkinson's catalyst), depicted in Figure 14.13. The process can be run at low H$_2$ pressures.

A similar catalyst can be generated by starting with [(diene)Rh(PPh$_3$)$_2$]$^+$. In this case, the diene is hydrogenated, leaving [Rh(PPh$_3$)$_2$(solvent)$_2$]$^+$, the active catalyst. In the *L*-dopa process, the catalyst precursor is [(diene)Rh(P—P)]$^+$, where (P—P) is the optically active chelating phosphine dipamp:

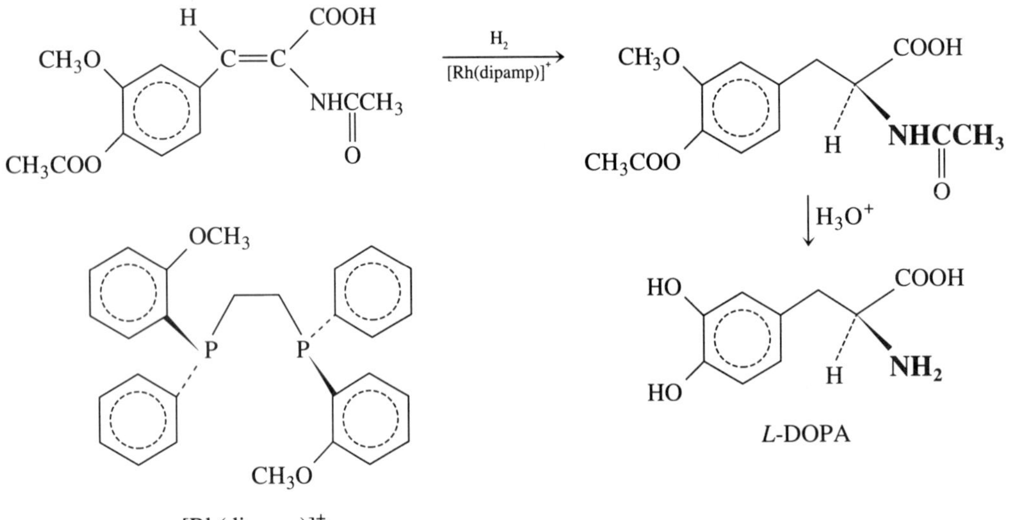

(14.68)

L-DOPA

[Rh(dipamp)]$^+$

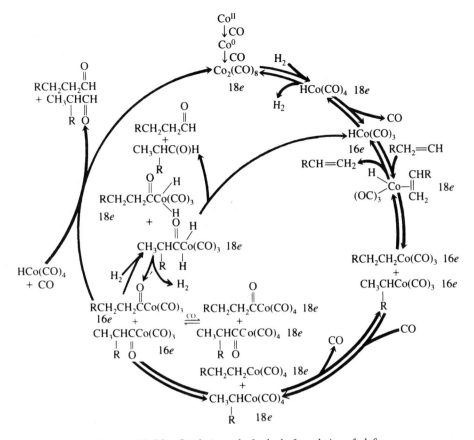

Figure 14.12 Catalytic cycle for hydroformylation of olefins.

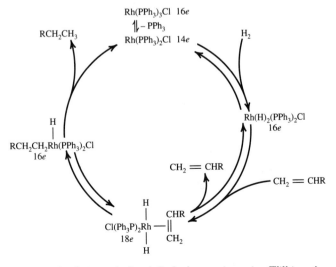

Figure 14.13 Catalytic cycle for olefin hydrogenation using Wilkinson's catalyst.

The optically active phosphine produces a "chiral hole" in the coordination sphere of the Rh, so that the prochiral olefin coordinates with one face preferentially to the complex. This leads to stereospecificity in the migratory insertion into the Rh—H bond, giving one optical isomer of the alkyl ligand, and subsequently, of the product 1-[3,4-$(OH)_2C_6H_3]CH_2C^*H(NH_2)COOH$.

Synthesis of Chiral Pharmaceuticals[36]

Biologically active compounds are usually chiral. Synthetic drugs produced in the past were usually racemic products. Commonly only one enantiomer is active. Production of the active enantiomer permits the use of lower dosage with possible reduction of side effects. Since the late 1960's there have been developments in metal complexes containing chiral ligands which react catalytically in highly enantioselective organic reactions. Small amounts of the chiral complexes as catalysts can produce much larger amounts of products of high enantiomeric purity. Applications of chiral metal complexes in enantioselective syntheses are increasing rapidly in the pharmaceutical industry. Production of L-DOPA was one of the early accomplishments. [Rh(pnnp)]$^+$ is the chiral catalyst for a similar enantioselective hydrogenation reaction used to produce L-phenylalanine for large scale production of the artificial sweetener aspartame.

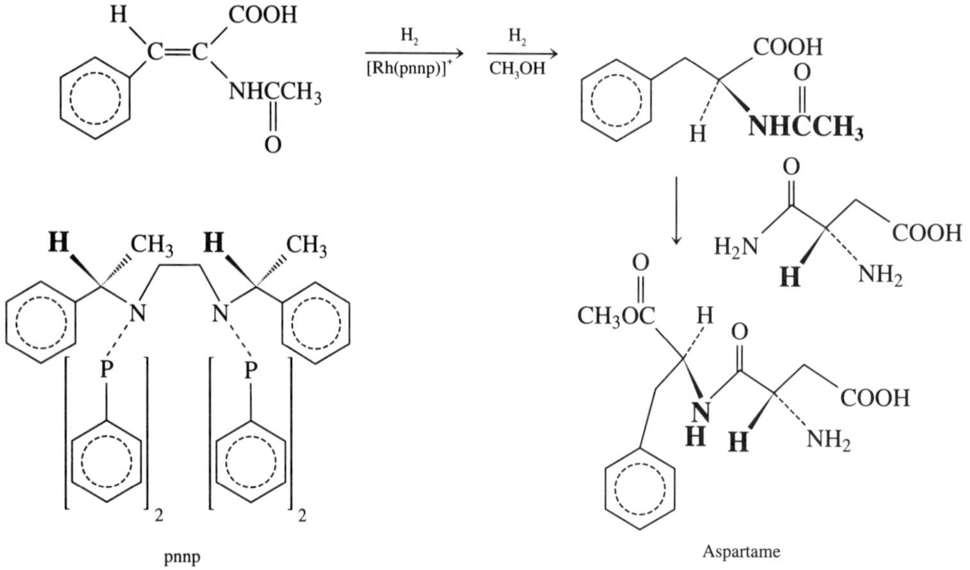

(14.69)

pnnp

Aspartame

In some cases such as [Rh(dipamp)]$^+$ and [Rh(pnnp)]$^+$ well defined complexes are used. In many cases the reaction is carried out in the presence of a chiral ligand and a metal compound. The many other enantioselective reactions catalyzed by chiral metal complexes include complexes of ligands of various chiral phosphines, tartrate ion, and others. Chiral

[36] R. Noyori, *Science* **1990**, *248*, 1194; **1992**, *258*, 584; J. W. Scott, *Topics in Stereochem.* **1989**, *19*, 209.

cyclopentadienyl metal complexes also are used. Chiral ligands are not limited to those containing an asymmetric carbon. $[Ru(binap)]^{2+}$ is a catalyst for enantioselective hydrogenation. Enzymes are usually specific for a particular reaction and a particular substrate. Some of the chiral metal catalysts, such as $[Ru(binap)]^{2+}$ as a hydrogenation catalyst, are quite general.

R-binap

▶ 14.2.4   Olefin Metathesis

Olefin metathesis involves exchange of two alkylidene groups by double-bond cleavage. Equation (14.70) shows the metathesis of ethylene with a symmetric olefin.

$$CH_2{=}CH_2 + RCH{=}CHR \longrightarrow 2\,RCH{=}CH_2 \qquad (14.70)$$

This interesting reaction can be catalyzed homogeneously in the laboratory by Schrock-type alkylidene complexes. The mechanism generally accepted involves formation of a metallacyclobutane intermediate and C—C bond cleavage to generate a new carbene which acts as a chain carrier:

$$(14.71)$$

In subsequent cycles, M=CHR reacts with ethylene, thereby generating more $CH_2{=}CHR$ and M=CH_2. In other words, the halves of olefin molecules are completely randomized in this mechanism.

In stoichiometric reactions, the presence of an O-containing ligand with a high-oxidation-state metal has been shown to result in effective metathesis. The π-donor properties of =O and —OR ligands presumably stabilize both the starting alkylidene and the metallacycle.

$$(PEt_3)_2Cl_2\overset{\overset{O}{\|}}{W}(=CHCMe_3) + CH_2=CHMe \xrightarrow{AlCl_3}$$

$$[Cl(PEt_3)_2\overset{\overset{O}{\|}}{W}(=CHCMe_3)]^+ [AlCl_4]^- \longrightarrow$$

$$\left[\begin{array}{c} \overset{O}{\|} \\ Cl(PEt_3)_2W=CHCMe_3 \\ | \\ CH_2=CHMe_3 \end{array}\right]^+ + \left[\begin{array}{c} \overset{O}{\|} \\ Cl(PEt_3)_2W=CHCMe_3 \\ | \\ MeHC=CH_2 \end{array}\right]^+ \qquad (14.72)$$

$$\left[\begin{array}{c} \overset{O}{\|} \\ Cl(PEt_3)_2W-CHCMe_3 \\ | \quad | \diagdown H \\ H_2C-C \\ \qquad \diagdown Me \end{array}\right]^+ + \left[\begin{array}{c} \overset{O}{\|} \\ Cl(PEt_3)_2W-CHCMe_3 \\ H \; | \quad | \\ \diagup C-CH_2 \\ Me \end{array}\right]^+$$

$$\downarrow +Cl^- \qquad\qquad\qquad \downarrow +Cl^-$$

$$Cl_2(PEt_3)_2W(O)(=CH_2) \quad + \quad Cl_2(PEt_3)_2W(O)(=CHMe)$$

$$+ \; MeCH=CHCMe_3 \qquad\qquad + \; CH_2=CHCMe_3$$

The function of the AlCl$_3$ co-catalyst is to remove a Cl$^-$ ligand, thereby vacating a position for olefin coordination. In general, Schrock alkylidenes are more effective at promoting metathesis than are Fischer-type carbenes, which are rendered so stable by the heteroatom on the carbene ligand that they do not react to form metallacycles with olefins. However, (CO)$_5$W=CPh$_2$, which lacks a heteroatom, *was* found to give metathesis with CH$_2$=CPh(C$_6$H$_4$*p*-OMe) when CO is removed by irradiation.[37]

Metathesis is much less significant in chemical industry than hydroformylation. However, internal olefins RCH=CHR$'$ containing around 20 C's are metathesized with CH$_2$=CH$_2$ in the Shell Higher Olefins Process (SHOP) to produce more commercially desirable C$_{11}$–C$_{14}$ terminal (α-) olefins.

Industrial processes involve catalyst precursors which are usually high-valent transition metal halides and main-group alkyls such as WCl$_6$/EtOH/EtAlCl$_2$ or WOCl$_4$/Me$_4$Sn. The way in which the presumed alkylidene active catalyst is produced is not clear. However, it is attractive to imagine production of an M(CH$_3$)$_2$ species followed by α-H abstraction to give M=CH$_2$ + CH$_4$ as in the synthesis of alkylidenes. (See Section 12.3.7.)

Cyclic olefins can be polymerized by repeated metathesis using catalyst systems such as R$_2$AlCl/MoCl$_5$. In the process, the rings are opened, accounting for the name *Ring*

---

[37] L. K. Fong and N. J. Cooper, *J. Am. Chem. Soc.* **1984**, *106*, 2595.

*O*pening *M*etathesis *P*olymerization (ROMP); these polymerizations exhibit high stereo-specificity. (See Problem 14.24 for an example.)

▶ 14.2.5   The Wacker Process

Another way of producing aldehydes (especially acetaldehyde) is by way of the Wacker process, in which olefins are oxidized by O_2 rather than being carbonylated. Hence the aldehyde product has the same number of carbons as the starting olefin, in contrast to hydroformylation. The overall process is given by Equation (14.73), and the catalytic cycle is diagramed in Figure 14.14.

$$CH_2{=}CH_2 + \tfrac{1}{2}O_2 \xrightarrow[\text{120°C, 4 atm}]{\text{PdCl}_2,\text{CuCl}_2} CH_3\overset{\overset{\textstyle O}{\|}}{C}H \qquad (14.73)$$

At the beginning of the cycle, square-planar $[PdCl_4]^{2-}$ undergoes ligand displacement by ethylene, giving $[PdCl_3(C_2H_4)]^-$, an analogue of Zeise's salt. Other ligands around Pd^{2+} also are replaced (since the reaction is inhibited by Cl^-), but the exact composition of the coordination sphere is uncertain and thus the catalytically active species is indicated as $Pd(C_2H_4)^{2+}$. Coordinated ethylene undergoes nucleophilic attack to give a β-hydroxyalkyl which undergoes β-elimination giving a coordinated enol; olefin insertion produces an isomeric hydroxy alkyl which undergoes reductive elimination leading to an aldehyde. Cu^{2+} in the solution reoxidizes the Pd^0 to Pd^{II}, which reenters the cycle. O_2 from air converts Cu^+ back to Cu^{2+}, making the process catalytic.

A similar cycle is used to manufacture acetone from propylene (see Problem 14.25). If OH^- is replaced by acetate, vinyl acetate can be made.

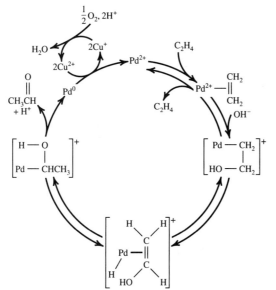

Figure 14.14   Catalytic cycle for the Wacker process.

▶ *14.2.6   The Monsanto Acetic Acid Synthesis*[38]

Chemists at Monsanto have developed a catalytic process for producing acetic acid by direct carbonylation of methanol in the presence of Rh and I^-:

$$CH_3OH + CO \xrightarrow[\text{180°C, 30–40 atm}]{\text{Rh, } I^-} CH_3COOH \qquad (14.74)$$

The acetic acid is employed in the manufacture of (a) vinyl acetate, (b) cellulose acetate, (c) other acetate esters, used as solvents, and (d) pesticides and other chemicals. An advantage is the relatively low pressure required. Figure 14.15 depicts the catalytic cycle. Rh can be added as the readily available $RhCl_3 \cdot 3H_2O$ at a concentration of 10^{-3} *M*. During an incubation period the Rh^{III} compound is reduced by CO to Rh^I and converted to the actual catalyst $[Rh(CO)_2I_2]^-$.

▶ *14.2.7   The 16- and 18-Electron*
Rule in Homogeneous Catalysis

Most mechanisms given in this section involve two-electron changes in metal oxidation state to give either 16- or 18-electron species, which the effective atomic number (EAN) rule predicts as most stable. There is considerable evidence for the importance[39] of species with these electron counts in homogeneous catalytic processes, and electron counts have

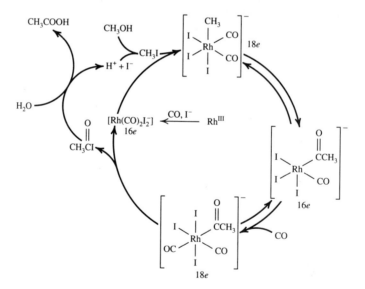

Figure 14.15   Catalytic cycle for the Monsanto acetic acid synthesis.

---

[38] See D. S. Forster, *Adv. Organomet. Chem.* **1979,** *17,* 255.

[39] C. A. Tolman, *Chem. Soc. Rev.* **1972,** *1,* 337.

been given on several catalytic cycles. Under vigorous catalytic reaction conditions, it may be that one-electron processes which give free-radical species are involved. Indeed, an increasing number of 17-*e* and 19-*e* species have been implicated in organometallic re-actions[40] (see Section 14.1.1).

GENERAL REFERENCES

J. P. Collman, L. S. Hegedus, J. R. Norton, and R. G. Finke, *Principles and Application of Organotransition Metal Chemistry,* University Science Books, Mill Valley, CA, 1987.

B. C. Gates, *Catalytic Chemistry,* John Wiley & Sons, New York, 1992. An excellent discussion of catalysis from a chemical engineering perspective.

F. R. Hartley and S. Patai, Eds., *The Chemistry of the Metal-Carbon Bond,* Wiley, Chichester, 1985-87. Four volumes of excellent reviews.

C. Masters, *Homogeneous Transition-Metal Catalysis, A Gentle Art,* Chapman & Hall, New York, 1981.

G. W. Parshall and S. D. Ittell, *Homogeneous Catalysis,* 2nd ed., Wiley-Interscience, New York, 1992.

F. G. A. Stone and R. West, Eds., *Advances in Organometallic Chemistry,* Academic Press, San Diego. A series of volumes published yearly containing authoritative reviews on organometallic chemistry.

G. Wilkinson, F. G. A. Stone and E. W. Abel, Eds., *Comprehensive Organometallic Chemistry,* Pergamon Press, Oxford, 1982. A nine-volume treatise with extensive treatment of all aspects of organometallic chemistry.

PROBLEMS

14.1 Activation parameters for CO exchange by $M(CO)_6$ (M = Cr, Mo, W) in Table 14.1 are for the gas-phase reactions. In decalin solution, the $\Delta H^{\ddagger}$ values are: Cr, 163.2 kJ/mol; Mo, 132.2 kJ/mol; and W, 169.9 kJ/mol. What is the significance of the agreement of gas-phase and solution activation enthalpies?

14.2 The substitution reaction $Mo(PPh_3)_2(CO)_4$ + bipy → $Mo(CO)_4(bipy)$ + $2\,PPh_3$ obeys the usual two-term rate law [Equation (14.2)]. Activation parameters are as follows: For the *cis* complex, $\Delta H_1^{\ddagger} = 109.5$ kJ/mol and $\Delta S_1^{\ddagger} = 25$ J/mol K; for the *trans* complex, $\Delta H_1^{\ddagger} = 99.9$ and $\Delta S_1^{\ddagger} = -33$. Comment on the significance of these numbers for the reaction mechanisms.

14.3 The following data refer to the substitution $(cod)Mo(CO)_4$ + $2\,L$ → $Mo(CO)_4L_2$ + cod. What do these data suggest about the common mechanism for these reactions?

---

[40] Many organometallic reactions are interpreted in terms of one-electron processes in J. K. Kochi, *Organometallic Mechanisms and Catalysis,* Academic Press, New York, 1978.

Ligand	$\Delta H_1^{\ddagger}$, kJ/mol	$\Delta S_1^{\ddagger}$, J/mol K	$\Delta H_2^{\ddagger}$, kJ/mol	$\Delta S_2^{\ddagger}$, J/mol K
PPh$_3$	104.5	9	87.7	−13
AsPh$_3$	104.5	9	71.0	−71
SbPh$_3$	104.5	9	75.2	− 4
py	100.3	13	83.6	−50
PCl$_2$Ph	100.3	13	36.8	−41

14.4 The salt $[Cp_2Co^+][Co(CO)_4^-]$ is dark red even though both ions are separately colorless; this is due to an intense charge-transfer band at 532 nm. When a solution of the salt is irradiated in the region of this band in thf in the presence of PPh$_3$, a 65% yield of $Co_2(CO)_6(PPh_3)_2$ is isolated. Explain this observation. (See T. M. Bockman and J. K. Kochi, *J. Am. Chem. Soc.* **1989**, *111*, 4669.)

14.5 Substitution of $(\eta^3\text{-}C_3H_5)Mn(CO)_4$ by PPh$_3$ could be envisioned as occurring by either associative or dissociative pathways.
(a) Write mechanisms for both pathways.
(b) The substitution reaction at 45°C is first-order in $[(\eta^3\text{-}C_3H_5)Mn(CO)_4]$, and $k_1 = 2.8 \times 10^{-4}$ s^{-1} at 45°C. The rate constant for decarbonylation of $(\eta^1\text{-}C_3H_5)Mn(CO)_5$ at 80°C was found to be 1.64×10^{-4} s^{-1}. How do these facts enable a choice between the two possible mechanisms?

14.6 With poor nucleophiles, $V(CO)_5(NO)$ undergoes substitution via a dissociative pathway. *trans*-$V(CO)_4(PPh_3)(NO)$ undergoes carbonyl substitution exclusively via a dissociative pathway 10^6 times slower than for $V(CO)_5(NO)$ and 10^4 times faster than dissociative substitution in isoelectronic $Cr(CO)_6$. Explain these observations. (See Q.-Z. Shi, T. G. Richmond, W. C. Trogler, and F. Basolo, *Inorg. Chem.* **1984**, *23*, 957.)

14.7 Figure 14.16 shows the cyclic voltammogram of $Cp^*W(CO)(p\text{-tolyl})_2$. Note that the voltage sweep is initially negative. What can you say about the electrochemical behavior of this complex? Why is such behavior reasonable?

14.8 Figure 14.17 shows the cyclic voltammogram of $[\{(\eta^5\text{-}Me_4C_5)SiMe_2(NMe_2)\}Fe(CO)_2]_2$. Coulometric measurements show that all electrochemical reactions involve two electrons. Write the equation which corresponds to the reaction occurring at each of the peaks.

14.9 Show that the product distribution for carbonylation depicted in Figure 14.4 would be different if the mechanism involved movement of coordinated CO which inserted itself into the Mn—C bond. (This is the so-called "CO insertion" mechanism for carbonylation.)

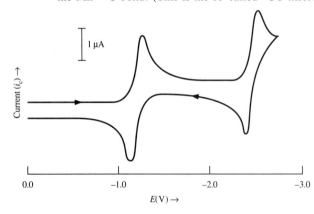

Figure 14.16 Cyclic voltammogram for $Cp^*W(NO)(p\text{-tolyl})_2$ in thf. Sweep rate is 400 mV/s. (Reprinted with permission from N. H. Dryden, P. Legzdins, S. J. Rettig, and J. E. Veltheer, *Organometallics*, **1992**, *11*, 2583. Copyright 1992, American Chemical Society.)

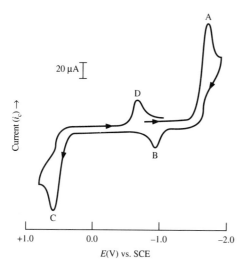

20 μA

Current (i_z) →

C

+1.0   0.0   −1.0   −2.0

E(V) vs. SCE

Figure 14.17   Cyclic voltammogram for $[\{(\eta^5\text{-}Me_4C_5)SiMe_2(NMe_2)\}Fe(CO)_2]_2$ in MeCN. Sweep rate is 200 mV/s. (Reprinted with permission from M. Moran, M. C. Pascual, I. Cuadrado, and J. Losada, *Organometallics*, **1993**, *12*, 811. Copyright 1992, American Chemical Society.)

14.10   Some data for the insertion reactions $Cp_2^*Nb(H)(\eta^2\text{-}CH_2{=}CHR) + CNMe \rightarrow Cp_2^*Nb(CH_2CH_2R)(CNMe)$ are given below. Rate $= k_2[Cp_2^*Nb(H)(\eta^2\text{-}CH_2{=}CHR)]\text{-}[CNMe]$.

R	k_1, s^{-1} (50°C in C_6D_6)
H	2.62
Me	890
Ph	3.18
$p\text{-}NMe_2C_6H_4$	6.80
$p\text{-}MeOC_6H_4$	4.81
$p\text{-}MeC_6H_4$	3.47
$p\text{-}CF_3C_6H_4$	0.91

(a) Recognizing that k is determined by the energy difference between the ground state and the transition state, what assumption needs to be made in comparing activation parameters as a function of R?

(b) If this assumption is made, what can be said about the nature of the transition state from these data? (See N. M. Doherty and J. E. Bercaw, *J. Am. Chem. Soc.* **1985,** *107,* 2670.)

14.11   Show how a reversible olefin insertion $\rightleftarrows$ elimination can lead to equilibration of D as shown below (dmpe $= Me_2PCH_2CH_2PMe_2$):

(dmpe) Pt$^+$ Me   (dmpe) Pt$^+$ Me   Me $\rightleftarrows$ Me   D D   D D

(See T. C. Flood and S. P. Bitler, *J. Am. Chem. Soc,* **1984,** *106,* 6076.)

14.12 The ease of oxidative addition and the tendency toward five-coordination for d^8 metals increase as shown in Figure 14.7. Explain these trends.

14.13 The rate of reaction of O_2 with *trans*-IrX(PPh₃)₂(CO) in benzene decreases in the order $X = NO_2 > I > ONO_2 > Br > Cl > N_3 > F$. Explain this observation. (See L. Vaska, L. S. Chen, and C. V. Senoff, *Science* **1971**, *174*, 587.)

14.14 Account for the following observations (*Hint:* β-elimination and/or reductive elimination are involved.)
(a) (dppe)Pt(CH₂CH₃)₂ affords C₂H₆ and C₂H₄ on heating in xylene at 150°C (dppe = Ph₂PCH₂CH₂PPh₂).
(b) (dppe)Pt(CH₂CH₃)(OMe) affords C₂H₆ and C₂H₄ in a 40 : 60 ratio as well as MeOH and H₂C=O on heating in toluene at 100°C.
(c) (dppe)Pt(OMe)₂ affords MeOH and H₂C=O on stirring in CH₂Cl₂ at 25°C. (See H. E. Bryndza, J. C. Calabrese, M. Marsi, D. C. Roe, W. Tam, and J. E. Bercaw, *J. Am. Chem. Soc.* **1986**, *108*, 4805.)

14.15 Propose a mechanism for the stoichiometric decarbonylation of C₆H₅CH₂C(O)Cl by Rh(PPh₃)₃Cl giving benzyl chloride. Keep in mind the 16- and 18-electron rule.

14.16 Using data given in Section 14.1.4:
(a) Show that the difference in activation energy for oxidative addition of benzene and propane to Cp*Rh(PMe)₃ is 2.5 kJ as depicted in Figure 14.11.
(b) Verify the value for K_{eq} at −17°C.

14.17 Write a cycle for the reaction of (CH₃)₂C=CHCH₂O₂CCH₃ with Na⁺CH(CO₂Et)₂⁻ catalyzed by Pd(PPh₃)₄. (*Hint:* The mechanism involves oxidative addition of the acetate to generate an η^3-allyl complex.)

14.18 Predict the products of the following attacks on coordinated ligands.
(a) *cis*-PtCl₂(PPh₃)(η^2-C₂H₄) + *xs* NMe₂H
(b) [CpMoCO)(NO)(η^3-C₈H₁₃)]⁺ + PhS⁻ (C₈H₁₃ = cyclooctadienyl)
(c) (η^4-C₄H₆)Fe(CO)₃ + HBF₄
(d) (η^4-C₆H₈)Fe(CO)₃ + Ph₃C⁺BF₄
(e) (η^4-C₆H₈)Fe(CO)₃ + Li⁺CH₂CN⁻
(f) [η^6-(6-phenylhexadienyl)Mn(CO)₂(NO)]⁺ + PPh₃
(g) [(η^5-C₆H₇)Fe(CO)₃]⁺ + I⁻
(h) [(η^6-C₆H₆)Mn(CO)₃]⁺ + NaBH₄
(i) [(η^7-C₇H₇)W(CO)₃]⁺ + MeO⁻

14.19 Exchange of ¹⁷O in the reaction [Re(CO)₆⁺] + H₂¹⁷O → [Re(CO)₅(C¹⁷O)]⁺ + H₂O obeys the rate law: rate = k[Re(CO)₆⁺][H₂O]². The experiments were conducted by adding measured amounts of ¹⁷O-enriched water to a MeCN solution of [Re(CO)₆]PF₆. Addition of OH⁻ does not appreciably increase the rate. Propose a mechanism for exchange. (See R. L. Kump and L. J. Todd, *Inorg. Chem.* **1981**, *20*, 3715.)

14.20 Give electron counts for all the species postulated to be involved in the catalytic cycle for hydroformylation (Figure 14.12).

14.21 Kinetic studies indicate that hydroformylation reaction rate is enhanced by an increase in H₂ pressure and retarded by an increase in CO pressure. How is the mechanism proposed in this chapter consistent with these observations?

14.22 What products would you expect from hydroformylation of C₃H₇C(CH₃)DCH=CH₂? Show how each is obtained.

14.23 Ni[P(OEt)$_3$]$_4$ reacts with H$^+$, affording HNi[P(OEt)$_3$]$_4^+$.

    (a) Formulate the electronic structure of this cation.

    (b) The cation is known to catalyze olefin isomerization. Write a cycle for the isomerization of 1-butene catalyzed by HNi[P(OEt)$_3$]$_4^+$.

14.24 Cyclopentene undergoes *Ring Opening Metathesis Polymerization* (ROMP) with a catalyst

of R$_3$Al/WCl$_6$, producing

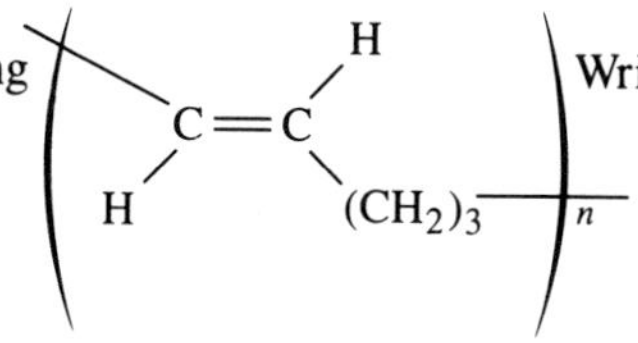

$$\left(\underset{H}{\overset{}{\diagdown}}C = C\underset{(CH_2)_3}{\overset{H}{\diagup}} \right)_n$$

Write the mechanism for the reaction.

14.25 Write a catalytic cycle for the production of acetone from propylene via the Wacker process.

14.26 Write a catalytic cycle for the production of methyl acetate via the Monsanto acetic acid synthesis.

14.27 Propose a mechanism for the following reaction: IrCl$_3$(PEt$_3$)$_3$ + C$_2$H$_5$OH + KOH → HIrCl$_2$(PEt$_3$)$_3$ + CH$_3$CHO + KCl + H$_2$O. (*Hint:* The mechanism involves β-elimination.) (See J. Chatt and B. L. Shaw, *Chem. Ind. (London)* **1960**, 931.)

PART VI

Selected Topics

·15·

Chemistry and Periodic Trends Among Metals

···························

Three-fourths of the elements are metals. Their melting points range from −38.9°C for Hg to 3380°C for W. Sodium is soft enough to be cut like cheese, whereas W and Cr are very hard. Metals with only one *s* electron in the outer shell include the most active metals (Li through Cs) and the least active (i.e., most noble) metal, namely, gold. Compounds of metals range from high-melting ionic compounds such as NaF and CaO to covalent volatile substances such as $TiCl_4$ and OsO_4. Even though properties of metals and their compounds cover a tremendous range, the variations follow periodic trends. The effects of size, charge, and electron configuration permit us to understand, and even predict, chemical and physical properties. We shall rely on the background of structures of ionic solids and interactions among ions from Chapter 5. There are now many examples of compounds involving metal–metal bonding. Simple compounds involving M—M bonding are included here, but metal cluster compounds are considered in Chapter 17, on cage and cluster compounds.

15.1 GENERAL PERIODIC TRENDS AMONG METALS

Periodic trends among the metals are rather regular. Melting points and hardness of the metals increase from Group 1 (IA) to the middle of the transition series (Group 6) and then decrease to the soft, low-melting metals of Group 12 (IIB). Melting points for metals of Groups 2–11 are shown in Figure 15.1. Metals of Groups 1, 12, and 13 have melting points too low for the scale used for the figure. Other post-transition-series metals also are rather soft and low-melting. The metals Sn, Pb, Sb, and Bi are used for low-melting al-

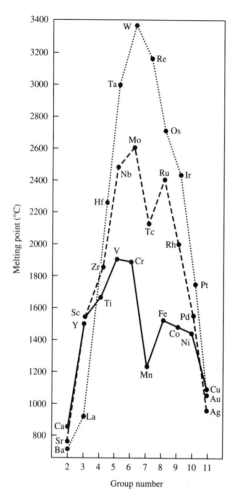

Figure 15.1   Melting points for metals of Groups 2–11.

loys. In looking for periodic trends among melting points, only large differences and broad trends are significant, since melting points depend on specific interactions in the solid. Thus Ga has the widest liquidus range of any element (mp 29.8°C, bp 2250°C), indicating that the attraction between Ga atoms is great (high bp), whereas the lattice energy is low. Ga is used in high-temperature thermometers for which the high vapor pressure of Hg makes it inappropriate.

We often analyze the activity of elements from a thermodynamic point of view by dissecting reactions into individual steps for which energy measurements have been made. (See, for example, the Born–Haber cycle on page 230.) Steps which are endoergic (such as sublimation of a metal or ionization of valence electrons) cause decreased reactivity, whereas exoergic steps (such as hydration of ions) increase reactivity. Entropy changes also affect activity. A balance must be achieved, and the very large increases in successive ionization energies (IEs) eliminate some possible compounds such as NaF_2 and CaF_3. Balance of lattice energies and successive IEs make $CaCl(s)$ unstable (page 233). As a result of these limitations and balances, there is only one important oxidation state for Groups 1,

2, and 3. Most of the transition metals have $+II$ oxidation states of varying stability and higher oxidation states, achieved by the stepwise removal of d electrons. For all but the largest metal atoms, the removal of four electrons requires too much energy, so ionic compounds are limited to those of M^+, M^{2+}, M^{3+}, and, in very few cases, M^{4+}, e.g., Ce^{4+} and Th^{4+}. In addition to the high IE(s) involved, ions of very high charge density are strongly polarizing, so compounds of metals in high oxidation states are covalent. Chemists insist on balancing redox equations with CrO_4^{2-} and MnO_4^-, rather than with the nonexistent Cr^{6+} and Mn^{7+} ions. These are covalent oxoanions. *The ease of formation of the highest oxidation states (same as the group number through Group 8) increases generally with increasing atomic radius within a transition metal group (family).* The electrons involved in forming oxidation states higher than II are from the d orbitals. These electrons are nonpenetrating and are well screened from the nuclear charge.

Main-group metals to the right of the transition series all show the positive oxidation number corresponding to the group number given in Roman numerals (III–VI) or the second digit of the IUPAC numbers (13–16), giving 18-electron (pseudo-noble-gas) configurations. *This highest oxidation state becomes more strongly oxidizing with increasing atomic number in each family* (the reverse is true for transition metal families). The stability of the highest oxidation state down through each family decreases because the last two electrons involved are ns^2. These become more penetrating with increasing n and are lost less easily. With the decrease in the stability of the highest oxidation state, the oxidation state two units lower than the maximum becomes increasingly more important (Tl^+, Pb^{2+}, Bi^{3+}). These ions have the $(18 + 2)$ configuration, corresponding to the retention of the ns^2 "inert" pair (not necessarily stereochemically inert). Group 11 (IB) metals show variation in oxidation states because the d electrons can participate. The characteristic oxidation number is II for Group 12 (IIB), but the $+I$ state involves the dimeric M_2^{2+} ions. This is important only for Hg_2^{2+}.

Size effects can be treated in a straightforward manner for ionic substances and in a qualitative way using Fajans' rules (Section 5.1.3) where polarization (covalence) is important. The Mn^{2+} and Fe^{2+} ions are comparable in size with Mg^{2+} and substitute for the more abundant Mg^{2+} ion in minerals. Such isomorphous substitution of transition metal ions for others of similar size is very common. Radii used for the ions of the transition metals are usually those of the high-spin ions, unless the metal ion is coordinated to strong-field ligands. Size trends through periodic columns and rows are rather regular, except for significant contraction following the first transition series (the transition series contraction) and the lanthanide series (the lanthanide contraction). The contraction through the first transition series has important effects in the chemistry of Ga through Br. Members of the third transition series have radii very similar to those elements of the same families in the second transition series because of the lanthanide contraction.

Metal ions with low oxidation numbers and 8-e configurations are hard cations, and their compounds with hard anions (F^- and O^{2-}) give typical ionic compounds. Softer cations [18-e or $(18 + 2)$-e configurations], particularly with soft anions (Cl^-, Br^-, I^-, and S^{2-}) give more covalent compounds. They form layer structures or others involving bridging anions. These covalent compounds usually have low solubilities in H_2O. Metal ions in high oxidation states form covalent compounds ($TiCl_4$, CrO_3, OsO_4, etc.).

Metal ions are commonly encountered as their complexes, with the metal ions coordinated by anions in the solid and solvent molecules or other ligands in solution. Metals to the far left of the periodic table with low charge densities of their ions and no ligand-field stabilization effects tend to form labile complexes with low formation constants. In solution, they form solvated ions, except in the presence of ligands that are stronger hard Lewis bases than the solvent or with multidentate ligands such as edta^{4-} (ethylenedi-aminetetraacetate ion) and cryptand ligands (Section 15.2.4). Main-group metals to the right of the transition series also have no ligand-field stabilization effects, but they are more highly polarizing and complexes of soft Lewis bases, particularly anionic ligands, are important—for example, complexes of Cl^- are more important than those of F^-. Transition metal complexes show the full range from labile to substitution-inert and from hard acid–base interaction to the complexes of zero-valent metals that are very soft Lewis acids. Colors of transition metal complexes make us very much aware of the presence of complexes. Often, we can observe striking color changes as substitution or redox reactions occur.

The most common ligand is H_2O, since metal ions in solution are aqua complexes (in the absence of other ligands). Hydrolysis of metal ions such as Fe^{3+} in H_2O does not involve OH^- addition, but involves simple ionization of the acid $[Fe(H_2O)_6]^{3+}$. Loss of H^+ decreases the positive charge of hydrated ions, and the acidity increases with increasing charge density of the metal ion. For high oxidation states, proton loss is complete, forming oxoanions.

15.2   GROUP 1 (IA)—THE ALKALI METALS

▶ *15.2.1 Group Trends*

The alkali metals are soft and low-melting because of weak bonding in the solid state. Only one electron beyond the noble-gas core is available for bonding. The melting points, boiling points, and hardnesses decrease with increasing atomic number, indicating a weakening in bonding between the atoms. The atoms are partly associated as diatomic molecules at temperatures just above the boiling points; the dissociation energies of the molecules decrease from 100.9 kJ/mol for Li_2 to 38.0 kJ/mol for Cs_2. The metals' densities, which are low because of the large radii, increase with increasing atomic number—except for K, which has lower density than that of Na. This irregularity results from differences in the magnitude of change in atomic masses as compared with the changes in the atomic radii. See Table 15.1 for a summary of some properties of the alkali metals.

Low IEs make the alkali metals the most active family of metals. They have bright luster, but except for Li, tarnish so rapidly in air that they are stored under hydrocarbons.

Because the alkali metals give only 1+ ions with noble-gas configurations, their properties vary more systematically than those of any other family in the periodic table. The very high second IEs indicate that oxidation numbers higher than I are ruled out for the alkali metals. Compounds such as Na_2O_2 (containing O_2^{2-}, peroxide ion), KO_2 (containing O_2^-, superoxide ion), and KI_3 (containing I_3^-, triiodide ion) contain only M^+ ions. Most alkali metal compounds are predominantly ionic in character, and because the ionic radii increase regularly down through the family, the Born treatment (Section 5.8.1) is particularly successful in dealing with trends. The cations have low charge and large radii (each has the largest radius of any cation from the same period), so that the lattice energies of

Table 15.1   The alkali metals

ns^1	$_3Li$	$_{11}Na$	$_{19}K$	$_{37}Rb$	$_{55}Cs$
Abundance (% of earth's crust)	0.0065	2.83	2.59	0.028	3.2×10^{-4}
Density (g/cm³)	0.534	0.97	0.87	1.53	1.873
Melting point (°C)	179	97.9	63.7	38.5	28.5
Boiling point (°C)	1317	883	760	668	705
Sublimation energy (kJ/mol 25°C)	155.1	108.7	90.00	85.81	78.78
Ionization energy (eV)					
1st	5.392	5.139	4.341	4.177	3.894
2nd	75.638	47.286	31.625	27.28	25.1
3rd	122.451	71.64	45.72	40	35
Atomic radius (pm)	152	185	231	246	263
Ionic radius (pm), C.N.6	90	116	152	166	188 (C.N.8)
Heat of hydration of $M^+(g)$ (kJ/mol)	515	406	322	293	264
E^0 for $M(aq)^+ + e = M(s)$	−3.040	−2.713	−2.924	−2.924	−2.92

their salts are relatively low. Consequently, most of the simple salts of the alkali metals are water-soluble. Low interionic attraction results in high conductance of the salts in solution or in the molten state. Most of the salts are dissociated completely in aqueous solution, and the hydroxides are among the strongest bases available in water.

▶ 15.2.2   Occurrence

Only Na and K of the alkali family are abundant in the earth's crust. Although abundances of light elements generally are high, the abundance of Li, like that of Be and B, is quite low. Not only is the terrestrial abundance of these elements low, but also the cosmic abundance is low. According to all current theories of the origin of the elements, these elements, with low nuclear charge, would be expected to undergo thermonuclear reactions to produce heavier elements and/or helium. The low nuclear charge gives a low potential barrier for proton- or alpha-capture reactions. These elements could not accumulate during the formation of the elements, because they were used up in such reactions.

Geochemically, the most important occurrence of the alkali metals is in the aluminosilicate minerals, which make up the bulk of the earth's crust. Because abundances of Rb and Cs are low, these elements rarely form independent minerals, but often are found in K minerals. In spite of its low abundance, Li forms independent minerals, because Li^+ is too small to replace the more abundant Na^+ or K^+ in their minerals. Lithium occurs in minerals along with Mg in aluminosilicates which separate in the very late stages of crystallization of a **magma.** In addition to the lithium ores obtained from **pegmatite** minerals, Searles Lake[1] in California is an important source, forming about half of the world's supply of lithium.

---

[1] Searles Lake is an almost dry lake in the Mohave Desert. It is strongly alkaline (pH 9.48) as is true of some other closed basins in volcanic areas. It contains almost no Mg^{2+} and only a trace of Ca^{2+}, although these are deposited in underlying mud. The brine is 4.70% KCl and 0.021% Li^+. In addition to Li, the brine produces K_2SO_4, bromine, and almost half of the world's output of borax ($Na_2B_4O_7 \cdot 10H_2O$).

A **magma** is the parent molten mass from which **igneous rocks** (those formed directly from magmas) can be considered to separate.

Pegmatites are formed during the last stages of cooling and solidification of a deep-seated igneous rock. Pegmatites are characterized by large and irregular grain size. Because pegmatites represent the end product of the crystallization of a magma, rare elements become concentrated and the conditions are right for further differentiation and growth of large crystals. The overall composition of pegmatites does not differ greatly from that of granites, but they are coarser-grained and are important sources of many rare elements.

The **sedimentary rocks** are secondary in origin, formed by such processes as weathering of other rocks. The ores found in sedimentary rocks have been concentrated to a great extent by chemical and physical changes caused by weathering. In the geological sense, a rock is any bed, layer, or mass of the material of the earth's crust. Natural waters are rocks composed mainly of the mineral water.

In geochemistry, elements are classified as **lithopile** (rock-liking), **chalcophile** (S-liking), **siderophile** (Fe-liking), and **atomophile** (gases in the atmosphere). Lithopile elements include the hard (8-*e*) cations in the aluminosilicates. Chalcophile elements include soft [18-*e* or (18 + 2)-*e*] cations often occurring as sulfides. Siderophiles are dense, not very active metals that occur as metallic alloys in the iron-rich core. Figure 15.2 shows the geochemical classification of the elements. Note that the different classes overlap. Fe, Co, and Ni are predominately siderophile, included in the box in the center of the periodic table. Dark with speckling indicates that Fe through Ge occur as oxides and sulfides in the crust. The overall groupings are very similar to the classification of metals as acceptors in Table 9.2 and as hard or soft acids. The figure includes acids or bases because nonmetals are listed also. Lithophiles occur as oxides or halides and are hard acids or bases. Soft acids or bases are chalcophiles or siderophiles.

Although the bulk of the Na and K present in the earth's crust occurs in the aluminosilicate minerals, the important ores are found in the **sedimentary rocks.** Sodium and potassium are leached from the parent rocks by weathering. Na^+ tends to concentrate in

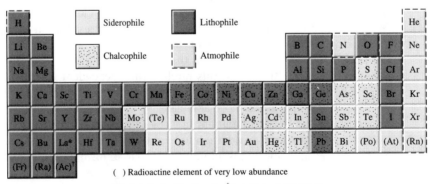

Figure 15.2 Geochemical classification of the elements. (Reproduced with permission from P. A. Cox, *The Elements*, Oxford University Press, Oxford, 1989, p. 13.)

the seas, but K^+ is absorbed strongly by clays. Some NaCl is recovered from sea water, but the most important sources are the extensive deposits of rock salt left by the evaporation of isolated bodies of water. Deposits of K^+ salts of marine origin are also important sources of K.

▶ 15.2.3 Preparation of Metals

The most important general process for the preparation of the alkali metals is the electrolysis of fused salts or hydroxides. Most of the sodium (the metal of this family produced in the greatest quantity) is obtained by the electrolysis of fused NaCl. Lithium also is obtained by electrolysis of fused LiCl with KCl added to lower the melting point of the mixture, permitting electrolysis to be carried out at a lower temperature (about 450°C). The heavier alkali metals are obtained by electrolysis or by reduction of molten MCl by Na vapor. Reduction of KCl with Na produces a K–Na alloy suitable for use as a heat exchanger. Potassium is obtained by reducing KF with CaC_2, known as **calcium carbide.** The heavier alkali metals also are obtained by reduction of the oxides with Al, Mg, Ca, Zr, or Fe:

$$2\,KF + CaC_2 \xrightarrow{\text{heat}} CaF_2 + 2\,K + 2\,C \tag{15.1}$$

$$3\,Cs_2O + 2\,Al \xrightarrow{\text{ignite}} Al_2O_3 + 6\,Cs(g) \tag{15.2}$$

The reduction by Al takes advantage of the high lattice energy of Al_2O_3 and the volatility of the alkali metals, which are vaporized in the process. Cesium can be obtained by reduction of Cs_2CO_3 with carbon

$$Cs_2CO_3 + 2\,C \longrightarrow 2\,Cs + 3\,CO \tag{15.3}$$

or by the thermal decomposition of cesium tartrate. The metals are best purified by distillation. Thermal decomposition of the azides yields all alkali metals except Li in a high state of purity. Lithium cannot be prepared from LiN_3 because it forms the very stable nitride, Li_3N. Sodium azide (NaN_3) is the source of N_2 for air bags in automobiles. Decomposition is ignited by a hot spark. Fe_2O_3 is present to react with Na and maintain the high temperature:

$$2\,NaN_3(s) \xrightarrow{\text{heat}} 2\,Na(l) + 3\,N_2(g) \tag{15.4}$$

$$6\,Na(l) + Fe_2O_3(s) \longrightarrow 3\,Na_2O(s) + 2\,Fe(s) \tag{15.5}$$

The strong reducing power of the alkali metals accounts for their extensive use in chemical synthesis. Sodium is the metal usually used for this purpose, because it is easily available and inexpensive. Lithium imparts toughness to alloys of Al and Cu. Cesium is used in photoelectric cells, since the absorption of radiant energy in the visible region of the spectrum can remove an electron because of its low IE. The photoelectric cells are widely used such as in automatic doors. Rb and Cs serve as "getters" in vacuum tubes, removing the last traces of O_2 or other reactive gases in the tube when it is first put into use.

▶ *15.2.4 Cryptates*[2]

Because of their large size and low charge density, the alkali metals show minimal tendency to form complexes. They form complexes with β-diketones, and some of the fluorinated β-diketone derivatives can be sublimed. The bonding is largely ionic, and the chelates are insoluble in nonpolar solvents.

An important development in the chemistry of the alkali metals has been the work on so-called **cryptate** complexes of polyethers (Figure 15.3) and nitrogen–oxygen macrocycles. (The name *cryptate* comes from the Greek word for "hidden.") The polyether macrocycles show selectivity among the alkali metal ions, depending on the size of the ring opening in the "crown." A cyclic polyether of four oxygens (crown-4) is selective for Li^+, whereas Na^+ prefers a crown-5 and K^+ prefers a crown-6 macrocycle. The organic linkages joining the oxygens are puckered to give the "crown" arrangement, whereas the oxygens, with their lone pairs, are arranged in a nearly planar fashion about the metal ion at the center of the ring. The ligand $N[CH_2CH_2OCH_2CH_2OCH_2CH_2]_3N$ (a *cryptand*, abbreviated C222) can enclose Rb^+ completely in a "cage." These cryptates owe their stabilities largely to entropy effects in displacing several solvent molecules. They are important because the complex presents a hydrocarbon exterior, so they are soluble in organic solvents. Also, they are interesting models of natural compounds that transport Na^+ or K^+ selectively across membranes in order to maintain the remarkable electrolyte balance inside and outside cells.

▶ *15.2.5 Solutions of the Alkali Metals in Ammonia*[3]

In the presence of catalysts, such as Fe, the alkali metals react with ammonia to form the metal amide and hydrogen. Without a catalyst and with no impurities present, the metals dissolve without liberation of H_2 and can be recovered by evaporating the ammonia. All of the alkali metals dissolve to give solutions that are bronze if concentrated and blue if di-

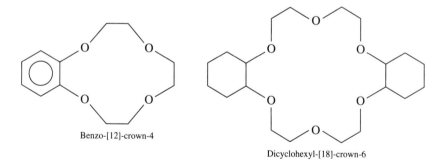

Benzo-[12]-crown-4

Dicyclohexyl-[18]-crown-6

Figure 15.3 Examples of crown ethers, using a simplified nomenclature that indicates the size of the macrocycle and the number of ligand atoms.

[2] C. J. Pedersen and H. K. Frensdorff, *Angew. Chem. Int. Ed. Engl.* **1972**, *11*, 16; D. Parker, *Adv. Inorg. Chem. Radiochem.* **1983**, *27*, 1; D. J. Cram, *Angew. Chem. Int. Ed. Engl.* **1986**, *25*, 1039.

[3] J. L. Dye, *Prog. Inorg. Chem.* **1984**, *32*, 327; D. Holton and P. P. Edwards, *Chem. Br.* **1985**, 1007.

lute. The bronze solutions conduct electricity about as well as many metals, and the blue solutions conduct electricity somewhat better than solutions of strong electrolytes (see Figure 15.4).

Characteristics of the solutions of the metals in ammonia are essentially the same for all alkali metals, for the more active alkaline earth metals, and for europium and ytterbium. Thus metals that dissolve in liquid ammonia are those with high negative reduction potentials—in other words, metals with low IEs, low sublimation energies, and high solvation energies. The solubilities of the alkaline earth metals in $NH_3(l)$ are much lower than those of the alkali metals. A saturated solution contains 25.1 moles of Cs per 1000 g of ammonia at $-50°C$.

The very dilute blue solutions are regarded as solutions of electrolytes: Metal atoms dissociate to give ammoniated cations and ammoniated electrons. Because there is a very marked increase in volume when the metals dissolve in ammonia, the electrons are thought to occupy cavities with the surrounding NH_3 molecules oriented such that the hydrogens are directed inward. The unusually high conductivity for an electrolyte results from the high mobility of the solvated electron.

In the polarographic reduction of alkali metal ions in liquid ammonia, amalgams are formed. In effect, alkali metal ions are captured by the electron-rich mercury electrode. A cathodic wave is observed polarographically for a solution of tetrabutylammonium iodide in ammonia, but this corresponds to the dissolution of electrons. Instead of the tetraalkylammonium ions dissolving in the mercury, the electrons are released as ammoniated electrons. The process is described by

$$e + x NH_3 \rightleftharpoons e(NH_3)_x \qquad E^0 = -1.89 \text{ V at } -34°C \qquad (15.6)$$

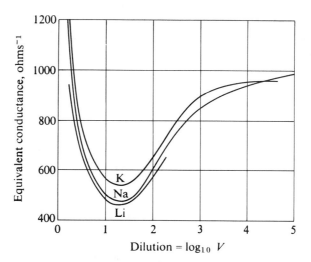

Figure 15.4 The equivalent conductance of solutions of K, Na, and Li in liquid ammonia at $-33.5°C$. V = liters of ammonia per gram-atom of metal. (Reproduced with permission from D. M. Yost and H. Russell, Jr., *Systematic Inorganic Chemistry*, Prentice–Hall, Englewood Cliffs, NJ, 1944, p. 137.)

When the electrolysis is carried out on a large scale, a blue solution results. The hydrated electron, in contrast, has a very short life ($t_{1/2} = 230$ μsec at pH 7).[4]

The decrease in conductance with increasing concentration apparently results from metal ions becoming bound together in clusters by electrons. At high concentrations, clusters become large enough to resemble liquid metals, with cationic sites occupied by ammoniated metal ions.

▶ 15.2.6   *Metal Anions*[5]

Because its electron affinity (EA) is higher than that of any element other than the halogens, gold might be expected to form Au⁻ compounds. Actually, CsAu is an ionic compound containing Au⁻, and Au dissolves in ammonia solutions by addition of K, Rb, or Cs, forming solvated Au⁻. Table 15.2 provides thermodynamic estimates to judge the stability of M_{solv}^- in ammonia relative to the solvated electron and the solid metal. Because estimates are involved, negative or even small positive ΔG^0 values indicate that the formation of ammoniated metal anions might be expected for all of the alkali metals, Au, Ag,

Table 15.2   Thermodynamic estimates to judge the stability of M_{solv}^- in ammonia (kJ/mol)[a]

Metal	Radius M^- est. (pm)	Electron affinity	ΔH^{0b}	ΔG^{0b}	$\Delta G^0(M_{solv}^-)^c$
Li	230	59.6	93.5	68.8	1
Na	263	52.9	48.3	25.7	−11
K	306	48.4	34.7	14.2	9
Rb	316	46.9	27.8	8.0	8
Cs	352	45.5	24.4	5.4	23
Au	201	222.8	137.2	105.5	0
Ag	195	125.6	153.1	122.2	8
Cu	180	118.5	212.9	181.1	41
Ba	342	−29	203	174	187
Pt	206	205.3	353.8	316.0	218
Te	219	190.2	0.4	−32.9	−113
Pb	236	35.1	153.9	123.8	62
Bi	231	91.3	109.6	76.5	9
Tl	242	19	157	130	74
Sb	223	103	152.9	118.5	42
Sn	227	120	179	147	75

[a] Data from S. G. Bratsch and J. J. Lagowski, *Polyhedron* **1986**, *5*, 1763.

[b] For $M_s + e_g \rightarrow M_g^-$ (sum of EA and sublimation energy).

[c] For $M_s + e_{solv}^- \rightarrow M_{solv}^-$ at 298 K.

[4] E. J. Hart, in *Survey of Progress in Chemistry*, Vol, 5, A. E. Scott, Ed., Academic Press, New York, 1969, p. 129.

[5] J. L. Dye, *Prog. Inorg, Chem.* **1984**, *32*, 327; S. G. Bratsch and J. J. Lagowski, *Polyhedron* **1986**, *5*, 1763

Bi, and Te. Because of the stabilization by metal–metal bonding, several main-group metals to the right of the periodic table form polyatomic anions such as Sn_5^{2-}, Sn_9^{4-}, Pb_5^{2-}, Pb_9^{4-}, Sb_7^{3-}, Bi_4^{2-}, Te_2^{2-}, and Te_3^{2-}. Even though Pt has a very high EA, its high sublimation energy makes the formation of Pt_{solv}^- very unfavorable, but anionic clusters might be formed.

Early studies of metal–ammonia solutions might have led to the discovery of M^-, but high solvation energies of M^+ and e in liquid ammonia favor the reaction

$$M_{solv}^- \rightleftharpoons M_{solv}^+ + 2e_{solv} \qquad (15.7)$$

in dilute solution. Solutions of Na in ethylenediamine and in methylamine contain Na^+ and Na^-, with electrical conductance of a normal 1:1 electrolyte. In these solvents the very high conductivity of dilute solutions of Cs indicates that the important species are Cs^+ and e_{solv}, but at higher concentrations the conductivity approaches that of Cs^+Cs^-.

The low solubility of alkali metals in amines has limited studies of M^- species. The crown ether (Figure 15.3) and cryptand ligands increase the solubilities of the metals and shift the equilibrium to the right, forming more M^-:

$$2\,M(s) \xrightleftharpoons{cry} M(cry)^+ + M^- \qquad (15.8)$$

Sodium and potassium form $M^+(C222)M^-$ in tetrahydrofuran and in simple amines. The crystal structure of $[Na^+(C222)]Na^-$ confirms the existence of the simple Na^- anion. Solids containing K^-, Rb^-, and Cs^- also have been obtained.

▶ *15.2.7 Francium*

In 1939, Mlle. Perey found element 87 in nature as a product of the decay of ^{227}Ac. The major product is ^{227}Th by β decay, but 1% of the ^{227}Ac undergoes α decay to produce ^{223}Fr. This nuclide undergoes β decay with a half-life of 21 min.

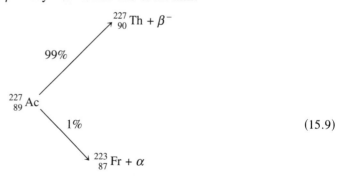

$$^{227}_{87}Fr \longrightarrow {}^{223}_{88}Ra + \beta^- \qquad t_{1/2} = 21\ min \qquad (15.10)$$

Enough tracer work has been done to confirm that the chemistry of Fr resembles that of Cs. Considering the great similarity between Rb and Cs, the properties of Fr are easily predictable.

▶ 15.2.8 *Diagonal Relationships—*
The Anomalous Behavior of Lithium

In many properties, Li is quite different from the other alkali metals. This behavior is not unusual, in that the first member of each main group of the periodic table shows marked deviations from the regular trends for the group as a whole. The deviations shown for Li can be explained on the basis of its small radius and its resulting high charge density. The nuclear charge of the Li^+ ion is screened only by two electrons. Because the ionic radius of Li^+ (90 pm) is closer to that of Mg^{2+} (86 pm) than to that of Na^+ (116 pm), Li^+ is more similar to Mg^{2+} than to Na^+ in a number of respects. Such **diagonal relationships** in the periodic table are of great importance among the active metals—for example, Li^+ and Mg^{2+}, Na^+ and Ca^{2+}, K^+ and Sr^{2+}, and Be^{2+} and Al^{3+}.

We can illustrate the diagonal relationship well by noting that the solubilities of Li compounds in water resemble those of Mg compounds to a greater extent than those of the other alkali metal compounds. Thus the fluoride, carbonate, and phosphate of Li or Mg are relatively insoluble, and those of the other alkali metals are reasonably soluble. Adding acid increases the solubility of the Li or Mg salts mentioned through the formation of acid salts (FHF^-, HCO_3^-, $H_2PO_4^-$). However, adding acid carefully to a solution of sodium carbonate precipitates $NaHCO_3$. LiOH is much less soluble than the other alkali metal hydroxides. Also, only LiOH forms Li_2O at red heat while the other hydroxides sublime as MOH.

Lithium is the only alkali metal that reacts with N_2 gas when heated. The expected high lattice energy of the ruby-red Li_3N is responsible for its great stability. On burning Mg in air, some Mg_3N_2 is formed along with MgO. Lithium carbide is the only alkali metal carbide fomed readily by direct reaction of the elements.

Lithium salts of small anions, like those of Mg, have exceptionally high lattice energies, accounting for their high stability and low solubility. The anomalous behavior of Li salts formed by large anions was discussed in the treatment of radius ratio effects (Section 5.2.2). For these salts the lattice energies are smaller than expected because of anion–anion repulsion. Only Li forms the oxide Li_2O on burning in air or oxygen at 1 atm. Na forms the peroxide Na_2O_2 while K, Rb, and Cs form the superoxide MO_2 as the principal product.

The hydration energy of Li exceeds that of any other alkali metal ion, because of the small radius of Li^+. Even though Li^+ is the smallest of the alkali metal ions, its ionic mobility is the lowest of the group because of the extensive hydration resulting from its high charge density. Lithium is the only alkali metal for which hydrolysis of salts is of any importance. In this respect also, it resembles Mg. Several Li compounds are useful because of properties appreciably different from those of Na compounds. Lithium alkyls or aryls are used in organic syntheses, where they undergo reactions similar to those of Grignard reagents (RMgX) (see Section 13.6.2). $LiAlH_4$ and $LiBH_4$ are employed extensively as reducing agents in organic reactions. Organolithium compounds and lithium hydride salts, soluble in nonpolar solvents, must be handled in the absence of air.

Ionization and sublimation energies decrease regularly down through the alkali metal family. However, although the IE and S are greatest for Li, the reduction potential of Li is slightly more negative than that of Cs. Lithium is a highly active reducing agent because of its very high hydration energy. The energy released in the hydration of the small Li^+ ion more than compensates for its high ionization and sublimation energies. On the basis of these effects alone, we would conclude that Li should be more active than Cs, but we have neglected the unfavorable entropy change for the formation of a well-ordered sheath of water molecules in $Li(aq)^+$. Although there are no outstanding anomalies among the electrode potentials of the remaining alkali metals, the order is not perfectly regular. Heat and entropy effects are rather delicately balanced, so the differences between the E^0 values are small and somewhat irregular.

Although we cannot measure the E_{cell} for a cell containing an alkali metal electrode, E^0 can be calculated from other thermodynamic data:

$$M(g) \xrightarrow{\;IE\;} M^+(g)$$

$$S \nearrow \qquad\qquad \searrow \Delta H_{M^+}$$

$$M(s) \xrightarrow{\qquad \Delta H^0 \qquad} M^+(aq)$$

$$\Delta H^0 = \Delta H^0_{M^+} \text{ (negative)} + IE + S$$

$$\Delta G^0 = \Delta H^0 - T\,\Delta S^0 = -n\mathscr{F}E^0$$

S is the sublimation energy of $M(s)$ and $\Delta H^0_{M^+}$ is the hydration energy of $M^+(g)$

New batteries use the high electrode potentials of the alkali metals. A lithium battery uses an Li anode, a polyvinylpyridine–I_2 cathode, and LiI as a solid electrolyte. The Na–S battery uses liquid Na at the anode, an S cathode, and β-Al_2O_3 as solid electrolyte.

▶ 15.2.9   *Compounds of the Other Alkali Metals*

The low solubility of $NaHCO_3$ in water serves as the basis for the production of Na_2CO_3 by the Solvay process. In this remarkably efficient process, CO_2 (from limestone, $CaCO_3$) is passed through a water solution of ammonia and salt. NH_4HCO_3 that is formed combines with NaCl to precipitate $NaHCO_3$:

$$NaCl + NH_4HCO_3 \longrightarrow NaHCO_3 + NH_4Cl \qquad (15.11)$$

Sodium hydrogen carbonate that precipitates is converted to Na_2CO_3 by heating. CO_2 is recycled.

$$2\,NaHCO_3 \longrightarrow Na_2CO_3 + H_2O + CO_2 \qquad (15.12)$$

NH_4HCO_3 in the mother liquor is decomposed by warming, and NH_3 is displaced by adding $Ca(OH)_2$ (from CaO produced on heating limestone). A solution of Na_2CO_3 is strongly basic, because of hydrolysis. For many large-scale uses calling for a strong base,

Na_2CO_3 can be substituted for more expensive bases such as NaOH. Because of the efficiency of the Solvay process and low-cost raw materials, Na_2CO_3, known as **soda ash,** is the least expensive strong base and consequently is produced in very large quantities. The $CaCl_2$ byproduct [from $Ca(OH)_2 + 2NH_4Cl$] is used for melting ice on roads and sidewalks.

Trends in properties of alkali metal compounds do not become really regular until we get to K, Rb, and Cs. Sodium does not deviate from regular trends as much as Li does, but it does differ from the heaviest alkali metals in properties such as solubilities and the extent of hydration of salts. Except for a few Li salts (mentioned above) and $NaHCO_3$, the slightly soluble salts of the alkali metals are those of K, Rb, and Cs—for example, the perchlorates and hexachloroplatinates. Because the ammonium ion is comparable in size (ionic radius 161 pm) to K^+ (152 pm), the solubilities of ammonium salts frequently parallel those of potassium. Actually, the radius of the NH_4^+ ion is closer to that of Rb^+, but comparison is usually made to more common K salts. We expect some differences between ammonium and alkali metal salts because of specific interactions between NH_4^+ ion and a solvent such as water through hydrogen bonding.

Similarities among the alkali metal salts often give no reason to prefer the salt of one alkali metal over another. Consequently when a soluble salt with a cation not likely to cause interference is needed, the Na salt is selected because of its lower cost and availability. Potassium salts sometimes are preferred, especially for analytical purposes, because they are less often hydrated and usually are not hygroscopic. More expensive Rb and Cs salts rarely offer advantages over those of Na or K. Organosodium and organopotassium compounds are ionic and not soluble in nonpolar solvents. They usually are not isolated, but are prepared *in situ* for carbanionic reactions.

The structure of Cs_2O is unique to metal oxides. The layer (*anti*-$CdCl_2$) structure has Cs—O—Cs sandwiches with only van der Waals interactions between adjacent layers of Cs^+. We think of alkali metal ions as hard cations, but this structure indicates extensive polarization of the large Cs^+ ions by O^{2-}.

15.3 GROUP 2 (IIA)—THE ALKALINE EARTH METALS

▶ *15.3.1 Group Trends*

We extend, as a convenience, the original definition of the alkaline earth metals to include Be, Mg, and Ra. The metals of this family are harder, more dense, and higher melting than the corresponding members of the group of alkali metals. Each of the metal atoms has two valence electrons beyond the noble-gas configuration. Hence the alkaline earth metal atoms are more strongly bonded in the solid state than are the alkali metals. Trends in properties dependent on bonding between atoms show a decrease in strength of bonding with increasing atomic radius, as is true for alkali metals.

First IEs of alkaline earth metals are higher than those of the corresponding alkali metals because of smaller radii and higher nuclear charge (compare Tables 15.1 and 15.3). Because the second IEs are considerably greater than the first, compounds with oxidation

Table 15.3 The alkaline earth metals

ns^2	$_4Be$	$_{12}Mg$	$_{20}Ca$	$_{38}Sr$	$_{56}Ba$	$_{88}Ra$
Abundance (% of earth's crust)	6×10^{-4}	2.09	3.63	0.015	0.040	1.3×10^{-10}
Density (g/cm³)	1.845	1.74	1.54	2.6	3.5	~5
Melting point (°C)	1284	651	851	770	710	(960)
Boiling point (°C)	2507	1103	1440	1380	1500	(1140)
Sublimation energy (kJ/mol, 25°C)	319.2	150	192.6	164	175.6	130
Ionization energy (eV)						
1st	9.322	7.646	6.113	5.695	5.212	5.279
2nd	18.211	15.035	11.871	11.030	10.004	10.147
3rd	153.893	80.143	50.908	43.6	35.5	—
Atomic radius (pm)	111	160	197	215	217	—
Ionic radius (pm) C.N. 6	41[a]	86	114	132	149	162[b]
Heat of hydration of $M^{2+}(g)$ (kJ/mol)	2385	1940	1600	1460	1320	—
E^0 for $M(aq)^{2+} + 2e = M(s)$	−1.97	−2.36	−2.84	−2.89	−2.92	−2.92

[a] C.N. 4.
[b] C.N. 8.

number I might be expected. Indeed, the formation of a compound such as MgF is an exothermic process [$\Delta H = -84$ kJ/mol for MgF(g)], and such molecules exist at high temperatures in the gaseous state. Although the heat of formation of MgF(s) would be considerably more negative than -84 kJ/mol, the *much* greater lattice energy of MgF$_2$ favors the disproportionation of MgF to give Mg and MgF$_2$ in the solid (see page 233). MgCl(g), CaCl(g), and CaF(g) have been reported also, but form MX$_2$ + M in the solid. In solution, monohalides are unstable because of the much higher hydration energy of divalent cations. Because removal of a third electron would break up a noble-gas configuration and energy required is prohibitively high, oxidation numbers higher than II are not encountered.

The large amount of energy necessary to sublime and to remove two electrons is compensated for by high hydration energies of divalent cations (or high lattice energies of the solids), so that the more active alkaline earth metals are comparable in activity to the alkali metals. Beryllium and magnesium metals are reasonably stable in air in spite of their high negative reduction potentials, presumably because oxidation is inhibited by the formation of a thin, adherent layer of metal oxide on the surface. The reduction potentials increase in magnitude with increasing atomic radius, following the order of decreasing IEs. The high hydration energy of Be^{2+} is not great enough to offset the trend established by the ionization and sublimation energies, because of the very large amount of energy required for these processes. Consequently, Be follows more regular trends for its group than does Li.

Salts of larger alkaline earth cations are typically ionic and conduct electricity well in aqueous solution and in the molten state. Conductances of fused salts of beryllium are low. Magnesium salts containing large, polarizable anions also are poor conductors of electricity in the fused state.

▶ 15.3.2 *Occurrence*

Magnesium and calcium are the only abundant alkaline earth metals in the earth's crust. The cosmic, as well as terrestrial, abundance of Be is low, as is true for Li for the same reasons (page 709). Most of the Be in the earth's crust is found in silicate minerals. The most important Be mineral is beryl, Be$_3$Al$_2$[Si$_6$O$_{18}$], in which Be^{2+} has C.N. 4 and Al^{3+} has C.N. 6. Emerald is crystalline beryl colored by a small amount of Cr^{3+}; aquamarine is also beryl of gem quality. Magnesium's abundant minerals should separate in the early stages of the crystallization of a magma because the concentration of Mg^{2+} is high. The **mantle** is believed to consist largely of forsterite, Mg$_2$SiO$_4$, and olivine, (Mg,Fe)$_2$SiO$_4$. The earth has a thin **crust** ($\sim$17 km) covering the mantle (thickness 2880 km) and inner core (radial thickness 3471 km). The important Mg ores are magnesite, MgCO$_3$; dolomite, (Mg,Ca)CO$_3$; and brucite, Mg(OH)$_2$—all found primarily in sedimentary rocks. Seawater is another important source of Mg.

Calcium is concentrated in minerals expected to separate during the main stage of crystallization of a magma. The bulk of the Ca in the earth's crust is found in the feldspars (Section 5.11.5), which make up about two-thirds of the crust. Ca^{2+}, leached from rocks by weathering, does not concentrate in the sea, because of the low solubility of $CaCO_3$. The $Ca^{2+} + CO_2 + H_2O \leftrightarrows CaCO_3 + 2H^+$ equilibrium is important in regulating the acidity of the sea and the CO_2 in the atmosphere. The most important Ca ore is $CaCO_3$ as limestone, marble, or sea shells. There are also extensive deposits of gypsum, $CaSO_4 \cdot 2H_2O$.

Strontium and barium form few independent minerals. Strontium frequently accompanies Ca and sometimes accompanies K, whereas Ba often replaces K in minerals such as potash feldspars. Heavy spar, $BaSO_4$, is an important Ba ore. Radium is found only as one of the daughters in the radioactive decay of heavier elements and occurs in all U minerals.

▶ 15.3.3   Preparation of the Metals

We obtain Be metal by electrolyzing fused BeF_2—NaF; adding NaF lowers the melting point, permitting the bath to operate at lower temperature, and improves the conductivity. Beryllium is used in alloys with Cu for nonsparking tools and is added to alloys to improve the resilience and fatigue-resisting properties of springs.

Magnesium, the most commercially important metal of the alkaline earth family, is obtained by electrolyzing a fused KCl–$MgCl_2$ mixture. $MgCl_2$ can be obtained from ores or from seawater: After $Mg(OH)_2$ is precipitated from partially evaporated seawater by addition of $Ca(OH)_2$, $Mg(OH)_2$ is converted to $MgCl_2$ for use in the electrolytic cell. We can also get Mg by reducing MgO with CaC_2, C, or Si. The reduction must be carried out at high temperature followed by quenching of the products to prevent reversal of the reaction. The low density of Mg makes it useful as a structural metal. Mg reacts rapidly with water at elevated temperatures, but slowly at ordinary temperatures. Some of its low-density alloys (Mg–Al and Mg–Mn) are much more resistant to air oxidation than pure Mg, presumably because the alloys form protective oxide layers to a greater extent.

The remaining alkaline earth metals can be obtained also by electrolyzing fused salts or by reducing the alkaline earth metal oxide or chloride with Al. Ca, Sr, and Ba are oxidized rapidly in air. They are produced in small amounts compared with Mg.

▶ 15.3.4   Anomalous Behavior of Beryllium

Like Li, Be stands apart from the remainder of the family in many respects. The hardness of Be (6–7 on Mohs' scale) is much greater than that of Mg (2.6) or.Ca (2.2–2.5). The melting points of all alkaline earth metals except for Be fall within a narrow range (Table 15.3 and Figure 15.1).

Because many properties depend greatly on charge density, there is some validity to the statement that Be is more similar to Al (diagonal relationship) than to Mg. The charge density (q/r) of Be [0.039, using r (C.N. 6) = 58 pm for comparison to Al^{3+} and Mg^{2+}] is closer to that for Al^{3+} (0.044) than to that for Mg^{2+} (0.023). Beryllium compounds are largely covalent, and fused $BeCl_2$ is a rather poor electrolyte. At high temperature, $BeCl_2$

is linear, Cl—Be—Cl. Beryllium oxide is hard and refractory, like Al_2O_3; and when fired it is practically insoluble in acids. Beryllium nitrate, sulfate, and the halides are soluble, but the hydroxide, oxide, and phosphate are all insoluble in water. Normal beryllium carbonate, $BeCO_3$, like that of Al, is unstable except in the presence of CO_2. Basic beryllium carbonate of variable composition is precipitated by the addition of Na_2CO_3 to a solution of a soluble Be salt. Be_2C and Al_4C_3 apparently are the only common saltlike metal carbides that yield methane on hydrolysis. Calcium carbide, CaC_2, contains the C_2^{2-} (acetylide) ion and gives acetylene on hydrolysis.

The extent of hydrolysis of Be salts in solution parallels much more that of Al salts than that of salts of the rest of the alkaline earth family. Beryllium halides cannot be obtained from hydrates by heating because HX is lost along with H_2O. Beryllium has a greater tendency than other Group 2 elements to form stable complexes, particularly with F^- or oxoanions. Tetrahedral complexes are formed usually. Salts of the type $[NaBeF_3]_n$ contain long chains of repeating BeF_3^- units, which are analogous to the metasilicates $(SiO_3^{2-})_n$ and metaphosphates $(PO_3^-)_n$. Bridging F^- ions give tetrahedral BeF_4 units. Fluorides are weakened models of oxides of the same structural type—for example, BeF_2 for SiO_2, MgF_2 for TiO_2, and NaF for CaO. Fluorides have lower lattice energies than do the corresponding oxides, because of the lower charges. Because of charge effects, solubilities, and melting points of salts of $(BeF_3^-)_n$ resemble more closely those of $(PO_3^-)_n$ salts than those of $(SiO_3^{2-})_n$ salts.

The remaining members of the family show little tendency to form complexes, except with rather exceptional ligands such as edta. The complexes of the alkaline earth metals other than Be usually are octahedral.

Beryllium forms an unusual compound by combining with acetic acid or other carboxylic acids. The compound is basic beryllium acetate, $Be_4O(O_2CCH_3)_6$, in which the four Be^{2+} are arranged tetrahedrally around a central oxide ion, with a carboxylate ion spanning each of the six edges of the tetrahedron (Figure 15.5). The molecule contains interlocking six-membered rings, satisfying the tetrahedral bonding requirements of Be^{2+}. Basic beryllium acetate is insoluble in water but soluble in organic solvents such as chloroform. It can be distilled under reduced pressure. Similar, though less stable, compounds of Zn^{2+} and ZrO^{2+} are known. The fact that the solubilities of both BeO and $BeSO_4$ are increased when *both* are present in the same solution further demonstrates Be's great tendency to form complexes involving oxide bridges. Individually, $BeSO_4$ is moderately soluble and BeO is virtually insoluble. Together they form a complex in which the Be^{2+} is "solvated" with four BeO molecules (Figure 15.6). The remaining coordination requirements of these BeO molecules can be satisfied by hydration.

The characteristic coordination number of Be^{2+} in crystals is four. The oxide, BeO, has the wurtzite structure (*hcp* O^{2-}), and BeS has the zinc blende structure (*ccp* S^{2-}). Most metal ions that do not replace Si^{4+} in silicates have C. N. 6; but in beryl, $Be_3Al_2[Si_6O_{18}]$, and phenacite, $Be_2[SiO_4]$, the Be^{2+} is in tetrahedral sites. The structure of BeF_2 is that of cristobalite (a polymorph of SiO_2—see Figure 5.32). Crystals of the more covalent $BeCl_2$ contain chains of $BeCl_4$ tetrahedra (distorted). The structure of solid $Be(CH_3)_2$ is similar, although $Be(CH_3)_2$ molecules are monomeric and linear in the vapor.

It is easy to separate Be^{2+} analytically from the remaining alkaline earth metals, but separation from Al^{3+} is difficult. We can take advantage of the formation of basic beryl-

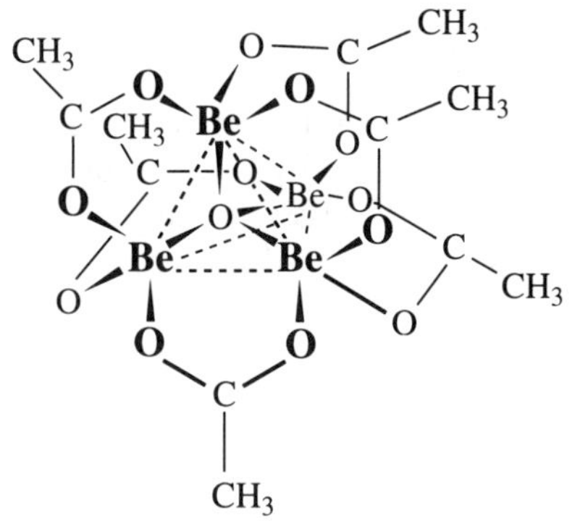

Figure 15.5   Basic beryllium acetate.

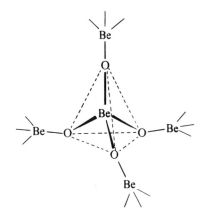

Figure 15.6   The $Be^{2+}(BeO)_4$ complex ion.

lium acetate, which can be extracted into chloroform, leaving Al^{3+} in the aqueous phase. Also, Al^{3+} can be precipitated as the hydroxide by adding an excess of ammonium carbonate, but Be^{2+} is soluble because of formation of a carbonato complex. We can precipitate Al^{3+} as $AlCl_3 \cdot 6H_2O$ by adding ether to a solution saturated with HCl gas. Beryllium chloride remains in solution.

Beryllium compounds are *extremely* toxic and should not be inhaled. Do *not* try to verify the sweet taste of beryllium compounds which led to the old name glucinium for Be. Be^{2+} can displace Mg^{2+} from important Mg-activated enzymes, blocking enzymatic activity.

▶ 15.3.5   Compounds of the Remaining Alkaline Earth Metals

Most compounds of the remaining members of Group 2 are rather typical ionic compounds for which trends are quite regular. The hydroxides are not very soluble, except for $Ba(OH)_2$ (a saturated solution is about $0.05\ M$). Barium hydroxide is a strong base and although slaked lime, $Ca(OH)_2$, is quite caustic, milk of magnesia, $Mg(OH)_2$, can be swallowed as a laxative because its solubility is low and it is a weaker base.

Magnesium bridges the gap between Be and the heavier members of the family in its slight tendency both to hydrolyze and to form complex ions. Compounds of Ca, Sr, and Ba are quite similar to one another in many respects. In an analytical separation scheme, these three metal ions usually are precipitated as carbonates and then separated from one another utilizing differences in solubilities of salts such as $MCrO_4$ at different acidities.

All alkaline earth metal compounds of the oxygen family except those of Ba (see above and Table 5.4) have the NaCl structure. The fluorite structure (C.N. 8 for M^{2+}) appears for the halides with the largest radius ratios: CaF_2, SrF_2, BaF_2, and $SrCl_2$. The rutile structure (C.N. 6 for M^{2+}) occurs for MgF_2, but layer structures (page 218) are encountered for MgCl$_2$ (CdCl$_2$ structure), CaI$_2$, MgBr$_2$, and MgI$_2$ (CdI$_2$ structure). A distorted rutile structure characterizes CaCl$_2$ and CaBr$_2$.

The Grignard reagents, usually given the general formula RMgX, have been widely used in organic reactions. The reagents are prepared by reaction of Mg metal with alkyl or aryl halides: For example, C$_2$H$_5$Br yields C$_2$H$_5$MgBr in anhydrous diethyl ether. See Section 13.6 for discussion of metal alkyls and Grignard reagents.

15.4   GROUP 11 (IB)—THE COINAGE METALS

▶ 15.4.1   *Group Trends*

Although K and Cu each contain a single electron in the fourth shell, their differences in properties are very great. The Cu atom has a smaller radius, because of its higher nuclear charge and significant contraction through the transition series. Both factors favor an increase in the IE and sublimation energy of Cu, resulting in a decrease in electropositive character and reactivity as a metal (see Table 15.4). Increases in sublimation and IEs are both quite large, and each contributes significantly to the nobility of the coinage metals. However, differences in the activities of the alkali and coinage metals are even greater than would be expected on the basis of the decrease in size and increase in nuclear charge

Table 15.4   The coinage metals

$(n-1)d^{10}ns^1$	$_{29}Cu$	$_{47}Ag$	$_{79}Au$
Abundance (% of earth's crust)	7×10^{-3}	1×10^{-5}	5×10^{-7}
Density (g/cm^3)	8.94	10.49	19.32
Melting point (°C)	1083	960.5	1063
Boiling point (°C)	2595	2212	2966
Sublimation energy (kJ/mol, 25°C)	341.1	299.2	344.3
Ionization energy (eV)			
1st	7.726	7.576	9.225
2nd	20.292	21.49	20.5
3rd	36.83	34.83	—
Atomic radius (pm)	128	144	144
Ionic radius (pm)			
C.N. 2	Cu$^+$ 60	Ag$^+$ 81	Au$^+$ 151 (C.N. 6)
C.N. 6	Cu^{2+} 87	Ag^{2+} 93 (C.N. 4)^{sq}	Au^{III} 82 (C.N. 4)^{sq}
Heat of hydration of M^{n+} (kJ/mol)	Cu$^+$ 581 Cu^{2+} 2120	Ag$^+$ 485	Au$^+$ 644
E^0 for M$^+(aq) + e \rightarrow$ M(s)	+0.520	+0.799	*ca.* + 1.83
E^0 for M$^{2+}(aq) + 2e \rightarrow$ M(s)	+0.340	+1.39	—

through the transition series. The additional factor is the difference in electron configuration. Beneath the outer shell, each of the alkali metals has a noble-gas core; but each of the coinage metals has an 18-electron ($18e$) shell, a pseudo-noble-gas configuration. The 18-e shell, which includes relatively nonpenetrating d electrons, does not screen the positive nuclear charge as effectively as an 8-e shell. Thus Z_{eff} becomes quite large (see Slater's rules, page 40). The contrast in reactivity is obvious from the fact that the alkali metals include the most active metals while the coinage metals include Au, the most noble metal. The alkali metals give hard cations, and the 18-e cations are soft.

The coinage metals offer a striking contrast to the alkali metals in another respect: They *decrease* in reactivity with increasing atomic number. Because of less effective shielding of the nuclear charge by the 18-e shell, the increase in Z_{eff} is more important than the increase in atomic radius in determining the attraction for electrons. Formation of $Ag(aq)^+$ from the metal requires more energy than for $Cu(aq)^+$, as indicated by the electromotive force (emf) values, even though the first IE of Ag is slightly *lower* than that of Cu and the sublimation energy is appreciably lower for Ag. Only the higher hydration energy of the smaller Cu^+ ion makes the oxidation process more favorable for Cu (see Table 15.4). Because the lanthanide series with the filling of $4f$ orbitals comes between Ag and Au, their atomic radii are the same. The resulting increase in the first IE (with higher nuclear charge) and the much higher sublimation energy account for the much greater nobility of Au.

The coinage metals are among the most dense metals (Table 15.4). Their melting points are very similar (Figure 15.1). There is a surprisingly large minimum in the sublimation energy for Ag. These metals have the highest electrical conductivities of any element and are the most malleable and ductile structural metals. Gold can be rolled into sheets as thin as 10^{-5} cm.

▶ 15.4.2   Occurrence

In spite of their low abundances, the coinage metals were among the earliest discovered metals. All three metals occur in the native state. Native Au usually is found in **placer deposits** (such as in the bed of a stream), left by the weathering of the original rocks. Because Au is dense and resistant to oxidation, it remains after most of the other minerals present have been removed by chemical attack or by the mechanical forces of weathering. The coinage metals also commonly occur as sulfide minerals (chalcophiles); early man could have obtained the metals in his campfires because their minerals are reduced easily by charcoal (carbon).

▶ 15.4.3   Preparation of the Metals

Copper sulfide ores are converted to the oxide by roasting in air. The oxide is reduced to Cu by carbon. If the supply of air is cut off before conversion to the oxide is complete, Cu can be obtained directly:

$$2\,Cu_2S + 3\,O_2 \xrightarrow{\text{heat}} 2\,Cu_2O + 2\,SO_2 \tag{15.13}$$

$$2\,Cu_2O + Cu_2S \xrightarrow{\text{heat}} SO_2 + 6\,Cu \tag{15.14}$$

Impure Cu is obtained directly from ores. Purification can be accomplished by heating crude Cu in a furnace to form some Cu_2O, which serves to oxidize the more reactive metals present. The remaining Cu_2O is reduced to the metal by wood charcoal or by stirring with wooden poles. The Cu obtained (blister Cu) is more than 99% pure.

Electrolytic refining is the important process for obtaining pure Cu. The crude blister Cu cast as a large anode is suspended in an acidified copper sulfate solution containing a small amount of chloride ion. Electrolysis oxidizes Cu and more active metals to their ions and deposits pure Cu at the cathode. Noble metals present in blister Cu are not oxidized and fall to the bottom of the cell along with other impurities as the **anode mud.** Any Pb that is oxidized precipitates as lead sulfate; any Ag that might be oxidized precipitates as AgCl. Recovery of precious metals from the anode mud pays a large share of the cost of refining Cu. More active metals remain in solution. Because of low operating voltage (a few tenths of a volt), many cells normally are operated in series.

Silver is obtained as a byproduct in the recovery of Pb, Zn, and Cu from their ores. Silver ores (Ag_2S, AgCl, or Ag) commonly are treated by the cyanide process. In the presence of oxygen (in air), Ag dissolves in aqueous NaCN solution as the complex ion $[Ag(CN)_2]^-$. Sulfide ion is oxidized to SO_4^{2-} or SCN^-. Ag is oxidized by oxygen, because of the great stability of $[Ag(CN)_2]^-$. The emf for oxidation is shifted by more than 1 V [compared with the formation of $Ag(aq)^+$].

$$Ag + 2\,CN^- \longrightarrow [Ag(CN)_2]^- + e \qquad E^0 = +0.31 \text{ V (as oxidation)} \tag{15.15}$$

Silver is recovered from the solution by addition of Zn dust or by electrolysis.

Gold-containing ore can be enriched by a washing process in which ore is broken up and suspended in water in order to separate less dense impurities from Au, which settles rapidly because of its great density. More complete recovery of Au is possible if the concentrated ore is treated by an amalgamation process. Au dissolves in the Hg to form an amalgam and then is recovered by distilling the Hg.

Extraction by the cyanide process is more important, particularly for very-low-grade ores. Ore is ground finely and treated with a dilute cyanide solution (0.1–0.2% KCN or NaCN). Oxygen in air oxidizes the Au to a soluble cyanide complex, to which Zn is added to yield the Au. Gold(I) is stabilized even more than silver(I) by coordination with CN^-, so that Au becomes a strong reducing agent. The emf is shifted by more than 2V compared with the formation of $Au(aq)^+$.

$$Au + 2\,CN^- \longrightarrow [Au(CN)_2]^- + e \qquad E^0 = +0.60 \text{ V (as oxidation)} \tag{15.16}$$

From this shift in emf, the formation constant for the complex is 2×10^{38}. (See Section 9.4 for the calculation of equilibrium constants from E^0 values.)

▶ *15.4.4   Compounds of the Coinage Metals*

Only one oxidation state is encountered for the alkali metals, because the first IEs are very low and the energy needed to remove additional electrons from the underlying noble-gas shell is prohibitively high (see Tables 15.1 and 15.4). The differences among the first three IEs are not so great for the coinage metals, involving *s* and *d* electrons. Formation of compounds in higher oxidation states (II and III) is favorable because of complex formation.

A number of subhalides of the type M_2X have been reported for the coinage metals, but most of them probably are mixtures. The only verified subhalide of the group is Ag_2F, a bronze-colored solid obtained by the cathodic reduction of AgF in aqueous solution. The solid has double layers of Ag "atoms" separated by layers of F^- ions. With the Ag "atoms" held together by metal–metal bonds, no discrete Ag^+ or Ag_2^+ ions can be identified.

The **oxidation number I** is common to all three of the coinage metals. Copper(I) is stable only in solid compounds or in a few complexes formed by ligands such as CN^-, SCN^-, $S_2O_3^{2-}$, or thiourea. In solid $K[Cu(CN)_2]$, each Cu^I is bonded to three CN^- in an almost planar trigonal arrangement. Two of the CN^- are bound through C, and the third (bridging) CN^- is bound through C and N. Water molecules doubtless are coordinated also in aqueous solution. Compounds of the type $M_2^ICuX_3$ and $M^ICu_2X_3$ contain single and double chains, respectively, of CuX_4 tetrahedra through sharing of X^- ions.

Copper(II) is reduced by I^- or CN^- with the formation of insoluble CuI or CuCN, but CuCN dissolves in the presence of excess CN^- with complex formation:

$$2\,Cu^{2+} + 4\,I^- \longrightarrow 2\,CuI(s) + I_2 \tag{15.17}$$

$$2\,Cu^{2+} + 4\,CN^- \longrightarrow 2\,CuCN(s) + (CN)_2 \tag{15.18}$$

$$CuCN(s) + CN^- \longrightarrow [Cu(CN)_2]^- \tag{15.19}$$

$$2\,Cu^+ \longrightarrow Cu + Cu^{2+} \qquad E^0 = +0.36\ V \tag{15.20}$$

The solids CuCl, CuBr, and CuI have the zinc blende structure (C.N. 4). The hydrated copper(I) ion and solid CuF are unstable with respect to disproportionation forming Cu and Cu^{2+}. Solutions of Cu^I complexes absorb olefins by forming Cu^I olefin complexes. This process has been used to separate olefins from other hydrocarbons and monoolefins from polyolefins. Lower oxidation states are generally more stable at high temperatures because all but strongest bonds are broken at high temperature. Only very simple molecules persist at very high temperature. CuO is converted is Cu_2O at elevated temperatures:

$$4\,CuO \xrightarrow{\text{heat}} 2\,Cu_2O + O_2 \tag{15.21}$$

Silver(I) is the most stable oxidation state of Ag. Silver nitrate, silver fluoride, and silver perchlorate are very soluble in water. The acetate, sulfate, chlorate, and bromate range from slightly soluble to moderately soluble, but most other common salts are quite insoluble. The solubilities of the silver halides decrease in the order AgCl ($K_{sp} = 1.8 \times 10^{-10}$), AgBr ($K_{sp} = 3.3 \times 10^{-13}$), AgI ($K_{sp} = 8.5 \times 10^{-17}$). The NaCl structure (C.N. 6) characterizes all the halides except AgI (wurtzite, C.N. 4), whose interesting polymorphism has been discussed (see Section 6.2.2). Chain structures are encountered for covalent AgCN and AgSCN. Many silver compounds containing colorless anions are themselves colored: For example, Ag_3PO_4, Ag_2CO_3, and AgI are yellow, and Ag_2S is black. Salts containing polarizable anions and highly polarizing cations—for example, those with 18-e or $(18 + 2)$-e configurations—often are colored because of charge-transfer absorption (see page 202). Silver(I) forms many stable complexes, particularly with N and S donors. In complexes, Ag^+ generally shows C.N. 2, whereas in solution, solvation likely increases its C.N. to 4.

Silver(I) halides (except the covalent AgI) show a higher C.N. in the solid than do copper(I) halides, but in its complexes, Ag^I demonstrates a pronounced preference, compared with Cu^I, for the lower C.N. 2. This indicates a greater tendency of Ag^I to form linear bonds. Whereas C.N. 2 also seems predominant for Au^I, in Group 12 (IIB) only Hg^{II} shows a preference for C.N. 2. Linear bonding involving sp hybridization can be enhanced, as a result of more directional character of the hybrid orbitals, by some mixing of the d_{z^2} orbital also. The energy differences between the ns and $(n - 1)$ d levels, as indicated by the (2nd IE − 1st IE), are about the same (12.6 ± 1.3 eV for Cu, Ag, and Au) although lowest for Au. These energy differences (3rd IE − 2nd IE) are much greater for Zn (21.7 eV) and Cd (20.6 eV), but comparable for Hg (15.4 eV). Planar trigonal coordination is encountered for Au^I in chlorobis(triphenylphosphine)gold(I).

Gold(I) chloride and bromide are obtained by gently heating the corresponding gold(III) halides [Reaction (15.22)]. These gold(I) halides disproportionate in aqueous solution to give the metal and the corresponding gold(III) halide [Reaction (15.23)].

$$AuCl_3 \xrightarrow{\text{heat}} AuCl + Cl_2 \tag{15.22}$$

$$3\,AuCl + H_2O \longrightarrow AuCl_3 + 2\,Au \tag{15.23}$$

$$AuCl_3 + 3\,I^- \longrightarrow AuI(s) + I_2 + 3\,Cl^- \tag{15.24}$$

Gold(I) iodide is precipitated from solutions containing gold(III) compounds and iodide ion [Reaction (15.24)]. Because of its low solubility, AuI is decomposed more slowly by water. Gold(I) cyanide, so insoluble that it is stable in the presence of water, dissolves in the presence of a cyanide salt to give a soluble complex, such as $K[Au(CN)_2]$. Gold(I) also gives soluble complexes with $S_2O_3^{2-}$ and SO_3^{2-}. Insoluble Au_2S can be precipitated from an aqueous solution of any soluble gold(I) complex by saturating the solution with H_2S.

Copper(II) occurs in most of the stable copper salts and complexes in the solid state and in solution. The copper(II) complexes are usually square planar, although there are often two additional ligands or solvent molecules at slightly greater distances—one above and one below the plane of the four closer ligands. Copper(II) fluoride has the fluorite structure typical of ionic salts. Copper(II) chloride and bromide have chain structures commonly encountered among salts with an appreciable amount of covalent character.

Silver(II) compounds are uncommon. Silver(II) fluoride, formed by the action of F_2 on AgF, is used as a fluorinating agent for organic compounds. An oxide, AgO, is obtained by anodic oxidation of aqueous solutions of silver salts. It is a diamagnetic silver(I)–silver(III) compound, $Ag^I[Ag^{III}O_2]$. The divalent state is stabilized by coordination with pyridine, $[Ag(py)_4](NO_3)_2$; pyridine derivatives such as picolinic acid; 1,10-phenanthroline; bipyridine; and a few other heterocyclic nitrogen ligands. The complexes presumably are planar, like those of copper(II).

Of the several gold(II) compounds reported, most have been shown to be mixed Au^I–Au^{III} compounds. The simple salts $AuSO_4$, AuO, and AuS have been reported, but whether they are authentic gold(II) compounds is questionable. No gold(II) complexes have been verified without Au–Au bonding.

Copper(III) (d^8) complexes are diamagnetic with square-planar coordination or five-coordination, except for the high-spin complex K_3CuF_6 as a green solid. Cu^{III} forms complexes with N_4—macrocyclic and tetradentate Schiff-base ligands. Oxidation of Cu^{2+} in alkaline solution by ClO^- in the presence of iodate or tellurate gives $[Cu(IO_5OH)_2]^{5-}$ or $[Cu\{TeO_4(OH)_2\}_2]^{5-}$. Silver(III) forms similar iodate and tellurate compounds. The fluorocomplexes $K[AgF_4]$ and even Cs_2KAgF_6 are formed by oxidation of $AgNO_3$ by F_2 in the presence of the alkali fluorides. Most commonly encountered compounds of Au contain gold(III). Gold is usually dissolved in *aqua regia* ($HCl + HNO_3$) to obtain chloroauric acid, $H[AuCl_4] \cdot 4H_2O$. Neither hydrochloric nor nitric acid alone will dissolve Au. Gold(III) is stabilized sufficiently by coordination with chloride ion to lower the emf required such that the oxidation can be accomplished by HNO_3. Copper(III) can be stabilized by coordination to peptides and seems to be biologically important (page 919).

All gold compounds are decomposed easily by heating. $AuCl_3$ and $AuBr_3$ give monohalides upon gentle heating, but the metal is formed when halides are heated more strongly. AuI_3 is unstable with respect to loss of I_2, even without heating. Except for a few stable solids, gold(III) compounds encountered are planar complexes. Fluorination of $Cs[AuF_4]$ yields $Cs[AuF_6]$ containing gold(V) (d^6) as the AuF_6^- ion.

15.5 GROUP 12 (IIB)—THE ZINC SUBGROUP

▶ 15.5.1 Group Trends

Zinc subgroup elements are more dense and less active than their counterparts in the alkaline earth family, as would be expected from their smaller radii and higher nuclear charge. Sublimation energies of the Group 12 metals are lower than for any other group of metals except the alkalis (Table 15.5). Correspondingly, the metals, with low melting points and boiling points, can be distilled. The $(18 + 2)$-e configuration characteristic of this family practically constitutes a closed configuration from the standpoint of metal–metal bonding in the free state. Increased activity of the Group 12 (IIB) family, compared with the Group 11 (IB) family, results from low sublimation energies of the metals and high hydration energies of the divalent ions. Obviously, it is not caused by a decrease in IEs. Radii of the Group 12 metals are slightly larger than those of the corresponding Group 11 metals. The Group 12 M^{2+} ions are soft cations compared to the hard cations of Group 2.

Table 15.5   The zinc family

$(n - 1)d^{10}ns^2$	$_{30}Zn$	$_{48}Cd$	$_{80}Hg$
Abundance (% of earth's crust)	8×10^{-3}	1.8×10^{-5}	5×10^{-5}
Density (g/cm³)	7.133	8.65	13.55
Melting point (°C)	419.5	320.9	−38.87
Boiling point (°C)	906	767	357
Vaporization energy (kJ/mol 25°C)	130.5	112.8	60.84
Ionization energy (eV)			
1st	9.394	8.993	10.437
2nd	17.964	16.908	18.756
3rd	39.722	37.48	34.2
Atomic radius (pm)	133	149	150
Ionic radius (pm) M^{2+} C.N. 6	88	109	116
Heat of hydration of M^{2+} (kJ/mol)	2060	1830	1840
E^0 for M^{2+} $(aq) + 2e \rightarrow M$	−0.763	−0.403	+0.853

The zinc family metals decrease in activity with increasing atomic number, like the Cu family metals. The increase in nuclear charge is more important than the increase in atomic radius. Zinc and cadmium are very active metals, and Hg is a distinctly noble metal—in fact, it has the highest first IE of *any* metal. Group 12 is the first group after Group 3 (IIIA) in which the members of the first and second transition series (Zn and Cd) are more similar than the members of the second and third transition series (Cd and Hg). Zn, Cd, and Hg are not strictly transition metals because the filled *d* subshells are not involved in bonding.

▶ 15.5.2   *Occurrence*

The most important zinc ores are smithsonite, $ZnCO_3$, and zinc blende, ZnS. Cadmium occurs along with Zn in the same minerals. Independent Cd minerals are rare. The principal Hg ore is cinnabar, HgS. Often, some native Hg is associated with cinnabar.

▶ 15.5.3   *Preparation of the Metals*

Roasting zinc ores (ZnS or $ZnCO_3$) yields the oxide, which is then reduced with C. Zn is distilled from the furnace. In the electrolytic process, which is becoming increasingly more important than C reduction, the ore is leached with sulfuric acid to give a solution of zinc sulfate, electrolysis of which produces Zn. Zinc is deposited even from acidic solution because the high H_2 overvoltage at the Zn cathode prevents H_2 evolution.

Cadmium can be recovered from Zn produced by C reduction, because of the greater volatility of Cd. Cd is obtained by repeated fractional distillation; it also can be separated from Zn in the electrolytic preparation, because of the lower deposition potential of Cd.

Mercury can be recovered by heating the sulfide in air—the Hg vapor distills. Metals more active can be removed by leaching the Hg produced with dilute nitric acid.

▶ 15.5.4   Compounds of the Group 12 (IIB) Metals

The characteristic oxidation number of Group 12 metals is II. Zinc forms four-coordinate tetrahedral complexes, with few examples of higher C.N. Cadmium forms six-coordinate complexes to a great extent. $Zn(OH)_2$ is amphoteric, dissolving in an excess of OH^- to form $[Zn(OH)_4]^{2-}$. $Cd(OH)_2$ is not amphoteric.

Zn, Cd, and Hg form organometallic compounds MR_2 with linear bonds. Often more ligands can be added to linear HgX_2 molecules to give tetrahedral complexes, but the formation constants for the two additional ligands are much lower (more weakly bonded). Mercury(II) shows a very marked preference for large, polarizable ligands. Although complexes containing nitrogen ligands are very stable, there is a great preference for S-ligands over O-ligands. The order of increasing stability of halide complexes is $Cl^- <$ $Br^- < I^-$, with little or no tendency to form fluoride complexes. Mercury(II) oxide is basic and does not tend to form hydroxide complexes. Zn, Cd, and Hg form complexes with *other metals bonded as ligands* as in $Zn[Co(CO)_4]_2$ and $[(en)_2Rh \rightarrow Hg \leftarrow Rh(en)_2]^{2+}$.

Except for CdO and HgO, the MX compounds of O, S, Se, and Te have the zinc blende or wurtzite structures (C.N. 4—see Table 5.5). Cadmium oxide has the NaCl structure (C.N. 6), whereas the orthorhombic form of HgO contains zigzag chains of essentially linear O—Hg—O units. A second form (the more common hexagonal form) of HgO is isostructural with cinnabar, HgS. Cinnabar is built of helical chains of S—Hg—S units (S—Hg—S angle 172°). Two Hg—S distances are 236 pm, two are 310 pm, and two are 330 pm.

ZnF_2 (rutile structure, C.N. 6 for Zn^{2+}) and CdF_2 (fluorite structure, C.N. 8 for Cd^{2+}) are the only ionic-type structures of the halides of Zn and Cd. Of the three polymorphic forms of $ZnCl_2$, two have structures built of $ZnCl_4$ tetrahedra and the third has the HgI_2 layer structure (Figure 5.18). Cadmium chloride and zinc bromide have a layer structure containing a cubic close-packed array of X^- with M^{2+} in octahedral holes in alternate layers (Figure 5.17). The CdI_2, $CdBr_2$, and ZnI_2 structures are similar except that the X^- ions are hexagonally close-packed.

The mercury(II) halides show interesting variations in lattice type (see page 220). HgF_2 has the fluorite structure common for ionic salts. Mercury(II) chloride has a molecular lattice consisting of discrete linear $HgCl_2$ molecules. (See Section 9.13.1 for discussion for the low force constants of linear molecules of the type HgX_2.) Although both $HgBr_2$ and red HgI_2 have layer lattices, $HgBr_2$ represents a transition between a molecular and layer lattice, since each Hg^{2+} has two nearest-neighbor Br atoms and two Br atoms at a greater distance. At 126°C, red HgI_2 undergoes a reversible phase transition to a yellow form that contains linear HgI_2 molecules.

Organozinc compounds, such as $(C_2H_5)_2Zn$, were used in organic reactions before the discovery of the Grignard reagents, which can be prepared and handled more conveniently. Organometallic compounds are formed easily by the reaction of Grignard reagents (organomagnesium compounds) with $HgBr_2$. Organomercury compounds are resistant to attack by oxygen or moisture and can be handled in contact with air (see Section 13.6.1).

Zinc is a biologically essential element that occurs at the active site of many enzymes (Section 18.5.1). At high levels, Zn is toxic, but its toxicity is obscured by the common presence of the much more toxic Cd^{II} in Zn^{II} compounds. The amount of highly toxic Cd in cigarette smoke is great enough for concern. Problems with Cd poisoning have surfaced in Sweden, in connection with the manufacture of alkaline Ni–Cd batteries, and in Japan, where the painful *itai itai* disease has been attributed to chronic Cd poisoning. Mercury also is highly toxic. The vapor pressure of Hg metal is sufficiently high so that spilled Hg is a laboratory hazard. Hg^{II} has a high affinity for S and inhibits metal enzymes, particularly those containing thiol groups. Microorganisms can transform Hg compounds into CH_3Hg^+, which concentrates in the food chain and poses a major threat of Hg poisoning in fish and other seafood. Cd, Hg, Pb, Sb, and Sn seem to have a cumulative effect, since they have no apparent biological function and hence there is no mechanism to regulate the levels of the toxic metals.

Zinc(I) and cadmium(I) compounds have been reported. The existence of Cd^I compounds has been demonstrated clearly, but they decompose in contact with oxygen or in water. Several Hg^I compounds are stable. The compounds are unusual in that they contain the only common diatomic metal cation, Hg_2^{2+}. The compounds are diamagnetic and hence cannot contain monomeric Hg^+. X-ray structure analysis has revealed that solid Hg^I compounds contain Hg—Hg bonds. Mercury(I) compounds can be prepared generally by the reaction of Hg with the corresponding Hg^{II} compound. The equilibrium constant for the reaction of Hg with Hg^{2+}, $Hg + Hg^{2+} \rightleftarrows Hg_2^{2+}$, is 166. Mercury(I) gives few stable complexes, so adding a complexing agent to a Hg^I salt usually results in disproportionation to give Hg and the Hg^{II} complex. The Cd^I and Zn^I ions have been shown to be M_2^{2+} from Raman spectra of melts containing MCl_2 and excess M. A yellow diamagnetic solid containing $Cd_2(AlCl_4)_2$ was isolated from melts containing excess $AlCl_3$. Unusual low oxidation states of metals are often obtained in a system in which the anion is $AlCl_4^-$ or some other large anion. Force constants for the M—M bonds demonstrate that the bond strengths increase in the order $Zn_2^{2+} < Cd_2^{2+} << Hg_2^{2+}$.

15.6 GROUP 3 (IIIA)—THE SCANDIUM FAMILY AND RARE EARTHS

► *15.6.1 Group Trends*

Trends in the family Sc, Y, La, and Ac are quite regular and follow the pattern expected from the properties of Group 1 and 2 metals. The metals are very active in spite of the large amount of energy required to remove three electrons and the fact that the sublimation energies are also appreciably higher than those of the alkaline earth metals. Lattice energies of compounds containing triply charged cations or hydration energies of the ions are sufficiently high to account for the high oxidation emfs of the metals (see Table 15.6). The trivalent state is most important for these metals.

Scandium, with a smaller radius than other members of the family, tends to form complexes. Its salts hydrolyze extensively in solution, and the tendency to hydrolyze decreases with increasing ionic radius of the metal ion through the family.

The rare earth metals are always found together in nature, because they all occur as compounds of the 3+ ions with very similar radii. The chemical similarities are so great

Table 15.6   Group 3 (IIIA) metals and the rare earths

	Abundance (% of earth's crust)	Density (g/cm³)	Melting point (°C)	Atomic radius (pm)	Ionic radius[a] M^{3+} (pm)	E^0 M^{3+}/M	E^0 M^{3+}/M^{2+}	E^0 M^{4+}/M^{3+}
$_{21}$Sc	5×10^{-4}	2.992	1539	164.1	88.5	-2.03		
$_{39}$Y	2.8×10^{-3}	4.472	1509	180.1	104.0	-2.37		
$_{57}$La	1.8×10^{-3}	6.174	920	187.7	117.2	-2.38		
$_{58}$Ce	4×10^{-3}	6.66	795	182	115	-2.34		$+1.72$
$_{59}$Pr	5.5×10^{-4}	6.782	935	182.8	113	-2.35		$+3.2$
$_{60}$Nd	2.4×10^{-3}	7.004	1024	182.1	112.3	-2.32		
$_{61}$Pm	—	—	1035	—	111	-2.29		
$_{62}$Sm	6.5×10^{-4}	7.536	1072	180.2	109.8	-2.30	-1.55	
$_{63}$Eu	1×10^{-4}	5.259	826	204.2	108.7	-1.99	-0.35	
$_{64}$Gd	6.5×10^{-4}	7.895	1312	180.2	107.8	-2.28		
$_{65}$Tb	9×10^{-5}	8.272	1356	178.2	106.3	-2.31		$+3.1$
$_{66}$Dy	4.5×10^{-4}	8.536	1407	177.3	105.2	-2.29		
$_{67}$Ho	1.1×10^{-4}	8.803	1461	176.6	104.1	-2.33		
$_{68}$Er	2.5×10^{-4}	9.051	1497	175.7	103.0	-2.32		
$_{69}$Tm	2×10^{-5}	9.332	1545	174.6	102.0	-2.32		
$_{70}$Yb	2.7×10^{-4}	6.977	824	194.0	100.8	-2.22	-1.05	
$_{71}$Lu	7.5×10^{-5}	9.843	1652	173.4	100.1	-2.30		

[a] C.N. 6.

Discovery of the Rare Earth Metals

The minerals gadolinite and cerite were discovered in Sweden in the latter part of the 18th century. The minerals were found to contain new "earths" (metal oxides), later shown to be mixtures and separated into other metal oxides. Berzelius characterized the mineral cerite as containing the new earth ceria. Later, ceria was resolved into oxides of lanthanum, cerium, and didymium. The name *didymium* ("twin") refers to the supposed new element as a twin of lanthanum because of their great similarities. Didymium oxide later was resolved into the oxides of praseodymium ("green twin") and neodymium ("new twin"). More careful separation and characterization of the earths obtained from these minerals ultimately yielded Sc, Y, La, and the 14 lanthanide metals. (The term *lanthanoid* is recommended by the IUPAC, to reserve the *-ide* ending for anions, but the well-established term *lanthanide* is commonly used.) Because of similar size, Y^{3+} occurs in minerals with the lanthanides, so the term rare earths includes Y_2O_3. Note in Table 15.6 that some of these "rare" earth metals (Y, La, Ce, and Nd) are more abundant than Ag, Cd, Sn, or Pb. The abundances exhibit a "sawtooth" pattern with higher abundances for elements with even Z and low abundances for those with odd Z. Generally, abundances are greater for elements with even Z than for neighbors with odd Z as seen in Figure 15.7. The noble gases with even Z have low abundances because they occur in the atmosphere and have been lost to space (see Section 16.5.2).

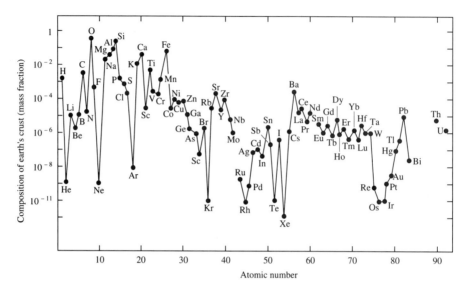

Figure 15.7   Abundances of elements in the continental crust, shown as mass fraction on a logarithmic scale. (Reproduced with permission from P. A. Cox, *The Elements*, Oxford University Press, Oxford, 1989, p. 9.)

that separating the metal ions is very difficult. The classical method involved fractional crystallization of salts from aqueous solution—a slow, tedious process not suitable for production of appreciable amounts of rare earths of high purity. Because rare earths are among the major products of the fission of uranium, their separation became a practical problem in the study of U fission. We now can separate rare earths using ion exchange procedures or a combination of solvent extraction and ion exchange procedures, so they currently are available in a high state of purity at only a small fraction of earlier costs.

Although the decrease in radius from one element to the next is very small within the rare earth series, the cumulative decrease is quite significant (the **lanthanide contraction**). The atomic radii decrease fairly regularly through the series, except for significant increases at Eu and Yb. Because the latter elements have only two electrons beyond the respective half-filled and completely filled *f* orbitals, apparently they have only two electrons available for metallic bonding, rather than the three available in the other rare earth metals. These irregularities disappear for ionic radii (M^{3+}) where three outer electrons have been removed. Although the ionic radius of La^{3+} is appreciably larger than that of Y^{3+}, the decrease in size through the rare earth series proceeds steadily until Ho^{3+} is only slightly larger than Y^{3+}. In the separation of the rare earths, Y concentrates with Dy and Ho.

The rare earth metals are very strong reducing agents, because of the high lattice energies of their compounds or their high hydration energies. Consequently, they are useful for the thermite-type reactions, taking advantage of the high ΔH_f of M_2O_3. [The thermite reaction uses Al to reduce metal oxides, e.g. Reaction (15.2)]. A mixture containing mostly Ce and La and smaller amounts of other rare earths and iron, known as **mischmetall,** is produced for technical uses. Mischmetall is pyrophoric when finely divided. When alloyed with more Fe to increase the hardness, it is used for flints in lighters.

We obtain the rare earth metals by reducing the anhydrous fluorides by using Ca at 800–1000°C [Reaction (15.25)]. Reduction proceeds only to the difluorides for Sm, Eu, and Yb [Reaction (15.26)]. These metals can be obtained by reduction of their oxides by La metal *in vacuo* [Reaction (15.27)], since they are more volatile than La. The anhydrous chlorides of La through Gd can be reduced to the metals with Ca, Mg, Li, or Na [Reaction (15.28)]. The chlorides are hygroscopic and those of the heavier earth metals are too volatile at high temperatures, so fluorides are used to obtain the metals [Reaction (15.29)].

$$2\,LaF_3 + 3\,Ca \longrightarrow 2\,La + 3\,CaF_2 \tag{15.25}$$

$$2\,EuF_3 + Ca \longrightarrow 2\,EuF_2 + CaF_2 \tag{15.26}$$

$$Eu_2O_3 + 2\,La \longrightarrow 2\,Eu(g) + La_2O_3 \tag{15.27}$$

$$2\,LaCl_3 + 3\,Ca \longrightarrow 2\,La + 3\,CaCl_2 \tag{15.28}$$

$$2\,LuF_3 + 3\,Ca \longrightarrow 2\,Lu + 3\,CaF_2 \tag{15.29}$$

► 15.6.2   *Crystal Structures*

Of the various crystal structures that characterize lanthanide compounds, many are less common than the structures considered in Chapter 5 and are not easily visualized from plane figures or descriptions. One striking characteristic of the lanthanide metal ions in crystal structures and in complexes is their range of coordination numbers (C.N.'s)—6, 7, 8, 9, 10, and 12. C.N. 6 is encountered in some crystals of divalent ions, in a few octahedral complexes of anionic ligands, and in the MCl_3 structures of the smaller ions forming the latter part of the series (Dy through Lu). C.N.'s 10 and 12 are encountered for the complexes of the didentate nitrate ligand $[Ce(NO_3)_5]^{2-}$ and $[Ce(NO_3)_6]^{3-}$ (page 431).

More common C.N.'s are 7 and, in particular, 8 and 9. C.N. 8 is represented by $La(acac)_3(H_2O)_2$ (distorted square antiprism, acac = $CH_3COCHCOCH_3^-$), $Ce^{IV}(acac)_4$ (Archimedian square antiprism), $Cs[Y(CF_3COCHCOCF_3)_4]$ (dodecahedron), CeO_2 (cube, CaF_2 structure), and LaI_3 (the $PuBr_3$ layer structure containing bicapped trigonal prisms). We see C.N. 9 in $[Nd(H_2O)_9]^{3+}$ (a tricapped trigonal prism—see Figure 15.8) and in the complex structures of $La(OH)_3$, LaF_3, and the trichlorides of the larger lanthanide ions (La through Gd).

► 15.6.3   *Oxidation States Other Than III*

Oxidation number III is characteristic for the rare earths. Other stable oxidation numbers occur just about as expected on the basis of the stability of a group of orbitals that is empty, half-filled, or completely filled (Table 15.7). Thus Ce^{4+} achieves a noble-gas configuration. Tb^{4+} and Eu^{2+} have the same configuration as Gd^{3+}: seven electrons in the $4f$ orbitals. Yb^{2+} has completely filled $4f$ orbitals, as does Lu^{3+}. There are a few other examples for metals that do not quite achieve one of the stable configurations of the f or-

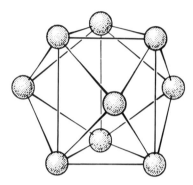

Figure 15.8   Positions of oxygen in $[Nd(H_2O)_9]^{3+}$.

bitals. Pr^{IV} is encountered in the mixed oxide Pr_6O_{11}, which can be obtained by heating Pr_2O_3 in the presence of O_2 at high pressure. The compounds PrF_4 and TbF_4 have been prepared. Tb_4O_7 has been reported as containing Tb^{IV}, but the composition does not agree with this formulation. Aqueous solutions of Eu^{2+}, obtained by reduction of $Eu^{3+}(aq)$ by Zn or Mg, can be handled easily in air. Samarium(II) salts can be obtained, but the blood-red Sm^{II} ion is unstable in aqueous solution, because it reduces water. Although thulium(II) can be obtained as the solid TmI_2, the ion is very unstable in water.

All of the rare earths have been obtained as M^{2+} ions in CaF_2 as the host lattice. When a rare earth fluoride, MF_3, is incorporated in solid CaF_2, the M^{3+} ions occupy the cation sites, with the "extra" F^- ions in interstitial sites. If electrolysis is carried out with graphite electrodes at each end of a crystal, electrons enter from the cathode to reduce M^{3+} to M^{2+} and the "extra" F^- ions migrate into the graphite anode.

The oxidation number of Eu in EuS clearly is II, since the magnetic and spectroscopic properties are those of $4f^7$ ion. The situation is not so clear for GdS, for which the magnetic properties are also those of a $4f^7$ cation, corresponding to Gd^{3+}. This can be interpreted as a Gd^{III} sulfide with the "extra" electron in a conduction band.

Table 15.7   Electronic configurations of some rare earth metals and ions

Metal	Electronic configuration	Configuration of M^{3+}	Configuration of M^{4+}	Configuration of M^{2+}
$_{57}La$	$[Xe]\,5d^1 6s^2$	$[Xe]$	—	—
$_{58}Ce$	$[Xe]\,5d^1 4f^1 6s^2$	$[Xe]\,4f^1$	$[Xe]$	—
	$\mid$ add $4f$ electrons			
$_{63}Eu$	$[Xe]\,4f^7 6s^2$	$[Xe]\,4f^6$	—	$[Xe]\,4f^7$
$_{64}Gd$	$[Xe]\,4f^7 5d^1 6s^2$	$[Xe]\,4f^7$	—	—
$_{65}Tb$	$[Xe]\,4f^9 6s^2$	$[Xe]\,4f^8$	$[Xe]\,4f^7$	—
	$\mid$ add $4f$ electrons			
$_{70}Yb$	$[Xe]\,4f^{14} 6s^2$	$[Xe]\,4f^{13}$	—	$[Xe]\,4f^{14}$
$_{71}Lu$	$[Xe]\,4f^{14} 5d^1 6s^2$	$[Xe]\,4f^{14}$	—	—

15.7   THE ACTINIDE METALS[6]

The metals Ac, Th, Pa, and U appeared to be members of the 3, 4, 5, and 6 transition element families, respectively, until the discovery of the transuranium metals. Each additional element discovered made it more obvious that these metals were members not of a transition series, but of an inner-transition series. The widely varying oxidation numbers of these metals made it difficult to determine which element was the first member of the series. Greater variation in oxidation numbers (compared with those of the rare earths) is to be expected among the actinide metals, because of their larger atomic radii and because the energy levels of the valence electrons are closer together. Several stable oxidation numbers are encountered for most of the actinide metals.

It was suggested that the second inner-transition series begins with thorium. Most of the metals following Th give dioxides that are isomorphous with ThO_2, and there are other similarities to Th. On proceeding through the series, however, the trivalent state becomes increasingly more important (Table 15.8). This is one of the important oxidation numbers of Pu and is the most stable state for americium and the remainder of the series. It is now clear that the second inner-transition series begins with actinium, to give a series similar to the lanthanides. Americium is the actinide counterpart of europium in the lanthanide series; and, as expected, it exhibits the divalent state, corresponding to the half-filled f^7 subshell. However, the $5f$ electrons are removed more easily than the $4f$ electrons; and consequently, for the actinide metals the lower oxidation numbers are less important and the higher oxidation numbers are more important, compared with the lanthanide metals. Even curium ($5f^7$) gives Cm^{IV} in solid CmO_2 and CmF_4, whereas only Gd^{III} compounds are known. The Cm^{4+} ion is not stable in solution.

With very short half-lives, the last few members of this series have been produced in very small quantities (a few atoms, in some cases). Studying elements that show a very high level of radioactivity is difficult, because drastic chemical changes occur in the sample. Even the water in which a compound of the element is dissolved is decomposed by the radioactive decay to give H_2 and O_2. Consequently, the chemical properties of the last few

Table 15.8   Oxidation numbers of the actinide elements (very stable states in boldface type; unstable states enclosed in parentheses; questionable states are indicated by "?")

Ac	*Th*	*Pa*	*U*	*Np*	*Pu*	*Am*	*Cm*	*Bk*	*Cf*	*Es*	*Fm*	*Md*	*No*	*Lr*
						(II)			(II)	(II)	II	II	**II**	
III	(III)	(III)	(III)	III	III	**III**	**III**	**III**	**III**	**III**	**III**	**III**	**III**	**III**
	IV	IV	IV	IV	IV	**IV**	IV	IV	(IV)	IV?				
		V	V	V	V	V	V?		V?					
			VI	VI	**VI**	VI	VI?							
				(VII)	(VII)	VII?								

[6] J. C. Spirlet, J. R. Peterson, and L. B. Asprey, *Adv. Inorg. Chem.* **1987,** *31,* 1; G. T. Seaborg and O. L. Keller, Jr., *The Chemistry of the Actinide Elements,* 2nd ed., Vols. 1 and 2, J. J. Katz, G. T. Seaborg, and L. R. Moss, Eds., Chapman & Hall, New York, 1986.

actinides have not been investigated in great detail. The last member of the actinide series should be element 103 (Lawrencium), which seems to have only III as an important oxidation number. Element 104 should be a member of the transition metal family 4 (IVA).

Elements beyond Element 103

Names proposed for element 104 are Rutherfordium (Rf) by Americans and Kurchatovium by Russians. Names proposed for 105 are Hahnium (Ha) by Americans and Neilsborium by Russians. Element 106 has been discovered. Germans have proposed the following names: 107 Nielsborium (Ns), 108 Hassium (Hs), and 109 Meitnerium (Mt). Although the half-lives and yields become progressively smaller for the elements beyond 104, it is predicted that there will be an "island of stability" for elements 108–120. Undiscovered element 112 should end the $6d$ transition series and element 168 should be a noble gas. After element 121 ($8s^2 7d^1$), electrons should be added to the $5g$ and $6f$ levels with completion at element 162 ($8s^2 7d^1 6f^{14} 5g^{18}$).

15.8 TRANSITION METALS, GROUPS 4–7 (IVA–VIIA)

Metals of Groups 11 (IB) and 12 (IIB) were discussed along with Groups I (IA) and 2 (IIA) in order to emphasize the great similarities and differences in the periodic relationships among these groups. Group 3 (IIIA) has been discussed separately because these metals, unlike most of the transition metals, vary little in oxidation numbers and also because the presence of the lanthanide and actinide metals makes this group unique. Only some of the overall trends will be discussed for Groups 4–7. Groups 8, 9, and 10 (Group VIII) have unique relationships and are treated separately.

▶ 15.8.1 Size Effects

The cumulative effect of the lanthanide contraction makes the radii of the members of the third transition series very similar to those of the corresponding members of the second transition series. The effect is great enough that Hf, which follows immediately after the rare earths, has a slightly *smaller* atomic radius than Zr (see Table 15.9). The almost identical ionic radii of Zr^{4+} and Hf^{4+} account for the fact that these metals always occur together in nature and are difficult to separate. There are no hafnium minerals. The less abundant Hf is too similar in chemical properties to those of Zr to become enriched. For most uses, a few percent of Hf in Zr does not matter. Separating them became important when Zr was found to be a good structural metal for atomic piles, because of its low-neutron cross section (it does not absorb neutrons greatly). In contrast, Hf has a very high cross section and can be used as a moderator to absorb neutrons. Early separation methods involved fractional crystallization or precipitation procedures, but solvent-extraction procedures are more effective.

Table 15.9   Properties of some transition metals

	$_{22}Ti$	$_{23}V$	$_{24}Cr$	$_{25}Mn$
Atomic radius (pm)	132	122	118	117
Ionic radius, C.N. 6 (pm)	M^{III} 81.0	M^{II} 93	M^{II} 94	M^{II} 97.0, 81[a]
	M^{IV} 74.5	M^{III} 78	M^{III} 75.5	M^{III} 78.5, 72[a]
Ionization energy (eV)				
1st	6.82	6.74	6.766	7.435
2nd	13.58	14.65	16.50	15.640
3rd	27.491	29.310	30.96	33.667
4th	43.266	46.707	49.1	51.2

	$_{40}Zr$	$_{41}Nb$	$_{42}Mo$	$_{43}Tc$
Atomic radius (pm)	145	134	130	127
Ionic radius, C.N. 6 (pm)	M^{IV} 86	M^{IV} 82	M^{IV} 79	M^{IV} 78.5
		M^{V} 78	M^{VI} 73	M^{VII} 70
Ionization energy (eV)				
1st	6.84	6.88	7.099	7.28
2nd	13.13	14.32	16.15	15.26
3rd	22.99	25.04	27.16	29.54
4th	34.34	38.3	46.4	

	$_{72}Hf$	$_{73}Ta$	$_{74}W$	$_{75}Re$
Atomic radius (pm)	144	134	130	128
Ionic radius, C.N. 6 (pm)	M^{IV} 85	M^{V} 78	M^{IV} 80	M^{IV} 77
			M^{VI} 74	M^{VII} 67
Ionization energy (eV)				
1st	7.0	7.89	7.98	7.88
2nd	14.9	16.2	17.7	16.6

[a] Low spin.

The atomic radius of Ta is the same as that of Nb, but the two elements are not as similar to one another as are Zr and Hf. They occur together in nature, but not always in the same proportions. The mineral $Fe(MO_3)_2$ is called **columbite** if niobium (formerly called columbium in America) predominates and **tantalite** if tantalum predominates. We can separate the two metals by crystallizing the less soluble K_2TaF_7 from a concentrated HF solution. In hot water, K_2NbF_7 dissolves as $K_2[NbOF_5]$, and the Ta complex is converted to a basic salt of low solubility. The chemical similarity between Mo and W is somewhat less: They form independent minerals, and their chemical separations are accomplished much more easily. The most important ore for Mo is molybdenite (MoS_2), but other ores are Pb or Mg molybdates. The tungstates of Fe, Co, and Pb are the important minerals. The effect of the lanthanide contraction diminishes for elements beyond W, but for each of the families of transition metals beyond Group 3, the second and third members are much more similar in properties than are the first and second members.

▶ 15.8.2 *Oxidation States*

Most transition metals show several oxidation numbers. The exclusive oxidation number III for the members of Group 3 arises because the energy required for the removal of all three outer electrons does not much exceed that needed for the removal of two electrons. The energy for the removal of three electrons is supplied by the high hydration energies of the 3^+ ions or the high lattice energies of their compounds.

The energy required for the removal of four or more electrons is prohibitively high for the formation of simple ions, except in the case of very large atoms with low IEs. We expect an appreciable amount of covalent character in higher oxidation states. Except for Groups 9 and 10 and Fe, the highest oxidation number for each transition metal corresponds to the group number. The lowest state usually encountered in simple salts is II, corresponding to the removal of the two *s* electrons in the outermost shell (Table 15.10). Compounds of second and third transition-series metals containing metal–metal bonds and metal atom clusters (see page 751) cloud comparisons of oxidation states. Thus, we can compare the oxidation number II for $CrCl_2$ with that of Mo in Mo_6Cl_{12}, with Mo—Mo bonding, only in a formal sense.

Although the first IEs do not differ greatly within most families of the first transition metals, the third and higher IEs generally are highest for the first member of each family. Differences between successive IEs also are greatest for the first member of each family. Consequently, the highest oxidation state is most stable, and corresponding compounds are poor oxidizing agents for the second or third member of each group (e.g., Zr^{4+} and Hf^{4+}). Lower oxidation states are relatively more stable for the first member of each group (e.g., Ti^{2+} and Ti^{3+}) as compared with the second and third members. Solid $ZrCl_2$, $ZrCl_3$, and $HfCl_3$ can be obtained by reducing MCl_4 with the free metal, but the lower oxidation states are not stable in solution. Several Ti^{II} and Ti^{III} compounds are known (see Table 15.10). Titanium(III) compounds can be obtained readily in solution by reducing Ti^{IV} salts with Zn and acid, or electrolytically. Titanium(III) chloride, which can be obtained by dissolving Ti metal in hydrochloric acid, finds some application as a reducing agent in analytical work.

Vanadium(II) compounds are strongly reducing, whereas V^{III} salts are stable in solution (see Section 8.8.2 for discussion of the Pourbaix diagram for V). Vanadium(V) compounds are moderately strong oxidizing agents. The chemistry of V^{IV} is dominated by blue VO^{2+} compounds containing a strong V=O bond. Niobium and tantalum most commonly show oxidation number V. Chromium(II) compounds also are strongly reducing, and most of the common salts are those of Cr^{III}. The oxide CrO_2, which until recently was virtually unknown, is now used as a magnetic oxide for high-quality audio recording tape. The oxidation number VI is encountered in CrO_3, K_2CrO_4, $K_2Cr_2O_7$, and so on. These compounds are strong oxidizing agents, being reduced to compounds of Cr^{III} in solution. As is generally true, oxidation to higher oxidation states encountered as oxoanions is accomplished much more easily in basic solution, and the oxoanions are stronger oxidizing agents in acidic solution. Molybdenum and its compounds are oxidized to MoO_3 by heating in air or by the reaction with HNO_3. Molybdenum trioxide dissolves in alkali hydroxides or ammonium hydroxide to form molybdates, $M_2^I[MoO_4]$. Molybdenum occurs as a sulfide, MoS_2 (molybdite), and ReS_2 sometimes occurs with the ore molybdenite. The layer structure of MoS_2 (see page 220) accounts for its uses as a solid lubricant, similar to graphite.

Table 15.10 Oxidation states of some binary compounds of transition and post-transition metals

	4	5	6	7	8	9	10	11	12	13	14
	Ti	V	Cr	Mn	Fe	Co	Ni	Cu	Zn	Ga[a]	Ge
F	3,4	2,3,4,5	2,3,4,5,6	2,3,4	2,3	2,3	2	2	2	3	2,4
Cl	2,3,4	2,3,4	2,3	2	2,3	2	2	1,2	2	1,3	2,4
Br	2,3,4	2,3,4	2,3	2	2,3	2	2	1,2	2	3	2,4
I	2,3,4	2,3	2,3	2	2	2	2	1	2	3	2,4
O	2,3,4	2,3,4,5	3,4, 6	2,3,4, 7	2,3	2,3	2,3	1,2	2	1,3	2,4

	4	5	6	7	8	9	10	11	12	13	14
	Zr	Nb	Mo	Tc	Ru	Rh	Pd	Ag	Cd	In[a]	Sn
F	4	3,4,5	3,4,5,6	6	3,4,5,6	3,4,5,6	2, 4	1,2	2	3	2,4
Cl	1,2,3,4	3,4,5	3,4,5	4, 6	3,4	3	2	1	2	1,3	2,4
Br	1,2,3,4	2,3,4,5	3,4	—	3	3	2	1	2	1,3	2,4
I	2,3,4	2,3,4,5	3	—	3	3	2	1	2	1,3	2,4
O	4	2, 4,5	4, 6	4, 6,7	4, 8	3,4	2	1, 3	2	1,3	2,4

	4	5	6	7	8	9	10	11	12	13	14
	Hf	Ta	W	Re	Os	Ir	Pt	Au	Hg	Tl	Pb
F	4	5	4,5,6	4,5,6,7	4,5,6	3,4,5,6	4,5,6	3	1,2	1,3	2,4
Cl	1, 3,4	3,4,5	3,4,5,6	3,4,5,6	3,4	3,4	2,3,4	1, 3	1,2	1,3	2,4
Br	1, 3,4	3,4,5	3,4,5,6	3,4,5,6	3	3,4	2,3,4	1, 3	1,2	1,3	2
I	3,4	3,4,5	3,4	2,3,4	3	3,4	2,3,4	1, 3	1,2	1	2
O	4	4,5	3,4	3,4, 6,7	5,6,7, 8	3,4	2,3,4	3	1,2	1,3	2

[a] Gallium and indium dihalides are diamagnetic compounds, which can be formulated as $M^I M^{III} X_4$.

741

Tungsten has an even greater tendency than Mo to form compounds with oxidation number VI. It forms a hexachloride, WCl_6, on heating freshly reduced W with Cl_2. The corresponding fluoride and even the bromide can be obtained. The hexahalides are hydrolyzed completely in water to give tungstic acid, H_2WO_4.

Manganese(I) forms the cyano complex, $K_5[Mn(CN)_6]$. The characteristic oxidation number of Mn in solution is II, although the III state is stabilized in complexes such as $K_3[Mn(CN)_6]$, $K_3[Mn(C_2O_4)_3]$, and $K_3[MnF_6]$. In nature, Mn occurs in the igneous rocks as the Mn^{2+} ion; but during the weathering process, Mn^{2+} is oxidized and finally deposited as MnO_2 (see Sections 7.3.2 and 8.6.2). Potassium permanganate, $KMnO_4$, the most important compound of Mn^{VII}, can be obtained by electrolytic oxidation or Cl_2 oxidation of K_2MnO_4, which is produced by strong oxidation of $Mn(OH)_2$ or $Mn(OH)_3$ in a basic solution. Freshly precipitated $Mn(OH)_2$ darkens quickly in air because of oxidation to $Mn(OH)_3$. Potassium permanganate is a strong oxidizing agent that finds wide analytical application because of the sharp color change from magenta to almost colorless Mn^{2+} on reduction. It usually is used in acidic solution, where it is a stronger oxidizing agent. In the absence of an excess of strong acid or strong base, MnO_4^- is reduced to MnO_2. The manganate(VI) ion is formed by reduction of MnO_4^- in concentrated NaOH or KOH solution. The bright green MnO_4^{2-} salts are unstable except in the solid or in strongly basic solution (see Section 8.6 for a detailed discussion of the redox chemistry of Mn in terms of Pourbaix diagrams). The oxide, Mn_2O_7, can be obtained by the reaction of $KMnO_4$ and concentrated H_2SO_4. Mn_2O_7 is a dark, oily liquid that decomposes explosively when heated. It is an acid anhydride, since it reacts with an excess of water to produce $HMnO_4$, a very strong acid.

The characteristic oxidation number of technetium and rhenium is VII. $KTcO_4$ and $KReO_4$ are not strong oxidizing agents. The volatile oxides, Tc_2O_7 and Re_2O_7, and sulfides, Tc_2S_7 and Re_2S_7, are stable. Properties of Tc are much closer to those of Re than to those of Mn.

The second most important oxidation number of Re is IV. Reducing Re_2O_7 in H_2 at about 300°C yields ReO_2, which combines with alkali hydroxides on fusion to produce the alkali metal rhenites, $M_2^I ReO_3$. TcO_2 also is stable. The reported Re(−I) turned out to be a hydride compound (ReH_9^{2-}), rather than a simple solvated Re^- ion.

Very high oxidation states of the transition metals usually are brought out in combination with F or O. The size of the anion, resulting in crowding for larger anions, is rarely the limiting factor. More important are the oxidizing power of the cation in higher oxidation states and the increasing ease of oxidation of the anions with increasing size within a group (e.g., F^-, Cl^-, Br^-, I^-). The oxides usually display the greatest number of oxidation states of a given metal, although this may be true in part because the oxides have been investigated most extensively (see Table 15.10). The emf diagrams given in Appendix E show the relative stabilities of the various oxidation states of the metals as hydrated cations or oxoanions. The relative stabilities of the various oxidation states can be altered greatly by complex formation—for example, the stabilization of high oxidation states in oxoanions—or by lattice forces in the solid state.

Basicity of metal ions in various oxidation states can be treated on the basis of charge/size relationships. For the same oxidation number (e.g., IV for Group 4), basicity increases markedly with increasing ionic radius, and the covalent character of resulting compounds decreases accordingly. Ti^{IV} compounds are much more covalent than those of larger Zr^{IV} or Hf^{IV} and are hydrolyzed much more extensively. Thus $TiCl_4$ is a liquid whereas $ZrCl_4$ is a solid, although it sublimes at about 300°C. In Group 6, Cr^{III} is ampho-teric and can be obtained as Cr^{3+} salts or salts of the chromite ion, $Cr(OH)_4^-$, whereas Mo^{III} and W^{III} are basic in character. The Cr^{2+} ion is basic because of the lower charge and larger radius (compared with Cr^{3+}). In general, members of the first transition series give ions that are predominately basic only in the lower oxidation states. The oxidation num-bers greater than IV for the first members of Groups 5–10 are generally encountered in oxides or oxoanions. The second and third members of these families give ions that are much more basic, showing a much greater tendency than the first members to form cationic species in higher oxidation states. Chromium trioxide is quite acidic and is very soluble in water, giving H_2CrO_4. Molybdenum trioxide is less acidic and is only slightly soluble in water, and WO_3 is very insoluble. All three oxides dissolve readily in bases. Be-cause differences between the radii of ions of Mo and W are small, basicities of their com-pounds are very similar.

Usually, zero and negative oxidation numbers are encountered for metal carbonyls (see Section 12.2). For elements in Groups 3–10, oxidation states lower than +II are en-countered only for π-acid-type ligands such as CO and CN^- or in organometallic com-pounds, except for Rh^I and Ir^I.

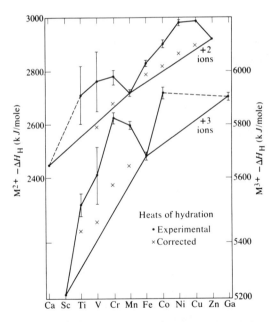

Figure 15.9 The hydration energies of the divalent and trivalent ions of the first transition series. (Repro-duced with permission from P. George and D. S. Mc-Clure, *Prog. Inorg. Chem.* **1959**, *1*, 418. Copyright 1959 by John Wiley & Sons, Inc.)

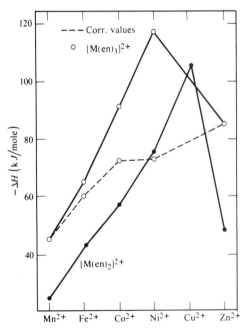

Figure 15.10 Heats of formation of $[M(en)_2]^{2+}$ and $[M(en)_3]^{2+}$ and the heats of formation of $[M(en)_3]^{2+}$ corrected for LFSE. (Adapted from M. Ciampolini, P. Paoletti, and L. Sacconi, in *Advances in the Chemistry of the Coordination Compounds*, S. Kirschner, Ed., Macmillan, New York, 1961, pp. 305 and 307.)

▶ *15.8.3 Ligand-Field Effects*

Hydration energies of first-transition-series metal ions might be expected to increase with decreasing ionic radius (and increasing nuclear charge) through the series. We see from plots of hydration energies for M^{2+} and M^{3+} (Figure 15.9) that there are two maxima with minima for ions with the configurations d^0 (Ca^{2+} and Sc^{3+}), d^5 (Mn^{2+} and Fe^{3+}), and d^{10} (Zn^{2+} and Ga^{3+}). These are cases with zero ligand-field stabilization energy (LFSE). For the other ions the hydration energy is increased by the LFSE (see Section 9.8.1), giving the two maxima. Subtracting the calculated LFSE from the observed hydration energies gives the "corrected" points, which fall close to the straight lines joining the d^0, d^5, and d^{10} ions.

Although increasing positive charge density through the transition series primarily determines the "natural order of stability" of metal complexes ($Mn^{2+} < Fe^{2+} < Co^{2+} < Ni^{2+} < Cu^{2+} > Zn^{2+}$), LFSE is also a factor, particularly for the decrease in stability of Zn^{2+} complexes as compared with those of Cu^{2+}. Figure 15.10 shows a plot of ΔH_f^0 for the formation of $[M(en)_2]^{2+}$ and $[M(en)_3]^{2+}$ complexes in aqueous solution. Subtracting the LFSE for $[M(en)_3]^{2+}$ gives values (dashed line) close to the straight line joining Mn^{2+} and Zn^{2+}. No value is plotted for $[Cu(en)_3]^{2+}$ because the C.N. of Cu^{2+} is uncertain in this case. Cu^{2+} shows strong tetragonal distortion in six-coordinate complexes (Jahn–Teller effect—see Section 9.8.2), lowering the stability of tris(didentate)copper(II) complexes. The stability trend is normal for $[M(en)_2]^{2+}$.

The "natural order of stability" applies to high-spin complexes and to cases in which the ligand does not impose some fixed stereochemistry, as do porphyrin or phthalocyanine ligands. Low-spin complexes have additional LFSE, but at the expense of the pairing energy. The ΔH_f^0 values given in Figure 15.10 are for formation of complexes in aqueous solution and involve replacement of coordinated H_2O by N of ethylenediamine. The LFSE from the plots is actually the difference between the coordination values for N and O, but this additional stabilization is proportional to the calculated LFSE.

▶ *15.8.4 Isopoly and Heteropoly Anions*

Isopoly Anions

One of the important characteristics of Mo^{VI} and W^{VI}—and, to a lesser extent, V^V, Nb^V, and Ta^V—is the tendency to form polymeric oxoanions containing MoO_6 octahedra. For tetrahedral oxoanions of the nonmetals, XO_4^{n-}, the tendency to form **polyanions** is greatest for Si^{IV} where the oxidation number is the same as the C.N. The oxidation number is also the same as the C.N. for the Mo^{VI} and W^{VI} polyanions. The smaller Cr atom shows C.N. 4 in oxoanions. The chromate ion CrO_4^{2-} forms $Cr_2O_7^{2-}$ in acidic solution and in concentrated H_2SO_4 or $HClO_4$, CrO_3 precipitates. The tetrahedral CrO_4 units share corners in $Cr_2O_7^{2-}$ and form chains in solid CrO_3. Simple tetrahedral anions of MoO_4^{2-}, WO_4^{2-}, VO_4^{3-}, NbO_4^{3-}, and TaO_4^{3-} are encountered, but in forming the polyacids or polyanions the C.N. increases to 6. Solid MoO_3 has a layer structure. Each layer consists of chains of

MoO_6 octahedra sharing adjacent edges with the chains joined by sharing apical O. Each MoO_6 has three triply shared O (in shared edges), one terminal O, and two doubly shared O (apical O).[7] The WO_3 structure involves WO_6 octahedra sharing all corners in a distorted version of the ReO_3 structure (Figure 5.7). Vanadium in V_2O_5 has a distorted octahedral environment with one V—O distance much longer than the others.

The **isopoly anions** can be regarded as fragments of oxide structures. Because discrete MO_6^{n-} ions have unfavorably high charges, the MO_4^{z-} ions are encountered instead. It is striking that we encounter only aggregates of well-defined size and that their sizes are different for different metals. The fact that the stable isopoly anions are compact rather than extended aggregates indicates that polymerization is favored by charge reduction and the highly favorable entropy change accompanying elimination of water molecules with the formation of compact structures.[8]

Sharing of edges between MO_6 octahedra causes repulsion between the metal ions. This repulsion is reduced partially by displacement of M^{n+} from the exact center of the octahedra. Repulsion should increase with increasing ionic radius ($V^{5+} < Mo^{6+} \leq W^{6+} < Nb^{6+} \leq Ta^{5+}$), in agreement with the degree of distortion observed and the sizes of the edge-shared isopoly anions: $V_{10}O_{28}^{6-}$, $Mo_8O_{26}^{4-}$, $Mo_7O_{24}^{6-}$, $HW_6O_{21}^{5-}$, $Nb_6O_{19}^{8-}$, and $Ta_6O_{19}^{8-}$. The choice of edges to be shared—and consequently, the shape of the aggregate—is determined also by the tendency to minimize charge repulsion. Larger aggregates require some extent of apex sharing, in order to reduce cation–cation repulsion.

As fragments of oxide structures, the isopoly anions contain cubic close-packed arrangements of oxide ions with metal ions in octahedral holes. The *hcp* arrangement involves face- and edge-shared octahedra. An important basic unit is the "super" octahedron, M_6O_{19} (Figure 15.11). The $M_{10}O_{28}$ unit contains two "super" octahedra sharing an edge. Fragments of this unit encountered include M_8O_{26}, from the removal of octahedra *a* and *b* in Figure 15.11; and M_7O_{24}, from the removal of octahedra *b*, *c*, and *d*.

Heteropoly Anions

As a qualitative test, phosphate ion is detected by adding $(NH_4)_2MoO_4$ to form a precipitate of ammonium 12-molybdophosphate, $(NH_4)_3[P(Mo_3O_{10})_4] \cdot 6H_2O$. This **heteropoly anion** has four groups of three edge-shared MoO_6 octahedra arranged tetrahedrally about P (Figure 15.12). A 2 : 18 ratio occurs in $K_6[P_2W_{18}O_{62}] \cdot 14H_2O$, where two PO_4 tetrahedra pointing in opposite directions are enclosed in a cage formed by two groups of three edge-shared octahedra (sharing the oxygens of the opposed apices of the PO_4 units) and six groups of two edge-shared octahedra. The edge-shared octahedral groupings are joined by shared apices. This structure would result from removing three octahedra from the 1 : 1 structure and then joining two of these fragments.

A second class of heteropoly anions involves edge-shared MO_6 units clustered about a different (hetero) octahedral atom. A 1 : 6 ratio—for example, $[Te^{VI}Mo_6O_{24}]^{6-}$ and $[M^{III}Mo_6O_{24}]^{9-}$—results from a ring of six octahedra about the hetero atom (Figure 15.13). Adding three edge-shared octahedra above and below this arrangement gives us a

[7] D. M. Adams, *Inorganic Solids*, Wiley, New York, 1974, p. 237.

[8] D. L. Kepert, *Inorg. Chem.* **1969,** *8,* 1556; K.-H. Tytoko and O. Glemser, *Adv. Inorg. Chem. Radiochem.* **1976,** *19,* 239.

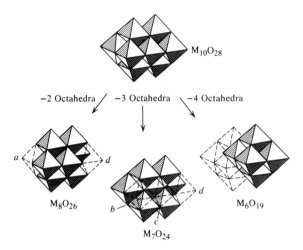

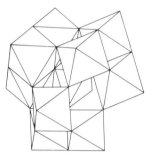

Figure 15.11 The structure of isopoly anions, showing their relationship to the $M_{10}O_{28}$ structure. (Reproduced with permission from D. L. Kepert, *Inorg. Chem.* **1969**, *8*, 1557. Copyright 1969, American Chemical Society.)

Figure 15.12. Structure of the 12-heteropoly anion $[P(Mo_3O_{10})_4]^{3-}$.

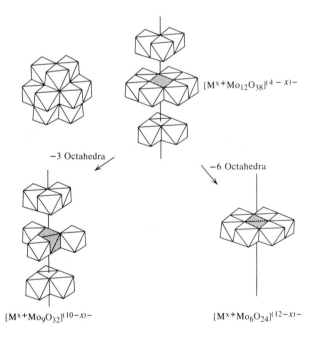

Figure 15.13 The structures of heteropolymolybdates and their relationship to the hypothetical $M^{x+}Mo_{12}O_{38}^{(4-x)-}$. (Reproduced with permission from D. L. Kepert, *Inorg. Chem.* **1969**, *8*, 1557. Copyright 1969, American Chemical Society.)

plausible 1 : 12 structure. Removing alternate octahedra from the ring of the 1 : 12 structure gives us the 1 : 9 ratio—for example, $[Ni^{IV}Mo_9O_{32}]^{6-}$. Other fragments are possible. Many hetero atoms capable of forming tetrahedral or octahedral oxoanions occur in heteropoly anions. The compound $(NH_4)_6[H_4Co_2Mo_{10}O_{38}] \cdot 7H_2O$ can be pictured as resulting from removal of an MoO_2 unit from the ring of each of two 1 : 6 heteropoly anions so as to expose the central CoO_6 groups, which are then joined with the elimination of six more oxygens along the shared edges. The dissymmetric $[H_4Co_2Mo_{10}O_{38}]^{6-}$ ion has been resolved to give stable optical isomers.[9]

The $[CeMo_{12}O_{42}]^{8-}$ ion does not have the 1 : 12 structure shown in Figure 15.13. There are six pairs of face-shared octahedra coordinated to Ce^{IV} to form an icosahedron.

We have taken only a broad structural view of isopoly and heteropoly anions. Many structures are unknown, and others do not fit into the simple pattern presented. Important questions remain to be answered in this not-fully-explored field. Isopoly and heteropoly acids are now important solid acids for catalysts (see Section 7.6).

15.9   GROUPS 8, 9, AND 10 (GROUP VIII)

▶ 15.9.1   *Periodic Classification*

Group VIII metals have been treated historically as a single periodic group, since they exhibit close similarities in the horizontal triads as well as in the vertical groups (Table 15.11). Properties of complexes are most closely related to the number of *d* electrons, which is similar for the ions of the metals in each of the vertical columns, Groups 8, 9, and 10. Not only are the radii very similar in each horizontal triad, but also the radii of the second- and third-transition-series metals in each vertical column are very similar, because of the lanthanide contraction. Hence, a useful classification places Fe, Co, and Ni in one group, known as the **iron triad.** The platinum metals might be divided into the light platinum metals (Ru, Rh, and Pd) and the heavy platinum metals (Os, Ir, and Pt), but their chemical similarities are more apparent in the vertical groups of two: Ru and Os, Rh and Ir, and Pd and Pt. In each of these pairs the metals have similar radii and similar electron configurations.

▶ 15.9.2   *The Iron Triad (Fe, Co, and Ni)*

Although the highest oxidation state encountered increases steadily through the first transition series for Groups 3–7 (see Table 15.10, particularly for the oxides), there is an abrupt change for Fe, Co, and Ni. The metals of the iron triad often occur together in nature as M^{2+}, and the simple salts of their M^{2+} ions are very similar. Only iron, as Fe^{III}, gives stable simple salts with an oxidation number other than II. Iron(II) salts are oxidized easily to

[9] T. Ama, J. Hidaka, and Y. Shimura, *Bull. Chem. Soc. Japan* **1970,** *43,* 2654.

Table 15.11   Properties of Group 8–10 metals

	₂₆ Fe	₂₇ Co	₂₈ Ni
Atomic radius (pm)	126	125	125
Ionic radius[a] (pm)	Fe^{II} 92, 75[b]	Co^{II} 88.5, 79[b]	Ni^{II} 83,71[c]
		72[c]	
	Fe^{III} 78.5, 69[b]	Co^{III} 75, 70[b]	Ni^{III} 68[b]
Ionization energy (eV)			
1st	7.870	7.86	7.635
2nd	16.18	17.06	18.168
3rd	30.651	33.50	35.17

	₄₄ Ru	₄₅ Rh	₄₆ Pd
Atomic radius (pm)	134	134	137
Ionic radius[a] (pm)	Ru^{III} 82	Rh^{III} 80.5	Pd^{II} 78[c]
			Pd^{IV} 75.5
Ionization energy (eV)			
1st	7.37	7.46	8.34
2nd	16.76	18.08	19.43
3rd	28.47	31.06	32.93

	₇₆ Os	₇₇ Ir	₇₈ Pt
Atomic radius (pm)	135	136	139
Ionic radius[a] (pm)	Os^{IV} 77.0	Ir^{III} 82	Pt^{II} 74[c]
		Ir^{IV} 76.5	Pt^{IV} 76.5
Ionization 1st	8.7	9.1	9.0
energy (eV) 2nd	17	17	18.563

[a] Ionic radii for C.N. 6, except as noted.

[b] Low-spin.

[c] Square-planar.

those of iron(III), whereas the hydrated Co^{3+} ion oxidizes water and the only simple salts of Ni are those of Ni^{II}. Strong oxidation of $Fe(OH)_3$ in strongly alkaline solution produces the ferrate(VI) ion—as in K_2FeO_4, a very strong oxidizing agent that forms lustrous black crystals.

Fe^{2+} ion is a d^6 ion, giving very stable low-spin octahedral complexes of the type $[Fe(CN)_6]^{4-}$ (ferrocyanide ion) and $[Fe(phen)_3]^{2+}$. Most ligands other than 1,10-phenanthroline (phen) and bipyridine give more stable complexes with Fe^{3+}, a d^5 ion—presumably because of the higher charge on the cation. $[Fe(CN)_6]^{3-}$ (ferricyanide ion) is a low-spin complex, but most others of Fe^{III}, such as $[FeF_6]^{3-}$, $[Fe(C_2O_4)_3]^{3-}$, and Fe(acetylactonate)$_3$, have high spin.

The cobalt(II) ion forms octahedral complexes, such as $[Co(NH_3)_6]^{2+}$; planar complexes, such as dimethylglyoximatocobalt(II),

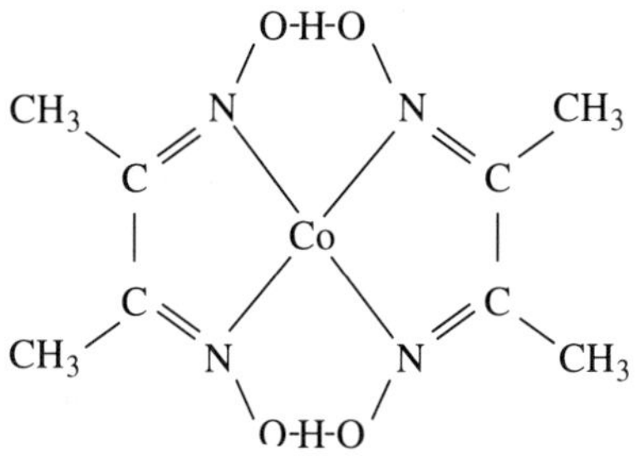

and tetrahedral complexes, such as $[CoCl_4]^{2-}$. However, in the presence of strong-field ligands, cobalt(II) is oxidized easily, even by O_2 in air, to cobalt(III) to give low-spin octahedral complexes such as $[Co(NH_3)_6]Cl_3$ and $K_3[Co(CN)_6]$. The cobalt(III) complexes are exceptionally stable and give stable geometrical isomers, such as *cis*- and *trans*-$[CoCl_2(en)_2]Cl$ (page 391), and optical isomers, such as $(+)$- and $(-)$-$[Co(en)_3]^{3+}$ (page 418) and $(+)$- and $(-)$- *cis*-$[CoCl_2(en)_2]^+$. There are very few high-spin octahedral Co^{III} complexes—for example, $[CoF_6]^{3-}$. Although octahedral complexes are almost always encountered for Co^{III}, a trigonal bipyramidal structure (C.N. 5) occurs for $[CoCl_3(Et_3P)_2]$.

Most nickel complexes are those of nickel(II). Planar complexes formed by very strong field ligands—$[Ni(CN)_4]^{2-}$, for example—are diamagnetic. The square-planar bis(dimethylglyoximato)nickel(II), with the structure shown above for the Co compound, is a bright red precipitate used in the analytical determination of Ni (see page 751). Octahedral complexes, such as $[Ni(H_2O)_6]^{2+}$, $[Ni(NH_3)_6]^{2+}$, and $[Ni(en)_3]^{2+}$, have two unpaired electrons. Halide complexes, such as $[NiCl_4]^{2-}$, or complexes containing halide ligands and bulky ligands, such as triphenylphosphine or triphenylphosphine oxide, are tetrahedral and paramagnetic. $Ni(acac)_2$ is a trimer with bridging oxygens to give C.N. 6 for Ni. Yellow diamagnetic bis($NH_2CHRCHRNH_2$) planar complexes are formed, but they add solvent molecules or other ligands to form blue octahedral complexes. Complexes of some Schiff bases, phosphines, and a few other ligands have shown planar-tetrahedral equilibria. Several phosphine and arsine ligands form five-coordinate trigonal bipyramidal complexes. Because tetradentate "tripod" ligands, such as $N[C_2H_4N(CH_3)_2]_3$ and $N[C_2H_4PPh_2]_3$, cannot span the apices of a regular tetrahedron, one face is left open for a fifth ligand, such as a halide ion, to form a trigonal bipyramidal complex. Nickel(III) complexes of the type $[Ni(PR_3)_2Br_3]$ also are trigonal bipyramidal.

▶ 15.9.3   Ruthenium and Osmium

Ru and Os form complexes similar to those of Fe^{II} and Fe^{III}. In addition to carbonyl derivatives of Ru^0 and Os^0, the elements show oxidation states I, II, III, IV, V, VI, VII, and VIII. The first complex involving elemental N_2 as a ligand was $[Ru(NH_3)_5N_2]^{2+}$. It can be prepared by substitution of N_2 for H_2O in $[Ru(NH_3)_5H_2O]^{2+}$ or by treating the common starting material $RuCl_3 \cdot 3H_2O$ with hydrazine, N_2H_4.

Oxidation of ruthenium metal produces RuO_2. Many of the M^{II} and M^{III} complexes of Ru and Ru can be oxidized to M^{IV} complexes. Strong oxidation of ruthenium salts in alkaline solution produces RuO_4, an orange liquid above 25°C. The stable oxoanions encountered for Ru are the ruthenate(VI) ion, RuO_4^{2-}, and ruthenate(VII) (perruthenate), RuO_4^-. Direct oxidation of the metal or compounds of Os produces colorless osmium tetraoxide, OsO_4. Like RuO_4, OsO_4 is volatile (mp 40°C, bp ~100°C) and highly toxic. Oxoanions of osmium are less stable than those of ruthenium, because of the ease of formation of OsO_4. Reduction of OsO_4 in CH_2Cl_2 by KI produces the perosmate ion, OsO_4^-. Mild reduction of OsO_4 in alkaline solution produces osmate salts such as K_2OsO_4. Osmates are normally octahedral such as the diamagnetic dark purple dihydroxodioxoosmate(VI) ion, $[OsO_2(OH)_4]^{2-}$. The diamagnetism results from π donation involving the *trans* Os=O bonds. This removes degeneracy of the octahedral e_g orbitals (d_{z^2} and $d_{x^2-y^2}$) and raises energy of the π bonding d_{xz} and d_{yz} orbitals of the octahedral t_{2g} set, leaving the filled d_{xy} orbital as lowest in energy. Hydroxide ions of $[OsO_2(OH)_4]^{2-}$ can be replaced by ligands such as Cl^-, Br^-, and NO_2^-. Even easily oxidized KCN reduces OsO_4 to only Os^{VI} in $K_2[OsO_2(CN)_4]$.

▶ *15.9.4 Rhodium and Iridium*

There are neutral and anionic carbonyl and PR_3 compounds of rhodium and iridium. Many compounds are known of Rh^I and Ir^I involving π-acceptor ligands such as CO, PR_3, alkenes, and arenes. The d^8 ions form square-planar complexes, but also tetrahedral $[Rh(PMe_3)_4]^+$ and trigonal bipyramidal $HRh(PF_3)_4$ complexes. M^I complexes are obtained by reduction of M^{III} complexes in the presence of the ligands. One of the most thoroughly studied examples of Ir^I are *trans*-$[IrCl(CO)(PPh_3)_2]$ (known as Vaska's compound) and its derivatives. Vaska's compound can give the five-coordinate $IrCl(CO)_2(PPh_3)_2$ by addition of CO, but commonly undergoes oxidative addition to give octahedral complexes of Ir^{III} (see Section 14.1.3). Rh^{II} and Ir^{II} complexes usually involve M—M bonding. M^{III} (d^6) complexes are most important. Octahedral complexes $[RhF_6]^{2-}$, $[RhCl_6]^{2-}$, and $[IrCl_6]^{2-}$ can be obtained by strong oxidation of the M^{III} complexes. $CsMF_6$ complexes are formed by strong fluorinating agents such as IF_5 and BrF_3.

▶ *15.9.5 Palladium and Platinum*

The important compounds of Pd and Pt are those of oxidation numbers II and IV, and primarily their complexes. Compounds containing M—M bonds occur with I and III oxidation states. Pd compounds generally are planar complexes of Pd^{II}, such as $K_2[PdCl_4]$ and $[Pd(NH_3)_4]Cl_2$. Even $PdCl_2$ in the solid state contains chains of planar $PdCl_4$ units. Oxidation of $K_2[PdCl_4]$ with Cl_2 or *aqua regia* produces $K_2[PdCl_6]$. One of the polymorphs of $PtCl_2$ is soluble in benzene, with Pt_6Cl_{12} units having the structure of the M_6Cl_{12} metal atom clusters (Figure 17.44), containing planar $PtCl_4$ units. Because the Pt orbitals are filled, the Pt—Pt bond order is zero, and the Pt_6Cl_{12} unit is held together primarily by Cl bridges. Pt complexes closely resemble those of Pd. Platinum(II) is the metal ion that gives the most stable planar complexes, giving rise to geometrical isomerism as exemplified by *cis-* and *trans*-[Pt(NH₃)₂Cl₂]. Pt^IV is much more stable than Pd^IV. Octahedral complexes of Pt^IV give particularly stable geometrical isomers of the type mentioned for Co^III.

15.10   COMPOUNDS CONTAINING METAL–METAL BONDS

The dimeric cation Hg_2^{2+} was long considered to be unique. Cd_2^{2+} and Zn_2^{2+} are known, but the bonding is much weaker than in Hg_2^{2+}. The compound $K_3W_2Cl_9$ has long been known to contain discrete $W_2Cl_9^{3-}$ ions consisting of two octahedra of Cl^- ions with one face in common (Figure 15.14). The octahedra are distorted because of the very short W—W distance (241 pm)—shorter than in the metal. Strong W—W bonding draws the positive ions together, and the compound is diamagnetic. The opposite distortion occurs in $[Cr_2Cl_9]^{3-}$, because of repulsion between the Cr^{3+} ions in the face-shared octahedra. The Cr—Cr distance is 312 pm, although the ionic radius of Cr^{3+} is much smaller than that of W^{3+}, and there is no spin pairing between the d^3 ions here. The bridging M—Cl—M bond angle is *less* than that for two regular face-shared octahedra for $[W_2Cl_9]^{3-}$, and *greater* for $[Cr_2Cl_9]^{3-}$. The isoelectronic $Re_2Cl_9^-$ has a structure similar to that of $[W_2Cl_9]^{3-}$, with a short Re—Re distance (~271 pm).

Metal–metal bonding occurs in many metal carbonyls. The structure of $Fe_2(CO)_9$ is similar to that of $[W_2Cl_9]^{3-}$, with three bridging carbonyls and an Fe—Fe bond. The dimeric carbonyls $Mn_2(CO)_{10}$ and $Re_2(CO)_{10}$ (Figure 12.2) contain metal–metal bonds without bridging carbonyls.

The high selectivity of dimethylglyoxime for the precipitation of Ni^II does not result from extraordinarily great stability of the discrete complex, but from the unusual structure

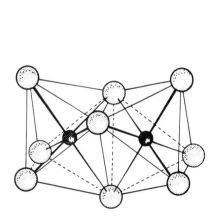

Figure 15.14   Structure of $[W_2Cl_9]^{3-}$.

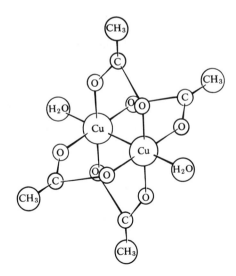

Figure 15.15   Structure of $[Cu(CH_3CO_2)_2 \cdot H_2O]_2$.

in the solid, which leads to low solubility. The planar units are stacked one on another so that the Ni "atoms" are bonded. In crystals of many Pd^{II} and Pt^{II} complexes the planar square units are stacked to form infinite chains of metal ions. The electronic interaction causes strong dichroism (different absorption or color along the direction of —M—M—M— interaction and the direction perpendicular to it). [$PdCl_2$(en)] is an example without counterions. If these complexes are partially oxidized they show high electrical conductivity—called "one-dimensional metals" (see page 280). In the stoichiometric $K_2[Pt(CN)_4] \cdot 3H_2O$ the Pt—Pt distance is 348 pm with low conductivity. Partially oxidized compounds have high conductivity because of short Pt—Pt distances: 296 pm for $K_{1.75}[Pt(CN)_4] \cdot 1.5H_2O$ and 287 pm for $K_2[Pt(CN)_4]Cl_{0.3} \cdot 3H_2O$. The Pt—Pt distance in Pt metal is 277.5 pm. There are also compounds with high electrical conductivity along one direction with halide ions bridging M^{II} and M^{IV} in —X—M^{II}—X—M^{IV}—. The M^{II} unit is square planar, and M^{IV} is octahedral with axial halides used for bridging.

The presence of M—M bonds is often inferred from low magnetic moments such as for the diamagnetic dimer $Cu_2(C_2H_3O_2)_4 \cdot 2H_2O$, which contains two d^9 Cu^{II} ions (Figure 15.15). The long Cu—Cu distance (264 pm) indicates very weak bonding interaction in this case. The corresponding Cr—Cr compound has a much shorter Cr—Cr distance (236 pm). The strong M—M bonding corresponds to a quadruple bond (see the next section). A very short bond length of 184.3 pm has been found for

$$\left(\begin{array}{c} C_6H_5 \\ | \\ H_3C \diagdown \; C \; \diagup CH_3 \\ N \quad N \\ | \quad | \\ Cr \text{——} Cr \end{array} \right)_4.$$

The "small bite" of the ligand favors a short Cr—Cr distance[10]. Magnetic interactions between metal ions might occur through overlap of diffuse orbitals without significant bonding, or through bridging groups. Lower-than-expected magnetic moments might also occur because of distortions, with the resulting removal of degeneracy of sets of orbitals in the lower symmetry; thus magnetic properties must be interpreted with caution.

Mercury forms compounds with transition metals bonded to Hg as *ligands*. In $Hg[Pt_3(2,6\text{-}Me_2C_6H_3NC)_6]_2$ the six Pt define a trigonal prism with Hg at the center. The Pt atoms in each triangular face are bridged by the C of three isocyanides, with six isocyanides bonded terminally, one to each Pt (Figure 15.16). In [η^5-$(CH_3C_5H_4)Mn(CO)_2Hg]_4$ there is an eight-membered ring of 4Mn and 4Hg with each Mn bonded to two Hg; the Hg—Hg distance corresponds to single bonding. The molecule has S_4 symmetry (two views are shown in Figure 15.17).

▶ 15.10.1 *Metal–Metal Quadruple and Triple Bonds*[11]

Whereas rhenium(III) compounds such as $CsReCl_4$ were presumed to contain low-spin tetrahedral $ReCl_4^-$ ions, Cotton and co-workers found instead a dimer, $[Re_2Cl_8]^{2-}$ (Figure 15.18a), and a trimer $[Re_3Cl_{12}]^{3-}$ (Figure 15.18b). The dimer has a surprising structure

[10] A. Bino, F. A. Cotton, and W. Kaim, *Inorg. Chem.* **1979**, *18*, 3566.
[11] F. A. Cotton and R. A. Walton, *Multiple Bonds Between Metal Atoms*, 2nd ed. Oxford: Clarendon Press, New York, 1993; M. H. Chisholm, *Angew. Chem. Int. Ed. Engl.* **1986**, *25*, 21.

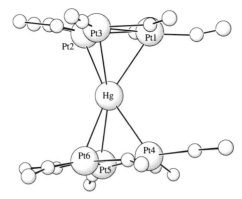

Figure 15.16 Crystal structure of $Hg[Pt_3(2,6-Me_2C_6H_3NC)_6]_2$. 2,6-Xylyl groups are omitted for clarity. (Y. Yamamoto, H. Hiroshi, and T. Sakurai, *J. Am. Chem. Soc.* **1982**, *104*, 2329. Copyright 1982, American Chemical Society.)

containing two *eclipsed* planar $ReCl_4^-$ units joined only by an Re—Re bond. The eclipsed configuration is unexpected, because of strong repulsion between adjacent Cl^- ions. This repulsion is evident from the distortion from planar $ReCl_4$ units with Cl—Cl distances between units much longer than the Re—Re distance. The unusually short Re—Re distance (224 pm, compared with 275 pm in the metal) and the eclipsed configuration led Cotton and co-workers to propose a *quadruple* bond. The nearly planar $ReCl_4$ units are assumed to use the $d_{x^2-y^2}$ sp^2 orbitals of Re for Re—Cl bonding. A σ Re—Re bond can be formed using d_{z^2} orbitals. Two equivalent π bonds are formed using d_{xz} and d_{yz} orbitals on each Re. The d_{xy} orbitals of the two Re atoms are parallel, forming a δ bond (see Figure 4.4 page 155). This bond *requires* the eclipsed configuration for the proper alignment of the d orbitals.

The very short Re—Re distance is retained in $[Re_2Br_8]^{2-}$, or by replacement of one halide ion on each Re by the tertiary phosphine ligand. Halide ions of $[Re_2X_8]^{2-}$ can be replaced by reaction with carboxylic acids to give dimeric carboxylates, $Re_2(O_2CR)_4X_2$, with the structure of the dimeric Cu^{II} acetate (Figure 15.15). The intermediate case $Re_2(O_2CR)_2X_4(H_2O)_2$ is also obtained with bridging carboxylates on opposite sides. Presumably the Re—Re quadruple bond is retained in these compounds.

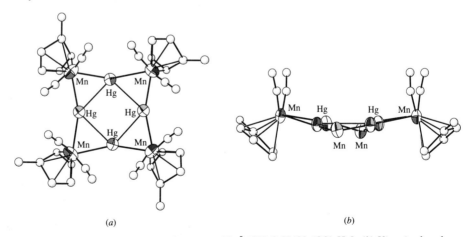

(a)   (b)

Figure 15.17 (a) Structure of the rings of $[\eta^5\text{-}(CH_3C_5H_4)Mn(CO)_2Hg]_4$ (b) View in the plane of the molecule. The Mn atoms projecting out and behind the figure are shown without ligands. (W. Gade and E. Weiss, *Angew. Chem. Int. Ed. Engl.* **1981**, *20*, 803.)

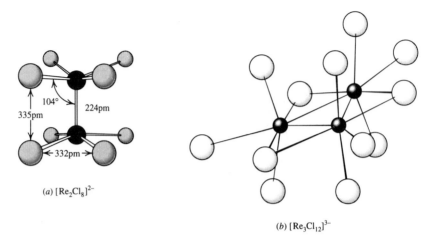

(a) [Re₂Cl₈]²⁻

(b) [Re₃Cl₁₂]³⁻

Figure 15.18 The forms of ReCl₄⁻. (a) The structure of the [Re₂Cl₈]²⁻ ion. (b) The structure of the [Re₃Cl₁₂]³⁻. (Reproduced with permission from F. A. Cotton, *Acc. Chem. Res.* **1969**, *2*, 242. Copyright 1969, American Chemical Society.)

The isoelectronic Mo^II compounds [Mo₂Cl₈]⁴⁻ and Mo₂(O₂CCH₃)₄ (without axial ligands) also have the eclipsed configuration (required by the bridging acetates) and exceedingly short Mo—Mo distances—214 and 211, respectively. Figure 15.19 shows the energy-level diagram for the highest occupied molecular orbital (HOMO) and lowest unoccupied molecular orbital (LUMO) involved in quadruple bonding in [Mo₂Cl₈]⁴⁻ and [Re₂Cl₈]²⁻. The $\sigma^2\pi^4\delta^2$ bonding is more clear for [Mo₂Cl₈]⁴⁻. In the case of [Re₂Cl₈]²⁻ the $\sigma\ a_{1g}$ orbital of about the same energy of σ for [Mo₂Cl₈]⁴⁻ has only 34% Re d orbital character. This orbital is involved to a great extent in Re—Cl $\sigma\ a_{1g}$ bonding. The lower-energy a_{1g} orbital (62% Re d character) is more involved in Re—Re bonding. Cotton and

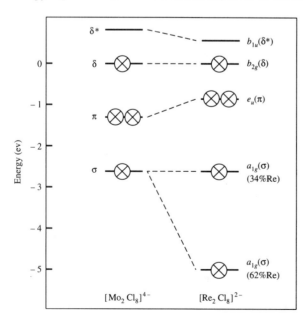

Figure 15.19 Portion of the energy-level diagrams for [Mo₂Cl₈]²⁻ and [Re₂Cl₈]²⁻ involving quadruple bonding. Energies are relative to that of the δ-orbital as zero. (Data from A. P. Mortola, J. W. Moskowitz, N. Rösch, D. D. Cowan, and H. B. Gray, *Chem. Phys. Lett.* **1975**, *32*, 283; and F. A. Cotton and R. A. Walton, *Multiple Bonds Between Metal Atoms*, Wiley, New York, 1982, p.358.)

Walton[12] gave a detailed comparison of MO energy levels for $[M_2X_8]^{n-}$ where M = Mo, W, Os, Re, and Tc.

Although many examples of compounds containing M—M bonds have been known for some time, they have been considered anomalies. Recent structural investigations indicate that this type of interaction is much more common than previously supposed. Quadruple bonds occur for Cr, Mo, W, Re, and Tc. There are fewer W compounds with quadruple bonds compared with Mo and Cr, but some of them are very stable, including $W_2(O_2CC_2H_5)_4$ (bond length 219 pm) and $[W_2Cl_8]^{4-}$ (bond length 225 pm). There are more compounds with M≡M (triple bonds) than with M≣M (quadruple bonds). Triple bonds (with bond lengths in pm) occur in $Mo_2(CH_2SiMe_3)_6$ (217), $W_2(CH_2SiMe_3)_6$ (226), $Mo_2(NMe_2)_6$ (221), and $W_2(NMe_2)_6$ (229). Other triply bonded compounds are obtained from $M_2(NR_2)_6$.

Electrons available for M—M bonding are the metal d electrons. Each Re^{III} in $[Re_2Cl_8]^{2-}$ and Mo^{II} in $[Mo_2Cl_8]^{4-}$ has 4 d electrons available for M—M bonding, or 8 electrons for a quadruple bond, $\sigma^2\pi^4\delta^2$. In $W_2(NMe_2)_6$ the two W^{III} (d^3) ions have 6 electrons forming a triple bond. In $Rh_2(O_2CCH_3)_4(H_2O)_2$ the two Rh^{II} (d^7) have 14 electrons for M—M bonding occupying $\sigma^2\pi^4\delta^2\delta^{*2}\pi^{*4}$ orbitals, giving a single bond.

▶ 15.10.2   Some Reactions of Compounds Involving M—M Multiple Bonds[13]

Substitution reactions occur readily for some Re≡Re and Mo≡Mo compounds:

$$[Re_2Cl_8]^{2-} + 8\,SCN^- \xrightarrow{\text{CH}_3\text{OH}} [Re_2(NCS)_8]^{2-} + 8\,Cl^- \qquad (15.30)$$

$$Mo_2(O_2CCH_3)_4 + 8\,HCl(aq) \xrightarrow{0°C} [Mo_2Cl_8]^{4-} + 4\,CH_3CO_2H + 4\,H^+ \qquad (15.31)$$

Many compounds containing Mo≡Mo bonds can be obtained from $Mo_2(O_2CCH_3)_4$ and $[Mo_2Cl_8]^{4-}$. At 60°C, adding HCl to $Mo_2(O_2CCH_3)_4$ results in oxidative addition to the Mo—Mo quadruple bond, to form a symmetrical compound (C_{2v}) containing an M—M bond (238 pm) with two bridging Cl and one symmetrical bridging H (Figure 15.20):

$$Mo_2^{II}(O_2CCH_3)_4 + 8\,HCl(aq) \xrightarrow{60°C} [Mo_2^{III}Cl_8H]^{3-} + 3\,H^+ + 4\,CH_3CO_2H \qquad (15.32)$$

The H is involved in a 3-center bond as in the diborane (see page 185). We might expect easy oxidation (removal of a δ electron) or reduction (addition of a $\delta*$ electron) to produce compounds of lower bond order, but redox processes usually involve structural changes. Oxidation of $[Re_2Cl_8]^{2-}$ by Cl_2 gives $[Re_2Cl_9]^-$ with the structure of $[W_2Cl_9]^{3-}$ (Figure 15.14). The $[Mo_2(SO_4)_4]^{4-}/[Mo_2(SO_4)_4]^{3-}$ case is an exception: Oxidation of $[Mo_2(SO_4)_4]^{4-}$ causes little structural change in the dimeric unit other than that expected for the decrease in bond order.

[12.] Reference 11, p. 658.

[13] M. H. Chisholm, Ed., *Reactivity of Metal–Metal Bonds*, ACS Symposium Series 155, American Chemical Society, Washington, D.C., 1981.

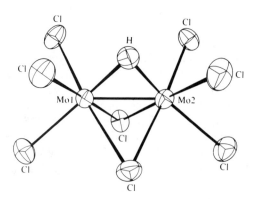

Figure 15.20   The structure of the [Cl₃Mo(μ-Cl)₂(μ-H)MoCl₃]³⁻ ion as it occurs in [C₅NH₆]₃[Mo₂Cl₈H]. (From A. Bono and F. A. Cotton, *Angew. Chem. Int. Ed. Engl.* **1979**, *18*, 332.)

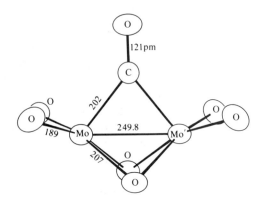

Figure 15.21   A view of the coordination geometry of Mo₂(O-*t*-Bu)₆(CO), showing the main internuclear distances. The tertiary butyl groups are omitted for clarity. (Reproduced with permission from M. H. Chisholm, R. L. Kelly, F. A. Cotton, and M. W. Extine, *J. Am. Chem. Soc.* **1978**, *100*, 2256. Copyright 1978, American Chemical Society.)

Unbridged triple-bonded Mo₂(O-*t*-Bu)₆ adds a molecule of CO reversibly to form Mo₂(O-*t*-Bu)₆CO, which contains two bridging alkoxides, a bridging CO, and a rare Mo≡Mo bond (249.8 pm) (Figure 15.21). Mo₂(OR)₆ compounds (R = Me₃Si, Me₃C, Me₂CH, and Me₃CCH₂) add two moles of CO₂ reversibly in hydrocarbon solvents to produce Mo₂(OR)₄(O₂COR)₂. The two carboxylates produced by the insertion reaction serve to bridge the Mo—Mo bond. Carbon monoxide can be inserted reversibly as a bridging carbonyl in Pd₂(dmp)₂X₂(X = Cl, Br; dmp = Ph₂PCH₂PPh₂) in dichloromethane to form an "A-frame" complex.

$$\text{(15.33)}$$

A-frame complexes have a vacant bridging site (X) that might serve as an active site for addition of a small molecule for possible catalytic activation (Figure 15.22). The complex ion [(CO)Rh(μ-Cl)(Ph₂PCH₂PPh₂)₂Rh(CO)]⁺ adds SO₂ reversibly at the active site. Carbon monoxide adds reversibly also, to give a compound with a bridging CO, but labeling studies show that CO attack occurs at a terminal position and then one of the coordinated CO molecules swings into the bridging site.

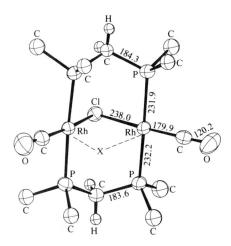

Figure 15.22   The inner coordination sphere of the $[Rh_2(CO)_2(\mu\text{-}Cl)(Ph_2PCH_2PPh_2)]^+$ ion, showing some relevant bond lengths. Only the first carbon atom of each phenyl ring is shown. (Reproduced with permission from M. Cowie and S. K. Dwight, *Inorg. Chem.* **1979,** *18,* 2700. Copyright 1979, American Chemical Society.)

GENERAL REFERENCES

J. C. Bailar, Jr., H. J. Emeléus, R. Nyholm, and A. F. Trotman-Dickerson, Eds., *Comprehensive Inorganic Chemistry,* Pergamon, Oxford, 1953. Five volumes. The most useful compilation for that date.

Coord. Chem. Rev. **1992,** *115;* **1993,** *127,* Transition Metals Chemistry Reviews.

F. A. Cotton and G. Wilkinson, *Advanced Inorganic Chemistry,* 5th ed., Wiley, New York, 1988.

S. A. Cotton and F. A. Hart, *The Heavy Transition Elements,* Wiley, New York, 1975. Good coverage of the second and third transition series, including the lanthanides and actinides.

N. N. Greenwood and A. Earnshaw, *Chemistry of the Elements,* Pergamon Press, Oxford, 1984.

K. A. Gschneider, Ed., *Industrial Applications of Rare Earth Elements,* ACS Symposium Series, No. 164, 1981.

J. J. Katz, G. T. Seaborg, and L. R Morss, Eds., *The Chemistry of the Actinide Elements,* Vols. 1 and 2, 2nd ed., Chapman & Hall, London, 1986.

S. P. Sinha, *Systematics and Properties of Lanthanides.* Riedel, Dordrecht, 1983.

A. F. Wells, *Structural Inorganic Chemistry,* 5th ed., Oxford, Oxford, 1984. The best single source for structural inorganic chemistry.

PROBLEMS

15.1   Write balanced equations of the preparation of
(**a**) Na_2CO_3 (**b**) MgF (**c**) $TiBr_3$ (**d**) $ZrCl_3$ (**e**) $VOCl_3$ (**f**) WCl_6

15.2   The bond dissociation energies for the alkali metal M_2 molecules decrease regularly from 100.9 kJ/mol for Li_2 to 38.0 kJ/mol for Cs_2. The bond dissociation energies are *greater* for the M_2^+ ions, decreasing from 138.9 kJ/mol for Li_2^+ to 59 kJ/mol for Cs_2^+. Explain why the dissociation energies are greater for the M_2^+ ions.

15.3   Predict the following for Fr.
(**a**) The product of the burning of Fr in air.
(**b**) An insoluble compound of Fr.
(**c**) The structure of FrCl.
(**d**) The relative heats of formation of FrF and FrI.

15.4 How could you remove unreacted Na (metal) in liquid ammonia safely?

15.5 Write equations for the preparation of the following metals.
(a) Na (b) K (c) Cs (d) Mg

15.6 How can you stabilize solutions containing Na^-?

15.7 What are laboratory and everyday uses of Li, K, and Cs?

15.8 Cite several properties that show the diagonal relationships between Li^+ and Mg^{2+}, and between Be^{2+} and Al^{3+}.

15.9 The element pairs Li–Mg, Na–Ca, and Be–Al are closely related because of the diagonal relationship. Would you expect Mg^{2+} to be more closely related to Sc^{3+} or to Ga^{3+}? Why?

15.10 Why is Au expected to form Au^-?

15.11 Write equations for the reduction of Cu^{2+} with a limited amount of CN^- and with an excess of CN^-.

15.12 Why is *aqua regia* (HCl–HNO_3) effective in oxidizing noble metals when neither HCl nor HNO_3 is effective alone?

15.13 Why are large anions effective in stabilizing unusually low oxidation states, each as Cd^I?

15.14 Sketch a linear combination of s, p_z, and d_{z^2} orbitals that would give very favorable overlap for bonding in a linear MX_2 molecule.

15.15 The structures of ZnO, CdO, and HgO are quite different. Describe the structures.

15.16 The structures of HgF_2, $HgCl_2$, $HgBr_2$, and HgI_2 show interesting variations. Describe the structures.

15.17 Sketch the structure of the silicate anion in beryl.

15.18 Give equations for reactions that could be used to separate a mixture of Zn^{2+}, Cd^{2+}, and Hg^{2+} present in solution.

15.19 Can one obtain
(a) Hg^{2+} salts free of Hg_2^{2+}? (b) Hg_2^{2+} salts free of Hg^{2+}?
Explain.

15.20 Discuss the factors involved in determining the following solubility patterns: LiF is much less soluble than LiCl, but AgF is much more soluble than AgCl.

15.21 The metal perchlorates have been referred to as *universal solutes*. What properties are important in causing most metal perchlorates to be quite soluble in water and several other solvents?

15.22 For which elements in the rare earth series are M^{II} and M^{IV} oxidation numbers expected?

15.23 Why should it have been expected that there would be more uncertainty concerning the identity of the first member of the second inner transition series compared to the lanthanide series?

15.24 One of the common M_2O_3 structures is that of α-alumina (Al_2O_3). Another is the La_2O_3 structure (C.N. 7). Check the La_2O_3 structure in Wells (1984) and describe the coordination about La^{3+}.

15.25 Ti is the ninth most abundant element in the earth's crust, and its minerals are reasonably concentrated in nature. Why is Ti less commonly used than much rarer metals?

15.26 Some elements are known as dispersed elements, forming no independent minerals, even though their abundances are not exceptionally low; whereas others of comparable or lesser abundance are highly concentrated in nature. Explain the following cases:
(a) *Dispersed:* Rb, Ga, Ge, Hf. (b) *Concentrated:* Li, Be, Au.

15.27 The occurrence of stable well-characterized oxidation states of the lanthanides other than +III can be explained in terms of empty, half-filled, and filled *f* orbitals. Attempt to apply a similar approach to the series Sc through Zn. Where does it work well? Why is it much less useful? Is this approach effective for the actinide elements?

15.28 Compare the syntheses of the highest and lowest stable oxidation states of Mn and Re. Which halide ions can be oxidized by MnO_4^- and by ReO_4^-? (See Appendix E for emf values.)

15.29 Would the removal of Hf from Zr be important in most applications of zirconium compounds? In the use of Zr metal in flash bulbs?

15.30 What properties of tungsten make it so suitable for filaments for light bulbs?

15.31 Given an example of
   (a) An acidic oxide of a metal.
   (b) An amphoteric oxide of a transition metal.
   (c) A diamagnetic rare earth metal ion.
   (d) A compound of a metal in the +VIII oxidation state.
   (e) A liquid metal chloride.
   (f) A compound of a transition metal in a negative oxidation state.

15.32 Describe the quadruple bonding in $[Re_2Cl_8]^{2-}$ in terms of the bond types (σ, π, etc.) and the atomic orbitals involved. What is significant about the eclipsed configuration?

15.33 Determine the bond order and oxidation number for $Mo_2(O_2CCH_3)_4$ (bridging acetate ions), $W_2Cl_4(PR_3)_4$ (no bridging ligands), and $[Mo_2(O\text{-}t\text{-}Bu)_6(CO)]$ (Figure 15.16).

15.34 Give an example of each type of Co complex.
   (a) Co^{II} tetrahedral   **(c)** Co^{III} octahedral, high-spin
   (b) Co^{II} square-planar   **(d)** Co^{III} optically active

15.35 Give one example of a nickel complex illustrating square-planar, tetrahedral, and octahedral coordination. What type of ligands favor each of these cases?

15.36 The structures of complexes with high coordination number involving didentate groups can be described in terms of the "average" positions of the didentate groups. Describe $[Ce(NO_3)_5]^{2-}$ and $[Ce(NO_3)_6]^{2-}$ in this way.

15.37 How can one account for the color of the following?
   (a) Fe_3O_4 **(c)** $KFeFe(CN)_6$ **(e)** $[Ti(H_2O)_6]^{3+}$
   (b) Ag_2S **(d)** $KMnO_4$ **(f)** $[Cu(NH_3)_4]^{2+}$

15.38 Give the bond order and show the occupancy of bonding and antibonding orbitals for M—M bonds in the following. (Ligands bridge metals except for **c** and **d**; there are axial ligands for **e** and **f**.)
   (a) $[Mo_2(SO_4)_4]^{4-}$ **(c)** $[Mo_2(NMe_2)_6]$ **(e)** $[Tc_2(O_2CCMe_3)_4Cl_2]$
   (b) $[Mo_2(SO_4)_4]^{3-}$ **(d)** $[Tc_2Cl_8]^{3-}$ **(f)** $[Ru_2(O_2CC_2H_5)_4(OCMe_2)_2]$

15.39 Count the valence shell electrons for the compounds in Problem 15.38.

·16·

The Chemistry of Some Nonmetals

·····································

Only about one-fifth of the elements are nonmetals and they are found in fewer than half of the periodic groups. Within the limited block of the periodic table occupied by nonmetals, we see great differences from one family to the next. As an example, F is the most active nonmetal, the neighboring element Ne is a noble gas, and the next element beyond Ne is Na, a very active metal.

Although this is the only chapter dealing exclusively with nonmetals, the chemistry of nonmetals appears in most of the book. The stereochemistry and bonding in compounds of nonmetals are treated in Chapters 2 and 4. Solids treated in Chapter 5 include metal compounds with nonmetals, covalent crystals, and metal silicates. Coordination chemistry (Chapters 9–11) and organometallic compounds (Chapters 12–14) deal with the chemistry of metals with nonmetal molecules or groups as ligands or substituents. Chapter 7, on acid–base chemistry, is concerned largely with compounds of nonmetals, and the oxidation–reduction reactions of Chapter 8 include many examples of nonmetals. The chemistry of boranes, carboranes, and other compounds of Group 13 elements is covered extensively in Chapter 17. Here we examine most of the nonmetals of N, O, F families and noble gases in the framework of the periodic table in order to observe and understand the similarities and differences in their physical and chemical properties.

16.1 GENERAL PERIODIC TRENDS FOR NONMETALS

Here we consider the chemistry of nonmetals of Groups 15–17, as well as that of the noble gases. We noted in Chapter 15 that Li and Be differ significantly from other members of their periodic groups. Boron is the only nonmetal of Group 13. The first member of Groups 14–17 also differs in many respects from other members of their families. These elements C–F have available for bonding only *s* and *p* orbitals of the second shell. The octet rule applies well to their compounds such as CF_4, NF_3, OF_2, and ClF. C, N, and O show a greater tendency to form multiple bonds compared to others of their families. The limitations of the number of σ bonds and/or the tendency to form multiple bonds accounts for the discrete gaseous molecules H_2, N_2, O_2, F_2, Cl_2, Br_2, and I_2. The noble gases are gases because of lack of bonding.

Nonmetals of the third and later periods have ns^2np^x configurations with empty $(n-1)d$ orbitals. The extent of bonding using these d orbitals is questionable, but a striking difference between N, O, and F and later members of their families is the formation of compounds such as PCl_5, PCl_6^-, SF_4, SF_6, ClF_3, ClF_4^-, BrF_5, and IF_7. There are no corresponding compounds of N, O, or F. In each family the reactivity as a nonmetal and the hardness decrease with increase in radius. There is a parallel increase in metallic character, passing from insulators for the nonmetals to conductors for As, Sb, and Po. Si and Ge are semiconductors.

H and He are unique in having only the $1s$ orbital available. H forms only one covalent bond and He forms none. The few compounds of noble gases are best represented by those of Xe corresponding to the halogen halides—for example, XeF_2 and XeF_4.

16.2   NITROGEN AND PHOSPHORUS

▶ 16.2.1   Family Trends

The elements N, P, and As are nonmetals, although the chemistry of As shows many characteristics of metals. Antimony and bismuth are metals. Except for N, the family trends are fairly regular. Whereas negative oxidation states are very important in the chemistry of N (NH_3, RNH_2, R_2NH, for example), the positive oxidation states are more important for P (and still more so for As). Nitrogen forms compounds involving very stable multiple bonds of the $p\pi p$ type, as in N_2, $C\equiv N^-$, and $RC\equiv NR$, whereas multiple bonds to P usually involve $p\pi d$ bonding, as in oxoanions. The availability of only s and p orbitals in the valence shell of N limits compound formation to those compounds satisfying the octet rule, whereas the octet is often exceeded in P compounds such as PCl_5 and $M^+PCl_6^-$.

The ionization energies (IEs) decrease by small amounts from one element to the next for P → Bi, but the IE of N is much higher than that of P (Table 16.1). The electron affinities (EAs) in general for the Group 15 (VB) elements are lower than those of the neighboring elements (see Table 1.8) in the same period, because an electron must be added to the half-filled p^3 configuration. Unlike the trend for most main-group families (see Tables 16.1, 16.3, and 16.5), the EAs for Group 15 *increase* (more energy is released on adding an electron) with increasing atomic number (Figure 16.1), except for Bi. The increase might be expected because of the decrease in pairing energy with increasing atomic radius (the p orbitals become larger and more diffuse with increasing n value). The EA of N is slightly negative (the addition of an electron to N is endothermic) as a result of the combination of three trends: (1) Addition of an electron is unfavorable compared with P because of the high pairing energy for the small N atom; (2) it is not as favorable compared with C and O because of the half-filled configuration of N; and (3) the first members

Table 16.1   **Some properties of the group 15 (VB) elements**

	N	*P*	*As*	*Sb*	*Bi*
Crystal radius (pm)	132(N^{3-})	212(P^{3-})	72(As^{3+})	94(Sb^{3+})	117(Bi^{3+})
Covalent radius (pm)	74	110	121	145	152
Electronegativity	3.0	2.2	2.2	2.1	2.0
First ionization (MJ/mol)	1.4023	1.0118	0.947	0.8337	0.7033
energy (eV)	14.534	10.486	9.81	8.641	7.289
Electron affinity (kJ/mol)	−7	72.0	78	103	91.3
Outer electron configuration	$2s^2 2p^3$	$3s^2 3p^3$	$3d^{10} 4s^2 4p^3$	$4d^{10} 5s^2 5p^3$	$4f^{14} 5d^{10} 6s^2 6p^3$

of Groups 15–17 are anomalous with respect to lower tendencies to increase further the high electron density for the small compact atoms, as seen in their low EAs and single-bond energies. From Table 1.8, the EA of N appears to deviate more from the family trend than does O or F. However, because of the opposite trends in EAs for Groups 15 and 16, plots of IE versus EA (Figure 16.1) reveal that N comes closer to the straight line determined by the rest of the family than does O.

► *16.2.2 The Elements and Their Occurrence*

Nitrogen occurs in organic matter in proteins and amino acids. Air is the major source for preparing nitrogen compounds because N_2 is the main constituent of air (78.09 vol %). The deposits of sodium nitrate found in a few arid regions, such as Chile, are important for fertilizers, but most of the nitrogen for fertilizers comes from the fixation of N_2. Dinitrogen (N_2) has accumulated in the atmosphere because the very stable triple-bond makes it quite chemically inert. N_2 is recovered on a large scale by the fractional distillation of liquid air, which also produces O_2 and the noble gases. For small-scale laboratory uses, pure N_2 can be obtained by the thermal decomposition of NaN_3 at 275°C:

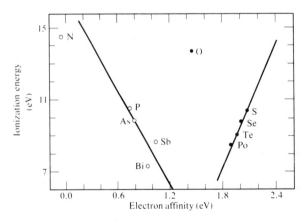

Figure 16.1   Ionization energy versus electron affinity plots for Groups 15 and 16. (Adapted from P. Politzer, in *Homoatomic Rings, Chains and Macromolecules of Main Group Elements*, A. L. Rheingold, Ed., Elsevier, Amsterdam, 1977, p. 95.)

$$2\,NaN_3 \xrightarrow{\text{heat}} 2\,Na + 3\,N_2 \tag{16.1}$$

The thermal decomposition of sodium azide inflates the air bags used as safety devices in cars. Na does not react directly with N_2. NaN_3 is prepared indirectly by the reaction of N_2O with $NaNH_2$ in $NH_3(l)$ or Na dissolved in $NH_3(l)$:

$$N_2O + 2\,NaNH_2 \longrightarrow NaN_3 + NaOH + NH_3 \tag{16.2}$$

$$3\,N_2O + 4\,Na + NH_3 \longrightarrow NaN_3 + 3\,NaOH + 2\,N_2 \tag{16.3}$$

With moderate heating, dinitrogen reacts only with Li of the alkali metals to form lithium nitride, Li_3N, reacts with Mg to form Mg_3N_2, and reacts with some transition metals giving nitrides.

Nitrogen-fixing bacteria, using the enzyme nitrogenase (Section 18.7), can reduce N_2 to NH_3 at ambient temperature and pressure. At elevated temperature, N_2 is more reactive. The **Haber process** is used in large-scale production of ammonia from the elements:

$$N_2 + 3\,H_2 \xrightarrow[\text{500°C, 500 atm}]{\text{Fe catalyst}} 2\,NH_3 \tag{16.4}$$

Most other N compounds are obtained from NH_3. Lightning converts N_2 to N oxides. In the atmosphere the N oxides are converted to HNO_3 and are then carried to the soil by rainfall.

Phosphorus occurs in sedimentary rocks (see page 710) as phosphates—primarily apatite, $Ca_5(PO_4)_3F$; and phosphorite, $Ca_5(PO_4)_3(OH,F,Cl)$, where there are variable relative proportions of OH^-, F^-, and Cl^-. Elemental phosphorus is obtained from phosphate rock by reduction with coke at 1300–1450°C with sand added to form slag:

$$Ca_3(PO_4)_2(s) + 5\,C(s) + 3\,SiO_2(s) \xrightarrow{\text{heat}} 3\,CaSiO_3(l) + 5\,CO(g) + P_2(g) \tag{16.5}$$

The P_2 vapor is condensed under water to form white phosphorus, P_4 (mp 44°C). Most phosphorus compounds are produced from the free element obtained in this way, with the notable exception of phosphate fertilizer. Phosphate rock can be used directly in pulverized form for fertilizer, or it can be treated with H_2SO_4 to give the more soluble **super-phosphate:**

$$2\,Ca_5(PO_4)_3F(s) + 7\,H_2SO_4(aq) + 10\,H_2O$$
$$\longrightarrow 3\,Ca(H_2PO_4)_2 \cdot H_2O(s) + 7\,CaSO_4 \cdot H_2O(s) + 2\,HF(g) \tag{16.6}$$

Superphosphate

Treatment of phosphate rock with aqueous phosphoric acid gives **triple superphosphate,** which has a higher P content because gypsum is not formed:

$$Ca_5(PO_4)_3F(s) + 7\,H_3PO_4(aq) + 5\,H_2O \longrightarrow 5\,Ca(H_2PO_4)_2 \cdot H_2O(s) + HF(g) \tag{16.7}$$

Triple
superphosphate

White phosphorus is a soft, waxy, highly toxic solid that, because it ignites spontaneously in air, is stored under water. It exists as P_4 molecules, with each P forming three single bonds:

When white P is heated about 250°C, or to lower temperature in the presence of light, it is converted to red phosphorus (mp ~600°C). Red P is polymeric and does not dissolve in solvents (diethyl ether and benzene) in which white P is soluble; it also is much less reactive than the white allotrope and does not ignite in air below about 400°C.

► 16.2.3 *Hydrogen Compounds*

Nitrogen forms the following volatile hydrogen compounds: ammonia, NH_3; hydrazine, H_2NNH_2; diimine, HNNH; tetrazene, N_4H_4; and hydrogen azide (hydrazoic acid), HN_3. Ammonia is the most important of these. Liquid ammonia can be sprayed directly into the soil as a fertilizer, or it can be converted to ammonium salts (nitrate, sulfate, or phosphate) for solid fertilizers. The catalytic oxidation of NH_3 produces NO (**Ostwald process**) as the first step in producing HNO_3 from NH_3:

$$4\,NH_3 + 5\,O_2 \xrightarrow{\text{Pt catalyst}} 4\,NO + 6\,H_2O \tag{16.8}$$

Heat is required to initiate the reaction, but then it becomes self-sustaining. The NO reacts with more O_2 to form NO_2 without a catalyst required. Nitric acid is produced by dissolving NO_2 in water: NO_2 disproportionates to form HNO_3 and HNO_2. The HNO_2 reacts with O_2 to form more HNO_3.

$$2\,NO_2 + H_2O \longrightarrow HNO_3 + HNO_2 \tag{16.9}$$

Interestingly, although NH_3 and amines, including $N(CH_3)_3$, are pyramidal, $N(SiH_3)_3$ is planar. This suggests $p\pi d$ bonding involving the empty d orbitals on Si and the filled p orbital on $N(sp^2$ hybridization). As this description implies, the basicity of $N(SiH_3)_3$ is very low. Because $P(SiH_3)_3$ is pyramidal, it appears that $p\pi d$ bonding is less favorable for P, because of poorer overlap with the more diffuse $3p$ orbitals of P.

As noted earlier (page 176), there is controversy over the extent of participation of the outer d orbitals in bonding, including $p\pi d$ bonding. Glidewell (*Inorg. Chim. Acta* **1975**, *12*, 219) prefers to explain the planarity of $N(SiH_3)_3$ in terms of stronger nonbonding interactions (for the larger Si substituents). Although his arguments are more profound, this is basically steric repulsion. The repulsion is less important for the larger P in $P(SiH_3)_3$.

Hydrazine, with a weak $N\text{—}N$ single bond (bond dissociation energy 247 kJ/mol), is a stronger reducing agent and a weaker base than ammonia. The **Raschig synthesis** of N_2H_4 uses NaOCl for oxidation of NH_3 to NH_2Cl, which then reacts with NH_3:

$$NH_3 + NaOCl(aq) \longrightarrow NH_2Cl + NaOH \tag{16.10}$$

$$NH_2Cl + NH_3 + NaOH \longrightarrow H_2NNH_2 + NaCl + H_2O \tag{16.11}$$

Competing reactions, catalyzed by traces of transition metal ions, are inhibited by gelatin or glue, added to sequester the metal ions, providing yields of ~70%.

Hydrogen azide, the only acidic uncharged $N\text{—}H$ compound, is a weak acid (Table 16.2). Anhydrous HN_3 (bp 36°C) decomposes explosively, but dilute aqueous solutions (up to ~20 wt %) can be handled safely. The N_3^- ion is linear and isoelectronic with CO_2. Ionic azides, such as NaN_3, are stable and decompose thermally (but controllably) to the elements. Covalent azides such as HN_3 and those of heavy metals (highly polarizing cations) are detonators. Heavy metal azides, such as those of lead and mercury, can be used as primers for explosives.

Derivatives of Ammonia

Amines have the general formulas RNH_2, R_2NH, and R_3N (where R is an akyl or aryl group). There are also derivatives in which R = OH, NH_2OH (hydroxylamine), and R = X—for example, NH_2Cl (chloroamine) and $NHCl_2$ (dichloroamine). The amide (NH_2^-), imide (NH^{2-}), and nitride (N^{3-}) ions yield metal salts. The formation of N^{3-} from $N(g)$ is highly endothermic (~2200 kJ/mol). The addition of one electron to N is endothermic, and the addition of a second and then a third electron is very unfavorable. Consequently, ionic nitrides are obtained only for the very electropositive metals.

Phosphine and Diphosphine

Phosphine, PH_3, is a toxic gas formed from the hydrolysis of metal phosphides or, on a larger scale, by the reaction of P_4 with NaOH solution:

$$Ca_3P_2 + 6H_2O \longrightarrow 3Ca(OH)_2 + 2PH_3(g) \tag{16.12}$$

Table 16.2   Acid dissociation constants for some NH compounds

Acid	K_a (25°C)
NH_4^+	5.5×10^{-10}
$N_2H_5^+$	1.2×10^{-8}
$N_2H_6^{2+}$	11.2
HN_3	1.8×10^{-5}

$$P_4 + 3\,NaOH + 3\,H_2O \longrightarrow PH_3(g) + 3\,NaH_2PO_2 \qquad (16.13)$$

Phosphine is not very soluble in water and is very weakly basic, but phosphonium salts, PH_4X, can be obtained. Pure phosphine is stable in air, but traces of P_2H_4 (diphosphine) commonly present cause the material prepared directly by Reactions (16.12) or (16.13) to burn spontaneously in air. Diphosphine can be separated from PH_3 by fractional distillation.

The fact that the bond angle for PH_3 (93.3°) is much smaller than that for NH_3 (107.3°) can be rationalized by the valence shell electron-pair repulsion (VSEPR) theory (Section 2.3.3) or interpreted as indicating greater p character (and less s character) for bonding in PH_3.

▶ *16.2.4 Halides*

The structures of the halides of N and P are shown in Figure 16.2. In addition, phosphorus forms PX_4^+ (T_d) and PX_6^- (O_h) ions. All the various N halides listed are encountered only for fluorine, since the stabilities of the N halides decrease with increasing atomic mass of the halogen. NF_3 is stable, but the other trihalides are explosive compounds. Of the N_2X_4 compounds, only N_2F_4 is known. Gaseous N_2F_4 contains both (C_2) *gauche* and (C_{2h}) *trans* conformers. Difluorodiazine (dinitrogen difluoride, N_2F_2) is a gaseous substance consisting of an equilibrium mixture of *cis* and *trans* isomers, with the *cis* isomer predominating. Only the *cis* isomer reacts with AsF_5 to form $[N_2F]^+[AsF_6]^-$, and this reacts with NaF/HF to yield pure *cis*-N_2F_2. The N—F bond length (121.7 pm) in $[N_2F]^+$ is the shortest known N—F bond, and the N—N bond length (109.9 pm) is almost as short as that in N_2 (109.76 pm). The short bonds are attributed to high s character of the sp σ bonds on the central N and the positive charge.[1] NF_3 can be oxidized[2] to NF_4^+ by F_2 in the presence of SbF_3 (or SbF_5) to form NF_4SbF_4 (or NF_4SbF_6). The N—F bond length is 130 pm in the tetrahedral ion. Structure and bonding of compounds of N with F or F and O have been reviewed by Peters and Allen.[3]

The symmetry of *cis*-N_2F_2 is C_{2v}, (*cis*-N_2X_2 is shown in Figure 16.2) and that of *trans*-N_2F_2 is C_{2h}. The C_2 *gauche*-N_2F_4 has one N—F bond in a plane bisecting the angle formed by the N—F bonds of the other NF_2. The eclipsed *cis*-N_2F_4(C_{2v}) isomer is unknown.

All phosphorus trihalides, formed by combination of the elements, are known. They hydrolyze easily and react with ammonia (ammonolyze) to produce $P(NH_2)_3$:

$$PCl_3 + 3\,H_2O \longrightarrow P(OH)_3 + 3\,HCl \qquad (16.14)$$

$$PCl_3 + 6\,NH_3 \longrightarrow P(NH_2)_3 + 3\,NH_4Cl \qquad (16.15)$$

[1] K. O. Christe, R. D. Wilson, W. W. Wilson, R. Bau, S. Sukermar, and D. A. Dixon, *J. Am. Chem. Soc.* **1991,** *113,* 3795.

[2] K. O. Christe, M. D. Lind, N. Thorup, D. R. Russell, J. Fawcett, and R. Bau, *Adv. Inorg. Chem.* **1988,** *27,* 2450; H. J. Emeléus, J. M. Shreeve, and R. D. Verma, *Adv. Inorg. Chem.* **1989,** *33,* 139.

[3] N. J. S. Peters and L. C. Allen, in *Fluorine-Containing Molecules,* J. F. Liebman, A. Greenberg, and W. R. Dolbier, Jr., Eds., VCH Publishers, New York, 1988, p. 199.

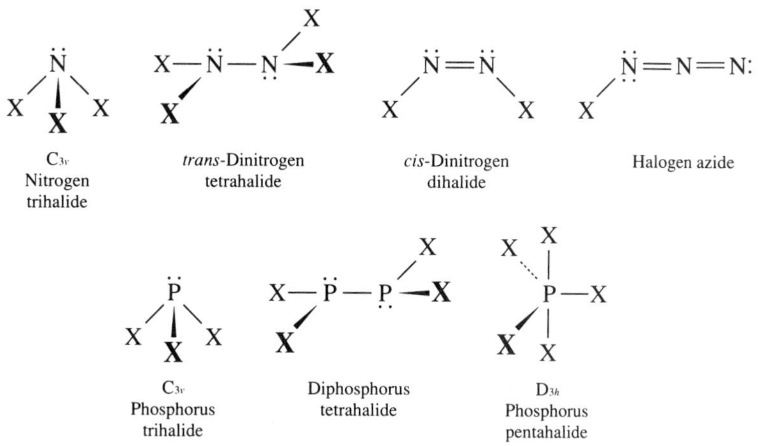

Figure 16.2   Nitrogen and phosphorus halides.

In addition, they react with O_2 to produce phosphoryl halides, POX_3 (also called phosphorus oxohalides). All P pentahalides are known except for PI_5, where steric factors probably are critical. PF_5 and PCl_5 have $\mathbf{D}_{3h}$ symmetry in the vapor phase, but PCl_5 exists as $[PCl_4]^+[PCl_6]^-$ in the solid. Phosphorus pentabromide exists as $[PBr_4]^+Br^-$ in the solid and dissociates into $PBr_3 + Br_2$ in the vapor phase.

Compounds of the type PR_5 are named as substituted phosphoranes. The fully hydrogenated hydrides are named as methane, CH_4; silane, SiH_4; and phosphorane, PH_5. The parent compound, phosphorane, is unknown. Chemists show no inclination to replace the names for water and ammonia by systematic names.

▶ *16.2.5   Oxygen Compounds*

Oxides of Nitrogen

Nitrogen forms an extensive series of oxides: N_2O (nitrous oxide), NO (nitric oxide), N_2O_3, NO_2, N_2O_4, N_2O_5, and NO_3. Scheme 16.1 demonstrates the complexity of the chemistry of the oxo compounds of N. The higher oxides are named systematically—dinitrogen trioxide, dinitrogen pentaoxide, and so on. Nitrous oxide is obtained by moderately heating NH_4NO_3:

$$NH_4NO_3 \xrightarrow{\sim 250°C} N_2O + 2\,H_2O \qquad (\textbf{Danger:} \text{ Explosive!}) \qquad (16.16)$$

or, more safely, by heating an acidic solution of NH_4NO_3. Nitrous oxide is not very reactive. Along with serving as an anesthetic (laughing gas), it is used as an aerosol propellant in whipped cream because it is soluble in fats.

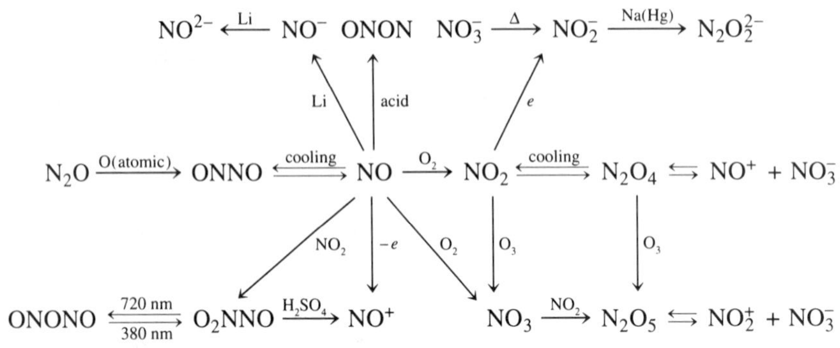

$$NO^{2-} \xleftarrow{Li} NO^- \quad ONON \quad NO_3^- \xrightarrow{\Delta} NO_2^- \xrightarrow{Na(Hg)} N_2O_2^{2-}$$

$$N_2O \xrightarrow{O(atomic)} ONNO \underset{cooling}{\rightleftharpoons} NO \xrightarrow{O_2} NO_2 \underset{cooling}{\rightleftharpoons} N_2O_4 \leftrightharpoons NO^+ + NO_3^-$$

$$ONONO \xrightleftharpoons[380\,nm]{720\,nm} O_2NNO \xrightarrow{H_2SO_4} NO^+ \qquad NO_3 \xrightarrow{NO_2} N_2O_5 \leftrightharpoons NO_2^+ + NO_3^-$$

Scheme 16.1 Chemistry of the oxo compounds of N. (Reproduced with permission from J. Laane and J. R. Ohlsen, *Prog. Inorg. Chem.* **1980,** 27, 465. Copyright 1980, John Wiley & Sons, Inc.)

Nitric oxide is obtained by reducing dilute HNO_3 by Cu:

$$8\,HNO_3 + 3\,Cu \longrightarrow 3\,Cu(NO_3)_2 + 4\,H_2O + 2\,NO \qquad (16.17)$$

The Ostwald process (page 764) for the catalytic oxidation of NH_3 to NO in the production of HNO_3 has been mentioned. NO, as an odd-electron molecule, dimerizes in the liquid or solid state to form ONNO; but in the presence of HCl or a Lewis acid, a red solid known to be the asymmetric dimer, ONON, is formed. Nitric oxide is oxidized easily to NO^+, the nitrosyl ion. Note that in the electron-counting scheme for metal carbonyl nitrosyl compounds (Section 12.2.3), NO is regarded as a three-electron donor. Reactions of nitrogen oxides are shown in Scheme 16.1.

Nitric oxide is very reactive, and it is surprising that it occurs in biological systems. Nevertheless, NO is produced in the brain as a messenger molecule of neurons. NO serves as a neurotransmitter in penile erection. Its function is related presumably to its role as a physiologic mediator of blood vessel relaxation.

Dinitrogen trioxide, $ONNO_2$, is blue as a liquid or solid. Obtained from the proper combination of NO and O_2 or NO and N_2O_4, it dissociates above $\sim-30°C$ to form $NO + NO_2$. The asymmetric $ONNO_2$ is converted into a symmetrical isomer $O{=}N\diagdown_{O}\diagup N{=}O$ by irradiation at 720 nm and is reconverted to the asymmetric isomer by irradiation at 380 nm. Nitrogen dioxide and dinitrogen tetraoxide exist in equilibrium:

$$2\,NO_2 \rightleftharpoons N_2O_4 \qquad (16.18)$$
$$\text{Brown} \qquad \text{Colorless}$$

The equilibrium is used for lecture demonstrations or laboratory experiments, since it is strongly dependent on temperature and pressure and only one species is colored. Liquid

N_2O_4 is useful as a nonaqueous solvent[4] in which NO^+ salts are acids and NO_3^- salts are bases. NO_2 is easily oxidized to NO_2^+ (nitryl ion) and reduced to NO_2^- (nitrite ion).

Dinitrogen pentaoxide, the anhydride of HNO_3, can be obtained by dehydrating HNO_3 with P_4O_{10}. The solid (mp 30°C, decomposes ~47°C) dissolves in H_2O to form HNO_3, as a true anhydride. Nitrogen trioxide is formed when N_2O_5 decomposes or by its reaction with ozone. The paramagnetic radical NO_3 is presumed to be planar.

Oxoacids of Nitrogen

Nitric acid, the most important acid of nitrogen, is synthesized from NH_3 [Reactions (16.8) and (16.9)] on a large scale as one of the common strong acids and for the production of N fertilizers. Concentrated HNO_3 is 15.7 M. It can be dehydrated and vacuum-distilled to give anhydrous HNO_3. Nitric acid is a strong oxidizing agent, commonly being reduced to NO in dilute solution and to NO_2 in concentrated solution. The E^0 values are given for comparison, but these refer to unit activities.

$$NO_3^- + 4H^+ + 3e \rightleftharpoons NO + 2H_2O \qquad E_A^0 = 0.96 \text{ V} \qquad (16.19)$$

$$NO_3^- + 2H^+ + e \rightleftharpoons NO_2 + H_2O \qquad E_A^0 = 0.803 \text{ V} \qquad (16.20)$$

NO_2 is the major reduction product only for concentrated HNO_3. A strong reducing agent such as Zn reduces NO_3^- in dilute acidic solution to NH_4^+. Nitrate salts are used frequently for studies of metal cations because most metal nitrates are water-soluble and the nitrate ion is a weak ligand for complex formation.

The orthonitrate ion, NO_4^{3-}, has been verified by crystal structure determination.[5] The N—O bond is very short (139 pm compared to ~150 pm expected for a single N—O bond). Because $d\pi p$ or $p\pi p$ bonding is not possible, the bond is shortened by polar interactions. The O—O distance (226 pm) is the shortest O—O nonbonding distance known. The van der Waals O—O distance is 304 pm.

Nitrous acid is unstable except in dilute solution, which can be obtained from $H_2SO_4 + Ba(NO_2)_2$ after removal of the $BaSO_4$ precipitate. Nitrous acid oxidizes Fe^{2+} or I^- and reduces MnO_4^-. Alkali-metal nitrite salts can be prepared by thermal decomposition of the nitrates or by reducing the nitrates with carbon, lead, or iron. Sodium nitrite is used as an additive in cured meats (hot dogs, hams, bacon, and cold cuts), to which the presence of some NO from decomposition of NO_2^- imparts an appealing red color, because of the formation of a bright red compound with hemoglobin. Also, the NO_2^- ion inhibits bacterial growth—particularly, that of *Clostridium botulinum*, which causes botulism. Nitrites are controversial food additives, since they can be converted during cooking to nitrosoamines ($R_2NN=O$), compounds considered to be carcinogens. However, nitrites have been approved as food additives.

$$HNO_2 + H^+ + e \rightleftharpoons NO + H_2O \qquad E^0 = 1.0 \text{ V} \qquad (16.21)$$

$$NO_3^- + 3H^+ + 2e \rightleftharpoons HNO_2 + H_2O \qquad E^0 = 0.94 \text{ V} \qquad (16.22)$$

[4] C. C. Addison, *Chem. Rev.* **1980**, *80*, 21.

[5] M. Jansen, *Angew. Chem. Int. Ed. Engl.* **1979**, *18*, 698.

Hyponitrous Acid

The hyponitrite ion, $N_2O_2^{2-}$, exists as *cis* and *trans* isomers. The more stable *trans* isomer is obtained by reducing NO in organic solvents, or by reducing $NaNO_2$ with sodium amalgam. The reaction of NO with Na or K in liquid ammonia yields the *cis* isomer. A solution of the free acid can be obtained from the silver salt of the *trans* isomer plus HCl. A variety of redox reactions is possible for hyponitrites, but they usually serve as reducing agents.

$$\begin{bmatrix} O & & & & O \\ & \diagdown & \ddot{N} = N & \diagup & \\ O & \diagup & & \ddot{} & \diagdown & \end{bmatrix}^{2-}$$

trans-Hyponitrite ion ($\mathbf{C}_{2h}$)

Oxides and Oxoacids of Phosphorus

The reaction of P_4 with a limited supply of O_2 yields P_4O_6; with an excess of O_2, P_4O_{10} is formed. Oxygen atoms are inserted into the P—P bonds of P_4 to form P_4O_6, and then terminal oxygen atoms are added to each P to form P_4O_{10}, with the retention of $\mathbf{T}_d$ symmetry (Figure 16.3). For the terminal P—O bonds we can write either single-bonded structures, corresponding to the addition of an O atom to the lone pair of P, or double-bonded structures. Figure 16.3 shows the formal charges. Taking into account the polarization effects that displace electron density toward the more electronegative O, the double-bonded structure is more favorable for charge distribution. There are also the intermediate oxides P_4O_7, P_4O_8, and P_4O_9, which correspond to the stepwise addition of terminal oxygens to P_4O_6.

Phosphorus(III) oxide is the anhydride of phosphorous acid, H_3PO_3, which features four covalent bonds to P. Only the hydrogens attached to O are acidic, so $H_2(HPO_3)$ is a dibasic acid, forming salts of $H_2PO_3^-$ and HPO_3^{2-}. Correspondingly, hypophorous acid is a

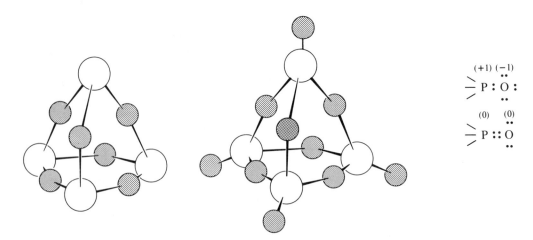

Figure 16.3 Structures of P_4O_6 and P_4O_{10}.

monobasic acid. It is obtained by reaction of P_4 with $Ba(OH)_2$ and subsequent treatment with a stoichiometric amount of H_2SO_4.

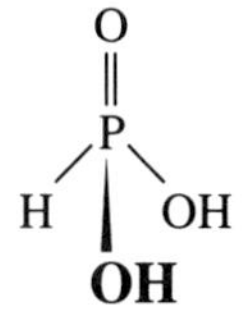

Phosphorous acid
(Phosphonic acid)

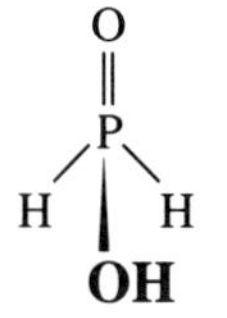

Hypophosphorous acid
(Phosphinic acid)

Phosphorus(V) oxide reacts with water stepwise to produce cyclic metaphosphoric acid, $(HPO_3)_4$, diphosphoric acid ($H_4P_2O_7$, also called pyrophosphoric acid), and, finally, orthophosphoric acid (H_3PO_4—the ortho prefix is usually omitted unless the distinction is important). Other polyphosphates are formed by dehydration of H_3PO_4.

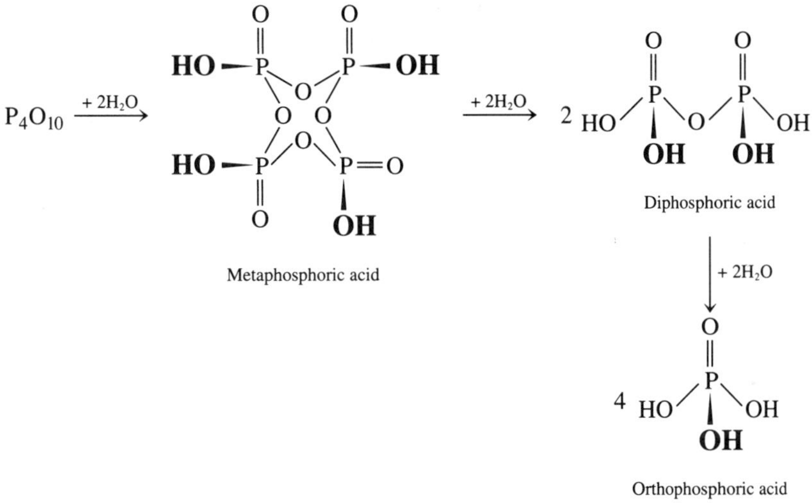

Metaphosphoric acid

Diphosphoric acid

Orthophosphoric acid

(16.23)

P_4O_{10} is used as a desiccant when a more powerful drying agent than $CaCl_2$ or other low-cost agent is needed. Reaction of P_4O_{10} with aqueous hydrogen peroxide gives H_3PO_5, peroxophosphoric acid. Salts of peroxodiphosphoric acid, $H_4P_2O_8$, are obtained by anodic oxidation of phosphate ion. Sometimes called (improperly) *per*phosphoric acids, these acids contain the O—O linkage, as H—O—O—P in H_3PO_5 and as P—O—O—P in peroxodiphosphates.

▶ 16.2.6   *Redox Chemistry of N and P*

As the potential diagrams for N reveal, the stable species are $NH_4^+(NH_3)$, N_2, and NO_3^-. For P the stable species are PH_3, $H_3PO_3(HPO_3^{2-})$, and $H_3PO_4(PO_4^{3-})$:

$$E_A^0 \qquad NO_3^- \xrightarrow{\;0.94\;} HNO_2 \xrightarrow{\;1.45\;} N_2 \xrightarrow{\;0.27\;} NH_4^+ \tag{16.24}$$

$$H_3PO_4 \xrightarrow{\;-0.276\;} H_3PO_3 \xrightarrow{\;-0.50\;} P_4 \xrightarrow{\;-0.06\;} PH_3(aq) \tag{16.25}$$

$$E_B^0 \qquad NO_3^- \xrightarrow{0.01} NO_2^- \xrightarrow{0.41} N_2 \xrightarrow{-0.74} NH_3(aq) \qquad (16.26)$$

$$PO_4^{3-} \xrightarrow{-1.12} HPO_3^{2-} \xrightarrow{-1.73} P_4 \xrightarrow{-0.89} PH_3 \qquad (16.27)$$

Elemental P disproportionates in basic solution. Although many oxides and oxoacids of N are kinetically stable, in aqueous solution the intermediate oxidation states are thermodynamically unstable. Note that HNO_3 is strongly oxidizing, but H_3PO_4 is not. Whereas HNO_2 is an even stronger oxidizing agent than HNO_3, H_3PO_3 is easily oxidized and not easily reduced.

Families of Pourbaix (E/pH) diagrams (in the form of the periodic table, *J. Chem. Educ.* **1969,** *46,* 90) are useful for comparing the range of stabilities of various species. (See Section 8.6 for a discussion of Pourbaix diagrams.) In this section the species referred to as stable are those that do not oxidize or reduce water (or H^+ or OH^-) or undergo disproportionation. These reactions limit redox stability in the absence of added oxidizing or reducing agents.

▶ *16.2.7   Phosphazenes*

The phosphazenes contain the $-P=N-$ linkage, most commonly as the chlorophosphazenes, $(-PCl_2=N-)_n$. There are cyclic chlorophosphazenes, with $n = 3$ to 8, and chain phosphazenes. Cyclic phosphazenes are formed by the reaction of PCl_5 with NH_4Cl at 120–135°C:

$$n\,PCl_5 + n\,NH_4Cl \longrightarrow (PCl_2N)_n + 4n\,HCl \qquad (16.28)$$

The oligomers can be separated by vacuum distillation. The Cl can be replaced partially or fully by other groups, such as F or organic groups, R. Hexachlorotriphosphazene has a considerable amount of π delocalization, although there is a sign mismatch for $p\pi d$ bonding going around the ring for the trimer (or for any case where n is odd). Puckering occurs for larger rings.

Polyphosphazenes

Chemists have known since 1897 that hexachlorotriphosphazene could be treated to form a rubbery material now known to be a cross-linked polymer that is hydrolytically unstable. Allcock[6] and co-workers, however, found the means to control the polymerization and substitute alcohol, phenol, and amine groups for Cl^-. By careful purification and control of conditions, they obtained soluble un-cross-linked polymers from $(PNCl_2)_3$. The Cl of the soluble polymers are replaced by organic groups (Scheme 16.2) and the polymers are cross-linked. The polyphosphazenes include polymers that are flame retardants. Other

---

[6] H. R. Allcock, *Chem. Eng. News,* March 18, 1985, p. 22. A review of polyphosphanes and other inorganic macromolecules. Also see H. R. Allcock, *Science* **1992,** *225,* 1106.

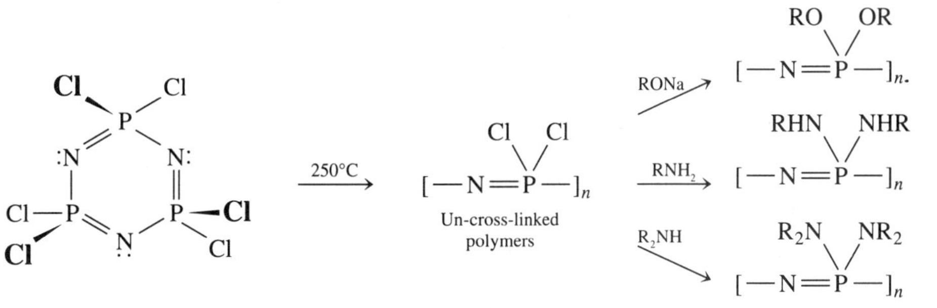

Scheme 16.2   Polyphosphazenes

polymers are (a) glasses, (b) elastomers that retain their elasticity at low temperature and are oil-resistant, (c) film-forming polymers, and (d) polymers that can be fabricated into fibers. Some of the polymers are water-repellant and compatible with living tissue for use in biomedical devices. Those derived from amino acid esters are biodegradable into harmless products through hydrolysis. The polymers can be modified by chemical reactions to produce adhesive, biocompatible, and even antibacterial surfaces.

16.3   OXYGEN, SULFUR, AND SELENIUM

► 16.3.1   Family Trends

Some properties of the elements of the oxygen family (chalcogens) are listed in Table 16.3. Oxygen stands apart in many physical and chemical properties, with close similarities among S, Se, and Te. A major difference is that third-period (and later) elements have d orbitals available for $p\pi d$ bonding or for the expansion of the valence shell beyond the octet. There are no oxygen counterparts of SF_4 and SF_6. Thus for Groups 15–17 we have a big change in properties from the first member of each group to the next, but a very small change from the second member to the third. Within each main-group family to the right of the transition series, the pairs Al and Ga, Si and Ge, P and As, S and Se, and Cl and Br display the greatest similarities, because of their equivalent sets of valence shell or-

Table 16.3   Some properties of group 16 (VIB) elements

	O	S	Se	Te	Po
Crystal radius (X^{2-}, pm)	126	170	184	207	—
Covalent radius (pm)	74	103	118	142	—
Electronegativity	3.4	2.6	2.6	2.1	2.0
First ionization (MJ/mol)	1.3140	0.99960	0.9409	0.8693	0.812
energy (eV)	13.62	10.360	9.752	9.009	8.42
Electron affinity (kJ/mol)	141.0	200.43	194.99	190.16	180
Outer electron configuration	$2s^2 2p^4$	$3s^2 3p^4$	$3d^{10} 4s^2 4p^4$	$4d^{10} 5s^2 5p^4$	$4f^{14} 5d^{10} 6s^2 6p^4$

bitals and similar radii. The corresponding similarity between Zr and Hf and between Nb and Ta is attributed to the lanthanide contraction (page 738). Actually, the decrease in atomic radius is greater for the addition of *d* electrons than for *f* electrons, so that the atomic radius of Ga (136 pm from the average Ga—Ga distance in the solid), following the first transition series, is *smaller* than that of Al (143 pm). This is the consequence of the "transition series contraction." As for the lanthanide contraction, the effect diminishes for each succeeding group, but the chemical consequences are profound for Ga through Br.

The IE and electronegativity of O are much higher than those of other members of the family. EAs decrease from S to Po, but that of O is the lowest of the group (see Table 16.3 and Figure 16.1). Nitrogen, oxygen, and fluorine have much lower EAs than expected from the group trends, because of their small size and very high electron densities. Removal of electrons from the neutral atoms requires a great deal of energy, but the addition of an electron is less favorable than for less compact atoms (see Section 1.6).

Oxygen shows negative oxidation states except in combination with F. The oxide ion is a hard Lewis base, whereas S^{2-} and Se^{2-} are very soft bases. Metal oxides commonly have typical ionic-type lattice structures, whereas metal sulfides and selenides more likely have layer-type (page 218) structures or other structures, such as NiAs (page 207). These structures are encountered only where polarization effects are great.

Family trends are quite regular for S, Se, and Te, but with greater importance of positive oxidation states and an increasing similarity to metals. The last member of the group, Po, is a radioactive metal. It is the only element with the simple cubic structure.

▶ 16.3.2   *The Elements and Their Occurrence*

Oxygen

Oxygen is the most abundant element (46.6 wt %) in the earth's crust, which consists primarily of Si, H, and metal oxides. We obtain the element most readily from the atmosphere (20.9 vol %) by fractional distillation of liquid air. Liquid-air plants are located at major usage sites, particularly in the steel industry. Liquid and solid O_2 are pale blue. Molecular orbital theory gives a description of bonding in O_2 consistent with a bond order of 2 and two unpaired electrons.

The earth is believed to have lost most of its volatile constituents during formation. The original atmosphere was reducing in character (H_2 and H compounds), the present O_2 atmosphere being produced by photosynthesis. Accordingly, elements in igneous rocks (rocks formed originally; see page 710) usually are in relatively low oxidation states and elements in sedimentary rocks (those subjected to the weathering process) commonly are in higher oxidation states.

Dioxygen is a rather strong oxidizing agent, as indicated by the reduction potentials, but the reactions are slow except at elevated temperature. In our remarkable environment, living organisms are dependent on O_2, but the organic molecules comprising the organisms are thermodynamically unstable in an O_2 atmosphere. These molecules eventually are converted to CO_2 and H_2O, and other oxides, but the process is slow except at elevated temperature. In living organisms, cells die, but they are replaced continuously as energy from metabolism brings about the thermodynamically unfavorable reactions. When an organism dies, the life-sustaining functions cease.

The usual intermediate in the reduction of O_2 in aqueous solution is H_2O_2 (or HO_2^-). The $O_2 \rightarrow H_2O_2$ couple gives the effective electromotive force (emf) available in determining whether a substance can be oxidized by O_2. The next step to form H_2O is more favorable because H_2O_2 is a stronger oxidizing agent than O_2.

$$E_A^0 \qquad O_2 \xrightarrow{\ 0.682\ } H_2O_2 \xrightarrow{\ 1.776\ } H_2O \tag{16.29}$$

with overall 1.229 above.

$$E_B^0 \qquad O_2 \xrightarrow{\ -0.065\ } HO_2^- \xrightarrow{\ 0.867\ } OH^- \tag{16.30}$$

with overall 0.401 below.

Ozone, O_3, is an allotropic form of oxygen produced by the absorption of ultraviolet light (sunlight) or electrical energy (lightning or a spark source). The ozone layer in the upper atmosphere (15- to 25-km altitude) absorbs most ultraviolet radiation (200 to 360 nm) harmful to life on earth. There is great concern about the depletion of the ozone layer by catalytic decomposition of O_3 to O_2 caused by chlorofluorocarbons (freons) used for refrigerants and aerosol propellants (hair sprays, etc.) and by NO and NO_2 produced by supersonic aircraft flying near to the ozone layer. The chlorofluorocarbons (e.g., $CFCl_3$ and CF_2Cl_2) decompose photochemically to form atomic Cl, which reacts with O_3 to form O_2. Freons are now prohibited as aerosol propellants in the United States, and they are being phased out as refrigerants.

The bond distance in O_2 is 121 pm and in O_3, 127.8 pm. The Lewis structure of O_3 is similar to that of NO_2^- and SO_2. Ozone is a powerful and highly reactive oxidant.

$$O_3 + 2H^+ + 2e \rightleftharpoons O_2 + H_2O \qquad E^0 = +2.076 \text{ V} \tag{16.31}$$

Sulfur and Selenium

Sulfur occurs in the crust of the earth as metal sulfides, many of which are important ores for metals. Sulfur is recovered in the smelting process for some of the metal sulfides. Gaseous H_2S and SO_2 are produced in volcanic activity, which is the source of some S deposits. Sulfates, such as anhydrite ($CaSO_4$), occur in sedimentary rocks as a result of complete oxidation of S. The most important sources of S are the remarkably pure deposits of elemental S in North and South America, Japan, Poland, the former Soviet Union, and Sicily. Many of these underground deposits were formed by the leaching of sulfide minerals, the eventual oxidation to SO_4^{2-}, and deposition of metal sulfates. Bacteria reduced the sulfates to H_2S, which reacted with more SO_4^{2-} to produce elemental S. Underground S is recovered by the **Frasch process,** in which superheated water melts the S and an admixture of air brings the molten S to the surface. The major use of S is in the production of H_2SO_4.

Sulfur in coal and petroleum is a source of pollution, because of combustion to SO_2; on the other hand, removal of SO_2 from flue gases is an increasingly important source of the element. The effective removal of S at reasonable cost is still a major problem in increasing the use of coal with high S content.

High-performance batteries that are smaller, cheaper, and have higher capacity than the lead storage cell are of interest for vehicles and for energy storage in off-peak hours in electric utility networks. The Na–S and Li–S batteries being developed for such applications operate at high temperature and use molten or solid electrolytes (see page 265).

Selenium and tellurium occur in nature as metal selenides and tellurides, usually accompanying metal sulfides. Se and Te, as the dioxides, concentrate in fly ash from the roasting of sulfide ores.

The many allotropic forms of S include S_6, S_8, S_{12}, and S_n rings, where n can be as high as 20. Rapid cooling of molten S yields metastable plastic S, consisting of zigzag chains. The S_8 rings are most important, as encountered in rhombic S, the stable room-temperature form. Monoclinic S crystallizes from the melt (119°C), but converts only slowly to the stable rhombic form below the transition temperature of 95.5°C. The stable metallic form of Se contains helical chains.

▶ 16.3.3   *Oxides, Peroxides, and Superoxides*

Although O_2 reacts with most elements to form oxides, the reactions are usually slow except at high temperature. The active metals and others in low oxidation states give ionic oxides. Nonmetals and metals in high oxidation states give covalent compounds—for example, CO_2, SO_3, CrO_3, Mn_2O_7, and OsO_4. Ionic oxides commonly have high heats of formation, even though energy is required to form the metal atom (heat of vaporization) and ion (IE), dissociate O_2, and add two electrons to O:

		ΔH (kJ/mol)
$O_2(g)$	$\longrightarrow$ $2\,O(g)$	493.59
$O(g) + e$	$\longrightarrow$ $O^-(g)$	−141.0
$O^-(g) + e$	$\longrightarrow$ $O^{2-}(g)$	744
$O(g) + 2e$	$\longrightarrow$ $O^{2-}(g)$	603

The favorable heats of formation result from the high lattice energies of ionic oxides. Less energy is required to form the peroxide ion from O_2 because it is not necessary to break the O—O bond, and the addition of a second electron is not so unfavorable, because of delocalization of charge over two atoms.

		ΔH (kJ/mol)
$O_2(g) + e$	$\longrightarrow$ $O_2^-(g)$	−42.5
$O_2^-(g) + e$	$\longrightarrow$ $O_2^{2-}(g)$	+512
$O_2^{2-}(g)$	$\longrightarrow$ $2\,O^-(g)$	+126

The lattice energies are smaller for peroxides because of the larger size of O_2^{2-}. The peroxides are highly reactive, because of the low O—O bond energy. The two electrons added to O_2 to form O_2^{2-} are antibonding.

Potassium superoxide is used in a self-contained breathing apparatus. The moisture from the user's breath reacts with KO_2 to release oxygen at a steady rate, and CO_2 is absorbed by reacting with K_2O to form K_2CO_3.

The superoxide ion, as a radical anion, is highly reactive. Anaerobic bacteria die in an O_2 atmosphere because O_2 is reduced to O_2^- and the bacteria lack the enzyme superoxidase to destroy the destructive O_2^-. The alkali metal hydroxides react with ozone to produce ozonides MO_3. As with superoxides, the ozonides are more stable for larger cations. For small cations, decomposition of MO_2 or MO_3 is favored because of the very high lattice energy of M_2O.

▶ 16.3.4   *The Dioxygenyl Cation*

The discovery that O_2 gas reacts with PtF_6 to form an orange solid, O_2PtF_6, together with recognition of the similar IEs of O_2 and Xe, led to the discovery of Xe compounds. The O_2^+ ion can be obtained by oxidation of O_2 with F_2, if a large anion is available for stabilization (see Section 5.8.5 for a discussion of the stabilizing effects of large counterions).

$$O_2 + \tfrac{1}{2}F_2 + GeF_4 \xrightarrow[-78°C]{\text{Photolysis}} O_2^+GeF_5^- \tag{16.32}$$

▶ 16.3.5   *Hydrogen Compounds of Oxygen*

The hydrogen compounds of O are H_2O, OH^-, H_2O_2, HO_2^-, and the unstable HO_2. H_2O_3 and H_2O_4 are formed in the decomposition of H_2O_2 in a glow discharge. Water is our most common reaction medium. Because it is so familiar, we tend to think of it as a typical solvent. In fact, it is an extraordinary solvent, because of its high dielectric constant, large dipole moment, and great tendency to form hydrogen bonds—both with itself and with other species capable of participating as H donors or acceptors.

Hydrogen peroxide is produced by the hydrolysis of the peroxodisulfate ion, which is obtained by electrolytic oxidation. The more important methods produce H_2O_2 as one of the products of oxidation of organic compounds such as a substituted 9, 10-anthracenediol. The quinone is reduced catalytically with H_2 to the diol, and H_2O_2 is extracted into water:

$$\tag{16.33}$$

Pure H_2O_2 has a greater liquidus range (mp $-0.43°C$, bp $150.2°C$) and greater density (1.44, 25°C) than H_2O. It is more acidic ($K_{20°} = 1.5 \times 10^{-12}$) than H_2O as expected, since the negative charge on HO_2^- can be delocalized to a greater extent. H_2O_2 is thermo-dynamically unstable with respect to disproportionation:

$$2\,H_2O_2 \longrightarrow 2\,H_2O + O_2 \qquad \Delta G^0 = -210 \text{ kJ/mol} \qquad (16.34)$$

The decomposition is catalyzed by traces of transition or heavy metal ions, so complexing agents are added to sequester the metal ions. Usually, it is used at concentrations of 30% or less.

The long O—O bond (148 pm) in H_2O_2 indicates a weak single bond. The H—O—O bond angle is 96.8°. If you imagine the O—O bond at the intersection of this page and the one opposite, with one O—H bond drawn in each page, you would open the pages to an angle of 93.8°. The barrier to rotation about the O—O bond is small.

The standard reduction potentials for H_2O_2 indicate that it is a strong oxidizing agent, particularly in acidic solution, and a mild reducing agent. It oxidizes NO_2^- to NO_3^- and Cr^{III} to chromate, but it is oxidized by strong oxidizing agents such as MnO_4^-, Cl_2, and Ce^{IV} salts.

$$H_2O_2 + 2\,H^+ + 2e \rightleftharpoons 2\,H_2O \qquad E^0 = 1.78 \text{ V} \qquad (16.35)$$

$$O_2 + 2\,H^+ + 2e \rightleftharpoons H_2O_2 \qquad E^0 = 0.68 \text{ V} \qquad (16.36)$$

$$HO_2^- + H_2O + 2e \rightleftharpoons 3\,OH^- \qquad E^0 = 0.87 \text{ V} \qquad (16.37)$$

EXAMPLE: What are the reactions expected of H_2O_2 in 1 M acid with U^{4+}? Write the overall equation and calculate E^0_{cell}.

$$UO_2^{2+} \xrightarrow{0.17} UO_2^+ \xrightarrow{0.380} U^{4+} \xrightarrow{-0.520} U^{3+} \qquad E^0_A$$

Solution: First consider whether H_2O_2 can reduce U^{4+} or oxidize U^{4+} to the next higher oxidation state (UO_2^+).

$$U^{4+} + e \longrightarrow U^{3+} \qquad E^0 = -0.520 \text{ V}$$

$$H_2O_2 \longrightarrow O_2 + 2\,H^+ + 2e \qquad E^0 = -0.68 \text{ V as a reducing agent}$$

Because both have negative E^0 values, the overall reaction (sum of the half-cells) cannot occur. The potential for oxidation of U^{4+} to UO_2^+ is -0.38 V, and that for reduction of H_2O_2 (+1.776 V) is very high, so this reaction will occur. However, note that the next oxidation step is even easier (-0.16 V), so UO_2^+ disproportionates; and if enough H_2O_2 is present, U^{4+} will be oxidized to UO_2^{2+}. See Section 8.5.3 for calculation of the emf for a third half-reaction (**3**) from **1** and **2**.

1. $U^{4+} + 2\,H_2O \longrightarrow UO_2^+ + 4\,H^+ + e \qquad\qquad E^0 = -0.38 \text{ V}$

2. $\qquad\qquad UO_2^+ \longrightarrow UO_2^{2+} + e \qquad\qquad E^0 = -0.17$

---

3. $U^{4+} + 2\,H_2O \longrightarrow UO_2^{2+} + 4\,H^+ + 2e \qquad E^0 = -0.28$

$H_2O_2 + 2\,H^+ + 2e \longrightarrow 2\,H_2O \qquad\qquad E^0 = 1.77$

$H_2O_2 + U^{4+} \longrightarrow UO_2^{2+} + 2\,H^+ \qquad\qquad E^0_{cell} = 1.49 \text{ V}$

Note that for a metal in an intermediate oxidation state, such as Mn^{3+}, the reactions for oxidation *and* reduction of H_2O_2 are favorable and should occur, decomposing H_2O_2 to H_2O and O_2 catalytically.

$$MnO_2 \xrightarrow{+0.95} Mn^{3+} \xrightarrow{1.51} Mn^{2+} \tag{16.38}$$

▶ 16.3.6   Oxygen Fluorides

Because O is less electronegative than F, OF compounds are fluorides rather than oxides. These include OF_2, O_2F_2, O_4F_2, the unstable HOF, and the radical O_2F obtained by matrix isolation. Oxygen difluoride is prepared by the reaction of F_2 with dilute (2%) NaOH solution:

$$2 F_2 + 2 NaOH \longrightarrow OF_2 + H_2O + 2 NaF \tag{16.39}$$

The product can be handled in the presence of reducing agents such as H_2 or CH_4, but it reacts violently with other halogens and oxidizes or fluorinates many elements. It hydrolyzes in basic solution, and more slowly in neutral solution.

$$OF_2 + 2 OH^- \longrightarrow O_2 + 2 F^- + H_2O \tag{16.40}$$

The compound HOF is obtained in low yield from the reaction of F_2 and H_2O at low temperature (page 802). For this unusual oxygen compound, assigning the most electronegative element (F) a $-I$ oxidation number and assigning H $+I$ gives O an oxidation number of zero.

Dioxygen difluoride, O_2F_2, is obtained as a yellow-orange solid from mixtures of O_2 and F_2 in a high-voltage electric discharge at low pressure and temperature. It decomposes into the elements at $-50°C$.

Earlier we noted that single bonds between highly electronegative atoms are weak: N—N, O—O, and F—F. The same effect is apparent for O—F bonds. The compound O_2F_2 is interesting because of the unusually short O—O distance and long O—F distances (Table 16.4). The structure resembles that of H_2O_2, but with O—O—F bond an-

Table 16.4   Bond distances for some oxygen and fluorine compounds

Molecule	O—H distance (pm)	O—O distance (pm)	O—F distance (pm)
H_2O	96		
HOF	96		144
OF_2			142
H_2O_2	96	148	
O_2F_2		122	158
O_2F (est.)		(122)	(158)
O_2		121	
Sum of covalent radii		148	145

gles of 109.5° and an 87.5° angle between the planes containing the O—O—F bonds. The O—O bond distance corresponds to a bond order of 2, and the O—F bond corresponds to a fractional bond order. This suggests that the O—O bond is much like that of O_2, with a singly occupied σ orbital of each F atom overlapping one of the the singly occupied π^* orbitals of oxygen to form a weak three-center electron-pair bond. Because of the high electronegativity of F, little electron density is transferred into the π^* orbital to weaken the O—O bond. In the case of H_2O_2, the π^* orbital is occupied fully, decreasing the bond order of the O—O bond to 1. These cases and the related FNO, HNO, and LiNO have been discussed.[7]

► 16.3.7   *Chemistry of Sulfur and Selenium Compounds*

The chemistry of S is very similar to that of Se, and both differ greatly from oxygen. One unique feature of S is its great tendency toward self-linkage (**catenation,** catena, Latin for chain)—for example, S_8 and S_x in polysulfides and in the thionates.

Chalcogen Cations

Sulfur, selenium, and tellurium dissolve in oleum (H_2SO_4/SO_3) to form colored cyclic cations. All three elements form square-planar cyclic X_4^{2+} cations, but S also gives S_8^{2+} (C_{2h}, blue) and S_{16}^{2+} (red), and Se gives Se_8^{2+} (green). The X_8^{2+} rings are folded (Figure 16.4) with one short transannular distance, suggesting a weak bond to give a bicyclic ion. A unique trigonal prismatic (D_{3h}) cluster cation is found for Te_6^{4+} (Figure 16.4).

Hydrogen Compounds of Sulfur and Selenium

The hydrides H_2X can be obtained from the elements at elevated temperature or by reaction of metal sulfides and selenides with acid. They are highly toxic, and the odor of H_2Se

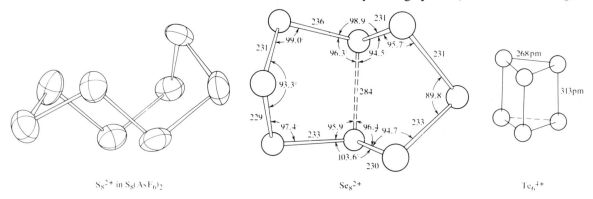

S_8^{2+} in $S_8(AsF_6)_2$ Se_8^{2+} Te_6^{4+}

Figure 16.4   Structures of S_8^{2+}, Se_8^{2+}, and Te_6^{4+} ions. (The Se_8^{2+} structure is reproduced with permission from R. K. McMullan, D. J. Prince, and J. D. Corbett, *Inorg. Chem.* **1971,** *10,* 1749. Copyright 1971, American Chemical Society.)

[7] R. D. Sprately and G. C. Pimintel, *J. Am. Chem. Soc.* **1966,** *88,* 2394.

is even worse than that of H_2S. The strengths of the acids H_2X increase in the order $H_2S < H_2Se < H_2Te$. Hydrogen sulfide dissolves in molten sulfur to form polysulfanes, H_2S_n. We also can produce polysulfanes by acidifying solutions of polysulfide salts at low temperature or by reaction of H_2S or H_2S_2 with SCl_2 or S_2Cl_2.

$$2\,H_2S + SCl_2 \longrightarrow H_2S_3 + 2\,HCl \tag{16.41}$$

$$2\,H_2S + S_2Cl_2 \longrightarrow H_2S_4 + 2\,HCl \tag{16.42}$$

The polysulfanes, H_2S_n, form zigzag chains up to $n = 8$.

Interest in the molten Na–S battery as a high-energy battery has led to the thorough study of the Na–S phase diagram. The species identified are Na_2S, Na_2S_2, Na_2S_3, Na_2S_4, and Na_2S_5. Potassium salts include K_2S_6.

Sulfur–Nitrogen Compounds

Disulfur dichloride reacts with NH_3 in CCl_4 to form S_4N_4 (mp 185°C). The crystals are thermochromic, changing from almost colorless at liquid N_2 temperature to yellow at room temperature and red above 100°C. Although the compound is the starting material for many S—N compounds, it explodes on grinding, by percussion, or by rapid heating. The S_4N_4 molecule consists of a cage (page 878) with an average S—N bond length of 161.6 pm and a long S—S distance (258 pm) that indicates weak bonding (Figure 16.5). Treatment of S_4N_4 with Ag_2F in CCl_4 gives $(NSF)_4$, breaking the weak S—S bond and forming an $(S—N)_4$ ring. Tetrasulfur tetranitride can be vaporized at about 80°C at low pressure and passed through silver wool to form S_2N_2. Disulfur dinitride, which is very nearly square-planar, can be obtained as a crystalline product that explodes on heating.

Disulfur dinitride vapor or single crystals can be polymerized to form crystals of $(SN)_x$, a shiny metallic solid. Polymeric sulfur nitride (polythiazyl) crystallizes as oriented fibers, each consisting of parallel zigzag $(SN)_x$ chains. Crystals of $(SN)_x$ are metallic conductors along the long axis of the crystal—a one-dimensional "metal" containing only S and N. There is great interest in $(SN)_x$ not only because of its one-dimensional conductivity, but also because it becomes superconducting at 0.26 K.

Crystals of high purity $(SN)_x$ are stable indefinitely *in vacuo* at room temperature. The crystals are relatively inert toward O_2 and H_2O, but heating in air causes explosions at ~240°C. In a vacuum, $(SN)_x$ sublimes at ~135°C to form a red-purple linear tetramer $(SN)_4$.

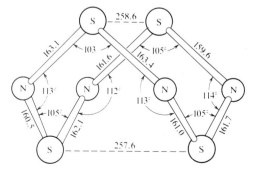

Figure 16.5   Structure of $S_4N_4(C_{2v})$. (From B. D. Sharma and J. Donohue, *Acta Cryst.* **1963,** *16,* 891.)

Sulfur and Selenium Halides

Sulfur forms S_2X_2 with all of the halogens, SF_2 (unstable), S_2F_4 (stable to $-75°C$), SCl_2, SF_4, SCl_4 (which dissociates above $-31°C$), SF_6, and S_2F_{10}. Selenium forms SeF_4, SeF_6, Se_2Cl_2, $SeCl_4$ (which dissociates in the vapor), Se_2Br_2 and $SeBr_2$ (both dissociate in the vapor), and $SeBr_4$. The important fluorides are SF_4 and SF_6. Sulfur dichloride is used as a solvent for S to form dichlorosulfanes, S_nCl_2, in the vulcanization of rubber. Chlorination of molten S gives S_2Cl_2; but in the presence of $FeCl_3$ or I_2 as a catalyst, SCl_2 is formed. The reaction of SCl_2 and NaF in acetonitrile yields SF_4:

$$3\,SCl_2 + 4\,NaF \xrightarrow[75°C]{CH_3CN} SF_4 + S_2Cl_2 + 4\,NaCl \qquad (16.43)$$

It hydrolyzes readily to form SO_2 and HF. It serves as a selective fluorinating agent, converting $\diagdown\!\!C\!=\!O$ into $\diagdown\!\!CF_2$, $-CO_2H$ into $-CF_3$, $-P\!=\!O$ into $-PF_2$, and $\diagdown\!\!P\!-\!OH$ (with $=O$) into $\diagdown\!\!PF_3$. The structure of the SF_4 molecule, as expected from the VSEPR treatment, is derived from a trigonal bipyramid with a lone pair in an equatorial position.

Direct fluorination of S or SO_2 yields SF_6, a colorless, odorless, nontoxic gas (bp $-64°C$) that is quite inert. The hydrolysis of SF_6 is thermodynamically favorable, so the stability toward hydrolysis must arise from kinetic factors—the effective screening of S from nucleophilic attack.

The S—F bond length (156 pm) in octahedral SF_6 is shorter than expected for an S—F single bond. Using a valence bond (VB) description, we would assume sp^3d^2 hybridization, with the possibility of some degree of S—F $d\pi p$ bonding to account for the short bond length. An equivalent molecular orbital (MO) description can be given also. There has been controversy[8] for some time over the participation of outer d orbitals in bonding, because of their high energy and diffuse character. The latter objection is met, in part, by the contraction of the d orbitals about a positive center, such as S bonded to very electronegative atoms. Kiang and Zare[9] determined the stepwise bond-dissociation energies of SF_6 and SF_4, finding that they are relatively high (~360 kJ/mol) for SF_5 and SF_3. The alternation was already known and had been explained in terms of a VB interpretation.

Zare prefers the MO description for SF_4 using p orbitals on S for three-center bonding, as described for XeF_2 (Figure 4.19). Sulfur difluoride is isoelectronic with OF_2, so bonding is localized with only two-center electron-pair bonds. An F atom can be added to form $F^-(SF_2)^+$ such that the F^- (on a line perpendicular to the SF_2^+ plane) is bonded by a two-center, three-electron bond; that is, the third electron is antibonding in the $\sigma*$ orbital. Another F can be added to $(SF_2)^+F^-$ to form a three-center F—S—F bond as shown in Figure 16.6. The bonding three-center MO is occupied by an electron pair, and the second electron pair occupies the nonbonding MO localized on the axial F atoms.

---

[8] K. A. Mitchell, *Chem. Rev.* **1969**, *69*, 157.

[9] T. Kiang and R. N. Zare, *J. Am. Chem. Soc.* **1980**, *102*, 4024.

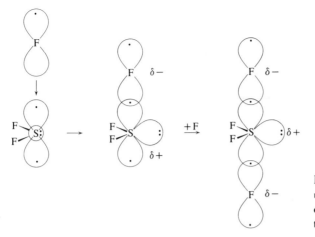

Figure 16.6 S—F bonding using only p orbitals showing the delocalization for SF_4. The electrons do not occupy separate lobes.

Similarly, a fifth F can be added to SF_4 to form a weak two-center, three-electron bond. SF_6 has three equivalent three-center, four-electron bonds. This description explains the low bond-dissociation energies of SF_3 and SF_5. Zare described the three-center, four-electron bonds as strong, but for SF_4, they are weaker than the two-center equatorial S—F bonds. Although the bond order is $\frac{1}{2}$ for each S—F bond of the three-center bond, there is an ionic contribution for SF_4 and SF_6.[10] Presumably, this is sufficiently important to enhance the strength of the S—F bond, with a covalent bond order of $\frac{1}{2}$, to account for a bond length somewhat *shorter* than that expected for a single bond.

Sulfur and Selenium Oxides

Unstable SO and $(SO)_2$ are known, but the important oxides are SO_2, SO_3, and SeO_2. Selenium trioxide is obtained as the anhydride of H_2SeO_4. The burning of S and the roasting of many sulfides in air produce SO_2 (bp $-10.1°C$). Many deposits of coal and oil contain enough S that their use as fuels produces objectionable amounts of SO_2. Selenium dioxide, a white volatile solid, can be obtained by burning Se in air or by the reaction of Se with HNO_3, followed by dehydration of the H_2SeO_3 formed.

Sulfur dioxide is oxidized slowly in air, but the reaction is catalyzed by V_2O_5, Pt sponge, or NO. In the gas phase, SO_3 is planar ($\mathbf{D}_{3h}$) with a bond order of about $1\frac{1}{3}$. In the γ-SO_3 form of solid SO_3 there are cyclic $(SO_3)_3$ molecules, involving three oxygen-bridged SO_4 tetrahedra. A more stable asbestoslike form consists of helical chains of SO_4 tetrahedra. Other forms also have been reported.

[10] R.Maclagan, *J. Chem. Educ.* **1980,** *57,* 428.

Oxoacids of Sulfur and Selenium

Sulfur forms an extensive series of oxoacids. Sulfur dioxide dissolves in water to produce solutions containing sulfurous acid. Although the free H_2SO_3 is not known, salts of HSO_3^- and SO_3^{2-} are stable. Sulfites are oxidized easily to sulfates.

$$SO_4^{2-} + 4H^+ + 2e \rightleftharpoons H_2SO_3 \text{ (or } SO_2 \cdot xH_2O) + H_2O \qquad E^0 = 0.16 \text{ V} \quad (16.44)$$

Adding an excess of SO_2 to $NaHSO_3$ solution gives the disulfite ion, which has an S—S bond and the unsymmetrical structure shown in Figure 16.7 for disulfurous acid.

Sulfur trioxide dissolves in water to produce sulfuric acid. When dissolved in sulfuric acid, it produces disulfuric acid, $H_2S_2O_7$, or higher polysulfuric acids, $H_2S_nO_{(3n+1)}$, involving O-bridged SO_4 tetrahedra. Replacing one oxygen of sulfuric acid by S gives thiosulfuric acid, known in the form of its salts (e.g., $Na_2S_2O_3$). In addition, there are polyacids involving S—S linkages with neither S in a terminal position, including dithionous acid, $H_2S_2O_4$; disulfurous acid, $H_2S_2O_5$; dithionic acid, $H_2S_2O_6$; and polythionic acids, $H_2S_{(n+2)}$ (Figure 16.7).

Anodic oxidation of solutions of HSO_4^- salts produces $S_2O_8^{2-}$, peroxodisulfate ion. Electrolysis is carried out at low temperature and high current density to diminish the extent of oxidation of H_2O to O_2. Hydrolysis of peroxodisulfate ion is one method for preparing H_2O_2. Peroxodisulfate salts are useful strong oxidizing agents,

$$S_2O_8^{2-} + 2e \rightleftharpoons 2SO_4^{2-} \qquad E^0 = 2.1 \text{ V} \qquad (16.45)$$

for example, oxidizing Mn^{2+} to MnO_4^-. Oxidations are often slow unless a catalyst, such as Ag^+, is added. Presumably Ag^{II} is the active oxidant in this case. Hydrolysis of $H_2S_2O_8$

produces [structure] OH (peroxomonosulfuric acid of Caro's acid), which is obtained also

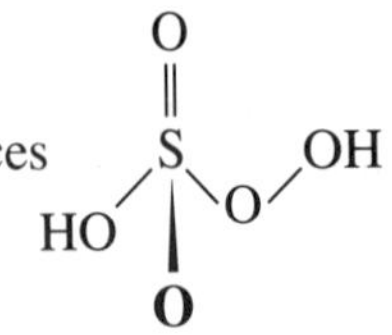

by the reaction of H_2O_2 with H_2SO_4 or HSO_3Cl (chlorosulfuric acid).

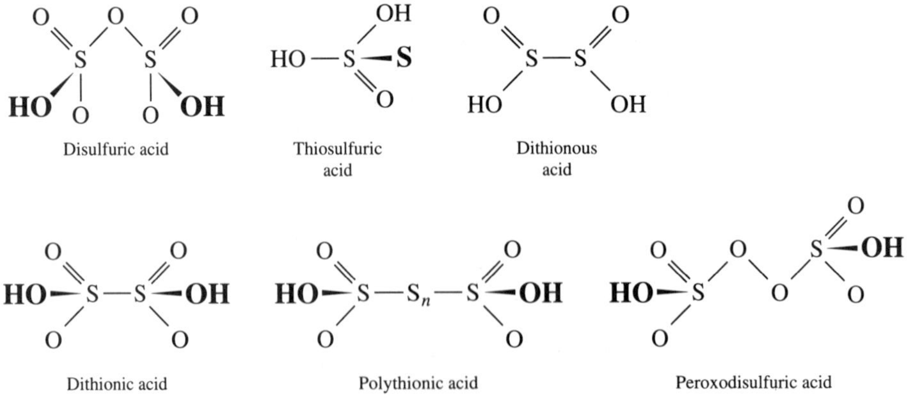

Figure 16.7   Oxoacids of sulfur.

Selenium and tellurium do not form such extensive series of oxoacids; those containing Se—Se or Te—Te bonds are unknown. Selenium dioxide dissolves in H_2O to form selenous acid, H_2SeO_3, which can be obtained as colorless crystals by evaporation of the solution. $H_2SeO_3(s)$ decomposes into SeO_2 and water on warming. Selenous acid is reduced easily to Se by mild reducing agents such as HI, SO_2, or N_2H_4. Concentrated solutions of $HSeO_3^-$ contain the $Se_2O_5^{2-}$ ion, with a symmetrical structure, unlike $S_2O_5^{2-}$ (Figure 16.7), with Se—O—Se bonding. Very strong oxidizing agents (O_3, MnO_4^-, and H_2O_2) oxidize H_2SeO_3 to H_2SeO_4. Crystalline selenic acid (mp 58°C) can be obtained from aqueous solution by evaporation at low pressure. Selenic acid is a strong acid and a rather strong oxidizing agent.

$$SeO_4^{2-} + 4H^+ + 2e \;\rightleftharpoons\; H_2SeO_3 + H_2O \qquad E^0 = 1.15 \text{ V} \qquad (16.46)$$

Sulfuric acid, the chemical produced in largest quantity (over 44 million tons annually in the United States), is obtained by the catalytic oxidation of SO_2 to SO_3 (contact process). Sulfur trioxide is water-soluble but tends to produce aerosols, so it is dissolved in concentrated H_2SO_4 to produce fuming sulfuric acid, oleum, which is diluted to produce H_2SO_4. In the manufacture of fertilizers and other chemicals, sulfuric acid is used whenever a strong acid is needed, except when its weak oxidizing character is a limitation. H_2SO_4 is a good drying agent—particularly for nonreactive gases, which can be bubbled through the liquid.

Thiosulfate salts are obtained by boiling aqueous solutions of sulfites with either H_2S or S:

$$8\,Na_2SO_3 + S_8 \;\longrightarrow\; 8\,Na_2S_2O_3 \qquad (16.47)$$

Sodium thiosulfate is used as a "fixer" in developing photographic film or prints, since it dissolves the AgBr not reduced by the developer:

$$AgBr + 2\,S_2O_3^{2-} \;\longrightarrow\; [Ag(S_2O_3)_2]^{3-} + Br^- \qquad (16.48)$$

Thiosulfates are useful as analytical reducing agents, particularly for the titration of I_2. The oxidation product is the tetrathionate ion:

$$2\,S_2O_3^{2-} + I_2 \;\longrightarrow\; S_4O_6^{2-} + 2I^- \qquad (16.49)$$

Dithionite ion ($S_2O_4^{2-}$) salts are obtained by using Zn to reduce solutions of sulfites, containing an excess of SO_2; they are oxidized easily by O_2. Oxidation of sulfites, by a mechanism not involving oxygen transfer, produces dithionates ($S_2O_6^{2-}$):

$$MnO_2 + 2\,SO_3^{2-} + 4H^+ \;\longrightarrow\; Mn^{2+} + S_2O_6^{2-} + 2H_2O \qquad (16.50)$$

Dithionate salts are stable with respect to reactions with water and with most oxidizing and reducing agents (kinetic, not thermodynamic, stability).

The polythionic acids, $H_2S_{(n+2)}O_6$, are unstable with respect to decomposition to form S and a variety of oxides and oxoacids of S. Polythionate salts are well characterized for $n = 1$ to 4, with much higher values of n reported. Polythionates are related to the dithionates only in the formal sense that the formula for dithionate ion, $S_2O_6^{2-}$, corresponds to $S_{(n+2)}O_6^{2-}$ with $n = 0$. Preparation of tetrathionate ion from $S_2O_3^{2-}$ and I_2 was described above. Oxidation of thiosulfate salts by H_2O_2 produces trithionate salts:

$$2\,S_2O_3^{2-} + 4\,H_2O_2 \longrightarrow S_3O_6^{2-} + SO_4^{2-} + 4\,H_2O \tag{16.51}$$

The reactions of $S_2O_3^{2-}(aq)$ with H_2S and SO_2 produces polythionates of varying chain length.

The rather complex chemistry of S provides many examples of the importance of reaction mechanisms in determining the stability of a compound (e.g., SF_6 and $S_2O_6^{2-}$) and the reaction products (e.g., oxidation of $S_2O_3^{2-}$ gives different products with I_2 compared with H_2O_2).[11]

► 16.3.8 *Redox Chemistry of Sulfur and Selenium*

Figures 16.8 and 16.9 show the Pourbaix (E/pH) diagrams for S and Se, respectively.

$$E_A^0 \qquad SO_4^{2-} \xrightarrow{\ 0.16\ } H_2SO_3 \xrightarrow{\ 0.45\ } S \xrightarrow{\ 0.14\ } H_2S \tag{16.52}$$

$$SeO_4^{2-} \xrightarrow{\ 1.1\ } H_2SeO_3 \xrightarrow{\ 0.74\ } Se \xrightarrow{\ -0.11\ } H_2Se \tag{16.53}$$

$$E_B^0 \qquad SO_4^{2-} \xrightarrow{\ -0.94\ } SO_3^{2-} \xrightarrow{\ -0.66\ } S \xrightarrow{\ -0.45\ } S^{2-} \tag{16.54}$$

$$SeO_4^{2-} \xrightarrow{\ 0.03\ } SeO_3^{2-} \xrightarrow{\ -0.36\ } Se \xrightarrow{\ -0.67\ } Se^{2-} \tag{16.55}$$

(For the discussion of Pourbaix diagrams, see Section 8.6.) In these diagrams a horizontal or sloping line represents the potential for two species in equilibrium as a function of pH. Solid lines separate thermodynamically stable species. The dashed lines are for the oxidation of H_2O (to O_2) and the reduction of H_2O (to H_2). The vertical lines separate the predominant species involved in acid–base equilibria. Figure 16.8 demonstrates that under reducing conditions, H_2S, HS^-, and S^{2-} are the stable species. H_2 is a strong enough reducing agent (thermodynamically) to reduce S or any other S compound. Under oxidizing conditions, H_2SO_4 and its salts are stable. $S_2O_8^{2-}$ is produced only at such high potentials that H_2O is oxidized. The element has a narrow range of stability, and H_2SO_3 does not appear because it disproportionates [see the potential diagrams (16.52 and 16.54)–(16.55)].

The diagram for Se (Figure 16.9) is more complex than that for S. H_2SeO_3 and its salts are stable, and elemental Se has a wide range of stability. A dashed line represents the reduction of H_2SeO_3 to H_2Se, because this is within the region of stability for Se and H_2SeO_3 will oxidize H_2Se to Se (see Problem 16.21). H_2Se is a strong enough reducing agent to reduce water, but the reaction is slow, because of the H_2 overvoltage.

---

[11] For thorough coverage of oxoacids of S see N. N. Greenwood and A. Earnshaw, *Chemistry of the Elements*, Pergamon, Oxford, 1984, pp. 834–854.

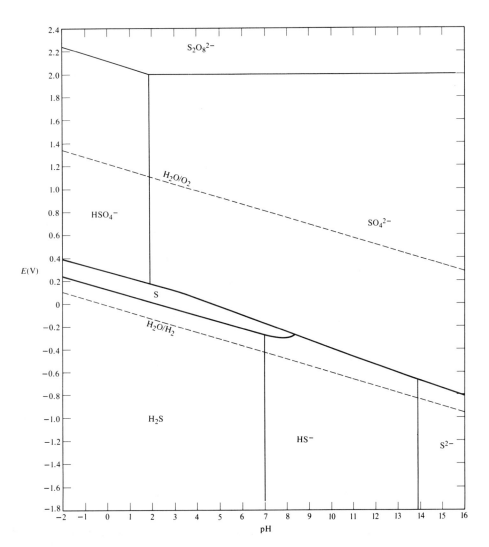

Figure 16.8 Pourbaix diagram for sulfur. [Adapted with permission from M. J. N. Pourbaix, in *Atlas of Electrochemical Equilibria in Aqueous Solution* (English translation, 2nd ed., by J. A. Franklin, National Association of Corrosion Engineering, Houston, TX, 1974, p. 551).]

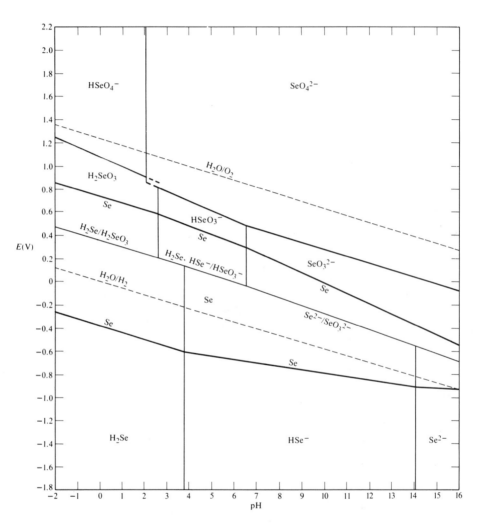

Figure 16.9 Pourbaix diagram for selenium. [Adapted with permission from M. J. N. Pourbaix, in *Atlas of Electrochemical Equilibria in Aqueous Solution* (English translation, 2nd ed., by J. A. Franklin, National Association of Corrosion Engineering, Houston, TX, 1974, p. 557).]

16.4   THE HALOGENS

► *16.4.1   Family Trends*

Table 16.5 lists some chemical properties of the halogens, and Table 16.6 gives physical properties for X_2. Note that the difference between F and Cl generally is greater than the differences between the succeeding heavier halogens. Of the halogens, Cl and Br are most similar in chemical properties. That the crystal radius of the Br^- ion is only slightly greater than that of Cl^- ion may be attributed to incomplete nuclear shielding that occurs as $3d$ electrons are added (the transition series contraction). The small difference in electronegativity between Br and Cl indicates that the polarity should be very similar for bonds formed by these elements with another element. The large chemical differences between F and Cl, and similarities between Cl and Br, are illustrated in pK_a values of HX (Table 16.7), the oxidizing ability of X_2 (Table 16.5), solubilities of the silver halides (page 238) and the alkali metal halides (page 237), oxidation states of known binary compounds (page 741), structures of compounds, and hydrogen bond strength of XHX^-.

The high oxidizing ability of F_2, together with the small atomic radius, permits it to form many compounds in which the elements exhibit their highest oxidation state. Thus there are 16 known binary hexafluorides, but no known hexaiodides. On the other hand, the lack of low-lying empty d orbitals prevents F from forming more than one covalent bond other than a dative bond with F the donor.

Crystal structures of the halides depend not only on the radius ratio of cation to anion, but also on the electronegativity difference of the metal and the halogen and on the polarizabilities of each. For compounds of the types MX, MX_2, or MX_3, fluorides have structures anticipated from the radius ratio, since F is sufficiently electronegative to form essentially ionic bonds with metals in low oxidation states, and F^- is not very polarizable. Chlorides, bromides, and iodides of MX_2- and MX_3-type compounds show the effect of lower electronegativities and greater polarizabilities by the formation of layer structures (page 218). Covalent bond formation or strong polarization of the halide ion is necessary to account for these structures involving adjacent layers of halide ions without cations between them.

We can also relate the colors of many of the halides and halide complexes to the polarizability of the anions. Thus, the colors of the TiX_6^{2-} ions vary from the fluoro complex (colorless) to the chloro complex (yellow), to the bromo complex (red), and to the iodo complex (deep red). The light absorption results from the excitation of an electron essentially in a halogen orbital to a metal orbital. The energy of this charge-transfer transition is lowest for the iodo complex (see page 463).

► *16.4.2   Occurrence*

The halogens occur in combined form in the earth's crust in the approximate amounts F, 0.08%; Cl, 0.05%; Br, 0.001%; I, 0.001%. Chlorides and bromides are concentrated in the oceans by leaching processes of natural waters (see Table 18.1, page 889). Large deposits of NaCl and, to a lesser extent, Ca, K, and Mg chlorides are found in dried-up beds

Table 16.5 Some properties of the halogens

	F	Cl	Br	I	At
Crystal radius (X^-, pm)	119	167	182	206	
Covalent radius (pm)	71	99.4	114.1	133.3	
Electronegativity	4.0	3.0	2.8	2.5	2.2
First ionization (MJ/mol)	1.6810	1.2511	1.1399	1.0084	0.92
energy (eV)	17.422	12.967	11.814	10.451	9.5
Outer electron configuration	$2s^2 2p^5$	$3s^2 3p^5$	$3d^{10} 4s^2 4p^5$	$4d^{10} 5s^2 5p^5$	$4f^{14} 5d^{10} 6s^2 6p^5$
$E^0 \frac{1}{2} X_2 + e \rightarrow X^-$ (V)	+2.87	+1.36	+1.07	+0.535	+0.2
$E_B^0 XO^- + e \rightarrow \frac{1}{2} X_2$ (V)	—	+0.40	+0.45	+0.45	0.0
Dissociation energy (kJ/mol, X_2)	154.6	239.23	190.15	148.82	115.9
Electron affinity (kJ/mol)	328.0	349.0	324.7	295.1	270
Polarizability (cm^3/atom)	1.04×10^{-24}	3.66×10^{-24}	4.77×10^{-24}	7.10×10^{-24}	

Table 16.6 Properties of the halogens, X_2

	Fluorine	Chlorine	Bromine	Iodine
Isotopes	19	35,37	79,81	127
Melting point (°C)	−223	−102.4	−7.3	+113.7
Boiling point (°C)	−187.9	−34.0	+58.8	+184.5
Density (g/cm³)	1.108[a]	1.57[a]	3.14	4.94
Color of gas	Light yellow	Greenish yellow	Reddish brown	Violet

[a]For the liquid at the boiling point.

of landlocked lakes, where Ca and Mg sulfates and mixed salts of KBr and $MgBr_2$ also occur. Because of differences in concentrations and temperature during deposition, different salts often are well separated. Limiting the fluoride content of seawater is the large concentration of Ca^{2+} present and the low solubility of CaF_2. The low solubility of CaF_2, compared with the other calcium halides, stems from the high lattice energy of CaF_2, resulting from the small radius of F^-. Natural deposits of CaF_2, called **fluorspar** or **fluorite,** serve as the primary source of fluorine. Cryolyte (Na_3AlF_6), another commercially important fluoride mineral, is used as the molten electrolyte in the production of Al by electrolysis.

Although the concentration of iodine in the ocean is small, it is absorbed selectively by, and can be obtained from, seaweed. More important sources are sodium iodate and sodium periodate, which occur in deposits in Chile. The natural occurrence of these oxoanions for I contrasts sharply with the lack of natural deposits of oxo salts of Cl and Br and illustrates the greater ease with which I attains the higher oxidation states, compared with the lighter halogens.

Astatine occurs naturally in very small amounts as the beta decay product of ^{215}Po, ^{216}Po, and ^{218}Po. Because polonium itself is rather rare, and α decay is the usual decay route giving Pb, At has not been isolated from natural sources. At is produced by α bombardment of bismuth, $^{209}Bi(\alpha,2n)^{211}At$. Astatine itself is radioactive: Its longest-lived isotope has a half-life of 8.3 hr.

Table 16.7 Properties of the hydrogen halides

	HF	HCl	HBr	HI
Melting point (°C)	−83.07	−114.19	−86.86	−50.79
Boiling point (°C)	19.9	−85.03	−66.72	−35.35
Density at bp (g/cm³)	0.991	1.187	2.160	2.799
Dielectric constant of liquid	66	9	6	3
Percent dissociation at 1000°C	—	3×10^{-7}	0.003	19
Composition of azeotrope with water at 1 atm (wt-%)	35.37	20.24	47	57.0
pK_a values (aqueous)	3.17	−7	−9	−10

▶ *16.4.3 Preparation of the Elements*

Because F, Cl, and Br occur in nature as halides, preparing the elements involves suitable oxidation reactions. Electrolytic oxidation of fused KF–HF adducts (page 98) is used to produce F_2. HF is removed from the F_2 by reaction with NaF. The cells are constructed of copper, monel, or steel with amorphous carbon anodes. Graphite electrodes cannot be used because graphite compounds are formed. A laboratory scale for the preparation of F_2 has been reported using the strong Lewis acid SbF_5 and K_2MnF_6 to liberate the unstable MnF_3 and F_2.[12]

$$K_2MnF_6 + 2\,SbF_5 \longrightarrow 2\,KSbF_6 + MnF_3 + \tfrac{1}{2}F_2 \qquad (16.56)$$

Chlorine is produced by electrolysis of either fused NaCl or concentrated aqueous solutions of NaCl, the latter being the more commonly used method:

$$Na^+ + Cl^- + H_2O \xrightarrow{\text{electr.}} Na^+ + OH^- + \tfrac{1}{2}Cl_2 + \tfrac{1}{2}H_2 \qquad (16.57)$$

In the laboratory, Cl_2 can be made by the action of oxidizing agents such as MnO_2 on HCl.

Bromine is produced commercially from the bromides in seawater by the oxidizing action of Cl_2. After the pH of the seawater is adjusted to a value between 1 and 4 with sulfuric acid and then treated with chlorine, the liberated Br_2 is blown out by an air stream and the bromine concentrated by a sequence of reactions, such as absorption in a carbonate solution and subsequent acidification. Br_2 disproportionates in the basic carbonate solution, and the reaction reverses in acidic solution.

$$Cl_2 + 2\,Br^- \longrightarrow Br_2 + 2\,Cl^- \qquad (16.58)$$

$$3\,Br_2 + 3\,CO_3^{2-} \longrightarrow 5\,Br^- + BrO_3^- + 3\,CO_2 \qquad (16.59)$$

$$5\,Br^- + BrO_3^- + 6\,H^+ \longrightarrow 3\,Br_2 + 3\,H_2O \qquad (16.60)$$

Cl_2 is removed from the Br_2 produced by reaction with a bromide such as iron(III) bromide.

Sulfite reduction of iodates yields I_2 which is purified by sublimation.

$$2\,NaIO_3 + 5\,NaHSO_3 \xrightarrow{\text{aq. soln.}} 3\,NaHSO_4 + 2\,Na_2SO_4 + H_2O + I_2 \qquad (16.61)$$

▶ *16.4.4 The Hydrogen Halides*

The hydrogen halides can be prepared by the action of a nonvolatile, nonoxidizing acid on a halide salt. Sulfuric acid serves well for the preparation of HF and HCl, but it oxidizes Br^- and I^-.

[12] K.O. Christe, *Inorg. Chem.* **1986**, *25*, 3721.

$$CaF_2 + H_2SO_4 \longrightarrow CaSO_4 + 2\,HF \qquad (16.62)$$

$$NaCl + H_2SO_4 \longrightarrow NaHSO_4 + HCl \qquad (16.63)$$

Phosphoric acid can be used for the preparation of HBr and HI. Commercially, the direct reaction of Cl_2 with H_2 is most important. Although all of the hydrogen halides can be obtained commercially, they usually contain appreciable amounts of impurities. HF, HCl, and HI usually contain H_2 from reaction with the metal cylinder. Hydrogen iodide usually contains I_2 in addition to H_2, from thermal decomposition of HI. For laboratory use, the best source of HF is the thermal decomposition of dry KHF_2. Hydrogen iodide is usually prepared by dropping the aqueous solution of HI on phosphorus pentaoxide.

Table 16.7 gives some of the properties of the hydrogen halides and their aqueous solutions. The following topics relating to the hydrogen halides have already been discussed: the effect of hydrogen bonding on the boiling point of HF (page 92), the ability to form hydrogen bonds with halide ions (page 97), and the acid strength of the hydrohalic acids (page 315). Further specific discussion will be limited to HF as a solvent system.

Hydrogen fluoride is highly toxic and should be handled only in a good fume hood. Because it reacts with glass, it is handled in metal systems, especially copper or monel. It is characterized by its dehydrating ability with its strong acidity. Because of its high dielectric constant and low viscosity, most of its solutions show high conductivity, although the pure liquid itself has a conductivity comparable with that of distilled water. Many of the reaction products formed upon dissolving substances in HF have been inferred from conductivities. Some typical reactions are

$$HNO_3 + HF \longrightarrow H_2NO_3^+ + F^- \qquad (16.64)$$

$$H_2SO_4 + 2\,HF \longrightarrow HSO_3F + H_3O^+ + F^- \qquad (16.65)$$

$$NaCl + HF \longrightarrow NaF + HCl \qquad (16.66)$$

$$ZnO + 3\,HF \longrightarrow ZnF_2 + H_3O^+ + F^- \qquad (16.67)$$

Although HF is a weak acid in water, anhydrous HF compares with anhydrous H_2SO_4 as a strong acid. Note that in Reaction (16.64), HNO_3 is a *base* in HF(*l*). The *only* known acids in anhydrous HF are fluoride acceptors such as NbF_5, SbF_5, AsF_5, and BF_3, the strongest of which is SbF_5. Adding metal fluorides to such solutions precipitates the metal hexafluoroantimonate(V).

▶ *16.4.5   Reactions of the Halogens*

When halogens react with other elements, the products often depend on the temperature and pressure at which the reaction is carried out. We can grasp the effect of these variables most readily by means of idealized phase diagrams.

Figure 16.10 shows generalized pressure–composition isotherms expected for a halogen–metal system, in which the MX_y compounds are assumed to be nonvolatile and

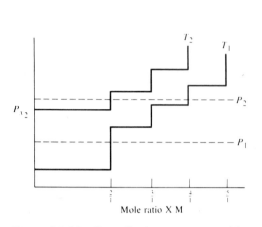

Figure 16.10 Generalized pressure–composition isotherms for the system $M(s)$–$X_2(g)$.

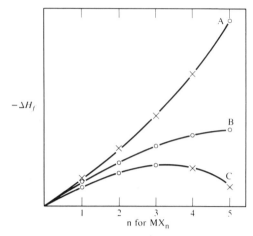

Figure 16.11 Stability curves from halides: $\times$, unstable compositions; $\circ$, stable compositions. (From L. H. Long, *Q. Rev.* **1953**, 7, 134.)

no solid solutions are formed. The vertical sections of the isotherms indicate the existence of a single solid compound (with the stoichiometry indicated on the abscissa) in equilibrium with gaseous X_2. The horizontal sections or plateaus indicate the coexistence of two solid phases in equilibrium with gaseous X_2. As the pressure of X_2 is increased, for a given equilibrium temperature a higher halide is produced. As the equilibrium temperature is increased, for a given pressure of X_2 a lower halide is produced. From the phase diagram, heating X_2 and M together at a pressure of P_2 produces MX_2 at the higher temperature T_2 and produces MX_4 at the lower temperature T_1. MX_2 may also be formed at the lower temperature T_1 by reducing the pressure of X_2 to P_1. Because the metal can coexist in equilibrium with only the lowest halide, heating the metal with an excess of a particular metal halide with no additional halogen produces some of the next-lower halides at equilibrium. If equilibrium is attained rapidly, it may be necessary to quench the products to keep other phases from forming.

The temperature range within which the various platinum chlorides may exist at 1 atm Cl_2 is as follows: $PtCl_4$, below 370°C; $PtCl_3$, 370–435°C; $PtCl_2$, 435–481°C; PtCl, 481–482°C; Pt, above 482°C.

Although the discussion here has used the systems M–X_2 as an example, similar considerations apply to other gas–solid systems in which the products are nonvolatile: $O_2(g)$–$M(s)$, $NH_3(g)$–$MX_y(s)$, $H_2O(g)$–$MO_y(s)$, and so on. When manganese oxides are ignited in a Bunsen flame in air, Mn_3O_4 is produced. If the partial pressure of O_2 is increased from 0.2 to 1 atm by using pure O_2, Mn_2O_3 is formed; at still higher O_2 pressures, MnO_2 is formed.

We can make some predictions as to whether a compound might be stable with respect to disproportionation from known and/or estimated heats of formation of the halides. The heat of the reaction

$$2\,MX \longrightarrow M + MX_2 \tag{16.68}$$

can be found from

$$\Delta H = \Delta H_f(MX_2) - 2\Delta H_f(MX) \tag{16.69}$$

If ΔH is negative, then ΔG probably is also negative and MX is unstable with respect to disproportionation. More generally, if we plot $-\Delta H_f$ against the number of halogen atoms in MX_y, as in Figure 16.11, we find that only the highest halide is stable if the curve has an increasing rate of ascent (A); that all halides are stable if the curve has a decreasing rate of ascent but no maximum (B); and that if a maximum does occur, only those halides at and preceding the maximum are stable (C). These situations are often approximated by the transition metal fluorides (A), the chlorides (B), and the bromides and iodides (C). (See Table 15.10.)

Preparation of the Binary Metal Halides

The binary halides of the transition metals are listed in Table 15.10. The metal halides formed in the solid state for Groups 1–3 are those expected—that is, MX, MX_2, and MX_3, respectively.

The preparation of MX or MX_2 compounds usually may be carried out by treating an oxide or carbonate of M with aqueous HX, evaporating the solution to obtain the metal halide hydrate, and then removing the water by means of a suitable desiccant—or if necessary to prevent hydrolysis, in a stream of HX. The dehydration step often is troublesome, and many specific procedures have been developed for individual preparations not involving aqueous solutions. Several preparations use the metal hydride as starting material:

$$2\,LiH(xs) + I_2 \xrightarrow{\text{ether}} 2\,LiI + H_2 \tag{16.70}$$

$$BaH_2(xs) + 2\,NH_4I \xrightarrow{\text{pyridine}} BaI_2 + 2\,NH_3 + H_2 \tag{16.71}$$

The excess hydride is insoluble in the solvent and may be removed by filtering.

Uranium(III) halides are made conveniently by reaction of the hydrogen halide with uranium hydride:

$$2\,U + 3\,H_2 \xrightarrow{250°C} 2\,UH_3 \tag{16.72}$$

$$UH_3 + 3\,HX \xrightarrow{250-300°C} UX_3 + 3\,H_2 \quad X = F, Cl \tag{16.73}$$

Anhydrous $FeCl_2$ has been obtained by the reaction

$$C_6H_5Cl + FeCl_3 \longrightarrow C_6H_4Cl_2 + FeCl_2 + HCl \tag{16.74}$$

Compounds of the type MX_3, MX_4, and MX_5 usually are obtained by one of the following reactions:

$$M + X_2 \longrightarrow MX_n \tag{16.75}$$

$$M + HX \longrightarrow MX_n + H_2 \tag{16.76}$$

$$MO + C + X_2 \longrightarrow MX_n + CO \tag{16.77}$$

$$MO + CCl_4 \longrightarrow MCl_n + CO + COCl_2 \tag{16.78}$$

$$MO + S_2Cl_2 \longrightarrow MCl_n + SO_2 \tag{16.79}$$

Because the direct reaction of the metal with the halogen is vigorous, an inert solvent medium often is used with Br_2 or I_2, the solvent serving as a diluent. Often, the vapor of the halogen is used, and for further safety a noble gas sometimes is used as a diluent. The **van Arkel process** for purifying metals (Ti, Hf, Zr, V, W, etc.) takes advantage of (1) the formation of a volatile halide by direct reaction of the halogen with the metal and (2) the decomposition of the metal halide to the metal and halogen at a higher temperature (depositing on a hot wire):

$$2M + nX_2 \xrightarrow{\ (1)\ } 2MX_n \xrightarrow{\ (2)\ } 2M + nX_2 \tag{16.80}$$

The reaction of the metal directly with anhydrous HX is useful for metals above hydrogen in the activity series (those with positive emfs for oxidation of the metal). $AlCl_3$ and $CrCl_2$ are made in this way. In the reactions involving C, S_2Cl_2, or CCl_4, we can presume that reduction to metal occurs, followed by halogenation.

Where a series of halides exists with the metal in different oxidation states, a lower halide can be produced by hydrogen reduction of the higher halide.

▶ 16.4.6 *Interhalogen Compounds and Ions*

The halogens form all combinations of XX′ compounds, of which only IF is unstable. IF is detectable spectroscopically, but it disproportionates to form I_2 and IF_5. Most of the XX′ compounds can be oxidized further to form XX_n' compounds, as shown in Table 16.8. Valence bond descriptions have been given (page 57), and these compounds were used as examples for the VSEPR approach. ICl_3 exists as a dimer, giving planar ICl_4 units. The planar structure can be predicted from VSEPR theory once we know that there are two bridging Cl.

The interhalogen compounds can serve as halogenating agents. The oxidizing power increases with increasing oxidation number in a series such as $BrF < BrF_3 < BrF_5$. Among the commercially available fluorinating agents the oxidizing power decreases with softer (less electronegative) partners:

Table 16.8   Interhalogen compounds and ions (boiling or melting points given for the neutral compounds)

	Interhalogens		Cations	Anions
ClF	−101°C	(bp)		ClF_2^-
ClF_3	11.8	(bp)	ClF_2^+	ClF_4^-
ClF_5	−14	(bp)	ClF_4^+	
BrF	20	(bp)		BrF_2^-
BrF_3	126	(bp)	BrF_2^+	BrF_4^-
BrF_5	41	(bp)	BrF_4^+	BrF_6^-
(IF)	—			IF_2^-
IF_3	—		IF_2^+	IF_4^-
IF_5	101	(bp)	IF_4^+	IF_6^-
IF_7	Subl. 40		IF_6^+	IF_8^-
BrCl	5	(bp)		$BrCl_2^-$
ICl	27	(mp)		ICl_2
$(ICl_3)_2$	—		ICl_2^+	ICl_4^-
IBr	41	(mp)		IBr_2^-

$$F_2 > ClF_5 > ClF_3 > BrF_3 > IF_5 > SF_4 > AsF_5 > SbF_5 > AsF_3 > SbF_3$$

Scheme 16.3

This is the thermodynamic order of oxidizing power, not the order of the rate of reaction. Thus SF_6 would probably be better as an oxidizing agent than IF_5, based on thermodynamic considerations, but SF_6 is inert, because of the lack of a low-energy path for the reaction. In similar fashion, ClF_3 is more reactive (greater *rate* of reaction) than F_2 with most substances, despite its lower oxidizing power. This may be attributed to the lower energy needed to break the F_2Cl—F bond compared with the F—F bond. ClF_3 reacts explosively with substances such as cotton, paper, wax, and stopcock grease! ClF_3 has been used as a fuel for short-range rockets. Where hydrazine (N_2H_4) is the fuel to be oxidized, both reactants may be stored without refrigeration. This mixture is a hypergolic fuel; that is, it ignites spontaneously on being mixed.

Interhalogen Halides

Almost all of the interhalogen compounds react with metal halides to form the $XX_n'^-$ ions (Table 16.8), as well as "mixed" anions such as I_2Cl^-, I_4Cl^-, ICl_3F^-, and so on. I_2Cl^- is described as a "mixed" anion because the more electropositive element (I) is the central atom, giving the unsymmetrical I—I—Cl$^-$ with nonequivalent iodines. The less stable anions can be stablized in the solid as salts of large cations (page 234). Because ions such as IF_2^- and IF_4^- have been known for a long time, and are isoelectronic with XeF_2 and XeF_4, respectively, it seems surprising, in retrospect, that the Xe compounds were not sought earlier. We have considered bonding in XeF_2 earlier (page 176). The $XX_2'^-$ ions are linear and $XX_4'^-$ ions are planar (page 81). The IF_6^- ion shows the expected distortion from octahedral symmetry, although for the smaller Br the lone pair is stereochemically inactive, giving a centrosymmetric BrF_6^- (isoelectronic with XeF_6; see Section 16.5.4).

In addition to the interhalogen halides, the halogens form polyhalide ions, X_n^-. The triiodide ion, I_3^-, is familiar for the use of KI to increase the amount of I_2 that can be dissolved in aqueous solution. Ions Cl_3^- and Br_3^- are known, but iodine gives MI_n salts with n as high as 9. I_n^- ions involve bonding between I_2 molecules and I^-. The I_5^- ion consists of a zigzag chain of two I_2 molecules bonded to I^- (C_{2v}).

Cations

Oxidation of Br_2 and I_2 by $S_2O_6F_2$ gives the salts $X_2^+(SO_3F^-)$. Also known are the cations Cl_3^+, Br_3^+, and I_3^+, which can be obtained in oleum (H_2SO_4/SO_3). Table 16.8 lists known interhalogen cations.

▶ *16.4.7   Pseudohalogens*[13]

Analogy plays an important role in descriptive chemistry. The ionic groups given in Table 16.9 behave chemically very much like halide ions and are often called **pseudohalides.** The approximate order of electronegativities of the pseudohalogens and the halogens is $F > N_3 > CN > Cl > NCS > Br > I$. The order of strength as oxidizing agents is approximately $F_2 > Cl_2 > Br_2 > (SCN)_2 > (CN)_2 > I_2 > (SeCN)_2$.

Electrolysis or chemical oxidation may be used to prepare many of the pseudohalogens from their ions.

$$4\,HSCN + MnO_2 \longrightarrow (SCN)_2 + Mn(SCN)_2 + 2\,H_2O \qquad (16.81)$$

Similarities to the halogens appear in reactions such as hydrolysis, addition to carbon–carbon double bonds, formation of complex ions and interpseudohalogen compounds, and insolubility of silver, lead, and Hg^I salts.

Table 16.9   The pseudohalogens

Ion	Dimeric molecule	Melting point of X_2, °C	pK_a of HX
N_3^- (azide)	—	—	4.55
CN^- (cyanide)	$(CN)_2$ (cyanogen)	mp -27.9 bp -21.17	9.21
OCN^- (cyanate)	$(OCN)_2$ (oxocyanogen)	—	3.92
SCN^- (thiocyanate)	$(SCN)_2$ (thiocyanogen)	-2 to $-3°$	(-0.74)
$SeCN^-$ (selenocyanate)	$(SeCN)_2$ (selenocyanogen)	Yellow powder	—

[13] A. M. Golub, H. Köhler, and V. V. Skopenko, Eds., *Chemistry of Polyhalides*, Elsevier, Amsterdam, 1986.

$$(CN)_2 + 2\,OH^- \longrightarrow OCN^- + CN^- + H_2O \qquad (16.82)$$

$$CH_2CH_2 + (SCN)_2 \longrightarrow NCSCH_2CH_2SCN \qquad (16.83)$$

$$Ag^+ + CN^- \longrightarrow AgCN(s) \qquad (16.84)$$

$$AgCN(s) + CN^- \longrightarrow [Ag(CN)_2]^- \qquad (16.85)$$

Because the pseudohalides, like the halides themselves, have more than one pair of electrons with which coordinate bonding can take place, they can serve as bridging groups. However, the bridging may result in a different degree of polymerization of the parent compound, because of the different geometric orientation of the available electron pairs. Thus, although dialkylgold chloride and bromide are dimeric, the cyanide is tetrameric.

In each case the geometry about the gold atom is essentially square-planar.

Other differences between the halogens and the pseudohalogens occur in the polymerization of many of the pseudohalogens under heating:

$$x(CN)_2 \xrightarrow{\;400^\circ C\;} (CN)_x \qquad (16.86)$$

$$x(SCN)_2 \xrightarrow{\;\text{room temp.}\;} (SCN)_x \qquad (16.87)$$

▶ 16.4.8   Chlorine Oxofluorides

The known chlorine oxofluorides form a nearly complete series of neutral, cationic, and anionic species (Figure 16.12). The Cl^{III} species ClF_2^+, ClF_3, and ClF_4^-, although not oxofluorides, are included to complete the series. Similarly, ClF, ClF_2^-, and Cl^+ (not known as a simple cation) could be added to complete the $FClO$, $FClO_2$, $FClO_3$ series. Note that here the formulas are written with the central atom (Cl) between the atoms bonded to it. This avoids possible confusion with isomers—for example, $FClO$ and $ClOF$, sometimes called chlorine hypofluorite.

The geometries of the species in Figure 16.12 are those expected from VSEPR considerations, with mutual repulsion decreasing in the order lone pair > double-bonded oxygen > fluorine. Detailed structures are known only for $FClO_2$ and $FClO_3$, but vibrational spectra are available for most of the compounds. Force constants reveal that the Cl—O bonds have varying degrees of double-bond character, whereas the Cl—F bonds appear to

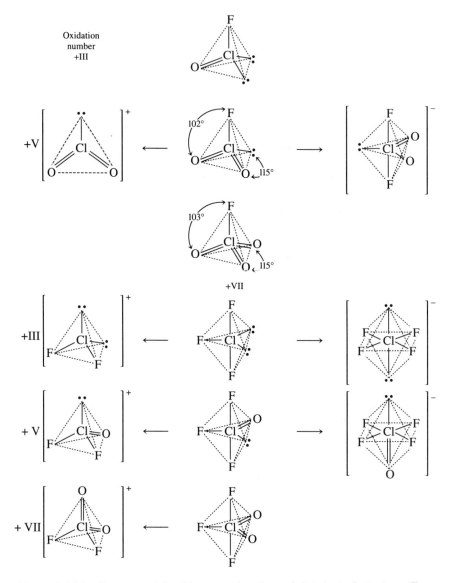

Figure 16.12 Structures of the chlorine oxofluorides and their ionic derivatives. (From K. O. Christe and C. J. Schack, *Adv. Inorg. Chem. Radiochem.* **1976,** *18,* 319.)

be of three types: (a) ordinary two-center covalent bonds; (b) three-center, four-electron bonds (as described for XeF_2; see page 176) for $F_2ClO_2^-$ and F_4ClO^- and for the axial F atoms in F_3ClO and F_3ClO_2; and (c) polar $p\sigma\pi^*$ three-center, two-electron bonding (as described for O_2F_2; see page 779) for $FClO_2$ and $FClO$.

Table 16.10 summarizes the properties of, as well as the syntheses for, the chlorine oxofluorides and their ionic derivatives. The oxofluorides are rather similar to the chlorine fluorides. As liquids, they undergo little self-ionization. They form stable adducts with strong Lewis acids (primarily fluorides such as BF_3, PF_5, and MF_5) or Lewis bases such as

Table 16.10   Synthesis and properties of chlorine oxofluorides and their cationic and anionic derivatives

Species	Properties	Synthesis
FClO	Unstable	Observed as intermediate and by matrix isolation
F_3ClO	Colorless mp −42 to −42.2°C bp 27 to 29°C Stable at room T	From $F_2 + Cl_2 + O_2$ and $2F_2 + ClONO_2 \xrightarrow{-35°C}$ $F_3ClO + FNO_2$
$FClO_2$ (chloryl fluoride)	Colorless mp ∼−120°C bp ∼−6°C Stable at room T	$6NaClO_3 + 4ClF_3 \xrightarrow{-196°C}$ $6NaF + 2Cl_2 + 3O_2 +$ $6FClO_2$
F_3ClO_2	Colorless mp −81.2°C bp −21.58°C Marginally stable at room T	From $FClO_2 + PtF_6$ followed by treatment with FNO_2
$FClO_3$ (perchloryl fluoride)	Colorless mp −147.75°C bp −46.67°C High thermal stability Not easily hydrolyzed	From $KClO_4 + SbF_5$
F_2ClO^+ (difluorooxochloronium cation)	Stable white solids with anions such as HF_2^-, BF_4^-, PF_6^-, and most MF_6^-	F_3ClO + Lewis acid
ClO_2^+ (chloryl cation)	White solids unless anion is colored Moderately stable	$FClO_2$ + Lewis acid
$F_2ClO_2^+$ (difluoroperchloryl cation)	React violently with H_2O or organic materials	$F_3ClO_2 + BF_3 \rightarrow [F_2ClO_2]BF_4$
F_4ClO^- (tetraflurooxochlorate ion)	Stable white solids	$MF + F_3ClO \rightarrow M[F_4ClO]$ $M = Cs^+, Rb^+, or K^+$
$F_2ClO_2^-$ (difluorochlorate ion)	White solid Stable at room T Reacts explosively with H_2O	$CsF + FClO_2 \rightarrow Cs[F_2ClO_2]$

CsF (a fluoride salt of a large nonoxidizable cation). The neutral molecules based upon a trigonal-bipyramid form adducts easily, giving stable pseudotetrahedral cations and pseudooctahedral anions. The tetrahedral $FClO_3$ is kinetically stable and does not form adducts.

▶ 16.4.9   Halogen Oxides

Oxygen fluorides (page 779) and ClO_2 (page 372) have been discussed. The other oxides of chlorine are Cl_2O, Cl_2O_4, Cl_2O_6, and Cl_2O_7. Two can be considered perchlorates: The unstable Cl_2O_4, with the structure $Cl-O-ClO_3$, is commonly called chlorine perchlorate; and Cl_2O_6 has an ionic structure, $ClO_2^+ClO_4^-$, in the solid. Chlorine monoxide, a yellow-red gas, is obtained by the reaction of Cl_2 with freshly prepared HgO. An angular molecule (C_{2v}), it decomposes explosively when heated. It reacts with water to produce HOCl. Even the most stable of chlorine oxides, Cl_2O_7, is explosive. It is obtained by dehydration of $HClO_4$ with P_4O_{10}. It can be isolated as a colorless liquid by low-pressure distillation.

All of the bromine oxides are unstable thermally: Br_2O (dark-brown liquid), Br_3O_8 (white solid), and BrO_2 (yellow solid). Two of the oxides of I—namely, I_2O_4 and I_4O_9—are not well-characterized. The important oxide is I_2O_5, a white crystalline product yielded by dehydration of HIO_4 at 250°C. I_2O_5 decomposes to the elements above 300°C at 170°C; and it is reduced to I_2 quantitatively by CO, providing the iodimetric determination of CO.

▶ *16.4.10 Oxoacids and Oxoanions of the Halogens*

We discussed the preparation of the oxoanions of Cl and some of their redox chemistry in Section 8.9. The emf values allow us to predict thermodynamically favorable reactions, but we cannot rely on a correlation of emf values with rates. In particular, the reactions of oxoanions (or acids) often involve O transfer mechanisms. The processes of breaking and reforming covalent bonds and the necessary molecular rearrangements often are slow. Thus the reduction potential for ClO_4^- indicates that it should be a strong oxidizing agent, and it is—in hot and/or concentrated acidic solutions. Yet cold dilute $HClO_4$ does not oxidize I^-, even though the E_{cell}^0 is 0.65 V. This does not make the potential diagrams useless; knowing which reactions are thermodynamically impossible is very valuable. With experience, you acquire a feeling for which reactions are likely to be slow even though thermodynamically favorable. Below, we examine some of the intercomparisons among and the chemistry of the halogen oxoanions.

The Hypohalites

The compound HOF, which has been prepared by the reaction of F_2 and H_2O at 0°C, is unstable, even oxidizing H_2O further to H_2O_2.

$$F_2 + H_2O \xrightarrow{0°C} HOF + HF \tag{16.88}$$

Although HOF has the same formula type as HOCl, F is assigned its usual $-I$ oxidation number, so it is hydrogen oxygen fluoride, or hydroxyl fluoride, not hypofluorous acid. Nevertheless, HOF should be acidic, and probably a stronger acid than HOCl. The $-OF$ group occurs in CF_3OF, O_2NOF, F_5SOF, and O_3ClOF—all of which are strongly oxidizing.

For the other halogens, the relative increase in ease of oxidation of the halide ion with increasing atomic number is demonstrated in the following potential diagrams for basic solution:

$$E_B^0 \quad ClO^- \xrightarrow[]{\;0.42\;} Cl_2 \xrightarrow[]{\;+1.36\;} Cl^- \qquad (16.89)$$

with the top bracket labeled $+0.89$ spanning ClO^- to Cl^-.

$$BrO^- \xrightarrow[]{\;+0.45\;} Br_2 \xrightarrow[]{\;+1.07\;} Br^- \qquad (16.90)$$

with the top bracket labeled $+0.76$ spanning BrO^- to Br^-.

$$IO^- \xrightarrow[]{\;+0.42\;} I_2 \xrightarrow[]{\;+0.535\;} I^- \qquad (16.91)$$

with the top bracket labeled $+0.48$ spanning IO^- to I^-.

All of the above hypohalite ions can be formed from the disproportionation of the corresponding halogen in base. The stability of the hypohalite ions with regard to further disproportionation to halite ions and halide ions decreases with increasing atomic number. Thus although hypoiodite solutions decompose within a few hours, hypochlorite solutions decompose only slightly over a period of weeks, with the following relative rates: ClO^-, 1, BrO^-, 100, and IO^-, 3×10^6. Accordingly, hypobromite is generated for analytical use from hypochlorite solutions at a pH of around 10:

$$ClO^- + Br^- \longrightarrow BrO^- + Cl^- \qquad (16.92)$$

Despite its lower emf, hypobromite is often a faster oxidizing agent than hypochlorite.

Hypoiodite ion is an oxidant used to determine β-keto functions through the "iodoform" reaction, so called because of one of the reaction products:

$$\underset{\underset{O}{\overset{\|}{}}{RCCH_3} + 3\,I_2 + 4\,OH^- \longrightarrow HCI_3 + RCO_2^- + 3\,I^- + 3\,H_2O \qquad (16.93)$$

The anhydrous hypohalous acids cannot be isolated. Their pK_a values are given in Table 7.3 (page 327). Hypoiodous acid also undergoes dissociation to produce hydroxide ion in solution, its pK_b being 10.

The Halites

Chlorite salts and chlorous acid, $HClO_2$, have been known for a long time. The acid cannot be isolated in anhydrous form. Earlier efforts to prepare bromites and iodites were unsuccessful, but bromite ion has been prepared by anodic oxidation of Br^- with very high current efficiency.[14] The bromite ion was isolated as a tetraalkylammonium salt, because of its insolubility in water. The salt was recrystallized from chloroform to give orange crystals.

[14] T. Kageyama and T. Yamamoto, *Bull, Chem. Soc. Japan* **1980,** *53,* 1175; *Chem. Lett.* **1980,** 671.

The Halates

The oxidizing strengths of chlorate and bromate solutions are similar, but the iodate is much weaker:

$$E_B^0 \qquad ClO_3^- \xrightarrow{\;+0.63\;} Cl^- \tag{16.94}$$

$$BrO_3^- \xrightarrow{\;+0.61\;} Br^- \tag{16.95}$$

$$IO_3^- \xrightarrow{\;+0.29\;} I^- \tag{16.96}$$

All the standard emf values are more positive by 0.83 V in acidic (1 M H^+) solution.

As oxidizing reagents, these ions have rates of reaction in the order $IO_3^- > BrO_3^- > ClO_3^-$. This order has been explained as stemming from the decreasing multiple-bond character in the O—X bond as the atomic number increases. The iodate thus has more negative charge at the oxygen surface than the bromate, which in turn has more surface charge than the chlorate. The increase in surface charge would allow O to coordinate more readily with other ions, and the low multiple-bond character also would allow I to expand its coordination sphere more readily through dative bonding to empty d orbitals. An increase in polarity for the O—I bond, because of the lower electronegativity of I, can explain the greater negative charge on O for IO_3^-.

Moreover, the low double-bond character or an increase in polarity and consequent higher negative charge on oxygen would account for the generally lower solubility of transition metal iodates compared with chlorates or bromates, and also for the greater viscosity of comparable solutions of iodates compared with chlorates. It might also help explain why acid salts containing the hydrogen-bonded units $(IO_3 \cdot HIO_3)^-$ and $(IO_3 \cdot 2HIO_3)^-$ are formed with iodic acid, but not with bromic or chloric acids. The high molecular weight of $KH(IO_3)_2$ and ease of purification through recrystallization make it suitable for standardizing base solutions for analytical work. Iodic acid also can be isolated in anyhdrous form, in contrast to chloric and bromic acids.

We can make both iodic and bromic acids in the laboratory by oxidizing the halogen or halide ion with Cl_2, removing the Cl^- by adding silver oxide:

$$X_2 + 5Cl_2 + 6H_2O \longrightarrow 2HXO_3 + 10HCl \tag{16.97}$$

$$2HCl + Ag_2O \longrightarrow 2AgCl + H_2O \tag{16.98}$$

Commercially, iodates and bromates are produced by electrolytic oxidation. As with Cl_2, reaction with hot basic solution produces the halate.

$$3X_2 + 6OH^- \longrightarrow XO_3^- + 5X^- + 3H_2O \qquad X = Cl, Br, I \tag{16.99}$$

In acidic solution, the above reaction is reversed for X = Br and I, and this serves as a convenient method of preparing Br_2 and I_2 in solution, usually using an excess of the halide ion:

$$6H^+ + 5Br^- + BrO_3^- \longrightarrow 3Br_2 + 3H_2O \tag{16.100}$$

The extent of unsaturation in olefins can be determined by the addition of the Br_2 produced to the double bond:

$$\text{C}=\text{C} + Br_2 \longrightarrow -\overset{|}{\underset{Br}{C}}-\overset{|}{\underset{Br}{C}}- \tag{16.101}$$

We can determine the excess Br_2 by adding KI and titrating with $Na_2S_2O_3$ solution:

$$2\,I^- + Br_2 \longrightarrow 2\,Br^- + I_2 \tag{16.102}$$

$$I_2 + S_2O_3^{2-} \longrightarrow 2\,I^- + S_4O_6^{2-} \tag{16.103}$$

Unlike the reactions of iodates and bromates with their ions, chlorates react with Cl^- in the presence of acid to produce ClO_2 and Cl_2:

$$2\,HClO_3 + 2\,HCl \longrightarrow 2\,ClO_2 + Cl_2 + 2\,H_2O \tag{16.104}$$

The Perhalates

The marked differences among the perhalic acids and their anions are not peculiar to this family, as can be illustrated by examining the properties of the following acids:

H_3PO_4	H_2SO_4	$HClO_4$
H_3AsO_4	H_2SeO_4	$HBrO_4$
$HSb(OH)_6{}^a$	H_6TeO_6	H_5IO_6

^{a}(in solution only)

Acid strength and oxidizing ability increase regularly on going from phosphoric acid to sulfuric acid to perchloric acid; the melting points decrease and the tendency toward formation of polyacid units decreases from phosphoric acid to perchlorate acid, which forms no polyacids. Likewise, the ability to form metal ion complexes decreases from phosphate through perchlorate. In acid strength, solubility of salts, and structure of the salts, arsenic and selenic acid resemble phosphoric and sulfuric acids more than they resemble antimonic and telluric acids. Properties of perbromic acid are more similar to those of perchloric acid than those of periodic acid, but $HBrO_4$ is a stronger oxidizing agent than $HClO_4$. Arsenic and selenic acids also are much stronger oxidizing agents than H_3PO_4 and H_2SO_4. Unlike H_2SO_4, H_2SeO_4 upon heating is unstable toward decomposition to form O_2, H_2O, and Se. Perbromic acid also is less stable thermally than $HClO_4$.

The coordination number of the central atom increases for antimonates and tellurates, which show only six-coordination. The smaller size of I^{VII} compared with Te^{VI} and Sb^V permits I to show C.N. 4 as well as 6 in periodates. Typical salts formed by these acids are $NaSb(OH)_6$, $LiSb(OH)_6$, $Na_3H_3TeO_6$, Ag_5TeO_6, Ag_5IO_6, $Na_3H_2IO_6$, and KIO_4.

$$E_A^0 \qquad ClO_4^- \xrightarrow{1.20} ClO_3^- \tag{16.105}$$

$$BrO_4^- \xrightarrow{1.85} BrO_3^- \tag{16.106}$$

$$H_5IO_6 \xrightarrow{1.60} IO_3^- \tag{16.107}$$

Perchloric Acid.   Perchloric acid is the only oxoacid of Cl that can be isolated in an anhydrous state. It is prepared most readily through the dehydration of the dihydrate using fuming sulfuric acid and removal of anhydrous $HClO_4$ by vacuum distillation in a grease-free system:

$$HClO_4 \cdot 2\,H_2O + 2\,H_2S_2O_7 \xrightarrow{25-80°C} HClO_4 + 2\,H_2SO_4 \qquad (16.108)$$

The anhydrous acid melts at $-102°C$. Molecular $HClO_4$ units exist in the gas, liquid, and solid. Contact with organic materials such as wood, paper, and rubber produces violent explosions, but it can be stored without explosive decomposition for $30-60$ days at liquid-air temperature ($-190°C$) or for $10-30$ days at room temperature. Normally colorless, it develops an amber color prior to detonation.

Perchloric acid and water form an azeotrope (72.5% $HClO_4$) that boils at approximately 203°C at 1 atm. Under these conditions, appreciable decomposition of the acid occurs. The commercially available concentrated acid is 70% $HClO_4$ (11.6 M).

Cold concentrated $HClO_4$ (70%) is a weak oxidizing agent. Hot concentrated $HClO_4$ has been used in the rapid "wet ashing" procedure for the complete oxidation of organic matter in samples, leaving the inorganic components for analysis—the so-called "liquid fire" reaction. Alcohols should not be present, because the perchlorate esters formed may explode, particularly ethyl perchlorate. This hazard can be eliminated by using a mixture of concentrated nitric acid and perchloric acid. The HNO_3 oxidizes the alcohols and other readily oxidizable groups, and is itself displaced on heating. The temperature of the reaction mixture slowly increases as the perchloric acid is concentrated, approaching the azeotropic conditions. As the temperature and perchloric acid concentration increase, the oxidizing ability of the solution increases. Completion of the reaction may often be determined by adding a small amount of $K_2Cr_2O_7$ to the reaction mixture. The dichromate reacts rapidly with the organic material, giving nearly colorless or green solutions containing Cr^{III} ion. At the end of the reaction, the Cr^{III} ion is oxidized to orange CrO_3. Wet-ashing in this fashion, which may be carried out in as little as $10-15$ minutes, has been used for analysis of samples of organic substances such as leather, wood, grain, and coal.

Perchlorates.   Ammonium perchlorate was used formerly as a nonfreezing blasting compound in mining operations. Today it is used as an oxidant in solid-fuel missiles.

The perchlorate ion is less extensively hydrated than the other oxo halogen anions and also shows very little tendency to form complexes with metal ions. Accordingly, it is used often as an inert anion in studies of metal ion complexes in aqueous solution; but in the absence of other ligands, ClO_4^- can function as a monodentate or didentate ligand.

Magnesium perchlorate is a very efficient desiccant. In this case the Mg^{2+} ions behave as if they were isolated in an inert matrix and accordingly form a very stable hexahydrate when in contact with water. Ammonium perchlorate absorbs sufficient ammonia to liquefy at room temperature, giving solutions resembling Diver's solution (NH_4NO_3, NH_3). Many perchlorate salts are isomorphous with permanganate, perrhenate, and tetrafluoroborate salts.

Perbromic Acid and Perbromates.   Efforts to prepare the perbromate ion were unsuccessful until it was obtained from the β-decay, $^{83}SeO_4^{2-} \rightarrow BrO_4^- + \beta^-$. This led to the discovery that perbromates can be prepared by the oxidation of bromates electrolytically, by XeF_2, or by oxidation by F_2 in basic solution. Solid perbromates are stable: For example, $KBrO_4$ is stable up to 275°C. Perbromate ion is a stronger oxidizing agent than ClO_4^- or IO_4^-, but in dilute solution reduction occurs slowly—it does not oxidize Cl^-. The free acid is stable in solutions up to 6 *M*. In concentrated solution, it is a vigorous oxidizing agent.

Periodic Acid and Periodates.   Periodates can be made in the laboratory by the oxidation of an iodate in basic solution using Cl_2. Commercially, periodates are prepared by electrolytic oxidation of iodates.

The equilibria involved in the dissociation of periodic acid in water are as follows:

$$H_5IO_6 \rightleftharpoons H^+(aq) + H_4IO_6^- \qquad K_1 = (10 \pm 4) \times 10^{-4} \qquad (16.109)$$

$$H_4IO_6^- \rightleftharpoons H^+(aq) + H_3IO_6^{2-} \qquad K_2 = (3 \pm 1) \times 10^{-7} \qquad (16.110)$$

$$H_4IO_6^- \rightleftharpoons 2 H_2O + IO_4^- \qquad K = 43 \pm 17 \qquad (16.111)$$

Periodic acid is thus a much weaker acid than $HClO_4$. The H_5IO_6 form is called "paraperiodic acid"—the form of the acid stable as a solid in contact with water. At 100°C, this form loses water to convert into *meta*-periodic acid, HIO_4.

Periodic acid is especially useful as an analytical reagent for the quantitative determination of α,β-dihydroxyorganic compounds:

$$-\overset{|}{\underset{|}{C}}-\overset{|}{\underset{|}{C}}- \;+\; H_5IO_6 \;\longrightarrow\; -\overset{|}{\underset{\|}{C}} \;+\; \overset{|}{\underset{\|}{C}}- \;+\; HIO_3 \;+\; 3 H_2O \qquad (16.112)$$
$$\quad OH \;\; OH \qquad\qquad\qquad\qquad O \quad O$$

An intermediate in this highly selective reaction is the complex

$$
\begin{array}{c}
-\overset{|}{C}-O \quad\; \overset{O}{\underset{\|}{\,}} \quad OH \\[2pt]
\underset{\diagup\;\;|\;\diagdown}{\overset{\diagdown\;\;\|\;\diagup}{\quad\;\; I \quad\;\;}} \\[2pt]
-\underset{|}{C}-O \quad\; OH \;\; OH
\end{array}
$$

Mercury(II) ion can be determined gravimetrically by precipitating it as $Hg_5(IO_6)_2$.

16.4   THE NOBLE GASES

► *16.5.1   Family Trends*

The elements of Group 18 (Group 0) have been called the rare gases or the inert gases, but both of these are misnomers. Argon constitutes almost 1% of air by volume, and helium is available readily from some natural gas deposits; moreover, the discovery of the xenon fluorides demonstrated that Xe is not truly inert. The term **noble gases** seems most appropriate, since we classify the least reactive metals as noble metals. The noble gases have "closed" eight-electron outer-shell configurations, ns^2np^6—except He, for which the first shell is complete, $1s^2$. The IEs of the noble gases are highest for each period following the regular trend. This increase across each period corresponds to the increase in Z_{eff}, with each electron added to the same outer shell not shielding the increased nuclear charge by a full unit (see page 40). The nonmetals become increasingly more reactive from left to right within a period until we reach the noble gases. The abrupt change in reactivity for the noble gases results from their very high IEs and low EAs, since an added electron would occupy a new shell. Actually, the EAs are zero or slightly negative, indicating that no bound state for the electron is achieved.

► 16.5.2 *Discovery, Occurrence, and Recovery*

Cavendish (1784) removed N_2 from air by sparking with O_2 to form NO_2. The NO_2 was absorbed in aqueous alkali and the excess O_2 was removed by burning S. After removal of the SO_2 in aqueous alkali, a small volume of unidentified, unreactive gas remained. Rayleigh (1892) prepared N_2 from the decomposition of ammonia and by removing the other known constituents of air (O_2, CO_2, and H_2O). The N_2 from ammonia was always 0.5% less dense than that from air. Rayleigh and Ramsay soon isolated a new gas, named argon ("inert"), with an emission spectrum different from that of any known element. Ramsay and Travers then isolated neon ("new"), krypton ("hidden"), and xenon ("stranger," all names are from Greek) by the fractional distillation of liquid air.

During an 1868 eclipse, Jansen identified a new yellow line in the spectrum of the sun. Because this line did not appear in the spectrum of any known element, Frankland and Lockyer concluded that the sun contained a new element, which they named helium (Greek *helios,* "the sun"). By heating clevite (a radioactive mineral related to uranite), Hillebrànd (1889) obtained an unreactive gas that he believed to be N_2, but Ramsay (1885) showed it to be a sample of Lockyer's helium. The study of the radioactive series led to the discovery of radon, the "emanation" or gas from the radioactive decay of radium. (Isotopes of radon from other radioactive decay series were called actinon and thoron for a time.)

The earth's atmosphere contains about 5 ppm He, but it would be very expensive to recover the He. Some natural gas deposits contain 0.5–0.8% He trapped from radioactive decay. These extensive deposits, which occur almost entirely in the southwestern United States, account for the present world production of He. They will be depleted in a few decades. Helium-lean (0.03–0.3%) natural gas deposits are found primarily in the former Soviet Union, the United States, Algeria, and Canada.

The U.S. government has had an on-again/off-again helium conservation program. Some He is stored in depleted gas fields, but most of the He in natural gas now being used is lost to the atmosphere. Helium can be obtained now at low cost from helium-rich gas, and the potential supply far exceeds the demand. The demand for He for the use of superconductors operating at liquid-He temperature for efficient power transmission could increase greatly just as the important supplies in natural gas are exhausted. Higher-temperature superconductors are anticipated to be available.

The other nonradioactive noble gases are recovered from the fractional distillation of liquid air. Only Ar is abundant in air (see Table 16.11). The greater abundance of Ar than

Table 16.11   Properties of the noble gases

	He	*Ne*	*Ar*	*Kr*	*Xe*	*Rn*
Atomic radius^a						
(van der Waals)	150 pm	160	190	200	220	—
First ionization (eV)	24.587	21.564	15.759	13.999	12.130	10.748
energy (MJ/mol)	2.3723	2.0806	1.5205	1.3507	1.1704	1.0370
Outer shell						
configuration	$1s^2$	$2s^2 2p^6$	$3s^2 3p^6$	$4s^2 4p^6$	$5s^2 5p^6$	$6s^2 6p^6$
Boiling point (K)	4.18	27.13	87.29	120.26	166.06	208.16
% by volume (in						
atmosphere)	5.2×10^{-4}	1.82×10^{-3}	0.934	1.14×10^{-3}	8.7×10^{-6}	—

^{a} Calculated from the cubic close-packed lattice parameters extrapolated to 0 K. Helium also has hexagonal close-packed and body-centered cubic phases. Lattice parameters are from J. Donohue, *The Structures of the Elements*, Wiley, New York, 1974, Chapter 2.

Ne is taken as evidence that the earth lost most of its atmosphere during its formation. There are three isotopes of Ar, but the one of overwhelming abundance (99.63%) is ^{40}Ar, formed from the positron decay of ^{40}K (^{40}K $\rightarrow$ ^{40}Ar $+ \beta^+$). Argon in the atmosphere has accumulated from this source. The decay of ^{40}K also is believed to have provided a major source of heat in bringing the earth to its present temperature.

► *16.5.3   Physical Properties*

Some of the physical properties of the noble gases are given in Table 16.11. The gases are all low-boiling, because of their spherical closed shells and the lack of any mutual interaction other than the very weak van der Waals (dispersion) attraction. As we might expect in the absence of any preferred bonding directions, the solids have close-packed structures.

Helium, the lowest-boiling substance known, is the only element that forms a solid only at high pressure (~25 atm). Liquid He (helium I) is a normal liquid down to a λ-point (~2 K, depending on pressure), where it is converted to helium II—a most extraordinary liquid, with zero viscosity and very high heat conductivity, among other unusual physical properties.

The major uses of He are for cryogenic work and for lighter-than-air vehicles. Neon is used for display signs and lights. Argon is used to provide an inert atmosphere, as in incandescent bulbs, and for chemical manipulations where N_2 is too reactive.

► *16.5.4   Chemical Properties*

Introductions to MO theory commonly deal with molecular ions such as He_2^+ and HeH^+. These ions, along with those of other noble gases—Ar_2^+, $HeNe^+$, and so on—are formed in gaseous discharge tubes, but the corresponding neutral molecules are unstable. Also, clathrate compounds of the noble gases have been known for a long time. Unlike ordinary compounds displaying typical chemical bonding, these compounds involve the trapping of atoms or molecules of the proper dimensions in cavities in a crystal. Aqueous hydroquinone solutions under several atmospheres pressure of a noble gas (G = Ar, Kr, or Xe) can be cooled to give crystalline solids with the approximate composition $[C_6H_4(OH)_2]_3G$. Noble gas hydrates are formed as clathrates, with the noble gas atoms occupying the cavities of regular pentagonal dodecahedra. The anesthetic effect of substances such as chloroform has been attributed to the formation of clathrate hydrate crystals in the brain, and this might explain also the anesthetic effect of xenon.

The first real noble-gas compound was obtained when Bartlett and Lohman oxidized Xe with platinum hexafluoride. Prior to this experiment, they had prepared $O_2^+ PtF_6^-$. Because the IE of both O_2 and Xe is 12.1 eV (1.17 MJ/mol), only lattice energy effects would prevent formation of the Xe derivative. They formulated the product obtained as $Xe^+[PtF_6]^-$, but it is now known to be more complex.[15] Bartlett described the experiment as follows:

> The predicted interaction of Xe and PtF_6 was confirmed in a simple and visually dramatic experiment. The deep red PtF_6, of known pressure, was mixed, by breaking a glass diaphragm, with the same volume of xenon, the pressure of which was greater than that of the hexafluoride. Combination, to produce a yellow solid, was immediate at room temperature, and the quantity of Xe which remained was commensurate with a combining ratio of 1:1.
>
> (From N. Bartlett, *Am. Sci.* **1963,** *51,* 114.)

Soon after publication of the preparation of "$XePtF_6$," XeF_4 was obtained by the direct reaction of Xe and F_2 at 400°C. XeF_4 is a colorless volatile solid that is isoelectronic (in terms of valence electrons) and isostructural with ICl_4^-. The parallel with interhalogen halides was apparent, and groups all over the world began the study of noble-gas compounds.

Table 16.12 lists the xenon compounds. Krypton forms a linear KrF_2, but the ionization energies of He and Ar are too high to expect similar compounds. Radon gives compounds similar to those of Xe and should have even more varied chemistry. The high level of radioactivity has limited the work on radon compounds; however, there is some evidence for the formation of $XeCl_2$, and $XeCl_4$ has been characterized by Mössbauer spectroscopy from the β-decay of ^{129}I in $KICl_4 \cdot 4H_2O$ to form $^{129}Xe^*Cl_4$.

Xenon Difluoride

XeF_2 is obtained by the direct reaction of F_2 with Xe at high pressure. XeF_2 is linear, as expected from VSEPR considerations. It is soluble in water, undergoing hydrolysis that is slow except in the presence of base:

$$XeF_2 + 4\,OH^- \longrightarrow Xe + O_2 + 2\,F^- + 2\,H_2O \qquad (16.113)$$

It is a very powerful oxidizing agent.

---

[15] The product at 25°C contains $[XeF^+]$ with $[PtF_6^-]$ and $[PtF_5^-]$. At 60°C, $[XeF^+][Pt_2F_{11}^-]$ is formed.

Table 16.12   Xenon compounds

Oxidation state	Compound	Melting point (°C)	Structure	Properties
II	XeF_2	129	Linear	Hydrolyzed to Xe + O_2
	$XeF^+AsF_6^-$		Nearly linear XeF_2 with 1 bridging F	
IV	XeF_4	117	Square-planar	Stable
	$XeOF_2$	31		Unstable
VI	XeF_6	49.6	Distorted octahedral	Stable
	$CsXeF_7$			dec. >50°C
	Cs_2XeF_8		Archimedes' antiprismatic	Stable to 400°C
	$XeOF_4$		Square-pyramidal	Stable
	XeO_2F_2			Unstable
	XeO_3		Pyramidal	Explosive
VIII	XeO_4		Tetrahedral	Explosive
	XeO_6^{4-}		Octahedral	Anion of a very weak acid

Estimated values of E^0

$$E_A^0 \quad H_4XeO_6 \xrightarrow{2.36} XeO_3 \xrightarrow{2.12} Xe$$

$$XeF_2 \xrightarrow{2.64} Xe$$

$$E_B^0 \quad HXeO_6^{3-} \xrightarrow{0.94} HXeO_4^- \xrightarrow{1.26} Xe$$

Xenon Tetrafluoride

XeF_4 is easily prepared from the elements, but accompanied by XeF_2 at low F_2/Xe ratios and by XeF_6 at high F_2/Xe ratios. It is a strong oxidizing agent and reacts violently with water to give XeO_3:

$$3\,XeF_4 + 6\,H_2O \longrightarrow XeO_3 + 2\,Xe + \tfrac{3}{2}O_2 + 12\,HF \qquad (16.114)$$

Xenon Hexafluoride

The reaction of Xe with excess F_2 at high pressure at about 250°C yields XeF_6, which is hydrolyzed rapidly to form XeO_3 and reacts with quartz:

$$2\,XeF_6 + SiO_2 \longrightarrow 2\,XeOF_4 + SiF_4 \qquad (16.115)$$

There has been controversy about the structure of XeF_6, in which the Xe has one lone pair of electrons. Although it now seems clear that the molecule is *not* octahedral, it does not correspond to a structure expected where the lone pair plays a dominant stereochemical role. In the solid, XeF_6 exists as XeF_5^+ (square pyramidal) and F^-, but the F^- ions form bridges. Electron diffraction studies indicate that XeF_6 is slightly distorted from an octahedron in the gas phase. It might involve fluxional behavior, passing from one nonoctahedral structure, such as one with the lone pair directed through an octahedral face or edge, to another. The isoelectronic ions $SbBr_6^{3-}$, $TeCl_6^{2-}$, and $TeBr_6^{2-}$ are octahedral. The lone pair is "stereochemically inactive," possibly in an unhybridized s orbital. BrF_6^- is centrosymmetric, whereas the lone pair lowers the symmetry of IF_6^- to C_{2v} for the larger central atom.

XeF_6 is a fluoride donor, forming complexes such as $2XeF_6 \cdot SbF_5$ and an adduct with PtF_5, $XeF_5^+ PtF_6^-$. The latter compound has a sixth, bridging F^- with an Xe—F distance longer than the others. Similarly, XeF_2 forms $XeF^+ AsF_6^-$ with AsF_5. An attempt to prepare the low-spin d^6 Au^V ion, AuF_6^-, yielded the compound $Xe_2F_{11}^+ AuF_6^-$. The cation consists of 2 XeF_5 bridged by a shared F^- at greater distance (223 pm) than for the other Xe—F bonds (184 pm). The geometry about each Xe suggests that the lone pair is stereochemically active. Coordination numbers of 7 and 8 are achieved in reactions of XeF_6 with alkali metal fluorides (M = Na, K, Rb, or Cs) to form $MXeF_7$ and M_2XeF_8. The $MXeF_7$ salts decompose above room temperature to form M_2XeF_8 and XeF_6. The Cs^+ and Rb^+ salts of XeF_8^{2-} are stable to about 400°C.

The oxofluorides of Xe are as follows: $XeOF_2$ (an unstable partial hydrolysis product of XeF_4), $XeOF_4$ (from the partial hydrolysis of XeF_6), and XeO_2F_2 (from the reaction of $XeOF_4$ and XeO_3):

$$XeOF_4 + XeO_3 \longrightarrow 2XeO_2F_2 \tag{16.116}$$

Bonding in Xenon Fluorides

Because XeF_2 and XeF_4 closely parallel ICl_2^- and ICl_4^-, respectively, and because of the success of the VSEPR approach in predicting their structures, these are commonly dealt with in terms of an expanded octet. Whether the $5d$ orbitals are sufficiently low in energy for effective bonding is questionable, however; and bonding can be explained without using d orbitals (see Section 4.4).

Xenon Oxides

Xenon trioxide is an explosive white solid that appears to be present as XeO_3 molecules in water but forms $HXeO_4^-$ in basic solution. The $HXeO_4^-$ disproportionates slowly to produce perxenate ion, XeO_6^{4-}:

$$2HXeO_4^- + 2OH^- \longrightarrow XeO_6^{4-} + Xe + O_2 + 2H_2O \tag{16.117}$$

The strongly oxidizing perxenate ion also can be made by oxidizing $HXeO_4^-$ with O_3. The alkali metal and barium salts of XeO_6^{4-} are stable. Xenon tetraoxide is formed as an explosive gas by the reaction of Ba_2XeO_6 with concentrated sulfuric acid.

GENERAL REFERENCES

F. A. Cotton and G. Wilkinson, *Advanced Inorganic Chemistry,* 5th ed., Wiley–Interscience, New York, 1988. An excellent single-volume source of inorganic descriptive chemistry.

Coord. Chem Rev. **1992,** *112* (Main Group Chemistry). Reviews of main group elements appear almost annually.

L. Gmelin, *Handbuch der anorganischen Chemie,* Verlag Chemie, Weinheim, Germany, 1924–1991.

N. N. Greenwood and A. Earnshaw, *Chemistry of the Elements,* Pergamon Press, Oxford, 1984. An excellent single-volume source of inorganic descriptive chemistry.

Inorganic Chemistry of the Main-Group Elements, A Specialist Periodical Report, Chemical Society, London, Vol. 5, 1978, and other volumes.

M. C. Sneed, J. L. Maynard, and R. C. Brasted, Eds., *Comprehensive Inorganic Chemistry.* Van Nostrand–Reinhold, New York, 1953–1961.

R. Steudel, *Chemistry of the Non-Metals,* F. C. Nachod and J. J Zuckerman, translators, deGruther, Berlin, 1977.

A. F. Wells, *Structural Inorganic Chemistry,* 5th ed., Oxford, Oxford, 1984. The best single-volume source for inorganic structures.

G. Wilkinson, R. D. Gillard, and J. A. McCleverty, Eds., *Comprehensive Coordination Chemistry.* Pergamon Press, Oxford, 1987, Vol. 2, Ligands. These volumes are the definitive source for coordination chemistry.

PROBLEMS

16.1   The electron affinity of N (-7 kJ) seems anomalous. Explain the order of electronegativities for Group 15 (Table 16.1) and why one can conclude that N is really more "regular" than O.

16.2   What compound is decomposed for inflating airbags in cars?

16.3   Write balanced equations for the preparation of HNO_3, starting with N_2.

16.4   When dilute nitric acid reacts with Cu turnings in a test tube, a colorless gas is formed that turns brown near the mouth of the tube. Explain the observations and write equations for the reactions involved.

16.5   Write equations for the preparation of five nitrogen oxides.

16.6   Sketch the *cis* and *trans* isomers of hyponitrous acid.

16.7   What properties of polymers of $(SN)_x$ and $(N\text{-}PR_2)_x$ make them of interest for practical uses?

16.8   Give the formula of a biodegradable polyphosphazene polymer that could be used for medical devices.

16.9 PI_5 is not known. What would be its likely structure in the solid?

16.10   (a) What N species are compatible with the H_3PO_4/PO_4^{3-} species from the Pourbaix diagrams given by J. A. Campbell and R. A. Whiteker (*J. Chem. Educ.* **1969,** *46,* 90)?
   (b) What P species are compatible with $NH_3(aq)$?

16.11   How can white P be separated from red P?

16.12   How are P_4, P_4O_6, and P_4O_{10} related structurally?

16.13   In what way are phosphine ligands in metal complexes similar to CO?

16.14   Write the formulas for the diethylester of phosphorous acid and the monoethylester of hypophosphorous acid. Would these be protonic acids?

16.15   Sketch the $p\pi d$ bonding in hexachlorotriphosphazene (Scheme 16.2) and in cyclic $(PCl_2N)_4$, assuming it to be planar.

16.16   Write balanced equations and calculate E^0_{cell} for the reaction in acidic solution of (**a**) H_2O_2 with NO_2^- and (**b**) H_2O_2 with MnO_4^-.

16.17 What thermodynamic factors are important in determining the relative stabilities of solid metal oxides, peroxides, and superoxides? What cation characteristics (size and charge) would you choose to prepare oxides, peroxides, superoxides, and ozonides?

16.18 Give a description of bonding in O_2F_2 to account for the very short O—O bond and very long O—F bonds.

16.19 Write the equations for the oxidation of $S_2O_3^{2-}$: (a) by I_2 and (b) by H_2O_2.

16.20 In the Pourbaix diagram for oxygen, H_2O_2 does not appear. Explain. (If you care to check the diagram, see reference in Problem 16.10.)

16.21 From Figures 16.8 and 16.9: (a) What S and Se species are stable in contact with O_2? (b) What S and Se species are stable in contact with H_2? (c) What S species are stable in contact with Se?

16.22 How are dithionates and polythionates alike, and in what respects do they differ?

16.23 What are the products of reaction of SF_4 with alkylcarbonyls, carboxylic acids, and phosphonic acids [R(RO)PO(OH)]?

16.24 Give valence bond descriptions for SF_3 and SF_5 and indicate why they have low stability.

16.25 Write balanced equations for the following preparations:
(a) Cl_2 (from NaCl)   (d) Br_2 (recovery from seawater)
(b) I_2 (from $NaIO_3$)   (e) HF (from CaF_2)
(c) HCl (from NaCl)

16.26 Draw the structures of the following species, indicating the approximate bond angles: IF_2^+, IF_4^+, and IF_6^+.

16.27 Indicate reactions which might be suitable for the preparation of:
(a) Anhydrous tetramethylammonium fluoride
(b) Aluminum bromide
(c) Barium iodide

16.28 The ΔH_f^0 of $BrF(g)$, $BrF_3(l)$, and $BrF_5(l)$ are -61.5, -314, and -533 kJ/mol, respectively. Which of these species should be predominant on reacting Br_2 and F_2 under standard conditions?

16.29 Bromites and perbromates have been obtained only recently. Give the preparation of each.

16.30 What are the expected structures of ClO_3^+, $F_2ClO_3^-$, and $F_4ClO_2^-$? Sketch the regular figure and indicate deviations from idealized bond angles.

16.31 Use the emf data in Appendix E to predict the results of mixing the following: (a) Cl^- and BrO_3^- in 1 M acid; (b) Cl_2 and IO_3^- in 1 M base; (c) At_2 and Cl_2 in 1 M base.

16.32 Compare the expected *rate* of reaction of AtO_3^- and IO_3^- as oxidizing agents. What formula is expected for perastatinate? Why?

16.33 Give practical uses of He, Ne, and Ar, and give the sources of each.

16.34 What known compounds related in bonding and structure to the Xe halides should have prompted a search for the Xe halides earlier?

16.35 Give the expected shape and approximate bond angles for ClF_3O, considering the effects of the lone pair on Cl and the directional effects of the $Cl{=}O$ π bond. (See K. O. Christe and H. Oberhammer, *Inorg. Chem.* **1981**, *20*, 297.)

16.36 Draw the structure expected for $XeOF_4$ and indicate the approximate bond lengths (the Xe—F bond length in XeF_4 is 195 pm).

16.37 XeF_6 gives solutions in HF which conduct electricity. How might one distinguish between the following possible modes of dissociation:

$$XeF_6 + HF \rightleftharpoons XeF_5^+ + HF_2^-$$

and

$$XeF_6 + 2\,HF \rightleftharpoons XeF_7^- + H_2F^+$$

16.38   How might one distinguish between Xe^+ ions and Xe_2^{2+} ions in the compound $XePtF_6$?

▶ 17 ◀

Cluster and Cage Compounds

∙∙∙∙∙∙∙∙∙∙∙∙∙∙∙∙∙∙∙∙∙∙∙∙∙∙∙∙∙∙∙

The preceding chapters have shown how to predict formulas and structures of several kinds of compounds with the aid of models provided by valence bond theory, molecular orbital theory, the octet rule, and the 18-electron rule. But the electronic and molecular structures of one large class of compounds—namely, cages and clusters—are not predictable in these terms. This chapter discusses a number of such compounds, starting with the boron hydride cages—their structures, syntheses, and chemistry. Subsequently, we study related cluster compounds containing transition metals, noting their remarkable behavior and structures as well as the possible relevance of these cluster compounds to heterogeneous catalysis. Finally, we deal briefly with inorganic chains, rings, and polyhedra whose structures are related to those of cages and clusters. We develop models for understanding the formulas and structures of these classes of compounds.

17.1 INTRODUCTION

The valence bond model dominated bonding theory for some time. Couper's lines between symbols in molecular formulas (1858) were interpreted as electron pair bonds by Lewis (1918). The stereochemistry postulated by van't Hoff and LeBel for organic compounds and by Werner for inorganic complexes was justified on a quantum-mechanical basis through hybridization by Pauling (1931). Along with the valence shell electron-pair repulsion (VSEPR) theory of Nyholm and Gillespie (1957), these considerations have accommodated the x-ray structures of thousands of ionic and molecular species. Using the valence bond (VB) model, Pauling related bond lengths to bond order, as well as electronegativity to bond polarity. Organic chemists developed "electronic" theories of reactivity to systematize the field.

Yet at the very time G. N. Lewis proposed the electron-pair bond, Alfred Stock was preparing a series of compounds whose formulas gave no hint as to their structures and whose structures, once determined, could not be accommodated by a simple VB model.

These remarkable compounds are the boron hydrides or boranes. Because they are so re-active toward O_2 (boron is found in nature as oxides), Stock and his co-workers devised vacuum line techniques to handle the volatile boron hydrides in an O_2-free and moisture-free environment. Even so, the experimental difficulties to be overcome were enormous. Yields of boranes typically were only 4–5%. Product mixtures often required tedious and laborious procedures for fractionation. Quantities of pure compounds attained were mil-limolar at maximum. In spite of these problems, Stock was able to prepare and character-ize B_2H_6, B_4H_{10}, B_5H_9, B_5H_{11}, B_6H_{10}, and $B_{10}H_{14}$. These compounds divide into a "hydrogen-rich" group (of general formula B_pH_{p+6}) and a "hydrogen-poor" group (of for-mula B_pH_{p+4}). Subsequently, other members of both series have been discovered. More recently, a third series of very stable anions, $B_pH_p^{2-}$ (which can be thought of as derived via deprotonation of B_pH_{p+2}), has been prepared for $p = 6-12$.

Figure 17.1 depicts structures of the hydrides discovered by Stock. Each B is bonded to at least one H, with some H's bridging two boron atoms. These are called **terminal** and **bridging** H, respectively. Physical properties of the boron hydrides are reported in Table 17.1. A very important chemical property of most boranes is their vigorous (often explo-sive) reactivity with O_2. Of the neutral boranes, only $B_{10}H_{14}$ can be handled in air. Table 17.2 records some known boron hydride species, ionic as well as neutral. Several poten-tial members of each series are missing—perhaps because no one has yet discovered a workable synthesis, or because the compounds themselves are not stable. A likely example of the latter is B_3H_9, which would have three very crowded bridging hydrogens.[1] The isoelectronic species resulting from its (hypothetical) deprotonation, $B_3H_8^-$, is stable. The different series of boranes are named *closo* (for species isoelectronic with B_pH_{p+2}), *nido* (isoelectronic with B_pH_{p+4}), *arachno* (isoelectronic with B_pH_{p+6}) and *hypho* (isoelectronic with B_pH_{p+8}).

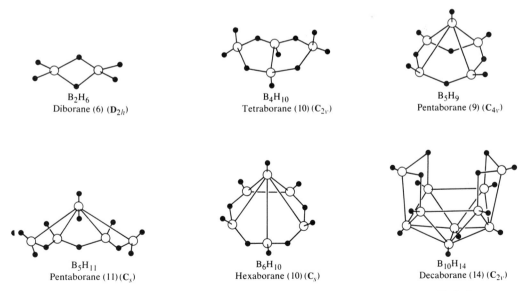

B_2H_6
Diborane (6) ($\mathbf{D}_{2h}$)

B_4H_{10}
Tetraborane (10) ($\mathbf{C}_{2v}$)

B_5H_9
Pentaborane (9) ($\mathbf{C}_{4v}$)

B_5H_{11}
Pentaborane (11) ($\mathbf{C}_s$)

B_6H_{10}
Hexaborane (10) ($\mathbf{C}_s$)

$B_{10}H_{14}$
Decaborane (14) ($\mathbf{C}_{2v}$)

Figure 17.1   Structures of Stock's boron hydrides. (From S. G. Shore in *Boron Hydride Chemistry*, E. L. Muetterties, Ed., Academic Press, New York, 1975.)

[1] R. E. Williams, *Adv. Inorg. Radiochem.* **1976,** *18,* 67.

Table 17.1 Physical properties[a] of boron hydrides prepared by Stock

Formula	Name	Melting point (°C)	Boiling point (°C)
B_2H_6	Diborane	−165.5	−92.5
B_4H_{10}	Tetraborane	−120	16
B_5H_9	Pentaborane(9)	−46.8	58.4
B_5H_{11}	Pentaborane(11)	−123.3	65
B_6H_{10}	Hexaborane(10)	−65.1	108
$B_{10}H_{14}$	Decaborane	99.5	213

[a] From S. G. Shore, in *Boron Hydride Chemistry,* E. Muetterties, Ed., Academic Press, New York, 1975.

Table 17.2 Some boron hydride species containing 12 or fewer borons[a]

Closo (B_pH_{p+2})	Nido (B_pH_{p+4})	Arachno (B_pH_{p+6})	Hypho (B_pH_{p+8})
	B_2H_6	$B_2H_7^-$	
		$B_3H_8^-$	
	$B_4H_7^-$	B_4H_{10}	
		$B_4H_9^-$	
	B_5H_9	B_5H_{11}	$B_5H_{12}^-$
	$B_5H_8^-$		$B_5H_{11}^{2-}$
$B_6H_6^{2-}$	B_6H_{10}	B_6H_{12}	
	$B_6H_9^-$	$B_6H_{11}^-$	
	$B_6H_{11}^+$		
$B_7H_7^{2-}$		$B_7H_{12}^-$	
$B_8H_8^{2-}$	B_8H_{12}	B_8H_{14}	
$B_9H_9^{2-}$	B_9H_{13}	n-B_9H_{15}	
	$B_9H_{12}^-$	i-B_9H_{15}	
		i-$B_9H_{14}^-$	
		$B_9H_{13}^{2-}$	
$B_{10}H_{10}^{2-}$	$B_{10}H_{14}$	$B_{10}H_{15}^-$	
	$B_{10}H_{13}^-$	$B_{10}H_{14}^{2-}$	
$B_{11}H_{11}^{2-}$	$B_{11}H_{15}(?)$		
	$B_{11}H_{14}^-$		
	$B_{11}H_{13}^{2-}$		
$B_{12}H_{12}^{2-}$			

[a] The general formula is given for the parent compound.

Nomenclature for Boranes[2]

Neutral boron hydrides are all named as *boranes;* a Greek prefix indicates the number of B atoms, and an Arabic number in parentheses gives the number of H atoms. Thus B_5H_{11} is pentaborane(11). The number is omitted if only one borane containing a particular B count is known. B_2H_6 is usually referred to as diborane.

Anionic species are named as hydroborates. Greek prefixes separately indicate the numbers of H and B; the charge on the anion is given in parentheses following. Thus $B_5H_8^-$ is octahydropentaborate(1-). The structural type sometimes is specified when naming anions. Thus $B_5H_8^-$ is also octahydro-*nido*-pentaborate(1-).

17.2 THE BONDING PROBLEM IN BORANES

▶ 17.2.1 Localized Bonding Model

Retaining the VB concept of the relationship between bond distance and bond order, we immediately encounter a problem on examining the known structures of boron hydrides (Figure 17.1): The coordination number of each B (and of some hydrogens) exceeds the number of low-energy orbitals—not to mention the number of electron pairs. Obviously, the lines in these structures cannot represent electron-pair bonds: They serve only to show which atoms are connected. Such structures are called **topological** structures. Boranes are examples of so-called **electron-deficient** compounds, in which the number of valence atomic orbitals exceeds the number of valence electrons.

Ideally, a bonding model for electron-deficient compounds would allow the same straightforward prediction of geometry, reactivity, stoichiometry, redox properties, acidity, and so on, that the VB model permits for "regular" compounds. Early attempts to account for the electronic structure of diborane, the simplest borane, included the observation that B_2H_6 is isoelectronic with ethylene, C_2H_4. In this view, we could regard the two bridging H's as protonating the double bond of $B_2H_4^{2-}$. Subsequent research has confirmed the acidic nature of bridging H's in the boranes; however, this bonding picture is difficult to extend to the higher boranes. A straightforward application of VB theory to the electronic structure of diborane requires some 20 resonance structures (see Figure 17.2). Plainly, as the task of describing more complex molecules begins to require unmanageable numbers of canonical structures, the simple VB approach loses its utility.

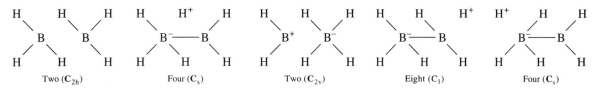

Figure 17.2 Resonance structures for B_2H_6.

[2] B. P. Block, W. H. Powell, and W. C. Fernelius, *Inorganic Chemical Nomenclature, Principles and Practice.* American Chemical Society, Washington, D.C., 1990, pp. 97–112.

The number of contributing structures is given by h/h', where h is the order of the point group of the parent molecule and h' is the order of the point group we would assign to a single contributing structure, assuming the bonds physically occupy space. You should verify this result for the canonical structures in Figure 17.2.

A VB model for simple molecules such as H_2O and NH_3 differs from a localized molecular orbital (MO) model mainly in terminology. A simple extension to include three-center, two-electron bonds accommodates many electron-deficient compounds. If each B forms at least one three-center bond, it achieves an octet of electrons (*vide infra*). Figure 17.3 shows possible ways of forming three-center B—B—B and B—H—B bonds, along with diagrammatic representations of each. The difference between open and closed B—B—B bonds lies in the choice of hybrid orbitals used. Because theoretical studies have not revealed evidence in favor of the open three-center bond, all valence structures in this book depict only closed ones. For B—H—B bonds there is no alternative choice of orbitals. In the older literature, these are depicted as open three-center bonds, but the current trend is to show them as closed.

We can provide a simple picture of the electronic structure of B_2H_6 by employing three-center bonds. The B atoms may be regarded as sp^3-hybridized. Each B forms two two-center, two-electron bonds with the terminal H's, thus accounting for eight of the 12 valence electrons. Two three-center, two-electron bonds involve both B atoms and a bridging H atom. The observed bond angles in B_2H_6 are far from tetrahedral angles, so the MOs probably have some of the character of sp^2 hybrids also. A valence structure for diborane is

$$\begin{array}{ccccc} & & H & & \\ H & & | & & H \\ & B & & B & \\ H & & | & & H \\ & & H & & \end{array}$$

An MO treatment of the B—H—B bridging bonds has been given on page 185.

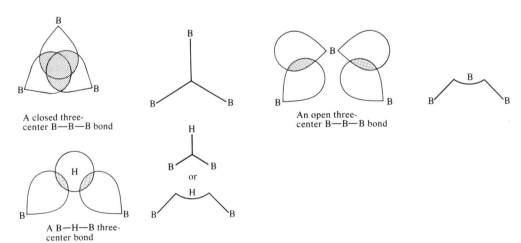

Figure 17.3   Representations of three-center bonds.

▶ *17.2.2   styx Numbers*

Lipscomb and co-workers established a systematic procedure for writing valence structures of more complex boron hydrides, incorporating three-center bonding. Valence structures drawn by Lipscomb's procedure may be employed in two ways: given the molecular formula, to predict a molecular structure; or, given the molecular structure, to describe the electronic structure. The procedure consists essentially of determining the total number of orbitals and electrons available for bonding; the number of B—H bonds and B—H—B three-center bonds is then counted and the requisite orbitals and electrons allotted. The remaining orbitals and electrons, considered to be available for framework bonding, are distributed among two-center B—B bonds and three-center B—B—B bonds. An algorithm (systematic prescription) for accomplishing this is outlined.

Consider a neutral borane whose formula can be written as B_pH_{p+q}. The molecule consists of p (BH) groups and q "extra" hydrogens distributed between bridging positions and BH groups. If we let

$$s = \text{number of B—H—B bonds}$$

$$t = \text{number of B—B—B three-center bonds}$$

$$y = \text{number of B—B bonds}$$

$$x = \text{number of "excess" H's attached to BH groups}$$

then we can formulate several relations between structural features and available orbitals and electrons, called **Equations of Balance.**

Hydrogen Balance

$$s + x = q$$

All "extra" hydrogens must be in B—H—B bonds or attached to BH forming BH_2 (or occasionally BH_3).

Orbital Balance

$$s + t = p$$

Each of the p B atoms must participate in one three-center bond if it is to attain an octet. This can be either a B—H—B or B—B—B bond.

Electron Balance.

The total number of electron pairs available for framework bonding is p from the BH groups (because each B uses one electron in bonding to the terminal H) plus $\frac{1}{2}q$ from the "extra" H's. These must be just enough to occupy the $s + t + y$ framework bonds and the x bonds to BH. Hence,

$$p + \tfrac{1}{2}q = s + t + y + x \tag{17.1}$$

Substituting from the hydrogen balance equation, we obtain

$$p = t + y + q/2 \qquad (17.2)$$

Applying the Equations of Balance to a compound of given composition allows us to determine a set of *styx* numbers that specify a valence structure. For example, for B_2H_6 [$(BH)_2H_4$], $p = 2$, $q = 4$, and we have $4 = s + x$, $2 = s + t$, $0 = t + y$. Because all numbers must be ≥ 0, the only possible solution is $s = 2$, $t = 0$, $y = 0$, $x = 2$ (written 2002); the structure corresponds to that already given for diborane:

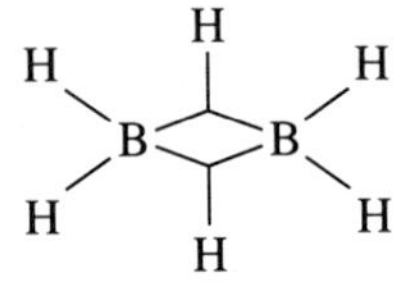

2002

You can write structures for higher boranes with this same procedure. For B_5H_9 [$(BH)_5H_4$], the *nido* species with a square-pyramidal framework is shown in Figure 17.1, where we have $4 = s + x$, $5 = s + t$, $5 - 2 = 3 = t + y$. Three sets of *styx* numbers are possible solutions 2302, 3211, and 4120. Obviously, the Equations of Balance do not always give unequivocal answers, but they do aid us by limiting the structures to be considered. In choosing the best structure, keep in mind these additional considerations.

1. Every pair of adjacent B's must be bonded to each other through a B—B, B—H—B, or B—B—B bond.

2. Pairs of B atoms bonded by a B—B bond may not also be bonded to one another by B—B—B or B—H—B bonds.

3. Nonadjacent pairs of B atoms may not be bonded by B—B framework bonds.

4. Other things being equal, the preferred structure is the one with the highest symmetry.

These considerations eliminate the first two solutions, leaving the 4120 structure (see Figure 17.4). Although no one structure shown exhibits the C_{4v} symmetry of the B_5 framework, the entire group of symmetry-equivalent structures taken together does. A C_4 operation yields each successive structure from the preceding one. Of course, the existence of several symmetry-equivalent structures corresponds to electron delocalization.

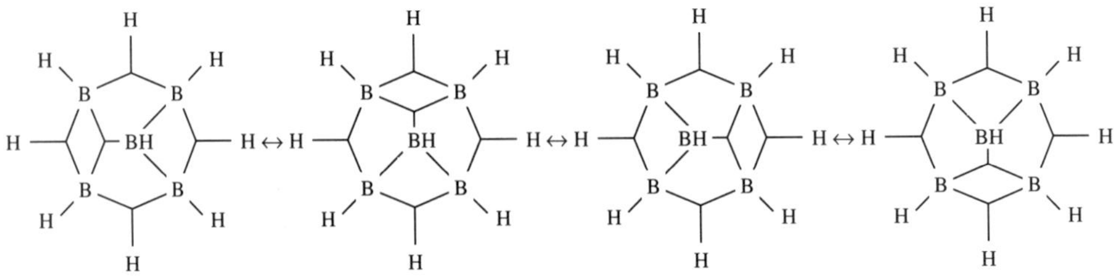

Figure 17.4   Symmetry-equivalent 4120 structures for B_5H_9.

We can modify the Equations of Balance to treat species of charge c. If we imagine that positive charge is created by protonation of neutral species, then the formulas can be written in the form $[(BH)_p H_{q+c}]^c$, and the Equations of Balance become

$$s + x = q + c$$

$$s + t = p + c$$

$$p = t + y + q/2 + c$$

Thus for $B_3 H_8^-$, $p = 3$, $q + c = 5$, $c = -1$, and $q = 6$; the two solutions are 1104 and 2013. The structures are

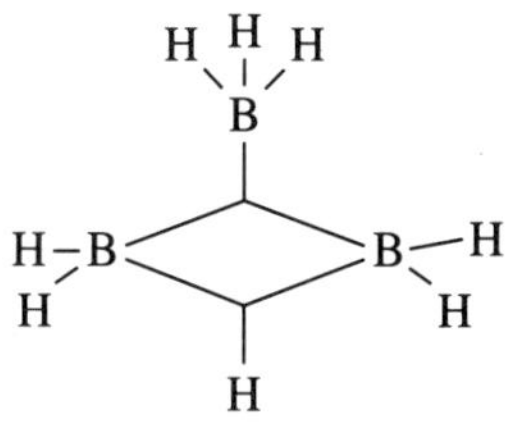

1104 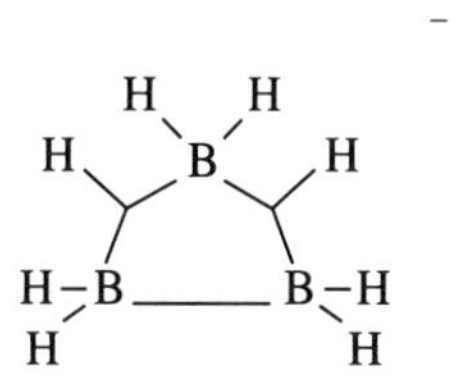   2013

The 2013 structure is found in the solid state. Proton exchange in solution may occur through the 1104 structure. Lipscomb[3] also has formulated rules for estimating formal charges on individual atoms from these VB structures.

The number of possible valence structures increases rapidly with the complexity of the species; this indicates considerable electron delocalization in the higher boron hydrides. Hence, a simpler approach to electronic structures lies in the MO model (also useful in treating delocalization in "regular" species). As usual, it is necessary to know something about the molecular geometry before the MO approach can be applied usefully.

17.3   STRUCTURES OF THE BORON HYDRIDES

X-ray diffraction, electron diffraction, and ^{11}B NMR techniques have allowed elucidation of the structures of many boron hydride species besides those prepared by Stock. Figure 17.5 shows idealized structures for boron hydrides. With the exception of $B_2 H_6$ and $B_3 H_8^-$ (not shown), these cage structures are based on polyhedra (Figure 17.6) or polyhedral fragments having triangular faces with BH groups at vertices. Such polyhedra are often called **deltahedra**, and they also form the structural basis for cluster compounds which contain metal–metal bonds. Boron hydrides adopt such structures because of the favorable possibilities for three-center bonding by borons sharing a triangular face.

The structures of several known hydrides result from joining deltahedral frameworks or fragments by sharing edges or vertices or by formation of B—B bonds with elimination of H_2. Some species of this type are shown in Figure 17.7.

---

[3] W. N. Lipscomb, *Boron Hydrides*, Benjamin, New York, 1963.

closo   nido   arachno   number of vertices

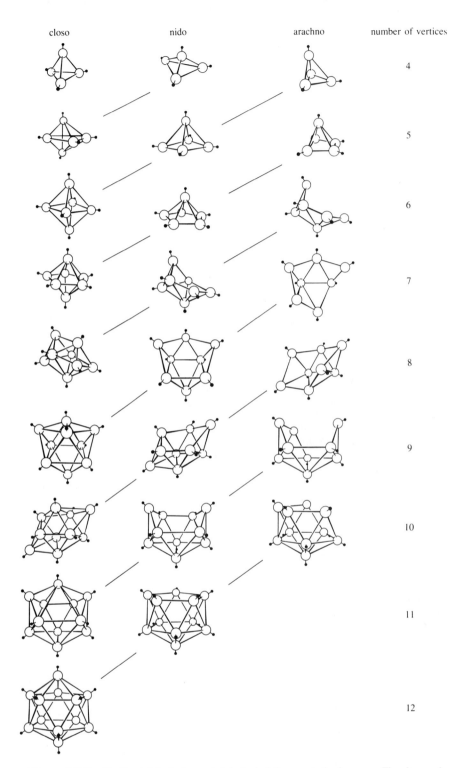

4

5

6

7

8

9

10

11

12

Figure 17.5 Idealized deltahedra and deltahedral fragments for boranes. The diagonal progressions represent excision of successive BH groups while retaining their electrons, thus generating *nido* from *closo* and *arachno* from *nido* species. (Reprinted with permission from R. W. Rudolph, *Acc. Chem. Res.* **1976**, *9*, 446. Copyright 1976, American Chemical Society.)

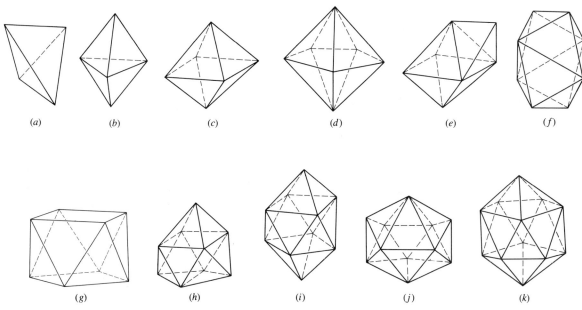

Figure 17.6   Closed triangulated polyhedra that form the structural bases for cage and cluster compounds. (*a*) Tetrahedron (**T**$_d$). (*b*) Trigonal bipyramid (**D**$_{3h}$). (*c*) Octahedron (**O**$_h$). (*d*) Pentagonal bipyramid (**D**$_{5d}$). (*e*) Capped octahedron (**C**$_s$). (*f*) Dodecahedron (**D**$_{2h}$). g) Archimedean (square) antiprism (**D**$_{4d}$). (*h*) Tricapped trigonal prism (**C**$_{3v}$). (*i*) Bicapped Archimedean antiprism (**D**$_{4d}$). (*j*) Octadecahedron (**C**$_{2v}$). (*k*) Icosahedron (**I**$_h$).

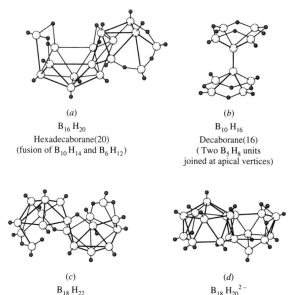

(*a*)
$B_{16}H_{20}$
Hexadecaborane(20)
(fusion of $B_{10}H_{14}$ and B_6H_{12})

(*b*)
$B_{10}H_{16}$
Decaborane(16)
(Two B_5H_8 units
joined at apical vertices)

(*c*)
$B_{18}H_{22}$
Octadecaborane(22)
(fusion of two
$B_{10}H_{14}$ units)

(*d*)
$B_{18}H_{20}{}^{2-}$
Octadecaborate(20) (2−)
(fusion of two $B_{10}H_{14}$ units
deprotonated)

Figure 17.7   Some structures containing fused and bonded boron cages. (Structures *a*, *b*, and *c* are from S. G. Shore, in *Boron Hydride Chemistry*, E. L. Muetterties, Ed., Academic Press, New York, 1975. Structure *d* is from X. L. R. Fontaine, N. N. Greenwood, J. D. Kennedy, and P. MacKinnon, *J. Chem. Soc. Dalton Trans.* **1988**, 1785.)

Conceptually, it is simplest to regard the structures of the *nido*-B_pH_{p+4} and *arachno*-B_pH_{p+6} series as derived from those of the *closo*-$B_pH_p^{2-}$ anions. The *closo* species (known for $p = 6-12$) adopt structures based on closed deltahedra having all vertices occupied by BH units. If one BH unit is removed from the most highly connected vertex and the two electrons it contributes are left behind, a series of hypothetical anions $B_pH_p^{4-}$ results. If these are protonated by distributing 4 H^+ in the most symmetrical fashion, each bridging two B's around the open face, the series of neutral *nido* compounds B_pH_{p+4} results. Removing a second BH unit of high connectivity on the open face of the B_pH_{p+4} molecule and leaving its pair of electrons gives the hypothetical $B_pH_{p+4}^{2-}$ series of anions. If these species are protonated, the *arachno* compounds B_pH_{p+6}, with a still more open structure, result. The placement of "extra" protons in the *arachno* species is not always easy to predict. At least one is always attached to a boron adjacent to the BH unit formally removed. This creates a BH_2 unit, a structural feature appearing for the first time in the *arachno* series. (The lone exception to this statement is B_2H_6.)

In summary, we can think of the known boranes as derived from parent *closo* anions $B_pH_p^{2-}$ by successive excision of BH units which leave behind their electron pairs followed by protonation of the resulting hypothetical anions to give B_pH_{p+4} and B_pH_{p+6} compounds featuring progressively more open structures. This feature is reflected in their names: *closo* = "closed"; *nido* is Greek "nest"; *arachno* and *hypho* mean "web" and "net", respectively. In the *nido* series, the "extra" H's are distributed around the open face in bridging positions. In the *arachno* series, at least one is placed so as to give a terminal BH_2 unit. The diagonal series in Figure 17.5 show this process. Clearly, the formulas of the hydrides have structural consequences: As the H/B ratio increases, the *closo*, *nido*, and *arachno* series exhibit progressively more open structures as well as increased reactivity.

17.4 MOLECULAR ORBITAL DESCRIPTION OF BONDING IN BORON HYDRIDES

The MO method provides a more satisfactory picture of electron delocalization in boranes than does the VB approach. A good illustration is[4] *closo*-$B_6H_6^{2-}$. Each BH group contributes a radially directed *sp* hybrid and two tangential *p* orbitals to framework bonding (Figure 17.8); each B also contributes two electrons. The other B *sp* hybrid and electron go to form the bond to the exopolyhedral (*exo*) H. Figure 17.9 shows the framework MOs from radial *sp* hybrids (a_{1g}, e_g, and t_{1u}) and tangential *p*'s (t_{1g}, t_{2g}, t_{1u}, t_{2u}). Their bonding or antibonding properties are obvious from the pictorial representations and the principle that energy increases with the number of nodes.[5] Interaction between radial and tangential t_{1u}'s lowers the energy of $1t_{1u}$, leading to the MO energy diagram of Figure 17.10. The a_{1g}, t_{2g}, and $1t_{1u}$ (the in-phase combination of the radial and tangential t_{1u}'s) orbitals are

---

[4] See Section 4.4 for B_2H_6.

[5] Here we are referring to the number of nodes related to the combining of atomic orbitals (AOs) into MOs because *p* orbitals all have an intrinsic node.

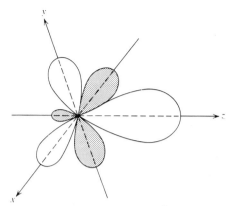

Figure 17.8   The radially directed *sp* hybrid along *z* and the tangentially directed p_x and p_y orbitals used for framework bonding in boron hydrides.

bonding, whereas the e_g, t_{1g}, $2t_{1u}$ (out-of-phase combination), and t_{2u} are antibonding. The total of seven electron pairs (one from each BH unit and one from the charge) exactly fills the bonding MOs. This *closo* species with six vertices has an MO energy diagram which accommodates seven pairs in bonding orbitals, leaving a large highest occupied molecular orbital (HOMO)–lowest unoccupied molecular orbital (LUMO) gap. In general, graph theoretical methods show that the *closo* anions $B_pH_p^{2-}$ can accommodate $(p + 1)$ electron pairs in bonding framework MOs. Thus, any species having *p* BH groups and $(p + 1)$ electron pairs for framework bonding should adopt a *closo* structure.

Electron-deficient molecules usually have only bonding orbitals filled, whereas most other stable molecules have both bonding and nonbonding orbitals filled. In constructing MO descriptions of other boranes, you may find it convenient to use hybrids whose geometry corresponds closely to bond directions in the structure—for example, sp^3 for B_4H_{10} and for the basal borons in B_5H_9.

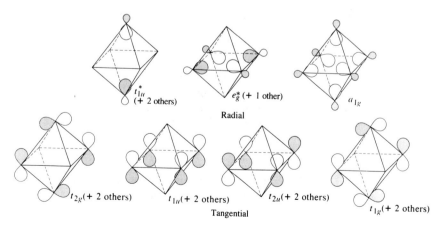

Figure 17.9   Framework orbitals for $B_6H_6^{2-}$.

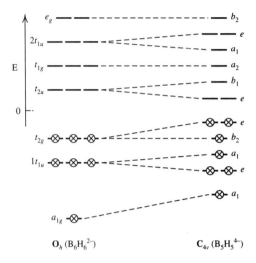

Figure 17.10 Energy-level diagrams for framework bonding orbitals in $B_6H_6^{2-}$ and $B_5H_5^{4-}$. The "exo" hydrogens involved in localized two-center bonds are not included here.

Suppose a BH unit is now removed from $B_6H_6^{2-}$ and its electron pair retained to generate $B_5H_5^{4-}$. The five BH groups occupy all but one of the six octahedral vertices, giving a C_{4v} square pyramid. Figure 17.10 depicts the fate of $B_6H_6^{2-}$ MOs on formation of $B_5H_5^{4-}$. The main change is that three framework orbitals are lost and that triply degenerate MOs split. Interactions between bonding and antibonding a_1 and e orbitals depress the energies of lower ones so that there are still seven bonding orbitals. The seven skeletal electron pairs (one from each BH unit and two from the overall charge) exactly fill the bonding MOs. A general feature[6] of *nido* compounds is that their energy-level diagrams provide stable levels for $(p + 2)$ skeletal electron pairs when the p BH units occupy all but one of the vertices of the $(p + 1)$ *closo* polyhedron. Thus we can expect any species having p BH units and $(p + 2)$ skeletal electron pairs to adopt a *nido* structure with the BH units occupying all but one vertex of the $(p + 1)$ *closo* polyhedron.

We can extend the same procedure to produce *arachno*-$B_4H_4^{6-}$. A general result is that the energy diagrams of $B_pH_p^{6-}$ anions provide stable orbitals for $(p + 3)$ skeletal electron pairs when the BH units occupy all but two adjacent vertices of the $(p + 2)$ *closo* polyhedron.

You should note that the framework orbitals for $B_pH_p^{2-}$ are the same as the ligand-group orbitals (LGOs) for a complex ML_p with the same symmetry. Thus, the combinations of the six radial sp orbitals for $B_6H_6^{2-}$ are the same as the σ LGOs for ML_6, and the tangential combinations correspond to the π LGOs for ML_6. The LGOs match in symmetry the orbitals of a *hypothetical* atom at the center of the cluster. Because the l quantum number of AOs dictates their symmetry properties and labels in the point group, the symmetry labels of "matching" framework orbitals also describe the overlap patterns and count the nodes of the cluster orbitals which increase with l. Thus, their relative energies are

[6] R. B. King and D. H. Rouvray, *J. Am. Chem. Soc.* **1977,** *99,* 7834.

specified. The set of O_h MOs is shown below along with their "matching" central AOs:

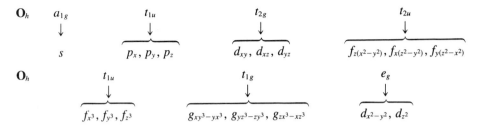

The radial cluster orbitals for O_h $B_6H_6^{2-}$ are $a_{1g} + t_{1u} + e_g$. These increase in energy in the order $a_{1g} < t_{1u} < e_g$ just as the energies of the matching AOs increase with l: $s < p < d$. a_{1g} is bonding, t_{1u} is nonbonding, and e_g is antibonding. Tangential orbitals have overlaps of π-symmetry (e.g., t_{1u}) or overlaps which can be resolved into $\sigma + \pi$ components (e.g., t_{2g}), and their energies thus lie between a_{1g}^b and e_g^*. Interaction of the radial and tangential t_{1u}'s depresses the energy of the former, making $1t_{1u}$ bonding. Hence, energies of tangential MOs lie between $1t_{1u}^b$ and e_g^* and increase in the order $t_{2g} < t_{2u} < t_{1u} < t_{1g}$, paralleling those of the matching AOs $d < f_{t_{2u}} < f_{t_{1u}} < g$. The difference in energy between parent $f_{t_{2u}}$ and $f_{t_{1u}}$ can be justified by crystal-field arguments. The actual energy order involves an inversion such that $t_{1g} < 2t_{1u}$ due to interaction of radial and tangential t_{1u}. The a_{2u} orbital of the f set (which would match f_{xyz}) is not needed since the reducible representation of the set of valence orbitals does not include a_{2u}.

The $B_5H_5^{4-}$ energy level diagram can also be derived using this general procedure which works best with close-to-spherical symmetry.

First, we obtain the symmetry species of the cluster orbitals. (We had already done this for the O_h case in Chapter 10.) The valence orbitals are s and p_x, p_y, p_z on each B. Only orbitals on atoms unshifted by a symmetry operation contribute to the character of the reducible representation for which the collection of valence orbitals serves as a basis. For each unshifted atom, the character for the p's is the same as that for x, y, and z. Because s always transforms as the totally symmetric representation, its character is $+1$ for every operation. For C_{4v} we have the following:

	E	$2C_4$	C_2	$2\sigma_v$	$2\sigma_d$
$\Gamma_{x,y,z}$	3	1	-1	1	1
Γ_s	1	1	1	1	1
$\Gamma_{\text{val. orbs.}}$ (sum)	4	2	0	2	2

The reducible representation characters are obtained by multiplying the number of unshifted atoms by the resulting $\Gamma_{\text{val.orbs}}$ for each atom:

Number of unshifted atoms:	5	1	1	3	1
Γ_{total} (product)	20	2	0	6	2

This reducible representation reduces to $5A_1 + A_2 + 3B_1 + B_2 + 5E$. Because we consider the cluster orbitals to be the radial and tangential orbitals, we must subtract out the irreducible representations for the *exo sp* hybrids which are $2A_1 + B_1 + E$. Thus, the cluster bonding orbitals span the remaining irreducible representations $3A_1 + A_2 + 2B_1 + B_2 + 4E$.

Next, we need to determine the relative orbital energies. We use arguments similar to those for the O_h case starting with the octahedral energies already derived:

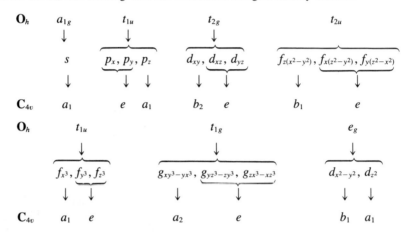

In C_{4v} $B_5H_5^{4-}$, three orbitals are lost because of removal of a BH from $B_6H_6^{2-}$; only the 15 lowest-energy orbitals are required from the O_h set of 18. We can see that the lowest-energy a_{1g} of $B_6H_6^{2-}$ correlates with a_1 of $B_5H_5^{4-}$. t_{1u} splits into $e + a_1$, whereas t_{2g} splits into $b_2 + e$. The "lost" orbitals are the high-energy e from t_{1g} and the a_1 from e_g because all needed orbitals of these symmetries have already been generated from AOs of lower l. The a_1 and e orbitals generated from bonding orbitals are depressed in energy by interaction with same-symmetry species whose parentage lies in antibonding orbitals. Thus, seven MOs remain bonding, and filling these with $7 = (5 + 2)$ electron pairs demonstrates the stability of $B_5H_5^{4-}$. This interaction also raises the energies of antibonding orbitals placing $a_1 + e$ from $2t_{1u}^*$ above a_2 which has no bonding counterpart with which to interact.

In viewing boranes as derived from anions, the "extra" hydrogens required to give neutral species from the (hypothetical) anions are thought of as arising by protonation (usually in the most symmetrical way) of B—B bonds around the open face (in *nido* and *arachno* compounds) or of BH groups to give BH_2 units (in *arachno* compounds).

A significant result of a generalized MO approach is the classification of boranes as *closo*, *nido*, or *arachno* via framework electron count as well as by molecular formula. Thus a link between electronic structure and molecular geometry is established for electron-deficient species.

Framework electron counting can be done from molecular formulas for species of the type $[(BH)_p H_{q+c}]^c$ by assuming the following contributions:

$$1 \text{ electron pair from each BH unit}$$
$$\tfrac{1}{2} \text{ electron pair from each "extra" H}$$
$$-\tfrac{1}{2} \text{ electron pair from each } + \text{ charge}$$

Hence, the total number of electron pairs is

$$p + \tfrac{1}{2}(q + c) - \tfrac{1}{2}c = p + \tfrac{1}{2}q$$

where p is the number of vertices. Thus, $B_9H_{14}^-$ has $p = 9$, $c = -1$, and $q + c = q - 1 = 5$, so that $q = 6$. The species has $p + \tfrac{1}{2}q = 9 + 3$ skeletal electron pairs and is expected to exhibit an *arachno* structure. Figure 17.11 shows the structure of isoelectronic $B_9H_{13}^{2-}$. The second B—H group excised from the related B_{11} *closo* structure is not the same as that shown in Figure 17.5. This shows that *arachno* structures are less certainly predicted than *nido* ones, although they can be rationalized as discussed above. The four "extra" H's are disordered over five B's in the crystal; however, all the H's are on the open face.

We also can use *styx* numbers to count framework electrons pairs.[7] The total number of bonding pairs is exactly that required to fill the $s + t + y + x$ orbitals, and from the Equations of Balance we have

$$s + t + y + x \;=\; q + c + p - q/2 - c \;=\; p + \tfrac{1}{2}q \qquad (17.3)$$

For hydrides having p vertices of formula $[(BH)_pH_{q+c}]^c$, the relation between structure and skeletal electron count may be summarized as follows:

Structural type	Skeletal electron pairs
	$s + t + y + x = p + q/2$
closo	$p + 1$
nido	$p + 2$
arachno	$p + 3$
hypho	$p + 4$

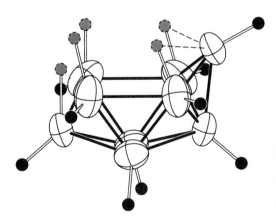

Figure 17.11 Structure of the $B_9H_{13}^{2-}$ dianion. Disordered H atoms are indicated by dashed circles. (Reprinted with permission from T. D. Getman, J. A. Krause, P. M. Niedenzu, and S. G. Shore, *Inorg. Chem.* **1989,** *28,* 1507. Copyright 1989, American Chemical Society.)

[7] W. N. Lipscomb, *Inorg. Chem.* **1979,** *18,* 2328.

Formula (17.3) is general. Note that in applying it to ions, the value of q is not straightforwardly given by the subscript of the "extra" H's and must be calculated. Thus, for $[B_6H_{11}]^+ = [(BH)_6H_5]^+$, $q + c = q + 1 = 5$ and $q = 4$. Hence, $p + q/2 = 6 + 2$ for a *nido* classification.

17.5   HETEROBORANES

In many known compounds, one or more framework BH groups are replaced by atoms of a different element. For example, CH^+, P^+, and S^{2+} are all equivalent to BH (having two electrons for framework bonding plus two more in a bond to an *exo*-H or as a lone pair). Hence, a series of neutral *closo* compounds exists of formula $C_2B_pH_{p+2}$ derived by substitution of CH^+ for BH in the $B_pH_p^{2-}$ series. These are called **carboranes.** Although $B_5H_5^{2-}$ has not yet been prepared, isoelectronic $C_2B_3H_5$ having the anticipated trigonal-bipyramid (TBP) structure is known. Likewise, other generic types of *closo* compounds include SB_pH_p, B_pCH_{p+2}, and B_pPH_{p+1}. Similar series of *nido* and *arachno* heteroboranes exist, although all members of any series are not always known. For example, $C_2B_4H_8$ is isoelectronic with B_6H_{10}, and $C_2B_7H_{13}$ is isoelectronic with B_9H_{15}; that is, both pairs of species have the same skeletal electron count. Structures of these species are related simply to those of the parent boron hydrides, except that atoms of differing size may cause some distortion of the polyhedral framework. (See Figure 17.12 for $Bi_2B_{10}H_{10}$.)

A frequently encountered problem is the reverse of the one above—namely, given the molecular formula, to predict the structure of the heteroborane. We can extend previous methods in a straightforward way to carboranes $[(CH)_a(BH)_pH_{q+c}]^c$. The number of electron pairs contributed by CH is $\frac{3}{2}$, so the total number of skeletal electron pairs is:

$$\tfrac{3}{2}a + p + \tfrac{1}{2}(q + c) - \tfrac{1}{2}c = (a + p) + \tfrac{1}{2}(a + q) = n + \tfrac{1}{2}(a + q) \qquad (17.4)$$

where n is the number of polyhedral vertices. Hence for $C_2B_8H_{12} = [(CH)_2(BH)_8H_2]$, $a = 2$, $p = 8$, $q = 2$, $c = 0$, and the number of electron pairs is $10 + \tfrac{1}{2}(2 + 2) =$

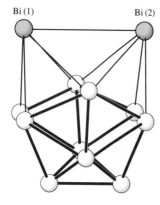

Bi (1)   Bi (2)

Figure 17.12   The framework of 1,2,-$Bi_2B_{10}H_{10}$ showing distortion of the icosahedron due to differences in B and Bi sizes. (Reprinted with permission from J. L. Little, M. A. Whitesell, J. G. Kester, K. Folting, and L. J. Todd, *Inorg. Chem.* **1990,** *29*, 804. Copyright 1990, American Chemical Society.)

10 + 2. A *nido* structure is expected. The "extra" hydrogens in bridging positions in carboranes normally are not bonded to C.

Nomenclature for Heteroboranes

Vertices of *closo, nido,* and *arachno* polyhedra are given numbers based by convention on planar projections of polyhedral structures. The numbering is by zones (planes) perpendicular to the major axis (see Figure 17.13). Numbering is shown for B_5, *closo*-B_{13}, and B_{12}. Interior vertices on the projection are numbered first, then peripheral ones. This corresponds to numbering apical vertices with lowest numbers. Numbering proceeds clockwise starting from the twelve o'clock position or at the first position clockwise. Heteroatom locations can be specified by numbers. Thus, the $C_2B_3H_7$ isomer shown in Figure 17.13 is 1,2-dicarba-*nido*-pentaborane(7). $PB_{11}H_{12}$ is phospha-*closo*-dodecaborane(12). There is no need to specify the P position because all icosahedral vertices are equivalent.

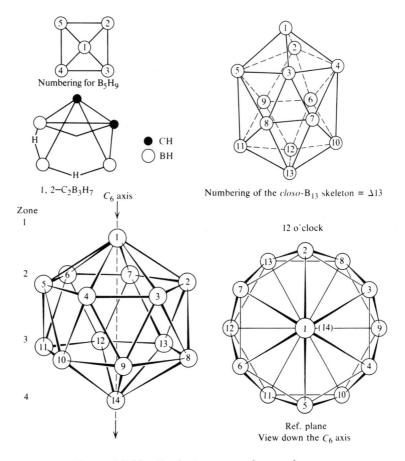

Figure 17.13   Numbering systems for some boranes.

Table 17.3 Isoelectronic equivalents of various groups

Group	1e	2e	3e	BH 4e	CH 5e	(BH + 2e) 6e	(BH + 3e) 7e
2	Be$^+$	Be/BeH$^+$	BeH	BeH$^-$ Be(NR$_3$)			
		Mg					
12		Zn		ZnH$^-$	ZnH$_2^{2-}$		
		Cd		CdH$^-$	CdH$_2^{2-}$		
13	B$^+$		B/BH$^+$	BH	BH$^-$ B(NR$_3$)		
	Al$^+$		Al/AlH$^+$	AlH	AlH$^-$		
	Ga$^+$		Ga/GaH$^+$	GaH	GaH$^-$		
	In$^+$		In/InH$^+$	InH	InH$^-$		
14			C$^+$	C/CH$^+$	CH	CH$^-$ C(NR$_3$)	
			Si$^+$	Si	SiH	SiH$^-$	
			Ge$^+$	Ge	GeH	GeH$^-$	
			Sn$^+$	Sn	SnH	SnH$^-$	
15				N$^+$	N	NH	NH$^-$
				P$^+$	P/PH$^+$	PH	PH$^-$
				As$^+$	As	AsH	AsH$^-$
				Sb$^+$	Sb	SbH	SbH$^-$
16					S$^+$	S/SH$^+$	SH
					Se$^+$	Se	SeH
					Te$^+$	Te	TeH

Table 17.3, which gives isoelectronic equivalents of various groups, can be used to predict and rationalize the structures of heteroboranes. Note especially that B : L (where L is a Lewis base) is isoelectronic with CH, and thus we expect the structures of Lewis base adducts of the boron hydrides to be related to those of the carboranes.

17.6 CHEMISTRY OF THE BORANES

Boron hydrides are extremely reactive with air, often explosively so. Their heats of combustion are quite large—up to 1.5 times greater than those of the most nearly comparable hydrocarbons. Accordingly, it is not surprising that they have been of interest as potential rocket fuels (a fact which provided impetus for several years for the investigation of their chemistry). Unfortunately, the combustion product is solid boric oxide which does not provide sufficient thrust and which coats engine parts.

Thermal stability of boron hydrides varies widely. Whereas B_8H_{12} decomposes within a few minutes at room temperature, B_5H_9 can be handled at temperatures above 250°C. $Cs_2B_{12}H_{12}$ is stable on heating to 810°C in a sealed, evacuated quartz tube, whereas $Cs_2B_{10}H_{10}$ is stable to 600°C. This behavior reflects the low reactivity of closed structures.

All boron hydrides are thermodynamically unstable to hydrolysis that affords $B(OH)_3$ and H_2. Some species, especially $B_{10}H_{10}^{2-}$ and $B_{12}H_{12}^{2-}$, are very kinetically stable, however. In general, the closed structures of $B_pH_p^{2-}$ species lead to lower reactivity than is the case for *nido* and *arachno* boranes. $B_{12}H_{12}^{2-}$ can be handled in open beakers in aqueous solutions. It has recently been reported[8] that pentaborane(9) solutions in thf are indefinitely stable under inert atmosphere and are sufficiently stable in dry air for synthetic work. $BH_3 \cdot$ thf and $BH_3 \cdot SMe_2$, which may be handled with syringe and septum techniques, are commercially available substitutes for B_2H_6.

▶ 17.6.1 *Reactions with Lewis Bases*

Not surprisingly, the electron-deficient boron hydrides react with reagents having unshared electron pairs (Lewis bases). Depending on the base and the reaction conditions, several types of reactions have been observed: **cleavage** (symmetric and unsymmetric), **addition,** and **bridge proton abstraction.**

Base Cleavage

Diborane undergoes **symmetric** and **unsymmetric bridge cleavage** reactions with Lewis bases:

$$\tfrac{1}{2}\ \text{[diborane]} + L \longrightarrow LBH_3 \quad \text{(Symmetric)} \qquad (17.5)$$

$$\text{[diborane]} + 2L \longrightarrow [L_2BH_2]^+BH_4^- \quad \text{(Unsymmetric)} \qquad (17.6)$$

[8] S. M. Cendrowski-Guillaume and J. T. Spencer, *Organometallics* **1992**, *11*, 969.

where L = amines, ethers, phosphines, H⁻, NCS⁻, or CN⁻. Some symmetric cleavage reactions of diborane are:

$$\tfrac{1}{2}\,B_2H_6 + Me_2O \longrightarrow Me_2O:BH_3 \tag{17.7}$$

$$\tfrac{1}{2}\,B_2H_6 + CO \longrightarrow H_3B:CO \tag{17.8}$$

$$\tfrac{1}{2}\,B_2H_6 + C_5H_5N \longrightarrow C_5H_5N:BH_3 \tag{17.9}$$

$$\tfrac{1}{2}\,B_2H_6 + Me_3N \longrightarrow Me_3N:BH_3 \tag{17.10}$$

$$\tfrac{1}{2}\,B_2H_6 + MH \longrightarrow MBH_4 \tag{17.11}$$

In all of the above reactions, diborane is split into BH_3 units, each of which accepts an electron pair from the attacking reagent, thereby completing its octet of electrons. The order of thermal stability of the products is generally that expected on the basis of strength of the Lewis base (see Chapter 7), with dimethyl ether-borane being the least stable of the above and the borohydride the most stable.

Reaction (17.7) has been used in the purification of diborane. When dimethyl ether is used, the unstable addition compound with diborane is formed at $-78.5°C$. Other volatile compounds that may be present are removed by pumping on the addition compound at $-100°C$. On warming to $-78.5°C$, the addition compound is dissociated sufficiently to allow vacuum-line fractionation of the dimethyl ether and the regenerated diborane.[9]

Reaction (17.11) proceeds only in the presence of a solvent, diethyl ether being satisfactory for the formation of $LiBH_4$ and diglyme $[(MeOCH_2CH_2)_2O]$ for $NaBH_4$. Borohydrides are quite stable thermally; $LiBH_4$ loses hydrogen at $275°C$, and $NaBH_4$ loses hydrogen above $400°C$. BH_4^- is reasonably stable in water at high pH. Lithium borohydride reacts slowly with water, apparently because hydrolysis of the Li^+ ions increases the hydrogen ion concentration (see Section 7.3.2). Other borohydrides may be prepared by metathesis with $LiBH_4$ and $NaBH_4$.

Ammonia reacts with diborane to afford the ionic product $[(NH_3)_2BH_2]^+BH_4^-$ of unsymmetrical cleavage, known as the **"diammoniate of diborane."** It is not completely clear which factors promote one type of cleavage over the other. Probably, the first step is the displacement of bridging H, giving a singly bridged species:

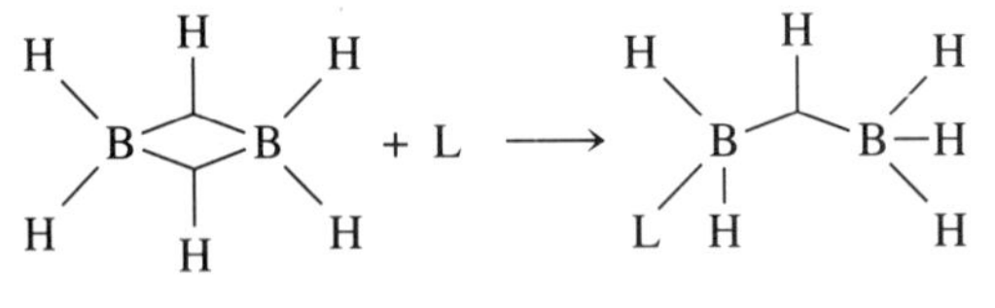

$$\tag{17.12}$$

---

[9] The technique of purifying a volatile compound through formation of an adduct which is stable at one temperature but dissociates at a higher temperature is a fairly general one. Other examples include the purification of HF through the KHF_2 adduct, of BF_3 through the anisole adduct, and low-molecular-weight ethers through the $LiBH_4$ adducts.

When L = H^-, we can prepare salts of the monobridged $B_2H_7^-$ anion that are stable in vaccum at room temperature. The second step involves another bridge displacement and leads to symmetric cleavage (via attack on the BH_3 boron) or unsymmetric cleavage (via attack on the BH_2L boron). When L is bulky, symmetric cleavage is more likely. Electronic factors may also be important. As in much of boron chemistry, solvent plays an important role, with more basic solvents giving a higher yield of symmetric cleavage products. Many adducts $L \cdot BH_3$ can be prepared via displacement of weaker Lewis bases by stronger ones, especially using $BH_3 \cdot thf$.

Reactions of higher boron hydrides resemble those of diborane, especially when a BH_2 group is present. Otherwise, fragmentation by Lewis bases occurs less readily. Tetraborane reacts with L (where L may be Me_3N, R_2O, Me_3P, or H^-) to give adducts of BH_3 and B_3H_7, with the exact structure of $B_3H_7 \cdot L$ dependent on L:

$$\text{(17.13)}$$

$B_3H_7 \cdot L$ is isoelectronic with the unknown B_3H_9. When an excess of L = Me_3N is used, $Me_3N : BH_3$ and $[HB : NMe_3]_x$ result from further attack.

The reaction of tetraborane with ammonia proceeds by unsymmetrical cleavage, as in the case of diborane:

$$B_4H_{10} + 2\,NH_3 \longrightarrow [H_2B(NH_3)_2]^+[B_3H_8]^- \tag{17.14}$$

thf affords an unsymmetric cleavage product at low temperature, $[(thf)_2BH_2]^+B_3H_8^-$, which rearranges to $BH_3(thf)$ and $B_3H_7(thf)$ on warming.

We can obtain an ammonia–triborane adduct in the same fashion as the ammonia–borane adduct—that is, by displacing the ether from an etherate with ammonia:

$$NH_3 + R_2O : B_3H_7 \longrightarrow H_3N : B_3H_7 + R_2O \tag{17.15}$$

Even though B_5H_9 has no BH_2 group, it is cleaved unsymmetrically by NH_3:

$$B_5H_9 + 2\,NH_3 \longrightarrow [(NH_3)_2BH_2]^+[B_4H_7]^- \tag{17.16}$$

Other higher hydrides are also cleaved by Lewis bases, with NH_3 usually yielding ionic products:

$$B_5H_{11} + 2\,NH_3 \longrightarrow [(NH_3)_2BH_2][B_4H_9] \tag{17.17}$$

$$B_5H_{11} + 2\,CO \longrightarrow BH_3CO + B_4H_8(CO) \tag{17.18}$$

$$B_6H_{12} + 2\,NH_3 \longrightarrow [(NH_3)_2BH_2][B_5H_{10}] \tag{17.19}$$

$$B_6H_{12} + PMe_3 \longrightarrow BH_3(PMe_3) + B_5H_9 \tag{17.20}$$

The amine boranes usually are much more stable in air than boron hydrides. Thus, a common way of disposing of reaction wastes is to condense a volatile amine into the reaction vessel and allow it to react on warming before exposing the contents to air.

Base Addition

Not all Lewis bases give cleavage products. Formation of the adduct $B_2H_7^-$ from B_2H_6 has been mentioned already. Evidence also exists for the monobridged $(H_3N)_2H_2B{\overset{\displaystyle H}{\diagup\diagdown}}BH_2$ which has not been isolated.

PMe$_3$ reacts with B_5H_9 to form the adduct $B_5H_9(PMe_3)_2$, whose structure is shown in Figure 17.14. The electron count of $p + 4$ makes this the first well-characterized *hypho* boron hydride.

Because $B_{10}H_{14}$ is air-stable, crystalline, and easily handled, its chemistry has been investigated extensively. Lewis bases such as CH_3CN, C_5H_5N, and Me_2S react with decaborane to liberate H_2:

$$2\,CH_3CN + B_{10}H_{14} \longrightarrow B_{10}H_{12}\cdot 2\,CH_3CN + H_2 \tag{17.21}$$

These have the ten-vertex *nido* structure shown in Figure 17.15 with base molecules bonded to the borons on the open face.

Triethylamine reacts with $B_{10}H_{12}\cdot 2CH_3CN$ to give two compounds of the formula

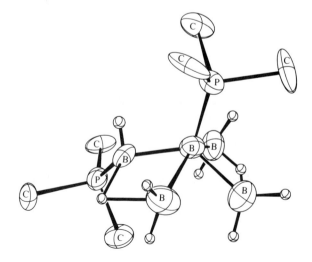

Figure 17.14 Structure of the *hypho* compound $B_5H_9(PMe_3)_2$. (Reprinted with permission from A. V. Fratini, G. W. Sullivan, M. L. Denniston, R. K. Hertz, and S. G. Shore, *J. Am. Chem. Soc.* **1974**, *96*, 3013. Copyright 1974, American Chemical Society.)

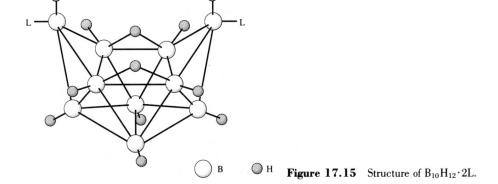

B ○ H ◉ **Figure 17.15** Structure of $B_{10}H_{12} \cdot 2L$.

$B_{10}H_{12} \cdot 2Et_3N$. One is assumed to be a covalent compound similar to those discussed above, since it is soluble in benzene and the amine readily may be displaced by triphenylphosphine. The other compound is ionic, containing triethylammonium ions and the *closo* anion $B_{10}H_{10}^{2-}$.

We also can obtain $B_{10}H_{10}^{2-}$ by direct reaction of triethylamine with decaborane [Equation (17.39)]. An interesting reaction of $B_{10}H_{12} \cdot 2Me_2S$ is that with $HC \equiv CH$, which affords 1,2-dicarbadodecahydro-*closo*-dodecaborane, $1,2\text{-}C_2B_{10}H_{12}$. This carborane also can be prepared directly from the reaction of Lewis base and acetylene with $B_{10}H_{14}$ without isolation of the intermediate (see Section 17.8).

Bridge Proton Abstraction

As early as 1956 it was shown that $B_{10}H_{14}$ behaved as an acid in aqueous media and could be titrated with NaOH. Exchange experiments revealed the exchange of the four bridging (μ) protons in decaborane and pointed to the acidic character of μ-H.

The bridging protons in *nido* and *arachno* boranes may be abstracted by (usually) very strong bases in ethers:

$$B_5H_9 + NaH \longrightarrow Na[B_5H_8] + H_2 \tag{17.22}$$

$$B_6H_{10} + KH \longrightarrow K[B_6H_9] + H_2 \tag{17.23}$$

$$B_{10}H_{14} + NH_3 \longrightarrow [NH_4][B_{10}H_{13}] \tag{17.24}$$

Metathesis to the tetraalkylammonium salts gives significantly more stable products.

The relative acidities of boron hydrides can be established by competition reactions such as

$$B_4H_9^- + B_{10}H_{14} \rightarrow B_4H_{10} + B_{10}H_{13}^- \tag{17.25}$$

showing that the acidity ordering is $B_4H_{10} < B_{10}H_{14}$. A more extensive list, in order of increasing acidity, is

nido $B_5H_9 < B_6H_{10} < B_{10}H_{14} < B_{16}H_{20}$

arachno $B_4H_{10} < B_5H_{11} < B_6H_{12}$

The experimental orderings $B_4H_{10} < B_{10}H_{14}$ and $B_5H_9 < B_6H_{12}$ provide a connection between the two series. Some idea of absolute acidity can be gained from the fact that NH_3 deprotonates B_4H_{10}, showing that $NH_4^+ < B_4H_{10}$ in ether solution. In general, larger boranes are more acidic than smaller ones, and *arachno* hydrides are more acidic than *nido* hydrides of comparable size.[10]

The borane conjugate bases contain electron-rich B—B bonds that can be reprotonated to give the parent boranes. Figure 17.16 shows the structure of $B_5H_8^-$ and $B_6H_9^-$ as elucidated by [11]B nuclear magnetic resonance (NMR).

These electron-rich B—B sites react with electrophiles to insert them into a bridging position. An isomer of $B_6H_8(CH_3)_2$ having a bridging $B(CH_3)_2$ group has been prepared from the reaction of LiB_5H_8 with $(CH_3)_2BCl$ at low temperature; its structure is shown in Figure 17.17. On warming, the compound undergoes cage expansion to give the more thermally stable 4,5-dimethylhexaborane(10).

▶ 17.6.2 *Electrophilic Substitution*

Terminal H atoms on all classes of boranes can be substituted by electrophiles. The most studied examples involve $B_{10}H_{10}^{2-}$, $B_{12}H_{12}^{2-}$, B_5H_9, and $B_{10}H_{14}$. Typical of these reactions are halogenations. $B_{10}H_{10}^{2-}$ has two different kinds of H, apical and equatorial, whereas in icosahedral $B_{12}H_{12}^{2-}$ all positions are equivalent. (See Figure 17.6.) Molecular orbital calculations indicate that negative charge on B atoms decreases in the order apical $B_{10}H_{10}^{2-} >$ equatorial $B_{10}H_{10}^{2-} \geq B_{12}H_{12}^{2-}$. As expected for electrophilic attack, the reactivity order is $B_{10}H_{10}^{2-} > B_{12}H_{12}^{2-}$ and $Cl_2 > Br_2 > I_2$. Halogenations proceed easily in the dark, and multiply substituted products are obtained. Fully substituted $B_{10}H_{10}^{2-}$ and $B_{12}X_{12}^{2-}$ (X = Cl, Br, I) can be prepared.

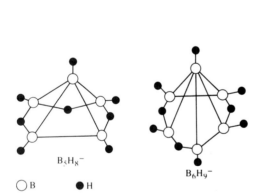

$B_5H_8^-$

○ B   ● H

Figure 17.16 Static structures of borane anions.

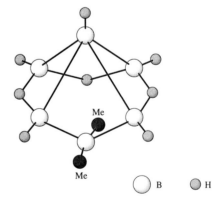

Me

Me

○ B   ● H

Figure 17.17 The low-temperature isomer of $B_6H_8Me_2$. (From S. G. Shore, in *Boron Hydride Chemistry*, E. L. Muetterties, Ed., Academic Press, New York, 1975.)

[10] R. J. Remmel, H. D. Johnson II, H. S. Jaworsky, and S. G. Shore, *J. Am. Chem. Soc.* **1975,** *97,* 5395.

B_5H_9 reacts with elemental halogens to give 1-XB_5H_8 in the presence of a Friedel–Crafts catalyst. In the absence of a catalyst, both 1- and 2-XB_5H_8 are obtained. The formation of the apical (1-) isomer is consistent with the assignment of greater negative charge to apical B and with the assumption that the ground-state charge distribution parallels that in the transition state for electrophilic attack. The following rearrangement occurs quantitatively, indicating that apical substitution is kinetically controlled:

$$1\text{-}ClB_5H_8 \xrightarrow{Et_2O} 2\text{-}ClB_5H_8 \tag{17.26}$$

Repeated substitution followed by isomerization allows the preparation of multiply substituted species.

Halogenation of $B_{10}H_{14}$ give both 1- and 2-$XB_{10}H_{13}$, with the latter predominating. Long treatment with Br_2 or I_2 gives 2,4- and 1,3-disubstituted derivatives. The reactivity toward electrophilic substitutions by several different reagents appears to be 2,4 > 1,3, paralleling the ground-state charge distribution.

Boranes undergo reactions with other electrophiles (alkyl halides, for example) that parallel those with halogens. Acids bring about H exchange except when an electron-rich B—B bond is available for protonation. An example is B_6H_{10} which reacts with HCl in BCl_3 to yield $B_6H_{11}^+$.

17.7 SYNTHESIS OF BORON HYDRIDES

Stock prepared boron hydrides from the hydrolysis of magnesium borides with aqueous acid. Because many boranes are susceptible to hydrolysis, he isolated those whose reaction with water were slowest—especially tetraborane (10). This compound then was pyrolyzed to give diborane and other boron hydrides:

$$2\,MgB_2 + 4\,Mg + 12\,HCl \longrightarrow B_4H_{10} + H_2 + 6\,MgCl_2 \tag{17.27}$$

$$B_4H_{10} \xrightarrow{\Delta} B_2H_6 + \text{higher boranes} \tag{17.28}$$

As Stock realized, protonic reagents are not good choices for preparing hydrides, since hydrolysis occurs with liberation of H_2. The metal hydrides now available are often employed—for example,

$$2\,KBH_4(s) + 2\,H_3PO_4 \longrightarrow B_2H_6(g) + 2\,KH_2PO_4(s) + 2\,H_2(g) \tag{17.29}$$

Historically, boranes often were prepared by pyrolysis of B_2H_6. We can write reaction schemes involving successive additions of BH_3 and H_2 eliminations to give most of the known *nido* and *arachno* boranes. In recent years, Shore has developed systematic procedures for making boron hydrides in good yield, taking advantage of the acidity of μ-H.

The anions resulting from H^+ abstraction may then react to add BH_3, presumably at electron-rich B—B bonds yielding species subject to cage expansion and to H_2 or MX elimination.

$$B_4H_{10} + KH \xrightarrow[Et_2O]{-78°C} K[B_4H_9] + H_2$$

$$K[B_4H_9] + \tfrac{1}{2}B_2H_6 \xrightarrow[Et_2O]{-35°C} K[B_5H_{12}] \qquad (17.30)$$

$$K[B_5H_{12}] + HCl(l) \xrightarrow{-110°C} B_5H_{11} + H_2 + KCl$$

$$1\text{-}BrB_5H_8 + KH \xrightarrow[Me_2O]{-78°C} K[1\text{-}BrB_5H_7] + H_2$$

$$K[1\text{-}BrB_5H_7] + \tfrac{1}{2}B_2H_6 \xrightarrow[Me_2O]{-78°C} K[B_6H_{10}Br] \qquad (17.31)$$

$$K[B_6H_{10}Br] \xrightarrow[Me_2O]{-35°C} B_6H_{10} + KBr$$

In another approach, H^- may be abstracted from anions with BX_3, giving species that transfer BH_3, sometimes with accompanying H_2 elimination.

$$[n\text{-}Bu_4N][BH_4] + BCl_3 \xrightarrow[CH_2Cl_2]{25°C} \tfrac{1}{2}B_2H_6 + [n\text{-}Bu_4N][HBCl_3] \qquad (17.32)$$

$$[n\text{-}Bu_4N][B_3H_8] + BBr_3 \xrightarrow{0°C} \tfrac{1}{2}B_4H_{10} + [n\text{-}Bu_4N][HBBr_3] + \tfrac{1}{x}(BH_2)_x \qquad (17.33)$$

$$[Me_4N][B_9H_{14}] + BCl_3 \xrightarrow{25°C} \tfrac{1}{2}B_{10}H_{14} + [Me_4N][HBCl_3] + \tfrac{1}{2}H_2 + \tfrac{1}{2}x(B_8H_{10})_x \qquad (17.34)$$

Commercially available B_5H_9 is a good starting material for preparation of higher boranes.[11] Deprotonation of B_5H_9 in the presence of excess B_5H_9 is a route to $B_9H_{14}^-$:

$$B_5H_9 + KH \xrightarrow{thf \text{ or glyme}} K[B_5H_8] + H_2$$

$$K[B_5H_{14}] + B_5H_9 \longrightarrow K[B_9H_{14}] + \tfrac{1}{2}B_2H_6 \qquad (17.35)$$

Because the reaction

$$K[B_5H_8] + BH_3 \longrightarrow K[B_6H_{11}] \qquad (17.36)$$

is known to occur, the yield of $K[B_5H_{14}]$ is maximized with a $1.8:1$ $B_5H_9:KH$ stoichiometry rather than the $2:1$ stoichiometry suggested by (17.35).

$B_9H_{14}^-$ can be protonated to give B_9H_{15} and can be used to make $B_{18}H_{22}$ and $B_9H_{13}\cdot MeCN$.

Hydride abstraction from $B_9H_{14}^-$ is a route to $B_{10}H_{14}$ in 55–60% yield:

$$[Me_4N][B_9H_{14}] + BCl_3 \xrightarrow[glyme]{25°C} \tfrac{1}{2}B_{10}H_{14} + [Me_4N][HBCl_3] + \tfrac{1}{2x}[B_8H_{10}]_x + \tfrac{1}{2}H_2 \qquad (17.37)$$

---

[11] S. H. Lawerence, J. R. Wermer, S. K. Boocock, M. A. Banks, P. C. Keller, and S. G. Shore, *Inorg. Chem.* **1986**, *25*, 367.

In a one-pot synthesis starting with B_5H_9, NaH, and $[Me_4N]Cl$, BCl_3 is generated *in situ*. Solvent, temperature, and cation choice are critical in these preparations.

Among the *closo* boranes, $B_{10}H_{10}^{2-}$ and $B_{12}H_{12}^{2-}$ are the most thermally stable and easiest to prepare. The condensation and closure reaction involving a *nido* boron hydride gives a nearly quantitative yield of $B_{12}H_{12}^{2-}$:

$$B_{10}H_{14} + 2\,Et_3N \cdot BH_3 \xrightarrow[\text{hydrocarbon solution}]{190°C} [Et_3NH]_2[B_{12}H_{12}] + 3\,H_2 \tag{17.38}$$

$B_{10}H_{10}^{2-}$ can be prepared nearly quantitatively by a simple base-promoted closure reaction:

$$B_{10}H_{14} + 2\,Et_3N \longrightarrow [Et_3NH]_2[B_{10}H_{10}] + H_2 \tag{17.39}$$

Other *closo* hydrides can be synthesized by related processes involving condensations and closures (both base-induced and pyrolytic) or degradation of larger *closo* species to smaller ones. The *closo* hydrides ordinarily are isolated as their salts. By using cationic exchange columns, the crystalline acid hydrate forms of more hydrolytically stable *closo* ions $(H_3O)_2(B_pH_p)(H_2O)_m$ can be prepared. These are strong acids comparable to H_2SO_4 in aqueous solution.

17.8 CARBORANES

The most important heteroboranes are the carboranes. These are much more stable to air and moisture than the boranes. An interesting feature of their structures is the appearance of carbon with coordination number (C.N.) up to 6. C's are the most electropositive centers in the frameworks. For species containing two or more C (many are C_2 species), isomers are possible. The most thermodynamically stable ones have maximum separation of the electropositive C's. The most common preparative route is the reaction of boranes with acetylenes:

$$B_5H_9 + C_2H_2 \xrightarrow{500-600°C} 2,4\text{-}C_2B_5H_7 + 1,6\text{-}C_2B_4H_6 + 1,5\text{-}C_2B_3H_5 \tag{17.40}$$

$$B_8H_{12} + MeC{\equiv}CMe \longrightarrow (MeC)_2B_7H_9 + \tfrac{1}{2}\,B_2H_6 \tag{17.41}$$

$$B_{10}H_{14} + C_2H_2 \longrightarrow 1,2\text{-}C_2B_{10}H_{12} + 2\,H_2 \tag{17.42}$$

The isomers of *closo* icosahedral $C_2B_{10}H_{12}$ exhibit extremely high kinetic and thermodynamic stability. The 1,2-, 1,7- and 1,12-isomers have the common names *o*-, *m*- and *p*-carborane, respectively. *o*-Carborane has been given the following symbol in the literature:

$$\begin{array}{c} HC{-}CH \\ \diagdown O \diagup \\ B_{10}H_{10} \end{array}$$

Reactions at B centers in carboranes parallel those of the boranes: bridge proton abstraction and electrophilic substitutions, including halogenation. The terminal H atoms attached to electropositive C are relatively acidic. Hence these C centers can be metallated:

$$1,7\text{-}C_2B_4H_6 + 2\, n\text{-BuLi} \longrightarrow Li_2[1,7\text{-}C_2B_4H_4] + 2\,C_4H_{10} \qquad (17.43)$$

$$HC\!-\!CH + RMgX \longrightarrow HC\!-\!CMgX + RH \qquad (17.44)$$
$$B_{10}H_{10} \qquad\qquad B_{10}H_{10}$$

The metallated products retain structural integrity and can react as nucleophiles to produce a large number of C-substituted derivatives. *o*-Carborane has a very extensive organic chemistry.

o-Carborane isomerizes to *m*-carborane at ~450°C. Lipscomb proposed that this reaction occurs via the so-called diamond–square–diamond (DSD) mechanism in which a pair of triangular faces open into a square by breaking a bond. Bond formation in an orthogonal direction leads to a rearranged icosahedron as shown in Figure 17.18*a*; the motion involved in forming a square stretches and breaks five other bonds, leading to the intermediate shown in Figure 17.18*b*. This model can account for the rearrangement of *o*- to *m*-carborane.

At still higher temperatures, *m*-carborane rearranges to the thermodynamically stable product *p*-carborane. The DSD mechanism cannot, however, produce *p*- from either *o*- or *m*-carborane. A more general mechanism (which also can account for the rearrangement products of substituted icosahedral boranes) is the extended triangle rotation (ETR) mechanism depicted in Figure 17.18*c*.[12] One triangular face rotates anticlockwise through a 30° angle, giving a transition state having three square faces resulting from cage bonds broken by the motion. A further 30° rotation gives an intermediate **Z** containing all triangular faces, but with the rotated triangle on the front now congruent with the triangle on the back. **Z** possesses a plane of symmetry passing through framework atoms 1, 5, 10, 12, 7, and 3. If the back triangle is now rotated anticlockwise through 30°, then another 30°, another icosahedron with interchanged vertices is reached. The result is equivalent to the DSD mechanism; however, the ETR mechanism has the advantage of preserving the maximum possible number of triangular faces and their attendant three-center bonding. An intermediate such as **Z** can be reached by rotating a front triangle clockwise (fc) or anticlockwise (fa) or by rotating the rear triangle directly behind the front one (ra or rc). (See Figure 17.18*d*.) Thus, for any choice of two parallel front and back faces, there are four ways of reaching **Z**. There are also four ways of relaxing from intermediate **Z** to an icosahedral product (fc, fa, rc, ra), for a total of 16 possible pathways connecting the starting and ending icosahedra. Because there is a choice of ten pairs of faces to rotate, a total of 160 rearrangement pathways is available, not all of which lead to different products. (For example, fcfa leads back to the starting configuration.)

The ETR mechanism in conjunction with the assumption that some triangular faces undergo preferential rotation is helpful in rationalizing the product distributions in thermal

[12] S.-h. Wu and M. Jones, Jr., *J. Am. Chem. Soc.* **1989**, *111*, 5373.

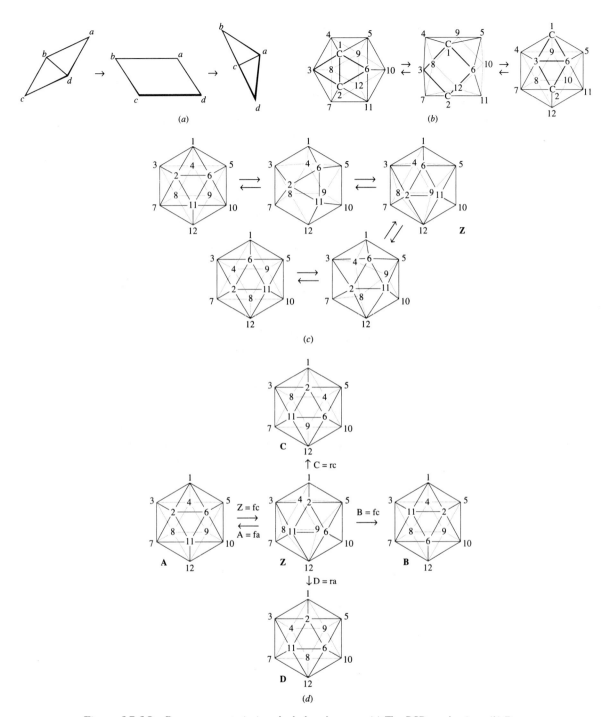

Figure 17.18 Rearrangements in icosahedral carboranes. (*a*) The DSD mechanism. (*b*) Rearrangement of *o*- to *m*-carborane via a cuboctahedral intermediate. (*c*) The ETR equivalent of DSD. (*d*) Possible fates of a Z-type intermediate showing possible rearrangement products. (Part *b* is from H. Beall in *Boron Hydride Chemistry*, E. L. Muetterties, Ed., Academic Press, New York, 1975. Parts *c* and *d* are reprinted with permission from S.-h. Wu and M. P. Jones, Jr., *J. Am. Chem. Soc.* **1989**, *111*, 5373. Copyright 1989, American Chemical Society.)

rearrangements of substituted icosahedral carboranes. In particular, BC_2 faces seem more likely to rotate by 30° than do B_3 faces, for reasons that are not entirely clear. The resulting **Z**-type intermediate is likely to rearrange by a further rotation of the favored BC_2 face.

EXAMPLE 17.1: Account for the experimental fact that the major product from thermal rearrangement of 9-bromo-*o*-carborane is 8-bromo-*o*-carborane by the ETR model.

Solution: Successive anticlockwise rotations of the front face in the drawing below lead to the 8-bromo product. The numbering in the third drawing corresponds to the initial positions of the B atoms, while that in the last structure is appropriate for the cage as depicted. Presumably, the triangle containing the substituted B must rotate preferentially.

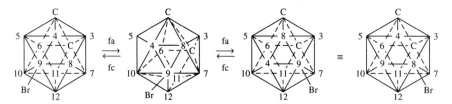

o-Carborane is attacked by base that excises a BH group, generating a *nido* anion that retains structural integrity. The 1,7 isomer can be obtained by thermal rearrangement of the anion or by starting with *m*-carborane:

$$1,2\text{-}C_2B_{10}H_{12} + OEt^- + 2\,EtOH \xrightarrow{85°C} 1,2\text{-}C_2B_9H_{12}^- + H_2 + B(OEt)_3 \qquad (17.45)$$

NaH in tetrahydrofuran deprotonates the anion, giving a dianion $C_2B_9H_{11}^{2-}$ whose structure is shown in Figure 17.19a. Assuming for convenience sp^3 hybridization of the five atoms on the open face, a set of MOs reminiscent of the Cp^- anion may be constructed. These MOs are occupied by six electrons. Hawthorne[13] has exploited the analogy between $C_2B_9H_{11}^{2-}$ (the dicarbollide ion) and Cp^- to prepare metal dicarbollide complexes related to the metallocenes:

$$2\,[1,2\text{-}C_2B_9H_{11}]^{2-} + Fe^{2+} \longrightarrow [(\eta^5\text{-}1,2\text{-}C_2B_9H_{11})_2Fe]^{2-} \qquad (17.46)$$

$$[1,7\text{-}C_2B_9H_{11}]^{2-} + Re(CO)_5Br \longrightarrow [(\eta^5\text{-}1,7\text{-}C_2B_9H_{11})Re(CO)_3]^- + 2CO + Br^- \qquad (17.47)$$

Representative structures appear in Figures 17.19b and 17.19c. Dicarbollide complexes undergo one-electron transfer processes, giving products having a range of metal oxidation numbers. For example, complexes of Ni^{II}, Ni^{III}, and Ni^{IV} can be prepared.

The structures in Figures 17.19b and 17.19c can also be regarded as metallacarboranes having a framework vertex occupied by a metal with its attached ligands. A general technique for preparing such complexes involves (a) excision of a BH group via base degradation and (b) reaction of the resulting anion with a metal halide.

---

[13] M. F. Hawthorne, *Pure Appl. Chem.* **1972**, *29*, 547; L. F. Warren and M. F. Hawthorne, *J. Am. Chem. Soc.* **1970**, *92*, 1157.

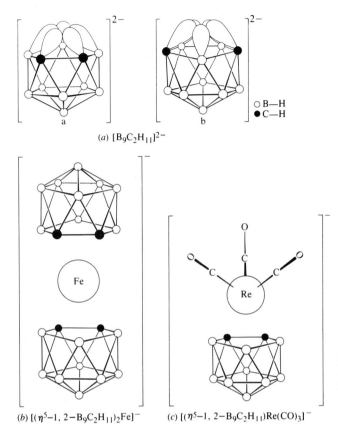

(a) $[B_9C_2H_{11}]^{2-}$

○ B—H
● C—H

(b) $[(\eta^5-1, 2-B_9C_2H_{11})_2Fe]^-$ (c) $[(\eta^5-1, 2-B_9C_2H_{11})Re(CO)_3]^-$

Figure 17.19 Structures of dicar-bollides. (Reprinted with permission from M. F. Hawthorne et. al., *J. Am. Chem. Soc.* **1968**, *90*, 879. Copyright 1968, American Chemical Society.)

17.9   SOME CHEMISTRY OF GROUP 13

▶ 17.9.1   *Borazine*[14] *and Boron Nitride*

Reaction of alkali metal borohydrides with NH_4Cl produces $B_3N_3H_6$, borazine (formerly called borazole). Borazine is isoelectronic with benzene, and because of its very similar physical properties (see Table 17.4) it has been referred to as "inorganic benzene".

The possible analogy with benzene is apparent from examination of its molecular structure parameters. In particular, borazine exhibits D_{3h} symmetry and displays B—N distances of 143.5 pm—significantly shorter than the distance in $H_3B:NH_3$ (156 pm).

---

[14] D. F. Gaines and J. Borlin, in *Boron Hydride Chemistry*, E. L. Muetterties, Ed., Academic Press, New York, 1975, p. 241.

Table 17.4   Physical properties of borazine and benzene

	Borazine	Benzene
Molecular weight	80.5	78.1
bp (°C)	55.0	80.10
mp (°C)	−56.2	5.51
Critical temperature (°C)	252	288.0
Liquid density at bp (g/cm³)	0.81	0.81
Crystal density at mp (g/cm³)	1.00	1.01
Trouton constant (J/K mol)	89.5	88.2

This is suggestive of a bond order greater than one. This information is represented by resonance structures showing aromatic character:

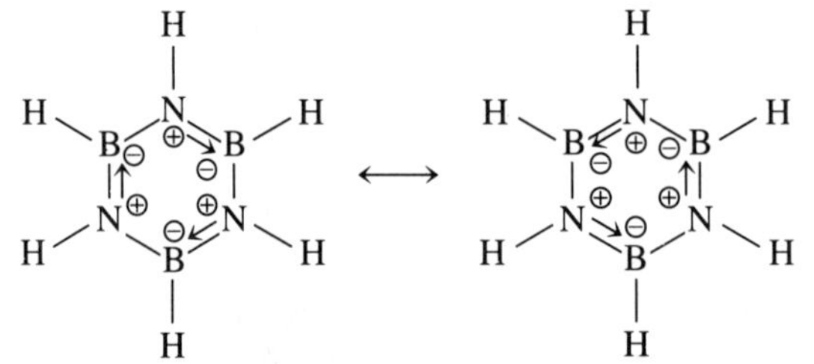

^{1}H NMR studies of substituted borazines show that the influence of Br, Cl, and Me is about equal on the chemical shifts of *ortho* and *para* NH protons. The *para* N—H chemical shifts can only be influenced by π (resonance) effects, whereas the *ortho* shifts result from both σ (inductive) and π effects. This behavior indicates the aromatic character of borazines.

The *chemical* behavior of borazines, however, is dominated by the polarity of the B—N bonds. The electrons required for aromaticity must come from the N. Whereas Lewis bases always add to the B's, Lewis acids add to N, suggesting an important contribution from the structure

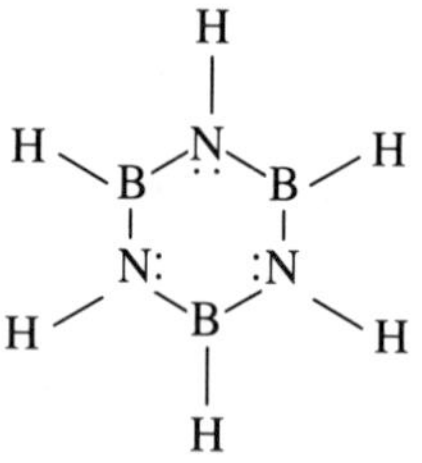

For example, *B*-substituted alkyl borazines can be prepared from borazine and nucleophilic reagents such as Li alkyls or Grignard reagents. *N*-Substitution can be achieved by reaction sequences involving *B*-alkylated species which have acidic *N*-hydrogens:

$$R_3B_3N_3H_3 + MeLi \longrightarrow R_3B_3N_3H_2Li + CH_4 \tag{17.48}$$

$$R_3B_3N_3H_2Li + R'X \longrightarrow R_3B_3N_3H_2R' + LiX \qquad (17.49)$$

(The first-listed substituents are those bound to B, and the last-listed ones are bound to N.) This reactivity contrasts with the alkylation of benzene, which is an electrophilic substitution process. Also in contrast to benzene chemistry is the facile addition of polar reagents across the B—N bonds. For example,

$$H_3B_3N_3H_3 + 3HCl \longrightarrow H_3Cl_3B_3N_3H_6 \qquad (17.50)$$

The saturated products (borazanes) are formal analogues of cyclohexanes (Figure 13.21). Thus, the chemistry of substituted borazines differs quite substantially from that of benzene. Figure 17.20 summarizes the chemistry of borazines.

Pure samples of borazine are light-sensitive and have been known to explode unless stored in the dark.

Substituted borazines can be made directly. Metal borohydrides react with monosubstituted alkylammonium salts to afford *N*-alkylated borazines:

$$3MBH_4 + 3RNH_3^+Cl^- \longrightarrow H_3B_3N_3R_3 + 3MCl + 9H_2 \qquad (17.51)$$

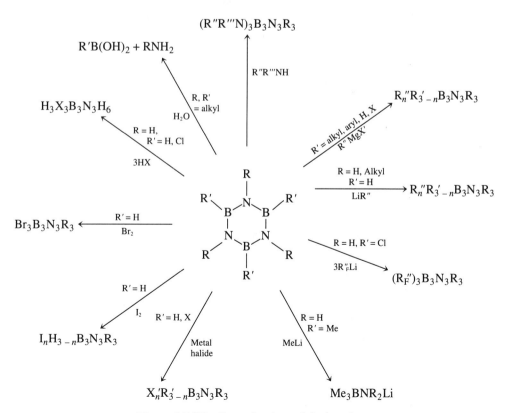

Figure 17.20 Some chemistry of the borazines.

Using a mixture of ammonium salts produces unsymmetrically substituted borazines. *B*-Halogenated borazines result from the reaction of boron trihalides with primary amines:

$$BX_3 + RNH_2 \longrightarrow RNH_2BX_3 \xrightarrow{-HX} RNHBX_2 \xrightarrow{-HX} \tfrac{1}{3}(X_3B_3N_3R_3) \quad (17.52)$$

Recently, P analogues of borazine have been prepared.[15] The crystal structure of $(MesBPPh)_3$ (Mes = mesityl) shows it to be planar with equivalent B—P distances of 184 pm, suggesting a high degree of aromaticity. Some distortion from ideal trigonal bond angles at B and P is seen presumably because of the difference in atomic size. ^{31}P NMR data indicate delocalization of electron density from P onto B. Consistent with these structural features, the boraphosphabenzenes are less reactive than borazine.

Complete pyrolysis of borazine, or any compound containing a 1 : 1 ratio of B to N, yields boron nitride. The crystal structure of the hexagonal form of BN is related to that of graphite and is shown in Figure 17.21. The stacking of layers differs from that in graphite in that B atoms in one layer lie directly over N atoms in another, and *vice versa*. This suggests the importance of polar interactions between the layers; however, interlayer bonding energy is only about 16 kJ/mol. Both graphite and BN are good lubricants, because of the ability of layers to slide over one another. In contrast to graphite, BN is a white insulator. High pressure converts BN to a cubic form displaying the diamond structure. The hardness of the cubic form surpasses that of diamond. Cubic BN can be used as an abrasive with metals prone to carbide formation. Like diamond, cubic BN is thermodynamically unstable at low pressures, but very stable kinetically. Recently a tubular form of BN has been prepared.[16]

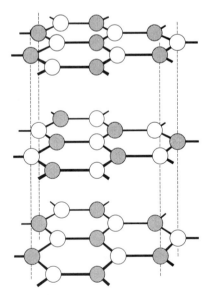

Figure 17.21   The layer structure of hexagonal BN. (From K. Niedenzu and J. W. Dawson, *Boron–Nitrogen Chemistry*, Academic Press, New York, 1965.)

[15] H. V. Rasika Dias and P. P. Power, *J. Am. Chem. Soc.* **1989**, *111*, 144.

[16] E. J. M. Hamilton, S. E. Dolan, C. M. Mann, H. O. Colijn, C. A. McDonald, and S. G. Shore, *Science* **1993**, *260*, 659.

► *17.9.2 The Borohydrides and*
Lithium Aluminum Hydride

Lithium borohydride was synthesized by applying the Lewis acid–base concept directly to the metal hydrides and boron compounds. The hydride ion is a very strong Lewis base, and it reacts even with the weak Lewis acid methyl borate:

$$\text{NaH}(s) + \text{B(OMe)}_3(l) \xrightarrow{68°C} \text{Na[BH(OMe)}_3] \tag{17.53}$$

The relative strengths of Lewis acids in the sequence $\text{BF}_3 > \text{(BH}_3)_2 > \text{Me}_3\text{B} > \text{B(OMe)}_3$ suggested that diborane might displace methyl borate from the trimethoxyborohydride ion:

$$2\,\text{Na[BH(OMe)}_3] + \text{B}_2\text{H}_6 \; \dashrightarrow \; 2\,\text{NaBH}_4 + 2\,\text{B(OMe)}_3 \tag{17.54}$$

This reaction does take place—so rapidly that it may be used to absorb diborane from a gas stream. The commercial synthesis of sodium borohydride is carried out without having to use diborane:

$$4\,\text{NaH} + \text{B(OMe)}_3 \xrightarrow{225–275°C} \text{NaBH}_4 + 3\,\text{NaOMe} \tag{17.55}$$

The reaction of lithium hydride with aluminum chloride in dry ether yields lithium aluminum hydride:

$$4\,\text{LiH} + \text{AlCl}_3 \xrightarrow{\text{Et}_2\text{O}} \text{LiAlH}_4 + 3\,\text{LiCl} \tag{17.56}$$

It has also been found possible to prepare the alkali-metal aluminum hydrides directly from the elements in tetrahydrofuran solvent:

$$\text{Na} + \text{Al} + \text{H}_2(5000\ \text{psi}) \xrightarrow[\text{thf}]{140°C} \text{NaAlH}_4 \tag{17.57}$$

Reactions of the Borohydrides

The borohydride ion has served as a starting material for preparing most of the boron hydrides and the hydroborate ions. However, B_5H_9 often provides cleaner reactions (*vide supra*).

The BH_4^- ion is stable in water above pH of ~9 and is a mild reducing agent, useful for selective organic reductions in water or alcohol solvents. It reduces aqueous solution of arsenites, antimonites, germanates, or stannates to the corresponding hydrides:

$$3\,\text{BH}_4^- + 4\,\text{H}_3\text{AsO}_3 + 3\,\text{H}^+ \longrightarrow 3\,\text{H}_3\text{BO}_3 + 4\,\text{AsH}_3 + 3\,\text{H}_2\text{O} \tag{17.58}$$

Many transition-metal ions are reduced quantitatively to lower oxidation states in aqueous solution, with the products being dependent on their half-cell values. Appendix E gives E^0 values for B half-reactions.

As indicated earlier, strong acids, protonic or Lewis, displace B_2H_6 from borohydrides, and this accounts for the success of the laboratory preparation of B_2H_6 given in Reaction (17.29). In dilute aqueous solution, complete hydrolysis of the borohydride yields boric acid and hydrogen.

Reactions of Lithium Aluminum Hydride

Lithium aluminum hydride, a much stronger reducing agent than the borohydrides, reacts violently with water. Its solutions in diethyl ether or tetrahydrofuran are used widely for organic reductions, and its reaction with binary halides provides a general method for preparing the volatile hydrides:

$$SiCl_4 + LiAlH_4 \xrightarrow{Et_2O} SiH_4 + LiCl + AlCl_3 \tag{17.59}$$

$$4\,PCl_3 + 3\,LiAlH_4 \xrightarrow{Et_2O} 4\,PH_3 + 3\,LiCl + 3\,AlCl_3 \tag{17.60}$$

Solvent plays an essential role in the above reaction; thus, reduction of PCl_3 does not take place in the absence of ether.

Lithium aluminum hydride reacts with diborane to produce liquid aluminum borohydride:

$$LiAlH_4 + 2\,B_2H_6 \longrightarrow LiBH_4 + \text{[structure]} \tag{17.61}$$

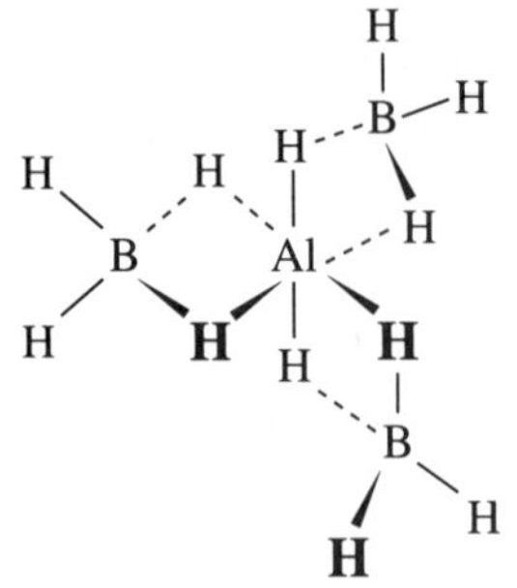

This borohydride has six bridge bonds about each aluminum. It is extremely hazardous, detonating violently on contact with air containing trace amounts of moisture, and slowly evolves hydrogen on standing at room temperature. $Al(BH_4)_3$ is the most volatile aluminum compound known.

► 17.9.3 *Hydroboration*

A new area of organic chemistry involving boron hydrides (a thorough treatment of which is beyond the scope of this text) was developed by H. C. Brown,[17] who received the 1979 Nobel Prize in Chemistry for this work.

Diborane adds across double bonds in olefins and has the further unique property of undergoing reversible addition and dissociation, during which time the double bond "walks" to the end of the chain! The process is kinetically controlled and permits contrathermodynamic isomerizations of internal to terminal olefins. Alkaline peroxide oxidation then produces predominately terminal alcohols from mixed olefins:

$$
\begin{aligned}
-C-C-C=C-C + \tfrac{1}{2}B_2H_6 &\longrightarrow -C-C-C-C-C-BH_2 \\
&\xrightarrow{[O]} -C-C-C-C-C-OH
\end{aligned} \tag{17.62}
$$

---

[17] H. C. Brown, *Organic Synthesis via Boranes,* Wiley–Interscience, New York, 1975.

The diborane is prepared *in situ* from the reaction of $LiBH_4$ or $NaBH_4$ with acids such as $Et_2O \cdot BF_3$, HCl, or H_2SO_4. An ether solvent, such as diethyl ether, tetrahydrofuran, or polyethers, must be used. $BH_3 \cdot thf$ is available commercially. Also, the oxidation usually is carried out *in situ*.

Addition reactions with 1-alkenes give 94% B placement on the terminal position. Branching of the alkyl on the 2-position does not change the distribution significantly:

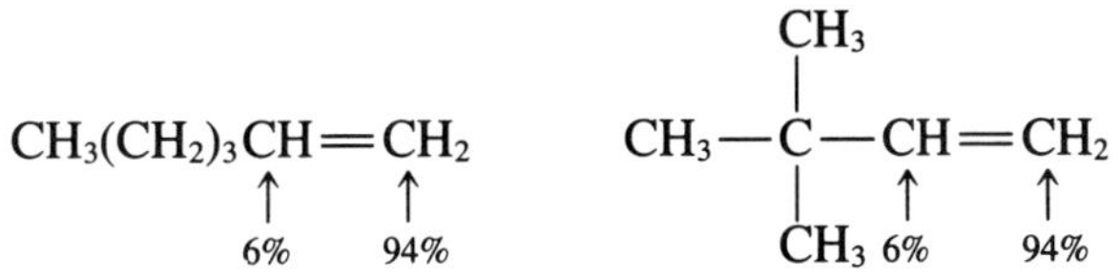

$$CH_3(CH_2)_3CH\!=\!CH_2 \qquad \underset{CH_3}{\overset{CH_3}{CH_3-C-CH\!=\!CH_2}}$$

$$\underset{6\%}{\uparrow}\quad\underset{94\%}{\uparrow}\qquad\qquad \underset{6\%}{\uparrow}\quad\underset{94\%}{\uparrow}$$

With internal olefins, B shows a similar preference to add at the less hindered site:

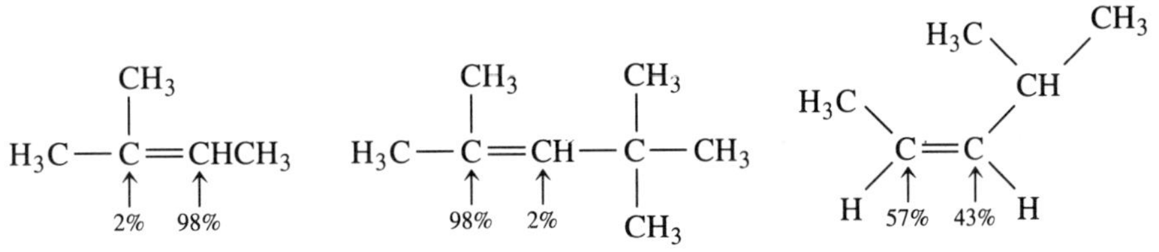

$$H_3C-\underset{\underset{2\%}{\uparrow}}{\overset{CH_3}{C}}\!=\!\underset{\underset{98\%}{\uparrow}}{CHCH_3} \qquad H_3C-\underset{\underset{98\%}{\uparrow}}{\overset{CH_3}{C}}\!=\!\underset{\underset{2\%}{\uparrow}}{CH}-\underset{CH_3}{\overset{CH_3}{C}}-CH_3 \qquad$$

$$\underset{H_3C}{\overset{H_3C}{\Big\rangle}}C\!=\!C\underset{\underset{57\%}{\uparrow}\,H}{\overset{CH}{\Big\langle}}\overset{H_3C\diagup\;\diagdown CH_3}{\underset{43\%\,H}{}}$$

However, no great preference exists for substituted positions containing alkyl groups of very different steric requirements. Steric effects do have some influence on the extent of hydroboration, as the following reactions show:

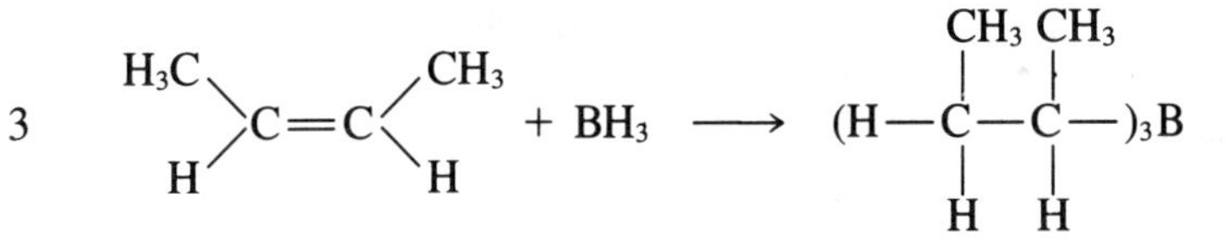

$$3 \quad \underset{H}{\overset{H_3C}{\Big\rangle}}C\!=\!C\underset{H}{\overset{CH_3}{\Big\langle}} + BH_3 \longrightarrow (H-\underset{\underset{H}{|}}{\overset{\overset{CH_3}{|}}{C}}-\underset{\underset{H}{|}}{\overset{\overset{CH_3}{|}}{C}}-)_3B \qquad (17.63)$$

$$2 \quad \underset{H_3C}{\overset{H_3C}{\Big\rangle}}C\!=\!C\underset{H}{\overset{CH_3}{\Big\langle}} + BH_3 \longrightarrow (H-\underset{\underset{CH_3}{|}}{\overset{\overset{CH_3}{|}}{C}}-\underset{\underset{H}{|}}{\overset{\overset{CH_3}{|}}{C}}-)_2BH \qquad (17.64)$$

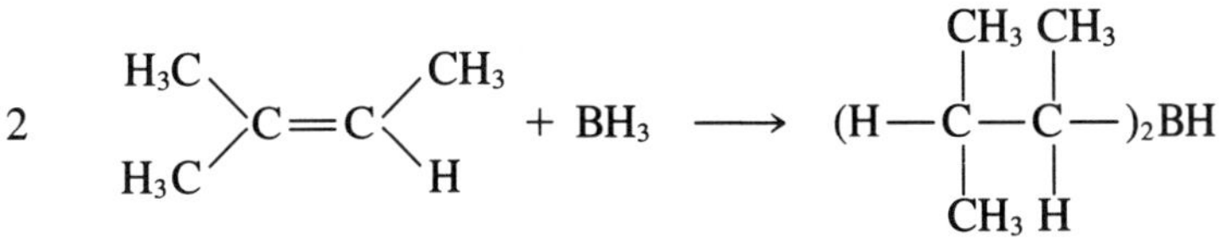

$$\underset{H_3C}{\overset{H_3C}{\Big\rangle}}C\!=\!C\underset{CH_3}{\overset{CH_3}{\Big\langle}} + BH_3 \longrightarrow (H-\underset{\underset{CH_3}{|}}{\overset{\overset{CH_3}{|}}{C}}-\underset{\underset{CH_3}{|}}{\overset{\overset{CH_3}{|}}{C}}-)BH_2 \qquad (17.65)$$

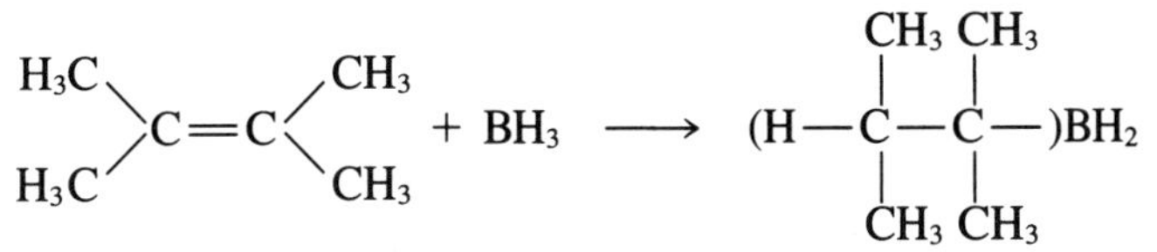

Under forcing conditions, Reactions (17.64) and (17.65) can proceed to the next stage, giving trialkyl and dialkyl boranes, respectively.

Electronic rather than steric effects seem to predominate in directing the site for B substitution. For example, in crotyl chloride, $CH_3CH\!=\!CHCH_2Cl$, the 2-position is substituted 90% of the time with B.

The *cis* stereochemistry of hydroboration permits stereoselective reactions such as the following:

$$(17.66)$$

17.10 CLUSTER COMPOUNDS

Inorganic chemistry encompasses numerous examples of compounds such as the boron hydrides, in which atoms are bonded in cage structures based on the triangulated polyhedra of Figure 17.6. We already have seen several examples in this book, including P_4, P_4O_6, P_4O_{10}, and trigonal prismatic Te_6^{2+} (Figure 16.4).

Atoms at cage vertices must have sufficient valence orbitals properly directed for three-dimensional bonding. Among the elements satisfying this requirement are metals, which often form **cluster compounds**—structures displaying polyhedral frameworks held together by two or more metal–metal bonds per metal. (Section 15.10 discusses the ability of metals to bond to each other.) Some clusters contain only metals. Examples include Bi_9^{5+} (a tricapped trigonal prism—Figure 17.6h), Sn_9^{4-} (a capped Archimedean antiprism—Figure 17.6g). These species are present in ammonia solutions of so-called Zintl alloys consisting of intermetallic phases of Na and the metal. Recently, the use of cryptate ligands (Section 15.2.4) to complex Na^+ has permitted crystallization of some of these strange species.[18]

Of more widespread recent interest are transition-metal clusters. Each metal generally is also bonded to other ligands that are π-acids, halides, hydrides, and/or organic π-donors. These are related structurally to the boron hydrides in that a transition-metal-containing group can often replace a BH unit in a borane framework. An example is $H_2Ru_6(CO)_{18}$ (Figure 17.22), which contains an octahedron of Ru atoms, each bonded to three terminal CO's. A pair of opposite octahedral faces is triply H-bridged. Regarding the compound as the diprotonated form of $[Ru_6(CO)_{18}]^{2-}$ gives a total of 54 Ru valence orbitals (the d, s, and p orbitals of six Ru's) and 86 electrons. Allotting 18 (6×3) orbitals and 36 electrons ($6 \times 3 \times 2$) for Ru—CO bonding, the remaining 36 orbitals and 50 electrons could be used to provide skeletal bonding of the Ru_6 octahedron. The effective atomic number (EAN) rule does not rationalize this electron count: If we regard the Ru's as being bonded by two-electron bonds along each octahedral edge, then each Ru gains four electrons from Ru—Ru bonds; because each neutral Ru atom has eight valence electrons and six are supplied by the three CO's, each Ru obeys the EAN rule with a total of $(6 \times 8) + (6 \times 3 \times 2) = 84$ electrons. However, the electron count of this cluster is 86.

---

[18] J. D. Corbett and P. A. Edwards, *J. Am. Chem. Soc.* **1977**, *99*, 3313 and references therein.

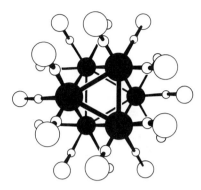

Figure 17.22   The structure of $H_2Ru_6(CO)_{18}$ projected onto one of the open faces. (H not shown.) (From M. R. Churchill, J. Wormald, J. Knight, and M. Mays, *J. Chem. Soc. Chem. Commun.* **1970**, 458.)

The situation seems just the reverse of that in the boranes: There are more electrons than valence orbitals. Thus, important questions emerge regarding transition-metal clusters: (a) How are their molecular structures related to electron count, and (b) how are the molecular structures related to those of the boranes?

▶ 17.10.1   Clusters and Catalysis

A principal reason for the extensive research on cluster compounds is their possible relevance to heterogeneous catalysis.[19] Industrially important syntheses of organic compounds rely heavily on catalysis (see Section 14.2).

Most industrial reactions are catalyzed heterogeneously, usually by transition metals or their solid oxides; reactions occur at the solid surface with reactants and products in solution. Practical considerations such as ease of product separation often dictate a preference for heterogeneous catalysts. Solid catalysts can be filtered off from a solution or can be suspended in a gaseous mixture. These solids also are stable at the high temperatures required for many reactions.

Directly combining the constituents of synthesis gas could afford such industrially useful products as CH_3OH, $HOCH_2CH_2OH$, and so on. Thus, it is desirable to develop suitable catalysts for the combination of CO and H_2 with high yield and selectivity. The Fischer–Tropsch process catalyzes such reactions heterogeneously. However, very high temperatures with attendant energy expenditure are required.

Investigation of reactions at solid catalyst surfaces presents a formidable problem; in particular, usual solution spectroscopic techniques are not applicable.[20] Instead, techniques such as low-energy electron diffraction (LEED), Auger spectroscopy, and scanning tunneling microscopy (STM), to name a few, are employed.[21] Nevertheless, knowledge of the course of reactions at surfaces is highly desirable as an aid in designing improved heterogeneous catalysts.

---

[19] E. L. Muetterties, *Bull. Soc. Chim. Belg.* **1975**, *84*, 959.

[20] J. C. Slater and K. H. Johnson, *Physics Today*, October 1974, 34.

[21] See, for example, *Chemical and Engineering News*, March 30, 1992, p. 22.

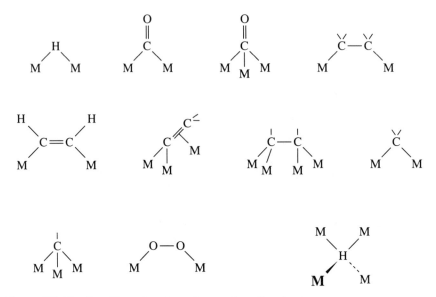

Figure 17.23 Possible modes of attachment of small molecules to metal surfaces or clusters. (After E. L. Muetterties, *Bull. Soc. Belg. Chem.* **1975,** *84,* 959.)

Interaction of small molecules with metal surfaces is a field of active investigation. We know that many small molecules are chemisorbed; that is, a bond is formed between surface metal atoms and the molecules that may even disrupt intramolecular bonding. Figure 17.23 indicates several modes of attachment for small molecules of catalytic interest to surfaces containing many metal atoms. The types of molecules bonded and the extent of surface coverage depend critically on the nature of the metal. Transition metals display the greatest activity. Chemisorbed species can migrate across surfaces. We can visualize the course of a chemical reaction at a metal surface as involving chemisorption of different molecules at different sites, possibly with bond breaking, migration of reactive species to adjacent sites, reaction, and, finally, product desorption. In addition, surface metal atoms themselves are known to move.

Two modeling approaches currently are employed for investigating the interaction of small molecules with surface metal atoms. The first[22] (which will not be discussed further here) involves vaporization of metals and their condensation at cryogenic temperatures in a matrix of CO, $CH_2 = CH_2$, or some other small molecule of interest on the window of an infrared (IR) cell. The IR spectra of the mixture, as relative numbers of metal atoms and small molecules are varied, give clues to the nature of the species present. The second modeling approach, and the one of interest here, involves transition-metal-containing clusters as models for metal surfaces with their chemisorbed species. The next sections point out several analogies between cluster compounds and metal surfaces.

It may be that cluster compounds ultimately will not prove very relevant to heterogeneous catalysis. However, this possible relation is an example of why the scientific com-

---

[22] G. A. Ozin, *Acc. Chem. Res.* **1977,** *10,* 21; K. Klabunde, *Acc. Chem. Res.* **1975,** *8,* 393.

munity focuses on a particular area. Whatever the ultimate result, much interesting chemistry has been and continues to be found.

► 17.10.2 Molecular Structures of Clusters

Although bare metal cluster skeletons generally are derived from the polyhedra of Figure 17.6, in some known examples the arrangement of metal atoms is related to that in bulk metals. The Rh framework in $[Rh_{13}(CO)_{24}H_3]^{2-}$ corresponds to a hexagonal close-packed (*hcp*) structure (Figure 17.24), whereas in $[Rh_{14}(CO)_{25}H_3]^{4-}$ a body-centered cubic (*bcc*) arrangement is evident (Figure 17.25). These metal frameworks can be regarded as "*round surfaces*". It is also significant that catalytically active metals have *hcp* or cubic close-packed (*ccp*) structures displaying the same triangular arrangements of metal atoms seen on the faces of cluster polyhedra.

In contrast to single metal centers, metal surfaces and clusters present possibilities for binding small molecules to more than one atom simultaneously. $[Ni_5(CO)_{12}]^{2-}$ (Figure 17.26) shows both terminal and μ_2-bridging carbonyls, and $[Rh_6(CO)_{15}I]^-$ (Figure 17.27) (derived from $Rh_6(CO)_{16}$ by carbonyl substitution) displays four μ_3-carbonyls [as does $Rh_6(CO)_{16}$]. As more negative charge accumulates in isoelectronic species, more carbonyls tend to be bridging. A case in point is $[Co_6(CO)_{15}]^{2-}$ (Figure 17.28), which is isoelectronic with $[Rh_6(CO)_{15}I]^-$ and has two more bridging CO's, whereas $[Co_6(CO)_{14}]^{4-}$

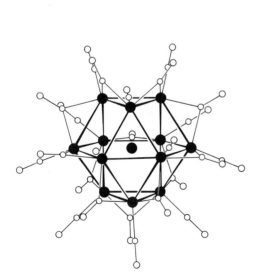

Figure 17.24 The structure of $[Rh_{13}(CO)_{24}H_3]^{2-}$ (H omitted). (Reprinted with permission from S. Martinengo, G. Ciani, A. Sironi, and P. Chini, *J. Am. Chem. Soc.* **1978**, *100*, 7096. Copyright 1978, American Chemical Society.)

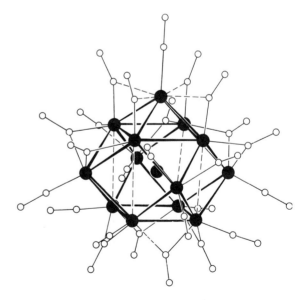

Figure 17.25 The Structure of $[Rh_{14}(CO)_{25}]^{4-}$. (Reprinted with permission from S. Martinengo, G. Ciani, A. Sironi, and P. Chini, *J. Am. Chem. Soc.* **1978**, *100*, 7096. Copyright 1978, American Chemical Society.)

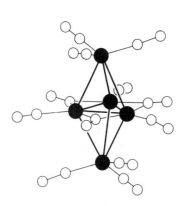

Figure 17.26   The structure of $[Ni_5(CO)_{12}]^{2-}$. (J. K. Ruff, R. P. White, and L. F. Dahl, *J. Am. Chem. Soc.* **1971,** *93,* 2159. Reprinted with permission from P. Chini, G. Longoni, and V. G. Albano, *Adv. Orgomet. Chem.* **1976,** *14,* 285.)

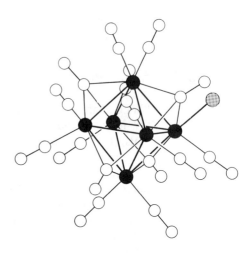

Figure 17.27   The structure of $[Rh_6(CO)_{15}I]^-$. (V. G. Albano, P. L. Bellon, and M. Sansoni, *J. Chem. Soc. A* **1971,** 678. Reprinted with permission from P. Chini, G. Longoni, and V. G. Albano, *Adv. Orgomet. Chem.* **1976,** *14,* 285.)

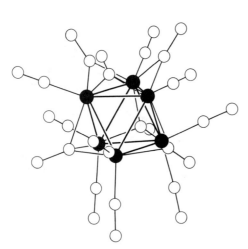

Figure 17.28   The structure of $[Co_6(CO)_{15}]^{2-}$. (V. G. Albano, P. Chini, and V. Scatturin, *J. Organomet. Chem.* **1968,** *15,* 423. Reprinted with permission from P. Chini, G. Longoni, and V. G. Albano, *Adv. Orgomet. Chem.* **1976,** *14,* 285.)

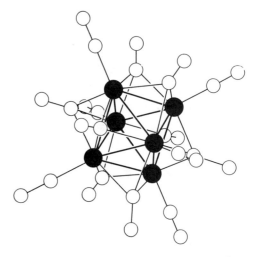

Figure 17.29   The structure of $[Co_6(CO)_{14}]^{4-}$. (V. G. Albano, P. L. Bellon, P. Chini, and V. Scatturin, *J. Organomet. Chem.* **1969,** *16,* 461. Reprinted with permission from P. Chini, G. Longoni, and V. G. Albano, *Adv. Orgomet.*

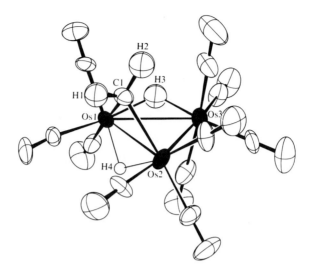

Figure 17.30   The structure of $H_2Os_3(CO)_{10}CH_2$. (Reprinted with permission from R. B. Calvert, J. R. Shapley, A. J. Schultz, J. M. Williams, S. L. Suib, and G. D. Stuckey, *J. Am. Chem. Soc.* **1978,** *100,* 6240. Copyright 1978, American Chemical Society.)

(Figure 17.29) (also isoelectronic) has eight μ_3-CO's. Other examples of coordination to multiple metal sites include μ_2-H and μ_2-CH$_2$ in $H_2Os_3(CO)_{10}CH_2$ (Figure 17.30). μ_3-H in $H_2Ru_6(CO)_{12}$ was mentioned previously. μ_3-CCH$_3$ is found in $Co_3(CO)_9CCH_3$ (Figure 17.31), whereas μ_4-C appears in $Fe_4C(CO)_{13}$ (Figure 17.32*a*). Coordination of a ligand to more than one metal may modify the reactivity of small molecules acting as cluster ligands in ways paralleling their reactivity on metal surfaces. Indeed, cluster compounds catalyze some reactions that mononuclear species do not. As examples, $Ir_4(CO)_{12}$ catalyzes the reaction of CO with H_2 to produce CH_3OH,[23] whereas $Ru_2Rh(CO)_{12}$ catalyzes the production of ethylene glycol $HOCH_2CH_2OH$[24] (although both these reactions are too slow to be of commercial importance). Coordination of CO to more than one metal possibly may weaken the triple bond and permit hydrogenation. Another similarity between the structures of clusters and those of metals is the incorporation of nonmetallic atoms in the framework. $Fe_4(CO)_{13}C$ (Figure 17.32*a*) and $Fe_5(CO)_{15}C$ (Figure 17.32*b*) are analogous to the interstitial carbides.

Cluster compounds containing six or more metals are often referred to as **high nuclearity cluster compounds** (HNCCs).

[23] M. G. Thomas, B. F. Beier, and E. L. Muetterties, *J. Am. Chem. Soc.* **1976,** *98,* 1296.
[24] J. F. Knifton, *J. Chem. Soc. Chem. Commun.* **1983,** 729.

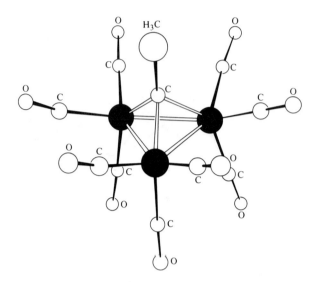

Figure 17.31   The structure of $Co_3(CO)_9CCH_3$. (Reprinted with permission from P. W. Sutton and L. F. Dahl, *J. Am. Chem. Soc.* **1967,** *89,* 261. Copyright 1967, American Chemical Society.)

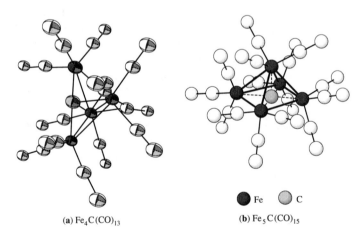

(a) $Fe_4C(CO)_{13}$ **(b)** $Fe_5C(CO)_{15}$

Figure 17.32   Structures of iron clusters. [(*a*) Reprinted with permission from J. S. Bradley, G. B. Ansell, M. E. Leonowicz, and E. W. Hall, *J. Am. Chem. Soc.* **1981,** *103,* 4968. Copyright 1981, American Chemical Society. (*b*) E. H. Braye, L. F. Dahl, W. Hubel, and D. Wampler, *J. Am. Chem. Soc.* **1962,** *84,* 4633. Reprinted with permission from P. Chini, G. Longoni, and V. G. Albano, *Adv. Organomet. Chem.* **1976,** *14,* 285.]

▶ *17.10.3 Stereochemical Nonrigidity in Clusters*

Rearrangements of cluster-bound ligands can occur via suitable vibrational motions that allow permutation of atomic positions; this is tantamount to ligand migration over the cluster surface. Structural techniques (e.g., NMR) that have a slower time scale than the vibrations permuting nuclear positions "see" a time-averaged structure (Section 9.15). In cluster compounds, vibrations may include motions of ligands or of the polyhedral metal structure.

An interesting case involving ligand mobility is $Cp_3Rh_3(CO)_3$, whose ^{13}C NMR spectrum in the CO region at $-65°C$ and $+26°C$ appears in Figure 17.33b. Because Rh has nuclear spin $I = \frac{1}{2}$, each equivalent bridging CO gives a triplet at low temperatures. At high T each CO visits each Rh and is thus coupled to all three; a ^{13}C quartet $[2 \times 3(\frac{1}{2}) + 1]$ results. Figure 17.33c also shows a possible idealized intermediate in the process which permutes CO's; it contains only terminal CO's.

The variable temperature ^{13}C NMR spectrum of $Rh_4(CO)_{12}$ appears in Figure 17.34. The low-temperature spectrum indicates a static structure containing four different kinds of CO's coupled to one (doublets) or two (triplets) Rh atoms. Coupling to two Rh's indicates μ-CO. The other chemically equivalent CO sets are coupled to a single Rh and must be terminal. The low-temperature structure depicted in Figure 17.34c agrees with the symmetry requirements of the NMR spectrum. When the sample is warmed, nuclei begin to permute, and each one spends less and less time in a particular environment. As the en-

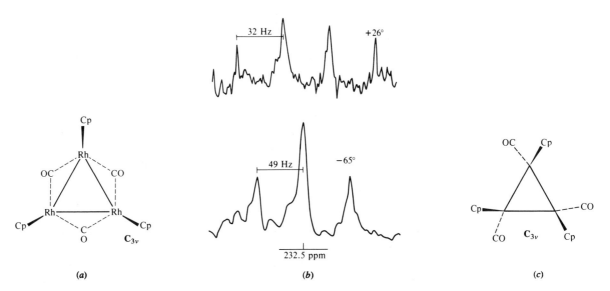

Figure 17.33 (*a*) Molecular structure of $Cp_3Rh_3(CO)_3$. (*b*) Carbonyl ^{13}C NMR spectra for $Cp_3Rh_3(CO)_3$ in the slow-exchange ($-65°C$) and fast-exchange ($26°C$) limits. (*c*) Proposed intermediate in CO exchange process. (Reprinted with permission form R. J. Lawson and J. R. Shapley, *J. Am. Chem. Soc.* **1976**, *98*, 7433. Copyright 1976, American Chemical Society.)

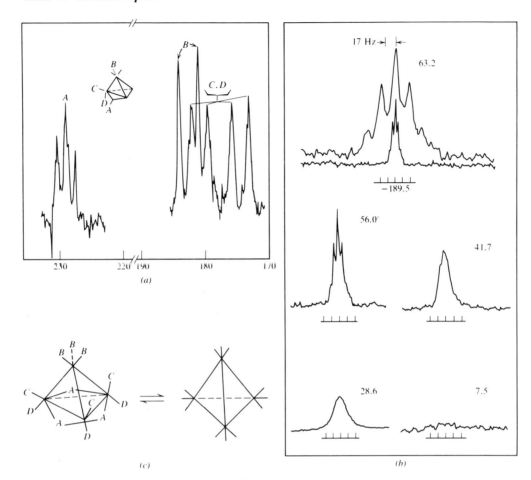

Figure 17.34   Variable-temperature ^{13}C NMR spectrum of $Rh_4(CO)_{12}$. (a) Slow-exchange limit. (From R. D. Adams and F. A. Cotton, in *Dynamic Nuclear Resonance Spectroscopy*, L. M. Jackman and F. A. Cotton, Eds., Academic Press, New York, 1975.) (b) At higher temperatures. (c) Proposed mechanism for CO exchange. (Parts b and c reprinted with permission from F. A. Cotton, L. Kruczynski, B. L. Shapiro, and L. F. Johnson, *J. Am. Chem. Soc.* **1972,** *94,* 6191. Copyright 1972, American Chemical Society.)

vironment becomes less and less well-defined over the time period required for a nuclear transition, the lines begin to broaden. When the time of residence is comparable to the time required for nuclear transition (at about 7.5°C), the CO's are not in any well-defined environment during the time period of the measurement and the signal is broadened into the baseline. On raising the temperature further, exchange becomes sufficiently rapid to reduce the residence time in any environment far below the time scale of the measurement. The radiation "sees" the CO's in their time-averaged environment. That ligand exchange is intramolecular is shown by the persistence of Rh—C coupling at high temperatures. Now each CO is coupled to all four Rh's, and a five-line $[2 \times 4(\frac{1}{2}) + 1]$ signal is seen. Again, bridge-terminal interchange of CO's provides a possible exchange mecha-

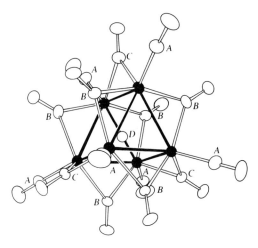

Figure 17.35 Structure of [Rh$_6$(CO)$_{15}$C]$^{2-}$. (Reprinted with permisson from V. G. Albano, P. Chini, S. Martinengo, D. J. A. McCaffrey, D. Strumolo, and B. T. Heaton, *J. Am. Chem. Soc.* **1974**, *96*, 8106. Copyright 1974, American Chemical Society.)

nism. This must be occurring over all six edges of the tetrahedron. [Rh$_6$(CO)$_{15}$C]$^{2-}$ (Figure 17.35), on the other hand, does not show carbonyl exchange below 298 K, perhaps because all edges are occupied by CO's and concerted motion of several CO's would be required to effect bridge–terminal interchange.

An example of H migration around a cluster framework is provided by [H$_4$(CpCo)Os$_3$(CO)$_9$] (Figure 17.36). At low temperature, two ^{1}H NMR signals are seen whose chemical shift values are in the range for μ-H, indicating two chemically equivalent pairs of H. The difference could arise from bridging two different types of tetrahedral edge: Os–Os or Co–Os. On warming, a single peak at the average chemical shift appears, indicating the equivalence of all the H's on the NMR time scale. Again, a bridge–terminal interchange involving all edges is likely.

Mobility in clusters is not confined to ligands. Relatively small cage motions can deform polyhedral structures with seven or more vertices into alternative structures. Figure 17.37 shows how stretching or compression of edges easily can lead to the traverse of

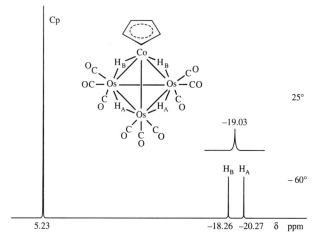

Figure 17.36 ^{1}H NMR spectrum of [H$_4$(CpCo)Os$_3$(CO)$_9$] at two temperatures and its static structure. (Reprinted with permission from S. G. Shore, W.-L. Hsu, C. L. Weisenberger, M. L. Caste, M. R. Churchill, and B. Bueno, *Organometallics* **1982**, *1*, 1405. Copyright 1982, American Chemical Society.)

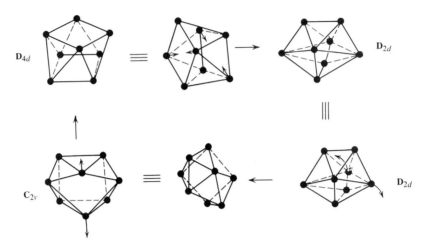

Figure 17.37 Interconversion of eight-vertex structures.

three different polytopal forms in eight-atom clusters. $Fe_3(CO)_{12}$ shows only one ^{13}C NMR signal even at $-150°C$ even though its C_{2v} solid-state structure (Figure 12.2) shows the presence of both terminal and bridging CO's. Cotton and Troup explained this result by postulating simultaneous opening of both bridges producing two $Fe(CO)_4$ groups, rotation around each Fe, and simultaneous bridge re-closing. This mechanism is shown in Figure 17.38a. Repeating the reclosing over the other two edges of the triangle scrambles all CO's. An alternative explanation is provided by the so-called Ligand Polyhedral Model[25] in which the ligands are viewed as occupying the vertices of a polyhedron circumscribing the polygon or polyhedron of metal atoms. Figure 17.38b shows the distorted icosahedron of ligands surrounding the Fe_3 triangle in $Fe_3(CO)_{12}$. Vertices 2 and 12 are bridging CO's, and vertex 9 is an axial CO on the $Fe(CO)_4$ group. Figure 17.38c shows how a DSD rearrangement (Section 17.8) coupled with a librational motion of the Fe_3 triangle effects the same ligand exchange as does the Cotton–Troup mechanism. A study of $Fe_3(\mu\text{-}CNCF_3)$-$(\mu\text{-}CO)(CO)_{10}$ showed that the bridging isocyanide and one equatorial CO (at positions 2 and 9, respectively, in the figure) did not scramble with the others. This was accounted for by proposing that rotation and libration of the Fe_3 triangle around the pseudo C_5 axis through vertices 2 and 9 is occurring inside the icosahedron.[26] (See Figure 17.38d.) This would involve distortion of the ligand polyhedron to a regular icosahedron. However, reference 25 shows that the Ligand Polyhedral Model can also account for this result.

Stereochemical nonrigidity in the ligands of mononuclear species can also be related to the polyhedra. For example, the CO's in $Fe(CO)_5$ can be considered to lie at the vertices of a circumscribed TBP. Down to the lowest experimentally accessible temperatures, only a single ^{13}C line is observed for $Fe(CO)_5$. This obviously indicates permutation of all CO's via a process with extremely low activation energy. Ligand exchange in such C.N. 5 species via Berry pseudorotation was discussed in Chapter 11.

NMR investigations of ligand and metal mobility in clusters have revealed aspects that should apply to species adsorbed on metal surfaces. The structures of boranes also are

[25] B. F. G. Johnson, E. Parisini, and Y. G. Roberts, *Organometallics* **1993**, *12*, 233 and references therein.

[26] D. Lentz and R. Marschall, *Organometallics* **1991**, *10*, 1467.

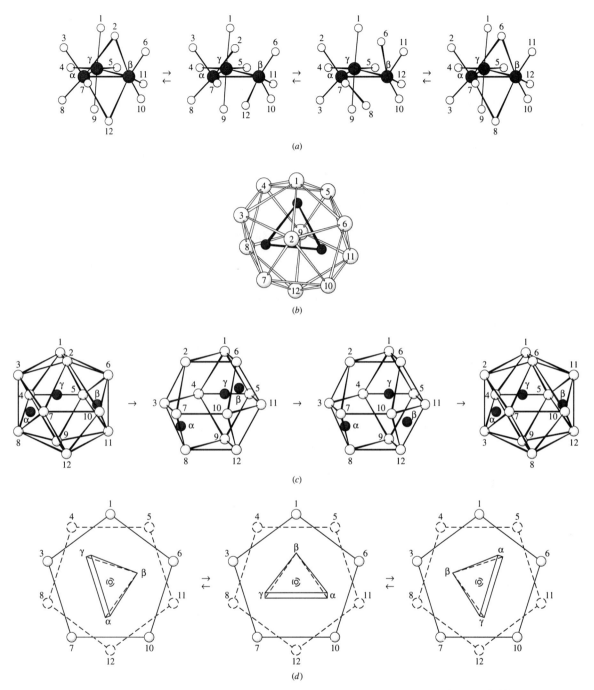

Figure 17.38 Ligand permutation in Fe₃(CO)₁₂. (*a*) Interconversion of μ- and terminal CO's. (*b*) Distorted icosahedron of CO's surrounding Fe₃ triangle. (*c*) Ligand Polyhedral Model for CO permutation. (*d*) Permutation via rotation of the Fe₃ triangle within the distorted icosahedron of CO's. (Parts *a* and *c* reprinted with permission from B. F. G. Johnson, E. Parisine, and Y. V. Roberts, *Organometallics* **1993**, *12*, 233. Copyright 1993, American Chemical Society. Parts *b* and *d* reprinted from D. Lentz and R. Marschall, *Organometallics* **1991**, *10*, 1487. Copyright 1991, American Chemical Society.)

very commonly nonrigid and temperature-dependent. ^{11}B and ^{1}H NMR are important structural techniques in boron chemistry.

▸ 17.10.4 *Electronic Structures of Clusters with π-Acid Ligands*

The relation between electron count and molecular structure for metal clusters with π-acid ligands displays an important analogy with that of boron hydrides, as pointed out by Wade, Mingos, and others.[27] The counting scheme is referred to as the **Polyhedral Skeletal Electron-Pair Theory** (PSEPT).

Returning to the problem of $H_2Ru_6(CO)_{18}$, recall that after reserving 18 Ru orbitals and 36 electrons for carbonyl bonding, 36 Ru orbitals and 50 electrons remain. Because transition-metal compounds usually contain nonbonding electrons, it is plausible to assign some electrons as nonbonding on Ru. Assuming six nonbonding electrons for each Ru and reserving three atomic orbitals per Ru to accommodate them leaves a total of 18 orbitals [three per $Ru(CO)_3$ unit] and 14 electrons (seven pairs) for bonding the Ru_6 framework. This same situation prevailed in $B_6H_6^{2-}$. The group at each vertex contributes three orbitals and two electrons to framework bonding. Seven electron pairs are available for bonding of the six-vertex framework; and by analogy with the $B_6H_6^{2-}$ case, we expect the octahedral *closo* structure.

This approach works because the three frontier orbitals of the $M(CO)_3$ fragment (Figure 17.39) consist of a radial (*sp*) and two tangential (*pd*) hybrids which have orientations and symmetry properties like those of the BH group (Figure 17.8). That is, they are isolobal with the BH set, and thus the $3n$ skeletal valence orbitals are expected to produce an MO scheme qualitatively like that for $B_6H_6^{2-}$ having seven $(n + 1)$ low-lying orbitals (which are just filled by the 7 electron pairs not involved in M—CO bonding or occupying nonbonding orbitals on the metal) and $(2n - 1)$ antibonding orbitals (which are empty). In general, conical fragments such as ML_3, CpM, and so on, are isolobal with BH, and their clusters are expected to be analogous to the boranes.

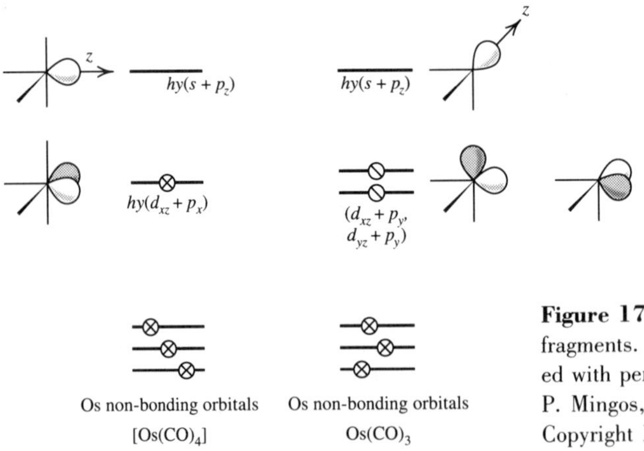
z
hy(s + p_z)
hy(s + p_z)
z
hy(d_xz + p_x)
(d_xz + p_y, d_yz + p_y)
Os non-bonding orbitals [Os(CO)_4]
Os non-bonding orbitals Os(CO)_3

Figure 17.39 Frontier orbitals of d^8 metal fragments. M—CO orbitals not shown. (Reprinted with permission from D. G. Evans and D. M. P. Mingos, *Organometallics* **1983**, *2*, 435. Copyright 1983, American Chemical Society.)

[27] See S. M. Owen, *Polyhedron* **1988,** *7,* 253; M. McPartlin, *Polyhedron* **1984,** *3,* 1279; M. McPartlin and D. M. P. Mingos, *Polyhedron* **1984,** *3,* 1321.

An alternative way of viewing the electron count for *closo* transition-metal clusters is that they require $(n + 1)$ skeletal bonding pairs as well as enough electrons to fill six orbitals on each metal with electrons which are either nonbonding or form ligand bonds. This leaves free three orbitals on each metal for skeletal bonding. Thus, a *closo* cluster is expected to contain $(n + 1) + 6n = (7n + 1)$ pairs of electrons or $(14n + 2)$ electrons.

Nido clusters will require $(n + 2)$ skeletal bonding pairs plus $6n$ nonbonding or ligand bonding pairs, for a total of $(7n + 2)$ pairs or $(14n + 4)$ electrons. A similar argument leads to the expectation of $(14n + 6)$ as the total electron count for *arachno* clusters. Table 17.5 summarizes the expected total electron counts for a variety of clusters containing π-acid ligands.

Heteroatoms in clusters such as $Fe_4(CO)_{13}C$ (Figure 17.32*a*) or $Fe_5(CO)_{15}C$ (Figure 17.32*b*) are located interstitially and do not occupy a vertex; all the heteroatom valence electrons are contributed to the skeletal electron count. Exposed heteroatoms in clusters often display interesting reactivity involving proton addition or reaction with methanol (Figure 17.40).

Table 17.5   PSEPT electron counts for some clusters with π-acid ligands

n (number of vertices)	*Geometry*	*Electron count*	*Example*
closo		**$(14n + 2)$**	
5	Trigonal bipyramid (TBP)	72	$Os_5(CO)_{16}$
6	Octahedron	86	$Rh_6(CO)_{16}$
7	Pentagonal bipyramid	100	—
8	Dodecahedron	114	—
9	Tricapped trigonal prism	128	—
10	Bicapped square antiprism	142	$[Rh_{10}S(CO)_{22}]^{2-}$
12	Icosahedron	170	$[Rh_{12}Sb(CO)_{27}]^{3-}$
nido		**$(14n + 4)$**	
4	Tetrahedron (TBP minus axial vertex)	60	$Ir_4(CO)_{12}$ $[Fe_4(CO)_{13}]^{2-}$
5	Square pyramid	74	$Fe_5C(CO)_{15}$
6	Pentagonal pyramid	88	—
8	Bicapped trigonal prism	116	—
9	Capped square antiprism	130	$[Rh_9P(CO)_{21}]^{2-}$
arachno		**$(14n + 6)$**	
3	Triangle (TBP minus both axial vertices)	48	$Os_3(CO)_{12}$
4	Butterfly (octahedron minus *cis* vertices)	62	$[Fe_4C(CO)_{13}]$ $[Fe_4C(CO)_{12}]^{2-}$
6	Trigonal prism	90	$[Rh_6C(CO)_{15}]^{2-}$

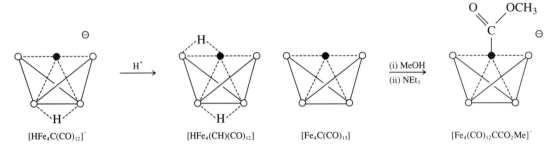

Figure 17.40 Reactions occurring at the carbido carbon atoms of Fe_4C clusters. (From J. N. Nicholls, *Polyhedron* **1984**, *3*, 1307.)

The treatment of tetrahedral clusters as the *nido* relatives of *closo*-TBP structures and of triangles as the *arachno* relatives in Table 17.5 may seem strange; however, it is otherwise impossible to fit them into the PSEPT scheme. These two structures can also be considered as **electron-precise;** that is, each metal obeys the EAN rule.

Note also that the simplest way to count electrons for clusters is to consider all atoms as neutral and to add or subtract from their total the number of electrons appropriate to the charge on the cluster. Some judgment is required to do this; for example, μ_2-Cl contributes three electrons, whereas terminal Cl contributes only one.

Structural predictions based on the PSEPT electron counts of Table 17.5 give the gross geometry of clusters, but not the details of their ligand arrangements. For example, $Rh_6(CO)_{16}$ is correctly predicted to have an octahedron of Rh's. However, the formula does not permit six $Rh(CO)_3$ units. The actual structure has $Rh(CO)_2$ fragments at the six vertices with μ_3-CO's on alternating octahedral faces; the PSEPT count does not predict the bridging CO's. (Other structural possibilities such as 4 $Rh(CO)_3$ and 2 $Rh(CO)_2$ correspond to the same electron count.)

The isolobal relationship between transition-metal fragments and BH units makes it reasonable that there should be clusters in which some BH units are replaced by transition-metal fragments.[28] (Such compounds may not correspond to the strict definition of clusters as containing two or more M—M bonds per metal atom, but are often called clusters in the literature. It is just as correct to think of them as cages derived from the boranes.) For these, it is convenient to calculate the number of electrons contributed to skeletal bonding by various metal-containing fragments. Because we require 12 electrons to fill the six orbitals which are nonbonding or ligand bonding for each transition metal, the remaining three orbitals and whatever electrons remain are contributed to skeletal bonding. Thus, each fragment contributes $(v + x - 12)$, where v is the number of metal valence electrons and x is the number supplied by ligands. Electron counts for some groups are given in Table 17.6. For example, B_5H_9, $B_4H_8Fe(CO)_3$, and $B_3H_7[Fe(CO)_3]_2$ all have five groups and seven skeletal electron pairs leading to *nido* structures as does the six-group, eight-skeletal-pair species $BC_4H_5Fe(CO)_3$.[29] Table 17.3 fulfills the same function for main-group heteroatoms as Table 17.6 does for transition-metal groups.

[28] C. E. Housecroft, *Adv. Organomet. Chem.* **1991**, *33*, 1.

[29] T. P. Fehlner, *J. Am. Chem. Soc.* **1978**, *100*, 3250.

Table 17.6   Number of skeletal bonding electrons ($= v + x - 12$) contributed by some transition-metal units[a]

| Metal group | M | Cluster unit | | | |
		$M(CO)_2$	MCp	$M(CO)_3$	$M(CO)_4$
6(VIA)	Cr, Mo, W	-2	-1	0	2
7(VIIA)	Mn, Tc, Re	-1	0	1	3
8 ⎫	Fe, Ru, Os	0	1	2	4
9 ⎬(VIII)	Co, Rh, Ir	1	2	3	5
10 ⎭	Ni, Pd, Pt	2	3	4	6

[a] From K. Wade, *Adv. Inorg. Radiochem.* **1975**, 18, 1.

▶ 17.10.5   The Capping Principle

A large number of structures is known in which triangular cluster faces are capped. An example is $[Rh_7(CO)_{16}]^{3-}$ (Figure 17.41a), where an $[Rh(CO)_3]^{3+}$ group caps an octahedral face on $[Rh_6(CO)_{13}]^{6-}$. Such structures can be rationalized (as Mingos showed) by the **capping principle**, which states that *the skeletal electron count for a capped structure is the same as that for the parent.* In the present example, the parent $[Rh_6(CO)_{13}]^{5-}$ has a count of 86 as expected for a *closo* octahedron. The $[Rh(CO)_3]^{3+}$ fragment has 12 electrons occupying nonbonding and Rh—CO bonding orbitals and thus contributes to skeletal bonding only its three empty orbitals of A and E symmetry (referred to the threefold axis of the capped triangular face). Figure 17.41b shows the result of interaction of these empty orbitals with filled cluster orbitals of suitable symmetry: one a and a pair of e

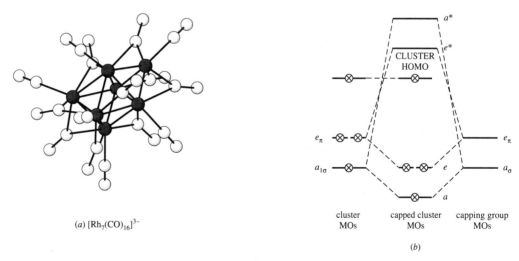

(a) $[Rh_7(CO)_{16}]^{3-}$

(b)

Figure 17.41   (a) The structure of $[Rh_7(CO)_{16}]^{3-}$. (V. G. Albano, P. L. Bellon, and G. F. Ciani, *J. Chem. Soc. Chem. Commun.* **1969**, 1024. Reprinted with permission from P. Chini, G. Longoni, and V. G. Albano, *Adv. Organomet. Chem.* **1976**, *14*, 285.) (b) Schematic diagram showing the result of capping a triangular cluster face. Symmetry labels refer to the pseudo-threefold axis passing through the face center.

Table 17.7 **Electron counts for some capped π-acid clusters**

Geometry	Electron count	Example
closo parent cluster	**$(14n + 2 + 12m)$**	
Capped trigonal bipyramid	$72 + 12 = 84$	$Os_6(CO)_{18}$
Monocapped octahedron	$86 + 12 = 98$	$Os_7(CO)_{21}$
Bicapped octahedron	$86 + 24 = 110$	$[Os_8(CO)_{22}]^{2-}$
Tetracapped octahedron	$86 + 48 = 134$	$[Os_{10}C(CO)_{24}]^{2-}$
nido parent cluster	**$(14n + 4 + 12m)$**	
Bicapped tetrahedron = capped trigonal bipyramid	$60 + 24 = 84$	$Os_6(CO)_{18}$
Monocapped square pyramid	$74 + 12 = 86$	$H_2Os_6(CO)_{18}$

bonding orbitals along with their antibonding counterparts result. Only the bonding orbitals are occupied, but these are more stable than those in the original cluster due to interaction with the empty capping group orbitals. Thus, for capped *closo* structure, the total electron count is expected to be $(14n + 2 + 12m)$, where m is the number of capping groups. Table 17.7 gives examples.

Figure 17.41*b* also provides insight into what happens as vertices of deltahedral clusters are excised while electrons are left behind as structures are converted from *closo* to *nido* to *arachno*. As the three valence orbitals on the vertex are removed, a set of three bonding and a set of three antibonding MOs collapses to a single set of three. Sufficient electrons remain to occupy this (less-bonding) set. The skeletal electron count remains the same, but the number of vertex groups decreases by one. In metal-containing clusters, this reduces the total electron count by 12 (as contrasted with boranes).

A two-dimensional capping principle can rationalize edge-bridged structures. $M(CO)_4$ fragments often play this role, and the frontier orbitals are depicted in Figure 17.39. $M(CO)_4$ is isolobal with CH_2 which often forms bridges. MO interactions analogous to

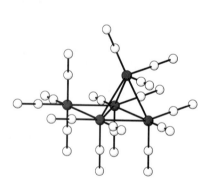

Figure 17.42 The structure of $H_2Os_5(CO)_{16}$. (From J. J. Guy and G. M. Sheldrick, *Acta, Cryst.* **1978**, *B34*, 1725.)

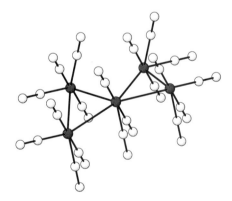

Figure 17.43 The "bow-tie" structure of $Os_5(CO)_{19}$. (From D. H. Farrar, B. F. G. Johnson, J. Lewis, J. N. Nicholls, P. R. Raithby, and M. J. Rosales, *J. Chem. Soc. Chem. Commun.* **1981**, 273.)

those in Figure 17.41*b* can be imagined, leading to stabilization via edge bridging. Leaving vacant the radial and one tangential orbital for interaction with cluster orbitals and filling the remaining $M(CO)_4$ orbitals with nonbonding or ligand bonding electrons leads to an electron count of 14 for the fragment. Thus, edge-bridged structures are expected to have a total electron count of $(N + 14)$, where N is the count for the parent. Thus, $[Os(CO)_4]^{2+}$ has 14 electrons and bridges one tetrahedral edge of $[H_2Os_4(CO)_{12}]^{2-}$; the electron count for $H_2Os_5(CO)_{16}$ is $(60 + 14) = 74$ (Figure 17.42). Likewise the "bow-tie" structure of $Os_5(CO)_{19}$ (Figure 17.43) can be rationalized as resulting of capping linear $[Os_3(CO)_{11}]^{4-}$ $(50e)$ with two $Os(CO)_4^{2+}$ fragments for a total count of 78.

► *17.10.6 Structures Not Rationalized by the PSEPT MODEL*

Although the PSEPT model correctly predicts a large number of cluster structures, many remain unrationalized. It is not yet completely clear how nonconical fragments such as $M(CO)_4$ (at vertices) fit into the scheme. Mingos has shown that their frontier orbitals are well-suited for forming planar clusters such as (electron precise) $Os_3(CO)_{12}$. One $M(CO)_4$ fragment can be incorporated into deltahedral clusters without disturbing the counting scheme, but M—M bonds are weakened.

Clusters containing Pt and Au conform to a different model requiring fewer skeletal bonding electrons as high-energy p orbitals are removed from the valence shell toward the end of the transition series.

Tetrahedral and planar triangular clusters can be viewed as electron-precise species having two-electron metal–metal bonds directed along edges. Consequently, there is often no advantage in using the PSEPT model for rationalization of their electronic structures.

A number of electron "hyperdeficient" clusters are known; an example with eight vertices is $(CpCo)_4B_4H_4$, which has only eight skeletal bonding pairs but adopts the *closo* dodecahedral structure (Figure 17.44). Such structures are frequently found among metallaboranes.

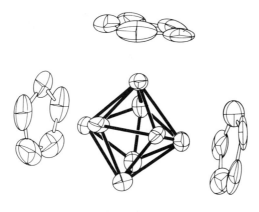

Figure 17.44 The structure of $(CpCo)_4B_4H_4$. (Reprinted with permission from J. R. Pipal and R. N. Grimes, *Inorg. Chem.* **1979**, *18*, 257. Copyright 1979, American Chemical Society.)

We have adopted the viewpoint of seeking to predict structural type from electron count in clusters. Although such an approach works fairly well for boron compounds, a number of important exceptions are known—such as B_4Cl_4, which is tetrahedral, and $B_pX_p(p = 6-12)$, which display *closo* structures. A possible alternative approach starts with the geometry given and develops a qualitative MO scheme for a metal framework.[30] (See Problem 17.15.) The framework electron count that will just fill the bonding and nonbonding levels is determined and used to rationalize the geometry adopted. For example, $[Ni_5(CO)_{12}]^{2-}$ has a total electron count of 76, leading to the prediction of an *arachno* structure derived from a dodecahedron. Instead, the complex is an elongated TBP (Figure 17.26). A regular TBP is predicted for a complex having 72 electrons. If the regular TBP is distorted by lengthening the axial distances, however, calculations show that two antibonding orbitals become stable enough to accommodate the two "extra" electron pairs. An MO calculation often serves to rationalize the structures of "nonconforming" complexes, but the results are without predictive value.

▶ *17.10.7 Halide Clusters*

Two long-known cluster compounds, $[Mo_6Cl_8]^{4+}$ and $[Ta_6Cl_{12}]^{2+}$, are depicted in Figure 17.45. As with other clusters of the types M_6X_8 and M_6X_{12}, both have octahedra of metal atoms. Mo is surrounded by a square-planar array of halide ligands, while Ta lies slightly below the halide plane. Their electronic structures can be rationalized in terms of localized bonds. Each metal uses dsp^2 hybrids to bond with its four square-planar ligands. A *pd* hybrid directed outward is used for bonding the additional *exo* ligands usually present. Four atomic orbitals for each metal are available for framework bonding. The Mo_6^{12+} and Ta_6^{14+} cores have 12 and 8 electron pairs, respectively. These could be used for M—M two-center bonds along the 12 octahedral edges or three-center bonds on the 8 octahedral faces.[31]

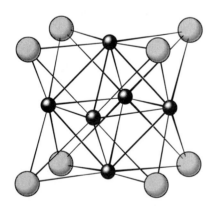 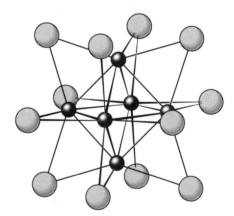

Figure 17.45   Structures of $[Mo_6Cl_8]^{4+}$ and $[Ta_6Cl_{12}]^{2+}$. (From L. Pauling, *The Nature of the Chemical Bond*, 3rd ed. Cornell University Press, Ithaca, NY, 1960, p. 440. Used with permission of Cornell University Press.)

[30] J. W. Lauher, *J. Am. Chem. Soc.* **1978**, *100*, 5305. See also Section 17.4 and Problem 17.15.
[31] F. A. Cotton and T. E. Haas, *Inorg. Chem.* **1964**, *3*, 10; S. F. A. Kettle, *Theo. Chim. Acta* **1965**, *3*, 211; R. F. Schneider and R. A. Mackay, *J. Chem. Phys.* **1968**, *48*, 843.

Condensation of octahedral halide (also OR^- and SR^-) clusters by ligand bridging and vertex-, edge- or face-sharing leads to a wide variety of two- and three-dimensional structures.[32]

▶ 17.10.8   Synthesis of Metal Clusters

A most active area in metal cluster chemistry is that of carbonyl-containing clusters. The syntheses of these species is not yet well understood. Unexpected products are obtained; for example, $H_2Ru_6(CO)_{18}$ is the product of the reaction in thf of $Ru_3(CO)_{12}$ with $[CpFe(CO)_2]^-$. Yields often are low. Single products seldom result, and separations are difficult. However, the redox condensation method (*vide infra*) does represent a considerable advance in systematic preparative strategies. Because of the large numbers of atoms, IR and NMR techniques are not always very informative for structure determination. Most progress depends on x-ray structure determinations (and, to a lesser extent, on mass spectra).

Thermochemical Considerations

The ability to prepare clusters and the conditions under which they are stable are related to the strengths of metal–metal and metal–ligand bonds. Table 17.8 gives bond energies determined from thermochemical experiments on the assumption that the M—CO terminal bond energy is invariant for all neutral carbonyls of the same metal. Also reported (in columns labeled "Set 2") are values calculated on the assumption that metal–metal bond energies are proportional to r^{-k}, where r is the M–M separation in the compound and k is a constant that can be evaluated from atomization enthalpies ΔH_{sub} of the pure metals.

Both methods indicate that metal–metal bond strengths increase on descending columns in the periodic table. Even so, the contribution of M–M bonds to the total bond energy is not very large. It accounts for only 6% of the total bond energy in dinuclear clusters, ~10% in trinuclear clusters, ~20% in tetranuclear clusters, and ~25% in hexanuclear clusters. Metal–CO bonds are energetically dominant.

Metal–ligand bond energies also increase on descending the periodic table (at about the same rate as M—M bond energies). In calculations that do not assume constancy for M—CO bond energies, we can see a significant increase in the average M—CO bond energy with increasing cluster nuclearity (as the CO/metal ratio decreases). The more electron-rich the metal, or the more negative the cluster charge, the more CO functions as a π-acid and the stronger the bonds become. In the set of calculations that differentiates bond energies by bond type, metal–terminal CO bonds have larger energies than metal–bridging CO bonds.

Thermal Condensation

Heating low-nuclearity clusters dissociates carbonyl ligands, creating coordinatively unsaturated species that condense:

$$3\,Rh_4(CO)_{12} \xrightleftharpoons[-19°C,\ 490\ atm\ CO]{25°C,\ 490\ atm\ CO} 2\,Rh_6(CO)_{16} + 4\,CO \qquad \Delta H \sim 15\ kJ \qquad (17.67)$$

[32] A. Simon, *Angew. Chem. Int. Ed. Engl.* **1988**, *27*, 159.

Table 17.8   Bond energies for some carbonyl complexes (kJ/mol)

Compound	d_{M-M} (pm)	E_{M-M} Set 1[a]	Set 2[b]	E_{CO} Set 1[a]	Set 2[b]
$Fe(CO)_5$				117	117
$Fe_2(CO)_9$	252	82	70	$117t^c$	123
				$64b^c$	
$Fe_3(CO)_{12}$	256	82	65	$117t^c$	126
				$64b^c$	
$Ru_3(CO)_{12}$	285	117	78	172	182
$Os_3(CO)_{12}$	268	130	94	190	201
$Co_2(CO)_8$	252	83	70	$136t^c$	136
				$68b^c$	
$Co_4(CO)_{12}$	249	83	74	136	140
$Rh_4(CO)_{12}$	273	114	86	166	178
$Rh_6(CO)_{16}$	278	114^d	80^d	166	182
$Ir_4(CO)_{12}$	268	130	117	190	196

[a] Assuming constant E_{M-M} and E_{CO} in series of compounds. From J. A. Connor, *Topics Curr. Chem.* **1977**, *71*, 71.

[b] Assuming $E_{M-M} = Ad^{-k}_{M-M}$. From C. E. Housecroft, *et. al., J. Chem. Soc., Chem. Commun.* **1978**, 765; C. E. Housecroft et al., *J. Organomet. Chem.* **1981**, *213*, 35.

[c] t = terminal; b = bridging

[d] Assuming 11 two-center, two-electron M—M bonds resonating among 12 octahedral edges.

Reaction (17.67) is endothermic, as we would expect for a process that breaks strong M—CO bonds and substitutes weaker M—M bonds.

The greater strength of M—CO bonds as compared with M—M bonds for first-row transition elements requires that cluster synthesis be carried out in the absence of CO, to avoid degrading the cluster. This dictates that a preformed carbonyl be used. With second- and third-row transition elements, reduction to carbonyls and condensation to clusters can be carried out in a single step; for example,

$$7\,K_3RhCl_6 + 48\,KOH + 28\,CO \xrightarrow[\text{MeOH}]{25°C,\ 1\ atm\ CO} K_3[Rh_7(CO)_{16}]$$
$$+ 42\,KCl + 12\,K_2CO_3 + 24\,H_2O \tag{17.68}$$

Buildup of negative charge renders even some third-row clusters unstable to CO-induced fragmentation to lower nuclearity species. Although $Rh_6(CO)_{16}$ and $[Rh_6(CO)_{15}]^{2-}$ are stable to CO at 25°C, isoelectronic $[Rh_6(CO)_{14}]^{4-}$ is not.

The major drawback to the thermal condensation method is lack of specificity; for example:

$$Os_3(CO)_{12} \xrightarrow{300°C} Os_6(CO)_{16} + Os_5(CO)_{16} + Os_7(CO)_{21} + Os_8(CO)_{23} \tag{17.69}$$

Redox Condensation

Redox condensation was first carried out by Hieber and Schubert and later by Chini. These reactions often take place under very mild conditions. Examples are

$$[Fe_3(CO)_{11}]^{2-} + Fe(CO)_5 \xrightarrow{25°C} [Fe_4(CO)_{13}]^{2-} + 3\,CO \qquad (17.70)$$

$$Ru_2Os(CO)_{12} + [Fe(CO)_4]^{2-} \xrightarrow{H^+} H_2OsRu_2Fe(CO)_{13} + 3\,CO \qquad (17.71)$$

The mechanisms may involve electron transfer to form two radicals which couple. Redox condensations offer particularly attractive possibilities for making mixed-metal clusters.

$$2\,Co_2(CO)_8 + Rh_2(CO)_4Cl_2 \longrightarrow Co_2Rh_2(CO)_{12} + 2\,CoCl_2 + 8\,CO \qquad (17.72)$$

CO Loss Promoted by Base

Alcoholic base solutions can cause CO loss (presumably through OH^- attack on CO to produce the metallacarboxylic acid which eliminates CO_2, leaving an anionic hydride cluster), and cluster condensation can occur (presumably by H_2 elimination involving hydrides):

$$Rh_4(CO)_{12} \xrightarrow[i\text{-PrOH, }H_2]{80°C,\ OH^-} [Rh_4(CO)_{11}(\overset{\overset{\displaystyle O}{\displaystyle \|}}{C}OH)]^- \xrightarrow{-CO_2} [Rh_4(CO)_{11}(H)]^-$$
$$\longrightarrow [Rh_{13}(CO)_{25}H_3]^{2-} \qquad (17.73)$$

$$Rh_4(CO)_{12} \xrightarrow[i\text{-PrOH, }N_2]{80°C,\ OH^-} [Rh_{15}(CO)_{27}]^{3-} \qquad (17.74)$$

► *17.10.9   Reactivity of Cluster Compounds*

Comparatively little is known of the reactivity of cluster species, and not much of reaction mechanisms. Only some rather general reactions types are mentioned here.

Reduction

Although in several isoelectronic series, carbonyl ligands are formally replaced by electron pairs (e.g., $Co_6(CO)_{16}$, $[Co_6(CO)_{15}]^{2-}$, $[Co_6(CO)_{14}]^{4-}$), these are usually not prepared cleanly by reduction of the first member with release of CO. The released carbon monoxide often leads to cluster degradation.

$$
\begin{array}{rcl}
11\,[Co_6(CO)_{15}]^{2-} + 22\,Na & \longrightarrow & 11\,[Co_6(CO)_{14}]^{4-} + 22\,Na^+ + 11CO \\
2\,[Co_6(CO)_{14}]^{4-} + 11\,CO & \longrightarrow & [Co_6(CO)_{15}]^{2-} + 6\,[Co(CO)_4]^- \\
\hline
10\,[Co_6(CO)_{15}]^{2-} + 22\,Na & \longrightarrow & 9\,[Co_6(CO)_{14}]^{4-} + 6\,[Co(CO)_4]^- + 22\,Na^+
\end{array} \qquad (17.75)
$$

Use of strong base oxidizes CO, thus providing a better yield of reduction product. (Compare treatment with OH^- mentioned above.)

$$Rh_4(CO)_{12} + OCH_3^- \xrightarrow[\text{MeOH}]{25°C} [Rh_4(CO)_{11}(\overset{\displaystyle O}{\overset{\displaystyle \|}{C}}OCH_3)]^-$$

$$\xrightarrow{3OH^-} [Rh_4(CO)_{11}]^{2-} + MeOH + CO_3^{2-} + H_2O \qquad (17.76)$$

In this reaction, the intermediate carboalkoxy complex has been isolated.

Oxidation

Electrochemical oxidation of $Cp_4Fe_4(CO)_4$ to the +1 and +2 ions is known; it probably indicates that the HOMO of the parent compound is slightly antibonding or nonbonding.

A second method is by protonation of anionic clusters to give intermediate hydrido compounds, which then react with an additional proton, eliminating H_2:

$$[Ir_6(CO)_{15}]^{2-} + 2H^+ \xrightarrow[\text{CH}_3\text{CO}_2\text{H}]{CO} Ir_6(CO)_{16} + H_2 \qquad (17.77)$$

$$[Ir_4(CO)_{11}H]^- + H^+ \xrightarrow{CO} Ir_4(CO)_{12} + H_2 \qquad (17.78)$$

Some chemical oxidations are effected with coordination of the oxidant.

$$[Rh_6(CO)_{15}]^{2-} + I_2 \xrightarrow{\text{thf}} [Rh_6(CO)_{15}I]^- + I^- \qquad (17.79)$$

Ligand Substitution

Carbonyl ligands often can be displaced by other Lewis bases. However, cluster fragmentation with first-row transition metals often occurs because of their relatively weak M–M bonds.

$$Co_4(CO)_{12} + MeNC \longrightarrow Co_4(CO)_{11}(CNMe) \xrightarrow{MeNC} Co_4(CO)_{10}(CNMe)_2$$

$$\xrightarrow{MeNC} Co_4(CO)_9(CNMe)_3 \xrightarrow{MeNC} Co_4(CO)_8(CNMe)_4 \qquad (17.80)$$

$$Ru_6(CO)_{17}C + PPh_3 \longrightarrow Ru_6(CO)_{16}(PPh_3)C + CO \qquad (17.81)$$

$$Fe_3(CO)_{12} + 3PPh_3 \longrightarrow Fe(CO)_4(PPh_3) + Fe(CO)_3(PPh_3)_2 \qquad (17.82)$$

Both **a** and **d** activation can occur. In fact, clusters present a new possibility for **a** mechanisms. For example, the closed tetrahedral cluster $(\mu_3\text{-PhP})Fe_3(\mu\text{-CO})(CO)_9$ (isoelectronic with $Fe_3(CO)_{12}$) adds L = $P(OMe)_3$ with breaking of an M—M bond to afford $(\mu_3\text{-PhP})(CO)_4FeFe(CO)_3Fe(CO)_3L$ which can be isolated but which, on heating, expels CO and re-forms the Fe—Fe bond, giving $(\mu_3\text{-PhP})Fe_3(\mu\text{-CO})(CO)_8L$. This addition–elimination involving bonded–opened intermediates can be repeated twice more to produce $(\mu_3\text{-PhP})(\mu_2\text{-CO})[Fe(CO)_2L]_3$.

A heterolytic bond fission mechanism is also in accord with the fact that $M_3(CO)_{11}L$ and $M_3(CO)_{10}L_2$ (M = Fe, Ru, Os) substitute faster than $M_3(CO)_{12}$. Heterolytic fission in

$M_3(CO)_{11}L$ could produce species such as $L(CO)_3M^+\!-\!M(CO)_4\!-\!M^-(CO)_4$ which would be stabilized by L's which are better donors than CO. Attack by another L followed by M—M bond closure with accompanying CO expulsion leads to $M_3(CO)_{10}L_2$. The initial site of L attack need not be the ultimate site of substitution because ligands can migrate on the metal framework.

Clusters containing nonmetal atoms may display reactivity at these centers as well. Figure 17.40 summarizes some reactions of carbide clusters.

17.11 ELECTRON-PRECISE COMPOUNDS AND THEIR RELATION TO CLUSTERS

The octet and 18-e rules define the very large set of **electron-precise** compounds. Some of these have structures related to the deltahedra of Figure 17.6. It is often useful to count the total number of electrons in chains, rings, and cages of such species, and this approach is developed in this section.

For a single main-group atom, the octet rule requires eight e; these can be either lone pairs or bonding pairs. A diatomic main-group compound requires a total electron count of 14 because the E—E bond supplies one electron for each E atom. For a chain of n main-group atoms, each of the $(n-2)$ "inner" atoms requires six electrons because the two E—E bonds supply two electrons; the two terminal atoms require seven. Thus, the total is $6(n-2) + (7 \times 2) = (6n+2)$. Because a transition-metal atom has five more valence orbitals, each requries ten more electrons so that an electron-precise chain of n transition-metal atoms requires $(16n+2)$ electrons.

An n-membered ring is formed by bonding the two end atoms of a chain; because these atoms now also form two E—E bonds, they require six electrons just like the "inner" atoms, for a total of $6n$. Again, the electron count is increased by ten for transition metals, giving a total of $16n$.

In a three-dimensional three-connected cage, each atom acquires three electrons from M—M bonding; thus, $5n$ electrons are required for electron-precise main-group cages and $15n$ for transition-metal cages. Table 17.9 gives examples.

Many cage compounds may be considered as electron-rich species derived from electron-precise ones. We can imagine each "extra" electron pair as occupying a localized M—M antibonding orbital, thereby breaking an M–M bond.[33] Thus $Co_4(CO)_{12}Sb_4$ (Figure 17.46) contains a tetrahedron of Co atoms bridged by μ_3-Sb's and has no Co—Co bonds. The distortion results from differing sizes of the groups. This is a hybrid compound containing half main-group and half transition-metal atoms, so the electron count for a cube should be $[5(4) + 15(4)] = 80$ as observed. A detailed MO treatment for molecules such as $Co_4Sb_4(CO)_{12}$ related to cubane has been worked out by Dahl and co-workers.[34]

[33] D. M. P. Mingos, *Nature Phys. Sci.* **1972**, *236*, 99.

[34] M. A. Neuman, Trinh-Toan, and L. F. Dahl, *J. Am. Chem. Soc.* **1972**, *94*, 3383; Trinh-Toan, W. P. Fehlhammer, and L. F. Dahl, *J. Am. Chem. Soc.* **1972**, *94*, 3389.

Table 17.9   Electron counts for some electron-precise molecules[a]

	Main group	Count	Transition metal	Count
Chains		**(6n + 2)**		**(16n + 2)**
$n = 2$	P_2Cl_4	14	$Mn_2(CO)_{10}$	34
$n = 3$	C_3H_8	20	$[Mn_3(CO)_{14}]^-$	50
Rings		**6n**		**16n**
$n = 3$	C_3H_6	18	$Os_3(CO)_{12}$, $Re_3(\mu\text{-}Cl)_9$	48
$n = 4$	$(CF_3P)_4$	24	$[Pt_4(\mu_2\text{-}CH_3CO_2)_8]$	64
$n = 6$	$[(FS)N]_3$	36	—	96
$n = 8$	S_8, $[(MeN)(PEt)]_4$	48	—	128
Cages		**5n**		**15n**
$n = 4$	P_4, $[Sn_2Bi_2]^{2-}$	20	$Ir_4(CO)_{12}$	60
$n = 6$	C_6H_6 (prismane)	30	$[Rh_6C(CO)_{15}]^{2-}$	90
$n = 8$	C_8H_8 (cubane)	40	$Ni_8(PPh)_6(CO)_8$	120

[a] Assuming only single bonds between vertex atoms.

EXAMPLE 17.2:   Predict the structure of $[Re_4(CO)_{16}]^{2-}$.

Solution: For a tetrahedral structure, 60 electrons are required. This ion has 62 electrons. We imagine the extra pair of electrons as occupying an antibonding orbital, breaking one Re—Re bond and giving the structure shown in Figure 17.47. Alternatively, we can view 48-electron triangular $[Re_3(CO)_{12}]^{3-}$ as being edge-bridged by 14-electron $[Re(CO)_4]^+$.

EXAMPLE 17.3:   Predict the structure of S_4N_4.

Solution: If we start by arranging the eight atoms at the corners of a cube, 40 electrons are required for an electron-precise structure. This compound has 44 electrons. The two "extra" pairs of electrons break two bonds. The actual structure of S_4N_4 shown in Figure 16.5 results from breaking two of the 12 bonds in the cage structure of cuneane (Figure 17.48), an isomer of the

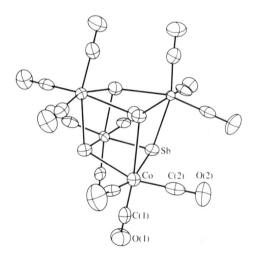

Figure 17.46   The structure of $Co_4(CO)_{12}Sb_4$. (Reprinted with permission from A. S. Foust and L. F. Dahl, *J. Am. Chem. Soc.* **1970**, *92*, 7337. Copyright 1970, American Chemical Society.)

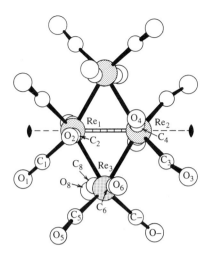

Figure 17.47 The structure of [Re$_4$(CO)$_{16}$]$^{2-}$. (Reprinted with permission from M. R. Churchill and R. Bau, *Inorg. Chem.* **1968,** *7,* 2606. Copyright 1968, American Chemical Society.)

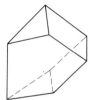

Figure 17.48 The structure of cuneane, an isomer of cubane (C$_8$H$_8$).

eight-carbon cubane. Hence, with some main-group compounds, the straightforward choice of the cube as the starting eight-vertex polyhedron must be modified.

EXAMPLE 17.4: Rationalize the structure of Se$_8^{2+}$ (Figure 16.4).

Solution: If we arrange the eight Se atoms at the corners of a cube, 40 electrons are required for an electron-precise structure. Neutral Se$_8$ has 48 electrons, and its eight-membered ring structure could be rationalized as resulting from breaking four alternating bonds with the four "extra" electron pairs. Se$_8^{2+}$ has only six "extra" electrons, so one trans-annular bond remains in the distorted skeleton remaining after three bonds are broken.

GENERAL REFERENCES

H. C. Brown, *Boranes in Organic Chemistry,* Cornell University Press, Ithaca, NY, 1972.

Chem. Rev. 1992, No. 2. An entire issue devoted to boron chemistry.

L. M. Jackman and F. A. Cotton, Eds., *Dynamic Nuclear Magnetic Resonance Spectroscopy,* Academic Press, New York, 1975. A comprehensive treatment of stereochemical nonrigidity.

D. M. P. Mingos and D. J. Wales, *Introduction to Cluster Chemistry,* Prentice–Hall, Englewood Cliffs, NJ, 1990. An excellent and up-to-date treatment.

E. L. Muetterties, Ed., *Boron Hydride Chemistry,* Academic Press, New York, 1975. Excellent chapters on the chemistry of boranes.

G. A. Olah, K. Wade, and R. E. Williams, Eds., *Electron Deficient Boron and Carbon Clusters,* Wiley, New York, 1991. Current viewpoints on the title subjects.

D. F. Shriver, H. D. Kaesz, and R. D. Adams, Eds., *The Chemistry of Metal Cluster Complexes,* VCH Publishers, New York, 1990. Excellent chapters summarizing the current state of knowledge about clusters written by experts in the field.

PROBLEMS

17.1 Classify the following species as *closo, nido, arachno,* or *hypho*.
(a) C_8H_8 (cyclooctatetraene)
(b) B_6H_{12} (f) $B_6H_{11}^+$
(c) B_9H_{15} (g) B_9H_9NH
(d) B_4H_{10} (h) $B_9H_{12}^-$
(e) B_4H_8 (i) $B_9H_{13} \cdot CH_3CN$

17.2 What structure do you expect for each species in Problem 17.1?

17.3 Calculate *styx* numbers and draw valence structures for the following.
(a) B_5H_{11} (c) $B_5H_5^{2-}$
(b) B_6H_{10} (d) B_8H_{12}

17.4 Solve the Equations of Balance for $B_3H_6^+$ (not known) and write a "reasonable" structure for such a hydride.

17.5 Show that the number of framework electrons contributed by the groups in Table 17.3 is given by the formula $(v + x - 2)$. v is the number of valence electrons in the vertex atom, and x is the number contributed by *exo* ligands. For example, for BH, $v = 3$ and $x = 1$.

17.6 Classify the following species as *closo, nido, arachno,* or *hypho*.
(a) $2\text{-}CB_5H_9$ (f) $B_{11}SH_{10}Ph$ (Ph attached to B)
(b) $2\text{-}CH_3\text{-}2,3\text{-}C_2B_4H_7$ (g) $B_9H_{12}NH^-$
(c) $1,2\text{-}C_2B_9H_{11}$ (h) $C_2B_6H_{10}$
(d) $B_9H_{11}S$ (i) $B_2H_4(PF_3)_2$'
(e) $1,7\text{-}B_{10}CPH_{11}$
Name and sketch the structure of the above species.

17.7 Use the Equations of Balance to obtain a reasonable bonding picture of 1,5-dicarba-*closo*-pentaborane(5).

17.8 Predict the products of the following reactions.
(a) $B_5H_{11} + KH$ (g) $Li_2(o\text{-}C_2B_{10}H_{10}) + R_3SiCl$
(b) $B_5H_9 + NMe_3$ (h) $B_6H_9^- + Me_2SiCl_2$
(c) $B_{10}H_{14} + SMe_2$ (i) $2,3\text{-}C_2B_4H_8 + NaH$
(d) $B_5H_9 + HCl(l)$ (j) $ZrCl_4 + 4LiBH_4$
(e) $B_6H_{10} + Br_2$ (k) $Ph_2PCl + LiAlH_4$
(f) $Li_2(o\text{-}C_2B_{10}H_{10}) + MeI$

17.9 Give a reasonable method for preparing and purifying B_2H_6. How might the purity of the sample be determined? How could one dispose of the diborane?

17.10 (a) How may $(CH_3)_2B_2H_4$ be prepared?
(b) Draw structural formulas for all isomers expected for $(CH_3)_2B_2H_4$.
(c) How might one identify these isomers if they were all separated?

17.11 (a) Show that the diamond–square–diamond mechanism (Figures 17.18*a*, and 17.18*b*) cannot account for the known thermal rearrangement of *o*-carborane to *p*-carborane.
(b) What experiments might be helpful in shedding light on the *o*- to *p*- rearrangement?
(c) Show how the ETR mechanism (Figures 17.18*c* and 17.18*d*) could lead to the *o*- to *p*-rearrangement.

17.12 Write an essay on the possible relevance of transition metal cluster compounds to heterogeneous catalysis. (You will want to consult some of the references.)

17.13 Classify the following as *closo*, *nido*, or *arachno*.
    (a) $[Co_4Ni_2(CO)_{14}]^{2-}$ **(e)** $[(Et_3P)_2Pt(H)]B_9H_{10}S$
    (b) $[Fe(CO)_3]B_4H_8$ **(f)** $Rh_6(CO)_{16}$
    (c) $(CpCo)C_2B_7H_{11}$ **(g)** $[Rh_9P(CO)_{21}]^{2-}$ (P is at cluster center)
    (d) $Os_5(CO)_{16}$

17.14 Sketch the predicted geometry of the species in Problem 17.13.

17.15 Consider $B_4H_4^{n-}$ in a tetrahedral geometry. Derive the MO energy-level diagram and predict the even values of $0 \leq n \leq 6$ for which this geometry will be stable.

17.16 Figure 17.49 shows the ^{13}C NMR spectrum of (cot)Fe(CO)₃ at several temperatures. The signals at 214 and 212 ppm are attributed to CO, and the others are attributed to cot. Explain the appearance of the spectrum at $-134°C$, and its change with T. (See F. A. Cotton and D. L. Hunter, *J. Am. Chem. Soc.* **1976**, *98*, 1413.)

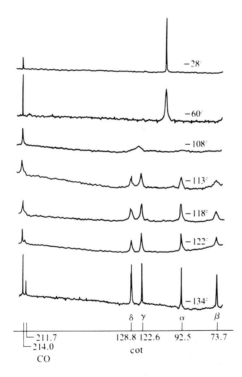

Figure 17.49 ^{13}C NMR spectrum of (cot)Fe(CO)₃ at several temperatures. (Reprinted with permission from F. A. Cotton and D. L. Hunter, *J. Am. Chem. Soc.* **1976**, *98*, 1413. Copyright 1976, American Chemical Society.)

17.17 Figure 17.50*a* shows the structure of $[Cr_3S(CO)_{12}]^{2-}$. Its ^{13}C NMR spectrum at several temperatures is shown in Figure 17.50*b*.
    (a) Give the Polyhedral Skeletal Electron-Pair Theory (PSEPT) electron count for the cluster and show how it is consistent with the structure shown.
    (b) What is the source of the inequivalence of the two types of terminal CO's.
    (c) Explain the changes in the NMR spectrum as the sample is warmed. (See D. J. Darensbourg and D. J. Zalewski, *Organometallics* **1984**, *3*, 1598.)

17.18 Figure 17.51*a* shows the structure of $[(\mu_3: \eta^2,\eta^2,\eta^2\text{-}C_6H_6)(\eta^2\text{-}CH_2CH_2)Os_3(CO)_8]$. Its ^1H NMR spectrum at several temperatures is shown in Figure 17.51*b* along with a labeling scheme for the protons.
    (a) Give the PSEPT electron count for the cluster and show how it is consistent with the structure shown.

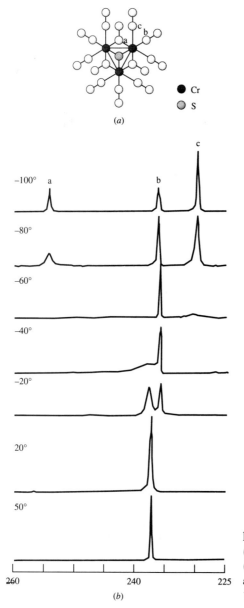

(a)

Figure 17.50 $[Cr_3(\mu_3\text{-}S)(CO)_{12}]^{2-}$. (a) Structure.
(b) Variable-temperature ^{13}C NMR spectrum.
(Reprinted with permission from D. J. Darensbourg
and D. J. Zalewski, *Organometallics* **1984**, *3*,
1598. Copyright 1984, American Chemical Society.)

(b) Account for the number of different kinds of protons observed at 188 K.

(c) As the temperature is raised, all the ring proton NMR signals coalesce to a single NMR signal around 4 ppm. What kind of molecular motion would account for this?

(d) As the temperature is raised, all the olefin proton signals coalesce first to two signals (not well-resolved) at around 230 K and finally to a single signal at 290 K. What kind of molecular motion would account for this?

(e) If the temperature were further raised beyond 290 K, would you expect the two signals to coalesce? Why or why not? (See M. A. Gallop, B. F. G. Johnson, J. Keeler, J. Lewis, S. J. Heyes, and C. M. Dobson, *J. Am. Chem. Soc.* **1992**, *114*, 2510.)

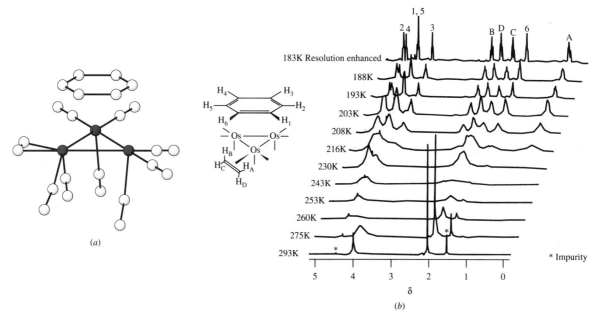

Figure 17.51 $[(\mu_3: \eta^2,\eta^2,\eta^2\text{-}C_6H_6)(\eta^2\text{-}C_2H_4)Os_3(CO)_9]$. (*a*) Static structure.
(*b*) Variable-temperature ^{1}H NMR spectrum. (Reprinted with permission from M. A. Gallop, B.
F. G. Johnson, J. Keeler, J. Lewis, S. J. Hayes, and C. M. Dobson, *J. Am. Chem. Soc.*
1992, *114*, 2510. Copyright 1992, American Chemical Society.)

17.19   The compound $Co_3Rh(CO)_{12}$ is a tetrahedral cluster.
  (a) At $-85°C$, its ^{13}C NMR spectrum displays seven signals in intensity ratio
  $1:2:2:2:3:1:1$. The second and last two signals are coupled to ^{103}Rh ($I = \frac{1}{2}$). What
  is the structure of the species "frozen out" at this temperature?
  (b) On warming to $+10°C$, two signals appear: a single line of relative intensity 2 showing
  coupling to ^{103}Rh, and one of relative intensity 10 which is somewhat broadened. At
  $+30°C$, only a single broad resonance is visible. Account for these observations. (See
  B. F. G. Johnson, J. Lewis and T. W. Matheson, *Chem. Commun.* **1974**, 441.)

17.20   Figure 17.52*a* shows the structure of $Cp^*Ru(acac)(\eta^2\text{-}CF_2{=}CF_2)$, and Figure 17.52*b*
  shows its ^{19}F NMR spectrum at various temperatures.
  (a) Account for the appearance of the spectrum at $-50°C$.
  (b) As the temperature increases, the signals coalesce to give *two* signals at 100°C. What
  does this indicate about the process which interchanges F's? (See O. J. Curnow, R. P.
  Hughes, and A. L Rheingold, *J. Am. Chem. Soc.* **1992**, *114*, 3153.)

17.21   Figure 17.53 shows the structures of some metal cluster compounds. Rationalize these observed structures by giving the PSEPT electron counts.

17.22   Figure 17.54 shows the structures of some metalloborane cluster compounds. Rationalize
  these observed structures by giving the PSEPT electron counts. With which borane is each
  cluster isoelectronic?

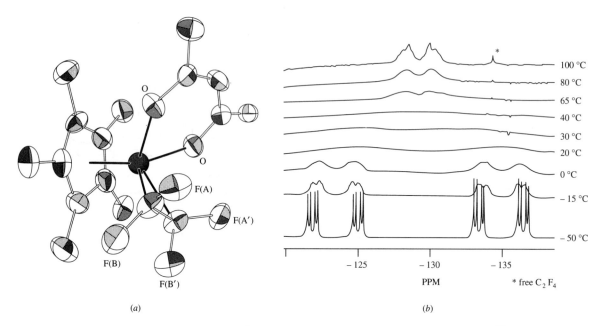

(a)   (b)

Figure 17.52   Cp*Ru(acac)(η^2-C_2F_4). (*a*) Static structure. (*b*) Variable-temperature ^{19}F NMR. (Reprinted with permission from O. J. Curnow, R. P. Hughes, and A. L. Rheingold, *J. Am. Chem. Soc.* **1992,** *114,* 3153.

17.23 (a) Give the electron count and sketch the geometry for 1,2-$B_9C_2H_{11}^{2-}$ and 1,7-$B_9C_2H_{11}^{2-}$.

      (b) Both of these ligands form a number of compounds with transition metals. Give the electron count and predict the structure for CpCo(1,2-$C_2B_9H_{11}$).

      (c) MO calculations indicate that six electrons are available on the open faces of the dicarbollide ligands for donation to transition metals, making them equivalent to Cp^-. Show that the following obey the EAN rule for the metal: (η^4-Ph_4C_4)Pd(1,2-$C_2B_9H_{11}$), [(1,2-$C_2B_9H_{11}$)Re(CO)$_3$]$^-$, [(1,2-$C_2B_9H_{11}$)Mo(CO)$_3$W(CO)$_5$]$^{2-}$.

17.24 What do the following experimental results suggest about the substitution mechanisms in the clusters?

(a)

Reaction	*Medium*	$\Delta V^{\ddagger}$ (cm^3 mol^{-1})
[HRu$_3$(CO)$_{11}$]$^-$ + PPh$_3$	thf	+21.2
[Ru$_3$(CO)$_{11}$(CO$_2$Me)]$^-$ + P(OMe)$_3$	90/10 thf/MeOH	+16
[Ru$_3$(CO)$_{11}$(CO$_2$Me)]$^-$ + P(OMe)$_3$	10/90 thf/MeOH	+2.5

(See D. J. Taube, R. van Eldik, and P. C. Ford, *Organometallics* **1987,** *6,* 125.)

(b) For the reaction

$$M_4(CO)_9[HC(PPh_2)_3] + L \rightarrow M_4(CO)_8L[HC(PPh_2)_3] + CO$$

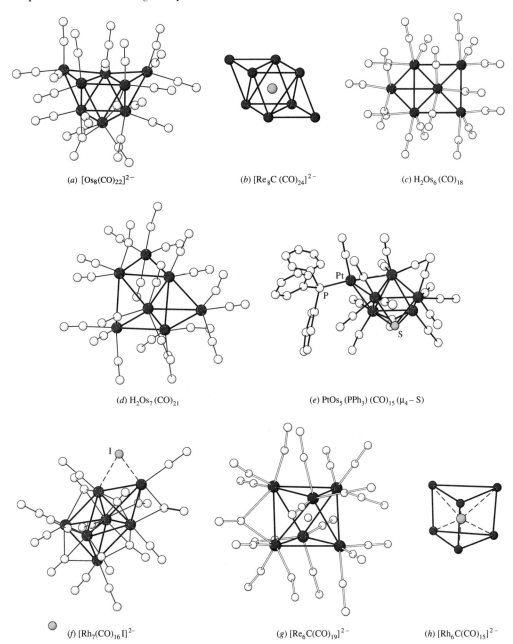

(a) $[Os_8(CO)_{22}]^{2-}$

(b) $[Re_8C(CO)_{24}]^{2-}$

(c) $H_2Os_6(CO)_{18}$

(d) $H_2Os_7(CO)_{21}$

(e) $PtOs_5(PPh_3)(CO)_{15}(\mu_4 - S)$

(f) $[Rh_7(CO)_{16}I]^{2-}$

(g) $[Re_6C(CO)_{19}]^{2-}$

(h) $[Rh_6C(CO)_{15}]^{2-}$

Figure 17.53   Structures for Problem 17.21. (Structure *a* is from P. F. Jackson, B. F. G. Johnson, J. Lewis, and P. R. Raithby, *J. Chem. Soc., Chem. Commun.* **1980,** 60. Structure *c* is from M. McPartlin, C. R. Eady, B. F. G. Johnson, and J. Lewis, *J. Chem. Soc. Chem. Commun.* **1976,** 883. Structure *d* is from B. F. G. Johnson, J. Lewis, M. McPartlin, J. Morris, G. L. Powell, P. R. Raithby, and M. D. Vargas, *J. Chem. Soc. Chem. Commun.* **1986,** 429. Structure *e* is reprinted with permission from R. D. Adams. J. E. Babin, R. Mathab, and S. Wang, *Inorg. Chem.* **1986,** *25,* 1623. Copyright 1986, American Chemical Society. Structure *f* is from V. G. Albano, G. Ciani, S. Martinengo, P. Chini, and G. Giordano, *J. Organomet. Chem.* **1975,** *88,* 381. Structure *g* is from J. Beringhelli, G. D'Alfonso, H. Molinari, and A. Sirone, *J. Chem. Soc. Dalton Trans.* **1992,** 689.)

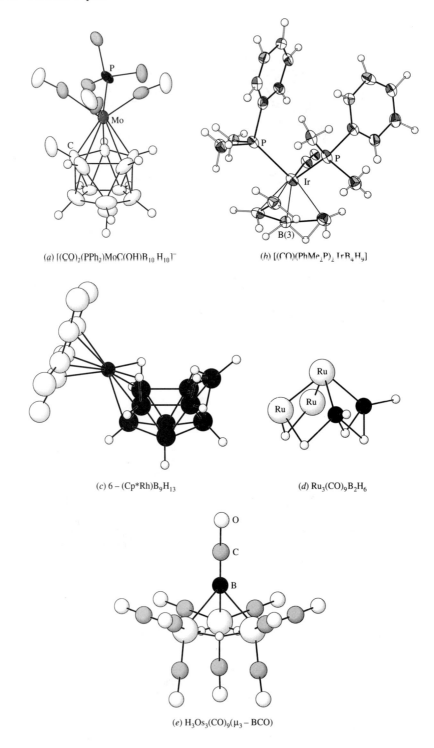

(a) $[(CO)_2(PPh_3)MoC(OH)B_{10}H_{10}]^-$

(b) $[(CO)(PhMe_2P)_2IrB_4H_9]$

(c) $6-(Cp*Rh)B_9H_{13}$

(d) $Ru_3(CO)_9B_2H_6$

(e) $H_3Os_3(CO)_9(\mu_3-BCO)$

(M = Co, Rh, Ir), rate = $(k_1 + k_2 [L])[M_4(CO)_9[HC(PPh_2)_3]]$ for M = Co and Rh, but the k_2 term dominates for Ir. Activation parameters are as follows:

M	L	$\Delta H_1^{\ddagger}$(kJ/mol)	$\Delta S_1^{\ddagger}$(kJ/mol)	$\Delta H_2^{\ddagger}$(kJ/mol)	$\Delta S_2^{\ddagger}$(kJ/mol)
Co	^{13}CO	101	−2.2		
	P(n-Bu)$_3$	105	31	43.1	−144
Rh	PPh$_3$	110	96	21.3	−121
	P(n-Bu)$_3$			25.5	−82.8
Ir	P(OMe)$_3$			59.4	−170
	P(n-Bu)$_3$			47.2	−192

(See J. R. Kennedy, P. Selz, A. L. Rheingold, W. C. Trogler, and F. Basolo, *J. Am. Chem. Soc.* **1991**, *111*, 3615.)

17.25   Using the approach developed by Mingos for cage and ring compounds (Section 17.11), predict plausible structures for the following.
(a)   S$_8$ (c)   P$_4$
(b)   P$_4$(C$_6$H$_{11}$)$_4$ (d)   [Fe(NO)$_2$]$_2$(SEt)$_2$

17.26   Figure 17.55 shows the structures of Os$_5$C(CO)$_{15}$ and its carbonylation product Os$_5$C(CO)$_{16}$. Rationalize these structures on the basis of their PSEPT electron counts. (See B. F. G. Johnson, J. Lewis, W. J. H. Nelson, J. N. Nicholls, J. Puga, P. R. Raithby, M. J. Rosales, M. Schröder, and M. D. Vargas, *J. Chem. Soc. Dalton Trans.* **1983**, 2447.)

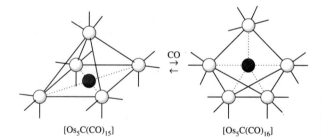

[Os$_5$C(CO)$_{15}$]   [Os$_5$C(CO)$_{16}$]

Figure 17.55   The reversible carbonylation of [Os$_5$C(CO)$_{15}$]. (From J. N. Nicholls, *Polyhedron*, **1984**, *3*, 1307.)

Figure 17.54   Structures for Problem 17.22. (Structure *a* is from X. L. R. Fontaine, N. N. Greenwood, J. D. Kennedy, P. I. MacKinnon, and I. Macpherson, *J. Chem. Soc. Dalton Trans.* **1987**, 2385. Structure *b* is reprinted with permission from S. K. Boocock, M. A. Toft, K. E. Inkrott, L.-Y. Hsu, J. C. Huffman, K. Folting, and S. G. Shore, *Inorg. Chem.* **1984**, *23*, 3084. Copyright 1984, American Chemical Society. Structure *c* is from X. L. R. Fontaine, H. Fowkes, N. N. Greenwood, J. D. Kennedy, and M. Thornton-Pett, *J. Chem. Soc. Dalton, Trans.* **1986**, 547. Structure *d* is from A. K. Chipperfield, C. E. Housecroft, and D. M. Matthews, *J. Organomet. Chem.* **1990**, *384*, C38. Reprinted from C. E. Housecroft, *Adv. Organomet. Chem.* **1991**, *33*, 1. Structure *e* is from D. P. Workman, D. -Y. Jan, and S. G. Shore, *Inorg. Chem.* **1990**, *29*, 3518. Reprinted from C. E. Housecroft, *Adv. Organomet. Chem.* **1991**, *33*, 1.)

·18·

Bioinorganic Chemistry

·····················

We now understand many biological processes at the molecular level. Complete sequences of amino acids, x-ray structures, and total syntheses have been achieved for large complex molecules of biological importance. The more complete our understanding of biological processes, the more we recognize the key roles of metal ions in the function of enzymes, oxidation–reduction reactions, and many other processes, in addition to their long-known roles in chlorophyll and hemoglobin. Bioinorganic chemistry is developing rapidly, and it illuminates the amazingly intricate and delicately balanced chemical processes in biological systems.

Obviously, life processes involve much more than just the reactions and syntheses of complex organic molecules. Biochemistry is a vast field. This chapter touches on important roles of metal ions in biochemical processes. The key role of coordination chemistry is evident, adapting a metal ion for a specific function.

Model studies have been very important in bioinorganic chemistry. The goal is not to improve on nature, but to understand it. Model building focuses on a simple system that we can understand, permitting us to control some factors and allowing us to evaluate alterations of the model and their effect on biological function.

18.1 ESSENTIAL ELEMENTS

▶ 18.1.1 Introduction

The essential biological roles of a few elements have been recognized for many years. Delicate electrolyte balances are achieved for Na^+ and Ca^{2+} as the major extracellular cations, for K^+ and Mg^{2+} as the major cellular cations, and for Cl^- as the major anion commonly found within and without the cell. The importance of calcium in bones was recognized before the development of biochemistry. Ca^{2+} is essential for formation of hydroxyapatite, $Ca_5(PO_4)_3OH$, the structural constituent of bone.

Quite early, iodine deficiencies were known to cause abnormal functioning of the thyroid gland. Iodine is an essential constituent of the thyroid hormones—for example, thyroxine:

Thyroxine

The polychlorinated biphenyls are structurally similar to thyroxine, and their toxicity might result from interference with the functions of thyroid hormones. The insecticide DDT [1,1,1-trichloro-2,2-bis(*p*-chlorophenyl)ethane, $Cl_3CCH(C_6H_4p\text{-}Cl)_2$] has been banned in the United States because of harmful effects on birds and other animals. These effects stem from biological amplification in food chains. Ultraviolet light—sunlight, for example—converts DDT in the vapor phase into toxic polychlorinated biphenyls.

About 30 elements[1] now recognized as essential for at least some forms of life (Table 18.1) include most of the elements through atomic number 35, except for Be, Al, Sc, Ti, Ga, Ge, and the noble gases. Molybdenum and iodine also are essential, and tungsten-containing enzymes have been found. In testing to determine if an element is essential at low concentrations, it is difficult to exclude an abundant element such as Al so completely that symptoms will appear as a result of the deficiency. Strikingly, all of the elements appreciably abundant in the human body, except for phosphorus, also are abundant in seawater—suggesting that our family tree is rooted in the sea. Figure 18.1*a* shows the abun-

Table 18.1 Chemical abundances
(percent of total number of atoms)

Composition of human body		*Composition of seawater*		*Composition of earth's crust*	
H	63%	H	66%	O	46.6%
O	25.5%	O	33%	Si	27.7%
C	9.5%	Cl	0.33%	Al	8.1%
N	1.4%	Na	0.28%	Fe	5.0%
Ca	0.31%	Mg	0.033%	Ca	3.6%
P	0.22%	S	0.017%	Na	2.8%
Cl	0.03%	Ca	0.006%	K	2.6%
K	0.06%	K	0.006%	Mg	2.1%
S	0.05%	C	0.0014%	Ti	0.44%
Na	0.03%	Br	0.0005%	H	0.14%
Mg	0.01%			C	0.20%

[1] E. Frieden, *J. Chem. Educ.* **1985,** *62,* 917.

dances of the elements in the oceans, and Figure 18.1*b* shows the abundance ratios of elements in seawater/abundances in the earth's crust. The elements giving water-soluble compounds occur as cations or simple anions and have high concentrations in seawater, as expected.

It is not at all unusual that elements essential at low concentrations, such as F, Se, As, and even Fe, are toxic at higher concentrations. Even NaCl is toxic at high concentrations because it upsets the essential electrolyte balance.

Cobalt Deficiency

Sheep raised in certain regions of Australia were found to suffer from an illness traced to cobalt deficiency resulting from cobalt–deficient soil. Adding cobalt salts to the soil remedies the problem, but it is expensive and must be repeated periodically. A more economical remedy is to force each sheep to swallow a cobalt pellet and a small screw. The pellet and screw stay in the rumen, adding sufficient cobalt salts to the diet as the mechanical action between them removes any coatings formed. The pellet is recovered when the animal dies, and used again. But if a little cobalt is good, more is not better. A higher rate of congestive heart failure among heavy beer drinkers was linked to the addition of small amounts of cobalt salts (1.2–1.5 ppm) to beer, to improve the foaming properties. It seems that the deleterious effect occurred only if there were a high alcohol level and a dietary deficiency of protein or thiamine, as often occurs among heavy drinkers.

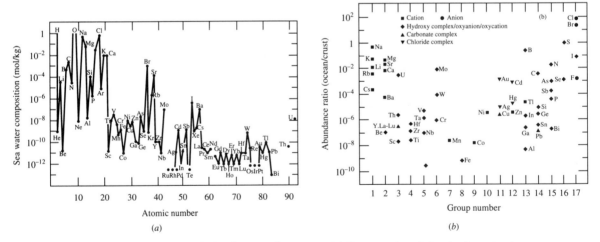

Figure 18.1   Elemental abundance in the oceans. (*a*) Concentration of dissolved species in seawater of average salinity. (*b*) Comparison of abundances between the ocean and the continental crust, showing the major species present. (Reproduced by permission from P. A. Cox, *The Elements*, Oxford University Press, Oxford, 1989, p. 153.)

▶ *18.1.2   Availability of Elements*

An organism must adapt to its environment, using the "raw materials" available to it and coping with unwanted or even toxic substances. Metal ions are made available and of proper compound type through the formation of metal complexes. In basic soils, iron, as $Fe(OH)_3$, is quite insoluble. The Fe is much less abundant in seawater than in the earth's crust (Figure 18.1*b*). Plants can synthesize chelating agents to form soluble Fe complexes (Sec. 18.3.2). Other chelate complexes are formed for the transport of Fe through cell membranes. Where the soil is deficient in Fe, a synthetic soluble Fe complex of ethylene-diaminetetraacetate ion (edta) is added. If the concentration of a metal ion is too high, the ions can be tied up as metal complexes or **sequestered.** The treatment for toxic metals, such as lead, is to inject a chelating agent (such as edta) to form a soluble complex that can be excreted.

Although phosphorus is not abundant in the earth's crust or in seawater (Table 18.1), it serves essential functions in plant and animal life. There was a controversy about detergents containing phosphates. They contribute to the pollution of lakes and streams by fostering the runaway growth of algae, often followed by its decay and the consequent depletion of oxygen. In the normal balance of nutrients, P is likely to be the limiting nutrient. The mean life of inorganic phosphate added to lake water is only a few minutes—it is used rapidly. Phosphate detergents thus upset the delicate natural balance.

The primary means for energy storage and release in cells involves the formation and hydrolysis of polyphosphates (see Figure 18.19). The low abundance of phosphorus makes its compounds good candidates for energy carriers: The molecules storing energy are less likely to get lost in a multitude. The nucleic acids **deoxyribonucleic acid** (DNA) and **ribonucleic acid** (RNA) consist of long chains of cyclic five-carbon sugars joined by phosphate linkages. Attached to the sugars along the chain are heterocyclic nitrogenous bases, the sequence of which constitutes the genetic code. The intertwined strands of DNA joined by hydrogen bonds to form the double helix permits the replication of the code. A portion of the double-helix structure of DNA in Figure 18.2 shows the four bases occurring in DNA. The uncoiled strands can generate complementary strands. In RNA the base uracil replaces thymine.

▶ *18.1.3   Heme and Chlorophyll*

Most early research into the role of metal ions in biological processes focused on magnesium in chlorophyll and iron in hemoglobin, two striking examples of the evolution of complex molecules for highly specialized roles. The heme of hemoglobin is the Fe complex of a substituted porphine (a phorphyrin). The substituents differ for chlorophyll, and the C_7—C_8 double bond of porphine is reduced. Chlorophyll *a*, shown in Figure 18.3, is the most abundant of the group of closely related pigments. The chlorophylls absorb light in the red region, making the energy available for photosynthesis. The appearance of hemoglobin in the evolutionary process might have depended on the availability of chlorophyll, or at least on the processes for its synthesis. Hemoglobin contains an Fe porphyrin

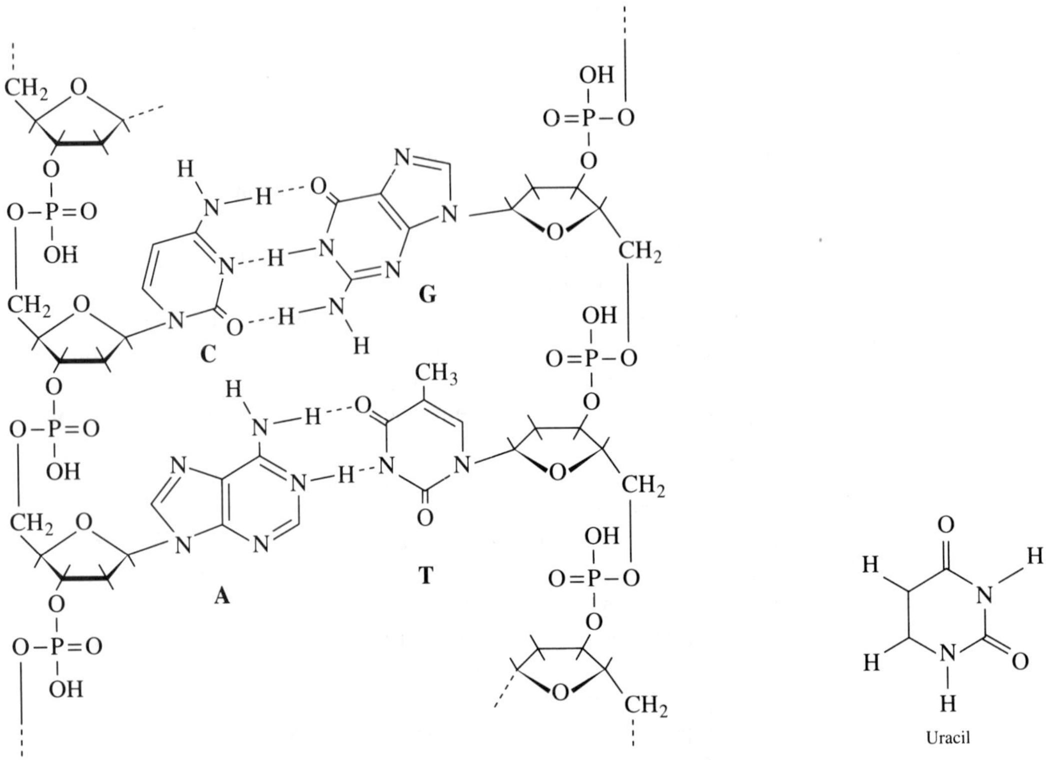

Figure 18.2 A small portion of the two strands of the double helix of DNA showing three hydrogen bonds between cytosine, C, and guanine, G, and two hydrogen bonds between adenine, A, and thymine, T. In RNA, uracil replaces thymine.

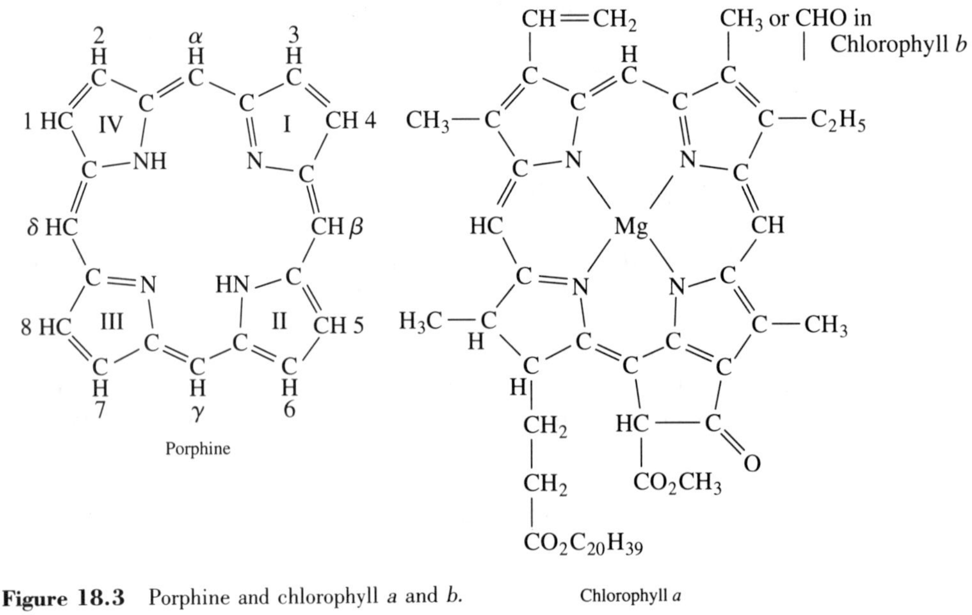

Figure 18.3 Porphine and chlorophyll *a* and *b*.

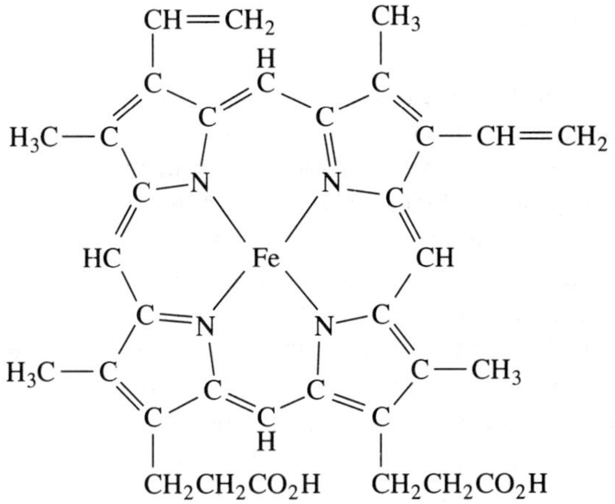

Figure 18.4   Heme *b* (protoheme).

with a large protein molecule, globin, coordinated to Fe on one side of the plane. The sixth coordination site is available for coordination of an O_2 molecule. One of the well-characterized hemes, heme *b* (shown in Figure 18.4), is isolated from hemoglobin of beef blood, or from other sources. All higher animals use iron porphyrins for O_2 transport (hemoglobins) and storage (myoglobins).

18.2   OXYGEN UTILIZATION

Life forms on earth are scientifically amazing, since they are made up of easily oxidizable organic molecules but depend on O_2 for their existence! The thermodynamically favored reaction of organomolecules with O_2 gives CO_2 and H_2O (plus oxides of N, etc.)—the products formed by decay or at high temperature (burning of fossil fuels). However, living organisms can use O_2 for controlled oxidation to supply the energy they need. Systems have evolved for the remarkable four-electron reduction of O_2 to form H_2O (cytochrome chain) or for dealing with the even stronger oxidizing agents produced by the more likely one-electron reduction product (O_2^-) or two-electron reduction product (O_2^{2-}). A strain of *micrococcus* that dies on contact with oxygen was found to survive in air if heme was added to their nutrient medium. Heme was taken up by the cells to provide the bacteria with respiratory enzymes and a catalyst for decomposing H_2O_2.

In a lighter vein we note that there is an inverse effect on animal life with decreasing concentration of O_2 in spite of the ultimate destruction of organic compounds by oxidation by O_2. Humans survive for decades at 20 vol % O_2, but die quickly at very low concentrations.

▶ *18.2.1 Oxygen Transport and Storage*

The best known function of Fe in biological systems is as an O_2 carrier in **hemoglobin.** The molar mass of hemoglobin is about 64,500. There are four subunits, each of which contains one heme group, an iron(II) complex of protoporphyrin IX (protoheme—see Figure 18.4), associated with the protein globin. In the common form of hemoglobin there are four heme sites. One heme with its protein chain is known as a **subunit.** Two of the subunit proteins form *alpha* chains of 141 amino acids, and two form *beta* chains of 146 amino acids. The chains are coiled so that a histidine side chain is coordinated to Fe on one side of the porphyrin ring. The sixth site is occupied by O_2 in oxyhemoglobin (Figure 18.5); in deoxyhemoglobin, it is vacant or a water molecule is weakly bonded.

The iron(II) is high-spin in deoxyhemoglobin. High-spin Fe^{II} has a larger radius than low-spin Fe^{II} because one electron occupies the $d_{x^2-y^2}$ orbital directed toward the N ligand atoms of the porphyrin. Because high-spin Fe^{II} is too large for the "hole" of the porphyrin, the Fe is above the plane of the nitrogens by about 70 pm, giving a square-pyramidal arrangement with the pyrrole rings tipped out-of-plane. When O_2 is coordinated, the ligand field is strong enough to cause spin-pairing, giving a low-spin d^6 (t_{2g}^6) complex. Release of the strain energy of the pyramidal complex helps balance the loss in spin-exchange energy (pairing energy) on going to the oxygenated hemoglobin, in much the same way that an overhead garage door spring and gravity counterbalance each other so that little energy is expended in changing states. If too stable an O_2 complex were formed, too much energy would be released in the lungs and less energy would be available when O_2 is released for use in the muscles. In the planar conjugated porphyrin ring, stable π and low-lying π^* orbitals are responsible for the characteristic charge-transfer electronic absorption spectrum of red blood. As heme adds O_2, Fe slips into the hole of the planar porphyrin ring, shifting with it the coordinated side chain and causing important conformational changes.[2]

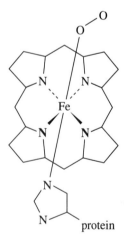

Figure 18.5   The coordination of Fe^{II} in oxyhemoglobin.

[2] M. F. Perutz, G. Fermi, B. Luisi, B. Shaanan, and R. C. Liddington, *Acc. Chem. Res.* **1987,** *20,* 309.

It is remarkable that O_2 does not oxidize hemoglobin, considering the reduction potentials for the reduction of O_2 and oxidation of $Fe^{2+}(aq)$. Fortunately, the reversible adduct is stabilized by the unique bonding features of the porphyrin ring system and the hydrophobic blocking by the large protein (globin). As we will see from model studies, an essential feature for reversible O_2 addition in iron porphyrins is the blocking of the irreversible formation of an oxidized dimer.

Cooperative binding of O_2 by the subunits of hemoglobin permits each successive O_2 molecule to be bound more strongly than the one before. In solution the tetrameric hemoglobin molecules ($\alpha_2\beta_2$) are in equilibrium with dimers ($\alpha\beta$). It has been proposed that upon addition of O_2 movement of the histidine in one chain of a dimer changes the conformation of the other chain, increasing its affinity for O_2 and affecting the dimer–tetramer equilibrium. Oxygenation of the second chain of the tetramer causes conformational changes in the other two chains, promoting addition of O_2. Although the four heme sites are well-separated, movement of one chain affects the conformations of the others significantly and charges the hydrophilic and hydrophobic surface area. This plausible explanation accounts for the cooperative binding of O_2 and for the differences in the O_2 saturation curves (Figure 18.6) of hemoglobin and myoglobin (see below), for which there is no cooperative effect. However, the effect is by no means completely understood, and we can anticipate further interesting developments. The affinity of hemoglobin for O_2 decreases with decreasing pH (Bohr effect); but blood is well-buffered, so the pH decreases only slightly with the accumulation of CO_2 in muscles (venous blood—see Figure 18.6).

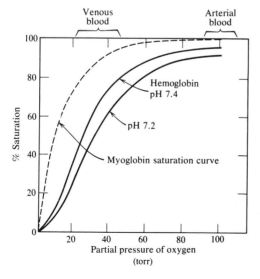

Figure 18.6 The oxygen saturation curves for myoglobin and hemoglobin. In muscles (venous blood, pH = 7.2), the partial pressure of oxygen varies considerably at work or at rest; but it varies little for arterial blood (pH = 7.4).

There has been controversy over whether the O_2 molecule is attached to Fe in a linear Fe—O—O group, an angular Fe—O $\overset{\text{O}}{\diagup}$, or a symmetrical group with sidewise interaction, Fe $\overset{\text{O}}{\underset{\text{O}}{\Big\langle}}$. X-ray structural studies indicate end-on angular binding for model compounds and for oxymyoglobin.

After being separated from air and transported from the lungs to the muscles by hemoglobin, O_2 is transferred to myoglobin for storage until needed for energetic processes. **Myoglobin** (Figure 18.7), although similar to one of the subunits of hemoglobin, binds O_2 more strongly than does hemoglobin, particularly at the low concentrations of O_2 and high concentrations of CO_2 (low pH) that exist in active muscles (Figure 18.6). After the first O_2 molecule is transferred by hemoglobin, the others are released even more easily, because of the cooperative effect (see above) in reverse. These effects result in efficient transfer of O_2 to myoglobin. Terminal amine groups bind CO_2 produced in muscles for transport to the lungs.

▶ *18.2.2 Oxygen Carriers* [3]

Model Compounds [4]

The question of central interest about hemoglobin is how it can bind O_2 reversibly. Unfortunately, the complexity of hemoglobin molecules makes identification of the relevant factors very difficult—a not-uncommon situation for biological systems. For example, even isolating hemoglobin from natural sources and purifying it is a tremendously difficult process. Preparing solutions of even moderate molarity for substances with very high molar mass is expensive, and the high molar masses often limit solubilities. (***Exercise:*** What is the maximum molarity of a substance with molar mass = 100,000 and density = 1 g/mL: *Answer:* 0.01 *M* neat—no solvent!) The tremendous number of atoms makes it impossible to determine by x-ray methods the mode of O_2 coordination—whether monodentate bent Fe $\overset{\text{O—O}}{\diagup}$ or didentate Fe $\overset{\text{O}}{\underset{\text{O}}{\Big\langle}}$ —in oxyhemoglobin.

Inorganic chemists have learned a good deal about the mode of functioning of hemoglobin by studying model compounds. These are simpler Fe-containing species that function like the hemoglobin prototype. Factors relevant to the biological functioning of hemoglobin can be understood by studying changes in chemical behavior of model compounds with planned structural differences.

The fact that free heme (without attached globin) is oxidized irreversibly by O_2 in aqueous solution to hematin—the Fe^{III} form of heme, which does not bind O_2—suggests that the globin part of the molecule somehow provides a structural feature required for reversibility. X-ray structure of hemoglobin reveals that the heme portion of the molecule is buried in the globin in such a way as to provide a hydrophobic "pocket" in the region of

---

[3] E. C. Niederhoffer, J. H. Timmons, and A. E. Martell, *Chem. Rev.* **1984,** *34,* 137.

[4] K. S. Suslick and T. J. Reinert, *J. Chem. Educ.* **1985,** *62,* 974.

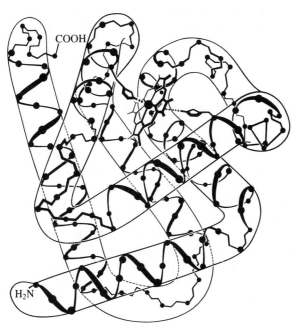

Figure 18.7   Structure of myoglobin.
(From R. E. Dickinson, in *The Proteins*,
Vol. 2. H. Neurath, Ed., Academic
Press, New York, 1964, p. 634.)

the Fe coordination site available for O_2 bonding. Studies on model compounds helped elucidate the role of this hydrophobic region. Possibilities include the importance of the low dielectric constant in the region in retarding possible oxidation processes, such as

$$Fe^{II}heme + O_2 \longrightarrow Fe^{III}heme + O_2^-$$

or   (18.1)

$$O_2 - Fe^{II}heme \longrightarrow Fe^{III}heme + O_2^-$$

the low acidity of the medium, which would retard

$$Fe^{II}heme + O_2 \xrightarrow{HA} Fe^{III}heme + HO_2$$

or   (18.2)

$$O_2 - Fe^{II}heme \xrightarrow{HA} Fe^{III}heme + HO_2$$

and the steric hindrance provided by the protein, which would retard oxidation with dimerization,

$$O_2 - Fe^{II}heme \longrightarrow Fe^{III} - O - Fe^{III} + \cdots$$ (18.3)

a process known for many simple Fe^{II} complexes.

For hemoglobin the Fe heme is the **active site.** In general, a good model compound for the active site should reproduce biological activity and approach or reproduce the bio unit in terms of composition, types of ligand, oxidation states, and physiochemical properties. The study of such models provides, by analogy, information on the role of proteins and the structure of the inside of the bio molecule. With hemoglobin, we are interested in clearing away much of the overwhelming mass of the molecule, in order to focus our attention on the active sites where the action is.

Iron Porphyrins

Early studies of hemoglobin models encountered problems with irreversible oxidation. Success in adding O_2 reversibly has been achieved more recently with synthetic iron(II) oxygen carriers. An iron porphyrin with a coordinated imidazole side chain adds O_2 reversibly at $-45°C$. The iron(II) complex of $\alpha,\beta,\gamma,\delta$-tetraphenylporphine with two pyridine ligands coordinated adds O_2 reversibly at $-78°C$ in methylene chloride solution. One pyridine is displaced by O_2. At higher temperature, these and other simple models, lacking the protective protein, form dimers, leading to the irreversible oxidation of Fe^{II}.

Because model studies have shown that irreversible oxidation can occur through dimerization, two approaches to obtaining O_2 carriers by preventing the formation of dimers have been devised. One involves attaching the Fe porphyrin to a surface so that the units cannot come together to form dimers. The other approach is illustrated by the "picket fence" porphyrins, in which bulky substituents on tetraphenylporphine form a "fence" around the O_2 bonding site (Figure 18.8). The O_2 is σ-bonded unsymmetrically,

$$Fe\text{---}O\overset{\displaystyle O}{\diagup}$$

. Cobalt(II) derivatives of the "picket fence" porphyrins also add O_2 reversibly at 25°C. Without globin or the "picket fence" protection, the cobalt(II) porphyrins have little tendency to bond O_2, except at very low temperature.

Baldwin achieved reversible O_2 addition with a macrocyclic Fe^{II} complex involving bulky ring substituents. He also prepared a porphyrin with a bridge involving a benzene ring as a "canopy" over the center of the porphyrin ring. With another ligand such as 1-methylimidazole coordinated below the porphyrin ring, O_2 could be added reversibly "beneath the canopy." After several hours, irreversible oxidation of the oxygenated capped porphyrin formed the oxo-bridged dimer. Thus, the ability of heme to bind O_2 without being irreversibly oxidized has been shown by model studies to be related to steric prevention of Fe—O—Fe dimer formation.

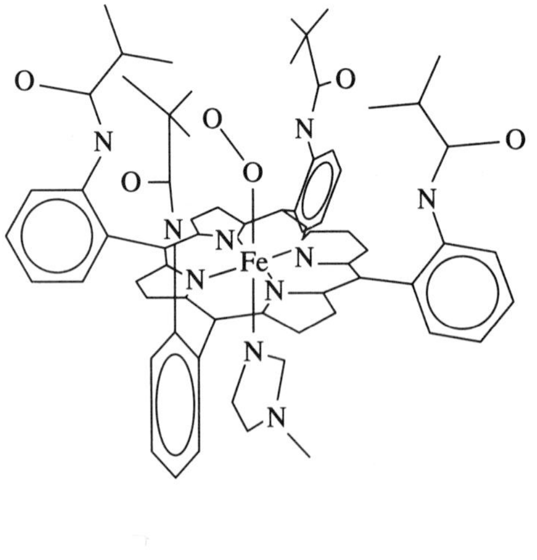

Figure 18.8 Dioxygen bonding to a "picket fence" porphyrin. (Adapted with permission from J. P. Collman et al., *J. Am. Chem. Soc.* **1975**, *97*, 1427. Copyright 1975, American Chemical Society.)

Other Synthetic Oxygen Carriers

Metals usually react irreversibly with O_2 to give metal oxides or other oxo compounds. Knowing which metals, in what oxidation states, and combined with what ligands can add O_2 reversibly is important not only for the production of "synthetic blood," but also for understanding the functioning of hemoglobin. The controlled activation of O_2 as achieved so delicately in biological systems is an important practical goal. During World War II, a considerable effort was made to synthesize O_2 carriers for storage of O_2, since shipment of O_2 at high pressure in heavy metal cylinders is expensive and hazardous.

The most successful models for O_2 carriers have been those of cobalt(II) and iron(II). Vaska, however, discovered that the complex $IrCl(CO)(PPh_3)_2$ can add O_2 (Figure 18.9).

The complex formed involves symmetrical $M{<}{\overset{O}{\underset{O}{|}}}$ bonding (see page 896). The coordination number is five if we consider the O_2 as a single ligand, perhaps π-bonded; it is six if the oxygen is regarded as a didentate ligand. The diamagnetic complex yields H_2O_2 on treatment with acid. Model compounds that more closely resemble natural O_2 carriers have been studied more extensively.

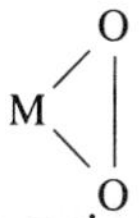

Until recently, the most successful O_2 carriers were cobalt(II) complexes. A ligand containing —NH_2 groups, such as ethylenediamine, can be condensed easily with salicylaldehyde through the formation of the Schiff-base linkages (C=N). The resulting ligand (salen) gives a square-planar complex with cobalt(II). The complex [Co(salen)] (Figure 18.10) and some related complexes react with O_2 in solution containing a base such as pyridine to give [(py)(salen)Co^{III}—O_2—Co^{III}(salen)(py)], a diamagnetic species involving an O_2^{2-} bridge. The bis(3-methoxy) derivative of the [Co(salen)] complex gives a monomeric complex with O_2 (1 : 1). The bis(dimethylglyoximato)cobalt(II) complex with a base such as pyridine coordinated (see Figure 18.30) can add O_2. Two molecules of the complex are bridged by O_2. The cobalt complex of the tetradentate macrocycle cyclam with an anion (X = Cl^-, NO_2^-, or NCS^-) coordinated also gives a bridged (2 : 1) complex with O_2 (Figure 18.11).

The cobalt(II) complex formed by *N,N'*-ethylene-bis(acetylacetonimine) (the Schiff base from acetylacetone and ethylenediamine) adds O_2 reversibly below 0°C to give a 1 : 1 complex. Oxidation of the ligand occurs at higher temperature. An x-ray crystallographic study reveals that the $Co^{III}O_2^-$ bonding is bent, end-on, $Co-O{\overset{O}{\diagup}}$. The five-coordinate complex formed by the ligand derived from $NH_2C_3H_6NHC_3H_6NH_2$ and salicylaldehyde adds O_2 to give a (1 : 1) complex (Figure 18.12). The similar cation complex of cobalt(II)

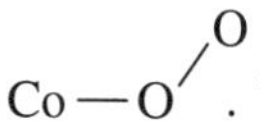

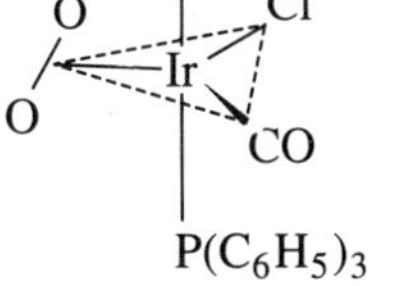

Figure 18.9 Vaska's oxygenated iridium complex.

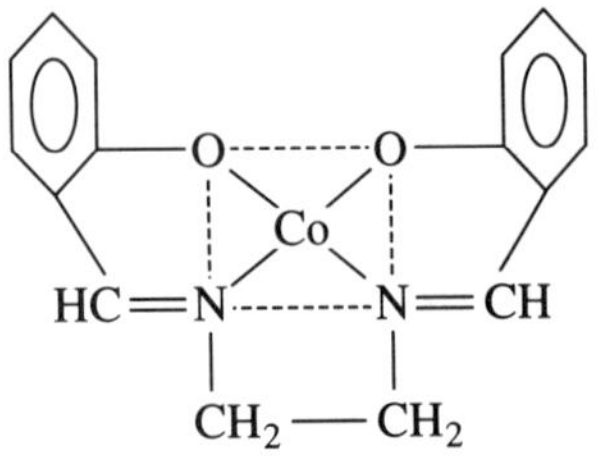

Figure 18.10 N,N'-ethylenebis(salicylaldiminato)cobalt(II) [Co(salen)].

with $(NH_2C_2H_4NHC_2H_4)_2NH$ gives a dimeric (2:1) complex with O_2. The 1:1 complex formed initially would contain Co^{III} and superoxide ion. The superoxide ion easily could oxidize another Co^{II} coordinated to a strong field ligand, to produce the dimer.

Synthetic **coboglobins,** prepared from cobalt(II) protoporphyrin IX (Figure 18.4) and globin from hemoglobin or myoglobin, function as O_2 carriers. The coordination of the Co^{II} and the orientation of the porphyrin in deoxy cobalt hemoglobin or myoglobin are the same as in hemoglobin or myoglobin, respectively.

▶ *18.2.3 Hemerythrin and Hemocyanin*

Hemerythrin,[5] a nonheme Fe protein, is the O_2 carrier for some marine worms. The typical and widely studied hemerythrin from the worm *Goldfingia gouldii* has a molar mass of 108,000, with eight subunits. The myohemerythrin found in the human muscles is very similar to just one of these subunits—as in the case for myoglobin in the heme series. Complete amino-acid sequences have been established for some hemerythrins. The long-chain protein has high helix content, with two coordinated high-spin Fe^{II} close together (344 pm) in deoxyhemerythrin. A coordinated water molecule in deoxyhemerythrin is replaced by O_2, presumably bridging the two irons in oxyhemerythrin. The irons in oxyhemerythrin are antiferromagnetically coupled Fe^{III}. Oxygen is bound as O_2^{2-}.

It is interesting that the blue blood of crabs, lobsters, snails, scorpions, and octopuses contains oxygenated **hemocyanins,** copper-containing proteins that bind one O_2 molecule

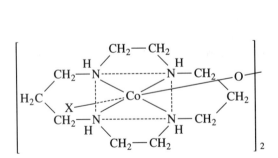

Figure 18.11   The oxygen complex of $[Co(cyclam)X]^+$.

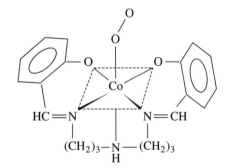

Figure 18.12   The 1:1 salicylaldimine-cobalt(II)–dioxygen complex.

[5] I. M. Klotz and D. M. Kurtz, Jr., *Acc. Chem. Res.* **1984,** *17,* 16.

for two Cu atoms. The name means "blue blood"; hemocyanin contains no heme. It appears that deoxygenated hemocyanins contain copper(I). The blood cells of sea squirts contain a vanadium protein that probably is an oxidation–reduction enzyme but might serve as an O_2 carrier.

18.3   SUPPLY AND STORAGE OF IRON

▶ 18.3.1   *Iron Utilization in Animals*

Animals absorb ingested iron in the gastrointestinal tract as Fe^{II}, but unlike bacteria, they do not require significant new supplies of Fe continually. Iron is recycled; little is absorbed from the diet or excreted. In humans, about 70% of Fe (4 g total) is present in hemoglobin (0.8 kg for an average person) and myoglobin. Most of the rest of the Fe is stored as **ferritin**.[6] Red-brown, water-soluble ferritin consists of a shell of protein [an **apoprotein**—the protein with the prosthetic group removed (a **prosthetic group** is the active group, for example, heme for hemoglobin or myoglobin)] surrounding a micelle (an aggregate whose surface bears a charge) of iron(III) hydroxide oxide phosphate. The micelle contains about 2000–4000 Fe atoms, or 12–20% Fe. The apoferritin shell has 24 subunits consisting of coiled polypeptide chains, each of molar mass about 18,500 (see Figure 18.13).

The micelle for horse ferritin is approximately spherical, with an outer diameter of 10,000–11,000 pm (~100 Å) and an inner diameter of ~6000 pm (~60 Å). Presumably the core is primarily iron(III) oxide, so we expect a close-packed oxide matrix with Fe^{3+} in octahedral holes. The hydroxide and phosphate groups probably serve only to balance charge and complete bonding at the surface. This view is supported by studies showing that careful neutralization of iron(III) nitrate solution yields a polymer with the formula $[Fe_4O_3(OH)_4(NO_3)_2 \cdot \frac{3}{2}H_2O]_n$. The micelles obtained have the same size as the ferritin core and have similar properties. This polymer and apoferritin yield a synthetic ferritin.

Ferritin stores Fe in spleen, liver, and bone marrow. Ingestion of Fe stimulates the synthesis of apoferritin. Iron salts are toxic at moderate concentrations. How ferritin releases iron is not known, but it can be achieved by reduction to Fe^{II} and/or by chelating a-

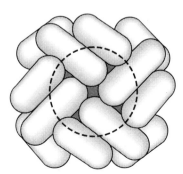

Figure 18.13   The arrangement of the subunits of the apoferritin shell. The iron(III) oxide core is shown by a dashed circle.

[6] P. M. Harrison et al, *Adv. Inorg. Chem.* **1991**, *36*, 449.

gents. The Fe is released to an iron-binding protein to form **transferrin.** Serum transferrin is a single-chain polypeptide with a molar mass of 76,000–80,000. There are two Fe^{III} binding

sites involving two or three tyrosyl residues (tyrosine is   HO—⟨◯⟩—CH_2CHCO_2H,

                                                                     NH_2

providing a phenolic oxygen for coordination) and one or two N atoms, perhaps from imidazole rings of histidyl residues. The Fe is bound only if a suitable anion also is bound. Several anions can promote the binding of Fe, but carbonate (or HCO_3^-) is the physiologically active one. Under physiological conditions, Fe^{III} is very strongly bound (binding constant $\sim 5 \times 10^{23}\ M^{-1}$), leaving less than one free Fe^{III} per liter of blood plasma. Carbonate ion could be displaced easily by acid, with the release of Fe^{III}, which has little affinity for the protein without the bound anion. The Fe is transported to bone marrow by transferrin, which can specifically recognize reticulocytes (immature red blood cells). Because Fe (~ 30 mg) used each day in building red cells in an adult is about 10 times the amount of Fe bound in transferrin, many cycles are required during the lifetime of a protein molecule (half-life ~ 7–8 days).

Cooley's anemia is a genetic blood disease. It results from the production of insufficient amounts of the beta chain of hemoglobin, lowering the ability to transport O_2. Patients require repeated blood transfusions. The increased intake of Fe into the body swamps the normal pathways for its removal. Chelating agents with a high affinity for Fe^{3+}, but not for other biologically important metal ions, are promising for treatment. Individuals with *atransferrinemia,* the genetic inability to synthesize transferrin, suffer from iron-deficiency anemia *and* an overload of Fe.

▶ 18.3.2   *Supplying Iron in Lower Organisms*

Because Fe plays a vital role in most organisms, several mechanisms have evolved for its storage and transport, depending on the availability and form of the Fe. Many organisms can use Fe^{3+} if it is available directly; if not, high-affinity mechanisms are activated.

    In our oxidizing environment, Fe is usually found as Fe^{III}. The Fe is present as very insoluble $Fe(OH)_3$ in basic soils, and thus is not utilized easily by plants. Such Fe deficiency can be remedied by supplying the soluble iron(II) complex of ethylenediaminetetraacetic acid, as is done on a large scale in Florida citrus groves.

Siderophores[7]

Red-brown iron-containing complexes for the transport of Fe in lower organisms are known as **siderophores.** The chelating agents are hydroxamates in fungi and yeast, and hydroxamates or substituted catechols in bacteria (Figure 18.14). Some bacteria can pro-

---

[7] R. C. Hider, *Struct. Bonding* **1984**, *58*, 25.

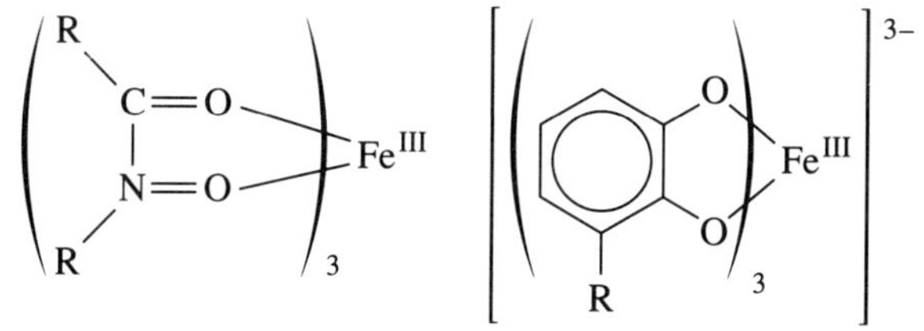

Iron(III) hydroxamate
complex

Iron(III) catechol
complex

Figure 18.14 Siderophores.

duce both types of chelating ligands, and other ligands that combine these functional groups. The production of the ligands is activated when the concentration of Fe is low. The siderophores that promote growth have been called **sideramines.** Those that are antibiotics (as Fe-free ligands) have been called **sideromycins;** these function as antibiotics because the ligands bind Fe strongly, making Fe unavailable for bacterial growth.

Siderophores also are known as **siderochromes,** because some of them are intensely colored. Older classifications as sideramines and sideromycins are not clear-cut.

These ligands give very stable Fe^{III} complexes (high formation constants, β_n), and they are rather specific for Fe^{III}. The formation constants for the Fe^{II} complexes are much lower, so that reduction of the Fe provides a mechanism for its release. Presumably, the ligands capture Fe^{3+} at the cell membrane and transport it into the cell. The mechanism of transfer through the cell membrane depends on the charge type of the complex. In cases of very low concentrations of Fe^{III}, overproduction of the ligands can result in their release to the surrounding medium to dissolve Fe^{III}.

Ferrichrome (Figure 18.15) is an hydroxamate complex in which the hydroxamate groups are side chains to a peptide ring. Crystallographic studies have shown that ferrichrome A, which differs in substituents, has the Λ absolute configuration, a left spiral of chelate rings about Fe^{III}. Other ferrichromes also differ in ring substituents. The **ferrioxamines** contain the hydroxamates as part of the peptide chain. Ferrioxamine D_1 (Figure 18.16) has a linear chain; there are cyclic ferrioxamines where the chain is closed. **Iron(III) enterobactin** (Figure 18.17) provides an example of a catechol-type siderophore in which the catechols are side chains to a cyclic ester. The siderophore has the Δ absolute configuration, a right spiral of chelate rings.

Iron is released from hydroxamate siderophores by reduction of Fe^{III} to Fe^{II}, which forms much less stable hydroxamate complexes. Then the ligand is available for reuse. Iron(III) enterobactin requires a potential for reduction (approximately -0.75 V at pH 7) that is too negative for physiological reductants. The Fe is released by the destruction of the ligand, a result of the hydrolysis of the ester linkages by an enzyme specific for the complex. The hydrolysis products are not used for ligand synthesis. Synthesis of a complex ligand for the transport of a single Fe^{III} seems very inefficient. However, the very great stability of the Fe^{III} complex ($\log K_f \sim 56$) allows the bacteria using enterobactin to acquire Fe present at very low concentrations, even at the expense of microorganisms which use other ligands.

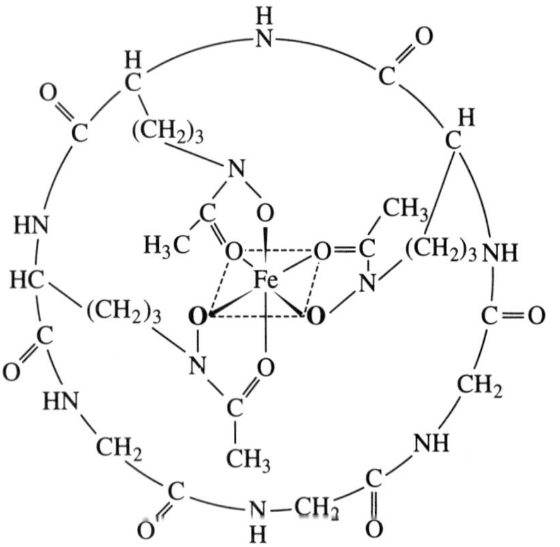

Figure 18.15   Ferrichrome.

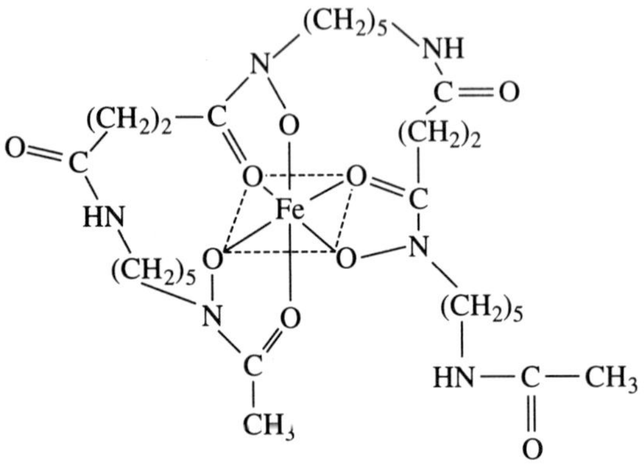

Figure 18.16   Ferrioxamine D_1.

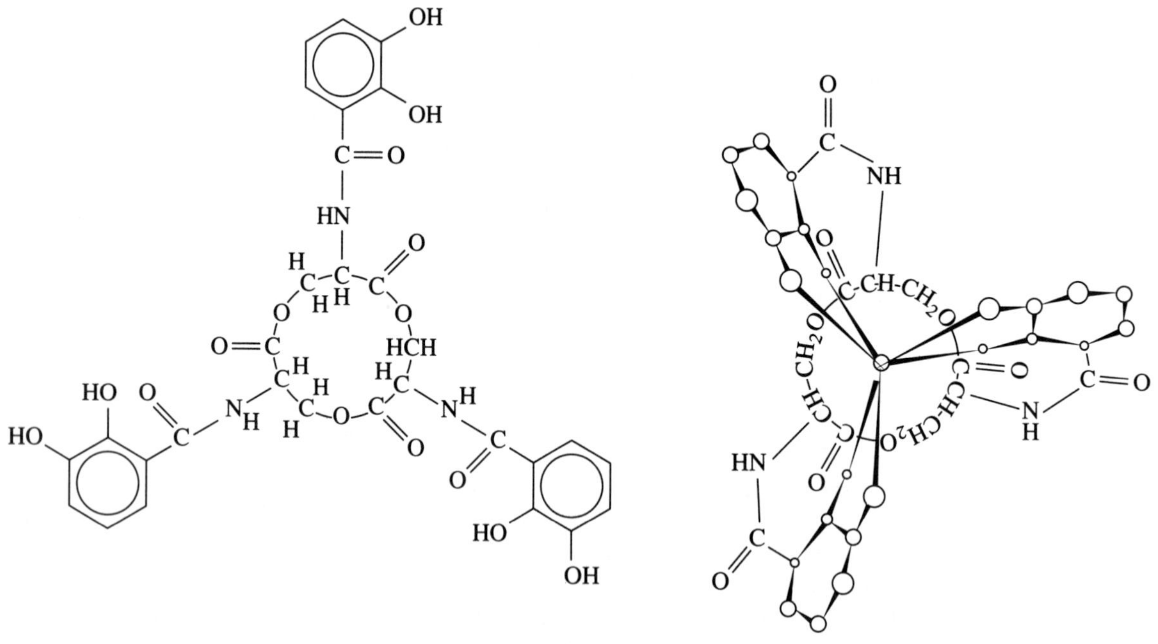

Figure 18.17   Enterobactin and the Fe^{III} complex.

18.4 OXIDATION–REDUCTION PROCESSES

► *18.4.1 Cytochromes*

Oxygen transported to cells is utilized for the reversible oxidation of cytochromes as a source of energy in mitochondria (energy-producing granules of aerobic cells). **Cytochromes,** which also are utilized in energy transfer in photosynthesis, are electron-transferring proteins that act in sequence to transfer electrons to O_2. The **prosthetic** group in all cytochromes is heme, which undergoes reversible Fe^{II}–Fe^{III} oxidation. For cases other than enzymes, the prosthetic group is the non-amino-acid portion of the protein. The cytochromes in a chain differ from one another in reduction potentials by about 0.2 V or less, with a total potential difference of about 1 V. Differences in potentials for the Fe^{II}–Fe^{III} oxidation result from changes in the porphyrin substituents, changes in the protein, and, in some cases, changes in axial ligands. Cytochrome a_3, the terminal member of a cytochrome chain, has water (instead of S) as a ligand. The H_2O can be replaced by O_2 to initiate electron transfer.

The structures of horse-heart and tuna-heart cytochrome c reveal that the protein chain is coordinated to Fe on one side of the porphyrin ring through a histidine, and on the other side through S of methionine [$H_3CSC_2H_4CH(NH_2)CO_2H$]. Higher-resolution (200 pm) structure determinations of tuna-heart cytochrome c show that the structures of the oxidized and reduced forms are very similar, with no great conformational changes. The heme group is enclosed in the hydrophobic interior of the protein except for one edge near the surface for substitution.

The sequence of reversible oxidations in a cytochrome chain is shown in Figure 18.18. At two stages in the cytochrome chain, where there are large differences in reduc-

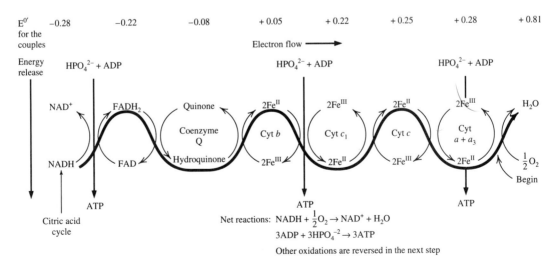

Figure 18.18 The respiratory chain, showing electron transport and oxidative phosphorylation. The $E^{0\prime}$ values for couples are shown above the figure. The potential ($+0.81$ V) is that for the reduction of gaseous (1 atm) O_2 in neutral solution. The potential for reduction of O_2 bound to cytochrome would differ.

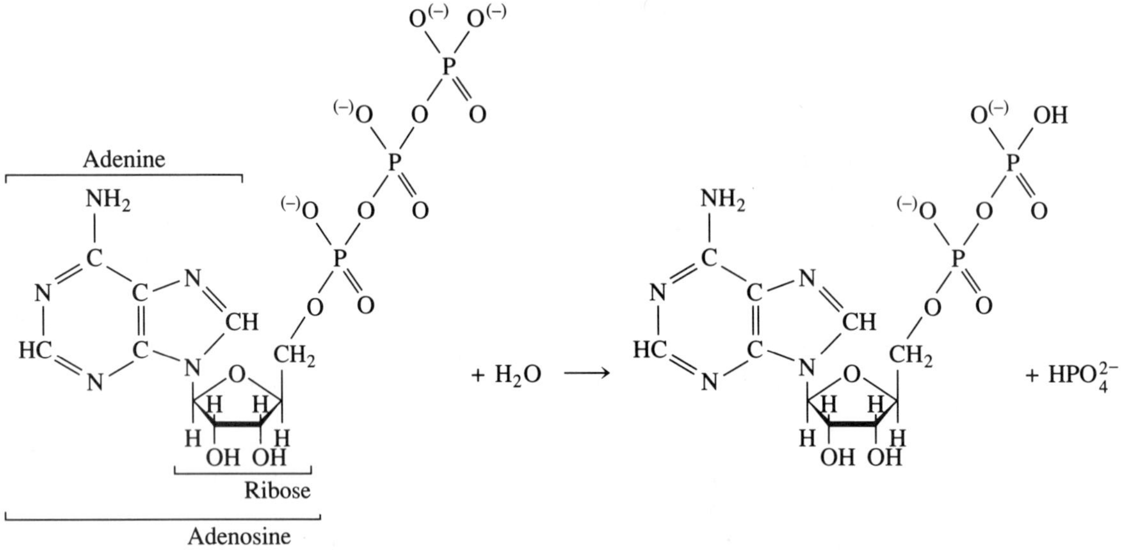

Figure 18.19 The energy-producing hydrolysis of adenosine triphosphate (ATP) to adenosine diphosphate (ADP).

tion potentials, enough free energy is supplied to drive the combination of inorganic PO_4^{3-} with **adenosine diphosphate** (ADP) to form **adenosine triphosphate** (ATP)—a conversion called **oxidative phosphorylation.** The reverse conversion of ATP to ADP (involving an Mg^{2+} complex) provides energy for muscle activity (Figure 18.19). The ΔG^0 for hydrolysis of ATP to give ADP + HPO_4^{2-} is ≈ -31 kJ/mol, but it has been estimated to be about -50 kJ/mol in a cell. The oxidized form of cytochrome b oxidizes the hydroquinone form of coenzyme Q (named ubiquinone, because it is ubiquitous in cells) to the quinone form (Figure 18.20). The quinone form oxidizes **flavin adenine dinucleotide hydride** (FADH$_2$) to **flavin adenine dinucleotide** (FAD). The last step in the cycle involves oxidation of **nicotinamide adenine dinucleotide hydride** (NADH) to **nicotinamide adenine dinucleotide** (NAD$^+$) (Figure 18.21), with the release of more energy for conversion of

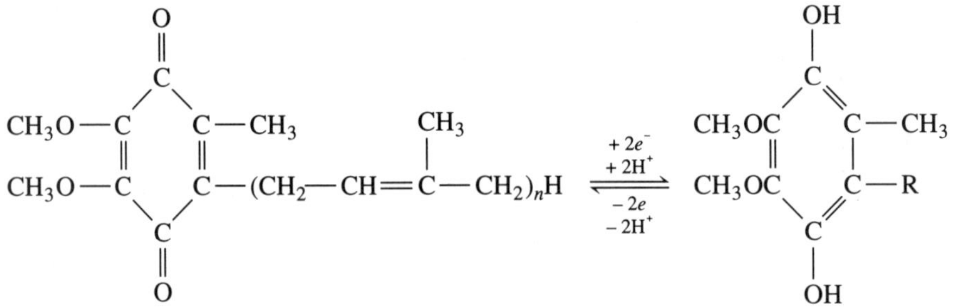

Figure 18.20 Ubiquinone (coenzyme Q). For most mammals, $m = 10$; for some microorganisms, $m = 6$.

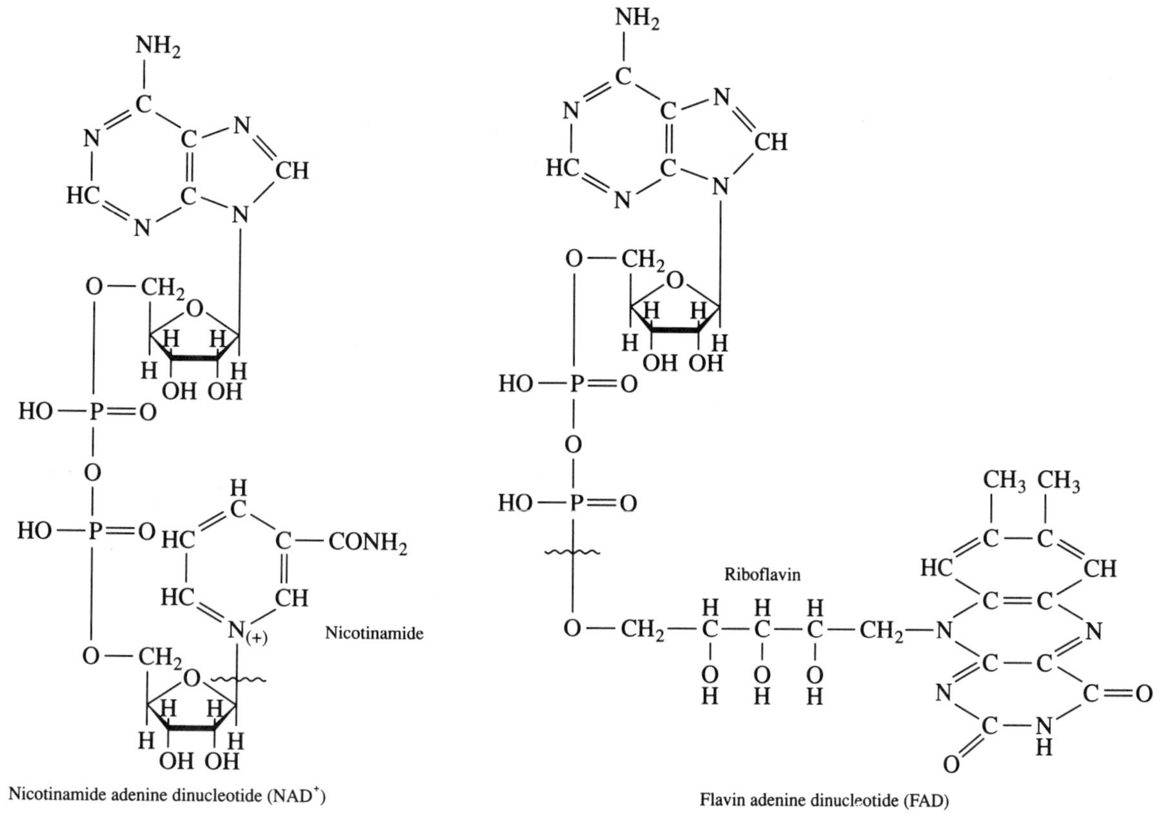

Figure 18.21   Structures of NAD⁺ and FAD.

ADP to ATP. Each oxidation is reversed in the next step, except for the last. The NADH is restored from the citric acid cycle. The complex sequence, bringing about the oxidation of NADH indirectly by O_2, is known as the **respiratory chain.**

The respiratory chain for the simple net oxidation of NADH by O_2 might seem unnecessarily complex, but the many steps serve to break down the large amount of energy involved in the reduction of O_2 into smaller units that can be stored as ATP and to keep the reactants (NADH and O_2) well-separated. The chain achieves high biological specificity, and H^+ is produced or consumed where needed. Photosynthesis in plants accomplishes the reverse reactions, with O_2 as the end product, by a similar cytochrome chain.

The electron-transfer proteins commonly are membrane-bound. This has the advantage of holding the links of the chain in the correct juxtaposition for reaction, and thus avoiding side reactions or short-circuiting of the chain. The coupled oxidative phosphorylation centers should not be bypassed because this is the mechanism for storing energy. Membrane binding also provides a nonaqueous environment for dehydration in the synthesis of ATP. Although the cytochromes have been characterized rather well, the links between them that provide efficient electron transfer along the chain offer an intriguing challenge.

▶ *18.4.2 Cytochrome P-450*[8]

Cytochrome P-450[9] represents a group of enzymes (containing heme) that act on O. P-450, unlike other cytochromes, transfers an O atom rather than transporting electrons. A typical reaction of a cytochrome P-450 enzyme converts a substrate R—H to R—OH:

$$R{-}H + O_2 \xrightarrow[\text{P-450}]{2e,\,2H^+} R{-}OH + H_2O \qquad (18.4)$$

The membrane-bound cytochrome P-450, found in plants, animals, and bacteria, serves several functions. In humans, P-450 catalyzes the hydroxylation of drugs, steroids, pesticides, and other foreign substances. The hydroxylated compounds are more water-soluble, favoring excretion in urine.

P-450, containing low-spin Fe^{III}, is converted to high-spin on binding a substrate, $R{-}H(P\text{-}450)Fe^{3+}$ (Scheme 18.1). A one-electron transfer produces $R{-}H(P\text{-}450)Fe^{2+}$ that binds O_2 to form $[R{-}H(P\text{-}450)Fe{-}O_2]^{2+}$. Another electron and two H^+ are added, and then H_2O leaves to produce $[R{-}H(P\text{-}450)Fe{=}O]^{3+}$. This presumably $Fe^V{=}O$ converts the substrate $(R{-}H)$ to $R{-}OH$ plus the starting P-450.

Various P-450 cytochromes are believed to have the same heme, but differ in the associated protein. The molar masses are ~50,000. The heme site is sheltered by the folded protein. The hydrocarbon substrate is bound by the protein near the heme site. On one side of the heme is a thiolate side chain of a cysteine residue. The other site is available for bonding O_2.

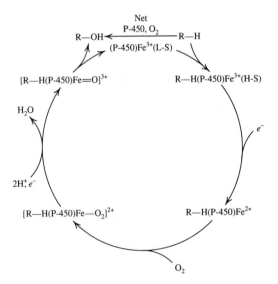

Scheme 18.1 Hydroxylation of a substrate R—H by cytochrome P-450.

---

[8] M. J. Gunter and P. Turner, *Coord. Chem. Rev.* **1991**, *108*, 115.

[9] The designation P-450 arises from the characteristic absorption at 450 nm of the CO adduct. This so-called Soret band usually occurs for shorter wavelengths by about 30 nm for other cytochromes as CO adducts.

▶ *18.4.3   Rubredoxins and Ferredoxins*[10]

There are nonheme iron proteins that participate in electron-transfer processes. They contain Fe bonded to S. **Rubredoxin** was first isolated from the bacterium *Clostridium pasteurianum,* but it also occurs in other anaerobic bacteria. The single Fe is bonded to four S atoms of cysteine [$HSCH_2CH(NH)_2CO_2H$] as part of a protein of about 54 amino acid residues (molar mass ~6000). The S atoms form a distorted tetrahedron around iron, with one very short Fe—S bond (Figure 18.22). The reduction potential of rubredoxin in neutral solution is close to 0 V. The oxidized protein is red, and the reduced form is colorless.

There are several types of **ferredoxin**[11]: One contains two Fe and two sulfide ions per unit, another contains four Fe and four S^{2-} per unit, and a third contains eight Fe and eight S^{2-}. The [2Fe:2S] proteins also contain cysteine groups bound to Fe, but the structure is not known. These proteins are important in photosynthesis and gain one electron per Fe upon reduction. Both the Fe^{III} and Fe^{II} are high-spin. The [2Fe:2S] ferredoxins were believed to occur only in plants, but Holm's core-extrusion technique, which permits the active Fe—S clusters to be removed by adding simple thiols to replace the S attached to the peptide chain, has shown that xanthine oxidase (easily available from milk) contains [2Fe:2S] units.[12]

Xanthine oxidase also contains Mo. In humans, this enzyme oxidizes purines to uric acid, which is excreted through the kidneys. An excess of uric-acid accumulation leads to gout, which can be treated with inhibitors of xanthine oxidase.

The ferredoxin from *Peptococcus aerogenes* 2[4Fe:4S], whose structure is known, has a molar mass of 6000. It contains two well-separated Fe_4S_4 cubes with a cysteine group attached to each Fe completing a tetrahedron of 4S about each Fe (Figure 18.23). The [4Fe:4S] ferredoxins contain just one cubane unit. Another group of [4Fe:4S] proteins are known as **high-potential iron–sulfur proteins** (HiPIPs). The HiPIPs, found in purple S bacteria and some other species, involve $[Fe_4S_4]^{2+}$ and $[Fe_4S_4]^{3+}$. The high-potential protein from *Chromatium vinosum* has a reduction potential of +0.3 V, whereas ferredoxins such as the one from *Peptococcus aerogenes* $[(Fe_4S_4)^{1+/2+}]$ have potentials of -0.3 to -0.7 V.

There are protein-bound Fe_3S_4 clusters in some enzymes and bacteria. $[Fe_3S_4]^{1+}$ is formed from $[Fe_4S_4]^{3+}$, and the reduced $[Fe_3S_4]^0$ can add Fe^{2+} to form $[Fe_4S_4]^{2+}$. $[Fe_3S_4]$ retains the cuboidal structure with a vacant corner.

Iron–sulfur proteins are easily oxidized by O_2, so they are usually found inside cells or in the membrane. They occur in plants, animals, and bacteria. The Fe_4S_4 clusters were used for redox functions in early forms of life. Before the evolution of an O_2 atmosphere, the $Fe/S/H_2$ system might have been the basis for energy production for life.

[10] "Iron-Sulfur Proteins," *Adv. Inorg. Chem.*, **1992**, *38;* J. J. R. Frausto Da Silva and R. J. P. Williams, *The Biological Chemistry of the Elements,* Oxford University Press, Oxford, 1991, p. 326.

[11] R. Cammack, *Adv. Inorg. Chem.* **1992**, *38*, 298.

[12] R. H. Holm, S. Ciurli, and J. A. Weigel, *Prog. Inorg. Chem.* **1990**, *38*, 1.

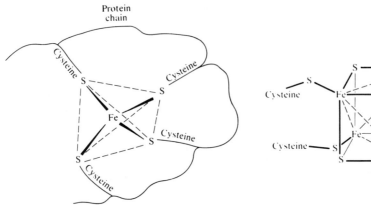

Figure 18.22 The FeS₄ flattened tetrahedron (smallest angle 101°) in rubredoxin. There is one short Fe—S bond (205 pm), with others at 224–235 pm.

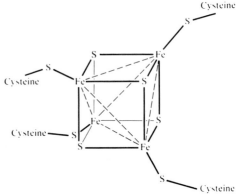

Figure 18.23 The cubane unit Fe₄S₄(cysteine)₄ found in *Peptococcus aerogenes*.

18.5 METALLOENZYMES

Enzymes are catalysts that greatly enhance the rates of specific reactions. The yields are often 100%, and reactions occur under the mildest conditions. High yields are achieved because the rate of the reaction catalyzed might be increased by a factor of as much as 10^{12}, making side reactions unimportant. Enzymes are proteins synthesized from amino acids by cells. The names of the over 1500 enzymes identified are obtained by adding the suffix *ase* to the name of the process catalyzed or to the name of the molecule on which the enzyme acts (the **substrate**). Some enzymes function alone, whereas others require the cooperation of a **cofactor,** a metal ion or an organic molecule. Some of the enzymes requiring metal ions as cofactors (*metalloenzymes*) are given in Table 18.2. Coenzymes

Table 18.2   Some metalloenzymes

Mg^{II}	Phosphohydrolases	Mn^{II}	Arginase
	Phosphotransferases		Oxaloacetate decarboxylase
			Phosphotransferases
Fe^{II} or Fe^{III}	Cytochromes	Zn^{II}	Alcohol dehydrogenase
	Peroxidase		Alkaline phosphatase
	Catalase		Carbonic anhydrase
	Ferredoxin		Carboxypeptidase
Cu^{II} or Cu^{I}	Tyrosinase	Fe and Mo	Nitrogenase
	Amine oxidases		
	Cytochrome oxidase		
	Ascorbate oxidase		
	Galactose oxidase		
	Dopamine-β-hydroxylase		

that are tightly bound to the enzyme are called prosthetic groups. The active centers of metalloenzymes often involve distortion (the *entatic state*)[13] of the normal stereochemistry of the metal ion with changes in stereochemistry occurring during the reaction cycle. We shall consider only a few metalloenzymes.

Catalysis is vitally important to the chemical industry, but the usual catalysts are very poor compared to enzymes. Imagine the frustration of a research director who is asked to devise a reaction for the synthesis of an organometallic compound with a molar mass of 50,000. The reaction is to be carried out in about 30 minutes total time, at 1 atm and at a temperature of less than 50°C, and to give a 100% yield! Such reactions occur with enzymes at low concentrations of reactants—in fact, the reactants are scavenged from the environment! Many complex compounds are being synthesized in your body as you read this. What a problem for the body of a student who subsists on snack foods for an extended period! We have a lot to learn about synthesis.

Most biologically active compounds are optically active. Synthetic drugs produced in the past were racemic products. Many of these as active enantiomers can now be produced by the use of chiral metal complexes in enantiomer-selective syntheses. The number of drugs as pure enantiomers is increasing rapidly. *L*-DOPA, used for Parkinson's disease, was one of the early successes (see Section 14.2.3). Enantiomer-selective catalysis can be accomplished in many cases with a catalyst only partially resolved. This process involving a partially resolved catalyst, called **amplification of chirality,** results from much greater turnover efficiency of the chiral catalytic system.

► *18.5.1   Nonredox Metalloenzymes*

Metalloenzymes not involving oxidation–reduction often serve as Lewis acids and/or they can alter the acidity or basicity of ligands. The active site may be protected from the aqueous surroundings, greatly altering its properties. Metal ions commonly utilized are Zn, Mg, or Mn. Zn^{2+} is a strong Lewis acid and generally available in nature. It is commonly used where a stronger acid than Mg^{2+} is required without involving oxidation–reduction.

Carbonic Anhydrase

Carbonic anhydrase is a zinc enzyme that catalyzes the hydration of CO_2 and dehydration of carbonic acid (and the hydrolysis of certain esters). The uncatalyzed dehydration of carbonic acid or hydration of CO_2 is too slow for respiration of animals. Carbonic anhydrase can hydrate 10^6 CO_2 molecules per second at 37°C, 10^7 times the uncatalyzed hydration rate. The enzyme (molar mass 30,000) occurs in blood and also in plants. The one Zn^{2+} per molecule lies inside a cavity associated with the active site. The Zn^{2+} has a tetrahedral environment in carbonic anhydrase C, with three imidazole ligands.

---

[13] R. J. P. Williams, *J. Mol. Catal.* **1985,** *30,* 1.

Carboxypeptidase

Yeast carboxypeptidase C belongs to a class of carboxypeptidases that are not metalloenzymes. They function at low pH. The second class of carboxypeptidases are zinc enzymes produced in the pancreas of animals. The enzyme aids protein digestion with maximum activity at neutral or slightly alkaline pH.

Carboxypeptidase A is a zinc enzyme that has been studied thoroughly.[14] The enzyme hydrolyzes (cleaves) the terminal peptide (amide) bond of a peptide chain for the carboxyl-terminal end.

$$\underset{\underset{CH}{|}}{\overset{R}{|}}-\underset{\overset{O}{\|}}{C}-NH-\underset{\underset{CH}{|}}{\overset{R'}{|}}-\underset{\overset{O}{\|}}{C}-NH-\underset{\underset{CH}{|}}{\overset{R''}{|}}-CO_2^-$$

The enzyme is selective for terminal peptides where the terminal amino acid has an aromatic or branched aliphatic substituent (R'') and the *L* configuration, but most terminal residues are hydrolyzed.

The structure of carboxypeptidase A has been studied extensively by x-ray crystallography. The one Zn^{2+} per molecule (molar mass 34,600) is bonded to two imidazoles of histidine, one glutamic acid group, and an H_2O molecule that can be replaced by the substrate. The Zn^{2+} sits in a depression in the surface of the molecule. This is the active site, from which extends a pocket into the interior of the molecule, for accommodating the substrate. The enzyme protects its own terminal end against self-digestion by tucking it inside. The Zn^{2+} site in Figure 18.24 shows coordination of histidine residues 69 and 196

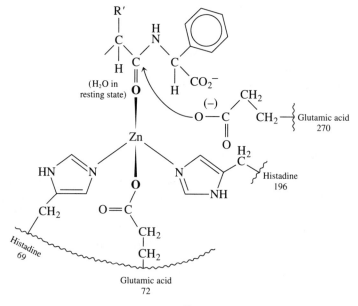

Figure 18.24   The active Zn^{2+} site of carboxypeptidase A.

---

[14] D. W. Christianson and W. N. Lipscomb, *Acc. Chem. Res.* **1989**, *22*, 62.

and glutamic acid residue 72.[15] Arginine-145 and tyrosine-248 seem to be involved in hydrogen bonding to the substrate, causing dramatic conformational changes of the protein. The glutamic acid residue 270 probably serves as a nucleophile aiding cleavage of the peptide linkage.

Zinc ion can be replaced by cobalt(II) in carboxypeptidase and in several other zinc enzymes, with retention of activity of the enzyme. In some cases there is even enhanced activity—perhaps Co was not available and nature had to settle for Zn. Replacement of Zn^{2+} by Co^{2+} has been particularly useful because Zn^{2+} (d^{10}) forms colorless complexes whereas Co^{2+} (d^7) complexes absorb in the visible region. Thus spectral studies (absorption, circular dichroism, and magnetic circular dichroism) of the Co-substituted enzyme provide valuable information about the metal ion environment at the active site. Cobalt serves as a "spectral probe" for the study of the active site. Other divalent metal ions that can be substituted for Zn with retention of some peptidase or esterase activity are Ni, Cd, Hg, and Pb.

A Model of Carboxypeptidase A

Co^{III} complexes of the type *cis*-$[Co(en)_2(H_2O)(OH)]^{2+}$ and *cis*-β-$[Co(trien)(H_2O)(OH)]^{2+}$ (en = $NH_2C_2H_4NH_2$ and trien = $NH_2C_2H_4NHC_2H_4NHC_2H_4NH_2$) promote the hydrolytic cleavage of *N*-terminal amino acids of a peptide chain. The reaction is not catalytic because a stable complex of the chelated amino acid is formed. The trien complex is more effective. The mechanism of enzymatic hydrolysis resembles that of hydrolysis of esters by such Co complexes. The first step is coordination of the terminal NH_2 in place of H_2O (Figure 18.25). By one path, the carbonyl oxygen coordinates next, replacing OH^-. Hydroxide attack on the carbonyl group leads to the chelated amino acid complex. By the other path, coordinated OH^- directly attacks the peptide carboxyl carbon, hydrolyzing the peptide bond and giving the amino acid complex. This case involves cleavage of the *N*-terminal amino acid, rather than the carboxyl-terminal amino acid cleaved by carboxypeptidase. However, metalloenzymes, such as leucine aminopeptidase [activated by Mg^{II} or Mn^{II}], catalyze hydrolysis of *N*-terminal peptide bonds through chelation.

▶ 18.5.2 Peroxidase and Catalase

The first forms of life probably emerged when the earth had a reducing atmosphere. The present O_2 atmosphere evolved later, from photosynthesis. Anaerobic bacteria still survive, but the dominant forms of life utilize O_2 for releasing energy stored primarily through photosynthesis. A crucial step in the evolution of aerobic life was the development of a means of coping with the highly toxic and reactive products of reduction of O_2—namely, peroxide (O_2^{2-}) or superoxide (O_2^-) ions.

---

[15] These amino acids are "residues" or part of a protein with the amino acids in the sequence as shown.

Histidine is [imidazole structure with N, HN] —$CH_2CH(NH_2)CO_2H$, glutamic acid is $HO_2CCH(NH_2)$-$C_2H_4CO_2H$, arginine is

$NH_2\underset{\underset{NH}{\|}}{C}NHC_3H_6\underset{\underset{NH_2}{|}}{C}HCO_2H$, and for tyrosine see page 902.

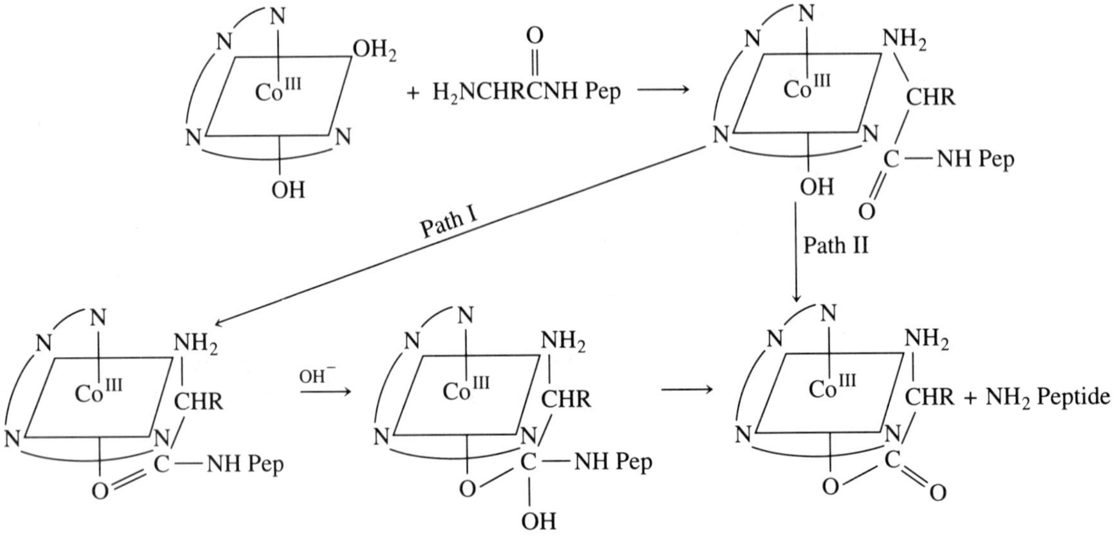

Figure 18.25   Mechanisms for peptide cleavage by *cis-β*-[Co(trien)(H$_2$O)OH]$^{2+}$.

The earliest Fe proteins utilized apparently were the ferredoxins and rubredoxins (nonheme Fe—S proteins—see Section 18.4.3) found in anaerobic bacteria. Some scientists have speculated that the porphyrins were formed geochemically and utilized before enzymes were developed for their synthesis.

The close link between the biological activity of Fe and Cu is apparent in *cytochrome oxidase* and in the role of *ceruloplasmin* in iron metabolism (Section 18.5.3). The enzyme *oxidase (cytochrome oxidase)* consists of cytochromes *a* and *a$_3$* together. The complex *a–a$_3$* system (molar mass ∼240,000) consists of six subunits, each containing a single heme *a* group and a copper atom. Two of the units form cytochrome *a*, and only the remaining four units of cytochrome *a$_3$* react directly with O$_3$. Oxidation of copper (Cu^I → Cu^{II}) and iron (Fe^{II} → Fe^{III}) occur in the reaction of *a$_3$* with oxygen. Cytochrome *c* receives an electron from cytochrome *c$_1$* and passes it along to cytochrome *a* + *a$_3$*. Figure 18.26*a* shows the transfers to accomplish O$_2$ + 4H$^+$ + 4e → 2H$_2$O. Also four protons are pumped across the membrane in which the enzyme is bound. The transformations involving cytochrome *a$_3$* are shown in Figure 18.26*b*. Electrons are provided by cytochrome *c* via cytochrome *a*.

The Cu proteins *erythocuprein* and *hepatocuprein* (molar mass 35,000; 2Cu/molecule), which occur in mammalian blood and liver, are believed to catalyze the disproportionation of superoxide to H$_2$O$_2$ + O$_2$.

$$2\,H^+ + 2\,O_2^- \xrightarrow[\text{dismutase}]{\text{superoxide}} H_2O_2 + O_2 \tag{18.5}$$

Dismutation is the term used by biochemists for disproportionation, so these are called *superoxide dismutases.*[16]

Bovine erthrocyte superoxide dismutase is one of the most thoroughly studied Cu proteins. The complete amino acid sequence and x-ray crystal structure are known. There

[16] J. S. Valentine and D. M. deFreitas, *J. Chem. Educ.* **1985,** *62,* 990.

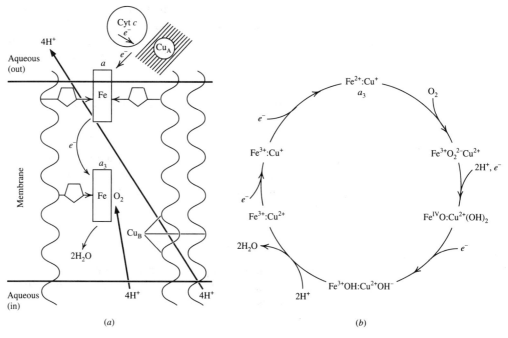

Figure 18.26 (*a*) The proposed outline structure of cytochrome oxidase bound in a membrane. (*b*) The intermediate steps proposed for a_3 of cytochrome oxidase. (Adapted with permission from J. J. R. Fraústo da Silva and R. J. P. Williams, *The Biological Chemistry of the Elements*, Oxford University Press, Oxford, 1991, p. 363.)

are two subunits with molar mass of 15,700, each containing one Cu and one Zn. At the active site a deprotonated imidazole ring bridges Cu and Zn.

Bacteria, algae, and other simple organisms without membrane-bound nuclei (prokaryotes) use Fe or Mn superoxide dismutase to get rid of O_2^-. Higher organisms (eukaryotes) use copper–zinc superoxide dismutase. The Fe and Mn enzymes dissociate to release Fe^{II} or Mn^{II} which could be harmful to DNA. Cu–Zn superoxide dismutase is very stable and does not release Cu^I or Cu^{II}. The Fe and Mn enzymes appeared early in evolution from the readily available cations. The Cu–Zn enzyme was needed with the accumulation of O_2 in the atmosphere and the emergence of higher organisms. The electrode potentials of the Cu–Zn, Mn, and Fe superoxide dismutases fall between the half-cell potentials for the couples (O_2/O_2^- and O_2^-/H_2O_2) involved in Reaction (18.5).

Hydrogen peroxide produced by reduction of O_2^- or O_2 can be handled by enzymes known as **peroxidases** by oxidizing a substrate. Peroxidases occur widely in plants and animals. The sap of the fig tree and the root of the horseradish are richest known sources. *Horseradish peroxidase* (molar mass ~40,000) contains the iron(III) derivative of heme *b* (see Figure 18.4) as the prosthetic group. A fifth coordination site is occupied by a protein of some 16 amino acid residues. The sixth coordination site can be occupied by H_2O or other ligands that exchange as the peroxidase functions. No single mechanism can explain all peroxidase oxidations. Horseradish peroxidase is oxidized by H_2O_2 to the extent of two equivalents. There is evidence for Fe^{IV}, suggesting that the second equivalent of oxidation involves the porphyrin or the protein. The oxidized enzyme oxidizes a substrate. The net reaction is

$$H_2O_2 + SubstrateH_2 \xrightarrow{\text{peroxidase}} 2\,H_2O + Substrate \qquad (18.6)$$

(reduced form)   (oxidized form)

The **catalases** are enzymes (molar mass ~240,000) with four iron(III) heme *b* groups per molecule. Each heme group has a separate peptide chain coordinated. Catalases promote the disproportionation of H_2O_2—similar to a peroxidase where the substrate oxidized is another H_2O_2 molecule:

$$2\,H_2O_2 \xrightarrow{\text{catalase}} 2\,H_2O + O_2 \qquad (18.7)$$

The heme of catalase is hidden in a deep pocket, presumably to prevent the escape of even more reactive intermediates such as $^{\bullet}OH$ and O_2^-. Catalases can also catalyze peroxide oxidation of substrates, acting as peroxidases.

A Catalase Model

The iron(III) complex of triethylenetetraamine, $[Fe(trien)(OH)_2]^+$, has catalase activity. In the proposed mechanism (Figure 18.27), the hydroxide ions are replaced by HO_2^- coordinated as a didentate group, stretching and weakening the O—O bond. A second HO_2^- can react to produce O_2 and $2\,OH^-$, regenerating the original complex. These reactions of H_2O_2 always seem to require substitution into the first coordination sphere of metal ions.

▶ *18.5.3 The Role of Copper*

Copper is vital in plants and animals, but some of its roles are less well understood than are those of Fe. Copper's roles are not as well-defined and consistent as are those of Zn and Fe. Biologically important characteristics are: redox processes involving Cu^+, Cu^{2+}, and even Cu^{3+}; significant differences in hardness of Cu^+ and Cu^{2+} (In the classification in Table 9.2 Cu^+ is a class *b* and Cu^{2+} is in the border region as Lewis acids and different stereochemistry of complexes of Cu^+ and Cu^{2+}.) The Cu protein hemocyanin has been

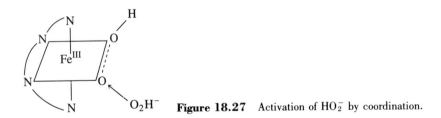

Figure 18.27   Activation of HO_2^- by coordination.

mentioned as the O_2^- carrier for a number of invertebrates. Cytochrome oxidase, an enzyme containing Co and Fe (heme), has been discussed as the terminal member of the respiratory chain (Section 18.4.1). Copper superoxidase dismutases and peroxidases were discussed (Section 18.5.2). Now we will consider some copper proteins and roles of Cu^{III}.

Blue Copper Proteins[17]

The intensely blue copper proteins *azurin* (from bacteria; molar mass 16,000; 1 Cu/molecule), *stellacyanin* (from the lacquer tree; molar mass 20,000; 1 Cu/molecule), and *plastocyanin* (from the chloroplasts of plants; molar mass 10,500; 1 Cu/molecule) are electron carriers. Plastocyanin is involved in electron transfer in photosynthesis. The charge-transfer absorption bands at 595–625 nm of these blue proteins are much more intense ($\varepsilon \sim 3000$–6000) than those associated with d–d (ligand-field) transitions. The Cu centers were believed to have similar highly distorted tetrahedral arrangements in azurin and plastocyanin. The blue Cu site in poplar plastocyanin is shown in Figure 18.28. The donors are N of imidazole rings of two histidines, S of methionine, and S of the thiol group of cysteine. The arrangement about the Cu center involves three short bonds in an almost planar trigonal arrangement with a fourth, longer bond to S of methionine.

An x-ray structure of azurin reported that the arrangement about Cu was approximately that of a trigonal bipyramid with O of a glycine residue (glycine, $NH_2CH_2CO_2H$) as the fifth donor. Here there are still three short bonds to the same donors as for plastocyanin with two very long bonds. These long bonds (312 and 316 pm) along the axis of the trigonal bipyramid are to S of methionine and O of glycine. The bond angles at Cu range from 85° to 132° for poplar plastocyanin and from 68° to 143° for azurin.

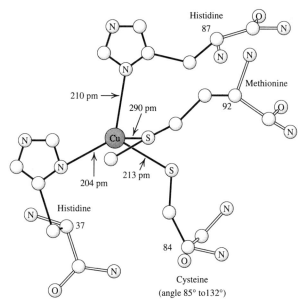

Figure 18.28   The blue copper site in poplar plastocyanin. Bond lengths are in pm.

[17] A. G. Sykes, *Adv. Inorg. Chem.* **1991**, *36*, 377; E. I. Solomon, K. W. Penfield, and D. E. Wilcox, *Struct. Bonding* **1983**, *53*, 1.

Recently, Lowery and Solomon[18] found that the O of glycine of azurin is too far away to be covalently bound. Plastocyanin and azurin have 2N and S in almost planar trigonal arrangement. Plastocyanin has a fourth weaker bond above the trigonal plane. Azurin has a similar arrangement with an even weaker bond to O of glycine opposite S of methionine. These sites are more suitable for Cu[I] than for Cu[II], giving high reduction potentials: 0.31 V for azurin and 0.38 V for plastocyanin at pH 7.5 ($E^0 = 0.16$ for $Cu^{2+} \rightarrow Cu^+$).

Azurin, stellacyanin, and plastocyanin do not have oxidase activity. Two related groups of Cu proteins do have oxidase activity, catalyzing the reduction of O_2 to H_2O and, in turn, oxidizing another substrate. These "blue" copper oxidases contain four Cu atoms per molecule; "nonblue" Cu oxidases contain only one or two Cu atoms per molecule. The "blue" oxidases contain one Cu[II], which is responsible for the very intense color. The other Cu atoms are referred to as nonblue, although one of them is believed to be Cu[II], but without the intense charge-transfer absorption. Two of the Cu atoms are diamagnetic. It has been shown for *laccase* that the removal of only the nonblue Cu[II] causes loss of catalytic activity, even though three atoms of Cu per molecule remain. Adding Cu to the sample restores the original Cu content and the oxidase activity. The ions N_3^-, CN^-, and F^- inhibit oxidase activity through the coordination to the nonblue Cu[II]. It has been proposed that the two diamagnetic Cu atoms are a spin-paired Cu[II]–Cu[II] pair capable of acting as a two-electron acceptor site in oxidizing a substrate. Only Cu[II] $\rightleftarrows$ Cu[I] conversion has been considered for Cu enzymes involved in oxidation–reduction processes.

Ceruloplasmin

Ceruloplasmin from human plasma is an intensely blue protein containing 0.3% Cu. Its molar mass is $\sim$134,000, with six or seven Cu atoms per molecule, and it has oxidase activity toward polyamines, polyphenols, and Fe[II]. The protein is found in the plasma of most animals. Interestingly, it occurs in the frog only after metamorphosis, accompanying changes in hemoglobin synthesis.

Ceruloplasmin apparently plays a role in Fe metabolism. Copper deficiency in animals results in reduced Fe mobilization and, eventually, anemia, even though there is still abundant Fe storage in the liver. Presumably, ceruloplasmin has ferroxidase activity (catalyzes Fe[II] $\rightarrow$ Fe[III]) in the uptake of Fe by transferrin. Although Fe is stored and transported as Fe[III], the transfer requires reduction to Fe[II] and then reoxidation. Other biological functions of ceruloplasmin have been suggested, including the storage and transport of Cu for other needs.

Several transition metals at trace levels can regulate the synthesis and degradation of heme. They inhibit γ-aminolevulinate ($H_2NCH_2\overset{\overset{\displaystyle O}{\|}}{C}C_2H_4CO_2H$) synthetase (the rate-limiting enzyme in heme synthesis) and induce heme oxygenase (the rate-limiting enzyme in heme degradation).

---

[18] M. D. Lowery and E. I. Solomon, *Inorg. Chim. Acta* **1992**, *198–200*, 233 and references cited.

A deficiency of ceruloplasmin occurs in individuals afflicted with the genetic Wilson's disease.[19] Copper accumulates in the liver, brain, and kidneys. The symptoms can be relieved by the use of the chelating agents edta [as $Na_2Ca(edta)$] or penicillamine for the mobilization of the accumulated Cu.

Nonblue Copper Proteins

Galactose oxidase catalyzes the oxidation of primary alcohols by O_2 to form the aldehyde plus H_2O_2. The enzyme with a molar mass 68,000 contains one Cu. There are probably two imidazole and H_2O as ligands. Several amine oxidases are Cu proteins using $O_2 \rightarrow H_2O$ for oxidation. Cytochrome oxidase containing Fe and Cu (page 914) and superoxide dismutase (page 914) are other examples of nonblue Cu enzymes.

Copper(III) in Nature

Iron and copper play many vital roles in life processes because of the availability of easily accessible $Fe^{II} \rightleftarrows Fe^{III}$ and $Cu^{I} \rightleftarrows Cu^{II}$ transformations. Copper(III) is stabilized so much in complexes formed with deprotonated amide or deprotonated peptide ligands that Cu^{II} can be oxidized to Cu^{III} by O_2. It was suggested that the biological role of Cu^{III}, involving a two-electron reduction $Cu^{III} \rightarrow Cu^{I}$, could be important in eliminating high-energy free radical intermediates.

18.6   VITAMIN B_{12} [20]

▶ 18.6.1   Structure

It has long been recognized that Co is one of the essential trace metals, but the discovery that vitamin B_{12} is a Co complex came as a great surprise. The vitamin was isolated from liver after it was found that eating raw liver would alleviate pernicious anemia. Vitamin B_{12} is now used for treatment (monthly 1-mg injections).

Figure 18.29a depicts the structure vitamin B_{12}. The Co is coordinated to a corrin ring. This macrocycle is similar to the porphyrin ring, except that rings **A** and **D** are joined directly in the corrin ring and there is less conjugation. On one side of the corrin ring the ligand bonded to Co is α-5,6-dimethylbenzimidazole nucleotide, which also is joined to the corrin ring. The sixth ligand is CN^-, but actually CN^- is introduced in the isolation procedure. The active form of the vitamin, called **coenzyme B_{12},** contains an adenosyl group (Figure 18.29b) as the sixth ligand. Solving the structure of B_{12} coenzyme required both chemical information and x-ray crystallography. Dorothy Crowfoot Hodgkin received the Nobel Prize in Chemistry in 1964 for her crystallographic work.

Vitamin B_{12} with CN^- removed is called **cobalamin,** so B_{12} is cyanocobalamin, the form used for treating pernicious anemia. Other ligands can substitute for CN^-. It was

[19] R. P. Csintalan and N. M. Senozan, *J. Chem. Educ.* **1991**, *68*, 365.
[20] D. Dolphin, Ed., *B_{12}*, Vols. 1 and 2, Wiley, New York, 1982.

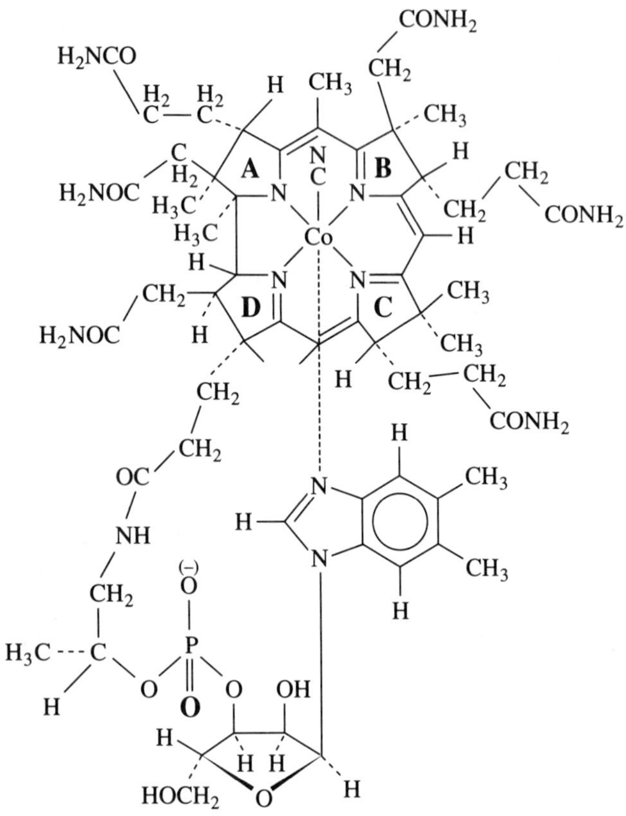

Figure 18.29   (*a*) Vitamin B_{12}. (*b*) The group present in place of CN^- in coenzyme B_{12}.

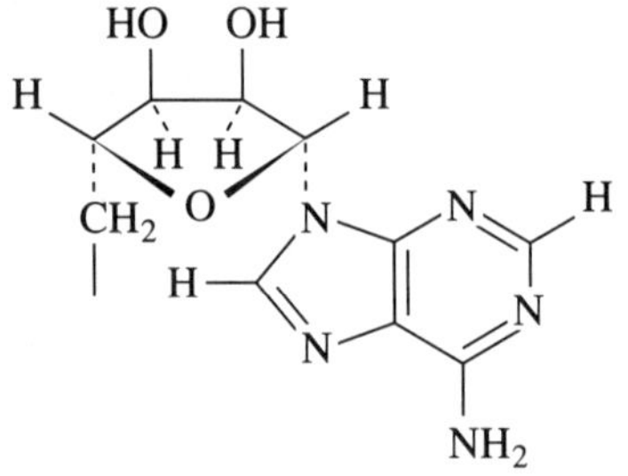

considered very unusual that the coenzyme contains a Co—C sigma bond, but —CH_3 or other organic groups can be so bonded as discussed in Section 13.6.1. Methylcobalamin reacts with Hg^{2+} via transmetallation to give CH_3Hg^+, whose accumulation in biological systems poses a problem. The vitamin and coenzyme are treated as cobalt(III) compounds. Hydroxocobalamin (and other derivatives) can be reduced (H_2/Pt) to Co^{II} cobalamins, designated B_{12r}, and by stronger reductants ($NaBH_4$) to Co^I cobalamins, designated B_{12s}.

► *18.6.2 Reactions*

Coenzyme B_{12} serves as a prosthetic group (see Section 18.3.1) to various enzymes in catalyzing reactions. In the case of enzymes the prosthetic group (see Section 18.4.1) is a coenzyme tightly bound to the enzyme. Many remarkable reactions have been discovered. We would expect a variety of reactions from the three possible ways of Co—C bond cleavage in alkylcobalamines:

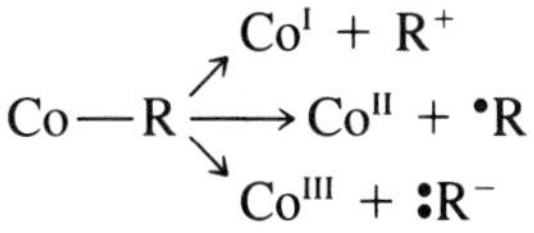

$$Co-R \nearrow\!\!\!\!\!\!\searrow\!\!\!\!\!\! \begin{array}{l} Co^I + R^+ \\ Co^{II} + {}^\bullet R \\ Co^{III} + :R^- \end{array} \qquad (18.8)$$

Among the very impressive reactions of the enzyme are:

One-Carbon Transfer. Coenzyme B_{12} readily accepts a methyl group (bound to Co) that can be transferfed to add a carbon to a substrate as shown in Reactions (18.9) and (18.10):

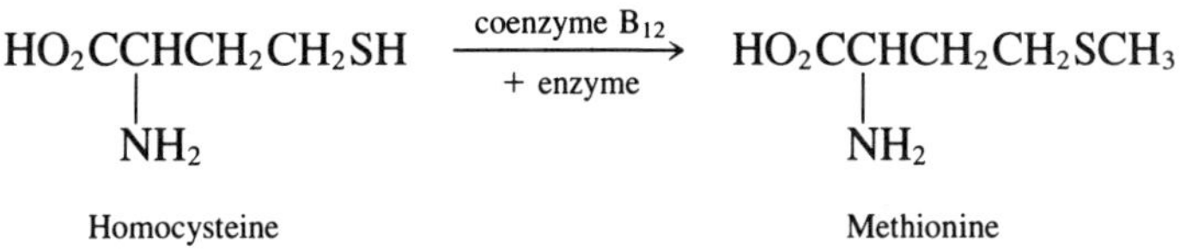

$$HO_2CCHCH_2CH_2SH \xrightarrow[+ \text{ enzyme}]{\text{coenzyme } B_{12}} HO_2CCHCH_2CH_2SCH_3 \qquad (18.9)$$
$$\quad\;| \qquad\qquad\qquad\qquad\qquad\qquad\qquad\quad |$$
$$\quad\;NH_2 \qquad\qquad\qquad\qquad\qquad\qquad\quad NH_2$$

Homocysteine   Methionine

$$H_2N-CH_2-CO_2H \xrightarrow[+ \text{ enzyme}]{\text{coenzyme } B_{12}} H_2N-CH-CO_2H \qquad (18.10)$$
$$\qquad\qquad\qquad\qquad\qquad\qquad\qquad\qquad\qquad\quad |$$
$$\qquad\qquad\qquad\qquad\qquad\qquad\qquad\qquad\quad CH_2OH$$

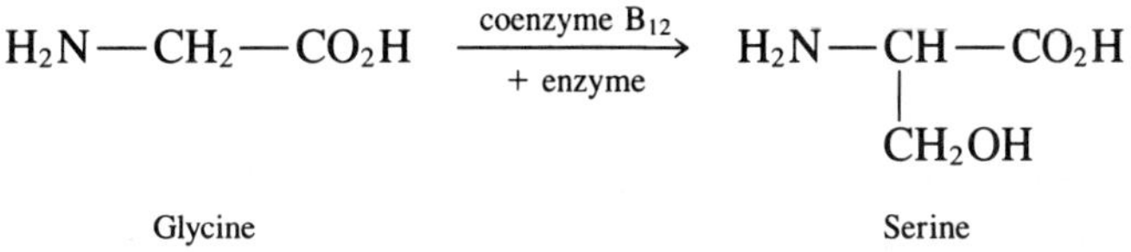

Glycine   Serine

Isomerization. The isomerization reactions that occur involve moving a substituent along a C chain.

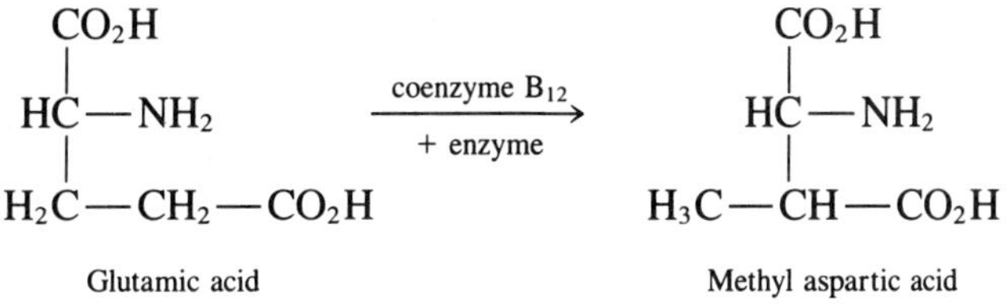

$$\begin{array}{c} CO_2H \\ | \\ HC-NH_2 \\ | \\ H_2C-CH_2-CO_2H \end{array} \xrightarrow[+ \text{ enzyme}]{\text{coenzyme } B_{12}} \begin{array}{c} CO_2H \\ | \\ HC-NH_2 \\ | \\ H_3C-CH-CO_2H \end{array} \qquad (18.11)$$

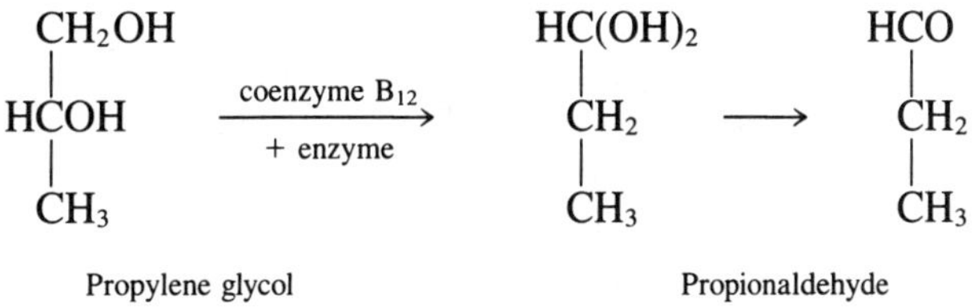

Glutamic acid   Methyl aspartic acid

$$\begin{array}{c} CH_2OH \\ | \\ HCOH \\ | \\ CH_3 \end{array} \xrightarrow[+ \text{ enzyme}]{\text{coenzyme } B_{12}} \begin{array}{c} HC(OH)_2 \\ | \\ CH_2 \\ | \\ CH_3 \end{array} \longrightarrow \begin{array}{c} HCO \\ | \\ CH_2 \\ | \\ CH_3 \end{array} \qquad (18.12)$$

Propylene glycol   Propionaldehyde

► **18.6.3   Models of B_{12}**

The Co complex of dimethylglyoxime (cobaloxime—see Figure 18.30) is an effective model for B_{12}. The cobaloximes can be reduced to cobaloximes(II) and cobaloximes(I). The reactions shown by adenosyl and alkyl derivatives of cobaloximes(I), which resemble those of B_{12s}, include methyl group transfer, reduction, and rearrangements. Presumably Co is used in B_{12} because of its versatility in showing three oxidation states achieved by one-electron transfers.

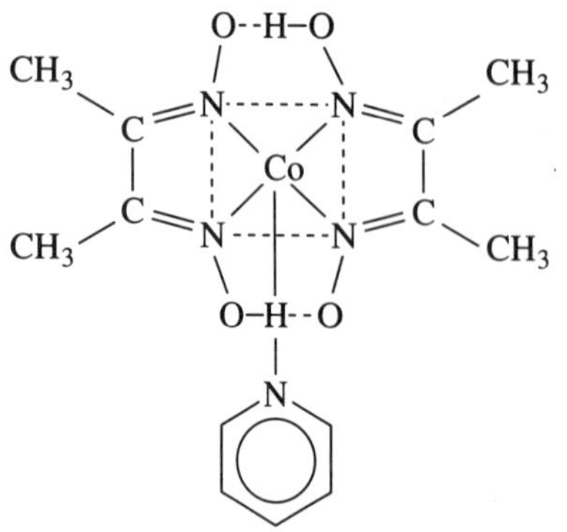

Figure 18.30   Pyridine cobaloxime.

18.7 NITROGENASE[21]

Nitrogen fertilizers have been in short supply because manufacturing them requires high energy consumption and so they are expensive. Over the period 1949–1974, consumption of N fertilizers increased more than tenfold. Usage is expected to increase still more, as undeveloped countries strive to improve their agriculture to meet the needs of population increases. In 1983 the industrial **nitrogen fixation** (conversion of N_2 to N compounds) was 4×10^7 tons compared with 1.2×10^8 tons fixed by biological fixation.

The N_2 molecule is so unreactive that N_2 commonly is used to provide an inert atmosphere. The commercial Haber process for N_2 fixation requires high temperature and pressure:

$$N_2 + 3\,H_2 \xrightarrow[\substack{200-600 \text{ atm} \\ 450-600°C}]{\text{catalyst}} 2\,NH_3 \tag{18.13}$$

Nevertheless, blue-green algae and some bacteria are able to fix N_2 at ambient temperature and pressure. *Rhizobium* bacteria living in the root nodules of certain legumes are an important source of fixed N.

A nitrogen-fixing enzyme, **nitrogenase,** was isolated from the anaerobic bacterium *Clostridium pasteurianum* in 1960. Nitrogenase has been found to contain two components, an Mo–Fe-containing protein and an Fe-containing protein. The Mo–Fe protein contains two Mo and about 30 atoms each of Fe and sulfide per molecule (molar mass ~220,000). The Fe protein contains two identical subunits, each containing an Fe_4S_4 cluster (molar mass ~60,000).

Extended x-ray absorption fine structure (EXAFS) spectroscopic studies of the Mo–Fe protein for two sources of nitrogenase indicate the presence of an Mo, Fe, S cluster, thought to be a constituent of all Mo–Fe nitrogenase systems. The cluster $[Mo_2Fe_6S_9(SC_2H_5)_8]^{3-}$ has been prepared by Holm et. al., and the structure has been determined (Figure 18.31). EXAFS studies of this cluster indicate that it is the closest model yet achieved for the active site of nitrogenase. It might differ from the core of the Mo–Fe site of nitrogenase only by the ligands bonded to Mo.

[21] R. R. Eady, *Adv. Inorg. Chem.* **1991,** *36,* 77; B. K. Burgess, *Chem. Rev.* **1990,** *90,* 1377.

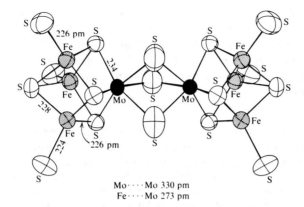

Figure 18.31 Structure of $[Mo_2Fe_6S_9(SC_2H_5)_8]^{3-}$. The ethyl groups of the six terminal thiolate ligands and the two bridging thiolates are omitted. (Reprinted with permission from T. E. Wolff, J. M. Berg, C. Warrick, K. O. Hodgson, and R. H. Holm, *J. Am. Chem. Soc.* **1978**, *100*, 4630. Copyright 1978, American Chemical Society.)

Nitrogenase reduces N_2 to NH_3 with no evidence for intermediates not bound by the enzyme. Even at high pressure of N_2, some H_2 is produced according to the limiting reaction:

$$N_2 + 8e + 8H^+ \rightarrow 2NH_3 + H_2 \tag{18.14}$$

Presumably, N_2 is complexed by the Mo and the Fe protein, probably of the ferredoxin type, bringing about the reduction through electron transfer. Adenosine triphosphate (ATP) is essential for nitrogenase activity. The catalytic cycle for the transformation of N_2 to NH_3 by the resting Mo–Fe protein E_0 according to the stoichiometry of Reaction (18.14) is shown in Scheme 18.2. The displacement of N_2 by H_2 from $E_3N_2(H^+)$ represents the competitive inhibition of N_2 reduction by H_2 present. The reduction of N_2 is carried out essentially under anaerobic conditions to avoid competition of O_2 for N_2 as a substrate.

Nitrogenase is effective in reducing N_3^-, N_2O, RCN, RNC, and HCCH. The reduction of acetylene to ethylene has been used for monitoring nitrogenase activity because the process is more efficient than the reduction of N_2.

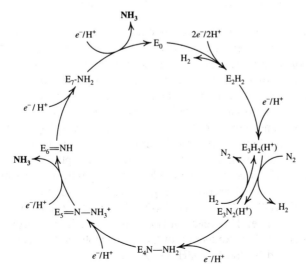

Scheme 18.2 The catalytic cycle for fixation of N_2 by the Mo-Fe protein E_0.

The discovery of stable complexes of N_2 led to intense study of model compounds and a possible new nitrogen fixation process. Ammonia has been obtained from metal complexes of N_2, and intermediate stages of reduction to complexes of N_2H^-, N_2H_2, and NNH_2^{2-} have been observed. A study of bonding in N_2 complexes favored end-on coordination and suggested that σ donation is more important than π back-donation from the metal to the π^* orbitals of N.

The metalloenzymes and Cu and Fe proteins are large molecules. The major organic components determine the polar/nonpolar compatibility with the environment, access to the metal ion site, selectivity of substrates, and the stereochemistry at the metal ion site. The stereochemical requirements of the metal ion are those of its coordination chemistry. These requirements can change significantly in redox reactions, particularly for Cu^{II}/Cu^{I}. The donor atoms and the stereochemistry imposed on the metal ion determine the electrode potential which can be fine-tuned by changing porphyrin substituents, axial ligands, or associated proteins.

18.8 PHOTOSYNTHESIS

Except for nuclear fuel and geothermal energy in the form of geysers, our energy sources ultimately are solar. Wood and fossil fuels (oil and coal) represent solar energy stored through photosynthesis. Hydroelectric power results from solar energy. The O_2 in our atmosphere is a byproduct of photosynthesis. We associate photosynthesis with plants, but perhaps half or more of the photosynthesis on earth occurs in algae, diatoms, and in certain red, green, brown, and purple bacteria.

The usual net reaction for photosynthesis converts H_2O and CO_2 to carbohydrates and O_2:

$$n\,H_2O + n\,CO_2 \xrightarrow{\text{light}} (CH_2O)_n + n\,O_2 \qquad (18.15)$$

In photosynthetic bacteria, however, the species oxidized is not H_2O, but instead an organic molecule or H_2S. In the latter case, elemental S is produced instead of O_2 by anaerobic bacteria that are poisoned by O_2. Plants can use NO_3^- as the electron acceptor instead of CO_2 (to form NH_3), and some organisms can use H^+ to form H_2, or N_2 to form NH_3.

Chlorophyll *a* (Figure 18.3) is just one of the many pigments involved in photosynthesis. A variety of colors of photosynthetic cells is possible from combinations of chlorophylls, yellow carotenoids, and blue or red pigments (*phycobilins*). Green plant cells contain chloroplasts, the membrane-surrounded units for photosynthesis, as well as mitochondria (page 905) for respiration—the reverse of the photosynthetic reactions. These O_2-producing chloroplasts contain two kinds of chlorophyll, one of which is chlorophyll *a*. O_2 is produced from H_2O accompanying the reduction of nicotinamide adenine dinucleotide phosphate ($NADP^+$). $NADP^+$ is NAD^+ (Figure 18.21) with one of the OH groups esterified with phosphate. This reaction is driven by the light energy absorbed and also produces ATP (Figure 18.19). Reduction of CO_2 to carbohydrates occurs in a dark reaction utilizing energy stored by the NADPH and ATP.

There are two light reaction paths in O_2-producing plants. The first of these (**Photosystem I**) involves chlorophyll *a* along with other pigments and does not produce O_2. This is the only photosystem for anaerobic bacteria and is believed to have been the first such system to evolve. Chlorophyll *a* absorbs at about 680 nm in cells. **Photosystem II,** which brings about the formation of O_2, involves chlorophyll *a*; chlorophyll *b*, which absorbs at shorter wavelengths; and other associated pigments. Typically, Photosystem I contains about 200 chlorophyll *a* molecules and about 50 carotenoid pigment molecules. Eight photons are needed, two photons for each electron transferred from H_2O to $NADP^+$, so energy must be accumulated. Photosystem I contains an energy trap, P700, that serves to accumulate the energy absorbed by the individual chlorophyll *a* molecules. P700 is thought to be a specialized chlorophyll that in the excited state can lose electrons to bring about the reduction of $NADP^+$. Photosystem II restores the electrons (from H_2O) to oxidized P700 and forms O_2 in the process. Manganese is involved in this process in an unknown manner. If this involves $Mn^{IV} \rightarrow Mn^{II}$ reduction, at least two Mn atoms are required. There are also cyclic dark reactions that produce ATP. These processes are represented in Figure 18.32.

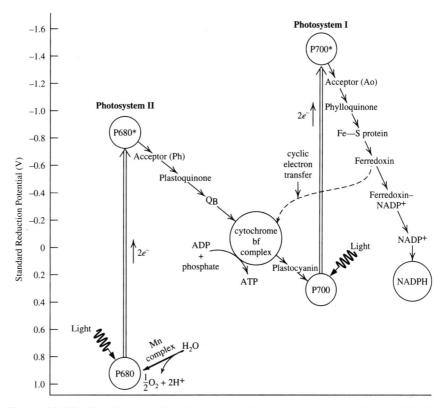

Figure 18.32 The "Z scheme" for the pathway of electron transfer from H_2O to $NADP^+$ in the two photosystems for green plants. Acceptor Ph is pheophytin, Q_B is a second quinone in Photosystem II, and Acceptor A_0 is an electron acceptor chlorophyll. The cytochrome bf complex contains a cytochrome chain and a Fe-S protein. (Adapted with permission from A. L. Lehninger, D. L. Nelson, and M. M. Cox, *Principles of Biochemistry*, 2nd ed., Worth, New York, 1993, p. 582.)

The net reaction is

$$2 H_2O + 2 NADP^+ + 8 \text{ photons} \longrightarrow 2 NADPH + 2 H^+ + O_2 \qquad (18.16)$$

The electron-transfer processes in Photosystem I utilize ferredoxin (Section 18.4.3). The electron transport between the photosystems involves a cytochrome chain similar to that for the respiratory chain (Figure 18.18) in reverse, converting $H_2O \rightarrow O_2$. A blue copper protein, plastocyanin (Section 18.5.3), also is involved in an electron-transfer process.

18.9   ROLES OF Na^+, K^+, Mg^{2+}, Ca^{2+}, AND ION PUMPS

So far we have dealt with the biological roles of trace metals (Table 18.1), as well as the role of P in the phosphorylation process. The pairs of major elements Na^+–K^+ and Ca^{2+}–Mg^{2+} are so similar chemically that it is surprising that they differ so greatly in their biological functions. The ions Ca^{2+} and Na^+ are concentrated in body fluids outside of cells. Calcium forms solid skeleton materials such as bones, stabilizes conformations of proteins and triggers muscle contraction and the release of hormones. The ions Mg^{2+} and K^+ are concentrated in cells. Magnesium ion forms a complex with ATP and is required for most enzymatic reactions involving ATP within the cell.

Cells are enclosed by a membrane about 7000 pm (70 Å) thick and composed of double layers of protein separated by lipids. Cations cannot pass through the lipid (fatty) layer without encapsulation, and thus the enclosed cation presents an organic, lipid-soluble surface to the membrane.

In most animal cells the concentration of K^+ is about 0.15 M and that of Na^+ is about 0.01 M. In the fluids outside the cells the concentration of Na^+ is about 0.15 M and that of K^+ is less than 0.004 M (concentrations rather close to those of seawater—see Table 18.1). Maintenance of these large-concentration gradients requires a "sodium pump" (really, an Na^+/K^+ pump). The energy of transport of the ions is provided by the hydrolysis of ATP. Kidney and brain cells use about 70% of the energy from ATP for this transport. In some cells, each ATP molecule hydrolyzed transports $3Na^+$ out of the cell and $2K^+$ ($+H^+$) into the cell. The K^+ is required in the cell for glucose metabolism, protein synthesis, and activation of some enzymes. The transport of glucose and amino acids into the cell is coupled with Na^+ transport, which is favored by the great concentration gradient. The Na^+ entering the cell in this way must be pumped out again.

The alkali metal ions are weak Lewis acids and form weak complexes. The encapsulation of a cation by crown ethers (Section 15.2.4) is selective, depending on the cation size. Some natural **ionophores** ("ion-bearers") for alkali metal ions have O donor atoms from carboxylates, cyclic peptides, and so on. Some of these, such as nonactin and valinomycin (Figure 18.33), are antibiotics. The formation constants for nonactin in acetone containing some H_2O for Na^+, K^+, and Cs^+ are 210, 2×10^4, and 400, respectively. Thus nonactin is selective for K^+.

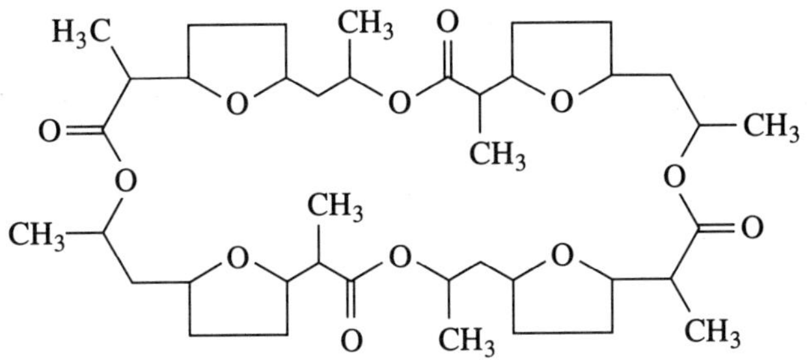

Nonactin

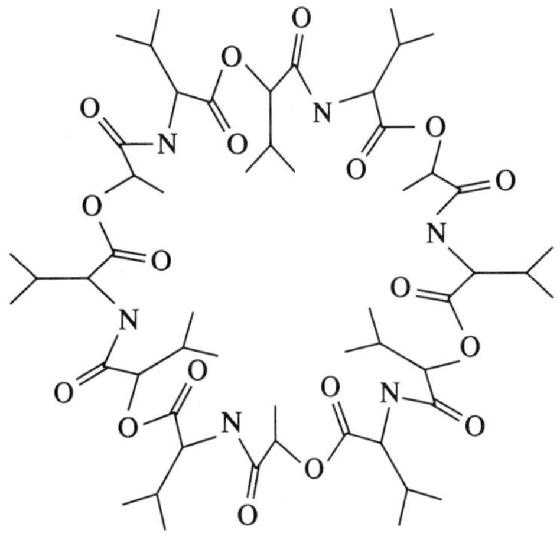

Valinomycin

Figure 18.33   Some macrocyclic antibiotics.

The membrane enzyme Na/K-ATPase serves as an Na^+/K^+ pump. It is thought that the large protein spans the membrane with internal and external sites for both Na^+ and K^+. When the outer sites are filled with K^+ and the inner ones with Na^+, a conformational change occurs rotating the sites and alternating the site preferences. This releases K^+ inside the cell and Na^+ outside the cell. This is a "revolving door" for transport of Na^+ and K^+ in opposite directions. The pump is reversible if there is a large enough concentration gradient for Na^+/K^+.

Initiation of muscle contraction results from the arrival of a nerve impulse at a motor ending in a fiber. This causes Ca^{2+} to be released from the *sarcoplasmic reticulum* (SR) of the muscle. The concentration of Ca^{2+} is increased by 100-fold in milliseconds by release of the SR. On relaxation of the muscle the Ca^{2+} is reabsorbed by SR. This process requires many cycles of a Ca pump identified as a Ca^{2+}/Mg^{2+}-ATPase similar to the sodium pump.

GENERAL REFERENCES

I. Bertini, H. B. Gray, S. J. Lippard, and J. S. Valentine, Eds., *Bioinorganic Chemistry,* University Science Books, Mill Valley, CA, 1993.

G. L. Eichorn and L. G. Marzilli, Eds., *Advances in Inorganic Biochemistry,* Elsevier North-Holland, New York. A series of volumes updating the treatise.

J. J. R. Frausto da Silva and R. J. R. Williams, *The Biological Chemistry of the Elements,* Oxford University Press, Oxford, 1992.

R. W. Hay, *Bio-Inorganic Chemistry,* Horwood (Wiley), Chichester (England), 1984.

M. N. Hughes, In *Comprehensive Coordination Chemistry,* G. Wilkinson, Ed., Pergamon Press, Oxford, 1987, Vol. 6, Chapter 62.1. Very thorough with 1500+ references.

J. A. Ibers and R. H. Holm, *Science,* **1980,** *209,* 223. An excellent article dealing with model studies of oxygen transport and storage, as well as of electron carriers.

S. J. Lippard and J. M. Berg, *Principles of Bioinorganic Chemistry*, University Science Books, Mill Valley, CA, 1993.

E. Ochiai, *General Principles of Biochemistry of the Elements*, Plenum, New York, 1987.

Science, **1993**, 261, 699–736. A series of articles on current activity in Bioinorganic Chemistry.

H. Sigel, Ed., *Metal Ions in Biological Systems*, Marcel Dekker, New York, Vols. 1–27 et seq.

PROBLEMS

18.1   For each of the following elements, identify one significant role in biological processes: Fe, Mn, Cu, Zn, I, Mg, Co, Ca, and K.

18.2   (a) Why is iron used as the edta complex rather than a simple salt for supplying iron for plants in basic soil? (b) Why are edta and other chelating ligands used for the removal of toxic and radioactive metal compounds?

18.3   High-spin iron(II) is too large for the opening of the porphyrin ring, but low-spin iron(II) can be accommodated in the opening. Why does high-spin iron(II) have a larger radius?

18.4   What prevents simple iron porphyrins from functioning as O_2 carriers? What features of successful models of Fe-porphyrin O_2 carriers have made it possible to avoid this problem?

18.5   How is iron stored and transported in mammals? What is the oxidation state of iron for storage and for transfer?

18.6   Identify two chemical types of siderophores.

18.7   The formation constant of Fe^{III} enterobactin is about 10^{52}. Calculate the concentration of Fe^{3+} in equilibrium with 10^{-4} M Fe^{III}enterobactin and 10^{-4} M ent^{6-}. This corresponds to how many liters per Fe^{3+}? The volume of the hydrosphere (all bodies of water, snow, and ice) is $\sim 1.37 \times 10^{21}$ liters (K. N. Raymond et. al., *J. Am. Chem. Soc.* **1979**, *101*, 6097).

18.8   What is the cytochrome chain? What are the advantages of such a complex system?

18.9   What are the prosthetic groups of cytochromes and hemoglobin?

18.10   What are the two important systems for biological electron-transfer processes?

18.11   What chemical properties of Fe and Cu make them suitable for redox processes in biological systems?

18.12   The conversion of carbonic acid to $CO_2 + H_2O$ is a spontaneous process; why is carbonic anhydrase needed?

18.13   Give an example of the substitution of Co^{II} for Zn^{II} in an enzyme to provide a "spectral probe" for study of the enzyme.

18.14   The direct reduction products of oxygen, H_2O_2 (or HO_2^-), and O_2^- are toxic. How are these products handled in biological systems?

18.15   Wilson's disease causes the accumulation of Cu in the body. How can symptoms be relieved?

18.16   Give an example of each of two types of reactions brought about by vitamin B_{12}.

18.17   What metals are at active centers of nitrogenase? Name some reduction processes other than that of N_2 that are accomplished by nitrogenase.

18.18   What electron transport systems are used in photosynthesis?

18.19   Magnesium has important roles as a cellular cation and in enzymes. What are likely reasons for the toxicity of Be?

APPENDIX A

Units

• •

By international convention the units employed for scientific measurement are those of the International System of Units (SI).[1] We use SI units in this book, except in a few cases (for example, we express pressure in atm.) Seven base units regarded as dimensionally independent are recognized. Other units are derived by combining base units according to algebraic relationships. Some of these derived units have special names. The SI base units are as follows.

	Name	*Symbol*
Mass	kilogram	kg
Length	meter	m
Time	second	s
Electric current	ampere	A
Thermodynamic temperature	kelvin	K
Amount of substance	mole	mol
Luminous intensity	candela	cd

[1] See NBS Special Publication 330, 1977 Edition, U.S. Government Printing Office, Washington, 1977.

Some derived units recognized in SI are listed below.

Quantity	Name	Symbol	Expression in Terms of Other Units	Expression in Terms of SI Base Units
Frequency	hertz	Hz		s^{-1}
Force	newton	N		$m \cdot kg \cdot s^{-2}$
Pressure, stress	pascal	Pa	N/m^2	$m^{-1} \cdot kg \cdot s^{-2}$
Energy, work, quantity of heat	joule	J	$N \cdot m$	$m^2 \cdot kg \cdot s^{-2}$
Power, radiant flux	watt	W	J/s	$m^2 \cdot kg \cdot s^{-3}$
Quantity of electricity, electric charge[a]	coulomb	C	$A \cdot s$	$A \cdot s$
Electric potential, potential difference, electromotive force	volt	V	J/c (W/A)	$m^2 \cdot kg \cdot s^{-3} \cdot A^{-1}$
Magnetic flux	weber	Wb	$V \cdot s$	$m^2 \cdot kg \cdot s^{-2} \cdot A^{-1}$
Magnetic flux density	tesla	T	Vs/m^2	$kg \cdot s^{-2} \cdot A^{-1}$
Celsius temperature	degree Celsius	°C		K
Heat capacity, entropy	joule per kelvin		J/K	$m^2 \cdot kg \cdot s^{-2} \cdot K^{-1}$
Specific heat capacity, specific entropy	joule per kilogram kelvin		$J/(kg \cdot K)$	$m^2 \cdot s^{-2} \cdot K^{-1}$
Thermal conductivity	watt per meter kelvin		$W/(m \cdot K)$	$m \cdot kg \cdot s^{-3} \cdot K^{-1}$
Energy density	joule per cubic meter		J/m^3	$m^{-1} \cdot kg \cdot s^{-2}$
Electric field strength	volt per meter		V/m	$m \cdot kg \cdot s^{-3} \cdot A^{-1}$
Electric charge density	coulomb per cubic meter		C/m^3	$m^{-3} \cdot s \cdot A$
Electric flux density	coulomb per square meter		C/m^2	$m^{-2} \cdot s \cdot A$
Molar energy	joule per mole		J/mol	$m^2 \cdot kg \cdot s^{-2} \cdot mol^{-1}$
Molar entropy, molar heat capacity	joule per mole kelvin		$J/(mol \cdot K)$	$m^2 \cdot kg \cdot s^{-2} \cdot K^{-1} \cdot mol^{-1}$

[a] In lattice energy calculations (see Section 5.8), dimensional analysis usually is simplified by expressing electrical charge in esu whose units are $dyne^{1/2}$ cm.

APPENDIX **B**

Nomenclature of Inorganic Chemistry[1]

· ·

B1 ATOMIC SYMBOLS, MASS, ATOMIC NUMBER, ETC.

The approved symbols and names of the elements are given in Table B.1. The mass number, atomic number, number of atoms, and atomic charges are to be represented as follows:

Mass number	Left upper index
Atomic number	Left lower index
Number of atoms	Right lower index
Ionic charge	Right upper index

EXAMPLE: $^{200}_{80}Hg_2^{2+}$ represents the doubly charged ion containing two mercury atoms, each of which has the mass number 200. The charge is to be written as Hg_2^{2+}, not Hg_2^{+2}.

The periodic A and B subgroups have become hopelessly confused by disparate use in different parts of the world. The IUPAC recommends numbering the groups 1–18 as shown in the table inside the front cover. Designation of families by name—such as the alkali metals, the halogens, or the carbon family—is unambiguous.

---

[1] For the full report see *Nomenclature of Inorganic Chemistry—Recommendations 1990,* International Union of Pure and Applied Chemistry, G. J. Leigh, Ed., Blackwell, Oxford, 1990. See also B. P. Block, W. H. Powell, and W. C. Fernelius, *Inorganic Chemical Nomenclature,* American Chemical Society, Washington, D.C., 1990.

B2   FORMULAS

The molecular formula is used for compounds that exist as discrete molecules—for example, S_2Cl_2, not SCl; $Co_2(CO)_8$, not $Co(CO)_4$. If the molar mass varies with changes in conditions, the simplest formula may be used unless it is desired to indicate the molecular complexity for given conditions, for example, S, P, and NO_2 may be used instead of S_8, P_4, and N_2O_4.

The electropositive constituent is placed first in the formula—for example, $NaCl$, $MgCO_3$. If the compound contains more than one electropositive or more than one electronegative constituent, the sequence within each class is in alphabetical order of their symbols. Acids are treated as hydrogen salts. In the case of binary compounds between nonmetals, that constituent should be placed first which appears earlier in the sequence:

$$Rn, Xe, Kr, B, Si, C, Sb, As, P, N, H, Te, Se, S, At, I, Br, Cl, O, F \qquad (B.1)$$

The sequence roughly follows the order of electronegativities without overlap of groups (except for O because of the great difference in electronegativities of O and S).

EXAMPLES: NH_3, H_2Te, $BrCl$, Cl_2O, OF_2, XeF_2.

Characteristic polyatomic ions (NH_4^+, CO_3^{2-}, SO_4^{2-}) are identified as units. Deviation from alphabetical order is allowed to emphasize similarities between compounds.

> $VOSO_4$(not VSO_5)
>
> $FeO(OH)$ (not $FeHO_2$)
>
> $Co_2NO_3(OH)_3$
>
> $CaTiO_3$ and $ZnTiO_3$ (not alphabetical to show similarity)

Exceptions to the above order are encountered in compounds in which the sequence of symbols is used to indicate the order in which the atoms are bonded in the molecule or ion—for example, $HOCN$ (cyanic acid), $HONC$ (fulminic acid), $HNCO$ (isocyanic acid).

If two or more different atoms or groups are attached to a single atom, the symbol of the central atom is placed first followed by the symbols of the remaining atoms or groups in alphabetical order.

Parentheses ( ) or square brackets [ ] are used to improve clarity—for example $[Co(NH_3)_6]_2(SO_4)_3$. Hydrates are written as follows: $Na_2SO_4 \cdot 10H_2O$.

The prefixes *cis* and *trans* are italicized and separated from the formula by a hyphen—for example *cis*-$[PtCl_2(NH_3)_2]$.

B3   SYSTEMATIC NAMES

The name of the electropositive constituent is not modified and is placed first. The name of the electronegative constituent [or the element later in sequence (B.1) for compounds of nonmetals] is modified to end in -ide if it is monoatomic or homopolyatomic (e.g., O_2^{2-} and N_3^-). If the compound contains more than one kind of electropositive constituent, the names should be spaced and cited in alphabetical order for the *name* of the element. Hydrogen is cited last among electropositive constituents and is separated from the following anion name(s) by a space unless it is known to be bound to an anion.

EXAMPLES: Sodium chloride, magnesium sulfide, lithium nitride, nickel arsenide, silicon carbide, sulfur hexafluoride, oxygen diflouride, potassium triiodide, and ammonium sodium hydrogen phosphate ($NaNH_4HPO_4$).

If the electronegative constituent is heteronuclear it should be designated by the termination -ate. Exceptions include OH^-, hydroxide ion; NH^{2-}, imide ion; NH_2^-, amide ion; and CN^-, cyanide ion. In case of two or more electronegative constituents, their sequence should be in alphabetical order. This order might differ in names and formulas.

Complex anions can be named using the name of the characteristic or central atom modified to end in -ate. Ligands attached to the central atom are indicated by the termination -o. The oxidation number of the central atom should be indicated by a Roman numeral (Stock system) or by making the charge on the anion clear by the use of prefixes.

EXAMPLES:

$Na_2[SO_4]$	sodium tetraoxosulfate(VI) or disodium tetraoxosulfate	$Na[SO_3F]$	sodium trioxofluorosulfate(VI)
$Na_2[SO_3]$	sodium trioxosulfate(IV) or disodium trioxosulfate	$Na[ICl_4]$	sodium tetrachloroiodate(III)
$Na_2[S_2O_3]$	disodium trioxothiosulfate	$Na[PCl_6]$	sodium hexachlorophosphate(V)
$[Fe(CO)_4]^{2-}$	tetracarbonylferrate(2−)		

Common names of many oxoanions and oxoacids may be used (see page B-7).

Stoichiometric proportions may be denoted by the use of numerical prefixes (mono, di, tri, tetra, penta, hexa, hepta, octa, nona, deca, undeca, dodeca) preceding without hyphen the names of the elements to which they refer. The prefix mono is omitted unless it is necessary to avoid confusion. Beyond 10, Arabic numerals may be used. The end vowels of numerical prefixes should not be elided—for example, tetr*a*oxide, with carbon monoxide an exception.

EXAMPLES:

N_2O	dinitrogen oxide	Fe_3O_4	triiron tetraoxide
N_2O_4	dinitrogen tetraoxide	U_3O_8	triuranium octaoxide
S_2Cl_2	disulfur dichloride		

The proportions of the constituents may also be indicated indirectly by the Stock system, in which Roman numerals are used to represent the oxidation number of an element or central atom. For zero the Arabic 0 is used. Latin names of the elements or Latin stems may be used with the Stock system; such usage is common for complex anions. The charge on the aggregate can be shown by an Arabic numeral followed by the charge in parentheses (Ewens and Bassett system) instead of giving the oxidation number of the central atom.

EXAMPLES:

$FeCl_2$	iron(II) chloride
$FeCl_3$	iron(III) chloride
$SnSO_4$	tin(II) sulfate
MnO_2	manganese(IV) oxide
BaO_2	barium(II) peroxide or barium peroxide
$Pb_2^{II}Pb^{IV}O_4$	dilead(II) lead(IV) oxide or trilead tetraoxide
$K_4[Ni(CN)_4]$	potassium tetracyanonickelate(0) or potassium tetracyanonickelate(4−)
$K_4[Fe(CN)_6]$	potassium hexacyanoferrate(II)
$K_2[Fe(CO)_4]$	potassium tetracarbonylferrate(−II) or potassium tetracarbonylferrate(2−)

The use of endings -ous and -ic for cations is not recommended and should not be used for elements exhibiting more than two oxidation numbers.

▶ B3.1 Hydrides

The following names are acceptable (parent hydrides used for derivatives named in substitutive nomenclature are given in parentheses):

H_2O	water (oxidane)	PH_3	phosphine (phosphane)
NH_3	ammonia (azane)	P_2H_4	diphosphane
N_2H_4	hydrazine (diazine)	AsH_3	arsine (arsane)
BH_3	borane	SbH_3	stibine (stibane)
B_2H_6	diborane	BiH_3	bismuthine (bismuthane)
SiH_4	silane (Si_2H_6,disilane, etc.)	H_2S_5	pentasulfane
GeH_4	germane	SnH_4	stannane

B4 NAMES FOR IONS AND RADICALS

▶ B4.1 Cations

Names of monoatomic cations are the same as the names of the elements. Oxidation numbers are designated by the use of the Stock system.

EXAMPLES: Cu^+, copper(I) ion; Cu^{2+}, copper(II) ion.
   Polyatomic cations formed from radicals which have special names use those names without change. Complex cations are discussed later.

EXAMPLES:

NO^+	nitrosyl cation
NO_2^+	nitryl cation
UO_2^{2+}	uranyl(VI) ion or uranyl(2+) ion

Polyatomic cations derived by the addition of protons to monoatomic anions are named by adding the ending -onium to the root of the name of the anion.

EXAMPLES: Phosphonium, arsonium, stibonium, oxonium, sulfonium, selenonium, telluronium, and iodonium ions.

EXCEPTIONS: NH_4^+, ammonium (or azanium) ion; $HONH_3^+$, hydroxylammonium ion; $NH_2NH_3^+$, hydrazinium (or diazanium) ion; $C_6H_5NH_3^+$, anilinium ion; $C_5H_5NH^+$, pyridinium; $HO_2CCH_2NH_3^+$, glycinium, etc.

The H_3O^+ ion is the oxonium ion. Hydrogen ion may be used for the indefinitely solvated proton or when the hydration is of no particular importance to the matter under consideration.

▶B4.2 Anions

Monoatomic anions are named by adding the ending -ide to the stem of the name of the element.

EXAMPLES: H^-, hydride ion; F^-, fluoride ion; Cl^-, chloride ion; Te^{2-}, telluride ion; N^{3-}, nitride ion; Sb^{3-}, antimonide ion; and so on.

Certain polyatomic anions have names ending in -ide. These are (systematic names are in brackets):

OH^- hydroxide ion	CN^- cyanide ion	N_3^- azide ion [trinitride(1−)]
O_2^{2-} peroxide ion [dioxide(2−)]	C_2^{2-} acetylide ion [dicarbide(2−)]	NH^{2-} imide ion
O_2^- hyperoxide ion or superoxide[2] ion [dioxide(1−)]	I_3^- triiodide ion	NH_2^- amide or azanide ion
O_3^- ozonide ion [trioxide(1−)]	HF_2^- hydrogendifluoride ion	$NHOH^-$ hydroxylamide ion
S_2^{2-} disulfide ion [disulfide(2−)]		$N_2H_3^-$ hydrazide ion

Ions such as HS^- and HO_2^- are named hydrogensulfide ion and hydrogenperoxide ion, respectively. The names of other polyatomic anions consist of the name of the central atom with the termination -ate in accordance with the naming of complex anions.

EXAMPLE: $[Sb(OH)_6]^-$ hexahydroxoantimonate(V) ion.

---

[2] The common name superoxide ion is used in this text.

Oxoacid Anions

The systematic names for oxoanions can always be used. Certain anions have names using prefixes (hypo-, per-, etc.) that are well-established. These are in accord with the names of the corresponding acids (see page B-7). The termination -ite has been used to denote a lower oxidation state and may be retained in trivial names in the following cases:

NO_2^- nitrite	SO_3^{2-} sulfite	ClO_2^- chlorite
AsO_3^{3-} arsenite	$S_2O_4^{2-}$ dithionite	ClO^- hypochlorite
	SeO_3^{2-} selenite	(and correspondingly for other halogens)

Other anions that have used the -ite ending—for example, antimonite, should be named according to the general rule—that is, trioxoantimonate($1-$) for SbO_3^- and tetraoxoantimonate($3-$) for SbO_4^{3-}.

▶ B4.3 Radicals

The names ending in -yl of the following radicals are approved:

HO hydroxyl	SO sulfinyl (thionyl)	ClO chloroxyl
CO carbonyl	SO_2 sulfonyl (sulfuryl)	ClO_2 chloryl
NO nitrosyl	S_2O_5 disulfuryl	ClO_3 perchloryl
NO_2 nitryl	SeO seleninyl	(and similarly for other halogens)
PO phosphoryl	SeO_2 selenonyl	
	CrO_2 chromyl	NpO_2 neptunyl
	UO_2 uranyl	PuO_2 plutonyl, etc.

The prefixes thio-, seleno-, and so on, are used for other chalcogens in place of oxygen.

EXAMPLES: PS, thiophosphoryl; CSe, selenocarbonyl, and so on. The oxidation number of the characteristic element is denoted by the Stock system, or the charge on the ion is indicated by the Ewens-Bassett system. UO_2^{2+} may be uranyl(VI) or uranyl($2+$) and UO_2^+ may be uranyl(V) or uranyl($1+$).

Radicals are treated as the positive part of a compound.

EXAMPLES: $COCl_2$, carbonyl dichloride; NOS, nitrosyl sulfide; POCl, phosphoryl chloride; $POCl_3$, phosphoryl trichloride, IO_2F, iodyl fluoride, and so on.

B5 ACIDS

Acids giving rise to -ide anions are named as binary and pseudobinary compounds of hydrogen—for example, hydrogen chloride, hydrogen cyanide, hydrogen azide, and so on.

Other acids may be named as pseudobinary compounds of hydrogen—for example, H_2SO_4, hydrogen sulfate; $H_4[Fe(CN)_6]$, hydrogen hexacyanoferrate(II).

► B5.1 Oxoacids

Most of the common acids are oxoacids commonly named using the -ic and -ous endings in place of the anion endings -ate and -ite, respectively. The acids using the -ous ending should be restricted to those listed above for the -ite anions (see page B-6).

The prefix hypo- is used to denote a lower oxidation state and the prefix per- is used to denote a higher oxidation state. These prefixes should be limited to the following cases:

$H_4P_2O_6$ hypophosphoric acid; HClO hypochlorous acid (also HBrO and HIO); and $HClO_4$ perchloric acid (also $HBrO_4$ and HIO_4)

$(HO)_2PHO$ is phosphonic acid rather than phosphorous acid in order to reserve the name phosphite for organic derivatives of $P(OH)_3$. Phosphinic acid is the name for $HOPH_2O$, rather than hypophosphorous acid.

The prefixes ortho- and meta- may be used to distinguish acids differing in the "content of water" in the following cases:

H_3BO_3	orthoboric acid	$(HBO_2)_n$	metaboric acids
H_4SiO_4	orthosilicic acid	$(H_2SiO_3)_n$	metasilicic acids
H_3PO_4	orthophosphoric acid	$(HPO_3)_n$	metaphosphoric acids
H_5IO_6	orthoperiodic acid		
H_6TeO_6	orthotelluric acid		

Acids obtained by removing water from H_5IO_6 or H_6TeO_6 and other acids not covered by specific names should be given systematic names—for example, H_2ReO_4, tetraoxorhenic(VI) acid; H_2NO_2, dioxonitric(II) acid.

The prefix di- should be used instead of the prefix pryo- for $H_2S_2O_7$, disulfuric acid; $H_2S_2O_5$, disulfurous acid; and $H_4P_2O_7$, diphosphoric acid.

► B5.2 Peroxoacids

The prefix peroxo- indicates the substitution of —O— by —O—O— (see coordination compounds for the use of μ for bridging groups).

EXAMPLES:

HNO_4	dioxoperoxonitric acid or peroxonitric acid
H_3PO_5	trioxoperoxophosphoric acid or peroxophosphoric acid
$H_4P_2O_8$	μ-peroxo-bis-trioxophosphoric acid or peroxodiphosphoric acid
H_2SO_5	trioxoperoxosulfuric acid or peroxosulfuric acid
$H_2S_2O_8$	μ-peroxo-bis-trioxosulfuric acid or peroxodisulfuric acid

▶ B5.3 *Thioacids*

The prefix thio- indicates the replacement of oxygen by sulfur. The prefixes seleno- and telluro- may be used in a similar manner.

EXAMPLES:

$H_2S_2O_2$	hydrogen dioxothiosulfate(2−)
$H_2S_2O_3$	hydrogen trioxothiosulfate(2−) or thiosulfuric acid
HSCN	thiocyanic acid
H_3PO_3S	hydrogen trioxothiophosphate(3−)
$H_3PO_2S_2$	hydrogen dioxodithiophosphate(3−)
H_2CS_3	hydrogen trithiocarbonate

Acids containing ligands other than O and S are generally named as complexes.

EXAMPLES:

$H[PF_6]$	hydrogen hexafluorophosphate(1−)
$H[AuCl_4]$	hydrogen tetrachloroaurate(1−)

B6 SALTS AND SALTLIKE COMPOUNDS

Simple salts are named as binary compounds using the names as prescribed for ions.

▶ B6.1 *Acid Salts*

Salts that contain acidic hydrogens are named by treating hydrogen as a positive constituent. No space is used if H is known to be bound.

EXAMPLES:

$NaHCO_3$	sodium hydrogencarbonate
NaH_2PO_4	sodium dihydrogenphosphate

▶ B6.2 *Double Salts, etc.*

Cations

Cations other than hydrogen are cited in alphabetical order, which may be different in formulas and names.

EXAMPLES:

$KMgF_3$	magnesium potassium fluoride (note orders)
$NaTl(NO_3)_2$	sodium thallium(I) nitrate
$KNaCO_3$	potassium sodium carbonate
$MgNH_4PO_4 \cdot 6H_2O$	ammonium magnesium phosphate hexahydrate [or water(1/6)]
$Na(UO_2)_3[Zn(H_2O)_6](C_2H_3O_2)_9$	hexaaquazinc sodium triuranyl(VI) nonaacetate
$NaNH_4HPO_4 \cdot 4H_2O$	ammonium sodium hydrogenphosphate tetrahydrate
$AlK(SO_4)_2 \cdot 12H_2O$	aluminum potassium sulfate water(1/12)

Anions

Anions are to be cited in alphabetical order which may be different in names and formulas.

EXAMPLES:

$NaCl \cdot NaF \cdot 2Na_2SO_4$ or $Na_6ClF(SO_4)_2$	hexasodium chloride fluoride bis(sulfate)[3]
$Ca_5F(PO_4)_3$	pentacalcium fluoride tris(phosphate)[3]

Basic salts should be treated as double salts, not as oxo or hydroxo salts.

EXAMPLES:

$MgCl(OH)$	magnesium chloride hydroxide
$BiClO$	bismuth chloride oxide (not bismuth oxychloride)
$ZrCl_2O \cdot 8H_2O$	zirconium dichloride oxide octahydrate
$CuCl_2 \cdot 3Cu(OH)_2$ or $Cu_2Cl(OH)_3$	dicopper chloride trihydroxide

B7 COORDINATION COMPOUNDS

The symbol for the central atom is placed first in the formula of coordination compounds followed by anionic ligands in the alphabetical order of the symbols and then neutral ligands in alphabetical order. The formula for the complex molecule or ion is enclosed in square brackets [ ]. In names the central atom is placed after the ligands. The ligands are listed in alphabetical order regardless of charge and regardless of the number of each. Thus diammine is listed under "a" whereas dimethylamine is listed under "d."

The oxidation number of the central atom is indicated by the Stock notation (Roman numbers). Alternatively, the proportion of constituents may be given by means of stoi-

---

[3] The prefixes bis, tris, tetrakis, etc., are used for anions to avoid confusion with disulfate, etc., and for complex expressions to avoid ambiguity.

chiometric prefixes, or the charge on the entire ion can be designated by the Ewens-Bassett number (Arabic numbers). Formulas and names may be supplemented by italicized prefixes *cis, trans, fac (facial), mer (meridional),* and so on. Names of complex anions end in -ate. Complex cations and neutral molecules are given no distinguishing ending.

▶ B7.1 Names of Ligands

The names for anionic ligands end in -o (-ido, -ito, and -ato commonly).

EXAMPLES:

$Li[AlH_4]$	lithium tetrahydridoaluminate
$K_2[OsCl_5N]$	potassium pentachloronitridoosmate(VI) or
	potassium pentachloronitridoosmate(2−)
$Na_3[Ag(S_2O_3)_2]$	sodium bis(thiosulfato)argentate(I) or (3−)
$[Ni(C_4H_7O_2N_2)_2]$	bis(2,3-butanedione dioximato)nickel(II) or omit (II)
$[Cu(C_5H_7O_2)_2]$	bis(2,4-pentanedionato)copper(II) or omit (II)
$K_2[Cr(CN)_2O_2(O_2)NH_3]$	potassium amminedicyanodioxoperoxochromate(VI) or (2−)

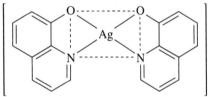

bis(8-quinolinolato)silver(II) or omit (II)

Some common anionic ligands (systematic names):

H^-	hydrido[4]		HS^-	mercapto (hydrogensulfido)
F^-	fluoro		S^{2-}	thio (sulfido)
Cl^-	chloro		CN^-	cyano
Br^-	bromo		CH_3O^-	methoxo or methanolato
I^-	iodo		$C_6H_5^-$	phenyl
O^{2-}	oxo (oxido)		$C_5H_5^-$	cyclopentadienyl
OH^-	hydroxo (hydroxido)			Other hydrocarbonanions are also given
O_2^{2-}	peroxo [dioxido(2−)]			radical names without the -o ending.

Neutral and cationic ligands are given no special endings. Water and ammonia are called aqua and ammine, respectively, in complexes. Groups such as NO and CO are named as radicals and treated as neutral ligands.

---

[4] Hydro is used in boron nomenclature.

EXAMPLES:

$Ba[BrF_4]_2$	barium tetrafluorobromate(III) or $(1-)$
$K[CrF_4O]$	potassium tetrafluorooxochromate(V) or $(1-)$
$Na[BH(OCH_3)_3]$	sodium hydrotrimethoxoborate(III) or $(1-)$
$[CuCl_2(CH_3NH_2)_2]$	dichlorobis(methylamine)copper(II) or omit (II)
$[Pt(py)_4][PtCl_4]$	tetrakis(pyridine)platinum(II) tetrachloroplatinate(II) or $(2+)$ and $(2-)$
$[Co(en)_3]_2(SO_4)_3$	tris(ethylenediamine)cobalt(III) sulfate or $(3+)$
$K[PtCl_3(C_2H_4)]$	potassium trichloro(ethylene)platinate(II) or $(1-)$
$[Al(OH)(H_2O)_5]^{2+}$	pentaaquahydroxoaluminum(III) or $(2+)$ ion
$K_3[Fe(CN)_5NO]$	potassium pentacyanonitrosylferrate(II) or $(3-)$
$[CoCl_3(NH_3)_2\{(CH_3)_2NH\}]$	diamminetrichloro(dimethylamine)cobalt(III)
$K[SbCl_5C_6H_5]$	potassium pentachloro(phenyl)antimonate(V) or $(1-)$
$[Fe(C_5H_5)_2]$	bis(cyclopentadienyl)iron(II) or omit (II)
$[Cr(C_6H_6)_2]$	bis(benzene)chromium(0) or omit (0)
$[Ru(NH_3)_5N_2]Cl_2$	pentaammine(dinitrogen)ruthenium(II) chloride or $(2+)$

▶ **B7.2   *Alternative Modes of Linkage***

A Ligand that may be attached through different atoms, for example SCN^-, may be distinguished as follows:

M—SCN thiocyanato-*S*   or M—NCS thiocyanato-*N*

EXAMPLES:

$K_2\left[Ni\left(\begin{matrix} S-C=O \\ \| \\ S-C=O \end{matrix} \right)_2 \right]$	potassium bis(dithiooxalato-*S,S′*)nickelate(II) or $(2-)$
$[Co(NO_2)_3(NH_3)_3]$	triamminetri(nitrito-*N*)cobalt(III)
$[Co(ONO)(NH_3)_5]SO_4$	pentaammine(nitrito-*O*)cobalt(III) sulfate or $(2+)$
$[Co(NCS)(NH_3)_5]Cl_2$	pentaammine(thiocyanato-*N*)cobalt(III) chloride

The Kappa Convention

The ligating atom of a ligand can be designated also by the Greek kappa, κ, preceding the italicized element symbol as thiocyanato-κS for SCN^- bonded through S. The κ convention is particularly helpful for polydentate ligands with not all donor atoms ligating and where the ligating atoms are not obvious. A right superscript to κ indicates the number of identically bound ligating atoms of a ligand.

EXAMPLES:

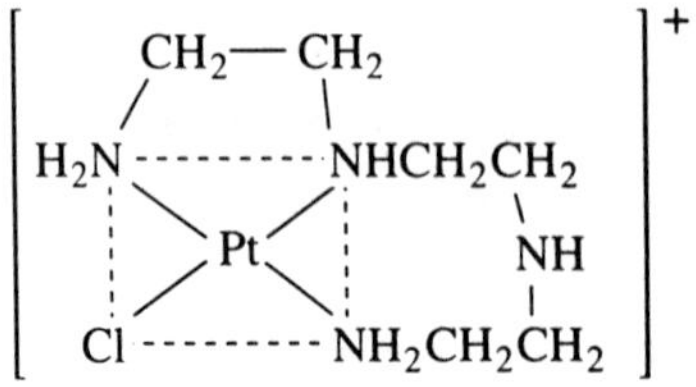

[*N,N'*-bis(2-amino-κN-ethyl)-1,2-
ethanediamine-κN]chloroplatinum(II)

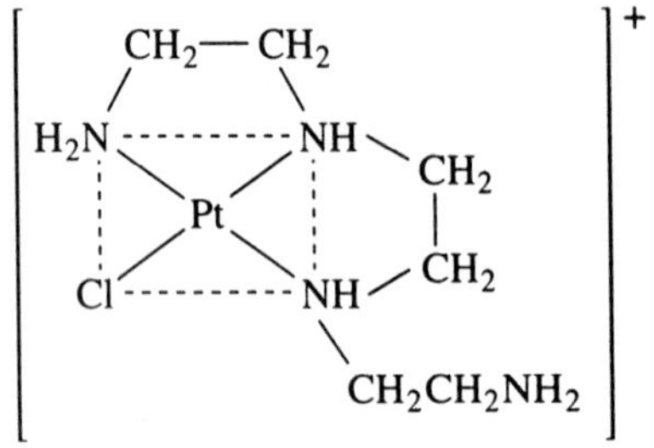

[*N*-(2-amino-κN-ethyl)-*N'*-(2-aminoethyl)-
1,2-ethanediamine-$\kappa^2 N,N'$]chloroplatinum(II)

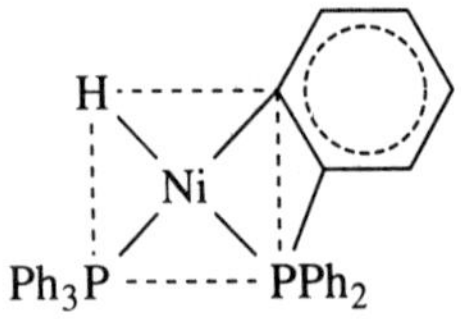

[2-(diphenylphosphino-κP)-phenyl-κC^1]hydrido-
(triphenylphosphine-κP)nickel(II)

► B7.3 Di- and Polynuclear Compounds

Bridging groups are indicated by adding the Greek letter μ immediately before the names of the groups. Two or more bridging groups of the same kind are indicated by di-μ, and so on. If a bridging group bridges more than two metals, use μ_3, μ_4, and so on. Bridging groups are listed followed by other groups in alphabetical order unless the symmetry of the molecule permits a simpler name (first example). If the same ligand is present in bridging and nonbridging roles, it is cited first as a bridging ligand.

EXAMPLES:

$[(NH_3)_5Cr-OH-Cr(NH_3)_5]^{5+}$	μ-hydroxo-bis(pentaamminechromium)(5+) ion or (III)
$[(CO)_3Fe(CO)_3Fe(CO)_3]$	tri-μ-carbonyl-bis(tricarbonyliron)
$[\{Au(CN)(C_3H_7)_2\}_4]$	*cyclo*-tetra-μ-cyano-tetrakis(dipropylgold)
$[Be_4O(CH_3COO)_6]$	hexa-μ-acetato(O,O')-μ_4-oxo-tetraberyllium(II) or omit (II)
$[(CH_3Hg)_4S]^{2+}$	μ_4-thio-tetrakis[methylmercury(II)] ion or (2+)

Extended structures may be indicated by the prefix *catena*-μ, for example, $CsCuCl_3$ contains the anion

$$
\cdots \text{Cl} \underset{\underset{\text{Cl}}{|}}{\overset{\overset{\text{Cl}}{|}}{\text{Cu}}} \text{Cl} \underset{\underset{\text{Cl}}{|}}{\overset{\overset{\text{Cl}}{|}}{\text{Cu}}} \text{Cl} \underset{\underset{\text{Cl}}{|}}{\overset{\overset{\text{Cl}}{|}}{\text{Cu}}} \text{Cl} \underset{\underset{\text{Cl}}{|}}{\overset{\overset{\text{Cl}}{|}}{\text{Cu}}} \cdots
$$

The compound may be named cesium *catena*-μ-chloro-dichlorocuprate(II). If the structure were in doubt, however, the substance would be called cesium copper(II) chloride (as a double salt).

Metal-metal bonding may be indicated by italicized atomic symbols of the appropriate metal atoms, separated by a long dash and enclosed in parentheses.

EXAMPLES:

$[(CO)_5Mn-Mn(CO)_5]$ bis(pentacarbonylmanganese)(*Mn—Mn*)
      or decarbonyldimanganese(*Mn—Mn*)

$[Br_4Re-ReBr_4]^{2-}$ bis(tetrabromorhenate)(*Re—Re*)(2−)

$Cs_3[(ReCl_4)_3]$ cesium *cyclo*-tris(tetrachlororhenate)(3*Re—Re*)(3−)
      (or *triangulo*, see Section B7.4)

Unsymmetrical Dinuclear Entities

Unsymmetrical dinuclear species may have different central atoms and/or different ligands or different numbers of ligands on the central atoms. They are named by citing the ligands and bridging groups in a *single* alphabetical ordered series, followed by the names of the central atoms in alphabetical order. The central atoms are numbered 1 and 2 to provide numerical locants for the ligands and the points of attachment of bridging groups.

If the central atoms are different, central 1 is the more electropositive element in the sequence going down group 17, then groups 16, 15, 14, and so on. It is the approximate sequence of electronegativities by groups. If the central atoms are identical, central atom 1 is designated with priority[5] decreasing from criteria 1 to 4:

1. Atom with higher oxidation number
2. Atom with higher coordination number
3. Atom with greater variety of bonding atoms
4. Atom to which is attached the side group with the name first in alphabetical order.

Each criterion is applied only if the previous criterion fails to make a distinction. The kappa convention is used to indicate the ligating atom(s) and its (their) role. A right superscript to κ indicates the number of equivalent ligating atoms bonded to a central atom. The numerical locant (1 or 2) of the central atom is placed before κ.

[5] B. P. Block, W. H. Powell, and W. C. Fernelius, *Inorganic Chemical Nomenclature*, American Chemical Society, Washington, D.C., 1990, pp. 50–54.

EXAMPLES:

$[(CO)_4Co—Re(CO)_5]$ nonacarbonyl-$1\kappa^5C$, $2\kappa^4C$-2-cobalt-1-rhenium($Co—Re$)
       2   1

$[(Ph_3As)Au—Mn(CO)_5]$ pentacarbonyl-$1\kappa^5C$-(triphenylarsine-$2\kappa As$)-2-gold-1-manganese($Au—Mn$)
       2   1

$[(H_3N)_5Cr(OH)Cr(NH_3)_4(NH_2Me)]Cl_5$ nonaammine-$1\kappa^5N$,$2\kappa^4N$-μ-hydroxo-
       1   2   (methylamine)$2\kappa N$-dichromium(5+) chloride

$[(Et_2PhP)_3(Cl)_2Re\text{-}N\text{-}Pt(PEt_3)(Cl)_2]$ tetrachloro-$1\kappa^2$,$2\kappa^2$-tris(diethylphenylphosphine-$1\kappa P$)-
       1   2 μ-nitrido-(triethylphosphine-$2\kappa P$)-2-platinum-1-rhenium

▶ B7.4 Metal Cluster Compounds

The geometrical shape of the cluster is designated by *triangulo, quadro, tetrahedro, octahedro,* and so on.

EXAMPLES:

$Os_3(CO)_{12}$ dodecacarbonyl-*triangulo*-triosmium($3Os—Os$)
               or *cyclo*-tris(tetracarbonylosmium)($3Os—Os$)

$Cs_3[Re_3Cl_{12}]$ cesium dodecachloro-*triangulo*-trirhenate($3Re—Re$)(3−)
               or cesium *cyclo*-tris(tetrachlororhenate)($3Re—Re$)(3−)

B_4Cl_4 tetrachloro-*tetrahedro*-tetraboron($6B—B$)
               or *tetrahedro*-tetrakis(chloroboron)

$[Nb_6Cl_{12}]^{2+}$ dodeca-μ-chloro-*octahedro*-hexaniobium(2+) ion

$[Mo_6Cl_8]^{4+}$ octa-μ_3-chloro-*octahedro*-hexamolybdenum(4+) ion

$Cu_4I_4(PEt_3)_4$ tetra-μ_3-iodo-tetrakis(triethylphosphine)-*tetrahedro*-tetracopper

Some aspects of the designation of geometrical and optical isomers are covered in Chapter 9, and the naming of boron compounds is discussed briefly in Chapter 17. For more details and for rules dealing with isopolyanions, heteropolyanions, and nonstoichiometric phases, and for tables of names of ions and radicals, see the original 283-page report.

Table B.1   The elements

Name	Symbol	Atomic number	Name	Symbol	Atomic number
Actinium	Ac	89	Astatine	At	85
Aluminum	Al	13	Barium	Ba	56
Americium	Am	95	Berkelium	Bk	97
Antimony	Sb	51	Beryllium	Be	4
Argon	Ar	18	Bismuth	Bi	83
Arsenic	As	33	Boron	B	5

(Continued)

Table B.1 The elements *(Continued)*

Name	Symbol	Atomic number	Name	Symbol	Atomic number
Bromine	Br	35	Neodymium	Nd	60
Cadmium	Cd	48	Neon	Ne	10
Calcium	Ca	20	Neptunium	Np	93
Californium	Cf	98	Nickel	Ni	28
Carbon	C	6	Niobium	Nb	41
Cerium	Ce	58	Nitrogen	N	7
Cesium	Cs	55	Nobelium	No	102
Chlorine	Cl	17	Osmium	Os	76
Chromium	Cr	24	Oxygen	O	8
Cobalt	Co	27	Palladium	Pd	46
Copper	Cu	29	Phosphorus	P	15
Curium	Cm	96	Platinum	Pt	78
Dysprosium	Dy	66	Plutonium	Pu	94
Einsteinium	Es	99	Polonium	Po	84
Erbium	Er	68	Potassium	K	19
Europium	Eu	63	Praseodymium	Pr	59
Fermium	Fm	100	Promethium	Pm	61
Fluorine	F	9	Protactinium	Pa	91
Francium	Fr	87	Radium	Ra	88
Gadolinium	Gd	64	Radon	Rn	86
Gallium	Ga	31	Rhenium	Re	75
Germanium	Ge	32	Rhodium	Rh	45
Gold	Au	79	Rubidium	Rb	37
Hafnium	Hf	72	Ruthenium	Ru	44
Helium	He	2	Samarium	Sm	62
Holmium	Ho	67	Scandium	Sc	21
Hydrogen	H	1	Selenium	Se	34
Indium	In	49	Silicon	Si	14
Iodine	I	53	Silver	Ag	47
Iridium	Ir	77	Sodium	Na	11
Iron	Fe	26	Strontium	Sr	38
Krypton	Kr	36	Sulfur	S	16
Lanthanum	La	57	Tantalum	Ta	73
Lawrencium	Lr	103	Technetium	Tc	43
Lead	Pb	82	Tellurium	Te	52
Lithium	Li	3	Terbium	Tb	65
Lutetium	Lu	71	Thallium	Tl	81
Magnesium	Mg	12	Thorium	Th	90
Manganese	Mn	25	Thulium	Tm	69
Mendelevium	Md	101	Tin	Sn	50
Mercury	Hg	80	Titanium	Ti	22
Molybdenum	Mo	42	Tungsten	W	74

(continued)

Table B.1   The elements *(Continued)*

Name	Symbol	Atomic number	Name	Symbol	Atomic number
Uranium	U	92	Yttrium	Y	39
Vanadium	V	23	Zinc	Zn	30
Xenon	Xe	54	Zirconium	Zr	40
Ytterbium	Yb	70			

APPENDIX C

Character Tables

· ·

THE NONAXIAL GROUPS

C_1	E
A	1

C_s	E	σ_h		
A'	1	1	x, y, R_z	x^2, y^2, z^2, xy
A''	1	-1	z, R_x, R_y	yz, xz

C_i	E	i		
A_g	1	1	R_x, R_y, R_z	$x^2, y^2, z^2, xy, xz, yz$
A_u	1	-1	x, y, z	

THE AXIAL GROUPS

▶ *The C_n Groups*

C_2	E	C_2		
A	1	1	z, R_z	x^2, y^2, z^2, xy
B	1	-1	x, y, R_x, R_y	yz, xz

C_3	E	C_3	C_3^2	$\varepsilon = \exp(2\pi i/3)$	
A	1	1	1	z, R_z	$x^2 + y^2, z^2$
E	$\begin{Bmatrix} 1 & \varepsilon & \varepsilon^* \\ 1 & \varepsilon^* & \varepsilon \end{Bmatrix}$			$(x, y), (R_x, R_y)$	$(x^2 - y^2, xy), (yz, xz)$

C_4	E	C_4	C_2	C_4^3		
A	1	1	1	1	z, R_z	$x^2 + y^2, z^2$
B	1	-1	1	-1		$x^2 - y^2, xy$
E	$\left\{\begin{matrix}1\\1\end{matrix}\right.$	$\begin{matrix}i\\-i\end{matrix}$	$\begin{matrix}-1\\-1\end{matrix}$	$\left.\begin{matrix}-i\\i\end{matrix}\right\}$	$(x, y), (R_x, R_y)$	(xz, yz)

C_5	E	C_5	C_5^2	C_5^3	C_5^4	$\varepsilon = \exp(2\pi i/5)$	
A	1	1	1	1	1	z, R_z	$x^2 + y^2, z^2$
E_1	$\left\{\begin{matrix}1\\1\end{matrix}\right.$	$\begin{matrix}\varepsilon\\\varepsilon^*\end{matrix}$	$\begin{matrix}\varepsilon^2\\\varepsilon^{2*}\end{matrix}$	$\begin{matrix}\varepsilon^{2*}\\\varepsilon^2\end{matrix}$	$\left.\begin{matrix}\varepsilon^*\\\varepsilon\end{matrix}\right\}$	$(x, y), (R_x, R_y)$	(yz, xz)
E_2	$\left\{\begin{matrix}1\\1\end{matrix}\right.$	$\begin{matrix}\varepsilon^2\\\varepsilon^{2*}\end{matrix}$	$\begin{matrix}\varepsilon^*\\\varepsilon\end{matrix}$	$\begin{matrix}\varepsilon\\\varepsilon^*\end{matrix}$	$\left.\begin{matrix}\varepsilon^{2*}\\\varepsilon^2\end{matrix}\right\}$		$(x^2 - y^2, xy)$

C_6	E	C_6	C_3	C_2	C_3^2	C_6^5	$\varepsilon = \exp(2\pi i/6)$	
A	1	1	1	1	1	1	z, R_z	$x^2 + y^2, z^2$
B	1	-1	1	-1	1	-1		
E_1	$\left\{\begin{matrix}1\\1\end{matrix}\right.$	$\begin{matrix}\varepsilon\\\varepsilon^*\end{matrix}$	$\begin{matrix}-\varepsilon^*\\-\varepsilon\end{matrix}$	$\begin{matrix}-1\\-1\end{matrix}$	$\begin{matrix}-\varepsilon\\-\varepsilon^*\end{matrix}$	$\left.\begin{matrix}\varepsilon^*\\\varepsilon\end{matrix}\right\}$	$(x, y),$ (R_x, R_y)	(xz, yz)
E_2	$\left\{\begin{matrix}1\\1\end{matrix}\right.$	$\begin{matrix}-\varepsilon^*\\-\varepsilon\end{matrix}$	$\begin{matrix}-\varepsilon\\-\varepsilon^*\end{matrix}$	$\begin{matrix}1\\1\end{matrix}$	$\begin{matrix}-\varepsilon^*\\-\varepsilon\end{matrix}$	$\left.\begin{matrix}-\varepsilon\\-\varepsilon^*\end{matrix}\right\}$		$(x^2 - y^2, xy)$

C_7	E	C_7	C_7^2	C_7^3	C_7^4	C_7^5	C_7^6	$\varepsilon = \exp(2\pi i/7)$	
A	1	1	1	1	1	1	1	z, R_z	$x^2 + y^2, z^2$
E_1	$\left\{\begin{matrix}1\\1\end{matrix}\right.$	$\begin{matrix}\varepsilon\\\varepsilon^*\end{matrix}$	$\begin{matrix}\varepsilon^2\\\varepsilon^{2*}\end{matrix}$	$\begin{matrix}\varepsilon^3\\\varepsilon^{3*}\end{matrix}$	$\begin{matrix}\varepsilon^{3*}\\\varepsilon^3\end{matrix}$	$\begin{matrix}\varepsilon^{2*}\\\varepsilon^2\end{matrix}$	$\left.\begin{matrix}\varepsilon^*\\\varepsilon\end{matrix}\right\}$	$(x, y),$ (R_x, R_y)	(xz, yz)
E_2	$\left\{\begin{matrix}1\\1\end{matrix}\right.$	$\begin{matrix}\varepsilon^2\\\varepsilon^{2*}\end{matrix}$	$\begin{matrix}\varepsilon^{3*}\\\varepsilon^3\end{matrix}$	$\begin{matrix}\varepsilon^*\\\varepsilon\end{matrix}$	$\begin{matrix}\varepsilon\\\varepsilon^*\end{matrix}$	$\begin{matrix}\varepsilon^3\\\varepsilon^{3*}\end{matrix}$	$\left.\begin{matrix}\varepsilon^{2*}\\\varepsilon^2\end{matrix}\right\}$		$(x^2 - y^2, xy)$
E_3	$\left\{\begin{matrix}1\\1\end{matrix}\right.$	$\begin{matrix}\varepsilon^3\\\varepsilon^{3*}\end{matrix}$	$\begin{matrix}\varepsilon^*\\\varepsilon\end{matrix}$	$\begin{matrix}\varepsilon^2\\\varepsilon^{2*}\end{matrix}$	$\begin{matrix}\varepsilon^{2*}\\\varepsilon^2\end{matrix}$	$\begin{matrix}\varepsilon\\\varepsilon^*\end{matrix}$	$\left.\begin{matrix}\varepsilon^{3*}\\\varepsilon^3\end{matrix}\right\}$		

C_8	E	C_8	C_4	C_2	C_4^3	C_8^3	C_8^5	C_8^7	$\varepsilon = \exp(2\pi i/8)$	
A	1	1	1	1	1	1	1	1	z, R_z	$x^2 + y^2, z^2$
B	1	-1	1	1	1	-1	-1	-1		
E_1	$\left\{\begin{matrix}1\\1\end{matrix}\right.$	$\begin{matrix}\varepsilon\\\varepsilon^*\end{matrix}$	$\begin{matrix}i\\-i\end{matrix}$	$\begin{matrix}-1\\-1\end{matrix}$	$\begin{matrix}-i\\i\end{matrix}$	$\begin{matrix}-\varepsilon^*\\-\varepsilon\end{matrix}$	$\begin{matrix}-\varepsilon\\-\varepsilon^*\end{matrix}$	$\left.\begin{matrix}\varepsilon^*\\\varepsilon\end{matrix}\right\}$	$(x, y),$ (R_X, R_y)	(xz, yz)
E_2	$\left\{\begin{matrix}1\\1\end{matrix}\right.$	$\begin{matrix}i\\-i\end{matrix}$	$\begin{matrix}-1\\-1\end{matrix}$	$\begin{matrix}1\\1\end{matrix}$	$\begin{matrix}-1\\-1\end{matrix}$	$\begin{matrix}-i\\i\end{matrix}$	$\begin{matrix}i\\-i\end{matrix}$	$\left.\begin{matrix}-i\\i\end{matrix}\right\}$		$(x^2 - y^2, xy)$
E_3	$\left\{\begin{matrix}1\\1\end{matrix}\right.$	$\begin{matrix}-\varepsilon\\-\varepsilon^*\end{matrix}$	$\begin{matrix}i\\-i\end{matrix}$	$\begin{matrix}-1\\-1\end{matrix}$	$\begin{matrix}-i\\i\end{matrix}$	$\begin{matrix}\varepsilon^*\\\varepsilon\end{matrix}$	$\begin{matrix}\varepsilon\\\varepsilon^*\end{matrix}$	$\left.\begin{matrix}-\varepsilon^*\\-\varepsilon\end{matrix}\right\}$		

▶ *The S_n Groups*

S_4	E	S_4	C_2	S_4^3		
A	1	1	1	1	R_z	$x^2 + y^2, z^2$
B	1	−1	1	−1	z	$x^2 − y^2, xy$
E	$\begin{Bmatrix}1 & i & −1 & −i \\ 1 & −i & −1 & i\end{Bmatrix}$				$(x, y), (R_x, R_y)$	(xz, yz)

S_6	E	C_3	C_3^2	i	S_6^5	S_6		$\varepsilon = \exp(2\pi i/3)$
A_g	1	1	1	1	1	1	R_z	$x^2 + y^2, z^2$
E_g	$\begin{Bmatrix}1 & \varepsilon & \varepsilon^* & 1 & \varepsilon & \varepsilon^* \\ 1 & \varepsilon^* & \varepsilon & 1 & \varepsilon^* & \varepsilon\end{Bmatrix}$						(R_x, R_y)	$(x^2 − y^2, xy),$ (xy, yz)
A_u	1	1	1	−1	−1	−1	z	
E_u	$\begin{Bmatrix}1 & \varepsilon & \varepsilon^* & −1 & −\varepsilon & −\varepsilon^* \\ 1 & \varepsilon^* & \varepsilon & −1 & −\varepsilon^* & −\varepsilon\end{Bmatrix}$						(x, y)	

S_8	E	S_8	C_4	S_8^3	C_2	S_8^5	C_4^3	S_8^7		$\varepsilon = \exp(2\pi i/8)$
A	1	1	1	1	1	1	1	1	R_z	$x^2 + y^2, z^2$
B	1	−1	1	−1	1	−1	1	−1	z	
E_1	$\begin{Bmatrix}1 & \varepsilon & i & −\varepsilon^* & −1 & −\varepsilon & −i & \varepsilon^* \\ 1 & \varepsilon^* & −i & −\varepsilon & −1 & −\varepsilon^* & i & \varepsilon\end{Bmatrix}$								$(x, y),$ (R_x, R_y)	
E_2	$\begin{Bmatrix}1 & i & −1 & −i & 1 & i & −1 & −i \\ 1 & −i & −1 & i & 1 & −i & −1 & i\end{Bmatrix}$									$(x^2 − y^2, xy)$
E_3	$\begin{Bmatrix}1 & −\varepsilon^* & −i & \varepsilon & −1 & \varepsilon^* & i & −\varepsilon \\ 1 & −\varepsilon & i & \varepsilon^* & −1 & \varepsilon & −i & −\varepsilon^*\end{Bmatrix}$									(xz, yz)

▶ *The C_{nv} Groups*

C_{2v}	E	C_2	$\sigma_v(xz)$	$\sigma_v'v(yz)$		
A_1	1	1	1	1	z	x^2, y^2, z^2
A_2	1	1	−1	−1	R_z	xy
B_1	1	−1	1	−1	x, R_y	xz
B_2	1	−1	−1	1	y, R_x	yz

C_{3v}	E	$2C_3$	$3\sigma_v$		
A_1	1	1	1	z	$x^2 + y^2, z^2$
A_2	1	1	−1	R_z	
E	2	−1	0	$(x, y), (R_x, R_y)$	$(x^2 − y^2, xy), (xz, yz)$

C_{4v}	E	$2C_4$	C_2	$2\sigma_v$	$2\sigma_d$		
A_1	1	1	1	1	1	z	$x^2 + y^2, z^2$
A_2	1	1	1	-1	-1	R_z	
B_1	1	-1	1	1	-1		$x^2 - y^2$
B_2	1	-1	1	-1	1		xy
E	2	0	-2	0	0	$(x, y), (R_x, R_y)$	(xz, yz)

C_{5v}	E	$2C_5$	$2C_5^2$	$5\sigma_v$		
A_1	1	1	1	1	z	$x^2 + y^2, z^2$
A_2	1	1	1	-1	R_z	
E_1	2	$2\cos 72°$	$2\cos 144°$	0	$(x, y), (R_x, R_y)$	(xz, yz)
E_2	2	$2\cos 144°$	$2\cos 72°$	0		$(x^2 - y^2, xy)$

C_{6v}	E	$2C_6$	$2C_3$	C_2	$3\sigma_v$	$3\sigma_d$		
A_1	1	1	1	1	1	1	z	$x^2 + y^2, z^2$
A_2	1	1	1	1	-1	-1	R_z	
B_1	1	-1	1	-1	1	-1		
B_2	1	-1	1	-1	-1	1		
E_1	2	1	-1	-2	0	0	$(x, y), (R_x, R_y)$	(xz, yz)
E_2	2	-1	-1	2	0	0		$(x^2 - y^2, xy)$

▶ The C_{nh} Groups

C_{2h}	E	C_2	i	σ_h		
A_g	1	1	1	1	R_z	x^2, y^2, z^2, xy
B_g	1	-1	1	-1	R_x, R_y	xz, yz
A_u	1	1	-1	-1	z	
B_u	1	-1	-1	1	x, y	

C_{3h}	E	C_3	C_3^2	σ_h	S_3	S_3^5	$\varepsilon = \exp(2\pi i/3)$	
A'	1	1	1	1	1	1	R_z	$x^2 + y^2, z^2$
E'	$\begin{cases} 1 \\ 1 \end{cases}$	$\begin{matrix}\varepsilon \\ \varepsilon^*\end{matrix}$	$\begin{matrix}\varepsilon^* \\ \varepsilon\end{matrix}$	$\begin{matrix}1 \\ 1\end{matrix}$	$\begin{matrix}\varepsilon \\ \varepsilon^*\end{matrix}$	$\begin{matrix}\varepsilon^* \\ \varepsilon\end{matrix}\Big\}$	(x, y)	$(x^2 - y^2, xy)$
A''	1	1	1	-1	-1	-1	z	
E''	$\begin{cases} 1 \\ 1 \end{cases}$	$\begin{matrix}\varepsilon \\ \varepsilon^*\end{matrix}$	$\begin{matrix}\varepsilon^* \\ \varepsilon\end{matrix}$	$\begin{matrix}-1 \\ -1\end{matrix}$	$\begin{matrix}-\varepsilon \\ -\varepsilon^*\end{matrix}$	$\begin{matrix}-\varepsilon^* \\ -\varepsilon\end{matrix}\Big\}$	(R_x, R_y)	(xz, yz)

C_{4h}	E	C_4	C_2	C_4^3	i	S_4^3	σ_h	S_4		
A_g	1	1	1	1	1	1	1	1	R_z	x^2+y^2, z^2
B_g	1	-1	1	-1	1	-1	1	-1		x^2-y^2, xy
E_g	$\begin{cases}1\\1\end{cases}$	$\begin{matrix}i\\-i\end{matrix}$	$\begin{matrix}-1\\-1\end{matrix}$	$\begin{matrix}-i\\i\end{matrix}$	$\begin{matrix}1\\1\end{matrix}$	$\begin{matrix}i\\-i\end{matrix}$	$\begin{matrix}-1\\-1\end{matrix}$	$\begin{matrix}-i\\i\end{matrix}$	(R_x, R_y)	(xz, yz)
A_u	1	1	1	1	-1	-1	-1	-1	z	
B_u	1	-1	1	-1	-1	1	-1	1		
E_u	$\begin{cases}1\\1\end{cases}$	$\begin{matrix}i\\-i\end{matrix}$	$\begin{matrix}-1\\-1\end{matrix}$	$\begin{matrix}-i\\i\end{matrix}$	$\begin{matrix}-1\\-1\end{matrix}$	$\begin{matrix}-i\\i\end{matrix}$	$\begin{matrix}1\\1\end{matrix}$	$\begin{matrix}i\\-i\end{matrix}$	(x, y)	

C_{5h}	E	C_5	C_5^2	C_5^3	C_5^4	σ_h	S_5	S_5^7	S_5^3	S_5^9		$\varepsilon = \exp(2\pi i/5)$
A'	1	1	1	1	1	1	1	1	1	1	R_z	x^2+y^2, z^2
E_1'	$\begin{cases}1\\1\end{cases}$	$\begin{matrix}\varepsilon\\\varepsilon^*\end{matrix}$	$\begin{matrix}\varepsilon^2\\\varepsilon^{2*}\end{matrix}$	$\begin{matrix}\varepsilon^{2*}\\\varepsilon^2\end{matrix}$	$\begin{matrix}\varepsilon^*\\\varepsilon\end{matrix}$	$\begin{matrix}1\\1\end{matrix}$	$\begin{matrix}\varepsilon\\\varepsilon^*\end{matrix}$	$\begin{matrix}\varepsilon^2\\\varepsilon^{2*}\end{matrix}$	$\begin{matrix}\varepsilon^{2*}\\\varepsilon^2\end{matrix}$	$\begin{matrix}\varepsilon^*\\\varepsilon\end{matrix}$	(x, y)	
E_2'	$\begin{cases}1\\1\end{cases}$	$\begin{matrix}\varepsilon^2\\\varepsilon^{2*}\end{matrix}$	$\begin{matrix}\varepsilon^*\\\varepsilon\end{matrix}$	$\begin{matrix}\varepsilon\\\varepsilon^*\end{matrix}$	$\begin{matrix}\varepsilon^{2*}\\\varepsilon^2\end{matrix}$	$\begin{matrix}1\\1\end{matrix}$	$\begin{matrix}\varepsilon^2\\\varepsilon^{2*}\end{matrix}$	$\begin{matrix}\varepsilon^*\\\varepsilon\end{matrix}$	$\begin{matrix}\varepsilon\\\varepsilon^*\end{matrix}$	$\begin{matrix}\varepsilon^{2*}\\\varepsilon^2\end{matrix}$		(x^2-y^2, xy)
A''	1	1	1	1	1	-1	-1	-1	-1	-1	z	
E_1''	$\begin{cases}1\\1\end{cases}$	$\begin{matrix}\varepsilon\\\varepsilon^*\end{matrix}$	$\begin{matrix}\varepsilon^2\\\varepsilon^{2*}\end{matrix}$	$\begin{matrix}\varepsilon^{2*}\\\varepsilon^2\end{matrix}$	$\begin{matrix}\varepsilon^*\\\varepsilon\end{matrix}$	$\begin{matrix}-1\\-1\end{matrix}$	$\begin{matrix}-\varepsilon\\-\varepsilon^*\end{matrix}$	$\begin{matrix}-\varepsilon^2\\-\varepsilon^{2*}\end{matrix}$	$\begin{matrix}-\varepsilon^{2*}\\-\varepsilon^2\end{matrix}$	$\begin{matrix}-\varepsilon^*\\-\varepsilon\end{matrix}$	(R_x, R_y)	(xz, yz)
E_2''	$\begin{cases}1\\1\end{cases}$	$\begin{matrix}\varepsilon^2\\\varepsilon^{2*}\end{matrix}$	$\begin{matrix}\varepsilon^*\\\varepsilon\end{matrix}$	$\begin{matrix}\varepsilon\\\varepsilon^*\end{matrix}$	$\begin{matrix}\varepsilon^{2*}\\\varepsilon^2\end{matrix}$	$\begin{matrix}-1\\-1\end{matrix}$	$\begin{matrix}-\varepsilon^2\\-\varepsilon^{2*}\end{matrix}$	$\begin{matrix}-\varepsilon^*\\-\varepsilon\end{matrix}$	$\begin{matrix}-\varepsilon\\-\varepsilon^*\end{matrix}$	$\begin{matrix}-\varepsilon^{2*}\\-\varepsilon^2\end{matrix}$		

C_{6h}	E	C_6	C_3	C_2	C_3^2	C_6^5	i	S_3^5	S_6^5	σ_h	S_6	S_3		$\varepsilon = \exp(2\pi i/6)$
A_g	1	1	1	1	1	1	1	1	1	1	1	1	R_z	x^2+y^2, z^2
B_g	1	-1	1	-1	1	-1	1	-1	1	-1	1	-1		
E_{1g}	$\begin{cases}1\\1\end{cases}$	$\begin{matrix}\varepsilon\\\varepsilon^*\end{matrix}$	$\begin{matrix}-\varepsilon^*\\-\varepsilon\end{matrix}$	$\begin{matrix}-1\\-1\end{matrix}$	$\begin{matrix}-\varepsilon\\-\varepsilon^*\end{matrix}$	$\begin{matrix}\varepsilon^*\\\varepsilon\end{matrix}$	$\begin{matrix}1\\1\end{matrix}$	$\begin{matrix}\varepsilon\\\varepsilon^*\end{matrix}$	$\begin{matrix}-\varepsilon^*\\-\varepsilon\end{matrix}$	$\begin{matrix}-1\\-1\end{matrix}$	$\begin{matrix}-\varepsilon\\-\varepsilon^*\end{matrix}$	$\begin{matrix}\varepsilon^*\\\varepsilon\end{matrix}$	(R_x, R_y)	(xz, yz)
E_{2g}	$\begin{cases}1\\1\end{cases}$	$\begin{matrix}-\varepsilon^*\\-\varepsilon\end{matrix}$	$\begin{matrix}-\varepsilon\\-\varepsilon^*\end{matrix}$	$\begin{matrix}1\\1\end{matrix}$	$\begin{matrix}-\varepsilon^*\\-\varepsilon\end{matrix}$	$\begin{matrix}-\varepsilon\\-\varepsilon^*\end{matrix}$	$\begin{matrix}1\\1\end{matrix}$	$\begin{matrix}-\varepsilon^*\\-\varepsilon\end{matrix}$	$\begin{matrix}-\varepsilon\\-\varepsilon^*\end{matrix}$	$\begin{matrix}1\\1\end{matrix}$	$\begin{matrix}-\varepsilon^*\\-\varepsilon\end{matrix}$	$\begin{matrix}-\varepsilon\\-\varepsilon^*\end{matrix}$		(x^2-y^2, xy)
A_u	1	1	1	1	1	1	-1	-1	-1	-1	-1	-1	z	
B_u	1	-1	1	-1	1	-1	-1	1	-1	1	-1	1		
E_{1u}	$\begin{cases}1\\1\end{cases}$	$\begin{matrix}\varepsilon\\\varepsilon^*\end{matrix}$	$\begin{matrix}-\varepsilon^*\\-\varepsilon\end{matrix}$	$\begin{matrix}-1\\-1\end{matrix}$	$\begin{matrix}-\varepsilon\\-\varepsilon^*\end{matrix}$	$\begin{matrix}\varepsilon^*\\\varepsilon\end{matrix}$	$\begin{matrix}-1\\-1\end{matrix}$	$\begin{matrix}-\varepsilon\\-\varepsilon^*\end{matrix}$	$\begin{matrix}\varepsilon^*\\\varepsilon\end{matrix}$	$\begin{matrix}1\\1\end{matrix}$	$\begin{matrix}\varepsilon\\\varepsilon^*\end{matrix}$	$\begin{matrix}-\varepsilon^*\\-\varepsilon\end{matrix}$	(x, y)	
E_{2u}	$\begin{cases}1\\1\end{cases}$	$\begin{matrix}-\varepsilon^*\\-\varepsilon\end{matrix}$	$\begin{matrix}-\varepsilon\\-\varepsilon^*\end{matrix}$	$\begin{matrix}1\\1\end{matrix}$	$\begin{matrix}-\varepsilon^*\\-\varepsilon\end{matrix}$	$\begin{matrix}-\varepsilon\\-\varepsilon^*\end{matrix}$	$\begin{matrix}-1\\-1\end{matrix}$	$\begin{matrix}\varepsilon^*\\\varepsilon\end{matrix}$	$\begin{matrix}\varepsilon\\\varepsilon^*\end{matrix}$	$\begin{matrix}-1\\-1\end{matrix}$	$\begin{matrix}\varepsilon^*\\\varepsilon\end{matrix}$	$\begin{matrix}\varepsilon\\\varepsilon^*\end{matrix}$		

THE DIHEDRAL GROUPS

▶ *The D_n Groups*

D_2	E	$C_2(z)$	$C_2(y)$	$C_2(x)$		
A	1	1	1	1		x^2, y^2, z^2
B_1	1	1	-1	-1	z, R_z	xy
B_2	1	-1	1	-1	y, R_y	xz
B_3	1	-1	-1	1	x, R_x	yz

D_3	E	$2C_3$	$3C_2$	(x axis is coincident with C_2)	
A_1	1	1	1		$x^2 + y^2, z^2$
A_2	1	1	-1	z, R_z	
E	2	-1	0	$(x, y), (R_x, R_y)$	$(x^2 - y^2, xy), (xz, yz)$

D_4	E	$2C_4$	$C_2(=C_4^2)$	$2C_2'$	$2C_2''$	(x axis coincident with C_2')	
A_1	1	1	1	1	1		$x^2 + y^2, z^2$
A_2	1	1	1	-1	-1	z, R_z	
B_1	1	-1	1	1	-1		$x^2 - y^2$
B_2	1	-1	1	-1	1		xy
E	2	0	-2	0	0	$(x, y), (R_x, R_y)$	(xz, yz)

D_5	E	$2C_5$	$2C_5^2$	$5C_2$	(x axis coincident with C_2)	
A_1	1	1	1	1		$x^2 + y^2, z^2$
A_2	1	1	1	-1	z, R_z	
E_1	2	$2 \cos 72°$	$2 \cos 144°$	0	$(x, y), (R_x, R_y)$	(xz, yz)
E_2	2	$2 \cos 144°$	$2 \cos 72°$	0		$(x^2 - y^2, xy)$

D_6	E	$2C_6$	$2C_3$	C_2	$3C_2'$	$3C_2''$	(x axis coincident with C_2')	
A_1	1	1	1	1	1	1		$x^2 + y^2, z^2$
A_2	1	1	1	1	-1	-1	z, R_z	
B_1	1	-1	1	-1	1	-1		
B_2	1	-1	1	-1	-1	1		
E_1	2	1	-1	-2	0	0	$(x, y), (R_x, R_y)$	(xz, yz)
E_2	2	-1	-1	2	0	0		$(x^2 - y^2, xy)$

▶ The D$_{nh}$ Groups

D$_{2h}$	E	C$_2$(z)	C$_2$(y)	C$_2$(x)	i	σ(xy)	σ(xz)	σ(yz)		
A$_g$	1	1	1	1	1	1	1	1		x^2, y^2, z^2
B$_{1g}$	1	1	-1	-1	1	1	-1	-1	R$_z$	xy
B$_{2g}$	1	-1	1	-1	1	-1	1	-1	R$_y$	xz
B$_{3g}$	1	-1	-1	1	1	-1	-1	1	R$_x$	yz
A$_u$	1	1	1	1	-1	-1	-1	-1		
B$_{1u}$	1	1	-1	-1	-1	-1	1	1	z	
B$_{2u}$	1	-1	1	-1	-1	1	-1	1	y	
B$_{3u}$	1	-1	-1	1	-1	1	1	-1	x	

D$_{3h}$	E	2C$_3$	3C$_2$	σ$_h$	2S$_3$	3σ$_v$	(x axis coincident with C$_2$)	
A$_1'$	1	1	1	1	1	1		$x^2 + y^2, z^2$
A$_2'$	1	1	-1	1	1	-1	R$_z$	
E$'$	2	-1	0	2	-1	0	(x, y)	$(x^2 - y^2, xy)$
A$_1''$	1	1	1	-1	-1	-1		
A$_2''$	1	1	-1	-1	-1	1	z	
E$''$	2	-1	0	-2	1	0	(R$_x$, R$_y$)	(xz, yz)

D$_{4h}$	E	2C$_4$	C$_2$	2C$_2'$	2C$_2''$	i	2S$_4$	σ$_h$	2σ$_v$	2σ$_d$	(x axis coincident with C$_2'$)	
A$_{1g}$	1	1	1	1	1	1	1	1	1	1		$x^2 + y^2, z^2$
A$_{2g}$	1	1	1	-1	-1	1	1	1	-1	-1	R$_z$	
B$_{1g}$	1	-1	1	1	-1	1	-1	1	1	-1		$x^2 - y^2$
B$_{2g}$	1	-1	1	-1	1	1	-1	1	-1	1		xy
E$_g$	2	0	-2	0	0	2	0	-2	0	0	(R$_x$, R$_y$)	(xz, yz)
A$_{1u}$	1	1	1	1	1	-1	-1	-1	-1	-1		
A$_{2u}$	1	1	1	-1	-1	-1	-1	-1	1	1	z	
B$_{1u}$	1	-1	1	1	-1	-1	1	-1	-1	1		
B$_{2u}$	1	-1	1	-1	1	-1	1	-1	1	-1		
E$_u$	2	0	-2	0	0	-2	0	2	0	0	(x, y)	

D$_{5h}$	E	2C$_5$	2C$_5^2$	5C$_2$	σ$_h$	2S$_5$	2S$_5^3$	5σ$_v$	(x axis coincident with C$_2$)	
A$_1'$	1	1	1	1	1	1	1	1		$x^2 + y^2, z^2$
A$_2'$	1	1	1	-1	1	1	1	-1	R$_z$	
E$_1'$	2	2 cos 72°	2 cos 144°	0	2	2 cos 72°	2 cos 144°	0	(x, y)	
E$_2'$	2	2 cos 144°	2 cos 72°	0	2	2 cos 144°	2 cos 72°	0		$(x^2 - y^2, xy)$
A$_1''$	1	1	1	1	-1	-1	-1	-1		
A$_2''$	1	1	1	-1	-1	-1	-1	1	z	
E$_1''$	2	2 cos 72°	2 cos 144°	0	-2	-2 cos 72°	-2 cos 144°	0	(R$_x$, R$_y$)	(xz, yz)
E$_2''$	2	2 cos 144°	2 cos 72°	0	-2	-2 cos 144°	-2 cos 72°	0		

D_{6h}	E	$2C_6$	$2C_3$	C_2	$3C_2'$	$3C_2''$	i	$2S_3$	$2S_6$	σ_h	$3\sigma_d$	$3\sigma_v$	(x axis coincident with C_2')	
A_{1g}	1	1	1	1	1	1	1	1	1	1	1	1		$x^2+y^2,\ z^2$
A_{2g}	1	1	1	1	-1	-1	1	1	1	1	-1	-1	R_z	
B_{1g}	1	-1	1	-1	1	-1	1	-1	1	-1	1	-1		
B_{2g}	1	-1	1	-1	-1	1	1	-1	1	-1	-1	1		
E_{1g}	2	1	-1	-2	0	0	2	1	-1	-2	0	0	(R_x, R_y)	(xz, yz)
E_{2g}	2	-1	-1	2	0	0	2	-1	-1	2	0	0		$(x^2 - y, xy)$
A_{1u}	1	1	1	1	1	1	-1	-1	-1	-1	-1	-1		
A_{2u}	1	1	1	1	-1	-1	-1	-1	-1	-1	1	1	z	
B_{1u}	1	-1	1	-1	1	-1	-1	1	-1	1	-1	1		
B_{2u}	1	-1	1	-1	-1	1	-1	1	-1	1	1	-1		
E_{1u}	2	1	-1	-2	0	0	-2	-1	1	2	0	0	(x, y)	
E_{2u}	2	-1	-1	2	0	0	-2	1	1	-2	0	0		

D_{8h}	E	$2C_8$	$2C_8^3$	$2C_4$	C_2	$4C_2'$	$4C_2''$	i	$2S_8^3$	$2S_8$	$2S_4$	σ_h	$4\sigma_v$	$4\sigma_d$	(x axis coincident with C_2')	
A_{1g}	1	1	1	1	1	1	1	1	1	1	1	1	1	1		$x^2 + y^2,\ z^2$
A_{2g}	1	1	1	1	1	-1	-1	1	1	1	1	1	-1	-1	R_z	
B_{1g}	1	-1	-1	1	1	1	-1	1	-1	-1	1	1	1	-1		
B_{2g}	1	-1	-1	1	1	-1	1	1	-1	-1	1	1	-1	1		
E_{1g}	2	$\sqrt{2}$	$-\sqrt{2}$	0	-2	0	0	2	$\sqrt{2}$	$-\sqrt{2}$	0	-2	0	0	(R_x, R_y)	(xz, yz)
E_{2g}	2	0	0	-2	2	0	0	2	0	0	-2	2	0	0		$(x^2 - y^2, xy)$
E_{3g}	2	$-\sqrt{2}$	$\sqrt{2}$	0	-2	0	0	2	$-\sqrt{2}$	$\sqrt{2}$	0	-2	0	0		
A_{1u}	1	1	1	1	1	1	1	-1	-1	-1	-1	-1	-1	-1		
A_{2u}	1	1	1	1	1	-1	-1	-1	-1	-1	-1	-1	1	1	z	
B_{1u}	1	-1	-1	1	1	1	-1	-1	1	1	-1	-1	-1	1		
B_{2u}	1	-1	-1	1	1	-1	1	-1	1	1	-1	-1	1	-1		
E_{1u}	2	$\sqrt{2}$	$-\sqrt{2}$	0	-2	0	0	-2	$-\sqrt{2}$	$\sqrt{2}$	0	2	0	0	(x, y)	
E_{2u}	2	0	0	-2	2	0	0	-2	0	0	2	-2	0	0		
E_{3u}	2	$-\sqrt{2}$	$\sqrt{2}$	0	-2	0	0	-2	$\sqrt{2}$	$-\sqrt{2}$	0	2	0	0		

► The D_{nd} Groups

D_{2d}	E	$2S_4$	C_2	$2C_2'$	$2\sigma_d$	(x axis coincident with C_2')	
A_1	1	1	1	1	1		$x^2 + y^2,\ z^2$
A_2	1	1	1	-1	-1	R_z	
B_1	1	-1	1	1	-1		$x^2 - y^2$
B_2	1	-1	1	-1	1	z	xy
E	2	0	-2	0	0	$(x, y), (R_x, R_y)$	(xz, yz)

D_{3d}	E	$2C_3$	$3C_2$	i	$2S_6$	$3\sigma_d$	(x axis coincident with C_2)	
A_{1g}	1	1	1	1	1	1		$x^2 + y^2,\ z^2$
A_{2g}	1	1	-1	1	1	-1	R_z	
E_g	2	-1	0	2	-1	0	(R_x, R_y)	$(x^2 - y^2, xy); (xz, yz)$
A_{1u}	1	1	1	-1	-1	-1		
A_{2u}	1	1	-1	-1	-1	1	z	
E_u	2	-1	0	-2	1	0	(x, y)	

$\mathbf{D_{4d}}$	E	$2S_8$	$2C_4$	$2S_8^3$	C_2	$4C_2'$	$4\sigma_d$	(x axis coincident with C_2')	
A_1	1	1	1	1	1	1	1		$x^2 + y^2, z^2$
A_2	1	1	1	1	1	-1	-1	R_z	
B_1	1	-1	1	-1	1	1	-1		
B_2	1	-1	1	-1	1	-1	1	z	
E_1	2	$\sqrt{2}$	0	$-\sqrt{2}$	-2	0	0	(x, y)	
E_2	2	0	-2	0	2	0	0		$(x^2 - y^2, xy)$
E_3	2	$-\sqrt{2}$	0	$\sqrt{2}$	-2	0	0	(R_x, R_y)	(xz, yz)

$\mathbf{D_{5d}}$	1	$2C_5$	$2C_5^2$	$5C_2$	i	$2S_{10}^3$	$2S_{10}$	$5\sigma_d$	(x axis coincident with C_2)	
A_{1g}	1	1	1	1	1	1	1	1		$x^2 + y^2, z^2$
A_{2g}	1	1	1	-1	1	1	1	-1	R_z	
E_{1g}	2	2 cos 72°	2 cos 144°	0	2	2 cos 72°	2 cos 144°	0	(R_x, R_y)	(xz, yz)
E_{2g}	2	2 cos 144°	2 cos 72°	0	2	2 cos 144°	2 cos 72°	0		$(x^2 - y^2, xy)$
A_{1u}	1	1	1	1	-1	-1	-1	-1		
A_{2u}	1	1	1	-1	-1	-1	-1	1	z	
E_{1u}	2	2 cos 72°	2 cos 144°	0	-2	-2 cos 72°	-2 cos 144°	0	(x, y)	
E_{2u}	2	2 cos 144°	2 cos 72°	0	-2	-2 cos 144°	-2 cos 72°	0		

$\mathbf{D_{6d}}$	E	$2S_{12}$	$2C_6$	$2S_4$	$2C_3$	$2S_{12}^5$	C_2	$6C'_2$	$6\sigma_d$	(x axis coincident with C_2)	
A_1	1	1	1	1	1	1	1	1	1		$x^2 + z^2, z^2$
A_2	1	1	1	1	1	1	1	-1	-1	R_z	
B_1	1	-1	1	-1	1	-1	1	1	-1		
B_2	1	-1	1	-1	1	-1	1	-1	1	z	
E_1	2	$\sqrt{3}$	1	0	-1	$-\sqrt{3}$	-2	0	0	(x, y)	
E_2	2	1	-1	-2	-1	1	2	0	0		$(x^2 - y^2, xy)$
E_3	2	0	-2	0	2	0	-2	0	0		
E_4	2	-1	-1	2	-1	-1	2	0	0		
E_5	2	$-\sqrt{3}$	1	0	-1	$\sqrt{3}$	-2	0	0	(R_x, R_y)	(xz, yz)

THE CUBIC GROUPS

▶ Tetrahedral Groups

$\mathbf{T}$	E	$4C_3$	$4C_3^2$	$3C_2$	$\varepsilon = \exp(2\pi i/3)$	
A	1	1	1	1		$x^2 + y^2 + z^2$
E	$\begin{cases} 1 \\ 1 \end{cases}$	$\begin{matrix} \varepsilon \\ \varepsilon^* \end{matrix}$	$\begin{matrix} \varepsilon^* \\ \varepsilon \end{matrix}$	$\begin{matrix} 1 \\ 1 \end{matrix}$		$(2z^2 - x^2 - y^2,$ $x^2 - y^2)$
T	3	0	0	-1	$(R, R_y, R_z), (x, y, z)$	(xy, xz, yz)

T_h	E	$4C_3$	$4C_3^2$	$3C_2$	i	$4S_6$	$4S_6^5$	$3\sigma_h$		$(\varepsilon = \exp(2\pi i/3))$
A_g	1	1	1	1	1	1	1	1		$x^2 + y^2 + z^2$
A_u	1	1	1	1	-1	-1	-1	-1		
E_g	$\begin{cases}1 \\ 1\end{cases}$	$\begin{matrix}\varepsilon \\ \varepsilon^*\end{matrix}$	$\begin{matrix}\varepsilon^* \\ \varepsilon\end{matrix}$	$\begin{matrix}1 \\ 1\end{matrix}$	$\begin{matrix}1 \\ 1\end{matrix}$	$\begin{matrix}\varepsilon \\ \varepsilon^*\end{matrix}$	$\begin{matrix}\varepsilon^* \\ \varepsilon\end{matrix}$	$\begin{matrix}1 \\ 1\end{matrix}\Big\}$		$(2z^2 - x^2 - y^2,$ $x^2 - y^2)$
E_u	$\begin{cases}1 \\ 1\end{cases}$	$\begin{matrix}\varepsilon \\ \varepsilon^*\end{matrix}$	$\begin{matrix}\varepsilon^* \\ \varepsilon\end{matrix}$	$\begin{matrix}1 \\ 1\end{matrix}$	$\begin{matrix}-1 \\ -1\end{matrix}$	$\begin{matrix}-\varepsilon \\ -\varepsilon^*\end{matrix}$	$\begin{matrix}-\varepsilon^* \\ -\varepsilon\end{matrix}$	$\begin{matrix}-1 \\ -1\end{matrix}\Big\}$		
T_g	3	0	0	-1	3	0	0	-1	(R_x, R_y, R_z)	(xz, yz, xy)
T_u	3	0	0	-1	-3	0	0	1	(x, y, z)	

T_d	E	$8C_3$	$3C_2$	$6S_4$	$6\sigma_d$		
A_1	1	1	1	1	1		$x^2 + y^2 + z^2$
A_2	1	1	1	-1	-1		
E	2	-1	2	0	0		$(2z^2 - x^2 - y^2, x^2 - y^2)$
T_1	3	0	-1	1	-1	(R_x, R_y, R_z)	
T_2	3	0	-1	-1	1	(x, y, z)	(xy, xz, yz)

► Octahedral Groups

O	E	$6C_4$	$3C_2(=C_4^2)$	$8C_3$	$6C_2$		
A_1	1	1	1	1	1		$x^2 + y^2 + z^2$
A_2	1	-1	1	1	-1		
E	2	0	2	-1	0		$(2z^2 - x^2 - y^2, x^2 - y^2)$
T_1	3	1	-1	0	-1	$(R_x, R_y, R_z), (x, y, z)$	
T_2	3	-1	-1	0	1		(xy, xz, yz)

O_h	E	$8C_3$	$6C_2$	$6C_4$	$3C_2(=C_4^2)$	i	$6S_4$	$8S_6$	$3\sigma_h$	$6\sigma_d$		
A_{1g}	1	1	1	1	1	1	1	1	1	1		$x^2 + y^2 + z^2$
A_{2g}	1	1	-1	-1	1	1	-1	1	1	-1		
E_g	2	-1	0	0	2	2	0	-1	2	0		$(2z^2 - x^2 - y^2, x^2 - y^2)$
T_{1g}	3	0	-1	1	-1	3	1	0	-1	-1	(R_x, R_y, R_z)	
T_{2g}	3	0	1	-1	-1	3	-1	0	-1	1		(xz, yz, xy)
A_{1u}	1	1	1	1	1	-1	-1	-1	-1	-1		
A_{2u}	1	1	-1	-1	1	-1	1	-1	-1	1		
E_u	2	-1	0	0	2	-2	0	1	-2	0		
T_{1u}	3	0	-1	1	-1	-3	-1	0	1	1	(x, y, z)	
T_{2u}	3	0	1	-1	-1	-3	1	0	1	-1		

► The Icosahedral Groups[a]

I_h	E	$12C_5$	$12C_5^2$	$20C_3$	$15C_2$	i	$12S_{10}$	$12S_{10}^3$	$20S_6$	15σ		
A_g	1	1	1	1	1	1	1	1	1	1		$x^2 + y^2 + z^2$
T_{1g}	3	$\frac{1}{2}(1+\sqrt{5})$	$\frac{1}{2}(1-\sqrt{5})$	0	-1	3	$\frac{1}{2}(1-\sqrt{5})$	$\frac{1}{2}(1+\sqrt{5})$	0	-1	(R_x, R_y, R_z)	
T_{2g}	3	$\frac{1}{2}(1-\sqrt{5})$	$\frac{1}{2}(1+\sqrt{5})$	0	-1	3	$\frac{1}{2}(1+\sqrt{5})$	$\frac{1}{2}(1-\sqrt{5})$	0	-1		
G_g	4	-1	-1	1	0	4	-1	-1	1	0		
H_g	5	0	0	-1	1	5	0	0	-1	1		$(2z^2-x^2-y^2,$ $x^2-y^2,$ $xy, yz, zx)$
A_u	1	1	1	1	1	-1	-1	-1	-1	-1		
T_{1u}	3	$\frac{1}{2}(1+\sqrt{5})$	$\frac{1}{2}(1-\sqrt{5})$	0	-1	-3	$-\frac{1}{2}(1-\sqrt{5})$	$-\frac{1}{2}(1+\sqrt{5})$	0	1	(x, y, z)	
T_{2u}	3	$\frac{1}{2}(1-\sqrt{5})$	$\frac{1}{2}(1+\sqrt{5})$	0	-1	-3	$-\frac{1}{2}(1+\sqrt{5})$	$-\frac{1}{2}(1-\sqrt{5})$	0	1		
G_u	4	-1	-1	1	0	-4	1	1	-1	0		
H_u	5	0	0	-1	1	-5	0	0	1	-1		

[a] For the pure rotation group **I**, the outlined section in the upper left is the character table; the g subscripts should, of course, be dropped and (x, y, z) assigned to the T_1 representation.

► The Groups $C_{\infty v}$ and $D_{\infty h}$ for Linear Molecules

$C_{\infty v}$	E	$2C_\infty^\phi$	$\cdots$	$\infty\sigma_v$		
$A_1 \equiv \Sigma^+$	1	1	$\cdots$	1	z	$x^2 + y^2, z^2$
$A_2 \equiv \Sigma^-$	1	1	$\cdots$	-1	R_z	
$E_1 \equiv \Pi$	2	$2\cos\phi$	$\cdots$	0	$(x, y); (R_x, R_y)$	(xz, yz)
$E_2 = \Delta$	2	$2\cos 2\phi$	$\cdots$	0		$(x^2 - y^2, xy)$
$E_3 \equiv \Phi$	2	$2\cos 3\phi$	$\cdots$	0		
$\vdots$	$\vdots$	$\vdots$	$\vdots$	$\vdots$		

$D_{\infty h}$	E	$2C_\infty^\phi$	$\cdots$	$\infty\sigma_v$	i	$2S_\infty^\phi$	$\cdots$	∞C_2		
$A_{1g} \equiv \Sigma_g^+$	1	1	$\cdots$	1	1	1	$\cdots$	1		$x^2 + y^2, z^2$
$A_{2g} \equiv \Sigma_g^-$	1	1	$\cdots$	-1	1	1	$\cdots$	-1	R_z	
$E_{1g} \equiv \Pi_g$	2	$2\cos\phi$	$\cdots$	0	2	$-2\cos\phi$	$\cdots$	0	(R_x, R_y)	(xz, yz)
$E_{2g} \equiv \Delta_g$	2	$2\cos 2\phi$	$\cdots$	0	2	$2\cos 2\phi$	$\cdots$	0		$(x^2 - y^2, xy)$
$\cdots$		$\cdots$		$\cdots$		$\cdots$		$\cdots$		
$A_{1u} \equiv \Sigma_u^+$	1	1	$\cdots$	1	-1	-1	$\cdots$	-1	z	
$A_{2u} \equiv \Sigma_u^-$	1	1	$\cdots$	-1	-1	-1	$\cdots$	1		
$E_{1u} \equiv \Pi_u$	2	$2\cos\phi$	$\cdots$	0	-2	$2\cos\phi$	$\cdots$	0	(x, y)	
$E_{2u} \equiv \Delta_u$	2	$2\cos 2\phi$	$\cdots$	0	-2	$-2\cos 2\phi$	$\cdots$	0		
$E_{3u} \equiv \Phi_u$	2	$2\cos 3\phi$	$\cdots$	0	-2	$2\cos 3\phi$	$\cdots$	0		
$\vdots$	$\vdots$	$\vdots$	$\vdots$	$\vdots$	$\vdots$	$\vdots$	$\vdots$	$\vdots$		

Tanabe–Sugano Diagrams

· ·

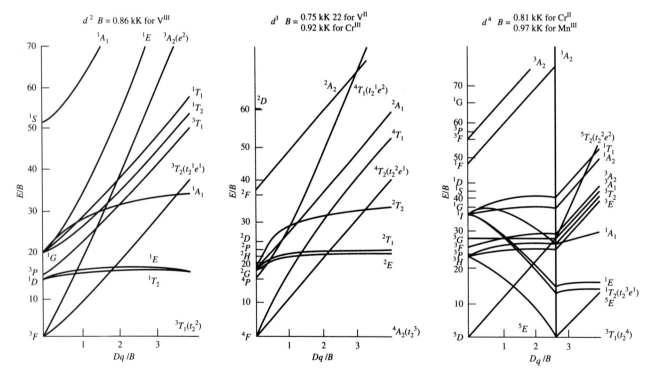

Semiquantitative energy-level diagrams for octahedral symmetry. (After Y. Tanabe and S. Sugano, *J. Phys. Soc. Japan* **1954**, *9*, 753.)

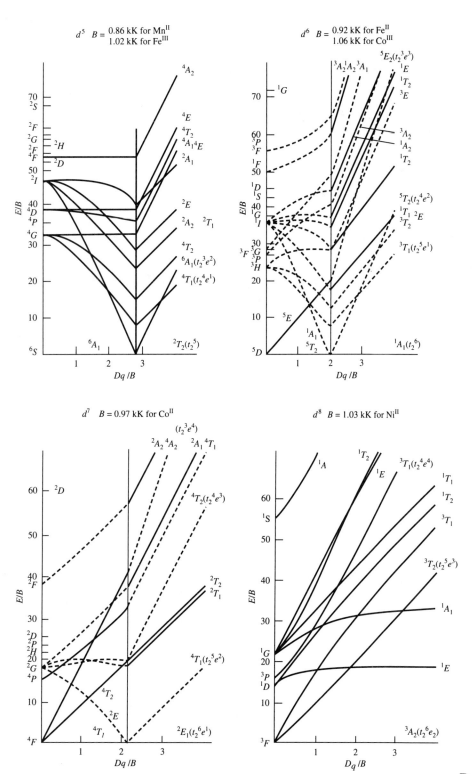

D-2

E

Standard Half-Cell EMF Data

. . . .

ACID SOLUTION (E_A^0)

$$H^+ \xrightarrow{\text{0.00}} H_2 \xrightarrow{\text{-2.25}} H^- \qquad\qquad \overset{\displaystyle\overset{1.229}{\longrightarrow}}{O_2 \xrightarrow{\text{0.682}} H_2O_2 \xrightarrow{\text{1.776}} H_2O}$$

Group 18

$$H_4XeO_6 \xrightarrow{\text{2.36}} XeO_3 \xrightarrow{\text{2.12}} Xe$$

Group 1		*Group 2*		*Group 3*	
$M^+ \longrightarrow M$		$M^{2+} \longrightarrow M$		$M^{3+} \longrightarrow M$	
Li	-3.040	Be	-1.97		
Na	-2.713	Mg	-2.36	Sc	-2.03
K	-2.924	Ca	-2.84	Y	-2.37
Rb	-2.924	Sr	-2.89	La	-2.38
Cs	-2.92	Ba	-2.92	Ac	-2.13
		Ra	-2.92		

Lanthanides

	IV/III	III/II	III/0
La	—		-2.38
Ce	1.7		-2.34
Pr	3.2		-2.35
Nd	4.9	-2.62	-2.32
Pm		-2.67	-2.29

Group 3 (continued)

	IV/III	III/II	III/0
Sm		−1.55	−2.30
Eu		−0.35	−1.99
Gd			−2.28
Tb	3.1		−2.31
Dy		−2.5	−2.29
Ho			−2.33
Er			−2.32
Tm		−2.3	−2.32
Yb		−1.05	−2.22
Lu		—	−2.30

Actinides

$$MO_2^{2+} \xrightarrow{VI/V} MO_2^+ \xrightarrow{V/IV} M^{4+} \xrightarrow{IV/III} M^{3+} \xrightarrow{III/II} M^{2+} \xrightarrow{II/0} M$$

(III/0 spanning M³⁺ → M)

	VI/V	V/IV	IV/III	III/II	III/0	II/0
Ac				(−4.9)	−2.13	(−0.7)
Th			(−3.8)	(−4.9)	−1.83	(0.7)
Pa		−0.05	(−1.4)	(−5.0)	−1.47	(0.3)
U	0.17	0.380	−0.52	(−4.7)	−1.66	(−0.1)
Np	1.24	0.64	0.15	(−4.7)	−1.79	(−0.3)
Pu	1.02	1.04	1.01	(−3.5)	−2.00	(−1.2)
Am	1.60	0.82	2.62	(−2.3)	−2.07	(−2.0)
Cm			3.1	(−3.7)	−2.06	(−1.2)
Bk			1.67	(−2.8)	−2.00	(−1.6)
Cf			(3.2)	(−1.6)	−1.91	(−2.1)
Es			(4.5)	(−1.6)	−1.98	(−2.2)
Fm			(5.2)	(−1.15)	−2.07	(−2.5)
Md				−0.15	−1.74	−2.53
No				1.45	−1.26	−2.6
Lr					−2.1	

Group 4

$$TiO^{2+} \xrightarrow{0.10} Ti^{3+} \xrightarrow{-0.37} Ti^{2+} \xrightarrow{-1.63} Ti$$

$$ZrO_2 \xrightarrow{-1.46} Zr \quad Zr^{4+} \xrightarrow{-1.55} Zr \quad HfO_2 \xrightarrow{-1.57} Hf \quad Hf^{4+} \xrightarrow{-1.70} Hf$$

Group 5

$$V(OH)_4^+ \xrightarrow{1.00} VO^{2+} \xrightarrow{0.34} V^{3+} \xrightarrow{-0.26} V^{2+} \xrightarrow{-1.13} V$$

$$Nb_2O_5 \xrightarrow{-0.1} Nb^{3+} \xrightarrow{-1.1} Nb$$

$$Ta_2O_5 \xrightarrow{-0.81} Ta$$

Group 6

$$Cr_2O_7^{2-} \xrightarrow{1.38} Cr^{3+} \xrightarrow{-0.42} Cr^{2+} \xrightarrow{-0.90} Cr$$

$$MoO_2^{2+} \xrightarrow{0.48} MoO_2^{+} \xrightarrow{0.089} Mo^{3+} \xrightarrow{-0.20} Mo$$

$$WO_3 \xrightarrow{-0.03} W_2O_5 \xrightarrow{-0.03} WO_2 \xrightarrow{-0.15} W^{3+} \xrightarrow{-0.11} W$$
$$WO_2 \xrightarrow{-0.119} W$$

Group 7

$$MnO_4^{-} \xrightarrow{1.507} Mn^{2+}$$
$$MnO_4^{-} \xrightarrow{0.564} MnO_4^{2-} \xrightarrow{0.274} MnO_4^{3-} \xrightarrow{4.27} MnO_2 \xrightarrow{0.95} Mn^{3+} \xrightarrow{1.51} Mn^{2+} \xrightarrow{-1.18} Mn$$
$$MnO_4^{-} \xrightarrow{1.70} MnO_2$$

$$TcO_4^{-} \xrightarrow{0.70} TcO_3 \xrightarrow{0.76} TcO_2 \xrightarrow{0.14} Tc^{2+} \xrightarrow{-0.40} Tc \xrightarrow{-1} TcH_9^{2-}$$

$$ReO_4^{-} \xrightarrow{0.768} ReO_3 \xrightarrow{0.63} ReO_2 \xrightarrow{0.22} Re \xrightarrow{-0.1} ReH_9^{2-}$$

Group VIII (8–10)

$$FeO_4^{2-} \xrightarrow{2.20} Fe^{3+} \xrightarrow{0.771} Fe^{2+} \xrightarrow{-0.44} Fe$$

$$RuO_4 \xrightarrow{1.40} RuO_2 \cdot x H_2O \xrightarrow{1.99} Ru^{3+} \xrightarrow{0.25} Ru^{2+} \xrightarrow{0.45} Ru$$
$$RuO_2 \cdot x H_2O \xrightarrow{0.68} Ru$$

$$OsO_4 \xrightarrow{0.85} Os$$
$$OsO_4 \xrightarrow{1.005} OsO_2 \cdot 2H_2O \xrightarrow{0.687} Os$$

$$CoO_2 \xrightarrow{1.416} Co^{3+} \xrightarrow{1.92} Co^{2+} \xrightarrow{-0.277} Co$$

$$RhO_4^{2-} \xrightarrow{1.5} RhO^{2+} \xrightarrow{1.4} Rh^{3+} \xrightarrow{1.1} Rh^{2+} \xrightarrow{0.6} Rh^{+} \xrightarrow{0.6} Rh$$

$$IrO_4^{2-} \xrightarrow{1.448} Ir^{3+}$$
$$Ir^{3+} \xrightarrow{1.156} Ir$$
$$IrO_4^{2-} \xrightarrow{2.056} IrO_2 \xrightarrow{0.223} Ir^{3+} \xrightarrow{1.73} IrO \xrightarrow{0.87} Ir$$
$$IrO_2 \xrightarrow{0.926} Ir$$

$$NiO_4^{2-} \xrightarrow{>1.8} NiO_2 \xrightarrow{1.59} Ni^{2+} \xrightarrow{-0.257} Ni$$
$$NiO_4^{2-} \xrightarrow{>1.6} Ni$$

$$Pd^{4+} \xrightarrow{1.26} Pd^{2+} \xrightarrow{0.915} Pd$$

$$PtO_3 \xrightarrow{2.0} PtO_2 \xrightarrow{0.84} Pt^{2+} \xrightarrow{1.188} Pt$$
$$PtO_2 \xrightarrow{1.05} PtO \xrightarrow{0.98} Pt$$

Group 11

$$CuO^+ \xrightarrow{1.8} Cu^{2+} \xrightarrow{0.159} Cu^+ \xrightarrow{0.521} Cu$$

(overall $Cu^{2+} \xrightarrow{0.340} Cu$)

$$AgO^+ \xrightarrow{2.1} Ag^{2+} \xrightarrow{1.98} Ag^+ \xrightarrow{0.799} Ag$$

$$Au^{3+} \xrightarrow{1.36} Au^+ \xrightarrow{1.83} Au$$

(overall $Au^{3+} \xrightarrow{1.52} Au$)

Group 12

$$Zn^{2+} \xrightarrow{-0.763} Zn$$

$$Cd^{2+} \xrightarrow{-0.403} Cd$$

$$Hg^{2+} \xrightarrow{0.911} Hg_2^{2+} \xrightarrow{0.796} Hg$$

Group 13

$$H_3BO_3 \xrightarrow{-0.890} B \xrightarrow{0.08} B_{12}H_{12}^{2-} \xrightarrow{0.32} B_{10}H_{10}^{2-} \xrightarrow{-0.50} B_{10}H_{14} \xrightarrow{0.16} B_2H_6 \xrightarrow{0.36} BH_4^-$$

(overall spans: 0.47, 0.69, 0.67, 0.24)

$$Al^{3+} \xrightarrow{-1.68} Al$$

$$Ga^{3+} \xrightarrow{-0.65} Ga^{2+} \xrightarrow{-0.45} Ga$$

(overall $Ga^{3+} \xrightarrow{-0.529} Ga$)

$$In^{3+} \xrightarrow{-0.49} In^{2+} \xrightarrow{-0.40} In^+ \xrightarrow{-0.126} In$$

(overall $In^{3+} \xrightarrow{-0.338} In$)

$$Tl^{3+} \xrightarrow{0.30} Tl^{2+} \xrightarrow{2.22} Tl^+ \xrightarrow{-0.34} Tl$$

(overall $Tl^{3+} \xrightarrow{1.25} Tl^+$)

Group 14

$$CO_2 \xrightarrow{-0.106} CO \xrightarrow{0.517} C \xrightarrow{0.132} CH_4$$

$$SiO_2 \xrightarrow{-0.848} Si \xrightarrow{-0.143} SiH_4$$

$$GeO_2 \xrightarrow{-0.50} Ge^{2+} \xrightarrow{0.25} Ge \xrightarrow{-0.42} GeH_4$$

(overall $GeO_2 \xrightarrow{-0.25} Ge$)

$$Sn^{4+} \xrightarrow{0.15} Sn^{2+} \xrightarrow{-0.137} Sn$$

$$PbO_2 \xrightarrow{1.46} Pb^{2+} \xrightarrow{-0.125} Pb$$

Group 15

$$NO_3^- \xrightarrow{0.803} N_2O_4 \xrightarrow{1.07} HNO_2 \xrightarrow{0.996} NO \xrightarrow{1.59} N_2O \xrightarrow{1.77} N_2 \xrightarrow{-1.87} NH_3OH^+ \xrightarrow{1.41} N_2H_5^+ \xrightarrow{1.275} NH_4^+$$

Over $N_2O_4 \to HNO_2$: 0.94 (spanning NO_3^- to HNO_2)
Over $NO \to N_2O$: 0.71 to $H_2N_2O_2$ $\xrightarrow{2.65}$
Over $N_2 \to NH_4^+$: 0.275

$$H_3PO_4 \xrightarrow{-0.276} H_3PO_3 \xrightarrow{-0.499} H_3PO_2 \xrightarrow{-0.365} P_4 \xrightarrow{-0.100} P_2H_4 \xrightarrow{-0.006} PH_3$$

Over $P_4 \to PH_3$: -0.063

$$H_3AsO_4 \xrightarrow{0.560} H_3AsO_3 \xrightarrow{0.240} As \xrightarrow{-0.225} AsH_3$$

$$Sb_2O_5 \xrightarrow{0.605} SbO^+ \xrightarrow{0.204} Sb \xrightarrow{-0.510} SbH_3$$

$$Bi_2O_5 \xrightarrow{1.60} BiO^+ \xrightarrow{0.317} Bi \xrightarrow{-0.97} BiH_3$$

Group 16

$$O_3 \xrightarrow{2.075} O_2\,(+H_2O)$$

$$O_2 \xrightarrow{-0.125} HO_2 \xrightarrow{1.51} H_2O_2 \xrightarrow{1.763} H_2O$$

Over $O_2 \to H_2O_2$: 0.695
Over $O_2 \to H_2O$: 1.229

$$SO_4^{2-} \xrightarrow{-0.25} S_2O_6^{2-} \xrightarrow{0.57} SO_2 \xrightarrow{0.51} S_4O_6^{2-} \xrightarrow{0.07} S_2O_3^{2-} \xrightarrow{0.60} S \xrightarrow{0.14} H_2S$$

Over $S_2O_6^{2-} \to SO_2$: 0.16
Over $SO_2 \to S$: 0.45

$$SeO_4^{2-} \xrightarrow{1.1} H_2SeO_3 \xrightarrow{0.74} Se \xrightarrow{-0.11} H_2Se$$

$$H_2TeO_4 \xrightarrow{1.00} TeO_2(s) \xrightarrow{0.53} Te \xrightarrow{-0.64} H_2Te$$

$$PoO_3 \xrightarrow{1.51} PoO_2 \xrightarrow{1.1} Po^{2+} \xrightarrow{0.37} Po \xrightarrow{-1.0} H_2Po$$

Over $PoO_2 \to Po$: 0.73

Group 17

$$F_2 \xrightarrow{2.87} F^-$$

$$ClO_4^- \xrightarrow{1.201} ClO_3^- \xrightarrow{1.18} HClO_2 \xrightarrow{1.70} HClO \xrightarrow{1.63} Cl_2 \xrightarrow{1.358} Cl^-$$

$$BrO_4^- \xrightarrow{1.85} BrO_3^- \xrightarrow{1.45} HBrO \xrightarrow{1.60} Br_2(l) \xrightarrow{1.065} Br^-$$

$$H_5IO_6 \xrightarrow{1.60} IO_3^- \xrightarrow{1.13} HIO \xrightarrow{1.44} I_2(s) \xrightarrow{-0.535} I^-$$

$$H_5AtO_6 \xrightarrow{1.6} AtO_3^- \xrightarrow{1.4} HAtO \xrightarrow{0.7} At \xrightarrow{0.2} At^-$$

BASE SOLUTION (E_B^0)

$$O_3 \xrightarrow{1.246} O_2 + OH^- \qquad\qquad O_2 \xrightarrow{-0.065} HO_2^- \xrightarrow{0.867} OH^-$$
$$\underset{0.401}{\underline{\qquad\qquad\qquad\qquad}}\uparrow$$

$$H_2O \xrightarrow{-0.828} H_2 + OH^-$$

Group 18

$$HXeO_6^{3-} \xrightarrow{0.94} HXeO_4^- \xrightarrow{1.24} Xe$$

Group 2

$$Be_2O_3^{2-} \xrightarrow{-2.62} Be \qquad Sr(OH)_2 \xrightarrow{-2.88} Sr$$

$$Mg(OH)_2 \xrightarrow{-2.69} Mg \qquad Ba(OH)_2 \xrightarrow{-2.81} Ba$$

$$Ca(OH)_2 \xrightarrow{-3.03} Ca$$

Group 3

$$Sc(OH)_3 \xrightarrow{-2.6} Sc \qquad Y(OH)_3 \xrightarrow{-2.85} Y \qquad La(OH)_3 \xrightarrow{-2.80} La$$

Lanthanides

$$M(OH)_3 \longrightarrow M$$

La	−2.80	Sm	−2.80	Ho	−2.85
Ce	−2.78	Eu	−2.51	Er	−2.84
Pr	−2.79	Gd	−2.82	Tm	−2.83
Nd	−2.78	Tb	−2.82	Yb	−2.74
Pm	−2.76	Dy	−2.80	Lu	−2.83

Actinides

$$Ac(OH)_3 \xrightarrow{-2.5} Ac \qquad ThO_2 \xrightarrow{-2.56} Th$$

$$UO_2(OH)_2 \xrightarrow{-0.3} UO_2 \xrightarrow{-2.6} U(OH)_3 \xrightarrow{-2.10} U$$

$$NpO_2(OH)_2 \xrightarrow{0.6} NpO_2OH \xrightarrow{0.3} Np(OH)_4 \xrightarrow{-2.1} Np(OH)_3 \xrightarrow{-2.23} Np$$

$$PuO_2(OH)_2 \xrightarrow{0.3} PuO_2OH \xrightarrow{0.9} Pu(OH)_4 \xrightarrow{-1.4} Pu(OH)_3 \xrightarrow{-2.46} Pu$$

$$AmO_2(OH)_2 \xrightarrow{0.9} AmO_2OH \xrightarrow{0.7} Am(OH)_4 \xrightarrow{0.5} Am(OH)_3 \xrightarrow{-2.53} Am$$

$$CmO_2 \xrightarrow{0.7} Cm(OH)_3 \xrightarrow{-2.53} Cm$$

Group 4

$$TiO_2 \xrightarrow{-1.90} Ti \qquad H_2ZrO_3 \xrightarrow{-2.36} Zr \qquad HfO(OH)_2 \xrightarrow{-2.50} Hf$$

Group 5

$$VO_4^{3-} \xrightarrow{0.120} V$$

Group 6

$$CrO_4^{2-} \xrightarrow{-0.11} Cr(OH)_3 \xrightarrow{-1.1} Cr(OH)_2 \xrightarrow{-1.4} Cr$$
$$CrO_2^- \xrightarrow{\hspace{3cm} -1.3 \hspace{3cm}}$$

$$MoO_4^{2-} \xrightarrow{-0.78} MoO_2 \xrightarrow{-0.98} Mo$$
$$WO_4^{2-} \xrightarrow{-1.074} W$$

Group 7

$$MnO_4^- \xrightarrow{0.564} MnO_4^{2-} \xrightarrow{0.27} MnO_4^{3-} \xrightarrow{0.96} MnO_2 \xrightarrow{+0.15} Mn_2O_3 \xrightarrow{-0.25} Mn(OH)_2 \xrightarrow{-1.56} Mn$$
$$TcO_4^- \xrightarrow{-0.322} TcO_2 \xrightarrow{-0.55} Tc$$
$$ReO_4^- \xrightarrow{-0.7} ReO_4^{3-} \xrightarrow{-0.5} ReO_2 \xrightarrow{-0.53} Re(OH)_3 \xrightarrow{-0.6} Re \xrightarrow{-0.4} ReH_9^{2-}$$

Group VIII (8–10)

$$FeO_4^{2-} \xrightarrow{0.72} Fe(OH)_3 \xrightarrow{-0.56} Fe(OH)_2 \xrightarrow{-0.887} Fe$$
$$RuO_4 \xrightarrow{1.00} RuO_4^- \xrightarrow{0.59} RuO_4^{2-} \xrightarrow{0.35} RuO_2 \cdot xH_2O \xrightarrow{-0.15} Ru$$
$$HOsO_5^- \xrightarrow{0.1} OsO_2 \cdot 2H_2O \xrightarrow{-0.12} Os$$
$$CoO_2 \xrightarrow{0.7} Co(OH)_3 \xrightarrow{0.17} Co(OH)_2 \xrightarrow{-0.73} Co$$
$$RhO_4^{2-} \xrightarrow{-0.1} Rh(OH)_3 \xrightarrow{0.0} Rh$$
$$IrO_4^{2-} \xrightarrow{0.4} IrO_2 \xrightarrow{0.1} Ir_2O_3 \xrightarrow{0.1} Ir$$
$$NiO_2^{2-} \xrightarrow{>0.4} NiO_2 \xrightarrow{0.490} Ni(OH)_2 \xrightarrow{-0.72} Ni$$
$$Pd(OH)_4 \xrightarrow{0.7} Pd(OH)_2 \xrightarrow{0.07} Pd$$
$$PtO_4^{2-} \xrightarrow{>0.4} Pt(OH)_6^{2-} \xrightarrow{\sim 0.2} Pt(OH)_2 \xrightarrow{0.15} Pt$$

Group 11

$$CuO \xrightarrow{-0.29} Cu_2O \xrightarrow{-0.365} Cu$$
$$Ag_2O_3 \xrightarrow{0.74} AgO \xrightarrow{0.60} Ag_2O \xrightarrow{0.34} Ag$$
$$HAuO_3^{2-} \xrightarrow{0.8} Au(OH)_2^- \xrightarrow{0.4} Au$$

Group 12

$$ZnO_2^{2-} \xrightarrow{-1.285} Zn$$

$$Cd(OH)_2 \xrightarrow{-0.824} Cd$$

$$HgO \xrightarrow{0.098} Hg$$

Group 13

$$B_4O_7^{2-} \xrightarrow{-1.76} B \xrightarrow{-1.04} B_{12}H_{12}^{2-} \xrightarrow{0.32} B_{10}H_{10}^{2-} \xrightarrow{1.15} B_{10}H_{14}(s) \xrightarrow{0.98} B_2H_6 \xrightarrow{0.78} BH_4^-$$

$$H_2AlO_3^- \xrightarrow{-2.31} Al \quad H_2GaO_3^- \xrightarrow{-1.22} Ga \quad In(OH)_3 \xrightarrow{1.0} In$$

$$Tl(OH)_3 \xrightarrow{-0.05} Tl(OH) \xrightarrow{-0.34} Tl$$

Group 14

$$CO_3^{2-} \xrightarrow{-1.01} HCO_2^- \xrightarrow{-0.52} C \xrightarrow{-0.70} CH_4$$

$$SiO_3^{2-} \xrightarrow{-1.69} Si \xrightarrow{-0.93} SiH_4$$

$$HGeO_3^- \xrightarrow{-0.89} Ge \xrightarrow{<-1.1} GeH_4$$

$$Sn(OH)_6^{2-} \xrightarrow{-0.90} HSnO_2^- \xrightarrow{-0.91} Sn$$

$$PbO_2 \xrightarrow{0.25} PbO \xrightarrow{-0.58} Pb$$

Group 15

$$NO_3^- \xrightarrow{-0.86} N_2O_4 \xrightarrow{0.867} NO_2^- \xrightarrow{-0.46} NO \xrightarrow{0.76} N_2O \xrightarrow{0.94} N_2 \xrightarrow{-3.04} NH_2OH \xrightarrow{0.73} N_2H_4 \xrightarrow{0.1} NH_3$$

$$PO_4^{3-} \xrightarrow{-1.12} HPO_3^{2-} \xrightarrow{-1.57} H_2PO_2^- \xrightarrow{-2.05} P_4 \xrightarrow{-0.89} PH_3$$

$$AsO_4^{3-} \xrightarrow{-0.67} H_2AsO_3^- \xrightarrow{-0.68} As \xrightarrow{-1.37} AsH_3$$

$$Sb(OH)_6^- \xrightarrow{-0.465} SbO_2^- \xrightarrow{-0.639} Sb \xrightarrow{-1.34} SbH_3$$

$$Bi_2O_5 \xrightarrow{0.78} Bi_2O_4 \xrightarrow{0.56} Bi_2O_3 \xrightarrow{-0.46} Bi \xrightarrow{<-1.6} BiH_3$$

Group 16

$$O_2 \xrightarrow{-0.0649} HO_2^- \xrightarrow{0.867} OH^-$$

$$\overset{-0.66}{\overbrace{\qquad\qquad\qquad\qquad}}$$

$$SO_4^{2-} \xrightarrow{-0.94} SO_3^{2-} \xrightarrow{0.79} S_4O_6^{2-} \xrightarrow{0.08} S_2O_3^{2-} \xrightarrow{-0.74} S \xrightarrow{-0.45} S^{2-}$$

$$SeO_4^{2-} \xrightarrow{0.03} SeO_3^{2-} \xrightarrow{-0.36} Se \xrightarrow{-0.67} Se^{2-}$$

$$TeO_2(OH)_4^{2-} \xrightarrow{0.07} TeO_3^{2-} \xrightarrow{-0.42} Te \xrightarrow{-1.14} Te^{2-}$$

$$P_oO_3 \xrightarrow{0.16} Po \xrightarrow{<-1.4} Po^{2-}$$

Group 17

$$F_2 \xrightarrow{2.87} F^-$$

$$\text{ClO}_4^- \xrightarrow{0.374} \text{ClO}_3^- \xrightarrow{0.295} \overset{\displaystyle \overset{-0.48}{\longrightarrow}\ \text{ClO}_2 \xrightarrow{1.07}}{\text{ClO}_2^-} \xrightarrow{0.681} \text{ClO}^- \xrightarrow{0.421} \text{Cl}_2 \xrightarrow{1.358} \text{Cl}^-$$

$$\text{BrO}_4^- \xrightarrow{1.025} \text{BrO}_3^- \xrightarrow{0.49} \text{BrO}^- \xrightarrow{0.455} \text{Br}_2 \xrightarrow{1.065} \text{Br}^-$$

$$\text{H}_3\text{IO}_6^{2-} \xrightarrow{0.65} \text{IO}_3^- \xrightarrow{0.15} \text{IO}^- \xrightarrow{0.42} \text{I}_2 \xrightarrow{0.535} \text{I}^-$$

$$\text{AtO}_3^- \xrightarrow{0.5} \text{AtO}^- \xrightarrow{0.0} \text{At}_2 \xrightarrow{0.2} \text{At}^-$$

REFERENCES

A. J. Bard, R. Parsons, and J. Jordan, Eds., *Standard Potentials in Aqueous Solution,* Marcel Dekker, New York, 1985. The primary source.

A. Kaczmarcyzk, W. C. Nichols, W. H. Stockmayer, and T. B. Ames, *Inorg. Chem.* **1968,** *7,* 1057. Data for boron hydrides.

L. R. Morss, in *The Chemistry of the Actinide Elements,* J. J. Katz, G. T. Seaborg, and L. R. Morss, Eds., Chapman and Hall, New York, 1986, Chapter 17. Data for the actinide metals.

M. Pourbaix, *Atlas of Electrochemical Equilibria in Aqueous Solution,* 2nd ed., translated by J. A. Franklin, National Association of Corrosion Engineers, Houston, TX, 1974. Source for species not listed in the other sources.

Index

. . . .

THE PERIODIC TABLE

Outer shell n	1 1A IA	2 II A II A	3 III A III B	4 IV A IV B	5 V A V B	6 VI A VI B	7 VII A VII B	8	9 VIII A VIII B
1	1 H 1.00794								
2	3 Li 6.941	4 Be 9.01218							
3	11 Na 22.98977	12 Mg 24.3050							
4	19 K 39.0983	20 Ca 40.078	21 Sc 44.95591	22 Ti 47.88	23 V 50.9415	24 Cr 51.9961	25 Mn 54.93805	26 Fe 55.847	27 Co 58.93320
5	37 Rb 85.4678	38 Sr 87.62	39 Y 88.90585	40 Zr 91.224	41 Nb 92.9064	42 Mo 95.94	43 Tc (98)	44 Ru 101.07	45 Rh 102.90550
6	55 Cs 132.9054	56 Ba 137.327	57–71 Lantha-nides	72 Hf 178.49	73 Ta 180.9479	74 W 183.84	75 Re 186.207	76 Os 190.23	77 Ir 192.22
7	87 Fr (223)	88 Ra 226.0254	89– Acti-nides	104 Rf (261)	105 Ha (262)	106 — (263)	107 Ns	108 Hs	109 Mt

Lanthanides

57 La 138.9055	58 Ce 140.115	59 Pr 140.90765	60 Nd 144.24	61 Pm (145)	62 Sm 150.36	63 Eu 151.965

Actinides

89 Ac 227.0278	90 Th 232.0381	91 Pa 231.0359	92 U 238.0289	93 Np 237.0482	94 Pu (244)	95 Am (243)

Atomic number	Name	Symbol
1	Hydrogen	H
2	Helium	He
3	Lithium	Li
4	Beryllium	Be
5	Boron	B
6	Carbon	C
7	Nitrogen	N
8	Oxygen	O
9	Fluorine	F
10	Neon	Ne
11	Sodium (Natrium)	Na
12	Magnesium	Mg
13	Aluminium	Al
14	Silicon	Si
15	Phosphorus	P
16	Sulfur	S
17	Chlorine	Cl
18	Argon	Ar
19	Potassium (Kalium)	K
20	Calcium	Ca

Atomic number	Name	Symbol
21	Scandium	Sc
22	Titanium	Ti
23	Vanadium	V
24	Chromium	Cr
25	Manganese	Mn
26	Iron	Fe
27	Cobalt	Co
28	Nickel	Ni
29	Copper	Cu
30	Zinc	Zn
31	Gallium	Ga
32	Germanium	Ge
33	Arsenic	As
34	Selenium	Se
35	Bromine	Br
36	Krypton	Kr
37	Rubidium	Rb
38	Strontium	Sr
39	Yttrium	Y
40	Zirconium	Zr

Atomic number	Name	Symbol
41	Niobium	Nb
42	Molybdenum	Mo
43	Technetium*	Tc
44	Ruthenium	Ru
45	Rhodium	Rh
46	Palladium	Pd
47	Silver	Ag
48	Cadmium	Cd
49	Indium	In
50	Tin	Sn
51	Antimony (Stibium)	Sb
52	Tellurium	Te
53	Iodine	I
54	Xenon	Xe
55	Caesium (Cesium)	Cs
56	Barium	Ba
57	Lanthanum	La
58	Cerium	Ce
59	Praseodymium	Pr
60	Neodymium	Nd